Lecture Notes in Computer Science 12346

More information about this series at http://www.springer.com/series/7412

Andrea Vedaldi · Horst Bischof ·
Thomas Brox · Jan-Michael Frahm (Eds.)

Computer Vision – ECCV 2020

16th European Conference
Glasgow, UK, August 23–28, 2020
Proceedings, Part I

Editors
Andrea Vedaldi
University of Oxford
Oxford, UK

Thomas Brox
University of Freiburg
Freiburg im Breisgau, Germany

Horst Bischof
Graz University of Technology
Graz, Austria

Jan-Michael Frahm
University of North Carolina at Chapel Hill
Chapel Hill, NC, USA

ISSN 0302-9743 ISSN 1611-3349 (electronic)
Lecture Notes in Computer Science
ISBN 978-3-030-58451-1 ISBN 978-3-030-58452-8 (eBook)
https://doi.org/10.1007/978-3-030-58452-8

LNCS Sublibrary: SL6 – Image Processing, Computer Vision, Pattern Recognition, and Graphics

This Springer imprint is published by the registered company Springer Nature Switzerland AG
The registered company address is: Gewerbestrasse 11, 6330 Cham, Switzerland

Foreword

Hosting the European Conference on Computer Vision (ECCV 2020) was certainly an exciting journey. From the 2016 plan to hold it at the Edinburgh International Conference Centre (hosting 1,800 delegates) to the 2018 plan to hold it at Glasgow's Scottish Exhibition Centre (up to 6,000 delegates), we finally ended with moving online because of the COVID-19 outbreak. While possibly having fewer delegates than expected because of the online format, ECCV 2020 still had over 3,100 registered participants.

Although online, the conference delivered most of the activities expected at a face-to-face conference: peer-reviewed papers, industrial exhibitors, demonstrations, and messaging between delegates. In addition to the main technical sessions, the conference included a strong program of satellite events with 16 tutorials and 44 workshops.

Furthermore, the online conference format enabled new conference features. Every paper had an associated teaser video and a longer full presentation video. Along with the papers and slides from the videos, all these materials were available the week before the conference. This allowed delegates to become familiar with the paper content and be ready for the live interaction with the authors during the conference week. The live event consisted of brief presentations by the oral and spotlight authors and industrial sponsors. Question and answer sessions for all papers were timed to occur twice so delegates from around the world had convenient access to the authors.

As with ECCV 2018, authors' draft versions of the papers appeared online with open access, now on both the Computer Vision Foundation (CVF) and the European Computer Vision Association (ECVA) websites. An archival publication arrangement was put in place with the cooperation of Springer. SpringerLink hosts the final version of the papers with further improvements, such as activating reference links and supplementary materials. These two approaches benefit all potential readers: a version available freely for all researchers, and an authoritative and citable version with additional benefits for SpringerLink subscribers. We thank Alfred Hofmann and Aliaksandr Birukou from Springer for helping to negotiate this agreement, which we expect will continue for future versions of ECCV.

August 2020

Vittorio Ferrari
Bob Fisher
Cordelia Schmid
Emanuele Trucco

Preface

Welcome to the proceedings of the European Conference on Computer Vision (ECCV 2020). This is a unique edition of ECCV in many ways. Due to the COVID-19 pandemic, this is the first time the conference was held online, in a virtual format. This was also the first time the conference relied exclusively on the Open Review platform to manage the review process. Despite these challenges ECCV is thriving. The conference received 5,150 valid paper submissions, of which 1,360 were accepted for publication (27%) and, of those, 160 were presented as spotlights (3%) and 104 as orals (2%). This amounts to more than twice the number of submissions to ECCV 2018 (2,439). Furthermore, CVPR, the largest conference on computer vision, received 5,850 submissions this year, meaning that ECCV is now 87% the size of CVPR in terms of submissions. By comparison, in 2018 the size of ECCV was only 73% of CVPR.

The review model was similar to previous editions of ECCV; in particular, it was double blind in the sense that the authors did not know the name of the reviewers and vice versa. Furthermore, each conference submission was held confidentially, and was only publicly revealed if and once accepted for publication. Each paper received at least three reviews, totalling more than 15,000 reviews. Handling the review process at this scale was a significant challenge. In order to ensure that each submission received as fair and high-quality reviews as possible, we recruited 2,830 reviewers (a 130% increase with reference to 2018) and 207 area chairs (a 60% increase). The area chairs were selected based on their technical expertise and reputation, largely among people that served as area chair in previous top computer vision and machine learning conferences (ECCV, ICCV, CVPR, NeurIPS, etc.). Reviewers were similarly invited from previous conferences. We also encouraged experienced area chairs to suggest additional chairs and reviewers in the initial phase of recruiting.

Despite doubling the number of submissions, the reviewer load was slightly reduced from 2018, from a maximum of 8 papers down to 7 (with some reviewers offering to handle 6 papers plus an emergency review). The area chair load increased slightly, from 18 papers on average to 22 papers on average.

Conflicts of interest between authors, area chairs, and reviewers were handled largely automatically by the Open Review platform via their curated list of user profiles. Many authors submitting to ECCV already had a profile in Open Review. We set a paper registration deadline one week before the paper submission deadline in order to encourage all missing authors to register and create their Open Review profiles well on time (in practice, we allowed authors to create/change papers arbitrarily until the submission deadline). Except for minor issues with users creating duplicate profiles, this allowed us to easily and quickly identify institutional conflicts, and avoid them, while matching papers to area chairs and reviewers.

Papers were matched to area chairs based on: an affinity score computed by the Open Review platform, which is based on paper titles and abstracts, and an affinity score computed by the Toronto Paper Matching System (TPMS), which is based on the paper's full text, the area chair bids for individual papers, load balancing, and conflict avoidance. Open Review provides the program chairs a convenient web interface to experiment with different configurations of the matching algorithm. The chosen configuration resulted in about 50% of the assigned papers to be highly ranked by the area chair bids, and 50% to be ranked in the middle, with very few low bids assigned.

Assignments to reviewers were similar, with two differences. First, there was a maximum of 7 papers assigned to each reviewer. Second, area chairs recommended up to seven reviewers per paper, providing another highly-weighed term to the affinity scores used for matching.

The assignment of papers to area chairs was smooth. However, it was more difficult to find suitable reviewers for all papers. Having a ratio of 5.6 papers per reviewer with a maximum load of 7 (due to emergency reviewer commitment), which did not allow for much wiggle room in order to also satisfy conflict and expertise constraints. We received some complaints from reviewers who did not feel qualified to review specific papers and we reassigned them wherever possible. However, the large scale of the conference, the many constraints, and the fact that a large fraction of such complaints arrived very late in the review process made this process very difficult and not all complaints could be addressed.

Reviewers had six weeks to complete their assignments. Possibly due to COVID-19 or the fact that the NeurIPS deadline was moved closer to the review deadline, a record 30% of the reviews were still missing after the deadline. By comparison, ECCV 2018 experienced only 10% missing reviews at this stage of the process. In the subsequent week, area chairs chased the missing reviews intensely, found replacement reviewers in their own team, and managed to reach 10% missing reviews. Eventually, we could provide almost all reviews (more than 99.9%) with a delay of only a couple of days on the initial schedule by a significant use of emergency reviews. If this trend is confirmed, it might be a major challenge to run a smooth review process in future editions of ECCV. The community must reconsider prioritization of the time spent on paper writing (the number of submissions increased a lot despite COVID-19) and time spent on paper reviewing (the number of reviews delivered in time decreased a lot presumably due to COVID-19 or NeurIPS deadline). With this imbalance the peer-review system that ensures the quality of our top conferences may break soon.

Reviewers submitted their reviews independently. In the reviews, they had the opportunity to ask questions to the authors to be addressed in the rebuttal. However, reviewers were told not to request any significant new experiment. Using the Open Review interface, authors could provide an answer to each individual review, but were also allowed to cross-reference reviews and responses in their answers. Rather than PDF files, we allowed the use of formatted text for the rebuttal. The rebuttal and initial reviews were then made visible to all reviewers and the primary area chair for a given paper. The area chair encouraged and moderated the reviewer discussion. During the discussions, reviewers were invited to reach a consensus and possibly adjust their ratings as a result of the discussion and of the evidence in the rebuttal.

After the discussion period ended, most reviewers entered a final rating and recommendation, although in many cases this did not differ from their initial recommendation. Based on the updated reviews and discussion, the primary area chair then made a preliminary decision to accept or reject the paper and wrote a justification for it (meta-review). Except for cases where the outcome of this process was absolutely clear (as indicated by the three reviewers and primary area chairs all recommending clear rejection), the decision was then examined and potentially challenged by a secondary area chair. This led to further discussion and overturning a small number of preliminary decisions. Needless to say, there was no in-person area chair meeting, which would have been impossible due to COVID-19.

Area chairs were invited to observe the consensus of the reviewers whenever possible and use extreme caution in overturning a clear consensus to accept or reject a paper. If an area chair still decided to do so, she/he was asked to clearly justify it in the meta-review and to explicitly obtain the agreement of the secondary area chair. In practice, very few papers were rejected after being confidently accepted by the reviewers.

This was the first time Open Review was used as the main platform to run ECCV. In 2018, the program chairs used CMT3 for the user-facing interface and Open Review internally, for matching and conflict resolution. Since it is clearly preferable to only use a single platform, this year we switched to using Open Review in full. The experience was largely positive. The platform is highly-configurable, scalable, and open source. Being written in Python, it is easy to write scripts to extract data programmatically. The paper matching and conflict resolution algorithms and interfaces are top-notch, also due to the excellent author profiles in the platform. Naturally, there were a few kinks along the way due to the fact that the ECCV Open Review configuration was created from scratch for this event and it differs in substantial ways from many other Open Review conferences. However, the Open Review development and support team did a fantastic job in helping us to get the configuration right and to address issues in a timely manner as they unavoidably occurred. We cannot thank them enough for the tremendous effort they put into this project.

Finally, we would like to thank everyone involved in making ECCV 2020 possible in these very strange and difficult times. This starts with our authors, followed by the area chairs and reviewers, who ran the review process at an unprecedented scale. The whole Open Review team (and in particular Melisa Bok, Mohit Unyal, Carlos Mondragon Chapa, and Celeste Martinez Gomez) worked incredibly hard for the entire duration of the process. We would also like to thank René Vidal for contributing to the adoption of Open Review. Our thanks also go to Laurent Charling for TPMS and to the program chairs of ICML, ICLR, and NeurIPS for cross checking double submissions. We thank the website chair, Giovanni Farinella, and the CPI team (in particular Ashley Cook, Miriam Verdon, Nicola McGrane, and Sharon Kerr) for promptly adding material to the website as needed in the various phases of the process. Finally, we thank the publication chairs, Albert Ali Salah, Hamdi Dibeklioglu, Metehan Doyran, Henry Howard-Jenkins, Victor Prisacariu, Siyu Tang, and Gul Varol, who managed to compile these substantial proceedings in an exceedingly compressed schedule. We express our thanks to the ECVA team, in particular Kristina Scherbaum for allowing open access of the proceedings. We thank Alfred Hofmann from Springer who again serve as the publisher. Finally, we thank the other chairs of ECCV 2020, including in particular the general chairs for very useful feedback with the handling of the program.

August 2020

Andrea Vedaldi
Horst Bischof
Thomas Brox
Jan-Michael Frahm

Organization

General Chairs

Vittorio Ferrari	Google Research, Switzerland
Bob Fisher	University of Edinburgh, UK
Cordelia Schmid	Google and Inria, France
Emanuele Trucco	University of Dundee, UK

Program Chairs

Andrea Vedaldi	University of Oxford, UK
Horst Bischof	Graz University of Technology, Austria
Thomas Brox	University of Freiburg, Germany
Jan-Michael Frahm	University of North Carolina, USA

Industrial Liaison Chairs

Jim Ashe	University of Edinburgh, UK
Helmut Grabner	Zurich University of Applied Sciences, Switzerland
Diane Larlus	NAVER LABS Europe, France
Cristian Novotny	University of Edinburgh, UK

Local Arrangement Chairs

Yvan Petillot	Heriot-Watt University, UK
Paul Siebert	University of Glasgow, UK

Academic Demonstration Chair

Thomas Mensink	Google Research and University of Amsterdam, The Netherlands

Poster Chair

Stephen Mckenna	University of Dundee, UK

Technology Chair

Gerardo Aragon Camarasa	University of Glasgow, UK

Tutorial Chairs

Carlo Colombo — University of Florence, Italy
Sotirios Tsaftaris — University of Edinburgh, UK

Publication Chairs

Albert Ali Salah — Utrecht University, The Netherlands
Hamdi Dibeklioglu — Bilkent University, Turkey
Metehan Doyran — Utrecht University, The Netherlands
Henry Howard-Jenkins — University of Oxford, UK
Victor Adrian Prisacariu — University of Oxford, UK
Siyu Tang — ETH Zurich, Switzerland
Gul Varol — University of Oxford, UK

Website Chair

Giovanni Maria Farinella — University of Catania, Italy

Workshops Chairs

Adrien Bartoli — University of Clermont Auvergne, France
Andrea Fusiello — University of Udine, Italy

Area Chairs

Lourdes Agapito — University College London, UK
Zeynep Akata — University of Tübingen, Germany
Karteek Alahari — Inria, France
Antonis Argyros — University of Crete, Greece
Hossein Azizpour — KTH Royal Institute of Technology, Sweden
Joao P. Barreto — Universidade de Coimbra, Portugal
Alexander C. Berg — University of North Carolina at Chapel Hill, USA
Matthew B. Blaschko — KU Leuven, Belgium
Lubomir D. Bourdev — WaveOne, Inc., USA
Edmond Boyer — Inria, France
Yuri Boykov — University of Waterloo, Canada
Gabriel Brostow — University College London, UK
Michael S. Brown — National University of Singapore, Singapore
Jianfei Cai — Monash University, Australia
Barbara Caputo — Politecnico di Torino, Italy
Ayan Chakrabarti — Washington University, St. Louis, USA
Tat-Jen Cham — Nanyang Technological University, Singapore
Manmohan Chandraker — University of California, San Diego, USA
Rama Chellappa — Johns Hopkins University, USA
Liang-Chieh Chen — Google, USA

Yung-Yu Chuang	National Taiwan University, Taiwan
Ondrej Chum	Czech Technical University in Prague, Czech Republic
Brian Clipp	Kitware, USA
John Collomosse	University of Surrey and Adobe Research, UK
Jason J. Corso	University of Michigan, USA
David J. Crandall	Indiana University, USA
Daniel Cremers	University of California, Los Angeles, USA
Fabio Cuzzolin	Oxford Brookes University, UK
Jifeng Dai	SenseTime, SAR China
Kostas Daniilidis	University of Pennsylvania, USA
Andrew Davison	Imperial College London, UK
Alessio Del Bue	Fondazione Istituto Italiano di Tecnologia, Italy
Jia Deng	Princeton University, USA
Alexey Dosovitskiy	Google, Germany
Matthijs Douze	Facebook, France
Enrique Dunn	Stevens Institute of Technology, USA
Irfan Essa	Georgia Institute of Technology and Google, USA
Giovanni Maria Farinella	University of Catania, Italy
Ryan Farrell	Brigham Young University, USA
Paolo Favaro	University of Bern, Switzerland
Rogerio Feris	International Business Machines, USA
Cornelia Fermuller	University of Maryland, College Park, USA
David J. Fleet	Vector Institute, Canada
Friedrich Fraundorfer	DLR, Austria
Mario Fritz	CISPA Helmholtz Center for Information Security, Germany
Pascal Fua	EPFL (Swiss Federal Institute of Technology Lausanne), Switzerland
Yasutaka Furukawa	Simon Fraser University, Canada
Li Fuxin	Oregon State University, USA
Efstratios Gavves	University of Amsterdam, The Netherlands
Peter Vincent Gehler	Amazon, USA
Theo Gevers	University of Amsterdam, The Netherlands
Ross Girshick	Facebook AI Research, USA
Boqing Gong	Google, USA
Stephen Gould	Australian National University, Australia
Jinwei Gu	SenseTime Research, USA
Abhinav Gupta	Facebook, USA
Bohyung Han	Seoul National University, South Korea
Bharath Hariharan	Cornell University, USA
Tal Hassner	Facebook AI Research, USA
Xuming He	Australian National University, Australia
Joao F. Henriques	University of Oxford, UK
Adrian Hilton	University of Surrey, UK
Minh Hoai	Stony Brooks, State University of New York, USA
Derek Hoiem	University of Illinois Urbana-Champaign, USA

Timothy Hospedales	University of Edinburgh and Samsung, UK
Gang Hua	Wormpex AI Research, USA
Slobodan Ilic	Siemens AG, Germany
Hiroshi Ishikawa	Waseda University, Japan
Jiaya Jia	The Chinese University of Hong Kong, SAR China
Hailin Jin	Adobe Research, USA
Justin Johnson	University of Michigan, USA
Frederic Jurie	University of Caen Normandie, France
Fredrik Kahl	Chalmers University, Sweden
Sing Bing Kang	Zillow, USA
Gunhee Kim	Seoul National University, South Korea
Junmo Kim	Korea Advanced Institute of Science and Technology, South Korea
Tae-Kyun Kim	Imperial College London, UK
Ron Kimmel	Technion-Israel Institute of Technology, Israel
Alexander Kirillov	Facebook AI Research, USA
Kris Kitani	Carnegie Mellon University, USA
Iasonas Kokkinos	Ariel AI, UK
Vladlen Koltun	Intel Labs, USA
Nikos Komodakis	Ecole des Ponts ParisTech, France
Piotr Koniusz	Australian National University, Australia
M. Pawan Kumar	University of Oxford, UK
Kyros Kutulakos	University of Toronto, Canada
Christoph Lampert	IST Austria, Austria
Ivan Laptev	Inria, France
Diane Larlus	NAVER LABS Europe, France
Laura Leal-Taixe	Technical University Munich, Germany
Honglak Lee	Google and University of Michigan, USA
Joon-Young Lee	Adobe Research, USA
Kyoung Mu Lee	Seoul National University, South Korea
Seungyong Lee	POSTECH, South Korea
Yong Jae Lee	University of California, Davis, USA
Bastian Leibe	RWTH Aachen University, Germany
Victor Lempitsky	Samsung, Russia
Ales Leonardis	University of Birmingham, UK
Marius Leordeanu	Institute of Mathematics of the Romanian Academy, Romania
Vincent Lepetit	ENPC ParisTech, France
Hongdong Li	The Australian National University, Australia
Xi Li	Zhejiang University, China
Yin Li	University of Wisconsin-Madison, USA
Zicheng Liao	Zhejiang University, China
Jongwoo Lim	Hanyang University, South Korea
Stephen Lin	Microsoft Research Asia, China
Yen-Yu Lin	National Chiao Tung University, Taiwan, China
Zhe Lin	Adobe Research, USA

Haibin Ling	Stony Brooks, State University of New York, USA
Jiaying Liu	Peking University, China
Ming-Yu Liu	NVIDIA, USA
Si Liu	Beihang University, China
Xiaoming Liu	Michigan State University, USA
Huchuan Lu	Dalian University of Technology, China
Simon Lucey	Carnegie Mellon University, USA
Jiebo Luo	University of Rochester, USA
Julien Mairal	Inria, France
Michael Maire	University of Chicago, USA
Subhransu Maji	University of Massachusetts, Amherst, USA
Yasushi Makihara	Osaka University, Japan
Jiri Matas	Czech Technical University in Prague, Czech Republic
Yasuyuki Matsushita	Osaka University, Japan
Philippos Mordohai	Stevens Institute of Technology, USA
Vittorio Murino	University of Verona, Italy
Naila Murray	NAVER LABS Europe, France
Hajime Nagahara	Osaka University, Japan
P. J. Narayanan	International Institute of Information Technology (IIIT), Hyderabad, India
Nassir Navab	Technical University of Munich, Germany
Natalia Neverova	Facebook AI Research, France
Matthias Niessner	Technical University of Munich, Germany
Jean-Marc Odobez	Idiap Research Institute and Swiss Federal Institute of Technology Lausanne, Switzerland
Francesca Odone	Università di Genova, Italy
Takeshi Oishi	The University of Tokyo, Tokyo Institute of Technology, Japan
Vicente Ordonez	University of Virginia, USA
Manohar Paluri	Facebook AI Research, USA
Maja Pantic	Imperial College London, UK
In Kyu Park	Inha University, South Korea
Ioannis Patras	Queen Mary University of London, UK
Patrick Perez	Valeo, France
Bryan A. Plummer	Boston University, USA
Thomas Pock	Graz University of Technology, Austria
Marc Pollefeys	ETH Zurich and Microsoft MR & AI Zurich Lab, Switzerland
Jean Ponce	Inria, France
Gerard Pons-Moll	MPII, Saarland Informatics Campus, Germany
Jordi Pont-Tuset	Google, Switzerland
James Matthew Rehg	Georgia Institute of Technology, USA
Ian Reid	University of Adelaide, Australia
Olaf Ronneberger	DeepMind London, UK
Stefan Roth	TU Darmstadt, Germany
Bryan Russell	Adobe Research, USA

Kwang Moo Yi	University of Victoria, Canada
Zhaozheng Yin	Stony Brook, State University of New York, USA
Chang D. Yoo	Korea Advanced Institute of Science and Technology, South Korea
Shaodi You	University of Amsterdam, The Netherlands
Jingyi Yu	ShanghaiTech University, China
Stella Yu	University of California, Berkeley, and ICSI, USA
Stefanos Zafeiriou	Imperial College London, UK
Hongbin Zha	Peking University, China
Tianzhu Zhang	University of Science and Technology of China, China
Liang Zheng	Australian National University, Australia
Todd E. Zickler	Harvard University, USA
Andrew Zisserman	University of Oxford, UK

Technical Program Committee

Sathyanarayanan N. Aakur
Wael Abd Almgaeed
Abdelrahman Abdelhamed
Abdullah Abuolaim
Supreeth Achar
Hanno Ackermann
Ehsan Adeli
Triantafyllos Afouras
Sameer Agarwal
Aishwarya Agrawal
Harsh Agrawal
Pulkit Agrawal
Antonio Agudo
Eirikur Agustsson
Karim Ahmed
Byeongjoo Ahn
Unaiza Ahsan
Thalaiyasingam Ajanthan
Kenan E. Ak
Emre Akbas
Naveed Akhtar
Derya Akkaynak
Yagiz Aksoy
Ziad Al-Halah
Xavier Alameda-Pineda
Jean-Baptiste Alayrac
Samuel Albanie
Shadi Albarqouni
Cenek Albl
Hassan Abu Alhaija
Daniel Aliaga
Mohammad S. Aliakbarian
Rahaf Aljundi
Thiemo Alldieck
Jon Almazan
Jose M. Alvarez
Senjian An
Saket Anand
Codruta Ancuti
Cosmin Ancuti
Peter Anderson
Juan Andrade-Cetto
Alexander Andreopoulos
Misha Andriluka
Dragomir Anguelov
Rushil Anirudh
Michel Antunes
Oisin Mac Aodha
Srikar Appalaraju
Relja Arandjelovic
Nikita Araslanov
Andre Araujo
Helder Araujo
Pablo Arbelaez
Shervin Ardeshir
Sercan O. Arik
Anil Armagan
Anurag Arnab
Chetan Arora
Federica Arrigoni
Mathieu Aubry
Shai Avidan
Angelica I. Aviles-Rivero
Yannis Avrithis
Ismail Ben Ayed
Shekoofeh Azizi
Ioan Andrei Bârsan
Artem Babenko
Deepak Babu Sam
Seung-Hwan Baek
Seungryul Baek
Andrew D. Bagdanov
Shai Bagon
Yuval Bahat
Junjie Bai
Song Bai
Xiang Bai
Yalong Bai
Yancheng Bai
Peter Bajcsy
Slawomir Bak

Mahsa Baktashmotlagh
Kavita Bala
Yogesh Balaji
Guha Balakrishnan
V. N. Balasubramanian
Federico Baldassarre
Vassileios Balntas
Shurjo Banerjee
Aayush Bansal
Ankan Bansal
Jianmin Bao
Linchao Bao
Wenbo Bao
Yingze Bao
Akash Bapat
Md Jawadul Hasan Bappy
Fabien Baradel
Lorenzo Baraldi
Daniel Barath
Adrian Barbu
Kobus Barnard
Nick Barnes
Francisco Barranco
Jonathan T. Barron
Arslan Basharat
Chaim Baskin
Anil S. Baslamisli
Jorge Batista
Kayhan Batmanghelich
Konstantinos Batsos
David Bau
Luis Baumela
Christoph Baur
Eduardo Bayro-Corrochano
Paul Beardsley
Jan Bednavr'ik
Oscar Beijbom
Philippe Bekaert
Esube Bekele
Vasileios Belagiannis
Ohad Ben-Shahar
Abhijit Bendale
Róger Bermúdez-Chacón
Maxim Berman
Jesus Bermudez-cameo
Florian Bernard
Stefano Berretti
Marcelo Bertalmio
Gedas Bertasius
Cigdem Beyan
Lucas Beyer
Vijayakumar Bhagavatula
Arjun Nitin Bhagoji
Apratim Bhattacharyya
Binod Bhattarai
Sai Bi
Jia-Wang Bian
Simone Bianco
Adel Bibi
Tolga Birdal
Tom Bishop
Soma Biswas
Mårten Björkman
Volker Blanz
Vishnu Boddeti
Navaneeth Bodla
Simion-Vlad Bogolin
Xavier Boix
Piotr Bojanowski
Timo Bolkart
Guido Borghi
Larbi Boubchir
Guillaume Bourmaud
Adrien Bousseau
Thierry Bouwmans
Richard Bowden
Hakan Boyraz
Mathieu Brédif
Samarth Brahmbhatt
Steve Branson
Nikolas Brasch
Biagio Brattoli
Ernesto Brau
Toby P. Breckon
Francois Bremond
Jesus Briales
Sofia Broomé
Marcus A. Brubaker
Luc Brun
Silvia Bucci
Shyamal Buch
Pradeep Buddharaju
Uta Buechler
Mai Bui
Tu Bui
Adrian Bulat
Giedrius T. Burachas
Elena Burceanu
Xavier P. Burgos-Artizzu
Kaylee Burns
Andrei Bursuc
Benjamin Busam
Wonmin Byeon
Zoya Bylinskii
Sergi Caelles
Jianrui Cai
Minjie Cai
Yujun Cai
Zhaowei Cai
Zhipeng Cai
Juan C. Caicedo
Simone Calderara
Necati Cihan Camgoz
Dylan Campbell
Octavia Camps
Jiale Cao
Kaidi Cao
Liangliang Cao
Xiangyong Cao
Xiaochun Cao
Yang Cao
Yu Cao
Yue Cao
Zhangjie Cao
Luca Carlone
Mathilde Caron
Dan Casas
Thomas J. Cashman
Umberto Castellani
Lluis Castrejon
Jacopo Cavazza
Fabio Cermelli
Hakan Cevikalp
Menglei Chai
Ishani Chakraborty
Rudrasis Chakraborty
Antoni B. Chan

Kwok-Ping Chan
Siddhartha Chandra
Sharat Chandran
Arjun Chandrasekaran
Angel X. Chang
Che-Han Chang
Hong Chang
Hyun Sung Chang
Hyung Jin Chang
Jianlong Chang
Ju Yong Chang
Ming-Ching Chang
Simyung Chang
Xiaojun Chang
Yu-Wei Chao
Devendra S. Chaplot
Arslan Chaudhry
Rizwan A. Chaudhry
Can Chen
Chang Chen
Chao Chen
Chen Chen
Chu-Song Chen
Dapeng Chen
Dong Chen
Dongdong Chen
Guanying Chen
Hongge Chen
Hsin-yi Chen
Huaijin Chen
Hwann-Tzong Chen
Jianbo Chen
Jianhui Chen
Jiansheng Chen
Jiaxin Chen
Jie Chen
Jun-Cheng Chen
Kan Chen
Kevin Chen
Lin Chen
Long Chen
Min-Hung Chen
Qifeng Chen
Shi Chen
Shixing Chen
Tianshui Chen
Weifeng Chen
Weikai Chen
Xi Chen
Xiaohan Chen
Xiaozhi Chen
Xilin Chen
Xingyu Chen
Xinlei Chen
Xinyun Chen
Yi-Ting Chen
Yilun Chen
Ying-Cong Chen
Yinpeng Chen
Yiran Chen
Yu Chen
Yu-Sheng Chen
Yuhua Chen
Yun-Chun Chen
Yunpeng Chen
Yuntao Chen
Zhuoyuan Chen
Zitian Chen
Anchieh Cheng
Bowen Cheng
Erkang Cheng
Gong Cheng
Guangliang Cheng
Jingchun Cheng
Jun Cheng
Li Cheng
Ming-Ming Cheng
Yu Cheng
Ziang Cheng
Anoop Cherian
Dmitry Chetverikov
Ngai-man Cheung
William Cheung
Ajad Chhatkuli
Naoki Chiba
Benjamin Chidester
Han-pang Chiu
Mang Tik Chiu
Wei-Chen Chiu
Donghyeon Cho
Hojin Cho
Minsu Cho
Nam Ik Cho
Tim Cho
Tae Eun Choe
Chiho Choi
Edward Choi
Inchang Choi
Jinsoo Choi
Jonghyun Choi
Jongwon Choi
Yukyung Choi
Hisham Cholakkal
Eunji Chong
Jaegul Choo
Christopher Choy
Hang Chu
Peng Chu
Wen-Sheng Chu
Albert Chung
Joon Son Chung
Hai Ci
Safa Cicek
Ramazan G. Cinbis
Arridhana Ciptadi
Javier Civera
James J. Clark
Ronald Clark
Felipe Codevilla
Michael Cogswell
Andrea Cohen
Maxwell D. Collins
Carlo Colombo
Yang Cong
Adria R. Continente
Marcella Cornia
John Richard Corring
Darren Cosker
Dragos Costea
Garrison W. Cottrell
Florent Couzinie-Devy
Marco Cristani
Ioana Croitoru
James L. Crowley
Jiequan Cui
Zhaopeng Cui
Ross Cutler
Antonio D'Innocente

Matthew Fisher
Boris Flach
Corneliu Florea
Wolfgang Foerstner
David Fofi
Gian Luca Foresti
Per-Erik Forssen
David Fouhey
Katerina Fragkiadaki
Victor Fragoso
Jean-Sébastien Franco
Ohad Fried
Iuri Frosio
Cheng-Yang Fu
Huazhu Fu
Jianlong Fu
Jingjing Fu
Xueyang Fu
Yanwei Fu
Ying Fu
Yun Fu
Olac Fuentes
Kent Fujiwara
Takuya Funatomi
Christopher Funk
Thomas Funkhouser
Antonino Furnari
Ryo Furukawa
Erik Gärtner
Raghudeep Gadde
Matheus Gadelha
Vandit Gajjar
Trevor Gale
Juergen Gall
Mathias Gallardo
Guillermo Gallego
Orazio Gallo
Chuang Gan
Zhe Gan
Madan Ravi Ganesh
Aditya Ganeshan
Siddha Ganju
Bin-Bin Gao
Changxin Gao
Feng Gao
Hongchang Gao
Jin Gao
Jiyang Gao
Junbin Gao
Katelyn Gao
Lin Gao
Mingfei Gao
Ruiqi Gao
Ruohan Gao
Shenghua Gao
Yuan Gao
Yue Gao
Noa Garcia
Alberto Garcia-Garcia
Guillermo Garcia-Hernando
Jacob R. Gardner
Animesh Garg
Kshitiz Garg
Rahul Garg
Ravi Garg
Philip N. Garner
Kirill Gavrilyuk
Paul Gay
Shiming Ge
Weifeng Ge
Baris Gecer
Xin Geng
Kyle Genova
Stamatios Georgoulis
Bernard Ghanem
Michael Gharbi
Kamran Ghasedi
Golnaz Ghiasi
Arnab Ghosh
Partha Ghosh
Silvio Giancola
Andrew Gilbert
Rohit Girdhar
Xavier Giro-i-Nieto
Thomas Gittings
Ioannis Gkioulekas
Clement Godard
Vaibhava Goel
Bastian Goldluecke
Lluis Gomez
Nuno Gonçalves
Dong Gong
Ke Gong
Mingming Gong
Abel Gonzalez-Garcia
Ariel Gordon
Daniel Gordon
Paulo Gotardo
Venu Madhav Govindu
Ankit Goyal
Priya Goyal
Raghav Goyal
Benjamin Graham
Douglas Gray
Brent A. Griffin
Etienne Grossmann
David Gu
Jiayuan Gu
Jiuxiang Gu
Lin Gu
Qiao Gu
Shuhang Gu
Jose J. Guerrero
Paul Guerrero
Jie Gui
Jean-Yves Guillemaut
Riza Alp Guler
Erhan Gundogdu
Fatma Guney
Guodong Guo
Kaiwen Guo
Qi Guo
Sheng Guo
Shi Guo
Tiantong Guo
Xiaojie Guo
Yijie Guo
Yiluan Guo
Yuanfang Guo
Yulan Guo
Agrim Gupta
Ankush Gupta
Mohit Gupta
Saurabh Gupta
Tanmay Gupta
Danna Gurari
Abner Guzman-Rivera

JunYoung Gwak
Michael Gygli
Jung-Woo Ha
Simon Hadfield
Isma Hadji
Bjoern Haefner
Taeyoung Hahn
Levente Hajder
Peter Hall
Emanuela Haller
Stefan Haller
Bumsub Ham
Abdullah Hamdi
Dongyoon Han
Hu Han
Jungong Han
Junwei Han
Kai Han
Tian Han
Xiaoguang Han
Xintong Han
Yahong Han
Ankur Handa
Zekun Hao
Albert Haque
Tatsuya Harada
Mehrtash Harandi
Adam W. Harley
Mahmudul Hasan
Atsushi Hashimoto
Ali Hatamizadeh
Munawar Hayat
Dongliang He
Jingrui He
Junfeng He
Kaiming He
Kun He
Lei He
Pan He
Ran He
Shengfeng He
Tong He
Weipeng He
Xuming He
Yang He
Yihui He
Zhihai He
Chinmay Hegde
Janne Heikkila
Mattias P. Heinrich
Stéphane Herbin
Alexander Hermans
Luis Herranz
John R. Hershey
Aaron Hertzmann
Roei Herzig
Anders Heyden
Steven Hickson
Otmar Hilliges
Tomas Hodan
Judy Hoffman
Michael Hofmann
Yannick Hold-Geoffroy
Namdar Homayounfar
Sina Honari
Richang Hong
Seunghoon Hong
Xiaopeng Hong
Yi Hong
Hidekata Hontani
Anthony Hoogs
Yedid Hoshen
Mir Rayat Imtiaz Hossain
Junhui Hou
Le Hou
Lu Hou
Tingbo Hou
Wei-Lin Hsiao
Cheng-Chun Hsu
Gee-Sern Jison Hsu
Kuang-jui Hsu
Changbo Hu
Di Hu
Guosheng Hu
Han Hu
Hao Hu
Hexiang Hu
Hou-Ning Hu
Jie Hu
Junlin Hu
Nan Hu
Ping Hu
Ronghang Hu
Xiaowei Hu
Yinlin Hu
Yuan-Ting Hu
Zhe Hu
Binh-Son Hua
Yang Hua
Bingyao Huang
Di Huang
Dong Huang
Fay Huang
Haibin Huang
Haozhi Huang
Heng Huang
Huaibo Huang
Jia-Bin Huang
Jing Huang
Jingwei Huang
Kaizhu Huang
Lei Huang
Qiangui Huang
Qiaoying Huang
Qingqiu Huang
Qixing Huang
Shaoli Huang
Sheng Huang
Siyuan Huang
Weilin Huang
Wenbing Huang
Xiangru Huang
Xun Huang
Yan Huang
Yifei Huang
Yue Huang
Zhiwu Huang
Zilong Huang
Minyoung Huh
Zhuo Hui
Matthias B. Hullin
Martin Humenberger
Wei-Chih Hung
Zhouyuan Huo
Junhwa Hur
Noureldien Hussein
Jyh-Jing Hwang
Seong Jae Hwang

Sung Ju Hwang
Ichiro Ide
Ivo Ihrke
Daiki Ikami
Satoshi Ikehata
Nazli Ikizler-Cinbis
Sunghoon Im
Yani Ioannou
Radu Tudor Ionescu
Umar Iqbal
Go Irie
Ahmet Iscen
Md Amirul Islam
Vamsi Ithapu
Nathan Jacobs
Arpit Jain
Himalaya Jain
Suyog Jain
Stuart James
Won-Dong Jang
Yunseok Jang
Ronnachai Jaroensri
Dinesh Jayaraman
Sadeep Jayasumana
Suren Jayasuriya
Herve Jegou
Simon Jenni
Hae-Gon Jeon
Yunho Jeon
Koteswar R. Jerripothula
Hueihan Jhuang
I-hong Jhuo
Dinghuang Ji
Hui Ji
Jingwei Ji
Pan Ji
Yanli Ji
Baoxiong Jia
Kui Jia
Xu Jia
Chiyu Max Jiang
Haiyong Jiang
Hao Jiang
Huaizu Jiang
Huajie Jiang
Ke Jiang
Lai Jiang
Li Jiang
Lu Jiang
Ming Jiang
Peng Jiang
Shuqiang Jiang
Wei Jiang
Xudong Jiang
Zhuolin Jiang
Jianbo Jiao
Zequn Jie
Dakai Jin
Kyong Hwan Jin
Lianwen Jin
SouYoung Jin
Xiaojie Jin
Xin Jin
Nebojsa Jojic
Alexis Joly
Michael Jeffrey Jones
Hanbyul Joo
Jungseock Joo
Kyungdon Joo
Ajjen Joshi
Shantanu H. Joshi
Da-Cheng Juan
Marco Körner
Kevin Köser
Asim Kadav
Christine Kaeser-Chen
Kushal Kafle
Dagmar Kainmueller
Ioannis A. Kakadiaris
Zdenek Kalal
Nima Kalantari
Yannis Kalantidis
Mahdi M. Kalayeh
Anmol Kalia
Sinan Kalkan
Vicky Kalogeiton
Ashwin Kalyan
Joni-kristian Kamarainen
Gerda Kamberova
Chandra Kambhamettu
Martin Kampel
Meina Kan
Christopher Kanan
Kenichi Kanatani
Angjoo Kanazawa
Atsushi Kanehira
Takuhiro Kaneko
Asako Kanezaki
Bingyi Kang
Di Kang
Sunghun Kang
Zhao Kang
Vadim Kantorov
Abhishek Kar
Amlan Kar
Theofanis Karaletsos
Leonid Karlinsky
Kevin Karsch
Angelos Katharopoulos
Isinsu Katircioglu
Hiroharu Kato
Zoltan Kato
Dotan Kaufman
Jan Kautz
Rei Kawakami
Qiuhong Ke
Wadim Kehl
Petr Kellnhofer
Aniruddha Kembhavi
Cem Keskin
Margret Keuper
Daniel Keysers
Ashkan Khakzar
Fahad Khan
Naeemullah Khan
Salman Khan
Siddhesh Khandelwal
Rawal Khirodkar
Anna Khoreva
Tejas Khot
Parmeshwar Khurd
Hadi Kiapour
Joe Kileel
Chanho Kim
Dahun Kim
Edward Kim
Eunwoo Kim
Han-ul Kim

Hansung Kim
Heewon Kim
Hyo Jin Kim
Hyunwoo J. Kim
Jinkyu Kim
Jiwon Kim
Jongmin Kim
Junsik Kim
Junyeong Kim
Min H. Kim
Namil Kim
Pyojin Kim
Seon Joo Kim
Seong Tae Kim
Seungryong Kim
Sungwoong Kim
Tae Hyun Kim
Vladimir Kim
Won Hwa Kim
Yonghyun Kim
Benjamin Kimia
Akisato Kimura
Pieter-Jan Kindermans
Zsolt Kira
Itaru Kitahara
Hedvig Kjellstrom
Jan Knopp
Takumi Kobayashi
Erich Kobler
Parker Koch
Reinhard Koch
Elyor Kodirov
Amir Kolaman
Nicholas Kolkin
Dimitrios Kollias
Stefanos Kollias
Soheil Kolouri
Adams Wai-Kin Kong
Naejin Kong
Shu Kong
Tao Kong
Yu Kong
Yoshinori Konishi
Daniil Kononenko
Theodora Kontogianni
Simon Korman
Adam Kortylewski
Jana Kosecka
Jean Kossaifi
Satwik Kottur
Rigas Kouskouridas
Adriana Kovashka
Rama Kovvuri
Adarsh Kowdle
Jedrzej Kozerawski
Mateusz Kozinski
Philipp Kraehenbuehl
Gregory Kramida
Josip Krapac
Dmitry Kravchenko
Ranjay Krishna
Pavel Krsek
Alexander Krull
Jakob Kruse
Hiroyuki Kubo
Hilde Kuehne
Jason Kuen
Andreas Kuhn
Arjan Kuijper
Zuzana Kukelova
Ajay Kumar
Amit Kumar
Avinash Kumar
Suryansh Kumar
Vijay Kumar
Kaustav Kundu
Weicheng Kuo
Nojun Kwak
Suha Kwak
Junseok Kwon
Nikolaos Kyriazis
Zorah Lähner
Ankit Laddha
Florent Lafarge
Jean Lahoud
Kevin Lai
Shang-Hong Lai
Wei-Sheng Lai
Yu-Kun Lai
Iro Laina
Antony Lam
John Wheatley Lambert
Xiangyuan lan
Xu Lan
Charis Lanaras
Georg Langs
Oswald Lanz
Dong Lao
Yizhen Lao
Agata Lapedriza
Gustav Larsson
Viktor Larsson
Katrin Lasinger
Christoph Lassner
Longin Jan Latecki
Stéphane Lathuilière
Rynson Lau
Hei Law
Justin Lazarow
Svetlana Lazebnik
Hieu Le
Huu Le
Ngan Hoang Le
Trung-Nghia Le
Vuong Le
Colin Lea
Erik Learned-Miller
Chen-Yu Lee
Gim Hee Lee
Hsin-Ying Lee
Hyungtae Lee
Jae-Han Lee
Jimmy Addison Lee
Joonseok Lee
Kibok Lee
Kuang-Huei Lee
Kwonjoon Lee
Minsik Lee
Sang-chul Lee
Seungkyu Lee
Soochan Lee
Stefan Lee
Taehee Lee
Andreas Lehrmann
Jie Lei
Peng Lei
Matthew Joseph Leotta
Wee Kheng Leow

Gil Levi
Evgeny Levinkov
Aviad Levis
Jose Lezama
Ang Li
Bin Li
Bing Li
Boyi Li
Changsheng Li
Chao Li
Chen Li
Cheng Li
Chenglong Li
Chi Li
Chun-Guang Li
Chun-Liang Li
Chunyuan Li
Dong Li
Guanbin Li
Hao Li
Haoxiang Li
Hongsheng Li
Hongyang Li
Houqiang Li
Huibin Li
Jia Li
Jianan Li
Jianguo Li
Junnan Li
Junxuan Li
Kai Li
Ke Li
Kejie Li
Kunpeng Li
Lerenhan Li
Li Erran Li
Mengtian Li
Mu Li
Peihua Li
Peiyi Li
Ping Li
Qi Li
Qing Li
Ruiyu Li
Ruoteng Li
Shaozi Li
Sheng Li
Shiwei Li
Shuang Li
Siyang Li
Stan Z. Li
Tianye Li
Wei Li
Weixin Li
Wen Li
Wenbo Li
Xiaomeng Li
Xin Li
Xiu Li
Xuelong Li
Xueting Li
Yan Li
Yandong Li
Yanghao Li
Yehao Li
Yi Li
Yijun Li
Yikang LI
Yining Li
Yongjie Li
Yu Li
Yu-Jhe Li
Yunpeng Li
Yunsheng Li
Yunzhu Li
Zhe Li
Zhen Li
Zhengqi Li
Zhenyang Li
Zhuwen Li
Dongze Lian
Xiaochen Lian
Zhouhui Lian
Chen Liang
Jie Liang
Ming Liang
Paul Pu Liang
Pengpeng Liang
Shu Liang
Wei Liang
Jing Liao
Minghui Liao
Renjie Liao
Shengcai Liao
Shuai Liao
Yiyi Liao
Ser-Nam Lim
Chen-Hsuan Lin
Chung-Ching Lin
Dahua Lin
Ji Lin
Kevin Lin
Tianwei Lin
Tsung-Yi Lin
Tsung-Yu Lin
Wei-An Lin
Weiyao Lin
Yen-Chen Lin
Yuewei Lin
David B. Lindell
Drew Linsley
Krzysztof Lis
Roee Litman
Jim Little
An-An Liu
Bo Liu
Buyu Liu
Chao Liu
Chen Liu
Cheng-lin Liu
Chenxi Liu
Dong Liu
Feng Liu
Guilin Liu
Haomiao Liu
Heshan Liu
Hong Liu
Ji Liu
Jingen Liu
Jun Liu
Lanlan Liu
Li Liu
Liu Liu
Mengyuan Liu
Miaomiao Liu
Nian Liu
Ping Liu
Risheng Liu

Sheng Liu
Shu Liu
Shuaicheng Liu
Sifei Liu
Siqi Liu
Siying Liu
Songtao Liu
Ting Liu
Tongliang Liu
Tyng-Luh Liu
Wanquan Liu
Wei Liu
Weiyang Liu
Weizhe Liu
Wenyu Liu
Wu Liu
Xialei Liu
Xianglong Liu
Xiaodong Liu
Xiaofeng Liu
Xihui Liu
Xingyu Liu
Xinwang Liu
Xuanqing Liu
Xuebo Liu
Yang Liu
Yaojie Liu
Yebin Liu
Yen-Cheng Liu
Yiming Liu
Yu Liu
Yu-Shen Liu
Yufan Liu
Yun Liu
Zheng Liu
Zhijian Liu
Zhuang Liu
Zichuan Liu
Ziwei Liu
Zongyi Liu
Stephan Liwicki
Liliana Lo Presti
Chengjiang Long
Fuchen Long
Mingsheng Long
Xiang Long
Yang Long
Charles T. Loop
Antonio Lopez
Roberto J. Lopez-Sastre
Javier Lorenzo-Navarro
Manolis Lourakis
Boyu Lu
Canyi Lu
Feng Lu
Guoyu Lu
Hongtao Lu
Jiajun Lu
Jiasen Lu
Jiwen Lu
Kaiyue Lu
Le Lu
Shao-Ping Lu
Shijian Lu
Xiankai Lu
Xin Lu
Yao Lu
Yiping Lu
Yongxi Lu
Yongyi Lu
Zhiwu Lu
Fujun Luan
Benjamin E. Lundell
Hao Luo
Jian-Hao Luo
Ruotian Luo
Weixin Luo
Wenhan Luo
Wenjie Luo
Yan Luo
Zelun Luo
Zixin Luo
Khoa Luu
Zhaoyang Lv
Pengyuan Lyu
Thomas Möllenhoff
Matthias Müller
Bingpeng Ma
Chih-Yao Ma
Chongyang Ma
Huimin Ma
Jiayi Ma
K. T. Ma
Ke Ma
Lin Ma
Liqian Ma
Shugao Ma
Wei-Chiu Ma
Xiaojian Ma
Xingjun Ma
Zhanyu Ma
Zheng Ma
Radek Jakob Mackowiak
Ludovic Magerand
Shweta Mahajan
Siddharth Mahendran
Long Mai
Ameesh Makadia
Oscar Mendez Maldonado
Mateusz Malinowski
Yury Malkov
Arun Mallya
Dipu Manandhar
Massimiliano Mancini
Fabian Manhardt
Kevis-kokitsi Maninis
Varun Manjunatha
Junhua Mao
Xudong Mao
Alina Marcu
Edgar Margffoy-Tuay
Dmitrii Marin
Manuel J. Marin-Jimenez
Kenneth Marino
Niki Martinel
Julieta Martinez
Jonathan Masci
Tomohiro Mashita
Iacopo Masi
David Masip
Daniela Massiceti
Stefan Mathe
Yusuke Matsui
Tetsu Matsukawa
Iain A. Matthews
Kevin James Matzen
Bruce Allen Maxwell
Stephen Maybank

Helmut Mayer
Amir Mazaheri
David McAllester
Steven McDonagh
Stephen J. Mckenna
Roey Mechrez
Prakhar Mehrotra
Christopher Mei
Xue Mei
Paulo R. S. Mendonca
Lili Meng
Zibo Meng
Thomas Mensink
Bjoern Menze
Michele Merler
Kourosh Meshgi
Pascal Mettes
Christopher Metzler
Liang Mi
Qiguang Miao
Xin Miao
Tomer Michaeli
Frank Michel
Antoine Miech
Krystian Mikolajczyk
Peyman Milanfar
Ben Mildenhall
Gregor Miller
Fausto Milletari
Dongbo Min
Kyle Min
Pedro Miraldo
Dmytro Mishkin
Anand Mishra
Ashish Mishra
Ishan Misra
Niluthpol C. Mithun
Kaushik Mitra
Niloy Mitra
Anton Mitrokhin
Ikuhisa Mitsugami
Anurag Mittal
Kaichun Mo
Zhipeng Mo
Davide Modolo
Michael Moeller
Pritish Mohapatra
Pavlo Molchanov
Davide Moltisanti
Pascal Monasse
Mathew Monfort
Aron Monszpart
Sean Moran
Vlad I. Morariu
Francesc Moreno-Noguer
Pietro Morerio
Stylianos Moschoglou
Yael Moses
Roozbeh Mottaghi
Pierre Moulon
Arsalan Mousavian
Yadong Mu
Yasuhiro Mukaigawa
Lopamudra Mukherjee
Yusuke Mukuta
Ravi Teja Mullapudi
Mario Enrique Munich
Zachary Murez
Ana C. Murillo
J. Krishna Murthy
Damien Muselet
Armin Mustafa
Siva Karthik Mustikovela
Carlo Dal Mutto
Moin Nabi
Varun K. Nagaraja
Tushar Nagarajan
Arsha Nagrani
Seungjun Nah
Nikhil Naik
Yoshikatsu Nakajima
Yuta Nakashima
Atsushi Nakazawa
Seonghyeon Nam
Vinay P. Namboodiri
Medhini Narasimhan
Srinivasa Narasimhan
Sanath Narayan
Erickson Rangel Nascimento
Jacinto Nascimento
Tayyab Naseer
Lakshmanan Nataraj
Neda Nategh
Nelson Isao Nauata
Fernando Navarro
Shah Nawaz
Lukas Neumann
Ram Nevatia
Alejandro Newell
Shawn Newsam
Joe Yue-Hei Ng
Trung Thanh Ngo
Duc Thanh Nguyen
Lam M. Nguyen
Phuc Xuan Nguyen
Thuong Nguyen Canh
Mihalis Nicolaou
Andrei Liviu Nicolicioiu
Xuecheng Nie
Michael Niemeyer
Simon Niklaus
Christophoros Nikou
David Nilsson
Jifeng Ning
Yuval Nirkin
Li Niu
Yuzhen Niu
Zhenxing Niu
Shohei Nobuhara
Nicoletta Noceti
Hyeonwoo Noh
Junhyug Noh
Mehdi Noroozi
Sotiris Nousias
Valsamis Ntouskos
Matthew O'Toole
Peter Ochs
Ferda Ofli
Seong Joon Oh
Seoung Wug Oh
Iason Oikonomidis
Utkarsh Ojha
Takahiro Okabe
Takayuki Okatani
Fumio Okura
Aude Oliva
Kyle Olszewski

Björn Ommer
Mohamed Omran
Elisabeta Oneata
Michael Opitz
Jose Oramas
Tribhuvanesh Orekondy
Shaul Oron
Sergio Orts-Escolano
Ivan Oseledets
Aljosa Osep
Magnus Oskarsson
Anton Osokin
Martin R. Oswald
Wanli Ouyang
Andrew Owens
Mete Ozay
Mustafa Ozuysal
Eduardo Pérez-Pellitero
Gautam Pai
Dipan Kumar Pal
P. H. Pamplona Savarese
Jinshan Pan
Junting Pan
Xingang Pan
Yingwei Pan
Yannis Panagakis
Rameswar Panda
Guan Pang
Jiahao Pang
Jiangmiao Pang
Tianyu Pang
Sharath Pankanti
Nicolas Papadakis
Dim Papadopoulos
George Papandreou
Toufiq Parag
Shaifali Parashar
Sarah Parisot
Eunhyeok Park
Hyun Soo Park
Jaesik Park
Min-Gyu Park
Taesung Park
Alvaro Parra
C. Alejandro Parraga
Despoina Paschalidou
Nikolaos Passalis
Vishal Patel
Viorica Patraucean
Badri Narayana Patro
Danda Pani Paudel
Sujoy Paul
Georgios Pavlakos
Ioannis Pavlidis
Vladimir Pavlovic
Nick Pears
Kim Steenstrup Pedersen
Selen Pehlivan
Shmuel Peleg
Chao Peng
Houwen Peng
Wen-Hsiao Peng
Xi Peng
Xiaojiang Peng
Xingchao Peng
Yuxin Peng
Federico Perazzi
Juan Camilo Perez
Vishwanath Peri
Federico Pernici
Luca Del Pero
Florent Perronnin
Stavros Petridis
Henning Petzka
Patrick Peursum
Michael Pfeiffer
Hanspeter Pfister
Roman Pflugfelder
Minh Tri Pham
Yongri Piao
David Picard
Tomasz Pieciak
A. J. Piergiovanni
Andrea Pilzer
Pedro O. Pinheiro
Silvia Laura Pintea
Lerrel Pinto
Axel Pinz
Robinson Piramuthu
Fiora Pirri
Leonid Pishchulin
Francesco Pittaluga
Daniel Pizarro
Tobias Plötz
Mirco Planamente
Matteo Poggi
Moacir A. Ponti
Parita Pooj
Fatih Porikli
Horst Possegger
Omid Poursaeed
Ameya Prabhu
Viraj Uday Prabhu
Dilip Prasad
Brian L. Price
True Price
Maria Priisalu
Veronique Prinet
Victor Adrian Prisacariu
Jan Prokaj
Sergey Prokudin
Nicolas Pugeault
Xavier Puig
Albert Pumarola
Pulak Purkait
Senthil Purushwalkam
Charles R. Qi
Hang Qi
Haozhi Qi
Lu Qi
Mengshi Qi
Siyuan Qi
Xiaojuan Qi
Yuankai Qi
Shengju Qian
Xuelin Qian
Siyuan Qiao
Yu Qiao
Jie Qin
Qiang Qiu
Weichao Qiu
Zhaofan Qiu
Kha Gia Quach
Yuhui Quan
Yvain Queau
Julian Quiroga
Faisal Qureshi
Mahdi Rad

Filip Radenovic
Petia Radeva
Venkatesh B. Radhakrishnan
Ilija Radosavovic
Noha Radwan
Rahul Raguram
Tanzila Rahman
Amit Raj
Ajit Rajwade
Kandan Ramakrishnan
Santhosh K. Ramakrishnan
Srikumar Ramalingam
Ravi Ramamoorthi
Vasili Ramanishka
Ramprasaath R. Selvaraju
Francois Rameau
Visvanathan Ramesh
Santu Rana
Rene Ranftl
Anand Rangarajan
Anurag Ranjan
Viresh Ranjan
Yongming Rao
Carolina Raposo
Vivek Rathod
Sathya N. Ravi
Avinash Ravichandran
Tammy Riklin Raviv
Daniel Rebain
Sylvestre-Alvise Rebuffi
N. Dinesh Reddy
Timo Rehfeld
Paolo Remagnino
Konstantinos Rematas
Edoardo Remelli
Dongwei Ren
Haibing Ren
Jian Ren
Jimmy Ren
Mengye Ren
Weihong Ren
Wenqi Ren
Zhile Ren
Zhongzheng Ren
Zhou Ren
Vijay Rengarajan
Md A. Reza
Farzaneh Rezaeianaran
Hamed R. Tavakoli
Nicholas Rhinehart
Helge Rhodin
Elisa Ricci
Alexander Richard
Eitan Richardson
Elad Richardson
Christian Richardt
Stephan Richter
Gernot Riegler
Daniel Ritchie
Tobias Ritschel
Samuel Rivera
Yong Man Ro
Richard Roberts
Joseph Robinson
Ignacio Rocco
Mrigank Rochan
Emanuele Rodolà
Mikel D. Rodriguez
Giorgio Roffo
Grégory Rogez
Gemma Roig
Javier Romero
Xuejian Rong
Yu Rong
Amir Rosenfeld
Bodo Rosenhahn
Guy Rosman
Arun Ross
Paolo Rota
Peter M. Roth
Anastasios Roussos
Anirban Roy
Sebastien Roy
Aruni RoyChowdhury
Artem Rozantsev
Ognjen Rudovic
Daniel Rueckert
Adria Ruiz
Javier Ruiz-del-solar
Christian Rupprecht
Chris Russell
Dan Ruta
Jongbin Ryu
Ömer Sümer
Alexandre Sablayrolles
Faraz Saeedan
Ryusuke Sagawa
Christos Sagonas
Tonmoy Saikia
Hideo Saito
Kuniaki Saito
Shunsuke Saito
Shunta Saito
Ken Sakurada
Joaquin Salas
Fatemeh Sadat Saleh
Mahdi Saleh
Pouya Samangouei
Leo Sampaio Ferraz Ribeiro
Artsiom Olegovich Sanakoyeu
Enrique Sanchez
Patsorn Sangkloy
Anush Sankaran
Aswin Sankaranarayanan
Swami Sankaranarayanan
Rodrigo Santa Cruz
Amartya Sanyal
Archana Sapkota
Nikolaos Sarafianos
Jun Sato
Shin'ichi Satoh
Hosnieh Sattar
Arman Savran
Manolis Savva
Alexander Sax
Hanno Scharr
Simone Schaub-Meyer
Konrad Schindler
Dmitrij Schlesinger
Uwe Schmidt
Dirk Schnieders
Björn Schuller
Samuel Schulter
Idan Schwartz

William Robson Schwartz
Alex Schwing
Sinisa Segvic
Lorenzo Seidenari
Pradeep Sen
Ozan Sener
Soumyadip Sengupta
Arda Senocak
Mojtaba Seyedhosseini
Shishir Shah
Shital Shah
Sohil Atul Shah
Tamar Rott Shaham
Huasong Shan
Qi Shan
Shiguang Shan
Jing Shao
Roman Shapovalov
Gaurav Sharma
Vivek Sharma
Viktoriia Sharmanska
Dongyu She
Sumit Shekhar
Evan Shelhamer
Chengyao Shen
Chunhua Shen
Falong Shen
Jie Shen
Li Shen
Liyue Shen
Shuhan Shen
Tianwei Shen
Wei Shen
William B. Shen
Yantao Shen
Ying Shen
Yiru Shen
Yujun Shen
Yuming Shen
Zhiqiang Shen
Ziyi Shen
Lu Sheng
Yu Sheng
Rakshith Shetty
Baoguang Shi
Guangming Shi
Hailin Shi
Miaojing Shi
Yemin Shi
Zhenmei Shi
Zhiyuan Shi
Kevin Jonathan Shih
Shiliang Shiliang
Hyunjung Shim
Atsushi Shimada
Nobutaka Shimada
Daeyun Shin
Young Min Shin
Koichi Shinoda
Konstantin Shmelkov
Michael Zheng Shou
Abhinav Shrivastava
Tianmin Shu
Zhixin Shu
Hong-Han Shuai
Pushkar Shukla
Christian Siagian
Mennatullah M. Siam
Kaleem Siddiqi
Karan Sikka
Jae-Young Sim
Christian Simon
Martin Simonovsky
Dheeraj Singaraju
Bharat Singh
Gurkirt Singh
Krishna Kumar Singh
Maneesh Kumar Singh
Richa Singh
Saurabh Singh
Suriya Singh
Vikas Singh
Sudipta N. Sinha
Vincent Sitzmann
Josef Sivic
Gregory Slabaugh
Miroslava Slavcheva
Ron Slossberg
Brandon Smith
Kevin Smith
Vladimir Smutny
Noah Snavely
Roger D. Soberanis-Mukul
Kihyuk Sohn
Francesco Solera
Eric Sommerlade
Sanghyun Son
Byung Cheol Song
Chunfeng Song
Dongjin Song
Jiaming Song
Jie Song
Jifei Song
Jingkuan Song
Mingli Song
Shiyu Song
Shuran Song
Xiao Song
Yafei Song
Yale Song
Yang Song
Yi-Zhe Song
Yibing Song
Humberto Sossa
Cesar de Souza
Adrian Spurr
Srinath Sridhar
Suraj Srinivas
Pratul P. Srinivasan
Anuj Srivastava
Tania Stathaki
Christopher Stauffer
Simon Stent
Rainer Stiefelhagen
Pierre Stock
Julian Straub
Jonathan C. Stroud
Joerg Stueckler
Jan Stuehmer
David Stutz
Chi Su
Hang Su
Jong-Chyi Su
Shuochen Su
Yu-Chuan Su
Ramanathan Subramanian
Yusuke Sugano

Masanori Suganuma
Yumin Suh
Mohammed Suhail
Yao Sui
Heung-Il Suk
Josephine Sullivan
Baochen Sun
Chen Sun
Chong Sun
Deqing Sun
Jin Sun
Liang Sun
Lin Sun
Qianru Sun
Shao-Hua Sun
Shuyang Sun
Weiwei Sun
Wenxiu Sun
Xiaoshuai Sun
Xiaoxiao Sun
Xingyuan Sun
Yifan Sun
Zhun Sun
Sabine Susstrunk
David Suter
Supasorn Suwajanakorn
Tomas Svoboda
Eran Swears
Paul Swoboda
Attila Szabo
Richard Szeliski
Duy-Nguyen Ta
Andrea Tagliasacchi
Yuichi Taguchi
Ying Tai
Keita Takahashi
Kouske Takahashi
Jun Takamatsu
Hugues Talbot
Toru Tamaki
Chaowei Tan
Fuwen Tan
Mingkui Tan
Mingxing Tan
Qingyang Tan
Robby T. Tan
Xiaoyang Tan
Kenichiro Tanaka
Masayuki Tanaka
Chang Tang
Chengzhou Tang
Danhang Tang
Ming Tang
Peng Tang
Qingming Tang
Wei Tang
Xu Tang
Yansong Tang
Youbao Tang
Yuxing Tang
Zhiqiang Tang
Tatsunori Taniai
Junli Tao
Xin Tao
Makarand Tapaswi
Jean-Philippe Tarel
Lyne Tchapmi
Zachary Teed
Bugra Tekin
Damien Teney
Ayush Tewari
Christian Theobalt
Christopher Thomas
Diego Thomas
Jim Thomas
Rajat Mani Thomas
Xinmei Tian
Yapeng Tian
Yingli Tian
Yonglong Tian
Zhi Tian
Zhuotao Tian
Kinh Tieu
Joseph Tighe
Massimo Tistarelli
Matthew Toews
Carl Toft
Pavel Tokmakov
Federico Tombari
Chetan Tonde
Yan Tong
Alessio Tonioni
Andrea Torsello
Fabio Tosi
Du Tran
Luan Tran
Ngoc-Trung Tran
Quan Hung Tran
Truyen Tran
Rudolph Triebel
Martin Trimmel
Shashank Tripathi
Subarna Tripathi
Leonardo Trujillo
Eduard Trulls
Tomasz Trzcinski
Sam Tsai
Yi-Hsuan Tsai
Hung-Yu Tseng
Stavros Tsogkas
Aggeliki Tsoli
Devis Tuia
Shubham Tulsiani
Sergey Tulyakov
Frederick Tung
Tony Tung
Daniyar Turmukhambetov
Ambrish Tyagi
Radim Tylecek
Christos Tzelepis
Georgios Tzimiropoulos
Dimitrios Tzionas
Seiichi Uchida
Norimichi Ukita
Dmitry Ulyanov
Martin Urschler
Yoshitaka Ushiku
Ben Usman
Alexander Vakhitov
Julien P. C. Valentin
Jack Valmadre
Ernest Valveny
Joost van de Weijer
Jan van Gemert
Koen Van Leemput
Gul Varol
Sebastiano Vascon
M. Alex O. Vasilescu

Subeesh Vasu
Mayank Vatsa
David Vazquez
Javier Vazquez-Corral
Ashok Veeraraghavan
Erik Velasco-Salido
Raviteja Vemulapalli
Jonathan Ventura
Manisha Verma
Roberto Vezzani
Ruben Villegas
Minh Vo
MinhDuc Vo
Nam Vo
Michele Volpi
Riccardo Volpi
Carl Vondrick
Konstantinos Vougioukas
Tuan-Hung Vu
Sven Wachsmuth
Neal Wadhwa
Catherine Wah
Jacob C. Walker
Thomas S. A. Wallis
Chengde Wan
Jun Wan
Liang Wan
Renjie Wan
Baoyuan Wang
Boyu Wang
Cheng Wang
Chu Wang
Chuan Wang
Chunyu Wang
Dequan Wang
Di Wang
Dilin Wang
Dong Wang
Fang Wang
Guanzhi Wang
Guoyin Wang
Hanzi Wang
Hao Wang
He Wang
Heng Wang
Hongcheng Wang
Hongxing Wang
Hua Wang
Jian Wang
Jingbo Wang
Jinglu Wang
Jingya Wang
Jinjun Wang
Jinqiao Wang
Jue Wang
Ke Wang
Keze Wang
Le Wang
Lei Wang
Lezi Wang
Li Wang
Liang Wang
Lijun Wang
Limin Wang
Linwei Wang
Lizhi Wang
Mengjiao Wang
Mingzhe Wang
Minsi Wang
Naiyan Wang
Nannan Wang
Ning Wang
Oliver Wang
Pei Wang
Peng Wang
Pichao Wang
Qi Wang
Qian Wang
Qiaosong Wang
Qifei Wang
Qilong Wang
Qing Wang
Qingzhong Wang
Quan Wang
Rui Wang
Ruiping Wang
Ruixing Wang
Shangfei Wang
Shenlong Wang
Shiyao Wang
Shuhui Wang
Song Wang
Tao Wang
Tianlu Wang
Tiantian Wang
Ting-chun Wang
Tingwu Wang
Wei Wang
Weiyue Wang
Wenguan Wang
Wenlin Wang
Wenqi Wang
Xiang Wang
Xiaobo Wang
Xiaofang Wang
Xiaoling Wang
Xiaolong Wang
Xiaosong Wang
Xiaoyu Wang
Xin Eric Wang
Xinchao Wang
Xinggang Wang
Xintao Wang
Yali Wang
Yan Wang
Yang Wang
Yangang Wang
Yaxing Wang
Yi Wang
Yida Wang
Yilin Wang
Yiming Wang
Yisen Wang
Yongtao Wang
Yu-Xiong Wang
Yue Wang
Yujiang Wang
Yunbo Wang
Yunhe Wang
Zengmao Wang
Zhangyang Wang
Zhaowen Wang
Zhe Wang
Zhecan Wang
Zheng Wang
Zhixiang Wang
Zilei Wang
Jianqiao Wangni

Anne S. Wannenwetsch
Jan Dirk Wegner
Scott Wehrwein
Donglai Wei
Kaixuan Wei
Longhui Wei
Pengxu Wei
Ping Wei
Qi Wei
Shih-En Wei
Xing Wei
Yunchao Wei
Zijun Wei
Jerod Weinman
Michael Weinmann
Philippe Weinzaepfel
Yair Weiss
Bihan Wen
Longyin Wen
Wei Wen
Junwu Weng
Tsui-Wei Weng
Xinshuo Weng
Eric Wengrowski
Tomas Werner
Gordon Wetzstein
Tobias Weyand
Patrick Wieschollek
Maggie Wigness
Erik Wijmans
Richard Wildes
Olivia Wiles
Chris Williams
Williem Williem
Kyle Wilson
Calden Wloka
Nicolai Wojke
Christian Wolf
Yongkang Wong
Sanghyun Woo
Scott Workman
Baoyuan Wu
Bichen Wu
Chao-Yuan Wu
Huikai Wu
Jiajun Wu
Jialin Wu
Jiaxiang Wu
Jiqing Wu
Jonathan Wu
Lifang Wu
Qi Wu
Qiang Wu
Ruizheng Wu
Shangzhe Wu
Shun-Cheng Wu
Tianfu Wu
Wayne Wu
Wenxuan Wu
Xiao Wu
Xiaohe Wu
Xinxiao Wu
Yang Wu
Yi Wu
Yiming Wu
Ying Nian Wu
Yue Wu
Zheng Wu
Zhenyu Wu
Zhirong Wu
Zuxuan Wu
Stefanie Wuhrer
Jonas Wulff
Changqun Xia
Fangting Xia
Fei Xia
Gui-Song Xia
Lu Xia
Xide Xia
Yin Xia
Yingce Xia
Yongqin Xian
Lei Xiang
Shiming Xiang
Bin Xiao
Fanyi Xiao
Guobao Xiao
Huaxin Xiao
Taihong Xiao
Tete Xiao
Tong Xiao
Wang Xiao
Yang Xiao
Cihang Xie
Guosen Xie
Jianwen Xie
Lingxi Xie
Sirui Xie
Weidi Xie
Wenxuan Xie
Xiaohua Xie
Fuyong Xing
Jun Xing
Junliang Xing
Bo Xiong
Peixi Xiong
Yu Xiong
Yuanjun Xiong
Zhiwei Xiong
Chang Xu
Chenliang Xu
Dan Xu
Danfei Xu
Hang Xu
Hongteng Xu
Huijuan Xu
Jingwei Xu
Jun Xu
Kai Xu
Mengmeng Xu
Mingze Xu
Qianqian Xu
Ran Xu
Weijian Xu
Xiangyu Xu
Xiaogang Xu
Xing Xu
Xun Xu
Yanyu Xu
Yichao Xu
Yong Xu
Yongchao Xu
Yuanlu Xu
Zenglin Xu
Zheng Xu
Chuhui Xue
Jia Xue
Nan Xue

Tianfan Xue
Xiangyang Xue
Abhay Yadav
Yasushi Yagi
I. Zeki Yalniz
Kota Yamaguchi
Toshihiko Yamasaki
Takayoshi Yamashita
Junchi Yan
Ke Yan
Qingan Yan
Sijie Yan
Xinchen Yan
Yan Yan
Yichao Yan
Zhicheng Yan
Keiji Yanai
Bin Yang
Ceyuan Yang
Dawei Yang
Dong Yang
Fan Yang
Guandao Yang
Guorun Yang
Haichuan Yang
Hao Yang
Jianwei Yang
Jiaolong Yang
Jie Yang
Jing Yang
Kaiyu Yang
Linjie Yang
Meng Yang
Michael Ying Yang
Nan Yang
Shuai Yang
Shuo Yang
Tianyu Yang
Tien-Ju Yang
Tsun-Yi Yang
Wei Yang
Wenhan Yang
Xiao Yang
Xiaodong Yang
Xin Yang
Yan Yang
Yanchao Yang
Yee Hong Yang
Yezhou Yang
Zhenheng Yang
Anbang Yao
Angela Yao
Cong Yao
Jian Yao
Li Yao
Ting Yao
Yao Yao
Zhewei Yao
Chengxi Ye
Jianbo Ye
Keren Ye
Linwei Ye
Mang Ye
Mao Ye
Qi Ye
Qixiang Ye
Mei-Chen Yeh
Raymond Yeh
Yu-Ying Yeh
Sai-Kit Yeung
Serena Yeung
Kwang Moo Yi
Li Yi
Renjiao Yi
Alper Yilmaz
Junho Yim
Lijun Yin
Weidong Yin
Xi Yin
Zhichao Yin
Tatsuya Yokota
Ryo Yonetani
Donggeun Yoo
Jae Shin Yoon
Ju Hong Yoon
Sung-eui Yoon
Laurent Younes
Changqian Yu
Fisher Yu
Gang Yu
Jiahui Yu
Kaicheng Yu
Ke Yu
Lequan Yu
Ning Yu
Qian Yu
Ronald Yu
Ruichi Yu
Shoou-I Yu
Tao Yu
Tianshu Yu
Xiang Yu
Xin Yu
Xiyu Yu
Youngjae Yu
Yu Yu
Zhiding Yu
Chunfeng Yuan
Ganzhao Yuan
Jinwei Yuan
Lu Yuan
Quan Yuan
Shanxin Yuan
Tongtong Yuan
Wenjia Yuan
Ye Yuan
Yuan Yuan
Yuhui Yuan
Huanjing Yue
Xiangyu Yue
Ersin Yumer
Sergey Zagoruyko
Egor Zakharov
Amir Zamir
Andrei Zanfir
Mihai Zanfir
Pablo Zegers
Bernhard Zeisl
John S. Zelek
Niclas Zeller
Huayi Zeng
Jiabei Zeng
Wenjun Zeng
Yu Zeng
Xiaohua Zhai
Fangneng Zhan
Huangying Zhan
Kun Zhan

Xiaohang Zhan
Baochang Zhang
Bowen Zhang
Cecilia Zhang
Changqing Zhang
Chao Zhang
Chengquan Zhang
Chi Zhang
Chongyang Zhang
Dingwen Zhang
Dong Zhang
Feihu Zhang
Hang Zhang
Hanwang Zhang
Hao Zhang
He Zhang
Hongguang Zhang
Hua Zhang
Ji Zhang
Jianguo Zhang
Jianming Zhang
Jiawei Zhang
Jie Zhang
Jing Zhang
Juyong Zhang
Kai Zhang
Kaipeng Zhang
Ke Zhang
Le Zhang
Lei Zhang
Li Zhang
Lihe Zhang
Linguang Zhang
Lu Zhang
Mi Zhang
Mingda Zhang
Peng Zhang
Pingping Zhang
Qian Zhang
Qilin Zhang
Quanshi Zhang
Richard Zhang
Rui Zhang
Runze Zhang
Shengping Zhang
Shifeng Zhang
Shuai Zhang
Songyang Zhang
Tao Zhang
Ting Zhang
Tong Zhang
Wayne Zhang
Wei Zhang
Weizhong Zhang
Wenwei Zhang
Xiangyu Zhang
Xiaolin Zhang
Xiaopeng Zhang
Xiaoqin Zhang
Xiuming Zhang
Ya Zhang
Yang Zhang
Yimin Zhang
Yinda Zhang
Ying Zhang
Yongfei Zhang
Yu Zhang
Yulun Zhang
Yunhua Zhang
Yuting Zhang
Zhanpeng Zhang
Zhao Zhang
Zhaoxiang Zhang
Zhen Zhang
Zheng Zhang
Zhifei Zhang
Zhijin Zhang
Zhishuai Zhang
Ziming Zhang
Bo Zhao
Chen Zhao
Fang Zhao
Haiyu Zhao
Han Zhao
Hang Zhao
Hengshuang Zhao
Jian Zhao
Kai Zhao
Liang Zhao
Long Zhao
Qian Zhao
Qibin Zhao
Qijun Zhao
Rui Zhao
Shenglin Zhao
Sicheng Zhao
Tianyi Zhao
Wenda Zhao
Xiangyun Zhao
Xin Zhao
Yang Zhao
Yue Zhao
Zhichen Zhao
Zijing Zhao
Xiantong Zhen
Chuanxia Zheng
Feng Zheng
Haiyong Zheng
Jia Zheng
Kang Zheng
Shuai Kyle Zheng
Wei-Shi Zheng
Yinqiang Zheng
Zerong Zheng
Zhedong Zheng
Zilong Zheng
Bineng Zhong
Fangwei Zhong
Guangyu Zhong
Yiran Zhong
Yujie Zhong
Zhun Zhong
Chunluan Zhou
Huiyu Zhou
Jiahuan Zhou
Jun Zhou
Lei Zhou
Luowei Zhou
Luping Zhou
Mo Zhou
Ning Zhou
Pan Zhou
Peng Zhou
Qianyi Zhou
S. Kevin Zhou
Sanping Zhou
Wengang Zhou
Xingyi Zhou

Yanzhao Zhou
Yi Zhou
Yin Zhou
Yipin Zhou
Yuyin Zhou
Zihan Zhou
Alex Zihao Zhu
Chenchen Zhu
Feng Zhu
Guangming Zhu
Ji Zhu
Jun-Yan Zhu
Lei Zhu
Linchao Zhu
Rui Zhu
Shizhan Zhu
Tyler Lixuan Zhu
Wei Zhu
Xiangyu Zhu
Xinge Zhu
Xizhou Zhu
Yanjun Zhu
Yi Zhu
Yixin Zhu
Yizhe Zhu
Yousong Zhu
Zhe Zhu
Zhen Zhu
Zheng Zhu
Zhenyao Zhu
Zhihui Zhu
Zhuotun Zhu
Bingbing Zhuang
Wei Zhuo
Christian Zimmermann
Karel Zimmermann
Larry Zitnick
Mohammadreza Zolfaghari
Maria Zontak
Daniel Zoran
Changqing Zou
Chuhang Zou
Danping Zou
Qi Zou
Yang Zou
Yuliang Zou
Georgios Zoumpourlis
Wangmeng Zuo
Xinxin Zuo

Additional Reviewers

Victoria Fernandez Abrevaya
Maya Aghaei
Allam Allam
Christine Allen-Blanchette
Nicolas Aziere
Assia Benbihi
Neha Bhargava
Bharat Lal Bhatnagar
Joanna Bitton
Judy Borowski
Amine Bourki
Romain Brégier
Tali Brayer
Sebastian Bujwid
Andrea Burns
Yun-Hao Cao
Yuning Chai
Xiaojun Chang
Bo Chen
Shuo Chen
Zhixiang Chen
Junsuk Choe
Hung-Kuo Chu
Jonathan P. Crall
Kenan Dai
Lucas Deecke
Karan Desai
Prithviraj Dhar
Jing Dong
Wei Dong
Turan Kaan Elgin
Francis Engelmann
Erik Englesson
Fartash Faghri
Zicong Fan
Yang Fu
Risheek Garrepalli
Yifan Ge
Marco Godi
Helmut Grabner
Shuxuan Guo
Jianfeng He
Zhezhi He
Samitha Herath
Chih-Hui Ho
Yicong Hong
Vincent Tao Hu
Julio Hurtado
Jaedong Hwang
Andrey Ignatov
Muhammad Abdullah Jamal
Saumya Jetley
Meiguang Jin
Jeff Johnson
Minsoo Kang
Saeed Khorram
Mohammad Rami Koujan
Nilesh Kulkarni
Sudhakar Kumawat
Abdelhak Lemkhenter
Alexander Levine
Jiachen Li
Jing Li
Jun Li
Yi Li
Liang Liao
Ruochen Liao
Tzu-Heng Lin
Phillip Lippe
Bao-di Liu
Bo Liu
Fangchen Liu

Hanxiao Liu
Hongyu Liu
Huidong Liu
Miao Liu
Xinxin Liu
Yongfei Liu
Yu-Lun Liu
Amir Livne
Tiange Luo
Wei Ma
Xiaoxuan Ma
Ioannis Marras
Georg Martius
Effrosyni Mavroudi
Tim Meinhardt
Givi Meishvili
Meng Meng
Zihang Meng
Zhongqi Miao
Gyeongsik Moon
Khoi Nguyen
Yung-Kyun Noh
Antonio Norelli
Jaeyoo Park
Alexander Pashevich
Mandela Patrick
Mary Phuong
Bingqiao Qian
Yu Qiao
Zhen Qiao
Sai Saketh Rambhatla
Aniket Roy
Amelie Royer
Parikshit Vishwas Sakurikar
Mark Sandler
Mert Bülent Sarıyıldız
Tanner Schmidt
Anshul B. Shah
Ketul Shah
Rajvi Shah
Hengcan Shi
Xiangxi Shi
Yujiao Shi
William A. P. Smith
Guoxian Song
Robin Strudel
Abby Stylianou
Xinwei Sun
Reuben Tan
Qingyi Tao
Kedar S. Tatwawadi
Anh Tuan Tran
Son Dinh Tran
Eleni Triantafillou
Aristeidis Tsitiridis
Md Zasim Uddin
Andrea Vedaldi
Evangelos Ververas
Vidit Vidit
Paul Voigtlaender
Bo Wan
Huanyu Wang
Huiyu Wang
Junqiu Wang
Pengxiao Wang
Tai Wang
Xinyao Wang
Tomoki Watanabe
Mark Weber
Xi Wei
Botong Wu
James Wu
Jiamin Wu
Rujie Wu
Yu Wu
Rongchang Xie
Wei Xiong
Yunyang Xiong
An Xu
Chi Xu
Yinghao Xu
Fei Xue
Tingyun Yan
Zike Yan
Chao Yang
Heran Yang
Ren Yang
Wenfei Yang
Xu Yang
Rajeev Yasarla
Shaokai Ye
Yufei Ye
Kun Yi
Haichao Yu
Hanchao Yu
Ruixuan Yu
Liangzhe Yuan
Chen-Lin Zhang
Fandong Zhang
Tianyi Zhang
Yang Zhang
Yiyi Zhang
Yongshun Zhang
Yu Zhang
Zhiwei Zhang
Jiaojiao Zhao
Yipu Zhao
Xingjian Zhen
Haizhong Zheng
Tiancheng Zhi
Chengju Zhou
Hao Zhou
Hao Zhu
Alexander Zimin

Contents – Part I

Quaternion Equivariant Capsule Networks for 3D Point Clouds

Yongheng Zhao[1,3], Tolga Birdal[2(✉)], Jan Eric Lenssen[4], Emanuele Menegatti[1], Leonidas Guibas[2], and Federico Tombari[3,5]

[1] University of Padova, Padova, Italy
[2] Stanford University, Stanford, USA
tbirdal@stanford.edu
[3] TU Munich, Munich, Germany
[4] TU Dortmund, Dortmund, Germany
[5] Google, Mountain View, USA

Abstract. We present a 3D capsule module for processing point clouds that is equivariant to 3D rotations and translations, as well as invariant to permutations of the input points. The operator receives a sparse set of local reference frames, computed from an input point cloud and establishes end-to-end transformation equivariance through a novel dynamic routing procedure on quaternions. Further, we theoretically connect dynamic routing between capsules to the well-known Weiszfeld algorithm, a scheme for solving *iterative re-weighted least squares* (IRLS) problems with provable convergence properties. It is shown that such group dynamic routing can be interpreted as robust IRLS rotation averaging on capsule votes, where information is routed based on the final inlier scores. Based on our operator, we build a capsule network that disentangles geometry from pose, paving the way for more informative descriptors and a structured latent space. Our architecture allows joint object classification and orientation estimation without explicit supervision of rotations. We validate our algorithm empirically on common benchmark datasets.

Keywords: 3D · Equivariance · Disentanglement · Rotation · Quaternion

1 Introduction

It is now well understood that in order to learn a compact and informative representation of the input data, one needs to respect the symmetries in the problem domain [17,73]. Arguably, one of the primary reasons for the success of 2D convolutional neural networks (CNN) is the *translation-invariance* of the 2D convolution acting on the image grid [29,36]. Recent trends aim to transfer

Electronic supplementary material The online version of this chapter (https://doi.org/10.1007/978-3-030-58452-8_1) contains supplementary material, which is available to authorized users.

A. Vedaldi et al. (Eds.): ECCV 2020, LNCS 12346, pp. 1–19, 2020.
https://doi.org/10.1007/978-3-030-58452-8_1

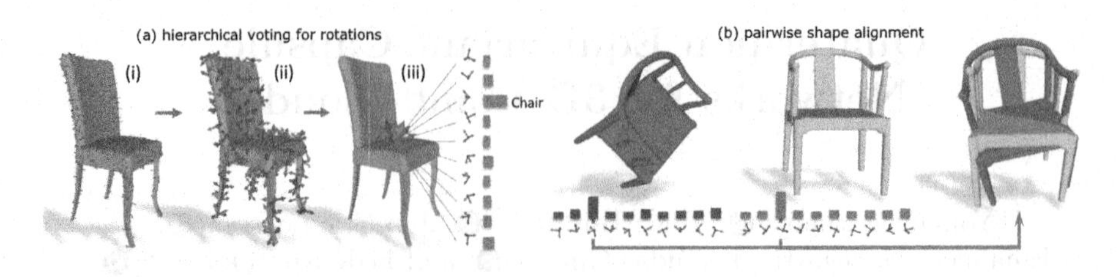

Fig. 1. (**a**) Our network operates on local reference frames (LRF) of an input point cloud (**i**). A hierarchy of quaternion equivariant capsule modules (QEC) then pools the LRFs to a set of latent capsules (**ii, iii**) disentangling the activations from poses. We can use activations in classification and the capsule (quaternion) with the highest activation in absolute (canonical) pose estimation without needing the supervision of rotations. (**b**) Our siamese variant can also solve for the relative object pose by aligning the capsules of two shapes with different point samplings. Our network directly consumes point sets and LRFs. Meshes are included only to ease understanding.

this success into the 3D domain in order to support many applications such as shape retrieval, shape manipulation, pose estimation, 3D object modeling and detection, etc. There, the data is naturally represented as sets of 3D points [55, 57]. Unfortunately, an extension of CNN architectures to 3D point clouds is non-trivial due to two reasons: 1) point clouds are irregular and unorganized, 2) the group of transformations that we are interested in is more complex as 3D data is often observed under arbitrary non-commutative $SO(3)$ rotations. As a result, learning appropriate embeddings requires 3D point-networks to be *equivariant* to these transformations, while also being invariant to point permutations.

In order to fill this gap, we present a quaternion equivariant point capsule network that is suitable for processing point clouds and is equivariant to $SO(3)$ rotations, compactly parameterized by quaternions, while also preserving translation and permutation invariance. Inspired by the local group equivariance [17,40], we efficiently cover $SO(3)$ by restricting ourselves to a sparse set of local reference frames (LRFs) that collectively determine the object orientation. The proposed *quaternion equivariant capsule (QEC) module* deduces equivariant latent representations by robustly combining those LRFs using the proposed *Weiszfeld dynamic routing* with inlier scores as activations, so as to route information from one layer to the next. Hence, our latent features specify to local orientations and activations, disentangling orientation from evidence of object existence. Such explicit and factored storage of 3D information is unique to our work and allows us to perform rotation estimation jointly with object classification. Our final architecture is a hierarchy of QEC modules, where LRFs are routed from lower level to higher level capsules as shown in Fig. 1. We use classification error as the only training cue and adapt a Siamese version for regression of the relative rotations. We neither explicitly supervise the network with pose annotations nor train by augmenting rotations. In summary, our contributions are:

1. We propose a novel, fully $SO(3)$-equivariant capsule module that produces invariant latent representations while explicitly decoupling the orientation

into capsules. Notably, equivariance results have not been previously achieved for $SO(3)$ capsule networks.
2. We connect dynamic routing between capsules [60] and generalized Weiszfeld iterations [4]. Based on this connection, we theoretically argue for the convergence of the included rotation estimation on votes and extend our understanding of dynamic routing approaches.
3. We propose a capsule network that is tailored for simultaneous classification and orientation estimation of 3D point clouds. We experimentally demonstrate the capabilities of our network on classification and orientation estimation on ModelNet10 and ModelNet40 3D shape data.

2 Related Work

Deep Learning on Point Sets. The capability to process raw, unordered point clouds within a neural network is introduced by the prosperous PointNet [55] thanks to the point-wise convolutions and the permutation invariant pooling functions. Many works have extended PointNet primarily to increase the local receptive field size [42,57,62,71]. Point-clouds are generally thought of as sets. This makes any permutation-invariant network that can operate on sets an amenable choice for processing points [58,81]. Unfortunately, common neural network operators in this category are solely equivariant to permutations and translations but to no other groups.

Equivariance in Neural Networks. Early attempts to achieve invariant data representations usually involved data augmentation techniques to accomplish tolerance to input transformations [49,55,56]. Motivated by the difficulty associated with augmentation efforts and acknowledging the importance of theoretically equivariant or invariant representations, the recent years have witnessed a leap in theory and practice of equivariant neural networks [6,37].

While laying out the fundamentals of the group convolution, G-CNNs [18] guaranteed equivariance with respect to finite symmetry groups. Similarly, Steerable CNNs [21] and its extension to 3D voxels [75] considered discrete symmetries only. Other works opted for designing filters as a linear combination of harmonic basis functions, leading to frequency domain filters [74,76]. Apart from suffering from the dense coverage of the group using group convolution, filters living in the frequency space are less interpretable and less expressive than their spatial counterparts, as the basis does not span the full space of spatial filters.

Achieving equivariance in 3D is possible by simply generalizing the ideas of the 2D domain to 3D by voxelizing 3D data. However, methods using dense grids [16,21] suffer from increased storage costs, eventually rendering the implementations infeasible. An extensive line of work generalizes the harmonic basis filters to $SO(3)$ by using *e.g.*, a spherical harmonic basis instead of circular harmonics [19,22,25]. In addition to the same downsides as their 2D counterparts, these approaches have in common that they require their input to be projected to the unit sphere [33], which poses additional problems for unstructured point clouds. A related line of research are methods which define a regular structure on the sphere to propose equivariant convolution operators [13,44].

To learn a rotation equivariant representation of a 3D shape, one can either act on the input data or on the network. In the former case, one either presents augmented data to the network [49,55] or ensures rotation-invariance in the input [23,24,34]. In the latter case one can enforce equivariance in the bottleneck so as to achieve an invariant latent representation of the input [50,63,66]. Further, equivariant networks for discrete sets of views [27] and cross-domain views [26] have been proposed. Here, we aim for a different way of embedding equivariance in the network by means of an explicit latent rotation parametrization in addition to the invariant feature.

Vector Field Networks. [47] followed by the 3D *Tensor Field Networks* (TFN) [66] are closest to our work. Based upon a geometric algebra framework, the authors did achieve localized filters that are equivariant to rotations, translations and permutations. Moreover, they are able to cover the continuous groups. However, TFN are designed for physics applications, are memory consuming and a typical implementation is neither likely to handle the datasets we consider nor can provide orientations in an explicit manner.

Capsule Networks. The idea of capsule networks was first mentioned by Hinton et al. [30], before Sabour et al. [60] proposed the *dynamic routing by agreement*, which started the recent line of work investigating the topic. Since then, routing by agreement has been connected to several well-known concepts, e.g. the EM algorithm [59], clustering with KL divergence regularization [68] and equivariance [40]. They have been extended to autoencoders [38] and GANs [32]. Further, capsule networks have been applied for specific kinds of input data, e.g. graphs [78], 3D point clouds [64,83] or medical images [1].

3 Preliminaries and Technical Background

We now provide the necessary background required for the grasp of the equivariance of point clouds under the action of quaternions.

3.1 Equivariance

Definition 1 (Equivariant Map). *For a $\mathcal{G}$-space acting on $\mathcal{X}$, the map $\Phi : \mathcal{G} \times \mathcal{X} \mapsto \mathcal{X}$ is said to be equivariant if its domain and co-domain are acted on by the same symmetry group [18,20]:*

$$\Phi(\mathbf{g}_1 \circ \mathbf{x}) = \mathbf{g}_2 \circ \Phi(\mathbf{x}) \tag{1}$$

where $\mathbf{g}_1 \in \mathcal{G}$ and $\mathbf{g}_2 \in \mathcal{G}$. Equivalently $\Phi(T(\mathbf{g}_1)\,\mathbf{x}) = T(\mathbf{g}_2)\,\Phi(\mathbf{x})$, where $T(\cdot)$ is a linear representation of the group $\mathcal{G}$. Note that $T(\cdot)$ does not have to commute. It suffices for $T(\cdot)$ to be a homomorphism: $T(\mathbf{g}_1 \circ \mathbf{g}_2) = T(\mathbf{g}_1) \circ T(\mathbf{g}_2)$. In this paper we use a stricter form of equivariance and consider $\mathbf{g}_2 = \mathbf{g}_1$.

Definition 2 (Equivariant Network). *An architecture or network is said to be equivariant if all of its layers are equivariant maps. Due to the transitivity of the equivariance, stacking up equivariant layers will result in globally equivariant networks e.g., rotating the input will produce output vectors which are transformed by the same rotation [37,40].*

3.2 The Quaternion Group $\mathbb{H}_1$

The choice of 4-vector quaternions as representation for $SO(3)$ has multiple motivations: (1) All 3-vector formulations suffer from infinitely many singularities as angle goes to 0, whereas quaternions avoid those, (2) 3-vectors also suffer from infinitely many redundancies (the norm can grow indefinitely). Quaternions have a single redundancy: $q = -q$ that is in practice easy to enforce [9], (3) Computing the actual 'manifold mean' on the Lie algebra requires iterative techniques with subsequent updates on the tangent space. Such iterations are computationally and numerically harmful for a differentiable GPU implementation.

Definition 3 (Quaternion). *A quaternion* $\mathbf{q}$ *is an element of Hamilton algebra* $\mathbb{H}_1$*, extending the complex numbers with three imaginary units* $\boldsymbol{i}$, $\boldsymbol{j}$, $\boldsymbol{k}$ *in the form:* $\mathbf{q} = q_1\boldsymbol{1} + q_2\boldsymbol{i} + q_3\boldsymbol{j} + q_4\boldsymbol{k} = (q_1, q_2, q_3, q_4)^T$, *with* $(q_1, q_2, q_3, q_4)^T \in \mathbb{R}^4$ *and* $\boldsymbol{i}^2 = \boldsymbol{j}^2 = \boldsymbol{k}^2 = \boldsymbol{ijk} = -\boldsymbol{1}$. $q_1 \in \mathbb{R}$ *denotes the scalar part and* $\boldsymbol{v} = (q_2, q_3, q_4)^T \in \mathbb{R}^3$, *the vector part. The* conjugate $\bar{\mathbf{q}}$ *of the quaternion* $\mathbf{q}$ *is given by* $\bar{\mathbf{q}} := q_1 - q_2\boldsymbol{i} - q_3\boldsymbol{j} - q_4\boldsymbol{k}$. *A* unit quaternion $\mathbf{q} \in \mathbb{H}_1$ *with* $1 \stackrel{!}{=} \|\mathbf{q}\| := \mathbf{q} \cdot \bar{\mathbf{q}}$ *and* $\mathbf{q}^{-1} = \bar{\mathbf{q}}$, *gives a compact and numerically stable parametrization to represent orientation of objects on the unit sphere* $\mathcal{S}^3$, *avoiding gimbal lock and singularities [15]. Identifying antipodal points* $\mathbf{q}$ *and* $-\mathbf{q}$ *with the same element, the unit quaternions form a double covering group of* $SO(3)$. $\mathbb{H}_1$ *is closed under the non-commutative multiplication or the Hamilton product:*

$$(\mathbf{p} \in \mathbb{H}_1) \circ (\boldsymbol{r} \in \mathbb{H}_1) = [p_1 r_1 - \mathbf{v}_p \cdot \mathbf{v}_r \,;\, p_1 \mathbf{v}_r + r_1 \mathbf{v}_p + \mathbf{v}_p \times \mathbf{v}_r]. \tag{2}$$

Definition 4 (Linear Representation of $\mathbb{H}_1$). *We follow [12] and use the parallelizable nature of unit quaternions (d* $\in \{1, 2, 4, 8\}$ *where d is the dimension of the ambient space) to define* $T : \mathbb{H}_1 \mapsto \mathbb{R}^{4\times 4}$ *as:*

$$\mathbf{T}(\mathbf{q}) \triangleq \begin{bmatrix} q_1 & -q_2 & -q_3 & -q_4 \\ q_2 & q_1 & -q_4 & q_3 \\ q_3 & q_4 & q_1 & -q_2 \\ q_4 & -q_3 & q_2 & q_1 \end{bmatrix}.$$

To be concise we will use capital letters to refer to the matrix representation of quaternions e.g. $\mathbf{Q} \equiv T(\mathbf{q})$, $\mathbf{G} \equiv T(\mathbf{g})$. *Note that* $T(\cdot)$, *the injective homomorphism to the orthonormal matrix ring, by construction satisfies the condition in Definition 1 [65]:* $\det(\mathbf{Q}) = 1, \mathbf{Q}^\top = \mathbf{Q}^{-1}, \|\mathbf{Q}\| = \|\mathbf{Q}_{i,:}\| = \|\mathbf{Q}_{:,i}\| = 1$ *and* $\mathbf{Q} - q_1\mathbf{I}$ *is skew symmetric:* $\mathbf{Q} + \mathbf{Q}^\top = 2q_1\mathbf{I}$. *It is easy to verify these properties. T linearizes the Hamilton product or the group composition:* $\mathbf{g} \circ \mathbf{q} \triangleq T(\mathbf{g})\mathbf{q} \triangleq \mathbf{G}\mathbf{q}$.

3.3 3D Point Clouds

Definition 5 (Point Cloud). *We define a 3D surface to be a differentiable 2-manifold embedded in the ambient 3D Euclidean space:* $\mathcal{M}^2 \in \mathbb{R}^3$ *and a point cloud to be a discrete subset sampled on* $\mathcal{M}^2$: $\mathbf{X} \in \{\mathbf{x}_i \in \mathcal{M}^2 \cap \mathbb{R}^3\}$.

Definition 6 (Local Geometry). *For a smooth point cloud* $\{\mathbf{x}_i\} \in \mathcal{M}^2 \subset \mathbb{R}^{N\times 3}$, *a* local reference frame (LRF) *is defined as an ordered basis of the tangent space at* $\mathbf{x}$, $\mathcal{T}_{\mathbf{x}}\mathcal{M}$, *consisting of orthonormal vectors:* $\mathcal{L}(\mathbf{x}) = [\boldsymbol{\partial}_1, \boldsymbol{\partial}_2, \boldsymbol{\partial}_3 \equiv \boldsymbol{\partial}_1 \times \boldsymbol{\partial}_2]$. *Usually the first component is defined to be the surface normal* $\boldsymbol{\partial}_1 \triangleq \mathbf{n} \in \mathcal{S}^2 : \|\mathbf{n}\| = 1$ *and the second one is picked according to a heuristic.*

Note that recent trends, *e.g.* as in Cohen et al. [17], acknowledge the ambiguity and either employ a *gauge* (tangent frame) equivariant design or propagate the determination of a certain direction until the last layer [54]. Here, we will assume that $\boldsymbol{\partial}_2$ can be uniquely and repeatably computed, a reasonable assumption for the point sets we consider [52]. For the cases where this does not hold, we will rely on the robustness of the iterative routing procedures in our network. We will explain our method of choice in Sect. 6 and visualize LRFs of an airplane object in Fig. 1.

4 $SO(3)$-Equivariant Dynamic Routing

Disentangling orientation from representations requires guaranteed equivariances and invariances. Yet, the original capsule networks of Sabour et al. [60] cannot achieve equivariance to general groups. To this end, Lenssen et al. [40] proposed a dynamic routing procedure that guarantees equivariance and invariance under $SO(2)$ actions, by applying a manifold-mean and the geodesic distance as routing operators. We will extend this idea to the non-abelian $SO(3)$ and design capsule networks that sparsely operate on a set of LRFs computed via [53] on local neighborhoods of points. The $SO(3)$ elements are paremeterized by quaternions similar to [82]. In the following, we begin by introducing our novel equivariant dynamic routing procedure, the main building block of our architecture. We show the connection to the well known Weiszfeld algorithm, broadening the understanding of dynamic routing by embedding it into traditional computer vision methodology. Then, we present an example of how to stack those layers via a simple aggregation, resulting in an $SO(3)$-equivariant 3D capsule network that yields invariant representations (or activations) as well as equivariant orientations (latent capsules).

4.1 Equivariant Quaternion Mean

To construct equivariant layers on the group of rotations, we are required to define a left-equivariant averaging operator $\mathcal{A}$ that is invariant under permutations of the group elements, as well as a distance metric δ that remains unchanged under the action of the group [40]. For these, we make the following choices:

Definition 7 (Geodesic Distance). *The Riemannian (geodesic) distance on the manifold of rotations lead to the following geodesic distance* $\delta(\cdot) \equiv d_{quat}(\cdot)$:

$$d(\mathbf{q}_1, \mathbf{q}_2) \equiv d_{quat}(\mathbf{q}_1, \mathbf{q}_2) = 2\cos^{-1}(|\langle \mathbf{q}_1, \mathbf{q}_2 \rangle|) \tag{3}$$

Algorithm 1: Quaternion Equivariant Dynamic Routing

1 **input** : Input points $\{\mathbf{x}_1, ..., \mathbf{x}_K\} \in \mathbb{R}^{K\times 3}$, input capsules (LRFs) $\mathcal{Q} = \{\mathbf{q}_1, \dots, \mathbf{q}_L\} \in {\mathbb{H}_1}^L$, with $L = N^c \cdot K$, N^c is the number of capsules per point, activations $\boldsymbol{\alpha} = (\alpha_1, \dots, \alpha_L)^T$, trainable transformations $\mathcal{T} = \{\mathbf{t}_{i,j}\}_{i,j} \in {\mathbb{H}_1}^{L\times M}$

2 **output**: Updated frames $\hat{\mathcal{Q}} = \{\hat{\mathbf{q}}_1, \dots, \hat{\mathbf{q}}_M\} \in {\mathbb{H}_1}^M$, updated activations $\hat{\boldsymbol{\alpha}} = (\hat{\alpha}_1, \dots, \hat{\alpha}_M)^T$

3 **for** *All primary (input) capsules i* **do**

4   **for** *All latent (output) capsules j* **do**

5         $\mathbf{v}_{i,j} \leftarrow \mathbf{q}_i \circ \mathbf{t}_{i,j}$ `// compute votes`

6 **for** *All latent (output) capsules j* **do**

7     $\hat{\mathbf{q}}_j \leftarrow \mathcal{A}(\{\mathbf{v}_{1,j} \dots \mathbf{v}_{K,j}\}, \boldsymbol{\alpha})$ `// initialize output capsules`

8   **for** *k iterations* **do**

9   **for** *All primary (input) capsules i* **do**

10             $w_{i,j} \leftarrow \alpha_i \cdot \text{sigmoid}\big(-\delta(\hat{\mathbf{q}}_j, \mathbf{v}_{i,j})\big)$ `// the current weight`

11         $\hat{\mathbf{q}}_j \leftarrow \mathcal{A}(\{\mathbf{v}_{1,j} \dots \mathbf{v}_{L,j}\}, \mathbf{w}_{:,j})$ `// see Eq (4)`

12     $\hat{\alpha}_j \leftarrow \text{sigmoid}\big(-\frac{1}{K}\sum_{1}^{L} \delta(\hat{\mathbf{q}}_j, \mathbf{v}_{i,j})\big)$ `// recompute activations`

Definition 8 (Quaternion Mean $\boldsymbol{\mu}(\cdot)$). *For a set of Q rotations $\mathbf{S} = \{\mathbf{q}_i\}$ and associated weights $\mathbf{w} = \{w_i\}$, the weighted mean operator $\mathcal{A}(\mathbf{S}, \mathbf{w}) : {\mathbb{H}_1}^n \times \mathbb{R}^n \mapsto {\mathbb{H}_1}^n$ is defined through the following maximization procedure [48]:*

$$\bar{\mathbf{q}} = \arg\max_{\mathbf{q}\in\mathbb{S}^3} \mathbf{q}^\top \mathbf{M} \mathbf{q} \tag{4}$$

where $\mathbf{M} \in \mathbb{R}^{4\times 4}$ is defined as: $\mathbf{M} \triangleq \sum_{i=1}^{Q} w_i \mathbf{q}_i \mathbf{q}_i^\top$.

The average quaternion $\bar{\mathbf{q}}$ is the eigenvector of $\mathbf{M}$ corresponding to the maximum eigenvalue. This operation lends itself to both analytic [46] and automatic differentiation [39]. The following properties allow $\mathcal{A}(\mathbf{S}, \mathbf{w})$ to be used to build an equivariant dynamic routing:

Theorem 1. *Quaternions, the employed mean $\mathcal{A}(\mathbf{S}, \mathbf{w})$ and geodesic distance $\delta(\cdot)$ enjoy the following properties:*

1. *$\mathcal{A}(\mathbf{g} \circ \mathbf{S}, \mathbf{w})$ is left-equivariant: $\mathcal{A}(\mathbf{g} \circ \mathbf{S}, \mathbf{w}) = \mathbf{g} \circ \mathcal{A}(\mathbf{S}, \mathbf{w})$.*
2. *Operator $\mathcal{A}$ is invariant under permutations:*

$$\mathcal{A}(\{\mathbf{q}_1, \dots, \mathbf{q}_Q\}, \mathbf{w}) = \mathcal{A}(\{\mathbf{q}_{\sigma(1)}, \dots, \mathbf{q}_{\sigma(Q)}\}, \mathbf{w}_\sigma). \tag{5}$$

3. *The transformations $\mathbf{g} \in \mathbb{H}_1$ preserve the geodesic distance $\delta(\cdot)$ given in Definition 7.*

Proof. The proofs are given in the supplementary material.

We also note that the above mean is closed form, differentiable and can be computed in a batch-wise fashion. We are now ready to construct the *dynamic routing* (DR) by agreement that is equivariant to $SO(3)$ actions, thanks to Theorem 1.

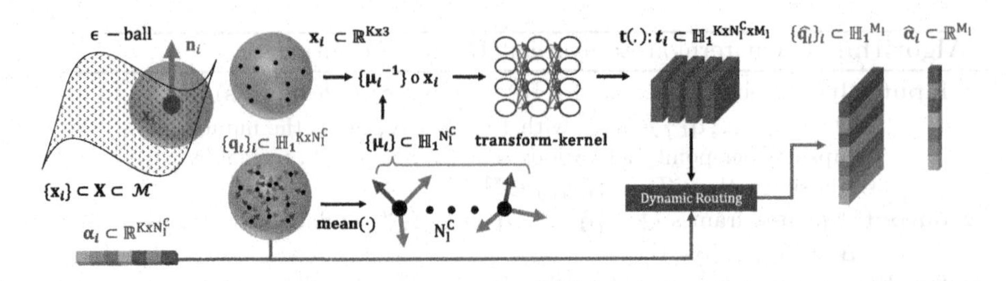

Fig. 2. Our quaternion equivariant capsule (QEC) layer for processing local patches: Our input is a 3D point set $\mathbf{X}$ on which we query local neighborhoods $\{\mathbf{x}_i\}$ with precomputed LRFs $\{\mathbf{q}_i\}$. Essentially, we learn the parameters of a fully connected network that continuously maps the canonicalized local point set to transformations $\mathbf{t}_i$, which are used to compute hypotheses (votes) from input capsules. By a special dynamic routing procedure that uses the activations determined in a previous layer, we arrive at latent capsules that are composed of a set of orientations $\hat{\mathbf{q}}_i$ and new activations $\hat{\boldsymbol{\alpha}}_i$. Thanks to the decoupling of local reference frames, $\hat{\boldsymbol{\alpha}}_i$ is invariant and orientations $\hat{\mathbf{q}}_i$ are equivariant to input rotations. All the operations and hence the entire QE-network are equivariant achieving a guaranteed disentanglement of the rotation parameters. *Hat symbol* ($\hat{\mathbf{q}}$) *refers to 'estimated'.*

4.2 Equivariant Weiszfeld Dynamic Routing

Our routing procedure extends previous work [40,60] for quaternion valued input. The core idea is to *route* from the *primary capsules* that constitute the input LRF set to the *latent capsules* by an iterative clustering of votes $\mathbf{v}_{i,j}$. At each step, we assign the weighted group mean of votes to the respective output capsules. The weights $w \leftarrow \sigma(\mathbf{x}, \mathbf{y})$ are inversely proportional to the distance between the vote quaternions and the new quaternion (cluster center). See Algorithm 1 for details. In the following, we analyze our variant of routing as an interesting case of the affine, Riemannian Weiszfeld algorithm [3,4].

Lemma 1. *For* $\sigma(\mathbf{x}, \mathbf{y}) = \delta(\mathbf{x}, \mathbf{y})^{q-2}$ *the equivariant routing procedure given in Algorithm 1 is a variant of the affine subspace Wieszfeld algorithm [3,4] that is a robust algorithm for computing the* L_q *geometric median.*

Proof (Proof Sketch). The proof follows from the definition of Weiszfeld iteration [3] and the mean and distance operators defined in Sect. 4.1. We first show that computing the weighted mean is equivalent to solving the normal equations in the iteratively reweighted least squares (IRLS) scheme [14]. Then, the inner-most loop corresponds to the IRLS or Weiszfeld iterations. We provide the detailed proof in supplementary material.

Note that, in practice one is quite free to choose the weighting function $\sigma(\cdot)$ as long as it is inversely proportional to the geodesic distance and concave [2]. The original dynamic routing can also be formulated as a clustering procedure with a KL divergence regularization. This holistic view paves the way to better routing algorithms [68]. Our perspective is akin yet more geometric due to the group structure of the parameter space. Thanks to the connection to Weiszfeld

Algorithm 2: Quaternion Equivariant Capsule Module

1 **input** : Input points of one patch $\{\mathbf{x}_1, ..., \mathbf{x}_K\} \in \mathbb{R}^{K\times3}$, input capsules (LRFs) $\mathcal{Q} = \{\mathbf{q}_1, \dots, \mathbf{q}_L\} \in \mathbb{H}_1{}^L$, with $L = N^c \cdot K$, N^c is the number of capsules per point, activations $\boldsymbol{\alpha} = (\alpha_1, \dots, \alpha_L)^T$
2 **output**: Updated frames $\hat{\mathcal{Q}} = \{\hat{\mathbf{q}}_1, \dots, \hat{\mathbf{q}}_M\} \in \mathbb{H}_1{}^M$, updated activations $\hat{\boldsymbol{\alpha}} = (\hat{\alpha}_1, \dots, \hat{\alpha}_M)^T$
3 **for** *Each input channel* n^c *of all the primary capsules channels* N^c **do**
4 $\quad \mu(n^c) \leftarrow \mathcal{A}(\mathcal{Q}(n^c))$ `// Input quaternion average, see Eq (4)`
5 $\quad$ **for** *Each point* $\mathbf{x}_i$ *of this patch* **do**
6 $\quad\quad \mathbf{x}'_i \leftarrow \mu(n^c)^{-1} \circ \mathbf{x}_i$ `// Rotate to a canonical orientation`
7 $\{\mathbf{x}'_i\} \in \mathbb{R}^{K\times N^c\times3}$`// Points in multiple(`$N^c$`) canonical frames`
8 **for** *Each point* $\mathbf{x}'_i$ *of this patch* **do**
9 $\quad \mathbf{t} \leftarrow t(\mathbf{x}'_i)$ `// Transform kernel,` $t(\cdot) : \mathbb{R}^{N^c\times3} \rightarrow \mathbb{R}^{N^c\times M\times4}$
10 $\mathcal{T} \equiv \{\mathbf{t}_i\} \in \mathbb{H}_1{}^{K\times N^c_i\times M} \leftarrow \{\mathbf{t}\} \in \mathbb{H}_1{}^{L\times M}$
11 $(\hat{\mathcal{Q}}, \hat{\boldsymbol{\alpha}}) \leftarrow \text{DynamicRouting}(X, \mathcal{Q}, \boldsymbol{\alpha}, \mathcal{T})$ `// See Alg. 1`

algorithm, the convergence behavior of our dynamic routing can be directly analyzed within the theoretical framework presented by [3,4].

Theorem 2 *Under mild assumptions provided in the appendix, the sequence of the DR-iterates generated by the inner-most loop almost surely converges to a critical point.*

Proof (Proof Sketch). Proof, given in the appendix, is a direct consequence of Lemma 1 and directly exploits the connection to the Weiszfeld algorithm.

In summary, the provided theorems show that our dynamic routing by agreement is in fact a variant of robust IRLS rotation averaging on the predicted votes, where refined inlier scores for combinations of input/output capsules are used to route information from one layer to the next.

5 Equivariant Capsule Network Architecture

In the following, we describe how we leverage the novel dynamic routing algorithm to build a capsule network for point cloud processing that is equivariant under $SO(3)$ actions on the input. The essential ingredient of our architecture, the *quaternion equivariant capsule (QEC) module* that implements a capsule layer with dynamic routing, is described in Sect. 5.1, before using it as building block in the full architecture, as described in Sect. 5.2.

5.1 QEC Module

The main module of our architecture, the QEC module, is outlined in Fig. 2. We also provide the corresponding pseudocode in Algorithm 2.

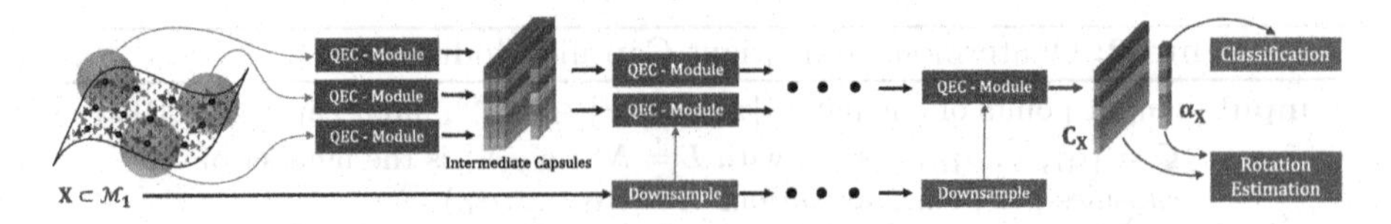

Fig. 3. Our entire capsule-network architecture. We hierarchically send all the local patches to our QEC-module as shown in Fig. 2. At each level the points are pooled in order to increase the receptive field, gradually reducing the LRFs into a single capsule per class. We use classification and orientation estimation (in the siamese case) as supervision cues to train the transform-kernels $\mathbf{t}(\cdot)$.

Input. The input to the module is a local patch of points with coordinates $\mathbf{x}_i \subset \mathbb{R}^{K\times 3}$, rotations (LRFs) attached to these points, parametrized as quaternions $\mathbf{q}_i \subset \mathbb{H}_1{}^{K\times N^c}$ and activations $\boldsymbol{\alpha}_i \subset \mathbb{R}^{K\times N^c}$. We also use $\mathbf{q}_i$ to denote the input capsules. N^c is the number of input capsule channels per point and it is equal to the number of output capsules (M) from the last layer.

Trainable Transformations. Recalling the original capsule networks of Sabour et al. [60], the trainable transformations $\mathbf{t}$, which are applied to the input rotations to compute the votes, lie in a grid kernel in the 2D image domain. Therefore, the procedure can learn to produce well-aligned votes if and only if the learned patterns in $\mathbf{t}$ match those in input capsule sets (agreement on evidence of object existence). Since our input points in the local receptive field lie in continuous $\mathbb{R}^3$, training a discrete set of pose transformations $\mathbf{t}_{i,j}$ based on discrete local coordinates is not possible. Instead, we use a similar approach as Lenssen et al. [40] and employ a continuous kernel $t(\cdot) : \mathbb{R}^{N^c\times 3} \rightarrow \mathbb{R}^{M\times N^c\times 4}$ that is defined on the continuous $\mathbb{R}^{N^c\times 3}$, instead of only a discrete set of positions. The network is shared over all points to compute the transformations $\mathbf{t}_{i,j} = (t(\mathbf{x}'_1),\ldots,t(\mathbf{x}'_K))_{i,j} \subset \mathbb{R}^{K\times M\times N^c\times 4}$, which are used to calculate the votes for dynamic routing with $\mathbf{v}_{i,j} = \mathbf{q}_i \circ \mathbf{t}_{i,j}$. The network $t(\cdot)$ consists of fully-connected layers that regresses the transformations, similar to common operators for continuous convolutions [28,61,70], just with quaternion output. The kernel is able to learn pose patterns in the 3D space, which align the resulting votes if certain pose sets are present. Note that $t(\cdot)$ predicts quaternions by unit-normalizing the regressed output: $\mathbf{t}_{i,j} \subset \mathbb{H}_1{}^{K\times M\times N^c}$. Although Riemannian layers [7] or spherical predictions [43] can improve the performance, the simple strategy works reasonably for our case.

In order for the kernel to be invariant, it needs to be aligned using an equivariant initial orientation candidate [40]. Given points $\mathbf{x}_i$ and rotations $\mathbf{q}_i$, we compute the mean $\boldsymbol{\mu}_i$ in a channel-wise manner like that of the initial candidates: $\boldsymbol{\mu}_i \subset \mathbb{H}_1{}^{N^c}$. These candidates are used to bring the kernels in canonical orientations by inversely rotating the input points: $\mathbf{x}'_i = (\boldsymbol{\mu}_i{}^{-1} \circ \mathbf{x}_i) \subset \mathbb{R}^{K\times N^c\times 3}$.

Computing the Output. After computing the votes, we utilize the input activation $\boldsymbol{\alpha}_i$ as initialization weights and iteratively refine the output capsule rotations (robust rotation estimation on votes) $\hat{\mathbf{q}}_i$ and activations $\hat{\boldsymbol{\alpha}}_i$ (final inlier scores) by our Weiszfeld routing by agreement as shown in Algorithm 1.

Table 1. Classification accuracy on ModelNet40 dataset [77] for different methods as well as ours. We also report the number of parameters optimized for each method. **X/Y** means that we train with **X** and test with **Y**.

	PN	PN++	DGCNN	KDTreeNet	Point2Seq	Sph.CNNs	PRIN	PPF	Ours (Var.)	Ours
NR/NR	88.45	89.82	**92.90**	86.20	92.60	-	80.13	70.16	85.27	74.43
NR/AR	12.47	21.35	29.74	8.49	10.53	43.92	68.85	70.16	11.75	**74.07**
#Params	3.5M	1.5M	2.8M	3.6M	1.8M	0.5M	1.5M	3.5M	0.4M	**0.4M**

5.2 Network Architecture

For processing point clouds, we use multiple QEC modules in a hierarchical architecture as shown in Fig. 3. In the first layer, the input primary capsules are represented by LRFs computed with FLARE algorithm [53]. Therefore, the number of input capsule channels N^c in the first layer is equal to 1 and activations are uniform. The output of a former layer is propagated to the input of the latter, creating the hierarchy.

In order to gradually increase the receptive field, we stack QEC modules creating a deep hierarchy, where each layer reduces the number of points and increases the receptive field. In our experiments, we use a two level architecture, which receives $N = 64$ patches as input. We call the centers of these patches *pooling centers* and compute them via a uniform farthest point sampling as in [11]. Pooling centers serve as the positions of output capsules of the current layer. Each of those centers is linked to their immediate vicinity leading to $K = 9$-star local connectivity from which serve as input to the first QEC module to compute rotations and activations of $64 \times 64 \times 4$ intermediate capsules. The second module connects those intermediate capsules to the output capsules, whose number corresponds to the number of classes. Specifically, for layer 1, we use $K = 9, N_l{}^c = 1, M_l = 64$ and for layer 2, $K = 64, N_l{}^c = 64, M_l = C = 40$. This way, the last QEC module receives only one input patch and pools all capsules into a single point with an estimated LRF. For further details, we refer to our source code, which we will make available online before publication and provide in the supplemental materials.

6 Experimental Evaluations

Implementation Details. We implement our network in PyTorch and use the ADAM optimizer [35] with a learning rate of 0.001. Our point-transformation mapping network (transform-kernel) is implemented by two FC-layers composed of 64 hidden units. We set the initial activation of the input LRF to 1.0. In each layer, we use 3 iterations of DR. For classification we use the spread loss [59] and the rotation loss is identical to $\delta(\cdot)$.

The first axis of the LRF is the surface normal computed by local plane fits [31]. We compute the second axis, $\boldsymbol{\partial}_2$, by FLARE [53], that uses the normalized projection of the point with the largest distance within the periphery of the support, onto the tangent plane of the center: $\boldsymbol{\partial}_2 = \frac{\mathbf{p}_{\max} - \mathbf{p}}{\|\mathbf{p}_{\max} - \mathbf{p}\|}$. Using other

Table 2. Relative angular error (RAE) of rotation estimation in different categories of ModelNet10. Right side of the table denotes the objects with rotational symmetry, which we include for completeness. PCA-S refers to running PCA only on a resampled instance, while PCA-SR applies both rotations and resampling.

Method	Avg.	No_Sym	Chair	Bed	Sofa	Toilet	Monitor	Table	Desk	Dresser	NS	Bathtub
Mean LRF	0.41	0.35	0.32	0.36	0.34	0.41	0.34	0.45	0.60	0.50	0.46	0.32
PCA-S	0.40	0.42	0.60	0.53	0.46	0.32	0.12	0.47	**0.23**	**0.33**	0.43	0.55
PCA-SR	0.67	0.67	0.69	0.70	0.67	0.68	0.61	0.67	0.67	0.67	0.66	0.70
PointNetLK [5]	0.37	0.38	0.43	0.31	0.40	0.40	0.31	0.40	0.33	0.39	0.38	0.34
IT-Net [80]	0.27	0.19	0.10	0.22	0.17	0.20	0.28	**0.31**	0.41	0.44	0.40	0.39
Ours	0.27	0.17	0.11	0.20	0.16	0.18	0.19	0.43	0.40	0.48	0.33	0.31
Ours (siamese)	**0.20**	**0.09**	**0.08**	**0.10**	**0.08**	**0.11**	**0.08**	0.40	0.35	0.34	**0.32**	**0.30**

choices such as SHOT [67] or GFrames [51] is possible. We found FLARE to be sufficient for our experiments. Prior to all operations, we flip all the LRF quaternions such that they lie on the northern hemisphere : $\{\mathbf{q}_i \in \mathbb{S}^3 : q_i^w > 0\}$.

3D Shape Classification. We use ModelNet40 dataset of [57,77] to assess our classification performance where each shape is composed of $10K$ points randomly sampled from the mesh surfaces of each shape [55,57]. We use the official split with 9,843 shapes for training and 2,468 for testing. We assign the LRFs to a subset of the uniformly sampled points, $N = 512$ [11].

During training, we do not augment the dataset with random rotations. All the shapes are trained with single orientation (well-aligned). We call this *trained with NR*. During testing, we randomly generate multiple arbitrary $SO(3)$ rotations for each shape and evaluate the average performance for all the rotations. This is called *test with AR*. This protocol is similar to [5]'s and is used both for our algorithms and for the baselines. Our results are shown in Table 1 along with that of PointNet (PN) [55], PointNet++ (PN++) [55], DGCNN [72], KD-treeNet [41], Point2Seq [45], Spherical CNNs [25], PRIN [79] and the theoretically invariant PPF-FoldNet (PPF) [23]. We also present a version of our algorithm (*Var*) that avoids the canonicalization within the QE-network. This is a non-equivariant network that we still train without data augmentation or orientation supervision. While this version gets comparable results to the state of the art for the NR/NR case, it cannot handle random $SO(3)$ variations (AR). Note that PPF uses the point-pair-feature [10] encoding and hence creates invariant input representations. For the scenario of NR/AR, our equivariant version outperforms all the other methods, including equivariant spherical CNNs [25] by a significant gap of at least 5% even when [25] exploits the 3D mesh. The object rotational symmetries in this dataset are responsible for a significant portion of the errors we make and we provide further details in supplementary material. It is worth mentioning that we also trained TFNs [66] for that task, but their memory demand made it infeasible to scale to this application.

Computational Aspects. As shown in Table 1 for ModelNet40 our network has $0.047M$ parameters. It incurs a computational cost in the order $O(MKL)$. The details are given in the supplementary material.

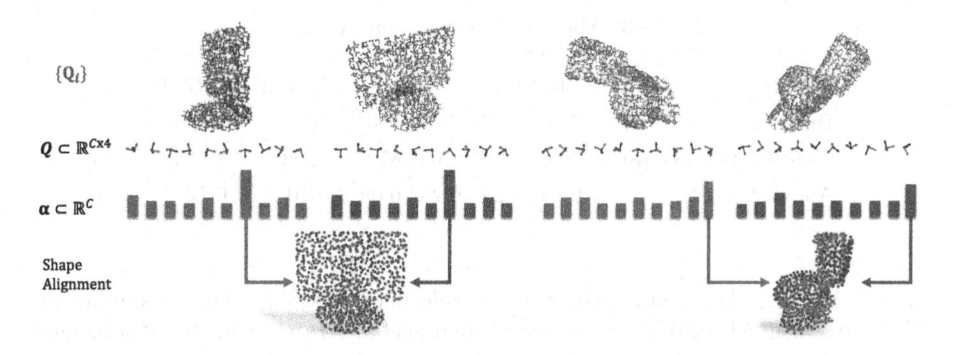

Fig. 4. Shape alignment on the **monitor** (left) and **toilet** (right) objects via our siamese equivariant capsule architecture. The shapes are assigned to the maximally activated class. The corresponding pose capsule provides the rotation estimate.

Rotation Estimation in 3D Point Clouds. Our network can estimate both the absolute and relative 3D object rotations without pose-supervision. To evaluate this desired property, we used the well classified shapes on ModelNet10 dataset, a sub-dataset of Modelnet40 [77]. This time, we use the official Modelenet10 dataset split with 3991 for training and 908 shapes for testing.

During testing, we generate multiple instances per shape by transforming the instance with five arbitrary $SO(3)$ rotations. As we are also affected by the sampling of the point cloud, we resample the mesh five times and generate different pooling graphs across all the instances of the same shape. Our QE-architecture can estimate the pose in two ways: 1) *canonical*: by directly using the output capsule with the highest activation, 2) *siamese*: by a siamese architecture that computes the relative quaternion between the capsules that are maximally activated as shown in Fig. 4. Both modes of operation are free of the data augmentation and we give further schematics of the latter in our appendix. It is worth mentioning that unlike regular pose estimation algorithms which utilize the same shape in both training and testing, our network never sees the test shapes during training. This is also known as *category-level* pose estimation [69].

Our results against the baselines including a naive averaging of the LRFs (Mean LRF) and principal axis alignment (PCA) are reported in Table 2 as the relative angular error (RAE). We further include results of PointNetLK [5] and IT-Net [80], two state of the art 3D networks that iteratively align two given point sets. These methods are in nature similar to iterative closest point (ICP) algorithm [8] but 1) do not require an initialization (*e.g.*first iteration estimates the pose), 2) learn data driven updates. Methods that use mesh inputs such as Spherical CNNs [25] cannot be included here as the random sampling of the same surface would not affect those. We also avoid methods that are just invariant to rotations (and hence cannot estimate the pose) such as Tensorfield Networks [66]. Finally, note that, IT-net [80] and PointNetLK [5] need to train for a lot of epochs (*e.g.* 500) with random $SO(3)$ rotation augmentation in order to obtain models with full coverage of $SO(3)$, whereas we train only for ~ 100

Table 3. Ablation study on point density.

LRF Input	LRF-10K				LRF-2K	LRF-1K
Dropout	50%	66%	75%	100%	100%	100%
Classification accuracy	77.8	83.3	83.4	87.8	85.46	79.74
Angular error	0.34	0.27	0.25	0.09	0.10	0.12

epochs. Finally, the recent geometric capsule networks [64] remains similar to PCA with an RAE of 0.42 on No_Sym when evaluated under identical settings. We include more details about the baselines in the appendix.

Relative Angle in Degrees (RAE) between the ground truth and the prediction is computed as: $d(\mathbf{q}_1, \mathbf{q}_2)/\pi$. Note that resampling and random rotations render the job of all methods difficult. However, both of our canonical and siamese versions which try to find a canonical and a relative alignment respectively, are better than the baselines. As pose estimation of objects with rotational symmetry is a challenging task due to inherent ambiguities, we also report results on the non-symmetric subset (No_Sym).

Robustness Against Point and LRF Resampling. Density changes in the local neighborhoods of the shape are an important cause of error for our network. Hence, we ablate by applying random resampling (patch-wise dropout) objects in ModelNet10 dataset and repeating the classification and pose estimation as described above. While we use all the classes in classification accuracy, we only consider the well classified non-symmetric (No_Sym) objects for ablating on the pose estimation. The first part (LRF-10K) of Table 3 shows our findings against gradual increases of the number of patches. Here, we sample 2K LRFs from the 10K LRFs computed on an input point set of cardinality 10K. 100% dropout corresponds to 2K points in all columns. On second ablation, we reduce the amount of points on which we compute the LRFs, to 2K and 1K respectively. As we can see from the table, our network is robust towards the changes in the LRFs as well as the density of the points.

7 Conclusion and Discussion

We have presented a new framework for achieving permutation invariant and $SO(3)$ equivariant representations on 3D point clouds. Proposing a variant of the capsule networks, we operate on a sparse set of rotations specified by the input LRFs thereby circumventing the effort to cover the entire $SO(3)$. Our network natively consumes a compact representation of the group of 3D rotations - quaternions. We have theoretically shown its equivariance and established convergence results for our Weiszfeld dynamic routing by making connections to the literature of robust optimization. Our network by construction disentangles the object existence that is used as global features in classification. It is among the few for having an explicit group-valued latent space and thus naturally estimates the orientation of the input shape, even without a supervision signal.

Limitations. In the current form our performance is severely affected by the shape symmetries. The length of the activation vector depends on the number of classes and for a sufficiently descriptive latent vector we need to have significant number of classes. On the other hand, this allows us to perform with merit on problems where the number of classes are large. The computation of LRFs are still sensitive to the point density changes and resampling. LRFs themselves can also be ambiguous and sometimes non-unique.

Future Work. Inspired by [17] and [54] our feature work will involve exploring the Lie algebra for equivariances, establishing invariance to the tangent directions, application of our network in the broader context of 6DoF object detection from point sets and looking for equivariances among point resampling.

References

1. Afshar, P., Mohammadi, A., Plataniotis, K.N.: Brain tumor type classification via capsule networks. In: 2018 25th IEEE International Conference on Image Processing (ICIP) (2018)
2. Aftab, K., Hartley, R.: Convergence of iteratively re-weighted least squares to robust m-estimators. In: Winter Conference on Applications of Computer Vision. IEEE (2015)
3. Aftab, K., Hartley, R., Trumpf, J.: Generalized Weiszfeld algorithms for Lq optimization. IEEE Trans. Pattern Anal. Mach. Intell. **37**(4), 728–745 (2014)
4. Aftab, K., Hartley, R., Trumpf, J.: l_q closest-point to affine subspaces using the generalized Weiszfeld algorithm. Int. J. Comput. Vis. **114**, 1–15 (2015)
5. Aoki, Y., Goforth, H., Srivatsan, R.A., Lucey, S.: PointNetLK: robust & efficient point cloud registration using PointNet. In: Proceedings of the IEEE Conference on Computer Vision and Pattern Recognition, pp. 7163–7172 (2019)
6. Bao, E., Song, L.: Equivariant neural networks and equivarification. arXiv preprint arXiv:1906.07172 (2019)
7. Becigneul, G., Ganea, O.E.: Riemannian adaptive optimization methods. In: International Conference on Learning Representations (2019)
8. Besl, P.J., McKay, N.D.: Method for registration of 3-D shapes. In: Sensor Fusion IV: Control Paradigms and Data Structures, vol. 1611, pp. 586–606. International Society for Optics and Photonics (1992)
9. Birdal, T., Arbel, M., Simsekli, U., Guibas, L.J.: Synchronizing probability measures on rotations via optimal transport. In: Proceedings of the IEEE/CVF Conference on Computer Vision and Pattern Recognition, pp. 1569–1579 (2020)
10. Birdal, T., Ilic, S.: Point pair features based object detection and pose estimation revisited. In: 2015 International Conference on 3D Vision, pp. 527–535. IEEE (2015)
11. Birdal, T., Ilic, S.: A point sampling algorithm for 3D matching of irregular geometries. In: IEEE/RSJ International Conference on Intelligent Robots and Systems (IROS). IEEE (2017)
12. Birdal, T., Simsekli, U., Eken, M.O., Ilic, S.: Bayesian pose graph optimization via Bingham distributions and tempered geodesic MCMC. In: Advances in Neural Information Processing Systems, pp. 308–319 (2018)
13. Boomsma, W., Frellsen, J.: Spherical convolutions and their application in molecular modelling. In: Advances in Neural Information Processing Systems, vol. 30, pp. 3433–3443 (2017)

14. Burrus, C.S.: Iterative reweighted least squares. OpenStax CNX (2012). http://cnx.org/contents/92b90377-2b34-49e4-b26f-7fe572db78a1
15. Busam, B., Birdal, T., Navab, N.: Camera pose filtering with local regression geodesics on the Riemannian manifold of dual quaternions. In: IEEE International Conference on Computer Vision Workshop (ICCVW) (October 2017)
16. Chakraborty, R., Banerjee, M., Vemuri, B.C.: H-CNNs: convolutional neural networks for Riemannian homogeneous spaces. arXiv preprint arXiv:1805.05487 (2018)
17. Cohen, T., Weiler, M., Kicanaoglu, B., Welling, M.: Gauge equivariant convolutional networks and the icosahedral CNN. In: Proceedings of the 36th International Conference on Machine Learning, pp. 1321–1330 (2019)
18. Cohen, T., Welling, M.: Group equivariant convolutional networks. In: International Conference on Machine Learning, pp. 2990–2999 (2016)
19. Cohen, T.S., Geiger, M., Köhler, J., Welling, M.: Spherical CNNs. In: 6th International Conference on Learning Representations (ICLR) (2018)
20. Cohen, T.S., Geiger, M., Weiler, M.: A general theory of equivariant CNNs on homogeneous spaces. In: Advances in Neural Information Processing Systems, pp. 9145–9156 (2019)
21. Cohen, T.S., Welling, M.: Steerable CNNs. In: International Conference on Learning Representations (ICLR) (2017)
22. Cruz-Mota, J., Bogdanova, I., Paquier, B., Bierlaire, M., Thiran, J.P.: Scale invariant feature transform on the sphere: theory and applications. Int. J. Comput. Vis. **98**(2), 217–241 (2012)
23. Deng, H., Birdal, T., Ilic, S.: PPF-FoldNet: unsupervised learning of rotation invariant 3D local descriptors. In: Ferrari, V., Hebert, M., Sminchisescu, C., Weiss, Y. (eds.) ECCV 2018. LNCS, vol. 11209, pp. 620–638. Springer, Cham (2018). https://doi.org/10.1007/978-3-030-01228-1_37
24. Deng, H., Birdal, T., Ilic, S.: PPFNet: global context aware local features for robust 3D point matching. In: Conference on Computer Vision and Pattern Recognition (2018)
25. Esteves, C., Allen-Blanchette, C., Makadia, A., Daniilidis, K.: Learning SO(3) equivariant representations with spherical CNNs. In: Ferrari, V., Hebert, M., Sminchisescu, C., Weiss, Y. (eds.) ECCV 2018. LNCS, vol. 11217, pp. 54–70. Springer, Cham (2018). https://doi.org/10.1007/978-3-030-01261-8_4
26. Esteves, C., Sud, A., Luo, Z., Daniilidis, K., Makadia, A.: Cross-domain 3D equivariant image embeddings. In: International Conference on Machine Learning (ICML) (2019)
27. Esteves, C., Xu, Y., Allen-Blanchette, C., Daniilidis, K.: Equivariant multi-view networks. In: Proceedings of the IEEE International Conference on Computer Vision, pp. 1568–1577 (2019)
28. Fey, M., Eric Lenssen, J., Weichert, F., Müller, H.: SplineCNN: fast geometric deep learning with continuous B-spline kernels. In: The IEEE Conference on Computer Vision and Pattern Recognition (CVPR) (June 2018)
29. Giles, C.L., Maxwell, T.: Learning, invariance, and generalization in high-order neural networks. Appl. Opt. **26**(23), 4972–4978 (1987)
30. Hinton, G.E., Krizhevsky, A., Wang, S.D.: Transforming auto-encoders. In: Honkela, T., Duch, W., Girolami, M., Kaski, S. (eds.) ICANN 2011. LNCS, vol. 6791, pp. 44–51. Springer, Heidelberg (2011). https://doi.org/10.1007/978-3-642-21735-7_6
31. Hoppe, H., DeRose, T., Duchamp, T., McDonald, J., Stuetzle, W.: Surface reconstruction from unorganized points. SIGGRAPH Comput. Graph. **26**(2), 71–78 (1992)

32. Jaiswal, A., AbdAlmageed, W., Wu, Y., Natarajan, P.: CapsuleGAN: generative adversarial capsule network. In: Leal-Taixé, L., Roth, S. (eds.) ECCV 2018. LNCS, vol. 11131, pp. 526–535. Springer, Cham (2019). https://doi.org/10.1007/978-3-030-11015-4_38
33. Jiang, C.M., Huang, J., Kashinath, K., Prabhat, Marcus, P., Niessner, M.: Spherical CNNs on unstructured grids. In: International Conference on Learning Representations (2019)
34. Khoury, M., Zhou, Q.Y., Koltun, V.: Learning compact geometric features. In: Proceedings of the IEEE International Conference on Computer Vision, pp. 153–161 (2017)
35. Kingma, D.P., Ba, J.: Adam: a method for stochastic optimization. arXiv preprint arXiv:1412.6980 (2014)
36. Kondor, R., Lin, Z., Trivedi, S.: Clebsch-Gordan Nets: a fully Fourier space spherical convolutional neural network. In: Advances in Neural Information Processing Systems (2018)
37. Kondor, R., Trivedi, S.: On the generalization of equivariance and convolution in neural networks to the action of compact groups. In: International Conference on Machine Learning, pp. 2747–2755 (2018)
38. Kosiorek, A., Sabour, S., Teh, Y.W., Hinton, G.E.: Stacked capsule autoencoders. In: Advances in Neural Information Processing Systems, pp. 15512–15522 (2019)
39. Laue, S., Mitterreiter, M., Giesen, J.: Computing higher order derivatives of matrix and tensor expressions. In: Advances in Neural Information Processing Systems (2018)
40. Lenssen, J.E., Fey, M., Libuschewski, P.: Group equivariant capsule networks. In: Advances in Neural Information Processing Systems, pp. 8844–8853 (2018)
41. Li, J., Chen, B.M., Hee Lee, G.: SO-Net: self-organizing network for point cloud analysis. In: Proceedings of the IEEE Conference on Computer Vision and Pattern Recognition (2018)
42. Li, Y., Bu, R., Sun, M., Wu, W., Di, X., Chen, B.: PointCNN: convolution on X-transformed points. In: Advances in Neural Information Processing Systems (2018)
43. Liao, S., Gavves, E., Snoek, C.G.: Spherical regression: learning viewpoints, surface normals and 3D rotations on n-spheres. In: Proceedings of the IEEE Conference on Computer Vision and Pattern Recognition. pp. 9759–9767 (2019)
44. Liu, M., Yao, F., Choi, C., Ayan, S., Ramani, K.: Deep learning 3D shapes using alt-az anisotropic 2-sphere convolution. In: International Conference on Learning Representations (ICLR) (2019)
45. Liu, X., Han, Z., Liu, Y.S., Zwicker, M.: Point2Sequence: learning the shape representation of 3D point clouds with an attention-based sequence to sequence network. Proc. AAAI Conf. Artif. Intell. **33**, 8778–8785 (2019)
46. Magnus, J.R.: On differentiating eigenvalues and eigenvectors. Econom. Theor. **1**(2), 179–191 (1985)
47. Marcos, D., Volpi, M., Komodakis, N., Tuia, D.: Rotation equivariant vector field networks. In: The IEEE International Conference on Computer Vision (ICCV) (October 2017)
48. Markley, F.L., Cheng, Y., Crassidis, J.L., Oshman, Y.: Averaging quaternions. J. Guid. Control Dyn. **30**(4), 1193–1197 (2007)
49. Maturana, D., Scherer, S.: VoxNet: A 3D convolutional neural network for real-time object recognition. In: Intelligent Robots and Systems (IROS). IEEE (2015)
50. Mehr, E., Lieutier, A., Sanchez Bermudez, F., Guitteny, V., Thome, N., Cord, M.: Manifold learning in quotient spaces. In: Proceedings of the IEEE Conference on Computer Vision and Pattern Recognition, pp. 9165–9174 (2018)

51. Melzi, S., Spezialetti, R., Tombari, F., Bronstein, M.M., Stefano, L.D., Rodola, E.: GFrames: gradient-based local reference frame for 3D shape matching. In: The IEEE Conference on Computer Vision and Pattern Recognition (CVPR) (June 2019)
52. Petrelli, A., Di Stefano, L.: On the repeatability of the local reference frame for partial shape matching. In: 2011 International Conference on Computer Vision. IEEE (2011)
53. Petrelli, A., Di Stefano, L.: A repeatable and efficient canonical reference for surface matching. In: 2012 2nd International Conference on 3D Imaging, Modeling, Processing, Visualization & Transmission, pp. 403–410. IEEE (2012)
54. Poulenard, A., Ovsjanikov, M.: Multi-directional geodesic neural networks via equivariant convolution. In: SIGGRAPH Asia 2018 Technical Papers, p. 236. ACM (2018)
55. Qi, C.R., Su, H., Mo, K., Guibas, L.J.: PointNet: deep learning on point sets for 3D classification and segmentation. In: Proceedings of the IEEE Conference on Computer Vision and Pattern Recognition, pp. 652–660 (2017)
56. Qi, C.R., Su, H., Nießner, M., Dai, A., Yan, M., Guibas, L.J.: Volumetric and multi-view CNNs for object classification on 3D data. In: Proceedings of the IEEE Conference on Computer Vision and Pattern Recognition, pp. 5648–5656 (2016)
57. Qi, C.R., Yi, L., Su, H., Guibas, L.J.: PointNet++: deep hierarchical feature learning on point sets in a metric space. In: Advances in Neural Information Processing Systems, pp. 5099–5108 (2017)
58. Rezatofighi, S.H., Milan, A., Abbasnejad, E., Dick, A., Reid, I., et al.: DeepSetNet: predicting sets with deep neural networks. In: 2017 IEEE International Conference on Computer Vision (ICCV), pp. 5257–5266. IEEE (2017)
59. Sabour, S., Frosst, N., Hinton, G.: Matrix capsules with EM routing. In: 6th International Conference on Learning Representations (ICLR) (2018)
60. Sabour, S., Frosst, N., Hinton, G.E.: Dynamic routing between capsules. In: Advances in Neural Information Processing Systems, pp. 3856–3866 (2017)
61. Schütt, K., Kindermans, P.J., Sauceda Felix, H.E., Chmiela, S., Tkatchenko, A., Müller, K.R.: SchNet: a continuous-filter convolutional neural network for modeling quantum interactions. In: Advances in Neural Information Processing Systems (2017)
62. Shen, Y., Feng, C., Yang, Y., Tian, D.: Mining point cloud local structures by kernel correlation and graph pooling. In: Proceedings of the IEEE Conference on Computer Vision and Pattern Recognition, pp. 4548–4557 (2018)
63. Spezialetti, R., Salti, S., Stefano, L.D.: Learning an effective equivariant 3D descriptor without supervision. In: Proceedings of the IEEE International Conference on Computer Vision, pp. 6401–6410 (2019)
64. Srivastava, N., Goh, H., Salakhutdinov, R.: Geometric capsule autoencoders for 3D point clouds. arXiv preprint arXiv:1912.03310 (2019)
65. Steenrod, N.E.: The Topology of Fibre Bundles, vol. 14. Princeton University Press, Princeton (1951)
66. Thomas, N., et al.: Tensor field networks: rotation-and translation-equivariant neural networks for 3D point clouds. arXiv preprint arXiv:1802.08219 (2018)
67. Tombari, F., Salti, S., Di Stefano, L.: Unique signatures of histograms for local surface description. In: Daniilidis, K., Maragos, P., Paragios, N. (eds.) ECCV 2010. Lecture Notes in Computer Science, vol. 6313. Springer, Heidelberg (2010)
68. Wang, D., Liu, Q.: An optimization view on dynamic routing between capsules (2018). https://openreview.net/forum?id=HJjtFYJDf

69. Wang, H., Sridhar, S., Huang, J., Valentin, J., Song, S., Guibas, L.J.: Normalized object coordinate space for category-level 6D object pose and size estimation. In: Proceedings of the IEEE Conference on Computer Vision and Pattern Recognition, pp. 2642–2651 (2019)
70. Wang, S., Suo, S., Ma, W.C., Pokrovsky, A., Urtasun, R.: Deep parametric continuous convolutional neural networks. In: The IEEE Conference on Computer Vision and Pattern Recognition (CVPR) (June 2018)
71. Wang, Y., Sun, Y., Liu, Z., Sarma, S.E., Bronstein, M.M., Solomon, J.M.: Dynamic graph CNN for learning on point clouds. ACM Trans. Graph. (TOG) **38**(5), 1–2 (2019)
72. Wang, Y., Sun, Y., Liu, Z., Sarma, S.E., Bronstein, M.M., Solomon, J.M.: Dynamic graph CNN for learning on point clouds. ACM Trans. Graph. (TOG) **38**(5), 1–12 (2019)
73. Weiler, M., Geiger, M., Welling, M., Boomsma, W., Cohen, T.: 3D Steerable CNNs: learning rotationally equivariant features in volumetric data. In: Advances in Neural Information Processing Systems, pp. 10381–10392 (2018)
74. Weiler, M., Hamprecht, F.A., Storath, M.: Learning steerable filters for rotation equivariant CNNs. In: The IEEE Conference on Computer Vision and Pattern Recognition (CVPR) (June 2018)
75. Worrall, D., Brostow, G.: CubeNet: equivariance to 3D rotation and translation. In: Ferrari, V., Hebert, M., Sminchisescu, C., Weiss, Y. (eds.) ECCV 2018. LNCS, vol. 11209, pp. 585–602. Springer, Cham (2018). https://doi.org/10.1007/978-3-030-01228-1_35
76. Worrall, D.E., Garbin, S.J., Turmukhambetov, D., Brostow, G.J.: Harmonic networks: deep translation and rotation equivariance. In: The IEEE Conference on Computer Vision and Pattern Recognition (CVPR) (July 2017)
77. Wu, Z., et al.: 3D ShapeNets: a deep representation for volumetric shapes. In: Proceedings of the IEEE Conference on Computer Vision and Pattern Recognition, pp. 1912–1920 (2015)
78. Xinyi, Z., Chen, L.: Capsule graph neural network. In: International Conference on Learning Representations (ICLR) (2019). openreview.net/forum?id=Byl8BnRcYm
79. You, Y., et al.: Pointwise rotation-invariant network with adaptive sampling and 3D spherical voxel convolution. In: AAAI, pp. 12717–12724 (2020)
80. Yuan, W., Held, D., Mertz, C., Hebert, M.: Iterative transformer network for 3D point cloud. arXiv preprint arXiv:1811.11209 (2018)
81. Zaheer, M., Kottur, S., Ravanbakhsh, S., Poczos, B., Salakhutdinov, R.R., Smola, A.J.: Deep sets. In: Advances in Neural Information Processing Systems (2017)
82. Zhang, X., Qin, S., Xu, Y., Xu, H.: Quaternion product units for deep learning on 3D rotation groups. In: Proceedings of the IEEE/CVF Conference on Computer Vision and Pattern Recognition, pp. 7304–7313 (2020)
83. Zhao, Y., Birdal, T., Deng, H., Tombari, F.: 3D point capsule networks. In: Conference on Computer Vision and Pattern Recognition (CVPR) (2019)

DeepFit: 3D Surface Fitting via Neural Network Weighted Least Squares

Yizhak Ben-Shabat(✉) and Stephen Gould(✉)

Australian Centre for Robotic Vision, Australian National University, Canberra, Australia
{yizhak.benshabat,stephen.gould}@anu.edu.au
https://github.com/sitzikbs/DeepFit

Abstract. We propose a surface fitting method for unstructured 3D point clouds. This method, called DeepFit, incorporates a neural network to learn point-wise weights for weighted least squares polynomial surface fitting. The learned weights act as a soft selection for the neighborhood of surface points thus avoiding the scale selection required of previous methods. To train the network we propose a novel surface consistency loss that improves point weight estimation. The method enables extracting normal vectors and other geometrical properties, such as principal curvatures, the latter were not presented as ground truth during training. We achieve state-of-the-art results on a benchmark normal and curvature estimation dataset, demonstrate robustness to noise, outliers and density variations, and show its application on noise removal.

Keywords: Normal estimation · Surface fitting · Least squares · Unstructured 3D point clouds · 3D point cloud deep learning

1 Introduction

Commodity 3D sensors are rapidly becoming an integral component of autonomous systems. These sensors, e.g., RGB-D cameras or LiDAR, provide a 3D point cloud representing the geometry of the scanned objects and surroundings. This raw representation, however, is challenging to process since it lacks connectivity information or structure, and is often incomplete, noisy and contains point density variations. In particular, processing it by means of convolutional neural networks (CNNs)—highly effective for images—is problematic because CNNs require structured, grid-like data as input.

When available, additional local geometric information, such as the surface normal and principal curvatures at each point, induces a partial local structure and improves performance of different tasks for interpreting the scene, such as over-segmentation [1], classification [19] and surface reconstruction [9].

Electronic supplementary material The online version of this chapter (https://doi.org/10.1007/978-3-030-58452-8_2) contains supplementary material, which is available to authorized users.

A. Vedaldi et al. (Eds.): ECCV 2020, LNCS 12346, pp. 20–34, 2020.
https://doi.org/10.1007/978-3-030-58452-8_2

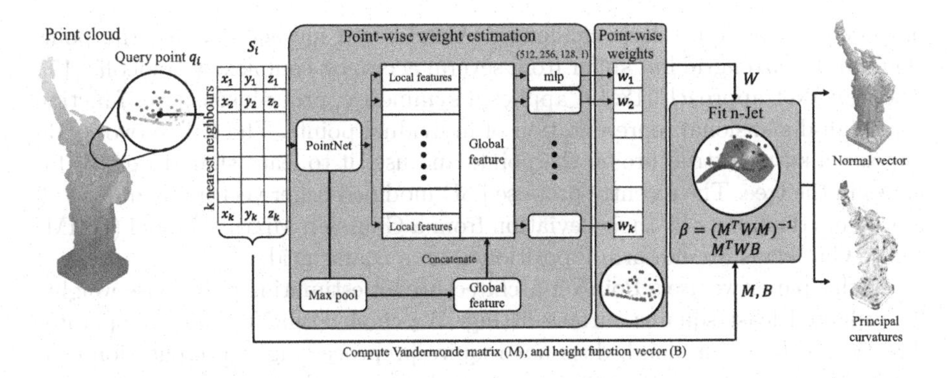

Fig. 1. DeepFit pipeline for normal and principal curvature estimation. For each point in a given point cloud, we compute a global and local representation and estimate a point-wise weight. Then, we fit an n-jet by solving a weighted least squares problem.

Estimating the normals and curvatures from a raw point cloud with no additional information is a challenging task due to difficulties associated with sampling density, noise, outliers, and detail level. The common approach is to specify a neighborhood around a point and then fit a local basic geometric surface (e.g., a plane) to the points in this neighborhood. The normal at the point under consideration is estimated from the fitted geometric surface. The chosen size (or scale) of the neighborhood introduces an unavoidable trade-off between robustness to noise and accuracy of fine details. A large neighborhood over-smooths sharp corners and small details but is otherwise robust to noise. A small neighborhood, on the other hand, may reproduce the normals more accurately around small details but is more sensitive to noise. Evidently, a robust, scale-independent, data-driven surface fitting approach should improve normal estimation performance.

We propose a surface fitting method for unstructured 3D point clouds. It features a neural network for point-wise weight prediction for weighted least squares fitting of polynomial surfaces. This approach removes the multi-scale requirement entirely and significantly increases robustness to different noise levels, outliers, and varying levels of detail. Moreover, the approach enables extracting normal vectors and additional geometric properties without the need for retraining or additional ground truth information. The main contribution of this paper is a method for per-point weight estimation for weighted least squares robust surface fitting that enables scale-free normal estimation, unsupervised principal curvature and geometric properties estimation using deep neural networks.

2 Background and Related Work

2.1 Deep Learning for Unstructured 3D Point Clouds

The point cloud representation of a 3D scene is challenging for deep learning methods because it is both unstructured and unordered. In addition, the number of points in the point cloud varies for different scenes. Several methods have been proposed to overcome these challenges. Voxel-based methods embed the point cloud into a voxel grid but suffer from several accuracy-complexity tradeoffs [15]. The PointNet approach [18,19] applies a symmetric, order-insensitive, function on a high-dimensional representation of individual points. The Kd-Network [13] imposes a kd-tree structure on the points and uses it to learn shared weights for nodes in the tree. The recently proposed 3D modified fisher vectors (3DmFV) [2] represents the points by their deviation from a Gaussian Mixture Model (GMM) whose Gaussians are uniformly positioned on a coarse grid.

In this paper we use a PoinNet architecture for estimating point-wise weights for weighted least squares surface fitting. We chose PointNet since it operates directly on the point cloud, does not require preprocessing, representation conversion or structure, and contains a relatively low number of parameters.

2.2 Normal Vector and Principal Curvature Estimation

A classic method for estimating normals uses principal component analysis (PCA) [10]. Here a neighborhood of points within some fixed scale is chosen and PCA regression is used to estimate a tangent plane. Variants that fit local spherical surfaces [8] or Jets [6] (truncated Taylor expansion) have also been proposed. Further detail on Jet fitting is given in Sect. 2.3. To be robust to noise, these methods usually choose a large-scale neighborhood, leading them to smooth sharp features and fail to estimate normals near 3D edges. Computing the optimal neighborhood size can decrease the estimation error [17] but requires the (usually unknown) noise standard deviation value and a costly iterative process to estimate the local curvature and additional density parameters.

A few deep learning approaches have been proposed to estimate normal vectors from unstructured point clouds. Boulch and Marlet proposed to transform local point cloud patches into a 2D Hough space accumulator by randomly selecting point triplets and voting for that plane's normal. Then, the normal is estimated from the accumulator by designing explicit criteria [4] for bin selection or, more recently, by training a 2D CNN [5] to estimate it continuously as a regression problem. This method does not fully utilize available 3D information since it loses information during the transformation stage. Another method, named PCPNet [9], uses a PointNet [18] architecture over local neighborhoods at multiple scales. It achieves good normal estimation performance and has been extended to estimating principal curvatures. However, it processes the multi-scale point clouds jointly and requires selecting a predefined set of scales. A more recent work, Nesti-Net [3] tries to predict the appropriate scale using a mixture of experts network and a local representation for different scales. It achieves high accuracy but suffers from high computation time due to the multiple scale computations. Nesti-Net shares PCPNet's drawback of requiring a predefined set of scales. A contemporary work [14] uses an iterative plane fitting approach which tries to predict all normals of a local neighborhood and iteratively adjusts the point weights to best fit the plane. The problem of estimating principal curvatures and principal directions is closely related to normal estimation. Meyer et al. [16] proposed a Voronoi-averaging finite element method for curvature estimation in arbitrary triangle meshes. Similarly, Rusinkiewicz [20] proposed a finite-differences approach for estimating curvatures on irregular triangle meshes. Kamberov and Kamberova [11] proposed to use a general orientability constraint that quantifies the confidence in estimating principal curvatures as well as other geometrical quantities on 3D point clouds directly. Recently, PCPNet [9] proposed to train a network to regress the principal curvatures simultaneously with the normal vector.

In this paper we propose a novel approach for normal estimation by learning to fit an n-order Jet while predicting informative points' weights. Our approach removes the need of predefined scales and optimal scale selection since the informative points are extracted at any given scale. Our method generalizes the contemporary method proposed by Lenssen et al. [14], avoids the iterative process, and enables the computation of additional geometric properties such as principal curvatures and principal directions.

2.3 Jet Fitting Using Least Squares and Weighted Least Squares

We now provide background and mathematical notation for truncated Taylor expansion surface fitting using least-squares (LS) and weighted least-squares (WLS). We refer the interested reader to Cazals and Pouget [6] for further detail.

Any regular embedded smooth surface can be locally written as the graph of a bi-variate "height function" with respect to any z-direction that does not belong to the tangent space [21]. We adopt the naming convention of Cazals and Pouget [6] and refer to the truncated Taylor expansion as a degree n jet or n-jet for short. An n-jet of the height function over a surface is given by:

$$f(x, y) = J_{\beta,n}(x, y) = \sum_{k=0}^{n} \sum_{j=0}^{k} \beta_{k-j,j} x^{k-j} y^{j} \tag{1}$$

Here β is the jet coefficients vector that consists of $N_n = (n+1)(n+2)/2$ terms.

In this work we wish to fit a surface to a set of N_p 3D points. For clarity, we move to the matrix notation and specify the Vandermonde matrix M as $(1, x_i, y_i, \ldots, x_i y_i^{n-1}, y_i^n)_{i=1,\ldots,N_p} \in \mathbb{R}^{N_p \times N_n}$ and the height function vector $B = (z_1, z_2, \ldots, z_{N_p})^T \in \mathbb{R}_p^N$ representing the sampled points. We require that every point satisfy Eq. 1, yielding the system of linear equations:

$$M\beta = B \tag{2}$$

When $N_n > N_p$ the system is over-determined and an exact solution may not exist. Therefore we use an LS approximation that minimizes the sum of square errors between the value of the jet and the height function over all points:

$$\beta = \arg \min_{z \in \mathbb{R}^{N_n}} \|Mz - B\|^2 \tag{3}$$

It is well known that the solution can be expressed in closed-form as:

$$\beta = (M^T M)^{-1} M^T B \tag{4}$$

Typically the sampled points include noise and outliers that heavily reduce the fitting accuracy. To overcome this, the formulation given in Eq. 4 can be extended to a weighted least square problem. In this setting, some points have more influence on the fitted model than others. Let $W \in \mathbb{R}^{N_p \times N_p}$ be a diagonal weight matrix $W = \text{diag}(w_1, w_2, \dots, w_{N_p})$. Each element in the matrix's diagonal w_i corresponds to the weight of that point.
The optimization problem becomes:

$$\beta = \arg \min_{z \in \mathbb{R}^{N_n}} \left\| W^{1/2}(Mz - B) \right\|^2$$
$$= \arg \min_{z \in \mathbb{R}^{N_n}} \sum_{i=1}^{N_p} w_i \left(\sum_{j=1}^{N_n} M_{ij} z_j - B_i \right)^2$$

and its solution:

$$\beta = (M^T W M)^{-1} M^T W B \tag{5}$$

In this work, we choose to focus on n-jet fitting because any order n differential quantity can be computed from the n-jet. This is one of the main advantages of our method. That is, our method is trained for estimating normal vectors but is then able to estimate other differential quantities, e.g., principal curvatures, depending on the jet order.

3 DeepFit

3.1 Learning Point-Wise Weights

The full pipeline for our method is illustrated in Fig. 1. Given a 3D point cloud S and a query point $q_i \in S$ we first extract a local subset of points S_i using k-nearest neighbors. We then use a neural network to estimate the weight of each point in the neighborhood, which will subsequently be used for weighted least squares surface fitting. Specifically, we feed S_i into a PointNet [18] network, which outputs a global point cloud representation $G(S_i)$. Additionally, we extract local representations from an intermediate layer for each of the points $p_j \in S_i$ separately to give $g(p_j)$. These representations are then concatenated and fed into a multi-layer perceptron $h(\cdot)$ followed by a sigmoid activation function. We choose a sigmoid in order to limit the output values to be between 0 and 1. The output of this network is a weight per point that is used to construct the diagonal point-weight matrix, $W = \text{diag}(w_j)$ with

$$w_j = \text{sigmoid}(h(G(S_i), g(p_{ij}))) + \epsilon \tag{6}$$

For numerical stability, we add a constant small ϵ in order to avoid the degenerate case of a zero or poorly conditioned matrix. This weight matrix is then used to solve the WLS problem of Eq. 5 and approximate the n-jet coefficients β. All parts of the network are differentiable and therefore it is trained end-to-end.

3.2 Geometric Quantities Estimation

Given the n-jet coefficients β several geometric quantities can be easily extracted:

Normal Estimation. The estimated normal vector is given by:

$$N_i = \frac{(-\beta_1, -\beta_2, 1)}{\sqrt{\beta_1^2 + \beta_2^2 + 1}} \tag{7}$$

Shape Operator and Principal Curvatures. For the second order information we compute the Weingarten map of the surface by multiplying the inverse of the first fundamental form and the second fundamental form. Its eigenvalues are the principal curvatures (k_1, k_2), and its eigenvectors are the principal directions. The computation is done in the tangent space associated with the parametrization.

$$M_{\text{Weingarten}} = -\frac{1}{\sqrt{\beta_1^2 + \beta_2^2 + 1}} \begin{bmatrix} 1+\beta_1^2 & \beta_1\beta_2 \\ \beta_1\beta_2 & 1+\beta_2^2 \end{bmatrix}^{-1} \begin{bmatrix} 2\beta_3 & \beta_4 \\ \beta_4 & 2\beta_5 \end{bmatrix} \tag{8}$$

Generally, the principal curvatures can be used as ground truth in training, however, due to the eigenvalue decomposition, with the high probability of outputting two zero principal curvatures (planes) it suffers from numerical issues when computing the gradients for backpropagation [7]. Therefore, we compute the curvatures only at test time. Note that Monge basis and higher order Monge coefficients can also be computed, similar to Cazals and Pouget [6].

3.3 Consistency Loss

In order to learn point-wise weights, we introduce a local consistency loss L_{con}. This loss is composed of two terms, the weighted normal difference term and a regularization term. The weighted normal difference term computes a weighted average of the sine of the angle between the ground truth normal and the estimated normal at every local neighborhood point. These normals are computed analytically by converting the n-jet to the implicit surface form of $F(x, y, z) = 0$. Therefore, for every query point q_i and its local neighborhood S_i we can compute the normal at each neighboring point $p_j \in S_i$ using:

$$N_j = \left.\frac{\nabla F}{\|\nabla F\|}\right|_{p_j} = \left.\frac{(-\beta \frac{\partial M}{\partial x}^T, \beta \frac{\partial M}{\partial y}^T, 1)}{\|\nabla F\|}\right|_{p_j} \tag{9}$$

Here β is the coefficient vector for subset S_i. Note that this formulation assumes all points to lie on the surface, for points that are not on the surface, the normal error will be large, therefore that points weight will be encouraged to be small. This term can easily converge to an undesired local minimum by setting all weights to zero. In order to avoid that, we add a regularization term which computes the negative average *log* of all weights. In summary, the consistency loss for a query point q_i is then given by:

$$L_{con} = \frac{1}{N_{q_i}} \left[-\sum_{j=1}^{N_{q_i}} log(w_j) + \sum_{j=1}^{N_{q_i}} w_j \left| N_{GT} \times N_j \right| \right] \quad (10)$$

In contrast to Lenssen et al. [14], this formulation allows us to avoid solving multiple linear systems iteratively for each point in the local neighborhood.

In total, to train the network, we sum several loss terms: The *sine* loss between the estimated unoriented normal and the ground truth normal at the query point, the consistency loss, and PointNet's transformation matrix regularization terms $L_{reg} = \left| I - AA^T \right|$.

$$L_{tot} = \left| N_{GT} \times N_i \right| + \alpha_1 L_{con} + \alpha_2 L_{reg} \quad (11)$$

Here, α_1, and α_2 are weighting factors, chosen empirically.

3.4 Implementation Notes

In our experiments we report results using DeepFit with the following configuration, unless otherwise stated. A four layer MLP with sizes 512, 256, 128, and 1; a neighborhood size of 256 points, and a 3-order jet. In order to avoid numerical issues, simplify the notation, and reduce the linear algebra operations, we perform the following pre-processing stages on every local point cloud:

1. Normalization: we translate the point cloud to position the query point in the origin and scale the point cloud to fit a unit sphere.
2. Basis extraction: we perform principal component analysis (PCA) on the point cloud. We then use the resulting three orthonormal eigenvectors as the fitting basis so that the vector associated with the smallest eigenvalue is the last vector of the basis.
3. Coordinate frame transformation: We perform a change of coordinates to move the points into the coordinate system of the fitting basis.
4. Preconditioning: we precondition the Vandermonde matrix by performing column scaling. Each monomial $x_i^k y_i^l$ is divided by h^{k+l}. That is, $M' = MD^{-1}$ with D the diagonal matrix $D = \text{diag}(1, h, h^2, \ldots, h^n)$. We use the mean of the norm $\|(x_i, y_i)\|$ as h. The new system is then $M'(D\beta) = B$ and $\beta = D^{-1}(M'^T W M')^{-1} M'^T W B$.

Note that after the normal is estimated we apply the inverse transform to output the result in the original coordinate frame.

4 Results

4.1 Dataset and Training Details

For training and testing we used the PCPNet shape dataset [9]. The training set consists of eight shapes: four CAD objects (fandisk, boxunion, flower, cup) and four high quality scans of figurines (bunny, armadillo, dragon and turtle). All shapes are given as triangle meshes and densely sampled with 100k points. The data is augmented by introducing i.i.d. Gaussian noise for each point's spatial location with a standard deviation of 0.012, 0.006, 0.00125 w.r.t the bounding box size. This yields a set with 3.2M training examples. The test set consists of 22 shapes, including figurines, CAD objects, and analytic shapes. For evaluation we use the same 5000 point subset per shape as in Guerrero et al. [9].

All variations of our method were trained using 32,768 (1024 samples, 32 shapes) random subsets of the 3.2M training samples at each epoch. We used a batch size of 256, Adam optimizer and a learning rate of 10^{-3}. The implementation was done in PyTorch and trained on a single Nvidia RTX 2080 GPU.

4.2 Normal Estimation Performance

We use the RMSE metric for comparing the proposed DeepFit to other deep learning based methods [3,9,14] and classical geometric methods [6,10]. Additionally, we analyze robustness for two types of data corruption:

- Point density—applying two sampling regimes for point subset selection: gradient, simulating effects of distance from the sensor, and stripes, simulating local occlusions.
- Point perturbations–adding Gaussian noise to the points coordinates with three levels of magnitude specified by σ, given as a percentage of the bounding box.

For the geometric methods, we show results for three different scales: small, medium and large, which correspond to 18, 112, 450 nearest neighbors. For the deep learning based methods we show the results for the single-scale (ss) and multi-scale (ms) versions.

Table 1 shows the unoriented normal RMSE results for the methods detailed above. It can be seen that our method slightly outperforms all other methods for low, medium and no noise augmentation and for gradient density augmentation. For high noise, and striped occlusion augmentation we are a close second to the contemporary work of Lenssen et al. [14] which only estimates the normal vectors while DeepFit also estimates other geometric properties, e.g., principal curvatures. The results also show that all method's performance deteriorate as the noise level rises. In this context, both PCA and Jet perform well for specific noise-scale pairs. In addition, for PCPNet, using a multiple scales only mildly improves performance. Nesti-Net's mixture of experts mitigate the scale-accuracy tradeoff well at the cost of computational complexity. DeepFit's soft point selection process overcomes this tradeoff. In the supplemental materials we perform additional evaluation using the percentage of good points (PGPα) metric.

Figure 2a depicts a visualization of DeepFit's results on three point clouds. Here the normal vectors are mapped to the RGB cube. It shows that for complex shapes (pillar, liberty) with high noise levels, the general direction of the normal

Table 1. Comparison of the RMSE angle error for unoriented normal vector estimation of our DeepFit method to classical geometric methods (PCA [10] and Jet [6] - for three scales small, med, and large corresponding to $k = 18, 122, 450$), and deep learning methods (PCPNet [9], Lenssen et al. [14], and Nesti-Net [3]).

Aug.	Our DeepFit	PCA [10]			Jet [6]			PCPNet [9]		Lenssen et al. [14]	Nesti-Net
scale	ss	small	med	large	small	med	large	ss	ms	ss	ms (MoE)
None	**6.51**	8.31	12.29	16.77	7.60	12.35	17.35	9.68	9.62	6.72	6.99
Noiseσ											
0.00125	**9.21**	12.00	12.87	16.87	12.36	12.84	17.42	11.46	11.37	9.95	10.11
0.006	**16.72**	40.36	18.38	18.94	41.39	18.33	18.85	18.26	18.87	17.18	17.63
0.012	23.12	52.63	27.5	23.5	53.21	27.68	23.41	22.8	23.28	**21.96**	22.28
Density											
Gradient	**7.31**	9.14	12.81	17.26	8.49	13.13	17.8	13.42	11.7	7.73	9.00
Stripes	7.92	9.42	13.66	19.87	8.61	13.39	19.29	11.74	11.16	**7.51**	8.47
average	**11.8**	21.97	16.25	18.87	21.95	16.29	19.02	14.56	14.34	11.84	12.41

vector is predicted correctly, but, the fine details and exact normal vector are not obtained. For a basic shape (Boxysmooth) the added noise does not affect the results substantially. Most notably, DeepFit shows robustness to point density corruptions. Figure 2 depicts a visualization of the angular error in each point for the different methods using a heat map. For the Jet method [6] we display the results for medium scale. For all methods, it can be seen that more errors occur in regions with small details, high curvature e.g. edges and corners, and complex geometry. DeepFit suffers the least from this effect due to its point-wise weight estimation, which allows it to adapt to the different local geometry and disregard irrelevant points in the fitting process.

Figure 3 qualitatively visualizes the performance of DeepFit's point-wise weight prediction network. The colors of the points correspond to weight magnitude, mapped to a heatmap ranging from 0 to 1 i.e. red points highly affect the fit while blue points have low influence. It shows that the network learns to adapt well to corner regions (column $n = 1$), assigning high weights to points on one plane and excluding points on the perpendicular one. Additionally, it shows how the network adapted the weight to achieve a good fit for complex geometries (column $n = 2, 3, 4$).

Figure 4 shows the unoriented normal RMSE results for different parameter choices of our method. We explore different Jet orders $n = 1, 2, 3, 4$, and a different number of neighboring points $k = 64, 128, 256$. It shows that using a large neighborhood size highly improves the performance in high noise cases while only minimally affecting the performance in low noise. It also shows that all jet orders are comparable with a small advantage for order 1-jet (plane) and order 3-jet which is an indication for a bias in the dataset towards low curvature geometry. Additional ablation results, including more augmentations and the PGPα metric are provided in the supplemental material.

Timing and efficiency performance are provided in the supplemental material. DeepFit is faster and has fewer parameters than PCPNet and Nesti-Net and has the potential of only being slightly slower than CGAL implementation of Jet

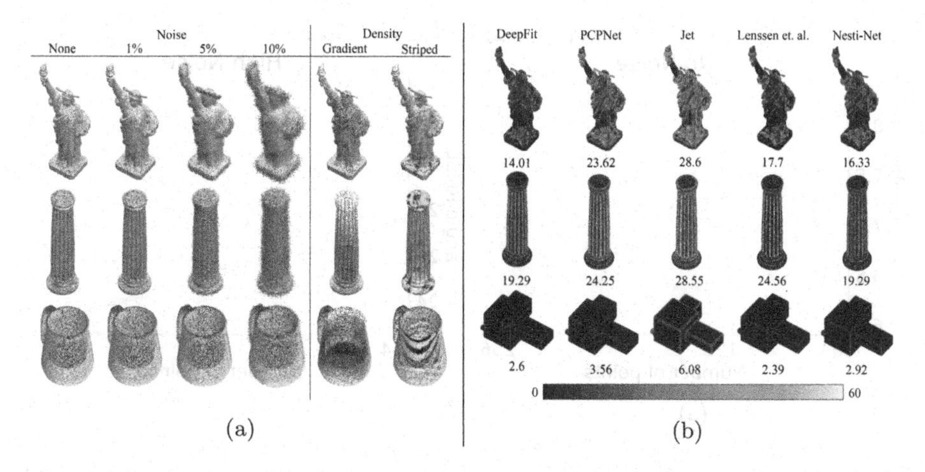

Fig. 2. (a) DeepFit's normal estimation results for different noise levels (columns 1–4), and density distortions (columns 5–6). The colors of the points are normal vectors mapped to RGB. (b) Normal estimation error visualization results of DeepFit compared to other methods for three types of point clouds without noise. The colors of the points correspond to angular difference, mapped to a heatmap ranging from 0–60°. (Color figure online)

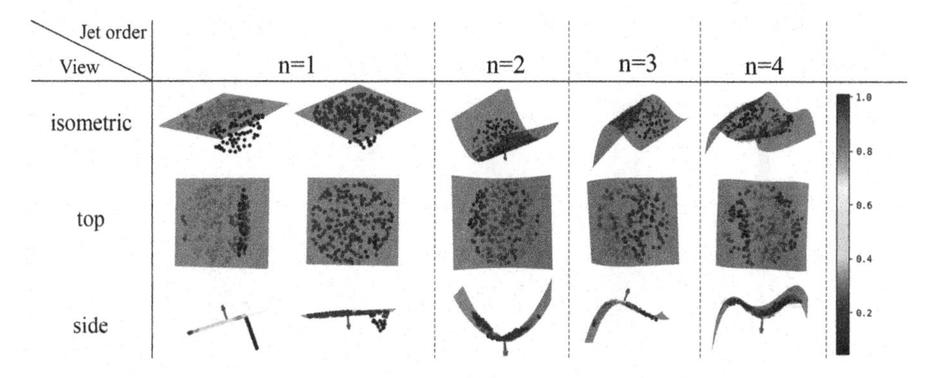

Fig. 3. DeepFit point-wise weight prediction. Three views of different n-jet surface fits. The colors of the points correspond to weight magnitude , mapped to a heatmap ranging from 0 to 1; see color bar on the right i.e. red points highly affect the fit while blue points have low influence. (Color figure online)

fitting because the forward pass for weight estimation is linear w.r.t the number of points and the network weights. Note that Lenssen et al. [14] is faster due to its lower number of parameters and its direct use of the graph structure.

4.3 Principal Curvature Estimation Performance

Figure 5 qualitatively depicts DeepFit's results on five point clouds. For visualization, the principal curvatures are mapped to RGB values according to the

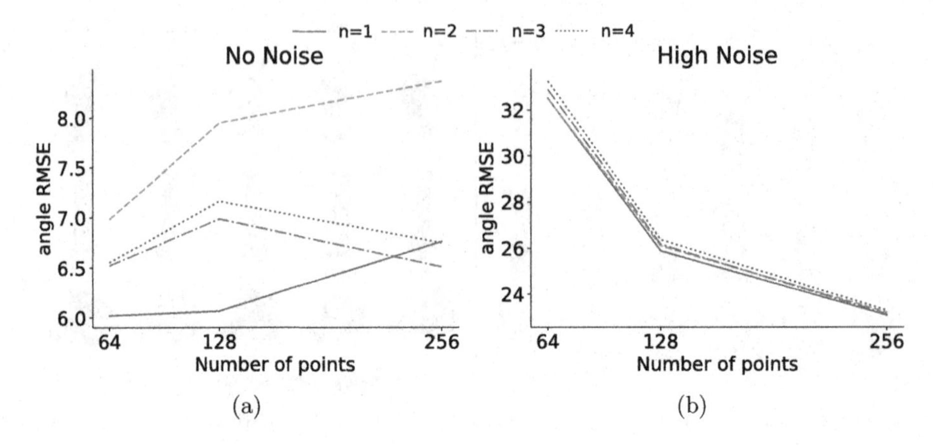

Fig. 4. Normal estimation RMSE results for DeepFit ablations for (a) no noise and (b) high noise augmentations. Comparing the effect of number of neighboring points and jet order.

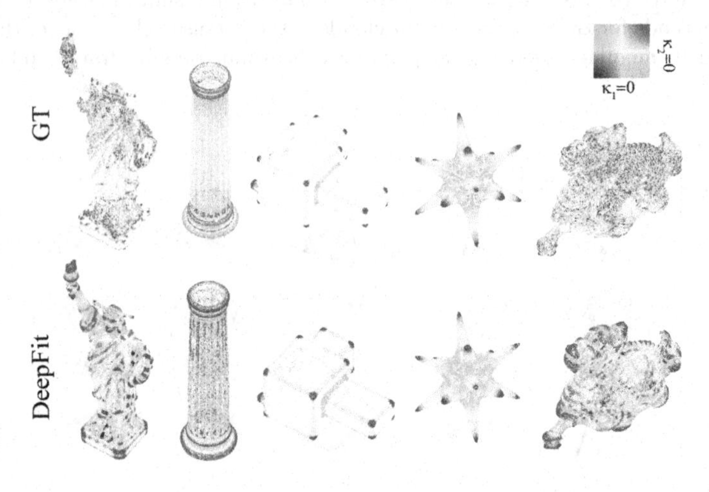

Fig. 5. Curvature estimation results visualization. The colors of the points corresponds to the mapping of k_1, k_2 to the color map given in the bottom right. Values in the range $[-(\mu(|k_i|) + \sigma(|k_i|)), \mu(|k_i|) + \sigma(|k_i|)]|_{i=1,2}$. (Color figure online)

commonly used mapping given in its bottom right corner i.e. both positive (dome) are red, both negative (bowl) are blue, one positive and one negative (saddle) are green, both zero (plane) are white, and one zero and one positive/negative (cylinder) are yellow/cyan. For consistency in color saturation we map each model differently according to the mean and standard deviation of the principal curvatures. Note that the curvature sign is determined by the ground truth normal orientation.

For quantitative evaluation we use the normalized RMSE metric curvature estimation evaluation proposed in Guerrero et al. [9] and given in Eq. 12, for comparing the proposed method to other deep learning based [9] and geometric methods [6]. Table 2 summarizes the results and shows an average error reduction of 35% and 13.7% for maximum and minimum curvatures respectively. We analyze robustness for the same types of data corruptions as in normal estimation i.e. point perturbation and density. DeepFit significantly outperforms all other methods for maximum principal curvature k_1. For the minimum principal curvature k_2 DeepFit outperforms all methods for low and no noise augmentation in addition to gradient and striped density augmentation, however PCPNet has a small advantage for medium and high noise levels. The results for the minimum curvature are very sensitive since most values are close to zero.

$$D_{k_j} = \left| \frac{k_j - k_{GT}}{\max\{|k_{GT}|, 1\}} \right|, \quad \text{for } j = 1, 2. \tag{12}$$

Table 2. Comparison of normalized RMSE for (left) maximal (k_1) and (right) minimal (k_2) principal curvature estimation of our DeepFit method to the classic Jet [6] with three scales, and PCPNet [9]

Aug.	Our DeepFit	PCPNet [9]	Jet [6]		
output scale	k_1+n ss	k_1+n ms	k_1 small	k_1 med.	k_1 large
None	**1.00**	1.36	2.19	6.55	2.97
Noise σ					
0.00125	**1.00**	1.48	57.35	6.68	2.90
0.006	**0.98**	1.46	60.91	9.86	3.30
0.012	**1.21**	1.59	49.40	10.78	3.58
Density					
Gradient	**0.59**	1.32	2.07	1.40	1.53
Stripes	**0.6**	1.09	2.04	1.54	1.89
average	**0.89**	1.38	28.99	6.13	2.69
reduc.	**35.5%**				

Aug.	Our DeepFit	PCPNet [9]	Jet [6]		
output scale	k_2+n ss	k_2+n ms	k_2 small	k_2 med.	k_2 large
None	**0.46**	0.54	1.61	2.91	1.59
Noise σ					
0.00125	**0.47**	0.53	25.83	2.98	1.53
0.006	0.57	**0.51**	22.27	4.88	1.73
0.012	0.68	**0.53**	18.17	5.22	1.84
Density					
Gradient	**0.31**	0.61	2.04	0.79	0.83
Stripes	**0.31**	0.55	1.92	0.89	1.09
average	**0.466**	0.54	11.97	2.94	1.43
reduc.	**13.7%**				

The normalized RMSE metric is visualized in Fig. 6 for DeepFit and PCPNet as the magnitude of the error vector mapped to a heatmap. It can be seen that more errors occur near edges, corners and small regions with a lot of detail and high curvature. These figures show that for both simple and complex geometric shapes DeepFit is able to predict the principal curvatures reliably.

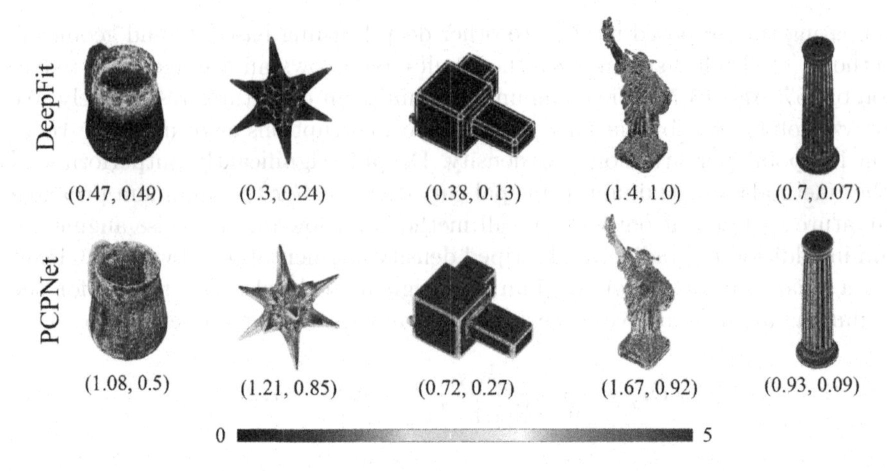

Fig. 6. Curvature estimation error results for DeepFit compared to PCPNet. The numbers under each point cloud are its normalized RMSE errors in the format (k_1, k_2). The color corresponds to the L2 norm of the error vector mapped to a heatmap ranging from 0–5. (Color figure online)

4.4 Surface Reconstruction and Noise Removal

We further investigate the effectiveness of our surface fitting in the context of two subsequent applications—Poisson surface reconstruction [12] and noise removal.

Surface Reconstruction. Figure 7a shows the results for the classical Jet fitting and our DeepFit approach. Since the reconstruction requires oriented normals, we orient the normals, in both methods, according to the ground truth normal. It shows that using DeepFit, the poisson reconstruction is moderately more satisfactory by being smoother overall, and crispier near corners. It also retains small details (liberty crown, cup rim).

Noise Removal. The point-wise weight prediction network enables a better fit by reducing the influence of neighboring points. This weight can also be interpreted as the network's confidence of that point to lie on the object's surface. Therefore, we can use the weight to remove points with low confidence. We first aggregate the weights by summing all of its weight prediction from all of its neighbors. Then we compute the mean and standard deviation of the aggregateed weights and remove points under a threshold of $\mu(\sum w_i) - \sigma(\sum w_i)$. The output point cloud contains less points than the original one and the removed points are mostly attributed to outliers or noise. The results are depicted in Fig. 7b.

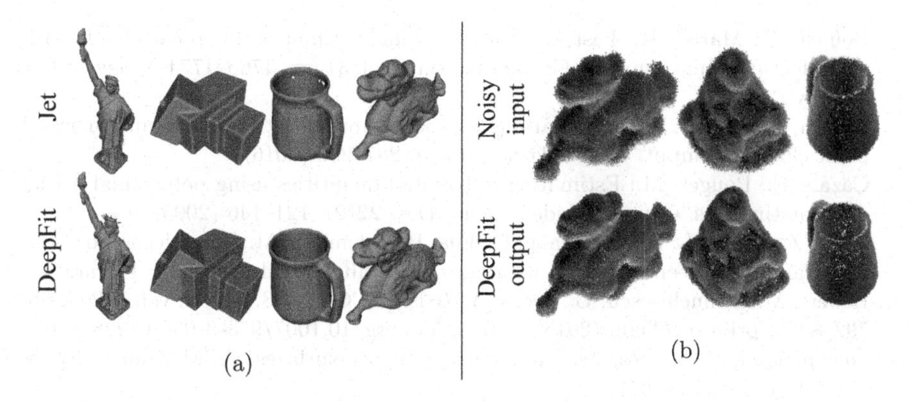

Fig. 7. DeepFit performance in two subsequent application pipelines: (a) Poisson surface reconstruction. (b) Noise removal results using DeepFit predicted weights.

5 Summary

In this paper we presented a novel method for deep surface fitting for unstructured 3D point clouds. The method consists of estimating point-wise weights for solving a weighted least square fitting of an n-jet surface. Our model is fully differentiable and can be trained end-to-end. The estimated weights (at test time) can be interpreted as a confidence measure for every point in the point cloud and used for noise removal. Moreover, the formulation enables the computation of normal vectors and higher order geometric quantities like principal curvatures. The approach demonstrates high accuracy, robustness and efficiency compared to state-of-the-art methods. This is attributed to it's ability to adaptively select the neighborhood of points through a learned model while leveraging classic robust surface fitting approaches, allowing the network to achieve high accuracy with a low number of parameters and computation time.

Acknowledgments. This research was conducted by the Australian Research Council Centre of Excellence for Robotic Vision (CE140100016).

References

1. Ben-Shabat, Y., Avraham, T., Lindenbaum, M., Fischer, A.: Graph based over-segmentation methods for 3D point clouds. Comput. Vis. Image Underst. **174**, 12–23 (2018)
2. Ben-Shabat, Y., Lindenbaum, M., Fischer, A.: 3DmFV: three-dimensional point cloud classification in real-time using convolutional neural networks. IEEE Robot. Autom. Lett. **3**(4), 3145–3152 (2018)
3. Ben-Shabat, Y., Lindenbaum, M., Fischer, A.: Nesti-Net: normal estimation for unstructured 3D point clouds using convolutional neural networks. In: Proceedings of the IEEE Conference on Computer Vision and Pattern Recognition, pp. 10112–10120 (2019)

4. Boulch, A., Marlet, R.: Fast and robust normal estimation for point clouds with sharp features. In: Computer Graphics Forum, vol. 31, pp. 1765–1774. Wiley Online Library (2012)
5. Boulch, A., Marlet, R.: Deep learning for robust normal estimation in unstructured point clouds. Comput. Graph. Forum **35**(5), 281–290 (2016)
6. Cazals, F., Pouget, M.: Estimating differential quantities using polynomial fitting of osculating jets. Comput. Aided Geom. Des. **22**(2), 121–146 (2005)
7. Dang, Z., Yi, K.M., Hu, Y., Wang, F., Fua, P., Salzmann, M.: Eigendecomposition-free training of deep networks with zero eigenvalue-based losses. In: Ferrari, V., Hebert, M., Sminchisescu, C., Weiss, Y. (eds.) ECCV 2018. LNCS, vol. 11209, pp. 792–807. Springer, Cham (2018). https://doi.org/10.1007/978-3-030-01228-1_47
8. Guennebaud, G., Gross, M.: Algebraic point set surfaces. ACM Trans. Graph. (TOG) **26**(3), 23 (2007)
9. Guerrero, P., Kleiman, Y., Ovsjanikov, M., Mitra, N.J.: PCPNet learning local shape properties from raw point clouds. Comput. Graph. Forum **37**(2), 75–85 (2018)
10. Hoppe, H., DeRose, T., Duchampt, T., McDonald, J., Stuetzle, W.: Surface reconstruction from unorganized points. Comput. Graph. **26**, 2 (1992)
11. Kamberov, G., Kamberova, G.: Geometric integrability and consistency of 3D point clouds. In 2007 IEEE 11th International Conference on Computer Vision, pp. 1–6. IEEE (2007)
12. Kazhdan, M., Bolitho, M., Hoppe, H.: Poisson surface reconstruction. In: Proceedings of the 4th Eurographics Symposium on Geometry processing, vol. 7 (2006)
13. Klokov, R., Lempitsky, V.: Escape from cells: deep Kd-networks for the recognition of 3D point cloud models. In The IEEE International Conference on Computer Vision (ICCV), pp. 863–872 (October 2017)
14. Lenssen, J.E., Osendorfer, C., Masci, J.: Differentiable iterative surface normal estimation. arXiv preprint arXiv:1904.07172 (2019)
15. Maturana, D., Scherer, S.: VoxNet: a 3D convolutional neural network for real-time object recognition. In: IEEE/RSJ International Conference on Intelligent Robots and Systems (IROS), pp. 922–928. IEEE (2015)
16. Meyer, M., Desbrun, M., Schröder, P., Barr, A.H.: Discrete differential-geometry operators for triangulated 2-manifolds. In: Hege, H.C., Polthier, K. (eds.) Visualization and Mathematics III. Mathematics and Visualization. Springer, Heidelberg (2003). https://doi.org/10.1007/978-3-662-05105-4_2
17. Mitra, N.J., Nguyen, A.: Estimating surface normals in noisy point cloud data. In: Proceedings of the 19th Annual Symposium on Computational Geometry, pp. 322–328. ACM (2003)
18. Qi, C.R., Su, H., Mo, K., Guibas, L.J.: PointNet: deep learning on point sets for 3D classification and segmentation. In: The IEEE Conference on Computer Vision and Pattern Recognition (CVPR) (July 2017)
19. Qi, C.R., Yi, L., Su, H., Guibas, L.J.: PointNet++: deep hierarchical feature learning on point sets in a metric space. In: Advances in Neural Information Processing Systems, pp. 5099–5108 (2017)
20. Rusinkiewicz, S.: Estimating curvatures and their derivatives on triangle meshes. In: 2004 Proceedings of the 2nd International Symposium on 3D Data Processing, Visualization and Transmission, 3DPVT 2004, pp. 486–493. IEEE (2004)
21. Spivak, M.D.: A Comprehensive Introduction to Differential Geometry. Publish or Perish, Inc., Houston (1970)

NSGANetV2: Evolutionary Multi-objective Surrogate-Assisted Neural Architecture Search

Zhichao Lu(✉), Kalyanmoy Deb, Erik Goodman, Wolfgang Banzhaf, and Vishnu Naresh Boddeti

Michigan State University, East Lansing, MI 48824, USA
{luzhicha,kdeb,goodman,banzhafw,vishnu}@msu.edu

Abstract. In this paper, we propose an efficient NAS algorithm for generating task-specific models that are competitive under multiple competing objectives. It comprises of two surrogates, one at the architecture level to improve sample efficiency and one at the weights level, through a supernet, to improve gradient descent training efficiency. On standard benchmark datasets (C10, C100, ImageNet), the resulting models, dubbed NSGANetV2, either match or outperform models from existing approaches with the search being orders of magnitude more sample efficient. Furthermore, we demonstrate the effectiveness and versatility of the proposed method on six diverse non-standard datasets, e.g. STL-10, Flowers102, Oxford Pets, FGVC Aircrafts etc. In all cases, NSGANetV2s improve the state-of-the-art (under mobile setting), suggesting that NAS can be a viable alternative to conventional transfer learning approaches in handling diverse scenarios such as small-scale or fine-grained datasets. Code is available at https://github.com/mikelzc1990/nsganetv2.

Keywords: NAS · Evolutionary algorithms · Surrogate-assisted search

1 Introduction

Neural networks have achieved remarkable performance on large scale supervised learning tasks in computer vision. A majority of this progress was achieved by architectures designed manually by skilled practitioners. Neural Architecture Search (NAS) [38] attempts to automate this process to find good architectures for a given dataset. This promise has led to tremendous improvements in convolutional neural network architectures, in terms of predictive performance, computational complexity and model size on standard large-scale image classification benchmarks such as ImageNet [27], CIFAR-10 [15], CIFAR-100 [15] etc. However, the utility of these developments, has so far eluded more widespread and

Electronic supplementary material The online version of this chapter (https://doi.org/10.1007/978-3-030-58452-8_3) contains supplementary material, which is available to authorized users.

A. Vedaldi et al. (Eds.): ECCV 2020, LNCS 12346, pp. 35–51, 2020.
https://doi.org/10.1007/978-3-030-58452-8_3

practical applications. These are cases where one wishes to use NAS to obtain high-performance models on custom non-standard datasets, optimizing possibly multiple competing objectives, and to do so without the steep computation burden of existing NAS methods.

The goal of NAS is to obtain both the optimal architecture and its associated optimal weights. The key barrier to realizing the full potential of NAS is the nature of its formulation. NAS is typically treated as a bi-level optimization problem, where an inner optimization loops over the weights of the network for a given architecture, while the outer optimization loops over the network architecture itself. The computational challenge of solving this problem stems from both the upper and lower level optimization. Learning the optimal weights of the network in the lower level necessitates costly iterations of stochastic gradient descent. Similarly, exhaustively searching the optimal architecture is prohibitive due to the discrete nature of the architecture description, size of search space and our desire to optimize multiple, possibly competing, objectives. Mitigating both of these challenges explicitly and simultaneously is the goal of this paper.

Many approaches have been proposed to improve the efficiency of NAS algorithms, both in terms of the upper level and the lower level. A majority of them focuses on the lower level, including weight sharing [1,18,25], proxy models [26,38], coarse training [30], etc. But these approaches still have to sample, explicitly or implicitly, a large number of architectures to evaluate in the upper level. In contrast, there is relatively little focus on improving the sample efficiency of the upper level optimization. A few recent approaches [9,17] adopt surrogates that predict the lower level performance with the goal of navigating the upper level search space efficiently. However, these surrogate predictive models are still very sample inefficient since they are learned in an offline stage by first sampling a large number of architectures that require full lower level optimization.

In this paper, we propose a practically efficient NAS algorithm, by adopting explicit surrogate models simultaneously at both the upper and the lower level. Our lower level surrogate adopts a fine-tuning approach, where the initial weights for fine-tuning are obtained by a supernet model, such as [1,4,5]. Our upper level surrogate adopts an online learning algorithm, that focuses on architectures in the search space that are close to the current trade-off front, as opposed to a random/uniform set of architectures used in the offline surrogate approaches [9,12,17]. Our online surrogate significantly improves the sample efficiency of the upper level optimization problem in comparison to the offline surrogates. For instance, OnceForAll [5] and PNAS [17] sample 16,000 and 1,160[1] architectures, respectively, to learn the upper level surrogate. In contrast, we only have to sample 350 architectures to obtain a model with similar performance.

An overview of our approach is shown in Fig. 1. We refer to the proposed NAS algorithm as MSuNAS and the resulting architectures as NSGANetV2. Our method is designed to provide a set of high-performance models on a custom dataset (large or small scale, multi-class or fine-grained) while optimizing possibly multiple objectives of interest. Our key contributions are:

[1] Estimate from # of models evaluated by PNAS, actual sample size is not reported.

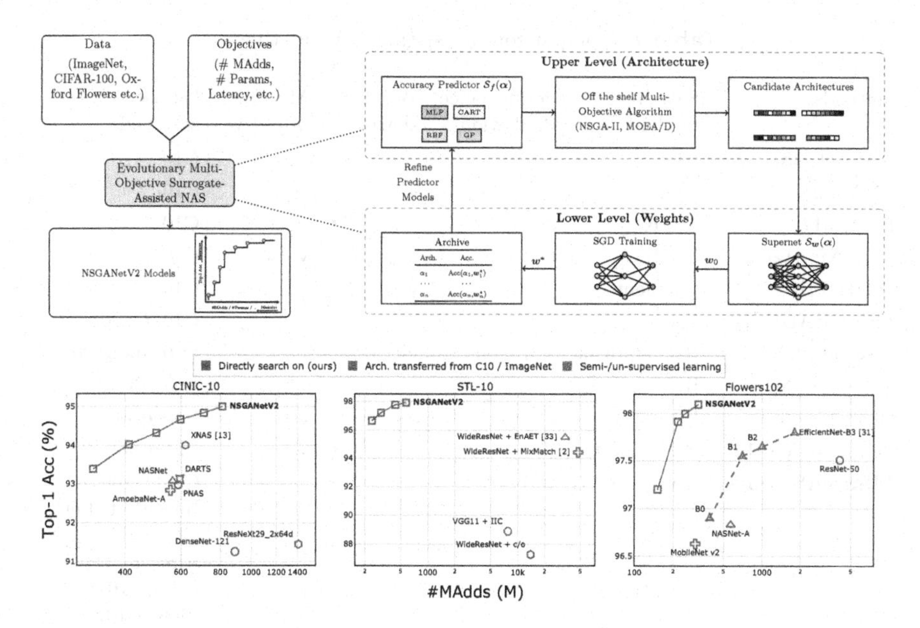

Fig. 1. (Top) **Overview:** Given a dataset and objectives, MSuNAS obtains a task-specific set of models that are competitive in all objectives with high search efficiency. It comprises of two surrogates, one at the upper level to improve sample efficiency and one at the lower level, through a supernet, to improve weight learning efficiency. (Bottom) Performance of the set of task-specific models, i.e. NSGANetV2s, on three different types of non-standard datasets, compared to SOTA from transfer learning [23,31] and semi-/un-supervised learning [2,33]

- An alternative approach to solve the bi-level NAS problem, i.e., simultaneously optimizing the architecture and learn the optimal model weights. However, instead of gradient based relaxations (e.g., DARTS), we advocate for surrogate models. Overall, given a dataset and a set of objectives to optimize, MSuNAS can design custom neural network architectures as efficiently as DARTS but with higher performance and extends to multiple, possibly competing objectives.
- A simple, yet highly effective, online surrogate model for the upper level optimization in NAS, resulting in a significant increase in sampling efficiency over other surrogate-based approaches.
- Scalability and practicality of MSuNAS on many datasets corresponding to different scenarios. These include standard datasets like ImageNet, CIFAR-10 and CIFAR-100, and six non-standard datasets like CINIC-10 [10] (multi-class), STL-10 [8] (small scale mutli-class), Oxford Flowers102 [24] (small scale fine-grained) etc. Under mobile settings ($\leq$ 600M MAdds), MSuNAS leads to SOTA performance.

Table 1. Comparison of existing NAS methods

Methods	Search method	Performance Prediction	Weight sharing	Multiple objective	Dataset searched
NASNet [38]	RL				C10
ENAS [25]	RL		✓		C10
PNAS [17]	SBMO	✓			C10
DPP-Net [12]	SBMO	✓		✓	C10
DARTS [18]	Gradient		✓		C10
LEMONADE [13]	EA		✓	✓	C10, C100
ProxylessNAS [6]	RL+gradient		✓	✓	C10, ImageNet
MnasNet [30]	RL			✓	ImageNet
ChamNet [9]	EA	✓		✓	ImageNet
MobileNetV3 [14]	RL+expert			✓	ImageNet
MSuNAS (ours)	EA	✓	✓	✓	C10, C100, ImageNet, Pets, STL-10, Aircraft, DTD, CINIC-10, Flowers102

2 Related Work

Lower Level Surrogate: Existing approaches [4,18,21,25] primarily focus on mitigating the computational overhead induced by SGD-based weight optimization in the lower level, as this process needs to be repeated for every architecture sampled by a NAS method in the upper level. A common theme among these methods involves training a supernet which contains all searchable architectures as its sub-networks. During search, accuracy using the weights inherited from the supernet becomes the metric to select architectures. However, completely relying on supernet as a substitute of actual weight optimization for evaluating candidate architectures is unreliable. Numerous studies [16,35,36] reported a weak correlation between the performance of the searched architectures (predicted by weight sharing) and the ones trained from scratch (using SGD) during the evaluation phase. MSuNAS instead uses the weights inherited from the supernet only as an initialization to the lower level optimization. Such a fine-tuning process affords the computation benefit of the supernet, while at the same time improving the correlation in the performance of the weights initialized from the supernet and those trained from scratch (Table 1).

Upper Level Surrogate: MetaQNN [1] uses surrogate models to predict the final accuracy of candidate architectures (as a time-series prediction) from the first 25% of the learning curve from SGD training. PNAS [17] uses a surrogate model to predict the top-1 accuracy of architectures with an additional branch added to the cell structure that are repeatedly stacked together. Fundamentally, both of these approaches seek to extrapolate rather than interpolate the performance of the architecture using the surrogates. Consequently, as we show later in the paper, the rank-order between the predicted accuracy and the true accuracy is very low[2] (0.476). OnceForAll [5] also uses a surrogate model to predict accuracy from architecture encoding. However, the surrogate model is trained offline for the entire search space, thereby needing a large number of samples for learning (16K samples -> 2 GPU-days -> 2x search cost of DARTS for just constructing the surrogate model). Instead of using uniformly sampled architectures and their validation accuracy to train the surrogate model to approximate the entire landscape, ChamNet [9] trains many architectures through full lower level optimization and selects only 300 samples with high accuracy with diverse efficiency (FLOPs, Latency, Energy) to train a surrogate model offline. In contrast, MSuNAS learns a surrogate model in an online fashion only on the samples that are close to the current trade-off front as we explore the search space. The online learning approach significantly improves the sample efficiency of our search, since we only need lower level optimization (full or surrogate assisted) for the samples near the current Pareto front.

Multi-objective NAS: Approaches that consider more than one objective to optimize the architecture can be categorized into two groups: (i) scalarization, and (ii) population based approaches. The former include, ProxylessNAS [6], MnasNet [30], FBNet [34], and MobileNetV3 [14] which use a scalarized objective that encourages high accuracy and penalizes compute inefficiency at the same time, e.g., maximize $Acc * (Latency/Target)^{-0.07}$. These methods require a pre-defined preference weighting of the importance of different objectives before the search, which typically requires a numbers of trials. Methods in the latter category include [7,12,13,19,20] and aim to approximate the entire Pareto-efficient frontier simultaneously. These approaches rely on heuristics (e.g., EA) to efficiently navigate the search space, which allows practitioners to visualize the trade-off between the objectives and to choose a suitable network *a posteriori* to the search. MSuNAS falls in the latter category using surrogate models to mitigate the computational overhead.

3 Proposed Approach

The neural architecture search problem for a target dataset $\mathcal{D} = \{\mathcal{D}_{trn}, \mathcal{D}_{vld}, \mathcal{D}_{tst}\}$ can be formulated as the following bilevel optimization problem [3],

$$\begin{aligned} \text{minimize} \quad & \mathbf{F}(\boldsymbol{\alpha}) = \left(f_1(\boldsymbol{\alpha}; \boldsymbol{w}^*(\boldsymbol{\alpha})), \dots, f_k(\boldsymbol{\alpha}; \boldsymbol{w}^*(\boldsymbol{\alpha})), f_{k+1}(\boldsymbol{\alpha}), \dots, f_m(\boldsymbol{\alpha})\right)^T, \\ \text{subject to} \quad & \boldsymbol{w}^*(\boldsymbol{\alpha}) \in \text{argmin } \mathcal{L}(\boldsymbol{w}; \boldsymbol{\alpha}), \\ & \boldsymbol{\alpha} \in \boldsymbol{\Omega}_\alpha, \quad \boldsymbol{w} \in \boldsymbol{\Omega}_w, \end{aligned} \tag{1}$$

[2] In the supplementary material we show that better rank-order correlation at the search stage ultimately leads to finding better performing architectures.

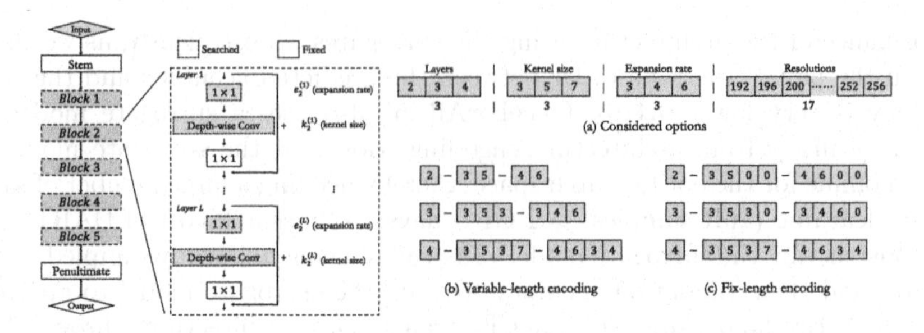

Fig. 2. Search Space: A candidate architecture comprises five computational blocks. Parameters we search for include image resolution, number of layers (L) in each block and the expansion rate (e) and the kernel size (k) in each layer.

where the upper level variable $\boldsymbol{\alpha}$ defines a candidate CNN architecture, and the lower level variable $\boldsymbol{w}(\boldsymbol{\alpha})$ defines the associated weights. $\mathcal{L}(\boldsymbol{w};\boldsymbol{\alpha})$ denotes the cross-entropy loss on the training data $\mathcal{D}_{trn}$ for a given architecture $\boldsymbol{\alpha}$. $\mathbf{F}:\boldsymbol{\Omega}\rightarrow\mathbb{R}^m$ constitutes m desired objectives. These objectives can be further divided into two groups, where the first group (f_1 to f_k) consists of objectives that depend on both the architecture and the weights—e.g., predictive performance on validation data $\mathcal{D}_{vld}$, robustness to adversarial attack, etc. The other group (f_{k+1} to f_m) consists of objectives that only depend on the architecture—e.g., number of parameters, floating point operations, latency etc.

3.1 Search Space

MSuNAS searches over four important dimensions of convolutional neural networks (CNNs), including depth (# of layers), width (# of channels), kernel size and input resolution. Following previous works [5,14,30], we decompose a CNN architecture into five sequentially connected blocks, with gradually reduced feature map size and increased number of channels. In each block, we search over the number of layers, where only the first layer uses stride 2 if the feature map size decreases, and we allow each block to have minimum of two and maximum of four layers. Every layer adopts the inverted bottleneck structure [28] and we search over the expansion rate in the first 1×1 convolution and the kernel size of the depth-wise separable convolution. Additionally, we allow the input image size to range from 192 to 256. We use an integer string to encode these architectural choices, and we pad zeros to the strings of architectures that have fewer layers so that we have a fixed-length encoding. A pictorial overview of this search space and encoding is shown in Fig. 2.

3.2 Overall Algorithm Description

The problem in Eq. 1 poses two main computational bottlenecks for conventional bi-level optimization methods. First, the lower level problem of learning the optimal weights $\boldsymbol{w}^*(\boldsymbol{\alpha})$ for a given architecture $\boldsymbol{\alpha}$ involves a prolonged training

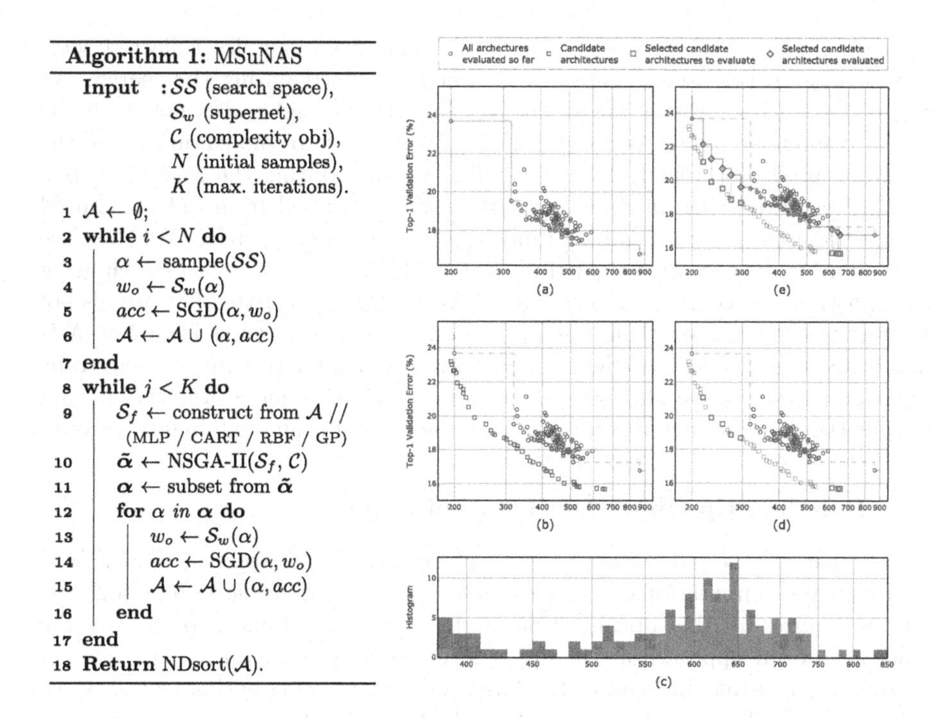

Algorithm 1: MSuNAS
Input : $\mathcal{SS}$ (search space), $\mathcal{S}_w$ (supernet), $\mathcal{C}$ (complexity obj), N (initial samples), K (max. iterations).
1 $\mathcal{A} \leftarrow \emptyset$;
2 while $i < N$ do
3 $\alpha \leftarrow$ sample($\mathcal{SS}$)
4 $w_o \leftarrow \mathcal{S}_w(\alpha)$
5 $acc \leftarrow$ SGD(α, w_o)
6 $\mathcal{A} \leftarrow \mathcal{A} \cup (\alpha, acc)$
7 end
8 while $j < K$ do
9 $\mathcal{S}_f \leftarrow$ construct from $\mathcal{A}$ // (MLP / CART / RBF / GP)
10 $\tilde{\boldsymbol{\alpha}} \leftarrow$ NSGA-II($\mathcal{S}_f$, $\mathcal{C}$)
11 $\boldsymbol{\alpha} \leftarrow$ subset from $\tilde{\boldsymbol{\alpha}}$
12 for α in $\boldsymbol{\alpha}$ do
13 $w_o \leftarrow \mathcal{S}_w(\alpha)$
14 $acc \leftarrow$ SGD(α, w_o)
15 $\mathcal{A} \leftarrow \mathcal{A} \cup (\alpha, acc)$
16 end
17 end
18 Return NDsort($\mathcal{A}$).
All architectures evaluated so far
Candidate architectures
Selected candidate architectures to evaluate
Selected candidate architectures evaluated
Top-1 Validation Error (%)
Histogram
(a)
(b)
(c)
(d)
(e)

Fig. 3. A sample run of MSuNAS on ImageNet: In each iteration, accuracy-prediction surrogate models $\mathcal{S}_f$ are constructed from an archive of previously evaluated architectures (a). New candidate architectures (brown boxes in (b)) are obtained by solving the auxiliary single-level multi-objective problem $\tilde{F} = \{\mathcal{S}_f, \mathcal{C}\}$ (line 10 in Algorithm 1). A subset of the candidate architectures is chosen to diversify the Pareto front (c)–(d). The selected candidate architectures are then evaluated and added to the archive (e). At the conclusion of search, we report the non-dominated architectures from the archive. The x-axis in all sub-figures is #MAdds. (Color figure online)

process—e.g., one complete SGD training on ImageNet dataset takes two days on an 8-GPU server. Second, even though there exist techniques like weight-sharing to bypass the gradient-descent-based weight learning process, extensively sampling architectures at the upper level can still render the overall process computationally prohibitive, e.g., 10,000 evaluations on ImageNet take 24 GPU hours, and for methods like NASNet, AmoebaNet that require more than 20,000 samples, it still requires days to complete the search even with weight-sharing.

Algorithm 1 and Fig. 3 show the pseudocode and corresponding steps from a sample run of MSuNAS on ImageNet, respectively. To overcome the aforementioned bottlenecks, we use surrogate models at both upper and lower levels to make our NAS algorithm practically useful for a variety of datasets and objectives. At the upper level, we construct a surrogate model that predicts the top-1 accuracy from integer strings that encode architectures. Previous approaches [5,9,29] that also used surrogate-modeling of the accuracy follow an offline approach, where the accuracy predictor is built from samples collected separately prior to the architecture search and not refined during the search. We argue that such a process makes the search outcome highly dependent on the initial training samples. As an alternative, we propose to model and refine the accuracy predictor iteratively in an online manner during the search. In particular, we start with an accuracy predictor constructed from only a limited number of architectures sampled randomly from the search space. We then use a standard multi-objective algorithm (NSGA-II [11], in our case) to search using the constructed accuracy predictor along with other objectives that are also of interest to the user. We then evaluate the outcome architectures from NSGA-II and refine the accuracy predictor model with these architectures as new training samples. We repeat this process for a pre-specified number of iterations and output the non-dominated solutions from the pool of evaluated architectures.

3.3 Speeding Up Upper Level Optimization

Recall that the nested nature of the bi-level problem makes the upper level optimization computationally very expensive, as every upper level function evaluation requires another optimization at the lower level. Hence, to improve the efficiency of our approach at the upper level, we focus on reducing the number of architectures that we send to the lower level for learning optimal weights. To achieve this goal, we need a surrogate model to predict the accuracy of an architecture before we actually train it. There are two desired properties of such a predictor: (1) high rank-order correlation between the predicted and the true performance; and (2) sample efficient such that the required number of architectures to be trained through SGD are minimized for constructing the predictor.

We first collected four different surrogate models for accuracy prediction from the literature, namely, Multi Layer Perceptron (MLP) [17], Classification And Regression Trees (CART) [29], Radial Basis Function (RBF) [1] and Gaussian Process (GP) [9]. From our ablation study, we observed that no one surrogate model is consistently better than others in terms of the above two criteria on all datasets (see Sect. 4.1). Hence, we propose a selection mechanism, dubbed Adaptive Switching (AS), which constructs all four types of surrogate models at every iteration and adaptively selects the best model via cross-validation.

With the accuracy predictor selected by AS, we apply the NSGA-II algorithm to simultaneously optimize for both accuracy (predicted) and other objectives of interest to the user (line 10 in Algorithm 1). For the purpose of illustration, we assume that the user is interested in optimizing #MAdds as the second objective. At the conclusion of the NSGA-II search, a set of non-dominated architectures is output, see Fig. 3(b). Often times, we cannot afford to train all architectures in the set. To select a subset, we first select the architecture with highest predicted accuracy. Then we project all other architecture candidates to the #MAdds axis, and pick the remaining architectures from the sparse regions that help in extending the Pareto frontier to diverse #MAdds regimes, see Fig. 3(c)–(d). The architectures from the chosen subset are then sent to the lower level for SGD training. We finally add these architectures to the training samples to refine our accuracy predictor models and proceed to next iteration, see Fig. 3(e).

3.4 Speeding Up Lower Level Optimization

To further improve the search efficiency of the proposed algorithm, we adopt the widely-used weight-sharing technique [4,21,22]. First, we need a supernet such that all searchable architectures are sub-networks of it. We construct such a supernet by taking the searched architectural hyperparameters at their maximum values, i.e., with four layers in each of the five blocks, with expansion ratio set to 6 and kernel size set to 7 in each layer (See Fig. 2). Then we follow the progressive shrinking algorithm [5] to train the supernet. This process is executed once before the architecture search. The weights inherited from the trained supernet are used as a warm-start for the gradient descent algorithm during architecture search.

4 Experiments and Results

In this section, we evaluate the surrogate predictor, the search efficiency and the obtained architectures on CIFAR-10 [15], CIFAR-100 [15], and ImageNet [27].

4.1 Performance of the Surrogate Predictors

To evaluate the effectiveness of the considered surrogate models, we uniformly sample 2,000 architectures from our search space, and train them using SGD for 150 epochs on each of the three datasets and record their accuracy on 5,000 held-out images from the training set. We then fit surrogate models with different number of samples randomly selected from the 2,000 collected. We repeat the process for 10 trials to compare the mean and standard deviation of the rank-order correlation between the predicted and true accuracy, see Fig. 4. In general, we observe that no single surrogate model consistently outperforms the others on all three datasets. Hence, at every iteration, we adopt an Adaptive Switching (AS) routine that compares the four surrogate models and chooses the best based on 10-fold cross-validation. It is evident from Fig. 4 that AS works better than any one of the four surrogate models alone on all three datasets. The construction time of the AS is negligible (relatively to the search cost).

4.2 Search Efficiency

In this section, we first compare the search efficiency of MSuNAS to other single-objective methods on both CIFAR-10 and ImageNet. To quantify the speedup, we compare the two governing factors, namely, the total number of architectures evaluated by each method to reach the reported accuracy and the number of epochs undertaken to train each sampled architecture during search. The results are provided in Table 2. We observe that MSuNAS is **20x faster** than methods

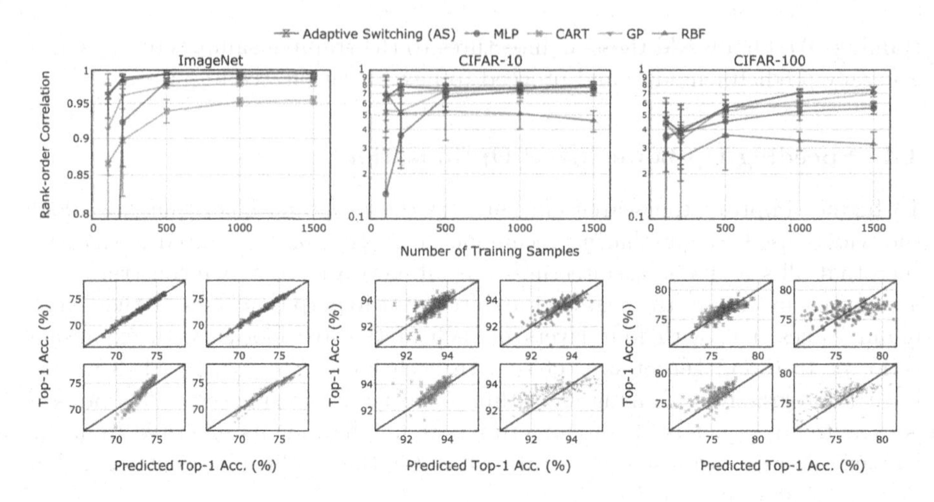

Fig. 4. Comparing the relative prediction performance of the proposed Adaptive Switching (AS) method to the existing four surrogate models. Top row compares Spearman rank-order correlation coefficient as number of training samples increases. Bottom row visualizes the true vs. predicted accuracy under 500 training samples (RBF method is omitted to conserve space).

that use RL or EA. When compared to PNAS [17], which also utilizes an accuracy predictor, MSuNAS is still at least **3x faster**.

We then compare the search efficiency of MSuNAS to NSGANet [20] and random search under a bi-objective setup: Top-1 accuracy and #MAdds. To perform the comparison, we run MSuNAS for 30 iterations, leading to 350 architectures evaluated in total. We record the cumulative hypervolume [37] achieved against the number of architectures evaluated. We repeat this process five times on both ImageNet and CIFAR-10 datasets to capture the variance in performance due to randomness in the search initialization. For a fair comparison to NSGANet, we apply the search code to our search space and record the number of architectures evaluated by NSGANet to reach a similar hypervolume than that achieved by MSuNAS. The random search baseline is performed by uniformly sampling from our search space. We plot the mean and the standard deviation of the hypervolume values achieved by each method in Fig. 5. Based on the incremental rate of hypervolume metric, we observe that MSuNAS is **2–5x faster**, on average, in achieving a better Pareto frontier in terms of number of architectures evaluated.

4.3 Results on Standard Datasets

Prior to the search, we train the supernet following the training hyperparameters setting from [5]. For each dataset, we start MSuNAS with 100 randomly sampled architectures and run for 30 iterations. In each iteration, we evaluate 8 architectures selected from the candidates recommended by NSGA-II according to the accuracy predictor. For searching on CIFAR-10 and CIFAR-100, we fine

Table 2. Comparing the relative search efficiency of MSuNAS to other single-objective methods: "#Model" is the total number of architectures evaluated during search, "#Epochs" is the number of epochs used to train each architecture during search. † and ‡ denote training epochs with and without a supernet to warm-start the weights, respectively.

	Method	Type	Top1 Acc	#MAdds	#Model	Speedup	#Epochs	Speedup
CIFAR-10	NASNet-A [38]	RL	97.4%	569M	20,000	57x	20	up to 4x
	AmoebaNet-B [26]	EA	97.5%	555M	27,000	77x	25	up to 5x
	PNASNet-5 [17]	SMBO	96.6%	588M	1,160	3.3x	20	up to 4x
	MSuNAS (ours)	EA	98.4%	468M	350	1x	$5^{\dagger}/20^{\ddagger}$	1x
ImageNet	MnasNet-A [30]	RL	75.2%	312M	8,000	23x	5	up to 5x
	OnceForAll [5]	EA	76.0%	230M	16,000	46x	0	-
	MSuNAS (ours)	EA	75.9%	225M	350	1x	$0^{\dagger}/5^{\ddagger}$	1x

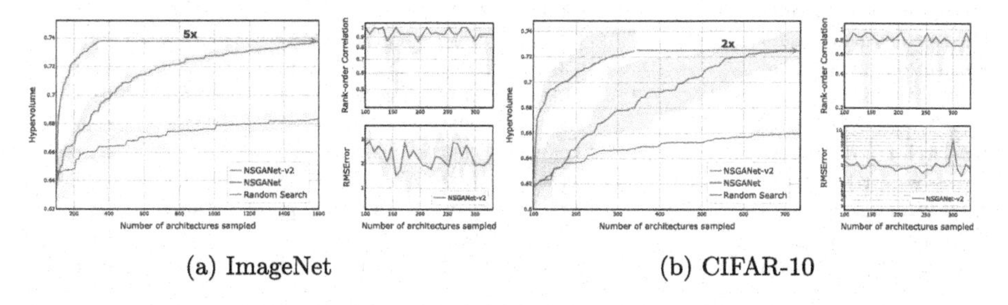

(a) ImageNet (b) CIFAR-10

Fig. 5. Comparing the relative search efficiency of MSuNAS to other methods under bi-objective setup on ImageNet (a) and CIFAR-10 (b). The left plots in each subfigure compares the hypervolume metric [37], where a larger value indicates a better Pareto front achieved. The right plots in each subfigure show the Spearman rank-order correlation (top) and the root mean square error (bottom) of MSuNAS. All results are averaged over five runs with standard deviation shown in shaded regions.

tune the weights inherited from the supernet for five epochs then evaluate on 5K held-out validation images from the original training set. For searching on ImageNet, we re-calibrate the running statistics of the BN layers after inheriting the weights from the supernet, and evaluate on 10K held-out validation images from the original training set. At the conclusion of the search, we pick the four architectures from the achieved Pareto front, and further fine-tune for additional 150–300 epochs on the entire training sets. For reference purpose, we name the obtained architectures as NSGANetV2-s/m/l/xl in ascending #MAdds order. Architectural details can be found in the supplementary materials.

Table 3 shows the performance of our models on the ImageNet 2012 benchmark [27]. We compare models in terms of predictive performance on the validation set, model efficiency (measured by #MAdds and latencies on different hardware), and associated search cost. Overall, NSGANetV2 consistently either matches or outperforms other models across different accuracy levels with highly competitive search costs. In particular, NSGANetV2-s is **2.2% more accurate** than MobileNetV3 [14] while being equivalent in #MAdds and latencies;

Table 3. ImageNet Classification [27]: comparing NSGANetV2 with manual and automated design of efficient networks. Models are grouped into sections for better visualization. Our results are underlined and best result in each section is in bold. CPU latency (batchsize = 1) is measured on Intel i7-8700K and GPU latency (batchsize = 64) is measured on 1080Ti. † The search cost excludes the supernet training cost. ‡ Estimated based on the claim that PNAS is 8x faster than NASNet from [17].

Model	Type	Search Cost (GPU days)	#Params	#MAdds	CPU Lat. (ms)	GPU Lat. (ms)	Top-1 Acc. (%)	Top-5 Acc. (%)
NSGANetV2-s	auto	1†	6.1M	225M	9.1	30	**77.4**	**93.5**
MobileNetV2 [28]	manual	0	3.4M	300M	**8.3**	**23**	72.0	91.0
FBNet-C [34]	auto	9	5.5M	375M	9.1	31	74.9	-
ProxylessNAS [6]	auto	8.3	7.1M	465M	8.5	27	75.1	92.5
MobileNetV3 [14]	combined	-	5.4M	219M	10.0	33	75.2	-
OnceForAll [5]	auto	2†	6.1M	230M	9.5	31	76.9	-
NSGANetV2-m	auto	1†	7.7M	312M	**11.4**	**37**	**78.3**	**94.1**
EfficientNet-B0 [31]	auto	-	5.3M	390M	14.4	46	76.3	93.2
MixNet-M [32]	auto	-	5.0M	360M	24.3	79	77.0	93.3
AtomNAS-C+ [22]	auto	1†	5.5M	329M	-	-	77.2	93.5
NSGANetV2-l	auto	1†	8.0M	400M	**12.9**	**52**	**79.1**	**94.5**
PNASNet-5 [17]	auto	250‡	5.1M	588M	35.6	82	74.2	91.9
NSGANetV2-xl	auto	1†	8.7M	593M	**16.7**	**73**	**80.4**	**95.2**
EfficientNet-B1 [31]	auto	-	7.8M	700M	21.5	78	78.8	94.4
MixNet-L [32]	auto	-	7.3M	565M	29.4	105	78.9	94.2

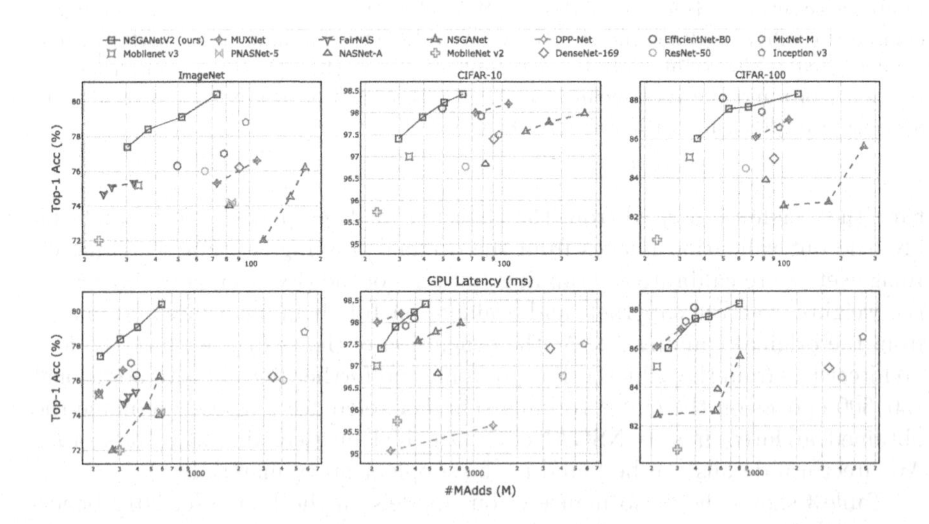

Fig. 6. Accuracy vs Efficiency: Top row compares predictive accuracy vs. GPU latency on a batch of 64 images. Bottom row compares predictive accuracy vs. number of multi-adds in millions. Models from multi-objective approaches are joined with lines. Our models are obtained by directly searching on the respective datasets. In most problems, MSuNAS finds more accurate solutions with fewer parameters.

NSGANetV2-xl achieves **80.4% Top-1 accuracy** under 600M MAdds, which is **1.5% more accurate** and **1.2x more efficient** than EfficientNet-B1 [31]. Additional comparisons to models from multi-objective approaches are provided in Fig. 6.

For CIFAR datasets, Fig. 6 compares our models with other approaches in terms of both predictive performance and computational efficiency. On CIFAR-10, we observe that NSGANetV2 dominates all previous models including (1) NASNet-A [38], PNASNet-5 [17] and NSGANet [20] that search on CIFAR-10 directly, and (2) EfficientNet [31], MobileNetV3 [14] and MixNet [32] that fine-tune from ImageNet.

5 Scalability of MSuNAS

5.1 Types of Datasets

Existing NAS approaches are rarely evaluated for their search ability beyond standard benchmark datasets, i.e., ImageNet, CIFAR-10, and CIFAR-100. Instead, they follow a conventional transfer learning setup, in which the architectures found by searching on standard benchmark datasets are transferred, with weights fine-tuned, to new datasets. We argue that such a process is conceptually contradictory to the goal of NAS, and the architectures identified under such a process are sub-optimal. In this section we demonstrate the scalability of MSuNAS to six[3] additional datasets with various forms of difficulties, in terms of diversity in classification classes (multi-classes vs. fine-grained) and size of training set (see Table 4). We adopt the settings of the CIFAR datasets as outlined in Sect. 3. For each dataset, one search takes less than one day on 8 GPU cards.

Table 4. Non-standard Datasets for MSuNAS

Datasets	Type	#Classes	#Train	#Test
CINIC-10 [10]	Multi-class	10	90,000	90,000
STL-10 [8]	Multi-class	10	5,000	8,000
Flowers102 [24]	Fine-grained	102	2,040	6,149

Figure 1 (Bottom) compares the performance of NSGANetV2 obtained by searching directly on the respective datasets to models from other approaches that transfer architectures learned from either CIFAR-10 or ImageNet. Overall, we observe that NSGANetV2 significantly outperforms other models on all three datasets. In particular, NSGANetV2 achieves a better performance than the currently known state-of-the-art on CINIC-10 [23] and STL-10 [2]. Furthermore, on Oxford Flowers102, NSGANetV2 achieves better accuracy to that of EfficientNet-B3 [31] while using **1.4B fewer** MAdds.

5.2 Number of Objectives

Single-Objective Formulation: Adding a hardware efficiency target as a penalty term to the objective of maximizing predictive performance is a common workaround to handle multiple objectives in the NAS literature [6,30,34].

[3] Due to space constraints, we report results from three datasets in the main paper and three more in the supplementary material.

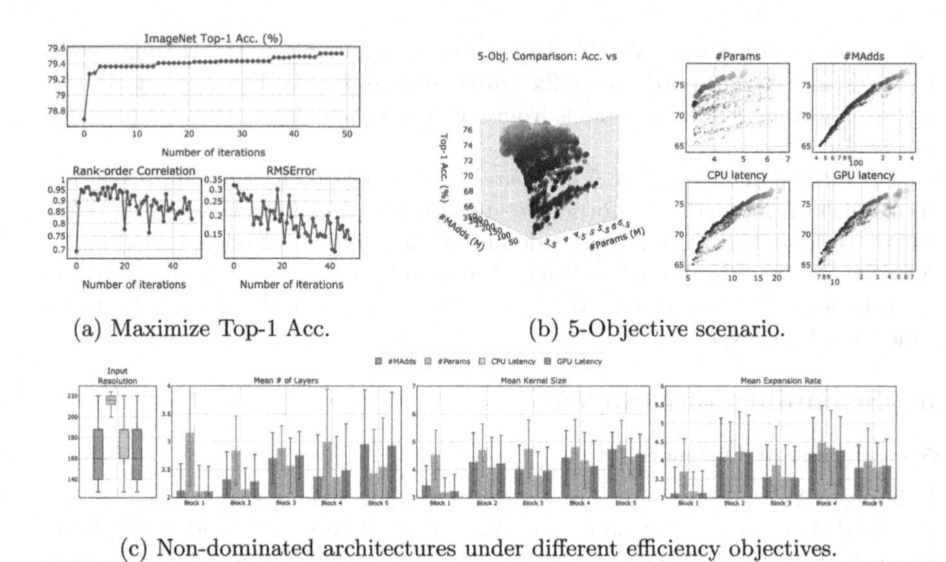

(a) Maximize Top-1 Acc.

(b) 5-Objective scenario.

(c) Non-dominated architectures under different efficiency objectives.

Fig. 7. Scalability of MSuNAS to different numbers and types of objectives: optimizing (a) a scalarized single-objective on ImageNet; (b) five objectives including accuracy, Params, MAdds, CPU and GPU latency, simultaneously. (c) Post-optimal analysis on the architectures that are non-dominated according to different efficiency objectives.

We demonstrate that our proposed algorithm can also effectively handle such a scalarized single-objective search. Following the scalarization method in [30], we apply MSuNAS to maximize validation accuracy on ImageNet with 600M MAdds as the targeted efficiency. The accumulative top-1 accuracy achieved and the performance of the accuracy predictor are provided in Fig. 7a. Without further fine-tuning, the obtained architecture yields 79.56% accuracy with 596M MAdds on the ImageNet validation set, which is more accurate and **100M fewer** MAdds than EfficientNet-B1 [31].

Many-Objective Formulation: Practical deployment of learned models are rarely driven by a single objective, and most often, seek to trade-off many different, possibly competing, objectives. As an example of one such scenario, we use MSuNAS to simultaneously optimize five objectives—namely, the accuracy on ImageNet, #Params, #MAdds, CPU and GPU latency. We follow the same search setup as in the main experiments and increase the budget to ensure a thorough search on the expanded objective space. We show the obtained Pareto-optimal (to five objectives) architectures in Fig. 7b. We use color and marker size to indicate CPU and GPU latency, respectively. We observe that a Pareto surface emerges, shown in the left 3D scatter plot, suggesting that trade-offs exist between objectives, i.e., #Params and #MAdds are not fully correlated. We then project all architectures to 2D, visualizing accuracy vs. each one of the four considered efficiency measurements, and highlight the architectures that are non-dominated in the corresponding two-objective cases. We observe that many architectures that are non-dominated in the five-objective case are now dominated when only considering two objectives. Empirically, we observe that accuracy is highly correlated with #MAdds, CPU and GPU latency, but not with #Params, to some extent.

6 Conclusion

This paper introduced MSuNAS, an efficient neural architecture search algorithm for rapidly designing task-specific models under multiple competing objectives. The efficiency of our approach stems from (i) online surrogate-modeling at the level of the architecture to improve the sample efficiency of search, and (ii) a supernet based surrogate-model to improve the weights learning efficiency via fine-tuning. On standard datasets (CIFAR-10, CIFAR-100 and ImageNet), NSGANetV2 matches the state-of-the-art with a search cost of one day. The utility and versatility of MSuNAS are further demonstrated on non-standard datasets of various types of difficulties and on different number of objectives. Improvements beyond the state-on-the-art on STL-10 and Flowers102 (under mobile setting) suggest that NAS is a more effective alternative to conventional transfer learning approaches.

References

1. Baker, B., Gupta, O., Raskar, R., Naik, N.: Accelerating neural architecture search using performance prediction. arXiv preprint arXiv:1705.10823 (2017)
2. Berthelot, D., Carlini, N., Goodfellow, I., Papernot, N., Oliver, A., Raffel, C.A.: Mixmatch: a holistic approach to semi-supervised learning. In: Advances in Neural Information Processing Systems (NeurIPS) (2019)
3. Bracken, J., McGill, J.T.: Mathematical programs with optimization problems in the constraints. Oper. Res. **21**(1), 37–44 (1973). http://www.jstor.org/stable/169087
4. Brock, A., Lim, T., Ritchie, J., Weston, N.: SMASH: one-shot model architecture search through hypernetworks. In: International Conference on Learning Representations (ICLR) (2018)
5. Cai, H., Gan, C., Wang, T., Zhang, Z., Han, S.: Once for all: train one network and specialize it for efficient deployment. In: International Conference on Learning Representations (ICLR) (2020)
6. Cai, H., Zhu, L., Han, S.: ProxylessNAS: direct neural architecture search on target task and hardware. In: International Conference on Learning Representations (ICLR) (2019)
7. Chu, X., Zhang, B., Xu, R., Li, J.: FairNAS: Rethinking evaluation fairness of weight sharing neural architecture search. arXiv preprint arXiv:1907.01845 (2019)
8. Coates, A., Ng, A., Lee, H.: An analysis of single-layer networks in unsupervised feature learning. In: Proceedings of the 14th International Conference on Artificial Intelligence and Statistics (2011)
9. Dai, X., et al.: ChamNet: towards efficient network design through platform-aware model adaptation. In: Proceedings of the IEEE/CVF Conference on Computer Vision and Pattern Recognition (CVPR) (2019)

10. Darlow, L.N., Crowley, E.J., Antoniou, A., Storkey, A.J.: CINIC-10 is not ImageNet or CIFAR-10. arXiv preprint arXiv:1810.03505 (2018)
11. Deb, K., Pratap, A., Agarwal, S., Meyarivan, T.: A fast and elitist multiobjective genetic algorithm: NSGA-II. IEEE Trans. Evol. Comput. **6**(2), 182–197 (2002). https://doi.org/10.1109/4235.996017
12. Dong, J.-D., Cheng, A.-C., Juan, D.-C., Wei, W., Sun, M.: DPP-Net: device-aware progressive search for pareto-optimal neural architectures. In: Ferrari, V., Hebert, M., Sminchisescu, C., Weiss, Y. (eds.) ECCV 2018. LNCS, vol. 11215, pp. 540–555. Springer, Cham (2018). https://doi.org/10.1007/978-3-030-01252-6_32
13. Elsken, T., Metzen, J.H., Hutter, F.: Efficient multi-objective neural architecture search via Lamarckian evolution. In: International Conference on Learning Representations (ICLR) (2019)
14. Howard, A., et al.: Searching for MobileNetV3. In: International Conference on Computer Vision (ICCV) (2019)
15. Krizhevsky, A., Hinton, G., et al.: Learning multiple layers of features from tiny images. Technical report. Citeseer (2009)
16. Li, L., Talwalkar, A.: Random search and reproducibility for neural architecture search. arXiv preprint arXiv:1902.07638 (2019)
17. Liu, C.: Progressive neural architecture search. In: Ferrari, V., Hebert, M., Sminchisescu, C., Weiss, Y. (eds.) ECCV 2018. LNCS, vol. 11205, pp. 19–35. Springer, Cham (2018). https://doi.org/10.1007/978-3-030-01246-5_2
18. Liu, H., Simonyan, K., Yang, Y.: DARTS: differentiable architecture search. In: International Conference on Learning Representations (ICLR) (2019)
19. Lu, Z., Deb, K., Boddeti, V.N.: MUXConv: information multiplexing in convolutional neural networks. In: Proceedings of the IEEE/CVF Conference on Computer Vision and Pattern Recognition (CVPR) (2020)
20. Lu, Z., et al.: NSGA-Net: neural architecture search using multi-objective genetic algorithm. In: Genetic and Evolutionary Computation Conference (GECCO) (2019)
21. Luo, R., Tian, F., Qin, T., Chen, E., Liu, T.Y.: Neural architecture optimization. In: Advances in Neural Information Processing Systems (NeurIPS) (2018)
22. Mei, J., et al.: AtomNAS: fine-grained end-to-end neural architecture search. In: International Conference on Learning Representations (ICLR) (2020)
23. Nayman, N., Noy, A., Ridnik, T., Friedman, I., Jin, R., Zelnik, L.: XNAS: neural architecture search with expert advice. In: Advances in Neural Information Processing Systems (NeurIPS) (2019)
24. Nilsback, M., Zisserman, A.: Automated flower classification over a large number of classes. In: 2008 6th Indian Conference on Computer Vision, Graphics Image Processing (2008)
25. Pham, H., Guan, M., Zoph, B., Le, Q., Dean, J.: Efficient neural architecture search via parameters sharing. In: International Conference on Machine Learning (ICML) (2018)
26. Real, E., Aggarwal, A., Huang, Y., Le, Q.V.: Regularized evolution for image classifier architecture search. In: AAAI Conference on Artificial Intelligence Conference on Artificial Intelligence (2019)
27. Russakovsky, O., et al.: ImageNet large scale visual recognition challenge. Int. J. Comput. Vision **115**(3), 211–252 (2015)
28. Sandler, M., Howard, A., Zhu, M., Zhmoginov, A., Chen, L.C.: MobileNetV2: inverted residuals and linear bottlenecks. In: Proceedings of the IEEE/CVF Conference on Computer Vision and Pattern Recognition (CVPR) (2018)

29. Sun, Y., Wang, H., Xue, B., Jin, Y., Yen, G.G., Zhang, M.: Surrogate-assisted evolutionary deep learning using an end-to-end random forest-based performance predictor. IEEE Trans. Evol. Comput. (2019). https://doi.org/10.1109/TEVC.2019.2924461
30. Tan, M., et al.: MnasNet: platform-aware neural architecture search for mobile. In: Proceedings of the IEEE/CVF Conference on Computer Vision and Pattern Recognition (CVPR) (2019)
31. Tan, M., Le, Q.V.: EfficientNet: rethinking model scaling for convolutional neural networks. In: International Conference on Machine Learning (ICML) (2019)
32. Tan, M., Le, Q.V.: MixConv: mixed depthwise convolutional kernels. In: British Machine Vision Conference (BMVC) (2019)
33. Wang, X., Kihara, D., Luo, J., Qi, G.J.: EnAET: Self-trained ensemble autoencoding transformations for semi-supervised learning. arXiv preprint arXiv:1911.09265 (2019)
34. Wu, B., et al.: FBNet: hardware-aware efficient ConvNet design via differentiable neural architecture search. In: Proceedings of the IEEE/CVF Conference on Computer Vision and Pattern Recognition (CVPR) (2019)
35. Xie, S., Kirillov, A., Girshick, R., He, K.: Exploring randomly wired neural networks for image recognition. In: Proceedings of the IEEE/CVF Conference on Computer Vision and Pattern Recognition (CVPR) (2019)
36. Yu, K., Sciuto, C., Jaggi, M., Musat, C., Salzmann, M.: Evaluating the search phase of neural architecture search. In: International Conference on Learning Representations (ICLR) (2020)
37. Zitzler, E., Thiele, L.: Multiobjective optimization using evolutionary algorithms - a comparative case study. In: Eiben, A.E., Bäck, T., Schoenauer, M., Schwefel, H.P. (eds.) PPSN 1998. LNCS, vol. 1498. Springer, Heidelberg (1998). https://doi.org/10.1007/BFb0056872
38. Zoph, B., Vasudevan, V., Shlens, J., Le, Q.V.: Learning transferable architectures for scalable image recognition. In: Proceedings of the IEEE/CVF Conference on Computer Vision and Pattern Recognition (CVPR) (2018)

Describing Textures Using Natural Language

Chenyun Wu(✉), Mikayla Timm, and Subhransu Maji

University of Massachusetts Amherst, Amherst, USA
{chenyun,mtimm,smaji}@cs.umass.edu

Abstract. Textures in natural images can be characterized by color, shape, periodicity of elements within them, and other attributes that can be described using natural language. In this paper, we study the problem of describing visual attributes of texture on a novel dataset containing rich descriptions of textures, and conduct a systematic study of current generative and discriminative models for grounding language to images on this dataset. We find that while these models capture some properties of texture, they fail to capture several compositional properties, such as the colors of dots. We provide critical analysis of existing models by generating synthetic but realistic textures with different descriptions. Our dataset also allows us to train interpretable models and generate language-based explanations of what discriminative features are learned by deep networks for fine-grained categorization where texture plays a key role. We present visualizations of several fine-grained domains and show that texture attributes learned on our dataset offer improvements over expert-designed attributes on the Caltech-UCSD Birds dataset.

1 Introduction

Texture is ubiquitous and provides useful cues for a wide range of visual recognition tasks. We rely on texture for estimating material properties of surfaces, for discriminating objects with a similar shape, for generating realistic imagery in computer graphics applications, and so on. Texture is localized and can be more easily modeled than shape that is affected by pose, viewpoint, or occlusion. The effectiveness of texture for perceptual tasks is also mimicked by deep networks trained on current computer vision datasets that have been shown to rely significantly on texture for discrimination (*e.g.*, [14,16,22,26]).

While there has been significant work in the last few decades on visual representations of texture, limited work has been done on describing detailed properties of textures using natural language. The ability to describe texture in rich detail can enable applications on domains such as fashion and graphics, as well as to interpret discriminative attributes of visual categories within a fine-grained

Electronic supplementary material The online version of this chapter (https://doi.org/10.1007/978-3-030-58452-8_4) contains supplementary material, which is available to authorized users.

A. Vedaldi et al. (Eds.): ECCV 2020, LNCS 12346, pp. 52–70, 2020.
https://doi.org/10.1007/978-3-030-58452-8_4

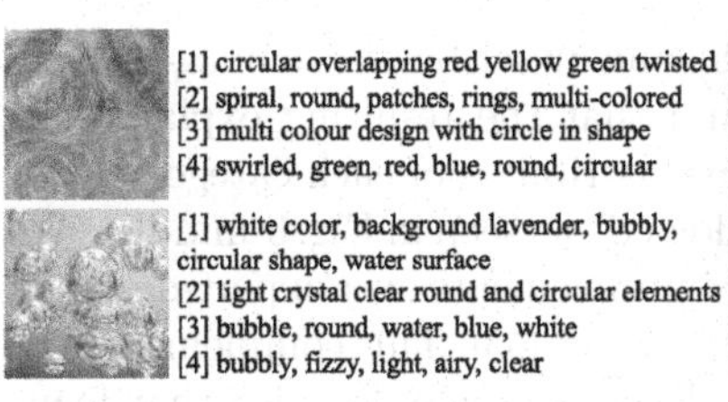

[1] circular overlapping red yellow green twisted
[2] spiral, round, patches, rings, multi-colored
[3] multi colour design with circle in shape
[4] swirled, green, red, blue, round, circular
[1] white color, background lavender, bubbly, circular shape, water surface
[2] light crystal clear round and circular elements
[3] bubble, round, water, blue, white
[4] bubbly, fizzy, light, airy, clear

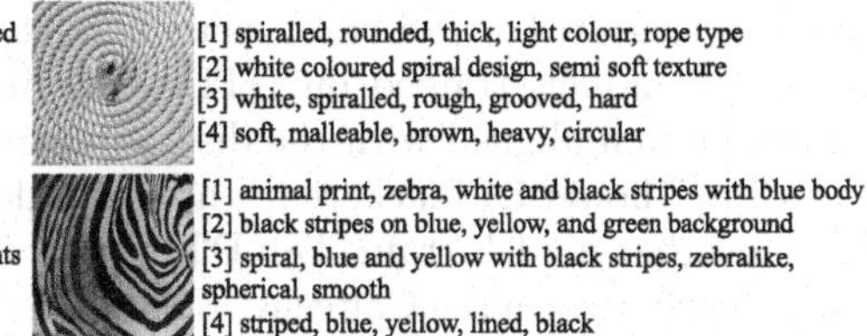

[1] spiralled, rounded, thick, light colour, rope type
[2] white coloured spiral design, semi soft texture
[3] white, spiralled, rough, grooved, hard
[4] soft, malleable, brown, heavy, circular
[1] animal print, zebra, white and black stripes with blue body
[2] black stripes on blue, yellow, and green background
[3] spiral, blue and yellow with black stripes, zebralike, spherical, smooth
[4] striped, blue, yellow, lined, black

Fig. 1. We introduce the Describable Textures in Detail Dataset (DTD^2) consisting of texture images from DTD [15] with natural language descriptions, which provide rich and fine-grained supervision for various aspects of texture such as color compositions, shapes, and materials. More examples are in the Supplementary material. (Color figure online)

taxonomy (*e.g.*, species of birds and flowers) where texture cues play a key role. However, existing datasets of texture (*e.g.*, [11,15]) are limited to a few binary attributes that describe patterns or materials, and do not describe detailed properties using the compositional nature of language (*e.g.*, descriptions of the color and shape of texture elements). At the same time, existing datasets of language and vision [7,27,32,38,44,48,58] primarily focus on objects and their relations with a limited treatment of textures (Sect. 2). Addressing this gap in the literature, we introduce a new dataset containing natural language descriptions of textures called the Describable Textures in Detail Dataset (DTD^2). It contains several manually annotated descriptions of each image from the Describable Texture Dataset (DTD) [15]. As seen in Fig. 1, these contain descriptions of colors of the structural elements within the texture (*e.g.* "circles" and "stripes"), their shape, and other high-level perceptual properties of the texture (*e.g.* "soft" and "protruding"). The resulting vocabulary vastly extends the 47 attributes present in the original DTD dataset (Sect. 3).

We argue that the domain of texture is rich and poses many challenges for compositional language modeling that are present in existing language and vision datasets describing objects and scenes. For example, to estimate the color of dots in a dotted texture the model must learn to associate the color to the dots and not to the background. Yet the domain of texture is simple enough that it allows us to analyze the robustness and generalization of existing vision and language models by synthetically generating variations of a texture. We conduct a systematic study of existing visual representations of texture, models of language, and methods for matching the two domains on this dataset (Sects. 4, 5 and 5.3). We find that adopting pre-trained language models significantly improve generalization. However, an analysis on synthetically generated variations of each texture by varying one attribute at a time (*e.g.*, foreground color and shape) shows that the representations fail to capture detailed properties.

We also present two novel applications of our dataset (Sect. 6). First, we visualize what discriminative texture properties are learned by existing deep networks for fine-grained classification on natural domains such as birds, flowers, and butterflies. To this end we generate "maximal images" for each category by "inverting" a texture-based classifier [35] and describe these images using captioning models trained on DTD^2. We find that the resulting explanations tend to be well aligned with the discriminative attributes of each category (*e.g.*, "Tiger Lily" flower is "black, red, white, and dotted" as seen in Fig. 6-middle). We also show that models trained on DTD^2 offer improvements over expert-designed binary attributes on the Caltech-UCSD Birds dataset [53]. This complements the capabilities of existing datasets for explainable AI on these domains that focus on shapes, parts, and their attributes such as color. Texture provides a domain-independent, albeit incomplete way of describing interpretable discriminative properties for several domains.

In summary, our contributions are:

- A novel dataset of texture descriptions (Sect. 3).
- An evaluation of existing models of grounding natural language to texture (Sects. 4 and 5).
- A critical analysis of these models using synthetic, but realistic variations of textures with their descriptions (Sect. 5.3).
- Application of our models for describing discriminative texture attributes and building interpretable models on fine-grained domains (Sect. 6).

Our dataset and code are at: https://people.cs.umass.edu/~chenyun/texture.

2 Related Work

Language and Texture. Describing textures using language has a long history. Early works [5,9,49] showed that textures can be categorized along a few semantic axes such as "coarseness", "contrast", "complexity" and "stochasticity". Bhusan et al. [12] systematically identified words in English that correspond to visual textures and analyzed their relationship to perceptual attributes of textures. This was the basis of the Describable Texture Dataset (DTD) [15] which consolidated a list of 47 texture attributes along with images downloaded from the Internet. The dataset captures attributes such as "dotted", "chequered", and "honeycombed". However, it does not capture properties such as the color of the structural elements ("red and green dots"), or the attributes that describe the background color. Our goal is to model the rich space of texture attributes in a compositional manner beyond these attributes.

Datasets of Images and Text. The vision and language community has put significant efforts into building large-scale datasets. Image captioning datasets such as MS-COCO [32], Flickr30K [56] and Conceptual Captions [48] contain sentences describing the general content of images. The Visual Question Answering dataset [7] provides language question and answer pairs for each image, which requires more detailed understanding of the image content. In visual grounding datasets such as RefClef [27], RefCOCO [38,58] and Flickr30K Entities [44], detailed descriptions of the target object instances are annotated to distinguish them from other objects. However, these tasks focus on recognizing object categories and descriptions of pose, viewpoint, and their relationships to other objects, and have a limited treatment of attributes related to texture.

Texture Representations. Representations based on orderless aggregations of local features originally developed for texture has had an significant influence on early computer vision (*e.g.*, "Textons" [30], "Bag-of-Visual-Words" [17], higher-order statistics [45], and Fisher vector [41,46]). Recent works (*e.g.*, [8,16,34]) have shown that combining texture representations with deep networks lead to better generalization on scene understanding and fine-grained categorization tasks. Even without explicit modeling, deep networks are capable of modeling texture through convolution, pooling, and non-linear encoding layers [21]. Indeed, several works have shown that deep networks trained on existing datasets tend to rely more on texture than shape for classification [14,22,26,33]. This motivates the need to develop techniques to describe texture properties using natural language as a way to explain the behavior of deep networks in an interpretable manner.

Methods for Vision and Language. There is a significant literature on techniques for various language and vision tasks. The Show-and-Tell [52] model was an early deep neural net based approach for captioning images that combined the convolutional image encoder followed by an LSTM [24] language decoder. Techniques for VQA are based on a joint encoding of the image and the question to retrieve or generate an answer [28,50,59]. For visual grounding, where the goal is to identify a region in the image given a "referring expression", a common approach is to learn a metric over expressions and regions [36,43,57]. The basic architectures for these tasks have been improved in a number of ways such as by incorporating attention mechanisms [6,20,28,37,54,59] and improved language models [19,47]. To model the relation between texture images and their descriptions we investigate a discriminative approach, a metric-learning based approach, and a generative modeling based approach [55] on our dataset.

3 Dataset and Tasks

We begin by describing how we collected DTD^2 in Sect. 3.1, followed by the tasks and evaluation metrics in Sect. 3.2. DTD^2 contains multiple crowdsourced descriptions for each image in DTD. Each image I contains k descriptions $\mathbf{S} = \{S_1, S_2, \ldots, S_k\}$ from k different annotators who are asked to describe the texture presented in the image. Instead of providing a grammatically coherent sentence, we found that it more effective for them to list a set of properties separated by commas. Thus each description S can be interpreted as a set of phrases $\{P_1, P_2, \ldots, P_n\}$. Figure 1 shows some examples of the collected data. We found that the ordering of phrases in a description is somewhat arbitrary, which motivates this annotation structure. Figure 2 shows the overall dataset statistics. DTD^2 contains 5,369 images and 24,697 descriptions. We split the images into 60% training, 15% validation, and 25% test. Below we describe details of the dataset collection pipeline and tasks.

Statistics	overall	frequent
#images	5369	-
#phrases	22,435	655
#words	7681	1673
#descriptions per image	4.60	-
#phrases per image	16.64	11.61
#words per description	7.13	6.69
#words per phrase	3.93	1.19

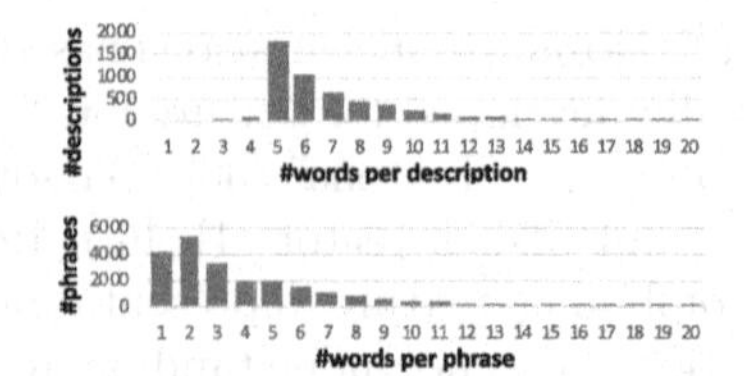

Fig. 2. Statistics of DTD². The "overall" column in the table shows the statistics of all data, while the "frequent" column only considers the phrases (or words) that occur at least 10 (or 5) times in the training split which forms our evaluation benchmark. The cloud of phrases has the font sizes proportional to square-root of frequencies in the dataset. The vocabulary significantly expands the 47 attributes of DTD.

3.1 Dataset Collection

Annotation. We present each DTD image and its corresponding texture category to 5 different annotators on Amazon Mechanical Turk, asking them to describe the texture using natural language with at least 5 words. Describable aspects of each image include texture, color, shape, pattern, style, and material (we provided description examples of several texture categories in the guidelines).

Verification. After collecting the raw annotations, we manually verified all of them and removed annotations that were irrelevant. For example, a breakfast waffle may have descriptions about the related food items such as strawberries instead of the texture which is our main goal. We also removed all images from "freckled" and "potholed" categories because they are primarily of human faces or scenes of roads with few texture-related terms in their descriptions. We also excluded images with fewer than 3 valid descriptions.

Post-processing. We found that the annotations (as seen in Fig. 1) describing aspects of texture are often expressed as a set of phrases separated by commas, instead of a fully grammatical sentence. We did find some users who provided long unbroken sentences, but these were few and far between. Therefore, we represent each description as a set of phrases indicated by commas (",") or semicolons (";"). For example, the first description of the top-right image in Fig. 1 is: "spiralled, rounded, thick, light colour, rope type", and it's split into 5 phrases: "spiralled", "rounded", "thick", "light colour", "rope type". For the purpose of evaluation, we consider words that appear at least 5 times and phrases that appear at least 10 times in the training split of the dataset, which results in 655 unique phrases. Although some long descriptions are lost in the process and most of the phrases are short (mostly within three words as seen in the lower histogram in Fig. 2), the collection of phrases captures a rich set of describable attributes for each image. Modeling the space of phrases poses significant challenges to existing techniques for language and vision (Sect. 5.3).

3.2 Tasks and Evaluation Metrics

The annotation for each image is in the form of a set of descriptions, with each description in the form of a set of phrases. A phrase is an ordered list of words. We consider several tasks and evaluation metrics on this dataset described next.

Phrase Retrieval. Given an image, the goal is to rank phrases $p \in \mathcal{P}$ that are relevant to the image. Here $\mathcal{P}$ is the set of all possible phrases, restricted to 655 frequent ones. For each image, the set of "true" relevant phrases are obtained by taking the union of phrases from all descriptions of the image. We can evaluate the ranked list using various metrics described as follows:

- Mean Average Precision (MAP): area under the precision-recall curve;
- Mean Reciprocal Rank (MRR): One over the ranking of first correct phrase;
- Precision at K (P@K): precision of the top K ranked phrases ($K \in \{5, 20\}$);
- Recall at K (R@K): recall of the top K ranked phrases ($K \in \{5, 20\}$).

Image Retrieval from a Phrase. The task is to retrieve images given a query phrase. When taking phrases as the query, we consider all phrases $p \in \mathcal{P}$ as before and ask the retrieval model to rank all images in the test or validation set. The "true" list is all images that contain the phrase (in any of its descriptions). We consider the same metrics as the phrase retrieval task.

Image Retrieval from a Description. When using descriptions the query, we consider all description $s \in \mathcal{S}$ as the input. Here $\mathcal{S}$ is the set of all descriptions in the test or validation set. We ask the retrieval model to rank all images in the corresponding set. We evaluate the rank of the image from which the description was collected (MRR metric). This metric allows us to evaluate the compositional properties of texture over phrases (*e.g.*, "red dots" + "white background"). While we only quantitatively evaluate phrases and descriptions in the dataset, the ranking models can potentially generalize to novel descriptions or phrases over the seen words. We present qualitative results and a detailed study of the models in Sects. 5 and 5.3.

Description Generation. The task is to generate a description for an input image. Given each image I, we compare the generated description against the set of its collected descriptions $\{S_1, S_2, \ldots, S_k\}$ using standard metrics for image captioning including BLEU-1,2,3,4 [40], METEOR [10], Rouge-L [31] and CIDEr [51]. However, we note that the task is open-ended and qualitative visualizations are just as important as these metrics.

4 Methods

We investigate three techniques to learn the mapping between visual texture and natural languages on our dataset—a discriminative approach, a metric learning approach, and a language generation approach. They are explained in detail in the next three sections.

4.1 A Discriminative Approach

A simple baseline is to treat each phrase $p \in \mathcal{P}$ as a binary attribute and train a multi-label classifier to map the images to phrase labels. Given a texture image I, let $\psi(I)$ be an embedding computed using a deep network. We use activations from layer 2 and layer 4 of ResNet101 [23] with mean-pooling over spatial locations as the image embedding. A comparison of features from different ResNet layers is included in the supplemental material. For the classification task, we attach a classifier head h to map the embeddings to a 655-dimensional space corresponding to each phrase in our frequent set $\mathcal{P}$. The function h is modeled as a two-layer network – the first is fully-connected layer with 512 units with BatchNorm and ReLU activation; the second is a linear layer with 655 units followed by sigmoid activation. Given a training set of $\{(I_i, Y_i)\}_{i=1}^N$ where Y_i is the ground-truth binary labels across 655 classes for image I_i, the model is trained to minimize the binary cross-entropy loss: $L_{BCE} = \sum_i \ell_{bce}(h \circ \psi(I_i), Y_i)$, where $\ell_{bce}(y, z) = \sum_i (z_i \log(y_i) - (1 - z_i) \log(1 - y_i))$.

Training Details. The ResNet101 is initialized with weights pre-trained on ImageNet [18] and fine-tuned on our training data for 75 epochs using the Adam optimizer [29] with an initial learning rate at 0.0001. We use image size 224×224 for all our experiments. The hyper-parameters are selected on the validation set.

Evaluation Setup. The classification scores over each phrase for each image are directly used to rank images or phrases for phrase retrieval or image retrieval with phrase input. Retrieving images given a description is more challenging since we need to aggregate the scores corresponding to different phrases, and the phrases in input descriptions may not be in $\mathcal{P}$. We found the following strategy works well: Given a description $S = \{P_1, P_2, \ldots, P_n\}$ and an image I, obtain the scores for each phrase $s(P_i) = \sigma(h \circ \psi(I))_k$ where k is the index of the phrase $P_i \in \mathcal{P}$. If the phrase is not in the set, we consider all its sub-sequences that are present in $\mathcal{P}$ and average the scores of them instead. For example, if the phrase "red maroon dot" is not present in $\mathcal{P}$, we consider all sub-sequences {red maroon, maroon dot, red, maroon, dot}, score each that is present in $\mathcal{P}$ separately and then average the scores. By concatenating the top 5 phrases for an image we can also use the classifier to generate a description for an image. The key disadvantage of the classification baseline is that it treats each phrase independently, and does not have a natural way to score novel phrases (our baseline using sub-sequences is an attempt to handle this).

4.2 A Metric Learning Approach

The metric learning approach aims to learn a common embedding over the images and phrases such that nearby image and phrase pairs in the embedding space are related. We adopt the standard metric learning approach based on triplet-loss [25]. Consider an embedding of an image $\psi(I)$ and of a phrase $\phi(P)$ in $\mathbb{R}^d$. Denote $||\psi(I) - \phi(P)||_2^2$ as the squared Euclidean distance between the two embeddings. Given an annotation (I, P) consisting of a positive (image, phrase) pair, we sample from the training set a negative image I' for P, and a negative phrase P' for I. We consider two losses; one from the negative phrase:

$$L_p(I,P,P') = \max(0, 1 + ||\psi(I) - \phi(P)||_2^2 - ||\psi(I) - \phi(P')||_2^2)$$

and another from the negative image:

$$L_i(P,I,I') = \max(0, 1 + ||\psi(I) - \phi(P)||_2^2 - ||\psi(I') - \phi(P)||_2^2)$$

The metric learning objective is to learn embeddings ψ and ϕ that minimize the loss $L = \mathbb{E}_{(I,P),(I',P')}\,(L_p + L_i)$ over the training set.

For embedding images, we use the same encoder as the classification approach with features from layer 2 and 4 from ResNet101. We add an additional linear layer with 256 units resulting in the embedding dimension $\psi(I) \in \mathbb{R}^{256}$. One advantage of the metric learning approach is that it allows us to consider richer embedding models for phrases. Specially we consider the following encoders:

- **Mean-pooling**: $\phi_{mean}(P) = \frac{1}{N_w}\sum_{w \in \texttt{tokenize}(P)} \texttt{embed}(w)$, where `tokenize(·)` splits the phrase into a list of words, `embed(·)` encodes each token into $\mathbb{R}^{300}$.
- **LSTM** [47]: $\phi_{lstm}(P) = \texttt{biLSTM}[\texttt{embed}(w) \text{ for } w \text{ in } \texttt{tokenize}(P)]$, with the same `tokenize(·)` and `embed(·)` as above. `biLSTM(·)` is a bi-directional LSTM with a single layer and hidden dimension 256 that returns the concatenation of the outputs on the last token from both directions.
- **ELMo** [42]: $\phi_{elmo}(P) = \texttt{ELMo}(P)$, where `ELMo(·)` uses pre-trained ELMo model [4] with its own tokenizer, and outputs the average embedding of all tokens in the phrase P.
- **BERT** [19]: $\phi_{bert}(P) = \texttt{BERT}(P)$, where `BERT(·)` uses pre-trained BERT model [3] with its own tokenizer, and outputs the average of last hidden states of all tokens in the phrase P.

To compute the final embedding of the phrase $\phi(P)$, we add a linear layer to map the embeddings to 256 dimensions compatible with the image embeddings.

Training Details. We trained this model on our training split using the Adam optimizer [29] with an initial learning rate at 0.0001. We found this model to be more prone to over-fitting than the classifier. Stopping the training when the image retrieval and phrase retrieval MAP on the validation set stops improving was effective. Same as the classifier, ResNet101 is initialized with ImageNet [18] weights and fine-tuned on our data. `embed(·)` in ϕ_{mean} and ϕ_{lstm} was initialized with FastText embeddings [1,13] and tuned end-to-end. Pre-trained encoders ϕ_{elmo} and ϕ_{bert} were fixed in our training.

Evaluation Setup. Given the joint embedding space, one can retrieve phrases for each image and images for each phrase based on the Euclidean distance. Similar to the classifier we concatenate the top 5 retrieved phrases as a baseline description generation model. We also investigate a metric learning approach over descriptions rather than phrases where the positive and negative triplets are computed over (image, description) pairs. The language embedding models are the same since they can handle descriptions of arbitrary length.

Table 1. Phrase retrieval and image retrieval on DTD². Metric learning models are trained with phrase input. Among the language encoders BERT works the best.

Data split	Task:	Phrase retrieval						Image retrieval					
	Model	MAP	MRR	P@5	P@20	R@5	R@20	MAP	MRR	P@5	P@20	R@5	R@20
Validation	MetricLearning: MeanPool	18.80	48.66	23.13	16.20	11.52	31.54	7.19	16.18	7.60	6.56	3.36	11.44
	MetricLearning: biLSTM	23.53	58.78	31.85	18.73	15.83	36.31	8.31	17.46	8.15	7.06	4.21	13.40
	MetricLearning: ELMo	28.13	68.46	37.02	21.11	18.44	41.12	11.25	24.05	12.79	10.27	5.85	18.57
	MetricLearning: BERT	**31.68**	**72.59**	**40.67**	**22.96**	**20.23**	**44.50**	**15.22**	**31.39**	**16.27**	**12.56**	**9.07**	**25.69**
Test	Classifier: Feat 2,4	27.12	61.28	33.50	21.71	16.07	41.48	**14.75**	**33.94**	**18.75**	**16.02**	**6.47**	**19.32**
	MetricLearning: BERT	**31.77**	**74.12**	**41.70**	**23.60**	**20.17**	**45.04**	13.50	31.12	16.52	14.57	5.24	17.32

Table 2. Retrieving texture images with descriptions as input.

Model	MRR
Classifier	12.40
MetricLearning(phrase)	12.92
MetricLearning(description)	**13.95**

Table 3. Description generation on textures. Synthesizing descriptions from phrases retrieved by the metric-learning based approach outperforms other baselines.

Model	Bleu-1	Bleu-2	Bleu-3	Bleu-4	METEOR	Rouge-L	CIDEr
Classifier: top 5	68.07	46.17	28.39	14.44	19.89	48.13	44.73
MetricLearning: top 5	**72.99**	**53.69**	**34.97**	**19.39**	**21.81**	**49.70**	**47.34**
Show-Attend-Tell	59.90	40.41	26.52	16.35	19.92	46.64	37.47

4.3 A Generative Language Approach

We adopt the Show-Attend-Tell model [55], a widely used model for image captioning. It combines a convolutional network to encode input images with an attention-based LSTM decoder to generate descriptions. Following the default setup, we encode images into the spatial features from the 4-th layer of ResNet101 (initialized with ImageNet [18] weights). The word embeddings are initialized from FastText [1,13]. The entire model is then trained end-to-end on the training set, using the Adam optimizer [29] with initial learning rate 0.0001 for the image encoder and 0.0004 for the language decoder. We apply early stopping based on the BLEU-4 score of generated descriptions on the validation images.

This model is primarily for the description generation task. For evaluation, we apply beam search with a beam-size of 5 to compute the best description.

5 Experiments and Analysis

5.1 Phrase and Image Retrieval

Tables 1 and 2 compare the classifier and the metric learning model on phrase and image retrieval tasks as described in Sect. 3.2. Figures 3 and 4 show examples of the top 5 retrieved images and phrases.

In Table 1 we first compare language encoders on the metric learning model. The performance of both phrase and image retrieval depends largely on the

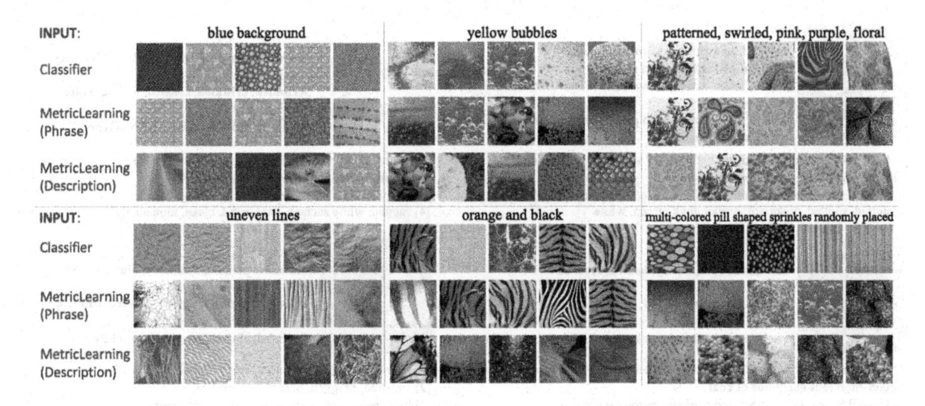

Fig. 3. Retrieve DTD^2 test images with language input. We show top 5 retrieved images from the classifier and the metric learning model (trained with phrase or description input). From left to right we show examples of (1) phrases the classifier has been trained on, (2) phrases beyond the frequent classes, and (3) full descriptions. (Color figure online)

Table 4. Image retrieval performance of R-Precision on synthetic tasks.

Model	Foreground	Background	Color+Pattern	Two-colors
Classifier	45.45 ± 20.34	$\mathbf{59.82 \pm 9.63}$	35.95 ± 21.48	26.82 ± 14.17
MetricLearning - phrase	46.55 ± 20.65	52.00 ± 6.32	$\mathbf{41.73 \pm 22.77}$	$\mathbf{27.45 \pm 15.13}$
MetricLearning - description	$\mathbf{47.64 \pm 18.97}$	53.64 ± 4.66	35.77 ± 21.12	21.59 ± 13.77
Random guess	50.00	50.00	7.40	5.26

language encoder, and BERT performs the best. The metric learning model is better at phrase retrieval while the classifier is slightly better at image retrieval.

Table 2 shows results of image retrieval from descriptions and here too the metric learning model outperforms the other two models. As shown in Fig. 3-right, although the models trained on phrases work reasonably well, the metric learning model trained on descriptions handles long queries better.

5.2 Description Generation

We compare the Show-Attend-Tell model [55] with a retrieval based approach. From the classifier and the metric learning model we retrieve the top 5 phrases and concatenate them in the order of their score to form a description. As shown in Table 3, the metric learning model reaches higher scores on the metrics. However, notice that in Fig. 4 the generative model's descriptions are more fluent and covers both the color and pattern of the images, while the retrieval baselines (especially the classifier) repeat phrases with similar meanings.

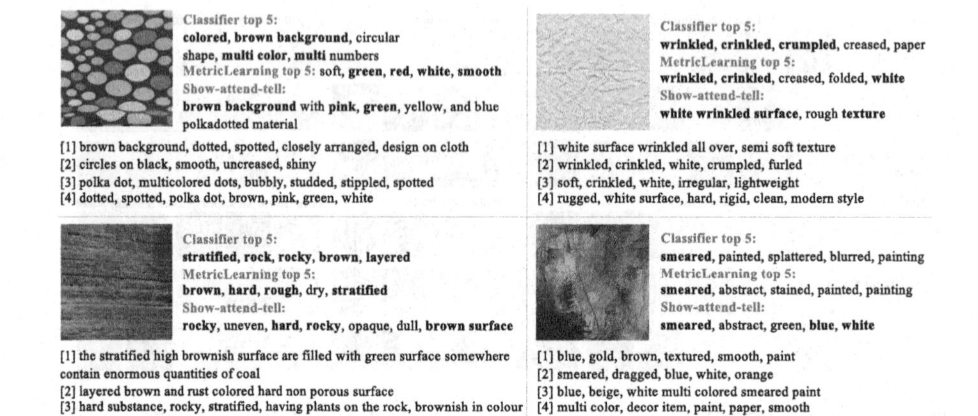

Fig. 4. Phrase retrieval and description generation on DTD^2 test images. For each input image, we list ground-truth descriptions beneath, and generated descriptions on the right. For the classifier and the metric learning model, we concatenate the top 5 retrieved phrases. Bold words are the ones included in ground-truth descriptions. (Color figure online)

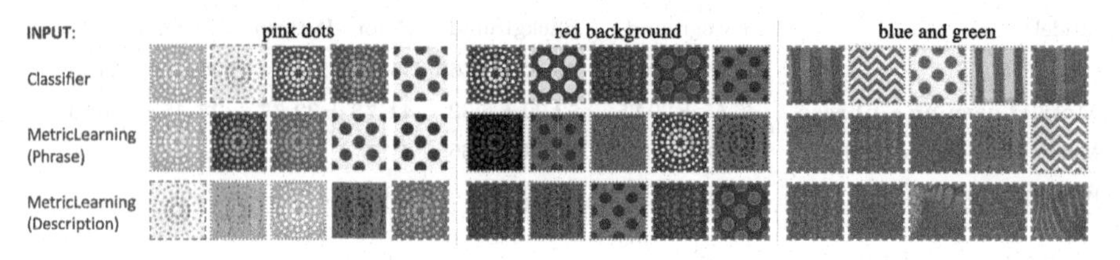

Fig. 5. Retrieval on synthetic images. Positive images are in dashed blue borders, hard negative ones are in dotted red borders. (Color figure online)

5.3 A Critical Analysis of Language Modeling

In this section, we evaluate the proposed models on tasks where we systematically vary the distribution of underlying texture attributes. This is relatively easy to do for textures than for natural images (*e.g.*, changing the color of dots) and allows us to understand the degree to which the models learn disentangled representations. We describe four tasks with varying degrees of difficulty to highlight the strengths and weaknesses of these models.

Automatically Generating Textures and Their Descriptions. To systematically generate textures with descriptions, we follow this procedure:

- Take the 11 most frequent colors in DTD^2 (white, black, brown, green, blue, red, yellow, pink, orange, gray, purple) and set their RGB values manually.
- Take 10 typical two-color images from ten different categories. We choose:
  - Type A: 5 images with "foreground on background": ['dots', 'polka-dots', 'swirls','web', 'lines' (thin lines on piece of paper)], and

  - Type B: 5 images with no clear distinction between the foreground and background: ['squares' (checkered), 'hexagon', 'stripes' (zebra-like), 'zigzagged', 'banded' (bands with similar width)].
- For each of these 10 images, we manually extract masks for the foreground and background(Type A), or two foreground colors(Type B).
- For each of the 10 images, generate a new image by picking 2 different colors from the 11 and modify pixel values of the two regions using the corresponding RGB value. This results in $10\times11\times10$=1,100 images.
- For each synthetic image, we construct the ground-truth description with the template as "[color1] [pattern], [color2] background" (such as "pink dots, white background") for Type A, and "[color1] and [color2] [pattern]" (such as "yellow and gray squares") for Type B.

Experiment 1: Foreground. On Type A set we construct:

- **Query:** A query of the form "[color=c] [pattern=p]" (*e.g.* "pink dots").
- **Positive set:** [color=c] [pattern=p] on randomly colored background (*e.g.* "pink dots, white background").
- **Negative set:** Randomly colored ($\neq$ c) [pattern=p] on [color=c] background (*e.g.* "blue dots, pink background").
- **Result:** Input the query description, we use the models to rank images from both the positive and negative set, and report R-Precision: the precision of top R predictions, where R is the number of positive images. The results are listed in Table 4 first column. Since half the images have the right attribute the chance performance is 50% and the various models are nearly at the chance level. Figure 5 shows that the model is unable to distinguish between "pink dots" and "dots on a pink background". This illustrates that the models are unable to associate color correctly with the foreground shapes.

Experiment 2: Background. This is similar to Experiment 1 but we focus on the background instead. On Type A set we construct: we know the name of its pattern (such as "dots", "squares", selected from the more frequent phrases that matches the category) and names of two colors (color1 and color2).

- **Query:** A query "[color=c] background" (*e.g.* "pink background").
- **Positive set:** Randomly colored pattern on [color=c] background (*e.g.* "red dots on pink background").
- **Negative set:** Random pattern of [color=c] on any [color$\neq$c] background (*e.g.* "pink dots on white background").
- **Result:** R-precision is shown in Table 4 second column. Once again the chance performance is 50% and the various models are nearly at the chance level. Fig. 5-middle shows that the model is unable to distinguish between "red background" and "red dots on random background".

Experiment 3: Color+Pattern. On both Type A and B images we construct:

- **Query:** A query "[color=c] [pattern=p]" (*e.g.* "pink dots").
- **Positive set:** [color=c] [pattern=p] on random colored background, or with another color (*e.g.* "pink dots, white background", "pink and blue squares").

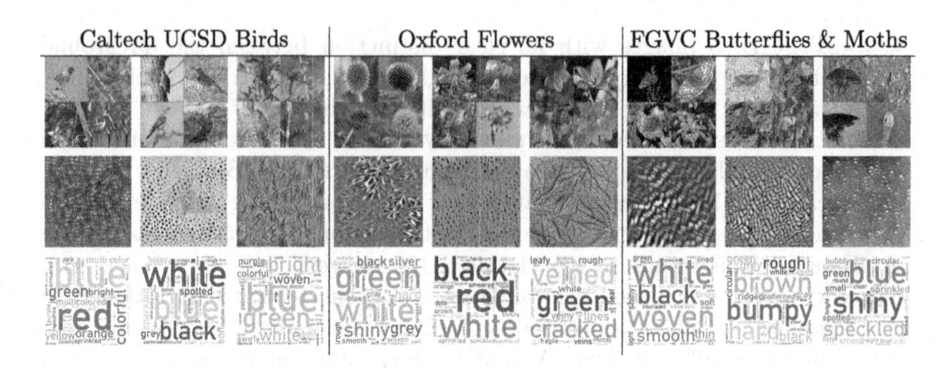

Fig. 6. Fine-grained categories visualized as their training images (top row), maximal texture images (middle row), and texture attributes (bottom row). The size of each phrase in the cloud is inversely decided by its Euclidean distance to the input maximal texture image calculated by the triplet model (Color figure online).

- **Negative set:** [color=c] [pattern≠p] or [color≠c] [pattern=p]. The negative set contains images with the correct pattern but wrong color or the wrong pattern with the right color (*e.g.*, "red dots" or "pink stripes"). Similar patterns (*e.g.*, "lines" *vs.* "banded") are not considered negative.
- **Result:** The positive and negative set is unbalanced which results in a chance performance of 7.4%. The models presented in the earlier section are able to rank the correct color and pattern combinations ahead of the negative set and achieve a considerably higher performance.

Experiment 4: Two colors. On both Type A and B images we construct:

- **Query:** A query "[color=c1] and [color=c2]" (*e.g.* "pink and green").
- **Positive set:** [color=c1] of random pattern on [color=c2] background (*e.g.* "pink dots on green background"), or [color=c1] and [color=c2] of random pattern (*e.g.* "pink and green squares").
- **Negative set:** pattern with one color from {c1, c2} and another color ≠{c1,c2} (*e.g.*, "pink dots on yellow background", "green and blue stripes").
- **Result:** The positive and negative set are unbalanced which results in a chance performance of 5.26%. The models once again are able to rank the two color combinations ahead of the negative set and achieve a considerably higher performance. Figure 5-right shows an example.

Summary. These experiments reveal that these models do indeed exhibit some high-level discriminative abilities (Exp. 3, 4), but they fail to disentangle properties such as the color of the foreground elements from background (Exp. 1, 2). This leaves much room for improvement, motivating future work, such as those that enforces spatial agreement between different attributes.

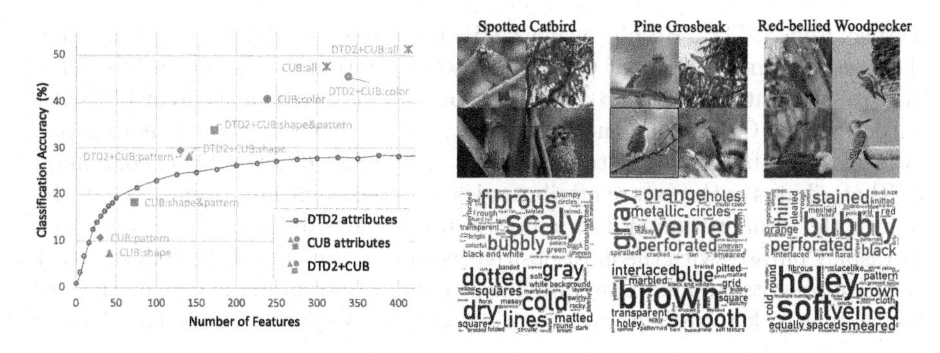

Fig. 7. Classification on CUB dataset with DTD^2 texture attributes. *Left:* classification accuracy *vs.* number of input features. Orange and green markers with the same shape are comparable with the same set of CUB attributes with or without the DTD^2 attributes. *Right:* The phrase clouds display important phrases for a few bird categories. Red phrases correspond to positive weights and blue are negative for a linear classifier for the category. Font sizes represent the absolute value of the coefficient. (Color figure online)

6 Applications

Describing Textures of Fine-Grained Categories. We analyze how the categories in fine-grained domains can be described by their texture. We consider categories from Caltech-UCSD Birds (CUB) [53], Oxford Flowers [39], and FGVC Butterflies and Moths [2] datasets. For each category, we follow the visualizing deep texture representations [33] to generate "maximal textures" – inputs that maximize the class probability using multi-layer bilinear CNN classifier [35]. These are provided as input to our metric learning model (with BERT encoder and phrase input) trained on DTD^2 to retrieve the top phrases. Figure 6 shows several categories with their maximal textures and a "phrase cloud" of the top retrieved phrases. These provide a qualitative description of each category.

Fine-Grained Classification with Texture Attributes. Here we apply models trained on our DTD^2 on the CUB dataset to show that embedding images into the space of texture attributes allows interpretable models for discriminative classification. Specifically, we input each image from the dataset to our phrase classifier (trained on DTD^2 and fixed) and obtain the log-likelihood over the 655 texture phrases as an embedding. We train a logistic regression model for the 200-way classification task. The dataset also comes with 312 binary attributes that describe the shape, pattern and color of specific parts of a bird, such as "has tail shape squared tail", "has breast pattern spotted", "has wing color yellow" (42 attributes for "shape", 31 for "pattern" and 239 for "color"). We also train a logistic regression classifier on top of these attributes.

Figure 7 shows the performance by varying the number of texture phrases ranked by their frequency on DTD^2 as the blue curve. It also shows a comparison of bird-specific attributes from CUB with generic texture attributes learned from DTD[2]. Results using CUB attributes are shown in green, while those using combinations of CUB and texture attributes are shown in orange. Texture attributes are able to distinguish bird species with an accuracy of 28.5%, outperforming CUB shape and pattern attributes. However, they do not outperform the part-based color attributes that are highly effective. Yet, combining CUB attributes with texture attributes lead to consistent improvements. On the right is the visualization of discriminative texture attributes for some categories: we display phrases with the most positive weights in red, and those with the most negative weights in blue. They provide a basis for interpretable explanations of discriminative features without requiring a category-specific vocabulary.

7 Conclusion

We presented a novel dataset of textures with natural language descriptions and analyzed the performance of several language and vision models. The domain of texture is poses challenges to existing models which fail to learn a sufficiently disentangled representation leading to poor generalization on synthetic tasks. Yet, the models show some generalization to novel domains and enabling us to provide interpretable models for describing some fine-grained domains. In particular they are complementary to existing domain-specific attributes on the CUB dataset.

Acknowledgements. We would like to thank Mohit Iyyer for helpful discussions and feedback. The project is supported in part by NSF grants #1749833 and #1617917. Our experiments were performed in the UMass GPU cluster obtained under the Collaborative Fund managed by the Mass. Technology Collaborative.

References

1. FastText pretrained embeddings. https://dl.fbaipublicfiles.com/fasttext/vectors-english/wiki-news-300d-1M.vec.zip
2. FGVC Butterflies and Moths Dataset. https://sites.google.com/view/fgvc6/competitions/butterflies-moths-2019
3. Pretrained BERT of version "bert-base-uncased". https://huggingface.co/transformers/pretrained_models.html
4. Pretrained ELMo. https://allennlp.s3.amazonaws.com/models/elmo/2x4096_512_2048cnn_2xhighway/elmo_2x4096_512_2048cnn_2xhighway_weights.hdf5
5. Amadasun, M., King, R.: Textural features corresponding to textural properties. IEEE Trans. Syst. Man Cybern. **19**(5), 1264–1274 (1989)
6. Anderson, P., et al.: Bottom-up and top-down attention for image captioning and visual question answering. In: IEEE Conference on Computer Vision and Pattern Recognition (CVPR), pp. 6077–6086 (2018)
7. Antol, S., et al.: VQA: visual question answering. In: IEEE International Conference on Computer Vision (ICCV) (2015)
8. Arandjelovic, R., Gronat, P., Torii, A., Pajdla, T., Sivic, J.: NetVLAD: CNN architecture for weakly supervised place recognition. In: IEEE Conference on Computer Vision and Pattern Recognition (CVPR), pp. 5297–5307 (2016)

9. Bajcsy, R.: Computer description of textured surfaces. In: Proceedings of the 3rd International Joint Conference on Artificial Intelligence, pp. 572–579 (1973)
10. Banerjee, S., Lavie, A.: METEOR: an automatic metric for MT evaluation with improved correlation with human judgments. In: Proceedings of the ACL Workshop on Intrinsic and Extrinsic Evaluation Measures for Machine Translation and/or Summarization, pp. 65–72. Association for Computational Linguistics, Ann Arbor (June 2005). https://www.aclweb.org/anthology/W05-0909
11. Bell, S., Upchurch, P., Snavely, N., Bala, K.: Material recognition in the wild with the materials in context database. In: IEEE Conference on Computer Vision and Pattern Recognition (CVPR), pp. 3479–3487 (2015)
12. Bhushan, N., Rao, A.R., Lohse, G.L.: The texture lexicon: understanding the categorization of visual texture terms and their relationship to texture images. Cogn. Sci. **21**(2), 219–246 (1997)
13. Bojanowski, P., Grave, E., Joulin, A., Mikolov, T.: Enriching word vectors with subword information. Trans. Assoc. Comput. Linguist. (TACL) **5**, 135–146 (2017)
14. Brendel, W., Bethge, M.: Approximating CNNs with Bag-of-Local-Features Models works surprisingly well on ImageNet. In: International Conference on Learning Representations (ICLR) (2019)
15. Cimpoi, M., Maji, S., Kokkinos, I., Mohamed, S., Vedaldi, A.: Describing textures in the wild. In: IEEE Conference on Computer Vision and Pattern Recognition (CVPR) (2014)
16. Cimpoi, M., Maji, S., Vedaldi, A.: Deep filter banks for texture recognition and segmentation. In: IEEE Conference on Computer Vision and Pattern Recognition (CVPR) (2015)
17. Csurka, G., Dance, C., Fan, L., Willamowski, J., Bray, C.: Visual categorization with bags of keypoints. In: Workshop on Statistical Learning in Computer Vision. ECCV, Prague, vol. 1, pp. 1–2 (2004)
18. Deng, J., Dong, W., Socher, R., Li, L.J., Li, K., Fei-Fei, L.: ImageNet: a large-scale hierarchical image database. In: IEEE Conference on Computer Vision and Pattern Recognition (CVPR) (2009)
19. Devlin, J., Chang, M.W., Lee, K., Toutanova, K.: BERT: Pre-training of deep bidirectional transformers for language understanding. arXiv preprint arXiv:1810.04805 (2018)
20. Gao, P., et al.: Dynamic fusion with intra-and inter-modality attention flow for visual question answering. In: IEEE Conference on Computer Vision and Pattern Recognition (CVPR), pp. 6639–6648 (2019)
21. Gatys, L.A., Ecker, A.S., Bethge, M.: Image style transfer using convolutional neural networks. In: IEEE Conference on Computer Vision and Pattern Recognition (CVPR), pp. 2414–2423 (2016)
22. Geirhos, R., Rubisch, P., Michaelis, C., Bethge, M., Wichmann, F.A., Brendel, W.: ImageNet-trained CNNs are biased towards texture; increasing shape bias improves accuracy and robustness. In: International Conference on Learning Representations (ICLR) (2018)
23. He, K., Zhang, X., Ren, S., Sun, J.: Deep residual learning for image recognition. In: IEEE Conference on Computer Vision and Pattern Recognition (CVPR), pp. 770–778 (2016)
24. Hochreiter, S., Schmidhuber, J.: Long short-term memory. Neural Comput. **9**(8), 1735–1780 (1997)
25. Hoffer, E., Ailon, N.: Deep metric learning using triplet network. In: Feragen, A., Pelillo, M., Loog, M. (eds.) SIMBAD 2015. LNCS, vol. 9370, pp. 84–92. Springer, Cham (2015). https://doi.org/10.1007/978-3-319-24261-3_7

26. Hosseini, H., Xiao, B., Jaiswal, M., Poovendran, R.: Assessing shape bias property of convolutional neural networks. In: The IEEE Conference on Computer Vision and Pattern Recognition (CVPR) Workshops (2018)
27. Kazemzadeh, S., Ordonez, V., Matten, M., Berg, T.: ReferItGame: referring to objects in photographs of natural scenes. In: Proceedings of the 2014 Conference on Empirical Methods in Natural Language Processing (EMNLP), pp. 787–798 (2014)
28. Kim, J.H., Jun, J., Zhang, B.T.: Bilinear attention networks. In: Advances in Neural Information Processing Systems, pp. 1564–1574 (2018)
29. Kingma, D.P., Ba, J.: Adam: A method for stochastic optimization. arXiv preprint arXiv:1412.6980 (2014)
30. Leung, T., Malik, J.: Representing and recognizing the visual appearance of materials using three-dimensional textons. Int. J. Comput. Vis. (IJCV) **43**(1), 29–44 (2001)
31. Lin, C.Y.: ROUGE: a package for automatic evaluation of summaries. In: Proceedings of the ACL Workshop: Text Summarization Braches Out 2004, p. 10 (January 2004)
32. Lin, T.Y., et al.: Microsoft COCO: common objects in context. In: Fleet, D., Pajdla, T., Schiele, B., Tuytelaars, T. (eds.) ECCV 2014. LNCS, vol. 8693, pp. 740–755. Springer, Cham (2014). https://doi.org/10.1007/978-3-319-10602-1_48
33. Lin, T.Y., Maji, S.: Visualizing and understanding deep texture representations. In: IEEE Conference on Computer Vision and Pattern Recognition (CVPR), pp. 2791–2799 (2016)
34. Lin, T.Y., RoyChowdhury, A., Maji, S.: Bilinear CNN models for fine-grained visual recognition. In: IEEE International Conference on Computer Vision (ICCV) (2015)
35. Lin, T.Y., RoyChowdhury, A., Maji, S.: Bilinear convolutional neural networks for fine-grained visual recognition. IEEE Trans. Pattern Anal. Mach. Intell. (PAMI) **40**(6), 1309–1322 (2018)
36. Liu, J., Wang, L., Yang, M.H.: Referring expression generation and comprehension via attributes. In: IEEE International Conference on Computer Vision (ICCV) (2017)
37. Liu, X., Wang, Z., Shao, J., Wang, X., Li, H.: Improving referring expression grounding with cross-modal attention-guided erasing. In: IEEE Conference on Computer Vision and Pattern Recognition (CVPR), pp. 1950–1959 (2019)
38. Mao, J., Huang, J., Toshev, A., Camburu, O., Yuille, A., Murphy, K.: Generation and comprehension of unambiguous object descriptions. In: IEEE Conference on Computer Vision and Pattern Recognition (CVPR) (2016)
39. Nilsback, M.E., Zisserman, A.: Automated flower classification over a large number of classes. In: Indian Conference on Computer Vision, Graphics and Image Processing (ICVGIP) (December 2008)
40. Papineni, K., Roukos, S., Ward, T., Zhu, W.J.: BLEU: a method for automatic evaluation of machine translation. In: Proceedings of the 40th Annual Meeting on Association for Computational Linguistics, pp. 311–318. Association for Computational Linguistics (2002)
41. Perronnin, F., Liu, Y., Sánchez, J., Poirier, H.: Large-scale image retrieval with compressed fisher vectors. In: IEEE Conference on Computer Vision and Pattern Recognition (CVPR), pp. 3384–3391. IEEE (2010)

42. Peters, M., et al.: Deep contextualized word representations. In: Proceedings of the 2018 Conference of the North American Chapter of the Association for Computational Linguistics: Human Language Technologies, Volume 1 (Long Papers), pp. 2227–2237 (June 2018)
43. Plummer, B.A., Kordas, P., Kiapour, M.H., Zheng, S., Piramuthu, R., Lazebnik, S.: Conditional image-text embedding networks. In: Ferrari, V., Hebert, M., Sminchisescu, C., Weiss, Y. (eds.) ECCV 2018. LNCS, vol. 11216, pp. 258–274. Springer, Cham (2018). https://doi.org/10.1007/978-3-030-01258-8_16
44. Plummer, B.A., Wang, L., Cervantes, C.M., Caicedo, J.C., Hockenmaier, J., Lazebnik, S.: Flickr30k entities: collecting region-to-phrase correspondences for richer image-to-sentence models. In: IEEE International Conference on Computer Vision (ICCV), pp. 2641–2649 (2015)
45. Portilla, J., Simoncelli, E.P.: A parametric texture model based on joint statistics of complex wavelet coefficients. Int. J. Comput. Vis. (IJCV) **40**(1), 49–70 (2000)
46. Sánchez, J., Perronnin, F., Mensink, T., Verbeek, J.: Image classification with the Fisher vector: theory and practice. Int. J. Comput. Vis. (IJCV) **105**(3), 222–245 (2013)
47. Schuster, M., Paliwal, K.K.: Bidirectional recurrent neural networks. IEEE Trans. Sig. Process. **45**(11), 2673–2681 (1997)
48. Sharma, P., Ding, N., Goodman, S., Soricut, R.: Conceptual captions: a cleaned, hypernymed, image alt-text dataset for automatic image captioning. In: Proceedings of the 56th Annual Meeting of the Association for Computational Linguistics (Volume 1: Long Papers), pp. 2556–2565 (2018)
49. Tamura, H., Mori, S., Yamawaki, T.: Textural features corresponding to visual perception. IEEE Trans. Syst. Man Cybern. **8**(6), 460–473 (1978)
50. Tan, H., Bansal, M.: LXMERT: Learning cross-modality encoder representations from transformers. arXiv preprint arXiv:1908.07490 (2019)
51. Vedantam, R., Lawrence Zitnick, C., Parikh, D.: CIDEr: consensus-based image description evaluation. In: IEEE Conference on Computer Vision and Pattern Recognition (CVPR), pp. 4566–4575 (2015)
52. Vinyals, O., Toshev, A., Bengio, S., Erhan, D.: Show and tell: a neural image caption generator. In: IEEE Conference on Computer Vision and Pattern Recognition (CVPR), pp. 3156–3164 (2015)
53. Wah, C., Branson, S., Welinder, P., Perona, P., Belongie, S.: The Caltech-UCSD Birds-200-2011 Dataset. Technical report, CNS-TR-2011-001, California Institute of Technology (2011)
54. Wang, P., Wu, Q., Cao, J., Shen, C., Gao, L., Hengel, A.: Neighbourhood watch: referring expression comprehension via language-guided graph attention networks. In: IEEE Conference on Computer Vision and Pattern Recognition (CVPR), pp. 1960–1968 (2019)
55. Xu, K., et al.: Show, attend and tell: neural image caption generation with visual attention. In: International Conference on Machine Learning (ICML), pp. 2048–2057 (2015)
56. Young, P., Lai, A., Hodosh, M., Hockenmaier, J.: From image descriptions to visual denotations: new similarity metrics for semantic inference over event descriptions. Trans. Assoc. Comput. Linguist. (TACL) **2**, 67–78 (2014)
57. Yu, L., et al.: MAttNet: modular attention network for referring expression comprehension. In: IEEE Conference on Computer Vision and Pattern Recognition (CVPR) (2018)

58. Yu, L., Poirson, P., Yang, S., Berg, A.C., Berg, T.L.: Modeling context in referring expressions. In: Leibe, B., Matas, J., Sebe, N., Welling, M. (eds.) ECCV 2016. LNCS, vol. 9906. Springer, Cham (2016). https://doi.org/10.1007/978-3-319-46475-6_5
59. Yu, Z., Yu, J., Cui, Y., Tao, D., Tian, Q.: Deep modular co-attention networks for visual question answering. In: IEEE Conference on Computer Vision and Pattern Recognition (CVPR), pp. 6281–6290 (2019)

Empowering Relational Network by Self-attention Augmented Conditional Random Fields for Group Activity Recognition

Rizard Renanda Adhi Pramono(✉), Yie Tarng Chen, and Wen Hsien Fang

National Taiwan University of Science and Technology, Taipei, Taiwan
{d10702801,ytchen,whf}@mail.ntust.edu.tw

Abstract. This paper presents a novel relational network for group activity recognition. The core of our network is to augment the conditional random fields (CRF), amenable to learning inter-dependency of correlated observations, with the newly devised temporal and spatial self-attention to learn the temporal evolution and spatial relational contexts of every actor in videos. Such a combination utilizes the global receptive fields of self-attention to construct a spatio-temporal graph topology to address the temporal dependency and non-local relationships of the actors. The network first uses the temporal self-attention along with the spatial self-attention, which considers multiple cliques with different scales of locality to account for the diversity of the actors' relationships in group activities, to model the pairwise energy of CRF. Afterward, to accommodate the distinct characteristics of each video, a new mean-field inference algorithm with dynamic halting is also addressed. Finally, a bidirectional universal transformer encoder (UTE), which combines both of the forward and backward temporal context information, is used to aggregate the relational contexts and scene information for group activity recognition. Simulations show that the proposed approach surpasses the state-of-the-art methods on the widespread Volleyball and Collective Activity datasets.

Keywords: Bidirectional universal transformer encoder · Self-attention mechanism · Conditional random field · Graph cliques · Group activity

1 Introduction

Group activity recognition has received much attention in view of numerous applications in abnormal event detection [1], sport tactical analysis [2], social

Electronic supplementary material The online version of this chapter (https://doi.org/10.1007/978-3-030-58452-8_5) contains supplementary material, which is available to authorized users.

A. Vedaldi et al. (Eds.): ECCV 2020, LNCS 12346, pp. 71–90, 2020.
https://doi.org/10.1007/978-3-030-58452-8_5

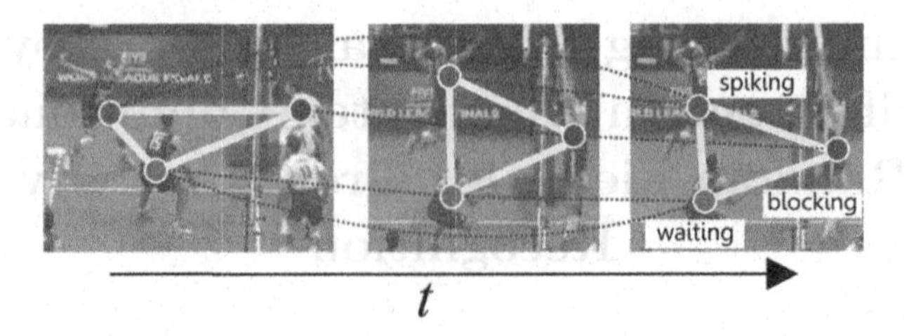

Fig. 1. A spatial-temporal graph learnt by the self-attention augmented CRF to model spatial relations and temporal evolution of actors in 'Left Spiking' activity.

behaviours [3], and *etc.* Understanding group activities requires reasoning on how interactions of every actor with different individual actions can lead to a collective activity. This is a challenging issue as the relations among the actors are dynamic [4] and, in addition, some individual actions may not be directly related to the group activity [1,5]. Therefore, it is of great importance to effectively learn the spatial relational contexts and temporal evolution of the actors in the group activity, as illustrated in Fig. 1.

A number of methods have been proposed to deal with group activity recognition. Earlier approaches [6–8] reasoned interactions of actors without a deep network architecture. However, these approaches do not properly address the temporal relationship of the actors and leverage complex semantic information from deep networks. To tackle this setback, [3,5,9] utilized recurrent neural network (RNN) to learn the dynamics of the individual actions. These methods, however, do not consider the spatial relational structure of the actors that is important in understanding complex group activities, where the actors' appearances and movements are dynamically changing. To learn the interactions of the actors, Deng *et al.* [10] introduced a structure inference machine to construct a graphical RNN model using a gating function. Shu *et al.* [11] proposed a graphical long short-term memory (LSTM), composed of an energy-based layer that can be optimized with a relatively small scale of data. Wang *et al.* [12] developed an efficient interaction model that combines person-level, group, and scene information. Biswas *et al.* [13] designed a grid pooling layer to aggregate the interaction information from the graphical RNN with a varying number of nodes and edges. Ibrahim *et al.* [2] proposed an autoencoder network comprising of multiple relational layers to learn multi-person interactions. Qi *et al.* [1] developed a soft attentive mechanism with message passing to model the interaction of relevant actors. However, the aforementioned approaches [1,2,10,11,13] are based on either RNN or LSTM, which generally requires a large variety of training data and may encounter the vanishing gradient problem [14]. Azar *et al.* [15] proposed a specific convolutional neural network (CNN) to learn group activities without explicitly detecting individual actions. Wu *et al.* [4] constructed relational graphs based on self-similarity of the actors. However, the relational graphs in [4] are limited to a few frames of observation without considering the diversity of actors' relationships at various spatial distances.

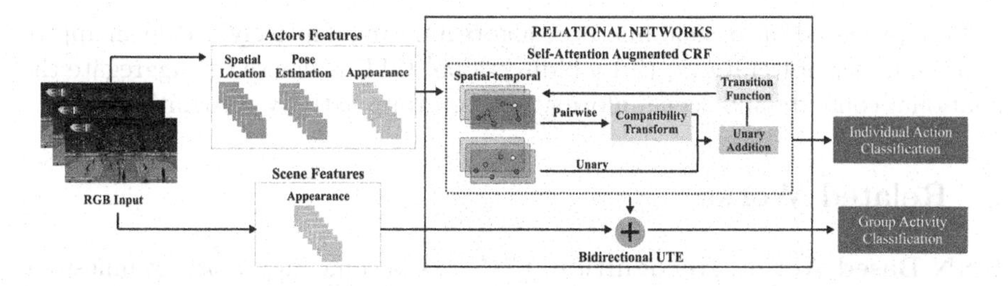

Fig. 2. Overview of the proposed network, which first uses a self-attention augmented CRF to produce the spatial relational contexts and temporal evolution of every actor. A bidirectional UTE is then used to aggregate the relational contexts and scene information.

In this paper, we propose a novel relational network for group activity recognition. The core of our network is to augment the mean-field conditional random fields (CRF) [16] with the newly devised temporal and spatial self-attention to learn the temporal evolution and spatial relational contexts of every actor in videos. In contrast to the convolutional or recurrent architectures in [16,17], self-attention, which calculates the response of every position in a sequence by relating it to all other positions, has global receptive fields across the whole data [18,19]. Such a combination thereby allows CRF, amenable to learning inter-dependency of correlated observations, to infer individual actions based on a spatio-temporal graph topology that considers temporal dependency of every actor across the frames and their non-local relationships. The network first employs the temporal self-attention along with the spatial self-attention, which considers multiple fully-connected sub-graphs, cliques [20], with different scales of locality to address the diversity of the actors' relationships in the group activities, in which their interactions can be local or non-local. As an illustration in Fig. 1, the interaction between 'blocking' and 'spiking' in the last frame of the 'Left Spiking' activity is local as those actions are close to each other while in the first frame it is non-local. Thereafter, to accommodate the distinct characteristics of each video, a new mean-field inference algorithm with dynamic halting is also addressed. Finally, a bidirectional universal transformer encoder (UTE) [21], which combines both of the forward and backward temporal context information to deal with videos with similar patterns in the first few frames, is utilized to aggregate the relational contexts with scene information. Simulations show that our network can achieve state-of-the-art performance on the widely adopted Volleyball and Collective Activity datasets.

The contributions of this paper include: (i) the mean-field CRF inference is reinforced by the temporal and spatial self-attention to facilitate the learning of the spatial relations and temporal evolution of the actors. To the best of the authors' knowledge, this is the first time CRF and self-attention are combined together to jointly model the spatial-temporal relations of multiple actors in action recognition; (ii) the CRF considers the pairwise energy with multi-scale cliques to deal with the diverse relationships of multiple actors in every frame;

(iii) the proposed mean-field inference algorithm can adaptively decide an appropriate number of iterations; (iv) a bidirectional UTE is devised to aggregate the relational contexts and scene information for group activity recognition.

2 Related Works

CNN Based Action Recognition. CNN has become an important milestone in video context understanding because of its effectiveness to extract meaningful image features. In action recognition, the majority of CNN architectures can be classified into two categories: two-stream networks [22–24], which are trained to capture appearance and motion information from RGB frames and optical flow images, and 3D networks [25,26], which are composed of spatial and temporal convolutional layers to process a number of consecutive video frames. Some recent methods [27–29] combined both architectures to attain a good trade-off between the accuracy and efficiency of the network in training and inference.

Attention Mechanism. Attention mechanism has been extensively used to improve the capability of CNN to extract fine-grained image features [30–33]. Li *et al.* [30] made use of an attentional masking scheme to filter noisy background information for more precise video object segmentation. Zhang *et al.* [31] proposed a generative adversarial network with a spatial attention mechanism that helps localize attribute-specific regions for face attribute editing. Zhao *et al.* [32] developed a pyramid feature attention network that can capture multi-level visual contexts for saliency detection. Fu *et al.* [33] introduced a dual attention network to learn correlation among channel and spatial feature maps.

Learning of Temporal Dependency. Temporal dependency is a core issue in video understanding as the past information can help infer the present and future behavior. For instance, a time-delayed graphical model was developed in [34] to detect global anomaly from multiple disjoint cameras. Swears *et al.* [35] took advantage of the Granger Causality to measure the temporal dependency between two time sequences. Meanwhile, the majority of the approaches [36–38] relied on RNN to learn temporal dependency from sequences of video frames. However, RNN and its variants such as LSTM have difficulty to generalize when the volume of training data is small and highly aperiodic [14]. Inspired by the impressive success of transformer based methods [18,19,21] in natural language processing, several recent works [39–43] made use of similar self-attention mechanisms in action recognition and detection.

Graphical Models. Graphical models [44–46] have been remarkably successful in various image and video analysis tasks. Intille *et al.* [45] modeled correlation of object trajectories to recognize complex activities. Morariu *et al.* [44] designed markov logic networks to recognize events in structured scenarios. Xu *et al.* [46] developed causal and-or graphs to learn the multiple person-object interaction for tracking humans in videos. However, [44–46] are based on non-deep networks, which can not capture spatial semantic information and not be integrated end-to-end with deeper networks. Several approaches [47,48] have addressed this setback by using CNN for graph models to learn complex semantic information. Li *et al.* [47] devised a deep relational network based on self-similarity to detect important persons in images. However, it is limited to low dimensional features from still images. Wang *et al.* [48] made use of graphical convolutional network to model local and long-term dependency for single human action recognition, but it is not devised for multiple-action scenario.

3 Feature Extraction Network

A feature extraction network, composed of a faster R-CNN on feature pyramid network (FPN) with ResNet-50 [49] and a two-stream inflated 3D network (I3D) [27] fine-tuned for individual action recognition, is employed to generate multi-scale and 3D features, respectively. The final fully-connected layer and the temporally-averaged last convolutional layer of FPN and I3D, respectively, are used to extract appearance features from actors. Also, scene features from the whole frame are generated using the same networks. The spatial location cues of each actor are obtained by concatenating its spatial position and the spatial distances with the other actors. In addition, the pose information of every actor, characterized by 17 keypoints, is extracted using AlphaPose [50]. The appearance, spatial location, and pose information are aggregated by a linear feed-forward layer to provide the final feature representation for every actor.

4 CRF for Individual Action Recognition

The graph representation for individual action recognition is illustrated in Fig. 1, where actors' interactions in every frame and temporal evolution of every actor are jointly modelled to infer individual action categories. Here, we define an actor's features in a particular frame obtained from the feature extraction network in Sect. 3 as a node. A node in each frame is fully connected to the other nodes in the same frame and a set of nodes from the same actor across a temporal sliding window of M frames is temporally interconnected. Denote a set of individual action labels $\mathcal{X} = \{x_1, \cdots, x_K\}$ and a set of random variables $\mathbf{z} = \{z_{1,1}, \cdots, z_{M,N}\}$, where $z_{i,j} \in \mathcal{X}$ is a random variable assigned to node i in frame j and N is the maximum number of actor in all frames.

The graph can be learnt by a conditional random fields (CRF) that abides by the Markov random fields conditioned on all actors' features $\mathbf{B}$. A CRF graph can be characterized by the Gibbs distribution of the form $P(\mathbf{z}|\mathbf{B}) = \frac{1}{n(\mathbf{B})}\exp(-E(\mathbf{z}|\mathbf{B}))$, where $E(\mathbf{z}|\mathbf{B})$ is the energy of the label assignment and $n(\mathbf{B})$ is the partition function [51]. In a fully-connected CRF model, the total energy can be expressed as a summation of the unary and pairwise energies [52]:

$$E(\mathbf{z}|\mathbf{B}) = \sum_{i,j} \phi_u(z_{i,j}|\mathbf{B}) + \sum_{(i',j') \neq (i,j)} \phi_p(z_{i,j}, z_{i',j'}|\mathbf{B}), \tag{1}$$

where the first term is the unary energy used to compute the cost of assigning a label $z_{i,j} \in \mathbf{z}$ to node i in frame j and the second term is the pairwise energy utilized to determine the cost of assigning labels to the same actor across the frames and to different actors in the same frame.

Minimizing the total energy can provide the most probable label assignment. However, in a dense pairwise graph, the exact minimization is intractable. So, a mean-field algorithm [52] can be adopted to approximate $P(\mathbf{z}|\mathbf{B})$ by a product of independent marginal distributions. However, the mean-field algorithm is not suitable for end-to-end training with deep CNN. Moreover, it is not designed for a spatio-temporal graph, in which the temporal dependency and non-local relationships of the actors are also considered.

5 Proposed Method

This section first introduces temporal and spatial self-attention in Subsect. 5.1, followed by a self-attention augmented CRF to generate the relational context and an individual action label for every actor in Subsect. 5.2. Next, reformulation of the mean-field inference as a self-attention network is described in Subsect. 5.3, followed by the bidirectional UTE for group activity recognition in Subsect. 5.4. For easy reference, the overall architecture of the proposed method is depicted in Fig. 2.

5.1 Temporal and Spatial Self-attention

Inspired by the success of self-attention to encode the structural information of a sequence of data, we use it to model the link between every pair of the nodes of a graph by their feature similarity. By self-attention, non-local edges can be constructed as it has global receptive fields that can simultaneously relate every node to the other nodes in the graph. Given an input sequence of nodes $\mathbf{V} = [\mathbf{v}_1, \cdots, \mathbf{v}_{T_s}] \in \mathbb{R}^{T_s \times F}$ with a feature length of F and a sequence length of T_s, the self-attention function $S(\mathbf{v}_m)$ can be defined as [18]:

$$S(\mathbf{v}_m) = \mathbf{v}_m + \sum_{\forall m' \in \{1, \cdots, T_s\}} p(\mathbf{v}_m, \mathbf{v}_{m'}) e_3(\mathbf{v}_m), \tag{2}$$

$$p(\mathbf{v}_m, \mathbf{v}_{m'}) = \frac{e_1(\mathbf{v}_m) e_2(\mathbf{v}_{m'})^T}{\sqrt{F}}, \tag{3}$$

where $e_1(\cdot)$, $e_2(\cdot)$, and $e_3(\cdot)$ are linear transformations implemented by matrix multiplication with trainable weights. Equation (3) computes the pairwise similarity between two different nodes. For our problem, we consider two types of self- attention, *i.e.* temporal self-attention and spatial self-attention.

The essence of **temporal self-attention** is to learn the temporal evolution of an actor by feature similarity of the same actor across the frames. The temporal evolution of actor i can be obtained by applying (2) to every node representation

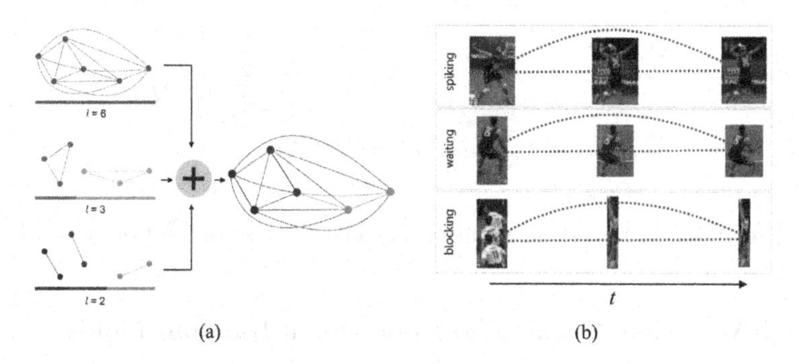

Fig. 3. (a) a pairwise fully connected graph constructed from three complete sub-graphs with the number of nodes $l = 2, 3, 6$; (b) temporal self-attention connecting the same actor across the frames.

of the actor i over M frames, $\{\mathbf{b}_{i,1}, \cdots, \mathbf{b}_{i,M}\}$. Suppose that the temporal self-attention output for node i in frame j is $\mathbf{r}_{i,j}^{t} = S(\mathbf{b}_{i,j})$. Thereby, the temporal self-attention applied to all actors in all frames, $\mathbf{B} = [\mathbf{b}_{1,1}, \cdots, \mathbf{b}_{N,M}]$, can be expressed as $S_T(\mathbf{B}) = [\mathbf{r}_{1,1}^{t}, \cdots, \mathbf{r}_{N,M}^{t}] \in \mathbb{R}^{(M \times N) \times F}$, where N is the maximum number of actors in all frames.

A new **spatial self-attention** mechanism is devised here to learn the statistical correlations of all actors in every frame by their pairwise similarity. It is designed to resolve the natural diversification of relationships among the actors as their interactions can be local or non-local, thereby making it difficult to learn the statistical correlations of all actors. The major idea behind the spatial self-attention is to prioritize the pairwise relationships of nodes with different scales of locality. A pair of local nodes is assigned a high priority edge while a pair of non-local nodes is assigned a low-priority edge. To compute the spatial self-attention, the actors in every frame are divided according to their horizontal positions into a set of cliques defined as fully-connected sub-graphs with l nodes, $2 \leq l \leq N$, where l represents the scale of locality of the cliques. For every clique, (3) is used to compute the edges based on the pairwise similarity of the nodes within that clique. The priority assignment for node i in frame j, $\mathbf{r}_{i,j}^{s}$, can thus be expressed as a linear combination of the self-attention outputs from all cliques enclosing node i:

$$\mathbf{r}_{i,j}^{s} = w_2 \cdot \mathbf{p}_{i,j}^{s,2} + \cdots + w_k \cdot \mathbf{p}_{i,j}^{s,k} + \cdots + w_N \cdot \mathbf{p}_{i,j}^{s,N}, \tag{4}$$

where $\mathbf{p}_{i,j}^{s,k} \in \mathbb{R}^{1 \times F}$ is the result of applying (2) to relate node i to the other k nodes in frame j and w_k is the corresponding trainable scalar weight. Equation (4) facilitates the learning of actors' diverse relationships by combining the edges from all cliques with different localities, as illustrated in Fig. 3. Likewise, the spatial self-attention applied to all nodes in all frames can be expressed as $S_S(\mathbf{B}) = [\mathbf{r}_{1,1}^{s}, \cdots, \mathbf{r}_{N,M}^{s}] \in \mathbb{R}^{(M \times N) \times F}$.

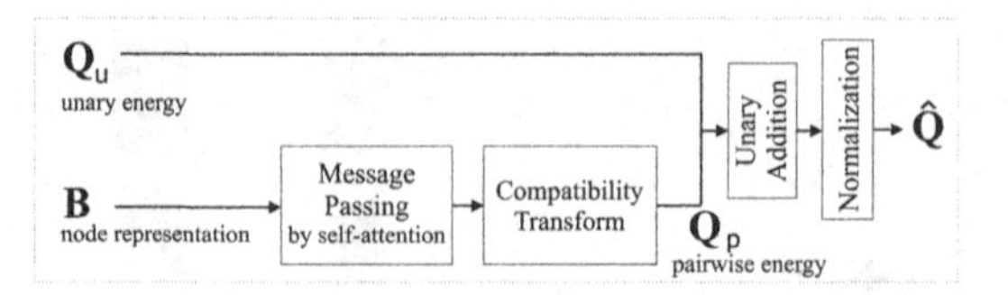

Fig. 4. A single mean-field iteration modelled as a self-attention network.

5.2 Self-Attention Augmented Conditional Random Fields

To resolve the setback of the conventional CRF, this section considers a self-attention augmented CRF to learn the temporal evolution and spatial relational contexts of every actor. Different from convolutional or recurrent architectures, self-attention can generate global receptive fields to facilitate CRF inference having non-local edges while addressing temporal dependency of the actors.

The **unary energy** of assigning an action label to an actor is now obtained by applying a linear feed-forward classifier, $f_u(\cdot)$, to the feature of each node, obtained from the feature extraction network described in Sect. 3:

$$\sum_{(i,j)} \phi_u(z_{i,j}|\mathbf{B}) = -\sum_{i,j} f_u(\mathbf{b}_{i,j}), \tag{5}$$

The **pairwise energy** of assigning different individual action labels to the same actor across the frames and to different actors in the same frame can be modelled by the spatial and temporal self-attention as follows:

$$\sum_{(i',j')\neq(i,j)} \phi_p(z_{i,j}, z_{i',j'}|\mathbf{B}) = -\sum_{i,j} f_p(\mathbf{r}^s_{i,j} + \mathbf{r}^t_{i,j}), \tag{6}$$

where $f_p(\cdot)$ is a linear transformation applied to the outputs of the spatial and temporal self-attention for each node, $\mathbf{r}^s_{i,j}$ and $\mathbf{r}^t_{i,j}$, respectively, discussed in Subsect. 5.1. Such a definition of pairwise energy can provide a measure of the cost for assigning the relevant labels to a pair of nodes based on their feature similarity. The new pairwise energy can be viewed as a degenerated version of the higher-order CRF [53–55], where several graph cliques are used to enforce consistency in labelling in image segmentation. However, learning the compatibility of several actors in a clique simultaneously is more difficult as the actors' interactions are dynamic. Therefore, in contrast to the higher order potentials, we restrict the problem to the minimization of the pairwise energy based on multi-scale cliques. As illustrated in Fig. 3, the edges of the pairwise graph are obtained by a combination of the spatial self-attention outputs at different scales of cliques.

The total energy formed by the unary and pairwise energy can be minimized by a new mean-field inference to be described in the next section.

5.3 Reformulation of Mean-Field Inference

The mean-field inference can be used to approximate Gibbs distribution of action labels by a product of independent marginal distribution of all actors, $Q(\mathbf{z}) = \prod_{i,j} Q_{z_{i,j}}$, each of which is obtained from the unary and pairwise energy [52]:

$$Q_{z_{i,j}} = \frac{1}{Z_{i,j}} \exp\big(- \mathbf{q}_u(i,j) - \mathbf{q}_p(i,j)\big), \tag{7}$$

in which $Z_{i,j}$ is the normalization constant [52], and $\mathbf{q}_u(i,j) = -f_u(\mathbf{b}_{i,j})$ and $\mathbf{q}_p(i,j) = -f_p(\mathbf{r}^s_{i,j} + \mathbf{r}^t_{i,j})$ are respectively the unary and pairwise energies of assigning a label $z_{i,j}$ to a node i in frame j and two labels at once to both the node and the other node connected to it. Such an inference can be implemented by iteratively stacking CNN kernels to refine the marginal distribution [16,17, 53], which, however, do not consider temporal dependency and non-local actors' interactions. To resolve the setback, we reformulate the mean-field inference algorithm as a self-attention network as summarized in Algorithm 1, described in details below.

Multiple mean-field iterations can be implemented by refining the marginal distribution $\hat{\mathbf{Q}} = [-\mathbf{q}_p(1,1), \cdots, -\mathbf{q}_p(M,N)]$ using self-attention networks, as depicted in Fig. 4. The marginal distribution is initialized by the unary energy $\mathbf{Q}_u = [-\mathbf{q}_u(1,1), \cdots, -\mathbf{q}_u(M,N)]$, as given in Step 5 of Algorithm 1. In each iteration, the pairwise energy, $\mathbf{Q}_p = [-\mathbf{q}_p(1,1), \cdots, -\mathbf{q}_p(M,N)]$, is calculated using the spatial and temporal self-attention, as given in Steps 7 to 8, followed by compatibility transform, in Step 9, to learn the penalty of assigning labels to a pair of nodes based on their correlation. Subsequently, in Steps 10 to 11, the marginal distribution $\hat{\mathbf{Q}}$ is refined by an addition of the pairwise energy and then normalized using a softmax layer [16]. Meanwhile, the node representation is updated by combining the outputs of the temporal and spatial self-attention, as given in Step 13. Finally, in Step 14, the probability of halting the iteration is computed using (8) based on the current node representation.

Message Passing. We employ the temporal and spatial self-attention in Subsect. 5.1 to connect every node by their feature similarity and address the non-local interactions of the nodes. Implementing the message passing within mean-field iterations is similar to increasing the depth of the self-attention network in [18,21]. First, denote the node feature representation at the vth mean-field iteration, $\mathbf{C}_v$, as the combination of the outputs of the temporal and spatial self-attention in the previous iteration, *i.e.*, $\mathbf{C}_v = S_T(\mathbf{C}_{v-1}) + S_S(\mathbf{C}_{v-1})$, where $\mathbf{C}_0 = \mathbf{B}$. The pairwise energy by the temporal and spatial self-attention $\mathbf{Q}^t_p, \mathbf{Q}^s_p \in \mathbb{R}^{(M \times N) \times K}$, where K is the number of individual action labels, is thus obtained by propagating the temporal and spatial self-attention outputs of the current node representation, $S_T(\mathbf{C}_v)$ and $S_S(\mathbf{C}_v)$, to linear feed forward layers. Thereafter, compatibility transform is performed by multiplying $\mathbf{U}^s$ and $\mathbf{U}^t \in \mathbb{R}^{K \times K}$ with $\mathbf{Q}^s_p$ and $\mathbf{Q}^t_p$, respectively, to learn the penalty of assigning labels to a pair of nodes based on their correlation. Instead of using fixed penalty, the compatibility matrices are trained to provide data-dependent penalty, provided that different labels are assigned to nodes with high correlation.

Algorithm 1 Mean-field inference of the self-attention augmented CRF

Input: $\mathbf{B}$, max_iter ▷ actors' features, maximum number of iterations
Output: $\hat{\mathbf{Q}}, \bar{\mathbf{C}}$ ▷ individual action scores, final context representation
1: halt = 0 ▷ initialization of the halting probability
2: $v = 0$
3: $\mathbf{C}_v = \mathbf{B}$ ▷ initialization of the node representation
4: $\mathbf{Q}_u = \mathrm{f_u}(\mathbf{B})$ ▷ feed forward classifier to obtain unary energy
5: $\hat{\mathbf{Q}} = \mathrm{softmax}(\mathbf{Q}_u)$ ▷ initialize the marginal distribution by the unary energy
6: **while** halt ≤ 1 and $v \leq$ max_iter **do**
7: Compute $S_T(\mathbf{C}_v)$ and $S_S(\mathbf{C}_v)$ ▷ message passing by temporal and spatial self-attention
8: Feed-forward layer applied to $S_T(\mathbf{B})$ and $S_S(\mathbf{B})$ to obtain $\mathbf{Q}_p^t, \mathbf{Q}_p^s$
9: $\mathbf{Q}_p = \mathbf{Q}_p^t\mathbf{U}^t + \mathbf{Q}_p^s\mathbf{U}^s$ ▷ compatibility transform
10: $\bar{\mathbf{Q}} = \mathbf{Q}_u + \mathbf{Q}_p$ ▷ unary addition
11: $\hat{\mathbf{Q}} = \mathrm{softmax}(\bar{\mathbf{Q}})$ ▷ update the marginal distribution of action labels
12: $v = v + 1$
13: $\mathbf{C}_v = S_T(\mathbf{C}_{v-1}) + S_S(\mathbf{C}_{v-1})$ ▷ update the node representation
14: halt = $\mathrm{f_s}(\mathbf{C}_v)$ ▷ adaptive halting probability by sigmoidal function in Eqn. (8)
15: **end while**
16: $\bar{\mathbf{C}} = \mathbf{C}_v$ ▷ final relational context information

Adaptive Mean-Field Iterations. Due to the diverse nature of actor interactions in group activities, the mean-field inference in some videos may require more iterations to converge. Consequently, instead of using a fixed number of iterations for all videos, we resort to a dynamic halting scheme, which computes the halting probability based on the current node representation. At the vth iteration, the halting probability given the current node representation, $\mathbf{C}_v$, is obtained by using the sigmoid function with the corresponding weight matrix $\mathbf{W}_h$ and bias B_h [21,56]:

$$f_s(\mathbf{C}_v) = f_s(\mathbf{C}_{v-1}) + \sigma(\mathbf{W}_h\mathbf{C}_v + B_h) \quad v \geq 1, \tag{8}$$

where $f_s(\mathbf{C}_0) = 0$. Equation (8) computes the accumulated halting probability up to the current iteration. Once the probability reaches one, the iteration stops and the node representation at the last iteration is considered as the final relational context information $\bar{\mathbf{C}}$. All inference parameters, including the trainable weights in (2)–(8), are updated by back propagation.

5.4 Bidirectional UTE for Group Activity Recognition

To recognize group activity in each frame, we aggregate the scene information and the relational context representation, $\bar{\mathbf{C}}$, obtained in Sect. 3 and Subsect. 5.2, respectively, by UTE [21]. UTE combines the advantages of recurrent neural networks and self-attention in modelling the temporal correlation at distant positions with more flexible network depth. To this end, the relational representation in each frame is first summarized as a weighted sum of feature vectors. Suppose $\bar{\mathbf{C}}_j \in \mathbb{R}^{N\times F}$ is the relational representation in frame j, it can be aggregated as:

$$\mathbf{g}_j = \mathbf{1}^T(\bar{\mathbf{C}}_j\mathbf{q}_j)^T\bar{\mathbf{C}}_j, \quad j = 1, \ldots, M. \tag{9}$$

where $\mathbf{1}$ is an all-one vector and $\mathbf{q}_j \in \mathbb{R}^{F \times N}$ is a trainable weight. Subsequently, for positional direction consideration, we employ the bidirectional self-attention encoding to combine both of the forward and backward temporal contexts to deal with videos with similar patterns in the first few frames. More specifically, denote a concatenation of scene and context information for frame j as $\mathbf{n}_j = [\mathbf{g}_j, \mathbf{f}_j] \in \mathbb{R}^{1 \times 2F}$. We can modify (2) to include positional masks as follows:

$$S_f(\mathbf{n}_j) = \mathbf{n}_j + \sum_{\forall j'} \left(p(\mathbf{n}_j, \mathbf{n}_{j'}) \odot M^f_{j,j'} \right) e_3(\mathbf{n}_j), \tag{10}$$

$$S_b(\mathbf{n}_j) = \mathbf{n}_j + \sum_{\forall j'} \left(p(\mathbf{n}_j, \mathbf{n}_{j'}) \odot M^b_{j,j'} \right) e_3(\mathbf{n}_j), \tag{11}$$

where $S_f(\cdot)$ and $S_b(\cdot)$ are the self-attention functions in the forward and backward directions, respectively, and

$$\begin{cases} \mathcal{M}^f_{j,j'} = 1, & j < j' \\ 0, & \text{otherwise} \end{cases} \quad \text{and} \quad \begin{cases} \mathcal{M}^b_{j,j'} = 1, & j > j' \\ 0, & \text{otherwise.} \end{cases} \tag{12}$$

The bidirectional representation of self-attention in UTE can be expressed as

$$S_{fb}(\mathbf{n}_j) = S_f(\mathbf{n}_j) + S_b(\mathbf{n}_j), \tag{13}$$

which is then propagated to a fully connected layer and a softmax layer to obtain the group activity distribution for every frame.

6 Experimental Results

We evaluate the performance of our proposed method on two popular group activity recognition datasets, Volleyball [5] and Collective Activity [6,57], which provide bounding boxes and tracking annotations.

Volleyball Dataset [5]**.** This dataset is a collection of 55 volleyball matches, which are trimmed into 4830 short videos. Every video is composed of 41 frames and categorized into one of eight group activity classes: 'right set', 'right spike', 'right pass', 'right winpoint', 'left winpoint', 'left pass', 'left spike' and 'left set' with nine possible individual action classes for every actor: 'waiting', 'setting', 'digging', 'falling', 'spiking', 'blocking', 'jumping', 'moving' and 'standing'.

Collective Activity Dataset [6,57]**.** This dataset contains 44 untrimmed video sequences captured by hand-held cameras from crowded daily environment. There are five collective activities: 'crossing', 'waiting', 'queueing', 'walking', and talking' and six individual action labels: 'NA', 'crossing', 'waiting', 'queueing', 'walking', and 'talking'. The group activity label in one frame is decided by the largest number of the existing individual actions.

Table 1. Impact of input features. The best results are bold-faced.

Appearance	Spatial location	Pose	Accuracy		
			Volleyball		Collective activity
			Group activity	Individual action	Group activity
✓	-	-	94.1	81.9	93.9
✓	✓	-	94.5	82.3	94.6
✓	✓	✓	**95.0**	**83.1**	**95.2**

Table 2. Performance comparison with various combination of modules. The best results are bold-faced.

Spatial max pooling [5]	Self-attention augmented CRF	UTE [21]	Bidirectional UTE	Accuracy		
				Volleyball		Collective activity
				Group activity	Individual action	Group activity
✓	-	-	-	87.1	78.4	84.6
✓	✓	-	-	94.0	82.5	93.9
-	✓	✓	-	94.6	82.8	94.8
-	✓	-	✓	**95.0**	**83.1**	**95.2**

Table 3. Impact of the number of iterations on our self-attention augmented CRF. The best results are bold-faced.

Number of iterations	Accuracy		
	Volleyball		Collective activity
	Group activity	Individual action	Group activity
1	92.1	81.0	92.3
5	93.4	82.1	94.3
8	94.2	82.5	93.5
10	93.1	81.8	92.7
Adaptive	**95.0 (5 iterations)**	**83.1 (5 iterations)**	**95.2 (3 iterations)**

6.1 Experimental Settings

Faster R-CNN is fine-tuned with RGB images using stochastic gradient descent (SGD) and a pre-trained COCO model with a learning rate of 0.0001, a momentum of 0.9, and a decay rate of 0.1 after 160 K iterations for a total of 240 K iterations. The two-stream I3D is fine-tuned with a volume of 64 RGB and optical flow images [58] using Adam and a pre-trained Kinetics with a learning rate of 0.0001, a batch of 6, and a decay rate of 0.1 in every 5 K iterations for a total of 20 K iterations. AlphaPose [50] with a pre-trained COCO model is used

Table 4. Comparison with the state-of-the-art methods on Volleyball and Collective Activity. The best results are bold-faced.

Method	Backbone	Accuracy		
		Volleyball		Collective activity
		Group activity	Individual action	Group activity
Discriminative latent models (RGB) [60]	-	-	-	79.1
Iterative belief propagation (RGB) [6]	-	-	-	79.6
Structure inference machine (RGB) [10]	AlexNet	-	-	81.2
HDTM (RGB) [5]	AlexNet	81.9	-	81.5
HiRF (RGB) [7]	-	-	-	83.1
Cardinality potential kernel (RGB) [8]	-	-	-	83.4
SBGAR (RGB + Flow) [9]	Inception-v3	67.6	-	86.1
SRNN (RGB) [13]	AlexNet	83.5	-	-
CERN (RGB) [11]	VGG16	83.3	69.1	87.2
StagNet w/o attention (RGB) [1]	VGG16	87.9	81.9	87.7
StagNet w/ attention (RGB) [1]	VGG16	89.3	-	89.1
SSU (RGB) [3]	Inception-v3	90.6	81.8	-
Recurrent modelling (RGB + Flow) [12]	AlexNet + GoogleNet	-	-	89.4
RCRG (RGB) [2]	VGG19	89.5	-	-
ARG (RGB) [4]	Inception-v3	92.5	83.0	91.0
CRM (RGB + Flow) [15]	I3D	93.0	-	85.8
Ours (RGB + Flow)	I3D + FPN	**95.0**	**83.1**	**95.2**

to generate the keypoints. The features of the faster R-CNN, two-stream I3D, spatial location, and pose information for each actor enclosed by a bounding box from the annotation are aggregated into a feature dimension of $F = 1024$. The self-attention augmented CRF and the bidirectional UTE are jointly trained for both individual action and group activity recognition using Adam with a learning rate of 0.0005, a decay rate of 0.1 after 60 epochs, for 100 epochs. The multi-task loss function for our relational network consists of the cross entropy for group activity and individual action recognition, L_2 regularization loss [59], and the pondering time penalty for the dynamic halting [21]. Same as [18], the number of the parallel self-attention heads is 8 and the drop out rate is 0.1. To simplify the pairwise graph complexity, we set $l = \{2, 4, 6, N\}$, where $N = 12$ and 13 for Volleyball and Collective Activity, respectively. The length of temporal sliding window, M, and the maximum number of iterations of the self-attention CRF are set as 10 for both datasets. The compatibility matrices are initialized with the Potts model [16]. The experiments mainly follow the protocols and evaluation metrics provided by Volleyball [5] and Collective Activity [6,57].

6.2 Ablation Studies

Input Features. We first scrutinize the performance with a different combination of input features as shown in Table 1, from which we can see that with the addition of spatial location information, the group activity recognition performance can be improved by 0.4% and 0.7% on Volleyball and Collective Activity, respectively. This is because some volleyball activities usually have distinct actors' positions as part of the game strategies. Also, for Collective Activity, the majority of the action classes, which determines the overall group activity, are located within a close distance to each other. The performance of individual action classification on Volleyball can also be boosted by 0.4%, as some individual action classes like 'spiking' and 'waiting' have different spatial positions throughout the video. Adding pose information helps bolster both group activity and individual action recognition performance on Volleyball by 0.5% and 0.8%, respectively. Also, the group activity classification on Collective Activity is improved by 0.6%. This is because our CRF can learn the dynamic changes of the actors' poses and their relation with the other actors that are crucial in determining the group activity and the individual action categories. Consequently, we employ all three features in the following simulations.

Functions of Modules. We assess the effect of the modules in our relational network, as shown in Table 2, from which we can see that compared with using only the node features by the feature extraction network, the proposed self-attention augmented CRF can improve the individual action recognition performance by 4.1% on Volleyball. The accuracy of group activity classification is also boosted by 6.9% and 9.3% on Volleyball and Collective Activity, respectively, compared to applying max pooling directly to aggregate the actors' features in every frame. This is because it can leverage the spatial relational context and temporal evolution of every actor to precisely infer the group activity category. Lastly, we assess the impact of bidirectional UTE. We can see that using the unidirectional UTE [21], the group activity recognition performance can be slightly improved by 0.6% and 0.9% on Volleyball and Collective Activity, respectively. The performance gain can be further enhanced respectively by 0.4% on both of the datasets with a replacement of the bidirectional UTE.

Number of Iterations. We inspect the significance of using the adaptive number of iterations by using the dynamic halting scheme instead of a fixed one. As shown in Table 3, the performance of the individual action and group activity recognition on Volleyball is improved with the number of iterations, but it begins to drop after it reaches 8 iterations. Similarly, on Collective Activity, the accuracy of group activity classification stops to improve after 5 iterations. This is because different videos require different numbers of iterations and increasing the number of iterations can lead to overfitting. Meanwhile, the proposed adaptive inference, which in average converges in 5 iterations and 3 iterations for Volleyball and Collective Activity, respectively, achieves the best performance.

6.3 Comparison with the State-of-the-Art Works

We first compare the proposed method with state-of-the-art works, including HDTM [5], SBGAR [9], SRNN [13], CERN [11], StagNet [1], Recurrent Modelling [12], SSU, [3], RCRG [2], ARG [4], and CRM [15], on Volleyball in terms of the group activity and individual action recognition accuracy, as shown in Table 4, from which we can see that SBGAR [9] is the worst as it relies on the high-level

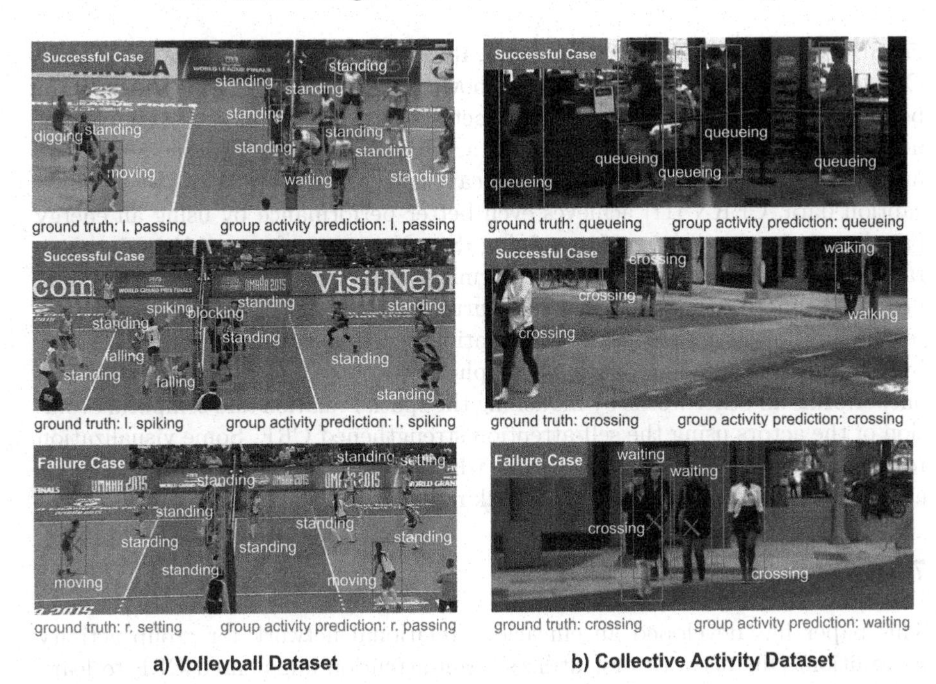

Fig. 5. Some group activity and individual action recognition results by our method, where the individual action classification is in white while the incorrect classification is marked by a red cross.

semantic caption data. This approach [9] is inferior to HDTM [5], CERN [11], and SRNN [13], which aggregate individual actions using two-level of RNN. Considerable improvement is attained by StagNet [1] as it models the relationships among the actors using semantic attentive graphs. RCRG [2] achieves similar performance by utilizing stacks of relational encoders. SSU [3] has even better performance by multi-task learning of action detection and group activity recognition. ARG [4] outperforms [3] by constructing an actor relational graph based on self similarity. Without explicitly learning individual actions, CRM [15] yields slightly better results by incorporating multi-stage refinement. Our work outperforms all of the aforementioned methods by learning the spatial relation and the temporal evolution of actors using the self-attention endowed CRF. Also, our approach excels the state-of-the-art works that reported their performance on the individual action recognition. This is because our self-attention augmented CRF can model the non-local relationships and temporal dependency of the actors. Some visualization results are shown in Fig. 5(a), from which we can observe that the error comes from differentiating between 'setting' and 'passing' that have similar actors' movements.

Next, we compare the group activity recognition with twelve baselines, Discriminative Latent Models [60], Iterative Belief Propagation (RGB) [6], Structure Inference Machine [10], HiRF [7], Cardinality Potential Kernel[8], HDTM

[5], SBGAR [9], CERN [11], StagNet [1], Recurrent Modelling [12], ARG [4], and CRM [15], on Collective Activity. As shown in Table 4, CRM [15] has superior performance compared with non-deep networks [6–8,60] and relatively shallow networks [5,10] by learning the group activity from multi-stage convolutional maps. SBGAR [9] outperforms [15] by capturing the dynamics of the semantic caption data. CERN [11] achieves even better performance by using an energy based optimization scheme for training. StagNet outperforms [11] by incorporating an attention mechanism to determine important actor features for more precise group activity recognition. Recurrent Modelling [12] yields even better results as it uses multi-context information. ARG [4] attains higher accuracy by constructing an actor relational graph based on self-similarity. Our method outperforms all other works by modelling the spatial relation and temporal evolution of the actors using the self-attention strengthened CRF. Some visualization results are depicted in Fig. 5(b), from which we can see that ‘waiting’ can be easily confused with ‘crossing’ and ‘walking’ as it has similar appearances.

7 Conclusions

This paper has developed an efficacious relational network for group activity recognition. Our network first utilizes a self-attention augmented CRF to learn the spatial relational context and temporal evolution of every actor. Such a combination explores actors’ interactions from a graph topology with different localities. Next, an adaptive mean-field inference is addressed. Finally, a bidirectional UTE is used to amass the actors’ relational context and scene information. Simulations show the effectiveness of the new approach on two common datasets.

Acknowledgement. This work was supported by the Ministry of Science and Technology and by ITRI, R.O.C., under contracts MOST 109-2221-E-011-131 and 109-2221-E-011-116, and ICL/ITRI B5-10903-HQ-07.

References

1. Qi, M., Qin, J., Li, A., Wang, Y., Luo, J., Van Gool, L.: stagNet: an attentive semantic RNN for group activity recognition. In: Ferrari, V., Hebert, M., Sminchisescu, C., Weiss, Y. (eds.) ECCV 2018. LNCS, vol. 11214, pp. 104–120. Springer, Cham (2018). https://doi.org/10.1007/978-3-030-01249-6_7
2. Ibrahim, M.S., Mori, G.: Hierarchical relational networks for group activity recognition and retrieval. In: Ferrari, V., Hebert, M., Sminchisescu, C., Weiss, Y. (eds.) ECCV 2018. LNCS, vol. 11207, pp. 742–758. Springer, Cham (2018). https://doi.org/10.1007/978-3-030-01219-9_44
3. Bagautdinov, T., Alahi, A., Fleuret, F., Fua, P., Savarese, S.: Social scene understanding: end-to-end multi-person action localization and collective activity recognition. In: Proceedings of the IEEE Conference on Computer Vision and Pattern Recognition, pp. 4315–4324 (2017)
4. Wu, J., Wang, L., Wang, L., Guo, J., Wu, G.: Learning actor relation graphs for group activity recognition. In: Proceedings of the IEEE Conference on Computer Vision and Pattern Recognition, pp. 9964–9974 (2019)

5. Ibrahim, M.S., Muralidharan, S., Deng, Z., Vahdat, A., Mori, G.: A hierarchical deep temporal model for group activity recognition. In: Proceedings of the IEEE Conference on Computer Vision and Pattern Recognition, pp. 1971–1980 (2016)
6. Choi, W., Savarese, S.: A unified framework for multi-target tracking and collective activity recognition. In: Fitzgibbon, A., Lazebnik, S., Perona, P., Sato, Y., Schmid, C. (eds.) ECCV 2012. LNCS, vol. 7575, pp. 215–230. Springer, Heidelberg (2012). https://doi.org/10.1007/978-3-642-33765-9_16
7. Amer, M.R., Lei, P., Todorovic, S.: HiRF: hierarchical random field for collective activity recognition in videos. In: Fleet, D., Pajdla, T., Schiele, B., Tuytelaars, T. (eds.) ECCV 2014. LNCS, vol. 8694, pp. 572–585. Springer, Cham (2014). https://doi.org/10.1007/978-3-319-10599-4_37
8. Hajimirsadeghi, H., Yan, W., Vahdat, A., Mori, G.: Visual recognition by counting instances: a multi-instance cardinality potential kernel. In: Proceedings of the IEEE Conference on Computer Vision and Pattern Recognition, pp. 2596–2605 (2015)
9. Li, X., Chuah, M.C.: SBGAR: semantics based group activity recognition. In: Proceedings of the IEEE International Conference on Computer Vision, pp. 2876–2885 (2017)
10. Deng, Z., Vahdat, A., Hu, H., Mori, G.: Structure inference machines: recurrent neural networks for analyzing relations in group activity recognition. In: Proceedings of the IEEE Conference on Computer Vision and Pattern Recognition, pp. 4772–4781 (2016)
11. Shu, T., Todorovic, S., Zhu, S.-C.: CERN: confidence-energy recurrent network for group activity recognition. In: Proceedings of the IEEE Conference on Computer Vision and Pattern Recognition, pp. 5523–5531 (2017)
12. Wang, M., Ni, B., Yang, X.: Recurrent modeling of interaction context for collective activity recognition. In: Proceedings of the IEEE Conference on Computer Vision and Pattern Recognition, pp. 3048–3056 (2017)
13. Biswas, S., Gall, J.: Structural recurrent neural network (SRNN) for group activity analysis. In: Proceedings of the IEEE Winter Conference on Applications of Computer Vision, pp. 1625–1632 (2018)
14. Butepage, J., Black, M.J., Kragic, D., Kjellstrom, H.: Deep representation learning for human motion prediction and classification. In: Proceedings of the IEEE Conference on Computer Vision and Pattern Recognition, pp. 6158–6166 (2017)
15. Azar, S.M., Atigh, M.G., Nickabadi, A., Alahi, A.: Convolutional relational machine for group activity recognition. In: Proceedings of the IEEE Conference on Computer Vision and Pattern Recognition, pp. 7892–7901 (2019)
16. Zheng, S., et al.: Conditional random fields as recurrent neural networks. In: Proceedings of the IEEE International Conference on Computer Vision, pp. 1529–1537 (2015)
17. Chu, X., Yang, W., Ouyang, W., Ma, C., Yuille, A.L., Wang, X.: Multi-context attention for human pose estimation. In: Proceedings of the IEEE Conference on Computer Vision and Pattern Recognition, pp. 1831–1840 (2017)
18. Vaswani, A., et al.: Attention is all you need. In: Proceedings of the Advances in Neural Information Processing Systems, pp. 5998–6008 (2017)
19. Devlin, J., Chang, M.-W., Lee, K., Toutanova, K.: BERT: pre-training of deep bidirectional transformers for language understanding. arXiv preprint arXiv:1810.04805 (2018)
20. Alba, R.D.: A graph-theoretic definition of a sociometric clique. J. Math. Sociol. **3**(1), 113–126 (1973)

21. Dehghani, M., Gouws, S., Vinyals, O., Uszkoreit, J., Kaiser, Ł.: Universal transformers. In: Proceedings of the International Conference on Learning Representations (2019)
22. Simonyan, K., Zisserman, A.: Two-stream convolutional networks for action recognition in videos. In: Proceedings of the Neural Information Processing Systems, pp. 568–576 (2014)
23. Feichtenhofer, C., Pinz, A., Zisserman, A.: Convolutional two-stream network fusion for video action recognition. In: Proceedings of the IEEE Conference on Computer Vision and Pattern Recognition, pp. 1933–1941 (2016)
24. Feichtenhofer, C., Pinz, A., Wildes, R.: Spatiotemporal residual networks for video action recognition. In: Proceedings of the Advances in Neural Information Processing Systems, pp. 3468–3476 (2016)
25. Tran, D., Bourdev, L., Fergus, R., Torresani, L., Paluri, M.: Learning spatiotemporal features with 3D convolutional networks. In: Proceedings of the IEEE International Conference on Computer Vision, pp. 4489–4497 (2015)
26. Ji, S., Wei, X., Yang, M., Kai, Yu.: 3D convolutional neural networks for human action recognition. IEEE Trans. Pattern Anal. Mach. Intell. **35**(1), 221–231 (2013)
27. Carreira, J., Zisserman, A.: Quo vadis, action recognition? A new model and the kinetics dataset. In: Proceedings of the IEEE Conference on Computer Vision and Pattern Recognition, pp. 6299–6308 (2017)
28. Crasto, N., Weinzaepfel, P., Alahari, K., Schmid, C.: MARS: motion-augmented RGB stream for action recognition. In: Proceedings of the IEEE Conference on Computer Vision and Pattern Recognition, pp. 7882–7891 (2019)
29. Xie, S., Sun, C., Huang, J., Tu, Z., Murphy, K.: Rethinking spatiotemporal feature learning: speed-accuracy trade-offs in video classification. In: Ferrari, V., Hebert, M., Sminchisescu, C., Weiss, Y. (eds.) ECCV 2018. LNCS, vol. 11219, pp. 318–335. Springer, Cham (2018). https://doi.org/10.1007/978-3-030-01267-0_19
30. Li, X., Loy, C.C.: Video object segmentation with joint re-identification and attention-aware mask propagation. In: Ferrari, V., Hebert, M., Sminchisescu, C., Weiss, Y. (eds.) ECCV 2018. LNCS, vol. 11207, pp. 93–110. Springer, Cham (2018). https://doi.org/10.1007/978-3-030-01219-9_6
31. Zhang, G., Kan, M., Shan, S., Chen, X.: Generative adversarial network with spatial attention for face attribute editing. In: Ferrari, V., Hebert, M., Sminchisescu, C., Weiss, Y. (eds.) ECCV 2018. LNCS, vol. 11210, pp. 422–437. Springer, Cham (2018). https://doi.org/10.1007/978-3-030-01231-1_26
32. Zhao, T., Wu, X.: Pyramid feature attention network for saliency detection. In: Proceedings of the IEEE Conference on Computer Vision and Pattern Recognition, pp. 3085–3094 (2019)
33. Fu, J., et al.: Dual attention network for scene segmentation. In: Proceedings of the IEEE Conference on Computer Vision and Pattern Recognition, pp. 3146–3154 (2019)
34. Loy, C.C., Xiang, T., Gong, S.: Modelling activity global temporal dependencies using time delayed probabilistic graphical model. In: Proceedings of the IEEE International Conference on Computer Vision, pp. 120–127 (2009)
35. Swears, E., Hoogs, A., Ji, Q., Boyer, K.: Complex activity recognition using granger constrained DBN (GCDBN) in sports and surveillance video. In: Proceedings of the IEEE Conference on Computer Vision and Pattern Recognition, pp. 788–795 (2014)
36. Lu, Y., Lu, C., Tang, C.-K.: Online video object detection using association LSTM. In: Proceedings of The IEEE International Conference on Computer Vision, pp. 2344–2352 (2017)

37. Luo, Y., et al.: LSTM pose machines. In: Proceedings of The IEEE Conference on Computer Vision and Pattern Recognition, pp. 7852–7861 (2018)
38. Perrett, T., Damen, D.: DDLSTM: dual-domain LSTM for cross-dataset action recognition. In: Proceedings of the IEEE Conference on Computer Vision and Pattern Recognition, pp. 5207–5215 (2019)
39. Purwanto, D., Pramono, R.R.A., Chen, Y.-T., Fang, W.-H.: Three-stream network with bidirectional self-attention for action recognition in extreme low resolution videos. IEEE Signal Process. Lett. **26**, 1 (2019)
40. Wang, X., Girshick, R., Gupta, A., He, K.: Non-local neural networks. In: Proceedings of the IEEE Conference on Computer Vision and Pattern Recognition, pp. 7794–7803 (2018)
41. Sun, C., Myers, A., Vondrick, C., Murphy, K., Schmid, C.: VideoBERT: a joint model for video and language representation learning. arXiv preprint arXiv:1904.01766 (2019)
42. Girdhar, R., Carreira, J., Doersch, C., Zisserman, A.: Video action transformer network. In: Proceedings of the IEEE Conference on Computer Vision and Pattern Recognition, pp. 244–253 (2019)
43. Pramono, R.R.A., Chen, Y.-T., Fang, W.-H.: Hierarchical self-attention network for action localization in videos. In: Proceedings of the IEEE International Conference on Computer Vision, pp. 61–70 (2019)
44. Morariu, V.I., Davis, L.S.: Multi-agent event recognition in structured scenarios. In: Proceedings of the IEEE Conference on Computer Vision and Pattern Recognition, pp. 3289–3296 (2011)
45. Intille, S.S., Bobick, A.F.: Recognizing planned, multiperson action. Comput. Vis. Image Underst. **81**(3), 414–445 (2001)
46. Xu, Y., Qin, L., Liu, X., Xie, J., Zhu, S.-C.: A causal and-or graph model for visibility fluent reasoning in tracking interacting objects. In: Proceedings of the IEEE Conference on Computer Vision and Pattern Recognition, pp. 2178–2187 (2018)
47. Li, W.-H., Hong, F.-T., Zheng, W.-S.: Learning to learn relation for important people detection in still images. In: Proceedings of the IEEE Conference on Computer Vision and Pattern Recognition, pp. 5003–5011 (2019)
48. Wang, X., Gupta, A.: Videos as space-time region graphs. In: Ferrari, V., Hebert, M., Sminchisescu, C., Weiss, Y. (eds.) ECCV 2018. LNCS, vol. 11209, pp. 413–431. Springer, Cham (2018). https://doi.org/10.1007/978-3-030-01228-1_25
49. Lin, T.-Y., Dollar, P., Girshick, R., He, K., Hariharan, B., Belongie, S.: Feature pyramid networks for object detection. In: Proceedings of the IEEE Conference on Computer Vision and Pattern Recognition, pp. 2117–2125 (2017)
50. Fang, H.-S., Xie, S., Tai, Y.-W., Lu, C.: RMPE: regional multi-person pose estimation, pp. 2334–2343 (2017)
51. Lafferty, J., McCallum, A., Pereira, F.C.N.: Conditional random fields: probabilistic models for segmenting and labeling sequence data. In: Proceedings of the International Conference on Machine Learning (2001)
52. Krähenbühl, P., Koltun, V.: Efficient inference in fully connected CRFs with Gaussian edge potentials. In: Proceedings of the Advances in Neural Information Processing Systems, pp. 109–117 (2011)
53. Arnab, A., Jayasumana, S., Zheng, S., Torr, P.H.S.: Higher order conditional random fields in deep neural networks. In: Leibe, B., Matas, J., Sebe, N., Welling, M. (eds.) ECCV 2016. LNCS, vol. 9906, pp. 524–540. Springer, Cham (2016). https://doi.org/10.1007/978-3-319-46475-6_33

54. Liu, B., He, X.: Learning dynamic hierarchical models for anytime scene labeling. In: Leibe, B., Matas, J., Sebe, N., Welling, M. (eds.) ECCV 2016. LNCS, vol. 9910, pp. 650–666. Springer, Cham (2016). https://doi.org/10.1007/978-3-319-46466-4_39
55. Xiong, Y., et al.: UPSNet: a unified panoptic segmentation network. In: Proceedings of the IEEE Conference on Computer Vision and Pattern Recognition, pp. 8818–8826 (2019)
56. Graves, A.: Adaptive computation time for recurrent neural networks. arXiv preprint arXiv:1603.08983 (2016)
57. Choi, W., Shahid, K., Savarese, S.: What are they doing?: Collective activity classification using spatio-temporal relationship among people. In: Proceedings of the International Conference on Computer Vision Workshops, pp. 1282–1289 (2009)
58. Zach, C., Pock, T., Bischof, H.: A duality based approach for realtime TV-L1 optical flow. In: Proceedings of the Joint Pattern Recognition Symposium, pp. 214–223 (2007)
59. Goodfellow, I., Bengio, Y., Courville, A.: Deep Learning. The MIT Press, Cambridge (2016)
60. Lan, T., Wang, Y., Yang, W., Robinovitch, S.N., Mori, G.: Discriminative latent models for recognizing contextual group activities. IEEE Trans. Pattern Anal. Mach. Intell. **34**(8), 1549–1562 (2011)

AiR: Attention with Reasoning Capability

Shi Chen, Ming Jiang, Jinhui Yang, and Qi Zhao(✉)

University of Minnesota, Minneapolis, MN 55455, USA
{chen4595,mjiang,yang7004,qzhao}@umn.edu

Abstract. While attention has been an increasingly popular component in deep neural networks to both interpret and boost performance of models, little work has examined how attention progresses to accomplish a task and whether it is reasonable. In this work, we propose an Attention with Reasoning capability (AiR) framework that uses attention to understand and improve the process leading to task outcomes. We first define an evaluation metric based on a sequence of atomic reasoning operations, enabling quantitative measurement of attention that considers the reasoning process. We then collect human eye-tracking and answer correctness data, and analyze various machine and human attentions on their reasoning capability and how they impact task performance. Furthermore, we propose a supervision method to jointly and progressively optimize attention, reasoning, and task performance so that models learn to look at regions of interests by following a reasoning process. We demonstrate the effectiveness of the proposed framework in analyzing and modeling attention with better reasoning capability and task performance. The code and data are available at https://github.com/szzexpoi/AiR.

Keywords: Attention · Reasoning · Eye-tracking dataset

1 Introduction

Recent progress in deep neural networks (DNNs) has resulted in models with significant performance gains in many tasks. Attention, as an information selection mechanism, has been widely used in various DNN models, to improve their ability of localizing important parts of the inputs, as well as task performances. It also enables fine-grained analysis and understanding of the black-box DNN models, by highlighting important information in their decision-making. Recent studies explored different machine attentions and showed varied degrees of agreement on where human consider important in various vision tasks, such as captioning [15,33] and visual question answering (VQA) [7].

S. Chen and M. Jiang–Equal contributions.

Electronic supplementary material The online version of this chapter (https://doi.org/10.1007/978-3-030-58452-8_6) contains supplementary material, which is available to authorized users.

A. Vedaldi et al. (Eds.): ECCV 2020, LNCS 12346, pp. 91–107, 2020.
https://doi.org/10.1007/978-3-030-58452-8_6

Similar to humans who look and reason actively and iteratively to perform a visual task, attention and reasoning are two intertwined mechanisms underlying the decision-making process. As shown in Fig. 1, answering the question requires humans or machines to make a sequence of decisions based on the relevant regions of interest (ROIs) (*i.e.,* to sequentially look for the jeans, the girl wearing the jeans, and the bag to the left of the girl). Guiding attention to explicitly look for these objects following the reasoning process has the potential to improve both interpretability and performance of a computer vision model.

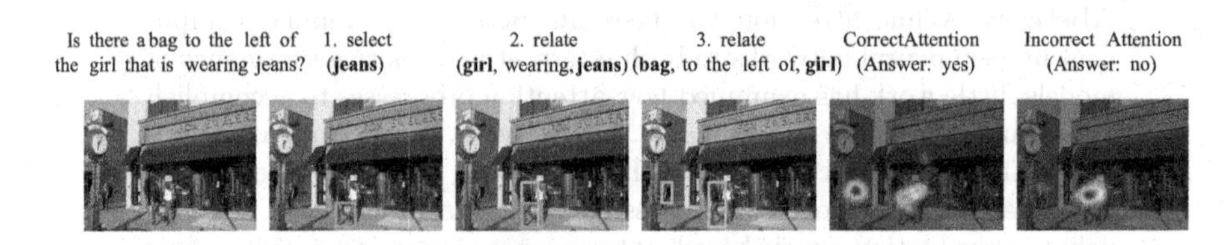

Fig. 1. Attention is an essential mechanism that affects task performances in visual question answering. Eye fixation maps of humans suggest that people who answer correctly look at the most relevant ROIs in the reasoning process (*i.e.,* jeans, girl, and bag), while incorrect answers are caused by misdirected attention

To understand the roles of visual attention in the visual reasoning context, and leverage it for model development, we propose an integrated Attention with Reasoning capability (AiR) framework. It represents the visual reasoning process as a sequence of atomic operations each with specific ROIs, defines a metric and proposes a supervision method that enables the quantitative evaluation and guidance of attentions based on the intermediate steps of the visual reasoning process. A new eye-tracking dataset is collected to support the understanding of human visual attention during the visual reasoning process, and is also used as a baseline for studying machine attention. This framework is a useful toolkit for research in visual attention and its interaction with visual reasoning.

Our work has three distinctions from previous attention evaluation [7,18,19,26] and supervision [28,29,44] methods: (1) We go beyond the existing evaluation methods that are either qualitative or focused only on the alignment with outputs, and propose a measure that encodes the progressive attention and reasoning defined by a set of atomic operations. (2) We emphasize the tight correlation between attention, reasoning, and task performance, conducting fine-grained analyses of the proposed method with various types of attention, and incorporating attention with the reasoning process to enhance model interpretability and performance. (3) Our new dataset with human eye movements and answer correctness enables more accurate evaluation and diagnosis of attention.

To summarize, the proposed framework makes the following contributions:

1. A new quantitative evaluation metric (AiR-E) to measure attention in the reasoning context, based on a set of constructed atomic reasoning operations.

2. A supervision method (AiR-M) to progressively optimize attention throughout the entire reasoning process.
3. An eye-tracking dataset (AiR-D) featuring high-quality attention and reasoning labels as well as ground truth answer correctness.
4. Extensive analyses of various machine and human attention with respect to reasoning capability and task performance. Multiple factors of machine attention have been examined and discussed. Experiments show the importance of progressive supervision on both attention and task performance.

2 Related Works

This paper is most closely related to prior studies on the evaluation of attention in visual question answering (VQA) [7,18,19,26]. In particular, the pioneering work by Das et al. [7] is the only one that collected human attention data on a VQA dataset and compared them with machine attention, showing considerable discrepancies in the attention maps. Our proposed study highlights several distinctions from related works: (1) Instead of only considering one-step attention and its alignment with a single ground-truth map, we propose to integrate attention with progressive reasoning that involves a sequence of operations each related to different objects. (2) While most VQA studies assume human answers to be accurate, it is not always the case [38]. We collect ground truth correctness labels to examine the effects of attention and reasoning on task performance. (3) The only available dataset [7], with post-hoc attention annotation collected on blurry images using a "bubble-like" paradigm and crowdsourcing, may not accurately reflect the actual attention of the task performers [32]. Our work addresses these limitations by using on-site eye tracking data and QA annotations collected from the same participants. (4) Das et al. [7] compared only spatial attention with human attention. Since recent studies [18,26] suggest that attention based on object proposals are more semantically meaningful, we conduct the first quantitative and principled evaluation of object-based attentions.

This paper also presents a progressive supervision approach for attention, which is related to the recent efforts on improving attention accuracy with explicit supervision. Several studies use different sources of attention ground truth, such as human attention [29], adversarial learning [28] and objects mined from textual descriptions [44], to explicitly supervise the learning of attentions. Similar to the evaluation studies introduced above, these attention supervision studies only consider attention as a single-output mechanism and ignores the progressive nature of the attention process or whether it is reasonable or not. As a result, they fall short of acquiring sufficient information from intermediate steps. Our work addresses these challenges with joint prediction of the reasoning operations and the desired attentions along the entire decision-making process.

Our work is also related to a collection of datasets for eye-tracking and visual reasoning. Eye tracking data is collected to study passive exploration [1,5,10,21,24,36] as well as task-guided attention [1,9,24]. Despite the less accurate and post-hoc mouse-clicking approximation [7], there has been no eye-tracking data recorded from human participants performing the VQA tasks.

To facilitate the analysis of human attention in VQA tasks, we construct the first dataset of eye-tracking data collected from humans performing the VQA tasks. A number of visual reasoning datasets [3,13,18,20,31,43] are collected in the form of VQA. Some are annotated with human-generated questions and answers [3,31], while others are developed with synthetic scenes and rule-based templates to remove the subjectiveness of human answers and language biases [13,18,20,43]. The one most closely related to this work is GQA [18], which offers naturalistic images annotated with scene graphs and synthetic question-answer pairs. With balanced questions and answers, it reduces the language bias without compromising generality. Their data efforts benefit the development of various visual reasoning models [2,8,11,16,23,27,30,35,39–42]. In this work, we use a selection of GQA data and annotations in the development of the proposed framework.

3 Method

Real-life vision tasks require looking and reasoning interactively. This section presents a principled framework to study attention in the reasoning context. It consists of three novel components: (1) a quantitative measure to evaluate attention accuracy in the reasoning context, (2) a progressive supervision method for models to learn where to look throughout the reasoning process, and (3) an eye-tracking dataset featuring human eye-tracking and answer correctness data.

3.1 Attention with Reasoning Capability

To model attention as a process and examine its reasoning capability, we describe reasoning as a sequence of atomic operations. Following the sequence, an intelligent agent progressively attends to the key ROIs at each step and reasons what to do next until eventually making a final decision. A successful decision-making relies on accurate attention for various reasoning operations, so that the most important information is not filtered out but passed along to the final step.

Table 1. Semantic operations of the reasoning process

Operation	Semantic
Select	Searching for objects from a specific category
Filter	Determining the targeted objects by looking for a specific attribute
Query	Retrieving the value of a specific attribute from the ROIs
Verify	Examining the targeted objects and checking if they have a given attribute
Compare	Comparing the values of an attribute between multiple objects
Relate	Connecting different objects through their relationships
And/Or	Serving as basic logical operations that combine the results of the previous operation(s)

To represent the reasoning process and obtain the corresponding ROIs, we define a vocabulary of atomic operations emphasizing the role of attention. These operations are grounded on the 127 types of operations of GQA [18] that completely represent all the questions. As described in Table 1, some operations require attention to a specific object (*query*, *verify*); some require attention to objects of the same category (*select*), attribute (*filter*), or relationship (*relate*); and others require attention to any (*or*) or all (*and*, *compare*) ROIs from the previous operations. The ROIs of each operation are jointly determined by the type of operation and the scene information (*i.e.*, object categories, attributes and relationships). Given the operation sequence and annotated scene information, we can traverse the reasoning process, starting with all objects in the scene, and sequentially apply the operations to obtain the ROIs at each step. Details of this method are described in the supplementary materials.

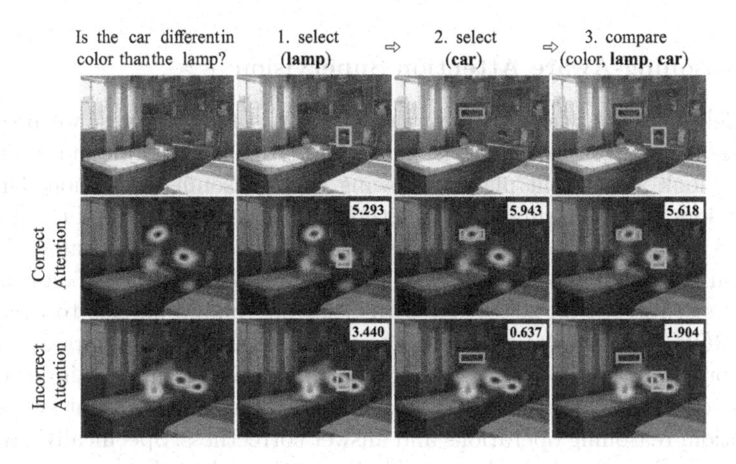

Fig. 2. AiR-E scores of Correct and Incorrect human attention maps, measuring their alignments with the bounding boxes of the ROIs

3.2 Measuring Attention Accuracy with ROIs

Decomposing the reasoning process into a sequence of operations allows us to evaluate the quality of attention (machine and human attentions) according to its alignment with the ROIs at each operation. Attention can be represented as a 2D probability map where values indicate the importance of the corresponding input pixels. To quantitatively evaluate attention accuracy in the reasoning context, we propose the AiR-E metric that measures the alignment of the attention maps with ROIs relevant to reasoning. As shown in Fig. 2, for humans, a better attention map leading to the correct answer has higher AiR-E scores, while the incorrect attention with lower scores fails to focus on the most important object (*i.e.*, car). It suggests potential correlation between the AiR-E and the task performance. The specific definition of AiR-E is introduced as follows:

Inspired by the Normalized Scanpath Saliency [6] (NSS), given an attention map $A(x)$ where each value represents the importance of a pixel x, we first standardize the attention map into $A^*(x) = (A(x) - \mu)/\sigma$, where μ and σ are the mean and standard deviation of the attention values in $A(x)$, respectively. For each ROI, we compute AiR-E as the average of $A^*(x)$ inside its bounding box B: AiR-E$(B) = \sum_{x \in B} A^*(x)/|B|$. Finally, we aggregate the AiR-E of all ROIs for each reasoning step:

1. For operations with one set of ROIs (*i.e.*, *select*, *query*, *verify*, and *filter*), as well as *or* that requires attention to one of multiple sets of ROIs, an accurate attention map should align well with at least one ROI. Therefore, the aggregated AiR-E score is the maximum AiR-E of all ROIs.
2. For those with multiple sets of ROIs (*i.e.*, *relate*, *compare*, *and*), we compute the aggregated AiR-E for each set, and take the mean across all sets.

3.3 Reasoning-Aware Attention Supervision

For models to learn where to look along the reasoning process, we propose a reasoning-aware attention supervision method (AiR-M) to guide models to progressively look at relevant places following each reasoning operation. Different from previous attention supervision methods [28,29,44], the AiR-M method considers the attention throughout the reasoning process and jointly supervises the prediction of reasoning operations and ROIs across the sequence of multiple reasoning steps. Integrating attention with reasoning allows models to accurately capture ROIs along the entire reasoning process for deriving the correct answers.

The proposed method has two major distinctions: (1) integrating attention progressively throughout the entire reasoning process and (2) joint supervision on attention, reasoning operations and answer correctness. Specifically, following the reasoning decomposition discussed in Sect. 3.1, at the t-th reasoning step, the proposed method predicts the reasoning operation $\boldsymbol{r}_t$, and generates an attention map $\boldsymbol{\alpha}_t$ to predict the ROIs. With the joint prediction, models learn desirable attentions for capturing the ROIs throughout the reasoning process and deriving the answer. The predicted operations and the attentions are supervised together with the prediction of answers:

$$L = L_{ans} + \theta \sum_t L_{\alpha_t} + \phi \sum_t L_{r_t} \tag{1}$$

where θ and ϕ are hyperparameters. We use the standard cross-entropy loss L_{ans} and L_{r_t} to supervise the answer and operation prediction, and a Kullback–Leibler divergence loss L_{α_t} to supervise the attention prediction. We aggregate the loss for operation and attention predictions over all reasoning steps.

The proposed AiR-M supervision method is general, and can be applied to various models with attention mechanisms. In the supplementary materials, we illustrate the implementation details for integrating AiR-M with different state-of-the-art models used in our experiments.

3.4 Evaluation Benchmark and Human Attention Baseline

Previous attention data collected under passive image viewing [21], approximations with post-hoc mouse clicks [7], or visually grounded answers [19] may not accurately or completely reflect human attention in the reasoning process. They also do not explicitly verify the correctness of human answers. To demonstrate the effectiveness of the proposed evaluation metric and supervision method, and to provide a benchmark for attention evaluation, we construct the first eye-tracking dataset for VQA. It, for the first time, enables the step-by-step comparison of how humans and machines allocate attention during visual reasoning.

Specifically, we (1) select images and questions that require humans to actively look and reason; (2) remove ambiguous or ill-formed questions and verify the ground truth answer to be correct and unique; (3) collect eye-tracking data and answers from the same human participants, and evaluate their correctness with the ground-truth answers.

Images and Questions. Our images and questions are selected from the balanced validation set of GQA [18]. Since the questions of the GQA dataset are automatically generated from a number of templates based on scene graphs [25], the quality of these automatically generated questions may not be sufficiently high. Some questions may be too trivial or too ambiguous. Therefore, we perform automated and manual screenings to control the quality of the questions. First, to avoid trivial questions, all images and questions are first screened with these criteria: (1) image resolution is at least 320×320 pixels; (2) image scene graph consists of at least 16 relationships; (3) total area of question-related objects does not exceed 4% of the image. Next, one of the authors manually selects 987 images and 1,422 questions to ensure that the ground-truth answers are accurate and unique. The selected questions are non-trivial and free of ambiguity, which require paying close attention to the scene and actively searching for the answer.

Eye-Tracking Experiment. The eye-tracking data are collected from 20 paid participants, including 16 males and 4 females from age 18 to 38. They are asked to wear a Vive Pro Eye headset with an integrated eye-tracker to answer questions from images presented in a customized Unity interface. The questions are randomly grouped into 18 blocks, each shown in a 20-min session. The eye-tracker is calibrated at the beginning of each session. During each trial, a question is first presented, and the participant is given unlimited time to read and understand it. The participant presses a controller button to start viewing the image. The image is presented in the center for 3 s. The image is scaled such that both the height and width occupy 30° of visual angle (DVA). After that, the question is shown again and the participant is instructed to provide an answer. The answer is then recorded by the experimenter. The participant presses another button to proceed to the next trial.

Human Attention Maps and Performances. Eye fixations are extracted from the raw data using the Cluster Fix algorithm [22], and a fixation map is computed for each question by aggregating the fixations from all participants. The fixation maps are scaled into 256×256 pixels, smoothed using a Gaussian kernel ($\sigma = 9$ pixels, $\approx$1 DVA) and normalized to the range of [0,1]. The overall accuracy of human answers is $77.64 \pm 24.55\%$ (M$\pm$SD). A total of 479 questions have consistently correct answers, and 934 have both correct and incorrect answers. The histogram of human answer accuracy is shown in Fig. 3a. We further separate the fixations into two groups based on answer correctness and compute a fixation map for each group. Correct and incorrect answers have comparable numbers of fixations per trial (10.12 *vs.*10.27), while the numbers of fixations for the correct answers have a lower standard deviation across trials (0.99 *vs.*1.54). Figure 3b shows the prior distributions of the two groups of fixations, and their high similarity (Pearson's $r = 0.997$) suggests that the answer correctness is independent of center bias. The correct and incorrect fixation maps are considered as two human attention baselines to compare with machine attentions, and also play a role in validating the effectiveness of the proposed AiR-E metric. More illustration is provided in the supplementary video.

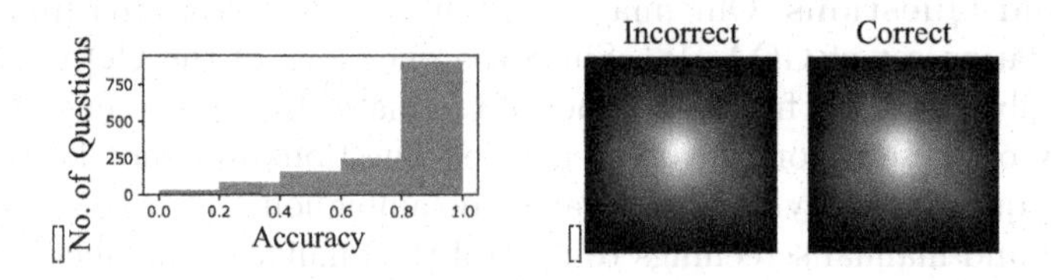

Fig. 3. Distributions of answer accuracy and eye fixations of humans. (a) Histogram of human answer accuracy (b) Center biases of the correct and incorrect attention

4 Experiments and Analyses

In this section, we conduct experiments and analyze various attention mechanisms of humans and machines. Our experiments aim to shed light on the following questions that have yet to be answered:

1. Do machines or humans look at places relevant to the reasoning process? How does the attention process influence task performances? (Sect. 4.1)
2. How does attention accuracy evolve over time, and what about its correlation with the reasoning process? (Sect. 4.2)
3. Does guiding models to look at places progressively following the reasoning process help? (Sect. 4.3)

4.1 Do Machines or Humans Look at Places Important to Reasoning? How Does Attention Influence Task Performances?

First, we measure attention accuracy throughout the reasoning process with the proposed AiR-E metric. Answer correctness is also compared, and its correlation with the attention accuracy reveals the joint influence of attention and reasoning operations to task performance. With these experiments, we

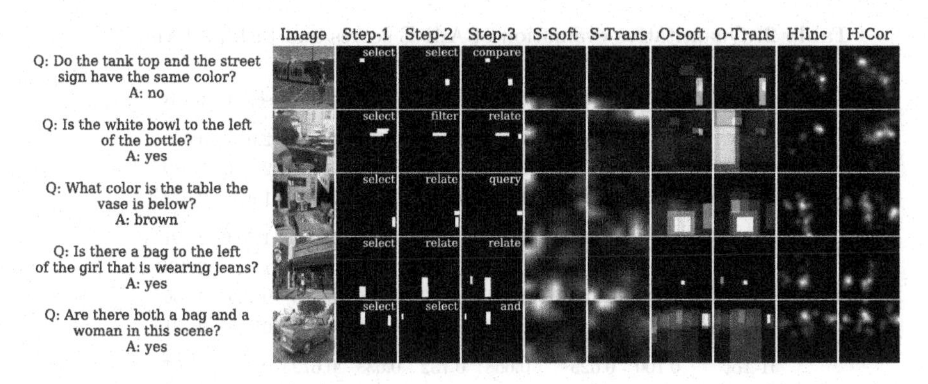

Fig. 4. Example question-answer pairs (column 1), images (column 2), ROIs at each reasoning step (columns 3–5), and attention maps (columns 6–11)

observe that humans attend more accurately than machines, and the correlation between attention accuracy and task performance is dependent on the reasoning operations.

We evaluate four types of attentions that are commonly used in VQA models, including spatial soft attention (S-Soft), spatial Transformer attention (S-Trans), object-based soft attention (O-Soft), and object-based Transformer attention (O-Trans). Spatial and object-based attentions differ in terms of their inputs (*i.e.,* image features or regional features), while soft and Transformer attention methods differ in terms of the computational methods of attention (*i.e.,* with convolutional layers or matrix multiplication). We use spatial features extracted from ResNet-101 [14] and object-based features from [2] as the two types of inputs, and follow the implementations of [2] and [12] for the soft attention [37] and Transformer attention [34] computation, respectively. We integrate the aforementioned attentions with different state-of-the-art VQA models as backbones. Our observations are general and consistent across various backbones. In the following sections, we use the results on UpDown [2] for illustration (results for the other backbones are provided in the supplementary materials). For human attentions, we denote the fixation maps associated with correct and incorrect answers as H-Cor and H-Inc, and the aggregated fixation map regardless of correctness is denoted as H-Tot. Figure 4 presents examples of ROIs for different reasoning operations and the compared attention maps.

Attention Accuracy and Task Performance of Humans and Models. Table 2 quantitatively compares the AiR-E scores and VQA task performance across humans and models with different types of attentions. The task performance for models is the classification score of the correct answer, while the task performance for humans is the proportion of correct answers. Three clear gaps can be observed from the table: (1) Humans who answer correctly have significantly higher AiR-E scores than those who answer incorrectly. (2) Humans consistently outperform models in both attention and task performance. (3) Object-based attentions attend much more accurately than spatial attentions.

Table 2. Quantitative evaluation of AiR-E scores and task performance

	Attention	and	compare	filter	or	query	relate	select	verify
AiR-E	H-Tot	2.197	2.669	2.810	2.429	3.951	3.516	2.913	3.629
	H-Cor	2.258	2.717	2.925	2.529	4.169	3.581	2.954	3.580
	H-Inc	1.542	1.856	1.763	1.363	2.032	2.380	1.980	2.512
	O-Soft	1.334	**1.204**	**1.518**	1.857	**3.241**	**2.243**	**1.586**	2.091
	O-Trans	**1.579**	1.046	1.202	**1.910**	3.041	1.839	1.324	**2.228**
	S-Soft	-0.001	-0.110	0.251	0.413	0.725	0.305	0.145	0.136
	S-Trans	0.060	-0.172	0.243	0.343	0.718	0.370	0.173	0.101
Accuracy	H-Tot	0.700	0.625	0.668	0.732	0.633	0.672	0.670	0.707
	O-Soft	0.604	**0.547**	0.603	0.809	**0.287**	0.483	0.548	0.605
	O-Trans	**0.606**	0.536	**0.608**	**0.832**	0.282	**0.487**	**0.550**	0.592
	S-Soft	0.592	0.520	0.558	0.814	0.203	0.427	0.511	0.544
	S-Trans	0.597	0.525	0.557	0.811	0.211	0.435	0.517	**0.607**

The low AiR-E of spatial attentions confirms the previous conclusion drawn from the VQA-HAT dataset [7]. By constraining the visual inputs to a set of semantically meaningful objects, object-based attention typically increases the probabilities of attending to the correct ROIs. Between the two object-based attentions, the soft attention slightly outperforms its Transformer counterpart. Since the Transformer attentions explicitly learn the inter-object relationships, they perform better for logical operations (*i.e., and, or*). However, due to the complexity of the scenes and fewer parameters used [34], they do not perform as well as soft attention. The ranks of different attentions are consistent with the intuition and literature, suggesting the effectiveness of the proposed AiR-E metric.

Attention Accuracy and Task Performance Among Different Reasoning Operations. Comparing the different operations, Table 2 shows that *query* is the most challenging operation for models. Even with the highest attention accuracy among all operations, the task performance is the lowest. This is probably due to the inferior recognition capability of models compared with humans. To humans, 'compare' is the most challenging in terms of task performance, largely because it often appears in complex questions that require close attention to multiple objects and thus take longer processing time. Since models can process multiple input objects in parallel, their performance is not highly influenced by the number of objects to look at.

Correlation Between Attention Accuracy and Task Performance. The similar rankings of AiR-E and task performance suggest a correlation between attention accuracy and task performance. To further investigate this correlation on a sample basis, for each attention and operation, we compute the Pearson's r between the attention accuracy and task performance across different questions.

Table 3. Pearson's r between attention accuracy (AiR-E) and task performance. Bold numbers indicate significant positive correlations ($p < 0.05$)

Attention	And	Compare	Filter	Or	Query	Relate	Select	Verify
H-Tot	0.205	**0.329**	0.051	0.176	**0.282**	**0.210**	**0.134**	**0.270**
O-Soft	0.167	**0.217**	−0.022	0.059	**0.331**	0.058	0.003	0.121
O-Trans	0.168	**0.205**	0.090	0.174	**0.298**	0.041	**0.063**	−0.027
S-Soft	0.177	**0.237**	−0.084	0.082	−0.017	−0.170	−0.084	0.066
S-Trans	0.171	**0.210**	−0.152	0.086	−0.024	−0.139	−0.100	**0.270**

As shown in Table 3, human attention accuracy and task performance are correlated for most of the operations (up to $r = 0.329$). The correlation is higher than most of the compared machine attentions, suggesting that humans' task performance is more consistent with their attention quality. In contrast, though commonly referred as an interface for interpreting models' decisions [7,19,26], spatial attention maps do not reflect the decision-making process of models. They typically have very low and even negative correlations (*e.g., relate, select*). By limiting the visual inputs to foreground objects, object-based attentions achieve higher attention-answer correlations.

The differences of correlations between operations are also significant. For the questions requiring focused attention to answer (*i.e.,* with *query* and *compare* operations), the correlations are relatively higher than the others.

4.2 How Does Attention Accuracy Evolve Throughout the Reasoning Process?

To complement our previous analysis on the spatial allocation of attentions, we move forward to analyze the spatiotemporal alignment of attentions. Specifically, we analyze the AiR-E scores according to the chronological order of reasoning operations. We show in Fig. 5a that the AiR-E scores peak at the 3rd or 4th steps, suggesting that human and machine attentions focus more on the ROIs closely related to the final task outcome, instead of the earlier steps. In the rest of this section, we focus our analysis on the spatiotemporal alignment between multiple attention maps and the ROIs at different reasoning steps. In particular, we study the change of human attention over time, and compare it with multi-glimpse machine attentions. Our analysis reveals the significant spatiotemporal discrepancy between human and machine attentions.

Do Human Attentions Follow the Reasoning Process? First, to analyze the spatiotemporal deployment of human attention in visual reasoning, we group the fixations into three temporal bins (0–1 s, 1–2 s and 2–3 s), and compute AiR-E scores for each fixation map and reasoning step (see Fig. 5b–c). Humans start exploration (0–1 s) with relatively low attention accuracy. After the initial exploration, human attention shows improved accuracy across all reasoning steps

(1–2 s), and particularly focuses on the early-step ROIs. In the final steps (2–3 s), depending on the correctness of answers, human attention either shifts to the ROIs at later stages (correct), or becomes less accurate with lowered AiR-E scores (incorrect). Such observations suggest high spatiotemporal alignments between human attention and the sequence of reasoning operations.

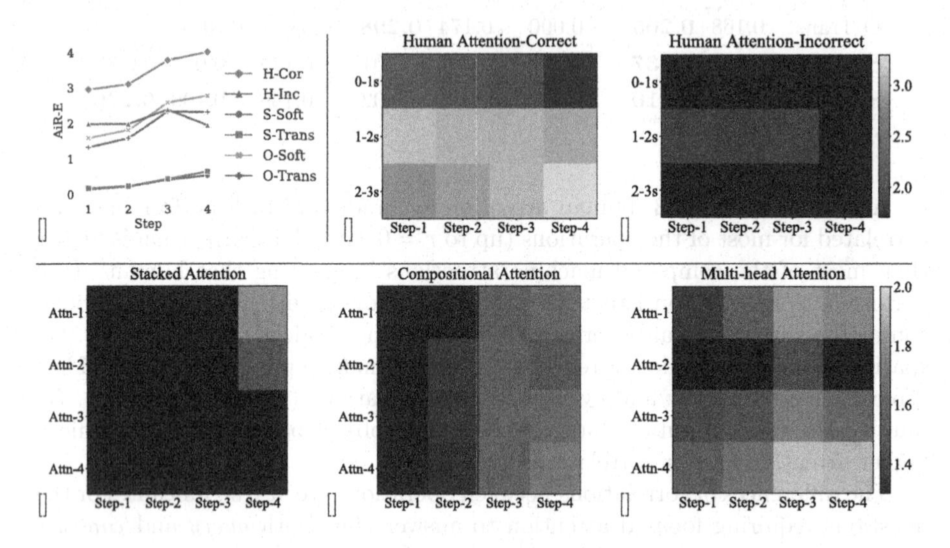

Fig. 5. Spatiotemporal accuracy of attention throughout the reasoning process. (a) shows the AiR-E of different reasoning steps for human aggregated attentions and single-glimpse machine attentions, (b)–(c) AiR-E scores for decomposed human attentions with correct and incorrect answers, (d)–(f) AiR-E for multi-glimpse machine attentions. For heat maps shown in (b)–(f), the x-axis denotes different reasoning steps while the y-axis corresponds to the indices of attention maps

Do Machine Attentions Follow the Reasoning Process? Similarly, we evaluate multi-glimpse machine attentions. We compare the stacked attention from SAN [39], compositional attention from MAC [17] and the multi-head attention [11,42], which all adopt the object-based attention. Figure 5d–f shows that multi-glimpse attentions do not evolve with the reasoning process. Stacked attention's first glimpse already attends to the ROIs at the 4th step, and the other glimpses contribute little to the attention accuracy. Compositional attention and multi-head attention consistently align the best with the ROIs at the 3rd or 4th step, and ignore those at the early steps.

The spatiotemporal correlations indicate that following the correct order of reasoning operations is important for humans to attend and answer correctly. In contrast, models tend to directly attend to the final ROIs, instead of shifting their attentions progressively.

Table 4. Comparative results on GQA test sets (test-dev and test-standard). We report the single-model performance trained on the balanced training set of GQA

	UpDown [2]		MUTAN [4]		BAN [23]	
	Dev	Standard	Dev	Standard	Dev	Standard
w/o Supervision	51.31	52.31	50.78	51.16	50.14	50.38
PAAN [28]	48.03	48.92	46.40	47.22	n/a	n/a
HAN [29]	49.96	50.58	48.76	48.99	n/a	n/a
ASM [44]	52.96	53.57	51.46	52.36	n/a	n/a
AiR-M	**53.46**	**54.10**	**51.81**	**52.42**	**53.36**	**54.15**

4.3 Does Progressive Attention Supervision Improve Attention and Task Performance?

Experiments in Sect. 4.1 and Sect. 4.2 suggest that attention towards ROIs relevant to the reasoning process contributes to task performance, and furthermore, the order of attention matters. Therefore, we propose to guide models to look at places important to reasoning in a progressive manner. Specifically, we propose to supervise machine attention along the reasoning process by jointly optimizing attention, reasoning operations, and task performance (AiR-M, see Sect. 3.3). Here we investigate the effectiveness of the AiR-M supervision method on three VQA models, *i.e.,* UpDown [2], MUTAN [4], and BAN [23]. We compare AiR-M with a number of state-of-the-art attention supervision methods, including supervision from human-like attention (HAN) [29], attention supervision mining (ASM) [44] and adversarial learning (PAAN) [28]. Note that while the other compared methods are typically limited to supervision on a single attention map, our AiR-M method is generally applicable to various VQA models with single or multiple attention maps (*e.g.,* BAN [23]).

According to Table 4, the proposed AiR-M supervision significantly improves the performance of all baselines and consistently outperforms the other attention supervision methods. Two of the compared methods, HAN and PAAN, fail to improve the performance of object-based attention. Supervising attention with knowledge from objects mined from language, ASM [44] is able to consistently improve the performance of models. However, without considering the intermediate steps of reasoning, it is not as effective as the proposed method.

Figure 6 shows the qualitative comparison between supervision methods. The proposed AiR-M not only directs attention to the ROIs most related to the answers (*i.e.,* freezer, wheel, chair, purse), but also highlights other important ROIs mentioned in the questions (*i.e.,* keyboard, man), thus reflecting the entire reasoning process, while attentions in other methods fail to localize these ROIs.

Table 5 reports the AiR-E scores across operations. It shows that the AiR-M supervision method significantly improves attention accuracy (attention aggregated across different steps), especially on those typically positioned in early steps (*e.g., select, compare*). In addition, the AiR-M supervision method also

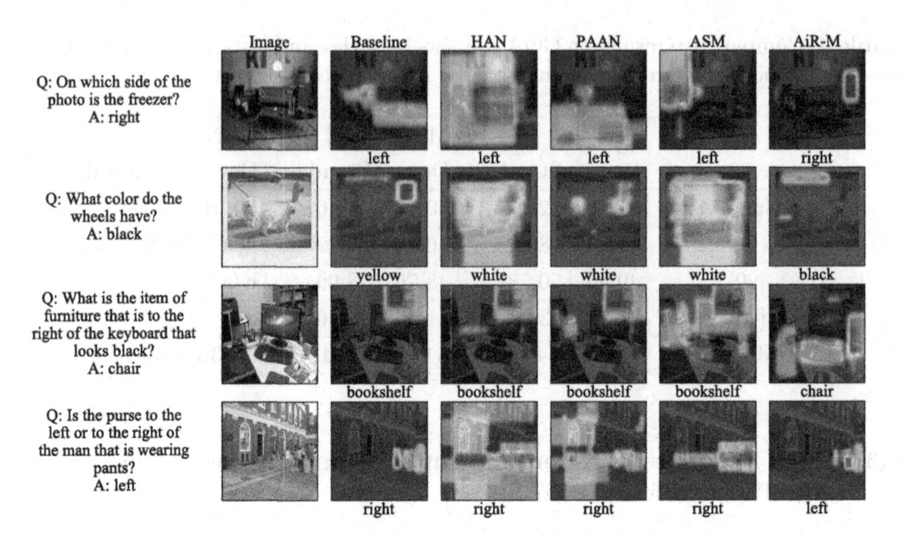

Fig. 6. Qualitative comparison between attention supervision methods, where Baseline refers to UpDown [2]. For each row, from left to right are the questions and the correct answers, input images, and attention maps learned by different methods. The predicted answers associated with the attentions are shown below its respective attention map

Table 5. AiR-E scores of the supervised attentions

Attention	And	Compare	Filter	Or	Query	Relate	Select	Verify
Human	2.197	2.669	2.810	2.429	3.951	3.516	2.913	3.629
AiR-M	**2.396**	**2.553**	**2.383**	**2.380**	3.340	**2.862**	**2.611**	**4.052**
Baseline [2]	1.859	1.375	1.717	2.271	**3.651**	2.448	1.796	2.719
ASM	1.415	1.334	1.443	1.752	2.447	1.884	1.584	2.265
HAN	0.581	0.428	0.468	0.607	1.576	0.923	0.638	0.680
PAAN	1.017	0.872	1.039	1.181	2.656	1.592	1.138	1.221

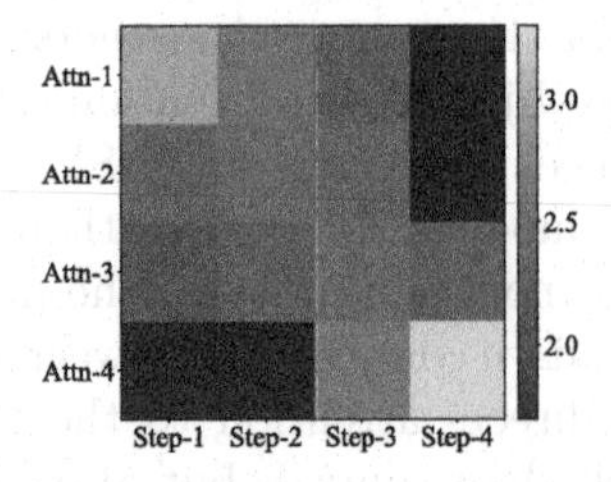

Fig. 7. Alignment between the proposed attention and reasoning process

aligns the multi-glimpse attentions better according to their chronological order in the reasoning process (see Fig. 7 and the supplementary video), showing progressive improvement of attention throughout the entire process.

5 Conclusion

We introduce AiR, a novel framework with a quantitative evaluation metric (AiR-E), a supervision method (AiR-M), and an eye-tracking dataset (AiR-D) for understanding and improving attention in the reasoning context. Our analyses show that accurate attention deployment can lead to improved task performance, which is related to both the task outcome and the intermediate reasoning steps. Our experiments also highlight the significant gap between models and humans on the alignment of attention and reasoning process. With the proposed attention supervision method, we further demonstrate that incorporating the progressive reasoning process in attention can improve the task performance by a considerable margin. We hope that this work will be helpful for future development of visual attention and reasoning method, and inspire the analysis of model interpretability throughout the decision-making process.

Acknowledgements. This work is supported by NSF Grants 1908711 and 1849107.

References

1. Alers, H., Liu, H., Redi, J., Heynderickx, I.: Studying the effect of optimizing the image quality in saliency regions at the expense of background content. In: Image Quality and System Performance VII, vol. 7529, p. 752907. International Society for Optics and Photonics (2010)
2. Anderson, P., et al.: Bottom-up and top-down attention for image captioning and visual question answering. In: CVPR (2018)
3. Antol, S., et al.: VQA: visual question answering. In: ICCV (2015)
4. Ben-Younes, H., Cadène, R., Thome, N., Cord, M.: MUTAN: multimodal tucker fusion for visual question answering. In: ICCV (2017)
5. Borji, A., Itti, L.: CAT 2000: a large scale fixation dataset for boosting saliency research. arXiv preprint arXiv:1505.03581 (2015)
6. Bylinskii, Z., Judd, T., Oliva, A., Torralba, A., Durand, F.: What do different evaluation metrics tell us about saliency models? IEEE Trans. Pattern Anal. Mach. Intell. **41**, 740–757 (2019)
7. Das, A., Agrawal, H., Zitnick, C.L., Parikh, D., Batra, D.: Human attention in visual question answering: do humans and deep networks look at the same regions? In: Conference on Empirical Methods in Natural Language Processing (EMNLP) (2016)
8. Do, T., Do, T.T., Tran, H., Tjiputra, E., Tran, Q.D.: Compact trilinear interaction for visual question answering. In: ICCV (2019)
9. Ehinger, K.A., Hidalgo-Sotelo, B., Torralba, A., Oliva, A.: Modelling search for people in 900 scenes: a combined source model of eye guidance. Vis. Cogn. **17**(6–7), 945–978 (2009)
10. Fan, S., et al.: Emotional attention: a study of image sentiment and visual attention. In: Proceedings of the IEEE Conference on Computer Vision and Pattern Recognition, pp. 7521–7531 (2018)
11. Fukui, A., Park, D.H., Yang, D., Rohrbach, A., Darrell, T., Rohrbach, M.: Multimodal compact bilinear pooling for visual question answering and visual grounding. In: Proceedings of the 2016 Conference on Empirical Methods in Natural Language Processing, pp. 457–468 (2016)

12. Gao, P., et al.: Dynamic fusion with intra- and inter-modality attention flow for visual question answering. In: CVPR (2019)
13. Goyal, Y., Khot, T., Summers-Stay, D., Batra, D., Parikh, D.: Making the V in VQA matter: elevating the role of image understanding in visual question answering. In: CVPR (2017)
14. He, K., Zhang, X., Ren, S., Sun, J.: Deep residual learning for image recognition. In: CVPR (2016)
15. He, S., Tavakoli, H.R., Borji, A., Pugeault, N.: Human attention in image captioning: dataset and analysis. In: ICCV (2019)
16. Hu, R., Andreas, J., Rohrbach, M., Darrell, T., Saenko, K.: Learning to reason: end-to-end module networks for visual question answering. In: ICCV (2017)
17. Hudson, D.A., Manning, C.D.: Compositional attention networks for machine reasoning (2018)
18. Hudson, D.A., Manning, C.D.: GQA: a new dataset for real-world visual reasoning and compositional question answering. In: CVPR (2019)
19. Huk Park, D., et al.: Multimodal explanations: justifying decisions and pointing to the evidence. In: CVPR (2018)
20. Johnson, J., Hariharan, B., van der Maaten, L., Fei-Fei, L., Lawrence Zitnick, C., Girshick, R.: CLEVR: a diagnostic dataset for compositional language and elementary visual reasoning. In: CVPR (2017)
21. Judd, T., Ehinger, K., Durand, F., Torralba, A.: Learning to predict where humans look. In: 2009 IEEE 12th International Conference on Computer Vision, pp. 2106–2113. IEEE (2009)
22. König, S.D., Buffalo, E.A.: A nonparametric method for detecting fixations and saccades using cluster analysis: removing the need for arbitrary thresholds. J. Neurosci. Methods **227**, 121–131 (2014)
23. Kim, J.H., Jun, J., Zhang, B.T.: Bilinear attention networks. In: NeurIPS, pp. 1571–1581 (2018)
24. Koehler, K., Guo, F., Zhang, S., Eckstein, M.P.: What do saliency models predict? J. Vis. **14**(3), 14 (2014)
25. Krishna, R., et al.: Visual genome: connecting language and vision using crowdsourced dense image annotations. Int. J. Comput. Vision **123**(1), 32–73 (2017)
26. Li, W., Yuan, Z., Fang, X., Wang, C.: Knowing where to look? Analysis on attention of visual question answering system. In: Leal-Taixé, L., Roth, S. (eds.) ECCV 2018. LNCS, vol. 11132, pp. 145–152. Springer, Cham (2018). https://doi.org/10.1007/978-3-030-11018-5_13
27. Mascharka, D., Tran, P., Soklaski, R., Majumdar, A.: Transparency by design: closing the gap between performance and interpretability in visual reasoning. In: CVPR (2018)
28. Patro, B.N., Anupriy, Namboodiri, V.P.: Explanation vs attention: a two-player game to obtain attention for VQA. In: AAAI (2020)
29. Qiao, T., Dong, J., Xu, D.: Exploring human-like attention supervision in visual question answering. In: AAAI (2018)
30. Selvaraju, R.R., et al.: Taking a hint: leveraging explanations to make vision and language models more grounded. In: ICCV (2019)
31. Tapaswi, M., Zhu, Y., Stiefelhagen, R., Torralba, A., Urtasun, R., Fidler, S.: MovieQA: understanding stories in movies through question-answering. In: CVPR (2016)
32. Tavakoli, H.R., Ahmed, F., Borji, A., Laaksonen, J.: Saliency revisited: analysis of mouse movements versus fixations. In: CVPR (2017)

33. Tavakoli, H.R., Shetty, R., Borji, A., Laaksonen, J.: Paying attention to descriptions generated by image captioning models. In: ICCV (2017)
34. Vaswani, A., et al.: Attention is all you need. In: NeurIPS, pp. 5998–6008 (2017)
35. Wu, J., Mooney, R.: Self-critical reasoning for robust visual question answering. In: NeurIPS (2019)
36. Xu, J., Jiang, M., Wang, S., Kankanhalli, M.S., Zhao, Q.: Predicting human gaze beyond pixels. J. Vis. **14**(1), 28 (2014)
37. Xu, K., et al.: Show, attend and tell: neural image caption generation with visual attention. In: ICML, pp. 2048–2057 (2015)
38. Yang, C.J., Grauman, K., Gurari, D.: Visual question answer diversity. In: Sixth AAAI Conference on Human Computation and Crowdsourcing (2018)
39. Yang, Z., He, X., Gao, J., Deng, L., Smola, A.: Stacked attention networks for image question answering. In: CVPR (2016)
40. Yi, K., Wu, J., Gan, C., Torralba, A., Kohli, P., Tenenbaum, J.: Neural-symbolic VQA: disentangling reasoning from vision and language understanding. In: NeurIPS, pp. 1031–1042 (2018)
41. Yu, Z., Yu, J., Cui, Y., Tao, D., Tian, Q.: Deep modular co-attention networks for visual question answering. In: CVPR (2019)
42. Yu, Z., Yu, J., Fan, J., Tao, D.: Multi-modal factorized bilinear pooling with co-attention learning for visual question answering. In: ICCV (2017)
43. Zellers, R., Bisk, Y., Farhadi, A., Choi, Y.: From recognition to cognition: visual commonsense reasoning. In: CVPR (2019)
44. Zhang, Y., Niebles, J.C., Soto, A.: Interpretable visual question answering by visual grounding from attention supervision mining. In: WACV, pp. 349–357 (2019)

Self6D: Self-supervised Monocular 6D Object Pose Estimation

Gu Wang[1,2](✉), Fabian Manhardt[2], Jianzhun Shao[1], Xiangyang Ji[1], Nassir Navab[2], and Federico Tombari[2,3]

[1] BNRist, Tsinghua University, Beijing, China
{wanggl6,sjz18}@mails.tsinghua.edu.cn, xyji@tsinghua.edu.cn
[2] Technical University of Munich, Munich, Germany
{fabian.manhardt,nassir.navab,}@tum.de, tombari@in.tum.de
[3] Google, Menlo Park, USA

Abstract. 6D object pose estimation is a fundamental problem in computer vision. Convolutional Neural Networks (CNNs) have recently proven to be capable of predicting reliable 6D pose estimates even from monocular images. Nonetheless, CNNs are identified as being extremely data-driven, and acquiring adequate annotations is oftentimes very time-consuming and labor intensive. To overcome this shortcoming, we propose the idea of monocular 6D pose estimation by means of self-supervised learning, removing the need for real annotations. After training our proposed network fully supervised with synthetic RGB data, we leverage recent advances in neural rendering to further self-supervise the model on unannotated real RGB-D data, seeking for a visually and geometrically optimal alignment. Extensive evaluations demonstrate that our proposed self-supervision is able to significantly enhance the model's original performance, outperforming all other methods relying on synthetic data or employing elaborate techniques from the domain adaptation realm.

Keywords: Self-supervised learning · 6D pose estimation

1 Introduction

While learning-based techniques have recently demonstrated great performance in estimating the 6D pose (*i.e.* the 3D translation and rotation), a huge amount of training data is required [30,44,53]. Furthermore, contrary to most 2D computer vision tasks such as classification, object detection and segmentation, acquiring real world 6D object pose annotations is much more labor intensive, time consuming, and error-prone [13,21].

G. Wang and F. Manhardt—Equal contribution.

Electronic supplementary material The online version of this chapter (https://doi.org/10.1007/978-3-030-58452-8_7) contains supplementary material, which is available to authorized users.

A. Vedaldi et al. (Eds.): ECCV 2020, LNCS 12346, pp. 108–125, 2020.
https://doi.org/10.1007/978-3-030-58452-8_7

Fig. 1. Abstract illustration of our proposed method. We visualize the 6D pose by overlaying the image with the corresponding transformed 3D bounding box. To circumvent the use of real 6D pose annotations, we firstly train our model purely on synthetic RGB data (*a*). Secondly, employing a large amount of unlabeled real RGB-D images (*b*), we significantly improve its performance (*right*). While *Blue* constitutes the ground truth pose, we demonstrate in *Red* and *Green* the results before and after applying our self-supervision, respectively. (Color figure online)

In order to deal with the lack of real annotations, one common approach is to simulate a large amount of synthetic images [49,51]. This is especially appealing for object pose estimation as one usually aims at estimating the 6D pose from an image *w.r.t.* the corresponding CAD model. Knowing the CAD model enables easy generation of enormous RGB images by randomly sampling 6D poses. Many approaches typically rely on rendering the models using OpenGL and placing them on random background images (drawn from large-scale 2D object datasets such as COCO [33]) in order to impose invariance to changing scenes [23,40]. Recent works propose to instead employ physically-based rendering to produce high quality renderings, and additionally enforce real physical constraints, as they can provide additional cues for the 6D pose [16,56].

Despite compelling results, these methods usually still exhibit inferior performance when inferring from real world data, due to the withstanding domain gap between real and synthetic data. Although techniques for domain adaption [2], domain randomization [52] and photorealistic rendering [16] can mitigate the problem to some extent, the performance is still far from satisfactory.

This motivated us to investigate the problem from an entirely different angle. Humans have the amazing ability to learn about the 3D world, whilst only perceiving it through 2D images. Moreover, they can even learn 3D world properties without supervision from another human or *labels* in a self-supervised fashion through making observations and validating if these observations are in accordance with the expected outcome [50]. In our context, while labeling the 6D pose is a severe bottleneck, recording unannotated data can be easily achieved at scale. Therefore, similar to learning for humans, we aim at teaching a neural network to reason about the 6D pose of an object by leveraging these unsupervised examples. As shown in Fig. 1, we first train our method fully-supervised with synthetic data. Afterwards, employing unannotaed RGB-D data, we make use of self-supervised learning to enhance the model's performance on real data.

To accomplish this, it is required to understand 3D properties solely from 2D images. The mechanism of experiencing the 3D world as images on the eye's retina is known as *rendering* and has been also extensively explored in Computer Graphics [41]. Unfortunately, rendering is also known to be non-differentiable due to the rasterization step, as gradients cannot be computed for the *argmax* function. Nevertheless, many approaches for differentiable rendering have been recently proposed. The real gradient is thereby either approximated [22,37], or computed analytically by approximating the rasterization function itself [6,35].

In summary, we make the following contributions. i) To the best of our knowledge, we are the first to conduct self-supervised 6D object pose estimation from real data, without the need of 6D labels. ii) Leveraging neural rendering, we formulate a self-supervised 6D pose estimation solution by means of visual and geometric alignment. iii) We experimentally show that the proposed method, which we dub Self6D, outperforms state-of-the-art methods for monocular 6D object pose estimation trained without real annotations by a large margin.

2 Related Work

We first introduce recent work in monocular 6D pose estimation. Afterwards, we discuss important methods from neural rendering as they form a core part of our (as well as other) self-supervised learning frameworks. We then outline other successful approaches grounded on self-supervised learning. Lastly, we take a brief look at domain adaptation in the field of 6D pose, since our method can be considered an implicit formulation to close the synthetic-to-real domain gap.

2.1 Monocular 6D Pose Estimation

Recently, monocular 6D pose estimation has received a lot of attention and several very promising works have been proposed [15].

One major branch is grounded on establishing 2D-3D correspondences between the image and the 3D CAD model. After estimating these correspondences, PnP is commonly employed to solve for the 6D pose. Inspired by [3,4], Rad *et al.* propose to employ a CNN to estimate the 2D projections of the 3D bounding box corners in image space [46]. Similarly, [17,44] also regress 2D projections of associated sparse 3D keypoints, however, both employ segmentation paired with voting to improve the reliability. In contrast, [30,43,61] ascertain dense 2D-3D correspondences, rather than sparse ones.

Another branch of work learns a pose embedding, which can be utilized for latter retrieval. In particular, inspired by [24,58], [52] employs an Augmented AutoEncoder (AAE) to learn latent representations for the 3D rotation.

A few methods also directly regress the 6D pose. For instance, while [23] extends [36] to also classify the viewpoint and in-plane rotation, [38] further adjusts [23] to implicitly deal with ambiguities via multiple hypotheses (MHP). In [59] and [29] the authors minimize a point matching loss.

The majority of these methods [17,43,46,53,59] exploit annotated real data to train their models. However, labeling real data commonly comes with a large cost in time and labor. Moreover, a shortage of sufficient real world annotations can lead to overfitting, regardless of exploiting strategies such as *crop&paste* [8, 21]. Other works, in contrast, fully rely on synthetic data to deal with these pitfalls [38,52]. Nonetheless, the performance falls far behind the methods based on real data. We, thus, harness the best of both worlds. While unannotated data can be easily obtained at scale, this combined with our self-supervision for pose is able to outperform all methods trained on synthetic data by a large margin.

2.2 Neural Rendering

Rasterization is a core part of all traditional rendering pipelines. Nonetheless, rasterization involves discrete assignment operations, preventing the flow of gradients throughout the rendering process. A series of work have been devoted to circumvent the hard assignment in order to reestablish the gradient flow.

Loper and Black introduce the first differentiable renderer by means of first-order Taylor approximation to calculate the derivative of pixel values [37]. In [22], the authors instead approximate the gradient as the potential change of the pixel's intensity *w.r.t.* the meshes' vertices. *SoftRas* [35] conducts rendering by aggregating the probabilistic contributions of each mesh triangle in relation to the rendered pixels. Consequently, the gradients can be calculated analytically, however, with the cost of extra computation. *DIB-R* [6] further extends [35] to render of a variety of different lighting conditions. In this work, we use *DIB-R* [6] since it can be considered state-of-the-art for neural rendering.

2.3 Recent Trends in Self-supervised Learning

Self-supervised learning, *i.e.* learning despite the lack of properly labeled data, has recently enabled a large number of applications ranging from 2D image understanding all the way down to depth estimation for autonomous driving. In the core, self-supervised learning approaches implicitly learn about a specific task through solving related proxy tasks. This is commonly achieved by enforcing different constraints such as pixel consistencies across multiple views or modalities.

One prominent approach in this area is *MonoDepth* [9], which conducts monocular depth estimation by warping the 2D image points into another view and enforcing a minimum reprojection loss. In the following many works to extend *MonoDepth* have been introduced [10,11,45]. In visual representation learning, consistency is ensured by solving pretext tasks [26]. Another line of works explore self-supervised learning for 3D human pose estimation, leveraging multi-view epipolar geometry [25] or imposing 2D-3D consistency after lifting and reprojection of keypoints [5]. Self-supervised learning approaches using neural rendering have also been proposed in the field of 3D object and human body reconstruction from single RGB images [1,20,42,57,64].

In the domain of 6D pose estimation, self-supervised learning is still a rather unexplored field. [7] proposes a novel self-labeling pipeline with an interactive robotic manipulator. Essentially, running several methods for 6D pose estimation, they can reliably generate precise annotations. Nonetheless, the final 6D pose estimation model is still trained fully-supervised using the acquired data. In this work, we propose to instead directly employ self-supervision for 6D pose by enforcing visual and geometric consistencies on top of neural rendering.

2.4 Domain Adaptation for 6D Pose Estimation

Bridging the domain gap between synthetic and real data is crucial in 6D pose estimation. Many works tackle this problem by learning a transformation to align the synthetic and real domains via Generative Adversarial Networks (GANs) [2, 28,60] or by means of feature mapping [47]. Exemplary, [28] uses a cross-cycle consistency loss based on disentangled representations to embed images onto a domain-invariant content space and a domain-specific attribute space. [47] instead maps the features of a color-based pose estimator to a depth-based pose estimator.

In contrast, works from domain randomization aim at learning domain-invariant attributes. For instance, harnessing random backgrounds and severe augmentations [23,52] or employing adversarial training to generate backgrounds and image augmentations [60].

3 Self-supervised 6D Pose Estimation

In this work we aim at conducting 6D pose estimation from monocular images via self-supervised learning. To this end, we propose a novel model that can learn monocular pose estimation from both synthetic RGB data and real world unannotated RGB-D data. Employing neural rendering, the model can be self-supervised by establishing coherence between real and rendered images *w.r.t.* the 6D pose. Since this requires good initial pose estimates, we rely on a two-stage approach. As shown in Fig. 1, we start by training our model using synthetic RGB data only. Afterwards, we further enhance the pose estimation performance by leveraging unlabeled real world RGB-D data.

We harness different visual and geometric constraints to seek the best alignment *w.r.t.* 6D pose. Unfortunately, while a 3D model contains information about the visible and invisible regions, the depth map only covers the visible surface. This complicates supervision since the invisible points would mistakenly contribute to the alignment. Therefore, we aim to extract only the model's visible surface given the current pose. This can be achieved in different ways: by culling the hidden points, or simply rendering the object in its current pose. Since we are required to render color for visual alignment, we resort to rendering depth for visible surface extraction, as it comes with no extra cost in computation.

We use the differentiable renderer *DIB-R* proposed by [6] to render 6D pose estimates from our model. Since *DIB-R* is only able to render RGB images and

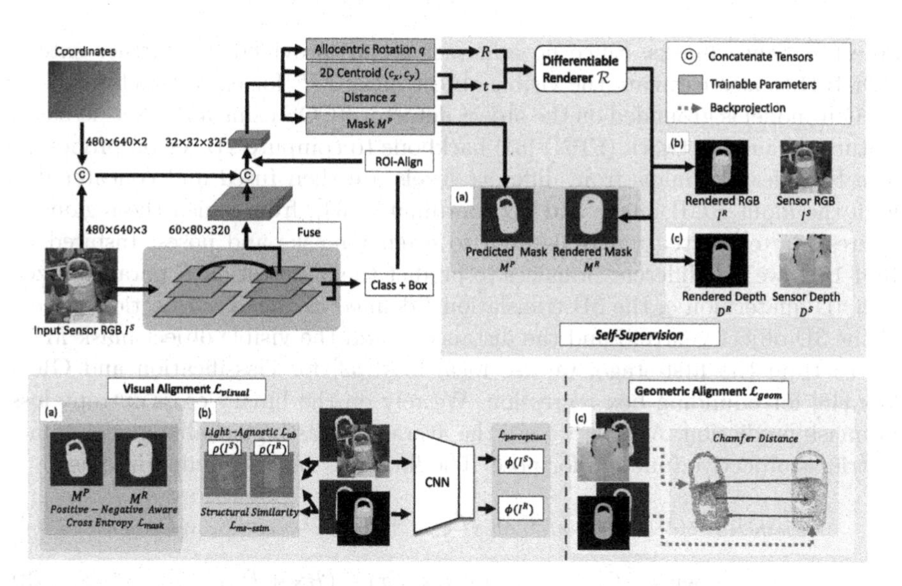

Fig. 2. Our self-supervised training pipeline. *Top*: We start training our model for 6D pose estimation purely on synthetic RGB data, to predict a 3D rotation R, translation t and object instance mask M^P. Using a large amount of unlabeled RGB-D images (I^S, D^S), we enhance the model's performance by means of self-supervised learning. We differentiably render $(\mathcal{R})$ the associated RGB-D image and mask (I^R, D^R, M^R). *Bottom*: We impose various constraints to visually (a and b), and geometrically (c) align the 6D pose.

object masks, we extend it to also provide the depth map fully differentiably. We additionally modify the camera projection to conduct a real perspective projection.[1] Given the estimated 6D pose as 3D rotation R, 3D translation t, together with the 3D CAD model $\mathcal{M}$ and the camera intrinsics matrix K, we render the triplet (I^R, D^R, M^R) consisting of the rendered RGB image I^R, the rendered depth map D^R and the rendered mask M^R

$$\mathcal{R}(R, t, K, \mathcal{M}) = (I^R, D^R, M^R). \tag{1}$$

Architecture Details. Besides rendering, also the prediction of the 3D rotation and translation has to be differentiable in order to allow backpropagation. While methods based on establishing 2D-3D correspondences are currently dominating the field, it is infeasible to resort to them as gradients cannot be computed for PnP. To this end, we rely on a similar network architecture as ROI-10D [39], since they directly estimate rotation and translation. Unfortunately, the predicted poses from ROI-10D are not accurate enough to match the demands of our self-supervision, thus, we base our method on the more recent FCOS [54] detector. Moreover, a crucial part of our subsequent self-supervision requires

[1] The code of our extended renderer is available at https://github.com/THU-DA-6D-Pose-Group/Self6D-Diff-Renderer.

object instance masks. Since no annotations are provided, we further extend ROI-10D to also estimate the visible object mask M^P for each detection.

Our model is grounded on the object detector FCOS using a ResNet-50 based feature pyramid network (FPN) [31] backbone to compute 2D region proposals. The FPN feature maps from different levels are then fused and concatenated with the input RGB image and 2D coordinates [34], from which the regions of interest are extracted via ROI-Align to predict masks and poses. Inspired by ROI-10D, we use different branches to predict the 3D rotation R parameterized as a 4D quaternion q, the 3D translation t defined as the 2D projection (c_x, c_y) of the 3D object centroid and the distance z, and the visible object mask M^P.

To train the first-stage, we use focal loss [32] for classification and GIoU loss [48] for bounding box regression. We rely on the binary cross entropy loss for mask prediction. As [29], we use the average of distinguishable model points metric as objective function for pose. The final loss can be summarized as

$$\mathcal{L}_{synthetic} := \lambda_{class}\mathcal{L}_{focal} + \lambda_{box}\mathcal{L}_{giou} + \lambda_{mask}\mathcal{L}_{bce} + \lambda_{pose}\mathcal{L}_{pose}, \tag{2}$$

$$\text{with} \quad \mathcal{L}_{pose} := \operatorname*{avg}_{x \in \mathcal{M}} \|(R\mathbf{x} + t) - (\bar{R}\mathbf{x} + \bar{t})\|_1, \tag{3}$$

where $\lambda_{class}, \lambda_{box}, \lambda_{mask}$ and λ_{pose} denote the balance factors for each task, $\mathcal{M}$ denotes the 3D model, and $[R|t]$, $[\bar{R}|\bar{t}]$ represent the predicted and ground truth poses, respectively. We kindly refer to the supplementary material for more details on the employed hyper-parameters.

For simplicity of the following, we define all foreground and background pixels as $N_+ := \{(i,j) \mid \forall M^P(i,j) = 1\}$ and $N_- := \{(i,j) \mid \forall M^P(i,j) = 0\}$. We further denote all pixels together as $N = N_+ \cup N_-$.

Neural Rendering for Visual Alignment. The most intuitive way is to simply align the rendered image I^R with the sensor image I^S, deploying directly a loss on both samples. However, as the domain gap between I^S and I^R turns out to be very large, this does not work well in practice. In particular, lightning changes as well as reflection and bad reconstruction quality (especially in terms of color) oftentimes cause a high error despite having good pose estimates, eventually leading to divergence in the optimization. Hence, in an effort to keep the domain gap as small as possible, we impose multiple constraints measuring different domain-independent properties. In particular, we assess different visual similarities *w.r.t.* mask, color, image structure, and high-level content.

Since object masks are naturally domain agnostic, they can provide a particularly strong supervision. As our data is unannotated we refer to our predicted masks M^P for a weak supervision. However, due to imperfect predicted masks, we utilize a modified cross-entropy loss [18], which recalibrates the weights of positive and negative regions

$$\mathcal{L}_{mask} := -\frac{1}{|N_+|}\sum_{j\in N_+} M_j^P \log M_j^R - \frac{1}{|N_-|}\sum_{j\in N_-} \log(1 - M_j^R). \tag{4}$$

Although masks are not suffering from the domain gap, they discard a lot of valuable information. In particular, color information is often the only guidance to disambiguate the 6D pose, especially for geometrically simple objects.

Since the domain shift is at least partially caused by light, we attempt to decouple light prior to measuring color similarity. Let ρ denote the transformation from RGB to LAB space, additionally discarding the light channel, we evaluate color coherence on the remaining two channels according to

$$\mathcal{L}_{ab} := \frac{1}{|N_+|}\sum_{j\in N} \|\rho(I^S)_j \cdot M_j^P - \rho(I^R)_j\|_1. \tag{5}$$

We also avail various ideas from image reconstruction and domain translation, as they succumb the same dilemma. We assess the structural similarity (SSIM) in the RGB space and additionally follow the common practice to use a multi-scale variant, namely MS-SSIM [63]

$$\mathcal{L}_{ms\text{-}ssim} := 1 - ms\text{-}ssim(I^S \odot M^P, I^R, s). \tag{6}$$

Thereby, $\odot$ denotes the element-wise multiplication and $s = 5$ is the number of employed scales. For more details on MS-SSIM, we kindly refer the readers to the supplement and [63].

Another common practice is to appraise the perceptual similarity [19,62] in the feature space. To this end, a pretrained deep neural network as AlexNet [27] is typically employed to ensure low- and high-level similarity. We apply the perceptual loss at different levels of the CNN. Specifically, we extract the feature maps of $L = 5$ layers and normalize them along the channel dimension. Then we compute squared L_2 distances of the normalized feature maps $\hat{\phi}^l(\cdot)$ for each layer l. We average the individual contributions spatially and sum across all layers [62]

$$\mathcal{L}_{perceptual} := \sum_{l=1}^{L} \frac{1}{|N^l|}\sum_{j\in N^l} \|\hat{\phi}_j^l(I^S \odot M^P) - \hat{\phi}_j^l(I^R)\|_2^2. \tag{7}$$

The visual alignment is then composed as the weighted sum over all four terms

$$\mathcal{L}_{visual} := \mathcal{L}_{mask} + \alpha\mathcal{L}_{ab} + \beta\mathcal{L}_{ms\text{-}ssim} + \gamma\mathcal{L}_{perceptual}, \tag{8}$$

where α, β and γ denote the balance factors for $\mathcal{L}_{ab}$, $\mathcal{L}_{ms\text{-}ssim}$, and $\mathcal{L}_{perceptual}$, respectively. We refer to the supplement for more details on the hyper-parameters.

Neural Rendering for Geometric Alignment. Since the depth map only provides information for the visible areas, aligning it with the transformed 3D Model similar to Eq. 3 harms performance. Therefore, we exploit the rendered depth map to enable comparison of the visible areas only. Nevertheless, employing a loss directly on both depth maps leads to bad correspondences as the points where the masks are not intersecting cannot be matched.

Hence, we operate on the visible surface in 3D to find the best geometric alignment. We first backproject D^S and D^R using the corresponding masks M^P and M^R to retrieve the visible pointclouds $\mathcal{P}^S$ and $\mathcal{P}^R$ in camera space with

$$\pi^{-1}(D, M, K) = \{\, K^{-1} \left[x_j \; y_j \; 1\right]^T \cdot D_j \mid \forall j \in M > 0 \,\}, \tag{9}$$

$$\mathcal{P}^S := \pi^{-1}(D^S, M^P, K), \qquad \mathcal{P}^R := \pi^{-1}(D^R, M^R, K). \tag{10}$$

Thereby, (x_j, y_j) denotes the 2D pixel location of j in M.

Since it is infeasible to estimate direct 3D-3D correspondences between $\mathcal{P}^S$ and $\mathcal{P}^R$, we refer to the chamfer distance to seek the best alignment in 3D

$$\mathcal{L}_{geom} := \frac{1}{|\mathcal{P}^S|} \sum_{p^S \in \mathcal{P}^S} \min_{p^R \in \mathcal{P}^R} \|p^S - p^R\|_2 + \frac{1}{|\mathcal{P}^R|} \sum_{p^R \in \mathcal{P}^R} \min_{p^S \in \mathcal{P}^S} \|p^S - p^R\|_2. \tag{11}$$

The overall self-supervision is $\mathcal{L}_{Self} := \mathcal{L}_{visual} + \eta \mathcal{L}_{geom}$, with η denoting the balance factor of $\mathcal{L}_{geom}$. An overview is also presented in Fig. 2. Noteworthy, while we require RGB-D data for self-supervision, we do not need any depth data during latter inference.

4 Evaluation

In this section, we first introduce our experimental setup. Afterwards, we present the analysis on the quality of predicted masks and different ablations to illustrate the effectiveness of our proposed self-supervised loss. We conclude by comparing our method with other state-of-the-art methods for 6D pose estimation and domain adaptation. For better understanding, in addition to the results of Self6D, we also evaluate our method using synthetic data only and additionally employing real 6D pose labels. Since they can be considered the lower and upper bound of our method, we refer to them Self6D-LB and Self6D-UB in the following.

Synthetic Training Data. [55] and [16] recently proposed to employ photorealistic and physically plausible renderings to improve 2D detection and 6D pose estimation, in contrast to simple OpenGL rendering [23]. In our experiments it turns out that a mixture of both approaches, together with a lot of augmentations (*e.g.* random Gaussian noise, intensity jitter), leads to best results.

Datasets. To evaluate our proposed method we leverage the commonly used *LineMOD* dataset [12], which consists of 15 sequences, Only 13 of these provide water-tight CAD models and we, therefore, remove the other two sequences. In [3], the authors propose to sample 15% of the real data for training to close the domain gap. We use the same split, however discarding the pose labels. As second dataset, we utilize the recent *HomebrewedDB* [21] dataset. However, we only employ the sequence which covers three objects from *LineMOD*, to depict that we can even self-supervise the same model in a new environment.

To also show generalization to other common datasets for 6D pose, we demonstrate the effectiveness of our self-supervision on 5 objects from *YCB-Video* [59] in the supplementary material. To compare with domain adaptation based methods, we refer to the usual *Cropped LineMOD* dataset [58] including center-cropped 64 × 64 patches of 11 different small objects in cluttered scenes imaged in various of poses.

Metrics for 6D Pose. We report our results *w.r.t.* the ADD metric [12], measuring whether the average deviation of the transformed model points is less than 10% of the object's diameter. For *symmetric* objects (*e.g.*, *Eggbox* and *Glue* in *LineMOD*) we rely on the ADD-S metric, which instead measures the error as the average distance to the *closest* model point [12,14].

$$\mathbf{ADD} = \operatorname*{avg}_{x \in \mathcal{M}} \|(Rx + t) - (\bar{R}x + \bar{t})\|_2, \tag{12}$$

$$\mathbf{ADD\text{-}S} = \operatorname*{avg}_{x_2 \in \mathcal{M}} \min_{x_1 \in \mathcal{M}} \|(Rx_1 + t) - (\bar{R}x_2 + \bar{t})\|_2. \tag{13}$$

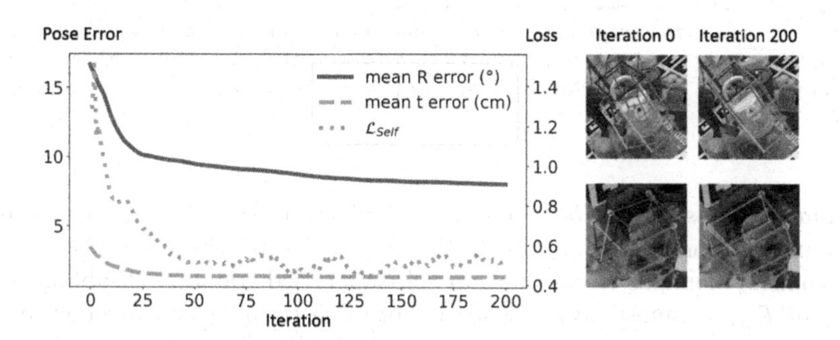

Fig. 3. Pose errors *v.s.* self-supervision. We optimize $\mathcal{L}_{self}$ on single images from *LineMOD* for 200 iterations and report the average over in total 100 images. We initialize the 6D poses with Self6D-LB.

4.1 Analysis on the Quality of Predicted Masks

Thanks to physically-based renderings, the predicted masks on the real data are very accurate, thus can be reliably used as a self-supervision signal. For instance, on the *LineMOD* test set, the average F1 score and mIoU between the predicted masks and the ground-truth masks are 89.63% and 90.38%. Please refer to the supplementary for detailed results and qualitative examples.

4.2 Ablation Study

Self-Supervision v.s. 6D Pose Error. We want to demonstrate that there is indeed a high correlation between our proposed $\mathcal{L}_{Self}$ and the actual 6D pose errors. To this end, we randomly draw 100 samples from *LineMOD* and optimize separately on each sample, always beginning from Self6D-LB. Figure 3 illustrates the average behavior *w.r.t.* loss *v.s.* 6D pose error at each iteration. As the loss decreases, also the pose error for both, rotation and translation, continuously declines until convergence. The accompanying qualitative images (Fig. 3, *right*) further support this observation, as the initial pose is significantly worse compared to the final optimized result. We refer to the supplementary material for more qualitative results.

Table 1. Ablation. We report the Average Recall of ADD(-S) on *LineMOD*.

	Ape	Bvise	Cam	Can	Cat	Drill	Duck	Eggbox	Glue	Holep	Iron	Lamp	Phone	Mean
w/o $\mathcal{L}_{mask}$	0.0	0.0	0.0	0.0	0.0	0.0	0.0	0.8	0.0	0.0	0.0	0.0	0.0	0.1
w/o $\mathcal{L}_{geom}$	0.0	10.1	3.1	0.0	0.0	7.5	0.1	33.0	0.2	0.0	5.9	20.7	2.4	6.4
w/o $\mathcal{L}_{ms\text{-}ssim}$	32.1	74.8	20.4	63.4	57.1	68.3	16.6	**99.0**	94.1	12.3	70.8	**68.5**	54.9	56.3
w/o $\mathcal{L}_{perceptual}$	34.9	74.4	33.5	64.8	55.3	**70.0**	17.2	98.7	**94.8**	10.7	76.3	68.1	**56.5**	58.1
w/o $\mathcal{L}_{ab}$	**40.9**	73.8	36.1	63.0	**58.1**	66.0	18.0	98.9	93.9	**16.2**	77.2	68.2	50.1	58.5
Self6D	38.9	**75.2**	**36.9**	**65.6**	57.9	67.0	**19.6**	**99.0**	94.1	15.5	**77.9**	68.2	50.1	**58.9**
Self6D-LB	14.8	68.9	17.9	50.4	33.7	47.4	18.3	64.8	59.9	5.2	68.0	35.3	36.5	40.1
Self6D-UB	62.3	95.3	86.5	93.0	80.7	93.7	63.4	99.7	99.4	73.6	96.0	96.6	90.0	86.9

Individual Loss Contributions. Table 1 illustrates the contribution of each individual loss component on *LineMOD*. Note that supervision from both visual and geometry domains is vital for our self-supervised training. Disabling either $\mathcal{L}_{mask}$ or $\mathcal{L}_{geom}$ almost always leads to unstable training and divergence (the average recall is only 0.1% and 6.4% *w.r.t.* ADD(-S)). The remaining three factors, measuring color similarity, have a comparably small impact. Concretely, we drop by more than 2% when disabling $\mathcal{L}_{ms\text{-}ssim}$, and about 1% referring to $\mathcal{L}_{ab}$ and $\mathcal{L}_{perceptual}$. Nonetheless, we still achieve the overall best results when applying all loss terms together. Most importantly, we can report a significant relative improvement of almost 50% from 40.1% to 58.9% leveraging the proposed self-supervision. Moreover, except for the *Duck* object, all other objects undergo a strong enhancement in ADD(-S). Noteworthy, we can almost halve the difference between training with and without real pose labels.

4.3 Comparison with State-of-the-Art

In the first part of this section we present a comparison with current state-of-the-art methods in 6D pose estimation. In the latter part, we present our results in the area of domain adaptation referring to *Cropped LineMOD*.

6D Pose Estimation

LineMOD Dataset. In line with other works, we distinguish between training with and without real pose labels, *i.e.* making use of annotated real training data. Despite exploiting real data, we do not employ any pose labels and must, therefore, be classified as the latter. We want to highlight that our model can produce state-of-the-art results for training with and without labels. Referring to Table 2, for training using only synthetic data, Self6D-LB reveals an average recall of 40.1%, which is deliberately better than AAE [52] with 31.4% and on par with MHP [38] and DPOD[2] [61] reporting 38.8% and 40.5%. On the other hand, as for training with real pose labels, we are again on par with other recently published methods such as PVNet [44] and CDPN [30] reporting a mean average recall of 86.9%. Furthermore, our proposed self-supervision Self6D achieves an overall average recall of 58.9%, which is more than 51% of relative improvement over all state-of-the-art methods using no real pose labels. Except for *Holep*, *Duck* and *Iron*, we can report a significant increase. Objects with little variation in color and geometry can become difficult to optimize. In addition, the 3D mesh of the *Holep* is rather different compared with the actual perceived object in the real images, which makes our visual alignment less meaningful.

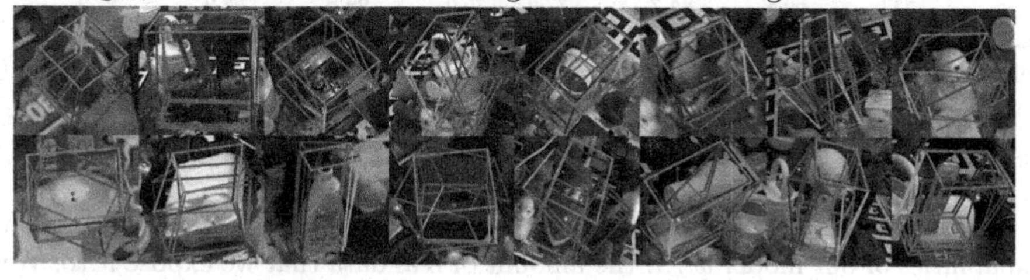

HomebrewedDB Dataset. In Fig. 4(*left*) we compare our method with DPOD [61] and SSD6D [23] after refinement using [40] (SSD6D+Ref.) on three objects of *HomebrewedDB*, which it shares with *LineMOD*.[2] Unfortunately, methods directly solving for the 6D pose always implicitly learn the camera intrinsics which degrades the performance when exposed to a new camera. 2D-3D correspondences based approaches are instead robust to camera changes as they simply run PnP using the new intrinsics. Therefore, the performance of our Self6D-LB is slightly outperformed by [61]. SSD6D+Ref. [40] employs contour-based pose refinement using renderings for the current hypotheses. Similarly, rendering the pose with the new intrinsics enables again easy adaptation and can even exceed [61] and our Self6D on the *Bvise* object. Nevertheless, we can easily adapt to the new domain and intrinsics by only leveraging 15% of unannotated data from [21]. In fact, we almost double their numbers for all other objects and reach a similar level as for *LineMOD* (Table 3).

[2] The numbers of [61] and [40] are different as in their paper since they used average precision instead. The authors provided us with their results for average recall.

Table 2. Results for *LineMOD*. *Top*: Qualitative results on unseen examples. The projected 3D bounding boxes with *blue*, *red* and *green* denote the poses of ground truth, Self6D-LB and Self6D, respectively. *Bottom*: Comparison with state-of-the-art. We present the results for the Average Recall(%) of ADD(-S) metric. *Real Pose Labels* refers to the 15% training split from [3] with pose labels. We use the same split for training, however, without employing labels.[2]

Train data	w/o Real Pose Labels				with Real Pose Labels			
Object	AAE [52]	MHP [38]	DPOD [61]	Self6D	Tekin [53]	DPOD [61]	PVNet [44]	CDPN [30]
Ape	4.0	11.9	35.1	**38.9**	21.6	53.3	43.6	**64.4**
Bvise	20.9	66.2	59.4	**75.2**	81.8	95.2	**99.9**	97.8
Cam	30.5	22.4	15.5	**36.9**	36.6	90.0	86.9	**91.7**
Can	35.9	59.8	48.8	**65.6**	68.8	94.1	95.5	**95.9**
Cat	17.9	26.9	28.1	**57.9**	41.8	60.4	79.3	**83.8**
Drill	24.0	44.6	59.3	**67.0**	63.5	**97.4**	96.4	96.2
Duck	4.9	8.3	**25.6**	19.6	27.2	66.0	52.6	**66.8**
Eggbox	81.0	55.7	51.2	**99.0**	69.6	99.6	99.2	**99.7**
Glue	45.5	54.6	34.6	**94.1**	80.0	93.8	95.7	**99.6**
Holep	17.6	15.5	**17.7**	16.2	42.6	64.9	81.9	**85.8**
Iron	32.0	60.8	**84.7**	77.9	75.0	**99.8**	98.9	97.9
Lamp	60.5	–	45.0	**68.2**	71.1	88.1	**99.3**	97.9
Phone	33.8	34.4	20.9	**50.1**	47.7	71.4	**92.4**	90.8
Mean	31.4	38.8	40.5	**58.9**	56.0	82.6	86.3	**89.9**

Based on this observation, we were curious to understand the adaptation capabilities of our model *w.r.t.* the amount of real data that we expose it to. We divided the samples from *HomebrewedDB* into 100 images for testing and 900 images for training. Afterwards, we repeatedly trained our model with increasing amount of data, however, always evaluating on the same test split. In Fig. 4 (*right*) we illustrate the corresponding results. When using only 15% (150 samples) of the real data for training, we can already almost double the mean average recall (mAR). Using $\approx$ 40% of the real data, the mAR can be improved by $\approx$ 130% from 31% to 71%. Afterwards, it slowly saturates at $\approx$ 74%.

LineMOD Occlusion Dataset. We also evaluate our method on *LineMOD Occlusion* which exhibits stronger occlusion. We follow the BOP [15] standard and evaluate on a subset of 200 samples. We compare Self6D with two state-of-the-art methods using synthetic data only, namely DPOD [61] and CDPN [30].[3] While our Self6D-LB can clearly outperform [61] with 15.1% compared to 6.3%, [30] exceeds our Self6D-LB by 5.4% and reports a mean average recall of 20.8%. 2D-3D correspondences based methods are more robust towards occlusion as they consider only the visible regions, while direct methods are less stable due to inferring poses from both visible and occluded regions. Nonetheless, after utilizing the remaining real RGB-D data via our self-supervision, we can easily surpass [30] (32.1% *v.s.* 20.8%), and double the performance of our Self6D-LB.

[3] The authors of [61] and [30] shared their results for the BOP 2019 challenge [15].

Method	Supervision		Object			
	Syn	Self	Bvise	Driller	Phone	Mean
DPOD [61]	✓		52.9	37.8	7.3	32.7
SSD6D+Ref. [40]	✓		**82.0**	22.9	24.9	43.3
Self6D-LB	✓		37.7	19.2	20.9	25.9
Self6D	✓	✓	72.1	**65.1**	**41.8**	**59.7**

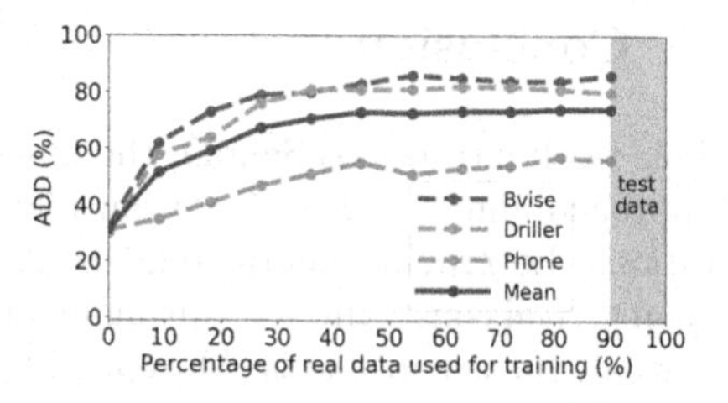

Fig. 4. Results for *HomebrewedDB*. *Left*: Comparison with [61] and [40].[2] While both train with synthetic data only, we report our results for synthetic data (Self6D-LB) and after self-supervision (Self6D) using 15% of real data from [21]. *Right*: Self-supervised training *w.r.t.* an increasing percentage of real training data. Results are always reported on the same unseen test split.

Table 3. Results for *LineMOD Occlusion*. Comparison with [61] and [30]. We evaluate the Average Recall(%) of ADD(-S) on the BOP [15] split.[3]

Method	Supervision			Object								
	Syn	Self	Real GT	Ape	Can	Cat	Driller	Duck	Eggbox	Glue	Holep	Mean
DPOD [61]	✓			2.3	4.0	1.2	10.5	7.2	4.4	12.9	7.5	6.3
CDPN [30]	✓			**20.0**	15.1	16.4	5.0	**22.2**	36.1	27.9	**24.0**	20.8
Self6D-LB	✓			7.4	14.1	7.6	18.0	12.2	18.3	31.4	11.5	15.1
Self6D	✓	✓		13.7	**43.2**	**18.7**	**32.5**	14.4	**57.8**	**54.3**	22.0	**32.1**
Self6D-UB	✓		✓	47.4	79.4	56.1	83.5	48.9	90.0	93.6	62.5	70.2

Table 4. Comparison with state of the art on *Cropped LineMOD*. We present the classification accuracy as well as mean angle error.

Method	PixelDA [2]	DRIT [28]	DeceptionNet [60]	Self6D-LB	Self6D
Classification accuracy (%)	99.9	98.1	95.8	100.0	100.0
Mean angle error (°)	23.5	34.4	51.9	19.8	**15.8**

Noteworthy, there is still plenty of room for all the methods trained without real labels, compared to our fully-supervised model Self6D-UB (70.2%).

Domain Adaptation for Pose Estimation Since our method is suitable for conducting synthetic to real domain adaptation, we assess transfer skills referring to the commonly used *Cropped LineMOD* scenario. We self-supervise the model with the real training set from *Cropped LineMOD*, and report the mean angle error on the real test set. As shown in Table 4, our synthetically trained model (Self6D-LB) slightly exceeds state-of-the-art methods as PixelDA [2]. Self6D can successfully surpass the original model on the target domain, reducing the mean angle error from 19.8° to 15.8°.

5 Conclusion

This work introduced Self6D, the first self-supervised 6D object pose estimation approach aimed at learning from real data without the need for 6D pose annotations. Leveraging neural rendering, we are able to enforce several visual and geometrical constraints, resulting in a remarkable leap forward compared to other state-of-the-art methods. Moreover, Self6D demonstrated to notably reduce the gap with the state of the art for pose estimation with real pose labels.

A main future direction is exploring how to overcome the need for depth data during self-supervision. Another interesting aspect is to incorporate also 2D detections into self-supervision, as this allows backpropagating the loss in an end-to-end fashion throughout the entire network.

Acknowledgments. This work was supported by China Scholarship Council (CSC) Grant #201906210393. This work was also supported by the National Key R&D Program of China under Grant 2018AAA0102801.

References

1. Alldieck, T., Magnor, M., Bhatnagar, B.L., Theobalt, C., Pons-Moll, G.: Learning to reconstruct people in clothing from a single RGB camera. In: CVPR, pp. 1175–1186 (2019)
2. Bousmalis, K., Silberman, N., Dohan, D., Erhan, D., Krishnan, D.: Unsupervised pixel-level domain adaptation with generative adversarial networks. In: CVPR, pp. 3722–3731 (2017)
3. Brachmann, E., Krull, A., Michel, F., Gumhold, S., Shotton, J., Rother, C.: Learning 6D object pose estimation using 3D object coordinates. In: Fleet, D., Pajdla, T., Schiele, B., Tuytelaars, T. (eds.) ECCV 2014. LNCS, vol. 8690, pp. 536–551. Springer, Cham (2014). https://doi.org/10.1007/978-3-319-10605-2_35
4. Brachmann, E., Michel, F., Krull, A., Ying Yang, M., Gumhold, S., Rother, C.: Uncertainty-driven 6D pose estimation of objects and scenes from a single RGB image. In: CVPR, pp. 3364–3372 (2016)
5. Chen, C.H., Tyagi, A., Agrawal, A., Drover, D., Stojanov, S., Rehg, J.M.: Unsupervised 3d pose estimation with geometric self-supervision. In: CVPR, pp. 5714–5724 (2019)
6. Chen, W., Ling, H., Gao, J., Smith, E., Lehtinen, J., Jacobson, A., Fidler, S.: Learning to predict 3d objects with an interpolation-based differentiable renderer. In: NeurIPS, pp. 9605–9616 (2019)
7. Deng, X., Xiang, Y., Mousavian, A., Eppner, C., Bretl, T., Fox, D.: Self-supervised 6d object pose estimation for robot manipulation. In: ICRA (2020)
8. Dwibedi, D., Misra, I., Hebert, M.: Cut, paste and learn: surprisingly easy synthesis for instance detection. In: ICCV, pp. 1301–1310 (2017)
9. Godard, C., Mac Aodha, O., Brostow, G.J.: Unsupervised monocular depth estimation with left-right consistency. In: CVPR, pp. 270–279 (2017)
10. Godard, C., Mac Aodha, O., Firman, M., Brostow, G.J.: Digging into self-supervised monocular depth estimation. In: ICCV, pp. 3828–3838 (2019)
11. Guizilini, V., Ambrus, R., Pillai, S., Gaidon, A.: Packnet-SFM: 3D packing for self-supervised monocular depth estimation. arXiv preprint arXiv:1905.02693 (2019)

12. Hinterstoisser, S., et al.: Model based training, detection and pose estimation of texture-less 3D objects in heavily cluttered scenes. In: ACCV, pp. 548–562 (2012)
13. Hodan, T., Haluza, P., Obdržálek, Š., Matas, J., Lourakis, M., Zabulis, X.: T-less: an RGB-D dataset for 6D pose estimation of texture-less objects. In: WACV, pp. 880–888 (2017)
14. Hodaň, T., Matas, J., Obdržálek, Š.: On evaluation of 6d object pose estimation. In: ECCVW, pp. 606–619 (2016)
15. Hodaň, T., et al.: BOP: benchmark for 6D object pose estimation. In: Ferrari, V., Hebert, M., Sminchisescu, C., Weiss, Y. (eds.) ECCV 2018. LNCS, vol. 11214, pp. 19–35. Springer, Cham (2018). https://doi.org/10.1007/978-3-030-01249-6_2
16. Hodaň, T., et al.: Photorealistic image synthesis for object instance detection. In: ICIP (2019)
17. Hu, Y., Hugonot, J., Fua, P., Salzmann, M.: Segmentation-driven 6D object pose estimation. In: CVPR, pp. 3385–3394 (2019)
18. Jiang, P.T., Hou, Q., Cao, Y., Cheng, M.M., Wei, Y., Xiong, H.K.: Integral object mining via online attention accumulation. In: ICCV, pp. 2070–2079 (2019)
19. Johnson, J., Alahi, A., Fei-Fei, L.: Perceptual losses for real-time style transfer and super-resolution. In: Leibe, B., Matas, J., Sebe, N., Welling, M. (eds.) ECCV 2016. LNCS, vol. 9906, pp. 694–711. Springer, Cham (2016). https://doi.org/10.1007/978-3-319-46475-6_43
20. Kanazawa, A., Tulsiani, S., Efros, A.A., Malik, J.: Learning category-specific mesh reconstruction from image collections. In: Ferrari, V., Hebert, M., Sminchisescu, C., Weiss, Y. (eds.) ECCV 2018. LNCS, vol. 11219, pp. 386–402. Springer, Cham (2018). https://doi.org/10.1007/978-3-030-01267-0_23
21. Kaskman, R., Zakharov, S., Shugurov, I., Ilic, S.: HomebrewedDB: RGB-D dataset for 6D pose estimation of 3d objects. In: ICCVW (2019)
22. Kato, H., Ushiku, Y., Harada, T.: Neural 3D mesh renderer. In: CVPR, pp. 3907–3916 (2018)
23. Kehl, W., Manhardt, F., Tombari, F., Ilic, S., Navab, N.: SSD-6D: Making RGB-based 3D detection and 6D pose estimation great again. In: ICCV, pp. 1521–1529 (2017)
24. Kehl, W., Milletari, F., Tombari, F., Ilic, S., Navab, N.: Deep learning of local RGB-D patches for 3D object detection and 6D pose estimation. In: Leibe, B., Matas, J., Sebe, N., Welling, M. (eds.) ECCV 2016. LNCS, vol. 9907, pp. 205–220. Springer, Cham (2016). https://doi.org/10.1007/978-3-319-46487-9_13
25. Kocabas, M., Karagoz, S., Akbas, E.: Self-supervised learning of 3D human pose using multi-view geometry. In: CVPR, pp. 1077–1086 (2019)
26. Kolesnikov, A., Zhai, X., Beyer, L.: Revisiting self-supervised visual representation learning. In: CVPR, pp. 1920–1929 (2019)
27. Krizhevsky, A., Sutskever, I., Hinton, G.E.: Imagenet classification with deep convolutional neural networks. In: NeurIPS, pp. 1097–1105 (2012)
28. Lee, H.-Y., Tseng, H.-Y., Huang, J.-B., Singh, M., Yang, M.-H.: Diverse image-to-image translation via disentangled representations. In: Ferrari, V., Hebert, M., Sminchisescu, C., Weiss, Y. (eds.) ECCV 2018. LNCS, vol. 11205, pp. 36–52. Springer, Cham (2018). https://doi.org/10.1007/978-3-030-01246-5_3
29. Li, Y., Wang, G., Ji, X., Xiang, Y., Fox, D.: DeepIM: deep iterative matching for 6d pose estimation. IJCV, 1–22 (2019)
30. Li, Z., Wang, G., Ji, X.: CDPN: coordinates-based disentangled pose network for real-time RGB-based 6-DoF object pose estimation. In: ICCV, pp. 7678–7687 (2019)

31. Lin, T.Y., Dollár, P., Girshick, R., He, K., Hariharan, B., Belongie, S.: Feature pyramid networks for object detection. In: CVPR, pp. 2117–2125 (2017)
32. Lin, T.Y., Goyal, P., Girshick, R., He, K., Dollar, P.: Focal loss for dense object detection. In: ICCV (2017)
33. Lin, T.-Y., et al.: Microsoft COCO: common objects in context. In: Fleet, D., Pajdla, T., Schiele, B., Tuytelaars, T. (eds.) ECCV 2014. LNCS, vol. 8693, pp. 740–755. Springer, Cham (2014). https://doi.org/10.1007/978-3-319-10602-1_48
34. Liu, R., et al.: An intriguing failing of convolutional neural networks and the coordconv solution. In: NeurIPS, pp. 9605–9616 (2018)
35. Liu, S., Li, T., Chen, W., Li, H.: Soft rasterizer: a differentiable renderer for image-based 3D reasoning. In: ICCV, pp. 7708–7717 (2019)
36. Liu, W.W., et al.: SSD: single shot multibox detector. In: Leibe, B., Matas, J., Sebe, N., Welling, M. (eds.) ECCV 2016. LNCS, vol. 9905, pp. 21–37. Springer, Cham (2016). https://doi.org/10.1007/978-3-319-46448-0_2
37. Loper, M.M., Black, M.J.: OpenDR: an approximate differentiable renderer. In: Fleet, D., Pajdla, T., Schiele, B., Tuytelaars, T. (eds.) ECCV 2014. LNCS, vol. 8695, pp. 154–169. Springer, Cham (2014). https://doi.org/10.1007/978-3-319-10584-0_11
38. Manhardt, F., et al.: Explaining the ambiguity of object detection and 6D pose from visual data. In: ICCV, pp. 6841–6850 (2019)
39. Manhardt, F., Kehl, W., Gaidon, A.: ROI-10D: monocular lifting of 2D detection to 6D pose and metric shape. In: CVPR, pp. 2069–2078 (2019)
40. Manhardt, F., Kehl, W., Navab, N., Tombari, F.: Deep model-based 6D pose refinement in RGB. In: Ferrari, V., Hebert, M., Sminchisescu, C., Weiss, Y. (eds.) Computer Vision – ECCV 2018. LNCS, vol. 11218, pp. 833–849. Springer, Cham (2018). https://doi.org/10.1007/978-3-030-01264-9_49
41. Marschner, S., Shirley, P.: Fundamentals of Computer Graphics. CRC Press (2015)
42. Omran, M., Lassner, C., Pons-Moll, G., Gehler, P., Schiele, B.: Neural body fitting: Unifying deep learning and model based human pose and shape estimation. In: 3DV. pp. 484–494 (2018)
43. Park, K., Patten, T., Vincze, M.: Pix2pose: pixel-wise coordinate regression of objects for 6D pose estimation. In: ICCV, pp. 7668–7677 (2019)
44. Peng, S., Liu, Y., Huang, Q., Zhou, X., Bao, H.: PVNet: pixel-wise voting network for 6DoF pose estimation. In: CVPR, pp. 4561–4570 (2019)
45. Pillai, S., Ambruş, R., Gaidon, A.: Superdepth: self-supervised, super-resolved monocular depth estimation. In: ICRA, pp. 9250–9256 (2019)
46. Rad, M., Lepetit, V.: BB8: A scalable, accurate, robust to partial occlusion method for predicting the 3D poses of challenging objects without using depth. In: ICCV, pp. 3828–3836 (2017)
47. Rad, M., Oberweger, M., Lepetit, V.: Domain transfer for 3D pose estimation from color images without manual annotations. In: ACCV, pp. 69–84 (2018)
48. Rezatofighi, H., Tsoi, N., Gwak, J., Sadeghian, A., Reid, I., Savarese, S.: Generalized intersection over union: a metric and a loss for bounding box regression. In: CVPR, pp. 658–666 (2019)
49. Richter, S.R., Vineet, V., Roth, S., Koltun, V.: Playing for data: ground truth from computer games. In: Leibe, B., Matas, J., Sebe, N., Welling, M. (eds.) ECCV 2016. LNCS, vol. 9906, pp. 102–118. Springer, Cham (2016). https://doi.org/10.1007/978-3-319-46475-6_7
50. Spelke, E.S.: Principles of object perception. Cogn. Sci. **14**(1), 29–56 (1990)

51. Su, H., Qi, C.R., Li, Y., Guibas, L.J.: Render for CNN: viewpoint estimation in images using CNNs trained with rendered 3D model views. In: ICCV, pp. 2686–2694 (2015)
52. Sundermeyer, M., Marton, Z.-C., Durner, M., Brucker, M., Triebel, R.: Implicit 3D orientation learning for 6D object detection from RGB images. In: Ferrari, V., Hebert, M., Sminchisescu, C., Weiss, Y. (eds.) ECCV 2018. LNCS, vol. 11210, pp. 712–729. Springer, Cham (2018). https://doi.org/10.1007/978-3-030-01231-1_43
53. Tekin, B., Sinha, S.N., Fua, P.: Real-time seamless single shot 6D object pose prediction. In: CVPR, pp. 292–301 (2018)
54. Tian, Z., Shen, C., Chen, H., He, T.: FCOS: fully convolutional one-stage object detection. In: ICCV, pp. 9627–9636 (2019)
55. Tremblay, J., To, T., Birchfield, S.: Falling things: a synthetic dataset for 3D object detection and pose estimation. In: CVPRW, pp. 2038–2041 (2018)
56. Tremblay, J., To, T., Sundaralingam, B., Xiang, Y., Fox, D., Birchfield, S.: Deep object pose estimation for semantic robotic grasping of household objects. In: Conference on Robot Learning (CoRL), pp. 306–316 (2018)
57. Tung, H.Y., Tung, H.W., Yumer, E., Fragkiadaki, K.: Self-supervised learning of motion capture. In: NeurIPS, pp. 5236–5246 (2017)
58. Wohlhart, P., Lepetit, V.: Learning descriptors for object recognition and 3D pose estimation. In: CVPR, pp. 3109–3118 (2015)
59. Xiang, Y., Schmidt, T., Narayanan, V., Fox, D.: PoseCNN: a convolutional neural network for 6D object pose estimation in cluttered scenes. In: RSS (2018)
60. Zakharov, S., Kehl, W., Ilic, S.: Deceptionnet: network-driven domain randomization. In: ICCV, pp. 532–541 (2019)
61. Zakharov, S., Shugurov, I., Ilic, S.: Dpod: 6D pose object detector and refiner. In: ICCV, pp. 1941–1950 (2019)
62. Zhang, R., Isola, P., Efros, A.A., Shechtman, E., Wang, O.: The unreasonable effectiveness of deep features as a perceptual metric. In: CVPR, pp. 586–595 (2018)
63. Zhao, H., Gallo, O., Frosio, I., Kautz, J.: Loss functions for image restoration with neural networks. IEEE Trans. Comput. Imaging **3**(1), 47–57 (2016)
64. Zuffi, S., Kanazawa, A., Berger-Wolf, T., Black, M.J.: Three-d safari: learning to estimate zebra pose, shape, and texture from images "in the wild". In: ICCV, pp. 5359–5368 (2019)

Invertible Image Rescaling

Mingqing Xiao[1], Shuxin Zheng[2(✉)], Chang Liu[2(✉)], Yaolong Wang[3], Di He[1], Guolin Ke[2], Jiang Bian[2], Zhouchen Lin[1], and Tie-Yan Liu[2]

[1] Peking University, Beijing, China
{mingqing_xiao,di_he,zlin}@pku.edu.cn
[2] Microsoft Research Asia, Beijing, China
{shuz,changliu,guoke,jiabia,tyliu}@microsoft.com
[3] University of Toronto, Toronto, Canada
yaolong.wang@mail.utoronto.ca

Abstract. High-resolution digital images are usually downscaled to fit various display screens or save the cost of storage and bandwidth, meanwhile the post-upscaling is adopted to recover the original resolutions or the details in the zoom-in images. However, typical image downscaling is a non-injective mapping due to the loss of high-frequency information, which leads to the ill-posed problem of the inverse upscaling procedure and poses great challenges for recovering details from the downscaled low-resolution images. Simply upscaling with image super-resolution methods results in unsatisfactory recovering performance. In this work, we propose to solve this problem by modeling the downscaling and upscaling processes from a new perspective, i.e. an invertible bijective transformation, which can largely mitigate the ill-posed nature of image upscaling. We develop an Invertible Rescaling Net (IRN) with deliberately designed framework and objectives to produce visually-pleasing low-resolution images and meanwhile capture the distribution of the lost information using a latent variable following a specified distribution in the downscaling process. In this way, upscaling is made tractable by inversely passing a randomly-drawn latent variable with the low-resolution image through the network. Experimental results demonstrate the significant improvement of our model over existing methods in terms of both quantitative and qualitative evaluations of image upscaling reconstruction from downscaled images. Code is available at https://github.com/pkuxmq/Invertible-Image-Rescaling.

1 Introduction

With exploding amounts of high-resolution (HR) images/videos on the Internet, image downscaling is quite indispensable for storing, transferring and sharing such large-sized data, as the downscaled counterpart can significantly save

M. Xiao and Y. Wang—Work done during an internship at Microsoft Research Asia.

Electronic supplementary material The online version of this chapter (https://doi.org/10.1007/978-3-030-58452-8_8) contains supplementary material, which is available to authorized users.

A. Vedaldi et al. (Eds.): ECCV 2020, LNCS 12346, pp. 126–144, 2020.
https://doi.org/10.1007/978-3-030-58452-8_8

the storage, efficiently utilize the bandwidth [12,34,38,49,56] and easily fit for screens with different resolution while maintaining visually valid information [27,50]. Meanwhile, many of these downscaling scenarios inevitably raise a great demand for the inverse task, i.e., upscaling the downscaled image to a higher resolution or its original size [19,47,58,59]. However, details are lost and distortions appear when users zoom in or upscale the low-resolution (LR) images. Such an upscaling task is quite challenging since image downscaling is well-known as a non-injective mapping, meaning that there could exist multiple possible HR images resulting in the same downscaled LR image. Hence, this inverse task is usually considered to be ill-posed [17,24,57].

Many efforts have been made to mitigate this ill-posed problem, but the gains fail to meet the expectation. For example, most of previous works choose super-resolution (SR) methods to upscale the downscaled LR images. However, mainstream SR algorithms [14,17,37,51,61,62] focus only on recovering HR images from LR ones under the guidance of a predefined and non-adjustable downscaling kernel (e.g., Bicubic interpolation), which omits its compatibility to the downscaling operation. Intuitively, as long as the target LR image is pre-downscaled from an HR image, taking the image downscaling method into consideration would be quite invaluable for recovering the high-quality upscaled image.

Instead of simply treating the image downscaling and upscaling as two separate and independent tasks, most recently, there have been efforts [27,34,50] attempting to model image downscaling and upscaling as a united task by an encoder-decoder framework. Specifically, they proposed to use an upscaling-optimal downscaling method as an encoder which is jointly trained with an upscaling decoder [27] or existing SR modules [34,50]. Although such an integrated training approach can significantly improve the quality of the HR images recovered from the corresponding downscaled LR images, neither can we do a perfect reconstruction. These efforts didn't tackle much on the ill-posedness since they link the two processes only through the training objectives and conduct no attempt to capture any feature of the lost information.

In this paper, with inspiration from the reciprocal nature of this pair of image rescaling tasks, we propose a novel method to largely mitigate the ill-posed problem of image upscaling. According to the Nyquist-Shannon sampling theorem, high-frequency contents are lost during downscaling. Ideally, we hope to keep all lost information to perfectly recover the original HR image, but storing or transferring the high-frequency information is unacceptable. In order to well address this challenge, we develop a novel invertible model called Invertible Rescaling Net (IRN) which captures some knowledge on the lost information in the form of distribution and embeds it into model's parameters to mitigate the ill-posedness. Given an HR image x, IRN not only downscales it into a visually-pleasing LR image y, but also embed the case-specific high-frequency content into an auxiliary case-agnostic latent variable z, whose marginal distribution obeys a fixed pre-specified distribution (e.g., isotropic Gaussian). Based on this model, we use a randomly drawn sample of z from this distribution for the inverse upscaling procedure, which holds the most information that one could have in upscaling.

Yet, there are still several great challenges needed to be addressed during the IRN training process. Specifically, it is essential to ensure the quality of reconstructed HR images, obtain visually pleasing downscaled LR ones, and accomplish the upscaling with a case-agnostic z, i.e., $z \sim p(z)$ instead of a case-specific $z \sim p(z|y)$. To this end, we design a novel compact and effective objective function by combining three respective components: an HR reconstruction loss, an LR guidance loss and a distribution matching loss. The last component is for the model to capture the true HR image manifold as well as for enforcing z to be case-agnostic. Neither the conventional adversarial training techniques of generative adversarial nets (GANs) [21] nor the maximum likelihood estimation (MLE) method for existing invertible neural networks [4,15,16,29] could achieve our goal, since the model distribution doesn't exist here, meanwhile these methods don't guide the distribution in the latent space. Instead, we take the pushed-forward empirical distribution of x as the distribution on y, which, in independent company with $p(z)$, is the actually used distribution to inversely pass our model to recover the distribution of x. We thus match this distribution with the empirical distribution of x (the data distribution). Moreover, due to the invertible nature of our model, we show that once this matching task is accomplished, the matching task in the (y, z) space is also solved, and z is made case-agnostic. We minimize the JS divergence to match the distributions, since the alternative sample-based maximum mean discrepancy (MMD) method [3] doesn't generalize well to the high dimension data in our task.

Our contributions are concluded as follows:

- To our best knowledge, the proposed IRN is the first attempt to model image downscaling and upscaling, a pair of mutually-inverse tasks, using an invertible (i.e., bijective) transformation. Powered by the deliberately designed invertibility, our proposed IRN can largely mitigate the ill-posed nature of image upscaling reconstruction from the downscaled LR image.
- We propose a novel model design and efficient training objectives for IRN to enforce the latent variable z, with embedded lost high-frequency information in the downscaling, to obey a simple case-agnostic distribution. This enables efficient upscaling based on the valuable samples of z drawn from the certain distribution.
- The proposed IRN can significantly boost the performance of upscaling reconstruction from downscaled LR images compared with state-of-the-art downscaling-SR and encoder-decoder methods. Moreover, the amount of parameters of IRN is significantly reduced, which indicates the light-weight and high-efficiency of the new IRN model.

2 Related Work

2.1 Image Upscaling After Downscaling

Super resolution (SR) is a widely-used image upscaling method and get promising results in low-resolution (LR) image upscaling task. Therefore, SR methods could be used to upscale downscaled images. Since the SR task is inherently ill-posed, previous SR works mainly focus on learning strong prior information by example-based strategy [18,20,28,47] or deep learning models [14,17,35,37,51,61–64]. However, if the targeted LR image is pre-downscaled from the corresponding high-resolution image, taking the image downscaling method into consideration would significantly help the upscaling reconstruction.

Traditional image downscaling approaches employ frequency-based kernels, such as Bilinear, Bicubic, etc. [42], as a low-pass filter to sub-sample the input HR images into target resolution. Normally, these methods suffer from resulting over-smoothed images since the high-frequency details are suppressed. Therefore, several detail-preserving or structurally similar downscaling methods [31,39,43,53,54] are proposed recently. Besides those perceptual-oriented downscaling methods, inspired by the potentially mutual reinforcement between downscaling and its inverse task, upscaling, increasing efforts have been focused on the upscaling-optimal downscaling methods, which aim to learn a downscaling model that is optimal to the post-upscaling operation. For instance, Kim et al. [27] proposed a task-aware downscaling model based on an auto-encoder framework, in which the encoder and decoder act as the downscaling and upscaling model, respectively, such that the downscaling and upscaling processes are trained jointly as a united task. Similarly, Li et al. [34] proposed to use a CNN to estimate downscaled compact-resolution images and leverage a learned or specified SR model for HR image reconstruction. More recently, Sun et al. [50] proposed a new content-adaptive-resampler based image downscaling method, which can be jointly trained with any existing differentiable upscaling (SR) models. Although these attempts have an effect of pushing one of downscaling and upscaling to resemble the inverse process of the other, they still suffer from the ill-posed nature of image upscaling problem. In this paper, we propose to model the downscaling and upscaling processes by leveraging the invertible neural networks.

Difference from SR. Note that image rescaling is a different task from super-resolution. In our scenario, the ground-truth HR image is available at the beginning but somehow we have to discard it and store/transmit the LR version instead. We hope that we can recover the HR image afterwards using the LR image. While for SR, the real HR is unavailable in applications and the task is to generate new HR images for LR ones.

2.2 Invertible Neural Network

The invertible neural network (INN) [8,13,15,16,22,29,32] is a popular choice for generative models, in which the generative process $x = f_\theta(z)$ given a latent variable z can be specified by an INN architecture f_θ. The direct access to the inverse mapping $z = f_\theta^{-1}(x)$ makes inference much cheaper. As it is possible to compute the density of the model distribution in INN explicitly, one can use the maximum likelihood method for training. Due to such flexibility, INN architectures are also used for many variational inference tasks [10,30,45]. INN is also used to learn representations without information loss [25], and has been applied for feature embedding in the super-resolution task [35,64].

INN is composed of invertible blocks. In this study, we employ the invertible architecture in [16]. For the l-th block, input h^l is split into h_1^l and h_2^l along the channel axis, and they undergo the additive affine transformations [15]:

$$\begin{aligned} h_1^{l+1} &= h_1^l + \phi(h_2^l), \\ h_2^{l+1} &= h_2^l + \eta(h_1^{l+1}), \end{aligned} \tag{1}$$

where ϕ, η are arbitrary functions. The corresponding output is $[h_1^{l+1}, h_2^{l+1}]$. Given the output, its inverse transformation is easily computed:

$$\begin{aligned} h_2^l &= h_2^{l+1} - \eta(h_1^{l+1}), \\ h_1^l &= h_1^{l+1} - \phi(h_2^l), \end{aligned} \tag{2}$$

To enhance the transformation ability, the identity branch is often augmented [16]:

$$\begin{aligned} h_1^{l+1} &= h_1^l \odot \exp(\psi(h_2^l)) + \phi(h_2^l), \\ h_2^{l+1} &= h_2^l \odot \exp(\rho(h_1^{l+1})) + \eta(h_1^{l+1}), \\ h_2^l &= (h_2^{l+1} - \eta(h_1^{l+1})) \odot \exp(-\rho(h_1^{l+1})), \\ h_1^l &= (h_1^{l+1} - \phi(h_2^l)) \odot \exp(-\psi(h_2^l)). \end{aligned} \tag{3}$$

Some prior works studied using INN for paired data (x, y). Ardizzone et al. [3] analyzed real-world problems from medicine and astrophysics. Compared to their tasks, image downscaling and upscaling bring more difficulties because of notably larger dimensionality, so that their losses do not work for our task. In addition, the ground-truth LR image y does not exist in our task. Guided image generation and colorization using INN is proposed in [4] where the invertible modeling between x and z is conditioned on a guidance y. The model cannot generate y given x thus is unsuitable for the image upscaling task. INN is also applied to the image-to-image translation task [44] where the paired domain (X, Y) instead of paired data is considered, thus is again not the case of image upscaling.

2.3 Image Compression

Image compression is a type of data compression applied to digital images, to reduce their cost for storage or transmission. Image compression may be lossy (e.g., JPEG, BPG) or lossless (e.g., PNG, BMP). Recently, deep learning based image compression methods [2,6,7,41,46,52] show promising results on both visual effect and compression ratio. However, the resolution of image won't be changed by compression, which means there is no visually meaningful low-resolution image but only bit-stream after compressing. Thus our task can't be served by image compression methods.

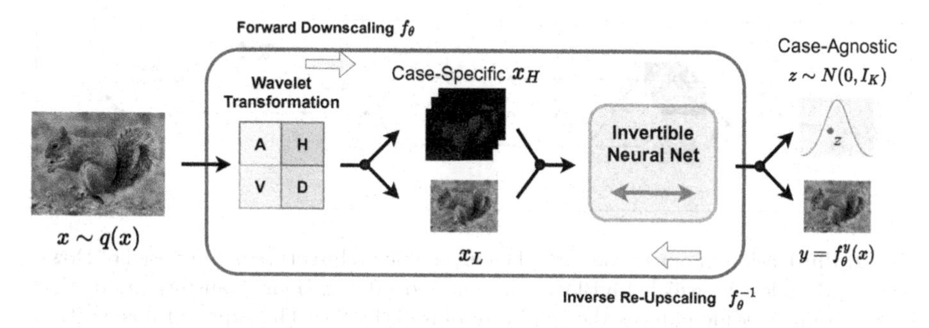

Fig. 1. Illustration of the problem formulation. In the forward downscaling procedure, HR image x is transformed to visually pleasing LR image y and case-agnostic latent variable z through a parameterized invertible function $f_\theta(\cdot)$; in the inverse upscaling procedure, a randomly drawn z combined with LR image y are transformed to HR image through the inverse function $f_\theta^{-1}(\cdot)$. I_K means identity matrix of variance in multivariate Gaussian distribution.

3 Methods

3.1 Model Specification

The sketch of our modeling framework is presented in Fig. 1. As explained in Introduction, we mitigate the ill-posed problem of the upscaling task by modeling the distribution of lost information during downscaling. We note that according to the Nyquist-Shannon sampling theorem [48], the lost information during downscaling an HR image amounts to high-frequency contents. Thus we firstly employ a wavelet transformation to decompose the HR image x into low and high-frequency component, denote as x_L and x_H respectively. Since the case-specific high-frequency information will be lost after downscaling, in order to best recover the original x as possible in the upscaling procedure, we use an invertible neural network to produce the visually-pleasing LR image y meanwhile model the distribution of the lost information by introducing an auxiliary latent variable z. In contrast to the case-specific x_H (i.e., $x_H \sim p(x_H|x_L)$), we force z to be case-agnostic (i.e., $z \sim p(z)$) and obey a simple specified distribution, e.g., an isotropic Gaussian distribution. In this way, there is no further need to preserve either x_H or z after downscaling, and z can be randomly sampled in the upscaling procedure, which is used to reconstruct x combined with LR image y by inversely passing the model.

3.2 Invertible Architecture

The general architecture of our proposed IRN is composed of stacked *Downscaling Modules*, each of which contains one *Haar Transformation* block and several invertible neural network blocks (*InvBlocks*), as illustrated in Fig. 2. We will show later that both of them are invertible, and thus the entire IRN model is invertible accordingly.

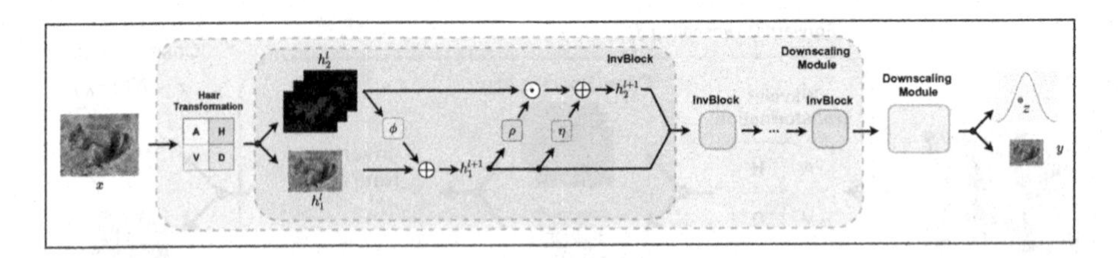

Fig. 2. Illustration of our framework. The invertible architecture is composed of Downscaling Modules, in which InvBlocks are stacked after a Haar Transformation. Each Downscaling Module reduces the spatial resolution by 2×. The exp(·) of ρ is omit.

The Haar Transformation. We design the model to contain certain inductive bias, which can efficiently learn to decompose x into the downscaled image y and case-agnostic high-frequency information embedded in z. To achieve this, we apply the Haar Transformation as the first layer in each downscaling module, which can explicitly decompose the input images into an approximate low-pass representation, and three directions of high-frequency coefficients [4,36,55,63]. More concretely, the Haar Transformation transforms the input raw images or a group of feature maps with height H, width W and channel C into a tensor of shape $(\frac{1}{2}H, \frac{1}{2}W, 4C)$. The first C slices of the output tensor are effectively produced by an average pooling, which is approximately a low-pass representation equivalent to the Bilinear interpolation downsampling. The rest three groups of C slices contain residual components in the vertical, horizontal and diagonal directions respectively, which are the high-frequency information in the original HR image. By such a transformation, the low and high-frequency information are effectively separated and will be fed into the following InvBlocks.

InvBlock. Taking the feature maps after the Haar Transformation as input, a stack of InvBlocks is used to further abstract the LR and latent representations. We leverage the general coupling layer architecture proposed in [15,16], i.e. Eqs. (1 and 3).

Utilizing the coupling layer is based on our considerations that (1) the input has already been split into low and high-frequency components by the Haar transformation; (2) we want the two branches of the output of a coupling layer to further polish the low and high-frequency inputs for a suitable LR image appearance and an independent and properly distributed latent representation of the high-frequency contents. So we match the low and high-frequency components respectively to the split of h_1^l, h_2^l in Eq. (1). Furthermore, as the shortcut connection is proved to be important in the image scaling tasks [37,51], we employ the additive transformation (Eq. 1) for the low-frequency part h_1^l, and the enhanced affine transformation (Eq. 3) for the high-frequency part h_2^l to increase the model capacity, as shown in Fig. 2.

Note that the transformation functions $\phi(\cdot), \eta(\cdot), \rho(\cdot)$ in Fig. 2 can be arbitrary. Here we employ a densely connected convolutional block, which is referred as Dense Block in [51] and demonstrated for its effectiveness of image upscaling task. Function $\rho(\cdot)$ is further followed by a centered sigmoid function and a scale term to prevent numerical explosion due to the exp(·) function. Note that Fig. 2 omits the exp(·) in function ρ.

Quantization. To save the output images of IRN as common image storage format such as RGB (8 bits for each R, G and B color channels), a quantization module is adopted which converts floating-point values of produced LR images to 8-bit unsigned int. We simply use rounding operation as the quantization module, store our output LR images by PNG format and use it in the upscaling procedure. There is one obstacle should be noted that the quantization module is nondifferentiable. To ensure that IRN can be optimized during training, we use Straight-Through Estimator [9] on the quantization module when calculating the gradients.

3.3 Training Objectives

Based on Sect. 3.1, our approach for invertible downscaling constructs a model that specifies a correspondence between HR image x and LR image y, as well as a case-agnostic distribution $p(z)$ of z. The goal of training is to drive these modeled relations and quantities to match our desiderata and HR image data $\{x^{(n)}\}_{n=1}^{N}$. This includes three specific goals, as detailed below.

LR Guidance. Although the invertible downscaling task does not pose direct requirements on the produced LR images, we do hope that they are valid visually pleasing LR images. To achieve this, we utilize the widely acknowledged Bicubic method [42] to guide the downscaling process of our model. Let $y_{\text{guide}}^{(n)}$ be the LR image corresponding to $x^{(n)}$ that is produced by the Bicubic method. To make our model follow the guidance, we drive the model-produced LR image $f_\theta^y(x^{(n)})$ to resemble $y_{\text{guide}}^{(n)}$:

$$L_{\text{guide}}(\theta) := \sum_{n=1}^{N} \ell_{\mathcal{Y}}(y_{\text{guide}}^{(n)}, f_\theta^y(x^{(n)})), \tag{4}$$

where $\ell_{\mathcal{Y}}$ is a difference metric on $\mathcal{Y}$, e.g., the L_1 or L_2 loss. We call it the LR guidance loss. This practice has also been adopted in the literature [27,50].

HR Reconstruction. Although f_θ is invertible, it is not for the correspondence between x and y when z is not transmitted. We hope that for a specific downscaled LR image y, the original HR image can be restored by the model using any sample of z from the case-agnostic $p(z)$. Inversely, this also encourages the forward process to produce a disentangled representation of z from y. As described in Sect. 3.1, given a HR image $x^{(n)}$, the model-downscaled LR image $f_\theta^y(x^{(n)})$ is to be upscaled by the model as $f_\theta^{-1}(f_\theta^y(x^{(n)}), z)$ with a randomly drawn $z \sim p(z)$. The reconstructed HR image should match the original one $x^{(n)}$, so we minimize the expected difference and traverse over all the HR images:

$$L_{\text{recon}}(\theta) := \sum_{n=1}^{N} \mathbb{E}_{p(z)}[\ell_{\mathcal{X}}(x^{(n)}, f_\theta^{-1}(f_\theta^y(x^{(n)}), z))], \tag{5}$$

where $\ell_{\mathcal{X}}$ measures the difference between the original image and the reconstructed one. We call $L_{\text{recon}}(\theta)$ the HR reconstruction loss. For practical minimization, we estimate the expectation w.r.t. z by one random draw from $p(z)$ for each evaluation.

Distribution Matching. The third part of the training goal is to encourage the model to catch the data distribution $q(x)$ of HR images, demonstrated by its sample cloud $\{x^{(n)}\}_{n=1}^N$. Recall that the model reconstructs a HR image $x^{(n)}$ by $f_\theta^{-1}(y^{(n)}, z^{(n)})$, where $y^{(n)} := f_\theta^y(x^{(n)})$ is the model-downscaled LR image, and $z^{(n)} \sim p(z)$ is the randomly drawn latent variable. When traversing over the sample cloud of true HR images $\{x^{(n)}\}_{n=1}^N$, $\{y^{(n)}\}_{n=1}^N$ also form a sample cloud of a distribution. We denote this distribution with the push-forward notation as $f_\theta^y{}_\#[q(x)]$, which represents the distribution of the transformed random variable $f_\theta^y(x)$ where the original random variable x obeys distribution $q(x)$, $x \sim q(x)$. Similarly, the sample cloud $\{f_\theta^{-1}(y^{(n)}, z^{(n)})\}_{n=1}^N$ represents the distribution of model-reconstructed HR images, and we denote it as $f_\theta^{-1}{}_\#\big[f_\theta^y{}_\#[q(x)]\, p(z)\big]$ since $(y^{(n)}, z^{(n)}) \sim f_\theta^y{}_\#[q(x)]\, p(z)$ (note that $y^{(n)}$ and $z^{(n)}$ are independent due to the generation process). The desideratum of distribution matching is to drive the model-reconstructed distribution towards data distribution, which can be achieved by minimizing their difference measured by some metric of distributions:

$$L_{\text{distr}}(\theta) := L_{\mathcal{P}}\big(f_\theta^{-1}{}_\#\big[f_\theta^y{}_\#[q(x)]\, p(z)\big], q(x)\big). \tag{6}$$

The distribution matching loss pushes the model-reconstructed HR images to lie on the manifold of true HR images so as to make the recovered images appear more realistic. It also drives the case-independence of z from y in the forward process. To see this, we note that if f_θ is invertible, then in the asymptotic case, the two distributions match on $\mathcal{X}$, i.e., $f_\theta^{-1}{}_\#\big[f_\theta^y{}_\#[q(x)]\, p(z)\big] = q(x)$, if and only if they match on $\mathcal{Y} \times \mathcal{Z}$, i.e., $f_\theta^y{}_\#[q(x)]\, p(z) = f_{\theta\#}[q(x)]$. The loss thus drives the coupled distribution $f_{\theta\#}[q(x)] = (f_\theta^y, f_\theta^z)_\#[q(x)]$ of (y, z) from the forward process towards the decoupled distribution $f_\theta^y{}_\#[q(x)]\, p(z)$. Neither effect can be fully guaranteed by the reconstruction and guidance losses.

As mentioned in Introduction, the minimization is generally hard since both distributions are high-dimensional and have unknown density function. We employ the JS divergence as the probability metric $L_{\mathcal{P}}$, and our distribution matching loss can be estimated in the following way:

$$\begin{aligned} L_{\text{distr}}(\theta) &= \text{JS}\big(f_\theta^{-1}{}_\#\big[f_\theta^y{}_\#[q(x)]\, p(z)\big], q(x)\big) \\ &\approx \frac{1}{2N} \max_T \sum_n \Big\{ \log \sigma(T(x^{(n)})) \\ &+ \log\Big(1 - \sigma\big[T\big(f_\theta^{-1}(f_\theta^y(x^{(n)}), z^{(n)})\big)\big]\Big)\Big\} + \log 2, \end{aligned} \tag{7}$$

where $\{z^{(n)}\}_{n=1}^N$ are i.i.d. samples from $p(z)$, σ is the sigmoid function, $T: \mathcal{X} \to \mathbb{R}$ is a function on $\mathcal{X}$ ($\sigma(T(\cdot))$ is regarded as a discriminator in GAN literatures), and "$\approx$" is due to Monte Carlo estimation. The appendix provides the details. For practical computation, the function T is parameterized as a neural network T_ϕ and $\max_T$ amounts to $\max_\phi$. The expression (7) is also suitable for estimating its gradient w.r.t. θ and ϕ, thus optimization is made practical.

Total Loss. We optimize our IRN model by minimizing the compact loss $L_{\text{total}}(\theta)$ with the combination of HR reconstruction loss $L_{\text{recon}}(\theta)$, LR guidance loss $L_{\text{guide}}(\theta)$ and distribution matching loss $L_{\text{distr}}(\theta)$:

$$L_{\text{total}} := \lambda_1 L_{\text{recon}} + \lambda_2 L_{\text{guide}} + \lambda_3 L_{\text{distr}}, \tag{8}$$

where $\lambda_1, \lambda_2, \lambda_3$ are coefficients for balancing different loss terms.

Loss Minimization in Practice. As an issue in practice, we find that directly minimizing the total loss $L_{\text{total}}(\theta)$ is difficult to train, due to the unstable training process of GANs [5]. We propose a pre-training stage that adopts a weakened but more stable surrogate of the distribution matching loss. Recall that the distribution matching loss $L_{\mathcal{P}}\big(f_\theta^{-1}{}_{\#}\big[f_\theta^y{}_{\#}[q(x)]\ p(z)\big], q(x)\big)$ on $\mathcal{X}$ has the same asymptotic effect as the loss $L_{\mathcal{P}}(f_\theta^y{}_{\#}[q(x)]\ p(z), (f_\theta^y, f_\theta^z)_{\#}[q(x)])$ on $\mathcal{Y} \times \mathcal{Z}$. The surrogate considers partial distribution matching on $\mathcal{Z}$, i.e., $L_{\mathcal{P}}(p(z), f_\theta^z{}_{\#}[q(x)])$. Since the density function of one of the distributions, $p(z)$, is now made available, we can choose more stable distribution metrics for minimization, such as the cross entropy (CE):

$$\begin{aligned} &L'_{\text{distr}}(\theta) := \mathrm{CE}(f_\theta^z{}_{\#}[q(x)], p(z)) \\ &= -\mathbb{E}_{f_\theta^z{}_{\#}[q(x)]}[\log p(z)] = -\mathbb{E}_{q(x)}[\log p(z = f_\theta^z(x))]. \end{aligned} \tag{9}$$

A related training method is the maximum likelihood estimation (MLE), i.e., $\max_\theta \mathbb{E}_{q(x)}[\log f_\theta^{-1}{}_{\#}[p(y,z)]]$, which is widely adopted by prevalent flow-based generative models [4,15,16,29]. It is equivalent to minimizing the Kullback-Leibler (KL) divergence $\mathrm{KL}(q(x), f_\theta^{-1}{}_{\#}[p(y,z)])$. The mentioned models explicitly specify the density function of $p(y,z)$, thus the density function of $f_\theta^{-1}{}_{\#}[p(y,z)]$ is made available together with the tractable Jacobian determinant computation of f_θ. However, the same objective cannot be leveraged for our model since we do not have the density function for $f_\theta^y{}_{\#}[q(x)]\ p(z)$; only that of $p(z)$ is known[1]. The invertible neural network (INN) [3] meets the same problem and cannot use MLE either.

We call IRN as our model trained by minimizing the following total objective:

$$L_{\text{IRN}} := \lambda_1 L_{\text{recon}} + \lambda_2 L_{\text{guide}} + \lambda_3 L'_{\text{distr}}. \tag{10}$$

After the pre-training stage, we restore the full distribution matching loss L_{distr} in the objective in place of L'_{distr}. Additionally, we also employ a perceptual

[1] MLEs corresponding to minimizing $\mathrm{KL}(q(x|y), f_\theta^{-1}(y,\cdot)_{\#}[p(z)])$ or $\mathrm{KL}\Big(q(x), \Big(\mathbb{E}_{f_\theta^y{}_{\#}[q(x)]}[f_\theta^{-1}(y,\cdot)]\Big)_{\#}[p(z)]\Big)$ are also impossible, since the pushed-forward distributions have a.e. zero density in $\mathcal{X}$ so the KL is a.e. infinite.

loss [26] L_{percp} on $\mathcal{X}$, which measures the difference of two images via their semantic features extracted by benchmarking models. It enhances the perceptual similarity between generated and true images thus helps to produce more realistic images. The perceptual loss has several slightly modified variants which mainly differ in the position of the objective features [33,51]. We adopt the variant proposed in [51]. We call IRN+ as our model trained by minimizing the following total objective:

$$L_{\text{IRN+}} := \lambda_1 L_{\text{recon}} + \lambda_2 L_{\text{guide}} + \lambda_3 L_{\text{distr}} + \lambda_4 L_{\text{percp}}. \tag{11}$$

4 Experiments

4.1 Dataset and Settings

We employ the widely used DIV2K [1] image restoration dataset to train our model, which contains 800 high-quality 2K resolution images in the training set, and 100 in the validation set. Besides, we evaluate our model on 4 additional standard datasets, i.e. the Set5 [11], Set14 [60], BSD100 [40], and Urban100 [23]. Following the setting in [37], we quantitatively evaluate the peak noise-signal ratio (PSNR) and SSIM [53] on the Y channel of images represented in the YCbCr (Y, Cb, Cr) color space. Due to space constraint, we leave training strategy details in the appendix.

4.2 Evaluation on Reconstructed HR Images

This section reports the quantitative and qualitative performance of HR image reconstruction with different downscaling and upscaling methods. We consider two kinds of reconstruction methods as our baselines: (1) downscaling with Bicubic interpolation and upscaling with state-of-the-art SR models [14,17,37,51, 61,62]; (2) downscaling with upscaling-optimal models [27,34,50] and upscaling with SR models. For the method of [51], we denote ESRGAN as their pre-trained model, and ESRGAN+ as their GAN-based model. We further investigate the influence of different z samples on the reconstructed image x. Finally, we empirically study the effectiveness of the different types of loss in the pre-training stage.

Quantitative Results. Table 1 summarizes the quantitative comparison results of different reconstruction methods where IRN significantly outperforms previous state-of-the-art methods regarding PSNR and SSIM in all datasets. We leave the results of IRN+ in the appendix because it is a visual-perception-oriented model. As shown in Table 1, upscaling-optimal downscaling models largely enhance the reconstruction of HR images by state-of-the-art SR models compared with downscaling with Bicubic interpolation. However, they still hardly achieve satisfying results due to the ill-posed nature of upscaling. In contract, with the invertibility, IRN significantly boosts the PSNR metric about 4–5 dB and 2–3 dB on each benchmark dataset in 2× and 4× scale downsampling and reconstruction, and the improvement goes as large as 5.94 dB compared with

Table 1. Quantitative evaluation results (PSNR/SSIM) of different downscaling and upscaling methods for image reconstruction on benchmark datasets: Set5, Set14, BSD100, Urban100, and DIV2K validation set. For our method, differences on average PSNR/SSIM from different z samples are less than 0.02. We report the mean result over 5 draws.

Downscaling & Upscaling	Scale	Param	Set5	Set14	BSD100	Urban100	DIV2K
Bicubic & Bicubic	2×	/	33.66/0.9299	30.24/0.8688	29.56/0.8431	26.88/0.8403	31.01/0.9393
Bicubic & SRCNN [17]	2×	57.3K	36.66/0.9542	32.45/0.9067	31.36/0.8879	29.50/0.8946	–
Bicubic & EDSR [37]	2×	40.7M	38.20/0.9606	34.02/0.9204	32.37/0.9018	33.10/0.9363	35.12/0.9699
Bicubic & RDN [62]	2×	22.1M	38.24/0.9614	34.01/0.9212	32.34/0.9017	32.89/0.9353	–
Bicubic & RCAN [61]	2×	15.4M	38.27/0.9614	34.12/0.9216	32.41/0.9027	33.34/0.9384	–
Bicubic & SAN [14]	2×	15.7M	38.31/0.9620	34.07/0.9213	32.42/0.9028	33.10/0.9370	–
TAD & TAU [27]	2×	–	38.46/ –	35.52/ –	36.68/ –	35.03/ –	39.01/ –
CNN-CR & CNN-SR [34]	2×	–	38.88/–	35.40/–	33.92/–	33.68/–	–
CAR & EDSR [50]	2×	51.1M	38.94/0.9658	35.61/0.9404	33.83/0.9262	35.24/0.9572	38.26/0.9599
IRN (ours)	2×	1.66M	43.99/0.9871	40.79/0.9778	41.32/0.9876	39.92/0.9865	44.32/0.9908
Bicubic & Bicubic	4×	/	28.42/0.8104	26.00/0.7027	25.96/0.6675	23.14/0.6577	26.66/0.8521
Bicubic & SRCNN [17]	4×	57.3K	30.48/0.8628	27.50/0.7513	26.90/0.7101	24.52/0.7221	–
Bicubic & EDSR [37]	4×	43.1M	32.62/0.8984	28.94/0.7901	27.79/0.7437	26.86/0.8080	29.38/0.9032
Bicubic & RDN [62]	4×	22.3M	32.47/0.8990	28.81/0.7871	27.72/0.7419	26.61/0.8028	–
Bicubic & RCAN [61]	4×	15.6M	32.63/0.9002	28.87/0.7889	27.77/0.7436	26.82/0.8087	30.77/0.8460
Bicubic & ESRGAN [51]	4×	16.3M	32.74/0.9012	29.00/0.7915	27.84/0.7455	27.03/0.8152	30.92/0.8486
Bicubic & SAN [14]	4×	15.7M	32.64/0.9003	28.92/0.7888	27.78/0.7436	26.79/0.8068	–
TAD & TAU [27]	4×	–	31.81/ –	28.63/ –	28.51/ –	26.63/ –	31.16/ –
CAR & EDSR [50]	4×	52.8M	33.88/0.9174	30.31/0.8382	29.15/0.8001	29.28/0.8711	32.82/0.8837
IRN (ours)	4×	4.35M	36.19/0.9451	32.67/0.9015	31.64/0.8826	31.41/0.9157	35.07/0.9318

the state-of-the-art downscaling and upscaling model. These results indicate an exponential improvement of IRN in the reduction of information loss, which also accords with the significant improvement in SSIM.

Moreover, the number of parameters of IRN is relatively small. When Bicubic downscaling and super-resolution methods require large model size (>15M) for better results, our IRN only has 1.66M and 4.35M parameters in scale 2× and 4× respectively. It indicates that our model is light-weight and efficient.

Qualitative Results. We then qualitatively evaluate IRN and IRN+ by demonstrating details of the upscaled images. As shown in Fig. 3, HR images reconstructed by IRN and IRN+ achieve better visual quality and fidelity than those of previous state-of-the-art methods. IRN recovers richer details, which contributes to the pleasing visual quality. IRN+ further produces sharper and more realistic images as the effect of the distribution matching objective. For the 'Comic' example, we observe that the IRN and IRN+ are the only models that can recover the complicated textures on the headwear and necklace, as well as the sharp and realistic fingers. Previous perceptual-driven methods such as ESRGAN [51] also claim that the sharpness and reality of their generated HR images are satisfied. However, the visually unreasonable and unpleasing details produced by their model often lead to dissimilarity to the original images. We leave the high-resolution version and more results in the appendix for spacing reason.

Visualisation on the Influence of z. As described in previous sections, we aim to let $z \sim p(z)$ focus on the randomness of high-frequency contents only. In Table 1, the PSNR difference is less than 0.02 dB for each image with different

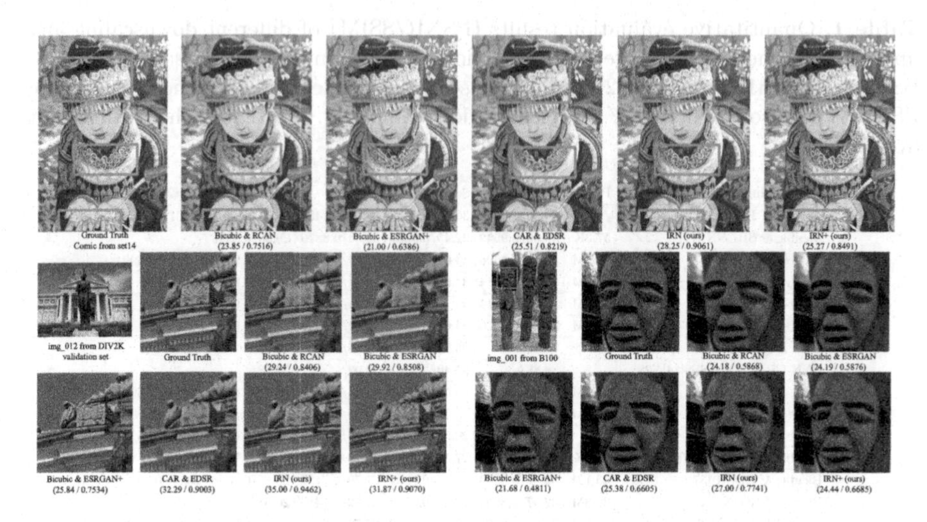

Fig. 3. Qualitative results of upscaling the 4× downscaled images. IRN recovers rich details, leading to both visually pleasing performance and high similarity to the original images. IRN+ produces even sharper and more realistic details. See the appendix for more results.

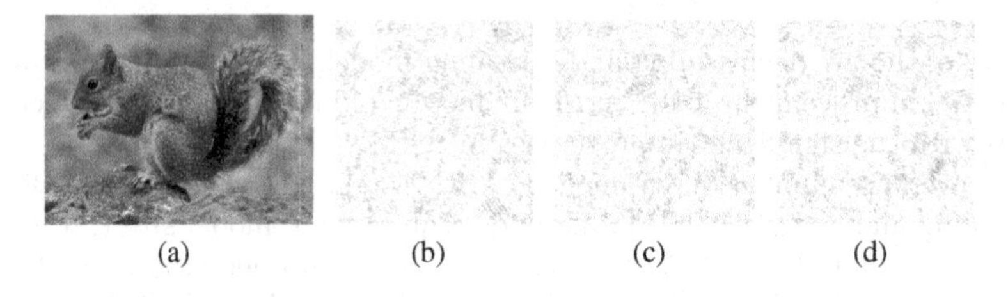

Fig. 4. Visualisation of the difference of upscaled HR images from multiple draws of z. (a): original image; (b–d): HR image differences of three z drawn from a common z sample. Darker color means larger difference. It shows that the differences are random noise in high-frequency regions without a typical texture.

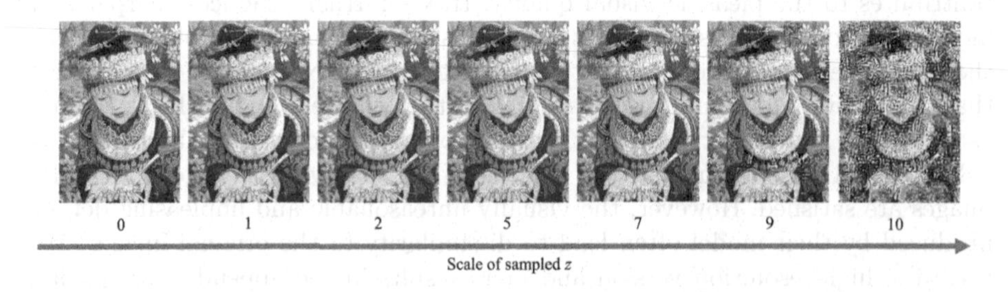

Fig. 5. Results of HR images by IRN+ with out-of-distribution samples of z. We train z with an isotropic Gaussian distribution, and illustrate upscaling results when scaling z sampled from the isotropic Gaussian distribution.

samples of z. In order to verify whether z has learned only to influence high-frequency information, we calculate and present the difference between different draws of z in Fig. 4. We can see in the figure that there is only a tiny noisy distinction in high-frequency regions without typical textures, which can hardly be perceived when combined with low-frequency contents. This indicates that our IRN has learned to reconstruct most meaningful high-frequency contents, while embedding senseless noise into randomness.

As mentioned above, we train the model to encourage $p(z)$ to obey a simple and easy-to-sample distribution, i.e., isotropic Gaussian distribution. In order to further verify the effectiveness of the learned model, we feed $(y, \alpha z)$ into our IRN+ to obtain x_α by controlling the scale of sampled z with different values of α. As shown in Fig. 5, a larger deviation to the original distribution results in more noisy textures and distortion. It demonstrates that our model transforms z faithfully to follow the specified distribution, and is also robust to slight distribution deviation.

Table 2. Analysis results (PSNR/SSIM) of training IRN with L_1 or L_2 LR guide and HR reconstruction loss, with/without partial distribution matching loss, on Set5, Set14, BSD100, Urban100 and DIV2K validation sets with scale 4×.

L_{guide}	L_{recon}	$L_{distr'}$	Set5	Set14	BSD100	Urban100	DIV2K
L_1	L_1	Yes	34.75/0.9296	31.42/0.8716	30.42/0.8451	30.11/0.8903	33.64/0.9079
L_1	L_2	Yes	34.93/0.9296	31.76/0.8776	31.01/0.8562	30.79/0.8986	34.11/0.9116
L_2	L_1	Yes	36.19/0.9451	32.67/0.9015	31.64/0.8826	31.41/0.9157	35.07/0.9318
L_2	L_2	Yes	35.93/0.9402	32.51/0.8937	31.64/0.8742	31.40/0.9105	34.90/0.9308
L_2	L_1	No	36.12/0.9455	32.18/0.8995	31.49/0.8808	30.91/0.9102	34.90/0.9308

Analysis on the Losses. We conduct experiments to analyze the components in the loss of Eqs. (4, 5, and 9). As shown in Table 2, IRN performs the best when the LR guidance loss is the L_2 loss and the HR reconstruction loss is the L_1 loss. The reason is that the L_1 loss encourages more pixel-wise similarity, while the L_2 loss is less sensitive to minor changes. In the forward procedure, we utilize the Bicubic-downscaled images as guidance, but we do not aim to exactly learn the Bicubic downscaling, which may harm the inverse procedure. The forward reconstruction loss only acts as a constraint to maintain visually pleasing downscaling, so the L_2 loss is more suitable. In the backward procedure, on the other hand, our goal is to reconstruct the ground truth image accurately. Therefore, the L_1 loss is more appropriate, as also identified by other super-resolution works. Table 2 also demonstrates the necessity of the partial distribution matching loss of Eq. (9), which restricts the marginal distributions on $\mathcal{Z}$, and benefits the forward distribution learning.

4.3 Evaluation on Downscaled LR Images

We also evaluate the quality of LR images downscaled by our IRN. We demonstrate the similarity index between our LR images and Bicubic-based LR images, and present similar visual perception of them, to show that IRN is able to perform as well as Bicubic.

Table 3. SSIM results between the images downscaled by IRN and by Bicubic on the Set5, Set14, BSD100, Urban100 and DIV2K validation sets.

Scale	Set5	Set14	BSD100	Urban100	DIV2K
2×	0.9957	0.9936	0.9936	0.9941	0.9945
4×	0.9964	0.9927	0.9923	0.9916	0.9933

As shown in Table 3, images downscaled by IRN are extremely similar to those by Bicubic. More figures in the appendix illustrate the visual similarity between them, which demonstrates the proper perception of our downscaled images.

5 Conclusion

In this paper, we propose a novel invertible network for the image rescaling task, with which the ill-posed nature of the task is largely mitigated. We explicitly model the statistics of the case-specific high-frequency information that is lost in downscaling as a latent variable following a specified case-agnostic distribution which is easy to sample from. The network models the rescaling processes by invertibly transforming between an HR image and an LR image with the latent variable. With the statistical knowledge of the latent variable, we draw a sample of it for upscaling from a downscaled LR image (whose specific high-frequency information was lost during downscaling, of course). We design a specific invertible architecture tailored for image rescaling, and an effective training objective to enforce the model to have desired downscaling and upscaling behavior, as well as to output the latent variable with the specified properties. Extensive experiments demonstrate that our model significantly improves both quantitative and qualitative performance of upscaling reconstruction from downscaled LR images, while being light-weighted.

References

1. Agustsson, E., Timofte, R.: NTIRE 2017 challenge on single image super-resolution: dataset and study. In: Proceedings of the IEEE Conference on Computer Vision and Pattern Recognition Workshops, pp. 126–135 (2017)

2. Agustsson, E., Tschannen, M., Mentzer, F., Timofte, R., Gool, L.V.: Generative adversarial networks for extreme learned image compression. In: Proceedings of the IEEE International Conference on Computer Vision, pp. 221–231 (2019)
3. Ardizzone, L., et al.: Analyzing inverse problems with invertible neural networks. In: Proceedings of the International Conference on Learning and Representations (2019)
4. Ardizzone, L., Lüth, C., Kruse, J., Rother, C., Köthe, U.: Guided image generation with conditional invertible neural networks. arXiv preprint arXiv:1907.02392 (2019)
5. Arjovsky, M., Bottou, L.: Towards principled methods for training generative adversarial networks. In: Proceedings of the International Conference on Learning and Representations (2017)
6. Ballé, J., Laparra, V., Simoncelli, E.P.: End-to-end optimized image compression. arXiv preprint arXiv:1611.01704 (2016)
7. Ballé, J., Minnen, D., Singh, S., Hwang, S.J., Johnston, N.: Variational image compression with a scale hyperprior. arXiv preprint arXiv:1802.01436 (2018)
8. Behrmann, J., Grathwohl, W., Chen, R.T., Duvenaud, D., Jacobsen, J.H.: Invertible residual networks. In: International Conference on Machine Learning, pp. 573–582 (2019)
9. Bengio, Y., Léonard, N., Courville, A.: Estimating or propagating gradients through stochastic neurons for conditional computation. arXiv preprint arXiv:1308.3432 (2013)
10. Berg, R., Hasenclever, L., Tomczak, J.M., Welling, M.: Sylvester normalizing flows for variational inference. In: Proceedings of the Conference on Uncertainty in Artificial Intelligence (2018)
11. Bevilacqua, M., Roumy, A., Guillemot, C., Alberi-Morel, M.L.: Low-complexity single-image super-resolution based on nonnegative neighbor embedding (2012)
12. Bruckstein, A.M., Elad, M., Kimmel, R.: Down-scaling for better transform compression. IEEE Trans. Image Process. **12**(9), 1132–1144 (2003)
13. Chen, R.T., Behrmann, J., Duvenaud, D., Jacobsen, J.H.: Residual flows for invertible generative modeling. arXiv preprint arXiv:1906.02735 (2019)
14. Dai, T., Cai, J., Zhang, Y., Xia, S.T., Zhang, L.: Second-order attention network for single image super-resolution. In: Proceedings of the IEEE Conference on Computer Vision and Pattern Recognition, pp. 11065–11074 (2019)
15. Dinh, L., Krueger, D., Bengio, Y.: NICE: non-linear independent components estimation. In: Workshop of the International Conference on Learning Representations (2015)
16. Dinh, L., Sohl-Dickstein, J., Bengio, S.: Density estimation using real NVP. In: Proceedings of the International Conference on Learning Representations (2017)
17. Dong, C., Loy, C.C., He, K., Tang, X.: Image super-resolution using deep convolutional networks. IEEE Trans. Pattern Anal. Mach. Intell. **38**(2), 295–307 (2015)
18. Freedman, G., Fattal, R.: Image and video upscaling from local self-examples. ACM Trans. Graph. (TOG) **30**(2), 12 (2011)
19. Giachetti, A., Asuni, N.: Real-time artifact-free image upscaling. IEEE Trans. Image Process. **20**(10), 2760–2768 (2011)
20. Glasner, D., Bagon, S., Irani, M.: Super-resolution from a single image. In: 2009 IEEE 12th International Conference on Computer Vision, pp. 349–356. IEEE (2009)
21. Goodfellow, I., et al.: Generative adversarial nets. In: Advances in Neural Information Processing Systems, Montréal, Canada, pp. 2672–2680. NIPS Foundation (2014)

22. Grathwohl, W., Chen, R.T., Betterncourt, J., Sutskever, I., Duvenaud, D.: FFJORD: free-form continuous dynamics for scalable reversible generative models. In: Proceedings of the International Conference on Learning and Representations (2019)
23. Huang, J.B., Singh, A., Ahuja, N.: Single image super-resolution from transformed self-exemplars. In: Proceedings of the IEEE Conference on Computer Vision and Pattern Recognition, pp. 5197–5206 (2015)
24. Glasner, D., Bagon, S., Irani, M.: Super-resolution from a single image. In: Proceedings of the IEEE International Conference on Computer Vision, Kyoto, Japan, pp. 349–356 (2009)
25. Jacobsen, J.H., Smeulders, A.W., Oyallon, E.: i-RevNet: deep invertible networks. In: International Conference on Learning Representations (2018)
26. Johnson, J., Alahi, A., Fei-Fei, L.: Perceptual losses for real-time style transfer and super-resolution. In: Leibe, B., Matas, J., Sebe, N., Welling, M. (eds.) ECCV 2016. LNCS, vol. 9906, pp. 694–711. Springer, Cham (2016). https://doi.org/10.1007/978-3-319-46475-6_43
27. Kim, H., Choi, M., Lim, B., Mu Lee, K.: Task-aware image downscaling. In: Ferrari, V., Hebert, M., Sminchisescu, C., Weiss, Y. (eds.) ECCV 2018. LNCS, vol. 11208, pp. 419–434. Springer, Cham (2018). https://doi.org/10.1007/978-3-030-01225-0_25
28. Kim, K.I., Kwon, Y.: Single-image super-resolution using sparse regression and natural image prior. IEEE Trans. Pattern Anal. Mach. Intell. **32**(6), 1127–1133 (2010)
29. Kingma, D.P., Dhariwal, P.: Glow: generative flow with invertible 1x1 convolutions. In: Advances in Neural Information Processing Systems, pp. 10215–10224 (2018)
30. Kingma, D.P., Salimans, T., Jozefowicz, R., Chen, X., Sutskever, I., Welling, M.: Improved variational inference with inverse autoregressive flow. In: Advances in Neural Information Processing Systems, pp. 4743–4751 (2016)
31. Kopf, J., Shamir, A., Peers, P.: Content-adaptive image downscaling. ACM Trans. Graph. (TOG) **32**(6), 173 (2013)
32. Kumar, M., et al.: VideoFlow: a flow-based generative model for video. arXiv preprint arXiv:1903.01434 (2019)
33. Ledig, C., et al.: Photo-realistic single image super-resolution using a generative adversarial network. In: Proceedings of the IEEE Conference on Computer Vision and Pattern Recognition, pp. 4681–4690 (2017)
34. Li, Y., Liu, D., Li, H., Li, L., Li, Z., Wu, F.: Learning a convolutional neural network for image compact-resolution. IEEE Trans. Image Process. **28**(3), 1092–1107 (2018)
35. Li, Z., Li, S., Zhang, N., Wang, L., Xue, Z.: Multi-scale invertible network for image super-resolution. In: Proceedings of the ACM Multimedia Asia, pp. 1–6 (2019)
36. Lienhart, R., Maydt, J.: An extended set of Haar-like features for rapid object detection. In: Proceedings of the International Conference on Image Processing, vol. 1, p. I. IEEE (2002)
37. Lim, B., Son, S., Kim, H., Nah, S., Mu Lee, K.: Enhanced deep residual networks for single image super-resolution. In: Proceedings of the IEEE Conference on Computer Vision and Pattern Recognition Workshops, pp. 136–144 (2017)
38. Lin, W., Dong, L.: Adaptive downsampling to improve image compression at low bit rates. IEEE Trans. Image Process. **15**(9), 2513–2521 (2006)
39. Liu, J., He, S., Lau, R.W.: $l\_\{0\}$-regularized image downscaling. IEEE Trans. Image Process. **27**(3), 1076–1085 (2017)

40. Martin, D., Fowlkes, C., Tal, D., Malik, J., et al.: A database of human segmented natural images and its application to evaluating segmentation algorithms and measuring ecological statistics. In: ICCV, Vancouver (2001)
41. Minnen, D., Ballé, J., Toderici, G.D.: Joint autoregressive and hierarchical priors for learned image compression. In: Advances in Neural Information Processing Systems, pp. 10771–10780 (2018)
42. Mitchell, D.P., Netravali, A.N.: Reconstruction filters in computer-graphics. ACM SIGGRAPH Comput. Graph. **22**(4), 221–228 (1988)
43. Oeztireli, A.C., Gross, M.: Perceptually based downscaling of images. ACM Trans. Graph. (TOG) **34**(4), 77 (2015)
44. van der Ouderaa, T.F., Worrall, D.E.: Reversible GANs for memory-efficient image-to-image translation. In: Proceedings of the IEEE Conference on Computer Vision and Pattern Recognition, pp. 4720–4728 (2019)
45. Rezende, D., Mohamed, S.: Variational inference with normalizing flows. In: Proceedings of the International Conference on Machine Learning, pp. 1530–1538 (2015)
46. Rippel, O., Bourdev, L.: Real-time adaptive image compression. In: Proceedings of the 34th International Conference on Machine Learning, vol. 70, pp. 2922–2930. JMLR.org (2017)
47. Schulter, S., Leistner, C., Bischof, H.: Fast and accurate image upscaling with super-resolution forests. In: Proceedings of the IEEE Conference on Computer Vision and Pattern Recognition, pp. 3791–3799 (2015)
48. Shannon, C.E.: Communication in the presence of noise. Proc. IRE **37**(1), 10–21 (1949)
49. Shen, M., Xue, P., Wang, C.: Down-sampling based video coding using super-resolution technique. IEEE Trans. Circ. Syst. Video Technol. **21**(6), 755–765 (2011)
50. Sun, W., Chen, Z.: Learned image downscaling for upscaling using content adaptive resampler. IEEE Trans. Image Process. **29**, 4027–4040 (2020)
51. Wang, X., et al.: ESRGAN: enhanced super-resolution generative adversarial networks. In: Leal-Taixé, L., Roth, S. (eds.) ECCV 2018. LNCS, vol. 11133, pp. 63–79. Springer, Cham (2019). https://doi.org/10.1007/978-3-030-11021-5_5
52. Wang, Y., Xiao, M., Liu, C., Zheng, S., Liu, T.Y.: Modeling lost information in lossy image compression. arXiv preprint arXiv:2006.11999 (2020)
53. Wang, Z., Bovik, A.C., Sheikh, H.R., Simoncelli, E.P., et al.: Image quality assessment: from error visibility to structural similarity. IEEE Trans. Image Process. **13**(4), 600–612 (2004)
54. Weber, N., Waechter, M., Amend, S.C., Guthe, S., Goesele, M.: Rapid, detail-preserving image downscaling. ACM Trans. Graph. (TOG) **35**(6), 205 (2016)
55. Wilson, P.I., Fernandez, J.: Facial feature detection using Haar classifiers. J. Comput. Sci. Coll. **21**(4), 127–133 (2006)
56. Wu, X., Zhang, X., Wang, X.: Low bit-rate image compression via adaptive down-sampling and constrained least squares upconversion. IEEE Trans. Image Process. **18**(3), 552–561 (2009)
57. Yang, J., Wright, J., Huang, T.S., Ma, Y.: Image super-resolution via sparse representation. IEEE Trans. Image Process. **19**(11), 2861–2873 (2010)
58. Yeo, H., Do, S., Han, D.: How will deep learning change internet video delivery? In: Proceedings of the 16th ACM Workshop on Hot Topics in Networks, pp. 57–64. ACM (2017)
59. Yeo, H., Jung, Y., Kim, J., Shin, J., Han, D.: Neural adaptive content-aware internet video delivery. In: 13th USENIX Symposium on Operating Systems Design and Implementation, OSDI 2018, pp. 645–661 (2018)

60. Zeyde, R., Elad, M., Protter, M.: On single image scale-up using sparse-representations. In: Boissonnat, J.-D., et al. (eds.) Curves and Surfaces 2010. LNCS, vol. 6920, pp. 711–730. Springer, Heidelberg (2012). https://doi.org/10.1007/978-3-642-27413-8_47
61. Zhang, Y., Li, K., Li, K., Wang, L., Zhong, B., Fu, Y.: Image super-resolution using very deep residual channel attention networks. In: Ferrari, V., Hebert, M., Sminchisescu, C., Weiss, Y. (eds.) ECCV 2018. LNCS, vol. 11211, pp. 294–310. Springer, Cham (2018). https://doi.org/10.1007/978-3-030-01234-2_18
62. Zhang, Y., Tian, Y., Kong, Y., Zhong, B., Fu, Y.: Residual dense network for image super-resolution. In: Proceedings of the IEEE Conference on Computer Vision and Pattern Recognition, pp. 2472–2481 (2018)
63. Zhong, Z., Shen, T., Yang, Y., Lin, Z., Zhang, C.: Joint sub-bands learning with clique structures for wavelet domain super-resolution. In: Advances in Neural Information Processing Systems, pp. 165–175 (2018)
64. Zhu, X., Li, Z., Zhang, X.Y., Li, C., Liu, Y., Xue, Z.: Residual invertible spatio-temporal network for video super-resolution. Proc. AAAI Conf. Artif. Intell. **33**, 5981–5988 (2019)

Synthesize Then Compare: Detecting Failures and Anomalies for Semantic Segmentation

Yingda Xia, Yi Zhang, Fengze Liu, Wei Shen(✉), and Alan L. Yuille

Johns Hopkins University, Baltimore, USA
shenwei1231@gmail.com

Abstract. The ability to detect failures and anomalies are fundamental requirements for building reliable systems for computer vision applications, especially safety-critical applications of semantic segmentation, such as autonomous driving and medical image analysis. In this paper, we systematically study failure and anomaly detection for semantic segmentation and propose a unified framework, consisting of two modules, to address these two related problems. The first module is an image synthesis module, which generates a synthesized image from a segmentation layout map, and the second is a comparison module, which computes the difference between the synthesized image and the input image. We validate our framework on three challenging datasets and improve the state-of-the-arts by large margins, *i.e.*, 6% AUPR-Error on Cityscapes, 7% Pearson correlation on pancreatic tumor segmentation in MSD and 20% AUPR on StreetHazards anomaly segmentation.

Keywords: Failure detection · Anomaly segmentation · Semantic segmentation

1 Introduction

Deep neural networks [15,19,31,45,47] have achieved great success in various computer vision tasks. However, when they come to real world applications, such as autonomous driving [22], medical diagnoses [5] and nuclear power plant monitoring [36], the safety issue [1] raises tremendous concerns particularly in conditions where failure cases have severe consequences. As a result, it is of enormous value that a machine learning system is capable of detecting the failures, *i.e.*, wrong predictions, as well as identifying the anomalies, *i.e.*, out-of-distribution (OOD) cases, that may cause these failures.

Y. Xia and Y. Zhang—The first two authors equally contributed to the work.

Electronic supplementary material The online version of this chapter (https://doi.org/10.1007/978-3-030-58452-8_9) contains supplementary material, which is available to authorized users.

A. Vedaldi et al. (Eds.): ECCV 2020, LNCS 12346, pp. 145–161, 2020.
https://doi.org/10.1007/978-3-030-58452-8_9

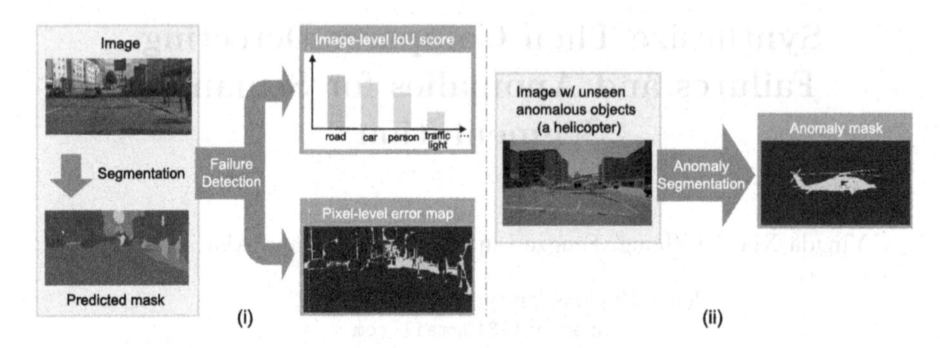

Fig. 1. We aim at addressing two tasks: (i) failure detection, *i.e.*, image-level per-class IoU prediction (top left) and pixel-level error map prediction (bottom left) (ii) anomaly segmentation i.e. segmenting anomalous objects (right middle).

Previous works on failure detection [7,17,24] and anomaly (OOD) detection [9,18,33–35] mainly focus on classifying small images. Although failure detection and anomaly detection for semantic segmentation have received little attention in the literature so far, they are more closely related to safety-critical applications, *e.g.*, autonomous driving and medical image analysis. The objective of failure detection for semantic segmentation is not only to determine whether there are failures in a segmentation result, but also to locate where the failures are. Anomaly detection for semantic segmentation, *a.k.a* anomaly segmentation, is related to failure detection, and its objective is to segment anomalous objects or regions in a given image.

In this paper, our goal is to build a reliable alarm system to address failure detection for semantic segmentation (Fig. 1(i)) and anomaly segmentation (Fig. 1(ii)). Unlike image classification outputs only a single image label, semantic segmentation outputs a structured semantic layout. Thus, this requires that the system should be able to provide more detailed analysis than those for image classification, *i.e.*, pixel-level error/confidence maps. Some previous works [7,16,26] directly applied the failure/anomaly detection strategies for image classification pixel by pixel to estimate a pixel-level error map, but they lack the consideration of the structured semantic layout of a segmentation result.

We propose a unified framework to address failure detection and anomaly detection for semantic segmentation. This framework consists of two components: an image synthesis module, which synthesizes an image from a segmentation result to reconstruct its input image, *i.e.*, a reverse procedure of semantic segmentation, and a comparison module which computes the difference between the reconstructed image and the input image. Our framework is motivated by the fact that the quality of semantic image synthesis [21,43,48] can be evaluated by the performance of segmentation network. Presumably the converse is also true, the better is the segmentation result, the closer a synthesized image generated from the segmentation result is to the input image. If a failure occurs during segmentation, for example, if a person is mis-segmented as a pole, the synthesized image generated from the segmentation result does not look like a person and an obvious difference between the synthesized image and the input image should occur. Similarly, when an anomalous (OOD) object occurs in a test image, it would be classified as any possible in-distribution objects in a segmentation result, and then appear as in-distribution objects in the synthesized image generated from the segmentation result. Consequently, the anomalous object can be identified by finding the differences between the test image and the synthesized image. We refer to our framework as SynthCP, for "synthesize then compare".

We model this synthesis procedure by a semantic-to-image conditional GAN (cGAN) [43], which is capable of modeling the mapping from the segmentation layout space to the image space. This cGAN is trained on label-image pairs. Given the segmentation result of an input image obtained by an semantic segmentation model, we apply the trained cGAN to the segmentation result to generate a reconstructed image. Then, the reconstructed image and the input image are fed into the comparison module to identify the failures/anomalies. The comparison module is designed task-specifically: For failure detection, the comparison module is modeled by a Siamese network, outputting both image-level confidences and pixel-level confidences; For anomaly segmentation, the comparison module is realized by computing the distance defined on the intermediate features extracted by the semantic segmentation model.

We validate SynthCP on the Cityscapes street scene dataset, a pancreatic tumor segmentation dataset in the Medical Segmentation Decathlon (MSD) challenge and the StreetHazards dataset, and show its superiority to other failure detection and anomaly segmentation methods. Specifically, we achieved improvements over the state-of-the-arts by approximately 6% AUPR-Error on Cityscapes pixel-level error prediction, 7% Pearson correlation on pancreatic tumor DSC prediction and 20% AUPR on StreetHazards anomaly segmentation.

We summarize our contribution as follows:

- To the best of our knowledge, we are the first to systematically study failure detection and anomaly detection for semantic segmentation
- We propose a unified framework, SynthCP, which enjoys the benefits of a semantic-to-image conditional GAN, to address both of the two tasks.
- SynthCP achieves state-of-the-art failure detection and anomaly segmentation results on three challenging datasets.

2 Related Work

In this section, we first review the topics closely related to failure detection and anomaly segmentation, such as uncertainty/confidence estimation, quality assessment and out-of-distribution (OOD) detection. Then, we review generative adversarial networks (GANs), which serves a key module in our framework.

Uncertainty estimation or confidence estimation has been a hot topic in the field of machine learning for years, and can be directly applied to the task of **failure detection**. Standard baselines was established in [17] for detecting failures in classification where maximum softmax probability (MSP) provides reasonable results. However, the main drawback of using MSP for confidence estimation is that deep networks tend to produce high confidence predictions [7]. Geifman *et al.* [12] controled the user specified risk-level by setting up thresholds on a pre-defined confidence function (e.g. MSP). Jiang *et al.* [24] measured the agreement between the classifier and a modified nearest-neighbor classifier on the test examples as a confidence score. A recent approach [7] proposed to direct regress "true class probability" which improved over MSP for failure detection. Additionally, Bayesian approaches have drawn attention in this field of study. Dropout based approaches [10,26] used Monte Carlo Dropout (MCDropout) for Bayesian approximation. Computing statistics such as entropy or variance is capable of indicating uncertainty. However, all these approaches mainly focus on small image classification tasks. When applied to semantic segmentation, they lack the information of semantic structures and contexts.

Segmentation quality assessment aims at estimating the overall quality of segmentation, without using ground-truth label, which is suitable to make alarms when model fails. Some approaches [25,32] utilize Bayesian CNNs to predict the segmentation quality of medical images. [29,44] regressed the segmentation quality from deep features computed from a pair of an image and its segmentation result. [20,23] plugged an extra IoU regression head into object detection or instance segmentation. [4,11] used unsupervised learning methods to estimate the segmentation quality using geometrical features. Recently, Liu *et al.* [38] proposed to use VAE [28] to capture a shape prior for segmentation quality assessment on 3D medical image. However, it is hardly applicable to natural images considering the complexity and large shape variance in 2D scenes and objects. Segmentation quality assessment will be referred to as image-level failure detection in the rest of the paper.

OOD detection aims at detecting out-of-distribution examples in testing data. Since the baseline MSP method [17] was brought up, many approaches have improved OOD detection from various aspects [9,18,33–35]. While these approaches mainly focus on image level OOD detection, *i.e.*, to determine whether an image is an OOD example, *e.g.*, [3,30] targeted at detecting hazardous scenes in the Wilddash dataset [49]. On the contrary, we focus on **anomaly segmentation**, *i.e.*, a pixel-level OOD detection task that aims at segmenting anomalous regions from an image. Pixel-wise reconstruction loss [2,14] with auto-encoders(AE) are the main stream approaches for anomaly segmentation. However, they can hardly model the complex street scenes in natural images and AEs can not guarantee to generate an in-distribution image from OOD regions. Recently, it was found that MSP surprisingly outperform AE and Bayesian network based approaches on a newly built larger scale street scene dataset StreetHazards [16] - with 250 types of anomalous objects and more than 6k high resolution images. Lis *et al.* [37] proposed to re-synthesize an image from the predicted semantic map to detect OOD objects in street scenes, which is the **pioneer** work for synthesis-based anomaly detection for semantic segmentation. SynthCP also follows this spirit, but we use a simple yet effective feature dis-

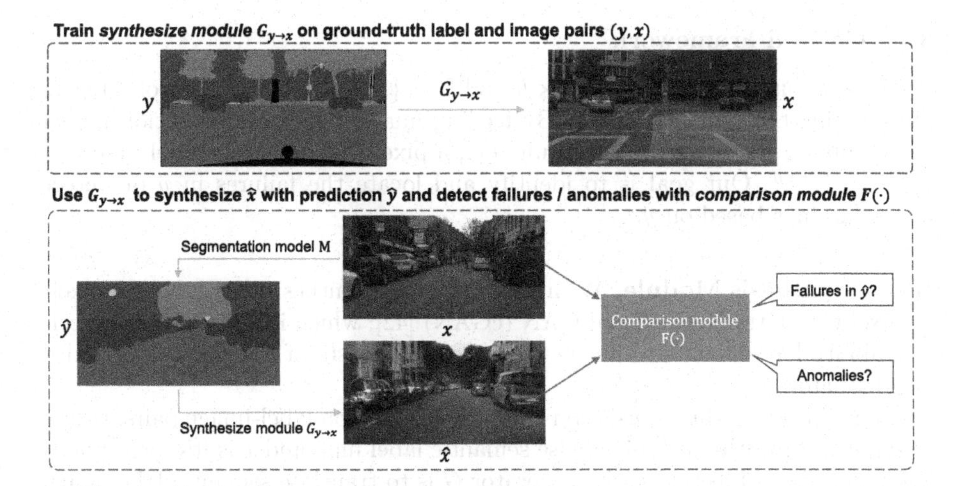

Fig. 2. We first train the synthesis module $G_{y \to x}$ on label-image pairs and then use this module to synthesize the image conditioning on the predicted segmentation mask $\hat{y}$. By comparing x and $\hat{x}$ with a comparison module $F(\cdot)$, we can detect failures as well as segment anomalous objects. $F(\cdot)$ is instantiated in Sect. 3.2 and Sect. 3.3.

tance measure rather than a discrepancy network to find anomalies. In addition, we extend this idea to do a systematic study of failure detection.

Generative adversarial networks [13] generate realistic images by playing a "min-max" game between a generator and a discriminator. GANs effectively minimize a Jensen-Shannon divergence, thus generating in-distribution images. SynthCP utilizes conditional GANs [42] (cGANs) for image translation [21], *a.k.a* pixel-to-pixel translation. Approaches designed for semantic image synthesis [40,43,48] improves pixel-to-pixel translation in synthesizing real images from semantic masks, which is the reverse procedure of semantic segmentation. Since semantic image synthesis is commonly evaluated by the performance of a segmentation model, reversely, we are motivated to use a semantic-to-image generator for failure detection for semantic segmentation.

3 Methodology

In this section, we introduce our framework, SynthCP, for failure detection and anomaly detection for semantic segmentation. SynthCP consists of two modules, an image synthesis module and a comparison module. We first introduce the general framework (shown in Fig. 2), then describe the details of the modules for failure detection and anomaly detection in Sect. 3.2 and Sect. 3.3, respectively. Unless otherwise specified, the notations in this paper follow this criterion: We use a lowercase letter, *e.g.*, x, to represent a tensor variable, such as a 1D array or a 2D map, and denote its i-th element as $x^{(i)}$; We use a capital letter, *e.g.*, F, to represent a function.

3.1 General Framework

Let x be an image with size of $w \times h$ and $\mathbb{L} = \{1, 2, \ldots, L\}$ be a set of integers representing the semantic labels. By feeding image x to a segmentation model M, we obtain its segmentation result, *i.e.*, a pixel-wise semantic label map $\hat{y} = M(x) \in \mathbb{L}^{w \times h}$. Our goal is to identify and locate the failures in $\hat{y}$ or detect anomalies in x based on $\hat{y}$.

Image Synthesis Module. We model this image synthesize module by a pixel-to-pixel translation conditional GAN (cGAN) [42], which is known for its excellent ability for semantic-to-image mapping. It consists of a generator G and a discriminator D.
Training. We train this translation conditional GAN on label-image pairs: (y, x), where y is a grouth-truth pixel-wise semantic label map and x is its corresponding image. The objective of the generator G is to translate semantic label maps to realistic-looking images, while the discriminator D aims to distinguish real images from the synthesized ones. This cGAN minimizes the conditional distribution of real images via the following min-max game:

$$\min_G \max_D \mathcal{L}_{GAN}(G, D), \tag{1}$$

where the objective function $\mathcal{L}_{GAN}(G, D)$ is defined as:

$$\mathbb{E}_{(y,x)}[\log D(y, x)] + \mathbb{E}_y[\log(1 - D(y, G(y))]. \tag{2}$$

Testing. After training, we fix the generator G. Given an image x and a segmentation model M, we feed the predicted segmentation mask $\hat{y} = M(x)$ into G, and obtain a synthesized (*i.e.*, reconstructed) image $\hat{x}$:

$$\hat{x} = G(\hat{y}). \tag{3}$$

$\hat{x}$ and x are then served as the input for the comparison module.

Comparison Module. We detect failures and anomalies in $\hat{y}$ by comparing $\hat{x}$ with x. Our assumption is that, if $\hat{x}$ is more similar to x, then $\hat{y}$ is more similar to y. However, since the optimization of G does not guarantee that the synthesized image $\hat{x}$ has the same style as the original image x, simple similarity measurements such as ℓ_1 distance between x and $\hat{x}$ is not accurate. In order to address this issue, we model the comparison module by a task-specific function F which estimates a trustworthy task-specific confidence measure $\hat{c}$ between x and $\hat{x}$:

$$\hat{c} = F(x, \hat{x}) = F(x, G(\hat{y})). \tag{4}$$

For the task of failure detection, the confidence measure $\hat{c} = (\hat{c}_{iu}, \hat{c}_m)$ includes an image-level per-class intersection over union (IoU) array $\hat{c}_{iu} \in [0, 1]^{|\mathbb{L}|}$ and a pixel-level error map $\hat{c}_m \in [0, 1]^{w \times h}$; For the task of anomaly segmentation, the confidence measure $\hat{c}$ is a pixel-level confidence map $\hat{c}_n \in [0, 1]^{w \times h}$ for anomalous objects.

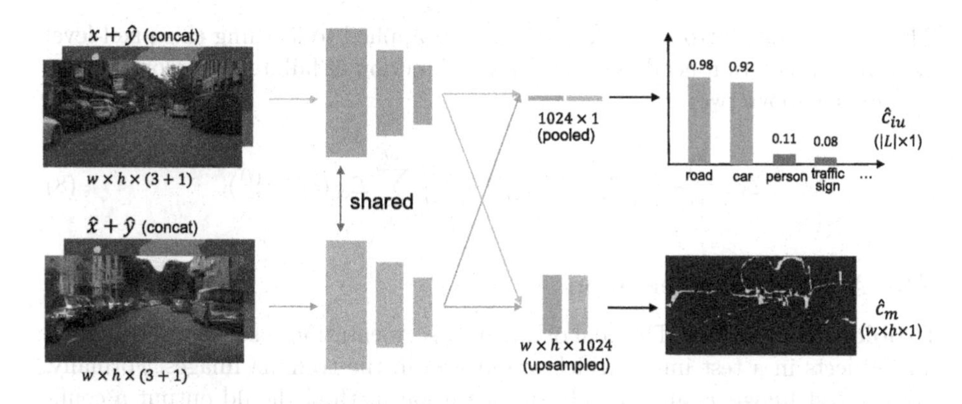

Fig. 3. We instantiate $F(\cdot)$ as a light-weighted siamese network $F(x, \hat{x}, \hat{y}; \theta)$ for joint image-level per-class IoU prediction and pixel-level error map prediction.

3.2 Failure Detection

Problem Definition. Our failure detection contains two tasks: 1) a per-class IoU prediction $\hat{c}_{iu} \in [0,1]^{|\mathbb{L}|}$, which is useful to indicate whether there are failures in the segmentation result $\hat{y}$, and 2) to locate the failures in $\hat{y}$, which needs to compute a pixel-level error map $\hat{c}_m \in [0,1]^{w \times h}$.

Instantiation of Comparison Module. We instantiate the comparison module $F(\cdot)$ as a light-weighted deep network. In practice, we use ResNet-18 [15] as the base network and follow a siamese-style design for learning the relationship between x and $\hat{x}$. As illustrated in Fig. 3, x and $\hat{x}$ are first concatenated with $\hat{y}$ and then separately encoded by a shared-weight siamese encoder. Then two heads are built upon the siamese encoder and output the image-level per-class IoU array $\hat{c}_{iu} \in [0,1]^{|\mathbb{L}|}$ and pixel-level error map $\hat{c}_m \in [0,1]^{w \times h}$, respectively. We rewrite the function F for failure detection as below:

$$\hat{c}_{iu}, \hat{c}_m = F(x, \hat{x}, \hat{y}; \theta) \tag{5}$$

where θ represents the network parameters.

In the training stage, the supervision of network training is obtained by computing the ground-truth confidence measure c from y and $\hat{y}$. For the ground-truth image-level per-class IoU array c_{iu}, we compute it by

$$c_{iu}^{(l)} = \frac{|\{i|\hat{y}^{(i)} = l\} \cap \{i|y^{(i)} = l\}|}{|\{i|\hat{y}^{(i)} = l\} \cup \{i|y^{(i)} = l\}|}, \tag{6}$$

where l is the l-th semantic class in label set $\mathbb{L}$. The ℓ_1 loss function $\mathcal{L}_{\ell_1}(c_{iu}^{(l)}, \hat{c}_{iu}^{(l)})$ is applied to learning this image-level per-class IoU prediction head. For the ground-truth pixel-level error map, we compute it by

$$c_m^{(i)} = \begin{cases} 1 & \text{if } y^{(i)} \neq \hat{y}^{(i)} \\ 0 & \text{if } y^{(i)} = \hat{y}^{(i)} \end{cases}. \tag{7}$$

The binary cross-entropy loss $\mathcal{L}_{ce}(c_m^{(i)}, \hat{c}_m^{(i)})$ is applied to learning this pixel-level error map prediction head. The overall loss function of failure detection $\mathcal{L}$ is the sum of the above two:

$$\mathcal{L} = \frac{1}{|\mathbb{L}|}\sum_{l}^{|\mathbb{L}|} \mathcal{L}_{\ell_1}(c_{iu}^{(l)}, \hat{c}_{iu}^{(l)}) + \frac{1}{wh}\sum_{i}^{wh} \mathcal{L}_{ce}(c_m^{(i)}, \hat{c}_m^{(i)}). \tag{8}$$

3.3 Anomaly Segmentation

Problem Definition. The goal of anomaly segmentation is segmenting anomalous objects in a test image which are unseen in the training images. Formally, given a test image x, an anomaly segmentation method should output a confidence score map $\hat{c}_n \in [0,1]^{w\times h}$ for the regions of the anomalous objects in the image, *i.e.*, $\hat{c}_n^{(i)} = 1$ and $\hat{c}_n^{(i)} = 0$ indicate the ith pixel belongs to an anomalous object and an in-distribution object (the object is seen in the training images), respectively.

Instantiation of Comparison Module. As the same as failure detection, we first train a cGAN generator G on the training images, which maps the in-distribution object labels to realistic images. Given a semantic segmentation model M, we feed its prediction $\hat{y} = M(x)$ into G and obtain $\hat{x} = G(\hat{y})$. Since $\hat{y}$ only contains in-distribution object labels, $\hat{x}$ also only contain in-distribution objects. Thus, we can compare x with $\hat{x}$ to find the anomalies. The pixel-wise semantic difference of x and $\hat{x}$ is a strong indicator of anomalous objects. Here, we simply instantiate the comparison function $F(\cdot)$ as the cosine distance defined on the intermediate features extracted by the segmentation model M:

$$\hat{c}_n^{(i)} = F(x, \hat{x}; M) = 1 - \langle \frac{\mathbf{f}_M^i(x)}{\|\mathbf{f}_M^i(x)\|_2}, \frac{\mathbf{f}_M^i(\hat{x})}{\|\mathbf{f}_M^i(\hat{x})\|_2} \rangle \tag{9}$$

where $\mathbf{f}_M^i$ is the feature vector at the ith pixel position outputted by the last layer of segmentation model M and $\langle\cdot,\cdot\rangle$ is the inner product of the two vectors.

Post-processing with MSP. Due to the artifacts and generalized errors of GANs, our approach may mis-classify an in-distribution object into an anomalous object (false positives). We use a simple post-processing to address this issue. We refine the result by maximum softmax probability (MSP) [17], which is known as an effective uncertainty estimation strategy: $\hat{c}_n^{(i)} \leftarrow \hat{c}_n^{(i)} \cdot \mathbb{1}\{p^{(i)} \le t\} + (1 - p^{(i)}) \cdot \mathbb{1}\{p^{(i)} > t\}$, where $p^{(i)}$ is the maximum soft-max probability at the i-th pixel outputted by the segmentation model M, $t \in [0,1]$ is a threshold and $\mathbb{1}\{\cdot\}$ is the indicator function.

3.4 Conceptual Explanation

We give conceptual explanations of SynthCP in Fig. 4, where $\mathcal{X}$ and $\mathcal{Y}$ correspond to image space and label space. $M_{x\to y}$ is the segmentation model and

$G_{y \to x}$ is a semantic-to-image generator. The left image shows when $M_{x \to y}$ correctly maps an image to its corresponding segmentation mask, the synthesized image generated from $G_{y \to x}$ is close to the original image. However, when $M_{x \to y}$ makes a failure (middle) or encounters an OOD case (right), the synthesized image should be far away from the original image. As a result, the synthesized image serves as a strong indicator for either failure detection or OOD detection.

4 Experiments

4.1 Failure Detection

Evaluation Metrics. Following [38], we evaluate the performance of image-level failure detection, *i.e.*, per-class IoU prediction, by four metrics: **MAE**, **STD**, **P.C** and **S.C**. MAE (mean absolute error) and its STD measure the average error between predicted IoUs and ground-truth IoUs. P.C (Pearson correlation) and S.C.(Spearman correlation) measures their correlation coefficients. For pixel-level failure detection, *i.e.*, pixel-level error map prediction, we use the metrics in literature [7,17]: **AUPR-Error**, **AUPR-Success**, **FPR at 95% TPR** and

AUROC. Following [7], AUPR-Error is our main metric, which computes the area under the Precision-Recall curve using errors as the positive class.

The Cityscapes Dataset. We validate SynthCP on the Cityscapes dataset [8], which contains 2975 high-resolution training images and 500 validation images. As far as we know, it is the largest one for failure detection for semantic segmentation.

Baselines. We compare SynthCP to MCDropout [10], VAE alarm [38], MSP [17], TCP [7] and "Direct Prediction". MCDropout, MSP and TCP output pixel-level confidence maps, serving as standard baselines for pixel-level failure prediction.

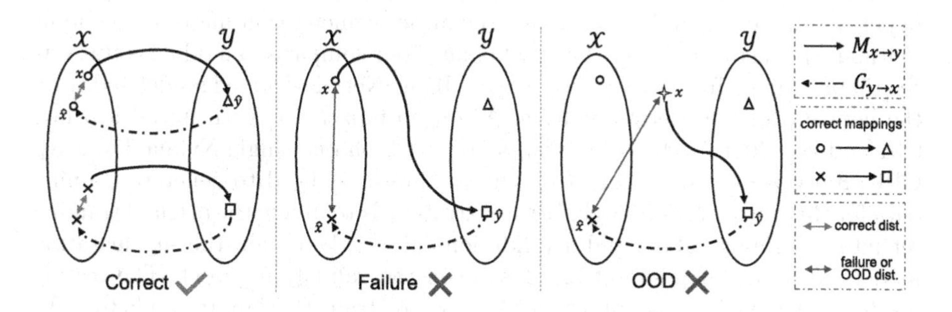

Fig. 4. An analysis of SynthCP. Left: $M_{x \to y}$ correctly maps x to $\hat{y}$, resulting in small distance between x and the synthesized $\hat{x}$. However, when there are failures in $\hat{y}$ (middle) or there are OOD examples in x (right), the distance between x and $\hat{x}$ is larger, given a reliable reverse mapping $G_{y \to x}$.

Table 1. Experiments on the Cityscapes dataset. We detect failures in the segmentation results of FCN-8 and Deeplab-v2. "SynthCP-separate" and "SynthCP-joint" mean training the image-level and pixel-level failure detection heads in our network separately and jointly, respectively.

	FCN-8				Deeplab-v2			
image-level	MAE↓	STD↓	P.C.↑	S.C.↑	MAE↓	STD↓	P.C.↑	S.C.↑
MCDropout [10]	17.28	13.33	3.62	5.97	19.31	12.86	4.55	1.37
VAE alarm [38]	16.28	11.88	21.82	18.26	16.78	12.21	17.92	19.63
Direct prediction	13.25	11.96	58.34	59.74	14.45	12.20	60.94	62.01
SynthCP-separate	**11.58**	11.50	**64.63**	**65.63**	**13.60**	12.32	62.51	63.41
SynthCP-joint	12.69	**11.29**	62.52	61.23	13.68	**11.60**	**64.05**	**65.42**
	FCN-8				Deeplab-v2			
pixel-level	AP-Err↑	AP-Suc↑	AUC↑	FPR95↓	AP-Err↑	AP-Suc↑	AUC↑	FPR95↓
MSP [17]	50.31	99.02	91.54	25.34	48.46	99.24	92.26	24.41
MCDropout [10]	49.23	99.02	91.47	25.16	47.85	99.23	92.19	24.68
TCP [7]	48.54	98.82	90.29	32.21	45.57	98.84	89.14	36.98
Direct prediction	52.17	99.15	92.55	**22.34**	48.76	99.34	92.94	**21.56**
SynthCP-separate	54.14	99.15	92.70	22.47	48.79	99.31	92.74	22.15
SynthCP-joint	**55.53**	**99.18**	**92.92**	22.47	**49.99**	**99.34**	**92.98**	21.69

VAE alarm [38] is the state-of-the-art in image-level failure prediction method. Following [38], we also use MCDropout to predict image-level failures. Direct Prediction is a method that directly uses a network to predict both image-level and pixel-level failures, by taking an image and its segmentation result as input. Note that, Direct Prediction shares the same experimental settings (backbone and training strategies) with SynthCP, which can be seen as an ablation study on the effectiveness of the synthesized image $\hat{x}$.

Implementation details. We use the state-of-the-art semantic-to-image cGAN - SPADE [43] in SynthCP. We re-trained SPADE from scratch following the same hyper-parameters as in [43] with only semantic segmentation maps as the input (without the instance maps). The backbone of our comparison module is ResNet-18 [15] pretrained from Image-Net. We use ImageNet pre-trained model and train the network for 20k iterations using Adam optimizer [27] with initial learning 0.01 and $\beta = (0.9, 0.999)$, which takes about 6 h on one single Nvidia Titan Xp GPU. Since we use a network to predict failures, we need to generate training data for this network. A straightforward strategy is to divide the original training set into a training subset and a validation subset, then train the segmentation model on the training subset and test it on the validation subset. The testing results on the validation subset can be used to train the failure predictor. We extend this strategy by doing 4-fold cross validation on the training set. Since the cross-validated results cover all samples in the original training set, we are able to generate sufficient training data to train our failure prediction network.

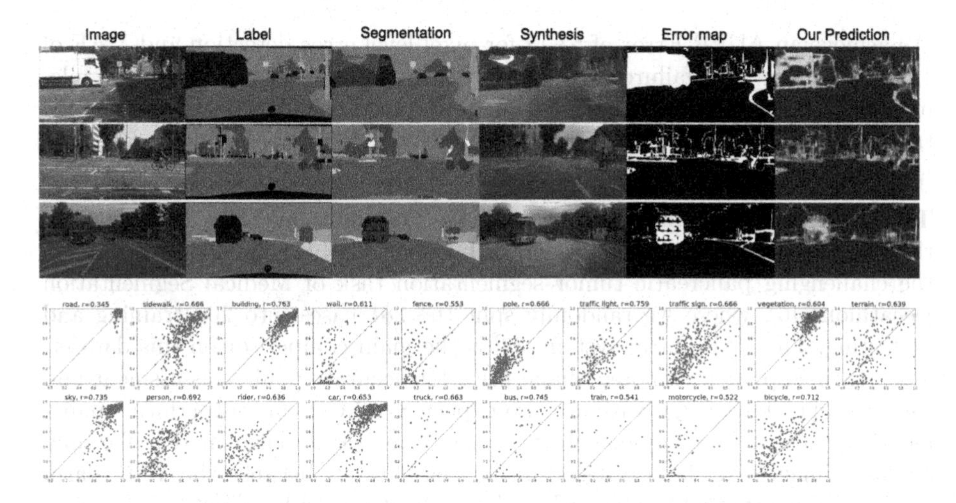

Fig. 5. Visualization on the Cityscapes dataset for pixel-level error map prediction (top) and image-level per-class IoU prediction (bottom). For each example from left to right (top), we show the original image, ground-truth label map, segmentation prediction, synthesized image conditioned on the segmentation prediction, (ground-truth) errors in the segmentation prediction and our pixel-level error prediction. The plots (bottom) show significant correlations between the ground-truth IoU and our predicted IoU on most of the classes.

Results. Experimental results are shown in Table 1 and visualizations are shown in Fig. 5. We use the well-known FCN8 [41] and Deeplab-v2 [6] as the segmentation models. For image-level failure detection, our approach consistently outperforms other methods on all metrics. Results are averaged over 19 classes for all four metrics (detailed results in supplementary). We find that VAE alarm does not perform well 2D images of street scenes, since small objects are easily missed in the VAE reconstruction. Without the synthesized images from the segmentation results, Direct Prediction performs worse than ours despite achieving better performance than the others.

For pixel-level failure detection, our approach achieves the state-of-the-art performance as well, especially for AP-Error metric where our approach outperforms other methods by a considerable margin. The comparison to Direct Prediction demonstrates that the improvements come from the image synthesis module in our framework. We hypothesis that TCP performs not as well because it is mainly designed for classification and it might be hard to fit the true class probability for dense predictions on large images in our settings. We find that our method produces slightly more false positives than "Direct Prediction" baseline (FPR95 is lower). We think the reason might be some correctly segmented regions are not synthesized well by the generative model.

We conducted another experiment to validate the **generalizability** on unseen segmentation models. We directly test our failure detection model, which is trained on Deeplab-v2 masks, on the segmentation masks produced by FCN8.

We achieve an AUPR-Error of 53.12 for pixel-level error detection and MAE of 12.91 for image-level failure prediction. Full results are available in the supplementary material. The results are comparable to those obtained by our model trained on FCN8 segmentation model, as shown in Table 1.

The Pancreatic Tumor Segmentation Dataset. We also validate SynthCP on medical images. Following VAE alarm [38], we applied SynthCP to the challenging pancreatic tumor segmentation task of Medical Segmentation Decathlon [46], where we randomly split the 281 cases into 200 training and 81 testing. The VAE alarm system [38] is the main competitor on this dataset. Since their approach explored shape prior for accurate quality assessment and tumor shapes have large variance, we expect SynthCP can outperform shape-based models or be complementary to the VAE-based alarm model. We only compare image-level failure detection in this dataset, because the VAE alarm system [38] is targeted to this task and sets up standard baselines.

We use the state-of-the-art network 3D AH-Net [39] as the segmentation model. Instead of IoU, the segmentation performance is measured by Dice coefficient (DSC), a standard evaluation metric used for medical image segmentation. Moving into 3D is challenging for SynthCP, since training 3D GANs is extremely hard, considering the limited GPU memory and high computational costs. In practice, we modify SPADE into 3D. Results and visualizations are shown in Table 2 and Fig. 6 respectively. In terms of baselines, we re-implement the VAE alarm system for a fair comparison in our settings, while the results of other methods are quoted from [38]. SynthCP achieved comparable performances as VAE alarm system. When combined with VAE alarm (a simple ensemble of the predicted DSC), all of the four metric improves significantly (P.C. and S.C correlation coefficient both improves by approximately 7% and 10% respectively), illustrating SynthCP which captures label-to-image information is complementary to the shape-based VAE approach.

Table 2. Failure detection results on the pancreatic tumor segmentation dataset in MSD [46]

	Tumor DSC prediction			
Method	MAE ↓	STD ↓	P.C. ↑	S.C. ↑
Direct prediction	23.20	29.81	45.50	45.36
Jungo *et al.* [25]	26.57	29.78	–23.87	–20.23
Kwon *et al.* [32]	26.14	29.24	14.61	14.70
VAE alarm [38]	20.21	23.60	60.24	63.30
VAE (our imple.)	18.60	13.73	63.42	58.47
SynthCP	18.13	13.77	61.11	62.66
SynthCP + VAE	**15.19**	**13.37**	**67.97**	**71.35**

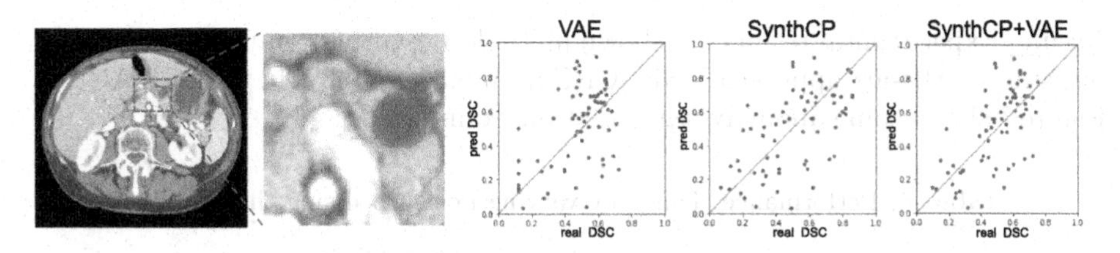

Fig. 6. Left two: an example of pancreatic tumor segmentation (in red). Right three: plots for tumor segmentation DSC score prediction by VAE alarm [38], SynthCP and the ensemble of SynthCP and VAE alarm. (Color figure online)

4.2 Anomaly Segmentation

Evaluation Metrics. We use the standard metrics for OOD detection and anomaly segmentation: area under the ROC curve (AUROC), false positive rate at 95% recall (FPR95), and area under the precision recall curve (AUPR).

The StreetHazards Dataset. We validate SynthCP on the StreetHazards dataset of CAOS Benchmark [16]. This dataset contains 5125 training images, 1000 validation images and 1500 test images. 250 types of anomaly objects appears only in the testing images.

Baselines. Baseline approaches include MSP [17], MSP+CRF [16], Dropout [10] and an auto-encoder (AE) based approach [2]. Except for AE, all the other three approaches require a segmentation model to provide either softmax probability or uncertainty estimation. AE is the only approach that requires extra training of an auto-encoder for the images and computes pixel-wise ℓ_1 loss for anomaly segmentation.

Implementation details. Following [16], we use two network backbones as the segmentation models: ResNet-101 [15] and PSPNet [50]. The cGAN is also SPADE [43] trained with the same training strategy as in Sect. 3.2. The post-processing threshold $t = 0.999$ is chosen for better AUPR and is discussed in detail in the following paragraph.

Table 3. Anomaly segmentation results on StreetHazards dataset [16]

Method	FPR95↓	AUROC↑	AUPR↑
AE [2]	91.7	66.1	2.2
Dropout [10]	79.4	69.9	7.5
MSP [17]	33.7	87.7	6.6
MSP + CRF [16]	29.9	88.1	6.5
SynthCP	**28.4**	**88.5**	**9.3**

Results Experimental results are shown in Table 3. SynthCP improves the previous state-of-the-art approach MSP+CRF from 6.5% to 9.3% in terms of AUPR. Figure 7 shows some anomaly segmentation examples.

Table 4. Performance change by varying post-processing threshold t

t	0.8	0.9	0.99	0.999	1.0
FPR95 ↓	**28.6**	28.5	28.2	28.4	46.0
AUROC ↑	88.3	88.4	**88.6**	88.5	81.9
AUPR ↑	7.4	7.7	8.8	**9.3**	8.1

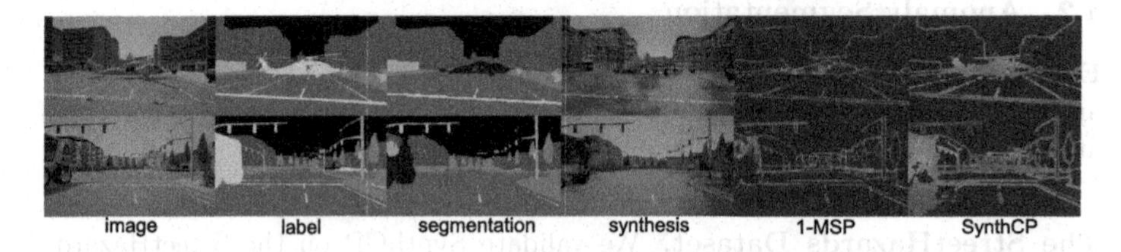

Fig. 7. Visualizations on the StreetHazards dataset. For each example, from left to right, we show the original image, ground-truth label map, segmentation prediction, synthesized image conditioned on segmentation prediction, MSP anomaly segmentation prediction and our anomaly segmentation prediction.

To study how much MSP post-processing contributes to SynthCP, we conduct experiments on different thresholds of t for post-processing. As shown in Table 4, without post-processing ($t = 1.0$), SynthCP achieves higher AUPR, but also produces more false positives, resulting in degrading FPR95 and AUROC. After pruning out false positives at high MSP positions ($p^{(i)} > 0.999$), we achieved the state-of-the-art performances under all three metrics.

5 Conclusions

We present a unified framework, SynthCP, to detect failures and anomalies for semantic segmentation, which consists of an image synthesize module and a comparison module. We model the image synthesize module with a semantic-to-image conditional GAN (cGAN) and train it on label-image pairs. We then use it to reconstruct the image based on the predicted segmentation mask. The synthesized image and the original image are fed forward to the comparison module and output either failure detection (both image-level and pixel-level) or the mask of anomalous objects, depending on the specific task. SynthCP achieved the state-of-the-art performances on three challenging datasets.

Acknowledgement. This work was supported by NSF BCS-1827427, the Lustgarten Foundation for Pancreatic Cancer Research and NSFC No. 61672336. We also thank the constructive suggestions from Dr. Chenxi Liu, Qing Liu and Huiyu Wang.

References

1. Amodei, D., et al.: Concrete problems in ai safety. arXiv preprint arXiv:1606.06565 (2016)
2. Baur, C., Wiestler, B., Albarqouni, S., Navab, N.: Deep autoencoding models for unsupervised anomaly segmentation in brain MR images. In: MICCAI Brainlesion Workshop (2018)
3. Bevandić, P., Krešo, I., Oršić, M., Šegvić, S.: Discriminative out-of-distribution detection for semantic segmentation. arXiv preprint arXiv:1808.07703 (2018)
4. Chabrier, S., Emile, B., Rosenberger, C., Laurent, H.: Unsupervised performance evaluation of image segmentation. EURASIP J. Adv. Signal Process. **2006**, 096306 (2006)
5. Challen, R., Denny, J., Pitt, M., Gompels, L., Edwards, T., Tsaneva-Atanasova, K.: Artificial intelligence, bias and clinical safety. BMJ Qual. Saf. **28**(3), 231–237 (2019)
6. Chen, L.C., Papandreou, G., Kokkinos, I., Murphy, K., Yuille, A.L.: DeepLab: semantic image segmentation with deep convolutional nets, atrous convolution, and fully connected crfs. IEEE Trans. Pattern Anal. Mach. Intell. (TPAMI) **40**(4), 834–848 (2018)
7. Corbière, C., Thome, N., Bar-Hen, A., Cord, M., Pérez, P.: Addressing failure prediction by learning model confidence. In: Advances in Neural Information Processing Systems (2019)
8. Cordts, M., et al.: The cityscapes dataset for semantic urban scene understanding. In: Proceedings of the IEEE Conference on Computer Vision and Pattern Recognition, CVPR (2016)
9. DeVries, T., Taylor, G.W.: Learning confidence for out-of-distribution detection in neural networks. arXiv preprint arXiv:1802.04865 (2018)
10. Gal, Y., Ghahramani, Z.: Dropout as a bayesian approximation: representing model uncertainty in deep learning. In: International Conference on Machine Learning, ICML (2016)
11. Gao, H., Tang, Y., Jing, L., Li, H., Ding, H.: A novel unsupervised segmentation quality evaluation method for remote sensing images. Sensors **17**(10), 2427 (2017)
12. Geifman, Y., El-Yaniv, R.: Selective classification for deep neural networks. In: Advances in Neural Information Processing Systems (2017)
13. Goodfellow, I., et al.: Generative adversarial nets. In: Advances in Neural Information Processing Systems (2014)
14. Haselmann, M., Gruber, D.P., Tabatabai, P.: Anomaly detection using deep learning based image completion. In: International Conference on Machine Learning and Applications, ICMLA. IEEE (2018)
15. He, K., Zhang, X., Ren, S., Sun, J.: Deep residual learning for image recognition. In: Proceedings of the IEEE Conference on Computer Vision and Pattern Recognition, CVPR (2016)
16. Hendrycks, D., Basart, S., Mazeika, M., Mostajabi, M., Steinhardt, J., Song, D.: A benchmark for anomaly segmentation. arXiv preprint arXiv:1911.11132 (2019)
17. Hendrycks, D., Gimpel, K.: A baseline for detecting misclassified and out-of-distribution examples in neural networks. In: International Conference on Learning Representations, ICLR (2017)
18. Hendrycks, D., Mazeika, M., Dietterich, T.: Deep anomaly detection with outlier exposure. In: International Conference on Learning Representations, ICLR (2019)

19. Huang, G., Liu, Z., Van Der Maaten, L., Weinberger, K.Q.: Densely connected convolutional networks. In: Proceedings of the IEEE Conference on Computer Vision and Pattern Recognition, CVPR (2017)
20. Huang, Z., Huang, L., Gong, Y., Huang, C., Wang, X.: Mask scoring R-CNN. In: Proceedings of the IEEE Conference on Computer Vision and Pattern Recognition, CVPR (2019)
21. Isola, P., Zhu, J.Y., Zhou, T., Efros, A.A.: Image-to-image translation with conditional adversarial networks. In: Proceedings of the IEEE Conference on Computer Vision and Pattern Recognition, CVPR (2017)
22. Janai, J., Güney, F., Behl, A., Geiger, A.: Computer vision for autonomous vehicles: problems, datasets and state-of-the-art. arXiv preprint arXiv:1704.05519 (2017)
23. Jiang, B., Luo, R., Mao, J., Xiao, T., Jiang, Y.: Acquisition of localization confidence for accurate object detection. In: Ferrari, V., Hebert, M., Sminchisescu, C., Weiss, Y. (eds.) Computer Vision – ECCV 2018. Lecture Notes in Computer Science, vol. 11218, pp. 816–832. Springer, Cham (2018). https://doi.org/10.1007/978-3-030-01264-9_48
24. Jiang, H., Kim, B., Guan, M., Gupta, M.: To trust or not to trust a classifier. In: Advances in Neural Information Processing Systems (2018)
25. Jungo, A., Meier, R., Ermis, E., Herrmann, E., Reyes, M.: Uncertainty-driven sanity check: Application to postoperative brain tumor cavity segmentation. arXiv preprint arXiv:1806.03106 (2018)
26. Kendall, A., Gal, Y.: What uncertainties do we need in Bayesian deep learning for computer vision? In: Advances in Neural Information Processing Systems (2017)
27. Kingma, D.P., Ba, J.: Adam: a method for stochastic optimization. In: International Conference on Learning Representations, ICLR (2015)
28. Kingma, D.P., Welling, M.: Auto-encoding variational bayes. In: International Conference on Learning Representations, ICLR (2014)
29. Kohlberger, T., Singh, V., Alvino, C., Bahlmann, C., Grady, L.: Evaluating segmentation error without ground truth. In: International Conference on Medical Image Computing and Computer-Assisted Intervention, MICCAI (2012)
30. Krešo, I., Oršić, M., Bevandić, P., Šegvić, S.: Robust semantic segmentation with ladder-densenet models. arXiv preprint arXiv:1806.03465 (2018)
31. Krizhevsky, A., Sutskever, I., Hinton, G.E.: Imagenet classification with deep convolutional neural networks. In: Advances in Neural Information Processing Systems (2012)
32. Kwon, Y., Won, J.H., Kim, B.J., Paik, M.C.: Uncertainty quantification using Bayesian neural networks in classification: application to biomedical image segmentation. Comput. Stat. Data Anal. **142**, 106816 (2020)
33. Lee, K., Lee, H., Lee, K., Shin, J.: Training confidence-calibrated classifiers for detecting out-of-distribution samples. In: International Conference on Learning Representations, ICLR (2018)
34. Lee, K., Lee, K., Lee, H., Shin, J.: A simple unified framework for detecting out-of-distribution samples and adversarial attacks. In: Advances in Neural Information Processing Systems (2018)
35. Liang, S., Li, Y., Srikant, R.: Enhancing the reliability of out-of-distribution image detection in neural networks. In: International Conference on Learning Representations, ICLR (2018)
36. Linda, O., Vollmer, T., Manic, M.: Neural network based intrusion detection system for critical infrastructures. In: International Joint Conference on Neural Networks, IJCNN (2009)

37. Lis, K., Nakka, K., Fua, P., Salzmann, M.: Detecting the unexpected via image Resynthesis. In: Proceedings of the IEEE International Conference on Computer Vision, pp. 2152–2161 (2019)
38. Liu, F., Xia, Y., Yang, D., Yuille, A.L., Xu, D.: An alarm system for segmentation algorithm based on shape model. In: Proceedings of the IEEE International Conference on Computer Vision, ICCV (2019)
39. Liu, S., et al.: 3D anisotropic hybrid network: transferring convolutional features from 2D images to 3D anisotropic volumes. In: International Conference on Medical Image Computing and Computer-Assisted Intervention, MICCAI (2018)
40. Liu, X., et al.: Learning to predict layout-to-image conditional convolutions for semantic image synthesis. In: Advances in Neural Information Processing Systems (2019)
41. Long, J., Shelhamer, E., Darrell, T.: Fully convolutional networks for semantic segmentation. In: Proceedings of the IEEE Conference on Computer Vision and Pattern Recognition, CVPR (2015)
42. Mirza, M., Osindero, S.: Conditional generative adversarial nets. arXiv preprint arXiv:1411.1784 (2014)
43. Park, T., Liu, M.Y., Wang, T.C., Zhu, J.Y.: Semantic image synthesis with spatially-adaptive normalization. In: Proceedings of the IEEE Conference on Computer Vision and Pattern Recognition, CVPR (2019)
44. Robinson, R., et al.: Real-time prediction of segmentation quality. In: International Conference on Medical Image Computing and Computer-Assisted Intervention, MICCAI (2018)
45. Simonyan, K., Zisserman, A.: Very deep convolutional networks for large-scale image recognition. In: International Conference on Learning Representations, ICLR (2014)
46. Simpson, A.L., et al.: A large annotated medical image dataset for the development and evaluation of segmentation algorithms. arXiv preprint arXiv:1902.09063 (2019)
47. Szegedy, C., et al.: Going deeper with convolutions. In: Proceedings of the IEEE Conference on Computer Vision and Pattern Recognition, CVPR (2015)
48. Wang, T.C., et al.: High-resolution image synthesis and semantic manipulation with conditional GANs. In: Proceedings of the IEEE Conference on Computer Vision and Pattern Recognition, CVPR (2018)
49. Zendel, O., Honauer, K., Murschitz, M., Steininger, D., Dominguez, G.F.: WildDash-creating hazard-aware benchmarks. In: Ferrari, V., Hebert, M., Sminchisescu, C., Weiss, Y. (eds.) Computer Vision – ECCV 2018. Lecture Notes in Computer Science, vol. 11210, pp. 407–421. Springer, Cham (2018). https://doi.org/10.1007/978-3-030-01231-1_25
50. Zhao, H., Shi, J., Qi, X., Wang, X., Jia, J.: Pyramid scene parsing network. In: Proceedings of the IEEE Conference on Computer Vision and Pattern Recognition, CVPR (2017)

House-GAN: Relational Generative Adversarial Networks for Graph-Constrained House Layout Generation

Nelson Nauata[1(✉)], Kai-Hung Chang[2(✉)], Chin-Yi Cheng[2(✉)], Greg Mori[1(✉)], and Yasutaka Furukawa[1(✉)]

[1] Simon Fraser University, Burnaby, Canada
{nnauata,furukawa}@sfu.ca, mori@cs.sfu.ca
[2] Autodesk Research, San Francisco, USA
{kai-hung.chang,chin-yi.cheng}@autodesk.com

Abstract. This paper proposes a novel graph-constrained generative adversarial network, whose generator and discriminator are built upon relational architecture. The main idea is to encode the constraint into the graph structure of its relational networks. We have demonstrated the proposed architecture for a new house layout generation problem, whose task is to take an architectural constraint as a graph (i.e., the number and types of rooms with their spatial adjacency) and produce a set of axis-aligned bounding boxes of rooms. We measure the quality of generated house layouts with the three metrics: the realism, the diversity, and the compatibility with the input graph constraint. Our qualitative and quantitative evaluations over 117,000 real floorplan images demonstrate that the proposed approach outperforms existing methods and baselines. We will publicly share all our code and data.

Keywords: GAN · Graph-constrained · Layout · Generation · Floorplan

1 Introduction

A house is the most important purchase one might make in life, and we all want to live in a safe, comfortable, and beautiful environment. However, designing a house that fulfills all the functional requirements with a reasonable budget is challenging. Only a small fraction of the residential building owners have enough budget to employ architects for customized house design.

House design is a time-consuming iterative process. A standard workflow is to 1) sketch a "bubble diagram" illustrating the number of rooms, their types, and

---

Electronic supplementary material The online version of this chapter (https://doi.org/10.1007/978-3-030-58452-8_10) contains supplementary material, which is available to authorized users.

A. Vedaldi et al. (Eds.): ECCV 2020, LNCS 12346, pp. 162–177, 2020.
https://doi.org/10.1007/978-3-030-58452-8_10

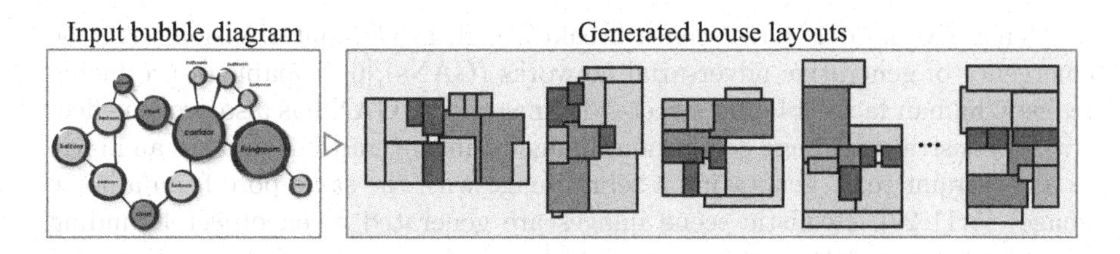

Fig. 1. House-GAN is a novel graph-constrained house layout generator, built upon a relational generative adversarial network. The bubble diagram (graph) is given as an input for automatically generating multiple house layout options.

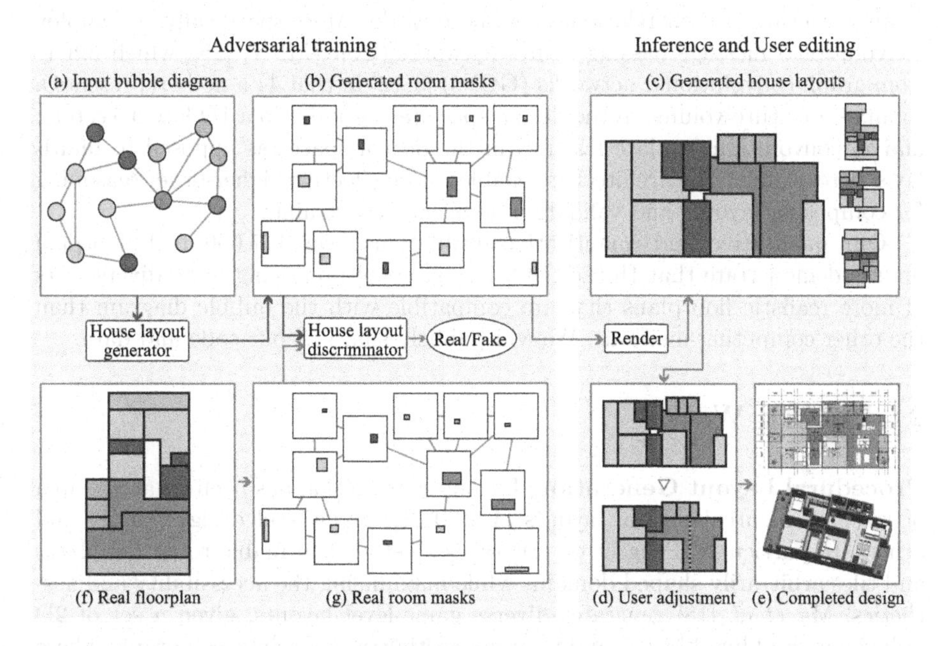

Fig. 2. Floorplan designing workflow with House-GAN. The input to the system is a bubble diagram encoding high-level architectural constraints. House-GAN learns to generate a diverse set of realistic house layouts under the bubble diagram constraint. Architects convert a layout into a real floorplan.

connections; 2) produce corresponding floorplans and collect clients feedback; 3) revert to the bubble diagram for refinement, and 4) iterate. Given limited budget, architects and their clients often compromise on the design quality. Automated floorplan generation techniques are in critical demand with immense potentials in the architecture, construction, and real-estate industries (See Fig. 2).

This paper proposes a novel house layout generation problem, whose task is to take a bubble diagram and generate a diverse set of realistic and compatible house layouts (See Fig. 1). A bubble diagram is represented as a graph where 1) nodes encode rooms with room types and 2) edges encode their spatial adjacency. A house layout is represented as a set of axis-aligned bounding boxes of rooms.

Generative models have seen a breakthrough in Computer Vision with the emergence of generative adversarial networks (GANs) [6], capable of producing realistic human faces [14] and street-side images [15]. GAN has also proven effective for constrained image generation. Image-to-image translation takes an image as a constraint (e.g., generating a zebra image with the same pose from a horse image) [5,11,27]. Realistic scene images are generated given object bounding boxes and placements as the constraint [3].

The house layout generation poses a new challenge: The graph is enforced as a constraint. We present a novel generative model called House-GAN that employs relational generator and discriminator, where the constraint is encoded into the graph structure of their relational neural networks. More specifically, we employ convolutional message passing neural networks (Conv-MPN) [26], which differs from graph convolutional networks (GCNs) [3,12] in that 1) a node represents a room as a feature volume in the design space (as opposed to a 1D latent vector), and 2) convolutions update features in the design space (as opposed to multi-layer perceptron). The architecture enables more effective higher-order reasoning for composing layouts and validating adjacency constraints.

Our qualitative and quantitative evaluations over 117,000 real floorplan images demonstrate that House-GAN is capable of producing more diverse sets of more realistic floorplans that are compatible with the bubble diagram than the other competing methods. We will publicly share all our code and data.

2 Related Work

Procedural Layout Generation: Layout composition has been an active area of research in architectural layouts [4,8,20,21], game-level design [9,18] and others. In particular, Peng *et al.* [21] takes a set of deformable room templates and tiles arbitrarily shaped domains while maximizing the accessibility and aesthetics. Ma *et al.* [18] generates diverse game-level layouts, given a set of 2D polygonal "building blocks" and their connectivity constraints as a graph. These methods are more traditional with hand-crafted energy minimization. Our approach exploits powerful data-driven techniques for robustness.

Data-Driven Space Planning: Data-driven sequential generative methods have been proposed for indoor scene synthesis by Wuang *et al.* [24] and Ritchie *et al.* [22], indoor plan generation by Wu *et al.* [25], and outdoor scene generation by Jyothi *et al.* [13]. In particular, Wu *et al.* [25] proposes a data-driven method for automatic floorplan generation for residential houses from a building footprint. The method starts from the living-room and sequentially adds rooms via an encoder-decoder network, followed by a final post-processing for the vectorization. Jyothi *et al.* [13] proposes a variational autoenconder (VAE), which iteratively predicts a diverse yet plausible counts and sets of bounding boxes, given a set of object labels as input. Li *et al.* [16] proposes a non-sequential adversarial generative method called LayoutGAN, which has a self-attention mechanism in the generator and a wireframe renderer in the discriminator. These methods produce impressive results but cannot take a graph as an input constraint.

Graph-Constrained Layout Generation: Graph-constrained layout generation has also been a focus of research. Wang *et al.* [23] plans an indoor scene as a relation graph and inserts a 3D model at each node via convolutional neural network (CNN) guided search. Merrel *et al.* [19] utilizes Bayesian Networks for retrieving candidate bubble diagrams, given high-level conditions such as the number of rooms, room types, and approximate square footage. These bubble diagrams are later converted to floorplans using the Metropolis algorithm. Jonhson *et al.* [12] and Ashual *et al.* [3] aim to generate image layouts and synthesize realistic images from input scene-graphs via GCNs. Our innovation is a relational generative adversarial network, where the input constraint is encoded into the graph structure of the relational generator and discriminator. The qualitative and quantitative evaluations demonstrate the effectiveness of our approach.

3 Graph-Constrained House Layout Generation Problem

We seek to generate a diverse set of realistic house layouts, compatible with a bubble diagram. The section explains our dataset, metrics, and limitations.

Dataset: LIFULL HOME's database offers five million real floorplans, from which we retrieved 117,587 [1] and rescaled uniformly to fit inside the 256×256 resolution (See Table 1). The database does not contain bubble diagrams. We used the floorplan vectorization algorithm [17] to generate the vector-graphics format, which is converted into bubble diagrams. A bubble diagram is a graph, where a node is a room with a room type as its property[1]. Two rooms are connected if the Manhattan distance between the bounding boxes is less than 8 pixels. An output house layout is axis-aligned bounding boxes (See Fig. 3).

Metrics: We divide the samples into five groups based on the number of rooms: (1–3, 4–6, 7–9, 10–12, and 13+). To test the generalization capability, we conduct k-fold validation (k = 5): When generating layouts in a group, we train a model while excluding samples in the same group so that a method cannot simply memorize. At test time, we randomly pick a house layout and generate X samples. $X = 10$ for measuring the realism and diversity, and $X = 1$ for measuring the compatibility whose evaluation is computationally expensive.

- The realism is measured by an average user rating. We present a generated house layout against a ground-truth or another method. A subject puts one of the four ratings: better (+1), worse (−1), equally-good (+1), or equally-bad (−1).
- The diversity is measured by the FID score [10] with the rasterized layout images. We rasterize a layout by 1) Setting the background to white; 2) Sorting the rooms in the decreasing order of the areas; and 3) Painting each room with a color based on its room type (e.g., orange for a bedroom) as shown in Fig. 3.

[1] Room types are "living room", "kitchen", "bedroom", "bathroom", "closet", "balcony", "corridor", "dining room", "laundry room", or "unkown".

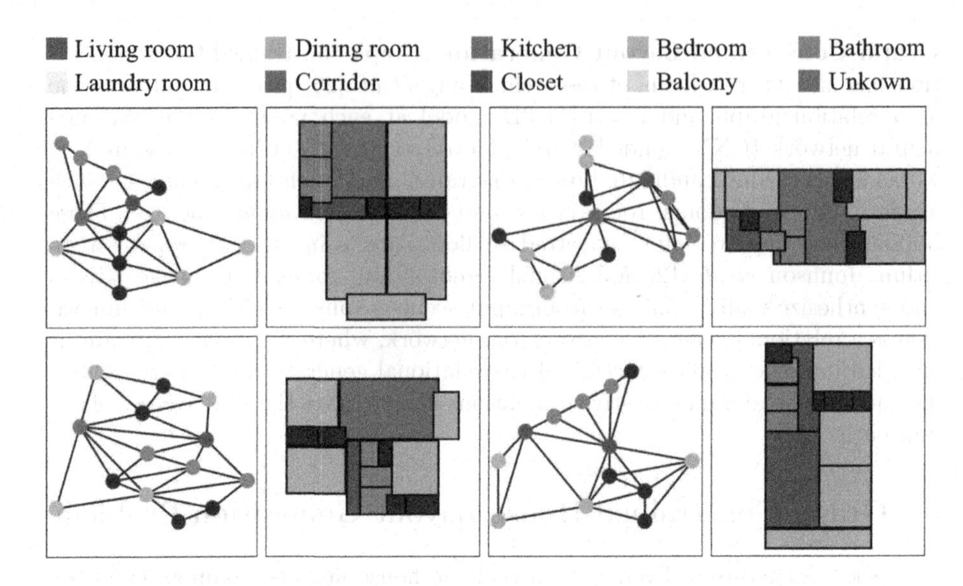

Fig. 3. Sample bubble diagrams and house layouts in our dataset.

Table 1. We divide the samples into five groups based on the room counts (1–3, 4–6, 7–9, 10–12, and 13+). The second row shows the numbers of samples. The remaining rows show the average number of rooms (left) and the average number of edge connections per room (right). The room counts are small for living-rooms, which are often interchangeable with dining-rooms and kitchens in the Japanese real-estate market.

	1–3	4–6	7–9	10–12	13+	All
(# of samples)	7,393	28,170	42,635	30,625	8,764	117,587
Living Room	0.0/1.2	0.0/3.1	0.1/4.8	0.3/5.6	0.3/5.9	0.1/5.1
Kitchen	0.6/1.3	1.0/3.3	1.2/4.5	1.1/5.4	1.3/5.3	1.1/4.4
Bedroom	0.4/1.3	0.8/2.8	1.3/3.5	2.0/3.9	2.8/4.1	1.4/3.6
Bathroom	0.7/1.2	1.6/2.4	2.6/2.9	3.0/3.3	3.4/3.5	2.4/3.0
Closet	0.3/1.2	1.0/2.2	1.6/2.6	2.4/3.1	3.6/3.2	1.7/2.8
Balcony	0.2/0.9	0.6/1.2	0.9/1.5	1.0/1.9	1.3/2.0	0.8/1.6
Corridor	0.1/1.1	0.1/2.6	0.4/3.7	1.0/4.6	1.4/5.0	0.5/4.3
Dining Room	0.0/1.5	0.0/3.0	0.0/3.6	0.0/3.2	0.0/1.9	0.0/2.9
Laundry Room	0.0/0.0	0.0/0.0	0.0/0.0	0.0/0.0	0.0/0.0	0.0/0.0
Unknown	0.0/0.0	0.0/0.0	0.0/0.0	0.0/0.0	0.0/0.0	0.0/0.0

- The compatibility with the bubble diagram is a graph editing distance [2] between the input bubble diagram and the bubble diagram constructed from the output layout in the same way as the GT preparation above.

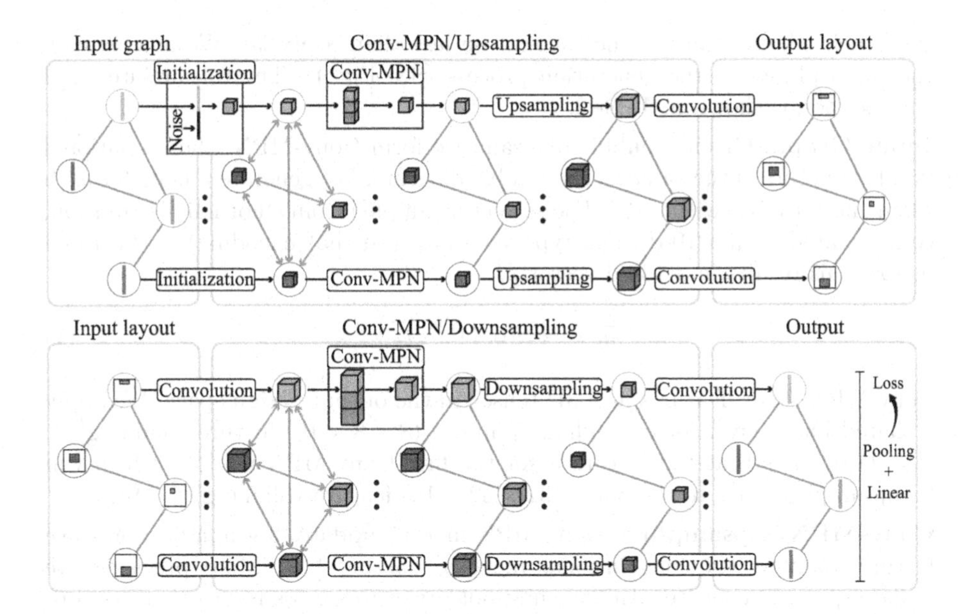

Fig. 4. Relational house layout generator (top) and discriminator (bottom). Conv-MPN is our backbone architecture [26]. The input graph constraint is encoded into the graph structure of their relational networks.

Assumptions: In contrast to the real design process, we make a few restrictive assumptions to simplify the problem setting: 1) A node property does not have a room size; 2) A room shape is always a rectangle; and 3) An edge property (i.e., room adjacency) does not reflect the presence of doors. This is the first research step in tackling the problem, where these extensions are our future work.

4 House-GAN

House-GAN is a relational generative adversarial network. The key specialization is our relational generator and discriminator, where the input graph constraint is encoded into the graph structure of the relational networks. In particular, we employ Conv-MPN [26], which differs from GCNs [3,12] in that a node stores a feature volume and convolutions update features in the design space (as opposed to a 1D latent vector space).

4.1 House Layout Generator

The generator takes a noise vector per room and a bubble diagram, then generates a house layout as an axis-aligned rectangle per room. The bubble diagram is represented as a graph, where a node represents a room with a room type, and an edge represents the spatial adjacency. More specifically, a rectangle should be generated for each room, and two rooms with an edge must be spatially adjacent (i.e., their Manhattan distance should be less than 8 pixels). We now explain the three phases of the generation process (See Fig. 4). The full architectural specification is shown in Table 2.

Input Graph: Given a bubble diagram, we form Conv-MPN whose relational graph structure is the same as the bubble diagram. We generate a node for each room and initialize with a 128-d noise vector sampled from a normal distribution, concatenated with a 10-d room type vector $\overrightarrow{t_r}$ (one-hot encoding). r is a room index. This results in a 138-d vector $\overrightarrow{g_r}$:

$$\overrightarrow{g_r} \leftarrow \left\{ \mathbb{N}(0,1)^{128}; \overrightarrow{t_r} \right\}. \quad (1)$$

Conv-MPN stores features as a 3D tensor in the output design space. We apply a shared linear layer for expanding $\overrightarrow{g_r}$ into a $(8 \times 8 \times 16)$ feature volume $\mathbf{g}_r^{l=1}$. $(l = 1)$ denotes that the feature is for the first Conv-MPN module, which will be upsampled twice to become a $(32 \times 32 \times 16)$ feature volume $\mathbf{g}_r^{l=3}$ later.

Conv-MPN/Upsampling: Conv-MPN module updates a graph of room-wise feature volumes via convolutional message passing [26]. More precisely, we update $\mathbf{g}_r^l$ by 1) concatenating a sum-pooled feature across rooms that are connected in the graph; 2) concatenating a sum-pooled feature across non-connected rooms; and 3) applying a CNN:

$$\mathbf{g}_r^l \leftarrow \text{CNN}\left[\mathbf{g}_r^l \quad ; \quad \underset{s \in \mathbf{N}(r)}{\text{Pool}}\, \mathbf{g}_s^l \quad ; \quad \underset{s \in \overline{\mathbf{N}}(r)}{\text{Pool}}\, \mathbf{g}_s^l\right]. \quad (2)$$

$\mathbf{N}(r)$ and $\overline{\mathbf{N}}(r)$ denote sets of rooms that are connected and not-connected, respectively. We upsample features by a factor of 2 using a transposed convolution (kernel = 4, stride = 2, padding = 1), while maintaining the number of channels. The generator has two rounds of Conv-MPN and upsampling, making the final feature volume $\mathbf{g}_r^{l=3}$ of size $(32 \times 32 \times 16)$.

Output Layout: A shared three-layer CNN converts a feature volume into a room segmentation mask of size $(32 \times 32 \times 1)$. This graph of segmentation masks will be passed to the discriminator during training. At test time, the room mask (an output of tanh function with the range $[-1, 1]$) is thresholded at 0.0, and we fit the tightest axis-aligned rectangle for each room to generate the house layout.

4.2 House Layout Discriminator

The discriminator performs a sequence of operations in the reverse order. The input is a graph of room segmentation masks either from the generator (before rectangle fitting) or a real floorplan (1.0 for foreground and −1.0 for background). A segmentation mask is of size $32 \times 32 \times 1$. To associate the room type information, we take a 10-d room type vector, apply a linear layer to expand to 8192-d, then reshape to a $(32 \times 32 \times 8)$ tensor, which is concatenated to the segmentation mask. A shared three-layer CNN converts the feature into a size $(32 \times 32 \times 16)$, followed by two rounds of Conv-MPN and downsampling. We

Table 2. House-GAN architectural specification. "s" and "p" denote stride and padding. "x", "z" and "t" denote the room mask, noise vector, and room type vector. "conv_mpn" layers have the same architecture in all occurrences. Convolution kernels and layer dimensions are specified as $(N_{in} \times N_{out} \times W \times H)$ and $(W \times H \times C)$.

Architecture	Layer	Specification	Output Size
House layout generator	$concat(z, t)$	N/A	1×138
	$linear\_reshape_1$	138×1024	$8 \times 8 \times 16$
	$conv\_mpn_1$	$\begin{bmatrix} 16 \times 16 \times 3 \times 3, (s = 1, p = 1) \\ 16 \times 16 \times 3 \times 3, (s = 1, p = 1) \\ 16 \times 16 \times 3 \times 3, (s = 1, p = 1) \end{bmatrix}$	$8 \times 8 \times 16$
	$upsample_1$	$16 \times 16 \times 4 \times 4, (s = 2, p = 1)$	$16 \times 16 \times 16$
	$conv\_mpn_2$	-	$16 \times 16 \times 16$
	$upsample_2$	$16 \times 16 \times 4 \times 4, (s = 2, p = 1)$	$32 \times 32 \times 16$
	$conv\_leaky\_relu_1$	$16 \times 256 \times 3 \times 3, (s = 1, p = 1)$	$32 \times 32 \times 256$
	$conv\_leaky\_relu_2$	$256 \times 128 \times 3 \times 3, (s = 1, p = 1)$	$32 \times 32 \times 128$
	$conv\_tanh_1$	$128 \times 1 \times 3 \times 3, (s = 1, p = 1)$	$32 \times 32 \times 1$
House layout discriminator	$linear\_reshape_1(t)$	10×8192	$32 \times 32 \times 8$
	$concat(t, x)$	N/A	$32 \times 32 \times 9$
	$conv\_leaky\_relu_1$	$9 \times 16 \times 3 \times 3, (s = 1, p = 1)$	$32 \times 32 \times 16$
	$conv\_leaky\_relu_2$	$16 \times 16 \times 3 \times 3, (s = 1, p = 1)$	$32 \times 32 \times 16$
	$conv\_leaky\_relu_3$	$16 \times 16 \times 3 \times 3, (s = 1, p = 1)$	$32 \times 32 \times 16$
	$conv\_mpn_1$	-	$32 \times 32 \times 16$
	$downsample_1$	$16 \times 16 \times 3 \times 3, (s = 2, p = 1)$	$16 \times 16 \times 16$
	$conv\_mpn_2$	-	$16 \times 16 \times 16$
	$downsample_2$	$16 \times 16 \times 3 \times 3, (s = 2, p = 1)$	$8 \times 8 \times 16$
	$conv\_leaky\_relu_1$	$16 \times 256 \times 3 \times 3, (s = 2, p = 1)$	$4 \times 4 \times 256$
	$conv\_leaky\_relu_2$	$256 \times 128 \times 3 \times 3, (s = 2, p = 1)$	$2 \times 2 \times 128$
	$conv\_leaky\_relu_3$	$128 \times 128 \times 3 \times 3, (s = 2, p = 1)$	$1 \times 1 \times 128$
	$pool\_reshape\_linear_1$	128×1	1

downsample by a factor of 2 each time by a convolution layer (kernel = 3, stride = 2, padding = 1). Lastly, we use a three layer CNN for converting a room feature into a 128-d vector ($\overrightarrow{d_r}$). We sum-pool over all the room vectors and add a single linear layer to output a scalar **d**, classifying ground-truth samples from generated ones.

$$\mathbf{d} \leftarrow \text{Linear}(\underset{r}{\text{Pool}}\ \overrightarrow{d_r}) \tag{3}$$

We use the WGAN-GP [7] loss with gradient penalty set to 10 and compute the gradient penalty as proposed by Gulrajani *et al.* [7]: Linearly and uniformly interpolating room segmentation masks pixel-wise between real samples and generated ones, while fixing the relational graph structure.

Table 3. The main quantitative evaluations. Realism is measured by a user study with graduate students and professional architects. Diversity is measured by the FID scores. Compatibility is measured by the graph edit distance. (↑) and (↓) indicate the-higher-the-better and the-lower-the-better metrics, respectively. We compare House-GAN against two baselines and two competing methods. The cyan, orange, and magenta colors indicate the first, the second, and third best results, respectively.

	Realism (↑)	Diversity (↓)					Compatibility (↓)				
Model	All groups	1–3	4–6	7–9	10–12	13+	1–3	4–6	7–9	10–12	13+
CNN-only	−0.49	13.2	26.6	43.6	54.6	90.0	0.4	3.1	8.1	15.8	34.7
GCN	0.11	18.6	17.0	18.1	22.7	31.5	0.1	0.8	2.3	3.2	3.7
Ashual *et al.* [3]	−0.61	64.0	92.2	87.6	122.8	149.9	0.2	2.7	6.2	19.2	36.0
Johnson *et al.* [12]	−0.62	69.8	86.9	80.1	117.5	123.2	0.2	2.6	5.2	17.5	29.3
House-GAN [Ours]	0.15	13.6	9.4	14.4	11.6	20.1	0.1	1.1	2.9	3.9	10.8

5 Implementation Details

The proposed architecture was implemented in PyTorch and utilized a workstation with dual Xeon CPUs and dual NVIDIA Titan RTX GPUs. Our model adopts WGAN-GP [7] with ADAM optimizer ($b_1 = 0.5$, $b_2 = 0.999$) and is trained for 200k iterations. The learning rates of the generator and the discriminator are 0.0001, respectively. The batch size is 32. We set the number of critics to 1 and use leaky-ReLUs ($\alpha = 0.1$) for all non-linearities except for the last one in the generator where we use hyperbolic tangent. We tried but do not use spectral normalization in the convolution layers and a per-room discriminator (before final sum-pooling), which did not lead to significant improvements.

6 Experimental Results

Realism, diversity, and compatibility metrics evaluate the performance of the proposed system against the two baselines and the two competing methods. We first introduce these methods, while referring to the supplementary document for the full architectural specification.

- **CNN-only**: We encode the bubble diagram into a fixed dimensional vector by assuming at most 40 rooms and sorting the rooms based on the room-center x-coordinate in the corresponding floorplan. To be precise, we concatenate a 128-d noise vector, a 10-d room type vector for 40 rooms, and $780 = \binom{40}{2}$ dimensional vector indicating the room connectivity, resulting in a 1308-d vector. We pad zeros for missing rooms. We convert a vector into a feature volume and apply two rounds of upsampling and CNN to produce room masks as a ($32 \times 32 \times 40$) feature volume. The discriminator takes the room masks, concatenates the room type and connectivity information (i.e. 1308-d vector) represented as a ($32 \times 32 \times 8$) feature volume and performs an inverse operation of the generator.

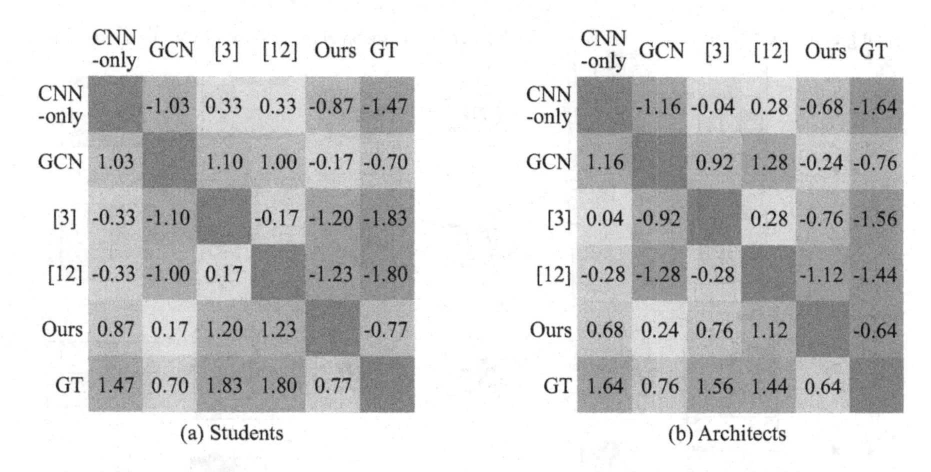

CNN -only GCN [3] [12] Ours GT
CNN -only -1.03 0.33 0.33 -0.87 -1.47
GCN 1.03 1.10 1.00 -0.17 -0.70
[3] -0.33 -1.10 -0.17 -1.20 -1.83
[12] -0.33 -1.00 0.17 -1.23 -1.80
Ours 0.87 0.17 1.20 1.23 -0.77
GT 1.47 0.70 1.83 1.80 0.77
CNN -only GCN [3] [12] Ours GT
CNN -only -1.16 -0.04 0.28 -0.68 -1.64
GCN 1.16 0.92 1.28 -0.24 -0.76
[3] 0.04 -0.92 0.28 -0.76 -1.56
[12] -0.28 -1.28 -0.28 -1.12 -1.44
Ours 0.68 0.24 0.76 1.12 -0.64
GT 1.64 0.76 1.56 1.44 0.64

(a) Students (b) Architects

Fig. 5. Realism evaluation. The user study score pairs of methods by graduate students (left) and professional architects (right). The tables should be read row-by-row. For example, the bottom row shows the results of GT against the other methods.

- **GCN**: The generator takes a 128-d noise vector concatenated with a 10-d room type vector per room. After 2 rounds of message passing as 1d vectors by GCN, a shared CNN module decodes the vector into a mask. The discriminator merges the room segmentation and type into a feature volume as in House-GAN. A shared CNN encoder converts it into a feature vector, followed by 2 rounds of message passing, sum-pooling, and a linear layer to produce a scalar.
- **Ashual *et al.*** [3] and **Johnson *et al.*** [12]: After converting our bubble diagram and floorplan data into their representation, we use their official code to train the models with two minor adaptations: 1) we limit scene-graphs to contain only two types of connections: "adjacent" and "not adjacent"; 2) we provide the rendered bounding boxes filled with their corresponding color during training.

Table 3 shows our main results. As explained in Sect. 3, we divide 117,587 samples into 5 groups based on their room counts. For the generation of layouts in each group, we exclude samples in the same group from the training so that methods cannot simply memorize layouts. House-GAN outperforms the competing methods and the baselines in all the metrics, except for the compatibility against GCN with a small margin. We now discuss each of the three metrics in more detail with more qualitative and quantitative evaluations.

Realism: We conducted a user study with 12 graduate students and 10 professional architectures. Each subject compared 75 pairs of layouts sampled from the five targeted methods and the ground-truth. Table 3 shows that House-GAN has the best overall user score. Figure 5 shows the direct pairwise comparisons. For each pair of methods, we look at the user scores when the 2 methods were compared, compute their average scores, and take the difference. If subjects

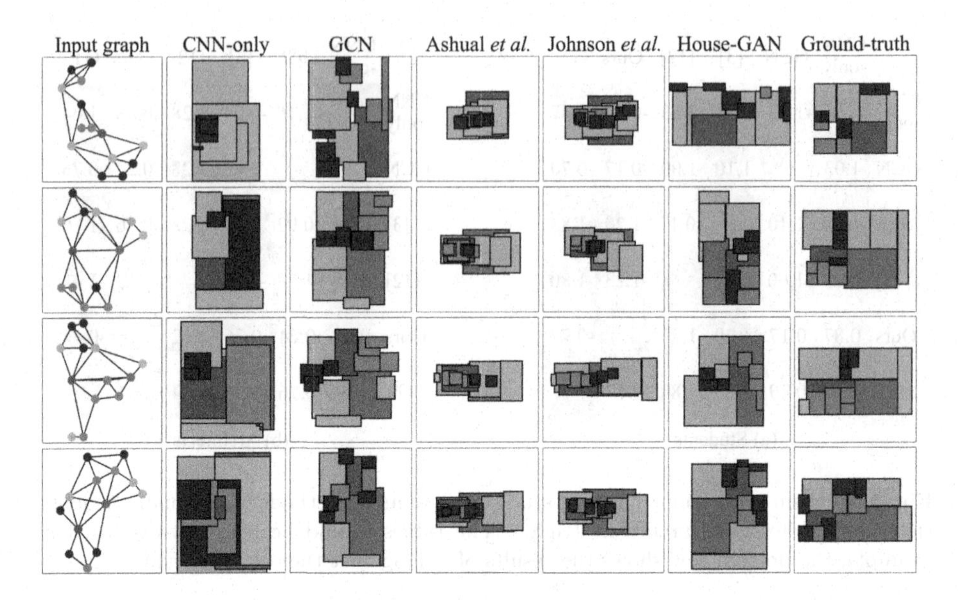

Fig. 6. Realism evaluation. We show one layout sample generated by each method from each input graph. Our approach (House-GAN) produces more realistic layouts whose rooms are well aligned and spatially distributed. See the supplementary document for the user study details.

always choose "better" for one method, the difference would be 2.0. Therefore, the score difference of 1.0 (e.g., GCN against [12] by students) could mean that the method was rated "better" half the time, and "equally X" half the time. We refer to the supplementary document for additional details on the user study.

The figure shows that both students and architects rate House-GAN the most realistic except of course the ground-truth. Figure 6 qualitatively supports the same conclusion. Ashual *et al.* did not produce compelling results, because they rather focus on realistic image generation and need rough locations of objects as input, which are not given in our problem. Johnson *et al.* also failed in our experiments. They produce more realistic results if samples from the same group are included in the training set, allowing the method to memorize answers. We believe that their network is not capable of generalizing to unseen cases.

Diversity: Diversity is another strength of our approach. For each group, we randomly sample 5,000 bubble diagrams and let each method generate 10 house layout variations. We rasterize the bounding boxes (sorted in a decreasing order of the areas) with the corresponding room colors, and compute the FID score. Ashual *et al.* generates variations by changing the input graph into an equivalent form (e.g., apple-right-orange to orange-left-apple). We implemented this strategy by changing the relation from room1-adjacent-room2 to room2-adjacent-room1. However, the method failed to create interesting variations. Johnson *et al.* also fails in the variation metric. Our observation is that they employ GCN but a noise vector is added after GCN near the end of the network.

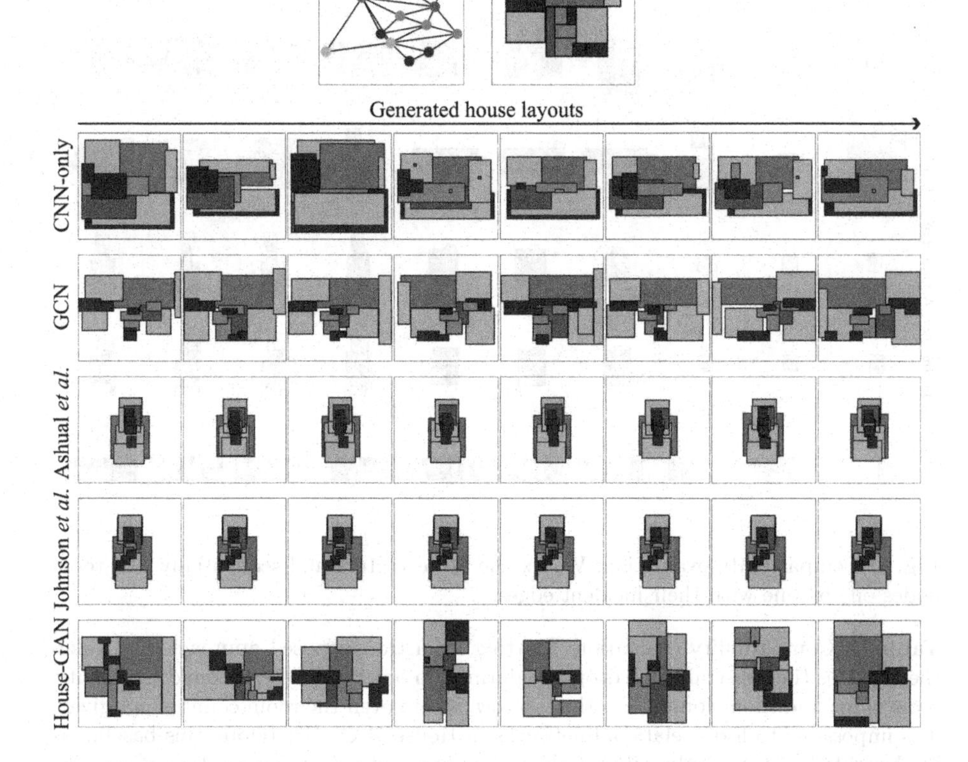

Fig. 7. Diversity evaluation: House layout examples generated from the same bubble diagram. House-GAN (ours) shows the most diversity/variations.

The system is not capable of generating variations. House-GAN has the best diversity scores except for the smallest group, where there is little diversity and the graph-constraint has little effect. Figure 7 qualitatively demonstrates the diversity of House-GAN, where the other methods tend to collapse into fewer modes.

Compatibility: All the methods perform fairly well on the compatibility metric, where many methods collapse to generating a few examples with high compatibility scores. The real challenge is to ensure compatibility while still keeping variations in the output, which House-GAN is the only method to achieve (See Fig. 8). To further validate the effectiveness of our approach, Table 4 shows the improvements of the compatibility scores as we increase the input constraint information (i.e., room count, room type, and room connectivity). The table demonstrates that House-GAN is able to achieve higher compatibility as we add more graph information. Figure 8 demonstrates another experiment, where we

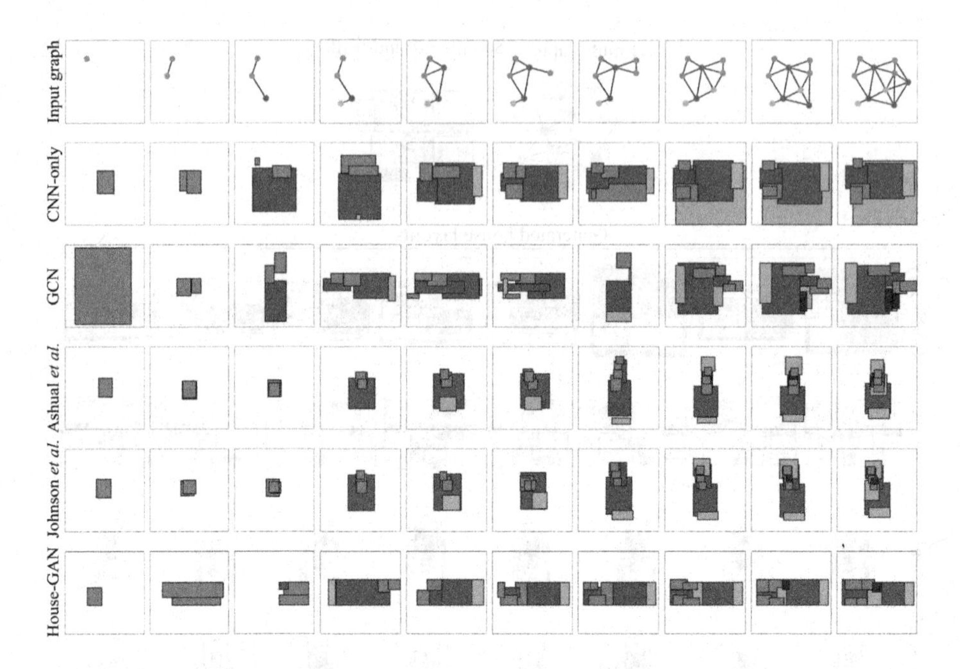

Fig. 8. Compatibility evaluation. We fix the noise vectors and sequentially add room nodes one by one with their incident edges.

Table 4. Compatibility evaluation: Starting from the proposed approach that takes the graph as the constraint, we drop its information one by one (i.e., room connectivity, room type, and room count). For the top row, where even the room count is not given, it is impossible to form relational networks in House-GAN. Therefore, this baseline is implemented as CNN-only without the room type and connectivity information. The second row is implemented as House-GAN while removing the room type information and making the graph fully-connected. Similarly, the third row is implemented as House-GAN while making the graph fully-connected. The last row is House-GAN.

Count	Type	Conn	1–3	4–6	7–9	10–12	13+
			28.5	28.8	19.0	26.4	32.2
✓			0.6	2.1	4.6	7.3	**37.3**
✓	✓		0.4	2.2	4.4	7.5	21.4
✓	✓	✓	0.1	1.1	2.9	3.9	10.8

fix the noise vectors and incrementally add room nodes one-by-one. It is interesting to see that House-GAN sometimes changes the layout dramatically to satisfy the connectivity constraint (e.g., from the 4th column to the 5th).

More Results and Discussion: Figure 9 shows interesting failure and success examples that were compared against the ground-truth in our user study. Professional architects rate the three success examples as "equally good" and

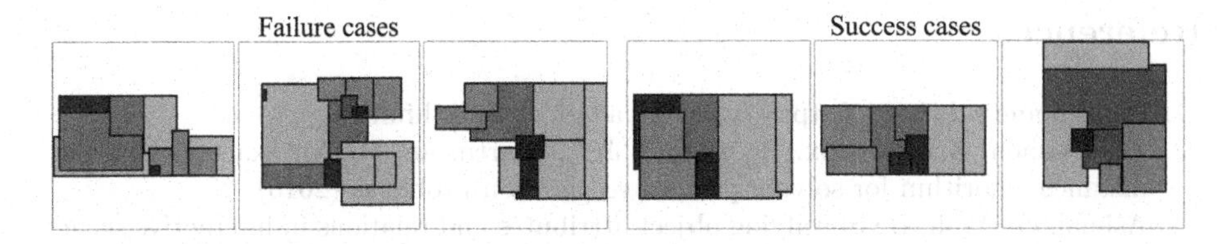

Fig. 9. Failure and success examples by House-GAN from the user study. Architects rate the success examples (right) as "equally good" and the failure examples (left) as "worse" against the ground-truth.

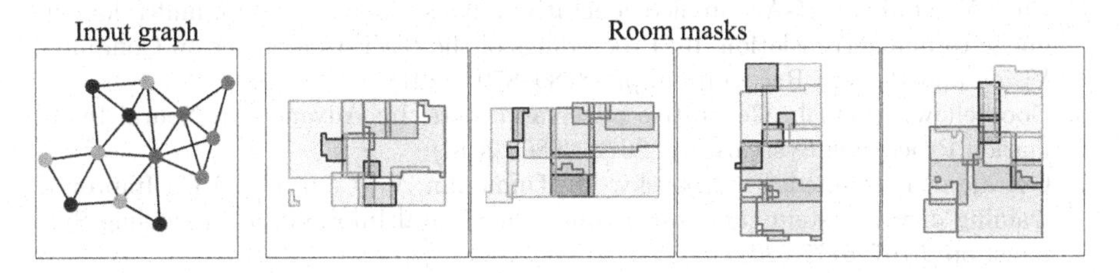

Fig. 10. Raw room segmentation outputs before the rectangle fitting.

the three failure cases as "worse" against the ground-truth. For instance, the first failure example looks strange because a balcony is reachabale only through bathrooms, and a closet is inside a kitchen. The second failure example looks strange, because a kitchen is separated into two disconnected spaces. Our major failure modes are 1) improper room size or shapes for a given room type (e.g., a bathroom is too big); 2) misalignment of rooms; and 3) inaccessible rooms (e.g., room entry is blocked by closets). Our future work is to incorporate room size information or door annotations to address these issues. Lastly, Fig. 10 illustrates the raw output of the room segmentation masks before the rectangle fitting. The rooms are often estimated as rectangular shapes, because rooms are represented as axis-aligned rectangles in our dataset, while the original floorplan contains non-rectangular rooms. Another future work is the generation of non-rectangular rooms. We refer to supplementary document for more results.

7 Conclusion

This paper proposes a house layout generation problem and a graph-constrained relational generative adversarial network as an effective solution. We define three metrics (realism, diversity, and compatibility) and demonstrate the effectiveness of the proposed system over competing methods and baselines. We believe this paper makes an important step towards computer aided design of house layouts.

Acknowledgement. This research is partially supported by NSERC Discovery Grants, NSERC Discovery Grants Accelerator Supplements, and DND/NSERC Discovery Grant Supplement. We would like to thank architects and students for participating in our user study.

References

1. Lifull home's dataset. https://www.nii.ac.jp/dsc/idr/lifull
2. Abu-Aisheh, Z., Raveaux, R., Ramel, J.Y., Martineau, P.: An exact graph edit distance algorithm for solving pattern recognition problems (2015)
3. Ashual, O., Wolf, L.: Specifying object attributes and relations in interactive scene generation. In: Proceedings of the IEEE International Conference on Computer Vision, pp. 4561–4569 (2019)
4. Bao, F., Yan, D.M., Mitra, N.J., Wonka, P.: Generating and exploring good building layouts. ACM Trans. Graph. (TOG) **32**(4), 1–10 (2013)
5. Choi, Y., et al.: StarGAN: unified generative adversarial networks for multi-domain image-to-image translation. In: Proceedings of the IEEE Conference on Computer Vision and Pattern Recognition, pp. 8789–8797 (2018)
6. Goodfellow, I., et al.: Generative adversarial nets. In: Advances in Neural Information Processing Systems, pp. 2672–2680 (2014)
7. Gulrajani, I., Ahmed, F., Arjovsky, M., Dumoulin, V., Courville, A.C.: Improved training of wasserstein gans. In: Advances in Neural Information Processing Systems, pp. 5767–5777 (2017)
8. Harada, M., Witkin, A., Baraff, D.: Interactive physically-based manipulation of discrete/continuous models. In: Proceedings of the 22nd Annual Conference on Computer Gand Interactive Techniques, pp. 199–208 (1995)
9. Hendrikx, M., Meijer, S., Van Der Velden, J., Iosup, A.: Procedural content generation for games: a survey. ACM Trans. Multimedia Comput. Commun. Appl. (TOMM) **9**(1), 1–22 (2013)
10. Heusel, M., Ramsauer, H., Unterthiner, T., Nessler, B., Hochreiter, S.: GANs trained by a two time-scale update rule converge to a local Nash equilibrium. In: Advances in Neural Information Processing Systems, pp. 6626–6637 (2017)
11. Isola, P., Zhu, J.Y., Zhou, T., Efros, A.A.: Image-to-image translation with conditional adversarial networks. In: Proceedings of the IEEE Conference on Computer Vision and Pattern Recognition, pp. 1125–1134 (2017)
12. Johnson, J., Gupta, A., Fei-Fei, L.: Image generation from scene graphs. In: Proceedings of the IEEE Conference on Computer Vision and Pattern Recognition, pp. 1219–1228 (2018)
13. Jyothi, A.A., Durand, T., He, J., Sigal, L., Mori, G.: LayoutVAE: stochastic scene layout generation from a label set. In: Proceedings of the IEEE International Conference on Computer Vision, pp. 9895–9904 (2019)
14. Karras, T., et al.: Analyzing and improving the image quality of styleGAN. arXiv preprint arXiv:1912.04958 (2019)
15. Kwon, Y.H., Park, M.G.: Predicting future frames using retrospective cycle GAN. In: Proceedings of the IEEE Conference on Computer Vision and Pattern Recognition, pp. 1811–1820 (2019)
16. Li, J., Yang, J., Hertzmann, A., Zhang, J., Xu, T.: LayoutGAN: generating graphic layouts with wireframe discriminators. arXiv preprint arXiv:1901.06767 (2019)
17. Liu, C., Wu, J., Kohli, P., Furukawa, Y.: Raster-to-vector: revisiting floorplan transformation. In: Proceedings of the IEEE International Conference on Computer Vision, pp. 2195–2203 (2017)
18. Ma, C., Vining, N., Lefebvre, S., Sheffer, A.: Game level layout from design specification. Comput. Graph. Forum **33**, 95–104 (2014). Wiley Online Library
19. Merrell, P., Schkufza, E., Koltun, V.: Computer-generated residential building layouts. ACM Trans. Graph. (TOG) **29**, 181 (2010). ACM

20. Müller, P., Wonka, P., Haegler, S., Ulmer, A., Van Gool, L.: Procedural modeling of buildings. In: ACM SIGGRAPH 2006 Papers, pp. 614–623 (2006)
21. Peng, C.H., Yang, Y.L., Wonka, P.: Computing layouts with deformable templates. ACM Trans. Graph. (TOG) **33**(4), 1–11 (2014)
22. Ritchie, D., Wang, K., Lin, Y.A.: Fast and flexible indoor scene synthesis via deep convolutional generative models. In: Proceedings of the IEEE Conference on Computer Vision and Pattern Recognition, pp. 6182–6190 (2019)
23. Wang, K., Lin, Y.A., Weissmann, B., Savva, M., Chang, A.X., Ritchie, D.: Planit: planning and instantiating indoor scenes with relation graph and spatial prior networks. ACM Trans. Graph. (TOG) **38**(4), 132 (2019)
24. Wang, K., Savva, M., Chang, A.X., Ritchie, D.: Deep convolutional priors for indoor scene synthesis. ACM Trans. Graph. (TOG) **37**(4), 1–14 (2018)
25. Wu, W., Fu, X.M., Tang, R., Wang, Y., Qi, Y.H., Liu, L.: Data-driven interior plan generation for residential buildings. ACM Trans. Graph. (TOG) **38**(6), 1–12 (2019)
26. Zhang, F., Nauata, N., Furukawa, Y.: Conv-MPN: convolutional message passing neural network for structured outdoor architecture reconstruction. arXiv preprint arXiv:1912.01756 (2019)
27. Zhu, J.Y., Park, T., Isola, P., Efros, A.A.: Unpaired image-to-image translation using cycle-consistent adversarial networks. In: Proceedings of the IEEE International Conference on Computer Vision, pp. 2223–2232 (2017)

Crowdsampling the Plenoptic Function

Zhengqi Li[✉], Wenqi Xian, Abe Davis, and Noah Snavely

Cornell Tech, Cornell University, Ithaca, USA
z1548@cornell.edu

Abstract. Many popular tourist landmarks are captured in a multitude of online, public photos. These photos represent a sparse and unstructured sampling of the plenoptic function for a particular scene. In this paper, we present a new approach to novel view synthesis under time-varying illumination from such data. Our approach builds on the recent *multi-plane image* (MPI) format for representing local light fields under fixed viewing conditions. We introduce a new *DeepMPI* representation, motivated by observations on the sparsity structure of the plenoptic function, that allows for real-time synthesis of photorealistic views that are continuous in both space and across changes in lighting. Our method can synthesize the same compelling parallax and view-dependent effects as previous MPI methods, while simultaneously interpolating along changes in reflectance and illumination with time. We show how to learn a model of these effects in an unsupervised way from an unstructured collection of photos without temporal registration, demonstrating significant improvements over recent work in neural rendering. More information can be found at crowdsampling.io.

1 Introduction

There is a thought experiment that goes something like this:

> Imagine a 'camera' with no optics or image sensor of any kind. Rather, it consists only of a box equipped with GPS, a radio for Internet access, a button for 'taking pictures', and a screen for displaying those pictures. When a user presses its button, the box searches the Internet for photos tagged with its current location, and from these selects a best match to display on the screen.

This thought experiment is perhaps best understood in the context of popular tourist attractions, of which one can often find countless images posted online (Fig. 1, second row). When pointed at such an attraction, one can imagine our box producing images very similar to those of a real camera, forcing us to consider

Electronic supplementary material The online version of this chapter (https://doi.org/10.1007/978-3-030-58452-8_11) contains supplementary material, which is available to authorized users.

A. Vedaldi et al. (Eds.): ECCV 2020, LNCS 12346, pp. 178–196, 2020.
https://doi.org/10.1007/978-3-030-58452-8_11

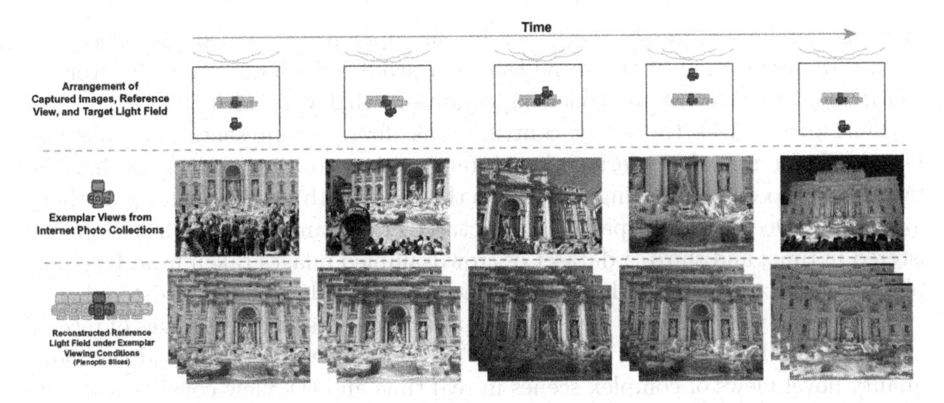

Fig. 1. Crowdsampled plenoptic slices. Given a large number of tourist photos taken at different times of day, our system learns to construct a continuous set of lightfields and to synthesize novel views capturing all-times-of-day scene appearance.

whether an image we capture ourselves is meaningfully different from a near-identical one captured by strangers. For many, the ensuing philosophical debate hinges on whether an image reflects the scene as they remember it. After all, appearance is not generally constant over time, even under a fixed geometry and viewpoint; in outdoor settings, for example, weather changes, shadows move, and day turns to night—all resulting in appearance changes that can be observed from a single view of the scene.

This poses an interesting challenge to the field of image-based rendering: can we use crowdsourced imagery to synthesize arbitrary views of a scene with viewing conditions that change over time? Without changing viewing conditions, this challenge would reduce to the more familiar problem of reconstructing a 4D light field $\mathcal{L}(u, v, x, y)$ that describes all light in our scene [33]. When we add time to our problem, it turns into a 5D reconstruction over what Adelson and Bergen [1] call the *plenoptic function.*[1]

In this paper, we propose a novel approach to neural image-based rendering from crowdsourced images that leverages the sparse structure of the plenoptic function to learn how scene appearance changes over space and time in an unsupervised manner. Our approach takes unstructured Internet photos spanning some range of time-varying appearance in a scene and learns how to reconstruct a *plenoptic slice*—a representation of the light field that respects temporal structure in the plenoptic function when interpolated over time—for each of the viewing conditions captured in our input data. By designing our model to preserve the structure of real plenoptic functions, we force it to learn time-varying phenomena like the motion of shadows according to sun position. This lets us, for example, recover plenoptic slices for images taken at different times of day (Fig. 1, bottom row) and interpolate between them to observe how shadows

[1] [1] describes the plenoptic function as 7D, but we can reduce this to a 4D color light field supplemented by time by applying the later observations of [33].

move as the day progresses (best seen in our supplemental video). In effect, we learn a representation of the scene that can produce high-quality views from a continuum of viewpoints *and* viewing conditions that vary with time.

Our work makes three key contributions: first, a representation, called a DeepMPI, for neural rendering that extends prior work on multiplane images (MPIs) [68] to model viewing conditions that vary with time; second, a method for training DeepMPIs on sparse, unstructured crowdsampled data that is unregistered in time; and third, a dataset of crowdsampled images taken from Internet photo collections, along with details on how it was collected and registered.

Compared with previous work, our approach inherits the advantages of recent methods based on MPIs [8,12,42,56,68], including the ability to produce high-quality novel views of complex scenes in real time and the view consistency that arises from a 3D scene representation (in contrast to neural rendering approaches that decode a separate view for each desired viewpoint). To these advantages we add the key ability to synthesize and interpolate continuous, photo-realistic, time-varying changes in appearance. We compare our approach both quantitatively and qualitatively to recent neural rendering methods, such as Neural Rerendering in the Wild [41], and show that our method produces superior results.

2 Related Work

The study of image-based rendering is motivated by a simple question: how do we use a finite set of images to reconstruct an infinite set of views? Different branches of research have explored this question from different angles and with different assumptions. Here we outline the space of approaches, highlighting work most closely related to our own.

Novel View Synthesis. Novel view synthesis has traditionally been approached through either explicit estimation of scene geometry and color [4,21,72], or using coarser estimates of geometry to guide interpolation between captured views [2, 10,55]. Light field rendering [3,17,33] pushes the latter strategy to an extreme by using dense structured sampling of the light field to make reconstruction guarantees independent of specific scene geometry. Subsequent works [9,32,44, 50,51,61] have leveraged observations on the structure of light fields to build on this approach. However, most IBR algorithms are designed to model static appearance, making them ill-suited for our problem.

Recently, deep learning techniques have been applied to this problem. Several works [22,59] rely on global meshes to guide view synthesis. However, such methods heavily rely on the accuracy of 3D models, and often fail to model complex scene components such as translucent and thin objects. Other works predict appearance flow [69], depth probabilities [13,65], or RGBD light fields [26,57]. However, many of these methods independently synthesize appearance for each view, leading to inconsistent renderings across views.

Our approach builds on the use of multiplane images (MPIs) [68] for novel view synthesis. Several recent methods have shown that MPIs are an effective and learnable representation for light fields [8,12,42,56]. We build on this representation by introducing the DeepMPI, which further captures viewing condition–dependent appearance. We are also inspired by recent work that poses view synthesis as decoding features from a learned latent space [7,11,36,53,54,59]. However, such work has been limited to synthetic environments or objects captured in controlled settings and is difficult to apply to crowdsampled images.

Appearance Modeling. Several works have modeled the time-varying appearance of outdoor scenes using physically-motivated approaches [19,29,48] or by combining data-driven methods and dense geometry [14,40,45,66]. Additionally, Martin-Brualla *et al.* [38,39] reconstruct time-lapses of urban scenes from Internet photos. However, their method relies on timestamps, and models appearance changes at much coarser granularity (scene dynamics across years). The recent work of Meshry *et al.* [41] is probably closest to our own. They model appearance changes across varying times of day by learning an appearance embedding. However, their method relies heavily on dense multi-view stereo geometry, and tends to produce temporal artifacts under complex appearance changes. In contrast, our approach is capable of rendering a more continuous range of photo-realistic views across diverse appearances, without relying on dense input geometry.

Deep Image Synthesis. Our work is also related to the problem of image-to-image translation [6,25,43,62], multi-model image-to-image translation [24,31,70,71] and style transfer [15,23,49,60]. Recently, Generative Adversarial Networks (GANs) [16,18,37] have successfully produced photo-realistic imagery, enabling a variety of applications in deep image synthesis [27,28,30,46,63,64]. However, there has been comparatively little investigation of 3D scene representations for deep image synthesis. Our method demonstrates the ability to learn a generative 3D scene representation and produce high-quality novel views of complex scenes.

3 Approach

Given a set $\mathcal{I} = \{I_1, I_2, ..., I_n\}$ of crowdsampled photos with corresponding camera viewpoints $\mathcal{C} = \{c_1, c_2, ..., c_n\}$ captured in a common scene, we formulate our problem as the reconstruction of *plenoptic slices* (local light fields parameterized by an appearance descriptor) around some reference view r conditioned on each of the scene appearances captured in $\mathcal{I}$ (see Fig. 1 for a geometric sketch of this setup). We present our approach in three parts: first, we describe how the input images $\mathcal{I}$ are collected and registered (Sect. 3.1); then we discuss our representation of the plenoptic function, which extends multiplane images (MPIs) to model appearance changes over time (Sect. 3.2); and finally we describe how to train this representation on our crowdsampled data (Sects. 3.3 and 3.4).

Note on Notation: Throughout the paper, we will use superscripts to denote camera viewpoints and subscripts to denote image or voxel indices.

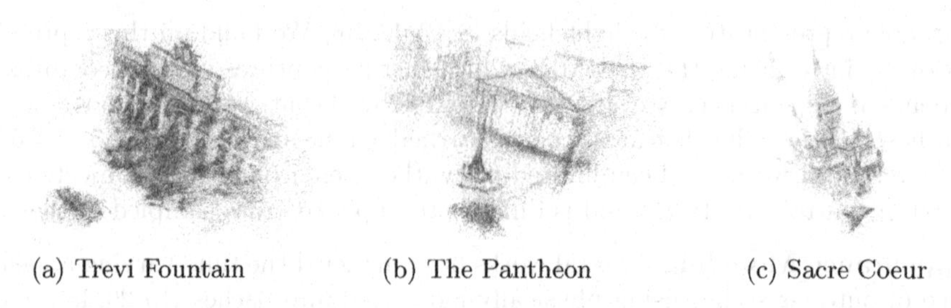

(a) Trevi Fountain (b) The Pantheon (c) Sacre Coeur

Fig. 2. Registered photo collections. Example SfM reconstructions of clusters of Internet photos sharing similar viewpoints, labeled as red dots. (Color figure online)

3.1 Collecting Crowdsampled Data

We selected a number of popular tourist sites and downloaded ∼50K photos from Flickr for each site. For each scene, we must then register these photos by solving for a camera pose and intrinsic parameters for each image. As running structure from motion (SfM) from scratch on such quantities of images is very expensive, we instead started with a existing SfM reconstruction of each site from the MegaDepth dataset [35], and performed camera relocalization to efficiently register each new image against the existing reconstruction [47].

For each landmark, we then identified a reference viewpoint r to center our reconstruction by using a canonical view selection algorithm similar to that of Simon *et al.*to find viewpoints with a high density of nearby views [52]. We then select all images captured from within a sphere centered at r for use in our method, randomly splitting the set gathered from each landmark into training and test data. We manually set the field of view of the reference viewpoint so that it has good coverage of the scene.

We found that the camera parameters estimated from relocalization are sometimes inaccurate, and so we reapply a global SfM and bundle adjustment to the smaller set of selected images near each scene's reference view to reestimate these images' camera parameters. We used this data pipeline to gather and register photos for eight locations, and will release this data to the research community. Figure 2 shows final SfM reconstructions for three of these landmarks.

3.2 The DeepMPI Scene Representation

We base our representation on the multiplane image (MPI) format [58,68], which represents light fields locally as a stack of fronto-parallel planar RGBα layers arranged at varying distances from the camera, akin to a stack of transparencies. Novel views are rendered from an MPI by warping the layers into a new view, then performing an *over* operation to composite the warped layers into a rendered image. Individual RGBα elements ("voxels") of an MPI are indexed by (x, y) position and plane depth d.

While MPIs have been remarkably effective for reconstructing fixed light fields from sparse views [12], they do not encode any information about how viewing conditions may vary with time. Furthermore, even if we were given a regular MPI corresponding to viewing conditions for each of our input images, directly interpolating between these MPIs would still fail to capture temporal structure in the plenoptic function. For example, interpolating between morning and afternoon MPIs would cause shadows cast by the sun to appear in duplicate when, in reality, a single shadow moved over time. This observation highlights the distinction between what we call a light field and what we call a plenoptic slice: we use the latter to describe a reparameterization of the light field that is better-suited for interpolation over time.

Inspired by DeepVoxels [53], we introduce *DeepMPIs* to help learn this reparameterization. DeepMPIs augment standard RGBα MPIs by appending a *learnable* latent feature vector at each MPI voxel (see Fig. 4). For a given scene, we position a DeepMPI at the reference viewpoint r, and denote this reference DeepMPI as $D^r = (B^r, \alpha^r, F^r)$. Each voxel of D^r at spatial location and depth $\mathbf{p} = (x, y, d)$ consists of a base RGB color $B^r_{\mathbf{p}}$, an alpha weight $\alpha^r_{\mathbf{p}}$, and a latent feature vector $F^r_{\mathbf{p}}$. We set the number of DeepMPI depth planes to 64 with uniform sampling in disparity space, and we adopt the method of Zhou *et al.* [68] to set the depth of the near and far planes of the DeepMPI.

In our supplemental document we relate the design of this representation and its training to priors on the sparse structure of the plenoptic function. At a high level, the α planes encode visibility information, which we expect to remain constant even as lighting and other viewing conditions change with time. The latent feature planes F^r are trained to capture correlations between different viewing conditions that arise from, for example, limited variation in material properties and correlation among surface normals within the scene. A plenoptic slice then consists of a DeepMPI and some exemplar image I_k. We can convert this to a standard RGBα MPI representing appearance under the specific conditions captured in I_k by using a decoder that is trained jointly with our DeepMPI, which we describe in Sect. 3.4.

To compute a DeepMPI from a collection of registered images, we use a two-stage process: first, we first estimate base color and α planes (Sect. 3.3), then optimize latent features F^r jointly with our neural rendering network (Sect. 3.4) to enable controllable, varying appearance.

3.3 Stage 1: Optimizing DeepMPI Color and α Planes

In the first stage of our method, we optimize base color planes B^r and alpha planes α^r in our DeepMPI as if it were a standard RGBα MPI. One simple approach would be to jointly optimize B^r and α^r from scratch so as to minimize a reconstruction loss over all images (i.e., the difference between a known image and an MPI-predicted image from that viewpoint, averaged over all input images). However, as described in [12], such a method exhibits slow convergence and can be prone to local minima. In addition, compared to [12], our setting is more challenging because Internet photos exhibit diversity in camera parameters and viewing conditions. Instead, we propose a simple yet effective approach to estimating B^r and α^r given a set of posed input views.

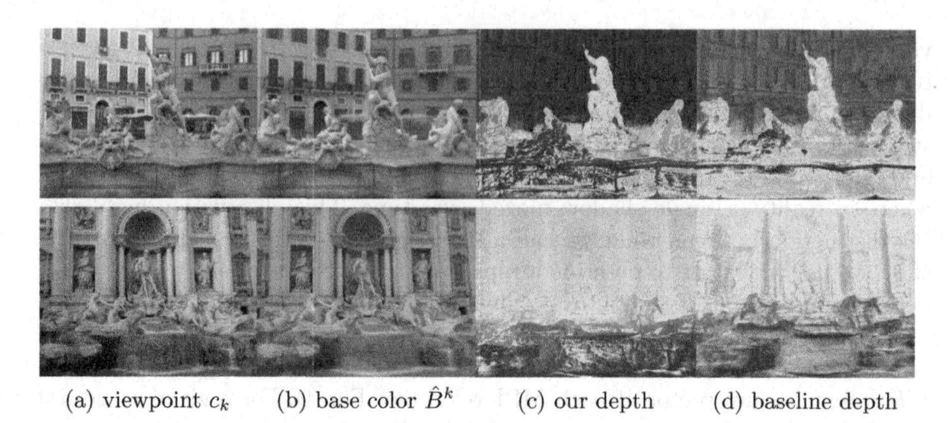

(a) viewpoint c_k (b) base color $\hat{B}^k$ (c) our depth (d) baseline depth

Fig. 3. Renderings of base color and alpha. From left to right: (a) original photos at target viewpoint c_k, (b) our estimated base color at c_k, (c) pseudo-depth computed from the RGBα MPI at c_k using our two-phase approach, (d) pseudo-depth from the baseline. For depth maps, red = close and blue = far. (Color figure online)

We start by creating a mean RGB plane sweep volume (PSV) at the reference viewpoint by reprojecting every image to the reference viewpoint via each depth plane, then averaging all reprojected images at each depth plane. We initialize the base color planes B^r to this mean RGB PSV. Keeping these color planes fixed, we optimize the alpha planes α^r to minimize reconstruction losses over the training photos. Specifically, given a photo I_k at viewpoint c_k, we project both B^r and α^r to c_k, then apply the over operation from back to front to render a base color image $\hat{B}^k$:

$$\hat{B}^k = \mathcal{O}\left(\mathcal{W}^k(B^r), \mathcal{W}^k(\alpha^r)\right), \tag{1}$$

where $\mathcal{O}$ is the over operation and $\mathcal{W}^k$ is the warping operation from the reference viewpoint r to the target viewpoint c_k. We compare the rendered base color image $\hat{B}^k$ and I_k using a reconstruction loss consisting of a pixel-wise l_1 loss and a multi-scale gradient consistency loss [34,35]. We observe that the gradient consistency loss leads to higher rendering quality and faster convergence.

Since the mean RGB PSV cannot accurately model scene content that is occluded in the reference view, after optimizing α^r with fixed B^r, we unfreeze B^r and jointly optimize B^r and α^r using the reconstruction loss described above. We observe that this two-phase training method leads to more accurate estimates of α^r than the alternative of optimizing B^r and α^r together from scratch. Figure 3, shows examples of input viewpoints and rendered base color images, as well as a comparison of pseudo-depths derived from alpha planes α^r computed by our two-phase training method and by the baseline. Once B^r and α^r are estimated, they are fixed for the subsequent stage of training, described below.

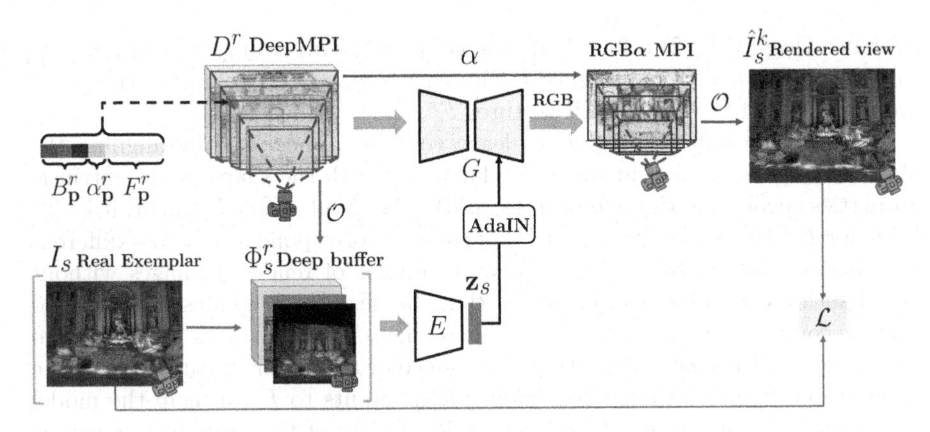

Fig. 4. Learning framework. Our method builds a reference DeepMPI D^r, consisting of base color, alpha, and latent feature components organized into planar layers. A rendering network G takes a DeepMPI projected to a target viewpoint c_k, and predicts corresponding RGB color layers. The appearance of these layers is modulated by an appearance vector $\mathbf{z}_s$ produced by encoder E. The over operation $\mathcal{O}$ is applied to the resulting RGBα MPI to render a view. We jointly train the encoder E, rendering network G, and latent features F^r in the DeepMPI by comparing a rendered view with an original exemplar image $I_k = I_s$. During inference, given an exemplar photo I_s, we can synthesize novel views close to the reference viewpoint, while also preserving the exemplar's appearance.

3.4 Stage 2: Learning How Appearance Changes with Time

Our method's second stage optimizes the latent features F^r in our DeepMPI, together with an appearance encoder E and rendering network G, to capture and render time-varying appearance. Our learning framework is summarized in Fig. 4.

Appearance Encoder. To model appearance variation, we devise a method wherein an encoder E learns to map an exemplar image I_s and an auxiliary *deep buffer* Φ_s^r to a latent appearance vector $\mathbf{z}_s$. Prior work, such as Meshry *et al.* [41], represents such variation by learning an appearance vector from the exemplar image and a deep buffer containing semantic and depth information. However, their deep buffer is aligned with the viewpoint of the exemplar image. This makes the encoding of exemplar data view-dependent when, under fixed conditions, the information (e.g., sun direction) it reflects should be largely view-independent. In contrast, we utilize a deep buffer *aligned with the reference viewpoint.*

In particular, our encoder E computes a latent appearance vector $\mathbf{z}_s$:

$$\mathbf{z}_s = E\left(I_s, \Phi_s^r\right) \tag{2}$$

where I_s is an exemplar image and Φ_s^r is a reference viewpoint–aligned deep buffer containing (1) a rectified RGB image over-composited from a PSV that reprojects exemplar I_s to the reference viewpoint via the depth planes of the reference DeepMPI, (2) a flattened base color image over-composited from base color layers B^r, and (3) a flattened latent feature map at the reference viewpoint over-composited from DeepMPI features F^r.

Such a deep buffer allows E to learn complex appearance by aligning the illumination information in the exemplar image with the shared scene intrinsic properties encoded in the reference DeepMPI. Without such alignment, it is difficult for E to consistently establish appearance correspondence across different viewpoints. Column (d) of Fig. 5 shows examples of rendered images without use of such a deep buffer. One can see that the deep buffer guides the model to capture complex illumination effects such as the realistic shadows highlighted in the first row. Moreover, integrating the base color and latent feature map at the reference viewpoint into Φ_s^r, and adding I_s as inputs to E can help the model to extrapolate appearance outside the field of view of the exemplar image, as shown in the last row of Fig. 6.

Neural Renderer. A plenoptic slice is now represented by the reference DeepMPI D^r and an appearance vector $\mathbf{z}_s$. Given these inputs, our neural renderer G predicts the corresponding RGB color planes. We could either predict these RGB planes at the reference viewpoint, or after first warping the DeepMPI to the target viewpoint. We choose the latter because it simplifies efficient implementation, as noted below. Let D^k denote the reference DeepMPI D^r after warping into target viewpoint c_k. Given D^k and $\mathbf{z}_s$, G predicts the RGB color planes C_s^k of a standard RGBα MPI at target viewpoint c_k:

$$C_s^k = G\left(D^k,\ \mathbf{z}_s\right) \tag{3}$$

In particular, G takes in each layer of D^k *independently* and predicts a corresponding RGB layer whose appearance is controlled by $\mathbf{z}_s$. A rendered RGB image with the appearance of I_s at viewpoint c_k can then be obtained using the over operation with precomputed alpha weights α^k in D^k:

$$\hat{I}_s^k = \mathcal{O}\left(C_s^k, \alpha^k\right) \tag{4}$$

As shown in Fig. 4, during training we set exemplar image $I_s = I_k$, i.e., we aim to reconstruct image I_k at viewpoint c_k. At inference, I_s is not necessarily I_k.

Our rendering network G is a U-Net variant with an encoder-decoder architecture. Prior methods [41,71] embed $\mathbf{z}$ in the bottleneck or input of G. Instead, we use Adaptive Instance Normalization (AdaIN) layers [23] whose parameters are dynamically generated from $\mathbf{z}$ via an MLP. AdaIN has been shown to be effective in capturing both global and spatially varying appearance of exemplar images. We find that AdaIN not only helps model natural scene appearance, but also stabilizes training. Column (b) of Fig. 5 shows examples of our rendered images without AdaIN; one can see the model using AdaIN preserves more faithful scene appearance including the style and color of exemplar images.

In practice, feeding a full-resolution DeepMPI into G and performing backpropagation is very memory intensive. Hence, during training, we operate on random 256×256 crops of training images, and only the necessary portion of D^r is warped to c_k and fed to G. At test time, any size input can be used.

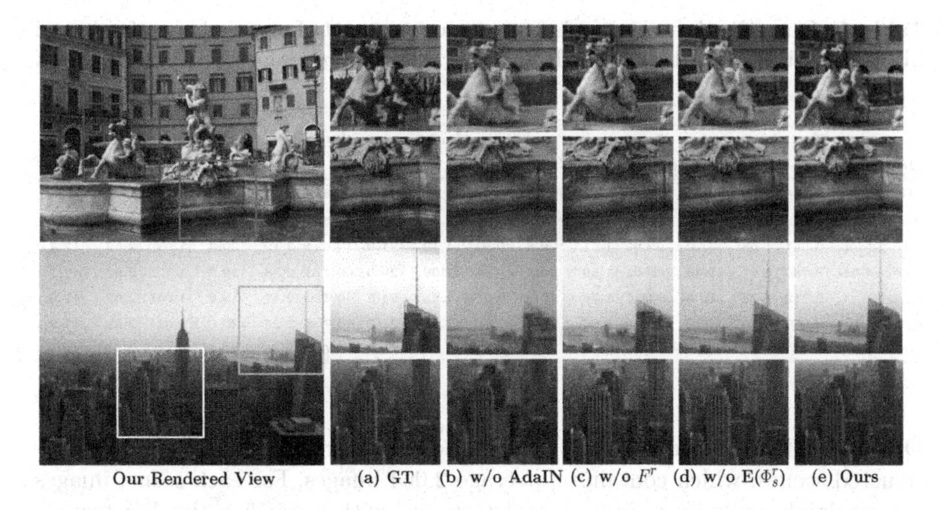

Fig. 5. Comparisons of images reconstructed with different configurations of our method. The images rendered from our full approach (e) are more similar to the ground truth images (a) than other configurations. In particular, the images rendered from the models without AdaIN (b) or the DeepMPI (c) are less realistic, and the model that does not feed the deep buffer Φ_s^r to the encoder (d) fails to capture accurate scene appearance, as indicated in the highlighted regions. (Color figure online)

Losses. To train G and E, we compute losses between output views and ground-truth exemplar views. Our training loss is composed of three terms:

$$\mathcal{L} = \mathcal{L}_{\mathsf{VGG}} + w_{\mathsf{GAN}}\mathcal{L}_{\mathsf{GAN}} + w_{\mathsf{style}}\mathcal{L}_{\mathsf{style}}, \tag{5}$$

where $\mathcal{L}_{\mathsf{VGG}}$, $\mathcal{L}_{\mathsf{GAN}}$, and $\mathcal{L}_{\mathsf{style}}$ denote VGG perceptual loss, adversarial loss, and style loss. For $\mathcal{L}_{\mathsf{VGG}}$, we adopt the formulation of [6,68]; $\mathcal{L}_{\mathsf{GAN}}$ is computed from multi-scale discriminators [62] with an objective similar to LSGAN [37].

To further enforce that the appearance of rendered images matches that of exemplar images, our style loss $\mathcal{L}_{\mathsf{style}}$ compares l_1 differences between Gram matrices constructed from VGG features at different layers. We empirically observe $\mathcal{L}_{\mathsf{style}}$ can guide our model to correctly capture the appearance of exemplar images, especially for rare photos such as those taken at sunset.

4 Experiments

We conduct extensive experiments to validate our proposed approach on our Internet photo dataset. We first compare with two baseline methods both quantitatively and qualitatively on the tasks of view synthesis, appearance transfer and appearance interpolation. We also present an ablation study to examine the impact of different configurations of our model. Finally, we perform a user study whose results demonstrate the quality of our synthesized novel views.

Table 1. Quantitative comparisons on our test set. Lower is better for l_1 and LPIPS and higher is better for PSNR. l_1 errors are scaled by 10 for ease of presentation.

	Trevi fountain			Sacre coeur			The pantheon			Top of the rock			Piazza navona		
Method	l_1	LPIPS	PSNR	l_1	LPIPS	PSNR	l_1	LPIPS	PSNR	l_1	LPIPS	PSNR	l_1	LPIPS	PSNR
MUNIT [24]	0.768	2.62	20.1	0.740	2.08	20.2	0.560	1.51	21.4	0.876	3.68	18.2	0.984	2.80	17.4
NRW [41]	0.779	2.07	20.0	0.808	1.90	19.6	0.592	1.35	21.1	0.802	2.76	19.3	1.050	2.64	17.1
w/o 2-phase	0.651	1.68	21.0	0.695	1.61	20.8	0.515	1.12	21.9	0.694	2.19	20.4	1.010	2.52	17.4
w/o AdaIN	0.780	1.87	19.8	0.801	1.89	19.6	0.609	1.30	20.9	0.773	2.58	19.3	1.150	2.97	17.1
w/o F^r	0.712	1.74	20.5	0.737	1.78	20.2	0.556	1.25	21.5	0.720	2.47	19.9	1.045	2.62	17.0
w/o $E(\Phi_s^r)$	0.670	1.70	20.9	0.715	1.66	20.5	0.549	1.16	21.5	0.703	2.24	20.0	1.017	2.52	17.2
Ours (full)	**0.618**	**1.56**	**21.8**	**0.676**	**1.57**	**21.0**	**0.495**	**1.08**	**22.5**	**0.642**	**2.48**	**20.7**	**0.933**	**2.32**	**17.6**

Data and Implementation. We evaluate our approach on five of our reconstructed scenes, which contain on average 2,064 images. For each scene, images are randomly split into training and test sets with a 85:15 ratio. We train a separate model for each scene. To mask out transient objects such as people and cars during training and evaluation, we adopt state-of-the-art semantic and instance segmentation algorithms [5,20] to create binary object masks. We set the dimension of the latent appearance vector to $\mathbf{z}_s \in \mathbb{R}^{16}$, and that of our latent DeepMPI features to $F^r_{\mathbf{p}} \in \mathbb{R}^8$. We refer readers to the supplemental material for scene statistics, network architectures, and other implementation details.

Baselines. We compare our approach to two state-of-the-art multi-modal image-to-image translation methods, adapted to our task: MUNIT [24] and Neural Rerendering in the Wild (NRW) [41]. To compare to MUNIT, we adapt their network G to predict an RGB image at the target viewpoint from a corresponding base color input, and train with a bidirectional reconstruction loss. For NRW, both E and G take as input base color, per-frame depth derived from the DeepMPI, and semantic segmentation at the target viewpoint. G then predicts a corresponding RGB image conditioned on the appearance vector extracted by E. We follow the same staged training strategy and use the same losses as in [41].

Error Metrics. Similar to [41], we report test image reconstruction errors using three error metrics: l_1 error, peak signal-to-noise ratio (PSNR), and perceptual similarity (via LPIPS [67]). Prior work has found the LPIPS metric to be better correlated with human visual perception than other metrics.

Quantitative Comparison. For fair comparison, we train and evaluate the baselines using the same data and hyperparameter settings as our method. Table 1 shows results of quantitative comparisons on our test set. Our proposed approach outperforms the two baseline methods by a large margin in terms of l_1 and PSNR, and is significantly better in terms of LPIPS, indicating that our method achieves higher rendering quality and realism.

Ablation Study. We perform an ablation study to analyze the effect of individual system components. In particular, we replace four components with simpler configurations: (1) using a train-from-scratch baseline to estimate alpha,

Fig. 6. Appearance transfer comparison. From left to right: (a) exemplar images used to extract appearance vectors, (b) predictions from MUNIT [24], (c) predictions from NRW [41], (d) predictions from our method. Compared to the baselines, our rendered images are more photo-realistic and are more faithful to the appearance of the exemplar images. Please zoom in to highlighted regions for better visual comparisons.

as described in Sect. 3.3 (w/o 2-phase), (2) including $\mathbf{z}$ as an input to G rather than using AdaIN (w/o AdaIN), (3) removing latent features from the DeepMPI (w/o F^r), and (4) encoding $\mathbf{z}$ only from the exemplar image and not additionally from the deep buffer (w/o $E(\Phi_s^r)$). Quantitative results are reported in Table 1. Latent DeepMPI features, as well as the use of AdaIN in our neural renderer, yield significant improvements, and lead to better rendering quality for thin structures and attached shadows, as highlighted in Fig. 5. Encoding the reference deep buffer also yields rendered images that better match the exemplar image.

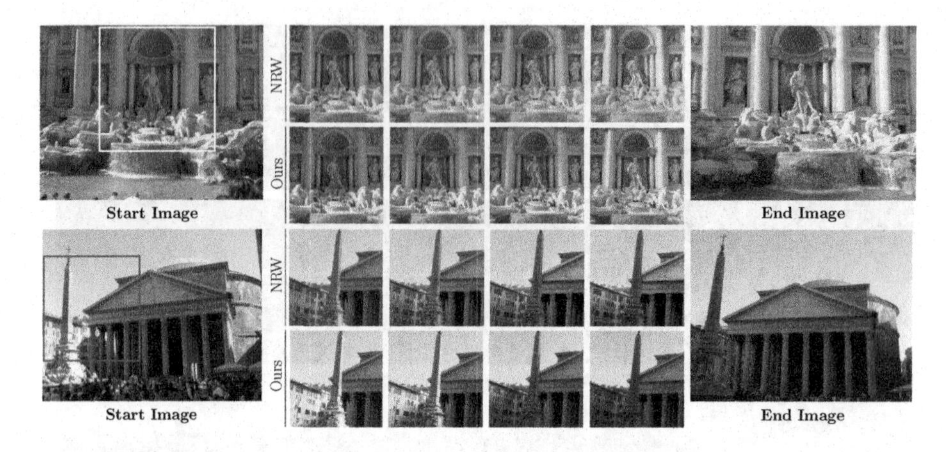

Fig. 7. Appearance interpolation. The left- and rightmost exemplar images indicate start and end appearance. Intermediate images are generated by linearly interpolating latent vectors from the two images. Odd rows show interpolation results from NRW [41], and even rows from our method. Moving shadows are indicated in highlighted regions.

Rendering with Appearance Transfer. Fig. 6 shows qualitative comparisons between our method and the two baselines on our test set in terms of rendering quality and appearance transferability (i.e., how well the model can transfer illumination and appearance of an exemplar image to a target viewpoint). We demonstrate compelling results in challenging cases such as sunset, which is a rare condition in the input photos. Compared to MUNIT, our rendered images are more realistic and exhibit fewer artifacts. For example, our rendered images successfully model specularities on glass windows, details on running water and droplets, cast shadows, and directional lighting effect as shown in the highlighted regions in Fig. 6. Our approach can also generate complex highlights and cast shadows from the sun. Compared with NRW, our rendered images are more faithful to the illumination in the exemplar image (e.g., for sunset appearance). Moreover, our approach can extrapolate appearance beyond the field of view of the exemplar image, as shown in the last row of Fig. 6. We refer readers to the supplemental video for visual comparisons with animated camera trajectories.

Appearance Interpolation. A key advantage of our method is the ability to interpolate between plenoptic slices in the latent appearance space. We conduct qualitative comparisons between our approach and NRW on appearance interpolation. We choose two images to define the start and end appearance, and linearly interpolate their latent vectors to produce in-between appearances. Figure 7 shows a comparison of interpolation results. In the first two rows of the figure, we observe that our method can simulate the progression of surfaces exposed to sunlight as the sun moves, while NRW fails to produce this effect. In the last row, our approach recovers the gradual motion of shadows throughout the day, while shadows in the NRW results tend to fade less naturally during

Fig. 8. 4D Photos. We demonstrate an application of creating *4D photos* by performing spatial-temporal interpolation in which both camera viewpoint and scene illumination change simultaneously. Results are best appreciated in the supplementary videos.

(a) insufficient view (b) MPI limits (c) exemplar for (d) (d) missing shadow

Fig. 9. Limitations. Some failure cases include: (a) input photo collections that do not span the full range of desired viewpoints, or (b) intrinsic limitations of MPI leading to poor extrapolation to large camera motions. In addition, as shown in (c) (exemplar image with strong shadow) and (d) (resulting rendering), our method can fail to model strong cast shadows produced by occluders outside the reference field of view.

interpolation. We refer readers to the supplemental videos for animated comparisons.

4D Photos. Figure 8 shows an application of our method to generating animated *4D photos* by animating the 3D viewpoints and simultaneously interpolating between latent appearance features. Our results achieve convincing changes across a variety of times of day and lighting conditions. The parallax effect of our results is best appreciated in the supplemental videos.

User Study. We ran a user study using 24 random sets of videos with camera movements and synthesized images from 5 different scenes. Each video is a sequence of novel views generated by our method, NRW [41], or MUNIT [24]. To quantify the performance of appearance transfer, we also show comparisons of results generated from different exemplar images selected from our test set. We invited 46 participants and asked them to rank the results of the three approaches. 88% of the time, participants responded that the videos produced by our system are the most temporally coherent. 82% of the time, they responded that the results from our method best reproduce the details of structure and illumination one would expect of a real-world scene. 77% of the time, they responded that the results from our method are the most faithful to the corresponding exemplar.

5 Discussion and Conclusion

Limitations. Our method inherits limitations from MPIs. For example, MPIs fail to generalize to viewpoints that are not well-sampled, or that are far from the reference view of the MPI (see Fig. 9(a–b)). In addition, our model can also sometimes fail to model cast shadows from occluders outside of the reference field of view, as shown in Fig. 9(c) and (d). Despite these limitations, we believe our work represents a significant advance towards photo-realistic capture and rendering of the world from crowd photography.

Conclusion. We presented a method for synthesizing novel views of scenes under time-varying appearance from Internet photos. We proposed a new DeepMPI representation and a method for optimizing and decoding DeepMPIs conditioned on viewing conditions present in different photos. Our method can synthesize plenoptic slices that can be interpolated to recover local regions of the full plenoptic function. In the future, we envision enabling even larger changes in viewpoint and illumination, including 4D walkthroughs of large-scale scenes.

Acknowledgements. We thank Kai Zhang, Jin Sun, and Qianqian Wang for helpful discussions. This research was supported in part by the generosity of Eric and Wendy Schmidt by recommendation of the Schmidt Futures program.

References

1. Adelson, E.H., Bergen, J.R.: The plenoptic function and the elements of early vision. In: Computational Models of Visual Processing, pp. 3–20. MIT Press (1991)
2. Buehler, C., Bosse, M., McMillan, L., Gortler, S., Cohen, M.: Unstructured lumigraph rendering. In: Proceedings of the 28th Annual Conference on Computer Graphics and Interactive Techniques, pp. 425–432 (2001)
3. Chai, J.X., Tong, X., Chan, S.C., Shum, H.Y.: Plenoptic sampling. In: Proceedings of the 27th Annual Conference on Computer Graphics and Interactive Techniques, SIGGRAPH 2000, pp. 307–318. ACM Press/Addison-Wesley Publishing Co., USA (2000). https://doi.org/10.1145/344779.344932
4. Chaurasia, G., Duchene, S., Sorkine-Hornung, O., Drettakis, G.: Depth synthesis and local warps for plausible image-based navigation. ACM Trans. Graph. **32**(3), 1–12 (2013)
5. Chen, L.C., Papandreou, G., Schroff, F., Adam, H.: Rethinking atrous convolution for semantic image segmentation. arXiv preprint arXiv:1706.05587 (2017)
6. Chen, Q., Koltun, V.: Photographic image synthesis with cascaded refinement networks. In: Proceedings of the International Conference on Computer Vision (ICCV), pp. 1511–1520 (2017)
7. Chen, Z., et al.: A neural rendering framework for free-viewpoint relighting. arXiv preprint arXiv:1911.11530 (2019)
8. Choi, I., Gallo, O., Troccoli, A., Kim, M.H., Kautz, J.: Extreme view synthesis. In: Proceedings of the International Conference on Computer Vision (ICCV), pp. 7781–7790 (2019)
9. Davis, A., Levoy, M., Durand, F.: Unstructured light fields. Comput. Graph. Forum **31**, 305–314 (2012)

10. Debevec, P.E., Taylor, C.J., Malik, J.: Modeling and rendering architecture from photographs: a hybrid geometry-and image-based approach. In: Proceedings of the 23rd Annual Conference on Computer Graphics and Interactive Techniques, pp. 11–20 (1996)
11. Eslami, S.A., et al.: Neural scene representation and rendering. Science **360**(6394), 1204–1210 (2018)
12. Flynn, J., et al.: DeepView: view synthesis with learned gradient descent. In: Proceedings of the Computer Vision and Pattern Recognition (CVPR), pp. 2367–2376 (2019)
13. Flynn, J., Neulander, I., Philbin, J., Snavely, N.: DeepStereo: learning to predict new views from the world's imagery. In: Proceedings of the Computer Vision and Pattern Recognition (CVPR), pp. 5515–5524 (2016)
14. Garg, R., Du, H., Seitz, S.M., Snavely, N.: The dimensionality of scene appearance. In: Proceedings of the International Conference on Computer Vision (ICCV), pp. 1917–1924. IEEE (2009)
15. Gatys, L.A., Ecker, A.S., Bethge, M.: Image style transfer using convolutional neural networks. In: Proc. Computer Vision and Pattern Recognition (CVPR). pp. 2414–2423 (2016)
16. Goodfellow, I., et al.: Generative adversarial nets. In: Neural Information Processing Systems, pp. 2672–2680 (2014)
17. Gortler, S.J., Grzeszczuk, R., Szeliski, R., Cohen, M.F.: The lumigraph. In: Proceedings of the 23rd Annual Conference on Computer Graphics and Interactive Techniques, pp. 43–54 (1996)
18. Gulrajani, I., Ahmed, F., Arjovsky, M., Dumoulin, V., Courville, A.C.: Improved training of Wasserstein GANs. In: Neural Information Processing Systems, pp. 5767–5777 (2017)
19. Hauagge, D.C., Wehrwein, S., Upchurch, P., Bala, K., Snavely, N.: Reasoning about photo collections using models of outdoor illumination. In: Proceedings of the British Machine Vision Conference (BMVC) (2014)
20. He, K., Gkioxari, G., Dollár, P., Girshick, R.: Mask R-CNN. In: Proceedings of the International Conference on Computer Vision (ICCV), pp. 2961–2969 (2017)
21. Hedman, P., Alsisan, S., Szeliski, R., Kopf, J.: Casual 3D photography. ACM Trans. Graph. **36**, 234:1–234:15 (2017)
22. Hedman, P., Philip, J., Price, T., Frahm, J.M., Drettakis, G., Brostow, G.: Deep blending for free-viewpoint image-based rendering. ACM Trans. Graph. **37**(6), 1–15 (2018)
23. Huang, X., Belongie, S.: Arbitrary style transfer in real-time with adaptive instance normalization. In: Proceedings of the International Conference on Computer Vision (ICCV), pp. 1501–1510 (2017)
24. Huang, X., Liu, M.-Y., Belongie, S., Kautz, J.: Multimodal unsupervised image-to-image translation. In: Ferrari, V., Hebert, M., Sminchisescu, C., Weiss, Y. (eds.) ECCV 2018. LNCS, vol. 11207, pp. 179–196. Springer, Cham (2018). https://doi.org/10.1007/978-3-030-01219-9_11
25. Isola, P., Zhu, J.Y., Zhou, T., Efros, A.A.: Image-to-image translation with conditional adversarial networks. In: Proceedings of the Computer Vision and Pattern Recognition (CVPR), pp. 1125–1134 (2017)
26. Kalantari, N.K., Wang, T.C., Ramamoorthi, R.: Learning-based view synthesis for light field cameras. ACM Trans. Graph. **35**(6), 1–10 (2016)
27. Karras, T., Aila, T., Laine, S., Lehtinen, J.: Progressive growing of GANs for improved quality, stability, and variation. arXiv preprint arXiv:1710.10196 (2017)

28. Karras, T., Laine, S., Aila, T.: A style-based generator architecture for generative adversarial networks. In: Proceedings of the Computer Vision and Pattern Recognition (CVPR), pp. 4401–4410 (2019)
29. Laffont, P.Y., Bousseau, A., Paris, S., Durand, F., Drettakis, G.: Coherent intrinsic images from photo collections. ACM Trans. Graph. **31**, 202:1–202:11 (2012)
30. Ledig, C., et al.: Photo-realistic single image super-resolution using a generative adversarial network. In: Proceedings of the Computer Vision and Pattern Recognition (CVPR), pp. 4681–4690 (2017)
31. Lee, H.-Y., Tseng, H.-Y., Huang, J.-B., Singh, M., Yang, M.-H.: Diverse image-to-image translation via disentangled representations. In: Ferrari, V., Hebert, M., Sminchisescu, C., Weiss, Y. (eds.) ECCV 2018. LNCS, vol. 11205, pp. 36–52. Springer, Cham (2018). https://doi.org/10.1007/978-3-030-01246-5_3
32. Levin, A., Durand, F.: Linear view synthesis using a dimensionality gap light field prior. In: Proceedings Computer Vision and Pattern Recognition (CVPR), pp. 1831–1838 (2010)
33. Levoy, M., Hanrahan, P.: Light field rendering. In: Proceedings of the 23rd Annual Conference on Computer Graphics and Interactive Techniques, pp. 31–42 (1996)
34. Li, Z., et al.: Learning the depths of moving people by watching Frozen people. In: Proceedings of the Computer Vision and Pattern Recognition (CVPR), pp. 4521–4530 (2019)
35. Li, Z., Snavely, N.: MegaDepth: learning single-view depth prediction from internet photos. In: Proceedings of the Computer Vision and Pattern Recognition (CVPR), pp. 2041–2050 (2018)
36. Lombardi, S., Simon, T., Saragih, J., Schwartz, G., Lehrmann, A., Sheikh, Y.: Neural volumes: learning dynamic renderable volumes from images. ACM Trans. Graph. **38**(4), 65 (2019)
37. Mao, X., Li, Q., Xie, H., Lau, R.Y., Wang, Z., Paul Smolley, S.: Least squares generative adversarial networks. In: Proceedings of the International Conference on Computer Vision (ICCV), pp. 2794–2802 (2017)
38. Martin-Brualla, R., Gallup, D., Seitz, S.M.: 3D time-lapse reconstruction from internet photos. In: Proceedings of the International Conference on Computer Vision (ICCV), pp. 1332–1340 (2015)
39. Martin-Brualla, R., Gallup, D., Seitz, S.M.: Time-lapse mining from internet photos. ACM Trans. Graph. **34**(4), 1–8 (2015)
40. Matzen, K., Snavely, N.: Scene chronology. In: Fleet, D., Pajdla, T., Schiele, B., Tuytelaars, T. (eds.) ECCV 2014. LNCS, vol. 8695, pp. 615–630. Springer, Cham (2014). https://doi.org/10.1007/978-3-319-10584-0_40
41. Meshry, M., et al.: Neural rerendering in the wild. In: Proceedings of the Computer Vision and Pattern Recognition (CVPR), pp. 6871–6880 (2019)
42. Mildenhall, B., et al.: Local light field fusion: practical view synthesis with prescriptive sampling guidelines. ACM Trans. Graph. **38**(4), 1–14 (2019)
43. Park, T., Liu, M.Y., Wang, T.C., Zhu, J.Y.: Semantic image synthesis with spatially-adaptive normalization. In: Proceedings of the Computer Vision and Pattern Recognition (CVPR), pp. 2337–2346 (2019)
44. Penner, E., Zhang, L.: Soft 3D reconstruction for view synthesis. ACM Trans. Graph. **36**(6), 1–11 (2017)
45. Philip, J., Gharbi, M., Zhou, T., Efros, A.A., Drettakis, G.: Multi-view relighting using a geometry-aware network. ACM Trans. Graph. **38**(4), 1–14 (2019)
46. Sangkloy, P., Lu, J., Fang, C., Yu, F., Hays, J.: Scribbler: controlling deep image synthesis with sketch and color. In: Proceedings of the Computer Vision and Pattern Recognition (CVPR), pp. 5400–5409 (2017)

47. Schonberger, J.L., Frahm, J.M.: Structure-from-motion revisited. In: Proceedings of the Computer Vision and Pattern Recognition (CVPR), pp. 4104–4113 (2016)
48. Shan, Q., Adams, R., Curless, B., Furukawa, Y., Seitz, S.M.: The visual turing test for scene reconstruction. In: International Conference on 3D Vision (3DV), pp. 25–32 (2013)
49. Sheng, L., Lin, Z., Shao, J., Wang, X.: Avatar-net: multi-scale zero-shot style transfer by feature decoration. In: Proceedings of the Computer Vision and Pattern Recognition (CVPR), pp. 1–9 (2018)
50. Shi, L., Hassanieh, H., Davis, A., Katabi, D., Durand, F.: Light field reconstruction using sparsity in the continuous Fourier domain. ACM Trans. Graph. **34**, 12:1–12:13 (2014)
51. Shi, L., Hassanieh, H., Davis, A., Katabi, D., Durand, F.: Light field reconstruction using sparsity in the continuous Fourier domain. ACM Trans. Graph. **34**(1) (2015). https://doi.org/10.1145/2682631
52. Simon, I., Snavely, N., Seitz, S.M.: Scene summarization for online image collections. In: Proceedings of the International Conference on Computer Vision (ICCV), pp. 1–8. IEEE (2007)
53. Sitzmann, V., Thies, J., Heide, F., Nießner, M., Wetzstein, G., Zollhofer, M.: DeepVoxels: learning persistent 3D feature embeddings. In: Proceedings of the Computer Vision and Pattern Recognition (CVPR), pp. 2437–2446 (2019)
54. Sitzmann, V., Zollhöfer, M., Wetzstein, G.: Scene representation networks: continuous 3D-structure-aware neural scene representations. In: Neural Information Processing Systems, pp. 1119–1130 (2019)
55. Snavely, N., Seitz, S.M., Szeliski, R.: Photo tourism: exploring photo collections in 3D. ACM Trans. Graph. (SIGGRAPH) (2006)
56. Srinivasan, P.P., Tucker, R., Barron, J.T., Ramamoorthi, R., Ng, R., Snavely, N.: Pushing the boundaries of view extrapolation with multiplane images. In: Proceedings of the Computer Vision and Pattern Recognition (CVPR), pp. 175–184 (2019)
57. Srinivasan, P.P., Wang, T., Sreelal, A., Ramamoorthi, R., Ng, R.: Learning to synthesize a 4D RGBD light field from a single image. In: Proceedings of the International Conference on Computer Vision (ICCV), pp. 2243–2251 (2017)
58. Szeliski, R., Golland, P.: Stereo matching with transparency and matting. Int. J. Comput. Vis. **32**, 45–61 (1998)
59. Thies, J., Zollhöfer, M., Nießner, M.: Deferred neural rendering: image synthesis using neural textures. ACM Trans. Graph. **38**(4), 1–12 (2019)
60. Ulyanov, D., Vedaldi, A., Lempitsky, V.: Improved texture networks: maximizing quality and diversity in feed-forward stylization and texture synthesis. In: Proceedings of the Computer Vision and Pattern Recognition (CVPR), pp. 6924–6932 (2017)
61. Vagharshakyan, S., Bregovic, R., Gotchev, A.P.: Light field reconstruction using shearlet transform. Trans. Pattern Anal. Mach. Intell. **40**, 133–147 (2015)
62. Wang, T.C., Liu, M.Y., Zhu, J.Y., Tao, A., Kautz, J., Catanzaro, B.: High-resolution image synthesis and semantic manipulation with conditional GANs. In: Proceedings of the Computer Vision and Pattern Recognition (CVPR), pp. 8798–8807 (2018)
63. Wang, X., Gupta, A.: Generative image modeling using style and structure adversarial networks. In: Leibe, B., Matas, J., Sebe, N., Welling, M. (eds.) ECCV 2016. LNCS, vol. 9908, pp. 318–335. Springer, Cham (2016). https://doi.org/10.1007/978-3-319-46493-0_20

64. Xian, W., et al.: TextureGAN: controlling deep image synthesis with texture patches. In: Proceedings of the Computer Vision and Pattern Recognition (CVPR), pp. 8456–8465 (2018)
65. Xu, Z., Bi, S., Sunkavalli, K., Hadap, S., Su, H., Ramamoorthi, R.: Deep view synthesis from sparse photometric images. ACM Trans. Graph. **38**(4) (2019)
66. Yu, Y., Smith, W.A.: InverseRenderNet: learning single image inverse rendering. In: Proceedings of the Computer Vision and Pattern Recognition (CVPR), pp. 3155–3164 (2019)
67. Zhang, R., Isola, P., Efros, A.A., Shechtman, E., Wang, O.: The unreasonable effectiveness of deep features as a perceptual metric. In: Proceedings of the Computer Vision and Pattern Recognition (CVPR), pp. 586–595 (2018)
68. Zhou, T., Tucker, R., Flynn, J., Fyffe, G., Snavely, N.: Stereo magnification: learning view synthesis using multiplane images. ACM Trans. Graph. **37**, 1–12 (2018)
69. Zhou, T., Tulsiani, S., Sun, W., Malik, J., Efros, A.A.: View synthesis by appearance flow. In: Leibe, B., Matas, J., Sebe, N., Welling, M. (eds.) ECCV 2016. LNCS, vol. 9908, pp. 286–301. Springer, Cham (2016). https://doi.org/10.1007/978-3-319-46493-0_18
70. Zhu, J.Y., Park, T., Isola, P., Efros, A.A.: Unpaired image-to-image translation using cycle-consistent adversarial networks. In: Proceedings of the International Conference on Computer Vision (ICCV), pp. 2223–2232 (2017)
71. Zhu, J.Y., et al.: Toward multimodal image-to-image translation. In: Neural Information Processing Systems, pp. 465–476 (2017)
72. Zitnick, C.L., Kang, S.B., Uyttendaele, M., Winder, S.A.J., Szeliski, R.: High-quality video view interpolation using a layered representation. In: SIGGRAPH 2004 (2004)

VoxelPose: Towards Multi-camera 3D Human Pose Estimation in Wild Environment

Hanyue Tu[1,2], Chunyu Wang[1(✉)], and Wenjun Zeng[1]

[1] Microsoft Research Asia, Beijing, China
chnuwa@microsoft.com, wezeng@microsoft.com
[2] University of Science and Technology of China, Hefei, China
tuhanyue@mail.ustc.edu.cn

Abstract. We present *VoxelPose* to estimate 3D poses of multiple people from multiple camera views. In contrast to the previous efforts which require to establish cross-view correspondence based on noisy and incomplete 2D pose estimates, *VoxelPose* directly operates in the 3D space therefore avoids making incorrect decisions in each camera view. To achieve this goal, features in all camera views are aggregated in the 3D voxel space and fed into *Cuboid Proposal Network* (CPN) to localize all people. Then we propose *Pose Regression Network* (PRN) to estimate a detailed 3D pose for each proposal. The approach is robust to occlusion which occurs frequently in practice. Without bells and whistles, it outperforms the previous methods on several public datasets.

Keyword: 3D human pose estimation

1 Introduction

Estimating 3D human pose from multiple cameras separated by wide baselines [1–7] has been a longstanding problem in computer vision. The goal is to predict 3D positions of the landmark joints for all people in a scene. The successful resolution of the task can benefit many applications such as intelligent sports [5] and retail analysis.

The previous works such as [4,5] propose to address the problem in three steps. They first estimate 2D poses in each camera view independently, for example, by Convolutional Neural Networks (CNN) [8,9]. Then, in the second step, the poses that correspond to the same person in different views are grouped into clusters according to appearance and geometry cues. The final step is to estimate a 3D pose for each person (*i.e.* each cluster) by standard methods such as triangulation [10] or pictorial structure models [11].

In spite of the fact that 2D pose estimation has quickly matured due to the development of CNN models [12,13], the estimation results are still unsatisfactory for challenging cases especially when occlusion occurs which is often the

This work is done when Hanyue Tu is an intern at Microsoft Research Asia.

A. Vedaldi et al. (Eds.): ECCV 2020, LNCS 12346, pp. 197–212, 2020.
https://doi.org/10.1007/978-3-030-58452-8_12

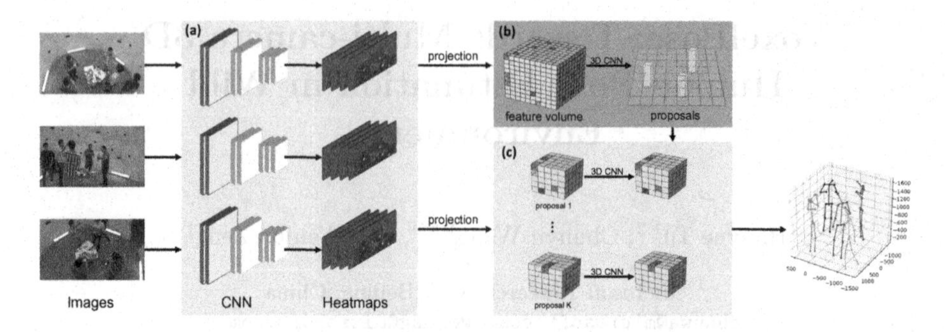

Fig. 1. Overview of our approach. It consists of three modules: (a) we first estimate 2D pose heatmaps for all views; (b) we warp the heatmaps to a common 3D space and construct a feature volume which is fed into a Cuboid Proposal Network to localize all people instances; (c) for each proposal, we construct a finer-grained feature volume and estimate a 3D pose.

case for natural scenes. See Fig. 5 for some example 2D poses estimated by the state-of-the-art method [13]. In addition, it is very difficult to establish cross-view correspondence when 2D poses are inaccurate. All these pose a serious challenge for 3D pose estimation in the wild.

To avoid making incorrect decisions for each camera view, we propose a 3D pose estimator which directly operates in the 3D space by gathering information from all camera views. Figure 1 shows an overview of our approach. It first estimates 2D heatmaps for each view to encode per-pixel likelihood of all joints as shown in Fig. 1(a). Different from the previous works, we do not determine the locations of joints (*e.g.*, by finding the maximum response) nor group the joints into different instances because estimated heatmaps are usually very noisy and incomplete. Instead, we project the heatmaps of all views to a common 3D space as in [6] and obtain a more complete feature volume which allows us to accurately estimate the 3D positions of all joints.

We first present Cuboid Proposal Network (CPN), as shown in Fig. 1(b), to coarsely localize all people in the scene by predicting a number of 3D cuboid proposals from the 3D feature volume. Then for each proposal, we construct a separate *finer-grained* feature volume centered at each proposal, and feed it into a Pose Regression Network (PRN) to estimate a detailed 3D pose. See Fig. 1(c) for illustration. The two networks are composed of basic 3D convolution layers and can be jointly trained.

It is worth noting that our approach implicitly accomplishes two types of association which was previously addressed by post-processing methods. Firstly, the joints of the same person *in a single camera view* are implicitly associated by the cuboid proposal. This was previously addressed in the 2D space either by bottom-up approaches [8,14] or by top-down approaches [13] which would suffer when occlusion occurs. Secondly, the joints that correspond to the same person *in different camera views* are also implicitly associated based on the fact that the 2D poses which overlap with the projections of a 3D pose belong to the same person. Our approach allows us to avoid the challenging association tasks therefore significantly improves the robustness.

We evaluate our approach on three public datasets including Campus [2], Shelf [2] and CMU Panoptic [15]. It outperforms the state-of-the-arts on the first two datasets. Since no work has reported numerical results on Panoptic, we conduct a series of ablation studies by comparing our approach to several baselines. In addition, we find that CPN and PRN can be accurately trained on automatically generated synthetic heatmaps. They achieve similar results as the models trained on realistic images. This is possible mainly because the heatmap based 3D feature volume representation is a high level abstraction that is disentangled from appearance/lighting, etc. This favorable property dramatically enhances the practical values of the approach.

2 Related Work

In this section, we briefly review the related works on 3D pose estimation for single and multiple people scenarios, respectively. We discuss their main difference from our work and summarize our contributions.

2.1 Single Person 3D Pose Estimation

We briefly classify the existing works into *analytical* and *predictive* approaches based on whether they have learnable parameters. Analytical methods [6,11,16,17] explicitly model the relationship between a 2D and 3D pose according to the camera geometry. **On one hand**, when multiple cameras are available, the 3D pose can be fully determined by simple geometry methods such as triangulation [10] based on the 2D poses in each view. So the bottleneck lies in the inaccuracy of 2D pose estimations. Some works [6,11] propose to model the conditional dependence between the joints and jointly infer their 3D positions to improve their robustness to errors in 2D poses. **On the other hand**, when only one camera is available, the problem is under-determined because multiple 3D poses may correspond to the same 2D pose. The previous works [16–18] propose to use low-dimensional pose representations to reduce ambiguities. They optimize the low-dimensional parameters of the representation such that the discrepancy between its projection and the 2D pose is minimized. The improvement in 2D pose estimation has boosted 3D accuracy.

The predictive models [19–27] are mainly proposed for the single camera setup aiming to address the ambiguity issue by powerful neural networks. The pioneer works [20,21] propose to regress 3D pose from 2D joint locations by various networks. Some recent works [6,19,26,28,29] also propose to regress a volumetric 3D pose representation from images. In particular, in [6,26], the authors project 2D features or pose heatmaps to 3D space and estimate 3D positions of the body joints. The approach achieves better performance than the triangulation and pictorial structure models on *single* person pose estimation. However, it requires to address the challenging association problem in order to apply to scenes with multiple people.

2.2 Multiple Person 3D Pose Estimation

There are two challenging association problems in this task. First, it needs to associate the joints of the same person by either top-down [9,30] or bottom-up [8,14,31] strategies. Second, it needs to associate the 2D poses of the same person in different views based on appearance features [4,5] which are unstable when people are occluded. The pictorial structure model is extended to deal with multiple people in [1,2]. The number of people is assumed to be known which is difficult by itself. Besides, the interactions between different people introduce loops into the graph model and complicate the optimization problem. These challenges limit the 3D pose estimation accuracy.

Our work differs from the previous methods [4,5] in that it elegantly avoids the two challenging association problems. This is because a 3D cuboid proposal already naturally associates the joints of the same person in the same and different views by projecting the proposals to image space. Different from the pictorial structure models [1,2], our approach does not suffer from the local optimum and does not need the number of people in each frame to be known as an input. We find in our experiments that our approach outperforms the previous methods on several public datasets.

3 Cuboid Proposal Network

The overview of our approach is shown in Fig. 1. It first estimates 2D pose heatmaps for every camera view independently by HRNet [13]. Then we introduce Cuboid Proposal Network (CPN) to localize all people by a number of cuboid proposals. Finally, we present Pose Regression Network (PRN) to regress a detailed 3D pose for each proposal. In this section, we focus on the details of CPN including the input, output and network structures.

3.1 Feature Volume

The input to CPN is a 3D feature volume which contains rich information for detecting people in the 3D space. The feature volume is constructed by projecting the 2D pose heatmaps in all camera views to a common discretized 3D space as will be detailed later. Since the 2D pose heatmaps encode location information of the joints, the resulting 3D feature volume also carries rich information for detecting 3D poses.

We discretize the 3D space, in which people can freely move, by $X \times Y \times Z$ discrete locations $\{\mathbf{G}^{x,y,z}\}$. Each location can be regarded as an anchor of people. In order to reduce the quantization error, we set the distance between the neighboring anchors to be small by adjusting the values of X, Y and Z, respectively. In general, the space is about $8 textm \times 8\,\mathrm{m} \times 2\,\mathrm{m}$ on the public datasets [2,15]. So we set X, Y and Z to be 80, 80 and 20, respectively, to strike a good balance between speed and precision. The distance between two neighboring bins is about 100 mm which is sufficiently accurate for coarsely localizing people. Note that we will obtain finer-grained 3D poses in PRN.

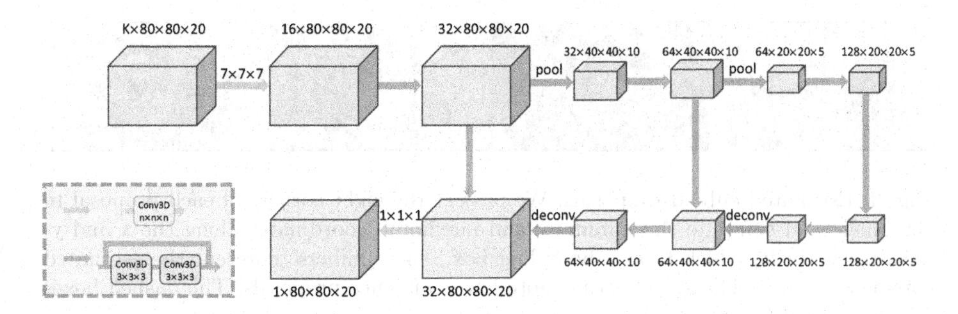

Fig. 2. Network structure of CPN. The input is a feature volume (see Sect. 3.1) and the output is the probability map **V** (see Sect. 3.2). The yellow arrow represents a standard 3D convolutional layer and the blue arrow represents a Residual Block of two 3D convolutional layers as shown in the legend.

We compute a feature vector for each anchor by sampling and fusing the 2D heatmap values at its projected locations in all camera views. Denote the 2D heatmap of view v as $\mathbf{M}_v \in \mathcal{R}^{K \times H \times W}$ where K is the number of body joints. For each anchor location $\mathbf{G}^{x,y,z}$, we compute its projected location in view v as $\mathbf{P}_v^{x,y,z}$. The heatmap values at $\mathbf{P}_v^{x,y,z}$ is denoted as $\mathbf{M}_v^{x,y,z} \in \mathcal{R}^K$. Then we compute a feature vector for the anchor as the average heatmap values in all camera views: $\mathbf{F}^{x,y,z} = \frac{1}{V} \sum_{v=1}^{V} \mathbf{M}_v^{x,y,z}$ where V is the number of cameras. More advanced fusion strategies, for example, assigning a data-dependent weight to reflect the heatmap estimation quality in each camera view, could be explored in the future work. In this work, we stick to the approach of computing average in order to keep the overall approach as simple as possible. We can see that $\mathbf{F}^{x,y,z}$ actually encodes the likelihood that the K joints are at $\mathbf{G}^{x,y,z}$ which is sufficient to infer people presence. In the following sections, we will describe how we estimate cuboid proposals from the feature volume $\mathbf{F}$.

3.2 Cuboid Proposals

We represent a cuboid proposal by a 3D bounding box whose orientation and size are fixed in our experiments. This is a reasonable simplification because the sizes of people in *3D space* have limited variations which differs from 2D proposals in object detection [32]. So the main task in CPN is to determine the people presence likelihood at each anchor location.

To generate proposals, we slide a small network over the feature volume $\mathbf{F}$. Each sliding window centered at an anchor is mapped to a lower-dimensional feature which is fed to a fully connected layer to regress a confidence score $\mathbf{V}^{x,y,z}$ representing the likelihood of people presence at the location. The likelihood at all anchors form a 3D heatmap $\mathbf{V} \in \mathcal{R}^{X,Y,Z}$. Since we use fixed box orientation and size, we do not estimate them as the 2D object detectors [32,33]. Neither do we estimate center offsets relative to the anchor locations for precise people locations because coarse locations are sufficient.

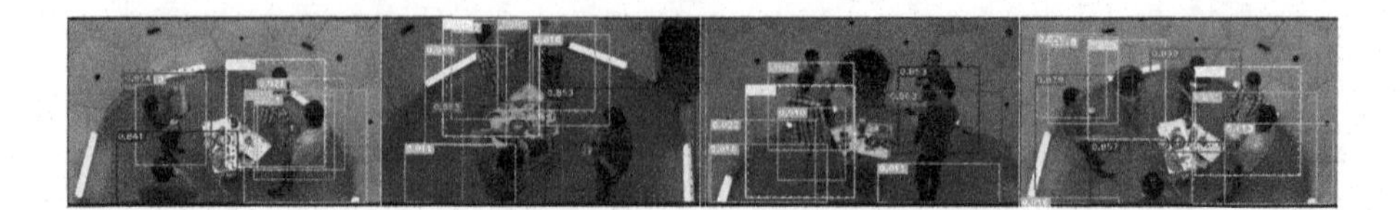

Fig. 3. Estimated cuboid proposals. We project the eight corners of each proposal to the image and compute the minimum and maximum coordinates along the x and y-axis, respectively, which form a bounding box. The numbers represent the estimated confidence scores. The gray boxes denote low confidence proposals. The dashed boxes are the ground-truth.

We compute a Ground-Truth (GT) heatmap score $\mathbf{V}_*^{x,y,z}$ for every anchor according to its distance to the GT poses. Specifically, for each pair of GT pose (root joint) and anchor, we compute a Gaussian score according to their distance. The score decreases exponentially when the distance increases. Note that there could be multiple scores for one anchor if there are multiple people in the scene and we simply keep the largest one. We train CPN by minimizing:

$$\mathcal{L}_{\mathrm{CPN}} = \sum_{x=1}^{X}\sum_{y=1}^{Y}\sum_{z=1}^{Z} \|\mathbf{V}_*^{x,y,z} - \mathbf{V}^{x,y,z}\|_2 \tag{1}$$

The edge length of every proposal is set to be 2000 mm which is sufficiently large to cover people in arbitrary poses. The orientations of the proposals are aligned with the world coordinate axes.

3.3 Non-maximum Suppression

We select the anchors with large regression confidence values as the proposals. On top of the 3D heatmap, we perform Non-Maximum Suppression (NMS) based on the heatmap scores to extract local peaks. Then, we keep the locations of peaks whose heatmap scores are larger than a threshold. Similar to [34], IOU based NMS is not needed for generating proposals because we only have one positive anchor per pose.

3.4 Network Structures of CPN

Inspired by the *voxel-to-voxel* prediction network in [35], we also adopt the 3D convolutions as the basic building blocks for CPN. Since the input feature volume is sparse and has clear semantic meanings, we propose a simpler structure than [35] which is shown in Fig. 2. In some scenarios such as football court, the motion capture space could be larger than that of the public datasets, which would lead to larger feature volume, thus notably reducing the inference speed. We solve the problem by using sparse 3D convolutions [36] because the feature volume only has a small number of non-zero values.

We visualize some estimated proposals in Fig. 3. We project the 3D proposals to 2D images using the camera parameters for the sake of simplicity. We can see that most people instances can be accurately retrieved even though some of them are severely occluded in the current view. This is mainly due to the effective fusion of multiview features in a common 3D space. We will numerically evaluate CPN in more details in the experiment section.

4 Pose Regression Network

In this section, we present the details of Pose Regression Network (PRN) which, for each proposal, predicts a complete 3D pose.

4.1 Constructing Feature Volume

Recall that we have already constructed a big feature volume in the previous CPN step, which covers the whole motion space, to coarsely localize people in the environment. However, we do NOT reuse it here in PRN because it is too coarse to accurately estimate the 3D positions of all joints. Instead, we construct a separate finer-grained feature volume centered at each proposal. The size of the feature volume is set to be 2000 mm × 2000 mm × 2000 mm, which is much smaller than that of CPN (8 m × 8 m × 2 m), but is still large enough to cover people in arbitrary poses. This volume is divided into a discrete grid with $X' \times Y' \times Z'$ bins where $X' = Y' = Z' = 64$. The edge length of a bin is about $\frac{2000}{64} = 31.25$ mm. Note that the precision of our approach is not bounded to 31.25 mm because we will use the integration trick [22] to reduce the impact of quantization error as will be described in more detail later. With these definitions, we compute the feature volume following the descriptions in Sect. 3.1.

4.2 Regression of Human Poses

We estimate a 3D heatmap $\mathbf{H}_k \in \mathcal{R}^{X' \times Y' \times Z'}$ for each joint k based on the constructed feature volume. Then the 3D location $\mathbf{J}_k$ of the joint can be obtained by computing the center of mass of $\mathbf{H}_k$ according to the following formula:

$$\mathbf{J}_k = \sum_{x=1}^{X'} \sum_{y=1}^{Y'} \sum_{z=1}^{Z'} (x, y, z) \cdot \mathbf{H}_k(x, y, z) \tag{2}$$

Note that we do not obtain the location $\mathbf{J}_k$ by finding the maximum of $\mathbf{H}_k$ because the quantization error of 31.25 mm is still large. Computing the expectation as in the above equation effectively reduces the error. This technique is frequently used in the previous works such as [22].

The estimated joint location is compared to the ground-truth location $\mathbf{J}_*$ to train PRN. Specifically, the L_1 loss is used:

$$\mathcal{L}_{\text{PRN}} = \sum_{k=1}^{K} \|\mathbf{J}_*^k - \mathbf{J}^k\|_1 \tag{3}$$

The network of PRN is kept the same as CPN as shown in Fig. 2 except that the input and output are different. The network weights are shared for different joints. We conducted experiments using different weights but that did not make much difference on current datasets.

4.3 Training Strategies

We first train the 2D pose estimation network for 20 epochs. The initial learning rate is 1e−4, and decreases to 1e−5 and 1e−6 at the 10_{th} and 15_{th} epochs, respectively. Then we jointly train the whole network including CPN and PRN for 10 epochs to convergence. The learning rate is set to be 1e−4. In some experiments (which will be described clearly), we directly use the backbone network learned on the COCO dataset without finetuning on target datasets.

5 Datasets and Metrics

The Campus Dataset [2]**.** This dataset captures three people interacting with each other in an outdoor environment by three cameras. We follow [2,4] to split the dataset into training and testing subsets. To avoid over-fitting to this small training data, we directly use the 2D pose estimator trained on the COCO dataset and only train CPN and PRN.

The Shelf Dataset [2]**.** It captures four people disassembling a shelf by five cameras. Similar to what we do for Campus, we use the 2D pose estimator trained on COCO and only train CPN and PRN.

The CMU Panoptic Dataset [15]**.** It captures people doing daily activities by dozens of cameras among which five HD cameras (3, 6, 12, 13, 23) are used in our experiments. We also report results for fewer cameras. Following [37], the training set consists of the following sequences: "160422_ultimatum1", "160224_haggling1", "160226_haggling1", "161202_haggling1", "160906_ian1", "160906_ian2", "160906_ian3", "160906_band1", "160906_band2", "16090 6_band3". The testing set consists of: "160906_pizza1", "160422_haggling1", "160906_ian5", and "160906_band4".

The Proposal Evaluation Metric. We compute recall of proposals at different proposal-groundtruth-distance. It is noteworthy that this metric is only loosely related to the 3D estimation accuracy. We keep ten proposals after NMS for evaluation on the three datasets.

The 3D Pose Estimation Metric. Following [4], we use the Percentage of Correct Parts (PCP3D) metric to evaluate the estimated 3D poses. Specifically, for each ground-truth 3D pose, it finds the closest pose estimation and computes the percentage of correct parts. This metric does not penalize false positive pose estimations. To overcome the limitation, we also extend the Average Precision (AP_K) metric [38] to the multi-person 3D pose estimation task which is more comprehensive than PCP3D. If the Mean Per Joint Position Error (MPJPE) of a pose is smaller than K millimeters, we think the pose is accurately estimated.

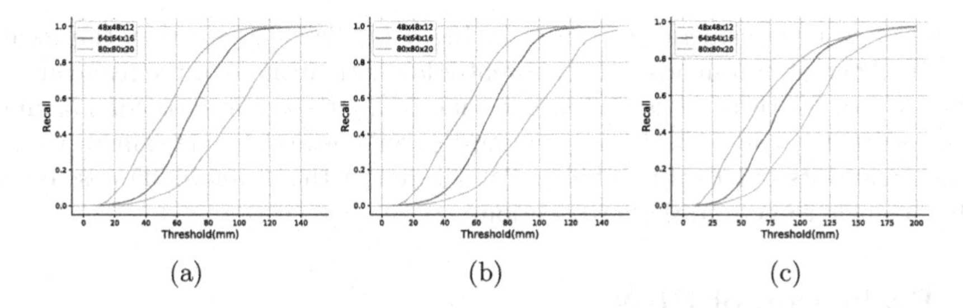

Fig. 4. Recall curves when the motion space is discretized with different numbers of bins on the Panoptic dataset. **(a)** CPN is trained/tested on the real images of five cameras from the Panoptic dataset. **(b)** CPN is trained on the synthetic heatmaps and tested on the real images of five cameras. **(c)** CPN is trained/tested on the real images of one camera from the Panoptic dataset.

6 Evaluation of CPN

We first study the impact of the space division granularity to the proposals by setting different values to the X, Y and Z parameters. The results are shown in Fig. 4(a). When we increase the number of bins from $48 \times 48 \times 12$ to $80 \times 80 \times 20$, the recall improves significantly for small thresholds. This is mainly because the quantization error is effectively reduced and the locations of the proposals become more precise. However, the gap becomes smaller for large thresholds. In our experiments, we use $80 \times 80 \times 20$ bins to strike a good balance between the accuracy and speed.

We also consider a practical situation where we do not have sufficient data to train CPN. We propose to address the problem by generating many synthetic heatmaps: we place a number of 3D poses (sampled from the motion capture datasets) at random locations in the space and project them to all views to get the respective 2D locations. Then we generate 2D heatmaps from the locations to train CPN. The experimental results are shown in Fig. 2 (b). We can see that the performance is on par with the model trained on the real images. This significantly improves the general applicability of CPN in the wild (we may also need to address the domain adaptation problem in 2D heatmap estimation but it is beyond the scope of this work).

Table 1. 2D pose estimation accuracy on the Panoptic dataset. Ours are obtained by projecting the estimated 3D poses to the images.

Methods	AP	AP^{50}	AP^{75}
HRNet-w48 [13]	55.8	67.4	59.0
Ours	**98.3**	**99.5**	**99.1**

Finally, we study the impact of the number of cameras to the proposals. In general, the recall decreases when fewer cameras are used. In particular, the results of a single camera are shown in Fig. 4 (c). We can see that the recall rates at different thresholds are consistently lower than those of the five-camera setup in (a). However, it is still larger than 95% at the threshold of 175 mm. It means that it can coarsely retrieve most people using a single camera which demonstrates its practical feasibility. We will report the ultimate 3D pose error using a single camera in the next section.

7 Evaluation of PRN

7.1 2D Pose Estimation Accuracy

We project the 3D poses estimated by our approach to 2D and compare them to the results of HRNet [13]. Since our approach also uses HRNet to estimate heatmaps, they are comparable. The two models are both trained on the Panoptic dataset [15]. The results are shown in Table 1. The AP of HRNet is only 55.8% because there is severe occlusion. Figure 5 shows some 2D poses estimated by HRNet and our approach, respectively. HRNet gets accurate estimates when there is no occlusion which validates its effectiveness. However, it gets inaccurate estimates when people are occluded. In addition, as a top-down method, it may generate false positives if object detector fails. For example, there are two poses mistakenly detected at the dome entrance area in the fourth example.

Fig. 5. Comparison of 2D poses estimated by HRNet [13] (top row) and our approach (bottom row). Ours are obtained by projecting the 3D poses to the images. Note that this is only proof-of-concept result rather than rigorous fair comparison as our approach uses multiview images as input.

7.2 Ablation Study on 3D Pose Estimation

We conduct ablation studies to evaluate a variety of factors of our approach. The results on the Panoptic dataset are shown in Table 2.

Space Division Granularity of CPN. By comparing the results of (a) and (b), we can see that increasing the number of bins from $64 \times 64 \times 16$ to $80 \times 80 \times 20$ improves accuracy in general. In particular, the AP_{25} metric improves most significantly whereas AP_{50} improves only marginally. The results represent that using finer-grained grids improves the precision but not accuracy which agrees with our expectation. Further increasing the grid size only slightly decreases the error but notably increases the computation time. To strike a good balance, we use $80 \times 80 \times 20$ for the rest of the experiments.

Number of Cameras. As shown in (b–d) of Table 2, reducing the number of cameras generally increases the 3D error because the information in the feature volume becomes less complete. In extreme cases, when there is only one camera, the 3D error increases dramatically to 66.95 mm as shown in row (d). This is mainly because there is severe ambiguity in monocular 3D pose estimation. If we align the pelvis joints of the estimated poses to the ground-truth (as the previous methods), the 3D pose error decreases to 51.14 mm as shown in Table 2 (j). This is comparable to the state-of-the-art monocular 3D pose estimation methods such as [20,26,39]. In addition, we find that AP_{25} drops dramatically but AP_{150} only drops slightly when we reduce the number of cameras from five to one. This means that it can estimate coarse 3D poses using a single camera although they are not as precise as the multiview setup.

Generalization to Different Cameras. We train and test our approach on different sets of cameras. Specifically, we randomly select a few cameras from the remaining HD cameras for training and test on the selected five cameras. The 3D error is about 25.51 mm (f) which is larger than the situation where training and testing are on the same cameras. But this is still a reasonably good result demonstrating that the approach has strong generalization capability.

Impact of Heatmaps. By comparing the results in (b) and (g), we can see that getting accurate 2D heatmaps is critical to the 3D accuracy. When the heatmaps are the ground-truth, the *AP*s at a variety of thresholds are very high suggesting that the estimated poses are accurate. The MPJPE remarkably decreases to 11.77 mm. The main reason for this remaining small error is the quantization error caused by space discretization.

Impact of Proposals. By comparing (b) and (h), we can see that replacing CPN by ground-truth proposals does not notably improve the results. The results suggest that the estimated proposals are already very accurate and more attention should be spent on improving the heatmaps and PRN. We do not compute APs when using ground-truth proposals because the confidence scores of all proposals are all set to be one.

Qualitative Study. We show the estimated 3D poses of three examples in Fig. 6. We can see that there are severe occlusions in the images of all camera views. However, by fusing the noisy and incomplete heatmaps from multiple cameras, our approach obtains more comprehensive features which allows us to successfully estimate the 3D poses without bells and whistles. It is noteworthy that we do not need to associate 2D poses in different views based on noisy observations by combining a number of sophisticated techniques. This significantly improves

Table 2. Ablation study on the Panoptic dataset. "*" means that CPN and PRN are trained on synthetic heatmaps. "+" means that CPN and PRN are trained and tested with different cameras. "rel" represents that we align the root joints of the estimated poses to the ground-truth.

	# Views	Backbone	CPN Size	AP_{25}	AP_{50}	AP_{100}	AP_{150}	MPJPE
(a)	5	ResNet-50	64 × 64 × 16	81.54	98.24	99.56	99.85	18.15 mm
(b)	5	ResNet-50	80 × 80 × 20	83.59	98.33	99.76	99.91	17.68 mm
(c)	3	ResNet-50	80 × 80 × 20	58.94	93.88	98.45	99.32	24.29 mm
(d)	1	ResNet-50	80 × 80 × 20	0.860	23.47	80.69	93.32	66.95 mm
(e)*	5	ResNet-50	80 × 80 × 20	71.26	96.96	99.12	99.52	20.31 mm
(f)+	5	ResNet-50	80 × 80 × 20	50.91	95.25	99.36	99.56	25.51 mm
(g)	5	GT Heatmap	80 × 80 × 20	98.61	99.82	99.98	99.99	11.77 mm
(h)	5	ResNet-50	GT Proposal	-	-	-	-	16.94 mm
(i)	5	GT Heatmap	GT Proposal	-	-	-	-	11.32 mm
	# Views	Backbone	CPN Size	AP^{rel}_{25}	AP^{rel}_{50}	AP^{rel}_{100}	AP^{rel}_{150}	$MPJPE^{rel}$
(j)	1	ResNet-50	80 × 80 × 20	1.520	39.86	92.37	96.98	51.14 mm

the robustness of the approach. Please see the supplementary video for more examples.[1]

Figure 7(B) shows two examples where our approach did not obtain accurate estimations in the three-camera setup. In the first example, most joints of the lady can be seen from two of the three cameras and our approach accurately estimates the 3D pose. However, the little child is only visible in the first view and, even in that view, many joints are actually occluded by its body. So the resulting 3D pose has large errors. The second example is also interesting. The person is only visible in one view but, fortunately, most joints are visible. We can see that our approach estimates a 3D pose which seems like a translated version of the ground-truth pose plotted in dashed lines. This is reasonable because there is ambiguity for 3D pose estimation from a single image.

Computational Complexity. It takes about 300 ms on a single Titan X GPU to estimate 3D poses in a five-camera setup. In particular, 93 ms is spent on estimating heatmaps and 24 ms is spent on generating proposals. The time spent on regressing poses depends on the number of proposals (people). In particular, it takes about 46 ms to process one proposal. The inference time has the potential to be further reduced by using sparse 3D convolutions [36].

7.3 Comparison to the State-of-the-Arts

Table 3 shows the results of the state-of-the-art methods on the Campus and the Shelf datasets in the top and bottom sections, respectively. On the Campus dataset, we can see that our approach improves PCP3D from 96.3% of [4] to

[1] https://youtu.be/qZAyHUzdpgw.

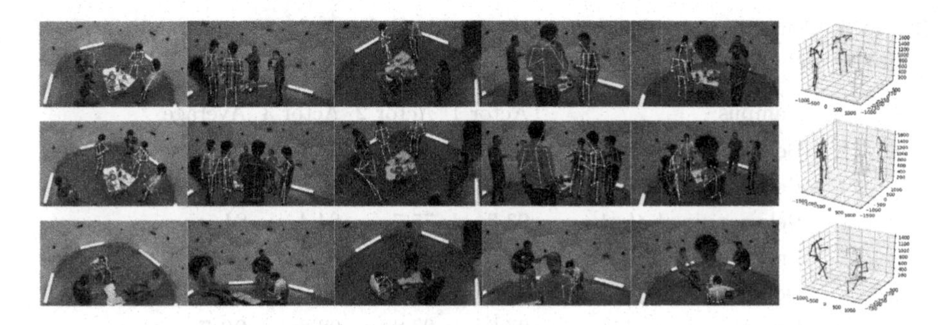

Fig. 6. Estimated 3D poses and their projections in images. The last column shows the estimated 3D poses.

96.7% which is a decent improvement considering the already very high accuracy. As discussed in Sect. 5, the PCP3D metric does not penalize false positive estimates. However, it is also meaningless to report AP scores because the GT pose annotations in this dataset are incomplete. So we propose to visualize and publish all of our estimated poses.[2] We find that our approach usually gets accurate estimates as long as joints are visible in at least two views.

Our approach also achieves better results than [4] on the Shelf dataset. In particular, it gets fewer false positives. For example, in Fig. 7 (A.2), there is a false positive pose in the pink dashed circle estimated by [4]. In contrast, our approach can suppress most false positives. We find that most errors of our approach are caused by inaccurate GT annotations. For example, as shown in the first column of Fig. 7 (A.1), the GT joint locations within the red circle are incorrect. In summary, 66 out of the 301 frames have completely correct annotations and our approach gets accurate estimates on them.

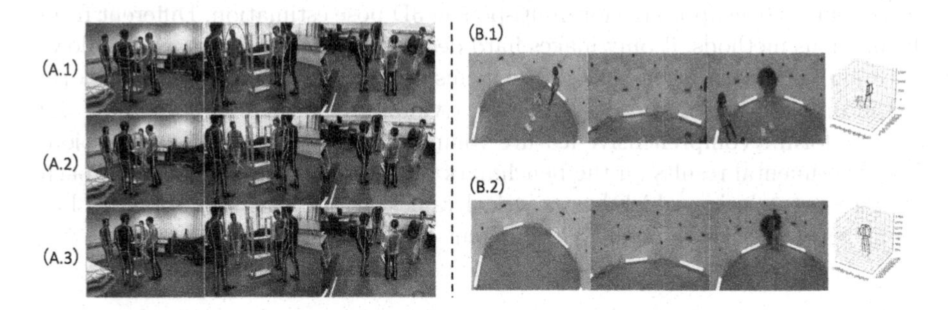

Fig. 7. (A) shows the 3D poses of ground-truth (A.1), estimated by [4] (A.2) and ours (A.3), respectively. The joints in the dashed circles represent the locations are incorrect. (B) shows two typical cases where our approach makes mistakes. The pose plotted by dashed lines in B.2 is the ground-truth.

[2] https://youtu.be/AgDQFIlL5IM.

Table 3. Comparison to the state-of-the-art methods on the Campus and the Shelf datasets. The metric is PCP3D.

Campus	Actor 1	Actor 2	Actor 3	Average
Belagiannis *et al.* [2]	82.0	72.4	73.7	75.8
Belagiannis *et al.* [3]	83.0	73.0	78.0	78.0
Belagiannis *et al.* [1]	93.5	75.7	84.4	84.5
Ershadi-Nasab *et al.* [40]	94.2	92.9	84.6	90.6
Dong *et al.* [4]	97.6	93.3	98.0	96.3
Ours	97.6	93.8	98.8	**96.7**
Shelf	Actor 1	Actor 2	Actor 3	Average
Belagiannis *et al.* [2]	66.1	65.0	83.2	71.4
Belagiannis *et al.* [3]	75.0	67.0	86.0	76.0
Belagiannis *et al.* [1]	75.3	69.7	87.6	77.5
Ershadi-Nasab *et al.* [40]	93.3	75.9	94.8	88.0
Dong *et al.* [4]	98.8	94.1	97.8	96.9
Ours	99.3	94.1	97.6	**97.0**

The previous works [1–4] did not report numerical results on the large scale Panoptic dataset. We encourage future works to do so as in Table 2 (b). We also evaluate our approach on the single person dataset Human3.6M [41]. The MPJPE of our approach is about 19mm which is comparable to [26]. We also visualize and publish our estimated poses.[3]

8 Conclusion

we present a novel approach for multi-person 3D pose estimation. Different from the previous methods, it only makes hard decisions in the 3D space which allows to avoid the challenging association problems in the 2D space. In particular, noisy and incomplete information of all camera views are warped to a common 3D space to form a comprehensive feature volume which is used for 3D estimation. The experimental results on the benchmark datasets validate that the approach is robust to occlusion which has practical values. In addition, the approach has strong generalization capability to different camera setups.

References

1. Belagiannis, V., Amin, S., Andriluka, M., Schiele, B., Navab, N., Ilic, S.: 3D pictorial structures revisited: multiple human pose estimation. TPAMI **38**(10), 1929–1942 (2015)

[3] https://youtu.be/S6G3TXaBukw.

2. Belagiannis, V., Amin, S., Andriluka, M., Schiele, B., Navab, N., Ilic, S.: 3D pictorial structures for multiple human pose estimation. In: CVPR, pp. 1669–1676 (2014)
3. Belagiannis, V., Wang, X., Schiele, B., Fua, P., Ilic, S., Navab, N.: Multiple human pose estimation with temporally consistent 3D pictorial structures. In: Agapito, L., Bronstein, M.M., Rother, C. (eds.) ECCV 2014. LNCS, vol. 8925, pp. 742–754. Springer, Cham (2015). https://doi.org/10.1007/978-3-319-16178-5_52
4. Dong, J., Jiang, W., Huang, Q., Bao, H., Zhou, X.: Fast and robust multi-person 3D pose estimation from multiple views. In: CVPR, pp. 7792–7801 (2019)
5. Bridgeman, L., Volino, M., Guillemaut, J.Y., Hilton, A.: Multi-person 3D pose estimation and tracking in sports. In: CVPRW (2019)
6. Qiu, H., Wang, C., Wang, J., Wang, N., Zeng, W.: Cross view fusion for 3D human pose estimation. In: ICCV, pp. 4342–4351 (2019)
7. Zhang, Y., An, L., Yu, T., Li, X., Li, K., Liu, Y.: 4D association graph for realtime multi-person motion capture using multiple video cameras. In: CVPR, pp. 1324–1333 (2020)
8. Cao, Z., Simon, T., Wei, S.E., Sheikh, Y.: Realtime multi-person 2D pose estimation using part affinity fields. In: CVPR, pp. 7291–7299 (2017)
9. He, K., Gkioxari, G., Dollár, P., Girshick, R.: Mask R-CNN. In: ICCV, pp. 2961–2969 (2017)
10. Hartley, R., Zisserman, A.: Multiple View Geometry in Computer Vision. Cambridge University Press, Cambridge (2003)
11. Amin, S., Andriluka, M., Rohrbach, M., Schiele, B.: Multi-view pictorial structures for 3D human pose estimation. In: BMVC. Citeseer (2013)
12. Newell, A., Yang, K., Deng, J.: Stacked hourglass networks for human pose estimation. In: Leibe, B., Matas, J., Sebe, N., Welling, M. (eds.) ECCV 2016. LNCS, vol. 9912, pp. 483–499. Springer, Cham (2016). https://doi.org/10.1007/978-3-319-46484-8_29
13. Sun, K., Xiao, B., Liu, D., Wang, J.: Deep high-resolution representation learning for human pose estimation. In: CVPR, pp. 5693–5703 (2019)
14. Newell, A., Huang, Z., Deng, J.: Associative embedding: End-to-end learning for joint detection and grouping. In: NIPS, pp. 2277–2287 (2017)
15. Joo, H., et al.: Panoptic studio: a massively multiview system for social interaction capture. IEEE Trans. Pattern Anal. Mach. Intell. **41**, 190–204 (2017)
16. Wang, C., Wang, Y., Lin, Z., Yuille, A.L., Gao, W.: Robust estimation of 3D human poses from a single image. In: CVPR, pp. 2361–2368 (2014)
17. Ramakrishna, V., Kanade, T., Sheikh, Y.: Reconstructing 3D human pose from 2D image landmarks. In: Fitzgibbon, A., Lazebnik, S., Perona, P., Sato, Y., Schmid, C. (eds.) ECCV 2012. LNCS, vol. 7575, pp. 573–586. Springer, Heidelberg (2012). https://doi.org/10.1007/978-3-642-33765-9_41
18. Zhou, X., Zhu, M., Leonardos, S., Daniilidis, K.: Sparse representation for 3d shape estimation: a convex relaxation approach. TPAMI **39**(8), 1648–1661 (2016)
19. Pavlakos, G., Zhou, X., Daniilidis, K.: Ordinal depth supervision for 3d human pose estimation. In: CVPR. (2018) 7307–7316
20. Martinez, J., Hossain, R., Romero, J., Little, J.J.: A simple yet effective baseline for 3D human pose estimation. In: ICCV (2017)
21. Moreno-Noguer, F.: 3D human pose estimation from a single image via distance matrix regression. In: CVPR, pp. 1561–1570. IEEE (2017)

22. Sun, X., Xiao, B., Wei, F., Liang, S., Wei, Y.: Integral human pose regression. In: Ferrari, V., Hebert, M., Sminchisescu, C., Weiss, Y. (eds.) ECCV 2018. LNCS, vol. 11210, pp. 536–553. Springer, Cham (2018). https://doi.org/10.1007/978-3-030-01231-1_33
23. Fang, H.S., Xu, Y., Wang, W., Liu, X., Zhu, S.C.: Learning pose grammar to encode human body configuration for 3D pose estimation. In: AAAI (2018)
24. Pavllo, D., Feichtenhofer, C., Grangier, D., Auli, M.: 3D human pose estimation in video with temporal convolutions and semi-supervised training. In: CVPR (2019)
25. Bogo, F., Kanazawa, A., Lassner, C., Gehler, P., Romero, J., Black, M.J.: Keep it SMPL: automatic estimation of 3D human pose and shape from a single image. In: Leibe, B., Matas, J., Sebe, N., Welling, M. (eds.) ECCV 2016. LNCS, vol. 9909, pp. 561–578. Springer, Cham (2016). https://doi.org/10.1007/978-3-319-46454-1_34
26. Iskakov, K., Burkov, E., Lempitsky, V., Malkov, Y.: Learnable triangulation of human pose. In: ICCV, pp. 7718–7727 (2019)
27. Remelli, E., Han, S., Honari, S., Fua, P., Wang, R.: Lightweight multi-view 3D pose estimation through camera-disentangled representation. In: CVPR, pp. 6040–6049 (2020)
28. Pavlakos, G., Zhou, X., Derpanis, K.G., Daniilidis, K.: Coarse-to-fine volumetric prediction for single-image 3d human pose. In: CVPR, pp. 1263–1272. IEEE (2017)
29. Zhou, X., Huang, Q., Sun, X., Xue, X., Wei, Y.: Towards 3d human pose estimation in the wild: a weakly-supervised approach. In: ICCV (2017)
30. Rogez, G., Weinzaepfel, P., Schmid, C.: LCR-Net++: multi-person 2D and 3D pose detection in natural images. TPAMI **42**, 1146–1161 (2019)
31. Kreiss, S., Bertoni, L., Alahi, A.: PifPaf: composite fields for human pose estimation. In: CVPR, pp. 11977–11986 (2019)
32. Ren, S., He, K., Girshick, R., Sun, J.: Faster R-CNN: towards real-time object detection with region proposal networks. In: NIPS, pp. 91–99 (2015)
33. Redmon, J., Farhadi, A.: YOLOv3: an incremental improvement. arXiv preprint arXiv:1804.02767 (2018)
34. Zhou, X., Wang, D., Krähenbühl, P.: Objects as points. arXiv preprint arXiv:1904.07850 (2019)
35. Moon, G., Yong Chang, J., Mu Lee, K.: V2V-PoseNet: voxel-to-voxel prediction network for accurate 3d hand and human pose estimation from a single depth map. In: CVPR, pp. 5079–5088 (2018)
36. Yan, Y., Mao, Y., Li, B.: SECOND: sparsely embedded convolutional detection. Sensors **18**(10), 3337 (2018)
37. Xiang, D., Joo, H., Sheikh, Y.: Monocular total capture: posing face, body, and hands in the wild. In: CVPR, pp. 10965–10974 (2019)
38. Pishchulin, L., et al.: DeepCut: joint subset partition and labeling for multi person pose estimation. In: CVPR, pp. 4929–4937 (2016)
39. Ci, H., Wang, C., Ma, X., Wang, Y.: Optimizing network structure for 3D human pose estimation. In: ICCV (2019)
40. Ershadi-Nasab, S., Noury, E., Kasaei, S., Sanaei, E.: Multiple human 3D pose estimation from multiview images. Multimedia Tools Appl. **77**(12), 15573–15601 (2018)
41. Ionescu, C., Papava, D., Olaru, V., Sminchisescu, C.: Human3.6M: large scale datasets and predictive methods for 3D human sensing in natural environments. T-PAMI **36**(7), 1325–1339 (2014)

End-to-End Object Detection with Transformers

Nicolas Carion[1,2](✉), Francisco Massa[2], Gabriel Synnaeve[2], Nicolas Usunier[2], Alexander Kirillov[2], and Sergey Zagoruyko[2]

[1] Paris Dauphine University, Paris, France
{alcinos,fmassa,gab,usunier,akirillov}@fb.com
[2] Facebook AI, Menlo Park, USA
szagoruyko@fb.com

Abstract. We present a new method that views object detection as a direct set prediction problem. Our approach streamlines the detection pipeline, effectively removing the need for many hand-designed components like a non-maximum suppression procedure or anchor generation that explicitly encode our prior knowledge about the task. The main ingredients of the new framework, called DEtection TRansformer or DETR, are a set-based global loss that forces unique predictions via bipartite matching, and a transformer encoder-decoder architecture. Given a fixed small set of learned object queries, DETR reasons about the relations of the objects and the global image context to directly output the final set of predictions in parallel. The new model is conceptually simple and does not require a specialized library, unlike many other modern detectors. DETR demonstrates accuracy and run-time performance on par with the well-established and highly-optimized Faster R-CNN baseline on the challenging COCO object detection dataset. Moreover, DETR can be easily generalized to produce panoptic segmentation in a unified manner. We show that it significantly outperforms competitive baselines. Training code and pretrained models are available at https://github.com/facebookresearch/detr.

1 Introduction

The goal of object detection is to predict a set of bounding boxes and category labels for each object of interest. Modern detectors address this set prediction task in an indirect way, by defining surrogate regression and classification problems on a large set of proposals [5,36], anchors [22], or window centers [45,52]. Their performances are significantly influenced by postprocessing steps to collapse near-duplicate predictions, by the design of the anchor sets and by the heuristics that assign target boxes to anchors [51]. To simplify these pipelines, we propose a direct set prediction approach to bypass the surrogate tasks. This

---

Electronic supplementary material The online version of this chapter (https://doi.org/10.1007/978-3-030-58452-8_13) contains supplementary material, which is available to authorized users.

A. Vedaldi et al. (Eds.): ECCV 2020, LNCS 12346, pp. 213–229, 2020.
https://doi.org/10.1007/978-3-030-58452-8_13

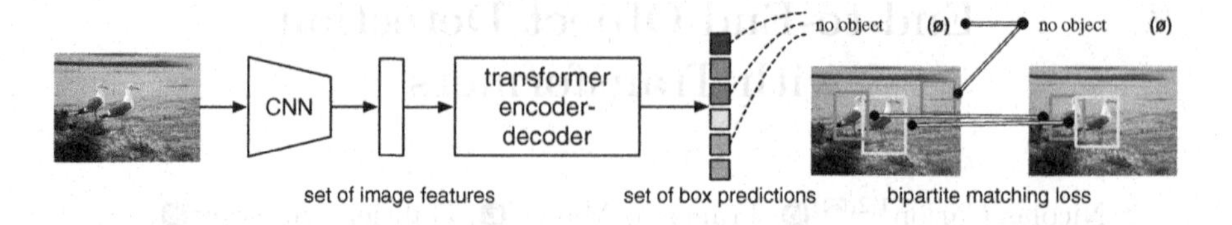

Fig. 1. DETR directly predicts (in parallel) the final set of detections by combining a common CNN with a transformer architecture. During training, bipartite matching uniquely assigns predictions with ground truth boxes. Prediction with no match should yield a "`no object`" (∅) class prediction.

end-to-end philosophy has led to significant advances in complex structured prediction tasks such as machine translation or speech recognition, but not yet in object detection: previous attempts [4,15,38,42] either add other forms of prior knowledge, or have not proven to be competitive with strong baselines on challenging benchmarks. This paper aims to bridge this gap.

We streamline the training pipeline by viewing object detection as a direct set prediction problem. We adopt an encoder-decoder architecture based on transformers [46], a popular architecture for sequence prediction. The self-attention mechanisms of transformers, which explicitly model all pairwise interactions between elements in a sequence, make these architectures particularly suitable for specific constraints of set prediction such as removing duplicate predictions.

Our DEtection TRansformer (DETR, see Fig. 1) predicts all objects at once, and is trained end-to-end with a set loss function which performs bipartite matching between predicted and ground-truth objects. DETR simplifies the detection pipeline by dropping multiple hand-designed components that encode prior knowledge, like spatial anchors or non-maximal suppression. Unlike most existing detection methods, DETR doesn't require any customized layers, and thus can be reproduced easily in any framework that contains standard ResNet [14] and Transformer [46] classes.

Compared to most previous work on direct set prediction, the main features of DETR are the conjunction of the bipartite matching loss and transformers with (non-autoregressive) parallel decoding [7,9,11,28]. In contrast, previous work focused on autoregressive decoding with RNNs [29,35,40–42]. Our matching loss function uniquely assigns a prediction to a ground truth object, and is invariant to a permutation of predicted objects, so we can emit them in parallel.

We evaluate DETR on one of the most popular object detection datasets, COCO [23], against a very competitive Faster R-CNN baseline [36]. Faster R-CNN has undergone many design iterations and its performance was greatly improved since the original publication. Our experiments show that our new model achieves comparable performances. More precisely, DETR demonstrates significantly better performance on large objects, a result likely enabled by the non-local computations of the transformer. It obtains, however, lower performances on small objects. We expect that future work will improve this aspect in the same way the development of FPN [21] did for Faster R-CNN.

Training settings for DETR differ from standard object detectors in multiple ways. The new model requires extra-long training schedule and benefits from auxiliary decoding losses in the transformer. We thoroughly explore what components are crucial for the demonstrated performance.

The design ethos of DETR easily extend to more complex tasks. In our experiments, we show that a simple segmentation head trained on top of a pre-trained DETR outperfoms competitive baselines on Panoptic Segmentation [18], a challenging pixel-level recognition task that has recently gained popularity.

2 Related Work

Our work build on prior work in several domains: bipartite matching losses for set prediction, encoder-decoder architectures based on the transformer, parallel decoding, and object detection methods.

2.1 Set Prediction

There is no canonical deep learning model to directly predict sets. The basic set prediction task is multilabel classification (see e.g., [32,39] for references in the context of computer vision) for which the baseline approach, one-vs-rest, does not apply to problems such as detection where there is an underlying structure between elements (i.e., near-identical boxes). The first difficulty in these tasks is to avoid near-duplicates. Most current detectors use postprocessings such as non-maximal suppression to address this issue, but direct set prediction are postprocessing-free. They need global inference schemes that model interactions between all predicted elements to avoid redundancy. For constant-size set prediction, dense fully connected networks [8] are sufficient but costly. A general approach is to use auto-regressive sequence models such as recurrent neural networks [47]. In all cases, the loss function should be invariant by a permutation of the predictions. The usual solution is to design a loss based on the Hungarian algorithm [19], to find a bipartite matching between ground-truth and prediction. This enforces permutation-invariance, and guarantees that each target element has a unique match. We follow the bipartite matching loss approach. In contrast to most prior work however, we step away from autoregressive models and use transformers with parallel decoding, which we describe below.

2.2 Transformers and Parallel Decoding

Transformers were introduced by Vaswani *et al.* [46] as a new attention-based building block for machine translation. Attention mechanisms [2] are neural network layers that aggregate information from the entire input sequence. Transformers introduced self-attention layers, which, similarly to Non-Local Neural Networks [48], scan through each element of a sequence and update it by aggregating information from the whole sequence. One of the main advantages of attention-based models is their global computations and perfect memory, which makes them more suitable than RNNs on long sequences. Transformers are now replacing RNNs in many problems in natural language processing, speech processing and computer vision [7,26,30,33,44].

Transformers were first used in auto-regressive models, following early sequence-to-sequence models [43], generating output tokens one by one. However, the prohibitive inference cost (proportional to output length, and hard to batch) lead to the development of parallel sequence generation, in the domains of audio [28], machine translation [9,11], word representation learning [7], and more recently speech recognition [6]. We also combine transformers and parallel decoding for their suitable trade-off between computational cost and the ability to perform the global computations required for set prediction.

2.3 Object Detection

Most modern object detection methods make predictions relative to some initial guesses. Two-stage detectors [5,36] predict boxes w.r.t. proposals, whereas single-stage methods make predictions w.r.t. anchors [22] or a grid of possible object centers [45,52]. Recent work [51] demonstrate that the final performance of these systems heavily depends on the exact way these initial guesses are set. In our model we are able to remove this hand-crafted process and streamline the detection process by directly predicting the set of detections with absolute box prediction w.r.t. the input image rather than an anchor.

Set-Based Loss. Several object detectors [8,24,34] used the bipartite matching loss. However, in these early deep learning models, the relation between different prediction was modeled with convolutional or fully-connected layers only and a hand-designed NMS post-processing can improve their performance. More recent detectors [22,36,52] use non-unique assignment rules between ground truth and predictions together with an NMS.

Learnable NMS methods [4,15] and relation networks [16] explicitly model relations between different predictions with attention. Using direct set losses, they do not require any post-processing steps. However, these methods employ additional hand-crafted context features like proposal box coordinates to model relations between detections efficiently, while we look for solutions that reduce the prior knowledge encoded in the model.

Recurrent Detectors. Closest to our approach are end-to-end set predictions for object detection [42] and instance segmentation [29,35,40,41]. Similarly to us, they use bipartite-matching losses with encoder-decoder architectures based on CNN activations to directly produce a set of bounding boxes. These approaches, however, were only evaluated on small datasets and not against modern baselines. In particular, they are based on autoregressive models (more precisely RNNs), so they do not leverage the recent transformers with parallel decoding.

3 The DETR Model

Two ingredients are essential for direct set predictions in detection: (1) a set prediction loss that forces unique matching between predicted and ground truth boxes; (2) an architecture that predicts (in a single pass) a set of objects and models their relation. We describe our architecture in detail in Fig. 2.

3.1 Object Detection Set Prediction Loss

DETR infers a fixed-size set of N predictions, in a single pass through the decoder, where N is set to be significantly larger than the typical number of objects in an image. One of the main difficulties of training is to score predicted objects (class, position, size) with respect to the ground truth. Our loss produces an optimal bipartite matching between predicted and ground truth objects, and then optimize object-specific (bounding box) losses.

Let us denote by y the ground truth set of objects, and $\hat{y} = \{\hat{y}_i\}_{i=1}^N$ the set of N predictions. Assuming N is larger than the number of objects in the image, we consider y also as a set of size N padded with $\varnothing$ (no object). To find a bipartite matching between these two sets we search for a permutation of N elements $\sigma \in \mathfrak{S}_N$ with the lowest cost:

$$\hat{\sigma} = \underset{\sigma \in \mathfrak{S}_N}{\arg\min} \sum_{i}^{N} \mathcal{L}_{\text{match}}(y_i, \hat{y}_{\sigma(i)}), \tag{1}$$

where $\mathcal{L}_{\text{match}}(y_i, \hat{y}_{\sigma(i)})$ is a pair-wise *matching cost* between ground truth y_i and a prediction with index $\sigma(i)$. This optimal assignment is computed efficiently with the Hungarian algorithm, following prior work (*e.g.* [42]).

The matching cost takes into account both the class prediction and the similarity of predicted and ground truth boxes. Each element i of the ground truth set can be seen as a $y_i = (c_i, b_i)$ where c_i is the target class label (which may be $\varnothing$) and $b_i \in [0,1]^4$ is a vector that defines ground truth box center coordinates and its height and width relative to the image size. For the prediction with index $\sigma(i)$ we define probability of class c_i as $\hat{p}_{\sigma(i)}(c_i)$ and the predicted box as $\hat{b}_{\sigma(i)}$. With these notations we define $\mathcal{L}_{\text{match}}(y_i, \hat{y}_{\sigma(i)})$ as $-\mathbb{1}_{\{c_i \neq \varnothing\}}\hat{p}_{\sigma(i)}(c_i) + \mathbb{1}_{\{c_i \neq \varnothing\}}\mathcal{L}_{\text{box}}(b_i, \hat{b}_{\sigma(i)})$.

This procedure of finding the matching plays the same role as the heuristic assignment rules used to match proposal [36] or anchors [21] to ground truth objects in modern detectors. The main difference is that we need to find one-to-one matching for direct set prediction without duplicates.

The second step is to compute the loss function, the *Hungarian loss* for all pairs matched in the previous step. We define the loss similarly to the losses of common object detectors, *i.e.* a linear combination of a negative log-likelihood for class prediction and a box loss $\mathcal{L}_{\text{box}}(\cdot,\cdot)$ defined later:

$$\mathcal{L}_{\text{Hungarian}}(y, \hat{y}) = \sum_{i=1}^{N} \left[-\log \hat{p}_{\hat{\sigma}(i)}(c_i) + \mathbb{1}_{\{c_i \neq \varnothing\}} \mathcal{L}_{\text{box}}(b_i, \hat{b}_{\hat{\sigma}}(i)) \right], \tag{2}$$

where $\hat{\sigma}$ is the optimal assignment computed in the first step (1). In practice, we down-weight the log-probability term when $c_i = \varnothing$ by a factor 10 to account for

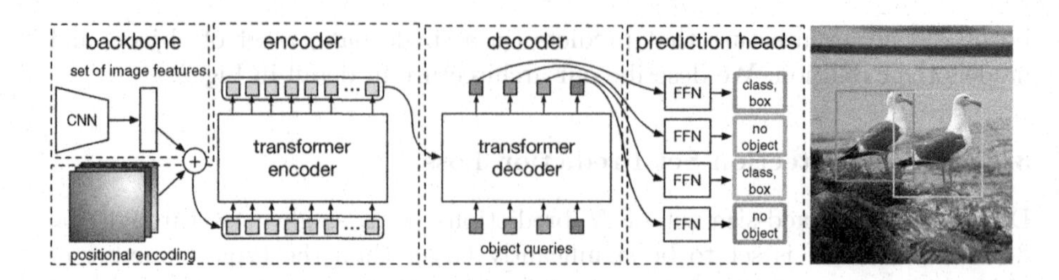

Fig. 2. DETR uses a conventional CNN backbone to learn a 2D representation of an input image. The model flattens it and supplements it with a positional encoding before passing it into a transformer encoder. A transformer decoder then takes as input a small fixed number of learned positional embeddings, which we call *object queries*, and additionally attends to the encoder output. We pass each output embedding of the decoder to a shared feed forward network (FFN) that predicts either a detection (class and bounding box) or a "`no object`" class.

class imbalance. This is analogous to how Faster R-CNN training procedure balances positive/negative proposals by subsampling [36]. Notice that the matching cost between an object and $\varnothing$ doesn't depend on the prediction, which means that in that case the cost is a constant. In the matching cost we use probabilities $\hat{p}_{\hat{\sigma}(i)}(c_i)$ instead of log-probabilities. This makes the class prediction term commensurable to $\mathcal{L}_{\rm box}(\cdot, \cdot)$, and we observed better empirical performances.

Bounding Box Loss. The second part of the matching cost and the Hungarian loss is $\mathcal{L}_{\rm box}(\cdot)$ that scores the bounding boxes. Unlike many detectors that do box predictions as a Δ w.r.t. some initial guesses, we make box predictions directly. While such approach simplify the implementation it poses an issue with relative scaling of the loss. The most commonly-used ℓ_1 loss will have different scales for small and large boxes even if their relative errors are similar. To mitigate this issue we use a linear combination of the ℓ_1 loss and the generalized IoU loss [37] $\mathcal{L}_{\rm iou}(\cdot, \cdot)$ that is scale-invariant. Overall, our box loss is $\mathcal{L}_{\rm box}(b_i, \hat{b}_{\sigma(i)})$ defined as $\lambda_{\rm iou}\mathcal{L}_{\rm iou}(b_i, \hat{b}_{\sigma(i)}) + \lambda_{\rm L1}||b_i - \hat{b}_{\sigma(i)}||_1$ where $\lambda_{\rm iou}, \lambda_{\rm L1} \in \mathbb{R}$ are hyperparameters. These two losses are normalized by the number of objects inside the batch.

3.2 DETR Architecture

The overall DETR architecture is surprisingly simple and depicted in Fig. 2. It contains three main components, which we describe below: a CNN backbone to extract a compact feature representation, an encoder-decoder transformer, and a simple feed forward network (FFN) that makes the final detection prediction.

Unlike many modern detectors, DETR can be implemented in any deep learning framework that provides a common CNN backbone and a transformer architecture implementation with just a few hundred lines. Inference code for DETR can be implemented in less than 50 lines in PyTorch [31]. We hope that the simplicity of our method will attract new researchers to the detection community.

Backbone. Starting from the initial image $x_{\rm img} \in \mathbb{R}^{3\times H_0\times W_0}$ (with 3 color channels[1]), a conventional CNN backbone generates a lower-resolution activation map $f \in \mathbb{R}^{C\times H\times W}$. Typical values we use are $C = 2048$ and $H, W = \frac{H_0}{32}, \frac{W_0}{32}$.

Transformer Encoder. First, a 1x1 convolution reduces the channel dimension of the high-level activation map f from C to a smaller dimension d. creating a new feature map $z_0 \in \mathbb{R}^{d\times H\times W}$. The encoder expects a sequence as input, hence we collapse the spatial dimensions of z_0 into one dimension, resulting in a $d\times HW$ feature map. Each encoder layer has a standard architecture and consists of a multi-head self-attention module and a feed forward network (FFN). Since the transformer architecture is permutation-invariant, we supplement it with fixed positional encodings [3,30] that are added to the input of each attention layer. We defer to the supplementary material the detailed definition of the architecture, which follows the one described in [46].

Transformer Decoder. The decoder follows the standard architecture of the transformer, transforming N embeddings of size d using multi-headed self- and encoder-decoder attention mechanisms. The difference with the original transformer is that our model decodes the N objects in parallel at each decoder layer, while Vaswani et al. [46] use an autoregressive model that predicts the output sequence one element at a time. We refer the reader unfamiliar with the concepts to the supplementary material. Since the decoder is also permutation-invariant, the N input embeddings must be different to produce different results. These input embeddings are learnt positional encodings that we refer to as *object queries*, and similarly to the encoder, we add them to the input of each attention layer. The N object queries are transformed into an output embedding by the decoder. They are then *independently* decoded into box coordinates and class labels by a feed forward network (described in the next subsection), resulting N final predictions. Using self- and encoder-decoder attention over these embeddings, the model globally reasons about all objects together using pair-wise relations between them, while being able to use the whole image as context.

Prediction Feed-Forward Networks (FFNs). The final prediction is computed by a 3-layer perceptron with ReLU activation function and hidden dimension d, and a linear projection layer. The FFN predicts the normalized center coordinates, height and width of the box w.r.t. the input image, and the linear layer predicts the class label using a softmax function. Since we predict a fixed-size set of N bounding boxes, where N is usually much larger than the actual number of objects of interest in an image, an additional special class label $\varnothing$ is used to represent that no object is detected within a slot. This class plays a similar role to the "background" class in standard object detection approaches.

Auxiliary Decoding Losses. We found helpful to use auxiliary losses [1] in the decoder during training, especially to help the model output the correct number of objects of each class. The output of each decoder layer is normalized with

[1] The input images are batched together, applying 0-padding adequately to ensure they all have the same dimensions (H_0, W_0) as the largest image of the batch.

a shared layer-norm then fed to the shared prediction heads (classification and box prediction). We then apply the Hungarian loss as usual for supervision.

4 Experiments

We show that DETR achieves competitive results compared to Faster R-CNN [36] and RetinaNet [22] in quantitative evaluation on COCO. Then, we provide a detailed ablation study of the architecture and loss, with insights and qualitative results. Finally, to show that DETR is a versatile model, we present results on panoptic segmentation, training only a small extension on a fixed DETR model.

Dataset. We perform experiments on COCO 2017 detection and panoptic segmentation datasets [17,23], containing 118k training images and 5k validation images. Each image is annotated with bounding boxes and panoptic segmentation. There are 7 instances per image on average, up to 63 instances in a single image in training set, ranging from small to large on the same images. If not specified, we report AP as bbox AP, the integral metric over multiple thresholds. For comparison with other models we report validation AP at the last training epoch, and in ablations we report the median over the last 10 epochs.

Technical Details. We train DETR with AdamW [25] setting the initial transformer's learning rate to 10^{-4}, the backbone's to 10^{-5}, and weight decay to 10^{-4}. All transformer weights are initialized with Xavier init [10], and the backbone is with ImageNet-pretrained ResNet model [14] from TORCHVISION with frozen batchnorm layers. We report results with two different backbones: a ResNet-50 and a ResNet-101. The corresponding models are called respectively DETR and DETR-R101. Following [20], we also increase the feature resolution by adding a dilation to the last stage of the backbone and removing a stride from the first convolution of this stage. The corresponding models are called respectively DETR-DC5 and DETR-DC5-R101 (dilated C5 stage). This modification increases the resolution by a factor of two, thus improving performance for small objects, at the cost of a 16x higher cost in the self-attentions of the encoder, leading to an overall 2x increase in computational cost. A full comparison of FLOPs of these models, Faster R-CNN and RetinaNet is given in Table 1.

We use scale augmentation, resizing the input images such that the shortest side is at least 480 and at most 800 pixels while the longest at most 1333 [49]. To help learning global relationships through the self-attention of the encoder, we also apply random crop augmentations during training, improving the performance by approximately 1 AP. Specifically, a train image is cropped with probability 0.5 to a random rectangular patch which is then resized again to 800–1333. The transformer is trained with default dropout of 0.1. At inference time, some slots predict empty class. To optimize for AP, we override the prediction of these slots with the second highest scoring class, using the corresponding confidence. This improves AP by 2 points compared to filtering out empty slots. Other training hyperparameters can be found in Appendix. For our ablation

Table 1. Comparison with RetinaNet and Faster R-CNN with a ResNet-50 and ResNet-101 backbones on the COCO validation set. The top section shows results for models in Detectron2 [49], the middle section shows results for models with GIoU [37], random crops train-time augmentation, and the long 9x training schedule. DETR models achieve comparable results to heavily tuned Faster R-CNN baselines, having lower AP_S but greatly improved AP_L. We use torchscript models to measure FLOPS and FPS. Results without R101 in the name correspond to ResNet-50.

Model	GFLOPS/FPS	#params	AP	AP_{50}	AP_{75}	AP_S	AP_M	AP_L
RetinaNet	205/18	38M	38.7	58.0	41.5	23.3	42.3	50.3
Faster RCNN-DC5	320/16	166M	39.0	60.5	42.3	21.4	43.5	52.5
Faster RCNN-FPN	180/26	42M	40.2	61.0	43.8	24.2	43.5	52.0
Faster RCNN-R101-FPN	246/20	60M	42.0	62.5	45.9	25.2	45.6	54.6
RetinaNet+	205/18	38M	41.1	60.4	43.7	25.6	44.8	53.6
Faster RCNN-DC5+	320/16	166M	41.1	61.4	44.3	22.9	45.9	55.0
Faster RCNN-FPN+	180/26	42M	42.0	62.1	45.5	26.6	45.4	53.4
Faster RCNN-R101-FPN+	246/20	60M	44.0	63.9	**47.8**	**27.2**	48.1	56.0
DETR	86/28	41M	42.0	62.4	44.2	20.5	45.8	61.1
DETR-DC5	187/12	41M	43.3	63.1	45.9	22.5	47.3	61.1
DETR-R101	152/20	60M	43.5	63.8	46.4	21.9	48.0	61.8
DETR-DC5-R101	253/10	60M	**44.9**	**64.7**	47.7	23.7	**49.5**	**62.3**

experiments we use training schedule of 300 epochs with a learning rate drop by a factor of 10 after 200 epochs, where a single epoch is a pass over all training images once. Training the baseline model for 300 epochs on 16 V100 GPUs takes 3 d, with 4 images per GPU (hence a total batch size of 64). For the longer schedule used to compare with Faster R-CNN we train for 500 epochs with learning rate drop after 400 epochs, which improves AP by 1.5 points.

4.1 Comparison with Faster R-CNN and RetinaNet

Transformers are typically trained with Adam or Adagrad optimizers with very long training schedules and dropout, and this is true for DETR as well. Faster R-CNN, however, is trained with SGD with minimal data augmentation and we are not aware of successful applications of Adam or dropout. Despite these differences we attempt to make our baselines stronger. To align it with DETR, we add generalized IoU [37] to the box loss, the same random crop augmentation and long training known to improve results [12]. Results are presented in Table 1. In the top section we show results from Detectron2 Model Zoo [49] for models trained with the 3x schedule. In the middle section we show results (with a "+") for the same models but trained with the 9x schedule (109 epochs) and the described enhancements, which in total adds 1–2 AP. In the last section of Table 1 we show the results for multiple DETR models. To be comparable in the number of parameters we choose a model with 6 transformer and 6 decoder layers of width 256 with 8 attention heads. Like Faster R-CNN with FPN this model

Fig. 3. Encoder self-attention for a set of reference points. The encoder is able to separate individual instances. Prediction made with baseline DETR on a validation image.

has 41.3M parameters, out of which 23.5M are in ResNet-50, and 17.8M are in the transformer. Even though both Faster R-CNN and DETR are still likely to further improve with longer training, we can conclude that DETR can be competitive with Faster R-CNN with the same number of parameters, achieving 42 AP on the COCO val subset. The way DETR achieves this is by improving AP_L (+7.8), however note that the model is still lagging behind in AP_S (-5.5). DETR-DC5 with the same number of parameters and similar FLOP count has higher AP, but is still significantly behind in AP_S too. Results on ResNet-101 backbone are comparable as well.

4.2 Ablations

Attention mechanisms in the transformer decoder are the key components which model relations between feature representations of different detections. In our ablation analysis, we explore how other components of our architecture and loss influence the final performance. For the study we choose ResNet-50-based DETR model with 6 encoder, 6 decoder layers and width 256. The model has 41.3M parameters, achieves 40.6 and 42.0 AP on short and long schedules respectively, and runs at 28 FPS, similarly to Faster R-CNN-FPN with the same backbone.

Number of Encoder Layers. We evaluate the importance of global image-level self-attention by changing the number of encoder layers. Without encoder layers, overall AP drops by 3.9 points, with a more significant drop of 6.0 AP on large objects. We hypothesize that, by using global scene reasoning, the encoder is important for disentangling objects. See results in appendix. In Fig. 3, we visualize the attention maps of the last encoder layer of a trained model, focusing on a few points in the image. The encoder seems to separate instances already, which likely simplifies object extraction and localization for the decoder.

Number of Decoder Layers. We apply auxiliary losses after each decoding layer (see Sect. 3.2), hence, the prediction FFNs are trained by design to predict objects out of the outputs of every decoder layer. We analyze the importance of each decoder layer by evaluating the objects that would be predicted at each

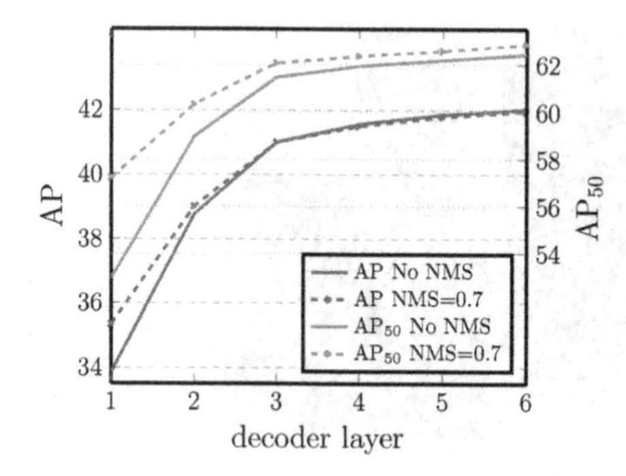

Fig. 4. AP and AP$_{50}$ performance after each decoder layer in a long schedule baseline model. DETR does not need NMS by design, which is validated by this figure. NMS lowers AP in the final layers, removing TP predictions, but improves it in the first layers, where DETR does not have the capability to remove double predictions.

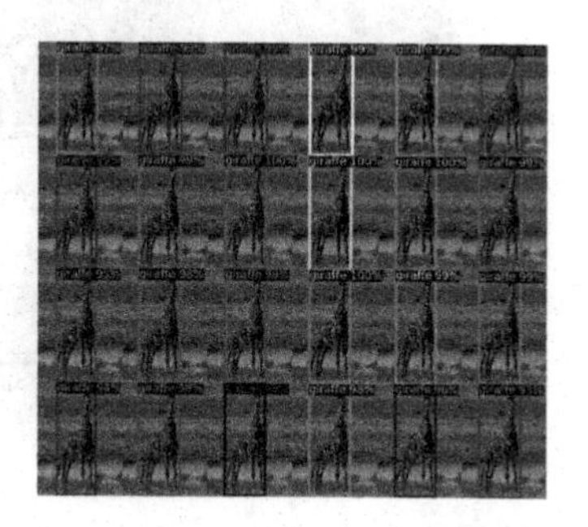

Fig. 5. Out of distribution generalization for rare classes. Even though no image in the training set has more than 13 giraffes, DETR has no difficulty generalizing to 24 and more instances.

stage of the decoding (Fig. 4). Both AP and AP$_{50}$ improve after every layer, totalling into a very significant +8.2/9.5 AP improvement between the first and the last layer. With its set-based loss, DETR does not need NMS by design. To verify this we run a standard NMS procedure with default parameters [49] for the outputs after each decoder. NMS improves performance for the predictions from the first decoder. This can be explained by the fact that a single decoding layer of the transformer is not able to compute any cross-correlations between the output elements, and thus it is prone to making multiple predictions for the same object. In the second and subsequent layers, the self-attention mechanism over the activations allows the model to inhibit duplicate predictions. We observe that the improvement brought by NMS diminishes as depth increases. It hurts AP in the last layers, as it incorrectly removes true positive predictions.

Similarly to visualizing encoder attention, we visualize decoder attentions in Fig. 6, coloring attention maps for each predicted object in different colors. We observe that decoder attention is fairly local, meaning that it mostly attends to object extremities such as heads or legs. We hypothesise that after the encoder has separated instances via global attention, the decoder only needs to attend to the extremities to extract the class and object boundaries.

Importance of FFN. FFN inside tranformers can be seen as 1×1 convolutional layers, making encoder similar to attention augmented convolutional networks [3]. We attempt to remove it completely leaving only attention in the transformer layers. By reducing the number of network parameters from 41.3M

Fig. 6. Visualizing decoder attention for every predicted object (images from COCO `val` set). Predictions are made with DETR-DC5 model. Decoder typically attends to object extremities, such as legs and heads.

to 28.7M, leaving only 10.8M in the transformer, performance drops by 2.3 AP, we thus conclude that FFN are important for achieving good results.

Importance of Positional Encodings. There are two kinds of positional encodings in our model: spatial positional encodings and output positional encodings (object queries). We experiment with various combinations of fixed and learned encodings, see results in appendix. Output positional encodings are required and cannot be removed, so we experiment with either passing them once at decoder input or adding to queries at every decoder attention layer. In the first experiment we completely remove spatial positional encodings and pass output positional encodings at input and, interestingly, the model still achieves more than 32 AP, losing 7.8 AP to the baseline. Then, we pass fixed sine spatial positional encodings and the output encodings at input once, as in the original transformer [46], and find that this leads to 1.4 AP drop compared to passing the positional encodings directly in attention. Learned spatial encodings passed to the attentions give similar results. Surprisingly, we find that not passing any spatial encodings in the encoder only leads to a minor AP drop of 1.3 AP. When we pass the encodings to the attentions, they are shared across all layers, and the output encodings (object queries) are always learned.

Given these ablations, we conclude that transformer components: the global self-attention in encoder, FFN, multiple decoder layers, and positional encodings, all significantly contribute to the final object detection performance.

Generalization to Unseen Numbers of Instances. Some classes in COCO are not well represented with many instances of the same class in the same image. For example, there is no image with more than 13 giraffes in the training set. We create a synthetic image[2] to verify the generalization ability of DETR (see Fig. 5). Our model is able to find all 24 giraffes on the image which is clearly out of distribution. This experiment confirms that there is no strong class-specialization in each object query.

[2] Base picture credit: https://www.piqsels.com/en/public-domain-photo-jzlwu.

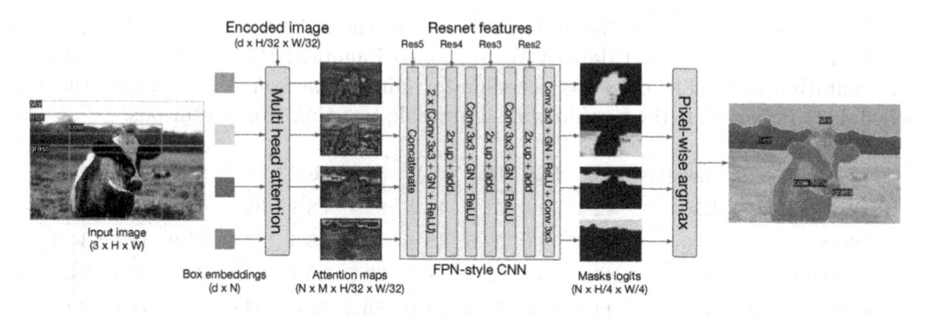

Fig. 7. Illustration of the panoptic head. A binary mask is generated in parallel for each detected object, then the masks are merged using pixel-wise argmax.

4.3 DETR for Panoptic Segmentation

Panoptic segmentation [18] has recently attracted a lot of attention from the computer vision community. Similarly to the extension of Faster R-CNN [36] to Mask R-CNN [13], DETR can be naturally extended by adding a mask head on top of the decoder outputs. In this section we demonstrate that such a head can be used to produce panoptic segmentation [18] by treating stuff and thing classes in a unified way. We perform our experiments on the panoptic annotations of the COCO dataset that has 53 stuff categories in addition to 80 things categories.

We train DETR to predict boxes around both *stuff* and *things* classes on COCO, using the same recipe. Predicting boxes is required for the training to be possible, since the Hungarian matching is computed using distances between boxes. We also add a mask head which predicts a binary mask for each of the predicted boxes, see Fig. 7. It takes as input the output of transformer decoder for each object and computes multi-head (with M heads) attention scores of this embedding over the output of the encoder, generating M attention heatmaps per object in a small resolution. To make the final prediction and increase the resolution, an FPN-like architecture is used. We refer to the supplement for more details. The final resolution of the masks has stride 4 and each mask is supervised independently using the DICE/F-1 loss [27] and Focal loss [22].

The mask head can be trained either jointly, or in a two steps process, where we train DETR for boxes only, then freeze all the weights and train only the mask head for 25 epochs. Experimentally, these two approaches give similar results, we report results using the latter method since it is less computationally intensive.

To predict the final panoptic segmentation we simply use an argmax over the mask scores at each pixel, and assign the corresponding categories to the resulting masks. This procedure guarantees that the final masks have no overlaps and thus DETR does not require a heuristic [18] to align different masks.

Training Details. We train DETR, DETR-DC5 and DETR-R101 models following the recipe for bounding box detection to predict boxes around stuff and things classes in COCO dataset. The new mask head is trained for 25 epochs (see supplementary for details). During inference we first filter out the detection

Table 2. Comparison with the state-of-the-art methods UPSNet [50] and Panoptic FPN [17] on the COCO `val` dataset We retrained PanopticFPN with the same data-augmentation as DETR, on a 18x schedule for fair comparison. UPSNet uses the `1x` schedule, UPSNet-M is the version with multiscale test-time augmentations.

Model	Backbone	PQ	SQ	RQ	PQ^{th}	SQ^{th}	RQ^{th}	PQ^{st}	SQ^{st}	RQ^{st}	AP
PanopticFPN++	R50	42.4	79.3	51.6	49.2	82.4	58.8	32.3	74.8	40.6	37.7
UPSnet	R50	42.5	78.0	52.5	48.6	79.4	59.6	33.4	75.9	41.7	34.3
UPSnet-M	R50	43.0	79.1	52.8	48.9	79.7	59.7	34.1	78.2	42.3	34.3
PanopticFPN++	R101	44.1	79.5	53.3	**51.0**	**83.2**	60.6	33.6	74.0	42.1	**39.7**
DETR	R50	43.4	79.3	53.8	48.2	79.8	59.5	36.3	78.5	45.3	31.1
DETR-DC5	R50	44.6	79.8	55.0	49.4	80.5	60.6	37.3	**78.7**	46.5	31.9
DETR	R101	45.1	79.9	55.5	50.5	80.9	61.7	37.0	78.5	46.0	33.0
DETR-DC5	R101	**45.6**	**80.0**	**56.1**	50.9	80.9	**62.2**	**37.5**	78.6	**46.8**	33.1

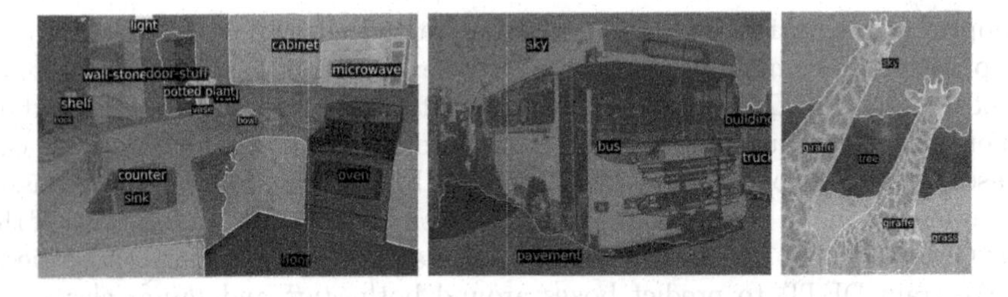

Fig. 8. Qualitative results for panoptic segmentation generated by DETR-R101. DETR produces aligned mask predictions in a unified manner for things and stuff.

with a confidence below 85%, then compute the per-pixel argmax to determine in which mask each pixel belongs. We then collapse different mask predictions of the same stuff category in one, and filter the empty ones (less than 4 pixels).

Main Results. Qualitative results are shown in Fig. 8. In Table 2 we compare our unified panoptic segmenation approach with several established methods that treat things and stuff differently. We report the Panoptic Quality (PQ) and the break-down on things (PQ^{th}) and stuff (PQ^{st}). We also report the mask AP (computed on the things classes), before any panoptic post-treatment (in our case, before taking the pixel-wise argmax). We show that DETR outperforms published results on COCO-val 2017, as well as our strong PanopticFPN baseline (trained with same data-augmentation as DETR, for fair comparison). The result break-down shows that DETR is especially dominant on stuff classes, and we hypothesize that the global reasoning allowed by the encoder attention is the key element to this result. For things class, despite a severe deficit of up to 8 mAP compared to the baselines on the mask AP computation, DETR obtains competitive PQ^{th}. We also evaluated our method on the test set of the COCO dataset, and obtained 46 PQ. We hope that our approach will inspire the exploration of fully unified models for panoptic segmentation in future work.

5 Conclusion

We presented DETR, a new design for object detection systems based on transformers and bipartite matching loss for direct set prediction. The approach achieves comparable results to an optimized Faster R-CNN baseline on the challenging COCO dataset. DETR is straightforward to implement and has a flexible architecture that is easily extensible to panoptic segmentation, with competitive results. In addition, it achieves significantly better performance on large objects, likely due to the processing of global information performed by the self-attention.

This new design for detectors also comes with new challenges, in particular regarding training, optimization and performances on small objects. Current detectors required several years of improvements to cope with similar issues, and we expect future work to successfully address them for DETR.

References

1. Al-Rfou, R., Choe, D., Constant, N., Guo, M., Jones, L.: Character-level language modeling with deeper self-attention. In: AAAI Conference on Artificial Intelligence (2019)
2. Bahdanau, D., Cho, K., Bengio, Y.: Neural machine translation by jointly learning to align and translate. In: ICLR (2015)
3. Bello, I., Zoph, B., Vaswani, A., Shlens, J., Le, Q.V.: Attention augmented convolutional networks. In: ICCV (2019)
4. Bodla, N., Singh, B., Chellappa, R., Davis, L.S.: Soft-NMS—improving object detection with one line of code. In: ICCV (2017)
5. Cai, Z., Vasconcelos, N.: Cascade R-CNN: high quality object detection and instance segmentation. PAMI (2019)
6. Chan, W., Saharia, C., Hinton, G., Norouzi, M., Jaitly, N.: Imputer: sequence modelling via imputation and dynamic programming. arXiv:2002.08926 (2020)
7. Devlin, J., Chang, M.W., Lee, K., Toutanova, K.: BERT: pre-training of deep bidirectional transformers for language understanding. In: NAACL-HLT (2019)
8. Erhan, D., Szegedy, C., Toshev, A., Anguelov, D.: Scalable object detection using deep neural networks. In: CVPR (2014)
9. Ghazvininejad, M., Levy, O., Liu, Y., Zettlemoyer, L.: Mask-predict: parallel decoding of conditional masked language models. arXiv:1904.09324 (2019)
10. Glorot, X., Bengio, Y.: Understanding the difficulty of training deep feedforward neural networks. In: AISTATS (2010)
11. Gu, J., Bradbury, J., Xiong, C., Li, V.O., Socher, R.: Non-autoregressive neural machine translation. In: ICLR (2018)
12. He, K., Girshick, R., Dollár, P.: Rethinking imagenet pre-training. In: ICCV (2019)
13. He, K., Gkioxari, G., Dollár, P., Girshick, R.B.: Mask R-CNN. In: ICCV (2017)
14. He, K., Zhang, X., Ren, S., Sun, J.: Deep residual learning for image recognition. In: CVPR (2016)
15. Hosang, J.H., Benenson, R., Schiele, B.: Learning non-maximum suppression. In: CVPR (2017)
16. Hu, H., Gu, J., Zhang, Z., Dai, J., Wei, Y.: Relation networks for object detection. In: CVPR (2018)

17. Kirillov, A., Girshick, R., He, K., Dollár, P.: Panoptic feature pyramid networks. In: CVPR (2019)
18. Kirillov, A., He, K., Girshick, R., Rother, C., Dollar, P.: Panoptic segmentation. In: CVPR (2019)
19. Kuhn, H.W.: The Hungarian method for the assignment problem (1955)
20. Li, Y., Qi, H., Dai, J., Ji, X., Wei, Y.: Fully convolutional instance-aware semantic segmentation. In: CVPR (2017)
21. Lin, T.Y., Dollár, P., Girshick, R., He, K., Hariharan, B., Belongie, S.: Feature pyramid networks for object detection. In: CVPR (2017)
22. Lin, T.Y., Goyal, P., Girshick, R.B., He, K., Dollár, P.: Focal loss for dense object detection. In: ICCV (2017)
23. Lin, T.-Y., et al.: Microsoft COCO: common objects in context. In: Fleet, D., Pajdla, T., Schiele, B., Tuytelaars, T. (eds.) ECCV 2014. LNCS, vol. 8693, pp. 740–755. Springer, Cham (2014). https://doi.org/10.1007/978-3-319-10602-1_48
24. Liu, W., et al.: SSD: single shot multibox detector. In: Leibe, B., Matas, J., Sebe, N., Welling, M. (eds.) ECCV 2016. LNCS, vol. 9905, pp. 21–37. Springer, Cham (2016). https://doi.org/10.1007/978-3-319-46448-0_2
25. Loshchilov, I., Hutter, F.: Decoupled weight decay regularization. In: ICLR (2017)
26. Lüscher, C., et al.: RWTH ASR systems for LibriSpeech: hybrid vs attention - w/o data augmentation. arXiv:1905.03072 (2019)
27. Milletari, F., Navab, N., Ahmadi, S.A.: V-net: Fully convolutional neural networks for volumetric medical image segmentation. In: 3DV (2016)
28. Oord, A., et al.: Parallel wavenet: fast high-fidelity speech synthesis. arXiv:1711.10433 (2017)
29. Park, E., Berg, A.C.: Learning to decompose for object detection and instance segmentation. arXiv:1511.06449 (2015)
30. Parmar, N., et al.: Image transformer. In: ICML (2018)
31. Paszke, A., et al.: Pytorch: an imperative style, high-performance deep learning library. In: NeurIPS (2019)
32. Pineda, L., Salvador, A., Drozdzal, M., Romero, A.: Elucidating image-to-set prediction: an analysis of models, losses and datasets. arXiv:1904.05709 (2019)
33. Radford, A., Wu, J., Child, R., Luan, D., Amodei, D., Sutskever, I.: Language models are unsupervised multitask learners (2019)
34. Redmon, J., Divvala, S., Girshick, R., Farhadi, A.: You only look once: unified, real-time object detection. In: CVPR (2016)
35. Ren, M., Zemel, R.S.: End-to-end instance segmentation with recurrent attention. In: CVPR (2017)
36. Ren, S., He, K., Girshick, R.B., Sun, J.: Faster R-CNN: towards real-time object detection with region proposal networks. PAMI **39** (2015)
37. Rezatofighi, H., Tsoi, N., Gwak, J., Sadeghian, A., Reid, I., Savarese, S.: Generalized intersection over union. In: CVPR (2019)
38. Rezatofighi, S.H., et al.: Deep perm-set net: learn to predict sets with unknown permutation and cardinality using deep neural networks. arXiv:1805.00613 (2018)
39. Rezatofighi, S.H., et al.: Deepsetnet: predicting sets with deep neural networks. In: ICCV (2017)
40. Romera-Paredes, B., Torr, P.H.S.: Recurrent instance segmentation. In: Leibe, B., Matas, J., Sebe, N., Welling, M. (eds.) ECCV 2016. LNCS, vol. 9910, pp. 312–329. Springer, Cham (2016). https://doi.org/10.1007/978-3-319-46466-4_19
41. Salvador, A., Bellver, M., Baradad, M., Marqués, F., Torres, J., Giró, X.: Recurrent neural networks for semantic instance segmentation. arXiv:1712.00617 (2017)

42. Stewart, R.J., Andriluka, M., Ng, A.Y.: End-to-end people detection in crowded scenes. In: CVPR (2015)
43. Sutskever, I., Vinyals, O., Le, Q.V.: Sequence to sequence learning with neural networks. In: NeurIPS (2014)
44. Synnaeve, G., et al.: End-to-end ASR: from supervised to semi-supervised learning with modern architectures. arXiv:1911.08460 (2019)
45. Tian, Z., Shen, C., Chen, H., He, T.: FCOS: fully convolutional one-stage object detection. In: ICCV (2019)
46. Vaswani, A., et al.: Attention is all you need. In: NeurIPS (2017)
47. Vinyals, O., Bengio, S., Kudlur, M.: Order matters: sequence to sequence for sets. In: ICLR (2016)
48. Wang, X., Girshick, R.B., Gupta, A., He, K.: Non-local neural networks. In: CVPR (2018)
49. Wu, Y., Kirillov, A., Massa, F., Lo, W.Y., Girshick, R.: Detectron2 (2019). https://github.com/facebookresearch/detectron2
50. Xiong, Y., et al.: Upsnet: a unified panoptic segmentation network. In: CVPR (2019)
51. Zhang, S., Chi, C., Yao, Y., Lei, Z., Li, S.Z.: Bridging the gap between anchor-based and anchor-free detection via adaptive training sample selection. arXiv:1912.02424 (2019)
52. Zhou, X., Wang, D., Krähenbühl, P.: Objects as points. arXiv:1904.07850 (2019)

DeepSFM: Structure from Motion via Deep Bundle Adjustment

Xingkui Wei[1], Yinda Zhang[2], Zhuwen Li[3], Yanwei Fu[1(✉)], and Xiangyang Xue[1]

[1] Fudan University, Shanghai, China
yanweifu@fudan.edu.cn
[2] Google Research, Menlo Park, USA
[3] Nuro, Inc, Mountain View, USA

Abstract. Structure from motion (SfM) is an essential computer vision problem which has not been well handled by deep learning. One of the promising trends is to apply explicit structural constraint, e.g. 3D cost volume, into the network. However, existing methods usually assume accurate camera poses either from GT or other methods, which is unrealistic in practice. In this work, we design a physical driven architecture, namely DeepSFM, inspired by traditional Bundle Adjustment (BA), which consists of two cost volume based architectures for depth and pose estimation respectively, iteratively running to improve both. The explicit constraints on both depth (structure) and pose (motion), when combined with the learning components, bring the merit from both traditional BA and emerging deep learning technology. Extensive experiments on various datasets show that our model achieves the state-of-the-art performance on both depth and pose estimation with superior robustness against less number of inputs and the noise in initialization.

1 Introduction

SfM is a fundamental human vision functionality which recovers 3D structures from the projected retinal images of moving objects or scenes. It enables machines to sense and understand the 3D world and is critical in achieving real-world artificial intelligence. Over decades of researches, there has been a lot of great success on SfM; however, the performance is far from perfect.

Conventional SfM approaches [1,5,8,46] heavily rely on Bundle-Adjustment (BA) [2,40], in which 3D structures and camera motions of each view are jointly optimized via Levenberg-Marquardt (LM) algorithm [32] according to the cross-view correspondence. Though successful in certain scenarios, conventional SfM based approaches are fundamentally restricted by the coverage of the provided

X. Wei, Y. Zhang and Z. Li—Equal contributions.

Electronic supplementary material The online version of this chapter (https://doi.org/10.1007/978-3-030-58452-8_14) contains supplementary material, which is available to authorized users.

A. Vedaldi et al. (Eds.): ECCV 2020, LNCS 12346, pp. 230–247, 2020.
https://doi.org/10.1007/978-3-030-58452-8_14

multiple views and the overlaps among them. They also typically fail to reconstruct textureless or non-lambertian (e.g. reflective or transparent) surfaces due to the missing of correspondence across views. As a result, selecting sufficiently good input views and the right scene requires excessive caution and is usually non-trivial to even experienced user.

Recent researches resort to deep learning to deal with the typical weakness of conventional SfM. Early effort utilizes deep neural network as a powerful mapping function that directly regresses the structures and motions [41,42,45,52]. Since the geometric constraints of structures and motions are not explicitly enforced, the network does not learn the underlying physics and prone to overfitting. Consequently, they do not perform as accurate as conventional SfM approaches and suffer from extremely poor generalization capability. Most recently, the 3D cost volume [22] has been introduced to explicit leveraging photo-consistency in a differentiable way, which significantly boosts the performance of deep learning based 3D reconstruction. However, the camera motion usually has to be known [20,50], which requires to run traditional methods on densely captured high resolution images or relies on extra calibration devices (Fig. 1 (b)). Some methods direct regress the motion [41,52], which still suffer from generalization issue (Fig. 1 (a)). Very rare deep learning approaches [38,39] can work well under noisy camera motion and improve both structure and motion simultaneously.

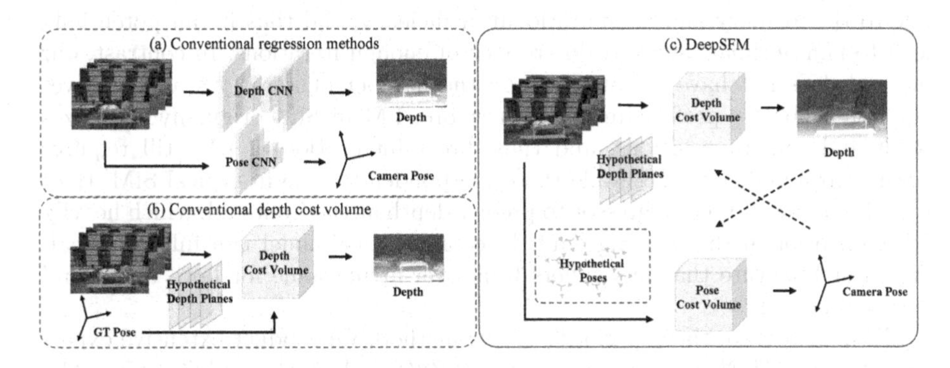

Fig. 1. DeepSFM refines the depth and camera pose of a target image given a few nearby source images. The network includes a depth based cost volume (D-CV) and a pose based cost volume (P-CV) which enforce photo-consistency and geometric-consistency into 3D cost volumes. The whole procedure is performed as iterations.

Inspired by BA and the success of cost volume for depth estimation, we propose a deep learning framework for SfM that iteratively improves both depth and camera pose according to cost volume explicitly built to measure photo-consistency and geometric-consistency. Our method does not require accurate pose, and a rough estimation is enough. In particular, our network includes a depth based cost volume (D-CV) and a pose based cost volume (P-CV). D-CV optimizes per-pixel depth values with the current camera poses, while P-CV optimizes camera poses with the current depth estimations (see Fig. 1 (c)). Conventional 3D cost volume enforces photo-consistency by unprojecting pixels into the discrete camera fronto-parallel planes and computing the photometric (i.e. image feature) difference as the cost. In addition to that, our D-CV further enforces geometric-consistency among cameras with their current depth estimations by adding the geometric (i.e. depth) difference to the cost. Note that the initial depth estimation can be obtained using the conventional 3D cost volume. When preparing this work, we notice that a concurrent work [49] which also utilizes this trick to build a better cost volume in their system. For pose estimation, rather than direct regression, our P-CV discretizes around the current camera positions, and also computes the photometric and geometric differences by hypothetically moving the camera into the discretized position. Note that the initial camera pose can be obtained by a rough estimation from the direct regression methods such as [41]. Our framework bridges the gap between the conventional and deep learning based SfM by incorporating explicit constraints of photo-consistency, geometric-consistency and camera motions all in the deep network.

The closest work in the literature is the recently proposed BA-Net [38], which also aims to explicitly incorporate multi-view geometric constraints in a deep learning framework. They achieve this goal by integrating the LM optimization into the network. However, the LM iterations are unrolled with few iterations due to the memory and computational inefficiency, and thus it can potentially lead to non-optimal solutions due to lack of enough iterations. In contrast, our method does not have a restriction on the number of iterations and achieves empirically better performance. Furthermore, LM in SfM originally optimizes point and camera positions, and thus direct integration of LM still requires good correspondences. To evade the correspondence issue in typical SfM, their models employ a direct regressor to predict depth at the front end, which heavily relies on prior in the training data. In contrast, our model is a fully physical-driven architecture that less suffers from over-fitting issue for both depth and pose estimation.

To demonstrate the superiority of our method, we conduct extensive experiments on *DeMoN datasets*, *ScanNet*, *ETH3D* and *Tanks and Temples*. The experiments show that our approach outperforms the state-of-the-art [33,38,41].

2 Related Work

There is a large body of work that focuses on inferring depth or motion from color images, ranging from single view, multiple views and monocular video. We discuss them in the context of our work.

Single-View Depth Estimation. While ill-posed, the emerging of deep learning technology enables the estimation of depth from a single color image. The early work directly formulates this into a per-pixel regression problem [7], and follow-up works improve the performance by introducing multi-scale network architectures [6,7], skip-connections [26,44], powerful decoder and post process [13,24–26,44], and new loss functions [10]. Even though single view based methods generate plausible results, the models usually resort heavily to the prior in the training data and suffer from generalization capability. Nevertheless, these methods still act as an important component in some multi-view systems [38].

Traditional Structure-from-Motion. Simultaneously estimating 3d structure and camera motion is a well studied problem which has a traditional toolchain of techniques [12,31,47]. Structure from Motion (SfM) has made great progress in many aspects. [15,28] aim at improving features and [35] introduce new optimization techniques. More robust structures and data representations are introduced by [14,33]. Simultaneous Localization and Sapping(SLAM) systems track the motion of the camera and build 3D structure from video sequence [9,29–31]. [9] propose the photometric bundle adjustment algorithm to directly minimize the photometric error of aligned pixels. However, traditional SfM and SLAM methods are sensitive to low texture region, occlusions, moving objects and lighting changes, which limit the performance and stability.

Deep Learning for Structure-from-Motion. Deep neural networks have shown great success in stereo matching and Structure-from-Motion problems. [41,42,45,52] regress depth and camera pose directly in a supervised manner or by introducing photometric constraints between depth and motion as a self-supervision signal. Such methods solve the camera motion as a regression problem, and the relation between camera motion and depth prediction is neglected.

Recently, some methods exploit multi-view photometric or feature-metric constraints to enforce the relationship between dense depth and the camera pose in network. The SE3 transformer layer is introduced by [39], which uses geometry to map flow and depth into a camera pose update. [43] propose the differentiable camera motion estimator based on the Direct Visual Odometry [36]. [4] using a LSTM-RNN [18] as the optimizer to solve nonlinear least squares in two-view SfM. [38] train a network to generate a set of basis depth maps and optimize depth and camera poses in a BA-layer by minimizing a feature-metric error.

3 Architecture

Our framework receives frames of a scene from different viewpoints, and produces accurate depth maps and camera poses for all frames. Similar to Bundle Adjustment (BA), we also assume initial structures (i.e depth maps) and motions (i.e. camera poses) are given. The initialization is not necessary to be accurate for the good performance using our framework and thus can be easily obtained from some direct regression based methods [41].

Now we introduce the overview of our model – DeepSFM. Without loss of generality, we describe our model taking two images as inputs, namely the target image and the source image, and all the technical components can be extended for multiple images straightforwardly. As shown in Fig. 2, we first extract feature maps from inputs through a shared encoder. We then sample the solution

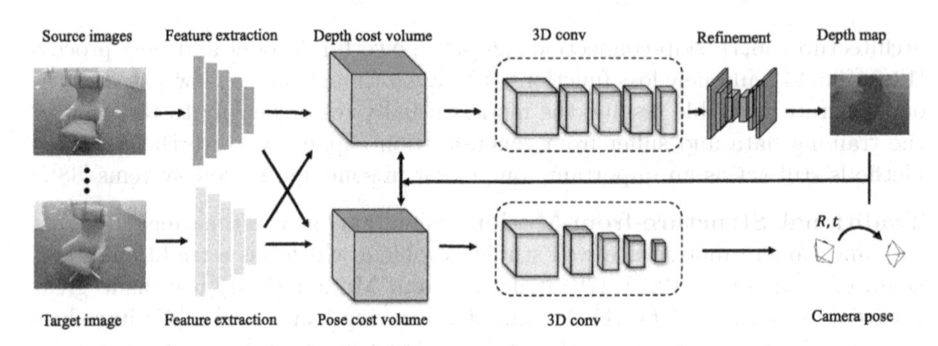

Fig. 2. Overview of DeepSFM. 2D CNN is used to extract photometric feature to construct cost volumes. Initial source depth maps and camera poses are used to introduce both photometric and geometric consistency. A series of 3D CNN layers are applied for D-CV and P-CV. Then a context network and depth regression operation are applied to produce predicted depth map of target image.

space for depth uniformly in the inverse-depth space between a predefined range, and camera pose around the initialization respectively. After that, we build cost volumes accordingly to reason the confidence of each depth and pose hypothesis. This is achieved by validating the consistency between the feature of the target view and the ones warped from the source image. Besides photo-metric consistency that measures the color image similarity, we also take into account the geometric consistency across warped depth maps. Note that depth and pose require different designs of cost volume to efficiently sample the hypothesis space. Gradients can back-propagate through cost volumes, and cost-volume construction does not affect any trainable parameters. The cost volumes are then fed into 3D CNN to regress new depth and pose. These updated values can be used to create new cost volumes, and the model improves the prediction iteratively.

For notations, we denote $\{\mathbf{I}_i\}_{i=1}^n$ as all the images in one scene, $\{\mathbf{D}_i\}_{i=1}^n$ as the corresponding ground truth depth maps, $\{\mathbf{K}_i\}_{i=1}^n$ as the camera intrinsics, $\{\mathbf{R}_i, \mathbf{t}_i\}_{i=1}^n$ as the ground truth rotations and translations of camera, $\{\mathbf{D}_i^*\}_{i=1}^n$ and $\{\mathbf{R}_i^*, \mathbf{t}_i^*\}_{i=1}^n$ as initial depth maps and camera pose parameters for constructing cost volumes, where n is the number of image samples.

3.1 2D Feature Extraction

Given the input sequences $\{\mathbf{I}_i\}_{i=1}^n$, we extract the 2D CNN feature $\{\mathbf{F}_i\}_{i=1}^n$ for each frame. Firstly, a 7 layers' CNN with kernel size 3×3 is applied to extract low contextual information. Then we adopt a spatial pyramid pooling (SPP) [21] module, which can extract hierarchical multi-scale features through 4 average pooling blocks with different pooling kernel size $(4 \times 4, 8 \times 8, 16 \times 16, 32 \times 32)$. Finally, we pass the concatenated features through 2D CNNs to get the 32-channel image features after upsampling these multi-scale features into the same resolution. These image sequence features are used by the building of both our depth based and pose based cost volumes.

3.2 Depth Based Cost Volume (D-CV)

Traditional plane sweep cost volume aims to back-project the source images onto successive virtual planes in the 3D space and measure photo-consistency error among the warped image features and target image features for each pixel. Different from the cost volume used in mainstream multi-view and structure-from-motion methods, we construct a D-CV to further utilize the local geometric consistency constraints introduced by depth maps. Inspired by the traditional plane sweep cost volumes, our D-CV is a concatenation of three components: the target image features, the warped source image features and the homogeneous depth consistency maps.

Hypothesis Sampling. To back-project the features and depth maps from source viewpoint to the 3D space in target viewpoint, we uniformly sample a set of L virtual planes $\{d_l\}_{l=1}^{L}$ in the inverse-depth space which are perpendicular to the forward direction (z-axis) of the target viewpoint. These planes serve as the hypothesis of the output depth map, and the cost volume can be built upon them.

Feature Warping. To construct our D-CV, we first warp source image features $\mathbf{F}_i$ (of size $CHannel \times Width \times Height$) to each of the hypothetical depth map planes d_l using camera intrinsic matrix $\mathbf{K}$ and initial camera poses $\{\mathbf{R}_i^*, \mathbf{t}_i^*\}$, according to:

$$\tilde{\mathbf{F}}_{il}(u) = \mathbf{F}_i\left(\tilde{u}_l\right), \tilde{u}_l \sim \mathbf{K}\left[\mathbf{R}_i^*|\mathbf{t}_i^*\right] \begin{bmatrix} \left(\mathbf{K}^{-1}u\right) d_l \\ 1 \end{bmatrix} \tag{1}$$

where u and $\tilde{u}_l$ are the homogeneous coordinates of each pixel in the target view and the projected coordinates onto the corresponding source view. $\tilde{\mathbf{F}}_{il}(u)$ denotes the warped feature of the source image through the l-th virtual depth plane. Note that the projected homogeneous coordinates $\tilde{u}_l$ are floating numbers, and we adopt a differentiable bilinear interpolation to generate the warped feature map $\tilde{\mathbf{F}}_{il}$. The pixels with no source view coverage are assigned with zeros. Following [20], we concatenate the target feature and the warped target feature together and obtain a $2CH \times L \times W \times H$ 4D feature volume.

Depth Consistency. In addition to photometric consistency, to exploit geometric consistency and promote the quality of depth prediction, we add two more channels on each virtual plane: the warped initial depth maps from the source views and the projected virtual depth plane from the perspective of the source view. Note that the former is the same as image feature warping, while the latter requires a coordinate transformation from the target to the source camera.

In particular, the first channel is computed as follows. The initial depth map of source image is first down-sampled and then warped to hypothetical depth planes similarly to the image feature warping as $\tilde{\mathbf{D}}_{il}^*(u) = \mathbf{D}_i^*\left(\tilde{u}_l\right)$, where the coordinates u and $\tilde{u}_l$ are defined in Eq. 1 and $\tilde{\mathbf{D}}_{il}^*(u)$ represents the warped one-channel depth map on the l-th depth plane. One distinction between depth warping and feature warping is that we adopt nearest neighbor sampling for depth warping, instead of bilinear interpolation. A comparison between the two methods is provided in the supplementary material.

The second channel contains the depth values of the virtual planes in the target view by seeing them from the source view. To transform the virtual planes to the source view coordinate system, we apply a T function on each virtual plane d_l in the following:

$$T(d_l) \sim [\mathbf{R}_i^*|\mathbf{t}_i^*] \begin{bmatrix} (\mathbf{K}^{-1}u)\, d_l \\ 1 \end{bmatrix} \tag{2}$$

We stack the warped initial depth maps and the transformed depth planes together, and get a depth volume of size $2 \times L \times W \times H$.

By concatenating the feature volume and depth volume together, we obtain a 4D cost tensor of size $(2CH + 2) \times L \times W \times H$. Given the 4D cost volume, our network learns a cost volume of size $L \times W \times H$ using several 3D convolutional layers with kernel size $3 \times 3 \times 3$. When there is more than one source image, we get the final cost volume by averaging over multiple input source views.

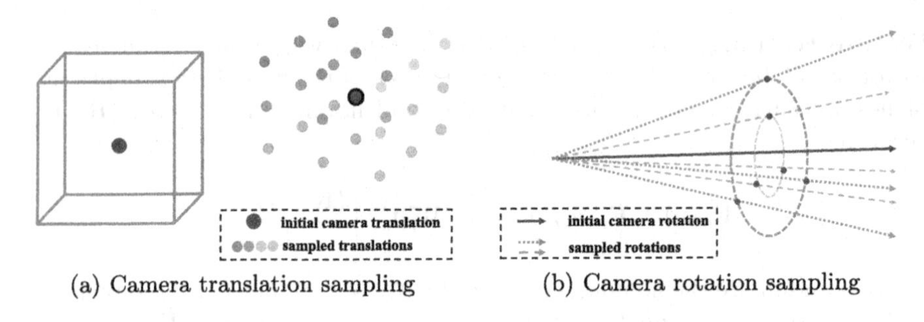

(a) Camera translation sampling (b) Camera rotation sampling

Fig. 3. Hypothetical camera pose sampling. (a) Camera translation sampling. We sample uniformly in the cubic space. (b) Camera rotation sampling. We sample around initial orientation vector in conical space.

3.3 Pose Based Cost Volume (P-CV)

In addition to the construction of D-CV, we also propose a P-CV, aiming at optimizing initial camera poses through both photometric and geometric consistency (see Fig. 3). Instead of building a cost volume based on hypothetical depth map planes, our novel P-CV is constructed based on a set of assumptive camera poses. Similar to D-CV, P-CV is also concatenated by three components: the target image features, the warped source image features and the homogeneous depth consistency maps. Given initial camera pose parameters $\{\mathbf{R}_i^*, \mathbf{t}_i^*\}$, we uniformly sample a batch of discrete candidate camera poses around. As shown in Fig. 3, we shift rotation and translation separately while keeping the other one unchanged. For rotation, we sample δR uniformly in the Euler angle space in a predefined range and multiply δR by the initial R. For translation, we sample δt uniformly and add δt to the initial t. In the end, a group of P

virtual camera poses noted as $\{\mathbf{R}^*_{ip}|\mathbf{t}^*_{ip}\}_{p=1}^P$ around input pose are obtained for cost volume construction.

The posed-based cost volume is also constructed by concatenating image features and homogeneous depth maps. However, source view features and depth maps are warped based on sampled camera poses. For feature warping, we compute $\tilde{u}_p$ as following equations:

$$\tilde{u}_p \sim \mathbf{K}\left[\mathbf{R}^*_{ip}|\mathbf{t}^*_{ip}\right]\begin{bmatrix}\left(\mathbf{K}^{-1}u\right)\mathbf{D}^*_i\\ 1\end{bmatrix} \tag{3}$$

where $\mathbf{D}^*_i$ is the initial target view depth. Similar to D-CV, we get warped source feature map $\tilde{\mathbf{F}}_{ip}$ after bilinear sampling and concatenate it with target view feature map. We also transform the initial target view depth and source view depth into one homogeneous coordinate system, which enhances the geometric consistency between camera pose and multi view depth maps.

After concatenating the above feature maps and depth maps together, we again build a 4D cost volume of size $(2CH+2)\times P\times W\times H$, where W and H are the width and height of feature map, CH is the number of channels. We get output of size $1\times P\times 1\times 1$ from the above 4-D tensor after eight 3D convolutional layers with kernel size $3\times 3\times 3$, three 3D average pooling layers with stride size $2\times 2\times 1$ and one global average pooling at the end.

3.4 Cost Aggregation and Regression

For depth prediction, we follow the cost aggregation technique introduced by [20]. We adopt a context network, which takes target image features and each slice of the coarse cost volume after 3D convolution as input and produce the refined cost slice. The final aggregated depth based volume is obtained by adding coarse and refined cost slices together. The last step to get depth prediction of target image is depth regression by soft-argmax as proposed in [20]. For camera poses prediction, we also apply a soft-argmax function on pose cost volume and get the estimated output rotation and translation vectors.

3.5 Training

The DeepSFM learns the feature extractor, 3D convolution, and the regression layers in a supervised way. We denote $\hat{\mathbf{R}}_i$ and $\hat{\mathbf{t}}_i$ as predicted rotation angles and translation vectors of camera pose. Then the pose loss $\mathcal{L}_{rotation}$ is defined as the $L1$ distance between prediction and groundtruth. We denote $\hat{D}^0_i$ and $\hat{D}_i$ as predicted coarse depth map and refined depth map, then the depth loss function is defined as $\mathcal{L}_{depth} = \sum_i \lambda H(\hat{D}^0_i, \mathbf{D}_i) + H(\hat{D}_i, \mathbf{D}_i)$, where λ is weight parameter and function H is Huber loss. Our final objective $\mathcal{L}_{final} = \lambda_r\mathcal{L}_{rotation} + \lambda_t\mathcal{L}_{translation} + \lambda_d\mathcal{L}_{depth}$. The λs are determined empirically, and are listed in the supplementary material.

The initial depth maps and camera poses are obtained from DeMoN. To keep correct scale, we multiply translation vectors and depth maps by the norm of the ground truth camera translation. The whole training and testing procedure are performed as four iterations. During each iteration, we take the predicted depth maps and camera poses of previous iteration as new initialization. More details are provided in the supplementary material.

4 Experiments

4.1 Datasets

We evaluate DeepSFM on widely used datasets and compare with state-of-the-art methods on accuracy, generalization capability and robustness to initialization.

DeMoN Datasets [41]. This dataset contains data from various sources, including SUN3D [48], RGB-D SLAM [37], and Scenes11 [3]. To test the generalization capability, we also evaluate on MVS [11] dataset but not use it for the training. In all four datasets, RGB image sequences and the ground truth depth maps are provided with the camera intrinsics and camera poses. Note that those datasets together provide a diverse set of both indoor and outdoor, synthetic and real-world scenes. For all the experiments, we adopt the same training and testing data split from DeMoN.

ETH3D Dataset [34]. It provides a variety of indoor and outdoor scenes with high-precision ground truth 3D points captured by laser scanners, which is a more solid benchmark dataset. Ground truth depth maps are obtained by projecting the point clouds to each camera view. Raw images are in high resolution but resized to 810×540 pixels for evaluation [20].

Tanks and Temples. [23] It is a benchmark for image-based large scale 3D reconstruction. The benchmark sequences are acquired in realistic conditions and of high quality. Point clouds captured using an industrial laser scanner are provided as ground truth. Again, our method are trained on DeMoN and tested on the dataset to show the robustness to noisy initialization.

4.2 Evaluation

DeMoN Datasets. Our results on DeMoN datasets and the comparison to other methods are shown in Table 1. We cite results of some strong baseline methods from DeMoN paper, named as Base-Oracle, Base-SIFT, Base-FF and Base-Matlab respectively [41]. Base-Oracle estimate depth with the ground truth camera motion using SGM [17]. Base-SIFT, Base-FF and Base-Matlab solve camera motion and depth using feature, optical flow, and KLT tracking correspondence from 8-pt algorithm [16]. We also compare to some most recent state-of-the-art methods LS-Net [4] and BA-Net [38]. LS-Net introduces the learned LSTM-RNN optimizer to minimizing photometric error for stereo reconstruction. BA-Net is the most recent work that minimizes the feature-metric error between

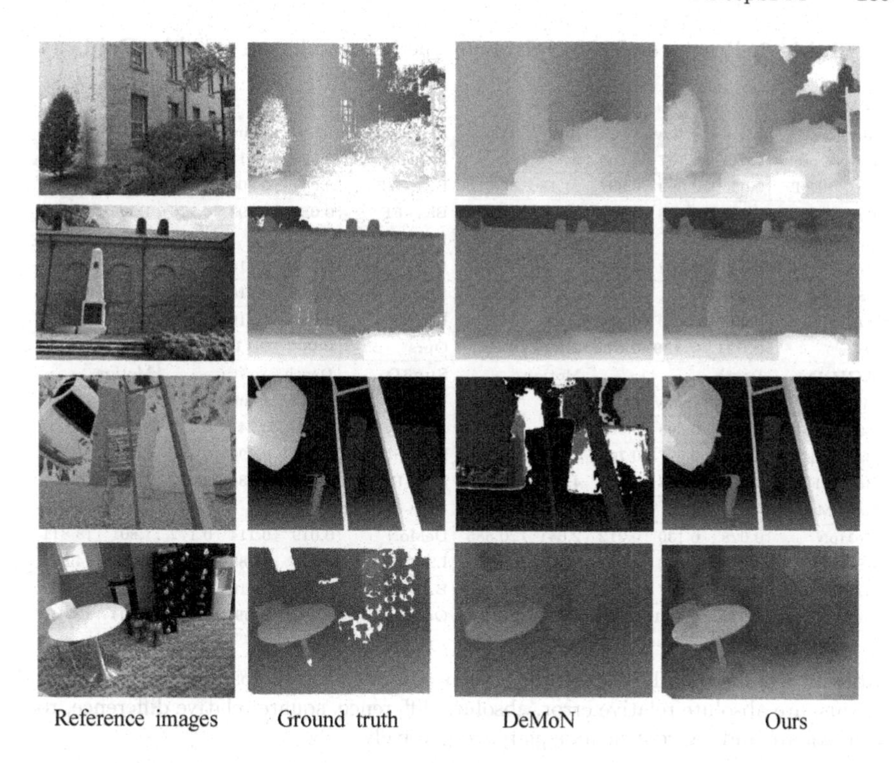

Fig. 4. Qualitative Comparisons with DeMoN [41] on DeMoN datasets. Results on more methods and examples are shown in the supplementary material.

multi-view via the differentiable Levenberg-Marquardt [27] algorithm. To make a fair comparison, we adopt the same metrics as DeMoN[41] for evaluation.

Our method outperforms all traditional baseline methods and DeMoN on both depth and camera poses. When compared with more recent LS-Net and BA-Net, our method produces better results in most metrics on four datasets. On RGB-D dataset, our performance is comparable to the state-of-the-art due to relatively higher noise in the RGB-D ground truth. LS-Net trains an initialization network which regresses depth and motion directly before adding the LSTM-RNN optimizer. The performance of the RNN optimizer is highly affected by the accuracy of the regressed initialization. The depth results of LS-Net are consistently poorer than BA-Net and our method, despite better rotation parameters are estimated by LS-Net on RGB-D and Sun3D datasets with very good initialization. Our method is slightly inferior to BA-Net on the L1-rel metric, which is probably due to that we sample 64 virtual planes uniformly as the hypothetical depth set, while BA-Net optimizes depth prediction based on a set of 128-channel estimated basis depth maps that are more memory consuming but have more fine-grained results empirically. Despite all that, it is shown that our learned cost volumes with geometric consistency work better than the photometric bundle adjustment (e.g. used in BA-Net) in most scenes. In particular,

Table 1. Results on DeMoN datasets, the best results are noted by **Bold**.

MVS	Depth			Motion		**Scenes11**	Depth			Motion	
Method	L1-inv	sc-inv	L1-rel	Rot	Trans	Method	L1-inv	sc-inv	L1-rel	Rot	Trans
Base-Oracle	0.019	0.197	0.105	0	0	Base-Oracle	0.023	0.618	0.349	0	0
Base-SIFT	0.056	0.309	0.361	21.180	60.516	Base-SIFT	0.051	0.900	1.027	6.179	56.650
Base-FF	0.055	0.308	0.322	4.834	17.252	Base-FF	0.038	0.793	0.776	1.309	19.426
Base-Matlab	-	-	-	10.843	32.736	Base-Matlab	-	-	-	0.917	14.639
DeMoN	0.047	0.202	0.305	5.156	14.447	DeMoN	0.019	0.315	0.248	0.809	8.918
LS-Net	0.051	0.221	0.311	4.653	11.221	LS-Net	0.010	0.410	0.210	4.653	8.210
BANet	0.030	0.150	0.080	3.499	11.238	BANet	0.080	0.210	0.130	3.499	10.370
Ours	**0.021**	**0.129**	**0.079**	**2.824**	**9.881**	Ours	**0.007**	**0.112**	**0.064**	**0.403**	**5.828**
RGB-D	Depth			Motion		**Sun3D**	Depth			Motion	
Method	L1-inv	sc-inv	L1-rel	Rot	Trans	Method	L1-inv	sc-inv	L1-rel	Rot	Trans
Base-Oracle	0.026	0.398	0.36	0	0	Base-Oracle	0.020	0.241	0.220	0	0
Base-SIFT	0.050	0.577	0.703	12.010	56.021	Base-SIFT	0.029	0.290	0.286	7.702	41.825
Base-FF	0.045	0.548	0.613	4.709	46.058	Base-FF	0.029	0.284	0.297	3.681	33.301
Base-Matlab	-	-	-	12.813	49.612	Base-Matlab	-	-	-	5.920	32.298
DeMoN	0.028	0.130	0.212	2.641	20.585	DeMoN	0.019	0.114	0.172	1.801	18.811
LS-Net	0.019	0.090	0.301	**1.010**	22.100	LS-Net	0.015	0.189	0.650	**1.521**	14.347
BANet	**0.008**	0.087	**0.050**	2.459	14.900	BANet	0.015	0.110	**0.060**	1.729	13.260
Ours	0.011	**0.071**	0.126	1.862	**14.570**	Ours	**0.013**	**0.093**	0.072	1.704	**13.107**

Table 2. Results on ETH3D (**Bold**: best; $\alpha = 1.25$). abs_rel, abs_diff, sq_rel, rms, and log_rms, are absolute relative error, absolute difference, square relative difference, root mean square and log root mean square, respectively.

Method	Error metric					Accuracy metric($\delta < \alpha^t$)		
	abs_rel	abs_diff	sq_rel	rms	log_rms	α	α^2	α^3
COLMAP	0.324	**0.615**	36.71	2.370	0.349	**86.5**	90.3	92.7
DeMoN	0.191	0.726	0.365	1.059	0.240	73.3	89.8	95.1
Ours	**0.127**	0.661	**0.278**	**1.003**	**0.195**	84.1	**93.8**	**96.9**

we improve mostly on the Scenes11 dataset, where the ground truth is perfect but the input images contain a lot of texture-less regions, which are challenging to photo-consistency based methods. The Qualitative Comparisons between our method and DeMoN are shown in Fig. 4.

ETH3D. We further test the generalization capability on ETH3D. We provide comparisons to COLMAP [33] and DeMoN on ETH3D. COLMAP is a state-of-the-art Structure-from-Motion method, while DeMoN introduces a classical deep network architecture that directly regress depth and motion in a supervised manner. Note that all the models are trained on DeMoN and then tested on the data provided by [19]. In the accuracy metric, the error δ s defined as $\max(\frac{y_i^*}{y_i}, \frac{y_i}{y_i^*})$, and the thresholds are typically set as $[1.25, 1.25^2, 1.25^3]$. In Table 2, our method shows the best performance overall among all the comparison methods. Our method produces better results than DeMoN consistently, since we impose geometric and physical constraints onto network rather than learning to regress

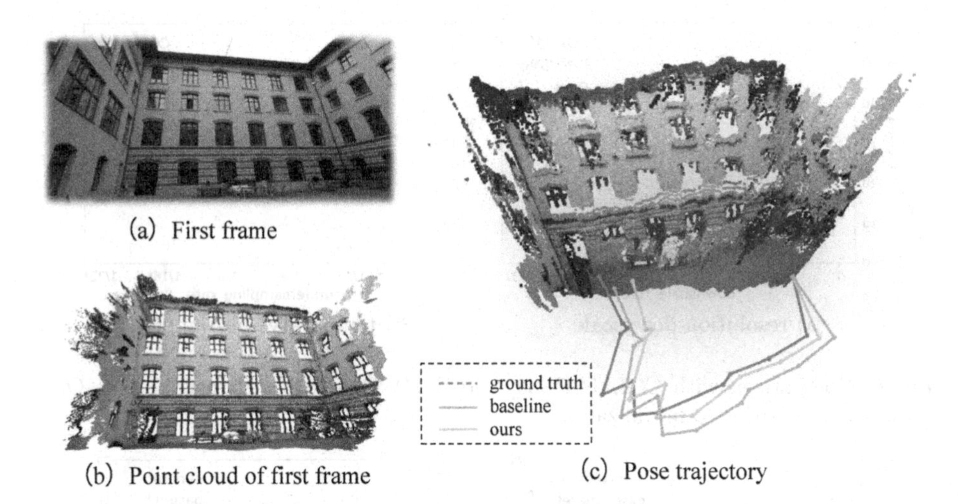

Fig. 5. Result on a sequence of the ETH3D dataset. (a) First frame of sequence. (b) The point cloud from estimated depth of the first frame. (c) Pose trajectories. Compared with the baseline method in Sect. 4.3, the accumulated pairwise pose trajectory predicted by our network (yellow) are more closely consistent with the ground truth (red) (Color figure online).

directly. When compared with COLMAP, our method performs better on most metrics. COLMAP behaves well in the accuracy metric (i.e. abs_diff). However, the presence of outliers is often observed in the predictions of COLMAP, which leads to poor performance in other metrics such as abs_rel and sq_rel, since those metrics are sensitive to outliers. As an intuitive display, we compute the motion of camera in a selected image sequence of ETH3D, as shown in Fig. 5c. The point cloud computed from the estimated depth map is showed in Fig. 5b, which is of good quality.

Tanks and Temples. To evaluate the robustness to initialization quality, we compare DeepSFM with COLMAP and the SOTA – R-MVSNet [51] on the Tanks and Temples [23] dataset as it contains densely captured high resolution images from which pose can be precisely estimated. To add noise on pose, we downscale the images and sub-sample temporal frames. For evaluation metrics, we adopt the F-score (higher is better) used in this dataset. The reconstruction qualities of Barn sequence are shown in Fig. 6. It is observed that the performance of R-MVSNet and COLMAP drops significantly as the input quality becomes lower, while our method maintains the performance in a certain range. It is worth noting that COLMAP completely fails when the number of images are sub-sampled to 1/16.

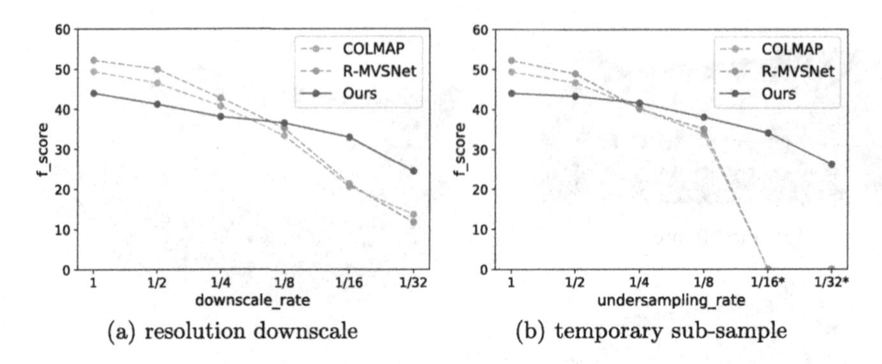

(a) resolution downscale (b) temporary sub-sample

Fig. 6. Comparison with COLMAP [33] and R-MVSNet [51] with noisy input. Our work is less sensitive to initialization.

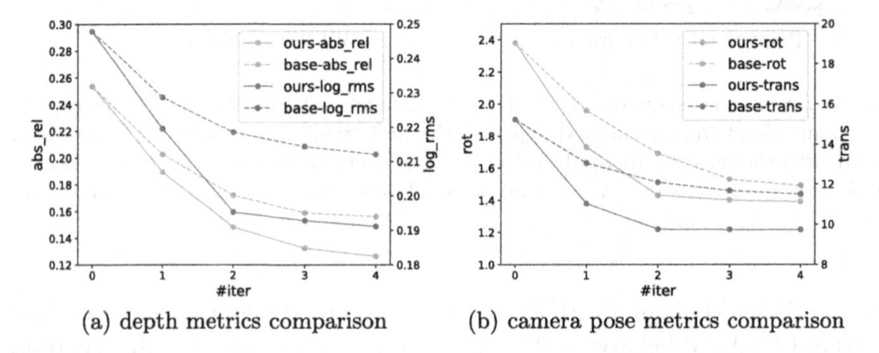

(a) depth metrics comparison (b) camera pose metrics comparison

Fig. 7. Comparison with baseline during iterations. Our work converges at a better position. (a) abs relative error and log RMSE. (b) rotation and translation error.

4.3 Model Analysis

In this section, we analyze our model on several aspects to verify the optimality and show advantages over previous methods. More ablation studies are provided in the supplementary material.

Iterative Improvement. Our model can run iteratively to reduce the prediction error. Figure 7 (solid lines) shows our performance over iterations when initialized with the prediction from DeMoN. As can be seen, our model effectively reduces both depth and pose errors upon the DeMoN output. Throughout the iterations, better depth and pose benefit each other by building more accurate cost volume, and both are consistently improved. The whole process is similar to coordinate descent algorithm, and finally converges at iteration 4.

Effect of P-CV. We compare DeepSFM to a baseline method for our P-CV. In this baseline, the depth prediction is the same as DeepSFM, but the pose prediction network is replaced by a direct visual odometry model [36], which updates camera parameters by minimizing pixel-wise photometric error between image features. Both methods are initialized with DeMoN results. As provided

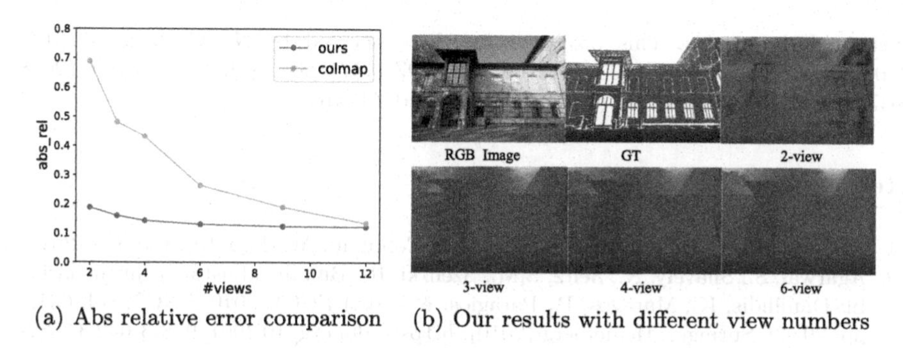

(a) Abs relative error comparison (b) Our results with different view numbers

Fig. 8. Depth map results w.r.t. the number of images. Our performance does not change much with varying number of views.

in Fig. 7, DeepSFM consistently produces lower errors on both depth and pose over all the iterations. This shows that our P-CV predicts more accurate pose and performs more robust against noise depth at early stages. Figure 5(c) shows the visualized pose trajectories which are estimated by baseline(cyan) and our method(yellow) on ETH3D.

View Number. DeepSFM works still reasonably well with fewer views due to the free from optimization based components. To show this, we compare to COLMAP with respect to the number of input views on ETH3D. As depicted in Fig. 8, more images yield better results for both methods as expected. However, our performance drops significantly slower than COLMAP with fewer number of inputs. Numerically, DeepSFM cuts the depth error by half under the same number of views as COLMAP, or achieves similar error with half number of views required by COLMAP. This clearly demonstrates that DeepSFM is more robust when fewer inputs are available.

5 Conclusions

We present a deep learning framework for Structure-from-Motion, which explicitly enforces photo-metric consistency, geometric consistency and camera motion constraints all in the deep network. This is achieved by two key components - namely D-CV and P-CV. Both cost volumes measure the photo-metric and geometric errors by hypothetically moving reconstructed scene points (structure) or camera (motion) respectively. Our deep network can be considered as an enhanced learning based BA algorithm, which takes the best benefits from both learnable priors and geometric rules. Consequently, our method outperforms conventional BA and state-of-the-art deep learning based methods for SfM.

Acknowledgements. This project is partly supported by NSFC Projects (6170 2108), STCSM Projects (19511120700, and 19ZR1471800), SMSTM Project (2018 SHZDZX01), SRIF Program (17DZ2260900), and ZJLab.

References

1. Agarwal, S., et al.: Building Rome in a day. Commun. ACM **54**(10), 105–112 (2011)
2. Agarwal, S., Snavely, N., Seitz, S.M., Szeliski, R.: Bundle adjustment in the large. In: Daniilidis, K., Maragos, P., Paragios, N. (eds.) ECCV 2010. LNCS, vol. 6312, pp. 29–42. Springer, Heidelberg (2010). https://doi.org/10.1007/978-3-642-15552-9_3
3. Chang, A.X., et al.: Shapenet: an information-rich 3D model repository. arXiv preprint arXiv:1512.03012 (2015)
4. Clark, R., Bloesch, M., Czarnowski, J., Leutenegger, S., Davison, A.J.: Learning to solve nonlinear least squares for monocular stereo. In: Ferrari, V., Hebert, M., Sminchisescu, C., Weiss, Y. (eds) Proceedings of the European Conference on Computer Vision (ECCV), vol. 11212, pp. 284–299. Springer, Cham (2018). https://doi.org/10.1007/978-3-030-01237-3_18
5. Delaunoy, A., Pollefeys, M.: Photometric bundle adjustment for dense multi-view 3d modeling. In: Proceedings of the IEEE Conference on Computer Vision and Pattern Recognition, pp. 1486–1493 (2014)
6. Eigen, D., Fergus, R.: Predicting depth, surface normals and semantic labels with a common multi-scale convolutional architecture. In: The IEEE International Conference on Computer Vision (ICCV), December 2015
7. Eigen, D., Puhrsch, C., Fergus, R.: Depth map prediction from a single image using a multi-scale deep network. In: Advances in Neural Information Processing Systems, pp. 2366–2374 (2014)
8. Engel, J., Koltun, V., Cremers, D.: Direct sparse odometry. IEEE Trans. Pattern Anal. Mach. Intell. **40**(3), 611–625 (2017)
9. Engel, J., Schöps, T., Cremers, D.: LSD-SLAM: large-scale direct monocular SLAM. In: Fleet, D., Pajdla, T., Schiele, B., Tuytelaars, T. (eds.) ECCV 2014. LNCS, vol. 8690, pp. 834–849. Springer, Cham (2014). https://doi.org/10.1007/978-3-319-10605-2_54
10. Fu, H., Gong, M., Wang, C., Batmanghelich, K., Tao, D.: Deep ordinal regression network for monocular depth estimation. In: The IEEE Conference on Computer Vision and Pattern Recognition (CVPR), June 2018
11. Fuhrmann, S., Langguth, F., Goesele, M.: Mve-a multi-view reconstruction environment. In: GCH, pp. 11–18 (2014)
12. Furukawa, Y., Curless, B., Seitz, S.M., Szeliski, R.: Towards internet-scale multi-view stereo. In: 2010 IEEE Computer Society Conference on Computer Vision and Pattern Recognition, pp. 1434–1441. IEEE (2010)
13. Garg, R., B.G., V.K., Carneiro, G., Reid, I.: Unsupervised CNN for single view depth estimation: geometry to the rescue. In: Leibe, B., Matas, J., Sebe, N., Welling, M. (eds.) ECCV 2016. LNCS, vol. 9912, pp. 740–756. Springer, Cham (2016). https://doi.org/10.1007/978-3-319-46484-8_45
14. Gherardi, R., Farenzena, M., Fusiello, A.: Improving the efficiency of hierarchical structure-and-motion. In: 2010 IEEE Computer Society Conference on Computer Vision and Pattern Recognition, pp. 1594–1600. IEEE (2010)

15. Han, X., Leung, T., Jia, Y., Sukthankar, R., Berg, A.C.: Matchnet: unifying feature and metric learning for patch-based matching. In: Proceedings of the IEEE Conference on Computer Vision and Pattern Recognition, pp. 3279–3286 (2015)
16. Hartley, R.I.: In defense of the eight-point algorithm. IEEE Trans. Pattern Anal. Mach. Intell. **19**(6), 580–593 (1997)
17. Hirschmuller, H.: Accurate and efficient stereo processing by semi-global matching and mutual information. In: 2005 IEEE Computer Society Conference on Computer Vision and Pattern Recognition (CVPR 2005), vol. 2, pp. 807–814. IEEE (2005)
18. Hochreiter, S., Younger, A.S., Conwell, P.R.: Learning to learn using gradient descent. In: Dorffner, G., Bischof, H., Hornik, K. (eds.) ICANN 2001. LNCS, vol. 2130, pp. 87–94. Springer, Heidelberg (2001). https://doi.org/10.1007/3-540-44668-0_13
19. Huang, P.H., Matzen, K., Kopf, J., Ahuja, N., Huang, J.B.: Deepmvs: larning multi-view stereopsis. In: Proceedings of the IEEE Conference on Computer Vision and Pattern Recognition, pp. 2821–2830 (2018)
20. Im, S., Jeon, H.G., Lin, S., Kweon, I.S.: Dpsnet: end-to-end deep plane sweep stereo. In: International Conference on Learning Representations (2019). https://openreview.net/forum?id=ryeYHi0ctQ
21. He, K., Zhang, X., Ren, S., Sun, J.: Spatial pyramid pooling in deep convolutional networks for visual recognition. In: Fleet, D., Pajdla, T., Schiele, B., Tuytelaars, T. (eds.) ECCV 2014. LNCS, vol. 8691, pp. 346–361. Springer, Cham (2014). https://doi.org/10.1007/978-3-319-10578-9_23
22. Kar, A., Häne, C., Malik, J.: Learning a multi-view stereo machine. In: Advances in Neural Information Processing Systems, pp. 365–376 (2017)
23. Knapitsch, A., Park, J., Zhou, Q.Y., Koltun, V.: Tanks and temples: benchmarking large-scale scene reconstruction. ACM Trans. Graph. **36**(4) (2017)
24. Kuznietsov, Y., Stuckler, J., Leibe, B.: Semi-supervised deep learning for monocular depth map prediction. In: The IEEE Conference on Computer Vision and Pattern Recognition (CVPR), July 2017
25. Laina, I., Rupprecht, C., Belagiannis, V., Tombari, F., Navab, N.: Deeper depth prediction with fully convolutional residual networks. In: 2016 Fourth International Conference on 3D Vision (3DV), pp. 239–248. IEEE (2016)
26. Liu, F., Shen, C., Lin, G., Reid, I.: Learning depth from single monocular images using deep convolutional neural fields. IEEE Trans. Pattern Anal. Mach. Intell. **38**(10), 2024–2039 (2016)
27. Lourakis, M., Argyros, A.A.: Is Levenberg-Marquardt the most efficient optimization algorithm for implementing bundle adjustment? In: Tenth IEEE International Conference on Computer Vision (ICCV 2005) Volume 1, vol. 2, pp. 1526–1531. IEEE (2005)
28. Lowe, D.G.: Distinctive image features from scale-invariant keypoints. Int. J. Comput. Vis. **60**(2), 91–110 (2004)
29. Mur-Artal, R., Montiel, J.M.M., Tardos, J.D.: ORB-SLAM: a versatile and accurate monocular SLAM system. IEEE Trans. Rob. **31**(5), 1147–1163 (2015)
30. Mur-Artal, R., Tardós, J.D.: ORB-SLAM2: an open-source slam system for monocular, stereo, and RGB-D cameras. IEEE Trans. Rob. **33**(5), 1255–1262 (2017)
31. Newcombe, R.A., Lovegrove, S.J., Davison, A.J.: DTAM: dense tracking and mapping in real-time. In: 2011 International Conference on Computer Vision, pp. 2320–2327. IEEE (2011)
32. Nocedal, J., Wright, S.: Numerical Optimization. Springer Science & Business Media, New York (2006). https://doi.org/10.1007/978-0-387-40065-5

33. Schonberger, J.L., Frahm, J.M.: Structure-from-motion revisited. In: Proceedings of the IEEE Conference on Computer Vision and Pattern Recognition, pp. 4104–4113 (2016)
34. Schöps, T., et al.: A multi-view stereo benchmark with high-resolution images and multi-camera videos. In: Conference on Computer Vision and Pattern Recognition (CVPR) (2017)
35. Snavely, N.: Scene reconstruction and visualization from internet photo collections: a survey. IPSJ Trans. Comput. Vis. Appl. **3**, 44–66 (2011)
36. Steinbrücker, F., Sturm, J., Cremers, D.: Real-time visual odometry from dense RGB-D images. In: 2011 IEEE International Conference on Computer Vision Workshops (ICCV Workshops), pp. 719–722. IEEE (2011)
37. Sturm, J., Engelhard, N., Endres, F., Burgard, W., Cremers, D.: A benchmark for the evaluation of RGB-D slam systems. In: 2012 IEEE/RSJ International Conference on Intelligent Robots and Systems, pp. 573–580. IEEE (2012)
38. Tang, C., Tan, P.: Ba-net: dense bundle adjustment network. arXiv preprint arXiv:1806.04807 (2018)
39. Teed, Z., Deng, J.: Deepv2d: video to depth with differentiable structure from motion. arXiv preprint arXiv:1812.04605 (2018)
40. Triggs, B., McLauchlan, P.F., Hartley, R.I., Fitzgibbon, A.W.: Bundle adjustment—a modern synthesis. In: Triggs, B., Zisserman, A., Szeliski, R. (eds.) IWVA 1999. LNCS, vol. 1883, pp. 298–372. Springer, Heidelberg (2000). https://doi.org/10.1007/3-540-44480-7_21
41. Ummenhofer, B., et al.: Demon: depth and motion network for learning monocular stereo. In: Proceedings of the IEEE Conference on Computer Vision and Pattern Recognition, pp. 5038–5047 (2017)
42. Vijayanarasimhan, S., Ricco, S., Schmid, C., Sukthankar, R., Fragkiadaki, K.: SFM-net: learning of structure and motion from video. arXiv preprint arXiv:1704.07804 (2017)
43. Wang, C., Miguel Buenaposada, J., Zhu, R., Lucey, S.: Learning depth from monocular videos using direct methods. In: Proceedings of the IEEE Conference on Computer Vision and Pattern Recognition, pp. 2022–2030 (2018)
44. Wang, P., Shen, X., Lin, Z., Cohen, S., Price, B., Yuille, A.L.: Towards unified depth and semantic prediction from a single image. In: The IEEE Conference on Computer Vision and Pattern Recognition (CVPR), June 2015
45. Wang, S., Clark, R., Wen, H., Trigoni, N.: Deepvo: towards end-to-end visual odometry with deep recurrent convolutional neural networks. In: 2017 IEEE International Conference on Robotics and Automation (ICRA), pp. 2043–2050. IEEE (2017)
46. Wu, C., Agarwal, S., Curless, B., Seitz, S.M.: Multicore bundle adjustment. In: CVPR 2011, pp. 3057–3064. IEEE (2011)
47. Wu, C., et al.: VisualSFM: a visual structure from motion system (2011). http://www.cs.washington.edu/homes/ccwu/vsfm
48. Xiao, J., Owens, A., Torralba, A.: Sun3d: a database of big spaces reconstructed using SFM and object labels. In: Proceedings of the IEEE International Conference on Computer Vision, pp. 1625–1632 (2013)

49. Xu, Q., Tao, W.: Multi-scale geometric consistency guided multi-view stereo. In: Proceedings of the IEEE Conference on Computer Vision and Pattern Recognition, pp. 5483–5492 (2019)
50. Yao, Y., Luo, Z., Li, S., Fang, T., Quan, L.: MVSNet: depth inference for unstructured multi-view stereo. In: Ferrari, V., Hebert, M., Sminchisescu, C., Weiss, Y. (eds.) ECCV 2018. LNCS, vol. 11212, pp. 785–801. Springer, Cham (2018). https://doi.org/10.1007/978-3-030-01237-3_47
51. Yao, Y., Luo, Z., Li, S., Shen, T., Fang, T., Quan, L.: Recurrent mvsnet for high-resolution multi-view stereo depth inference. In: Proceedings of the IEEE Conference on Computer Vision and Pattern Recognition, pp. 5525–5534 (2019)
52. Zhou, T., Brown, M., Snavely, N., Lowe, D.G.: Unsupervised learning of depth and ego-motion from video. In: Proceedings of the IEEE Conference on Computer Vision and Pattern Recognition, pp. 1851–1858 (2017)

Ladybird: Quasi-Monte Carlo Sampling for Deep Implicit Field Based 3D Reconstruction with Symmetry

Yifan Xu[1], Tianqi Fan[2,3], Yi Yuan[1(✉)], and Gurprit Singh[3]

[1] Netease Fuxi AI Lab, Hangzhou, China
{xuyifan,yuanyi}@corp.netease.com
[2] Saarland Informatics Campus, Saarbruecken, Germany
[3] MPI for Informatics, Saarbruecken, Germany
{tfan,gsingh}@mpi-inf.mpg.de

Abstract. Deep implicit field regression methods are effective for 3D reconstruction from single-view images. However, the impact of different sampling patterns on the reconstruction quality is not well-understood. In this work, we first study the effect of *point set discrepancy* on the network training. Based on Farthest Point Sampling algorithm, we propose a sampling scheme that theoretically encourages better generalization performance, and results in fast convergence for SGD-based optimization algorithms. Secondly, based on the reflective symmetry of an object, we propose a feature fusion method that alleviates issues due to self-occlusions which makes it difficult to utilize local image features. Our proposed system *Ladybird* is able to create high quality 3D object reconstructions from a single input image. We evaluate Ladybird on a large scale 3D dataset (ShapeNet) demonstrating highly competitive results in terms of Chamfer distance, Earth Mover's distance and Intersection Over Union (IoU).

Keywords: 3D reconstruction · Deep learning · Sampling · Symmetry

1 Introduction

Due to the under-constrained nature of the problem, 3D object reconstruction from a single-view image has been a challenging task. Large shape and structure variations among objects make it difficult to define one dedicated parameterized model. Methods based on template deformation are often restricted by the initial topology of the template, and are not able to recover holes for instance. Recently,

Y. Xu and T. Fan—These two authors contribute equally.

Electronic supplementary material The online version of this chapter (https://doi.org/10.1007/978-3-030-58452-8_15) contains supplementary material, which is available to authorized users.

A. Vedaldi et al. (Eds.): ECCV 2020, LNCS 12346, pp. 248–263, 2020.
https://doi.org/10.1007/978-3-030-58452-8_15

deep learning based implicit fields regression methods have shown great potential in monocular 3D reconstruction. Mescheder et al. [20] and DISN [32] create visually pleasing smooth shape reconstruction, with consistent normal and complex topology using implicit fields.

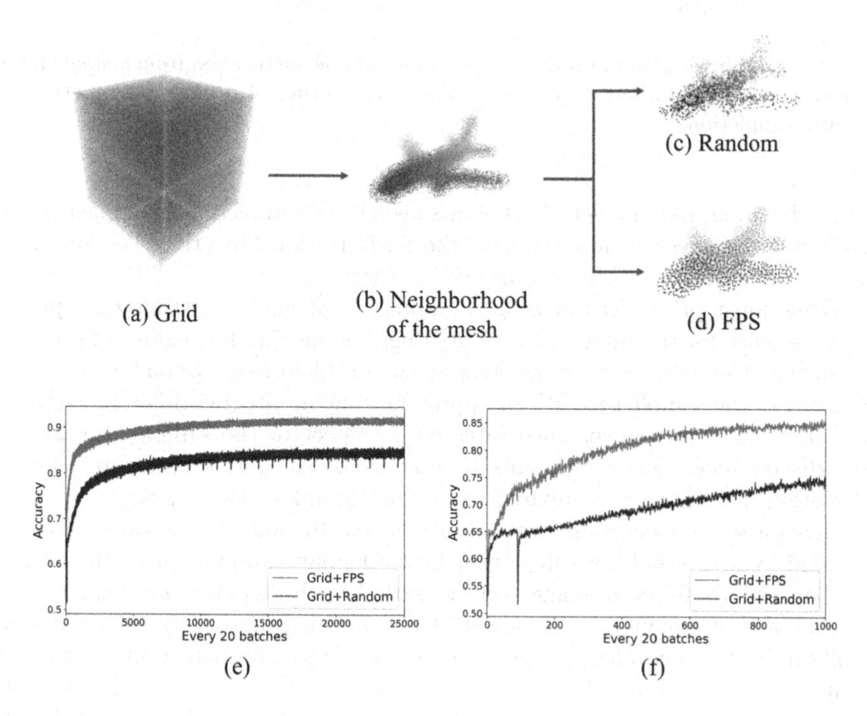

Fig. 1. Top: Demonstration of our sampling strategy for implicit field regression network training. A neighborhood of the mesh (b) is sampled from a set of dense grid points (a). A sparse set of points is sampled from (b) uniformly at random (c) or through FPS (d). Bottom: comparison of the training accuracy between Grid+FPS and Grid+Random sampling for the same network architecture during training. (e) is the plot of the training accuracy for the first 25 epoch. (f) is the plot of the training accuracy of the first epoch. Sampling with lower discrepancy results in faster convergence and better accuracy during training.

An implicit field is a real-valued function defined on $\mathbb{R}^3$ whose iso-surface recovers the mesh of interest. Common choices of implicit field are signed distance field, truncated signed distance field, or occupancy probability field. A network $g_w(I, p)$ is trained to predict the implicit field of point $p \in \mathbb{R}^3$, based on the input image I, where w are the parameters which are optimized with stochastic gradient descent (SGD) type algorithms. This is followed by post-processing methods like marching cube and sphere tracing to reconstruct the mesh.

The loss function for the implicit field regression problem is the L_2 distance between the ground truth implicit field and the network g_w predicted output.

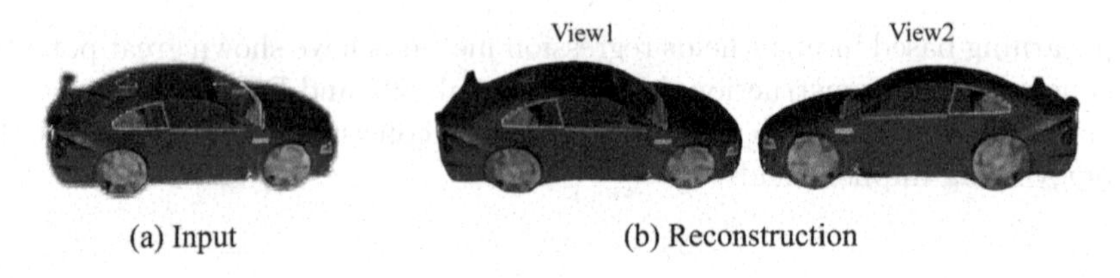

Fig. 2. Ladybird is able to produce high quality 3D reconstruction from a single input image. The consideration of symmetry allows recovering of occluded geometry and texture completion.

During training, a sparse set of 3D points need to be sampled in a compact region containing the mesh to approximate the optimization objective. We formulate this empirical loss as a Monte Carlo estimator.

While most prior discussion on sampling [20] focuses on designing a probability measure for the integral that puts different weights for regions of different distance to the mesh surface, we look at the problem from a point view of discrepancy of the sample sets. When approximating an integral, different samplers have different error convergence rates with respect to the sample size [22,24]. Low discrepancy sequences/points or blue noise (in 2D) samples give better estimation, for instance, compared to random samples (white noise).

Given a set of locally uniform samples whose distance to the target mesh is bounded by a threshold, we show that farthest point sampling algorithm (FPS) can be used to select a sparse subset with low discrepancy for training g_w. An overview of our method is shown in Fig. 1. Our proposed sampling scheme results in better generalization performance as it provides better approximation to the expected loss, thanks to the Koksma-Hlawka inequality [15]. Empirically our sampling scheme also results in faster convergence for SGD-based optimization algorithms, which speeds up the training process significantly as shown in Fig. 1(e, f).

Many deep 3D implicit field reconstruction works [5,20] explore the use of global shape encoding. While being good at capturing the general shape and obtaining interesting interpolation in the latent space, sometimes it is difficult to recover fine geometric details with only global features. Local features found via aligning image to mesh by modeling the camera are used to address the issue. However, for occluded points, it is ambiguous what local features should be used. Usually all the sampled points are projected to the images [32], and hence points in the back use features of the points that occlude them.

As most man-made objects are symmetric about a plane, we observe that this problem can be alleviated via the consideration of reflective symmetry. For a symmetric pair of points p and q, the implicit fields at p and q are the same, and often at least one of them is visible in the image. Hence we can use the local features of q to improve the implicit field predication of p, which can also be understood as utilizing two-view information. Our feature fusion method imposes a symmetry prior on the network g_w, which gives significant improvement of the reconstruction quality as shown in Fig. 2. Unlike previous works [29,33] that focus on the design of loss function, detection or encoding of symmetry, our method naturally integrates into the pixel-to-mesh alignment framework.

The advantage of spatially aligning the image to mesh and utilizing the corresponding local features is that the fine shape details and textures can be better recovered. However, when p is occluded, the feature obtained by such alignment no longer has an intuitive meaning. Recently Front2Back [33] addresses such issues by detecting reflective symmetries from the data and synthesizing the opposite orthographic view. Our approach is simpler and does not depend on symmetry detection.

Mesh. AtlasNet [11] represents a mesh as a locally parameterized surface and predicts the local patches from a latent shape representation learned via reconstruction objectives. Mitchell et al. [21] proposes to represent 3D shapes using higher order functions.

Pixel2Mesh [28] uses graph CNN to progressively deform an ellipsoid template mesh to fit the target. Features from different layers in the CNN are used to generate different resolution of details. 3DN [29] infers vertex offsets from a template mesh according to the image object's category, and proposes differentiable mesh sampling operator to compute the loss function. SDM-NET [9] uses VAE to generate a spatial arrangement of deformable parts of an object. Pan et al. [23] proposes a progressive method that alternates between deforming the mesh and modifying the topology. Mesh R-CNN [10] unifies object detection and shape reconstruction, with a mesh prediction branch that first produces coarse cubified meshes which are refined with a graph convolution network.

DIB-R [5], Soft Rasterizer [19] design differentiable rasterization layers that enable unsupervised training for reconstruction tasks. DIST [18] proposes an optimized differentiable sphere tracing layer for differentiable SDF rendering.

Point Cloud and Voxel. Fan et al. [8] proposes a conditional shape sampler to predict multiple plausible point clouds from an input image. Lin et al. [17] uses an auto-encoder to synthesize partial point clouds from multiple views, which is combined as a dense point cloud. Then the loss is computed via rendering the depth images from multiple views. Li et al. [16] uses a CNN to predict multiple depth maps and corresponding deformation fields, which are fused to form the full 3D shape.

3D-R2N2 [6] uses recurrent neural networks to generate voxelized 3D reconstruction. Pixel2Vox [30] uses encoder-decoder structures to generate a coarse 3D voxel.

1.1 Sampling Methods in Monte Carlo Integration

Realistic image synthesis involves evaluating very high-dimensional light transport integrals. (Quasi-)Monte Carlo (MC) numerical methods are traditionally employed to approximate these integrals which is highly error prone. This error

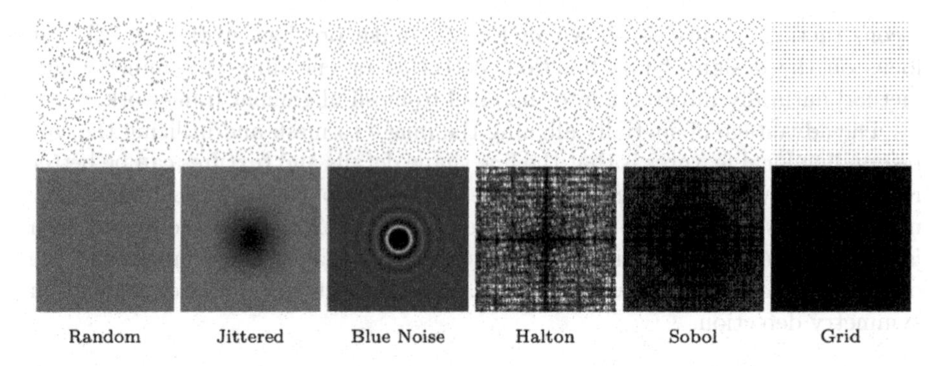

Fig. 3. Top row shows different point patterns for different samplers for $N = 1024$ samples. Bottom row shows the corresponding *expected* power spectra. Random samples are completely decorrelated which results in a flat power spectrum. 2D stratification (jittered) results in a power spectrum with small dark region around the center (DC frequency). For blue noise (Poisson Disk) sampler, this dark (no energy low-frequency) region is larger. However, for Halton and Sobol samplers, the corresponding power spectrum shows some spikes, but it preserves well the underlying stratification along dimensions which is characterized as a dark cross in the middle of the spectrum. Finally, a simple regular (Grid) pattern has a grid like power spectrum (zoom-in to the right-most bottom image to see the grid structure).

directly depends on the sampling pattern used to estimate the underlying integral [25]. These sampling patterns can be highly correlated. Fourier power spectra are commonly employed to characterize these correlations among samples (Fig. 3).

Blue noise samplers [25] are well-known to show good improvements for low-dimensional integration problems whereas low-discrepancy [22] samplers like Halton [12] and Sobol [13] are more effective for higher dimensional problems. In this work, we use farthest point selection strategy [7] from any given pointset to select our samples.

2 Our Approach

We first start with a theoretical motivation for our sampling methods. This is followed by the proposed symmetric feature fusion module and our 3D reconstruction pipeline (illustrated in Fig. 4).

2.1 Preliminary

In Quasi-Monte Carlo integration literature, the equidistribution of a point set is tested by calculating the discrepancy of the set. This approach assigns a single quality number, the discrepancy, to every point set. The lower the discrepancy, the better (uniform) the underlying point set would be. We focus on the *star*

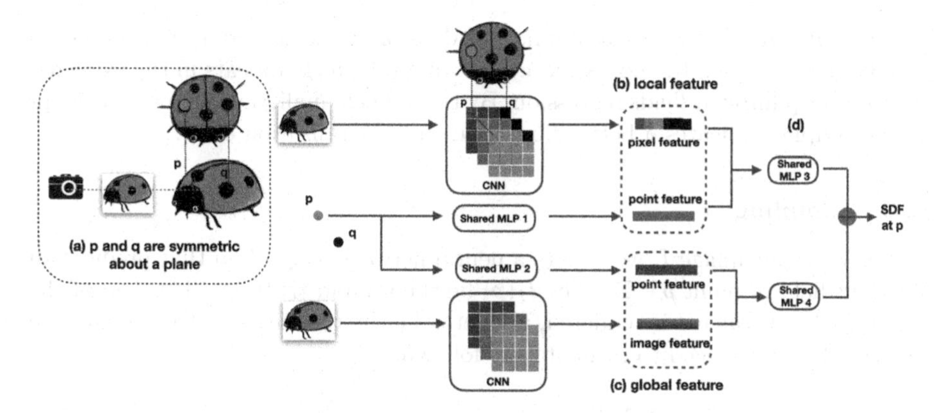

Fig. 4. Overview of Ladybird. a) $p, q \in \mathbb{R}$ are symmetric about a plane. Their projections to the image are found via a camera model. b) Local feature consists of point feature of p and local image feature from pixels corresponding to p and q. c) Global feature consists of point feature of p and global image feature. d) Local feature and global feature are encoded through two MLPs whose parameters are shared among all $p \in \mathbb{R}$. In the end, marching cube is used to extract iso-surface.

discrepancy of a point set, which computes discrepancy with respect to rectangular axis-aligned sub-regions with one of their corners fixed to the origin. Mathematically, the star discrepancy can be defined as follows:

Definition 1. *Let $P = \{x_1, x_2, ..., x_N\}$ be a set of points in $\mathbb{R}^d$, then the star discrepancy of P is*

$$D_N^*(P) = \sup_{B \in J} \|\frac{A(B;P)}{N} - \lambda_d(B)\|, \tag{1}$$

where λ_d is the Lebesgue measure on $\mathbb{R}^d$, $A(B;P)$ is the number of points in P that are in B, and $J = \{\prod_{i=1}^{d}[0, u_i)|0 < u_i \leq 1\}$.

For a given point set or a sequence P (stochastic or deterministic), the error due to sampling is directly related to the star discrepancy $D_N^*(P)$ of the point set P. This relation is given by the Koksma-Hlawka inequality [15] as described below:

Theorem 1. *Let $I = [0,1]^d$ and f is a function on I with bounded variation $V(f)$. Then for any $x_1, x_2, ..., x_N \in I$,*

$$\|\frac{1}{N}\sum_{i=1}^{N} f(x_i) - \int_I f(x)dx\| \leq V(f)D_N^*(\{x_1, ..., x_N\}) \tag{2}$$

The above inequality states that for f with bounded variation, a point set with lower discrepancy gives less error when numerically integrating f.

The distance between two implicit fields is an integral, and a set of points in needs to be sampled to approximate such integral which appears in the expected loss for deep implicit fields regression. By triangle inequality, using lower discrepancy sampler indicates a better bound on the generalisation error.

2.2 Sampling

Given an input image I, we denote a neural network as $g_w(I, p)$ that predicts an implicit field of point $p \in \mathbb{R}^3$. Let $f_I(p)$ be the ground truth implicit field of the mesh M from which I is rendered, and let S be the training set. To estimate the expected loss, we need to estimate the following:

$$\sum_{I \in S} \int_{\mathbb{R}^3} (f_I(p) - g_w(I, p))^2 \, m(p) dp \tag{3}$$

where $m(p)$ is a probability density function in $\mathbb{R}^3$ supported in a compact region near the mesh M.

Instead of studying different choices for $m(p)$ and their effects on training, we study the impact of different sampling patterns on the integral estimation.

The error convergence rate of an estimator is greatly influenced by the sampling pattern [22,24]. Sparse sampling could result in aliasing following the Nyquist-Shannon theorem. A better sampling strategy would allow faster convergence to the true integral resulting in better generalisation performance. Following the Koksma-Hlawka inequality, in order to better approximate the L_2 distance between g_w and f_M—which indicates better generalisation of the network on different input points p—sample sets of lower discrepancy should be preferred.

In consideration of the time efficiency, usually we pre-compute the implicit field of a dense set of points around the mesh surface, where a sparse subset is chosen uniformly during training. Hence we consider the following problem: given a set of points A, how to select a subset $B \subseteq A$ consisting of N points with low discrepancy. It is natural to consider farthest point sampling algorithm(FPS): initially $x_1 \in A$ is selected uniformly at random. Then iteratively,

$$x_i = \arg\max_{x \in A \setminus B}(\min_{y \in B} d(x, y)) \tag{4}$$

is added to B. In Sect. 3.3, we show that compared to randomly selecting a sample subset B from A, sampling using the FPS approach results in lower discrepancy.

2.3 Feature Fusion Based on Symmetry

For a fixed camera model, let π be the corresponding projection that maps 3D points to the image plane. Assume that the target mesh M is symmetric about xy plane[1], and A is the rigid transformation such that the input image is formed

[1] ShapeNet data set is aligned, and most objects are symmetric about xy plane.

via the composition $\pi \circ A$. In practice, either A is known or A is predicted via a camera network from input image.

For a point p not too far from M, let I_p be the pixel in the image that corresponds to $\pi(p)$. A convolution neural network (CNN) is used to extract features from the input image I. Let F_p be the concatenation of feature vectors at I_p in different layers of the CNN.

We can use F_p to guide the regression of the implicit field at p. However when p is occluded, the pixel value of I_p is not determined by p but by $r \in M$ with smallest z-buffer value whose projection $\pi(r)$ also lies in the pixel I_p. There is no clear relation between the implicit field at p and that at r.

For a point $v = (x, y, z)$, such that $p = Av$, the symmetric point q of p is $A\bar{v}$ where $\bar{v} = (x, y, -z)$. The implicit field at p should equal to that at q. Hence it is reasonable to include F_q as part of the local feature of p, which we call feature fusion. One straight-forward and effective way to implement feature fusion is to concatenate F_p and F_q.

3 Experiments

To show the effectiveness of our proposed system Ladybird, we provide quantitative as well qualitative comparisons to other methods. Our backbone network architecture is based on DISN [32]. Our implementation of Ladybird is in Tensorflow 1.9 [2], and the system is tested on Nvidia GTX 1080Ti with Cuda 9.0. In all our experiments, Adam optimizer [14] is used with $\beta = 0.5$ and an initial learning rate of 1e−4.

3.1 Data Processing

For dataset, we use ShapeNet Core v1 [3], and use the official train/test split. There are 13 categories of objects. For each object, 24 views are rendered as in 3D-R2N2 [6]. We randomly select 6000 images from the training set as the validation set, and our training set contains 726,600 images. The data is aligned and most objects (about 80%) are symmetric about xy plane. We normalize the object mesh such that its center of mass is at the origin and the mesh lies in the unit sphere.

To efficiently and accurately compute the SDF values, we use polygon soup algorithm [31] to compute the SDF on 256^3 grid points. After that, non-grid point SDF values are obtained through tri-linear interpolation.

For each mesh object, first we sample 256^3 points P_1 using Grid, Jitter, or Sobol sampler [1] and compute the corresponding SDF values. In Jitter, each grid point jitters with Gaussian noise of mean 0 and standard deviation 0.02. We then sample a subset $P_2 \subset P_1$ consisting of 32,768 points from P_1 in the following way: from each SDF range $[-0.10, -0.03]$, $[-0.03, 0.00]$, $[0.00, 0.03]$, and $[0.03, 0.10]$, 1/4-th of points are sampled uniformly at random. During training time, a subset $P_3 \subset P_2$ consisting of 2048 points are sampled from P_2 uniformly at random or

Table 1. Mean (×0.01) and standard deviation (×0.01) of star discrepancy of different samplers. A+B means we first sample 256^2 points using sampling method A and then select a subset of size $n = 1024, 2048, 4096$ with method B.

Sample size	Metric	Grid+Random	Grid+FPS	Jitter+FPS	Sobol+FPS
1024	Mean	4.51	3.86	6.49	4.35
	Std	0.66	0.19	0.49	1.43
2048	Mean	2.98	2.48	6.07	2.51
	Std	0.26	0.16	0.77	0.47
4096	Mean	2.34	1.96	6.10	1.65
	Std	0.33	0.08	0.30	0.37

through FPS at each epoch. Depending on the sampling pattern used to sample P_1 (say A) and P_3 (say B), the resulting sampling pattern is denoted by $A + B$.

At test time, the SDF of 256^3 grid points are predicated and marching cube is used to extract the iso-surface.

3.2 Network Details

We use a pre-trained camera pose estimation network from DISN [32], to predict a rigid transformation matrix A described in Sect. 2.3. VGG-16 is used as a CNN module to extract features form the input image. For a given point p, F_p is the concatenation of features (Sect. 2.3) at different layers of VGG-16 at pixel I_p that p projects under the known or predicted camera intrinsics. Assuming q and p being symmetric about a plane, the pixel feature of p is one of the following:

1. Base: F_p which is of dimension 1472.
2. Symm(Near): F_p or F_q depending on the one having smaller z-buffer value.
3. Symm(Avg): The average of F_p and F_q.
4. Symm(Concat): The concatenation of F_p and F_q.

As shown in Fig. 4, the image feature is the output of VGG-16 (of dimension 1024). Two stream of point features are processed with two MLPs, each of parameters (64, 256, 512). Each stream is concatenated with pixel feature and image feature respectively, to form a local and a global feature. These global and local features are encoded through two MLPs, each of parameters (512, 256, 1), and the encoded values are added as the predicted SDF at p.

3.3 Samplers Impact on Training

To assess the effect of different samplers on training, we set our pixel features to Base (see Sect. 3.2), use ground truth camera parameters and keep the batch size to 20. In Table 1, we report the star discrepancy of different samplers in 2D. We first sample 256^2 points using Grid, Jitter, Sobol sampler in $[0,1]^2$, then selecting 1024, 2048, 4096 points uniformly at random or through FPS.

Table 2. Effect of different samplers on the reconstruction results on ShapeNet test set. Method A+B means P_1 is sampled with A and P_3 is sampled with B. Metrics are class mean of CD ($\times 0.001$), and class mean of EMD ($\times 100$), computed on 2048 points. Grid+FPS outperforms other methods.

Metric	Grid+Random	Grid+FPS	Jitter+FPS	Sobol+FPS
CD	10.17	8.43	19.88	11.33
EMD	2.71	2.57	2.92	2.84

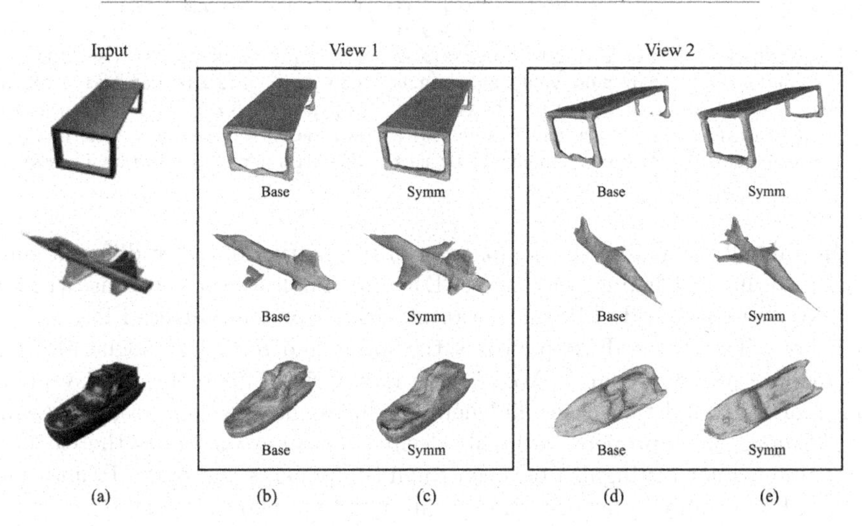

Fig. 5. Impact of feature fusion based on reflective symmetry. (a) indicates the input images. (b) and (d) are the reconstruction results using Base in two different views. (c) and (e) are the reconstruction results using Symm(Concat). We see that Symm(Concat) helps to improve the reconstruction quality.

In Jitter, each grid point jitters with Gaussian noise of mean 0 and standard deviation 0.01. We experimentally verify in 2D that Grid+FPS sampling has lower discrepancy and lower variance compared to Grid+Random.

Grid vs. Sobol: The SDF validation accuracy of Sobol+FPS (0.914) is similar to that of Grid+FPS (0.917), which is higher than Grid+Random (0.825). However, SDF prediction is an intermediate step for the reconstruction task. Marching Cube is used to recover the mesh from the SDF, which requires SDF values at grid points. Due to this grid restriction imposed by Marching Cube, Grid sampling ensures better training/test data consistency. In addition, Grid+FPS and Grid+Random leads to more stable training results (cf. Sobol+FPS) due to lower std. Our work advocates that Grid+FPS is suitable for 3D reconstruction based on deep implicit fields and marching cube.

Table 3. Comparison between different feature fusion operation evaluated on ShapeNet test set. Metrics are CD ($\times 0.001$), and EMD ($\times 100$), computed on 2048 points. Groud truth camera parameters are used.

Metric	Local image feature	Plane	Bench	Box	Car	Chair	Display	Lamp	Speaker	Rifle	Sofa	Table	Phone	Boat	Mean
CD	Base	5.33	5.37	9.33	4.42	7.73	7.07	24.36	13.65	**3.32**	5.78	9.37	8.13	5.79	8.43
	Symm(Avg)	7.27	17.00	12.29	4.97	14.83	15.83	58.77	23.76	6.72	11.15	12.06	61.73	5.96	19.41
	Symm(Near)	4.73	5.50	9.13	4.12	6.70	7.05	18.43	12.26	3.62	6.70	11.49	4.49	5.37	7.66
	Symm(Concat)	**3.86**	**4.30**	**8.04**	**4.11**	**5.43**	**6.09**	**14.10**	**10.53**	3.51	**5.05**	**8.13**	**4.16**	**4.92**	**6.33**
EMD	Base	2.35	2.30	2.91	2.47	2.66	2.44	4.21	3.19	1.69	2.29	2.78	1.95	2.14	2.57
	Symm(Avg)	2.14	2.36	2.98	2.42	2.56	2.54	4.69	3.41	1.71	2.45	2.77	3.25	2.12	2.72
	Symm(Near)	2.24	2.22	2.95	2.40	2.53	2.42	4.11	3.14	**1.65**	2.38	2.85	1.89	2.11	2.53
	Symm(Concat)	**2.07**	**2.06**	**2.80**	**2.38**	**2.32**	**2.28**	**3.59**	**2.98**	1.73	**2.18**	**2.57**	**1.85**	**2.07**	**2.38**
IoU	Base	63.4	56.3	52.0	77.8	58.1	60.2	41.7	58.4	70.4	71.3	53.8	75.7	66.0	61.9
	Symm(Near)	64.7	56.3	54.3	79.0	60.2	61.1	43.4	58.5	71.6	69.8	52.8	76.4	67.5	62.7
	Symm(Concat)	**66.6**	**60.3**	**56.4**	**80.2**	**64.7**	**63.7**	**48.5**	**61.5**	**71.9**	**73.5**	**58.1**	**78.1**	**68.8**	**65.6**

In Table 2, we report the comparison of reconstruction using different samplers in terms of Chamfer distance (CD)[2] and Earth Mover's distance (EMD) [27]. We see that Grid+FPS outperforms Grid+Random, Jitter+FPS, as well as Sobol+FPS. Jitter+FPS performs the worst and its 2D analogue also has the highest star discrepancy. We observe that Grid+FPS reduces noisy phantom blocks around the mesh, and hence reduces the need for post-processing and cleaning. This property is highly desired, because sometimes the cleaning algorithm cannot distinguish between small components and noise. In addition, Grid+FPS encourages faster training convergence as shown in Fig. 1.

3.4 Effect of Feature Fusion Based on Symmetry

To analyze the effect of symmetry-based feature fusion, we choose Grid+FPS sampling method. The corresponding batch size for this experiment is kept 16.

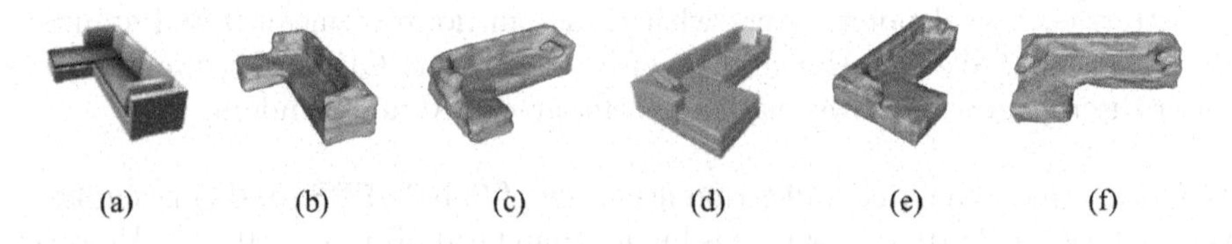

Fig. 6. Symm(Concat) can produce good reconstruction result for non-symmetrical object, without ground truth camera parameters. (a) and (d) are input images. (b) and (c) are reconstruction result of (a) rendered from 2 different views. (e) and (f) are reconstruction result of (d) from two different views.

In Table 3 and Fig. 5, we compare the effects of different feature fusion operations that are defined in Sect. 3.2 on the reconstruction result from ShapeNet.

[2] For two point set S_1 and S_2, CD is defined to be $\sum_{x \in S_1} \min_{y \in S_2} \|x-y\|_2^2 + \sum_{y \in S_2} \min_{x \in S_1} \|x-y\|_2^2$.

Ablation study shows that Symm(Near) and Symm(Concat) improve the reconstruction results. We see that concatenation of features from symmetrical pair performs the best. The reason is that Symm(Concat) better utilizes additional information comparing to Symm(Near) and Symm(Avg). When both p and its symmetry point q are visible in the image, the pixel features of p and q are both helpful for recovering the local shape at p. We observe that Symm(Concat) is able to produce reconstruction result for non-symmetrical object as shown in Fig. 6. It has the interpretation of adding the most promising additional local feature based on a symmetry prior.

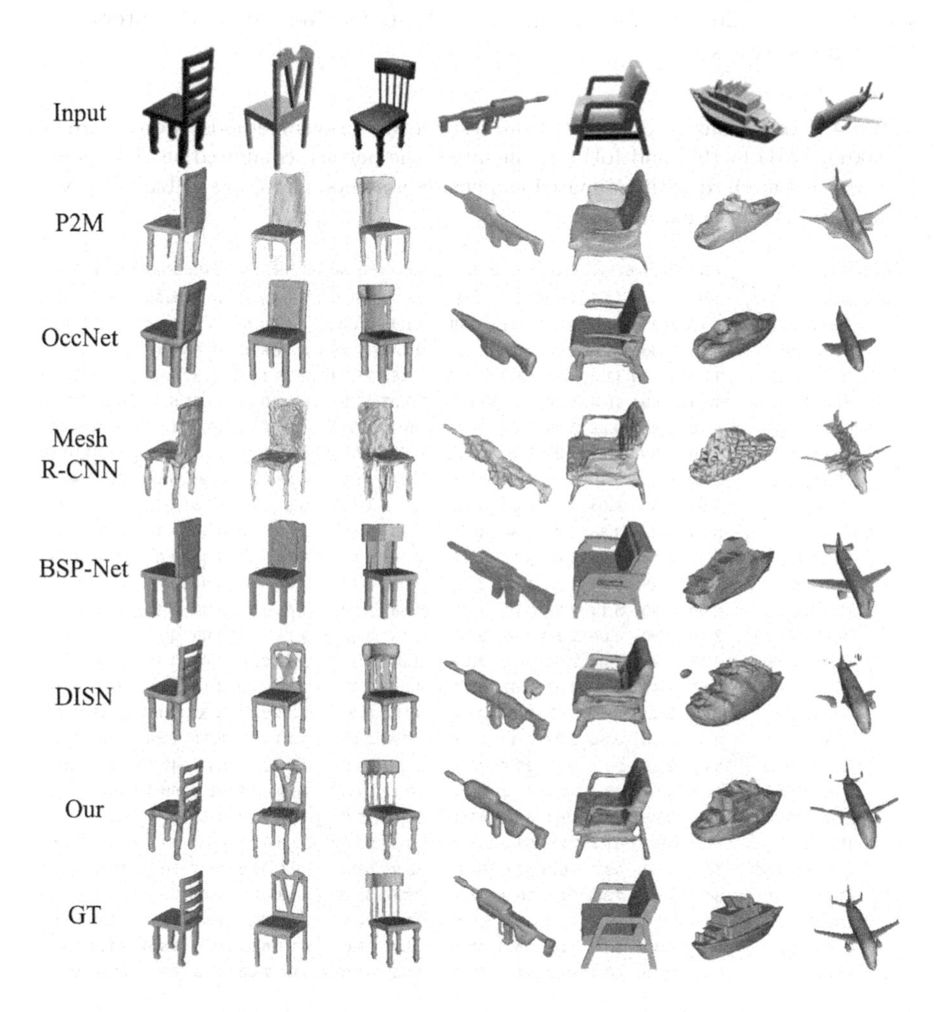

Fig. 7. Qualitative comparison with other methods. The first row contains the input image. The released model of P2M (Pixel2Mesh) [28], OccNet [20], Mesh-R-CNN [10], BSP-NET [4], DISN [32] are used to generate the results. The last row GT contains the ground truth meshes.

3.5 Comparison with Other Methods

In this subsection, the sampling method is Grid+FPS. The pixel feature is Symm(Concat). Camera parameters are estimated using the network mentioned in Sect. 4.2.

We report comparison with other state-of-the-art methods in terms of CD, EMD, and IoU. From Table 4, we see that Ladybird outperforms other methods. Figure 7 shows qualitative comparison of Ladybird with other methods. We see that Ladybird is able to reconstruct high quality mesh with fine geometric details from a single input image. Note that due to the difference between the train/test split of OccNet [20] and that of ours, we evaluate OccNet [20] on the intersection between two test sets.

Table 4. Evaluations on ShapeNet Core test set for various methods. Metrics are CD ($\times 0.001$), EMD ($\times 100$) and IoU (%, the larger the better), computed on 2048 points. $Ours_{cam}$ is Ladybird with estimated camera parameters, and $Ours$ is Ladybird with ground truth camera parameters.

Metric	Method	Plane	Bench	Box	Car	Chair	Display	Lamp	Speaker	Rifle	Sofa	Table	Phone	Boat	Mean
CD	AtlasNet [11]	5.98	6.98	13.76	17.04	13.21	7.18	38.21	15.96	4.59	8.29	18.08	6.35	15.85	13.19
	Pixel2Mesh [28]	6.10	6.20	12.11	13.45	11.13	**6.39**	31.41	**14.52**	4.51	**6.54**	15.61	6.04	12.66	11.28
	3DN [29]	6.75	7.96	**8.34**	7.09	17.53	8.35	**12.79**	17.28	**3.26**	8.27	14.05	5.18	10.20	9.77
	IMNET [5]	12.65	15.10	11.39	8.86	11.27	13.77	63.84	21.83	8.73	10.30	17.82	7.06	13.25	16.61
	3DCNN [32]	10.47	10.94	10.40	5.26	11.15	11.78	35.97	17.97	6.80	9.76	13.35	6.30	9.80	12.30
	OccNet [20]	7.70	6.43	9.36	5.26	7.67	7.54	26.46	17.30	4.86	6.72	10.57	7.17	9.09	9.70
	DISN [32]	9.96	8.98	10.19	5.39	7.71	10.23	25.76	17.90	5.58	9.16	13.59	6.40	11.91	10.98
	$Ours_{cam}$	**5.85**	**6.12**	9.10	**5.13**	**7.08**	8.23	21.46	14.75	5.53	6.78	**9.97**	**5.06**	**6.71**	**8.60**
	Ours	3.86	4.30	8.04	4.11	5.43	6.09	14.10	10.53	3.51	5.05	8.13	4.16	4.92	6.33
EMD	AtlasNet [11]	3.39	3.22	3.36	3.72	3.86	3.12	5.29	3.75	3.35	3.14	3.98	3.19	4.39	3.67
	Pixel2Mesh [28]	2.98	2.58	3.44	3.43	3.52	2.92	5.15	3.56	3.04	2.70	3.52	2.66	3.94	3.34
	3DN [29]	3.30	2.98	3.21	3.28	4.45	3.91	3.99	4.47	2.78	3.31	3.94	2.70	3.92	3.56
	IMNET[5]	2.90	2.80	3.14	2.73	3.01	2.81	5.85	3.80	2.65	2.71	3.39	2.14	2.75	3.13
	3DCNN [32]	3.36	2.90	3.06	**2.52**	3.01	2.85	4.73	3.35	2.71	2.60	3.09	2.10	2.67	3.00
	OccNet [20]	2.75	2.43	3.05	2.56	2.70	**2.58**	**3.96**	3.46	2.27	**2.35**	2.83	2.27	2.57	2.75
	DISN [32]	2.67	2.48	3.04	2.67	2.67	2.73	4.38	3.47	2.30	2.62	3.11	2.06	2.77	2.84
	$Ours_{cam}$	**2.48**	**2.29**	**3.03**	2.65	**2.60**	2.61	4.20	**3.32**	**2.22**	2.42	**2.82**	**2.06**	**2.46**	**2.71**
	Ours	2.07	2.06	2.80	2.38	2.32	2.28	3.59	2.98	1.73	2.18	2.57	1.85	2.07	2.38
IoU	AtlasNet [11]	39.2	34.2	20.7	22.0	25.7	36.4	21.3	23.2	45.3	27.9	23.3	42.5	28.1	30.0
	Pixel2Mesh [28]	51.5	40.7	43.4	50.1	40.2	55.9	29.1	52.3	50.9	60.0	31.2	69.4	40.1	47.3
	3DN [29]	54.3	39.8	49.4	59.4	34.4	47.2	35.4	45.3	57.6	60.7	31.3	71.4	46.4	48.7
	IMNET [5]	55.4	49.5	51.5	74.5	52.2	56.2	29.6	52.6	52.3	64.1	45.0	70.9	56.6	54.6
	3DCNN [32]	50.6	44.3	52.3	**76.9**	52.6	51.5	36.2	58.0	50.5	67.2	50.3	70.9	57.4	55.3
	OccNet [20]	54.7	45.2	**73.2**	73.1	50.2	47.9	**37.0**	**65.3**	45.8	67.1	**50.6**	70.9	52.1	56.4
	DISN [32]	57.5	52.9	52.3	74.3	54.3	56.4	34.7	54.9	59.2	65.9	47.9	72.9	55.9	56.9
	$Ours_{cam}$	**60.0**	**53.4**	50.8	74.5	**55.3**	**57.8**	36.2	55.6	**61.0**	**68.5**	48.6	**73.6**	**61.3**	**58.2**
	Ours	66.6	60.3	56.4	80.2	64.7	63.7	48.5	61.5	71.9	73.5	58.1	78.1	68.8	65.6

Since ShapeNet is a synthesized dataset, we further provide quantitative evaluation on Pix3D [26] (Table 5), and some qualitative examples of in-the-wild images which are randomly selected from the internet (Fig. 8). These results

Table 5. Evaluations on Pix3D [26] test set. Metrics are CD (×0.001), and EMD (×100), computed on 2048 points. Groud truth camera parameters are used.

Metric	Method	Bed	Bookcase	Chair	Desk	Misc	Sofa	Table	Tool	Wardrobe	Mean
CD	DISN [32]	12.74	35.29	23.82	18.70	31.18	3.85	18.46	46.00	**4.23**	18.51
	Ours	**5.73**	**15.89**	**13.03**	**10.38**	**30.34**	**3.28**	**8.38**	**28.39**	5.58	**10.02**
EMD	DISN [32]	2.84	4.65	3.97	4.04	**4.53**	1.99	3.85	5.66	2.11	3.53
	Ours	**2.35**	**3.07**	**3.23**	**2.77**	4.96	**1.84**	**2.42**	**3.68**	**1.99**	**2.75**
IoU	DISN [32]	71.2	43.0	59.0	53.7	48.8	89.4	57.8	37.3	85.6	64.4
	Ours	**78.2**	**67.8**	**66.5**	**67.5**	**49.5**	**91.8**	**74.2**	**58.4**	**86.8**	**73.3**

show that Ladybird generalizes well to natural images. For the experiment on Pix3D, we fine-tune Ladybird and DISN [32] (both pre-trained on ShapeNet) on Pix3D train set, and use the ground truth camera poses and the segmentation masks.

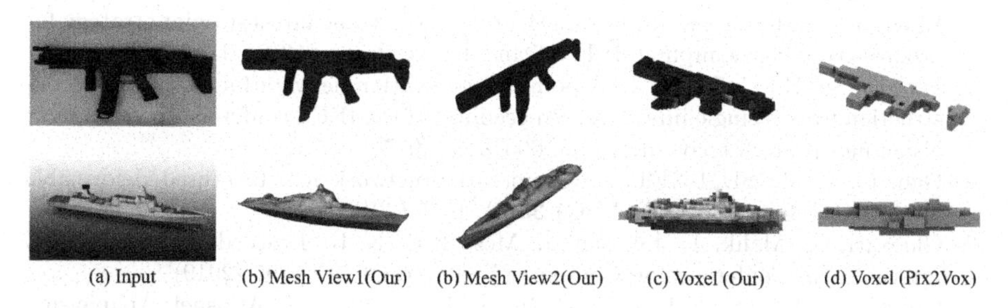

Fig. 8. Reconstruction results for online images. (a) indicates input images. (b) and (c) are our reconstruction results in mesh and voxel representation respectively. (d) shows the reconstruction results of Pixel2Vox [30]. Ladybird naturally produces accurate uv-map for texturing.

4 Conclusion

We study the impact of sample set discrepancy on the training efficiency of implicit field regression networks, and proposes to use FPS instead of Random sampling to select training points. We also propose to explore local feature fusion based on reflective symmetry to improve the reconstruction quality. Qualitatively and quantitatively we verify the efficiency of our methods through extensive experiments on large-scale dataset ShapeNet.

Acknowledgement. We would like to thank the anonymous reviewers for their helpful feedback and suggestions. We would like to thank Zilei Huang for his help in accelerating the data processing and debugging.

References

1. Uni(corn—form) tool kit. https://utk-team.github.io/utk/
2. Abadi, M., et al.: Tensorflow: a system for large-scale machine learning. In: 12th {USENIX} Symposium on Operating Systems Design and Implementation ({OSDI} 16), pp. 265–283 (2016)
3. Chang, A.X., et al.: Shapenet: An information-rich 3d model repository. arXiv preprint arXiv:1512.03012 (2015)
4. Chen, Z., Tagliasacchi, A., Zhang, H.: BSP-Net: generating compact meshes via binary space partitioning. In: Proceedings of the IEEE/CVF Conference on Computer Vision and Pattern Recognition, pp. 45–54 (2020)
5. Chen, Z., Zhang, H.: Learning implicit fields for generative shape modeling. In: Proceedings of the IEEE Conference on Computer Vision and Pattern Recognition, pp. 5939–5948 (2019)
6. Choy, C.B., Xu, D., Gwak, J.Y., Chen, K., Savarese, S.: 3D-R2N2: a unified approach for single and multi-view 3D object reconstruction. In: Leibe, B., Matas, J., Sebe, N., Welling, M. (eds.) ECCV 2016. LNCS, vol. 9912, pp. 628–644. Springer, Cham (2016). https://doi.org/10.1007/978-3-319-46484-8_38
7. Eldar, Y., Lindenbaum, M., Porat, M., Zeevi, Y.Y.: The farthest point strategy for progressive image sampling. IEEE Trans. Image Process. **6**(9), 1305–1315 (1997)
8. Fan, H., Su, H., Guibas, L.J.: A point set generation network for 3D object reconstruction from a single image. In: Proceedings of the IEEE conference on computer vision and pattern recognition, pp. 605–613 (2017)
9. Gao, L., et al.: SDM-NET: deep generative network for structured deformable mesh. ACM Trans. Graph. (TOG) **38**(6), 1–15 (2019)
10. Gkioxari, G., Malik, J., Johnson, J.: Mesh R-CNN. In: Proceedings of the IEEE International Conference on Computer Vision, pp. 9785–9795 (2019)
11. Groueix, T., Fisher, M., Kim, V.G., Russell, B.C., Aubry, M.: Atlasnet: Apapier-m\^ach\'e approach to learning 3d surfacegeneration. arXiv preprint arXiv:1802.05384 (2018)
12. Halton, J.H.: Algorithm 247: radical-inverse quasi-random point sequence. Commun. ACM **7**(12), 701–702 (1964)
13. Joe, S., Kuo, F.Y.: Constructing Sobol sequences with better two-dimensional projections. SIAM J. Sci. Comput. **30**(5), 2635–2654 (2008)
14. Kingma, D.P., Ba, J.: Adam: A method for stochastic optimization. arxiv:1412.6980 (2014)
15. Kuipers, L., Niederreiter, H.: Uniform Distribution of Sequences. Courier Corporation, North Chelmsford (2012)
16. Li, K., Pham, T., Zhan, H., Reid, I.: Efficient dense point cloud object reconstruction using deformation vector fields. In: Proceedings of the European Conference on Computer Vision (ECCV), pp. 497–513 (2018)
17. Lin, C.H., Kong, C., Lucey, S.: Learning efficient point cloud generation for dense 3D object reconstruction. In: Thirty-Second AAAI Conference on Artificial Intelligence (2018)
18. Liu, S., Zhang, Y., Peng, S., Shi, B., Pollefeys, M., Cui, Z.: Dist: Rendering deep implicit signed distance function with differentiable sphere tracing. arXiv preprint arXiv:1911.13225 (2019)
19. Liu, S., Li, T., Chen, W., Li, H.: Soft rasterizer: a differentiable renderer for image-based 3D reasoning. In: Proceedings of the IEEE International Conference on Computer Vision, pp. 7708–7717 (2019)

20. Mescheder, L., Oechsle, M., Niemeyer, M., Nowozin, S., Geiger, A.: Occupancy networks: Learning 3D reconstruction in function space. In: Proceedings of the IEEE Conference on Computer Vision and Pattern Recognition, pp. 4460–4470 (2019)
21. Mitchell, E., Engin, S., Isler, V., Lee, D.D.: Higher-order function networks for learning composable 3d object representations. arXiv preprint arXiv:1907.10388 (2019)
22. Niederreiter, H.: Low-discrepancy and low-dispersion sequences. J. Number Theor. **30**(1), 51–70 (1988)
23. Pan, J., Han, X., Chen, W., Tang, J., Jia, K.: Deep mesh reconstruction from single RGB images via topology modification networks. In: Proceedings of the IEEE International Conference on Computer Vision, pp. 9964–9973 (2019)
24. Pilleboue, A., Singh, G., Coeurjolly, D., Kazhdan, M., Ostromoukhov, V.: Variance analysis for Monte Carlo integration. ACM Trans. Graph. (Proc. SIGGRAPH) **34**(4), 124:1–124:14 (2015)
25. Singh, G., et al.: Analysis of sample correlations for Monte Carlo rendering. Comput. Graph. Forum **38**(2), 473–491 (2019)
26. Sun, X., et al.: Pix3D: Dataset and methods for single-image 3D shape modeling. In: Proceedings of the IEEE Conference on Computer Vision and Pattern Recognition, pp. 2974–2983 (2018)
27. Villani, C.: Optimal Transport: Old and New, vol. 338. Springer, Heidelberg (2008). https://doi.org/10.1007/978-3-540-71050-9
28. Wang, N., Zhang, Y., Li, Z., Fu, Y., Liu, W., Jiang, Y.G.: Pixel2mesh: generating 3D mesh models from single RGB images. In: Proceedings of the European Conference on Computer Vision (ECCV), pp. 52–67 (2018)
29. Wang, W., Ceylan, D., Mech, R., Neumann, U.: 3DN: 3D deformation network. In: Proceedings of the IEEE Conference on Computer Vision and Pattern Recognition, pp. 1038–1046 (2019)
30. Xie, H., Yao, H., Sun, X., Zhou, S., Zhang, S.: Pix2Vox: context-aware 3D reconstruction from single and multi-view images. In: Proceedings of the IEEE International Conference on Computer Vision, pp. 2690–2698 (2019)
31. Xu, H., Barbič, J.: Signed distance fields for polygon soup meshes. Proc. Graph. Interface **2014**, 35–41 (2014)
32. Xu, Q., Wang, W., Ceylan, D., Mech, R., Neumann, U.: Disn: Deep implicit surface network for high-quality single-view 3d reconstruction. arXiv preprint arXiv:1905.10711 (2019)
33. Yao, Y., Schertler, N., Rosales, E., Rhodin, H., Sigal, L., Sheffer, A.: Front2back: single view 3d shape reconstruction via front to back prediction. arXiv preprint arXiv:1912.10589 (2019)

Segment as Points for Efficient Online Multi-Object Tracking and Segmentation

Zhenbo Xu[1,2], Wei Zhang[2], Xiao Tan[2], Wei Yang[1](✉), Huan Huang[1], Shilei Wen[2], Errui Ding[2], and Liusheng Huang[1]

[1] University of Science and Technology of China, Hefei, China
qubit@ustc.edu.cn
[2] Department of Computer Vision Technology (VIS), Baidu Inc., Beijing, China

Abstract. Current multi-object tracking and segmentation (MOTS) methods follow the tracking-by-detection paradigm and adopt convolutions for feature extraction. However, as affected by the inherent receptive field, convolution based feature extraction inevitably mixes up the foreground features and the background features, resulting in ambiguities in the subsequent instance association. In this paper, we propose a highly effective method for learning instance embeddings based on segments by converting the compact image representation to un-ordered 2D point cloud representation. Our method generates a new tracking-by-points paradigm where discriminative instance embeddings are learned from randomly selected points rather than images. Furthermore, multiple informative data modalities are converted into point-wise representations to enrich point-wise features. The resulting online MOTS framework, named PointTrack, surpasses all the state-of-the-art methods including 3D tracking methods by large margins (5.4% higher MOTSA and 18 times faster over MOTSFusion) with the near real-time speed (22 FPS). Evaluations across three datasets demonstrate both the effectiveness and efficiency of our method. Moreover, based on the observation that current MOTS datasets lack crowded scenes, we build a more challenging MOTS dataset named APOLLO MOTS with higher instance density. Both APOLLO MOTS and our codes are publicly available at https://github.com/detectRecog/PointTrack.

Keywords: Motion and tracking · Tracking · Vision for robotics

1 Introduction

Multi-object tracking (MOT) is a fundamental task in computer vision with broad applications such as autonomous driving and video surveillance. Recent MOT methods [4,6,41] mainly adopt the tracking-by-detection paradigm which

---

Electronic supplementary material The online version of this chapter (https://doi.org/10.1007/978-3-030-58452-8_16) contains supplementary material, which is available to authorized users.

A. Vedaldi et al. (Eds.): ECCV 2020, LNCS 12346, pp. 264–281, 2020.
https://doi.org/10.1007/978-3-030-58452-8_16

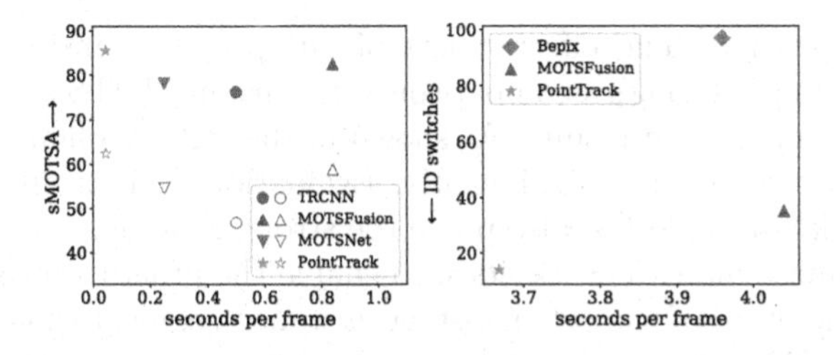

Fig. 1. Comparison between our PointTrack and the state-of-the-art MOTS methods on sMOTSA (Left) and id switches (Right). On the left subfigure, the filled symbols and the hollow symbols denote the results for cars and for pedestrians respectively. On the right subfigure, all methods perform tracking on the same segmentation result, which takes 3.66 s.

links detected bounding boxes across frames via data association algorithms. Since the performance of association highly depends on robust similarity measurements, which is widely noticed difficult due to the frequent occlusions among targets, challenges remain in MOT especially for crowded scenes [2]. More recently, the task of multi-object tracking and segmentation (MOTS) [34] extends MOT by jointly considering instance segmentation and tracking. As instance masks precisely delineate the visible object boundaries and separate adjacency naturally, MOTS not only provides pixel-level analysis, but more importantly encourages to learn more discriminative instance features to facilitate robust similarity measurements than bounding box (bbox) based methods.

Unfortunately, how to extract instance feature embeddings from segments have rarely been tackled by current MOTS methods. TRCNN [34] extends Mask RCNN by 3D convolutions and adopts ROI Align to extract instance embeddings in bbox proposals. To focus on the segment area in feature extraction, Porzi *et al.* [27] propose mask pooling to replace ROI Align. Nevertheless, as affected by the receptive field of convolutions, the foreground features and the background features are still mixed up, which is harmful for learning discriminative feature. Therefore, though current MOTS methods adopt advanced segmentation networks to extract image features, they fail to learn discriminative instance embeddings which are essential for robust instance association, resulting in limited tracking performances.

In this paper, we propose a simple yet highly effective method to learn instance embeddings on segments. Inspired by the success of PointNet [28] which enables feature aggregations directly from irregular formatted 3D point clouds, we regard 2D image pixels as un-ordered 2D point clouds and learn instance embeddings in a point cloud processing manner. Concretely, for each instance, we build two separate point clouds for the foreground segment and the surrounding area respectively. In each point cloud, we further propose to combine different modalities of point-wise features to realize a unified and context-aware instance embedding. In this way, the novel tracking-by-points paradigm can be easily established by combining our proposed instance embedding with any instance segmentation method. The effectiveness of our proposed instance embedding method is examined through a comparison with current MOTS approaches based on the same segmentation results. As shown in the right subfigure of Fig. 1, our method reduces id switches significantly. Evaluations across different datasets (see PointTrack* in Table 3,5) also prove the strong generalization ability of our proposed instance embedding. Besides, to enable the practical utility of MOTS, we enhance the state-of-the-art one-stage instance segmentation method SpatialEmbedding [23] for temporal coherence and build up a novel MOTS framework named PointTrack. Our proposed framework first achieves nearly real-time performance while out-performs all the state-of-the-art methods including 3D tracking methods on KITTI MOTS by large margins (see the left subfigure of Fig. 1).

Moreover, to facilitate better evaluations, we construct a more crowded thus more challenging MOTS dataset named APOLLO MOTS based on the public ApolloScape dataset [13]. APOLLO MOTS has a similar number of frames with KITTI MOTS but two times more tracks and car annotations (see Table 1). We believe APOLLO MOTS can further help promote researches in MOTS.

We summarize our main contributions as follows:

- We propose a highly effective method for learning discriminative instance embeddings on segments by breaking the compact image representation into un-ordered 2D point clouds.
- A novel online MOTS framework named PointTrack is introduced, which is more efficient and more effective than the state-of-the-art methods.
- We build APOLLO MOTS, a more challenging dataset with 68% higher instance density over KITTI MOTS.
- Evaluations across three datasets show that PointTrack outperforms all existing MOTS methods by large margins. Also, PointTrack can reduce id switches significantly and generalizes well on instance embedding extraction.

2 Related Work

Tracking-by-Detection. Detection based MOT approaches first detect objects of interests and then link objects into trajectories via data association. The data association can be accomplished on either the 2D image plane [4,6,7,14,32,37, 41] or the 3D world space [1,8,10,20,24,38]. ATOM [7] introduces a novel tracking architecture, which consists of dedicated target estimation and classification components, by predicting the overlap between the target object and an estimated bounding box. FAMNet [6] develops an end-to-end tracking architecture where feature extraction, affinity estimation and multi-dimensional assignment are jointly optimized. Most 3D tracking methods [24,31] merge track-lets based on 3D motion clues. Other approaches [10,18,22] further perform 3D reconstruction for objects to improve the tracking performance.

Tracking-by-Segmentation. Unlike 2D bounding boxes which might overlap heavily in crowded scenes, per-pixel segments locate objects precisely. Recently instance segments have been exploited for improving the tracking performance [12,19,25–27]. In [25], Osep *et al.* present a model-free multi-object tracking approach that uses a category-agnostic image segmentation method to track objects. Track-RCNN [34] extends Mask-RCNN with 3D convolutions to incorporate temporal information and extracts instance embeddings for tracking by ROI Align. MOTSNet [27] proposes a mask pooling layer to Mask-RCNN to improve object association over time. STE [12] introduces a new spatial-temporal embedding loss to generate temporally consistent instance segmentation and regard the mean embeddings of all pixels on segments as the instance embedding for data association. As features obtained by 2D or 3D convolutions are harmful for learning discriminative instance embeddings, different from previous methods, our PointTrack regards 2D image pixels as un-ordered 2D point clouds and learn instance embeddings in a point cloud processing manner.

MOTS Datasets. KITTI MOTS [34] extends the popular KITTI MOT dataset with dense instance segment annotations. Except for KITTI MOTS, popular datasets (like the ApolloScape dataset [13]) also provide video instance segmentation labels, but the instances are not consistent in time. Compared with KITTI MOTS, ApolloScape provides more crowded scenes which are more challenging for tracking. Based on this observation, we build Apollo MOTS in a semi-automatic annotation manner with the same metric as KITTI MOTS.

3 Method

In this section, we first formulate how PointTrack converts different data modalities into a unified per-pixel style and learns context-aware instance embeddings M on 2D segments. Then, details about instance segmentation are introduced.

3.1 Context-Aware Instance Embeddings Extraction

For an instance C with its segment C_s and its smallest circumscribed rectangle C_b, we enlarge C_b to $\hat{C}_b$ by extending its border in all four directions (top, down, left, and right) by a scale factor k ($k = 0.2$ by default). Both C_s and $\hat{C}_b$ are visualized in dark green in the lower-left corner of Fig. 2. Then, we regard the foreground segment as a 2D point cloud and denote it as F. Similarly, we regard the other area in $\hat{C}_b$ the environment point cloud and denote it as E. Each point inside $\hat{C}_b$ has six dimensional data space $(u, v, R, G, B, \mathcal{C})$ that contains the coordinate (u, v) in the image plane, the pixel color (R, G, B), and which class $\mathcal{C}$ the pixel belongs to.

For the foreground point cloud F, we uniformly random sample N_F points ($N_F = 1000$ by default) for feature extraction. As shown in Fig. 2, N_F points are enough to evenly cover a relatively large instance. For the environment point cloud E, N_E points ($N_E = 500$ by default) are randomly selected. The coordinate of the foreground point F_i is denoted as (u_i^F, v_i^F) and the coordinate of the environment point E_i is denoted as (u_i^E, v_i^E). The center point $P(u_c^F, v_c^F)$ is computed by averaging the coordinates of selected foreground points $\{F_i | i =$

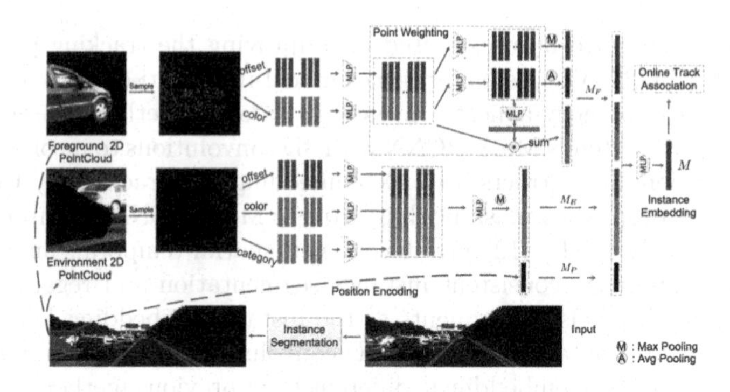

Fig. 2. Overview of PointTrack. For an input image, PointTrack obtains instance segments by an instance segmentation network. Then, PointTrack regards the segment and its surrounding environment as two 2D point clouds and learn features on them separately. MLP stands for multi-layer perceptron with Leaky ReLU. (Color figure online)

$1, ..., N_F\}$ in the image plane. P is highlighted in blue in the foreground point cloud (see Fig. 2).

Previous works [8,31,36] have demonstrated that features concerning position, appearance, scale, shape, and nearby objects, are useful for tracking. Intuitively, PointTrack can summarize all the above features by learning the following data modalities: (i) Offset; (ii) Color; (iii) Category; (iv) Position. In the following, we formulate these data modalities and show how PointTrack learns context-aware embeddings from them.

Offset. We define the offset data of each foreground point F_i and each environment point E_i as follows:

$$O_{F_i} = (u_i^F - u_c^F, v_i^F - v_c^F), O_{E_i} = (u_i^E - u_c^F, v_i^E - v_c^F) \tag{1}$$

The offset data, which are formulated as vectors from the instance center P to themselves, represent the relative locations inside the segment. Offset vectors of foreground points provide essential information concerning both the scale and the shape of instances.

Color. We consider RGB channels and formulate the color data as follows:

$$C_{F_i} = (R_i^F, G_i^F, B_i^F), C_{E_i} = (R_i^E, G_i^E, B_i^E) \tag{2}$$

When the color data combine with the offset data, the discriminative appearance features can be learned from foreground points and the surrounding color distribution can be learned from environment points. The ablation study (see Table 6) shows that the color data are critical for accurate instance association.

Category. To further incorporate the environmental context into point-wise features, we encode all semantic class labels including the background class (suppose Z classes include the background) into fixed-length one-hot vectors

$\{H_j | j = 1, ..., Z\}$. Then, for selected environment points E_i, the one-hot category vector are also gathered for feature extraction. Suppose that E_i belongs to the category C_i, the category data are formulated as follows:

$$Y_{E_i} = H_{C_i}, C_i \in [1, Z] \quad (3)$$

Strong context features can be learned by PointTrack by jointly learning from the category data and the offset data. When the current instance is adjacent to other instances, for E_i lying on the nearby instances, the category data Y_{E_i} together with the offset data O_{E_i} tell PointTrack both the relative position and the semantic class of nearby instances, which serve as strong clues for instance association. Visualizations (see Fig. 5) also confirm that environment points on nearby instances matter in learning discriminative instance embeddings.

Position. Since previous three data modalities focus on extracting features around C_b regardless of the position of C_b in the image plane, we encode the position of C_b into the position embedding M_P. Following [33], we embed the position of C_b (4-dim) into a high-dimensional vector (64-dim) to make it easier for learning by computing *cosine* and *sine* functions of different wavelengths.

Based on the above four data modalities, PointTrack learns the foreground embeddings M_F and the environment embeddings M_E in separate branches. As shown in Fig. 2, the environment embeddings M_E are learned by first fusing (O_E, C_E, Y_E) for all E_i and then applying the max pooling operation to the fused features. As aforementioned, by fusing (O_E, Y_E), PointTrack learns strong context clues concerning nearby instances from M_E. For the foreground point cloud F, M_F is learned by fusing (O_F, C_F). Based on the intuition that more prominent points should have higher weights for differentiating instances, and other points should also be considered, but have lower weights, we introduce the point weighting layer to actively weight all foreground points and sum the features of all points. Different from Max-Pooling which only selects features of prominent points and Average-Pooling which blindly averages features of all points, the point weighting layer learns to summarize the foreground features by learning to weight points. Visualizations (see Fig. 5) demonstrate that the point weighting layer learns to give informative areas higher weights. Afterward, as shown in Fig. 2, M_F, M_E, and the position embeddings M_P are concatenated for predicting the final instance embeddings M as follows:

$$M = \mathbf{MLP}(M_F + M_E + M_P) \quad (4)$$

where + represents concatenation and **MLP** denotes multi-layer perceptron.

Instance Association. To produce the final tracking result, we need to perform instance association based on similarities. Given segment C_{s_i} and segment C_{s_j}, and their embeddings M_i and M_j, the similarity S is formulated as follows:

$$S(C_{s_i}, C_{s_j}) = -D(M_i, M_j) + \alpha * U(C_{s_i}, C_{s_j}) \quad (5)$$

where D denotes the Euclidean distance and U represents the mask IOU. α is set to 0.5 by default. If an active track does not update for the recent β frames, we

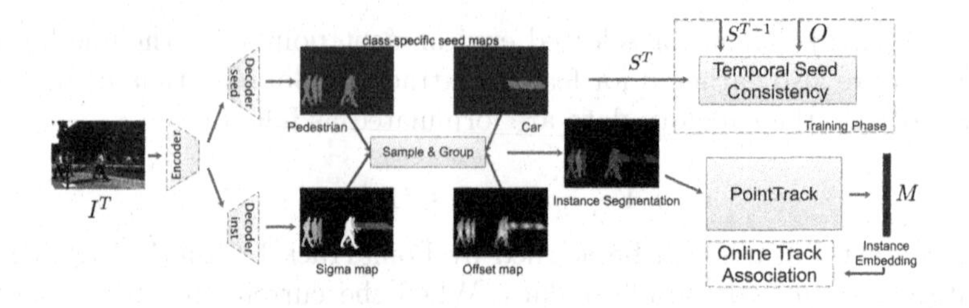

Fig. 3. Segmentation network of PointTrack.

end this track automatically. For each frame, we compute the similarity between the latest embeddings of all active tracks and embeddings of all instances in the current frame according to Eq. (5). Following [34], we set a similarity threshold γ for instance association and instance association is allowed only when the similarity is greater than γ. The Hungarian algorithm [17] is exploited to perform instance matching. After instance association, unassigned segments will start new tracks. By default, β and γ are set to 30 and -8.0 respectively.

3.2 Instance Segmentation with Temporal Seed Consistency

Different from previous methods [27,34] which put great efforts to adapt mask RCNN into the MOTS frameworks, PointTrack builds on a state-of-the-art one-stage instance segmentation method named SpatialEmbedding [23]. SpatialEmbedding performs instance segmentation without bbox proposals, and thus runs much faster than two-stage methods. As shown in Fig. 3, SpatialEmbedding follows an encoder-decoder structure with two separate decoders: (i) the seed decoder; (ii) the inst decoder. Given an input image I^T at time T, the seed decoder predicts seed maps S^T for all semantic classes. Besides, the inst decoder predicts a sigma map denoting the pixel-wise cluster margin and an offset map representing the vector pointing to the corresponding instance center. Afterward, instance centers are sampled from S^T and pixels are grouped into segments according to the learned clustering margin for each instance. When applied to MOTS, by studying the segmentation failure cases, we find that the seed map predictions are not consistent between consecutive frames, which results in many false positives and false negatives. Therefore, we introduce the temporal consistency loss in the training phase to improve the quality of seed map prediction as follows. First, we also feed the input image I^{T-1} at time $T-1$ to SpatialEmbedding to predict the seed maps S^{T-1}. Then, optical flow O between I^{T-1} and I^T is estimated by VCN[1] [39]. Subsequently, we synthesize the warped seed maps $\hat{S^T} = O(S^{T-1})$ by exploiting O to warp S^{T-1}. Our temporal consistency loss is formulated as: $L_{tc} = \frac{1}{N}\sum_i^N ||\hat{S_i^T} - S_i^T||^2$. where N is the number of foreground

[1] We exploit the pre-trained model provided at https://github.com/gengshan-y/VCN for optical flow estimation.

Table 1. Comparison between APOLLO MOTS and KITTI MOTS on their respective train/validation sets.

	Frames	Tracks	Annotations	Car density	Crowd cars	Frames per second
APOLLO MOTS	11488	1530	64930	5.65	36403	10
KITTI MOTS (Car)	8008	582	26899	3.36	14509	7

pixels and i denotes the i-th foreground pixel. Evaluations (see Table 2) demonstrate that our temporal consistency loss improves the instance segmentation performance.

4 Apollo MOTS Dataset

Tracking becomes more challenging with the increase of instances. However, for KITTI MOTS, the instance density is limited (only 3.36 cars per frame on average) and crowded scenes are also insufficient. Based on these observations, we present our Apollo MOTS dataset. We first briefly overview the dataset. Thereafter, the annotation procedures are introduced.

4.1 Overview

We build APOLLO MOTS on the public ApolloScape dataset [13] which contains video instance segment labels for 49,287 frames. As there are barriers on both sides of the road where data were collected, pedestrians are much fewer than cars in the ApolloScape dataset. Therefore, we focus on cars. As the ApolloScape dataset calibrates the camera for each frame, APOLLO MOTS can serve as a challenging MOTS dataset for both 2D tracking and 3D tracking.[2]

Detailed comparisons between our APOLLO MOTS and KITTI MOTS on their respective train/validation sets are shown in Table 1. APOLLO MOTS contains 22480 video frames including the testing set. We divide the train set, validation set, and test set according to the proportions of 30%, 20%, and 50%. Scenes in these three sets have similar tracking difficulties. The original image resolution in the ApolloScape dataset is 3384 (width) $\times$ 2710 (height). We crop it to 3384×1604 to remove the sky area and down-sample it to a lower and more suitable resolution 1692×802. As shown in Table 1, though APOLLO MOTS has a similar number of frames, we have two times more tracks and car annotations. We define the car density as the number of cars per frame. The average car density of APOLLO MOTS is 5.65, which is much higher than that of KITTI MOTS. Moreover, as tracking becomes more challenging when cars are overlapped, we count the number of crowded cars for both APOLLO MOTS and KITTI MOTS. A car is considered crowded if and only if its segment is adjacent to any other car. Our APOLLO MOTS has 2.5 times more crowded cars than KITTI MOTS.

[2] Sample videos are provided in the supplementary material.

4.2 Annotation

We annotate all video frames in the ApolloScape dataset. If consecutive frames contain no cars or are too easy for tracking, the entire video is removed. The resulting 22480 frames represent the most difficult tracking scenes in the ApolloScape dataset. We annotate APOLLO MOTS in the following three steps.

(1) In-complete instance segment removal. For cars occluded by the fence, the ground-truth instance segment is always in-complete in the ApolloScape dataset. We manually traverse all frames to remove these in-complete instances by setting the bounding box area compassing this instance to the 'Dontcare' category. The 'Dontcare' area will be ignored in the evaluation process. After this step, only instances with complete segments will be preserved.
(2) Semi-automatic tracking annotation. We incorporate PointTrack trained on KITTI MOTS into our data annotation tool for automatic instance association. For each frame, the tracking results generated by PointTrack are manually reviewed and corrected. Moreover, we subjectively assign different difficulty levels to all videos ($0 \sim 4$, from the easiest to the hardest) by jointly considering the crowded level, the rotation of the camera, the overlap level, etc..
(3) Simple video removal and dataset partitioning. All videos with difficulty level 0 are discarded. For each difficulty level from 1 to 4, we divide videos of this level to the train, validation, and test sets according to the aforementioned percentages to ensure that these sets share similar difficulties.

5 Experiments

Experiments are divided into four parts[3]. Firstly, we evaluate PointTrack across three datasets: the KITTI MOTS dataset [9], the MOTSChallenge dataset [21], and our proposed Apollo MOTS dataset. Secondly, we show the ablation study on data modalities. Thirdly, to investigate what PointTrack learns from 2D point clouds, we visualize both predicted instance embeddings and critical tracking points. Lastly, we provide our results on the official KITTI MOTS test set.

Metric. Following previous works [12,18], we focus on sMOTSA, MOTSA, and id switches (IDS). As an extension of MOTA, MOTSA measure segmentation as well as tracking accuracy. sMOTSA [34] is a soft version of MOTSA which weights the contribution of each true positive segment by its mask IoU with the corresponding ground truth segment.

Experimental Setup. Following previous works [12,34], we pre-train the segmentation network on the KINS dataset [29] due to the limit of training data in KITTI MOTS (only 1704 frames contains Pedestrian where merely 1957 masks are manually annotated). Afterward, SpatialEmbedding is fine-tuned on KITTI

[3] More details and ablation studies are provided in the supplementary material.

Table 2. Results on the KITTI MOTS validation. Speed is measured in seconds per frame. TC denotes the temporal consistency loss. BS represents BB2SegNet [19].

Type	Method	Det. & Seg.	Speed	Cars			Pedestrians		
				sMOTSA	MOTSA	IDS	sMOTSA	MOTSA	IDS
2D	TRCNN [34]	TRCNN	0.5	76.2	87.8	93	46.8	65.1	78
3D	BePix [31]	RRC[30]+TRCNN	3.96	76.9	89.7	88	-	-	-
2D	MOTSNet [27]	MOTSNet	-	78.1	87.2	-	54.6	69.3	-
3D	MOTSFusion [18]	TRCNN+BS	0.84	82.6	90.2	51	58.9	71.9	36
3D	BePix	RRC+BS	3.96	84.9	93.8	97	-	-	-
3D	MOTSFusion	RRC+BS	4.04	**85.5**	94.6	35	-	-	-
2D	PointTrack	PointTrack	**0.045**	**85.5**	**94.9**	**22**	**62.4**	**77.3**	**19**
2D	PointTrack (without TC)	PointTrack	**0.045**	82.9	92.7	25	61.4	76.8	21
2D	PointTrack (on Bbox)	PointTrack	**0.045**	85.3	94.8	36	61.8	76.8	36

Fig. 4. Quantitative results on KITTI MOTS. Instances of the same track id are plotted in the same color.

MOTS with our proposed seed consistency loss for 50 epochs at a learning rate of $5 \cdot 10^{-6}$. For MOTSChallenge, we fine-tune the model trained on KITTI MOTS for 50 epochs at a learning rate of $5 \cdot 10^{-6}$. For APOLLO MOTS, we train SpatialEmbedding from scratch following [23]. Besides, our PointTrack is trained from scratch in all experiments by margin based hard triplet loss [40]. An instance database is constructed from the train set by extracting all crops $\hat{C}_b$ of all track ids. Unlike previous method [34] which samples T frames as a batch, we sample D track ids as a batch, each with three crops. These three crops are selected from three equally spaced frames rather than three consecutive frames to increase the intra-track-id discrepancy. The space between frames is randomly chosen between 1 and 10. Empirically we find a smaller D ($16 \sim 24$) is better for training PointTrack, because a large D (more than 40) leads to a quick over-fitting. In addition, to test the generalization ability of instance association, we test PointTrack*, whose instance embeddings extraction is only fine-tuned on KITTI MOTS, on both MOTSChallenge and Apollo MOTS.

We compare recent works on MOTS: TRCNN [34], MOTSNet [27], BePix [31], and MOTSFusion (online) [18]. TRCNN and MOTSNet perform 2D tracking while BePix and MOTSFusion track on 3D.

Results on KITTI MOTS. Following MOTSFusion, we compare different methods on different segmentation results. The main results are summarized in Table 2, where our method outperforms all the state-of-the-art methods, especially for pedestrians. Quantitative results are shown in Fig. 4. On the 'Speed' column, we show the total time of detection, segmentation, and tracking[4]. On KITTI MOTS, our PointTrack takes 0.037s per frame for instance segmentation, 8ms per frame for tracking, and 3ms per instance for embedding extraction.

For cars, the 3D tracking method MOTSFusion adopts a time-consuming detector RRC [30] which takes 3.6s per frame to perform detection. MOTSFusion builds up short tracklets using 2D optical flow and segments. Afterward, 3D world-space motion consistency is used to merge tracklets together into accurate long-term tracks while recovering missed detections. By contrast, though tracking objects purely on 2D images with a light-weight instance segmentation network, PointTrack achieves comparable performance to the 3D tracking method MOTSFusion (0.3% gains on MOTSA) with significant speed improvement (0.045s VS. 4.04s). For pedestrians, PointTrack surpasses current approaches by 3.5% and 5.4% on sMOTSA and MOTSA respectively. It is worth noting that, though only small improvements over MOTSFusion are observed for cars on the KITTI MOTS validation set, PointTrack surpasses MOTSFusion by large margins on the official test set (see Table 8), which demonstrates the good generalization ability. Besides, when the temporal consistency loss is removed (see the last but one row in Table 2), the performance drops are observable (by 2.6% and 0.5% on sMOTSA for cars and pedestrians, respectively). This demonstrates the effectiveness of our temporal consistency loss.

The Effectiveness of Segment C_s. To investigate the effectiveness of segment C_s, we ignore C_s and instead sample points inside the inmodal bbox C_b. $N_E + N_F$ points are randomly sampled and the network branch for environment embedding is removed. As shown in the last row in Table 2, for cars, IDS increase by 64% (from 22 to 36) after the segments are removed. For pedestrians, more significant performance drops (89.5% IDS increase and 0.6% sMOTSA) are observed. The increase in IDS demonstrates that segment matters for better tracking performances. Moreover, the gap between the performance drop in cars and in pedestrians demonstrates that segments are more effective in improving the tracking performance for non-rigid objects where bbox level feature extraction introduces more ambiguities.

Results on APOLLO MOTS. We show comparisons on APOLLO MOTS in Table 3. All models are trained under the same setting as KITTI MOTS. We also train DeepSort [35] to tracks instances on inmodal bboxes surrounding segments as a baseline to PointTrack. DeepSort extends SORT [3] by incorporating convolution layers to extract appearance features for instance association. Different from DeepSort, our PointTrack extracts features from 2D point clouds rather than images. When applied to the same segmentation results (see the third row and the fifth row), PointTrack achieves 6% higher sMOTSA than DeepSort and reduces IDS from 1692 to 292. Also, compared with the performance on KITTI MOTS, the sMOTSA of TRCNN and PointTrack decreases by 14.7% and 26.4%

[4] Our calculated speed is different from MOTSFusion because in [18], the detection time of the RRC detector which takes 3.6s per frame is ignored. The speed of MOTSNet [27] is not mentioned in their work.

Table 3. Results on APOLLO MOTS validation.

	Seg.	sMOTSA	MOTSA
DeepSort [35]	TRCNN	45.71	57.06
TRCNN [34]	TRCNN	49.84	61.19
DeepSort	PointTrack	64.69	73.97
PointTrack*	PointTrack	70.58	79.87
PointTrack	PointTrack	**70.76**	**80.05**

Table 4. Comparisons of IDS on KITTI MOTS and APOLLO MOTS.

Dataset	Seg.	Method	IDS (car)	IDS (Ped.)
KITTI MOTS Val	TRCNN	TRCNN	93	78
		PointTrack	**46**	**30**
	RRC+BS	BePix	97	–
		MOTSFusion	35	–
		PointTrack	**14**	–
APOLLO MOTS	TRCNN	DeepSort	1263	–
		TRCNN	312	–
		PointTrack	**241**	–
	PointTrack	DeepSort	1692	–
		PointTrack	**292**	–
KITTI MOTS Test	MOTSFusion	MOTSFusion	201	279
		PointTrack	**187**	**150**

when evaluated on APOLLO MOTS. As the training and testing settings are the same, the significant performance drop shows that APOLLO MOTS is more challenging than KITTI MOTS.

The Significant IDS Reduction by PointTrack. As shown in Table 4, when applied to the same segmentation results on different datasets, PointTrack can effectively reduce IDS. The steady IDS reduction across different datasets and different segmentation results demonstrate the effectiveness of PointTrack.

Results on MOTSChallenge. Compared with KITTI MOTS, MOTSChallenge has more crowded scenarios and more different viewpoints. Following previous work [27,34], we train PointTrack in a leaving-one-out fashion and show comparisons on MOTSChallenge in Table 5. Our PointTrack outperforms the state-of-the-art methods by more than 1.1% on all three metrics. It's worth noting that, though the instance embeddings extraction is only fine-tuned on KITTI MOTS (see PointTrack* in Table 3,5), PointTrack* also achieves similar high performance on both APOLLO MOTS and MOTSChallenge, demonstrating the good generalization ability on instance embedding extraction.

Table 5. Results on MOTSChallenge. +MG denotes mask generation with a domain fine-tuned Mask R-CNN.

	sMOTSA	MOTSA
MOTDT [5]+ MG	47.8	61.1
MHT-DAM [16]+ MG	48.0	62.7
jCC [15]+ MG	48.3	63.0
FWT [11]+ MG	49.3	64.0
TrackRCNN [34]	52.7	66.9
MOTSNet [27]	56.8	69.4
PointTrack*	57.98	70.47
PointTrack	**58.09**	**70.58**

Ablation Study on the Impact of Data Modalities. We remove four data modalities in turn to examine their impacts on performance. As shown in Table 6, the largest performance drop occurs when the color data are removed. By contrast, the performance drop is minimal when the position data are removed. This difference in performance gap demonstrates that our PointTrack focuses more on the appearance features and the environment features while relies less on the bounding box position to associate instances, leading to higher tracking performances and much lower IDS than previous approaches.

Impact of the Point Weighting Layer, M_F, M_E, and the mask IOU. We remove the point weighting layer (P_W), the foreground embeddings M_F, the environment embeddings M_E, and the mask IOU (M_I) in turn to examine their impacts on performance. When we remove the mask IOU, we set α to zero in Eq. (5). As shown in Table 7, when the foreground embeddings M_F is removed, the performance drops a lot, demonstrating that the foreground point cloud in the segment area matters most in the instance association. By contrast, when the mask IOU is removed, the performance drop is minimal, especially for Pedestrians. Therefore, for instances with rigid shapes, considering the mask IOU in computing similarity is more beneficial than instances with non-rigid shapes like Pedestrians.

Visualizing Critical Points. We visualize critical foreground points as well as critical environment points in Fig. 5. For each instance, to validate the temporal consistency of critical points, we select crops from three consecutive frames.

For foreground points, points with 10% top weights predicted by the point weighting layer are plotted in red. As shown in Fig. 5, critical foreground points gather around car glasses and around car lights. We believe that the offsets of these points are essential for learning the shape and the pose of the vehicle. Also, their colors are important to outline the instance appearance and light distribution. Moreover, we find that PointTrack keeps the consistency of weighting points in consecutive frames even when different parts are occluded (the second and the fifth in the first row), or the car is moving to the image boundary (the

Table 6. Ablation study on the impact of different data modalities.

				Cars			Pedestrians		
Color	Offset	Category	Position	sMOTSA	MOTSA	IDS	sMOTSA	MOTSA	IDS
√	√	√	√	**85.51**	**94.93**	**22**	**62.37**	**77.35**	**19**
x				83.65	93.08	171	61.15	76.13	60
	x			85.32	94.74	37	62.16	77.14	26
		x		85.33	94.40	38	62.13	77.11	27
			x	85.35	94.77	35	62.31	77.29	21

Table 7. Experiments on impact of the point weighting layer, M_F, M_E, and the mask IOU.

				Cars		Pedestrians	
P_W	M_F	M_E	M_I	sMOTSA	MOTSA	sMOTSA	MOTSA
√	√	√	√	**85.51**	**94.93**	**62.37**	**77.35**
x				85.37	94.79	62.04	77.02
	x			83.59	93.01	61.27	76.25
		x		85.30	94.72	61.98	76.96
			x	85.33	94.76	62.31	77.29

fourth in the first row). The consistency in point weighting across frames shows the effectiveness of our point weighting layer.

For environment points, we visualize the five most critical points in yellow. These points are selected by first fetching the tensor with size of $256 * N_E$ before the max-pooling layer in the environment branch and then gathering the index with the max value for all 256-dimensions. Among these 256 indexes, points belonging to the five most common indexes are selected. As shown in Fig. 5, when instances are adjacent to any other instances, yellow points usually gather on nearby instances. As aforementioned, when combining the category data with the offset data, strong context clues are provided from environment points for instance association. The distribution of critical environment points validate that PointTrack learns discriminative context features from environment points.

Results on KITTI MOTS Testset. To further demonstrate the effectiveness of PointTrack, we report the evaluation results on the official KITTI test set in Table 8 where our PointTrack currently ranks first. It is worth noting that, on MOTSA, PointTrack surpasses MOTSFusion by 6.8% for cars and 3.6% for pedestrians. Also, PointTrack is the most efficient framework among current approaches. More detailed comparisons can be found online[5].

[5] The official leader-board: http://www.cvlibs.net/datasets/kitti/eval_mots.php.

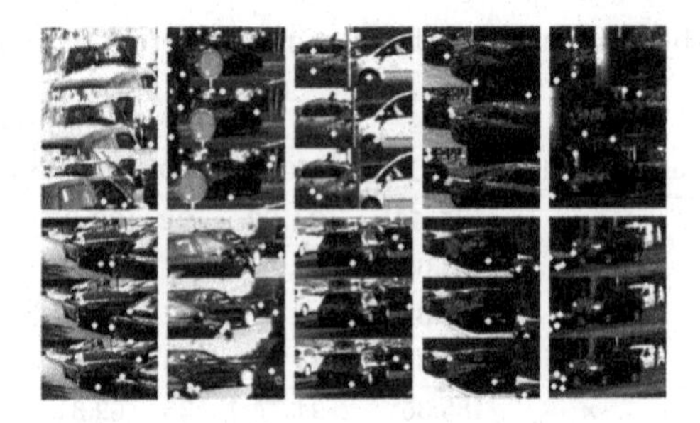

Fig. 5. Visualizations of critical points. Red points and yellow points represent the critical foreground points and the critical environment points respectively.

Table 8. Results on KITTI MOTS test set.

	Cars		Pedestrians	
	sMOTSA	MOTSA	sMOTSA	MOTSA
TRCNN	67.00	79.60	47.30	66.10
MOTSNet	71.00	81.70	48.70	62.00
MOTSFusion	75.00	84.10	58.70	72.90
PointTrack	**78.50**	**90.90**	**61.50**	**76.50**

6 Conclusions

In this paper, we presented a new tracking-by-points paradigm together with an efficient online MOTS framework named PointTrack, by breaking the compact image representation into 2D un-ordered point clouds for learning discriminative instance embeddings. Different informative data modalities are converted into point-level representations to enrich point cloud features. Evaluations across three datasets demonstrate that PointTrack surpasses all the state-of-the-art methods by large margins. Moreover, we built APOLLO MOTS, a more challenging MOTS dataset over KITTI MOTS with more crowded scenes.

Acknowledgement. This work was supported by the Anhui Initiative in Quantum Information Technologies (No. AHY150300).

References

1. Baser, E., Balasubramanian, V., Bhattacharyya, P., Czarnecki, K.: Fantrack: 3d multi-object tracking with feature association network. In: 2019 IEEE Intelligent Vehicles Symposium (IV), pp. 1426–1433. IEEE (2019)
2. Bergmann, P., Meinhardt, T., Leal-Taixe, L.: Tracking without bells and whistles. In: Proceedings of the IEEE International Conference on Computer Vision, pp. 941–951 (2019)

3. Bewley, A., Ge, Z., Ott, L., Ramos, F., Upcroft, B.: Simple online and realtime tracking. In: 2016 IEEE International Conference on Image Processing (ICIP), pp. 3464–3468. IEEE (2016)
4. Bhat, G., Danelljan, M., Gool, L.V., Timofte, R.: Learning discriminative model prediction for tracking. In: Proceedings of the IEEE International Conference on Computer Vision, pp. 6182–6191 (2019)
5. Chen, L., Ai, H., Zhuang, Z., Shang, C.: Real-time multiple people tracking with deeply learned candidate selection and person re-identification. In: 2018 IEEE International Conference on Multimedia and Expo (ICME), pp. 1–6. IEEE (2018)
6. Chu, P., Ling, H.: Famnet: joint learning of feature, affinity and multi-dimensional assignment for online multiple object tracking. In: Proceedings of the IEEE International Conference on Computer Vision, pp. 6172–6181 (2019)
7. Danelljan, M., Bhat, G., Khan, F.S., Felsberg, M.: Atom: accurate tracking by overlap maximization. In: Proceedings of the IEEE Conference on Computer Vision and Pattern Recognition, pp. 4660–4669 (2019)
8. Geiger, A., Lauer, M., Wojek, C., Stiller, C., Urtasun, R.: 3d traffic scene understanding from movable platforms. IEEE Trans. Pattern Anal. Mach. Intell. **36**(5), 1012–1025 (2013)
9. Geiger, A., Lenz, P., Urtasun, R.: Are we ready for autonomous driving? the kitti vision benchmark suite. In: Conference on Computer Vision and Pattern Recognition (CVPR) (2012)
10. Held, D., Levinson, J., Thrun, S.: Precision tracking with sparse 3D and dense color 2D data. In: 2013 IEEE International Conference on Robotics and Automation, pp. 1138–1145. IEEE (2013)
11. Henschel, R., Leal-Taixé, L., Cremers, D., Rosenhahn, B.: Fusion of head and full-body detectors for multi-object tracking. In: Proceedings of the IEEE Conference on Computer Vision and Pattern Recognition Workshops, pp. 1428–1437 (2018)
12. Hu, A., Kendall, A., Cipolla, R.: Learning a spatio-temporal embedding for video instance segmentation (2019). arXiv preprint arXiv:1912.08969
13. Huang, X., et al.: The apolloscape dataset for autonomous driving. In: Proceedings of the IEEE Conference on Computer Vision and Pattern Recognition Workshops, pp. 954–960 (2018)
14. Karunasekera, H., Wang, H., Zhang, H.: Multiple object tracking with attention to appearance, structure, motion and size. IEEE Access **7**, 104423–104434 (2019)
15. Keuper, M., Tang, S., Andres, B., Brox, T., Schiele, B.: Motion segmentation & multiple object tracking by correlation co-clustering. IEEE Trans. Pattern Anal. Mach. Intell. **42**(1), 140–153 (2018)
16. Kim, C., Li, F., Ciptadi, A., Rehg, J.M.: Multiple hypothesis tracking revisited. In: Proceedings of the IEEE International Conference on Computer Vision, pp. 4696–4704 (2015)
17. Kuhn, H.W.: The hungarian method for the assignment problem. Naval Res. Logistics Q. **2**(1–2), 83–97 (1955)
18. Luiten, J., Fischer, T., Leibe, B.: Track to reconstruct and reconstruct to track. IEEE Rob. Autom. Lett. **5**, 1803–10810 (2020)
19. Luiten, J., Voigtlaender, P., Leibe, B.: PReMVOS: proposal-generation, refinement and merging for video object segmentation. In: Jawahar, C.V., Li, H., Mori, G., Schindler, K. (eds.) ACCV 2018. LNCS, vol. 11364, pp. 565–580. Springer, Cham (2019). https://doi.org/10.1007/978-3-030-20870-7_35

20. Luo, W., Yang, B., Urtasun, R.: Fast and furious: Real time end-to-end 3d detection, tracking and motion forecasting with a single convolutional net. In: Proceedings of the IEEE conference on Computer Vision and Pattern Recognition, pp. 3569–3577 (2018)
21. Milan, A., Leal-Taixé, L., Reid, I., Roth, S., Schindler, K.: MOT16: a benchmark for multi-object tracking. arXiv:1603.00831 [cs] (Mar 2016). http://arxiv.org/abs/1603.00831, arXiv: 1603.00831
22. Mitzel, D., Leibe, B.: Taking mobile multi-object tracking to the next level: people, unknown objects, and carried items. In: Fitzgibbon, A., Lazebnik, S., Perona, P., Sato, Y., Schmid, C. (eds.) ECCV 2012. LNCS, vol. 7576, pp. 566–579. Springer, Heidelberg (2012). https://doi.org/10.1007/978-3-642-33715-4_41
23. Neven, D., Brabandere, B.D., Proesmans, M., Gool, L.V.: Instance segmentation by jointly optimizing spatial embeddings and clustering bandwidth. In: The IEEE Conference on Computer Vision and Pattern Recognition (CVPR), June 2019
24. Osep, A., Mehner, W., Mathias, M., Leibe, B.: Combined image-and world-space tracking in traffic scenes. In: 2017 IEEE International Conference on Robotics and Automation (ICRA), pp. 1988–1995. IEEE (2017)
25. Ošep, A., Mehner, W., Voigtlaender, P., Leibe, B.: Track, then decide: category-agnostic vision-based multi-object tracking. In: 2018 IEEE International Conference on Robotics and Automation (ICRA), pp. 1–8. IEEE (2018)
26. Payer, C., Štern, D., Neff, T., Bischof, H., Urschler, M.: Instance segmentation and tracking with cosine embeddings and recurrent hourglass networks. In: Frangi, A.F., Schnabel, J.A., Davatzikos, C., Alberola-López, C., Fichtinger, G. (eds.) MICCAI 2018. LNCS, vol. 11071, pp. 3–11. Springer, Cham (2018). https://doi.org/10.1007/978-3-030-00934-2_1
27. Porzi, L., Hofinger, M., Ruiz, I., Serrat, J., Bulò, S.R., Kontschieder, P.: Learning multi-object tracking and segmentation from automatic annotations (2019). arXiv preprint arXiv:1912.02096
28. Qi, C.R., Su, H., Mo, K., Guibas, L.J.: Pointnet: deep learning on point sets for 3D classification and segmentation. In: Proceedings of the IEEE Conference on Computer Vision and Pattern Recognition, pp. 652–660 (2017)
29. Qi, L., Jiang, L., Liu, S., Shen, X., Jia, J.: Amodal instance segmentation with kins dataset. In: Proceedings of the IEEE Conference on Computer Vision and Pattern Recognition, pp. 3014–3023 (2019)
30. Ren, J., et al.: Accurate single stage detector using recurrent rolling convolution. In: CVPR (2017)
31. Sharma, S., Ansari, J.A., Murthy, J.K., Krishna, K.M.: Beyond pixels: leveraging geometry and shape cues for online multi-object tracking. In: 2018 IEEE International Conference on Robotics and Automation (ICRA), pp. 3508–3515. IEEE (2018)
32. Tian, W., Lauer, M., Chen, L.: Online multi-object tracking using joint domain information in traffic scenarios. IEEE Trans. Intell. Transp. Syst. **21**, 374–384 (2019)
33. Vaswani, A., et al.: Attention is all you need. In: Advances in neural information processing systems, pp. 5998–6008 (2017)
34. Voigtlaender, P., et al.: MOTS: multi-object tracking and segmentation. In: Proceedings of the IEEE Conference on Computer Vision and Pattern Recognition, pp. 7942–7951 (2019)
35. Wojke, N., Bewley, A., Paulus, D.: Simple online and realtime tracking with a deep association metric. In: 2017 IEEE International Conference on Image Processing (ICIP), pp. 3645–3649. IEEE (2017)

36. Xu, J., Cao, Y., Zhang, Z., Hu, H.: Spatial-temporal relation networks for multi-object tracking. In: Proceedings of the IEEE International Conference on Computer Vision, pp. 3988–3998 (2019)
37. Xu, Z., et al.: Towards end-to-end license plate detection and recognition: a large dataset and baseline. In: Proceedings of the European Conference on Computer Vision (ECCV), pp. 255–271 (2018)
38. Xu, Z., et al.: ZoomNet: part-aware adaptive zooming neural network for 3D object detection. In: AAAI, pp. 12557–12564 (2020)
39. Yang, G., Ramanan, D.: Volumetric correspondence networks for optical flow. In: Advances in Neural Information Processing Systems, pp. 793–803 (2019)
40. Yuan, Y., Chen, W., Yang, Y., Wang, Z.: In defense of the triplet loss again: learning robust person re-identification with fast approximated triplet loss and label distillation (2019). arXiv preprint arXiv:1912.07863
41. Zhang, W., Zhou, H., Sun, S., Wang, Z., Shi, J., Loy, C.C.: Robust multi-modality multi-object tracking. In: Proceedings of the IEEE International Conference on Computer Vision, pp. 2365–2374 (2019)

Conditional Convolutions for Instance Segmentation

Zhi Tian, Chunhua Shen(✉), and Hao Chen

The University of Adelaide, Adelaide, Australia
chunhua.shen@adelaide.edu.au

Abstract. We propose a simple yet effective instance segmentation framework, termed CondInst (conditional convolutions for instance segmentation). Top-performing instance segmentation methods such as Mask R-CNN rely on ROI operations (typically ROIPool or ROIAlign) to obtain the final instance masks. In contrast, we propose to solve instance segmentation from a new perspective. Instead of using instance-wise ROIs as inputs to a network of fixed weights, we employ dynamic instance-aware networks, conditioned on instances. CondInst enjoys two advantages: (1) Instance segmentation is solved by a fully convolutional network, eliminating the need for ROI cropping and feature alignment. (2) Due to the much improved capacity of dynamically-generated conditional convolutions, the mask head can be very compact (*e.g.*, 3 conv. layers, each having only 8 channels), leading to significantly faster inference. We demonstrate a simpler instance segmentation method that can achieve improved performance in both accuracy and inference speed. On the COCO dataset, we outperform a few recent methods including well-tuned Mask R-CNN baselines, without longer training schedules needed. Code is available: https://git.io/AdelaiDet.

Keywords: Conditional convolutions · Instance segmentation

1 Introduction

Instance segmentation is a fundamental yet challenging task in computer vision, which requires an algorithm to predict a per-pixel mask with a category label for each instance of interest in an image. Despite a few works being proposed recently, the dominant framework for instance segmentation is still the two-stage method Mask R-CNN [15], which casts instance segmentation into a two-stage detection-and-segmentation task. Mask R-CNN first employs an object detector Faster R-CNN to predict a bounding-box for each instance. Then for each

Electronic supplementary material The online version of this chapter (https://doi.org/10.1007/978-3-030-58452-8_17) contains supplementary material, which is available to authorized users.

A. Vedaldi et al. (Eds.): ECCV 2020, LNCS 12346, pp. 282–298, 2020.
https://doi.org/10.1007/978-3-030-58452-8_17

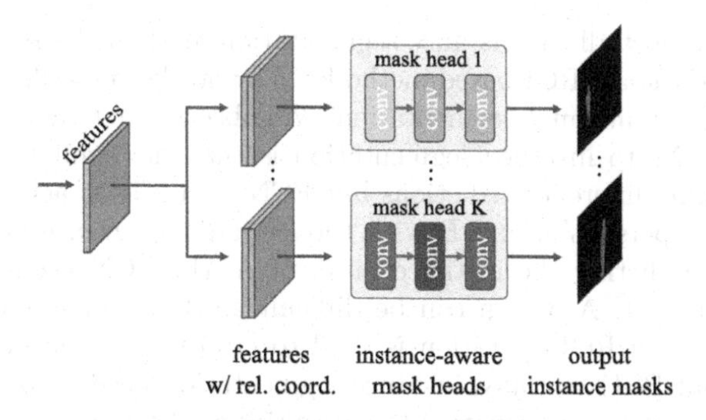

Fig. 1. CondInst uses instance-aware mask heads to predict the masks for each instance. K is the number of instances to be predicted. The filters in the mask head vary with different instances, which are dynamically-generated and conditioned on the target instance. ReLU is used as the activation function (excluding the last conv. layer).

instance, regions-of-interest (ROIs) are cropped from the networks' feature maps using the ROIAlign operation. To predict the final masks for each instance, a compact fully convolutional network (FCN) (*i.e.*, mask head) is applied to these ROIs to perform foreground/background segmentation. However, this ROI-based method may have the following drawbacks. (1) Since ROIs are often axis-aligned bounding-boxes, for objects with irregular shapes, they may contain an excessive amount of irrelevant image content including background and other instances. This issue may be mitigated by using rotated ROIs, but with the price of a more complex pipeline. (2) In order to distinguish between the foreground instance and the background stuff or instance(s), the mask head requires a relatively larger receptive field to encode sufficiently large context information. As a result, a stack of 3×3 convolutions is needed in the mask head (*e.g.*, four 3×3 convolutions with 256 channels in Mask R-CNN). It considerably increases computational complexity of the mask head, resulting that the inference time significantly varies in the number of instances. (3) ROIs are typically of different sizes. In order to use effective batched computation in modern deep learning frameworks [1,29], a resizing operation is often required to resize the cropped regions into patches of the same size. For instance, Mask R-CNN resizes all the cropped regions to 14×14 (upsampled to 28×28 using a deconvolution), which restricts the output resolution of instance segmentation, as large instances would require higher resolutions to retain details at the boundary.

In computer vision, the closest task to instance segmentation is semantic segmentation, for which fully convolutional networks (FCNs) have shown dramatic success [8,18,27,28,37]. FCNs also have shown excellent performance on many other per-pixel prediction tasks ranging from low-level image processing such as denoising, super-resolution; to mid-level tasks such as optical flow estimation and contour detection; and high-level tasks including recent single-shot object detection [38], monocular depth estimation [2,3,25,43,44] and counting

[5]. However, almost all the instance segmentation methods based on FCNs[1] lag behind state-of-the-art ROI-based methods. Why do the versatile FCNs perform unsatisfactorily on instance segmentation? We observe that the major difficulty of applying FCNs to instance segmentation is that the similar image appearance may require different predictions but FCNs struggle at achieving this. For example, if two persons A and B with the similar appearance are in an input image, when predicting the instance mask of A, the FCN needs to predict B as background w.r.t. A, which can be difficult as they look similar in appearance. Therefore, an ROI operation is used to crop the person of interest, *i.e.*, A; and filter out B. Essentially, instance segmentation needs two types of information: (1) *appearance* information to categorize objects; and (2) *location* information to distinguish multiple objects belonging to the same category. Almost all methods rely on ROI cropping, which explicitly encodes the location information of instances. In contrast, CondInst exploits the location information by using *instance-sensitive convolution filters* as well as relative coordinates that are appended to the feature maps.

Thus, we advocate a new solution that uses instance-aware FCNs for instance mask prediction. In other words, instead of using a standard ConvNet with a fixed set of convolutional filters as the mask head for predicting all instances, the network parameters are adapted according to the instance to be predicted. Inspired by dynamic filtering networks [21] and CondConv [42], for each instance, a controller sub-network (see Fig. 3) dynamically generates the mask FCN network parameters (conditioned on the center area of the instance), which is then used to predict the mask of this instance. It is expected that the network parameters can encode the characteristics (*e.g.*, relative position, shape and appearance) of this instance, and only fires on the pixels of this instance, which thus bypasses the difficulty mentioned above. These conditional mask heads are applied to the whole feature maps, *eliminating the need for ROI operations*. At the first glance, the idea may not work well as instance-wise mask heads may incur a large number of network parameters provided that some images contain as many as dozens of instances. However, we show that a very compact FCN mask head with dynamically-generated filters can already outperform previous ROI-based Mask R-CNN, resulting in much reduced computational complexity per instance than that of the mask head in Mask R-CNN.

We summarize our main contributions as follow.

- We attempt to solve instance segmentation from a new perspective. To this end, we propose the CondInst instance segmentation framework, which achieves improved instance segmentation performance than existing methods such as Mask R-CNN while being faster. To our knowledge, this is the first time that a new instance segmentation framework outperforms recent state-of-the-art both in accuracy and speed.
- CondInst is fully convolutional and avoids the aforementioned resizing operation used in many existing methods, as CondInst does not rely on ROI

[1] By FCNs, we mean the vanilla FCNs in [28] that only involve convolutions and pooling.

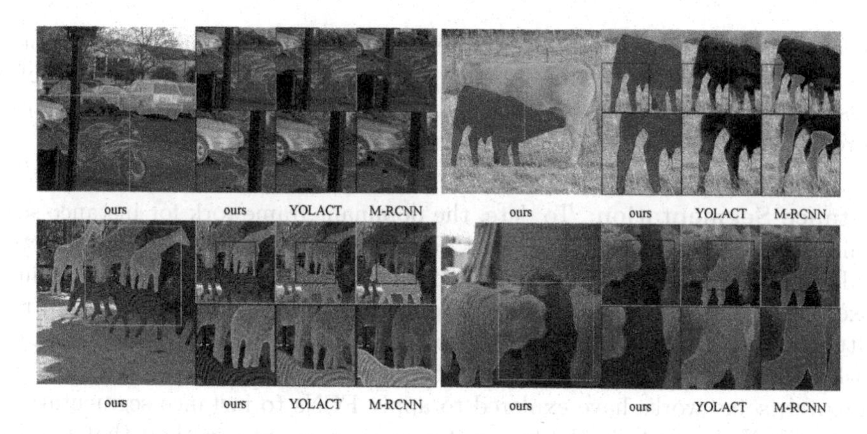

Fig. 2. Qualitative comparisons with other methods. We compare the proposed CondInst against YOLACT [4] and Mask R-CNN [15]. Our masks are generally of higher quality (*e.g.*, preserving more details). Best viewed on screen.

operations. Without having to resize feature maps leads to high-resolution instance masks with more accurate edges.
- Unlike previous methods, in which the filters in its mask head are fixed for all the instances once trained, the filters in our mask head are dynamically generated and conditioned on instances. As the filters are only asked to predict the mask of only one instance, it largely eases the learning requirement and thus reduces the load of the filters. As a result, the mask head can be extremely light-weight, significantly reducing the inference time per instance. Compared with the bounding box detector FCOS, CondInst needs only $\sim$10% more computational time, even processing the maximum number of instances per image (*i.e.*, 100 instances).
- Without resorting to longer training schedules as needed in recent works [4,9], CondInst achieves state-of-the-art performance while being faster in inference. We hope that CondInst can be a new strong alternative to popular methods such as Mask R-CNN for the instance segmentation task.

Moreover, CondInst can be immediately applied to panoptic segmentation due to its flexible design. We believe that with minimal re-design effort, the proposed CondInst can be used to solve all instance-level recognition tasks that were previously solved with an ROI-based pipeline.

1.1 Related Work

Here we review some work that is most relevant to ours.

Conditional Convolutions. Unlike traditional convolutional layers, which have fixed filters once trained, the filters of conditional convolutions are conditioned on the input and are dynamically generated by another network (*i.e.*, a controller). This idea has been explored previously in dynamic filter networks [21] and CondConv [42] mainly for the purpose of increasing the capacity of a classification network. In this work, we extend this idea to solve the significantly more challenging task of instance segmentation.

Instance Segmentation. To date, the dominant framework for instance segmentation is still Mask R-CNN. Mask R-CNN first employs an object detector to detect the bounding-boxes of instances (*e.g.*, ROIs). With these bounding-boxes, an ROI operation is used to crop the features of the instance from the feature maps. Finally, a compact FCN head is used to obtain the desired instance masks. Many works [7,19,26] with top performance are built on Mask R-CNN. Moreover, some works have explored to apply FCNs to instance segmentation. InstanceFCN [11] may be the first instance segmentation method that is fully convolutional. InstanceFCN proposes to predict position-sensitive score maps with vanilla FCNs. Afterwards, these score maps are assembled to obtain the desired instance masks. Note that InstanceFCN does not work well with overlapping instances. Others [13,31,32] attempt to first perform segmentation and the desired instance masks are formed by assembling the pixels of the same instance. Novotny et al. [33] propose semi-convolutional operators to make FCNs applicable to instance segmentation. To our knowledge, thus far none of these methods can outperform Mask R-CNN both in accuracy and speed on the public COCO benchmark dataset.

The recent YOLACT [4] and BlendMask [6] may be viewed as a reformulation of Mask RCNN, which decouple ROI detection and feature maps used for mask prediction. Wang et al. developed a simple FCN based instance segmentation method, showing competitive performance [39]. PolarMask developed a new simple mask representation for instance segmentation [41], which extends the bounding box detector FCOS [38]. EmbedMask [45] learns instance and pixel embedding, and then assigns pixels to an instance based on the similarity of their embedding.

Recently AdaptIS [36] proposes to solve panoptic segmentation with FiLM [34]. The idea shares some similarity with CondInst in that information about an instance is encoded in the coefficients generated by FiLM. Since only the batch normalization coefficients are dynamically generated, AdaptIS needs a large mask head to achieve good performance. In contrast, CondInst directly encodes them into conv. filters of the mask head, thus having much stronger capacity. As a result, even with a very compact mask head, we believe that CondInst can achieve instance segmentation accuracy that would not be possible for AdaptIS to attain.

2 Instance Segmentation with CondInst

2.1 Overall Architecture

Given an input image $I \in \mathbb{R}^{H \times W \times 3}$, the goal of instance segmentation is to predict the pixel-level mask and the category of each instance of interest in the

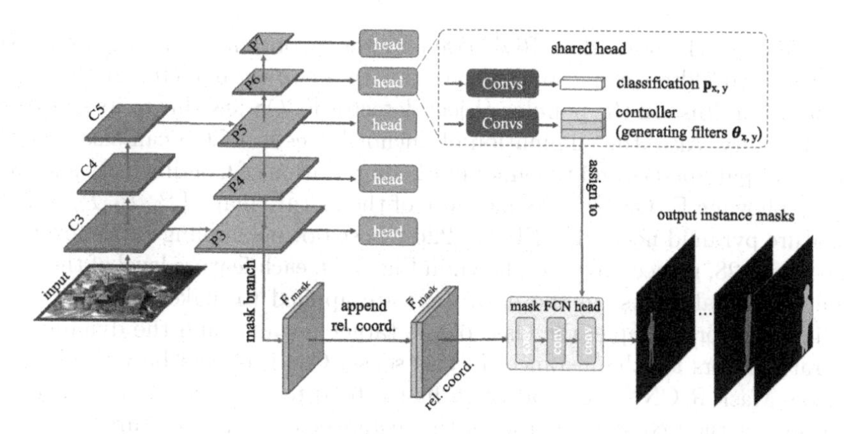

Fig. 3. The overall architecture of CondInst. C_3, C_4 and C_5 are the feature maps of the backbone network (*e.g.*, ResNet-50). P_3 to P_7 are the FPN feature maps as in [22,38]. $\mathbf{F}_{mask}$ is the mask branch's output and $\tilde{\mathbf{F}}_{mask}$ is obtained by concatenating the relative coordinates to $\mathbf{F}_{mask}$. The classification head predicts the class probability $\boldsymbol{p}_{x,y}$ of the target instance at location (x, y), same as in FCOS. The controller generates the filter parameters $\boldsymbol{\theta}_{x,y}$ of the mask head for the instance. Similar to FCOS, there are also center-ness and box heads in parallel with the controller (not shown in the figure for simplicity). Note that the heads in the dashed box are repeatedly applied to $P_3 \cdots P_7$. The mask head is instance-aware, and is applied to $\tilde{\mathbf{F}}_{mask}$ as many times as the number of instances in the image (refer to Fig. 1).

image. The ground-truths are defined as $\{(M_i, c_i)\}$, where $M_i \in \{0, 1\}^{H \times W}$ is the mask for the i-th instance and $c_i \in \{1, 2, ..., C\}$ is the category. C is 80 on MS-COCO [24]. Unlike semantic segmentation, which only requires to predict one mask for an input image, instance segmentation needs to predict a variable number of masks, depending on the number of instances in the image. This poses a challenge when applying traditional FCNs [28] to instance segmentation. In this work, our core idea is that for an image with K instances, K different mask heads will be dynamically generated, and each mask head will contain the characteristics of its target instance in their filters. As a result, when the mask is applied to an input, it will only fire on the pixels of the instance, thus producing the mask prediction of the instance. We illustrate the process in Fig. 1.

Recall that Mask R-CNN employs an object detector to predict the bounding-boxes of the instances in the input image. The bounding-boxes are actually the way that Mask R-CNN represents instances. Similarly, CondInst employs the instance-aware filters to represent the instances. In other words, instead of encoding the instance concept into the bounding-boxes, CondInst implicitly encodes it into the parameters of the mask heads, which is a much more flexible way. For example, it can easily represent the irregular shapes that are hard to be tightly enclosed by a bounding-box. This is one of CondInst's advantages over the previous ROI-based methods.

Similar to the way that ROI-based methods obtain bounding-boxes, the instance-aware filters can also be obtained with an object detector. In this work, we build CondInst on the popular object detector FCOS [38] due to its simplicity and flexibility. Also, the elimination of anchor-boxes in FCOS can also save the number of parameters and the amount of computation of CondInst. As shown in Fig. 3, following FCOS [38], we make use of the feature maps $\{P_3, P_4, P_5, P_6, P_7\}$ of feature pyramid networks (FPNs) [22], whose down-sampling ratios are 8, 16, 32, 64 and 128, respectively. As shown in Fig. 3, on each feature level of the FPN, some functional layers (in the dash box) are applied to make instance-related predictions. For example, the class of the target instance and the dynamically-generated filters for the instance. In this sense, CondInst can be viewed as the same as Mask R-CNN, both of which first attend to instances in an image and then predict the pixel-level masks of the instances (*i.e.*, instance-first).

Besides the detector, as shown in Fig. 3, there is also a mask branch, which provides the feature maps that our generated mask heads take as inputs to predict the desired instance mask. The feature maps are denoted by $\mathbf{F}_{mask} \in \mathbb{R}^{H_{mask} \times W_{mask} \times C_{mask}}$. The mask branch is connected to FPN level P_3 and thus its output resolution is $\frac{1}{8}$ of the input image resolution. The mask branch has four 3×3 convolutions with 128 channels before the last layer. Afterwards, in order to reduce the number of the generated parameters, the last layer of the mask branch reduces the number of channels from 128 to 8 (*i.e.*, $C_{mask} = 8$). Surprisingly, using $C_{mask} = 8$ can already achieve superior performance and using a larger C_{mask} here (*e.g.*, 16) cannot improve the performance, as shown in our experiments. Even more aggressively, using $C_{mask} = 2$ only degrades the performance by ∼0.3% in mask AP. Moreover, as shown in Fig. 3, $\mathbf{F}_{mask}$ is combined with a map of the coordinates, which are relative coordinates from all the locations on $\mathbf{F}_{mask}$ to the location (x, y) (*i.e.*, where the filters of the mask head are generated). Then, the combination is sent to the mask head to predict the instance mask. The relative coordinates provide a strong cue for predicting the instance mask, as shown in our experiments. Moreover, a single sigmoid is used as the final output of the mask head, and thus the mask prediction is class-agnostic. The class of the instance is predicted by the classification head in parallel with the controller, as shown in Fig. 3.

The resolution of the original mask prediction is same as the resolution of F_{mask}, which is $\frac{1}{8}$ of the input image resolution. In order to produce high-resolution instance masks, a bilinear upsampling is used to upsample the mask prediction by 4, resulting in 400×512 mask prediction (if the input image size is 800×1024). We will show that the upsampling is crucial to the final instance segmentation performance of CondInst in experiments. Note that the mask's resolution is much higher than that of Mask R-CNN (only 28×28 as mentioned before).

2.2 Network Outputs and Training Targets

Similar to FCOS, each location on the FPN's feature maps P_i either is associated with an instance, thus being a positive sample, or is considered a negative sample.

The associated instance and label for each location are determined as follows. Let us consider the feature maps $P_i \in \mathbb{R}^{H \times W \times C}$ and let s be its down-sampling ratio. As shown in previous works [16,35,38], a location (x, y) on the feature maps can be mapped back onto the input image as $(\lfloor \frac{s}{2} \rfloor + xs, \lfloor \frac{s}{2} \rfloor + ys)$. If the mapped location falls in the center region of an instance, the location is considered to be responsible for the instance. Any locations outside the center regions are labeled as negative samples. The center region is defined as the box $(c_x - rs, c_y - rs, c_x + rs, c_y + rs)$, where (c_x, c_y) denotes the mass center of the instance, s is the down-sampling ratio of P_i and r is a constant scalar being 1.5 as in FCOS [38]. As shown in Fig. 3, at a location (x, y) on P_i, CondInst has the following output heads.

Classification Head. The classification head predicts the class of the instance associated with the location. The ground-truth target is the instance's class c_i or 0 (*i.e.*, background). As in FCOS, the network predicts a C-D vector $\boldsymbol{p}_{x,y}$ for the classification and each element in $\boldsymbol{p}_{x,y}$ corresponds to a binary classifier, where C is the number of categories.

Controller Head. The controller head, which has the same architecture as the above classification head, is used to predict the parameters of the mask head for the instance at the location. The mask head predicts the mask of this particular instance. This is the core contribution of our work. To predict the parameters, we concatenate all the parameters of the filters (*i.e.*, weights and biases) together as an N-D vector $\boldsymbol{\theta}_{x,y}$, where N is the total number of the parameters. Accordingly, the controller head has N output channels. The mask head is a very compact FCN architecture, which has three 1×1 convolutions, each having 8 channels and using ReLU as the activation function except for the last one. No normalization layer such as batch normalization [20] is used here. The last layer has 1 output channel and uses sigmoid to predict the probability of being foreground. The mask head has 169 parameters in total ($\#weights = (8 + 2) \times 8(conv1) + 8 \times 8(conv2) + 8 \times 1(conv3)$ and $\#biases = 8(conv1) + 8(conv2) + 1(conv3)$). As mentioned before, the generated filters contain information about the instance at the location, and thus, ideally, the mask head with the filters will only fire on the pixels of the instance, even taking as the input the whole feature maps.

Center-ness and Box Heads. The center-ness and box heads are the same as that in FCOS. We refer readers to FCOS [38] for the details.

Conceptually, CondInst can eliminate the box head since CondInst needs no ROIs. However, we find that if we make use of box-based NMS, the inference time will be much reduced. Thus, we still predict boxes in CondInst. We would like to highlight that the predicted boxes are *only* used in NMS and do not involve any ROI operations. Moreover, as shown in Table 5, the box prediction can be removed if the NMS using no box used (*e.g.*, mask-based NMS). This is fundamentally different from previous ROI-based methods, in which the box prediction is mandatory.

2.3 Loss Function

Formally, the overall loss function of CondInst can be formulated as,

$$L_{overall} = L_{fcos} + \lambda L_{mask}, \tag{1}$$

where L_{fcos} and L_{mask} denote the original loss of FCOS and the loss for instance masks, respectively. λ being 1 in this work is used to balance the two losses. We refer readers to FCOS for the details of L_{fcos}. L_{mask} is defined as,

$$L_{mask}(\{\boldsymbol{\theta}_{x,y}\}) = \frac{1}{N_{\text{pos}}} \sum_{x,y} \mathbb{1}_{\{c^*_{x,y}>0\}} L_{dice}(MaskHead(\tilde{\mathbf{F}}_{x,y}; \boldsymbol{\theta}_{x,y}), \mathbf{M}^*_{x,y}), \tag{2}$$

where $c^*_{x,y}$ is the classification label of location (x, y), which is the class of the instance associated with the location or 0 (*i.e.*, background) if the location is not associated with any instance. N_{pos} is the number of locations where $c^*_{x,y} > 0$. $\mathbb{1}_{\{c^*_{x,y}>0\}}$ is the indicator function, being 1 if $c^*_{x,y} > 0$ and 0 otherwise. $\boldsymbol{\theta}_{x,y}$ is the generated filters' parameters at location (x, y). $\tilde{\mathbf{F}}_{x,y} \in \mathbb{R}^{H_{mask} \times W_{mask} \times (C_{mask}+2)}$ is the combination of $\mathbf{F}_{mask}$ and a map of coordinates $\mathbf{O}_{x,y} \in \mathbb{R}^{H_{mask} \times W_{mask} \times 2}$. As described before, $\mathbf{O}_{x,y}$ is the relative coordinates from all the locations on $\mathbf{F}_{mask}$ to (x, y) (*i.e.*, the location where the filters are generated). $MaskHead$ denotes the mask head, which consists of a stack of convolutions with dynamic parameters $\boldsymbol{\theta}_{x,y}$. $\mathbf{M}^*_{x,y} \in \{0, 1\}^{H \times W \times C}$ is the mask of the instance associated with location (x, y). L_{dice} is the dice loss as in [30], which is used to overcome the foreground-background sample imbalance. We do not employ focal loss here as it requires special initialization, which cannot be realized if the parameters are dynamically generated. Note that, in order to compute the loss between the predicted mask and the ground-truth mask $\mathbf{M}^*_{x,y}$, they are required to have the same size. As mentioned before, the prediction is upsampled by 4 and thus the resolution of the final prediction is half of that of the ground-truth mask $\mathbf{M}^*_{x,y}$. We downsample $\mathbf{M}^*_{x,y}$ by 2 to make the sizes equal. These operations are omitted in Eq. (2) for clarification.

Moreover, as shown in YOLACT [4], the instance segmentation task can benefit from a joint semantic segmentation task. Thus, we also conduct experiments with the joint semantic segmentation task. However, unless explicitly specified, all the experiments in the paper are *without* the semantic segmentation task. If used, the semantic segmentation loss is added to $L_{overall}$.

2.4 Inference

Given an input image, we forward it through the network to obtain the outputs including classification confidence $\boldsymbol{p}_{x,y}$, center-ness scores, box prediction $\boldsymbol{t}_{x,y}$ and the generated parameters $\boldsymbol{\theta}_{x,y}$. We first follow the steps in FCOS to obtain the box detections. Afterwards, box-based NMS with the threshold being 0.6 is used to remove duplicated detections and then the top 100 boxes are used to compute masks. Different from FCOS, these boxes are also associated with

Table 1. Instance segmentation results with different architectures of the mask head on MS-COCO `val2017` split. "depth": the number of layers in the mask head. "width": the number of channels of these layers. "time": the milliseconds that the mask head takes for processing 100 instances.

(a) Varying the depth (width = 8).

Depth	Time	AP	AP_{50}	AP_{75}	AP_S	AP_M	AP_L
1	**2.2**	30.9	52.9	31.4	14.0	33.3	45.1
2	3.3	35.5	56.1	37.8	17.0	38.9	50.8
3	4.5	**35.7**	**56.3**	37.8	17.1	**39.1**	50.2
4	5.6	**35.7**	56.2	**37.9**	**17.2**	38.7	**51.5**

(b) Varying the width (depth = 3).

Width	Time	AP	AP_{50}	AP_{75}	AP_S	AP_M	AP_L
2	**2.5**	34.1	55.4	35.8	15.9	37.2	49.1
4	2.6	35.6	**56.5**	**38.1**	17.0	**39.2**	**51.4**
8	4.5	**35.7**	**56.3**	37.8	17.1	39.1	50.2
16	4.7	35.6	56.2	37.9	**17.2**	38.8	50.8

the filters generated by the controller. Let us assume that K boxes remain after the NMS, and thus we have K groups of the generated filters. The K groups of filters are used to produce K instance-specific mask heads. These instance-specific mask heads are applied, in the fashion of FCNs, to the $\tilde{\mathbf{F}}_{x,y}$ (*i.e.*, the combination of $\mathbf{F}_{mask}$ and $\mathbf{O}_{x,y}$) to predict the masks of the instances. Since the mask head is a very compact network (three 1×1 convolutions with 8 channels and 169 parameters in total), the overhead of computing masks is extremely small. For example, even with 100 detections (*i.e.*, the maximum number of detections per image on MS-COCO), only less 5 ms in total are spent on the mask heads, which only adds ~10% computational time to the base detector FCOS. In contrast, the mask head of Mask R-CNN has four 3×3 convolutions with 256 channels, thus having more than 2.3M parameters and taking longer computational time.

3 Experiments

We evaluate CondInst on the large-scale benchmark MS-COCO [24]. Following the common practice [15,23,38], our models are trained with split `train2017` (115K images) and all the ablation experiments are evaluated on split `val2017` (5K images). Our main results are reported on the `test-dev` split (20K images). Experiments on Cityscapes [10] can be found in our supplementary material.

3.1 Implementation Details

Unless specified, we make use of the following implementation details. Following FCOS [38], ResNet-50 [17] is used as our backbone network and the weights pre-trained on ImageNet [12] are used to initialize it. For the newly added layers, we initialize them as in [38]. Our models are trained with stochastic gradient descent (SGD) over 8 V100 GPUs for 90K iterations with the initial learning rate being 0.01 and a mini-batch of 16 images. The learning rate is reduced by a factor of 10 at iteration $60K$ and $80K$, respectively. Weight decay and momentum are set as 0.0001 and 0.9, respectively. Following Detectron2 [40], the input images are resized to have their shorter sides in [640, 800] and their longer sides less or equal to 1333 during training. Left-right flipping data augmentation is also used

Table 2. The instance segmentation results by varying the number of channels of the mask branch output (*i.e.*, C_{mask}) on MS-COCO val2017 split. The performance keeps almost the same if C_{mask} is in a reasonable range, which suggests that CondInst is robust to the design choice.

C_{mask}	AP	AP_{50}	AP_{75}	AP_S	AP_M	AP_L
1	34.8	55.9	36.9	16.7	38.0	50.1
2	35.4	56.2	37.6	16.9	38.9	50.4
4	35.5	56.2	**37.9**	17.0	39.0	50.8
8	**35.7**	**56.3**	37.8	**17.1**	**39.1**	50.2
16	35.5	56.1	37.7	16.4	**39.1**	**51.2**

Table 3. Ablation study of the input to the mask head on MS-COCO val2017 split. As shown in the table, without the relative coordinates, the performance drops significantly from 35.7% to 31.4% in mask AP. Using the absolute coordinates cannot improve the performance remarkably. If the mask head only takes as input the relative coordinates (*i.e.*, no appearance in this case), CondInst also achieves modest performance.

w/abs. coord	w/rel. coord	w/ $\mathbf{F}_{mask}$	AP	AP_{50}	AP_{75}	AP_S	AP_M	AP_L	AR_1	AR_{10}	AR_{100}
		✓	31.4	53.5	32.1	15.6	34.4	44.7	28.4	44.1	46.2
	✓		31.3	54.9	31.8	16.0	34.2	43.6	27.1	43.3	45.7
✓		✓	32.0	53.3	32.9	14.7	34.2	46.8	28.7	44.7	46.8
	✓	✓	**35.7**	**56.3**	**37.8**	**17.1**	**39.1**	**50.2**	**30.4**	**48.8**	**51.5**

during training. When testing, we do not use any data augmentation and only the scale of the shorter side being 800 is used. The inference time in this work is measured on a single V100 GPU with 1 image per batch.

3.2 Architectures of the Mask Head

In this section, we discuss the design choices of the mask head in CondInst. To our surprise, the performance is insensitive to the architectures of the mask head. Our baseline is the mask head of three 1×1 convolutions with 8 channels (*i.e.*, width $= 8$). As shown in Table 1 (3rd row), it achieves 35.7% in mask AP. Next, we first conduct experiments by varying the depth of the mask head. As shown in Table 1a, apart from the mask head with depth being 1, all other mask heads (*i.e.*, depth $= 2$, 3 and 4) attain similar performance. The mask head with depth being 1 achieves inferior performance as in this case the mask head is actually a linear mapping, which has overly weak capacity. Moreover, as shown in Table 1b, varying the width (*i.e.*, the number of the channels) does not result in a remarkable performance change either as long as the width is in a reasonable range. We also note that our mask head is extremely light-weight as the filters in our mask head are dynamically generated. As shown in Table 1, our baseline mask head only takes 4.5 ms per 100 instances (the maximum number

Table 4. The instance segmentation results on MS-COCO `val2017` split by changing the factor used to upsample the mask predictions. "resolution" denotes the resolution ratio of the mask prediction to the input image. Without the upsampling (*i.e.*, factor = 1), the performance drops significantly. Almost the same results are obtained with ratio 2 or 4.

Factor	Resolution	AP	AP_{50}	AP_{75}	AP_S	AP_M	AP_L
1	1/8	34.4	55.4	36.2	15.1	38.4	50.8
2	1/4	**35.8**	**56.4**	**38.0**	17.0	**39.3**	**51.1**
4	1/2	35.7	56.3	37.8	**17.1**	39.1	50.2

of instances on MS-COCO), which suggests that our mask head only adds small computational overhead to the base detector. Moreover, our baseline mask head only has 169 parameters in total. In sharp contrast, the mask head of Mask R-CNN [15] has more than 2.3M parameters and takes ~2.5× computational time (11.4 ms per 100 instances).

3.3 Design Choices of the Mask Branch

We further investigate the impact of the mask branch. We first change C_{mask}, which is the number of channels of the mask branch's output feature maps (*i.e.*, $\mathbf{F}_{mask}$). As shown in Table 2, as long as C_{mask} is in a reasonable range (*i.e.*, from 2 to 16), the performance keeps almost the same. $C_{mask} = 8$ is optimal and thus we use $C_{mask} = 8$ in all other experiments by default.

As mentioned before, before taken as the input of the mask heads, the mask branch's output $\mathbf{F}_{mask}$ is concatenated with a map of relative coordinates, which provides a strong cue for the mask prediction. As shown in Table 3 (2nd row), the performance drops significantly if the relative coordinates are removed (35.7% vs. 31.4%). The significant performance drop implies that the generated filters not only encode the appearance cues but also encode the shape (and relative position) of the target instance. It can also be evidenced by the experiment only using the relative coordinates. As shown in Table 3 (2rd row), only using the relative coordinates can also obtain decent performance (31.3% in mask AP). We would like to highlight that unlike Mask R-CNN, which encodes the shape of the target instance by a bounding-box, CondInst implicitly encodes the shape into the generated filters, which can easily represent any shapes including irregular ones and thus is much more flexible. We also experiment with the absolute coordinates, but it cannot largely boost the performance as shown in Table 3 (32.0%). This suggests that the generated filters mainly carry translation-invariant cues such as shapes and relative position, which is preferable.

3.4 How Important to Upsample Mask Predictions?

As mentioned before, the original mask prediction is upsampled and the upsampling is of great importance to the final performance. We confirm this in the

Table 5. Instance segmentation results with different NMS algorithms. Mask-based NMS can obtain the same overall performance as box-based NMS, which suggests that CondInst can totally eliminate the box detection.

NMS	AP	AP_{50}	AP_{75}	AP_S	AP_M	AP_L
Box	**35.7**	56.3	**37.8**	17.1	39.1	50.2
Mask	**35.7**	**56.7**	37.7	**17.2**	**39.2**	**50.5**

experiment. As shown in Table 4, without using the upsampling (1st row in the table), in this case CondInst can produce the mask prediction with $\frac{1}{8}$ of the input image resolution, which merely achieves 34.4% in mask AP because most of the details (*e.g.*, the boundary) are lost. If the mask prediction is upsampled by factor = 2, the performance can be significantly improved by 1.4% in mask AP (from 34.4% to 35.8%). In particular, the improvement on small objects is large (from 15.1% to 17.0), which suggests that the upsampling can greatly retain the details of objects. Increasing the upsampling factor to 4 slightly worsens the performance (from 35.8% to 35.7% in mask AP), probably due to the relatively low-quality annotations of MS-COCO. We use factor = 4 in all other models as it has the potential to produce high-resolution instance masks.

3.5 CondInst without Bounding-Box Detection

Although we still keep the bounding-box detection branch in CondInst, it is conceptually feasible to totally eliminate it if we make use of the NMS using no bounding-boxes. In this case, all the foreground samples (determined by the classification head) will be used to compute instance masks, and the duplicated masks will be removed by mask-based NMS. As shown in Table 5, with the mask-based NMS, the same overall performance can be obtained as box-based NMS (35.7% vs. 35.7% in mask AP).

3.6 Comparisons with State-of-the-Art Methods

We compare CondInst against previous state-of-the-art methods on MS-COCO `test-dev` split. As shown in Table 6, with 1× learning rate schedule (*i.e.*, 90K iterations), CondInst outperforms the original Mask R-CNN by 0.8% (35.4% vs. 34.6%). CondInst also achieves a much faster speed than the original Mask R-CNN (49 ms vs. 65 ms per image on a single V100 GPU). To our knowledge, it is the first time that a new and simpler instance segmentation method, without any bells and whistles outperforms Mask R-CNN both in accuracy and speed. CondInst also obtains better performance (35.9% vs. 35.5%) and on-par speed (49 ms vs 49 ms) than the well-engineered Mask R-CNN in `Detectron2` (*i.e.*, Mask R-CNN* in Table 6). Furthermore, with a longer training schedule (*e.g.*, 3×) or a stronger backbone (*e.g.*, ResNet-101), a consistent improvement is achieved as well (37.8% vs. 37.5% with ResNet-50 3× and 39.1% vs. 38.8% with

Table 6. Comparisons with state-of-the-art methods on MS-COCO `test-dev`. "Mask R-CNN" is the original Mask R-CNN [15] and "Mask R-CNN*" is the improved Mask R-CNN in `Detectron2` [40]. "aug.": using multi-scale data augmentation during training. "sched.": the used learning rate schedule. 1× is 90K iterations, 2× is 180K iterations and so on. The learning rate is changed as in [14]. "w/sem": using the auxiliary semantic segmentation task.

Method	Backbone	Aug.	Sched.	AP	AP_{50}	AP_{75}	AP_S	AP_M	AP_L
Mask R-CNN [15]	R-50-FPN		1×	34.6	**56.5**	36.6	15.4	36.3	**49.7**
CondInst	R-50-FPN		1×	**35.4**	56.4	**37.6**	**18.4**	**37.9**	46.9
Mask R-CNN*	R-50-FPN	✓	1×	35.5	57.0	37.8	19.5	37.6	46.0
Mask R-CNN*	R-50-FPN	✓	3×	37.5	59.3	40.2	21.1	39.6	48.3
TensorMask [9]	R-50-FPN	✓	6×	35.4	57.2	37.3	16.3	36.8	49.3
CondInst	R-50-FPN	✓	1×	35.9	56.9	38.3	19.1	38.6	46.8
CondInst	R-50-FPN	✓	3×	37.8	59.1	40.5	21.0	40.3	48.7
CondInst w/sem.	R-50-FPN	✓	3×	38.8	60.4	41.5	21.1	41.1	51.0
Mask R-CNN	R-101-FPN	✓	6×	38.3	61.2	40.8	18.2	40.6	**54.1**
Mask R-CNN*	R-101-FPN	✓	3×	38.8	60.9	41.9	21.8	41.4	50.5
YOLACT-700 [4]	R-101-FPN	✓	4.5×	31.2	50.6	32.8	12.1	33.3	47.1
TensorMask	R-101-FPN	✓	6×	37.1	59.3	39.4	17.4	39.1	51.6
CondInst	R-101-FPN	✓	3×	39.1	60.9	42.0	21.5	41.7	50.9
CondInst w/sem.	R-101-FPN	✓	3×	**40.1**	**62.1**	**43.1**	**21.8**	**42.7**	52.6

ResNet-101 3×). Moreover, as shown in Table 6, with the auxiliary semantic segmentation task, the performance can be boosted from 37.8% to 38.8% (ResNet-50) or from 39.1% to 40.1% (ResNet-101), without increasing the inference time. For fair comparisons, all the inference time here is measured by ourselves on the same hardware with the official codes.

We also compare CondInst with the recently-proposed instance segmentation methods. Only with half training iterations, CondInst surpasses TensorMask [9] by a large margin (38.8% vs. 35.4% for ResNet-50 and 39.1% vs. 37.1% for ResNet-101). CondInst is also ~8× faster than TensorMask (49 ms vs 380 ms per image on the same GPU) with similar performance (37.8% vs. 37.1%). Moreover, CondInst outperforms YOLACT-700 [4] by a large margin with the same backbone ResNet-101 (40.1% vs. 31.2% and both with the auxiliary semantic segmentation task). Moreover, as shown in Fig. 2, compared with YOLACT-700 and Mask R-CNN, CondInst can preserve more details and produce higher-quality instance segmentation results.

4 Conclusions

We have proposed a new and simpler instance segmentation framework, named CondInst. Unlike previous method such as Mask R-CNN, which employs the mask head with fixed weights, CondInst conditions the mask head on instances and dynamically generates the filters of the mask head. This not only reduces the parameters and computational complexity of the mask head, but also eliminates the ROI operations, resulting in a faster and simpler instance segmentation framework. To our knowledge, CondInst is the first framework that can outperform Mask R-CNN both in accuracy and speed, without longer training schedules needed. We believe that CondInst can be a new strong alternative to Mask R-CNN for instance segmentation.

Acknowledgments. Correspondence should be addressed to CS. CS was in part supported by ARC DP 'Deep learning that scales'.

References

1. Paszke, A., et al.: PyTorch: an imperative style, high-performance deep learning library. In: Proceedings of the Advances in Neural Information Processing Systems, pp. 8024–8035 (2019)
2. Bian, J.W., Zhan, H., Wang, N., Chin, T.J., Shen, C., Reid, I.: Unsupervised depth learning in challenging indoor video: weak rectification to rescue. arXiv preprint arXiv:2006.02708 (2020)
3. Bian, J., et al.: Unsupervised scale-consistent depth and ego-motion learning from monocular video. In: Advances in Neural Information Processing Systems, pp. 35–45 (2019)
4. Bolya, D., Zhou, C., Xiao, F., Lee, Y.J.: YOLACT: real-time instance segmentation. In: Proceedings of the IEEE International Conference on the Computer Vision, pp. 9157–9166 (2019)
5. Boominathan, L., Kruthiventi, S., Babu, R.V.: CrowdNet: a deep convolutional network for dense crowd counting. In: Proceedings of the ACM International Conference on Multimedia, pp. 640–644. ACM (2016)
6. Chen, H., Sun, K., Tian, Z., Shen, C., Huang, Y., Yan, Y.: BlendMask: top-down meets bottom-up for instance segmentation. In: Proceedings of the IEEE Conference on Computer Vision and Pattern Recognition (2020)
7. Chen, K., et al.: Hybrid task cascade for instance segmentation. In: Proceedings of the IEEE Conference on Computer Vision and Pattern Recognition, pp. 4974–4983 (2019)
8. Chen, L.C., Papandreou, G., Kokkinos, I., Murphy, K., Yuille, A.: DeepLab: semantic image segmentation with deep convolutional nets, atrous convolution, and fully connected CRFs. IEEE Trans. Pattern Anal. Mach. Intell. **40**(4), 834–848 (2017)
9. Chen, X., Girshick, R., He, K., Dollár, P.: TensorMask: a foundation for dense object segmentation. In: Proceedings of the IEEE International Conference on Computer Vision, pp. 2061–2069 (2019)
10. Cordts, M., et al.: The cityscapes dataset for semantic urban scene understanding. In: Proceedings of the IEEE Conference on Computer Vision and Pattern Recognition (CVPR) (2016)
11. Dai, J., He, K., Li, Y., Ren, S., Sun, J.: Instance-sensitive fully convolutional networks. In: Leibe, B., Matas, J., Sebe, N., Welling, M. (eds.) ECCV 2016. LNCS, vol. 9910, pp. 534–549. Springer, Cham (2016). https://doi.org/10.1007/978-3-319-46466-4_32

12. Deng, J., Dong, W., Socher, R., Li, L.J., Li, K., Fei-Fei, L.: ImageNet: a large-scale hierarchical image database. In: Proceedings of the IEEE Conference on Computer Vision Pattern Recognition, pp. 248–255. IEEE (2009)
13. Fathi, A., et al.: Semantic instance segmentation via deep metric learning. arXiv: Comp. Res. Repository (2017)
14. He, K., Girshick, R., Dollár, P.: Rethinking ImageNet pre-training. In: Proceedings of the IEEE International Conference on Computer Vision, pp. 4918–4927 (2019)
15. He, K., Gkioxari, G., Dollár, P., Girshick, R.: Mask R-CNN. In: Proceedings of the IEEE International Conference on Computer Vision, pp. 2961–2969 (2017)
16. He, K., Zhang, X., Ren, S., Sun, J.: Spatial pyramid pooling in deep convolutional networks for visual recognition. IEEE Trans. Pattern Anal. Mach. Intell. **37**(9), 1904–1916 (2015)
17. He, K., Zhang, X., Ren, S., Sun, J.: Deep residual learning for image recognition. In: Proceedings of the IEEE Conference on Computer Vision and Pattern Recognition, pp. 770–778 (2016)
18. He, T., Shen, C., Tian, Z., Gong, D., Sun, C., Yan, Y.: Knowledge adaptation for efficient semantic segmentation. In: Proceedings of the IEEE Conference on Computer Vision and Pattern Recognition, pp. 578–587 (2019)
19. Huang, Z., Huang, L., Gong, Y., Huang, C., Wang, X.: Mask scoring R-CNN. In: Proceedings of the IEEE Conference Computer Vision Pattern Recognition, pp. 6409–6418 (2019)
20. Ioffe, S., Szegedy, C.: Batch normalization: accelerating deep network training by reducing internal covariate shift. arXiv preprint arXiv:1502.03167 (2015)
21. Jia, X., De Brabandere, B., Tuytelaars, T., Gool, L.V.: Dynamic filter networks. In: Proceedings of the Advances in Neural Information Processing System, pp. 667–675 (2016)
22. Lin, T.Y., Dollár, P., Girshick, R., He, K., Hariharan, B., Belongie, S.: Feature pyramid networks for object detection. In: Proceedings of the IEEE Conference on Computer Vision Pattern Recognition, pp. 2117–2125 (2017)
23. Lin, T.Y., Goyal, P., Girshick, R., He, K., Dollár, P.: Focal loss for dense object detection. In: Proceedings of the IEEE Conference on Computer Vision Pattern Recognition, pp. 2980–2988 (2017)
24. Lin, T.-Y., et al.: Microsoft COCO: common objects in context. In: Fleet, D., Pajdla, T., Schiele, B., Tuytelaars, T. (eds.) ECCV 2014. LNCS, vol. 8693, pp. 740–755. Springer, Cham (2014). https://doi.org/10.1007/978-3-319-10602-1_48
25. Liu, F., Shen, C., Lin, G., Reid, I.: Learning depth from single monocular images using deep convolutional neural fields. IEEE Trans. Pattern Anal. Mach **38**, 2024–2039 (2016)
26. Liu, S., Qi, L., Qin, H., Shi, J., Jia, J.: Path aggregation network for instance segmentation. In: Proceedings of the IEEE Conference on Computer Vision Pattern Recognition, pp. 8759–8768 (2018)
27. Liu, Y., Shu, C., Wang, J., Shen, C.: Structured knowledge distillation for dense prediction. IEEE Trans. Pattern Anal. Mach. Intell. (2020)
28. Long, J., Shelhamer, E., Darrell, T.: Fully convolutional networks for semantic segmentation. In: Proceedings of the IEEE Conference on Computer Vision Pattern Recognition, pp. 3431–3440 (2015)
29. Abadi, M., et al.: TensorFlow: a system for large-scale machine learning. In: USENIX Symposium Operating Systems Design & Implementation (OSDI), pp. 265–283 (2016)

30. Milletari, F., Navab, N., Ahmadi, S.A.: V-Net: fully convolutional neural networks for volumetric medical image segmentation. In: Proceedings of the International Conference on 3D Vision (3DV), pp. 565–571. IEEE (2016)
31. Neven, D., Brabandere, B.D., Proesmans, M., Gool, L.V.: Instance segmentation by jointly optimizing spatial embeddings and clustering bandwidth. In: Proceedings of the IEEE Conference on Computer Vision Pattern Recognition, pp. 8837–8845 (2019)
32. Newell, A., Huang, Z., Deng, J.: Associative embedding: End-to-end learning for joint detection and grouping. In: Proceedings of the Advances in Neural Information Processing System, pp. 2277–2287 (2017)
33. Novotny, D., Albanie, S., Larlus, D., Vedaldi, A.: Semi-convolutional operators for instance segmentation. In: Ferrari, V., Hebert, M., Sminchisescu, C., Weiss, Y. (eds.) ECCV 2018. LNCS, vol. 11205, pp. 89–105. Springer, Cham (2018). https://doi.org/10.1007/978-3-030-01246-5_6
34. Perez, E., Strub, F., De Vries, H., Dumoulin, V., Courville, A.: Film: visual reasoning with a general conditioning layer. In: Proceedings of the AAAI Conference on Artificial Intelligent (2018)
35. Ren, S., He, K., Girshick, R., Sun, J.: Faster R-CNN: towards real-time object detection with region proposal networks. In: Proceedings of the Advances in Neural Information Processings System, pp. 91–99 (2015)
36. Sofiiuk, K., Barinova, O., Konushin, A.: AdaptIS: adaptive instance selection network. In: Proceedings of the IEEE International Conference on Computer Vision, pp. 7355–7363 (2019)
37. Tian, Z., He, T., Shen, C., Yan, Y.: Decoders matter for semantic segmentation: data-dependent decoding enables flexible feature aggregation. In: Proceedings of the IEEE Conference on Computer Vision and Pattern Recognition, pp. 3126–3135 (2019)
38. Tian, Z., Shen, C., Chen, H., He, T.: FCOS: fully convolutional one-stage object detection. In: Proceedings of the IEEE International Conference on Computer Vision, pp. 9627–9636 (2019)
39. Wang, X., Kong, T., Shen, C., Jiang, Y., Li, L.: SOLO: segmenting objects by locations. arXiv: Comp. Res. Repository (2019)
40. Wu, Y., Kirillov, A., Massa, F., Lo, W.Y., Girshick, R.: Detectron2. https://github.com/facebookresearch/detectron2 (2019)
41. Xie, E., et al.: PolarMask: single shot instance segmentation with polar representation. In: Proceedings of the IEEE Conference on Computer Vision Pattern Recognition (2020)
42. Yang, B., Bender, G., Le, Q.V., Ngiam, J.: Condconv: conditionally parameterized convolutions for efficient inference. In: Proceedings of the Advances in Neural Information Processing System, pp. 1305–1316 (2019)
43. Yin, W., Liu, Y., Shen, C., Yan, Y.: Enforcing geometric constraints of virtual normal for depth prediction. In: The IEEE International Conference on Computer Vision (ICCV) (2019)
44. Yin, W., et al.: DiverseDepth: affine-invariant depth prediction using diverse data. arXiv preprint arXiv:2002.00569 (2020)
45. Ying, H., Huang, Z., Liu, S., Shao, T., Zhou, K.: EmbedMask: embedding coupling for one-stage instance segmentation. arXiv preprint arXiv:1912.01954 (2019)

MutualNet: Adaptive ConvNet via Mutual Learning from Network Width and Resolution

Taojiannan Yang[1], Sijie Zhu[1], Chen Chen[1](✉), Shen Yan[2], Mi Zhang[2], and Andrew Willis[1]

[1] University of North Carolina at Charlotte, Charlotte, USA
{tyang30,szhu3,chen.chen,arwillis}@uncc.edu
[2] Michigan State University, East Lansing, USA
{yanshen6,mizhang}@msu.edu

Abstract. We propose the width-resolution mutual learning method (MutualNet) to train a network that is executable at dynamic resource constraints to achieve adaptive accuracy-efficiency trade-offs at runtime. Our method trains a cohort of sub-networks with different widths (i.e., number of channels in a layer) using different input resolutions to mutually learn multi-scale representations for each sub-network. It achieves consistently better ImageNet top-1 accuracy over the state-of-the-art adaptive network US-Net under different computation constraints, and outperforms the best compound scaled MobileNet in EfficientNet by 1.5%. The superiority of our method is also validated on COCO object detection and instance segmentation as well as transfer learning. Surprisingly, the training strategy of MutualNet can also boost the performance of a single network, which substantially outperforms the powerful AutoAugmentation in both efficiency (GPU search hours: 15000 vs. 0) and accuracy (ImageNet: 77.6% vs. 78.6%). *Code is available at* https://github.com/taoyang1122/MutualNet.

1 Introduction

Deep neural networks have triumphed over various perception tasks. However, deep networks usually require large computational resources, making them hard to deploy on mobile devices and embedded systems. This motivates research in reducing the redundancy in deep neural networks by designing efficient convolutional blocks [14,25,29,39] or pruning unimportant network connections [1,20,21]. However, these works ignore the fact that the computational cost is determined by both the network scale and input scale. Only focusing on reducing network scale cannot achieve the optimal accuracy-efficiency trade-off. EfficientNet [33] has acknowledged the importance of balancing among network depth,

Electronic supplementary material The online version of this chapter (https://doi.org/10.1007/978-3-030-58452-8_18) contains supplementary material, which is available to authorized users.

A. Vedaldi et al. (Eds.): ECCV 2020, LNCS 12346, pp. 299–315, 2020.
https://doi.org/10.1007/978-3-030-58452-8_18

Table 1. Comparison between our framework and previous works.

Model	Adaptive	Network Scale	Input Scale	Mutual Learning (NS&IS)
MobileNet [14,29]	✗	✓	✗	✗
ShuffleNet [25,39]	✗	✓	✗	✗
EfficientNet [33]	✗	✓	✓	✗
US-Net [36]	✓	✓	✗	✗
MutualNet (Ours)	✓	✓	✓	✓

width and resolution. But it considers network scale and input scale separately. The authors conduct grid search over different configurations and choose the best-performed one, while we argue that *network scale and input scale should be considered jointly in learning* to take full advantage of the information embedded in different configurations.

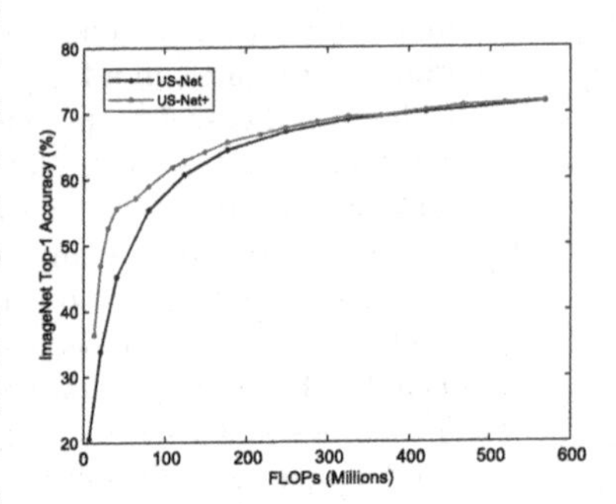

Fig. 1. Accuracy-FLOPs curves of US-Net+ and US-Net.

Another issue that prevents deep networks from practical deployment is that the resource budget (e.g., battery condition) varies in real-world applications, while traditional networks are only able to run at a specific constraint (e.g., FLOP). To address this issue, SlimNets [36,37] are proposed to train a single model to meet the varying resource budget at runtime. They only reduce the network width to meet lower resource budgets. As a result, the model performance drops dramatically as computational resource goes down. Here, we provide a concrete example to show the importance of balancing between input resolution and network width for achieving better accuracy-efficiency trade-offs. Specifically, to meet a dynamic resource constraint from 13 to 569 MFLOPs on MobileNet v1 backbone, US-Net [36] needs a network width range of $[0.05,1.0]\times$ given a 224×224 input, while this constraint can also be met via a network width of $[0.25,1.0]\times$ by adjusting the input resolution from {224, 192, 160, 128} during test time. We denote the latter model as US-Net+. As shown in Fig. 1, simply combining different resolutions with network widths *during inference* can achieve a better accuracy-efficiency trade-off than US-Net without additional efforts.

Inspired by the observations above, we propose a *mutual learning* scheme which incorporates both network width and input resolution into a unified learning framework. As depicted in Fig. 2, our framework feeds different sub-networks with different input resolutions. Since sub-networks share weights with each other, each sub-network can learn the knowledge shared by other sub-networks, which enables them to capture *multi-scale representations from both network level and input level.* Table 1 provides a comparison between our framework and previous works. In summary, we make the following contributions:

- We highlight the importance of input resolution for efficient network design. Previous works either ignore it or treat it independently from network structure. In contrast, we embed network width and input resolution in a unified mutual learning framework to learn a deep neural network (MutualNet) that can achieve adaptive accuracy-efficiency trade-offs at runtime.
- We carry out extensive experiments to demonstrate the effectiveness of our MutualNet. It significantly outperforms independently-trained networks and US-Net on various network structures, datasets and tasks under different constraints. To the best of our knowledge, we are the first to benchmark arbitrary-constraint adaptive networks on object detection and instance segmentation.
- We conduct comprehensive ablation studies to analyze the proposed mutual learning scheme. We further demonstrate that our framework is promising to serve as a plug-and-play strategy to boost the performance of a single network, which substantially outperforms the popular performance-boosting methods, e.g., data augmentations [7,9,22], SENet [15] and knowledge distillation [26].
- The proposed framework is a general training scheme and model-agnostic. It can be applied to any networks without making any adjustments to the structure. This makes it compatible with other state-of-the-art techniques (e.g., Neural Architecture Search (NAS) [13,32], AutoAugmentation [7,22]).

2 Related Work

Light-Weight Network. There has recently been a flurry of interest in designing light-weight networks. MobileNet [14] factorizes the standard 3×3 convolution into a 3×3 depthwise convolution and a 1×1 pointwise convolution which reduce computation cost by several times. ShuffleNet [39] separates the 1×1 convolution into group convolutions to further boost computation efficiency. MobileNet v2 [29] proposes the inverted residual and linear block for low-complexity networks. ShiftNet [35] introduces a zero-flop shift operation to reduce computation cost. Most recent works [13,32,34] also apply neural architecture search methods to search efficient networks. However, none of them considers the varying resource constraint during runtime in real-world applications. *To meet different resource budgets, these methods need to deploy several models and switch among them, which is not scalable.*

Adaptive Neural Networks. To meet the dynamic constraints in real-world applications, MSDNet [16] proposes a multi-scale and coarse to fine densenet framework. It has multiple classifiers and can make early predictions to meet varying resource demands. NestedNet [18] uses a nested sparse network which consists of multiple levels to enable nested learning. S-Net [37] introduces a 4-width framework to incorporate different complexities into one network, and proposes the switchable batch normalization for slimmable training. [27] leverages

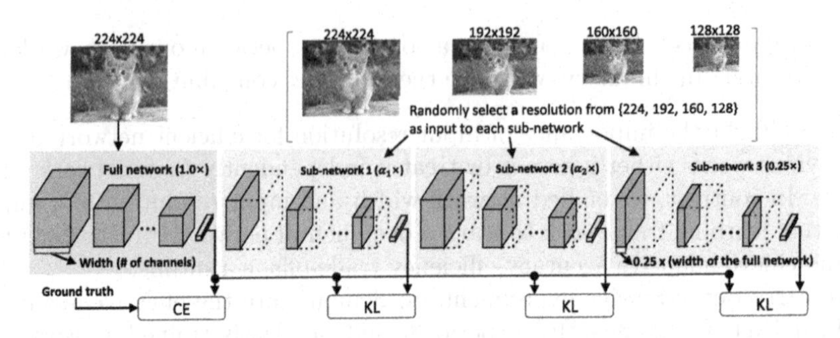

Fig. 2. The training process of our proposed MutualNet. The network width range is $[0.25, 1.0]\times$, input resolution is chosen from $\{224, 192, 160, 128\}$. This can achieve a computation range of [13, 569] MFLOPs on MobileNet v1 backbone. We follow the *sandwich rule* [36] to sample 4 networks, i.e., upper-bound full width network (1.0×), lower-bound width network (0.25×), and **two random width ratios** $\alpha_1, \alpha_2 \in (0.25, 1)$. For the full-network, we constantly choose 224×224 resolution. For the other three sub-networks, we randomly select its input resolution. The full-network is optimized with the ground-truth label. Sub-networks are optimized with the prediction of the full-network. Weights are shared among different networks to facilitate mutual learning. CE: Cross Entropy loss. KL: Kullback–Leibler Divergence loss.

knowledge distillation to train a multi-exit network. However, these approaches can only execute at a limited number of constraints. US-Net [36] can instantly adjust the runtime network width for arbitrary accuracy-efficiency trade-offs. But its performance degrades significantly as the budget lower bound goes down. [3] proposes progressive shrinking to finetune sub-networks from a well-trained large network, but the training process is complex and expensive.

Multi-scale Representation Learning. The effectiveness of multi-scale representation has been explored in various tasks. FPN [23] fuses pyramid features for object detection and segmentation. [17] proposes a multi-grid convolution to pass message across the scale space. HRNet [31] designs a multi-branch structure to exchange information across different resolutions. However, these works resort to the multi-branch fusion structure which is unfriendly to parallelization [25]. Our method does not modify the network structure, and the learned multi-scale representation is not only from image scale but also from network scale.

3 Methodology

3.1 Preliminary

Sandwich Rule. US-Net [36] trains a network that is executable at any resource constraint. The solution is to randomly sample several network widths for training and accumulate their gradients for optimization. However, the performance of the sub-networks is bounded by the smallest width (e.g., 0.25×) and the largest width (e.g., 1.0×). Thus, the *sandwich rule* is introduced to sample the smallest and largest widths plus two random ones for each training iteration.

Inplace Distillation. Knowledge distillation [12] is an effective method to transfer knowledge from a teacher network to a student network. Following the *sandwich rule*, since the largest network is sampled in each iteration, it is natural to use the largest network, which is supervised by the ground truth labels, as the teacher to guide smaller sub-networks in learning. This gives a better performance than training all sub-networks with ground truth labels.

Post-statistics of Batch Normalization (BN). US-Net proposes that each sub-network needs their own BN statistics (mean and variance), but it is insufficient to store the statistics of all the sub-networks. Therefore, US-Net collects BN statistics for the desired sub-network after training. Experimental results show that 2,000 samples are sufficient to get accurate BN statistics.

3.2 Rethinking Efficient Network Design

The computation cost of a vanilla convolution is $C_1 \times C_2 \times K \times K \times H \times W$, where C_1 and C_2 are the number of input and output channels, K is the kernel size, H and W are output feature map sizes. Most previous works only focus on reducing $C_1 \times C_2$. The most widely used group convolution decomposes standard convolution into groups to reduce the computation to $C_1 \times (C_2/g) \times K \times K \times H \times W$, where g is the number of groups. A larger g gives a lower computation but leads to higher memory access cost (MAC) [25], making the network inefficient in practical applications. Pruning methods [1,10,21] also only consider reducing structure redundancies.

In our approach, we shift the attention to reducing $H \times W$, i.e., lowering input resolution for the following reasons. First, as demonstrated in Fig. 1, balancing between width and resolution achieves better accuracy-efficiency trade-offs. Second, downsampling input resolution does not necessarily hurt the performance. It even sometimes benefits the performance. [6] points out that lower image resolution may produce better detection accuracy by reducing focus on redundant details. Third, different resolutions contain different information [5]. Lower resolution images may contain more global structures while higher resolution ones may encapsulate more fine-grained patterns. Learning multi-scale representations from different scaled images and features has been proven effective in previous works [17,23,31]. But these methods resort to a multi-branch structure which is unfriendly to parallelization [25]. Motivated by these observations, we propose a mutual learning framework to consider network scale and input resolution simultaneously for effective network accuracy-efficiency trade-offs.

3.3 Mutual Learning Framework

Sandwich Rule and Mutual Learning. As discussed in Sect. 3.2, different resolutions contain different information. We want to take advantage of this attribute to learn robust representations and better width-resolution trade-offs. The *sandwich rule* in US-Net can be viewed as a scheme of mutual learning [40] where an ensemble of networks are learned collaboratively. Since the sub-networks share weights with each other and are optimized jointly, they can transfer their knowledge to each other. Larger networks can take advantage of the features captured by smaller networks. Also, smaller networks can benefit from the stronger representation ability of larger networks. In light of this, we feed each sub-network with different input resolutions. By sharing knowledge, each sub-network is able to capture multi-scale representations.

Model Training. We present an example to illustrate our framework in Fig. 2. We train a network where its width ranges from 0.25× to 1.0×. We first follow the *sandwich rule* to sample four sub-networks, i.e., the smallest (0.25×), the largest (1.0×) and *two random width ratios* $\alpha_1, \alpha_2 \in (0.25, 1)$. Then, unlike traditional ImageNet training with 224×224 input, we resize the input image to four resolutions {224, 196, 160, 128} and feed them into different sub-networks. We denote the weights of a sub-network as $W_{0:w}$, where $w \in (0, 1]$ is the width of the sub-network and $0:w$ means the sub-network adopts the first $w \times 100\%$ weights of each layer of the full network. $I_{R=r}$ represents a $r \times r$ input image. Then $N(W_{0:w}, I_{R=r})$ represents the output of a sub-network with width w and input resolution $r \times r$. For the largest sub-network (i.e., the full-network in Fig. 2), we always train it with the highest resolution (224×224) and ground truth label y. The loss for the full network is

$$loss_{full} = CrossEntropy(N(W_{0:1}, I_{R=224}), y). \tag{1}$$

For the other sub-networks, we randomly pick an input resolution from {224, 196, 160, 128} and train it with the output of the full-network. The loss for the i-th sub-network is

$$loss_{sub_i} = KLDiv(N(W_{0:w_i}, I_{R=r_i}), N(W_{0:1}, I_{R=224})), \tag{2}$$

where $KLDiv$ is the Kullback-Leibler divergence. The total loss is the summation of the full-network and sub-networks, i.e.,

$$loss = loss_{full} + \sum_{i=1}^{3} loss_{sub_i}. \tag{3}$$

The reason for training the full-network with the highest resolution is that the highest resolution contains more details. Also, the full-network has the strongest learning ability to capture the discriminatory information from the image data.

Mutual Learning from Width and Resolution. In this part, we explain why the proposed framework can mutually learn from different widths and resolutions. For ease of demonstration, we only consider two network widths 0.4× and 0.8×, and two resolutions 128 and 192 in this example. As shown in Fig. 3, sub-network 0.4× selects input resolution 128, sub-network 0.8× selects input resolution 192. Then we can define the gradients for sub-network 0.4× and 0.8× as $\frac{\partial l_{W_{0:0.4}, I_{R=128}}}{\partial W_{0:0.4}}$ and $\frac{\partial l_{W_{0:0.8}, I_{R=192}}}{\partial W_{0:0.8}}$, respectively. Since sub-network 0.8× shares weights with 0.4×, we can decompose its gradient as

$$\frac{\partial l_{W_{0:0.8}, I_{R=192}}}{\partial W_{0:0.8}} = \frac{\partial l_{W_{0:0.8}, I_{R=192}}}{\partial W_{0:0.4}} \oplus \frac{\partial l_{W_{0:0.8}, I_{R=192}}}{\partial W_{0.4:0.8}} \tag{4}$$

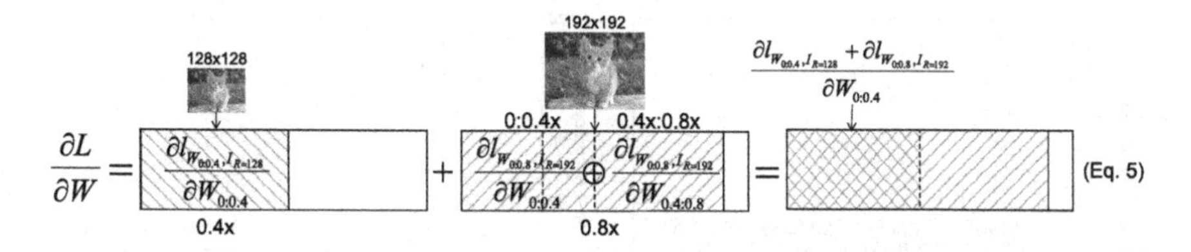

Fig. 3. Illustration of the mutual learning from network width and input resolution.

where $\oplus$ is vector concatenation. Since the gradients of the two sub-networks are accumulated during training, the total gradients are computed as

$$\begin{aligned}\frac{\partial L}{\partial W} &= \frac{\partial l_{W_{0:0.4},I_{R=128}}}{\partial W_{0:0.4}} + \frac{\partial l_{W_{0:0.8},I_{R=192}}}{\partial W_{0:0.8}} \\ &= \frac{\partial l_{W_{0:0.4},I_{R=128}}}{\partial W_{0:0.4}} + \left(\frac{\partial l_{W_{0:0.8},I_{R=192}}}{\partial W_{0:0.4}} \oplus \frac{\partial l_{W_{0:0.8},I_{R=192}}}{\partial W_{0.4:0.8}}\right) \\ &= \frac{\partial l_{W_{0:0.4},I_{R=128}} + \partial l_{W_{0:0.8},I_{R=192}}}{\partial W_{0:0.4}} \oplus \frac{\partial l_{W_{0:0.8},I_{R=192}}}{\partial W_{0.4:0.8}}\end{aligned} \tag{5}$$

Therefore, the gradient for sub-network 0.4$\times$ is $\frac{\partial l_{W_{0:0.4},I_{R=128}} + \partial l_{W_{0:0.8},I_{R=192}}}{\partial W_{0:0.4}}$, and it consists of two parts. The first part is derived from itself (0 : 0.4$\times$) with 128 input resolution. The second part is derived from sub-network 0.8$\times$ (i.e., 0 : 0.4$\times$ portion) with 192 input resolution. Thus the sub-network is able to capture multi-scale representations from different input resolutions and network scales. Due to the random sampling of network width, every sub-network is able to learn multi-scale representations in our framework.

Model Inference. The trained model is executable at various width-resolution configurations. The goal is to find the best configuration under a particular resource constraint. A simple way to achieve this is via a query table. For example, in MobileNet v1, we sample network width from 0.25$\times$ to 1.0$\times$ with a step-size of 0.05$\times$, and sample network resolution from {224, 192, 160, 128}. We test all these width-resolution configurations on a validation set and choose the best one under a given constraint (FLOPs or latency). Since there is no re-training, the whole process is once for all.

4 Experiments

In this section, we first present our results on ImageNet [8] classification to illustrate the effectiveness of MutualNet. Next, we conduct extensive ablation studies to analyze the mutual learning scheme. Finally, we apply MutualNet to transfer learning datasets and COCO [24] object detection and instance segmentation to demonstrate its robustness and generalization ability.

4.1 Evaluation on ImageNet Classification

We compare our MutualNet with US-Net and independently-trained networks on the ImageNet dataset. We evaluate our framework on two popular light-weight

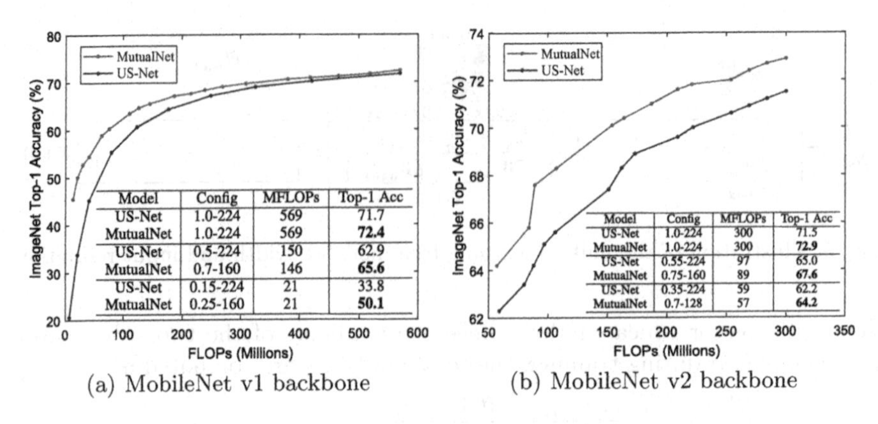

Model	Config	MFLOPs	Top-1 Acc
US-Net	1.0-224	569	71.7
MutualNet	1.0-224	569	**72.4**
US-Net	0.5-224	150	62.9
MutualNet	0.7-160	146	**65.6**
US-Net	0.15-224	21	33.8
MutualNet	0.25-160	21	**50.1**

(a) MobileNet v1 backbone

Model	Config	MFLOPs	Top-1 Acc
US-Net	1.0-224	300	71.5
MutualNet	1.0-224	300	**72.9**
US-Net	0.55-224	97	65.0
MutualNet	0.75-160	89	**67.6**
US-Net	0.35-224	59	62.2
MutualNet	0.7-128	57	**64.2**

(b) MobileNet v2 backbone

Fig. 4. Accuracy-FLOPs curves of our proposed MutualNet and US-Net. (a) is based on MobileNet v1 backbone. (b) is based on MobileNet v2 backbone.

structures, MobileNet v1 [14] and MobileNet v2 [29]. These two networks also represent non-residual and residual structures respectively.

Implementation Details. We compare with US-Net under the **same** dynamic FLOPs constraints ([13, 569] MFLOPs on MobileNet v1 and [57, 300] MFLOPs on MobileNet v2). US-Net uses width scale $[0.05, 1.0]\times$ on MobileNet v1 and $[0.35, 1.0]\times$ on MobileNet v2 based on the 224×224 input resolution. **To meet the same dynamic constraints**, our method uses width scale $[0.25, 1.0]\times$ on MobileNet v1 and $[0.7, 1.0]\times$ on MobileNet v2 with downsampled input resolutions {224, 192, 160, 128}. Due to the lower input resolutions, our method is able to use higher width lower bounds (i.e., $0.25\times$ and $0.7\times$) than US-Net. The other training settings are the same as US-Net.

Comparison with US-Net. We first compare our framework with US-Net on MobileNet v1 and MobileNet v2 backbones. The Accuracy-FLOPs curves are shown in Fig. 4. We can see that our framework consistently outperforms US-Net on both MobileNet v1 and MobileNet v2 backbones. Specifically, we achieve significant improvements under small computation costs. This is because our framework considers both network width and input resolution, and can find a better balance between them. For example, if the resource constraint is 150 MFLOPs, US-Net has to reduce the width to $0.5\times$ given its constant input resolution 224, while our MutualNet can meet this budget by a balanced configuration of (0.7×-160), leading to a better accuracy (65.6% (Ours) vs. 62.9% (US-Net) as listed in the table of Fig. 4(a)). On the other hand, our framework is able to learn multi-scale representations which further boost the performance of each sub-network. We can see that even for the same configuration (e.g., 1.0×-224) our approach clearly outperforms US-Net, i.e., 72.4% (Ours) vs. 71.7% (US-Net) on MobileNet v1, and 72.9% (Ours) vs. 71.5% (US-Net) on MobileNet v2 (Fig. 4).

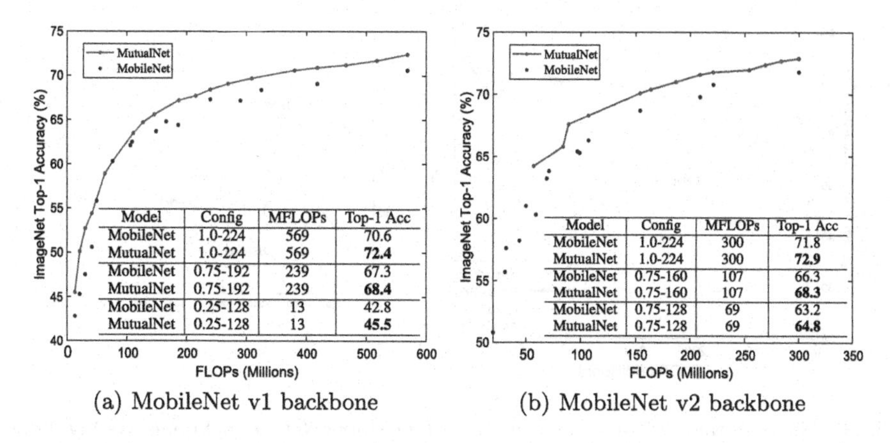

Model	Config	MFLOPs	Top-1 Acc
MobileNet	1.0-224	569	70.6
MutualNet	1.0-224	569	**72.4**
MobileNet	0.75-192	239	67.3
MutualNet	0.75-192	239	**68.4**
MobileNet	0.25-128	13	42.8
MutualNet	0.25-128	13	**45.5**

(a) MobileNet v1 backbone

Model	Config	MFLOPs	Top-1 Acc
MobileNet	1.0-224	300	71.8
MutualNet	1.0-224	300	**72.9**
MobileNet	0.75-160	107	66.3
MutualNet	0.75-160	107	**68.3**
MobileNet	0.75-128	69	63.2
MutualNet	0.75-128	69	**64.8**

(b) MobileNet v2 backbone

Fig. 5. Accuracy-FLOPs curves of our MutualNet and independently-trained MobileNets. (a) is MobileNet v1 backbone. (b) is MobileNet v2 backbone. The results for different MobileNets configurations are taken from the papers [14,29].

Comparison with Independently Trained Networks. Different scaled MobileNets are trained separately in [14,29]. The authors consider width and resolution as independent factors, thus cannot leverage the information contained in different configurations. We compare the performance of MutualNet with independently-trained MobileNets under different width-resolution configurations in Fig. 5. For MobileNet v1, widths are selected from {1.0×, 0.75×, 0.5×, 0.25×}, and resolutions are selected from {224, 192, 160, 128}, leading to 16 configurations in total. Similarly, MobileNet v2 selects configurations from {1.0×, 0.75×, 0.5×, 0.35×} and {224, 192, 160, 128}. From Fig. 5, our framework consistently outperforms MobileNets. Even for the same width-resolution configuration (although it may not be the best configuration MutualNet finds at that specific constraint), MutualNet can achieve much better performance. This demonstrates that MutualNet not only finds the better width-resolution balance but also learns stronger representations by the mutual learning scheme.

4.2 Ablation Study

Balanced Width-Resolution Configuration via Mutual Learning. As evident in Fig. 1, we can apply different resolutions to US-Net *during inference* to yield improvement over the original US-Net. However, this cannot achieve the optimal width-resolution balance due to lack of width-resolution mutual learning. In the experiment, we test US-Net at width scale [0.25, 1.0]× with input resolutions {224, 192, 160, 128} and denote this improved model as US-Net+. In Fig. 6, we plot the Accuracy-FLOPs curves of our method and US-Net+ based on MobileNet v1 backbone, and highlight the selected input resolutions with different colors. As we decrease the FLOPs (569 → 468 MFLOPs), our MutualNet first reduces network width to meet the constraint while keeping the 224 × 224 resolution (red line in Fig. 6). After 468 MFLOPs, MutualNet

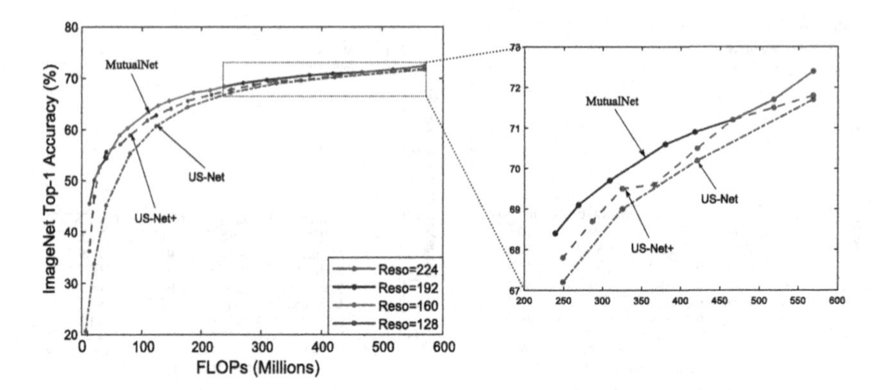

Fig. 6. The Accuracy-FLOPs curves are based on MobileNet v1 backbone. We highlight the selected resolution under different FLOPs with different colors. For example, the solid green line indicates when the constraint range is [41, 215] MFLOPs, our method constantly selects input resolution 160 but reduces the width to meet the resource constraint. Best viewed in color. (Color figure online)

selects lower input resolution (192) and continues reducing the width to meet the constraint. On the other hand, US-Net+ cannot find such balance. It always slims the network width and uses the same (224) resolution as the FLOPs decreasing until it goes to really low. This is because US-Net+ does not incorporate input resolution into the learning framework. Simply applying different resolutions during inference cannot achieve the optimal width-resolution balance.

Difference with EfficientNet. EfficientNet acknowledges the importance of balancing among network width, depth and resolution. But they are considered as independent factors. The authors use grid search over these three dimensions and train each configuration independently to find the optimal one under certain constraint, while our MutualNet incorporates width and resolution in a unified framework. We compare with the best model scaling EfficientNet finds for MobileNet v1 at 2.3 BFLOPs (scale up baseline by 4.0×). Similar to this scale setting, we train our framework with a width range of [1.0×, 2.0×] (scaled by [1.0×,4.0×]), and select resolutions from {224, 256, 288, 320}. This makes MutualNet executable in the range of [0.57, 4.5] BFLOPs. We pick the best performed width-resolution configuration at 2.3 BFLOPs. The results are compared in Table 2. MutualNet achieves significantly better performance than EfficientNet because it can capture multi-scale representations for each model scaling due to the width-resolution mutual learning.

Difference with Multi-scale Data Augmentation. In multi-scale data augmentation, the network may take images of different resolutions in different iterations. But within each iteration, the network weights are still optimized in the same resolution direction. While our method randomly samples several subnetworks which share weights with each other. Since sub-networks can select different image resolutions, the weights are optimized in a mixed resolution

Table 2. ImageNet Top-1 accuracy on MobileNet v1 backbone. d: depth, w: width, r: resolution.

Model	Best model scaling	FLOPs	Top-1 Acc
EfficientNet [33]	$d = 1.4, w = 1.2, r = 1.3$	2.3B	75.6%
MutualNet	$w = 1.6, r = 1.3$	2.3B	**77.1%**

direction in each iteration as illustrated in Fig. 3. This enables each sub-network to effectively learn multi-scale representations from both network width and resolution. To validate the superiority of our mutual learning scheme, we apply multi-scale data augmentation to MobileNet and US-Net and explain the difference with MutualNet.

MobileNet + Multi-scale Data Augmentation. We train MobileNet v2 (1.0× width) with multi-scale images. To have a fair comparison, input images are randomly sampled from scales {224, 192, 160, 128} and the other settings are the same as MutualNet. As shown in Table 3, multi-scale data augmentation only marginally improves the baseline (MobileNet v2) while our MutualNet (MobileNet v2 backbone) clearly outperforms both of them by considerable margins.

Table 3. Comparison between MutualNet and multi-scale data augmentation.

Model	ImageNet Top-1 Acc
MobileNet v2 (1.0× - 224) - Baseline	71.8%
Baseline + Multi-scale data augmentation	72.0%
MutualNet (MobileNet v2 backbone)	**72.9%**

US-Net + Multi-scale Data Augmentation. Different from our framework which feeds different scaled images to different sub-networks, in this experiment, we *randomly* choose a scale from {224, 192, 160, 128} and feed the *same* scaled image to all sub-networks in each iteration. That is each sub-network takes the same image resolution. In this way, the weights are still optimized towards a single resolution direction in each iteration. For example, as illustrated in Fig. 3, the gradient of the sub-network 0.4× in MutualNet is $\frac{\partial l_{W_{0:0.4}, I_{R=128}} + \partial l_{W_{0:0.8}, I_{R=192}}}{\partial W_{0:0.4}}$, while in US-Net + multi-scale it would be $\frac{\partial l_{W_{0:0.4}, I_{R=128}} + \partial l_{W_{0:0.8}, I_{R=128}}}{\partial W_{0:0.4}}$. With more sub-networks and input scales, the difference between their gradient flows becomes more distinct. As shown in Fig. 7, our method clearly performs better than *US-Net + multi-scale data augmentation* over the entire FLOPs spectrum. The experiment is based on MobileNet v2 with the same settings as in Sect. 4.1. *These experiments demonstrate that the improvement comes from our mutual learning scheme rather than the multi-scale data augmentation.*

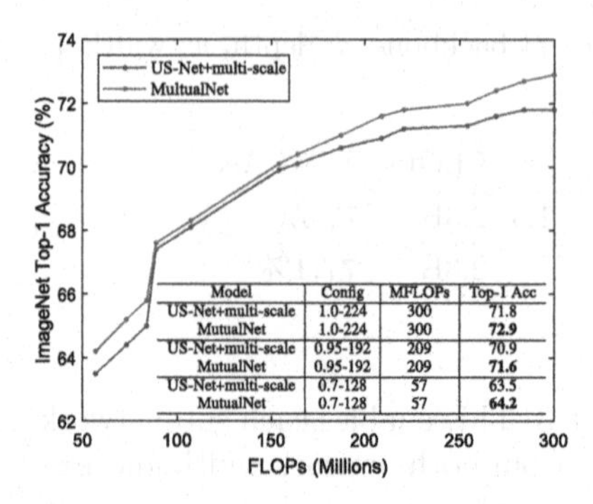

Fig. 7. MutualNet and US-Net + multi-scale data augmentation.

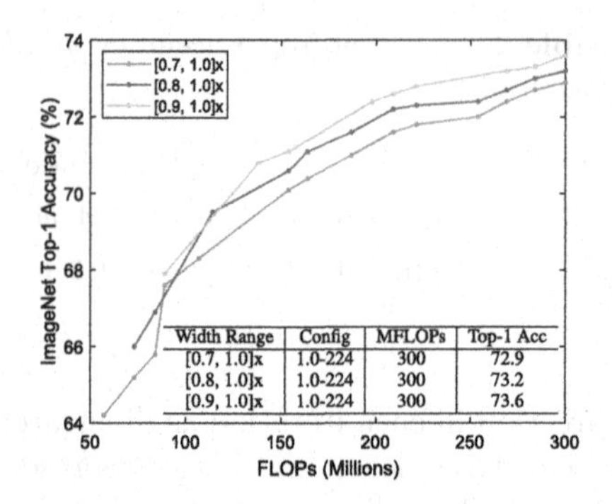

Fig. 8. Accuracy-FLOPs curves of different width lower bounds.

Table 4. Top-1 Accuracy (%) on Cifar-10 and Cifar-100.

WideResNet-28-10	GPU search hours	C-10	C-100
Baseline	0	96.1	81.2
Cutout [9]	0	96.9	81.6
AA [7]	5000	**97.4**	82.9
Fast AA [22]	3.5	97.3	82.7
MutualNet	0	97.2	**83.8**

Table 5. Top-1 Accuracy (%) on ImageNet.

ResNet-50	Additional cost	Top-1 Acc
Baseline	\	76.5
Cutout [9]	\	77.1
KD [26]	Teacher network	76.5
SENet [15]	SE block	77.6
AA [7]	15000 GPU hours	77.6
Fast AA [22]	450 GPU hours	77.6
MutualNet	\	**78.6**

Effects of Width Lower Bound. The executable constraint range and model performance are affected by the width lower bound. To study its effects, we conduct experiments with three different lower bounds (0.7×, 0.8×, 0.9×) on MobileNet v2. The results in Fig. 8 show that a higher lower bound gives better overall performance, but the executable range is narrower. One interesting observation is that the performance of the full-network (1.0×-224) is also largely improved as the width lower bound increases from 0.7× to 0.9×. This property is not observed in US-Net. We attribute this to the robust and well-generalized multi-scale representations which can be effectively re-used by the full-network, while in US-Net, the full-network cannot effectively benefit from sub-networks.

Boosting Single Network Performance. As discussed above, the performance of the full-network is greatly improved as we increase the width lower bound. Therefore, we apply our training framework to improve the performance of a single full network. We compare our method with the popular performance-boosting techniques (e.g., AutoAugmentation (AA) [7,22], SENet [15] and Knowledge Distillation (KD) [26]) to show its superiority. We conduct experiments using Wide-ResNet [38] on Cifar-10 and Cifar-100 and ResNet-50 on ImageNet. MutualNet adopts the width range [0.9, 1.0]× as it achieves the

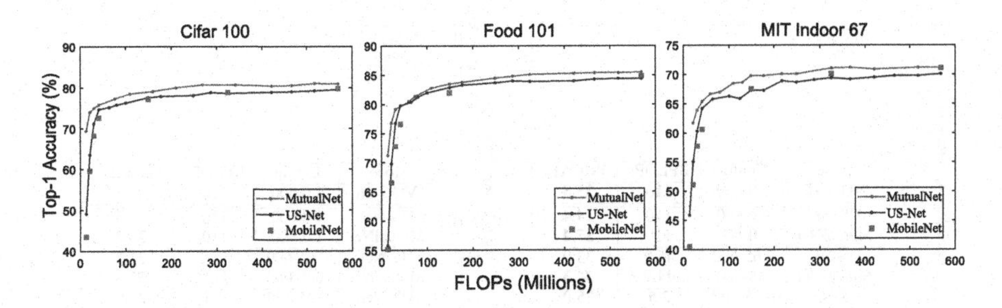

Fig. 9. Accuracy-FLOPs curves on different transfer learning datasets.

best-performed full-network in Fig. 8. The resolution is sampled from {32, 28, 24, 20} on Cifar-10 and Cifar-100 and {224, 192, 160, 128} on ImageNet. Wide-ResNet is trained for 200 epochs following [38]. ResNet is trained for 120 epochs. The results are compared in Table 4 and Table 5. MutualNet achieves substantial improvements over other techniques. It is important to note that MutualNet is a **general training scheme** which does not need the expensive searching procedure or additional network blocks or stronger teacher networks. Moreover, MutualNet training is as easy as the regular training process, and is orthogonal to other performance-boosting techniques, e.g., AutoAugmentation [7,22]. Therefore, it can be easily combined with those methods.

4.3 Transfer Learning

To evaluate the representations learned by our method, we further conduct experiments on three popular transfer learning datasets, Cifar-100 [19], Food-101 [2] and MIT-Indoor67 [28]. Cifar-100 is superordinate-level object classification, Food-101 is fine-grained classification and MIT-Indoor67 is scene classification. Such large variety is suitable to evaluate the robustness of the learned representations. We compare our approach with US-Net and MobileNet v1. We fine-tune ImageNet pre-trained models with a batch size of 256, initial learning rate of 0.1 with cosine decay schedule and a total of 100 epochs. Both MutualNet and US-Net are trained with width range [0.25, 1.0]× and tested with resolutions from {224, 192, 160, 128}. The results are shown in Fig. 9. Again, our MutualNet achieves consistently better performance than US-Net and MobileNet. This verifies that MutualNet is able to learn well-generalized representations.

4.4 Object Detection and Instance Segmentation

We also evaluate our method on COCO object detection and instance segmentation [24]. The experiments are based on Mask-RCNN-FPN [11,23] and MMDetection [4] toolbox on VGG-16 [30] backbone. We first pre-train VGG-16 on ImageNet following US-Net and MutualNet respectively. Both methods are trained with width range [0.25, 1.0]×. Then we fine-tune the pre-trained models on COCO. The FPN neck and detection head are shared among different

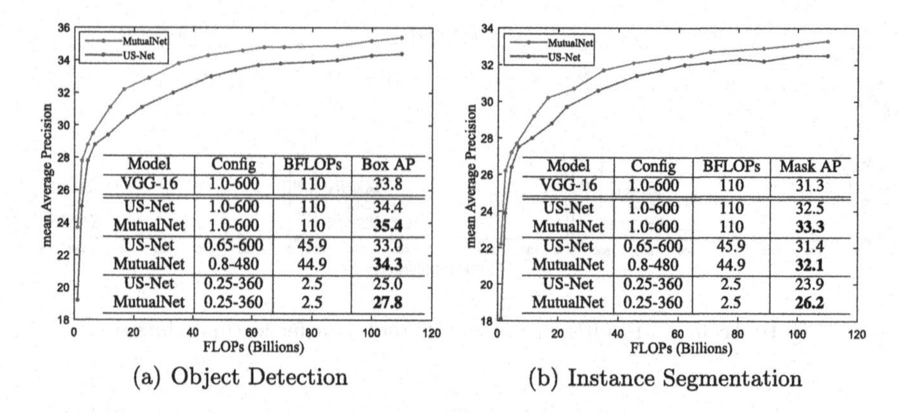

Model	Config	BFLOPs	Box AP
VGG-16	1.0-600	110	33.8
US-Net	1.0-600	110	34.4
MutualNet	1.0-600	110	**35.4**
US-Net	0.65-600	45.9	33.0
MutualNet	0.8-480	44.9	**34.3**
US-Net	0.25-360	2.5	25.0
MutualNet	0.25-360	2.5	**27.8**

Model	Config	BFLOPs	Mask AP
VGG-16	1.0-600	110	31.3
US-Net	1.0-600	110	32.5
MutualNet	1.0-600	110	**33.3**
US-Net	0.65-600	45.9	31.4
MutualNet	0.8-480	44.9	**32.1**
US-Net	0.25-360	2.5	23.9
MutualNet	0.25-360	2.5	**26.2**

Fig. 10. Average Precision - FLOPs curves of MutualNet and US-Net.

Fig. 11. Object detection and instance segmentation examples.

sub-networks. For simplicity, we don't use *inplace distillation*. Rather, each sub-network is trained with the ground truth. The other training procedures are the same as training ImageNet classification. Following common settings in object detection, US-Net is trained with image resolution 1000×600. Our method randomly selects resolutions from $1000 \times \{600, 480, 360, 240\}$. All models are trained with $2\times$ schedule for better convergence and tested with different image resolutions. The mean Average Precision (AP at IoU = 0.50:0.05:0.95) are presented in Fig. 10. These results reveal that our MutualNet significantly outperforms US-Net under all resource constraints. Specifically, for the full network (1.0×-600), MutualNet significantly outperforms both US-Net and independent network. This again validates the effectiveness of our width-resolution mutual learning scheme. Figure 11 provides some visual examples which reveal that MutualNet is more robust to small-scale and large-scale objects than US-Net.

5 Conclusion and Future Work

This paper highlights the importance of simultaneously considering network width and input resolution for efficient network design. A new framework namely

MutualNet is proposed to mutually learn from network width and input resolution for adaptive accuracy-efficiency trade-offs. Extensive experiments have shown that it significantly improves inference performance per FLOP on various datasets and tasks. The mutual learning scheme is also an effective training strategy for boosting single network performance. The generality of the proposed framework allows it to translate well to generic problem domains.

Acknowledgements. This material is based upon work supported by the National Science Foundation under Grant CNS-1910844.

References

1. Anwar, S., Hwang, K., Sung, W.: Structured pruning of deep convolutional neural networks. ACM J. Emerg. Technolo. Comput. Syst. (JETC) **13**(3), 32 (2017)
2. Bossard, L., Guillaumin, M., Van Gool, L.: Food-101 – mining discriminative components with random forests. In: Fleet, D., Pajdla, T., Schiele, B., Tuytelaars, T. (eds.) ECCV 2014. LNCS, vol. 8694, pp. 446–461. Springer, Cham (2014). https://doi.org/10.1007/978-3-319-10599-4_29
3. Cai, H., Gan, C., Wang, T., Zhang, Z., Han, S.: Once for all: train one network and specialize it for efficient deployment. In: International Conference on Learning Representations (2020). https://openreview.net/forum?id=HylxE1HKwS
4. Chen, K., et al.: MMDetection: open MMLab detection toolbox and benchmark. arXiv preprint arXiv:1906.07155 (2019)
5. Chen, Y., et al.: Drop an octave: reducing spatial redundancy in convolutional neural networks with octave convolution. arXiv preprint arXiv:1904.05049 (2019)
6. Chin, T.W., Ding, R., Marculescu, D.: AdaScale: towards real-time video object detection using adaptive scaling. arXiv preprint arXiv:1902.02910 (2019)
7. Cubuk, E.D., Zoph, B., Mane, D., Vasudevan, V., Le, Q.V.: AutoAugment: learning augmentation strategies from data. In: Proceedings of the IEEE Conference on Computer Vision and Pattern Recognition, pp. 113–123 (2019)
8. Deng, J., Dong, W., Socher, R., Li, L.J., Li, K., Fei-Fei, L.: ImageNet: a large-scale hierarchical image database. In: 2009 IEEE Conference on Computer Vision and Pattern Recognition, pp. 248–255. IEEE (2009)
9. DeVries, T., Taylor, G.W.: Improved regularization of convolutional neural networks with cutout. arXiv preprint arXiv:1708.04552 (2017)
10. Han, S., Pool, J., Tran, J., Dally, W.: Learning both weights and connections for efficient neural network. In: Advances in Neural Information Processing Systems, pp. 1135–1143 (2015)
11. He, K., Gkioxari, G., Dollár, P., Girshick, R.: Mask R-CNN. In: Proceedings of the IEEE International Conference on Computer Vision, pp. 2961–2969 (2017)
12. Hinton, G., Vinyals, O., Dean, J.: Distilling the knowledge in a neural network. arXiv preprint arXiv:1503.02531 (2015)
13. Howard, A., et al.: Searching for mobileNetV3. In: Proceedings of the IEEE International Conference on Computer Vision, pp. 1314–1324 (2019)
14. Howard, A.G., et al.: MobileNets: efficient convolutional neural networks for mobile vision applications. arXiv preprint arXiv:1704.04861 (2017)
15. Hu, J., Shen, L., Sun, G.: Squeeze-and-excitation networks. In: Proceedings of the IEEE Conference on Computer Vision and Pattern Recognition, pp. 7132–7141 (2018)

16. Huang, G., Chen, D., Li, T., Wu, F., van der Maaten, L., Weinberger, K.Q.: Multi-scale dense networks for resource efficient image classification. arXiv preprint arXiv:1703.09844 (2017)
17. Ke, T.W., Maire, M., Yu, S.X.: Multigrid neural architectures. In: Proceedings of the IEEE Conference on Computer Vision and Pattern Recognition, pp. 6665–6673 (2017)
18. Kim, E., Ahn, C., Oh, S.: NestedNet: learning nested sparse structures in deep neural networks. In: Proceedings of the IEEE Conference on Computer Vision and Pattern Recognition, pp. 8669–8678 (2018)
19. Krizhevsky, A.: Learning multiple layers of features from tiny images. Technical report (2009)
20. Lemaire, C., Achkar, A., Jodoin, P.M.: Structured pruning of neural networks with budget-aware regularization. arXiv preprint arXiv:1811.09332 (2018)
21. Li, H., Kadav, A., Durdanovic, I., Samet, H., Graf, H.P.: Pruning filters for efficient ConvNets. arXiv preprint arXiv:1608.08710 (2016)
22. Lim, S., Kim, I., Kim, T., Kim, C., Kim, S.: Fast AutoAugment. In: Advances in Neural Information Processing Systems, pp. 6662–6672 (2019)
23. Lin, T.Y., Dollár, P., Girshick, R., He, K., Hariharan, B., Belongie, S.: Feature pyramid networks for object detection. In: Proceedings of the IEEE Conference on Computer Vision and Pattern Recognition, pp. 2117–2125 (2017)
24. Lin, T.Y., et al.: Microsoft COCO: common objects in context. In: Fleet, D., Pajdla, T., Schiele, B., Tuytelaars, T. (eds.) ECCV 2014. LNCS, vol. 8693, pp. 740–755. Springer, Cham (2014). https://doi.org/10.1007/978-3-319-10602-1_48
25. Ma, N., Zhang, X., Zheng, H.-T., Sun, J.: ShuffleNet V2: practical guidelines for efficient CNN architecture design. In: Ferrari, V., Hebert, M., Sminchisescu, C., Weiss, Y. (eds.) Computer Vision – ECCV 2018. LNCS, vol. 11218, pp. 122–138. Springer, Cham (2018). https://doi.org/10.1007/978-3-030-01264-9_8
26. Mishra, A., Marr, D.: Apprentice: using knowledge distillation techniques to improve low-precision network accuracy. In: International Conference on Learning Representations (2018). https://openreview.net/forum?id=B1ae1lZRb
27. Phuong, M., Lampert, C.H.: Distillation-based training for multi-exit architectures. In: The IEEE International Conference on Computer Vision (ICCV), October 2019
28. Quattoni, A., Torralba, A.: Recognizing indoor scenes. In: 2009 IEEE Conference on Computer Vision and Pattern Recognition, pp. 413–420 (2009)
29. Sandler, M., Howard, A., Zhu, M., Zhmoginov, A., Chen, L.C.: MobileNetV2: inverted residuals and linear bottlenecks. In: Proceedings of the IEEE Conference on Computer Vision and Pattern Recognition, pp. 4510–4520 (2018)
30. Simonyan, K., Zisserman, A.: Very deep convolutional networks for large-scale image recognition. arXiv preprint arXiv:1409.1556 (2014)
31. Sun, K., Xiao, B., Liu, D., Wang, J.: Deep high-resolution representation learning for human pose estimation. In: Proceedings of the IEEE Conference on Computer Vision and Pattern Recognition, pp. 5693–5703 (2019)
32. Tan, M., et al.: MnasNet: platform-aware neural architecture search for mobile. In: Proceedings of the IEEE Conference on Computer Vision and Pattern Recognition, pp. 2820–2828 (2019)
33. Tan, M., Le, Q.: EfficientNet: rethinking model scaling for convolutional neural networks. In: Chaudhuri, K., Salakhutdinov, R. (eds.) Proceedings of the 36th International Conference on Machine Learning. Proceedings of Machine Learning Research, vol. 97, pp. 6105–6114. PMLR, Long Beach, 09–15 June 2019. http://proceedings.mlr.press/v97/tan19a.html

34. Wu, B., et al.: FBNet: hardware-aware efficient convnet design via differentiable neural architecture search. In: Proceedings of the IEEE Conference on Computer Vision and Pattern Recognition, pp. 10734–10742 (2019)
35. Wu, B., et al.: Shift: a zero flop, zero parameter alternative to spatial convolutions. In: Proceedings of the IEEE Conference on Computer Vision and Pattern Recognition, pp. 9127–9135 (2018)
36. Yu, J., Huang, T.: Universally slimmable networks and improved training techniques. arXiv preprint arXiv:1903.05134 (2019)
37. Yu, J., Yang, L., Xu, N., Yang, J., Huang, T.: Slimmable neural networks. In: International Conference on Learning Representations (2019). https://openreview.net/forum?id=H1gMCsAqY7
38. Zagoruyko, S., Komodakis, N.: Wide residual networks. In: Richard, C., Wilson, E.R.H., Smith, W.A.P. (eds.) Proceedings of the British Machine Vision Conference (BMVC), pp. 87.1–87.12. BMVA Press, September 2016. https://doi.org/10.5244/C.30.87. https://dx.doi.org/10.5244/C.30.87
39. Zhang, X., Zhou, X., Lin, M., Sun, J.: ShuffleNet: an extremely efficient convolutional neural network for mobile devices. In: Proceedings of the IEEE Conference on Computer Vision and Pattern Recognition, pp. 6848–6856 (2018)
40. Zhang, Y., Xiang, T., Hospedales, T.M., Lu, H.: Deep mutual learning. In: Proceedings of the IEEE Conference on Computer Vision and Pattern Recognition, pp. 4320–4328 (2018)

Fashionpedia: Ontology, Segmentation, and an Attribute Localization Dataset

Menglin Jia[1(✉)], Mengyun Shi[1,4], Mikhail Sirotenko[3], Yin Cui[3], Claire Cardie[1], Bharath Hariharan[1], Hartwig Adam[3], and Serge Belongie[1,2]

[1] Cornell University, Ithaca, USA
mj493@cornell.edu
[2] Cornell Tech, New York, USA
[3] Google Research, New York, USA
[4] Hearst Magazines, New York, USA

Abstract. In this work we explore the task of *instance segmentation with attribute localization*, which unifies instance segmentation (detect and segment each object instance) and fine-grained visual attribute categorization (recognize one or multiple attributes). The proposed task requires both localizing an object and describing its properties. To illustrate the various aspects of this task, we focus on the domain of fashion and introduce *Fashionpedia* as a step toward mapping out the visual aspects of the fashion world. Fashionpedia consists of two parts: (1) an ontology built by fashion experts containing 27 main apparel categories, 19 apparel parts, 294 fine-grained attributes and their relationships; (2) a dataset with everyday and celebrity event fashion images annotated with segmentation masks and their associated per-mask fine-grained attributes, built upon the Fashionpedia ontology. In order to solve this challenging task, we propose a novel Attribute-Mask R-CNN model to jointly perform instance segmentation and localized attribute recognition, and provide a novel evaluation metric for the task. Fashionpedia is available at: https://fashionpedia.github.io/home/.

Keywords: Dataset · Ontology · Instance segmentation · Fine-grained · Attribute · Fashion

1 Introduction

Recent progress in the field of computer vision has advanced machines' ability to recognize and understand our visual world, showing significant impacts in fields including autonomous driving [52], product recognition [14,32], *etc*. These real-world applications are fueled by various visual understanding tasks with the goals of *naming*, *describing (attribute recognition)*, or *localizing* objects within an image.

M. Jia, M. Shi, M. Sirotenko and Y. Cui—Equal Contribution.

Electronic supplementary material The online version of this chapter (https://doi.org/10.1007/978-3-030-58452-8_19) contains supplementary material, which is available to authorized users.

A. Vedaldi et al. (Eds.): ECCV 2020, LNCS 12346, pp. 316–332, 2020.
https://doi.org/10.1007/978-3-030-58452-8_19

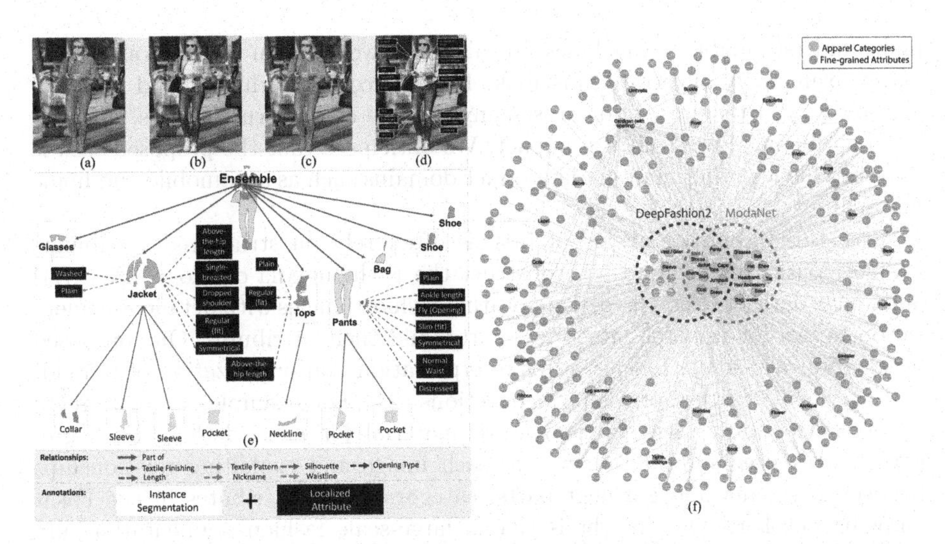

Fig. 1. An illustration of the Fashionpedia dataset and ontology: (a) main garment masks; (b) garment part masks; (c) both main garment and garment part masks; (d) fine-grained apparel attributes; (e) an exploded view of the annotation diagram: the image is annotated with both instance segmentation masks *(white boxes)* and per-mask fine-grained attributes *(black boxes)*; (f) visualization of the Fashionpedia ontology: we created Fashionpedia ontology and separate the concept of categories *(yellow nodes)* and attributes [38] *(blue nodes)* in fashion. It covers pre-defined garment categories used by both Deepfashion2 [11] and ModaNet [53]. Mapping with DeepFashion2 also shows the versatility of using attributes and categories. We are able to present all 13 garment classes in DeepFashion2 with 11 main garment categories, 1 garment part, and 7 attributes

Naming and localizing objects is formulated as an object detection task (Fig. 1(a–c)). As a hallmark for computer recognition, this task is to identify and indicate the boundaries of objects in the form of bounding boxes or segmentation masks [12,17,37]. *Attribute recognition* [6,9,29,36] (Fig. 1(d)) instead focuses on describing and comparing objects, since an object also has many other properties or attributes in addition to its category. Attributes not only provide a compact and scalable way to represent objects in the world, as pointed out by Ferrari and Zisserman [9], attribute learning also enables transfer of existing knowledge to novel classes. This is particularly useful for fine-grained visual recognition, with the goal of distinguishing subordinate visual categories such as birds [46] or natural species [43].

In the spirit of mapping the visual world, we propose a new task, *instance segmentation with attribute localization*, which unifies object detection and fine-grained attribute recognition. As illustrated in Fig. 1(e), this task offers a structured representation of an image. Automatic recognition of a rich set of attributes for each segmented object instance complements category-level object detection and therefore advance the degree of complexity of images and scenes we can make understandable to machines. In this work, we focus on the fashion domain as an example to illustrate this task. Fashion comes with rich and complex apparel with attributes, influences many aspects of modern societies, and has a strong financial and cultural impact. We anticipate that the proposed task is also suitable for other man-made product domains such as automobile and home interior.

Structured representations of images often rely on structured vocabularies [28]. With this in mind, we construct the Fashionpedia ontology (Fig. 1(f)) and image dataset (Fig. 1(a–e)), annotating fashion images with detailed segmentation masks for apparel categories, parts, and their attributes. Our proposed ontology provides a rich schema for interpretation and organization of individuals' garments, styles, or fashion collections [24]. For example, we can create a knowledge graph (see supplementary material for more details) by aggregating structured information within each image and exploiting relationships between garments and garment parts, categories, and attributes in the Fashionpedia ontology. Our insight is that a large-scale fashion segmentation and attribute localization dataset built with a fashion ontology can help computer vision models achieve better performance on fine-grained image understanding and reasoning tasks.

The contributions of this work are as follows:

A *novel task* of fine-grained instance segmentation with attribute localization. The proposed task unifies instance segmentation and visual attribute recognition, which is an important step toward structural understanding of visual content in real-world applications.

A *unified fashion ontology* informed by product descriptions from the internet and built by fashion experts. Our ontology captures the complex structure of fashion objects and ambiguity in descriptions obtained from the web, containing 46 apparel objects (27 main apparels and 19 apparel parts), and 294 fine-grained attributes (spanning 9 super categories) in total. To facilitate the development of related efforts, we also provide a mapping with categories from existing fashion segmentation datasets, see Fig. 1(f).

A *dataset* with a total of 48,825 clothing images in daily-life, street-style, celebrity events, runway, and online shopping annotated both by crowd workers for segmentation masks and fashion experts for localized attributes, with the goal of developing and benchmarking computer vision models for comprehensive understanding of fashion.

A new *model*, Attribute-Mask R-CNN, is proposed with the aim of jointly performing instance segmentation and localized attribute recognition; a novel *evaluation metric* for this task is also provided.

2 Related Work

The combined task of fine-grained instance segmentation and attribute localization has not received a great deal of attention in the literature. On one hand, COCO [31] and LVIS [15] represent the benchmarks of object detection for common objects. Panoptic segmentation is proposed to unify both semantic and

Table 1. Comparison of fashion-related datasets: (Cls. = Classification, Segm. = Segmentation, MG = Main Garment, GP = Garment Part, A = Accessory, S = Style, FGC = Fine-Grained Categorization). To the best of our knowledge, we include all fashion-related datasets focusing on visual recognition

Name	Category Annotation Type			Attribute Annotation Type		
	Cls.	BBox	Segm.	Unlocalized	Localized	FGC
Clothing Parsing [50]	MG, A	-	-	-	-	-
Chic or Social [49]	MG, A	-	-	-	-	-
Hipster [26]	MG, A, S	-	-	-	-	-
Ups and Downs [19]	MG	-	-	-	-	-
Fashion550k [23]	MG, A	-	-	-	-	-
Fashion-MNIST [48]	MG	-	-	-	-	-
Runway2Realway [44]	-	-	MG, A	-	-	-
ModaNet [53]	-	MG, A	MG, A	-	-	-
Deepfashion2 [11]	-	MG	MG	-	-	-
Fashion144k [40]	MG, A	-	-	✓	-	-
Fashion Style-128 Floats [41]	S	-	-	✓	-	-
UT Zappos50K [51]	A	-	-	✓	-	-
Fashion200K [16]	MG	-	-	✓	-	-
FashionStyle14 [42]	S	-	-	✓	-	-
Main Product Detection [39]	-	MG	-	✓	-	-
StreetStyle-27K [35]	-	-	-	✓	-	✓
UT-latent look [21]	MG, S	-	-	✓	-	✓
FashionAI [7]	MG, GP, A	-	-	✓	-	✓
iMat-Fashion Attribute [14]	MG, GP, A, S	-	-	✓	-	✓
Apparel classification-Style [4]	-	MG	-	✓	-	✓
DARN [22]	-	MG	-	✓	-	✓
WTBI [25]	-	MG, A	-	✓	-	✓
Deepfashion [32]	S	MG	-	✓	-	✓
Fashionpedia	-	**MG, GP, A**	**MG, GP, A**	-	✓	✓

instance segmentation, addressing both stuff and thing classes [27]. In spite of the domain differences, Fashionpedia has comparable mask qualities with LVIS and the similar total number of segmentation masks as COCO. On the other hand, we have also observed an increasing effort to curate datasets for fine-grained visual recognition, evolved from CUB-200 Birds [46] to the recent iNaturalist dataset [43]. The goal of this line of work is to advance the state-of-the-art in automatic image classification for large numbers of real world, fine-grained categories. A rather unexplored area of these datasets, however, is to provide a structured representation of an image. Visual Genome [28] provides dense annotations of object bounding boxes, attributes, and relationships in the general domain, enabling a structured representation of the image. In our work, we instead focus on fine-grained attributes and provide segmentation masks in the fashion domain to advance the clothing recognition task.

Clothing recognition has received increasing attention in the computer vision community recently. A number of works provide valuable apparel-related datasets [4,7,11,16,19,21–23,25,26,32,35,39–42,44,48–51,53]. These pioneering works enabled several recent advances in clothing-related recognition and knowledge discovery [10,34]. Table 1 summarizes the comparison among different fashion datasets regarding annotation types of clothing categories and attributes. Our dataset distinguishes itself in the following three aspects.

Exhaustive Annotation of Segmentation Masks: Existing fashion datasets [11,44,53] offer segmentation masks for the main garment (e.g., jacket, coat, dress) and the accessory categories (e.g., bag, shoe). Smaller garment objects such as collars and pockets are not annotated. However, these small objects could be valuable for real-world applications (e.g., searching for a specific collar shape during online-shopping). Our dataset is not only annotated with the segmentation masks for a total of 27 main garments and accessory categories but also 19 garment parts (e.g., collar, sleeve, pocket, zipper, embroidery).

Localized Attributes: The fine-grained attributes from existing datasets [14, 22,32,39] tend to be noisy, mainly because most of the annotations are collected by crawling fashion product attribute-level descriptions directly from large online shopping websites. Unlike these datasets, fine-grained attributes in our dataset are annotated manually by fashion experts. To the best of our knowledge, ours is the only dataset to annotate localized attributes: fashion experts are asked to annotate attributes associated with the segmentation masks labeled by crowd workers. Localized attributes could potentially help computational models detect and understand attributes more accurately.

Fine-grained Categorization: Previous studies on fine-grained attribute categorization suffer from several issues including: (1) repeated attributes belonging to the same category (e.g., zip,zipped and zipper) [21,32]; (2) basic level categorization only (object recognition) and lack of fine-grained categorization [4,11,16,23,25,26,41,42,44,48–50,53]; (3) lack of fashion taxonomies with the needs of real-world applications for the fashion industry, possibly due to the research gap in fashion design and computer vision; (4) diverse taxonomy structures from different sources in fashion domain. To facilitate research in the areas of fashion and computer vision, our proposed ontology is built and verified by fashion experts based on their own design experiences and informed by the following four sources: (1) world-leading e-commerce fashion websites (e.g., ZARA, H&M, Gap, Uniqlo, Forever21); (2) luxury fashion brands (e.g., Prada, Chanel, Gucci); (3) trend forecasting companies (e.g., WGSN); (4) academic resources [3,8].

Fig. 2. Image examples with annotated segmentation masks (a–f) and fine-grained attributes (g–i)

3 Dataset Specification and Collection

3.1 Ontology Specification

We propose a unified fashion ontology (Fig. 1(f)), a structured vocabulary that utilizes the basic level categories and fine-grained attributes [38]. The Fashionpedia ontology relies on similar definitions of object and attributes as previous well-known image datasets. For example, a Fashionpedia object is similar to "item" in Wikidata [45], or "object" in COCO [31] and Visual Genome [28]). In the context of Fashionpedia, objects represent common items in apparel (e.g., jacket, shirt, dress). In this section, we break down each component of the Fashionpedia ontology and illustrate the construction process. With this ontology and our image dataset, a large-scale fashion knowledge graph can be built as an extended application of our dataset (more details can be found in the supplementary material).

Apparel Categories. In the Fashionpedia dataset, all images are annotated with one or multiple main garments. Each main garment is also annotated with its garment parts. For example, general garment types such as jacket, dress, pants are considered as main garments. These garments also consist of several garment parts such as collars, sleeves, pockets, buttons, and embroideries. Main garments are divided into three main categories: outerwear, intimate and accessories. Garment parts also have different types: garment main parts (e.g., collars, sleeves), bra parts, closures (e.g., button, zipper) and decorations (e.g., embroidery, ruffle). On average, each image consists of 1 person, 3 main garments, 3 accessories, and 12 garment parts, each delineated by a tight segmentation mask (Fig. 1(a–c)). Furthermore, each object is assigned to a synset ID in our Fashionpedia ontology.

Fine-grained Attributes. Main garments and garment parts can be associated with apparel attributes (Fig. 1(e)). For example, "button" is part of the main garment "jacket"; "jacket" can be linked with the silhouette attribute "symmetrical"; the garment part "button" could contain the attribute "metal" with a relationship of material. The Fashionpedia ontology provide attributes for 13 main outerwear garments categories, and 5 out of 19 garments parts ("sleeve", "neckline", "pocket", "lapel", and "collar"). Each image has 16.7 attributes on average (max 57 attributes). As with the main garments and garment parts, we canonicalize all attributes to our Fashionpedia ontology.

Relationships. Relationships can be formed between categories and attributes. There are three main types of relationships (Fig. 1(e)): (1) outfits to main garments, main garments to garment parts: meronymy (part-of) relationship; (2) main garments to attributes or garment parts to attributes: these relationship types can be garment silhouette (e.g., peplum), collar nickname (e.g., peter pan collars), garment length (e.g., knee-length), textile finishing (e.g., distressed), or textile-fabric patterns (e.g., paisley), etc.; (3) within garments, garment parts or attributes: there is a maximum of four levels of hyponymy (is-an-instance-of) relationships. For example, weft knit is an instance of knit fabric, and the fleece is an instance of weft knit.

3.2 Image Collection and Annotation Pipeline

Image Collection. A total of 50,527 images were harvested from Flickr and free license photo websites, which included Unsplash, Burst by Shopify, Freestocks, Kaboompics, and Pexels. Two fashion experts verified the quality of the collected images manually. Specifically, the experts checked a scenes' diversity and made sure clothing items were visible in the images. Fdupes [33] is used to remove duplicated images. After filtering, 48,825 images were left and used to build our Fashionpedia dataset.

Annotation Pipeline. Expert annotation is often a time-consuming process. In order to accelerate the annotation process, we decoupled the work between crowd workers and experts. We divided the annotation process into the following two phases.

First, segmentation masks with apparel objects are annotated by 28 crowd workers, who were trained for 10 days before the annotation process (with prepared annotation tutorials of each apparel object, see supplementary material for details). We collected high-quality annotations by having the annotators follow the contours of garments in the image as closely as possible (See Sect. 4.2 for annotation analysis). This polygon annotation process was monitored daily and verified weekly by a supervisor and by the authors.

Second, 15 fashion experts (graduate students in the apparel domain) were recruited to annotate the fine-grained attributes for the annotated segmentation masks. Annotators were given one mask and one attribute super-category ("textile pattern", "garment silhouette" for example) at a time. An additional two options, "not sure" and "not on the list" were added during the annotation. The option "not on the list" indicates that the expert found an attribute that is not on the proposed ontology. If "not sure" is selected, it means the expert can not identify the attribute of one mask. Common reasons for this selection include occlusion of the masks and viewing angles of the image; (for example, a top underneath a closed jacket). More details can be found in Fig. 2. Each attribute supercategory is assigned to one or two fashion experts, depending on the number of masks. The annotations are also checked by another expert annotator before delivery.

We split the data into training, validation and test sets, with 45,623, 1158, 2044 images respectively. More details of the dataset creation can be found in the supplementary material.

4 Dataset Analysis

This section will discuss a detailed analysis of our dataset using the training images. We begin by discussing general image statistics, followed by an analysis of segmentation masks, categories, and attributes. We compare Fashionpedia with four other segmentation datasets, including two recent fashion datasets DeepFashion2 [11] and ModaNet [53], and two general domain datasets COCO [31] and LVIS [15].

4.1 Image Analysis

We choose to use images with high resolutions during the curating process since Fashionpedia ontology includes diverse fine-grained attributes for both garments and garment parts. The Fashionpedia training images have an average dimension of 1710 (width) $\times$ 2151 (height). Images with high resolutions are able to show apparel objects in detail, leading to more accurate and faster annotations for both segmentation masks and attributes. These high resolution images can also benefit downstream tasks such as detection and image generation. Examples of detailed annotations can be found in Fig. 2.

4.2 Mask Analysis

We define "masks" as one apparel instance that may have more than one separate components (the jacket in Fig. 1 for example), "polygon" as a disjoint area.

Mask Quantity. On average, there are 7.3 (median 7, max 74) number of masks per image in the Fashionpedia training set. Figure 3(a) shows that the Fashionpedia has the largest median value in the 5 datasets used for comparison. Fashionpedia also has the widest range among three fashion datasets, and a comparable range with COCO, which is a dataset in a general domain. Compared to ModaNet and Deepfashion2 datasets, Fashionpedia has the widest range of mask count distribution. However, COCO and LVIS maintain a wider distribution over Fashionpedia owing to their more common objects in their dataset. Figure 3(d) illustrates the distribution within the Fashionpedia dataset. One image usually contains more garment parts and accessories than outerwears.

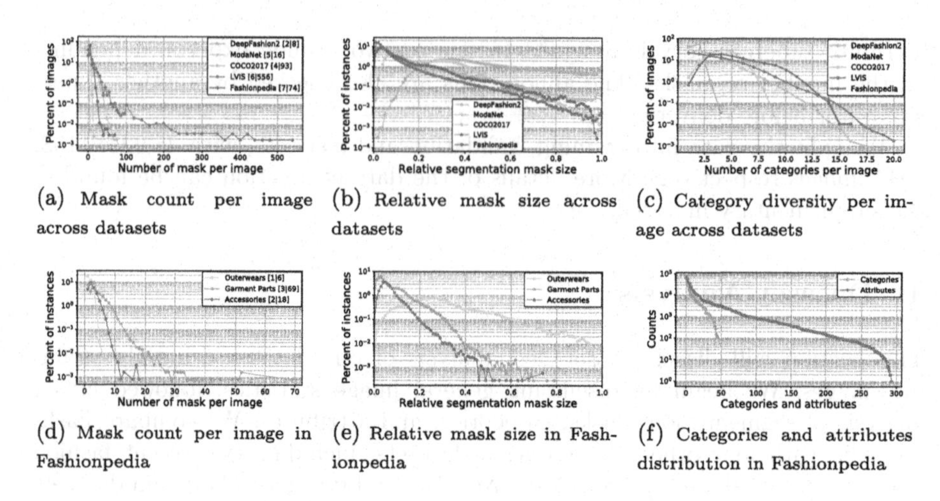

(a) Mask count per image across datasets (b) Relative mask size across datasets (c) Category diversity per image across datasets

(d) Mask count per image in Fashionpedia (e) Relative mask size in Fashionpedia (f) Categories and attributes distribution in Fashionpedia

Fig. 3. Dataset statistics: First row presents comparison among datasets. Second row presents comparison within Fashionpedia. Y-axes are in log scale. Relative segmentation mask size were calculated same as [15]. Relative mask size was rounded to precision of 2. For mask count per image comparisons (Fig. 3(a) and (d)), legends follow [*median* | *max*] format. Values in X-axis for Fig. 3(a) was discretized for better visual effect

Mask Sizes. Fig. 3(b) and (e) compares relative mask sizes within Fashionpedia and against other datasets. Ours has a similar distribution as COCO and LVIS, except for a lack of larger masks (area > 0.95). DeepFashion2 has a heavier tail, meaning it contains a larger portion of garments with a zoomed-in view. Unlike DeepFashion2, our images mainly focused on the whole ensemble of clothing. Since ModaNet focuses on outwears and accessories, it has more masks with relative area between 0.2 and 0.4. whereas ours has an additional 19 apparel parts categories. As illustrated in Fig. 3(e), garment parts and accessories are relatively small compared to the outerwear (e.g., "dress", "coat").

Mask Quality. Apparel categories also tend to have complex silhouettes. Table 2 shows that the Fashionpedia masks have the most complex boundaries amongst the five datasets (according to the measurement used in [15]). This suggests that our masks are better able to represent complex silhouettes of apparel categories more accurately than ModaNet and DeepFashion2. We also report the number of vertices per polygons, which is a measurement representing how granularity of the masks produced. Table 2 shows that we have the second-highest average number of vertices among five datasets, next to LVIS.

4.3 Category and Attributes Analysis

There are 46 apparel categories and 294 attributes presented in the Fashionpedia dataset. On average, each image was annotated with 7.3 instances, 5.4 categories, and 16.7 attributes. Of all the masks with categories and attributes,

Table 2. Comparison of segmentation mask complexity amongst segmentation datasets in both fashion and general domain (COCO 2017 instance training data was used). Each statistic (mean and median) represents a bootstrapped 95% confidence interval following [15]. Boundary complexity was calculated according to [2,15]. Reported mask boundary complexity for COCO and LVIS was different compared with [15] due to different image resolution and image sets. The number of vertices per polygon is calculated as the number of vertices in one polygon. Polygon is defined as one disjoint area. Masks (and polygons) with zero area were ignored

Dataset	Boundary complexity		No. of vertices per polygon		Images count
	mean	median	mean	median	
COCO [31]	6.65 - 6.66	6.07 - 6.08	21.14 - 21.21	15.96 - 16.04	118,287
LVIS [15]	6.78 - 6.80	5.89 - 5.91	**35.77 - 35.95**	**22.91 - 23.09**	57,263
ModaNet [53]	5.87 - 5.89	5.26 - 5.27	22.50 - 22.60	18.95 - 19.05	52,377
Deepfashion2 [11]	4.63 - 4.64	4.45 - 4.46	14.68 - 14.75	8.96 - 9.04	**191,960**
Fashionpedia	**8.36 - 8.39**	**7.35 - 7.37**	31.82 - 32.01	20.90 - 21.10	45, 623

each mask has 3.7 attributes on average (max 14 attributes). Fashionpedia has the most diverse number of categories within one image among three fashion datasets, while comparable to COCO (Fig. 3(c)), since we provide a comprehensive ontology for the annotation. In addition, Fig. 3(f) shows the distributions of categories and attributes in the training set, and highlights the long-tailed nature of our data.

During the fine-grained attributes annotation process, we also ask the experts to choose "not sure" if they are uncertain to make a decision, "not on the list" if they find another attribute that not provided. the majority of "not sure" comes from three attributes superclasses, namely "Opening Type", "Waistline", "Length". Since there are masks only show a limited portion of apparel (a top inside a jacket for example), the annotators are not sure how to identify those attributes due to occlusion or viewpoint discrepancies. Less than 15% of masks for each attribute superclasses account for "not on the list", which illustrates the comprehensiveness of our proposed ontology (see supplementary material for more details of the extra dataset analysis).

5 Evaluation Protocol and Baselines

5.1 Evaluation Metric

In object detection, a true positive (TP) for each category c is defined as a single detected object that matches a ground truth object with a Intersection over Union (IoU) over a threshold τ_{IoU}. COCO's main evaluation metric uses average precision averaged across all 10 IoU thresholds $\tau_{\text{IoU}} \in [0.5 : 0.05 : 0.95]$ and all 80 categories. We denote such metric as AP_{IoU}.

In the case of instance segmentation and attribute localization , we extend standard COCO metric by adding one more constraint: the macro F_1 score for predicted attributes of single detected object with category c (see supplementary material for the average choice of f1-score). We denote the F_1 threshold as τ_{F_1}, and it has the same range as τ_{IoU} ($\tau_{F_1} \in [0.5 : 0.05 : 0.95]$). The main metric AP_{IoU+F_1} reports averaged precision score across all 10 IoU thresholds, all 10 macro F_1 scores, and all the categories. Our evaluation API, code and trained models are available at: https://fashionpedia.github.io/home/Model_and_API.html.

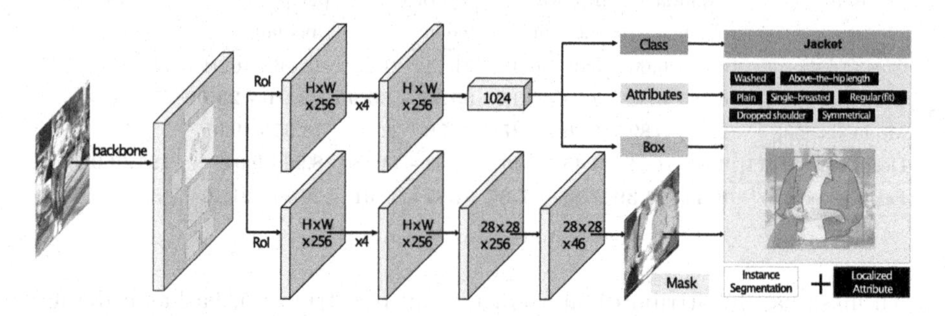

Fig. 4. Attribute-Mask R-CNN adds a multi-label attribute prediction head upon Mask R-CNN for instance segmentation with attribute localization

5.2 Attribute-Mask R-CNN

We perform two tasks on Fashionpedia: (1) apparel instance segmentation (ignoring attributes); (2) instance segmentation with attribute localization. To better facilitate research on Fashionpedia, we design a strong baseline model named Attribute-Mask R-CNN that is built upon Mask R-CNN [17]. As illustrated in Fig. 4, we extend Mask R-CNN heads to include an additional multi-label attribute prediction head, which is trained with sigmoid cross-entropy loss. Attribute-Mask R-CNN can be trained end-to-end for jointly performing instance segmentation and localized attribute recognition.

We leverage different backbones including ResNet-50/101 (R50/101) [18] with feature pyramid network (FPN) [30] and SpineNet-49/96/143 [5]. The input image is resized to 1024 of the longer edge to feed all networks except SpineNet-143. For SpineNet-143, we instead use the input size of 1280. During training, other than standard random horizontal flipping and cropping, we use large scale jittering that resizes an image to a random ratio between (0.5, 2.0) of the target input image size, following Zoph *et al.* [54]. We use an open-sourced Tensorflow [1] code base[1] for implementation and all models are trained with a batch size of 256. We follow the standard training schedule of 1× (5625 iterations), 2× (11250 iterations), 3× (16875 iterations) and 6× (33750 iterations) in Detectron2 [47], with linear learning rate scaling suggested by Goyal *et al.* [13].

[1] https://github.com/tensorflow/tpu/tree/master/models/official/detection.

Table 3. Baseline results of Mask R-CNN and Attribute-Mask R-CNN on Fashionpedia. The big performance gap between AP_{IoU} and AP_{IoU+F_1} suggests the challenging nature of our proposed task

Backbone	Schedule	FLOPs(B)	params(M)	$\mathbf{AP^{box}_{IoU/IoU+F_1}}$	$\mathbf{AP^{mask}_{IoU/IoU+F_1}}$
R50-FPN	1×	296.7	46.4	38.7 / 26.6	34.3 / 25.5
	2×			41.6 / 29.3	38.1 / 28.5
	3×			43.4 / 30.7	39.2 / 29.5
	6×			42.9 / 31.2	38.9 / 30.2
R101-FPN	1×	374.3	65.4	41.0 / 28.6	36.7 / 27.6
	2×			43.5 / 31.0	39.2 / 29.8
	3×			44.9 / 32.8	40.7 / 31.4
	6×			44.3 / 32.9	39.7 / 31.3
SpineNet-49	**6**×	267.2	40.8	43.7 / 32.4	39.6 / 31.4
SpineNet-96		314.0	55.2	46.4 / 34.0	41.2 / 31.8
SpineNet-143		**498.0**	**79.2**	**48.7 / 35.7**	**43.1 / 33.3**

Table 4. Per super-category results (for masks) using Attribute-Mask R-CNN with SpineNet-143 backbone. We follow the same COCO sub-metrics for overall and three super-categories for apparel objects. Result format follows [AP_{IoU} / AP_{IoU+F_1}] or [AR_{IoU} / AR_{IoU+F_1}] (see supplementary material for per-class results)

Category	AP	AP50	AP75	APl	APm	APs
overall	43.1 / 33.3	60.3 / 42.3	47.6 / 37.6	50.0 / 35.4	40.2 / 27.0	17.3 / 9.4
outerwear	64.1 / 40.7	77.4 / 49.0	72.9 / 46.2	67.1 / 43.0	44.4 / 29.3	19.0 / 4.4
Parts	19.3 / 13.4	35.5 / 20.8	18.4 / 14.4	28.3 / 14.5	23.9 / 16.4	12.5 / 9.8
Accessory	56.1 / -	77.9 / -	63.9 / -	57.5 / -	60.5 / -	25.0 / -

5.3 Results Discussion

Attribute-Mask R-CNN. From results in Table 3, we have the following observations: (1) Our baseline models achieve promising performance on challenging Fashionpedia dataset. (2) There is a significant drop (*e.g.*, from 48.7 to 35.7 for SpineNet-143) in box AP if we add τ_{F_1} as another constraint for true positive. This is further verified by per super-category mask results in Table 4. This suggests that joint instance segmentation and attribute localization is a significantly more difficult task than instance segmentation, leaving much more room for future improvements.

Main Apparel Detection Analysis. We also provide in-depth detector analysis following COCO detection challenge evaluation [31] inspired by Hoiem *et al.* [20]. Figure 5 illustrates a detailed breakdown of bounding boxes false positives produced by the detectors.

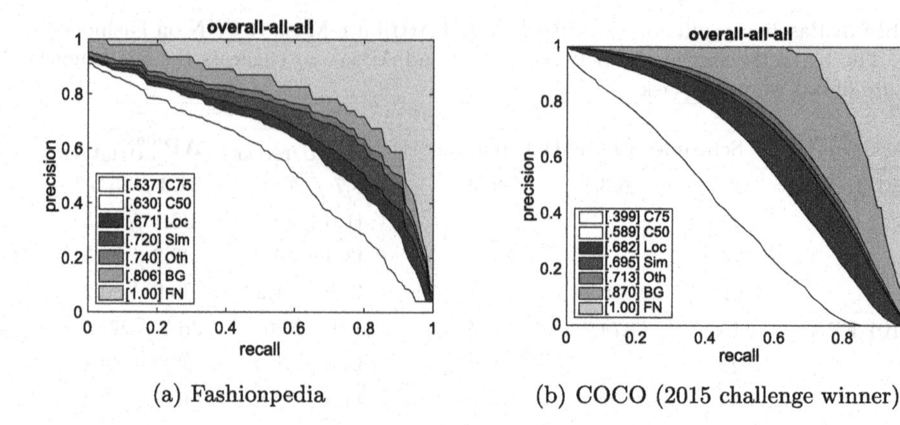

(a) Fashionpedia (b) COCO (2015 challenge winner)

Fig. 5. Main apparel detectors analysis. Each plot shows 7 precision recall curves where each evaluation setting is more permissive than the previous. Specifically, **C75**: strict IoU ($\tau_{\text{IoU}} = 0.75$); **C50**: PASCAL IoU ($\tau_{\text{IoU}} = 0.5$); **Loc**: localization errors ignored ($\tau_{\text{IoU}} = 0.1$); **Sim**: supercategory False Positives (FPs) removed; **Oth**: category FPs removed; **BG**: background (and class confusion) FPs removed; **FN**: False Negatives are removed. Two plots are a comparison between two detectors trained on Fashionpedia and COCO respectively. The results are averaged over all categories. Legends present the area under each curve (corresponds to AP metric) in brackets as well

Figure 5(a) and (b) compare two detectors trained on Fashionpedia and COCO. Errors of the COCO detector are dominated by imperfect localization (AP is increased by 28.3 from overall AP at $\tau_{\text{IoU}} = 0.75$) and background confusion (+15.7) (5(b)). Unlike the COCO detector, no mistake in particular dominates the errors produced by the Fashionpedia detector. Figure 5(a) shows that there are errors from localization (+13.4.0), classification (+6.9), background confusions (+6.6). Due to the space constraint, we leave super-category analysis in the supplementary material.

Prediction Visualization. Baseline outputs (with both segmentation masks and localized attributes) are also visualized in Fig. 6. Our Attribute-Mask R-CNN achieves good results even for small objects like shoes and glasses. Model can correctly predict fine-grained attributes for some masks on one hand (e.g., Fig. 6 second image at top row Sleeve 1). On the other hand, it also incorrectly predict the wrong nickname (welt) to pocket (e.g., Fig. 6 third image at top row Pocket 1). These results show that there is headroom remaining for future development of more advanced computer vision models on this task (see supplementary material for more details of the baseline analysis).

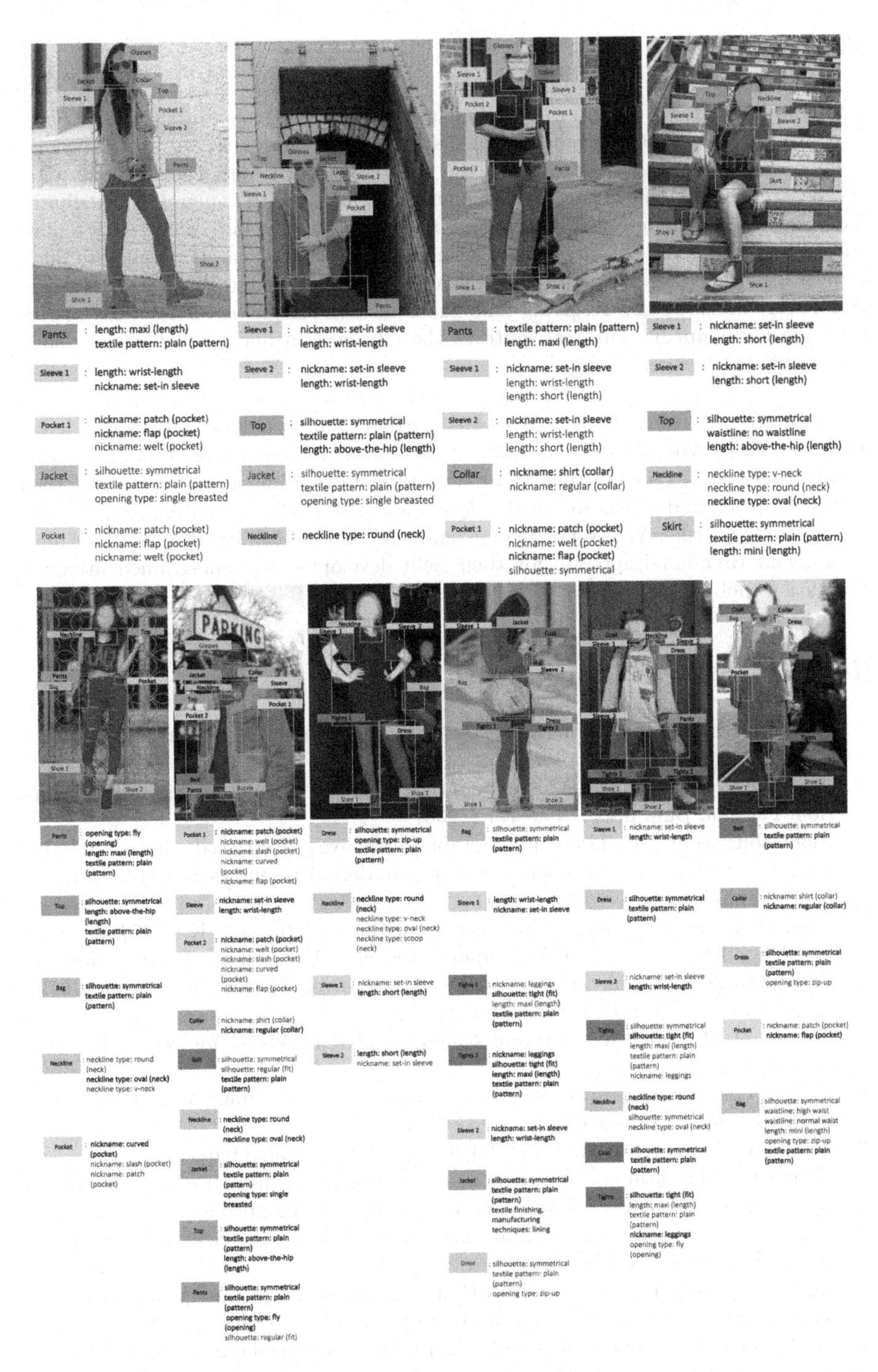

Fig. 6. Attribute-Mask R-CNN results on the Fashionpedia validation set. Masks, bounding boxes, and apparel categories (category score > 0.6) are shown. The localized attributes from the top 5 masks (that contain attributes) on each image are also shown. Correctly predicted categories and localized attributes are bolded

6 Conclusion

In this work, we focus on a new task that unifies instance segmentation and attribute recognition. To solve challenging problems entailed in this task, we introduced the Fashionpedia ontology and dataset. To the best of our knowledge, Fashionpedia is the first dataset that combines part-level segmentation masks with fine-grained attributes. We presented Attribute-Mask R-CNN, a novel model for this task, along with a novel evaluation metric. We expect models trained on Fashionpedia can be applied to many applications including better product recommendation in online shopping, enhanced visual search results, and resolving ambiguous fashion-related words for text queries. We hope Fashionpedia will contribute to the advances in fine-grained image understanding in the fashion domain.

Acknowledgements. This research was partially supported by a Google Faculty Research Award. We thank Kavita Bala, Carla Gomes, Dustin Hwang, Rohun Tripathi, Omid Poursaeed, Hector Liu, and Nayanathara Palanivel, Konstantin Lopuhin for their helpful feedback and discussion in the development of Fashionpedia dataset. We also thank Zeqi Gu, Fisher Yu, Wenqi Xian, Chao Suo, Junwen Bai, Paul Upchurch, Anmol Kabra, and Brendan Rappazzo for their help developing the fine-grained attribute annotation tool.

References

1. Abadi, M., et al.: TensorFlow: a system for large-scale machine learning. In: OSDI (2016)
2. Attneave, F., Arnoult, M.D.: The quantitative study of shape and pattern perception. Psychol. Bull. **53**(6), 452–471 (1956)
3. Bloomsbury.com: Fashion photography archive. Retrieved from 9 May 2019. https://www.bloomsbury.com/dr/digital-resources/products/fashion-photography-archive/
4. Bossard, L., et al.: Apparel classification with style. In: Lee, K.M., Matsushita, Y., Rehg, J.M., Hu, Z. (eds.) Computer Vision – ACCV 2012. Lecture Notes in Computer Science, vol. 7727, pp. 321–335. Springer, Berlin, Heidelberg (2012). https://doi.org/10.1007/978-3-642-37447-0_25
5. Du, X., et al.: SpineNet: learning scale-permuted backbone for recognition and localization. In: CVPR (2020)
6. Farhadi, A., Endres, I., Hoiem, D., Forsyth, D.: Describing objects by their attributes. In: CVPR (2009)
7. FashionAI: Retrieved from 9 May 2019 . http://fashionai.alibaba.com/
8. Fashionary.org: Fashionpedia – the visual dictionary of fashion design. Retrieved from 9 May 2019. https://fashionary.org/products/fashionpedia
9. Ferrari, V., Zisserman, A.: Learning visual attributes. In: Advances in Neural Information Processing Systems (2008)
10. Fu, C.Y., Berg, T.L., Berg, A.C.: IMP: instance mask projection for high accuracy semantic segmentation of things. arXiv preprint arXiv:1906.06597 (2019)
11. Ge, Y., Zhang, R., Wang, X., Tang, X., Luo, P.: Deepfashion2: a versatile benchmark for detection, pose estimation, segmentation and re-identification of clothing images. In: CVPR (2019)

12. Girshick, R., Donahue, J., Darrell, T., Malik, J.: Rich feature hierarchies for accurate object detection and semantic segmentation. In: CVPR (2014)
13. Goyal, P., et al.: Accurate, large minibatch SGD: training imageNet in 1 hour. arXiv preprint arXiv:1706.02677 (2017)
14. Guo, S., et al.: The imaterialist fashion attribute dataset. In: ICCV Workshops (2019)
15. Gupta, A., Dollar, P., Girshick, R.: LVIS: a dataset for large vocabulary instance segmentation. In: CVPR (2019)
16. Han, X., et al.: Automatic spatially-aware fashion concept discovery. In: ICCV (2017)
17. He, K., Gkioxari, G., Dollár, P., Girshick, R.: Mask R-CNN. In: ICCV (2017)
18. He, K., Zhang, X., Ren, S., Sun, J.: Deep residual learning for image recognition. In: CVPR (2016)
19. He, R., McAuley, J.: Ups and downs: modeling the visual evolution of fashion trends with one-class collaborative filtering. In: WWW (2016)
20. Hoiem, D., Chodpathumwan, Y., Dai, Q.: Diagnosing error in object detectors. In: Fitzgibbon, A., Lazebnik, S., Perona, P., Sato, Y., Schmid, C. (eds.) Computer Vision – ECCV 2012. Lecture Notes in Computer Science, vol. 7574, pp. 340–353. Springer, Berlin, Heidelberg (2012). https://doi.org/10.1007/978-3-642-33712-3_25
21. Hsiao, W.L., Grauman, K.: Learning the latent "look": unsupervised discovery of a style-coherent embedding from fashion images. In: ICCV (2017)
22. Huang, J., Feris, R., Chen, Q., Yan, S.: Cross-domain image retrieval with a dual attribute-aware ranking network. In: ICCV (2015)
23. Inoue, N., Simo-Serra, E., Yamasaki, T., Ishikawa, H.: Multi-label fashion image classification with minimal human supervision. In: ICCV (2017)
24. Kendall, E.F., McGuinness, D.L.: Ontology Engineering (Synthesis Lectures on The Semantic Web: Theory and Technology), pp. 1–136. Morgan & Claypool, San Rafael (2019)
25. Kiapour, M.H., Han, X., Lazebnik, S., Berg, A.C., Berg, T.L.: Where to buy it: matching street clothing photos in online shops. In: ICCV (2015)
26. Kiapour, M.H., Yamaguchi, K., Berg, A.C., Berg, T.L.: Hipster wars: discovering elements of fashion styles. In: Fleet, D., Pajdla, T., Schiele, B., Tuytelaars, T. (eds.) Computer Vision – ECCV 2014. Lecture Notes in Computer Science, vol. 8689, pp. 472–488. Springer, Cham (2014). https://doi.org/10.1007/978-3-319-10590-1_31
27. Kirillov, A., He, K., Girshick, R., Rother, C., Dollár, P.: Panoptic segmentation. In: CVPR (2019)
28. Krishna, R., et al.: Visual genome: connecting language and vision using crowdsourced dense image annotations. Int. J. Comput. Vis. (IJCV) **123**, 32–73 (2017)
29. Kumar, N., Berg, A.C., Belhumeur, P.N., Nayar, S.K.: Attribute and simile classifiers for face verification. In: ICCV (2009)
30. Lin, T.Y., et al.: Feature pyramid networks for object detection. In: CVPR (2017)
31. Lin, T.Y., et al.: Microsoft COCO: common objects in context. In: Fleet, D., Pajdla, T., Schiele, B., Tuytelaars, T. (eds.) Computer Vision – ECCV 2014. Lecture Notes in Computer Science, vol. 8693, pp. 740–755. Springer, Cham (2014). https://doi.org/10.1007/978-3-319-10602-1_48
32. Liu, Z., Luo, P., Qiu, S., Wang, X., Tang, X.: Deepfashion: Powering robust clothes recognition and retrieval with rich annotations. In: CVPR (2016)
33. Lopez, A.: Fdupes is a program for identifying or deleting duplicate files residing within specified directories. Retrieved from 9 May 2019. https://github.com/adrianlopezroche/fdupes

34. Mall, U., Matzen, K., Hariharan, B., Snavely, N., Bala, K.: Geostyle: discovering fashion trends and events. In: ICCV (2019)
35. Matzen, K., Bala, K., Snavely, N.: StreetStyle: exploring world-wide clothing styles from millions of photos. arXiv preprint arXiv:1706.01869 (2017)
36. Parikh, D., Grauman, K.: Relative attributes. In: ICCV (2011)
37. Ren, S., He, K., Girshick, R., Sun, J.: Faster R-CNN: towards real-time object detection with region proposal networks. In: Advances in Neural Information Processing Systems (2015)
38. Rosch, E.: Cognitive representations of semantic categories. J. Exp. Psychol. Gen. **104**(3), 192–233 (1975)
39. Rubio, A., Yu, L., Simo-Serra, E., Moreno-Noguer, F.: Multi-modal embedding for main product detection in fashion. In: ICCV (2017)
40. Simo-Serra, E., Fidler, S., Moreno-Noguer, F., Urtasun, R.: Neuroaesthetics in fashion: modeling the perception of fashionability. In: CVPR (2015)
41. Simo-Serra, E., Ishikawa, H.: Fashion style in 128 floats: joint ranking and classification using weak data for feature extraction. In: CVPR (2016)
42. Takagi, M., Simo-Serra, E., Iizuka, S., Ishikawa, H.: What makes a style: experimental analysis of fashion prediction. In: ICCV (2017)
43. Van Horn, G., et al.: The iNaturalist species classification and detection dataset. In: CVPR (2018)
44. Vittayakorn, S., Yamaguchi, K., Berg, A.C., Berg, T.L.: Runway to Realway: visual analysis of fashion. In: WACV (2015)
45. Vrandečić, D., Krötzsch, M.: Wikidata: a free collaborative knowledge base (2014)
46. Wah, C., Branson, S., Welinder, P., Perona, P., Belongie, S.: The Caltech-UCSD Birds-200-2011 Dataset. Technical report CNS-TR-2011-001, California Institute of Technology (2011)
47. Wu, Y., Kirillov, A., Massa, F., Lo, W.Y., Girshick, R.: Detectron2. https://github.com/facebookresearch/detectron2 (2019)
48. Xiao, H., Rasul, K., Vollgraf, R.: Fashion-MNIST: a novel image dataset for benchmarking machine learning algorithms. arXiv preprint arXiv:1708.07747 (2017)
49. Yamaguchi, K., Berg, T.L., Ortiz, L.E.: Chic or social: visual popularity analysis in online fashion networks. In: ACM MM (2014)
50. Yamaguchi, K., Kiapour, M.H., Ortiz, L.E., Berg, T.L.: Parsing clothing in fashion photographs. In: CVPR (2012)
51. Yu, A., Grauman, K.: Semantic jitter: dense supervision for visual comparisons via synthetic images. In: ICCV (2017)
52. Yu, F., et al.: Bdd100k: a diverse driving dataset for heterogeneous multitask learning. In: CVPR (2020)
53. Zheng, S., Yang, F., Kiapour, M.H., Piramuthu, R.: Modanet: a large-scale street fashion dataset with polygon annotations. In: ACM MM (2018)
54. Zoph, B., et al.: Rethinking pre-training and self-training. arXiv preprint arXiv:2006.06882 (2020)

Privacy Preserving Structure-from-Motion

Marcel Geppert[1(✉)], Viktor Larsson[1], Pablo Speciale[2], Johannes L. Schönberger[2], and Marc Pollefeys[1,2]

[1] Department of Computer Science, ETH Zurich, Zurich, Switzerland
marcel.geppert@inf.ethz.ch
[2] Microsoft MR & AI, Zurich, Switzerland

Abstract. Over the last years, visual localization and mapping solutions have been adopted by an increasing number of mixed reality and robotics systems. The recent trend towards cloud-based localization and mapping systems has raised significant privacy concerns. These are mainly grounded by the fact that these services require users to upload visual data to their servers, which can reveal potentially confidential information, even if only derived image features are uploaded. Recent research addresses some of these concerns for the task of image-based localization by concealing the geometry of the query images and database maps. The core idea of the approach is to lift 2D/3D feature points to random lines, while still providing sufficient constraints for camera pose estimation. In this paper, we further build upon this idea and propose solutions to the different core algorithms of an incremental Structure-from-Motion pipeline based on random line features. With this work, we make another fundamental step towards enabling privacy preserving cloud-based mapping solutions. Various experiments on challenging real-world datasets demonstrate the practicality of our approach achieving comparable results to standard Structure-from-Motion systems.

1 Introduction

Driven by the quickly growing mixed reality and robotics markets, there has been significant commercial interest in image-based localization and mapping solutions. Over the last years, several companies have launched their cloud services, including Microsoft Azure Spatial Anchors [7], Google's Visual Positioning System [49] underlying the Google Maps AR navigation [6], 6D.AI [40], and Scape Technologies [35]. For these services to function, they require users to upload image information to their servers, which can reveal potentially private user information to the service provider or an adversary. As Dosovitskiy *et al.* [14] and Pittaluga *et al.* [26] strikingly demonstrated, this is the case even when only uploading local features extracted from the image to the cloud.

Electronic supplementary material The online version of this chapter (https://doi.org/10.1007/978-3-030-58452-8_20) contains supplementary material, which is available to authorized users.

A. Vedaldi et al. (Eds.): ECCV 2020, LNCS 12346, pp. 333–350, 2020.
https://doi.org/10.1007/978-3-030-58452-8_20

In this emerging field, privacy concerns have been initially widely ignored by both consumers and the industry, while recent motions in the community [20,24, 31,47] spurred fundamental research to find technical solutions to address these concerns [41,42]. In their pioneering work, Speciale *et al.* propose an approach to enable privacy preserving image-based localization services. The core idea of their method is to obfuscate the geometry of the query images and maps in a way that hides private user information but still provides sufficient geometric constraints to enable camera localization. Specifically, they lift 2D image features to random 2D lines to preserve the privacy in query images sent by the client for server-based localization. In addition, they also show how the same concept can be applied in the 3D domain to obfuscate the geometry of maps [41], where they lift the 3D points of Structure-from-Motion maps to random 3D lines to enable the sharing of private maps with a service provider or another client for the purpose of localization.

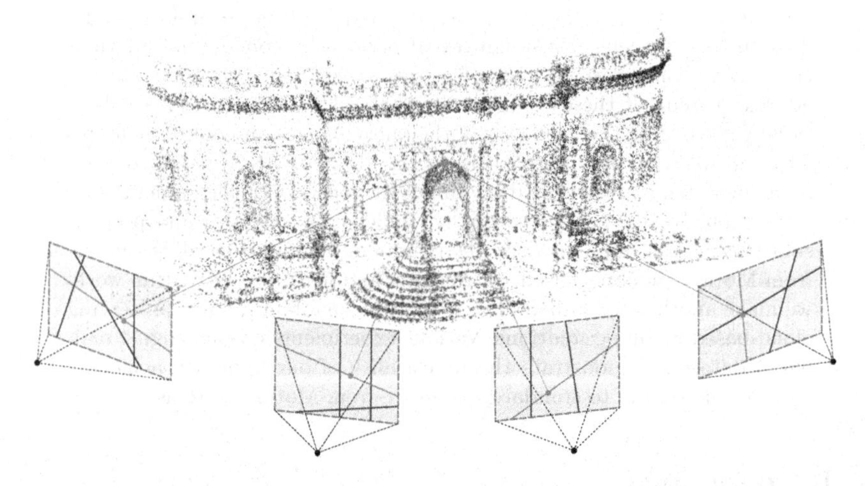

Fig. 1. Privacy Preserving SfM. Our proposed method takes privacy preserving feature lines as input and seeds the reconstruction from four views with at least eight corresponding line features. The intersection of at least three line features produces a point triangulation. Camera resectioning requiring at least six 2D line to 3D point correspondences is based on Speciale *et al.* [42].

The fundamental limitation with their approach is that the map reconstruction stage must be performed on the client side, thereby prohibiting server-based mapping solutions. In the context of mixed reality and robotics, client devices typically have low compute capabilities and thus offloading the map reconstruction stage to the cloud is often required. Furthermore, crowd-based mapping solutions become increasingly relevant, especially as an increasing number of heterogeneous clients navigate through the same space and mapping large-scale spaces becomes infeasible for a single agent alone. In these scenarios, a server-based mapping solution is required to merge the visual data from multiple clients into a single consistent map.

In this paper, we address this limitation and propose an approach to perform privacy preserving Structure-from-Motion (SfM). Our approach is the first to enable cloud-based mapping solutions which do not sacrifice the privacy of users by hiding the privacy concerning contents of the input images. Equivalent to Speciale *et al.*, only consistently observed, triangulated and therefore static scene structure is revealed during our reconstruction process, while privacy concerning transient structure only consistently visible in less than three views (*e.g.*, moving people) are concealed and cannot be reconstructed. The proposed method is based on the fundamental ideas from Speciale *et al.* in that we derive the necessary geometric constraints to perform end-to-end SfM from input images with obfuscated random 2D feature lines (see Fig. 1). This representation hides the appearance information in the extracted 2D features from feature inversion techniques such as [26]. Only in combination with the 3D structure some of the scene appearance can be recovered. In detail, we make the following contributions: (1) We present an end-to-end privacy preserving SfM pipeline based on line features. Our pipeline builds upon COLMAP [36] and the entire source code of our system will be released as open source. (2) For each of the main processing stages of an incremental SfM system (initialization, camera resectioning, triangulation, bundle adjustment), we propose its equivalent counterpart in the privacy preserving setting. (3) We derive a practical minimal solver to initialize our incremental SfM pipeline from four views. The underlying geometric constraints are based on the theory of trifocal tensors and, by exploiting gravity information, we are able to decompose the problem into feasible subproblems. (4) We demonstrate robust and efficient performance of our system on challenging datasets and achieve comparable results with the traditional point feature based baseline.

2 Related Work

In this section, we review related works on privacy preserving methods with a focus on privacy preserving localization approaches. In addition, we also discuss adversarial methods to reconstruct images from its features. Background on SfM in general will be discussed more broadly in the following section.

Privacy Preserving Databases. Querying data in databases without leaking side information has been studied in [12]. Furthermore, various works focused on the specific problems of location privacy [3,5,17,44], differential privacy [15], k-anonymity [34], or learning data-driven models [1,18,54]. However, these approaches are not applicable to geometric computer vision problems, in particular to SfM and image-based localization.

Privacy-Aware Vision Recognition and Hardware Sensors. Privacy-aware techniques were also explored for image retrieval [38], video surveillance [45], biometric verification [46], and face recognition [16]. Learning methods on encrypted data [1,18,54] and adversarial training for vision/action recognition [25,53] have also received much attention. Recent works on privacy in vision include anonymization for activity recognition [30,32]. Furthermore, there are works on privacy preserving optics [27,28] and lens-less coded aperture camera systems that preserve privacy by making the images/video incomprehensible while still allowing action recognition [50]. Zhao *et al.* [57] demonstrated how to localize objects indoors using active cameras and Shariati *et al.* [37] performed ego-motion estimation using low-resolution cameras. However, these privacy preserving techniques cannot be used for SfM from regular images, since they focus either on recognition tasks or rely on special hardware to achieve privacy.

Recovering Images from Features. The main privacy concern we aim to avoid in our pipeline is the reconstruction of images using only its features. Weinzaepfel *et al.* [51] were the first ones to try to invert SIFT features, followed by Vondrick *et al.* [48] interpreting HOG features, bag-of-words features [19], and finally CNN features [23,55,56]. Currently, there are methods that can recover remarkably accurate images from extracted SIFT features [13,14]. More recently, and to raise awareness in the community about the privacy implications, Pittaluga *et al.* [26] demonstrated that 3D point clouds of scenes reconstructed using SfM techniques retain enough information to reconstruct detailed images of the scene, even after the source images are discarded. This enables an adversary to recover confidential content, emphasizing the privacy risks of transmitting and permanently storing such data.

Privacy Preserving Localization. In response to such an attack, there have been a series of papers presented by Speciale *et al.* [41,42] investigating new camera pose estimation algorithms that can safeguard the user's privacy. While there were some existing approaches for recognizing objects in images and videos in a privacy-aware manner [9,10,30,32], those methods cannot be used for camera pose estimation or other geometric computer vision tasks used in mixed reality and robotics. The fundamental limitation of their approach is that the mapping stage (*i.e.*, Structure-from-Motion) must be performed on the client side. This prevents deploying their solution to a wide range of practical applications, where a server-based mapping solution is required. To address this limitation, we build upon their idea of obfuscating 2D feature points by lifting them to 2D random lines [42]. Instead of just enabling privacy aware image-based localization, our goal is to extend this task to create privacy preserving SfM models from the obfuscated images.

3 Method

In this section, we describe our proposed solution to privacy preserving SfM. Our approach is similar to other incremental methods for SfM that operate by alternating between registering new images to the reconstruction (so-called *resectioning*) and triangulating new 3D points. The main steps of these pipelines are **initialization** (Sect. 3.1), **triangulation** (Sect. 3.2), **camera resectioning**

(Sect. 3.3) and **bundle adjustment** (Sect. 3.4). In the following sections, we detail how we adapt each of these steps to the privacy preserving setting. The main difficulty is in dealing with the weaker geometric constraints induced by lifting 2D feature points $\boldsymbol{x}$ to random 2D feature lines $\boldsymbol{\ell}$ analogous to Speciale *et al.* [42], which renders the initialization stage especially challenging.

3.1 Initialization

From only corresponding 2D line features it is not possible to perform initialization from two views. To see this, note that two backprojected 2D lines become 3D planes and will always intersect in a 3D line, regardless of how the cameras are posed, thus the line-line correspondences provide no constraint on the two-view relative pose. Even with three views, the 3D planes will always intersect in a point and therefore not provide any constraints. In fact, the first constraints appear in four views. The relative poses of four images are described by the quadrifocal tensor (see *e.g.* [39]). While it is in theory possible to estimate the quadrifocal tensor from line correspondences, there is currently no tractable method for doing so due to the high complexity of the quadrifocal tensor's internal constraints.

Instead, we present an alternative approach for performing robust initialization from line-based correspondences. The method is based on the assumption of knowing the gravity direction of the images used for initialization, which is reasonable in practice as virtually any device nowadays comes with an intertial measurement unit. Furthermore, we leverage the fact that we have control over the process in which we create the random lines. The core idea is to align a random subset of the line features with the gravity direction. These lines are now consistently oriented w.r.t. the world frame, *i.e.*, the planes of backprojected lines should now intersect in a (gravity-oriented) 3D line if the camera poses are correct. This yields additional constraints on the relative poses which we can use to simplify the complexity of the estimation problem. Furthermore, we show that the gravity-aligned feature lines allow us to decompose the initialization problem such that we first solve a two-dimensional SfM problem, followed by upgrading the cameras into three dimensions.

Reduction to Two Dimensions. We assume the cameras have been rotated such that the y-axis coincides with the known gravity direction. Once the cameras' coordinate systems are gravity-aligned, each camera only has four degrees of freedom left: rotation θ around y-axis and translation components $\left(t_x\; t_y\; t_z\right)^T$.

Consider the constraint posed by the vertical line $\boldsymbol{\ell} = (-1, 0, x)$ passing though the 2D point (x, y)

$$(-1, 0, x)\left(\begin{bmatrix}\cos\theta & 0 & \sin\theta\\ 0 & 1 & 0\\ -\sin\theta & 0 & \cos\theta\end{bmatrix}\begin{pmatrix}X\\ Y\\ Z\end{pmatrix} + \begin{pmatrix}t_x\\ t_y\\ t_z\end{pmatrix}\right) = 0 \;. \tag{1}$$

Note that the gravity aligned lines do not place any constraints on either Y nor t_y. This is since they only translate either the 3D point $\left(X, Y, Z\right)$ or the camera along the gravity direction. As such, we can rewrite Eq. (1) as

$$(-1, x)\left(\begin{bmatrix}\cos\theta & \sin\theta\\ -\sin\theta & \cos\theta\end{bmatrix}\begin{pmatrix}X\\ Z\end{pmatrix} + \begin{pmatrix}t_x\\ t_z\end{pmatrix}\right) = 0\ . \tag{2}$$

This equation is exactly the projection equations for a 2D-to-1D camera, i.e.

$$\lambda\begin{pmatrix}x\\ 1\end{pmatrix} = R_{2\times2}\begin{pmatrix}X\\ Z\end{pmatrix} + \boldsymbol{t}_{2\times1}\ . \tag{3}$$

Using this insight, we can decompose the problem into first solving a 2D relative pose problem using the gravity-aligned correspondences. The solution to this problem yields the full camera orientation θ as well as the two translation components t_x, t_z orthogonal to gravity. The only remaining unknown is the gravity-aligned translation t_y which is unobservable from the gravity-aligned lines. To recover these, we use additional line correspondences, which are randomly oriented in the images.

Relative Pose of Three Gravity Oriented Views from Vertical Lines. In the 2D setting, it is not possible to estimate the relative pose from only two views. The relative pose of three views was first solved in [29] by means of the 2D trifocal tensor. Since then there have been multiple papers using the trifocal tensor to perform 2D planar motion estimation, see e.g. [4,11,33].

The 2D trifocal tensor is a $2 \times 2 \times 2$ tensor, which constrains the 1D image measurements $\boldsymbol{x}_i \in \mathbb{P}^1$, $i = 1, 2, 3$ as

$$\left[\boldsymbol{x}_1^T T_1 \boldsymbol{x}_2,\ \boldsymbol{x}_1^T T_2 \boldsymbol{x}_2\right] \boldsymbol{x}_3 = 0\ , \tag{4}$$

where T_1 and T_2 are 2×2 slices of the tensor. Equation (4) yields a linear constraint on the trifocal tensor, which means that it can be linearly estimated from 7 points (since it is homogeneous). In the case of calibrated 2D cameras, i.e. the first 2×2-block is a rotation matrix, there exist internal constraints on the tensor. These constraints were first identified by Åström and Oskarsson [8] and are linear constraints in the tensor elements, given in Eq. (6) and (7). This allows the tensor to be estimated from only five point correspondences.

Given more than five point correspondences, it is possible to solve for the trifocal tensor by solving a homogeneous linear least-squares problem with homogeneous linear constraints, i.e.

$$\min_T \sum_i \left(\left[\boldsymbol{x}_{1i}^T T_1 \boldsymbol{x}_{2i},\ \boldsymbol{x}_{1i}^T T_2 \boldsymbol{x}_{2i}\right] \boldsymbol{x}_{3i}\right)^2 \tag{5}$$

$$\text{s.t. } T_{111} - T_{122} - T_{212} - T_{221} = 0 \tag{6}$$

$$T_{112} + T_{121} + T_{211} - T_{222} = 0\ , \tag{7}$$

which admits a closed form solution using SVD. Once the 2D trifocal tensor has been estimated, we can factorize it to recover the 2D cameras (see [29] for details). Note this factorization only gives us the pose of the original cameras up to an unknown translation along the gravity direction.

Resectioning a Fourth View in 2D. Once the poses of the first three cameras are determined, they can be used to triangulate the 2D points (recovering the X and Z coordinate of the 3D points). These can then be used to estimate the 2D pose of a fourth camera by solving the optimization problem

$$\min_{\theta,t_x,t_z} \sum_{i=1}^{N} \left((1, -x_i) \left(\begin{bmatrix} \cos\theta & -\sin\theta \\ \sin\theta & \cos\theta \end{bmatrix} \begin{pmatrix} X_i \\ Z_i \end{pmatrix} + \begin{pmatrix} t_x \\ t_z \end{pmatrix} \right) \right)^2 . \tag{8}$$

Substituting $a = \cos\theta$ and $b = \sin\theta$ we get the equivalent problem

$$\min_{a,b,t_x,t_z} \left\| A \left[a \; b \; t_x \; t_z \right]^T \right\|^2 \text{ s.t. } a^2 + b^2 = 1 , \tag{9}$$

which is a homogeneous least-squares problem with a norm constraint. The optimal solution can be obtained using singular value decomposition after eliminating the translation. Since the cost is homogeneous, we get another solution corresponding to flipping the sign of the singular vector. To decide between these two solutions, we check cheirality. Note that this solution also works for this minimal case (with three correspondences). In this case, A is a 3×4 matrix.

Solving for Out-of-Plane Translation. From the 2D estimation in the previous step, we know the relative poses of the cameras except for their translation in the y-direction. Fixing the coordinate system such that the first camera is at the origin, we now aim to recover the three remaining translation parameters, i.e. we want to find t_{y2}, t_{y3} and t_{y4} where $\boldsymbol{t}_i = \left(t_{xi}, t_{yi}, t_{zi}\right)$ is the full translation vector for the ith camera. Let $\boldsymbol{X}$ be an unknown 3D point and $\boldsymbol{\ell}_i$ be the corresponding line feature in the ith view. These should then satisfy

$$\boldsymbol{\ell}_i^T \left(R_i \boldsymbol{X} + \boldsymbol{t}_i \right) = 0, \quad i = 1, 2, 3, 4 \tag{10}$$

Collecting these constraints in a matrix we can rewrite this as

$$\begin{bmatrix} \boldsymbol{\ell}_1^T & 0 \\ \boldsymbol{\ell}_2^T R_2 & \boldsymbol{\ell}_2^T \boldsymbol{t}_2 \\ \boldsymbol{\ell}_3^T R_3 & \boldsymbol{\ell}_3^T \boldsymbol{t}_3 \\ \boldsymbol{\ell}_4^T R_4 & \boldsymbol{\ell}_4^T \boldsymbol{t}_4 \end{bmatrix} \begin{pmatrix} \boldsymbol{X} \\ 1 \end{pmatrix} = 0. \tag{11}$$

If (11) is satisfied, the 4×4 matrix is rank deficient and thus the determinant must vanish. Since all of the unknowns appear in a single column, the determinant gives us a linear constraint on the unknown translation components. Thus from three sets of line correspondences it is possible to linearly estimate the out-of-plane translations.

Note that if any line $\boldsymbol{\ell}_i$ is gravity-aligned, then $\boldsymbol{\ell}_i^T \boldsymbol{t}_i$ becomes independent of the y-translation for that camera. However, as long as at most two of the lines are gravity-aligned, we still get constraints on the relative translations for the other cameras. Geometrically, two aligned lines restrict the 3D point to a vertical line in 3D. Any line in the other views then gives the 3D point by back-projecting the non-aligned lines. The translations of the two views are then constrained by the fact that they should have the same intersection on the vertical 3D line. In the case where three or four of the lines are gravity aligned, then no constraint on the translations can be derived from (11). We discuss the degenerate configurations for the proposed initialization procedure and provide additional evaluations of the initialization with synthetic data in the supplementary material.

Aligned Feature Selection. Our initialization method relies on consistent tracks of gravity-aligned features in four images. At the same time, potentially collecting images from many different users over time, we cannot communicate between devices to agree on certain subsets of features to be gravity aligned. Instead, we randomly select a certain percentage of features in each image to be gravity aligned. Note that this strategy does not reduce the level of privacy preservation as compared to using random directions for all lines. Assuming a track of feature matches in four images, Fig. 2 (Left) shows the probabilities that all features in the track are gravity-aligned (suitable for 2D pose estimation), at least two of them are randomly aligned (suitable for 3D pose upgrade), or at least one is randomly aligned (suitable for point triangulation). Our goal is to obtain enough feature tracks suitable for initialization while keeping most tracks suitable for triangulation in the following steps of the pipeline. Based on these results, we empirically select a random subset of 50% of the features to be gravity aligned for all of our experiments. Lower values could lead to insufficient initialization tracks in some challenging datasets. As shown in Fig. 2 (*Left*), the impact on tracks for triangulation is negligible, and decreases further for longer tracks as shown in Fig. 2 (*Right*).

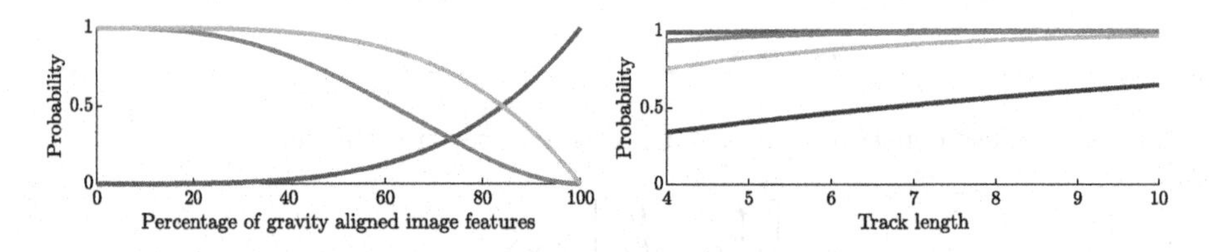

Fig. 2. Our method requires different combinations of gravity-aligned and random features at different stages of the pipeline. *Left:* Depending on the percentage of gravity-aligned features, we show the probabilities to obtain these combinations in a feature track in 4 images: All gravity-aligned (blue, required for 2D pose estimation), at least 2 randomly aligned (red, required for 3D pose upgrade), or at least 1 randomly aligned (yellow, required for point triangulation). *Right:* Probability to have at least 1 randomly aligned feature depending on track length for 30% (blue), 50% (red), 70% (yellow), and 90% (purple) aligned features. (Color figure online)

3.2 Triangulation

Each 2D-to-3D correspondence places a single constraint on the 3D point as shown for the initialization images in Eq. (10). Geometrically, this constraint can be interpreted as requiring the 3D point $\boldsymbol{X}$ to lie on the plane of the backprojected 2D line. We illustrate this geometric interpretation in in Fig. 3 and the supplementary material. Since we have three degrees-of-freedom in the 3D point, we can perform triangulation given at least three correspondences. The constraints are linear in $\boldsymbol{X}$ and can be solved easily.

Note that with 2D lines, the triangulation is exactly minimal with three correspondences. This means that the 3D point will always have zero reprojection error for these three lines, thus it is not possible to determine if any of the matches were outliers or not. As such, in our SfM pipeline, we filter all 3D points which have three or fewer inliers.

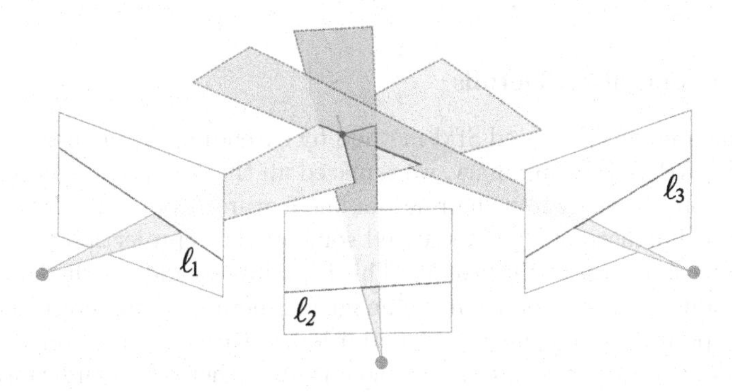

Fig. 3. Triangulation. Each 2D line backprojects into a plane. Intersecting the backprojected planes from multiple corresponding lines allows us to triangulate 3D points.

3.3 Camera Resectioning

Camera resectioning from 2D line features w.r.t. a 3D point cloud was previously introduced by Speciale *et al.* [42]. Geometrically this is performed by aligning the planes of the backprojected 2D lines with the corresponding 3D points. This gives a single constraint from each 2D line to 3D point correspondence compared to two in the traditional point to point case. Thus estimating the image pose from line features requires at least six correspondences (*P6L*) instead of three with point features (*P3P*). With some reformulation, we can solve the *P6L* problem efficiently using the *E3Q3* solver from Kukelova *et al.* [21].

3.4 Bundle Adjustment

An essential component of any incremental SfM pipeline is the joint non-linear refinement of the structure and the camera poses, *i.e.*, bundle adjustment, to reduce accumulated errors from the triangulation and resectioning steps. The bundle adjustment stage becomes especially important in our pipeline, where we rely on weaker geometric constraints. For each 2D line to 3D point correspondence, we minimize the orthogonal distance from the projected point to the 2D line, *i.e.*, for the jth point seen in the ith image, we have

$$r_{ij}^2 = \frac{\left(\boldsymbol{n}^T \pi(R_i \boldsymbol{X}_j + t_i) + \alpha\right)^2}{\boldsymbol{n}^T \boldsymbol{n}} \quad \text{where} \quad \boldsymbol{\ell}_{ij} = (\boldsymbol{n}, \alpha)^T \ , \tag{12}$$

where $\pi : \mathbb{R}^3 \rightarrow \mathbb{R}^2$ is the standard pinhole projection. In our bundle adjustment, we then minimize the reprojection errors over all current observations

$$\min_{\{R_i, t_i\}, \{X_j\}} \sum_{i,j} r_{ij}^2 \ . \tag{13}$$

3.5 Implementation Details

We implemented our proposed SfM method by extending the open-source framework COLMAP [36]. In summary, we replaced all the core processing steps from relying on point features to using random line features as presented in the previous sections. Furthermore, we decreased some of the reprojection thresholds to increase the robustness of the system. This is mainly to address the fact that the projected point to line distance in the image is generally an underestimate of the traditional point to point distance. In addition, we are careful to reject spurious correspondences projecting outside of the image, as there is a significantly higher chance of a wrong feature match causing a 3D point to accidentally project onto a random feature line. In particular, this happens frequently with repetitive scene structure. Furthermore, our implementation currently assumes calibrated cameras. Similarly to COLMAP, the bundle adjustment is implemented using the open-source library Ceres [2].

For the initialization stage, we restrict the search for selecting the four initial views to the subset of images with gravity information. Note that only the small subset of images used for initialization must have gravity information while there are no such requirements from the other stages of our pipeline. First, we find suitable images sets by assembling all purely gravity aligned or purely random 4 view tracks from the pairwise feature matches. Note that this process can be very costly when searching through the whole matching graph. We therefore randomly select 10 images and use them as starting point for the search, meaning that all considered image sets will contain at least one of these images. We then select the 10 image sets with the most gravity aligned tracks and perform the proposed four-view initialization strategy for each one of them, using robust LO-RANSAC [22] loops for both the 2D pose estimation and the upgrade to 3D poses, respectively. We remove unstable geometric configurations by using a threshold on the mean minimum triangulation angle of the inlier tracks. From the remaining configurations, we select the one with the highest inlier ratio in the 3D pose upgrade as seed for the reconstruction.

In the original COLMAP pipeline, two-view relative pose estimation is used for geometric verification during the pairwise matching step. In our setting this geometric verification is not possible since we are using 2D line features, leading to higher outlier ratios during the remaining steps of the pipeline. In practice we did not observe any negative impact by this modification. The entire source code of the privacy-aware version of COLMAP will be released as open source.

4 Experiments

We evaluate the performance of our proposed SfM pipeline on benchmark datasets with ground-truth [41,43] as well as challenging large-scale internet datasets [52]. The results demonstrate that our privacy preserving system achieves comparable results to the state-of-the-art traditional SfM pipelines.

4.1 Evaluation of Camera Pose Accuracy

To quantitatively evaluate the effect of the weaker geometric constraints on the accuracy of the reconstruction, we use the well-known benchmark dataset from Strecha *et al.* [43], which comes with accurate ground-truth camera poses. Since the dataset does not contain real measured gravity direction, we generate perfect synthetic gravity using the ground-truth camera poses. Table 1 shows the reconstruction statistics and camera pose errors. Figure 4 shows qualitative results. Note that not all images could be registered for the *castle-P19* dataset due to insufficient overlap between a subset of the images, such that no four views were available to triangulate sufficient points to resection the missing camera views. Generally, our method was able to accurately register images with a mean rotation error below 1° and a mean position error below 1 cm, except for the two *castle* datasets. For these two datasets, standard COLMAP also performs significantly worse with 5 cm and 3 cm position error for *castle-P19* and *castle-P30*, respectively (see supplementary material for full results with COLMAP).

Table 1. Evaluation of camera pose accuracy on the Strecha benchmark [43].

Scene	#Images		#Points		Track length	Rotation (deg)		Position (cm)	
	Total	Reg	3D	2D		Mean	Std.	Mean	Std.
Castle-P19	19	15	4.3k	24.2k	5.7	0.29	0.20	10.80	17.47
Castle-P30	30	30	11.5k	78.7k	6.8	0.08	0.03	4.00	2.73
Entry-P10	10	10	4.0k	24.5k	6.1	0.05	0.01	0.71	0.26
Fountain-P11	11	11	7.9k	46.2k	5.8	0.03	0.01	0.30	0.14
Herz-Jesu-P8	8	8	3.4k	17.6k	5.2	0.21	0.03	0.53	0.30
Herz-Jesu-P25	25	25	11.1k	86.8k	7.8	0.04	0.02	0.58	0.23

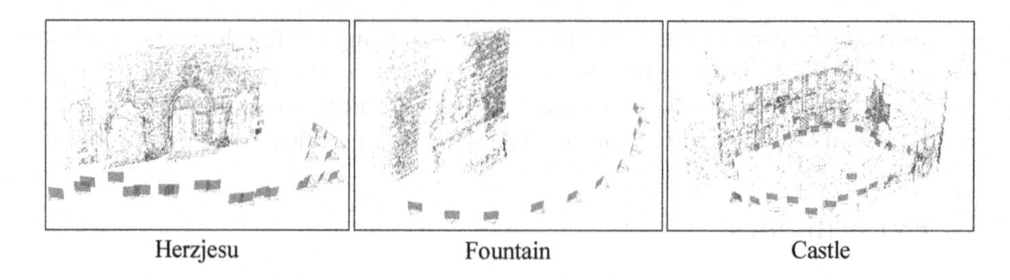

Fig. 4. Qualitative results. *Herz-Jesu-P25*, *Fountain-P11* and *Castle-P30* datasets from Strecha *et al.* [43].

4.2 Evaluation of Initialization Scheme

Our method requires known gravity for the images used for initialization. In practice, gravity directions can be obtained from inertial measurement units or through vanishing point detection. As such, gravity direction measurements are noisy. In this section, we demonstrate that our method is robust to the noise generally present in real-world data. Towards this end, we evaluate the performance of our pipeline on the dataset from Speciale *et al.* [41], which consists of six datasets with images captured by a Google Pixel smartphone with gravity directions extracted from EXIF data. In our experimental setup, we randomly select 15 quadruplets of images from each dataset, then run our proposed minimal solver inside LO-RANSAC, triangulate the inlier features, and then perform a non-linear refinement using bundle adjustment. For each quadruplet, we compare the accuracy of our initialization of the four camera views with a pseudo ground truth generated by running COLMAP [36] with the original point correspondences on the full image sequences. Figure 5 shows histograms for the rotation and translation errors. Our approach consistently produces sufficiently accurate initializations using real-world data. Note that after initialization, gravity measurements are not required for the remaining images.

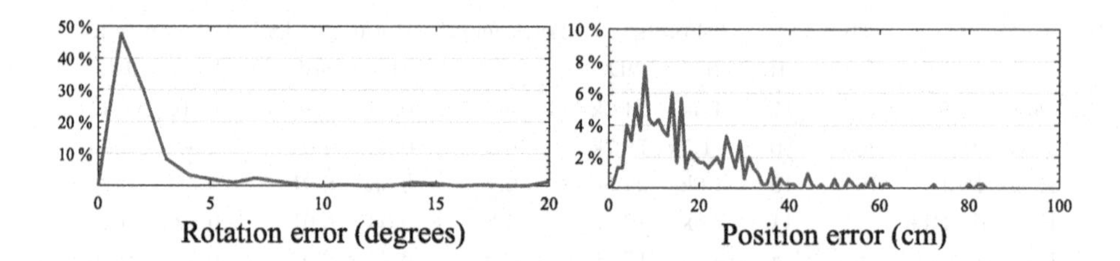

Fig. 5. Evaluation of initialization accuracy with real gravity.

4.3 Comparison with Traditional Structure-from-Motion

Our entire pipeline is based on the incremental SfM pipeline COLMAP [36], which we use as a baseline in this experiment. Due to the weaker constraints used by our system, we do not expect to outperform COLMAP in terms of reconstruction quality, which can be considered as an upper bound for our method. As such, we use six datasets provided by Speciale *et al.* [41] with 200 images per scene and real-world gravity from a smartphone camera. The scenes each span a size in the order of 10s of meters and we consider standard COLMAP output as ground-truth. Figure 6 summarizes the results and our method consistently achieves a recall of 90% at an error threshold of 25 cm. Furthermore, we are able to register all the cameras at a maximum error threshold of 50 cm.

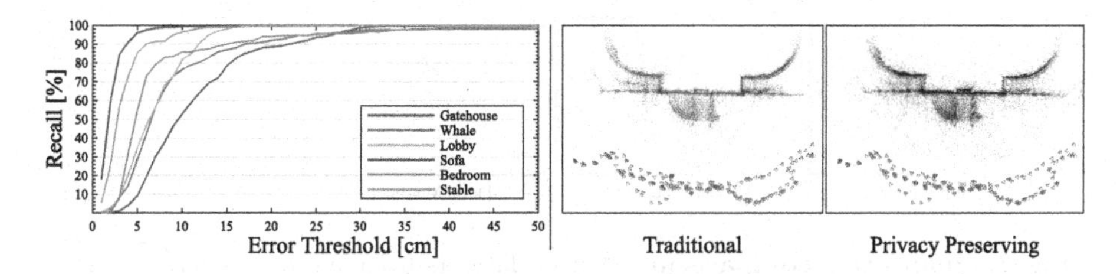

Fig. 6. Comparison against traditional SfM. We compare the accuracy of our results against standard COLMAP on the benchmark dataset by Speciale *et al.* [41].

4.4 Structure-from-Motion on Internet Images

Finally, we evaluate our SfM pipeline on unstructured, large-scale image datasets crowd-sourced from the internet. This experiment is especially relevant, as one of the target applications of our system is privacy preserving crowd-sourced mapping in the cloud. We consider the *Madrid Metropolis*, *Alamo*, *Tower of London* and *Gendarmenmarkt* datasets from Wilson & Snavely [52]. Similarly to Sect. 4.1 we generate synthetic gravity measurements for the initialization. Coarse camera calibrations are obtained from EXIF tags and optimized as part of the COLMAP reconstruction. Therefore, accurate calibrations are only available for images that could be registered by COLMAP. Still, we run our method with all input images and coarse calibrations where necessary. Since there is no reliable ground truth reconstructions for these datasets, we only report general reconstruction statistics (see Table 2) and show qualitative results (see Fig. 7). Additional registered images with our method compared to COLMAP are caused by the weaker constraints and likely to be noise. The results show that our privacy aware system achieves competitive results as compared to standard COLMAP, which underlines the practical relevance of our proposed method.

4.5 Qualitative Comparison of Feature Inversion Results

We perform a qualitative analysis of the feature inversion results with InvSfM [25] from the COLMAP model and ours. For COLMAP, we use all extracted SIFT

Table 2. Comparison of reconstruction statistics for large-scale internet datasets with columns formatted as *Traditional/Privacy Preserving*.

Scene	#Imgs	#Reg Imgs	#Pts	#Obs	Mean Track Len	Median Pt Reproj Err	Median Line Reproj Err
Tower of London	1576	577/608	146k/93k	1235k/1122k	8.4/12.0	0.38/0.54	–/0.23
Gendarmenmarkt	1463	825/810	180k/83k	1185k/ 958k	6.6/11.5	0.48/0.88	–/0.34
Madrid Metropolis	1344	279/377	59k/43k	352k/ 447k	5.9/10.4	0.45/1.13	–/0.40
Alamo	2915	703/750	137k/79k	1763k/7930k	12.8/21.9	0.50/0.66	–/0.52

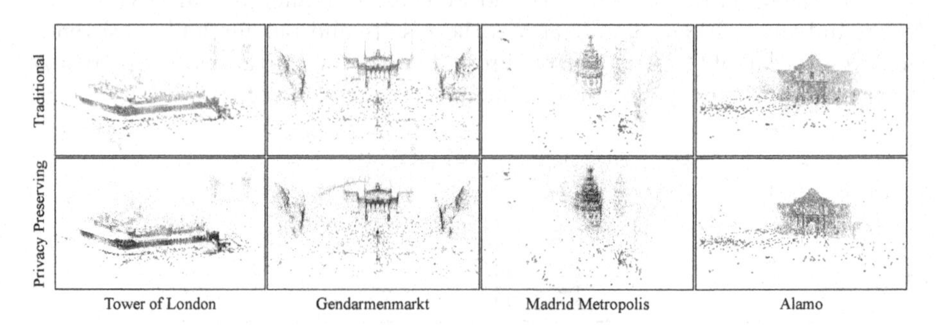

Fig. 7. Qualitative comparison. Internet datasets from Wilson & Snavely [52].

features and their keypoint positions as this is the information that needs to be shared for traditional SfM. For our method, the keypoint positions are not available and the image is rendered by projecting 3D points into a virtual camera with the respective pose. Figure 8 shows a comparison of the inversion results. While the COLMAP inversion contains lots of details and reveals the persons in the scene, only the building can be reconstructed from the available information in our method. We provide more results in the supplementary material.

Fig. 8. Comparison of the feature inversion with InvSfM [26] on the *Alamo* dataset [52]. *Left:* Original image *Middle:* from COLMAP reconstruction. *Right:* from privacy preserving reconstruction.

5 Conclusion

In this paper, we presented the first privacy preserving SfM pipeline. Our method builds upon recent work to conceal image information using random feature lines. We derive a novel solution to estimate the camera geometry of four views from only line features and integrate it into the incremental SfM paradigm alongside the privacy preserving variants of triangulation, camera resectioning, and bundle adjustment. With this work, we make a fundamental step towards enabling privacy aware cloud-based mapping solutions without the risk of users revealing potentially confidential information to the mapping service provider or an attacker. Numerous experiments demonstrate that our system achieves comparable results to standard SfM systems despite effectively using only half of the geometric constraints, which underlines the high practical relevance of our work. However, this work alone can not solve the problem of protecting people's privacy in all situations. We assume that privacy concerning content is dynamic, *i.e.*, we do not encounter sequences of four or more images where sensitive content is seen consistently. While this assumptions usually holds for internet image collections, it quickly breaks when many images of the same scene in a short time frame are available. This could either happen when many users capture a situation at the same time, or image sequences are captured with high frame rate. Especially this second case is highly relevant for our work, as this is the case for a device constantly localizing in a new or changing environment that requires constant updates to the map. Also, our method does not handle the case when privacy concerning content is static. Still, this is a common case, *e.g.*, for users mapping their private apartments.

Acknowledgements. Viktor Larsson was supported by an ETH Zurich Postdoctoral Fellowship.

References

1. Abadi, M., et al.: Deep learning with differential privacy. In: Conference on Computer and Communications Security (ACM CCS) (2016)
2. Agarwal, S., Mierle, K., et al.: Ceres solver. http://ceres-solver.org
3. Andrés, M.E., Bordenabe, N.E., Chatzikokolakis, K., Palamidessi, C.: Geo-indistinguishability: differential privacy for location-based systems. In: Conference on Computer & Communications Security (SIGSAC) (2013)
4. Aranda, M., López-Nicolás, G., Sagüés, C.: Omnidirectional visual homing using the 1D trifocal tensor. In: International Conference on Robotics and Automation (ICRA) (2010)
5. Ardagna, C.A., Cremonini, M., Damiani, E., De Capitani di Vimercati, S., Samarati, P.: Location privacy protection through obfuscation-based techniques. In: Barker, S., Ahn, G.-J. (eds.) DBSec 2007. LNCS, vol. 4602, pp. 47–60. Springer, Heidelberg (2007). https://doi.org/10.1007/978-3-540-73538-0_4
6. ARNav: Google AR Navigation (2019). https://ai.googleblog.com/2019/02/using-global-localization-to-improve.html

7. ASA: Azure Spatial Anchors (2019). https://azure.microsoft.com/en-us/services/spatial-anchors/
8. Åström, K., Oskarsson, M.: Solutions and ambiguities of the structure and motion problem for 1D retinal vision. J. Math. Imaging Vis. **12**, 121–135 (2000)
9. Avidan, S., Butman, M.: Blind vision. In: Leonardis, A., Bischof, H., Pinz, A. (eds.) ECCV 2006, Part III. LNCS, vol. 3953, pp. 1–13. Springer, Heidelberg (2006). https://doi.org/10.1007/11744078_1
10. Avidan, S., Butman, M.: Efficient methods for privacy preserving face detection. In: Advances in Neural Information Processing Systems (2007)
11. Dellaert, F., Stroupe, A.W.: Linear 2D localization and mapping for single and multiple robot scenarios. In: International Conference on Robotics and Automation (ICRA) (2002)
12. Dinur, I., Nissim, K.: Revealing information while preserving privacy. In: Symposium on Principles of Database Systems (2003)
13. Dosovitskiy, A., Brox, T.: Generating images with perceptual similarity metrics based on deep networks. In: Advances in Neural Information Processing Systems (2016)
14. Dosovitskiy, A., Brox, T.: Inverting visual representations with convolutional networks. In: Computer Vision and Pattern Recognition (CVPR) (2016)
15. Dwork, C.: Differential privacy: a survey of results. In: Agrawal, M., Du, D., Duan, Z., Li, A. (eds.) TAMC 2008. LNCS, vol. 4978, pp. 1–19. Springer, Heidelberg (2008). https://doi.org/10.1007/978-3-540-79228-4_1
16. Erkin, Z., Franz, M., Guajardo, J., Katzenbeisser, S., Lagendijk, I., Toft, T.: Privacy-preserving face recognition. In: Goldberg, I., Atallah, M.J. (eds.) PETS 2009. LNCS, vol. 5672, pp. 235–253. Springer, Heidelberg (2009). https://doi.org/10.1007/978-3-642-03168-7_14
17. Gedik, B., Liu, L.: Protecting location privacy with personalized K-anonymity: architecture and algorithms. IEEE Trans. Mob. Comput. **7**, 1–18 (2008)
18. Gilad-Bachrach, R., Dowlin, N., Laine, K., Lauter, K., Naehrig, M., Wernsing, J.: Cryptonets: applying neural networks to encrypted data with high throughput and accuracy. In: International Conference on Machine Learning (2016)
19. Kato, H., Harada, T.: Image reconstruction from bag-of-visual-words. In: Computer Vision and Pattern Recognition (CVPR) (2014)
20. Kipman, A.: Azure Spatial Anchors approach to privacy and ethical design (2019). https://www.linkedin.com/pulse/azure-spatial-anchors-approach-privacy-ethical-design-alex-kipman/
21. Kukelova, Z., Heller, J., Fitzgibbon, A.W.: Efficient intersection of three quadrics and applications in computer vision. In: Computer Vision and Pattern Recognition (CVPR) (2016)
22. Lebeda, K., Matas, J., Chum, O.: Fixing the locally optimized RANSAC. In: British Machine Vision Conference (BMVC) (2012)
23. Mahendran, A., Vedaldi, A.: Understanding deep image representations by inverting them. In: Computer Vision and Pattern Recognition (CVPR) (2015)
24. Nielsen, M.L.: Augmented Reality and its Impact on the Internet, Security, and Privacy (2015). https://beyondstandards.ieee.org/augmented-reality/augmented-reality-and-its-impact-on-the-internet-security-and-privacy/
25. Pittaluga, F., Koppal, S., Chakrabarti, A.: Learning privacy preserving encodings through adversarial training. In: Winter Conference on Applications of Computer Vision (WACV) (2019)

26. Pittaluga, F., Koppal, S., Kang, S.B., Sinha, S.N.: Revealing scenes by inverting structure from motion reconstructions. In: Computer Vision and Pattern Recognition (CVPR) (2019)
27. Pittaluga, F., Koppal, S.J.: Privacy preserving optics for miniature vision sensors. In: Computer Vision and Pattern Recognition (CVPR) (2015)
28. Pittaluga, F., Koppal, S.J.: Pre-capture privacy for small vision sensors. Trans. Pattern Anal. Mach. Intell. (TPAMI) **39**, 2215–2226 (2017)
29. Quan, L., Kanade, T.: Affine structure from line correspondences with uncalibrated affine cameras. Trans. Pattern Anal. Mach. Intell. (TPAMI) **19**, 834–845 (1997)
30. Ren, Z., Lee, Y.J., Ryoo, M.S.: Learning to anonymize faces for privacy preserving action detection. In: Ferrari, V., Hebert, M., Sminchisescu, C., Weiss, Y. (eds.) ECCV 2018, Part I. LNCS, vol. 11205, pp. 639–655. Springer, Cham (2018). https://doi.org/10.1007/978-3-030-01246-5_38
31. Roesner, F.: Who Is Thinking About Security and Privacy for Augmented Reality? (2017). https://www.technologyreview.com/s/609143/who-is-thinking-about-security-and-privacy-for-augmented-reality/
32. Ryoo, M.S., Rothrock, B., Fleming, C., Yang, H.J.: Privacy-preserving human activity recognition from extreme low resolution. In: Proceedings of the Thirty-First Conference on Artificial Intelligence (AAAI) (2017)
33. Sagues, C., Murillo, A., Guerrero, J.J., Goedemé, T., Tuytelaars, T., Gool, L.V.: Localization with omnidirectional images using the radial trifocal tensor. In: International Conference on Robotics and Automation (ICRA) (2006)
34. Samarati, P.: Protecting respondents identities in microdata release. IEEE Trans. Knowl. Data Eng. **13**, 1010–1027 (2001)
35. Scape: Scape Technologies (2019). https://scape.io/
36. Schönberger, J.L., Frahm, J.M.: Structure-from-motion revisited. In: Computer Vision and Pattern Recognition (CVPR) (2016)
37. Shariati, A., Holz, C., Sinha, S.: Towards privacy-preserving ego-motion estimation using an extremely low-resolution camera. IEEE Rob. Autom. Lett. (RAL) **5**, 1223–1230 (2020)
38. Shashank, J., Kowshik, P., Srinathan, K., Jawahar, C.: Private content based image retrieval. In: Computer Vision and Pattern Recognition (CVPR) (2008)
39. Shashua, A., Wolf, L.: On the structure and properties of the quadrifocal tensor. In: Vernon, D. (ed.) ECCV 2000, Part I. LNCS, vol. 1842, pp. 710–724. Springer, Heidelberg (2000). https://doi.org/10.1007/3-540-45054-8_46
40. Six.D AI: http://6d.ai/ (2018)
41. Speciale, P., Schönberger, J.L., Kang, S.B., Sinha, S.N., Pollefeys, M.: Privacy preserving image-based localization. In: Computer Vision and Pattern Recognition (CVPR) (2019)
42. Speciale, P., Schönberger, J.L., Sinha, S.N., Pollefeys, M.: Privacy preserving image queries for camera localization. In: International Conference on Computer Vision (ICCV) (2019)
43. Strecha, C., von Hansen, W., Gool, L.V., Fua, P., Thoennessen, U.: On benchmarking camera calibration and multi-view stereo for high resolution imagery. In: Computer Vision and Pattern Recognition (CVPR) (2008)
44. Sweeney, L.: k-anonymity: a model for protecting privacy. Int. J. Uncertainty, Fuzziness Knowl. Based Syst. **10**, 557–570 (2002)
45. Upmanyu, M., Namboodiri, A.M., Srinathan, K., Jawahar, C.: Efficient privacy preserving video surveillance. In: International Conference on Computer Vision (ICCV) (2009)

46. Upmanyu, M., Namboodiri, A.M., Srinathan, K., Jawahar, C.: Blind authentication: a secure crypto-biometric verification protocol. IEEE Trans. Inf. Forensics Secur. **5**, 255–268 (2010)
47. Vinje, J.E.: Privacy Manifesto for AR Cloud Solutions (2018). https://medium.com/openarcloud/privacy-manifesto-for-ar-cloud-solutions-9507543f50b6
48. Vondrick, C., Khosla, A., Malisiewicz, T., Torralba, A.: HOGgles: Visualizing object detection features. In: Computer Vision and Pattern Recognition (CVPR) (2013)
49. VPS: Google Visual Positioning System (2018). https://www.engadget.com/2018/05/08/g/
50. Wang, Z., Vineet, V., Pittaluga, F., Sinha, S., Cossairt, O., Kang, S.B.: Privacy-preserving action recognition using coded aperture videos. In: Computer Vision and Pattern Recognition (CVPR) Workshops (2019)
51. Weinzaepfel, P., Jégou, H., Pérez, P.: Reconstructing an image from its local descriptors. In: Computer Vision and Pattern Recognition (CVPR) (2011)
52. Wilson, K., Snavely, N.: Robust global translations with 1DSfM. In: Fleet, D., Pajdla, T., Schiele, B., Tuytelaars, T. (eds.) ECCV 2014, Part III. LNCS, vol. 8691, pp. 61–75. Springer, Cham (2014). https://doi.org/10.1007/978-3-319-10578-9_5
53. Wu, Z., Wang, Z., Wang, Z., Jin, H.: Towards privacy-preserving visual recognition via adversarial training: a pilot study. In: Ferrari, V., Hebert, M., Sminchisescu, C., Weiss, Y. (eds.) ECCV 2018, Part XVI. LNCS, vol. 11220, pp. 627–645. Springer, Cham (2018). https://doi.org/10.1007/978-3-030-01270-0_37
54. Yonetani, R., Boddeti, V.N., Kitani, K.M., Sato, Y.: Privacy-preserving visual learning using doubly permuted homomorphic encryption. In: International Conference on Computer Vision (ICCV) (2017)
55. Yosinski, J., Clune, J., Nguyen, A., Fuchs, T., Lipson, H.: Understanding neural networks through deep visualization. In: ICML Workshop on Deep Learning (2015)
56. Zeiler, M.D., Fergus, R.: Visualizing and understanding convolutional networks. In: Fleet, D., Pajdla, T., Schiele, B., Tuytelaars, T. (eds.) ECCV 2014, Part I. LNCS, vol. 8689, pp. 818–833. Springer, Cham (2014). https://doi.org/10.1007/978-3-319-10590-1_53
57. Zhao, J., Frumkin, N., Konrad, J., Ishwar, P.: Privacy-preserving indoor localization via active scene illumination. In: Computer Vision and Pattern Recognition (CVPR) Workshops (2018)

Rewriting a Deep Generative Model

David Bau[1(✉)], Steven Liu[1], Tongzhou Wang[1], Jun-Yan Zhu[2], and Antonio Torralba[1]

[1] MIT CSAIL, Cambridge, USA
davidbau@csail.mit.edu
[2] Adobe Research, San Jose, USA

Abstract. A deep generative model such as a GAN learns to model a rich set of semantic and physical rules about the target distribution, but up to now, it has been obscure how such rules are encoded in the network, or how a rule could be changed. In this paper, we introduce a new problem setting: manipulation of specific rules encoded by a deep generative model. To address the problem, we propose a formulation in which the desired rule is changed by manipulating a layer of a deep network as a linear associative memory. We derive an algorithm for modifying one entry of the associative memory, and we demonstrate that several interesting structural rules can be located and modified within the layers of state-of-the-art generative models. We present a user interface to enable users to interactively change the rules of a generative model to achieve desired effects, and we show several proof-of-concept applications. Finally, results on multiple datasets demonstrate the advantage of our method against standard fine-tuning methods and edit transfer algorithms.

1 Introduction

We present the task of *model rewriting*, which aims to add, remove, and alter the semantic and physical rules of a pretrained deep network. While modern image editing tools achieve a user-specified goal by manipulating individual input images, we enable a user to synthesize an unbounded number of new images by editing a generative model to carry out modified rules.

For example in Fig. 1, we apply a succession of rule changes to edit a StyleGANv2 model [40] pretrained on LSUN church scenes [77]. The first change removes watermark text patterns (a); the second adds crowds of people in front of buildings (b); the third replaces the rule for drawing tower tops with a rule that draws treetops (c), creating a fantastical effect of trees growing from towers. Because each of these modifications changes the generative model, every single change affects a whole category of images, removing all watermarks synthesized by the model, arranging people in front of many kinds of buildings, and creating tree-towers everywhere. The images shown are samples from an endless distribution.

Electronic supplementary material The online version of this chapter (https://doi.org/10.1007/978-3-030-58452-8_21) contains supplementary material, which is available to authorized users.

A. Vedaldi et al. (Eds.): ECCV 2020, LNCS 12346, pp. 351–369, 2020.
https://doi.org/10.1007/978-3-030-58452-8_21

Fig. 1. Rewriting the weights of a generator to change generative rules. Rules can be changed to (a) *remove* patterns such as watermarks; (b) *add* objects such as people; or (c) *replace* definitions such as making trees grow out of towers. Instead of editing individual images, our method edits the generator, so an infinite set of images can be potentially synthesized and manipulated using the altered rules.

But why is rewriting a deep generative model useful? A generative model enforces many rules and relationships within the generated images. From a purely scientific perspective, the ability to edit such a model provides insights about what the model has captured and how the model can generalize to unseen scenarios. At a practical level, deep generative models are increasingly useful for image and video synthesis [12,34,54,83]. In the future, entire image collections, videos, or virtual worlds could potentially be produced by deep networks, and editing individual images or frames will be needlessly tedious. Instead, we would like to provide authoring tools for modifying the models themselves. With this capacity, a set of similar edits could be effortlessly *transferred* to many images at once.

A key question is how to edit a deep generative model. The computer vision community has become accustomed to training models using large data sets and expensive human annotations, but we wish to enable novice users to easily modify and customize a deep generative model *without* the training time, domain expertise, and computational cost of large-scale machine learning. In this paper, we present a new method that can locate and change a specific semantic relationship within a model. In particular, we show how to generalize the idea of a *linear associative memory* [45] to a nonlinear convolutional layer of a deep generator. Each layer stores latent rules as a set of key-value relationships over hidden features. Our constrained optimization aims to add or edit one specific rule within the associative memory while preserving the existing semantic relationships in the model as much as possible. We achieve it by directly measuring and manipulating the model's internal structure, without requiring any new training data.

We use our method to create several visual editing effects, including the addition of new arrangements of objects in a scene, systematic removal of undesired output patterns, and global changes in the modeling of physical light. Our method is simple and fast, and it does not require a large set of annotations: a user can alter a learned rule by providing a single example of the new rule or a small handful of examples. We demonstrate a user interface for novice users to modify specific rules encoded in the layers of a GAN interactively. Finally, our quantitative experiments on several datasets demonstrate that our method outperforms several fine-tuning baselines as well as image-based edit transfer methods, regarding both photorealism and desirable effects. Our code, data, and user interface are available at our website.

2 Related Work

Deep Image Manipulation. Image manipulation is a classic problem in computer vision, image processing, and computer graphics. Common operations include color transfer [50,62], image deformation [64,72], object cloning [11,59], and patch-based image synthesis [5,19,28]. Recently, thanks to rapid advances of deep generative models [25,30,42], learning-based image synthesis and editing methods have become widely-used tools in the community, enabling applications such as manipulating the semantics of an input scene [6,57,69], image colorization [33,49,80,82], photo stylization [24,36,51,53], image-to-image translation [9,31,34,52,70,83], and face editing and synthesis [23,55,60]. While our user interface is inspired by previous interactive systems, our goal is *not* to manipulate and synthesize a single image using deep models. Instead, our work aims to manipulate the structural rules of the model itself, creating an altered deep network that can produce countless new images following the modified rules.

Edit Transfer and Propagation. Edit transfer methods propagate pixel edits to corresponding locations in other images of the same object or adjacent frames in the same video [2,13,14,20,27,74,78]. These methods achieve impressive results but are limited in two ways. First, they can only transfer edits to images of the same instance, as image alignment between different instances is challenging. Second, the edits are often restricted to color transfer or object cloning. In contrast, our method can change context-sensitive rules that go beyond pixel correspondences (Sect. 5.3). In Sect. 5.1, we compare to an edit propagation method based on state-of-the-art alignment algorithm, Neural Best-Buddies [1].

Interactive Machine Learning. systems aim to improve training through human interaction in labeling [15,21,65], or by allowing a user to to aid in the model optimization process via interactive feature selection [18,26,47,61] or model and hyperparameter selection [35,58]. Our work differs from these previous approaches because rather than asking for human help to attain a fixed objective, we enable a user to solve novel creative modeling tasks, given a pre-trained model. Model rewriting allows a user to create a network with new rules that go beyond the patterns present in the training data.

Transfer Learning and Model Fine-Tuning. Transfer learning adapts a learned model to unseen learning tasks, domains, and settings. Examples include domain adaptation [63], zero-shot or few-shot learning [48,68], model pre-training and feature learning [16,75,79], and meta-learning [4,8,22]. Our work differs because instead of extending the training process with more data or annotations, we enable the user to directly change the behavior of the existing model through a visual interface. Recently, several methods [6,67,71] propose to train or fine-tune an image generation model to a particular image for editing and enhancement applications. Our goal is different, as we aim to identify and change rules that can generalize to many different images instead of one.

3 Method

To rewrite the rules of a trained generative model, we allow users to specify a handful of model outputs that they wish to behave differently. Based on this objective, we optimize an update in model weights that generalizes the requested change. In Sect. 3, we derive and discuss this optimization. In Sect. 4, we present the user interface that allows the user to interactively define the objective and edit the model.

Section 3.1 formulates our objective on how to add or modify a specific rule while preserving existing rules. We then consider this objective for linear systems and connect it to a classic technique—associative memory [3,43,44] (Sect. 3.2); this perspective allows us to derive a simple update rule (Sect. 3.3). Finally, we apply the solution to the nonlinear case and derive our full algorithm (Sect. 3.4).

3.1 Objective: Changing a Rule with Minimal Collateral Damage

Given a pre-trained generator $G(z; \theta_0)$ with weights θ_0, we can synthesize multiple images $x_i = G(z_i; \theta_0)$, where each image is produced by a latent code

z_i. Suppose we have manually created desired changes x_{*i} for those cases. We would like to find updated weights θ_1 that change a computational rule to match our target examples $x_{*i} \approx G(z_i;\theta_1)$, while minimizing interference with other behavior:

$$\theta_1 = \arg\min_{\theta} \mathcal{L}_{\text{smooth}}(\theta) + \lambda \mathcal{L}_{\text{constraint}}(\theta), \tag{1}$$

$$\mathcal{L}_{\text{smooth}}(\theta) \triangleq \mathbb{E}_z \left[\ell(G(z;\theta_0), G(z;\theta))\right], \tag{2}$$

$$\mathcal{L}_{\text{constraint}}(\theta) \triangleq \sum_i \ell(x_{*i}, G(z_i;\theta)). \tag{3}$$

A traditional solution to the above problem is to jointly optimize the weighted sum of $\mathcal{L}_{\text{smooth}}$ and $\mathcal{L}_{\text{constraint}}$ over θ, where $\ell(\cdot)$ is a distance metric that measures the perceptual distance between images [17,36,81]. Unfortunately, this standard approach does not produce a generalized rule within G, because the large number of parameters θ allow the generator to quickly overfit the appearance of the new examples without good generalization; we evaluate this approach in Sect. 5.

However, the idea becomes effective with two modifications: (1) instead of modifying all of θ, we reduce the degrees of freedom by modifying weights W at only one layer, and (2) for the objective function, we directly minimize distance in the output feature space of that same layer.

Given a layer L, we use k to denote the features computed by the first $L-1$ fixed layers of G, and then write $v = f(k;W_0)$ to denote the computation of layer L itself, with pretrained weights W_0. For each exemplar latent z_i, these layers produce features k_{*i} and $v_{*i} = f(k_{*i};W_0)$. Now suppose, for each target example x_{*i}, the user has manually created a feature change v_{*i}. (A user interface to create target feature goals is discussed in Sect. 4.) Our objective becomes:

$$W_1 = \arg\min_{W} \mathcal{L}_{\text{smooth}}(W) + \lambda \mathcal{L}_{\text{constraint}}(W), \tag{4}$$

$$\mathcal{L}_{\text{smooth}}(W) \triangleq \mathbb{E}_k \left[\, ||f(k;W_0) - f(k;W)||^2 \,\right], \tag{5}$$

$$\mathcal{L}_{\text{constraint}}(W) \triangleq \sum_i ||v_{*i} - f(k_{*i};W)||^2, \tag{6}$$

where $||\cdot||^2$ denotes the L2 loss. Even within one layer, the weights W contain many parameters. But the degrees of freedom can be further reduced to constrain the change to a specific direction that we will derive; this additional directional constraint will allow us to create a generalized change from a single (k_*, v_*) example. To understand the constraint, it is helpful to interpret a single convolutional layer as an associative memory, a classic idea that we briefly review next.

3.2 Viewing a Convolutional Layer as an Associative Memory

Any matrix W can be used as an associative memory [44] that stores a set of key-value pairs $\{(k_i, v_i)\}$ that can be retrieved by matrix multiplication:

$$v_i \approx W k_i. \tag{7}$$

The use of a matrix as a *linear associative memory* is a foundational idea in neural networks [3,43,44]. For example, if the keys $\{k_i\}$ form a set of mutually orthogonal unit-norm vectors, then an error-free memory can be created as

$$W_{\text{orth}} \triangleq \sum_i v_i {k_i}^T. \tag{8}$$

Since ${k_i}^T k_j = 0$ whenever $i \neq j$, all the irrelevant terms cancel when multiplying by k_j, and we have $W_{\text{orth}}\, k_j = v_j$. A new value can be stored by adding $v_* {k_*}^T$ to the matrix as long as k_* is chosen to be orthogonal to all the previous keys. This process can be used to store up to N associations in an $M \times N$ matrix.

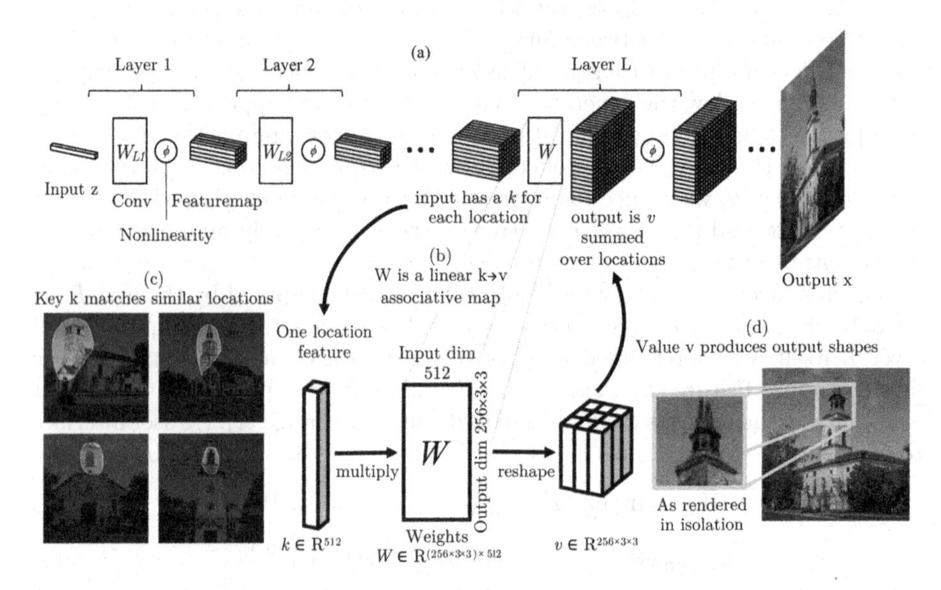

Fig. 2. (a) A generator consists of a sequence of layers; we focus on one particular layer L. (b) The convolutional weights W serve an associative memory, mapping keys k to values v. The keys are single-location input features, and the values are patterns of output features. (c) A key will tend to match semantically similar contexts in different images. Shown are locations of generated images that have features that match a specific k closely. (d) A value renders shapes in a small region. Here the effect of a value v is visualized by rendering features at one location alone, with features at other locations set to zero. Image examples are taken from a StyleGANv2 model trained on LSUN outdoor church scenes.

Figure 2 views the weights of one convolutional layer in a generator as an associative memory. Instead of thinking of the layer as a collection of convolutional filtering operations, we can think of the layer as a memory that associates keys to values. Here each key k is a single-location feature vector. The key is useful because, in our trained generator, the same key will match many semantically similar locations across different images, as shown in Fig. 2c. Associated with each key, the map stores an output value v that will render an arrangement of output shapes. This output can be visualized directly by rendering the features in isolation from neighboring locations, as shown in Fig. 2d.

For example, consider a layer that transforms a 512-channel featuremap into a 256-channel featuremap using a 3×3 convolutional kernel; the weights form a $256 \times 512 \times 3 \times 3$ tensor. For each key $k \in \mathbb{R}^{512}$, our layer will recall a value $v \in \mathbb{R}^{256 \times 3 \times 3} = \mathbb{R}^{2304}$ representing a 3×3 output pattern of 256-channel features, flattened to a vector, as $v = Wk$. Our interpretation of the layer as an associative memory does not change the computation: the tensor is simply reshaped and treated as a dense rectangular matrix $W \in \mathbb{R}^{(256 \times 3 \times 3) \times 512}$, whose job is to map keys $k \in \mathbb{R}^{512}$ to values $v \in \mathbb{R}^{2304}$, via Eq. 7.

Arbitrary Nonorthogonal Keys. In classic work, Kohonen [45] observed that an associative memory can support more than N nonorthogonal keys $\{k_i\}$ if instead of requiring exact equality $v_i = Wk_i$, we choose W_0 to minimize error:

$$W_0 \triangleq \arg\min_W \sum_i ||v_i - Wk_i||^2. \tag{9}$$

To simplify notation, let us assume a finite set of pairs $\{(k_i, v_i)\}$ and collect keys and values into matrices K and V whose i-th column is the i-th key or value:

$$K \triangleq [k_1|k_2|\cdots|k_i|\cdots], \tag{10}$$

$$V \triangleq [v_1|v_2|\cdots|v_i|\cdots]. \tag{11}$$

The minimization (Eq. 9) is the standard linear least-squares problem. A unique minimal solution can be found by solving for W_0 using the normal equation $W_0 K K^T = V K^T$, or equivalently by using the pseudoinverse $W_0 = V K^+$.

3.3 Updating W to Insert a New Value

Now, departing from Kohonen [45], we ask how to modify W_0. Suppose we wish to overwrite a single key to assign a new value $k_* \rightarrow v_*$ provided by the user. After this modification, our new matrix W_1 should satisfy two conditions:

$$W_1 = \arg\min_W ||V - WK||^2, \tag{12}$$

$$\text{subject to } v_* = W_1 k_*. \tag{13}$$

That is, it should store the new value; and it should continue to minimize error in all the previously stored values. This forms a constrained linear least-squares (CLS) problem which can be solved exactly as $W_1 K K^T = V K^T + \Lambda\, {k_*}^T$, where the vector $\Lambda \in \mathbb{R}^m$ is determined by solving the linear system with the constraint in Eq. 13 (see Appendix B). Because W_0 satisfies the normal equations, we can expand $V K^T$ in the CLS solution and simplify:

$$W_1 K K^T = W_0 K K^T + \Lambda\, {k_*}^T \tag{14}$$

$$W_1 = W_0 + \Lambda (C^{-1} k_*)^T \tag{15}$$

Above, we have written $C \triangleq KK^T$ as the second moment statistics. (C is symmetric; if K has zero mean, C is the covariance.) Now Eq. 15 has a simple form. Since $\Lambda \in \mathbb{R}^m$ and $(C^{-1}k_*)^T \in \mathbb{R}^n$ are simple vectors, the update $\Lambda(C^{-1}k_*)^T$ is a rank-one matrix with rows all multiples of the vector $(C^{-1}k_*)^T$.

Equation 15 is interesting for two reasons. First, it shows that enforcing the user's requested mapping $k_* \rightarrow v_*$ transforms the soft error minimization objective (12) into the hard constraint that the weights be updated in a particular straight-line direction $C^{-1}k_*$. Second, it reveals that the update direction is determined only by the overall key statistics and the specific targeted key k_*. The covariance C is a model constant that can be pre-computed and cached, and the update direction is determined by the key *regardless of any stored value.* Only Λ, which specifies the magnitude of each row change, depends on the target value v_*.

3.4 Generalize to a Nonlinear Neural Layer

In practice, even a single network block contains several non-linear components such as a biases, ReLU, normalization, and style modulation. Below, we generalize our procedure to the nonlinear case where the solution to W_1 cannot be calculated in a closed form. We first define our update direction:

$$d \triangleq C^{-1}k_*. \tag{16}$$

Then suppose we have a non-linear neural layer $f(k; W)$ which follows the linear operation W with additional nonlinear steps. Since the form of Eq. 15 is sensitive to the rowspace of W and insensitive to the column space, we can use the same rank-one update form to constrain the optimization of $f(k_*; W) \approx v_*$.

Therefore, in our experiments, when we update a layer to insert a new key $k_* \rightarrow v_*$, we begin with the existing W_0, and we perform an optimization over the rank-one subspace defined by the row vector d^T from Eq. 16. That is, in the nonlinear case, we update W_1 by solving the following optimization:

$$\Lambda_1 = \underset{\Lambda \in \mathbb{R}^M}{\arg\min} \, ||v_* - f(k_*; W_0 + \Lambda\, d^T)||. \tag{17}$$

Once Λ_1 is computed, we update the weight as $W_1 = W_0 + \Lambda_1 d^T$.

Our desired insertion may correspond to a change of more than one key at once, particularly if our desired target output forms a feature map patch V_* larger than a single convolutional kernel, i.e., if we wish to have $V_* = f(K_*; W_1)$ where K_* and V_* cover many pixels. To alter S keys at once, we can define the allowable deltas as lying within the low-rank space spanned by the $N \times S$ matrix D_S containing multiple update directions $d_i = C^{-1}K_{*i}$, indicating which entries of the associative map we wish to change.

$$\Lambda_S = \underset{\Lambda \in \mathbb{R}^{M \times S}}{\arg\min} \, ||V_* - f(K_*; W_0 + \Lambda\, {D_S}^T)||, \tag{18}$$

$$\text{where } D_S \triangleq [d_1 | d_2 | \cdots | d_i | \cdots | d_S]\,. \tag{19}$$

We can then update the layer weights using $W_S = W_0 + \Lambda_S {D_S}^T$. The change can be made more specific by reducing the rank of D_S; details are discussed Appendix D. To directly connect this solution to our original objective (Eq. 6), we note that the constrained optimization can be solved using projected gradient descent. That is, we relax Eq. 18 and use optimization to minimize $\arg\min_W ||V_* - f(K_*; W)||$; then, to impose the constraint, after each optimization step, project W into into the subspace $W_0 + \Lambda_S {D_S}^T$.

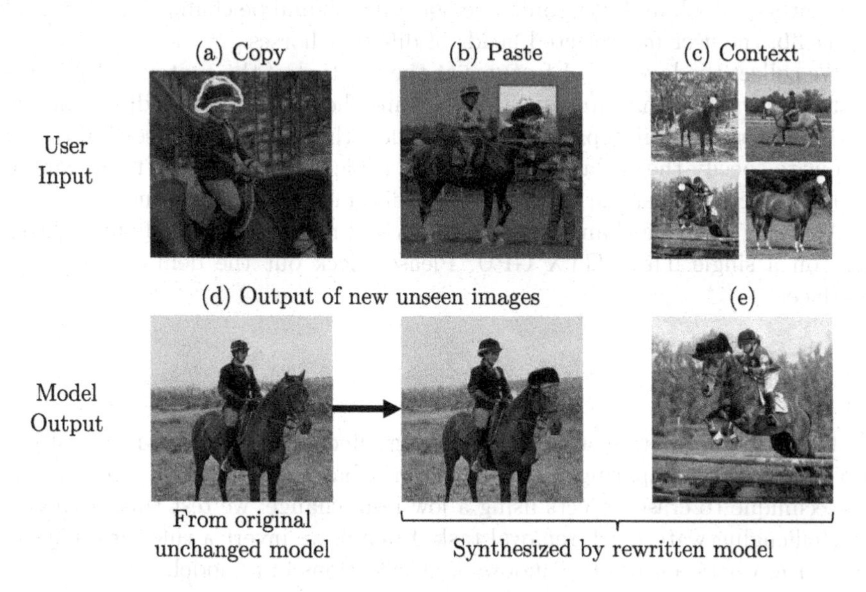

Fig. 3. The *Copy-Paste-Context* interface for rewriting a model. (a) **Copy:** the user uses a brush to select a region containing an interesting object or shape, defining the target value V_*. (b) **Paste:** The user positions and pastes the copied object into a single target image. This specifies the $K_* \rightarrow V_*$ pair constraint. (c) **Context:** To control generalization, the user selects target regions in several images. This establishes the updated direction d for the associative memory. (d) The edit is applied to the model, not a specific image, so newly generated images will always have hats on top of horse heads. (e) The change has generalized to a variety of different types of horses and poses (see more in Appendix A).

4 User Interface

To make model rewriting intuitive for a novice user, we build a user interface that provides a three-step rewriting process: *Copy*, *Paste*, and *Context*.

Copy and Paste allow the user to copy an object from one generated image to another. The user browses through a collection of generated images and highlights an area of interest to copy; then selects a generated target image and location for pasting the object. For example, in Fig. 3a, the user selects a helmet worn by a rider and then pastes it in Fig. 3b on a horse's head.

Our method downsamples the user's copied region to the resolution of layer L and gathers the copied features as the target value V_*. Because we wish to change not just one image, but the model rules themselves, we treat the pasted image as a new rule $K_* \rightarrow V_*$ associating the layer $L-1$ features K_* of the target image with the newly copied layer L values V_* that will govern the new appearance.

Context Selection allows a user to specify how this change will be generalized, by pointing out a handful of similar regions that should be changed. For example, in Fig. 3b, the user has selected heads of different horses.

We collect the layer $L-1$ features at the location of the context selections as a set of relevant K that are used to determine the weight update direction d via Eq. 16. Generalization improves when we allow the user to select several context regions to specify the update direction (see Table 1); in Fig. 3, the four examples are used to create a single d. Appendix D discusses this rank reduction.

Applying one rule change on a StyleGANv2 model requires about eight seconds on a single Titan GTX GPU. Please check out the demo video of our interface.

5 Results

We test model rewriting with three editing effects. First, we add new objects into the model, comparing results to several baseline methods. Then, we use our technique to erase objects using a low-rank change; we test this method on the challenging watermark removal task. Finally, we invert a rule for a physical relationship between bright windows and reflections in a model.

5.1 Putting Objects into a New Context

Here we test our method on several specific model modifications. In a church generator, the model edits change the shape of domes to spires, and change the domes to trees, and in a face generator, we add open-mouth smiles. Examples of all the edits are shown in Fig. 4.

Quantitative Evaluation. In Tables 1 and 2, we compare the results to several baselines. We compare our method to the naive approach of fine-tuning all weights according to Eq. 3, as well as the method of optimizing all the weights of a layer without constraining the direction of the change, as in Eq. 6, and to a state-of-the-art image alignment algorithm, *Neural Best-Buddies* (NBB [1]), which is used to propagate an edit across a set of similar images by compositing pixels according to identified sparse correspondences. To transfer an edit from a target image, we use NBB and Moving Least Squares [64] to compute a dense correspondence between the source image we would like to edit and the original target image. We use this dense correspondence field to warp the masked target into the source image. We test both direct copying and Laplace blending.

Fig. 4. Adding and replacing objects in three different settings. (a) Replacing domes with an angular peaked spire causes peaked spires to be used throughout the model. (b) Replacing domes with trees can generate images unlike any seen in a training set. (c) Replacing closed lips with an open-mouth smile produces realistic open-mouth smiles. For each case, we show the images generated by an unchanged model, then the edit propagation results, with and without blending. Our method is shown in the last row.

For each setting, we measure the efficacy of the edits on a sample of 10, 000 generated images, and we also quantify the undesired changes made by each method. For the smiling edit, we measure efficacy by counting images classified as smiling by an attribute classifier [66], and we also quantify changes made in the images outside the mouth region by masking lips using a face segmentation model [84] and using LPIPS [81] to quantify changes. For the dome edits, we

Table 1. Editing a StyleGANv2 [40] FFHQ [39] model to produce smiling faces in $n = 10,000$ images. To quantify the efficacy of the change, we show the percentage of smiling faces among the modified images, and we report the LPIPS distance on masked images to quantify undesired changes. For realism, workers make $n = 1,000$ pairwise judgements comparing images from other methods to ours.

	% smiling images ↑	LPIPS (masked) ↓	% more realistic than ours ↑
Our method (projected gradient descent)	84.37	**0.04**	–
With direct optimization of Λ	87.44	0.14	43.0
With single-image direction constraint	82.12	0.05	47.3
With single-layer, no direction constraint	90.94	0.30	6.8
Finetuning all weights	85.78	0.40	8.7
NBB + Direct copying	94.81	0.32	9.8
NBB + Laplace blending	**93.51**	0.32	8.6
Unmodified model	78.37	–	50.9

Table 2. We edit a StyleGANv2 [40] LSUN church [77] model to replace domes with spires/trees in $n = 10,000$ images. To quantify efficacy, we show the percentage of `dome` category pixels changed to the target category, determined by a segmenter [73]. To quantify undesired changes, we report LPIPS distance between edited and unchanged images, in non-dome regions. For realism, workers make $n = 1,000$ pairwise judgements comparing images from other methods to ours.

	Dome → Spire			Dome → Tree	
	% dome pixels correctly modified ↑	LPIPS (masked) ↓	% more realistic than ours ↑	% dome pixels correctly modified ↑	LPIPS (masked) ↓
Our method (projected gradient descent)	**92.03**	**0.02**	–	48.65	**0.03**
With direct optimization of Λ	80.03	0.10	53.7	**59.43**	0.13
With single-image direction constraint	90.14	0.04	48.8	39.72	0.03
With single-layer, no direction constraint	80.69	0.29	38.1	41.32	0.45
Finetuning all weights	41.16	0.36	27.1	10.16	0.31
NBB + Direct copying	69.99	0.08	8.9	46.44	0.09
NBB + Laplace blending	69.63	0.08	12.2	31.18	0.09
Unmodified model	–	–	63.8	–	–

measure how many dome pixels are judged to be changed to non-domes by a segmentation model [73], and we measure undesired changes outside dome areas using LPIPS. We also conduct a user study where users are asked to compare the realism of our edited output to the same image edited using baseline methods. We find that our method produces more realistic outputs that are more narrowly targeted than the baseline methods. For the smile edit, our method is not as aggressive as baseline methods at introducing smiles, but for the dome edits, our method is more effective than baseline methods at executing the change. Our metrics are further discussed in Appendix C.

5.2 Removing Undesired Features

Here we test our method on the removal of undesired features. Figure 5a shows several examples of images output by a pre-trained StyleGANv2 church model. This model occasionally synthesizes images with text overlaid in the middle and the bottom resembling stock-photo watermarks in the training set.

The GAN Dissection study [7] has shown that some objects can be removed from a generator by zeroing the units that best match those objects. To find these units, we annotated the middle and bottom text regions in ten generated images, and we identified a set of 60 units that are most highly correlated with features in these regions. Zeroing the most correlated 30 units removes some of the text, but leaves much bottom text unremoved, as shown in Fig. 5b. Zeroing all 60 units reduces more of the bottom text but begins to alter the main content of the images, as shown in Fig. 5c.

For our method, we use the ten user-annotated images as a context to create a rank-one constraint direction d for updating the model, and as an optimization target $K_* \rightarrow V_*$, we use one successfully removed watermark from the setting

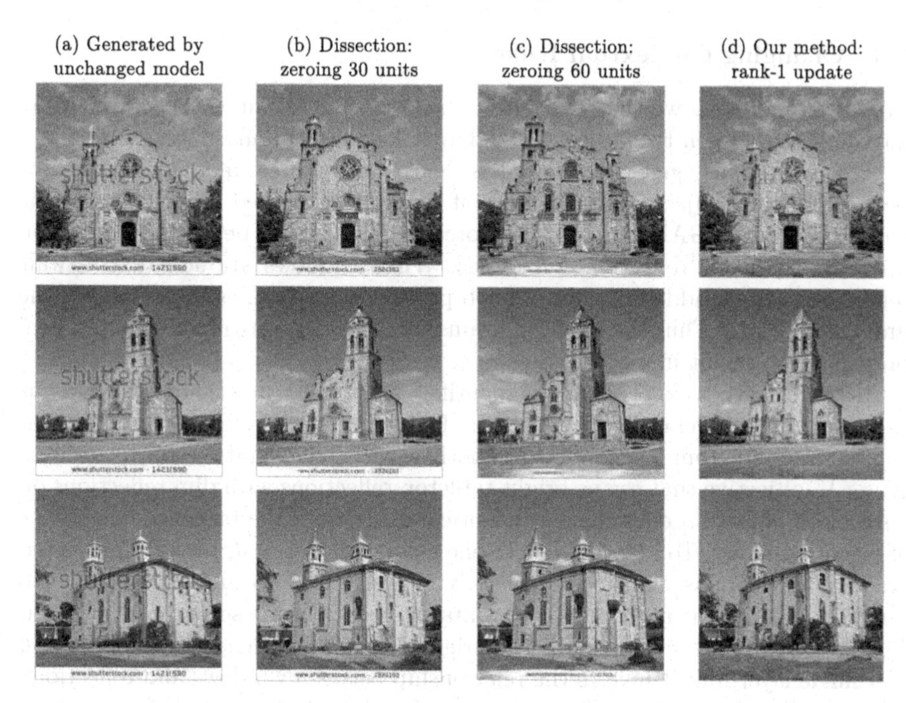

Fig. 5. Removing watermarks from StyleGANv2 [40] LSUN church [77] model. (a) Many images generated by this model include transparent watermarks in the center or text on the bottom. (b) Using GAN Dissection [7] to zero 30 text-specific units removes middle but not bottom text cleanly. (c) Removing 60 units does not fully remove text, and distorts other aspects of the image. (b) Applying our method to create a rank-1 change erases both middle and bottom text cleanly.

Table 3. Visible watermark text produced by StyleGANv2 church model in $n = 1000$ images, without modification, with sets of units zeroed (using the method of GAN Dissection), and using our method to apply a rank-one update.

Count of visible watermarks	Middle	Bottom
Zeroing 30 units (GAN Dissection)	**0**	6
Zeroing 60 units (GAN Dissection)	**0**	4
Rank-1 update (our method)	**0**	**0**
Unmodified model	64	26

shown in Fig. 5b. Since our method applies a narrow rank-1 change constraint, it would be expected to produce a loose approximation of the rank-30 change in the training example. Yet we find that it has instead improved specificity and generalization of watermark removal, removing both middle and bottom text cleanly while introducing few changes in the main content of the image. We repeat the process for 1000 images and tabulate the results in Table 3.

5.3 Changing Contextual Rules

In this experiment, we find and alter a rule that determines the illumination interactions between two objects at different locations in an image.

State-of-the-art generative models learn to enforce many relationships between distant objects. For example, it has been observed [6] that a kitchen-scene Progressive GAN model [38] enforces a relationship between windows on walls and specular reflections on tables. When windows are added to a wall, reflections will be added to shiny tabletops, and vice-versa, as illustrated in the first row of Fig. 6. Thus the model contains a rule that approximates the physical propagation of light in a scene.

In the following experiment, we identified an update direction that allows us to change this model of light reflections. Instead of specifying an objective that copies an object from one context to another, we used a similar tool to specify a $K_* \rightarrow V_*$ objective that swaps bright tabletop reflections with dim reflections on a set of 15 pairs of scenes that are identical other than the presence or absence of bright windows. To identify a rank-one change direction d, we used projected gradient descent, as described in Sect. 3.4, using SVD to limit the change to rank one during optimization. The results are shown in the second row of Fig. 6. The modified model differs from the original only in a single update direction of a single layer, but it inverts the relationship between windows and reflections: when windows are added, reflections are reduced, and vice-versa.

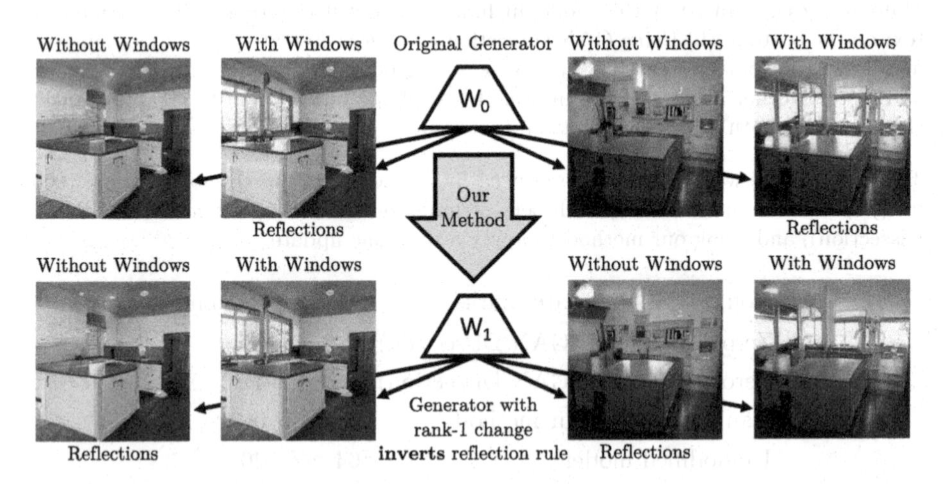

Fig. 6. Inverting a single semantic rule within a model. At the top row, a Progressive GAN [38] trained on LSUN kitchens [77] links windows to reflections: when windows are added by manipulating intermediate features identified by GAN Dissection [7], reflections appear on the table. In the bottom row, one rule has been changed within the model to *invert* the relationship between windows and reflections. Now adding windows *decreases* reflections and vice-versa.

6 Discussion

Machine learning requires data, so how can we create effective models for data that do not yet exist? Thanks to the rich internal structure of recent GANs, in this paper, we have found it feasible to create such models by rewriting the rules within existing networks. Although we may never have seen a tree sprouting from a tower, our network contains rules for both trees and towers, and we can easily create a model that connects those compositional rules to synthesize an endless distribution of images containing the new combination.

The development of sophisticated generative models beyond the image domain, such as the GPT-3 language model [10] and WaveNet for audio synthesis [56], means that it will be increasingly attractive to rewrite rules within other types of models as well. After training on vast datasets, large-scale deep networks have proven to be capable of representing an extensive range of different styles, sentiments, and topics. Model rewriting provides an avenue for using this structure as a rich medium for creating novel kinds of content, behavior, and interaction.

Acknowledgements. We thank Jonas Wulff, Hendrik Strobelt, Aaron Hertzman, Taesung Park, William Peebles, Gerald Sussman, and William T. Freeman for their vision, encouragement, and many valuable discussions. We are grateful for the support of DARPA XAI FA8750-18-C-0004, DARPA SAIL-ON HR0011-20-C-0022, NSF 1524817 on Advancing Visual Recognition with Feature Visualizations, NSF BIGDATA 1447476, and a hardware donation from NVIDIA.

References

1. Aberman, K., Liao, J., Shi, M., Lischinski, D., Chen, B., Cohen-Or, D.: Neural best-buddies: sparse cross-domain correspondence. ACM TOG **37**(4), 69 (2018)
2. An, X., Pellacini, F.: AppProp: all-pairs appearance-space edit propagation. ACM TOG **27**(3), 40 (2008)
3. Anderson, J.A.: A simple neural network generating an interactive memory. Math. Biosci. **14**(3–4), 197–220 (1972)
4. Andrychowicz, M., et al.: Learning to learn by gradient descent by gradient descent. In: NeurIPS, pp. 3981–3989 (2016)
5. Barnes, C., Shechtman, E., Finkelstein, A., Goldman, D.B.: PatchMatch: a randomized correspondence algorithm for structural image editing. ACM TOG **28**(3), 24 (2009)
6. Bau, D., et al.: Semantic photo manipulation with a generative image prior. ACM TOG **38**(4), 1–11 (2019)
7. Bau, D., et al.: Gan dissection: visualizing and understanding generative adversarial networks. In: ICLR (2019)
8. Bengio, S., Bengio, Y., Cloutier, J., Gecsei, J.: On the optimization of a synaptic learning rule. In: Optimality in Artificial and Biological Neural Networks, pp. 6–8. University of Texas (1992)
9. Bousmalis, K., Silberman, N., Dohan, D., Erhan, D., Krishnan, D.: Unsupervised pixel-level domain adaptation with generative adversarial networks. In: CVPR (2017)

10. Brown, T.B., et al.: Language models are few-shot learners. arXiv preprint arXiv:2005.14165 (2020)
11. Burt, P., Adelson, E.: The Laplacian pyramid as a compact image code. IEEE Trans. Commun. **31**(4), 532–540 (1983)
12. Chan, C., Ginosar, S., Zhou, T., Efros, A.A.: Everybody dance now. In: ICCV (2019)
13. Chen, X., Zou, D., Li, J., Cao, X., Zhao, Q., Zhang, H.: Sparse dictionary learning for edit propagation of high-resolution images. In: CVPR (2014)
14. Chen, X., Zou, D., Zhao, Q., Tan, P.: Manifold preserving edit propagation. ACM TOG **31**(6), 1–7 (2012)
15. Cohn, D., Atlas, L., Ladner, R.: Improving generalization with active learning. Mach. Learn. **15**(2), 201–221 (1994)
16. Donahue, J., et al.: DeCAF: a deep convolutional activation feature for generic visual recognition. In: ICML (2014)
17. Dosovitskiy, A., Brox, T.: Generating images with perceptual similarity metrics based on deep networks. In: NeurIPS (2016)
18. Dy, J.G., Brodley, C.E.: Visualization and interactive feature selection for unsupervised data. In: SIGKDD, pp. 360–364 (2000)
19. Efros, A.A., Freeman, W.T.: Image quilting for texture synthesis and transfer. In: SIGGRAPH. ACM (2001)
20. Endo, Y., Iizuka, S., Kanamori, Y., Mitani, J.: DeepProp: extracting deep features from a single image for edit propagation. Comput. Graph. Forum **35**(2), 189–201 (2016)
21. Fails, J.A., Olsen Jr, D.R.: Interactive machine learning. In: ACM IUI, pp. 39–45 (2003)
22. Finn, C., Abbeel, P., Levine, S.: Model-agnostic meta-learning for fast adaptation of deep networks. In: ICML, pp. 1126–1135 (2017). JMLR.org
23. Fried, O., et al.: Text-based editing of talking-head video. ACM TOG **38**(4), 1–14 (2019)
24. Gatys, L.A., Ecker, A.S., Bethge, M.: Image style transfer using convolutional neural networks. In: CVPR (2016)
25. Goodfellow, I., et al.: Generative adversarial nets. In: NeurIPS (2014)
26. Guo, D.: Coordinating computational and visual approaches for interactive feature selection and multivariate clustering. Inf. Vis. **2**(4), 232–246 (2003)
27. Hasinoff, S.W., Jóźwiak, M., Durand, F., Freeman, W.T.: Search-and-replace editing for personal photo collections. In: 2010 IEEE International Conference on Computational Photography (ICCP), pp. 1–8. IEEE (2010)
28. Hertzmann, A., Jacobs, C.E., Oliver, N., Curless, B., Salesin, D.H.: Image analogies. In: SIGGRAPH (2001)
29. Heusel, M., Ramsauer, H., Unterthiner, T., Nessler, B., Hochreiter, S.: GANs trained by a two time-scale update rule converge to a local Nash equilibrium. In: NeurIPS (2017)
30. Hinton, G.E., Osindero, S., Teh, Y.W.: A fast learning algorithm for deep belief nets. Neural Comput. **18**, 1527–1554 (2006)
31. Huang, X., Liu, M.-Y., Belongie, S., Kautz, J.: Multimodal unsupervised image-to-image translation. In: Ferrari, V., Hebert, M., Sminchisescu, C., Weiss, Y. (eds.) ECCV 2018. LNCS, vol. 11207, pp. 179–196. Springer, Cham (2018). https://doi.org/10.1007/978-3-030-01219-9_11
32. Huh, M., Zhang, R., Zhu, J.Y., Paris, S., Hertzmann, A.: Transforming and projecting images to class-conditional generative networks. In: ECCV (2020)

33. Iizuka, S., Simo-Serra, E., Ishikawa, H.: Let there be color!: Joint end-to-end learning of global and local image priors for automatic image colorization with simultaneous classification. ACM TOG **35**(4), 1–11 (2016)
34. Isola, P., Zhu, J.Y., Zhou, T., Efros, A.A.: Image-to-image translation with conditional adversarial networks. In: CVPR (2017)
35. Jiang, B., Canny, J.: Interactive machine learning via a gpu-accelerated toolkit. In: ACM IUI, pp. 535–546 (2017)
36. Johnson, J., Alahi, A., Fei-Fei, L.: Perceptual losses for real-time style transfer and super-resolution. In: Leibe, B., Matas, J., Sebe, N., Welling, M. (eds.) ECCV 2016. LNCS, vol. 9906, pp. 694–711. Springer, Cham (2016). https://doi.org/10.1007/978-3-319-46475-6_43
37. Karras, T.: FFHQ dataset (2019). https://github.com/NVlabs/ffhq-dataset
38. Karras, T., Aila, T., Laine, S., Lehtinen, J.: Progressive growing of GANs for improved quality, stability, and variation. In: ICLR (2018)
39. Karras, T., Laine, S., Aila, T.: A style-based generator architecture for generative adversarial networks. In: CVPR (2019)
40. Karras, T., Laine, S., Aittala, M., Hellsten, J., Lehtinen, J., Aila, T.: Analyzing and improving the image quality of styleGAN. In: CVPR (2020)
41. Kingma, D., Ba, J.: Adam: a method for stochastic optimization. In: ICLR (2015)
42. Kingma, D.P., Welling, M.: Auto-encoding variational Bayes. In: ICLR (2014)
43. Kohonen, T.: Correlation matrix memories. IEEE Trans. Comput. **100**(4), 353–359 (1972)
44. Kohonen, T.: Associative Memory: A System-Theoretical Approach, vol. 17. Springer, Heidelberg (2012). https://doi.org/10.1007/978-3-642-96384-1
45. Kohonen, T., Ruohonen, M.: Representation of associated data by matrix operators. IEEE Trans. Comput. **100**(7), 701–702 (1973)
46. Kokiopoulou, E., Chen, J., Saad, Y.: Trace optimization and eigenproblems in dimension reduction methods. Numer. Linear Algebra Appl. **18**(3), 565–602 (2011)
47. Krause, J., Perer, A., Bertini, E.: INFUSE: interactive feature selection for predictive modeling of high dimensional data. IEEE Trans. Vis. Comput. Graph. **20**(12), 1614–1623 (2014)
48. Lake, B.M., Salakhutdinov, R., Tenenbaum, J.B.: Human-level concept learning through probabilistic program induction. Science **350**(6266), 1332–1338 (2015)
49. Larsson, G., Maire, M., Shakhnarovich, G.: Learning representations for automatic colorization. In: Leibe, B., Matas, J., Sebe, N., Welling, M. (eds.) ECCV 2016. LNCS, vol. 9908, pp. 577–593. Springer, Cham (2016). https://doi.org/10.1007/978-3-319-46493-0_35
50. Levin, A., Lischinski, D., Weiss, Y.: Colorization using optimization. ACM TOG **23**(3), 689–694 (2004)
51. Liao, J., Yao, Y., Yuan, L., Hua, G., Kang, S.B.: Visual attribute transfer through deep image analogy. arXiv preprint arXiv:1705.01088 (2017)
52. Liu, M.Y., Breuel, T., Kautz, J.: Unsupervised image-to-image translation networks. In: NeurIPS (2017)
53. Luan, F., Paris, S., Shechtman, E., Bala, K.: Deep photo style transfer. In: CVPR, pp. 4990–4998 (2017)
54. Mathieu, M., Couprie, C., LeCun, Y.: Deep multi-scale video prediction beyond mean square error. In: ICLR (2016)
55. Nagano, K., et al.: paGAN: real-time avatars using dynamic textures. In: SIGGRAPH Asia, p. 258 (2018)
56. Oord, A.v.d., et al.: Wavenet: a generative model for raw audio. arXiv preprint arXiv:1609.03499 (2016)

57. Park, T., Liu, M.Y., Wang, T.C., Zhu, J.Y.: Semantic image synthesis with spatially-adaptive normalization. In: CVPR (2019)
58. Patel, K., Drucker, S.M., Fogarty, J., Kapoor, A., Tan, D.S.: Using multiple models to understand data. In: IJCAI (2011)
59. Pérez, P., Gangnet, M., Blake, A.: Poisson image editing. In: SIGGRAPH, pp. 313–318 (2003)
60. Portenier, T., Hu, Q., Szabó, A., Bigdeli, S.A., Favaro, P., Zwicker, M.: FaceShop: deep sketch-based face image editing. ACM Trans. Graph. (TOG) **37**(4), 99:1–99:13 (2018)
61. Raghavan, H., Madani, O., Jones, R.: Active learning with feedback on features and instances. JMLR **7**(Aug), 1655–1686 (2006)
62. Reinhard, E., Adhikhmin, M., Gooch, B., Shirley, P.: Color transfer between images. IEEE Comput. Graphics Appl. **21**(5), 34–41 (2001)
63. Saenko, K., Kulis, B., Fritz, M., Darrell, T.: Adapting visual category models to new domains. In: Daniilidis, K., Maragos, P., Paragios, N. (eds.) ECCV 2010. LNCS, vol. 6314, pp. 213–226. Springer, Heidelberg (2010). https://doi.org/10.1007/978-3-642-15561-1_16
64. Schaefer, S., McPhail, T., Warren, J.: Image deformation using moving least squares. ACM TOG **25**(3), 533–540 (2006)
65. Settles, B., Craven, M.: An analysis of active learning strategies for sequence labeling tasks. In: EMNLP, pp. 1070–1079 (2008)
66. Sharma, A., Foroosh, H.: Slim-CNN: a light-weight CNN for face attribute prediction. arXiv preprint arXiv:1907.02157 (2019)
67. Shocher, A., Cohen, N., Irani, M.: "zero-shot" super-resolution using deep internal learning. In: CVPR, pp. 3118–3126 (2018)
68. Socher, R., Ganjoo, M., Manning, C.D., Ng, A.: Zero-shot learning through cross-modal transfer. In: NeurIPS (2013)
69. Suzuki, R., Koyama, M., Miyato, T., Yonetsuji, T., Zhu, H.: Spatially controllable image synthesis with internal representation collaging. arXiv preprint arXiv:1811.10153 (2018)
70. Taigman, Y., Polyak, A., Wolf, L.: Unsupervised cross-domain image generation. In: ICLR (2017)
71. Ulyanov, D., Vedaldi, A., Lempitsky, V.: Deep image prior. In: CVPR (2018)
72. Wolberg, G.: Digital image warping. IEEE Computer Society Press, Los Alamitos (1990)
73. Xiao, T., Liu, Y., Zhou, B., Jiang, Y., Sun, J.: Unified perceptual parsing for scene understanding. In: Ferrari, V., Hebert, M., Sminchisescu, C., Weiss, Y. (eds.) ECCV 2018. LNCS, vol. 11209, pp. 432–448. Springer, Cham (2018). https://doi.org/10.1007/978-3-030-01228-1_26
74. Xu, K., Li, Y., Ju, T., Hu, S.M., Liu, T.Q.: Efficient affinity-based edit propagation using KD tree. ACM TOG **28**(5), 1–6 (2009)
75. Yosinski, J., Clune, J., Bengio, Y., Lipson, H.: How transferable are features in deep neural networks? In: NeurIPS, pp. 3320–3328 (2014)
76. Yu, C., Wang, J., Peng, C., Gao, C., Yu, G., Sang, N.: BiSeNet: bilateral segmentation network for real-time semantic segmentation. In: Ferrari, V., Hebert, M., Sminchisescu, C., Weiss, Y. (eds.) ECCV 2018. LNCS, vol. 11217, pp. 334–349. Springer, Cham (2018). https://doi.org/10.1007/978-3-030-01261-8_20
77. Yu, F., Seff, A., Zhang, Y., Song, S., Funkhouser, T., Xiao, J.: LSUN: construction of a large-scale image dataset using deep learning with humans in the loop. arXiv preprint arXiv:1506.03365 (2015)

78. Yücer, K., Jacobson, A., Hornung, A., Sorkine, O.: Transfusive image manipulation. ACM TOG **31**(6), 1–9 (2012)
79. Zeiler, M.D., Fergus, R.: Visualizing and understanding convolutional networks. In: Fleet, D., Pajdla, T., Schiele, B., Tuytelaars, T. (eds.) ECCV 2014. LNCS, vol. 8689, pp. 818–833. Springer, Cham (2014). https://doi.org/10.1007/978-3-319-10590-1_53
80. Zhang, R., Isola, P., Efros, A.A.: Colorful image colorization. In: Leibe, B., Matas, J., Sebe, N., Welling, M. (eds.) ECCV 2016. LNCS, vol. 9907, pp. 649–666. Springer, Cham (2016). https://doi.org/10.1007/978-3-319-46487-9_40
81. Zhang, R., Isola, P., Efros, A.A., Shechtman, E., Wang, O.: The unreasonable effectiveness of deep features as a perceptual metric. In: CVPR (2018)
82. Zhang, R., et al.: Real-time user-guided image colorization with learned deep priors. ACM TOG **9**(4), 11 (2017)
83. Zhu, J.Y., Park, T., Isola, P., Efros, A.A.: Unpaired image-to-image translation using cycle-consistent adversarial networks. In: ICCV (2017)
84. ZLL: Face-parsing PyTorch (2019). https://github.com/zllrunning/face-parsing.PyTorch

Compare and Reweight: Distinctive Image Captioning Using Similar Images Sets

Jiuniu Wang[1,2,3], Wenjia Xu[2,3,4], Qingzhong Wang[1], and Antoni B. Chan[1](✉)

[1] Department of Computer Science, City University of Hong Kong, Kowloon, Hong Kong
{jiuniwang2-c,qingzwang2-c}@my.cityu.edu.hk, abchan@cityu.edu.hk
[2] Aerospace Information Research Institute, Chinese Academy of Sciences, Beijing, China
[3] University of Chinese Academy of Sciences, Beijing, China
xuwenjia16@mails.ucas.ac.cn
[4] Max Planck Institute for Informatics, Saarbrücken, Germany

Abstract. A wide range of image captioning models has been developed, achieving significant improvement based on popular metrics, such as BLEU, CIDEr, and SPICE. However, although the generated captions can accurately describe the image, they are generic for similar images and lack distinctiveness, *i.e.*, cannot properly describe the uniqueness of each image. In this paper, we aim to improve the distinctiveness of image captions through training with sets of similar images. First, we propose a distinctiveness metric—between-set CIDEr (CIDErBtw) to evaluate the distinctiveness of a caption with respect to those of similar images. Our metric shows that the human annotations of each image are not equivalent based on distinctiveness. Thus we propose several new training strategies to encourage the distinctiveness of the generated caption for each image, which are based on using CIDErBtw in a weighted loss function or as a reinforcement learning reward. Finally, extensive experiments are conducted, showing that our proposed approach significantly improves both distinctiveness (as measured by CIDErBtw and retrieval metrics) and accuracy (*e.g.*, as measured by CIDEr) for a wide variety of image captioning baselines. These results are further confirmed through a user study. Project page: https://wenjiaxu.github.io/ciderbtw/.

1 Introduction

Image captioning is attracting increasing attention from researchers in the fields of computer vision and natural language processing. It is promising in various applications such as human-computer interaction and medical image understanding [3,11,24,36,37,41]. Currently, the limitation of image captioning models is that the generated captions tend to consist of common words so that many

Electronic supplementary material The online version of this chapter (https://doi.org/10.1007/978-3-030-58452-8_22) contains supplementary material, which is available to authorized users.

A. Vedaldi et al. (Eds.): ECCV 2020, LNCS 12346, pp. 370–386, 2020.
https://doi.org/10.1007/978-3-030-58452-8_22

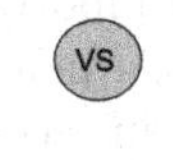

Target Image

CIDErBtw	Human Ground-truth Captions:
53.5	1: A living room with a big table next to a book shelf.
40.2	2: The large room has a wooden table with chairs and a couch.
54.2	3: A living room decorated with a modern theme.
73.0	4: A living room with wooden floors and furniture.
	Machine generated captions:
141.5	Baseline: A living room with a couch and a table.
68.5	Ours: A living room filled with wooden table and a large window.

Similar Image

CIDErBtw	Human Ground-truth Captions:
55.6	1: A living room filled with nice furniture and a persian rug.
55.9	2: An image of a living room setting with furniture and curtains.
38.9	3: An open living room with brown walls and beige carpeting.
31.2	4: A large tan living room bathed in sunlight.
	Machine generated captions:
174.2	Baseline: A living room with a couch and a table.
88.3	Ours: A living room with a white couch and a painting.

Fig. 1. The human ground-truth captions of a target image and a semantically similar image contain both common words (highlighted in green) and distinctive words (highlighted in red for the target, and blue for the similar image). The baseline model, Transformer [33] trained with MLE and SCST, generates the same caption for both images, while our model generates distinctive captions with words unique to each image. The distinctiveness is measured using CIDErBtw, the CIDEr metric between the target caption and the GT captions of the similar images set, where lower values mean more distinctive. (Color figure online)

images have similar or even the same captions (see Fig. 1). The distinctive concepts in images are ignored, which limits the application of image captioning. Although, auxiliary information such as where, when and who takes the picture could be used to generate personalized captions [4,26], many images do not have such information. In terms of the quality of generated captions, [21] summarizes four attributes that encourage auto-generated captions to resemble human language: fluency, relevance, diversity, and descriptiveness. Various models and metrics have been proposed to improve the fluency and relevance of the captions so as to obtain accurate results. However, these captions are poor at mimicking the inherent characteristics of human language: *distinctiveness*, which refers to the specific and detailed aspects of the image that distinguish it from other similar images.

Some recent works have focused on generating more diverse and descriptive captions, with techniques such as conditional generative adversarial networks (GANs) [5,30], self-retrieval [6,23,35] and two-stage LSTM [21]. Some works propose metrics for evaluating the *diversity* of a set of generated captions for a single image, based on the percentage of unique n-grams or novel sentences [30] or the similarity between pairs of captions [38]. However, only encouraging the diversity, such as using synonyms or changing word order, may not help with generating distinctive captions among multiple similar images. For instance, the human caption in Fig. 1 "*an image of a living room setting with furniture and curtains*" is telling the same story as "*a living room with furniture and curtains*". Although the two sentences have different syntax and the first sentence is more diverse according to some metrics, the distinctiveness is not improved. In this paper, we mainly focus on promoting the *distinctiveness* of image captioning, where the caption should describe the important and specific aspects of an image that can distinguish it from other similar images. To evaluate distinctiveness, the retrieval metric is generally employed in recent works [6,21–23]. However, using self-retrieval in captioning models could lead to repetition problem [35,38], *i.e.*, the generated captions could repeat distinct words, which hurts language fluency. Also, its result may vary when choosing different retrieval models or candidate images pool. In this work, we propose a general metric for distinctiveness, Between-Set CIDEr (CIDErBtw), by measuring the semantic distance between an image's caption and captions from a set of similar images. If the caption is distinct, i.e., captures unique concepts in its image, then it should have less overlap with its similar image set, i.e., lower CIDErBtw. We found that the human annotations of each image are not equivalent based on distinctiveness. Consider the example image and caption pairs shown in Fig. 1, some ground-truth captions contain more distinct concepts (e.g., *bathed in sunlight*) and detailed description that can distinguish the image from its similar image (e.g., *wooden floor* and *brown walls*). However, traditional training objectives such as maximum likelihood estimation (MLE) and reinforcement learning (RL) treat every ground-truth caption equally. Thus, one possible method for improving distinctiveness is to give more weight to the distinctive ground-truth captions during training. In this way, the captioning model learns to focus on important visual objects or properties, and generate distinctive words instead of generic ones. In summary, the contributions of our paper are three-fold:

- We propose a novel metric CIDErBtw to evaluate the distinctiveness of captions within similar image sets. Experiments show that our metric aligns with human judgment for distinctiveness.
- We use CIDErBtw as guidance for training, encouraging the model to learn from more distinctive captions. Experiments show that training with CIDErBtw is generic and yields consistent improvement for many baseline models.
- Based on the transformer network trained with SCST (self-critical sequence training) [29] and CIDErBtw strategies, we generate distinctive captions while maintaining state-of-the-art performance according to evaluation metrics such as CIDEr and BLEU. Both automatic metrics and human evaluation demonstrate that our captions are more accurate and more distinctive.

2 Related Work

Captioning Models. A wide range of image captioning models have been developed [3,11,24,36,37,41], achieving satisfying results as measured by popular metrics, such as BLEU [25], CIDEr [34] and SPICE [1]. Generally, an image captioning model is composed of three modules: 1) visual feature extractor, 2) language generator, and 3) the connection between vision and language. Convolutional neural networks (CNNs) [14,31] are widely used as visual feature extractors. Recently, object-level features extracted by Faster-RCNN [28] have also been introduced into captioning models [2], significantly improving the performance of image captioning models. [42] proposed a hierarchy parsing model to fuse multi-level image features extracted by mask-RCNN [13], which improves the performance of the baseline models. In terms of language generators, LSTMs [15] and its variants are the most popular, while some works [3,37] use CNNs as the decoder since LSTMs cannot be trained in parallel. More recently, transformers [9,27,33] show improved performance in both language generation and language understanding, where the multi-head attention plays the most important role and the receptive field is much larger than CNNs. Stacking multi-head attention layers could mitigate the long-term dependency problem in LSTMs. Hence, the transformer model could handle much longer texts. For vision-language connection, attention mechanisms [2,16,29,41] are used to reveal the co-occurrence between concepts and objects in the images.

Distinctive Image Captioning. Previous works [5,6,38] reveal that training the captioning model with MLE loss or CIDEr reward result in over-generic captions, since the captioning models try to predict an "average" caption that is close to all ground-truth captions. These captions lack distinctiveness, *i.e.*, they describe images with similar semantic content using the same caption. Recently, various works aim to solve this problem. In summary, they propose three aspects to consider: (1) *diversity*: describe one image with notably different expressions every time like humans [5], or use rich and diverse wording [38] to generate captions; (2) *discriminability*: describe an image by referring to the important and detailed aspects of the image, which is accurate, and informative [21–23,35]; (3) *distinctiveness*: describe the important and specific aspects of an image that can distinguish the image from other similar images [6,21]. In our paper, we focus on the last aspect, distinctiveness.

To promote diversity, some works [5,30] employ GANs, where an evaluator distinguishes the generated captions from human annotations, encouraging the captions to be similar to human annotations. Instead of using generative models, VisPara-Cap [21] employs two-stage LSTM and visual paraphrases to improve diversity and discriminability, where the two-stage model is trained with a pair of ground-truth image captions from an image—the first caption is less complex, and the next one with rich information is more distinctive. In contrast, our method is based on weighting all the ground-truth captions according to their distinctiveness, which retains more information for training. During inference, VisPara-Cap [21] first generates a simple caption and then paraphrases it into a more distinctive caption, which is a two-stage model and time-consuming. Another drawback of the model is that it cannot be trained in SCST [29] manner, and therefore the performance based on BLEU [25], CIDEr [34], and SPICE [1] is limited. In contrast, our method is able to improve both traditional metric scores and distinctiveness, and it can be applied to any image captioning model.

Contrastive learning [6] and self-retrieval [22,23,35] are introduced into captioning models to improve the distinctiveness of the generated captions. DiscCap [23], CL [6] and PSST [35] employ image retrieval to optimize the contrastive loss, which aims at pushing the generated caption far from other images in the training batch. On one hand, image retrieval encourages a model to generate distinctive words, while on the other hand, it hurts the accuracy and caption quality—weighting too much on image retrieval could lead a model to repeat the distinctive words [38]. In contrast, we encourage the generated caption to learn from its own ground-truth captions, giving more weights to captions that are distinct from other similar images, and disregard those generic captions. Thus both accuracy and distinctiveness are promoted in our model.

Metrics for Distinctiveness. Traditional metrics such as BLEU [25], METEOR [7], ROUGE-L [19], CIDEr [34] and SPICE [1] normally consider the overlap between a generated caption and the ground-truth captions. These metrics treat all ground-truth equally, and thus a generated caption that only uses common words could obtain high scores, reflecting the statistics of human annotations. Some works aim to generate multiple captions to cover more concepts in an image [5,8,30,39] and several diversity metrics are proposed, such as the number of novel captions, the number of distinct n-grams [40], mBLEU [30], local and global word recall [32], and self-CIDEr [38]. However, these metrics only encourage the diversity and discriminability and do not explicitly evaluate distinctiveness. Although generating multiple captions could cover distinctive concepts, it is difficult to summarize them into one human-like description.

Currently, the retrieval approach is the most popular evaluation metric for distinctiveness. A generated caption is used as the query and a pre-trained image-text embedding model, *e.g.*, VSE++ [10], is employed to rank the given images, with recall at K (R@K) normally used to measure the distinctiveness of captions. Ideally, a correct and distinctive caption should retrieve the image that was used to generate the caption. The drawback of retrieval-based metrics is that they are time-consuming, since it requires using a deep retrieval model. Moreover, different trained models could result in different R@K. In contrast, our proposed CIDErBtw metric for distinctiveness is fast and easy to implement, allowing it to be incorporated into various training protocols and captioning models.

3 Methodology

In this paper, we aim to obtain a distinct caption that describes the important, specific, and detailed aspects of an image. To achieve this goal, we train the captioning model to focus on important details that would distinguish the target image from semantically similar images. Our work involves two main components, the Between-Set CIDEr (CIDErBtw) that measures the distinctiveness of an image caption from those of similar images, and several strategies for training distinctive models based on CIDErBtw.

The image captioning model aims to generate a sentence c^* to describe the semantics of the target image I_0. In the image caption dataset, the image I_0 is provided with N annotated ground-truth captions $C^0 = \{c_1^0, c_2^0, \ldots, c_N^0\}$. We first find K similar images $\{I_1, I_2, \ldots, I_K\}$ that are semantically similar to I_0, and then calculate the CIDErBtw values of C^0 using these similar images. During training process, CIDErBtw can be used as an indicator of which ground-truth captions deserve more attention, or as a part of the reward in reinforcement learning (RL). This will train the model to generate a caption different from those of the similar images. Moreover, CIDErBtw can work as an evaluation metric to measure distinctiveness.

3.1 Similar Images Set

According to the split of the training, validation, and testing dataset, we measure the similarity of the target image I_0 to every image within the same split. For each image I_0 in the dataset, we find the top K images $\{I_1, I_2, \ldots, I_K\}$ with the highest semantic similarity to form a *similar images set*. Similar images sets in the training split are used when calculating the loss and the reward during training, while those in the validation and test split are used to evaluate the distinctiveness of generated captions.

Given every target image, we generate its similar images set according to an image-to-caption retrieval process. We use VSE++ [10] to encode images and captions into a joint semantic space, and obtain similar images sets via retrieval. Given target image I_0, we obtain a set of closest captions $\{c'_1, c'_2, \ldots, c'_{N'}\}$ in the joint space by image-to-caption retrieval, where $N' = N(K+1)$ to ensure that at least $K+1$ images are obtained to construct the similar images set. The top K images corresponding to this caption set are considered as similar to the target image I_0. When using the retrieval method, the similarity of I_i to I_j denoted as $S(I_i, I_j)$ can be expressed like

$$S(I_i, I_j) = \max_{k \in \{1, \cdots, N\}} g_r(I_i, c_k^j), \quad g_r(I_i, c_k^j) = \frac{\phi(I_i)^T \theta(c_k^j)}{\|\phi(I_i)\| \|\theta(c_k^j)\|}, \tag{1}$$

where $g_r(I_i, c_k^j)$ represents the retrieval score between the target image I_i and the k-th ground-truth caption of I_j, and $\phi(\cdot)$ and $\theta(\cdot)$ are the image and caption encoders.

3.2 Between-Set CIDEr (CIDErBtw)

Next, we introduce the definition of Between-set CIDEr (CIDErBtw) and its applications. In this paper, we mainly apply CIDErBtw in the following three aspects. During training, CIDErBtw is used to reweight the cross entropy (XE) loss and the reinforcement learning (RL) reward for each ground-truth caption. The CIDErBtw metric is also used directly as part of the reward to guide RL. During inference, CIDErBtw is used as a metric to measure the distinctiveness of a generated caption.

CIDErBtw Definition. CIDErBtw reflects the distinctiveness of a caption c by measuring the similarity of c to the captions of similar images $C^{(s)}$. Specifically, given a caption c for image I_0, the similar images set $\{I_1, I_2, \ldots, I_K\}$ retrieved in Section 3.1 and their ground-truth captions $C^{(s)} = \{c_n^k\}_{n=1,k=1}^{N,K}$, we define the CIDErBtw score of c as

$$CIDErBtw(c) = \frac{1}{KN} \sum_{k=1}^{K} \sum_{n=1}^{N} g_c(c, c_n^k), \tag{2}$$

where N is the number of ground-truth captions provided for each image, $g_c(c, c_n^k)$ represents the CIDEr value between c and c_n^k. Actually, the methodology could be extended to use any caption metric to measure between-set similarity. Here we use CIDEr because it focuses more on low frequency words (through TF-IDF vectors) that could be more distinctive, is efficient to compute, and is the most frequently used metric to evaluate performance of image captioning models.

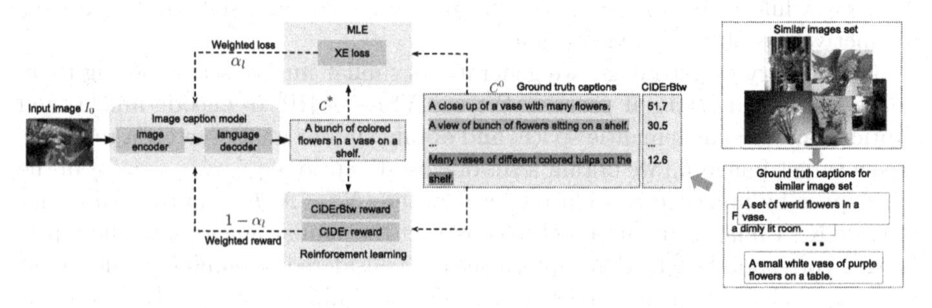

Fig. 2. The framework of our CIDErBtw image captioning model. α_l is a hyperparameter that controls the weight of the two optimization modules. The solid and dashed lines represent the forward and backward process. c^* and C^0 indicate the generated caption and the ground-truth captions. With CIDErBtw, we reweight the ground-truth captions when calculating the XE loss and reward. The shade of blue shows the CIDErBtw weight w_i for each caption.

CIDErBtw Weight. For conventional training strategies such as MLE and reinforcement learning, we maximize the likelihood or reward for the given ground-truth captions $C^0 = \{c_1^0, c_2^0, \ldots, c_N^0\}$. In previous methods, each ground-truth caption c_i^0 is treated equally, whereas these ground-truth might have different distinctiveness. In this work, we focus more attention to distinctive ground-truth captions by reweighting the training loss. For every training image I_0, we provide its N ground-truth captions C^0 with different weights $W = \{w_1, w_2, \ldots, w_N\}$, according to their CIDErBtw scores $V = \{v_1, v_2, \ldots, v_N\}$,

$$v_i = CIDErBtw(c_i^0), \quad w_i = \lambda_w - \alpha_w \frac{v_i}{\max_i(v_i)}, \tag{3}$$

where λ_w and α_w are hyperparameters. Here w_i indicates the contribution of the i-th ground-truth caption during model training. More distinctive captions will have lower v_i, leading to higher weight w_i.

3.3 CIDErBtw Training Strategies

Figure 2 shows the overall framework of our CIDErBtw Image Caption model. The model is composed of a image encoder and language decoder. These two modules can generate a caption c^* for input image I_0. There are two criteria to update the parameters of our image captioning model, the XE loss $\mathcal{L}_{XE}$ and RL reward $\mathcal{L}_{RL}$. We apply a hyperparameter α_l to control the weight of these two criteria,

$$\mathcal{L} = \alpha_l \mathcal{L}_{XE} + (1 - \alpha_l)\mathcal{L}_{RL}. \tag{4}$$

Following SCST (self-critical sequence training) [29], the training process of our model can be divided into two steps. The first step only trains with $\mathcal{L}_{XE}$, setting $\alpha_l = 1$, and the second step only trains with $\mathcal{L}_{RL}$, setting $\alpha_l = 0$.

Reweighting XE Loss. Given the words in a ground-truth caption $c_i^0 = \{d_1, d_2, \dots, d_T\}$, XE loss can be expressed as

$$L_{XE}(c_i^0) = -\sum_{t=1}^{T} \log p_\theta(d_t|d_{1:t-1}, I_0), \tag{5}$$

where $p_\theta(d_t|d_{1:t-1}, I_0)$ denotes the probability of the word d_t given the word sequence $d_1, \dots, d_{t-1}$ and image I_0. The CIDErBtw weighted XE loss is then

$$\mathcal{L}_{XE} = \sum_{i=1}^{N} w_i L_{XE}(c_i^0). \tag{6}$$

Reweighting RL Reward. For RL, we reweight the CIDEr reward according to the CIDErBtw to focus more on distinctive captions, resulting in a new reward,

$$\tilde{R}(c^*) = \frac{1}{N}\sum_{i=1}^{N} w_i g_c(c^*, c_i^0), \tag{7}$$

where $g_c(c^*, c_i^0)$ is the CIDEr value between c^* and ground-truth c_i^0.

CIDErBtw Reward. Finally, when performing RL, our CIDErBtw can also be used as a part of the reward related to distinctiveness. We combine the CIDErBtw score with the previous reward $\tilde{R}(c^*)$ and obtain the final RL reward $R(c^*)$ and RL loss $\mathcal{L}_{RL}$ as

$$R(c^*) = \tilde{R}(c^*) - \alpha_r CIDErBtw(c^*), \quad \mathcal{L}_{RL} = -\mathbb{E}_{c^* \sim p_\theta}[R(c^*)], \tag{8}$$

where $CIDErBtw(c^*)$ represents CIDErBtw score of the generated caption c^* defined in (2), α_r is a hyperparameter controlling the relative contributions, and the greedy sampling is used as the RL policy p_θ.

CIDErBtw Evaluation Metric. CIDEr measures the similarity between the generated caption c^* and its ground-truth captions C^0, and has become an important evaluation metric in image captioning. We believe the distinctiveness should also be measured when evaluating the quality of generated captions. Thus we propose to use CIDErBtw as a complementary evaluation metric for image captioning models. We hope that the caption c^* generated by the model is closer to the semantics of target image I_0, while far from the semantics of other K similar images $\{I_1, I_2, \ldots, I_K\}$. Therefore, the c^* generated by a more distinctive image captioning model will have a lower CIDErBtw. Note that for evaluation, the similar image sets are computed using the validation or test split. Note that CIDEtBtw requires human annotations to evaluate the generated captions, which is similar to other captioning evaluations, e.g., CIDEr [34], BLEU [25], METEOR [7], and ROUGE [19]. Although VSE++ does not require human annotation for evaluation, it still needs ground-truth captions in the training phase, and the performance is highly related to the training data.

4 Experiments

In this section, we conduct extensive experiments to evaluate the effectiveness of CIDErBtw in generating distinctive captions. Note that our motivation is to generate distinctive captions as well as achieve high caption quality.

4.1 Implementation Details

Dataset. We use the MSCOCO dataset [20] with Karpathy splitting [17]. The numbers of images are 113,287 for training, 5,000 for validation, and 5,000 for testing. There are five annotated captions for each image.

Models. For the image encoder, following Luo *et al.* [23], we use two types of features in the experiments, *i.e.*, the FC features and the spatial features. The FC features are extracted from Resnet-101 [14], and each image is encoded as a vector of dimension 2, 048. The spatial features are extracted from the output of a Faster-RCNN [28] following UpDown [2].

Our experiments are performed using four baseline models, *i.e.*, FC [29], Att2in [29], UpDown [2], and Transformer [33]. FC model only uses the FC features, Att2in and Transformer only use the spatial features, and UpDown uses both types of features. Each model is trained using four methods: 1) MLE with standard XE loss, denoted as "*model*"; 2) MLE with CIDErBtw-weighted XE loss in (6), denoted as "*model*+CIDErBtw"; 3) SCST [29], which trains with standard XE loss first, and then switches to RL with CIDEr reward, denoted as "*model*+SCST"; 4) SCST using weighted XE loss and weighted RL reward in (7), denoted as "*model*+SCST+CIDErBtw".

Training Details. We set λ_w as 1.5, α_w between 0.25 to 1.25 when reweighting the loss and the reward. α_r is set to 0.4 when using CIDErBtw reward, and 0 otherwise. We use Adam [18] to optimize the training parameters with an initial learning rate 5×10^{-4} and a decay factor 0.8 every three epochs. During test time, we apply beam search with size five to generate captions.

Metrics. For evaluation we consider two groups of metrics. The first group includes language quality metrics CIDEr, BLEU3, BLEU4, METEOR, ROUGE-L, and SPICE for evaluating the accuracy and quality of generated captions. The second group assesses the distinctiveness of captions, and includes our CIDErBtw metric and retrieval metrics (*i.e.*, R@1, R@5, R@10). When calculating CIDErBtw, we collect $K = 5$ similar images for each target image, so the CIDErBtw score measures the similarity between the generated caption and 25 captions from the similar images set, with lower values indicating more distinctiveness. Similar images sets are generated using a pre-trained VSE++ [10] to perform the caption-to-image retrieval (see Sect. 3.1). For the retrieval metrics, we follow the protocol in [6,21,23]. Given a generated caption, images are retrieved in the joint semantic space of the pre-trained VSE++, with the goal to retrieve the original image. Recall at K (R@K) is used to measure the retrieval performance, where a higher recall represents a better distinctiveness.

4.2 Experiment Results

In this section, we present the experiment results to show the effectiveness of CIDErBtw training strategies at improving caption distinctiveness. Due to space constraints, the ablation study is presented in the supplemental.

Effect of CIDErBtw Strategies. The main results are presented in the top and middle of Table 1. All baseline models obtain better performances when using CIDErBtw weighting in training process, for both MLE or SCST, which suggests that our method is widely applicable to many existing models. Specifically, our method both reduces the CIDErBtw score and improves other accuracy metrics, such as CIDEr. This shows that the generated captions become more similar to ground-truth captions, while more distinctive from other images' captions since redundancy is suppressed. Among the four baseline models, CIDErBtw reweighted loss and reward have the largest effect on Transformer [33]. Most likely the multi-head attention and larger receptive field of Transformer allow it better extract details and context from the image that is distinctive.

Next we apply all three of our CIDErBtw reward strategies together on Transformer+SCST, which is denoted as "+CIDErBtwReward" in Table 1. Compared to only using reweighted loss and reward (Transformer+SCST+CIDErBtw), adding the CIDErBtw reward in RL improves both the CIDErBtw and retrieval metrics significantly (i.e., improves distinctiveness), at the expense of a small decrease in accuracy (CIDEr).

Finally, we examine the disadvantage of SCST that directly optimizing CIDEr reward improves the fluency of captions but also leads to common and generic words. Consistent with [21,38], the baseline models trained with SCST obtain higher CIDEr but also perform worse in CIDErBtw and R@K, compared with models trained only with MLE. Optimizing the model with CIDErBtw weighted reward will relieve this problem, and the distinctness of captions will be promoted, while maintaining or even improving the overall quality of the captions.

Reasons for Improving CIDEr. Results in Table 1 show that models trained with CIDErBtw obtain better performance for *both* distinctiveness metrics and accuracy metrics. Given that our training method puts more weight on distinct ground-truth captions, it is expected that we will obtain lower CIDErBtw and higher R@K scores. However, the reason why our method also improves caption accuracy (CIDEr) is less obvious, especially for SCST, which *directly* optimizes CIDEr using RL. Note that CIDEr is based on the cosine similarity between TFIDF vectors, and thus low-frequency words (with higher IDF weights) will have higher impact on the CIDEr score. Since rare words are also distinct, their usage in a caption should increase CIDEr. If using distinct words can increase CIDEr, then why does RL with CIDEr reward not use distinct words? We speculate that RL gets stuck in a local minimum of models that only use frequent words because of two reasons: 1) equal weighting of an image's ground-truth captions encourages the model to predict the common words that match all captions; and 2) regularization encourages models to use smaller vocabularies – using less words means less non-zero weights in the network, and lower model complexity. By reweighting the reward with CIDErBtw, more reward is obtained when using diverse words, which effectively moves the learning process out of this local minimum.

Comparison with State-of-the-art. We list the performance of state-of-the-art captioning models that focus on distinctiveness at the bottom of Table 1. Compared to these models, our model (Transformer+SCST+CIDErBtw, and +CIDErBtwReward) generally achieves superior results in both accuracy and distinctiveness—our model obtains both a high CIDEr score and low CIDErBtw score (or high retrieval score) at the same time. Specifically, Stack-Cap [12] and DiscCap [23] have lower accuracy (CIDEr 120) and less distinctiveness (CIDErBtw 89, R@1 22), compared to our model. VisPara-Cap [21] has high distinctiveness by using visual paraphrases, slightly worse than our model (+CIDErBtwReward), while the accuracy (CIDEr 86.9) is much lower than our model. CL-Cap [6] and PSST [35] directly optimize the retrieval loss, aiming to identify the input image among a set of randomly-chosen distractor images, which improves the distinctiveness. CL-Cap has similar distinctiveness as our method, obtaining worse R@K than ours, but better CIDErBtw.[1] However,

[1] We could not compare distinctiveness with PSST since their captions are not publicly available, and they use a different evaluation protocol for R@K.

Table 1. Comparison of caption accuracy and distinctiveness on MSCOCO test split: (top) baseline models trained with MLE using standard or our weighted XE loss; (middle) models trained with SCST using standard or our weighted loss/reward; (bottom) SOTA methods for generating distinctive/discriminative captions. CIDEr, BLEU3/4, METEOR, ROUGE-L, and SPICE measure caption accuracy, while CIDErBtw and R@K measure distinctiveness. ↑ or ↓ show whether higher or lower scores are better for each metric. CIDErBtw could not be computed for some models because the captions are not publicly available. Our self-retrieval results (R@K) and those of [6,12,21,23] use the pre-trained VSE++ model and the same protocol. † Note that [35] reports self-retrieval results using a different retrieval model/protocol – they use their own model for retrieval – which makes it not directly comparable.

Method	CIDEr↑	CIDErBtw↓	BLEU3↑	BLEU4↑	METEOR↑	ROUGE-L↑	SPICE↑	R@1↑	R@5 ↑	R@10↑
FC [29]	97.90	83.35	41.81	31.58	25.22	53.34	17.99	15.44	40.36	55.08
FC+CIDErBtw (ours)	98.82	83.22	42.03	31.79	25.46	53.48	18.29	16.24	41.54	56.64
Att2in [29]	110.04	83.19	46.36	35.75	26.79	56.18	19.91	17.44	43.88	58.02
Att2in+ CIDErBtw (ours)	110.97	82.42	46.63	36.0	27.03	56.30	20.01	17.98	44.72	58.62
UpDown [2]	111.25	79.46	45.64	35.93	27.54	56.24	20.54	20.10	47.58	61.92
UpDown+ CIDErBtw (ours)	112.77	78.34	46.35	36.10	27.69	56.36	20.68	20.92	49.72	63.98
Transformer [33]	110.13	80.98	44.80	34.46	26.98	55.30	20.18	21.52	49.88	64.70
Transformer+ CIDErBtw (ours)	112.44	**75.35**	45.44	35.01	27.59	55.66	20.74	21.84	50.48	65.04
FC+SCST [29]	104.43	90.09	43.10	31.59	25.46	54.33	18.67	11.44	33.16	48.04
FC+SCST+ CIDErBtw (ours)	104.76	89.41	43.25	31.72	25.60	54.35	18.58	11.74	33.62	48.32
Att2in+SCST [29]	117.96	87.40	47.22	35.31	27.17	56.92	20.57	16.00	41.55	56.66
Att2in+SCST+ CIDErBtw (ours)	118.48	87.21	47.33	35.41	27.27	56.94	20.77	16.82	42.26	57.72
UpDown+SCST [2]	121.94	86.82	48.82	36.12	27.95	57.61	21.29	18.50	46.34	61.70
UpDown+SCST+ CIDErBtw (ours)	123.02	86.42	48.98	36.39	28.12	57.78	21.44	19.68	47.30	62.78
Transformer+ SCST [33]	125.13	86.68	50.26	38.04	27.96	58.60	22.30	23.38	54.34	68.44
Transformer+ SCST+ CIDErBtw (ours)	**128.11**	84.70	**51.29**	**39.0**	**29.12**	**59.24**	22.92	24.46	55.22	69.02
+CIDErBtw Reward (ours)	127.78	82.74	50.97	38.52	29.09	58.82	**22.96**	**26.46**	**57.98**	**71.28**
Stack-Cap [12]	120.4	88.7	47.9	36.1	27.4	56.9	20.9	21.9	49.7	63.7
DiscCap [23]	120.1	89.2	48.5	36.1	27.7	57.8	21.4	21.6	50.3	65.4
VisPara-Cap [21]	86.9	–	–	27.1	–	–	21.1	26.3	57.2	70.8
CL-Cap [6]	114.2	81.3	46.0	35.3	27.1	55.9	19.7	24.1	52.5	67.5
PSST [35]	111.9	–	–	32.2	26.4	54.4	20.6	45.3†	79.4†	89.9†

directly optimizing the training parameter with retrieval loss results in low-quality captions, lowering the accuracy (CIDEr 114.2 and 111.9) compared to our model.

4.3 User Study

To fairly evaluate the quality of generated sentences and verify the consistency between the metrics and human perspective, we conducted two user studies. Firstly, we performed a user study on image retrieval to assess distinctiveness, following the protocol in [23]. The task involves displaying the target image, a semantically similar image which is retrieved following the method in Sect. 3.1, and a generated caption describing the target image. The users are asked to choose the image that more closely matches the caption.

In the second experiment, we compare captions generated from a baseline model trained with and without CIDErBtw. In each trial, an image and two captions are displayed, and the user is asked to choose the better caption with respect to two criteria: distinctiveness and accuracy. In each experiment, we randomly sample 50 similar images pair from the test split. We perform the experiment on four captioning models: UpDown [2] and Transformer [33] trained by SCST with and without CIDErBtw (denoted as UD, UD+CIDErBtw, TF and TF+CIDErBtw). Twenty people participated in the user study, and we collected about $6,000$ responses in total. See the supplemental for more details.

The results for the image retrieval user study are shown in Table 2. Compared to the baseline model, our method increases the accuracy of image retrieval by 5.6% and 14.4%. This user study is consistent with the automatic image retrieval results (R@K), and indicates that captions generated by our model are more distinctive in terms of both machine and human perception, than those of the baseline models.

The result for the distinctiveness/accuracy user study are shown in Figure 3. From human perspective, captions from our models are more distinctive than the baseline models (our captions are selected 69% and 72% of the time). The improvement of accuracy is not as much (our model selected 59% and 63% of the time), since the baseline models already generate captions that are accurate. Again this is consistent with the observations from the machine-based metrics (CIDErBtw and CIDEr).

4.4 Qualitative Results

We next show qualitative results for the baseline model Transformer+SCST, and our model Transformer+SCST+CIDErBtw in Fig. 4. The baseline model generates captions that accurately describe the main object, but are quite generic and monotonous. Intuitively, in order to increase a caption's distinctiveness, the model should focus on more properties that would distinguish the image from others, such as color, numbers, or other objects/background in the image. Our method focuses on more of these aspects and generates accurate results. Our captions describe more properties of the main object, such as "*black suit*", "*red tie*" and "*a man and a child*". We also describe backgrounds that are distinctive, such as "*pictures on the wall*" and "*city street at night*".

In order to show the distinctiveness of our model, we present a similar images set with the same semantic meaning in Fig. 5. The baseline model generates

Table 2. User study on image retrieval to assess caption distinctiveness. Our models trained with CIDErBtw generated more distinctive captions, enabling the user to more accurately select the correct image, compared with the baselines (2-sample z-test on proportions, * p<0.05, ** p<0.01).

Method	Image retrieval
UD	68.7% **
UD+CIDErBtw	**74.3%** **
TF	75.2% *
TF+CIDErBtw	**79.6%** *

Distinctiveness
UD+CIDErBtw 69% 31% UD
TF+CIDErBtw 72% 28% TF
Accuracy
UD+CIDErBtw 59% 41% UD
TF+CIDErBtw 63% 37% TF

Fig. 3. User study comparing captions generated from models trained with and without our CIDErBtw. Users selected our models trained with CIDErBtw more frequently when assessing accuracy and distinctiveness (Chi-Square test, $p < 0.001$ for each pair).

Baseline: (86.7) A man in a suit and tie.
(59.0) A living room with a television and on the television
(66.8) A stop sign on the side of a road.
(90.2) A man standing on the beach with a surfboard.
Ours: (49.0) A man in a black suit wearing a red tie.
(56.6) A living room with a television and pictures on the wall.
(58.3) A stop sign on the side of a city street at night.
(69.7) A man and a child standing on the beach with a surfboard.

Fig. 4. Example captions from the baseline model and our model. The distinctive words are highlighted. The number in parenthesis is the CIDErBtw score, with lower values meaning more distinctive.

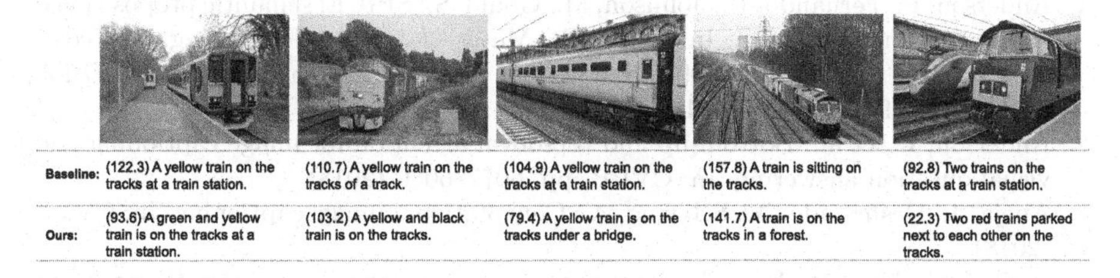

Baseline: (122.3) A yellow train on the tracks at a train station.
(110.7) A yellow train on the tracks of a track.
(104.9) A yellow train on the tracks at a train station.
(157.8) A train is sitting on the tracks.
(92.8) Two trains on the tracks at a train station.
Ours: (93.6) A green and yellow train is on the tracks at a train station.
(103.2) A yellow and black train is on the tracks.
(79.4) A yellow train is on the tracks under a bridge.
(141.7) A train is on the tracks in a forest.
(22.3) Two red trains parked next to each other on the tracks.

Fig. 5. Example captions for a set of similar images. (Color figure online)

captions that follow generic templates, e.g. "*train on the track*" or "*at a train station*". Although the captions are correct, it is hard to tell the images apart according to the captions. Our model enriches the description by mentioning the colors, e.g. the "*green and yellow*" and "*yellow and black*" distinguishing the first two images, and the background environment, e.g. "*under a bridge*" and "*in a forest*". Furthermore, our model is more sensitive to the relative positions of objects, e.g. "*next to each other on the tracks*". However, a more descriptive caption may also lead to some errors. For instance, the train in the third image is not actually "*under a bridge*". More details are in the supplementary material.

5 Conclusion

In this paper, we consider an important property, *distinctiveness* of image captions, and proposed a metric CIDErBtw to evaluate distinctiveness, which can be calculated quickly and easily implemented. We found that human annotations for each image vary in distinctiveness based on CIDErBtw. To improve the distinctiveness of generated captions, we developed a novel training strategy, where each human ground-truth annotation is assigned a weight based on its distinctiveness computed by CIDErBtw. Thus, during training the model pays more attention to the captions that are more distinctive. We also consider using CIDErBtw directly as part of the reward in RL. Extensive experiments were conducted, and we showed that our method is widely applicable to many captioning models. Experimental results demonstrate that our training strategy is able to improve both accuracy and distinctiveness, achieving state-of-the-art performance on CIDEr, CIDErBtw and retrieval metrics (R@K).

Acknowledgments. This work was supported by grants from the Research Grants Council of the Hong Kong Special Administrative Region, China (Project No. CityU 11212518) and from City University of Hong Kong (Strategic Research Grant No. 7005218).

References

1. Anderson, P., Fernando, B., Johnson, M., Gould, S.: SPICE: semantic propositional image caption evaluation. In: Leibe, B., Matas, J., Sebe, N., Welling, M. (eds.) ECCV 2016. LNCS, vol. 9909, pp. 382–398. Springer, Cham (2016). https://doi.org/10.1007/978-3-319-46454-1_24
2. Anderson, P., et al.: Bottom-up and top-down attention for image captioning and visual question answering. In: CVPR, pp. 6077–6086 (2018)
3. Aneja, J., Deshpande, A., Schwing, A.: Convolutional image captioning. In: CVPR (2018)
4. Park, C.C., Kim, B., Kim, G.: Attend to you: personalized image captioning with context sequence memory networks. In: CVPR, pp. 895–903 (2017)
5. Dai, B., Fidler, S., Urtasun, R., Lin, D.: Towards diverse and natural image descriptions via a conditional GAN. In: ICCV (2017)
6. Dai, B., Lin, D.: Contrastive learning for image captioning. In: NeurIPS (2017)
7. Denkowski, M., Lavie, A.: Meteor universal: language specific translation evaluation for any target language. In: EACL Workshop (2014)
8. Deshpande, A., Aneja, J., Wang, L., Schwing, A.G., Forsyth, D.: Fast, diverse and accurate image captioning guided by part-of-speech. In: CVPR (2019)
9. Devlin, J., Chang, M.W., Lee, K., Toutanova, K.: BERT: pre-training of deep bidirectional transformers for language understanding. In: NAACL (2018)
10. Faghri, F., Fleet, D.J., Kiros, J.R., Fidler, S.: VSE++: improving visual-semantic embeddings with hard negatives. In: BMVC (2018)

11. Farhadi, A., et al.: Every picture tells a story: generating sentences from images. In: Daniilidis, K., Maragos, P., Paragios, N. (eds.) ECCV 2010. LNCS, vol. 6314, pp. 15–29. Springer, Heidelberg (2010). https://doi.org/10.1007/978-3-642-15561-1_2
12. Gu, J., Cai, J., Wang, G., Chen, T.: Stack-captioning: coarse-to-fine learning for image captioning. In: AAAI (2018)
13. He, K., Gkioxari, G., Dollár, P., Girshick, R.: Mask R-CNN. In: CVPR, pp. 2961–2969 (2017)
14. He, K., Zhang, X., Ren, S., Sun, J.: Deep residual learning for image recognition. In: CVPR, pp. 770–778 (2016)
15. Hochreiter, S., Schmidhuber, J.: Long short-term memory. Neural Comput. **9**(8), 1735–1780 (1997)
16. Huang, L., Wang, W., Chen, J., Wei, X.Y.: Attention on attention for image captioning. In: ICCV, pp. 4634–4643 (2019)
17. Karpathy, A., Fei-Fei, L.: Deep visual-semantic alignments for generating image descriptions. In: CVPR, pp. 3128–3137 (2015)
18. Kingma, D.P., Ba, J.: Adam: a method for stochastic optimization. In: ICLR (2015)
19. Lin, C.Y.: ROUGE: a package for automatic evaluation of summaries. In: ACL Workshop (2004)
20. Lin, T.-Y., et al.: Microsoft COCO: common objects in context. In: Fleet, D., Pajdla, T., Schiele, B., Tuytelaars, T. (eds.) ECCV 2014. LNCS, vol. 8693, pp. 740–755. Springer, Cham (2014). https://doi.org/10.1007/978-3-319-10602-1_48
21. Liu, L., Tang, J., Wan, X., Guo, Z.: Generating diverse and descriptive image captions using visual paraphrases. In: CVPR, pp. 4240–4249 (2019)
22. Liu, X., Li, H., Shao, J., Chen, D., Wang, X.: Show, tell and discriminate: image captioning by self-retrieval with partially labeled data. In: Ferrari, V., Hebert, M., Sminchisescu, C., Weiss, Y. (eds.) ECCV 2018. LNCS, vol. 11219, pp. 353–369. Springer, Cham (2018). https://doi.org/10.1007/978-3-030-01267-0_21
23. Luo, R., Price, B., Cohen, S., Shakhnarovich, G.: Discriminability objective for training descriptive captions. In: CVPR, pp. 6964–6974 (2018)
24. Mao, J., Xu, W., Yang, Y., Wang, J., Huang, Z., Yuille, A.: Deep captioning with multimodal recurrent neural networks (m-RNN). In: ICLR (2015)
25. Papineni, K., Roukos, S., Ward, T., Zhu, W.J.: BLEU: a method for automatic evaluation of machine translation. In: ACL (2002)
26. Park, C.C., Kim, B., Kim, G.: Towards personalized image captioning via multimodal memory networks. In: IEEE TPAMI (2018)
27. Radford, A., Wu, J., Child, R., Luan, D., Amodei, D., Sutskever, I.: Language models are unsupervised multitask learners. OpenAI Blog **1**(8), 9 (2019)
28. Ren, S., He, K., Girshick, R., Sun, J.: Faster R-CNN: towards real-time object detection with region proposal networks. In: NeurIPS, pp. 91–99 (2015)
29. Rennie, S.J., Marcheret, E., Mroueh, Y., Ross, J., Goel, V.: Self-critical sequence training for image captioning. In: CVPR, pp. 7008–7024 (2017)
30. Shetty, R., Rohrbach, M., Hendricks, L.A.: Speaking the same language: matching machine to human captions by adversarial training. In: ICCV (2017)
31. Simonyan, K., Zisserman, A.: Very deep convolutional networks for large-scale image recognition. In: ICLR (2015)
32. Van Miltenburg, E., Elliott, D., Vossen, P.: Measuring the diversity of automatic image descriptions. In: COLING, pp. 1730–1741 (2018)
33. Vaswani, A., et al.: Attention is all you need. In: NeurIPS, pp. 5998–6008 (2017)
34. Vedantam, R., Lawrence Zitnick, C., Parikh, D.: CIDEr: consensus-based image description evaluation. In: CVPR, pp. 4566–4575 (2015)

35. Vered, G., Oren, G., Atzmon, Y., Chechik, G.: Joint optimization for cooperative image captioning. In: CVPR, pp. 8898–8907 (2019)
36. Vinyals, O., Toshev, A., Bengio, S., Erhan, D.: Show and tell: a neural image caption generator. In: CVPR (2015)
37. Wang, Q., Chan, A.B.: CNN+CNN: convolutional decoders for image captioning. In: CVPR Workshop (2018)
38. Wang, Q., Chan, A.B.: Describing like humans: on diversity in image captioning. In: CVPR (2019)
39. Wang, Q., Chan, A.B.: Towards diverse and accurate image captions via reinforcing determinantal point process. arXiv (2019)
40. Xu, J., Ren, X., Lin, J., Sun, X.: Diversity-promoting GAN: a cross-entropy based generative adversarial network for diversified text generation. In: EMNLP, pp. 3940–3949 (2018)
41. Xu, K., et al.: Show, attend and tell: neural image caption generation with visual attention. In: ICML (2015)
42. Yao, T., Pan, Y., Li, Y., Mei, T.: Hierarchy parsing for image captioning. In: ICCV, pp. 2621–2629 (2019)

Long-Term Human Motion Prediction with Scene Context

Zhe Cao[1(✉)], Hang Gao[1], Karttikeya Mangalam[1], Qi-Zhi Cai[2], Minh Vo[1,2], and Jitendra Malik[1]

[1] UC Berkeley, Berkeley, USA
zhecao@eecs.berkeley.edu

[2] Nanjing University, Nanjing, China

Abstract. Human movement is goal-directed and influenced by the spatial layout of the objects in the scene. To plan future human motion, it is crucial to perceive the environment – imagine how hard it is to navigate a new room with lights off. Existing works on predicting human motion do not pay attention to the scene context and thus struggle in long-term prediction. In this work, we propose a novel three-stage framework that exploits scene context to tackle this task. Given a single scene image and 2D pose histories, our method first samples multiple human motion goals, then plans 3D human paths towards each goal, and finally predicts 3D human pose sequences following each path. For stable training and rigorous evaluation, we contribute a synthetic dataset with clean annotations. In both synthetic and real datasets, our method shows consistent quantitative and qualitative improvements over existing methods. Project page: https://people.eecs.berkeley.edu/~zhecao/hmp/index.html (Please refer to our arXiv for a longer version of the paper with more visualizations.)

1 Introduction

Figure 1 shows the image of a typical indoor scene. Overlaid on this image is the pose trajectory of a person, depicted here by renderings of her body skeleton over time instants, where Frames 1–3 are in the past, Frame 4 is the present, and Frames 5–12 are in the future. In this paper, we study the following problem: *Given the scene image and the person's past pose and location history in 2D, predict her future poses and locations.*

Human movement is goal-directed and influenced by the spatial layout of the objects in the scene. For example, the person may be heading towards the window, and will find a path through the space avoiding collisions with various objects that might be in the way. Or perhaps a person approaches a chair with the intention to sit on it, and will adopt an appropriate path and pose sequence

Electronic supplementary material The online version of this chapter (https://doi.org/10.1007/978-3-030-58452-8_23) contains supplementary material, which is available to authorized users.

A. Vedaldi et al. (Eds.): ECCV 2020, LNCS 12346, pp. 387–404, 2020.
https://doi.org/10.1007/978-3-030-58452-8_23

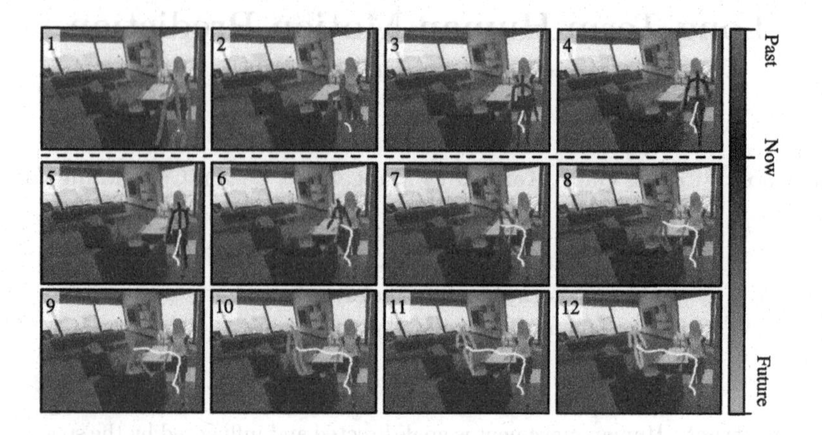

Fig. 1. Long-term 3D human motion prediction. Given a single scene image and 2D pose histories (the 1st row), we aim to predict long-term 3D human motion (projected on the image, shown in the 2–3 rows) influenced by scene. The human path is visualized as a yellow line. (Color figure online)

to achieve such a goal efficiently. We seek to understand such goal-directed, spatially contextualized human behavior, which we have formalized as a pose sequence and location prediction task.

With the advent of deep learning, there has been remarkable progress on the task of predicting human pose sequences [14,36,60,63]. However, these frameworks do not pay attention to scene context. As a representative example, Zhang et al. [63] detect the human bounding boxes across multiple time instances and derive their predictive signal from the evolving appearance of the human figure, but do not make use of the background image. Given this limitation, the predictions tend to be short-term (around 1 s), and local in space, e.g., walking in the same spot without global movement. If we want to make predictions that encompass bigger spatiotemporal neighborhoods, we need to make predictions conditioned on the scene context.

We make the following philosophical choices: (1) To understand long term behavior, we must reason in terms of goals. In the setting of moving through space, the goals could be represented by the destination points in the image. We allow multi-modality by generating multiple hypotheses of human movement "goals", represented by 2D destinations in the image space. (2) Instead of taking a 3D path planning approach as in the classical robotics literature [6,26], we approach the construction of likely human motions as a learning problem by constructing a convolutional network to implicitly learn the scene constraints from lots of human-scene interaction videos.

Specifically, we propose a learning framework that factorizes this task into three sequential stages as shown in Fig. 2. Our model sequentially predicts the motion goals, plans the 3D paths following each goal and finally generates the

(a) predicted goals (b) planned paths (c) final poses

Fig. 2. Overall pipeline. Given a single scene image and 2D pose histories, our method first samples (a) multiple possible future 2D destinations. We then predict the (b) 3D human path towards each destination. Finally, our model generates (c) 3D human pose sequences following paths, visualized with the ground-truth scene point cloud.

3D poses. In Sect. 5, we demonstrate our model outperforms existing methods quantitatively and generates more visually plausible 3D future motion.

To train such a learning system, we contribute a large-scale synthetic dataset focusing on human-scene interaction. Existing real datasets on 3D human motion have either contrived environment [21,58], relatively noisy 3D annotations [45], or limited motion range due to the depth sensor [17,45]. This motivates us to collect a diverse synthetic dataset with clean 3D annotations. We turn the Grand Theft Auto (GTA) gaming engine into an automatic data pipeline with control over different actors, scenes, cameras, lighting conditions, and motions. We collect over one million HD resolution RGB-D frames with 3D annotations which we discuss in detail in Sect. 4. Pre-training on our dataset stabilizes training and improves prediction performance on real dataset [17].

In summary, our key contributions are the following: (1) We formulate a new task of long-term 3D human motion prediction with scene context in terms of 3D poses and 3D locations. (2) We develop a novel three-stage computational framework that utilizes scene context for goal-oriented motion prediction, which outperforms existing methods both quantitatively and qualitatively. (3) We contribute a new synthetic dataset with diverse recordings of human-scene interaction and clean annotations.

2 Related Work

Predicting future human motion under real-world social context and scene constraints is a long-standing problem [4,16,18,24,44]. Due to its complexity, most of the current approaches can be classified into global trajectory prediction and local pose prediction. We connect these two components in a single framework for long-term scene-aware future human motion prediction.

Global Trajectory Prediction: Early approaches in trajectory prediction model the effect of social-scene interactions using physical forces [18], continuum dynamics [49], Hidden Markov model [24], or game theory [33]. Many of these approaches achieve competitive results even on modern pedestrian datasets [29,43]. With the resurgence of neural nets, data-driven prediction paradigm that captures multi-modal interaction between the scene and its agents becomes more dominant [4,5,8,16,34,44,47,62]. Similar to our method, they model the influence of the scene implicitly. However, unlike our formulation that considers images from diverse camera viewpoints, they make the key assumption of the bird-eye view image or known 3D information [4,16,24,44].

Local Pose Prediction: Similar to trajectory prediction, there has been plenty of interest in predicting future pose from image sequences both in the form of image generation [52,64], 2D pose [9,54], and 3D pose [11,15,61,63]. These methods exploit the local image context around the human to guide the future pose generation but do not pay attention to the background image or the global scene context. Approaches that focus on predicting 3D pose from 3D pose history also exist and are heavily used in motion tracking [12,53]. The goal is to learn 3D pose prior conditioning on the past motion using techniques such as Graphical Models [7], linear dynamical systems [42], trajectory basis [2,3], or Gaussian Process latent variable models [48,50,55,56], and more recently neural networks such as recurrent nets [14,22,32,36,41], temporal convolution nets [19,20,30], or graph convolution net in frequency space [60]. However, since these methods completely ignore the image context, the predicted human motion may not be consistent with the scene, i.e, waling through the wall. In contrast, we propose to utilize the scene context for future human motion prediction. This is similar in spirit to iMapper [37]. However, this approach relies on computationally expensive offline optimization to jointly reason about the scene and the human motion. Currently, there is no learning-based method that holistically models the scene context and human pose for more than a single time instance [10,28,31,57,59].

3D Human Motion Dataset. Training high capacity neural models requires large-scale and diverse training data. Existing human motion capture datasets either contain no environment [1], contrive environment [21,58], or in the outdoor setting without 3D annotation [35]. Human motion datasets with 3D scenes are often much smaller and have relatively noisy 3D human poses [17,45] due to the limitations of the depth sensor. To circumvent such problems, researchers exploit the interface between the game engine and the graphics rendering system to collect large-scale synthetic datasets [13,25]. Our effort on synthetic training data generation is a consolidation of such work to the new task of future human motion prediction with scene context.

3 Approach

In this work, we focus on long-term 3D human motion prediction that is goal-directed and is under the influence of scene context. We approach this problem by constructing a learning framework that factorizes long-term human motions into modeling their potential goal, planing 3D path and pose sequence, as shown in Fig. 3. Concretely, given a N-step 2D human pose history $\mathbf{X}_{1:N}$ and an 2D

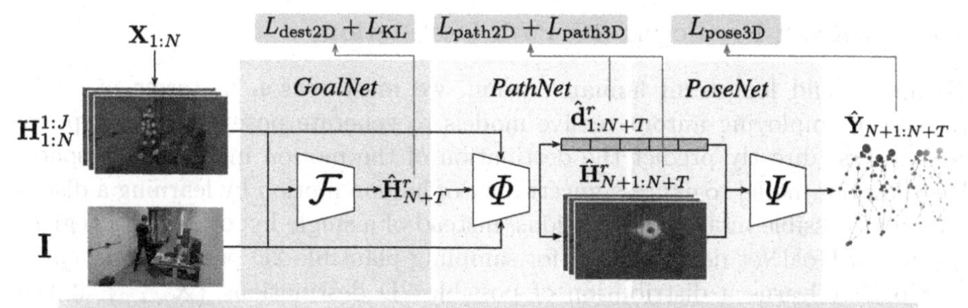

Fig. 3. Network architecture. Our pipeline contains three stages: *GoalNet* predicts 2D motion destinations of the human based on the reference image and 2D pose heatmaps (Sect. 3.1); *PathNet* plans the 3D global path of the human with the input of 2D heatmaps, 2D destination, and the image (Sect. 3.2); *PoseNet* predicts 3D global human motion, i.e., the 3D human pose sequences, following the predicted path (Sect. 3.3).

image[2] of the scene $\mathbf{I}$ (the Nth video frame in our case), we want to predict the next T-step 3D human poses together with their locations, denoted by a sequence $\mathbf{Y}_{N+1:N+T}$. We assume a known human skeleton consists of J keypoints, such that $\mathbf{X} \in \mathbb{R}^{J\times 2}, \mathbf{Y} \in \mathbb{R}^{J\times 3}$. We also assume a known camera model parameterized by its intrinsic matrix $\mathbf{K} \in \mathbb{R}^3$. To denote a specific keypoint position, we use the superscript of its index in the skeleton, e.g., $\mathbf{X}^r$ refers to the 2D location of the human center (torso) indexed by $r \in [1, J]$.

We motivate and elaborate our modular design for each stage in the rest of the section. Specifically, *GoalNet* learns to predict multiple possible human motion goals, represented as 2D destinations in the image space, based on a 2D pose history and the scene image. Next, *PathNet* learns to plan a 3D path towards each goal – the 3D location sequence of the human center (torso) – in conjunction with the scene context. Finally, *PoseNet* predicts 3D human poses at each time step following the predicted 3D path. In this way, the resulting 3D human motion has global movement and is more plausible considering the surrounding scene.

Thanks to this modular design, our model can have either deterministic or stochastic predictions. When deploying GoalNet, our model can sample multiple destinations, which results in stochastic prediction of future human motion. If not deploying GoalNet, our model generates single-mode prediction instead. We discuss them in more detail in the rest of the section and evaluate both predictions in our experiments.

[2] We choose to represent the scene by RGB images rather than RGBD scans because they are more readily available in many practical applications.

3.1 *GoalNet*: Predicting 2D Path Destination

To understand long-term human motion, we must reason in terms of goals. Instead of employing autoregressive models to generate poses step-by-step, we seek to first directly predict the destination of the motion in the image space. We allow our model to express uncertainty of human motion by learning a distribution of possible motion destinations, instead of a single hypothesis. This gives rise to our GoalNet denoted as $\mathcal{F}$ for sampling plausible 2D path destination.

GoalNet learns a distribution of possible 2D destinations $\{\hat{\mathbf{X}}^r_{N+T}\}$ at the end of the time horizon conditioned on the 2D pose history $\mathbf{X}_{1:N}$ and the scene image $\mathbf{I}$. We parametrize each human keypoint $\mathbf{X}^j$ by a heatmap channel $\mathbf{H}^j$ which preserves spatial correlation with the image context.

We employ GoalNet as a conditional variational auto-encoder [23]. The model encodes the inputs into a latent $\mathbf{z}$-space, from which we sample a random $\mathbf{z}$ vector for decoding and predicting the target destination positions. Formally, we have

$$\mathbf{z} \sim \mathcal{Q}(\mathbf{z}|\mathbf{H}^{1:J}_{1:N}, \mathbf{I}) \equiv \mathcal{N}(\boldsymbol{\mu}, \boldsymbol{\sigma}), \text{ where } \boldsymbol{\mu}, \boldsymbol{\sigma} = \mathcal{F}_{\text{enc}}(\mathbf{H}^{1:J}_{1:N}, \mathbf{I}). \tag{1}$$

In this way, we estimate a variational posterior $\mathcal{Q}$ by assuming a Gaussian information bottleneck using the decoder. Next, given a sampled $\mathbf{z}$ latent vector, we learn to predict our target destination heatmap with our GoalNet decoder,

$$\hat{\mathbf{H}}^r_{N+T} = \mathcal{F}_{\text{dec}}(\mathbf{z}, \mathbf{I}), \tag{2}$$

where it additionally conditions on the scene image. We use soft-argmax [46] to extract the 2D human motion destination $\hat{\mathbf{X}}^r_{N+T}$ from this heatmap $\hat{\mathbf{H}}^r_{N+T}$. We choose to use soft-argmax operation because it is differentiable and can produce sub-pixel locations. By constructing GoalNet, we have

$$\hat{\mathbf{H}}^r_{N+T} = \mathcal{F}(\mathbf{I}, \mathbf{H}^{1:J}_{1:N}). \tag{3}$$

We train GoalNet by minimizing two objectives: (1) the destination prediction error and (2) the KL-divergence between the estimated variational posterior $\mathcal{Q}$ and a normal distribution $\mathcal{N}(\mathbf{0}, \mathbf{1})$:

$$\begin{aligned} L_{\text{dest2D}} &= \|\mathbf{X}^r_{N+T} - \hat{\mathbf{X}}^r_{N+T}\|_1, \\ L_{\text{KL}} &= \text{KL}\big[\mathcal{Q}(\mathbf{z}|\mathbf{H}^{1:J}_{1:N}, \mathbf{I})\,||\,\mathcal{N}(0,1)\big], \end{aligned} \tag{4}$$

where we weigh equally between them. During testing, our GoalNet is able to sample a set of latent variables $\{\mathbf{z}\}$ from $\mathcal{N}(\mathbf{0}, \mathbf{1})$ and map them to multiple plausible 2D destinations $\{\hat{\mathbf{H}}^r_{N+T}\}$.

3.2 *PathNet*: Planning 3D Path towards Destination

With predicted destinations in the image space, our method further predicts 3D paths (human center locations per timestep) towards each destination. The destination determines where to move while the scene context determines how to move. We design a network that exploits both the 2D destination and the image for future 3D path planning. A key design choice we make here is that, instead of directly regressing 3D global coordinate values of human paths, we represent the 3D path as a combination of 2D path heatmaps and the depth values of the human center over time. This 3D path representation facilitates training as validated in our experiments (Sect. 5.3).

As shown in Fig. 3, our PathNet Φ takes the scene image $\mathbf{I}$, the 2D pose history $\mathbf{H}^{1:J}_{1:N}$, and the 2D destination assignment $\hat{\mathbf{H}}^r_{N+T}$ as inputs, and predicts global 3D path represented as $(\hat{\mathbf{H}}^r_{N+1:N+T}, \hat{\mathbf{d}}^r_{1:N+T})$, where $\hat{d}^r_t \in \mathbb{R}$ denotes the depth of human center at time t:

$$\hat{\mathbf{H}}^r_{N+1:N+T}, \hat{\mathbf{d}}^r_{1:N+T} = \Phi(\mathbf{I}, \mathbf{X}^{1:J}_{1:N}, \mathbf{X}^r_{N+T}). \tag{5}$$

We use soft-argmax to extract the resulting 2D path $\hat{\mathbf{X}}^r_{N+1:N+T}$ from predicted heatmaps $\hat{\mathbf{H}}^r_{N+1:N+T}$. Finally, we obtain the 3D path $\hat{\mathbf{Y}}^r_{1:N+T}$ by back-projecting the 2D path into the 3D camera coordinate frame using the human center depth $\hat{\mathbf{d}}^r_{1:N+T}$ and camera intrinsics $\mathbf{K}$.

We use Hourglass54 [27,38] as the backbone of PathNet to encode both the input image and 2D pose heatmaps. The network has two branches where the first branch predicts 2D path heatmaps and the second branch predicts the depth of the human torso.

We train our PathNet using two supervisions. We supervise our path predictions with ground-truth 2D heatmaps:

$$L_{\text{path2D}} = \|\mathbf{X}^r_{N+1:N+T} - \hat{\mathbf{X}}^r_{N+1:N+T}\|_1. \tag{6}$$

We also supervise path predictions with 3D path coordinates, while encouraging smooth predictions by penalizing large positional changes between consecutive frames:

$$L_{\text{path3D}} = \|\mathbf{Y}^r_{1:N+T} - \hat{\mathbf{Y}}^r_{1:N+T}\|_1 + \|\hat{\mathbf{Y}}^r_{1:N+T-1} - \hat{\mathbf{Y}}^r_{2:N+T}\|_1. \tag{7}$$

These losses are summed together with equal weight as the final training loss. During training, we use the ground-truth destination to train our PathNet, while during testing, we can use predictions from the GoalNet.

The GoalNet and PathNet we describe so far enable sampling multiple 3D paths during inference. We thus refer to it as the stochastic mode of the model. The modular design of GoalNet and PathNet is flexible. By removing GoalNet and input $\mathbf{X}^r_{N+T}$ from Eq. 5, we can directly use PathNet to produce deterministic 3D path predictions. We study these two modes, deterministic and stochastic mode, in our experiments.

3.3 *PoseNet*: Generating 3D Pose following Path

With the predicted 3D path $\hat{\mathbf{Y}}^r_{1:N+T}$ and 2D pose history $\mathbf{X}_{1:N}$, we use the transformer network [51] as our PoseNet Ψ to predict 3D poses following such path. Instead of predicting the 3D poses from scratch, we first lift 2D pose history into 3D to obtain a noisy 3D human pose sequence $\bar{\mathbf{Y}}_{1:N+T}$ as input, and further use Ψ to refine them to obtain the final prediction. Our initial estimation consists of two steps. We first obtain a noisy 3D poses $\bar{\mathbf{Y}}_{1:N}$ by back-projecting 2D pose history $\mathbf{X}_{1:N}$ into 3D using the human torso depth $\hat{\mathbf{d}}^r_{1:N}$ and camera intrinsics $\mathbf{K}$. We next replicate the present 3D pose $\bar{\mathbf{Y}}_N$ to each of the predicted future 3D path location for an initial estimation of future 3D poses $\bar{\mathbf{Y}}_{N+1:N+T}$. We then concatenate both estimations together to form $\bar{\mathbf{Y}}_{1:N+T}$ as input to our PoseNet:

$$\hat{\mathbf{Y}}_{N+1:N+T} = \Psi(\bar{\mathbf{Y}}_{1:N+T}). \tag{8}$$

The training objective for PoseNet is to minimize the distance between the 3D pose prediction and the ground-truth defined as:

$$L_{\text{pose3D}} = \|\mathbf{Y}_{N+1:N+T} - \hat{\mathbf{Y}}_{N+1:N+T}\|_1. \tag{9}$$

During training, ground-truth 3D path $\mathbf{Y}^r_{1:N+T}$ is used for estimating coarse 3D pose input. During testing, we use the predicted 3D path $\hat{\mathbf{Y}}^r_{1:N+T}$ from PathNet.

4 GTA Indoor Motion Dataset

We introduce the GTA Indoor Motion dataset (GTA-IM)[3] that emphasizes human-scene interactions. Our motivation for this dataset is that existing real datasets on human-scene interaction [17,45] have relatively noisy 3D human pose annotations and limited long-range human motion limited by depth sensors. On the other hand, existing synthetic human datasets [13,25] focus on the task of human pose estimation or parts segmentation and sample data in wide-open outdoor scenes with limited interactable objects.

To overcome the above issues, we spend extensive efforts in collecting a synthetic dataset by developing an interface with the game engine for controlling characters, cameras, and action tasks in a fully automatic manner. For each character, we randomize the goal destination inside the 3D scene, the specific task to do, the walking style, and the movement speed. We control the lighting condition by changing different weather conditions and daytime. We also diversify the camera location and viewing angle over a sphere around the actor such that it points towards the actor. We use in-game ray tracing API and synchronized human segmentation map to track actors. The collected actions include climbing the stairs, lying down, sitting, opening the door, and etc. – a set of basic activities within indoor scenes. For example, the character has 22 walking styles including 10 female and 12 male walking styles. All of these factors enable us to collect a diverse and realistic dataset with accurate annotations for our challenging task.

[3] Dataset available in https://github.com/ZheC/GTA-IM-Dataset

Fig. 4. Sample RGBD images from GTA-IM dataset. Our dataset contains realistic RGB images (visualized with the 2D pose), accurate depth maps, and clean 3D human pose annotations.

In total, we collect one million RGBD frames of 1920×1080 resolution with the ground-truth 3D human pose (98 joints), human segmentation, and camera pose. Some examples are shown in Fig. 4. The dataset contains 50 human characters acting inside 10 different large indoor scenes. Each scene has several floors, including living rooms, bedrooms, kitchens, balconies, and etc., enabling diverse interaction activities.

5 Evaluation

We perform extensive quantitative and qualitative evaluations of our future 3D human path and motion predictions. The rest of this section is organized as follows: We first describe the datasets we use in Sect. 5.1. We then elaborate on our quantitative evaluation metrics and strong baselines in Sect. 5.2. Further, we show our quantitative and qualitative improvement over previous methods in Sect. 5.3. Finally, we evaluate our long-term predictions and show qualitative results of destination samples and final 3D pose results in Sect. 5.4. We discussed some failure cases in Sect. 5.5.

5.1 Datasets

GTA-IM: We train and test our model on our collected dataset as described in Sect. 4. We split 8 scenes for training and 2 scenes for evaluation. We choose 21 out of 98 human joints provided from the dataset. We convert both the 3D path and the 3D pose into the camera coordinate frame for both training and evaluation.

PROX: Proximal Relationships with Object eXclusion (PROX) is a new dataset captured using the Kinect-One sensor by Hassan et al. [17]. It comprises of 12 different 3D scenes and RGB sequences of 20 subjects moving in and interacting with the scenes. We split the dataset with 52 training sequences and 8 sequences for testing. Also, we extract 18 joints from the SMPL-X model [39] from the provided human pose parameters.

5.2 Evaluation Metric and Baselines

Metrics: We use the Mean Per Joint Position Error (MPJPE) [21] as a metric for quantitatively evaluating both the 3D path and 3D pose prediction. We report the performance at different time steps (seconds) in millimeters (mm).

Baselines: To the best of our knowledge, there exists no prior work that predicts 3D human pose with global movement using 2D pose sequence as input. Thus, we propose three strong baselines for comparison with our method. For the first baseline, we combine the recent 2D-to-3D human pose estimation method [40] and 3D human pose prediction method [60]. For the second baseline, we use Transformer [51], the state-of-the-art sequence-to-sequence model, to perform 3D prediction directly from 2D inputs treating the entire problem as a single-stage sequence to sequence task. For the third baseline, we compare with is constructed by first predicting the future 2D human pose using [51] from inputs and then lifting the predicted pose into 3D using [40]. Note that none of these baselines consider scene context or deal with uncertainty in their future predictions. We train all models on both datasets for two-second-long prediction based on 1-second-long history and report their best performance for comparison.

5.3 Comparison with Baselines

In this section, we perform quantitative evaluations of our method in the two datasets. We also show some qualitative comparisons in Fig. 5. We evaluate the two modes of our model: the stochastic mode that generates multiple predictions by sampling different 2D destinations from the GoalNet; and the deterministic mode that can generate one identical prediction without deploying GoalNet.

GTA-IM: The quantitative evaluation of 3D path and 3D pose prediction in GTA-IM dataset is shown in Table 1. Our deterministic model with image input can outperform the other methods by a margin, i.e., with an average error of 173 mm vs. 193 mm. When using sampling during inference, the method can generate multiple hypotheses of the future 3D pose sequence. We evaluate different numbers of samples and select the predictions among all samples that best matches ground truth to report the error. We find using four samples during inference can match the performance of our deterministic model (173 mm error), while using ten samples, we further cut down the error to 165 mm. These results validate that our stochastic model can help deal with the uncertainty of future human motion and outperform the deterministic baselines with few samples.

As an ablation, we directly regress 3D coordinates ("Ours w/ xyz." in the Table 1) and observe an overall error that is 18 mm higher than the error of our deterministic model (191 mm vs. 173 mm). This validates that representing the 3D path as the depth and 2D heatmap of the human center is better due to its

Table 1. Evaluation results in GTA-IM dataset. We compare other baselines in terms of 3D path and pose error. The last column (All) is the mean average error of the entire prediction over all time steps. VP denotes Pavllo et al. [40], TR denotes transformer [51] and LTD denotes Wei et al. [60]. GT stands for ground-truth, xyz. stands for directly regressing 3D coordinates of the path. We report the error of our stochastic predictions with varying number of samples.

Time step (second)	3D path error (mm)				3D pose error (mm)				
	0.5	1	1.5	2	0.5	1	1.5	2	**All** ↓
TR [51]	277	352	457	603	291	374	489	641	406
TR [51] + VP [40]	157	240	358	494	174	267	388	526	211
VP [40] + LTD [60]	124	194	276	367	121	180	249	330	193
Ours (deterministic)	**104**	**163**	**219**	**297**	**91**	**158**	**237**	**328**	**173**
Ours (samples = 4)	114	162	227	310	94	161	236	323	173
Ours (samples = 10)	110	154	213	289	90	154	224	306	165
Ours w/ xyz. output	122	179	252	336	101	177	262	359	191
Ours w/o image	128	177	242	320	99	179	271	367	196
Ours w/ masked image	120	168	235	314	96	170	265	358	189
Ours w/ RGBD input	100	138	193	262	93	160	235	322	172
Ours w/ GT destination	104	125	146	170	85	133	178	234	137

strong correlation to the image appearance. We also ablates different types of input to our model. Without image input, the average error is 23 mm higher. With only masked images input, i.e., replacing pixels outside human crop by ImageNet mean pixel values, the error is 16 mm high. This validates that using full image to encode scene context is more helpful than only observing cropped human image, especially for long-term prediction. Using both color and depth image as input ("Ours w/ RGBD input"), the average error is 172 mm which is similar to the model with RGB input. This indicates that our model implicitly learns to reason about depth information from 2D input. If we use ground-truth 2D destinations instead of predicted ones, and the overall error decreases down to 137 mm. It implies that the uncertainty of the future destination is the major source of difficulty in this problem.

PROX: The evaluation results in Table 2 demonstrate that our method outperforms the previous state of the art in terms of mean MPJPE of all time steps, 270 mm vs. 282 mm. Overall, we share the same conclusion as the comparisons in GTA-IM dataset. When using sampling during inference, three samples during inference can beat the performance of our deterministic model (264 mm vs. 270 mm), while using ten samples, the error decreases to 249 mm. Note that these

Table 2. Evaluation results in PROX dataset. We compare other baselines in terms of 3D future path and pose prediction. VP denotes Pavllo et al. [40], TR denotes transformer [51] and LTD denotes Wei et al. [60]. GT stands for ground-truth. We rank all methods using mean average error of the entire prediction (last column).

Time step (second)	3D path error (mm)				3D pose error (mm)				
	0.5	1	1.5	2	0.5	1	1.5	2	**All ↓**
TR [51]	487	583	682	783	512	603	698	801	615
TR [51] + VP [40]	262	358	461	548	297	398	502	590	326
VP [40] + LTD [60]	194	263	332	394	216	274	335	**394**	282
Ours w/o GTA-IM pretrain	192	258	336	419	192	273	352	426	280
Ours (deterministic)	**189**	**245**	**317**	**389**	**190**	**264**	**335**	406	**270**
Ours (samples = 3)	192	245	311	398	187	258	328	397	264
Ours (samples = 6)	185	229	285	368	184	249	312	377	254
Ours (samples = 10)	181	222	273	354	182	244	304	367	249
Ours w/ gt destination	193	223	234	237	195	235	276	321	237

improvements are more prominent than what we observe on GTA-IM benchmark. This is because the uncertainty of future motion in the real dataset is larger. Therefore, stochastic predictions have more advantage.

Moreover, we find that pre-training in GTA-IM dataset can achieve better performance (270 mm vs. 280 mm). Our method exploits the image context and tends to overfit in PROX dataset because it is less diverse in terms of camera poses and background appearance (both are constant per video sequence). Pre-training in our synthetic dataset with diverse appearance and clean annotations can help prevent overfitting and boost the final performance.

Qualitative comparison: In Fig. 5, we show qualitative comparison with the baseline of VP [40] and LTD [60]. This baseline is quantitatively competitive as shown in Table 1 and 2. However, without considering scene context, their predicted results may not be feasible inside the 3D scene, e.g., the person cannot go through a table or sofa. In contrast, our model implicitly learns the scene constraints from the image and can generate more plausible 3D human motion.

5.4 Evaluation and Visualization on Longer-Term Predictions

To demonstrate our method can predict future human motion for more than 2 s, we train another model to produce the 3-second-long future prediction. In Fig. 6, we show the self-comparisons between our stochastic predictions and deterministic predictions. Our stochastic models can beat their deterministic counterpart using 5 samples. With increasing number of samples, the testing error decreases accordingly. The error gap between deterministic results and stochastic results becomes larger at the later stage of the prediction, i.e., more

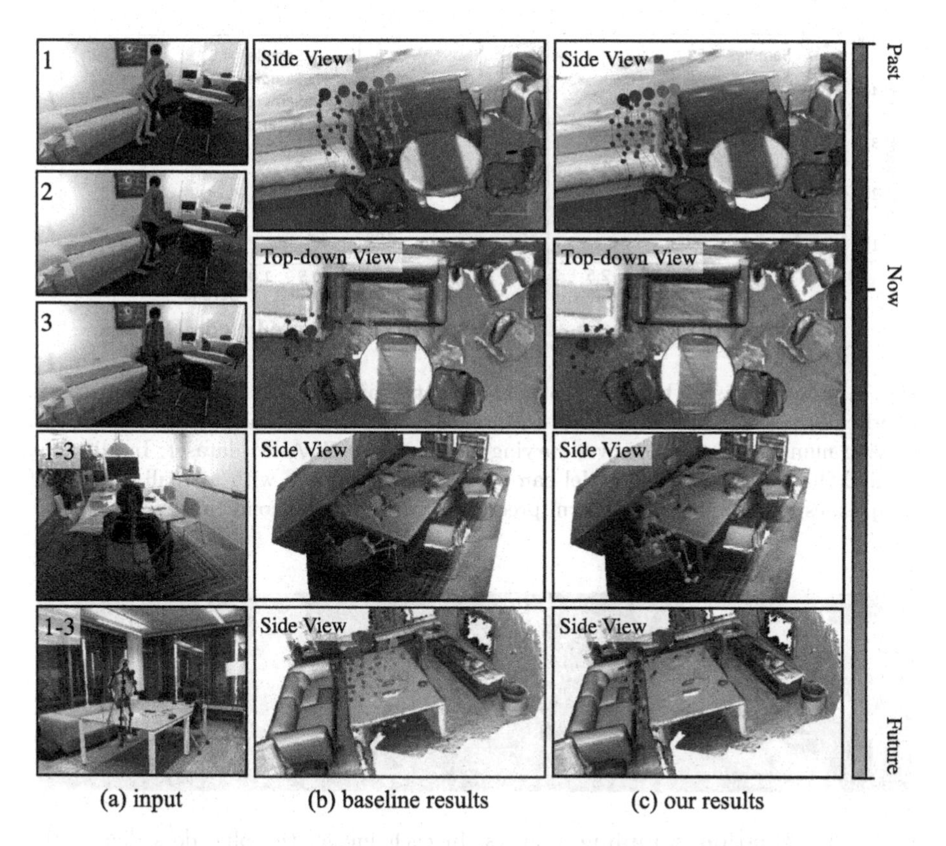

Fig. 5. Qualitative comparison. We visualize the input (a), the results of VP [40] and LTD [60] (b) and our results (c) in the ground-truth 3D mesh. The color of pose is changed over timesteps according to the color bar. The first example includes both top-down and side view. From the visualization, we can observe some collisions between the baseline results and the 3D scene, while our predicted motion are more plausible by taking the scene context into consideration.

than 100 mm difference at 3 second time step. This indicates the advantage of the stochastic model in long-term future motion prediction.

We show qualitative results of our stochastic predictions on movement destinations in Fig. 7. We include more visualizations of our long-term stochastic future motion prediction in the supplement and our arXiv paper. Our method can generate diverse human movement destinations, and realistic 3D future human motion considering the environment context, e.g., turning left/right, walking straight, taking a U-turn, climbing stairs, standing up from sitting, and laying back on the sofa.

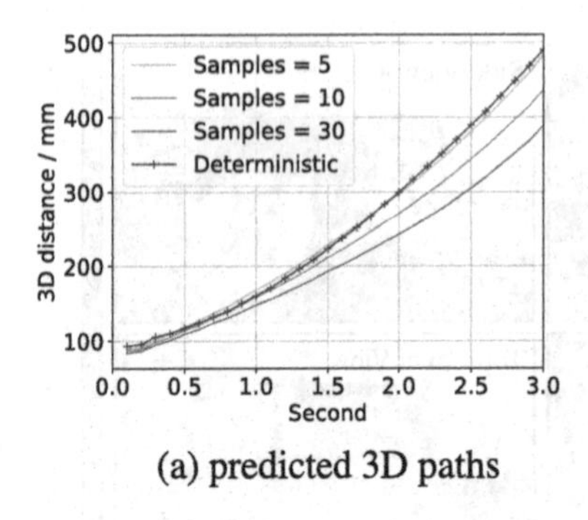

(a) predicted 3D paths

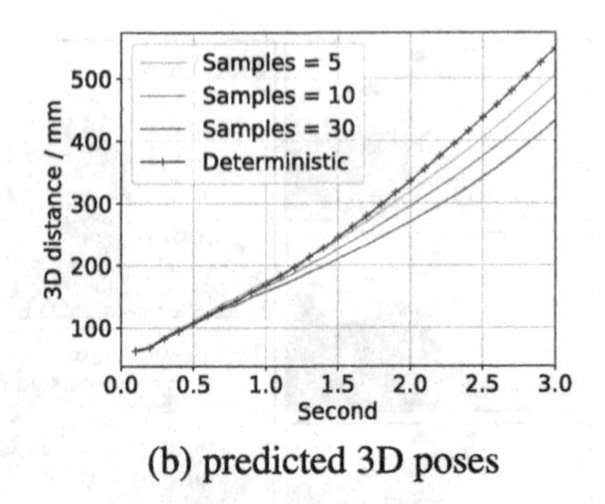

(b) predicted 3D poses

Fig. 6. Comparison between our stochastic predictions and deterministic predictions. We show error curves of predicted (a) 3D paths and (b) 3D poses with varying numbers of samples over varying timesteps on GTA-IM dataset. In all plots, we find that our stochastic model can achieve better results with a small number of samples, especially in the long-term prediction (within 2–3 second time span).

Fig. 7. Destination sampling results. In each image, the blue dots denote the path history, the green dots are ground-truth future destination, red dots are sample destinations from the GoalNet which we draw 30 times from the standard Gaussian. Our method can generate diverse plausible motion destination samples which leads to different activities, i.e., sitting still or standing up.

5.5 Discussion of Failure Cases

Our model implicitly learns scene constraints in a data-driven manner from large amounts of training data, and can produce consistent 3D human paths without serious collision comparing to previous methods which do not take scene context into consideration as shown in Fig. 5. However, without assuming we have access to the pre-reconstructed 3D mesh and using expensive offline optimization as [17], the resulting 3D poses may not strictly meet all physical constraints of the 3D scene geometry. Some failure cases are shown in Fig. 8. In the red circled area, we observe small intersections between the human feet and the 3D scene mesh, e.g., the ground floor or the bed. This issue could be relieved by integrating multi-view/temporal images as input to the learning system to recover the 3D scene better. The resulting 3D scene could be further used to enforce explicit scene geometry constraints for refining the 3D poses. We leave this to the future work.

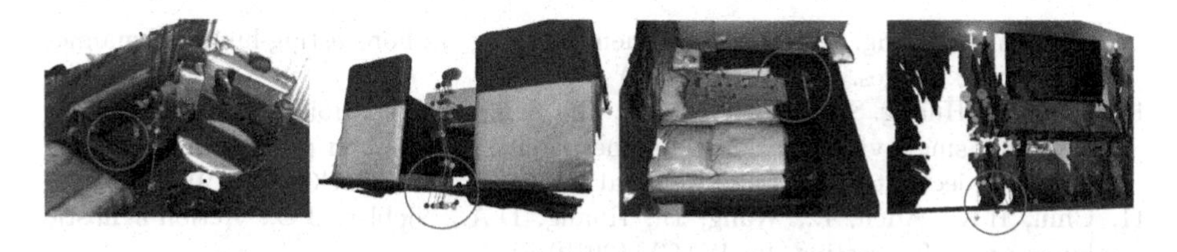

Fig. 8. Visualization of failure cases. In each red circle area, we observe the intersection between the human feet and the 3D mesh, e.g., the bed.

6 Conclusion

In this work, we study the challenging task of long-term 3D human motion prediction with only 2D input. This research problem is very relevant to many real-world applications where understanding and predicting feasible human motion considering the surrounding space is critical, e.g., a home service robot collaborating with the moving people, AR glass providing navigation guide to visually impaired people, and autonomous vehicle planning the action considering the safety of pedestrians. We present an initial attempt in attacking the problem by contributing a new dataset with diverse human-scene interactions and clean 3D annotations. We also propose the first method that can predict long-term stochastic 3D future human motion from 2D inputs, while taking the scene context into consideration. There are still many aspects in this problem that can be explored in the future, such as how to effectively evaluate the naturalness and feasibility of the stochastic human motion predictions, and how to incorporate information of dynamic objects and multiple moving people inside the scene.

Ackownledgement. We thank Carsten Stoll and Christoph Lassner for the helpful feedback. We are also very grateful for the discussion within the BAIR community.

References

1. CMU Motion Capture Database. http://mocap.cs.cmu.edu
2. Akhter, I., Sheikh, Y., Khan, S., Kanade, T.: Nonrigid structure from motion in trajectory space. In: NIPS (2009)
3. Akhter, I., Simon, T., Khan, S., Matthews, I., Sheikh, Y.: Bilinear spatiotemporal basis models. SIGGRAPH (2012)
4. Alahi, A., Goel, K., Ramanathan, V., Robicquet, A., Fei-Fei, L., Savarese, S.: Social LSTM: human trajectory prediction in crowded spaces. In: CVPR (2016)
5. Alahi, A., Ramanathan, V., Fei-Fei, L.: Socially-aware large-scale crowd forecasting. In: CVPR (2014)
6. Alexopoulos, C., Griffin, P.M.: Path planning for a mobile robot. IEEE Trans. Syst. Man Cybern. (1992)
7. Brand, M., Hertzmann, A.: Style machines. SIGGRAPH (2000)
8. Chai, Y., Sapp, B., Bansal, M., Anguelov, D.: Multipath: multiple probabilistic anchor trajectory hypotheses for behavior prediction. In: CoRL (2019)

9. Chao, Y.W., Yang, J., Price, B., Cohen, S., Deng, J.: Forecasting human dynamics from static images. In: CVPR (2017)
10. Chen, Y., Huang, S., Yuan, T., Qi, S., Zhu, Y., Zhu, S.C.: Holistic++ scene understanding: single-view 3D holistic scene parsing and human pose estimation with human-object interaction and physical commonsense. In: ICCV (2019)
11. Chiu, H.K., Adeli, E., Wang, B., Huang, D.A., Niebles, J.C.: Action-agnostic human pose forecasting. In: WACV (2019)
12. Elhayek, A., Stoll, C., Hasler, N., Kim, K.I., Seidel, H.P., Theobalt, C.: Spatio-temporal motion tracking with unsynchronized cameras. In: CVPR (2012)
13. Fabbri, M., Lanzi, F., Calderara, S., Palazzi, A., Vezzani, R., Cucchiara, R.: Learning to detect and track visible and occluded body joints in a virtual world. In: Ferrari, V., Hebert, M., Sminchisescu, C., Weiss, Y. (eds.) ECCV 2018. LNCS, vol. 11208, pp. 450–466. Springer, Cham (2018). https://doi.org/10.1007/978-3-030-01225-0_27
14. Fragkiadaki, K., Levine, S., Felsen, P., Malik, J.: Recurrent network models for human dynamics. In: ICCV (2015)
15. Ghosh, P., Song, J., Aksan, E., Hilliges, O.: Learning human motion models for long-term predictions. In: 3DV (2017)
16. Gupta, A., Johnson, J., Fei-Fei, L., Savarese, S., Alahi, A.: Social GAN: socially acceptable trajectories with generative adversarial networks. In: CVPR (2018)
17. Hassan, M., Choutas, V., Tzionas, D., Black, M.J.: Resolving 3D human pose ambiguities with 3D scene constraints. In: ICCV (2019)
18. Helbing, D., Molnar, P.: Social force model for pedestrian dynamics. Phys. Rev. E (1995)
19. Hernandez, A., Gall, J., Moreno-Noguer, F.: Human motion prediction via spatio-temporal inpainting. In: CVPR (2019)
20. Holden, D., Saito, J., Komura, T., Joyce, T.: Learning motion manifolds with convolutional autoencoders. In: SIGGRAPH Asian Technical Briefs (2015)
21. Ionescu, C., Papava, D., Olaru, V., Sminchisescu, C.: Human3.6M: large scale datasets and predictive methods for 3D human sensing in natural environments. TPAMI (2013)
22. Jain, A., Zamir, A.R., Savarese, S., Saxena, A.: Structural-RNN: deep learning on spatio-temporal graphs. In: CVPR (2016)
23. Kingma, D.P., Welling, M.: Auto-encoding variational bayes. ICLR (2014)
24. Kitani, K.M., Ziebart, B.D., Bagnell, J.A., Hebert, M.: Activity forecasting. In: Fitzgibbon, A., Lazebnik, S., Perona, P., Sato, Y., Schmid, C. (eds.) ECCV 2012. LNCS, vol. 7575, pp. 201–214. Springer, Heidelberg (2012). https://doi.org/10.1007/978-3-642-33765-9_15
25. Krähenbühl, P.: Free supervision from video games. In: CVPR (2018)
26. LaValle, S.M.: Planning Algorithms. Cambridge University Press (2006)
27. Law, H., Teng, Y., Russakovsky, O., Deng, J.: CornerNet-Lite: efficient keypoint based object detection. arXiv preprint arXiv:1904.08900 (2019)
28. Lee, D., Liu, S., Gu, J., Liu, M.Y., Yang, M.H., Kautz, J.: Context-aware synthesis and placement of object instances. In: NIPS (2018)
29. Lerner, A., Chrysanthou, Y., Lischinski, D.: Crowds by example. In: CGF (2007)
30. Li, C., Zhang, Z., Sun Lee, W., Hee Lee, G.: Convolutional sequence to sequence model for human dynamics. In: CVPR (2018)
31. Li, X., Liu, S., Kim, K., Wang, X., Yang, M.H., Kautz, J.: Putting humans in a scene: learning affordance in 3D indoor environments. In: CVPR (2019)
32. Li, Z., Zhou, Y., Xiao, S., He, C., Huang, Z., Li, H.: Auto-conditioned recurrent networks for extended complex human motion synthesis. In: ICLR (2018)

33. Ma, W.C., Huang, D.A., Lee, N., Kitani, K.M.: Forecasting interactive dynamics of pedestrians with fictitious play. In: CVPR (2017)
34. Makansi, O., Ilg, E., Cicek, O., Brox, T.: Overcoming limitations of mixture density networks: a sampling and fitting framework for multimodal future prediction. In: CVPR (2019)
35. von Marcard, T., Henschel, R., Black, M.J., Rosenhahn, B., Pons-Moll, G.: Recovering accurate 3D human pose in the wild using IMUs and a moving camera. In: Ferrari, V., Hebert, M., Sminchisescu, C., Weiss, Y. (eds.) ECCV 2018. LNCS, vol. 11214, pp. 614–631. Springer, Cham (2018). https://doi.org/10.1007/978-3-030-01249-6_37
36. Martinez, J., Black, M.J., Romero, J.: On human motion prediction using recurrent neural networks. In: CVPR (2017)
37. Monszpart, A., Guerrero, P., Ceylan, D., Yumer, E., Mitra, N.J.: iMapper: interaction-guided joint scene and human motion mapping from monocular videos. SIGGRAPH (2019)
38. Newell, A., Yang, K., Deng, J.: Stacked hourglass networks for human pose estimation. In: Leibe, B., Matas, J., Sebe, N., Welling, M. (eds.) ECCV 2016. LNCS, vol. 9912, pp. 483–499. Springer, Cham (2016). https://doi.org/10.1007/978-3-319-46484-8_29
39. Pavlakos, G., et al.: Expressive body capture: 3D hands, face, and body from a single image. In: CVPR (2019)
40. Pavllo, D., Feichtenhofer, C., Grangier, D., Auli, M.: 3D human pose estimation in video with temporal convolutions and semi-supervised training. In: CVPR (2019)
41. Pavllo, D., Grangier, D., Auli, M.: QuaterNet: a quaternion-based recurrent model for human motion. In: BMVC (2018)
42. Pavlovic, V., Rehg, J.M., MacCormick, J.: Learning switching linear models of human motion. In: NIPS (2001)
43. Pellegrini, S., Ess, A., Schindler, K., Van Gool, L.: You'll never walk alone: modeling social behavior for multi-target tracking. In: CVPR (2009)
44. Sadeghian, A., Kosaraju, V., Sadeghian, A., Hirose, N., Rezatofighi, H., Savarese, S.: SoPhie: an attentive GAN for predicting paths compliant to social and physical constraints. In: CVPR (2019)
45. Savva, M., Chang, A.X., Hanrahan, P., Fisher, M., Nießner, M.: PiGraphs: Learning Interaction Snapshots from Observations. TOG (2016)
46. Sun, X., Xiao, B., Wei, F., Liang, S., Wei, Y.: Integral human pose regression. In: Ferrari, V., Hebert, M., Sminchisescu, C., Weiss, Y. (eds.) ECCV 2018. LNCS, vol. 11210, pp. 536–553. Springer, Cham (2018). https://doi.org/10.1007/978-3-030-01231-1_33
47. Tai, L., Zhang, J., Liu, M., Burgard, W.: Socially compliant navigation through raw depth inputs with generative adversarial imitation learning. In: ICRA (2018)
48. Tay, M.K.C., Laugier, C.: Modelling smooth paths using gaussian processes. In: Laugier, C., Siegwart, R. (eds.) Field and Service Robotics, pp. 381–390. Springer, Heidelberg (2008). https://doi.org/10.1007/978-3-540-75404-6_36
49. Treuille, A., Cooper, S., Popović, Z.: Continuum crowds. TOG (2006)
50. Urtasun, R., Fleet, D.J., Geiger, A., Popović, J., Darrell, T.J., Lawrence, N.D.: Topologically-constrained latent variable models. In: ICML (2008)
51. Vaswani, A., et al.: Attention is all you need. In: NIPS (2017)
52. Villegas, R., Yang, J., Zou, Y., Sohn, S., Lin, X., Lee, H.: Learning to generate long-term future via hierarchical prediction. In: ICML (2017)
53. Vo, M., Narasimhan, S.G., Sheikh, Y.: Spatiotemporal bundle adjustment for dynamic 3D reconstruction. In: CVPR (2016)

54. Walker, J., Marino, K., Gupta, A., Hebert, M.: The pose knows: video forecasting by generating pose futures. In: CVPR (2017)
55. Wang, J.M., Fleet, D.J., Hertzmann, A.: Gaussian process dynamical models for human motion. TPAMI (2007)
56. Wang, J.M., Fleet, D.J., Hertzmann, A.: Multifactor gaussian process models for style-content separation. In: ICML (2007)
57. Wang, X., Girdhar, R., Gupta, A.: Binge watching: scaling affordance learning from sitcoms. In: CVPR (2017)
58. Wang, Z., Chen, L., Rathore, S., Shin, D., Fowlkes, C.: Geometric pose affordance: 3D human pose with scene constraints. arXiv preprint arXiv:1905.07718 (2019)
59. Wang, Z., Shin, D., Fowlkes, C.C.: Predicting camera viewpoint improves cross-dataset generalization for 3d human pose estimation. arXiv preprint arXiv:2004.03143 (2020)
60. Wei, M., Miaomiao, L., Mathieu, S., Hongdong, L.: Learning trajectory dependencies for human motion prediction. In: ICCV (2019)
61. Weng, C.Y., Curless, B., Kemelmacher-Shlizerman, I.: Photo wake-up: 3D character animation from a single photo. In: CVPR (2019)
62. Yu, T., et al.: One-shot imitation from observing humans via domain-adaptive meta-learning. IROS (2018)
63. Zhang, J.Y., Felsen, P., Kanazawa, A., Malik, J.: Predicting 3D human dynamics from video. In: ICCV (2019)
64. Zhao, L., Peng, X., Tian, Yu., Kapadia, M., Metaxas, D.: Learning to forecast and refine residual motion for image-to-video generation. In: Ferrari, V., Hebert, M., Sminchisescu, C., Weiss, Y. (eds.) ECCV 2018. LNCS, vol. 11219, pp. 403–419. Springer, Cham (2018). https://doi.org/10.1007/978-3-030-01267-0_24

NeRF: Representing Scenes as Neural Radiance Fields for View Synthesis

Ben Mildenhall[1(✉)], Pratul P. Srinivasan[1], Matthew Tancik[1], Jonathan T. Barron[2], Ravi Ramamoorthi[3], and Ren Ng[1]

[1] UC Berkeley, Berkeley, USA
bmild@berkeley.edu123460409
[2] Google Research, New York, USA
[3] UC San Diego, San Diego, USA

Abstract. We present a method that achieves state-of-the-art results for synthesizing novel views of complex scenes by optimizing an underlying continuous volumetric scene function using a sparse set of input views. Our algorithm represents a scene using a fully-connected (non-convolutional) deep network, whose input is a single continuous 5D coordinate (spatial location (x, y, z) and viewing direction (θ, ϕ)) and whose output is the volume density and view-dependent emitted radiance at that spatial location. We synthesize views by querying 5D coordinates along camera rays and use classic volume rendering techniques to project the output colors and densities into an image. Because volume rendering is naturally differentiable, the only input required to optimize our representation is a set of images with known camera poses. We describe how to effectively optimize neural radiance fields to render photorealistic novel views of scenes with complicated geometry and appearance, and demonstrate results that outperform prior work on neural rendering and view synthesis. View synthesis results are best viewed as videos, so we urge readers to view our supplementary video for convincing comparisons.

Keywords: Scene representation · View synthesis · Image-based rendering · Volume rendering · 3D deep learning

1 Introduction

In this work, we address the long-standing problem of view synthesis in a new way by directly optimizing parameters of a continuous 5D scene representation to minimize the error of rendering a set of captured images.

We represent a static scene as a continuous 5D function that outputs the radiance emitted in each direction (θ, ϕ) at each point (x, y, z) in space, and a

B. M. Pratul, P. Srinivasan and M. Tancik: Authors contributed equally to this work.

Electronic supplementary material The online version of this chapter (https://doi.org/10.1007/978-3-030-58452-8_24) contains supplementary material, which is available to authorized users.

A. Vedaldi et al. (Eds.): ECCV 2020, LNCS 12346, pp. 405–421, 2020.
https://doi.org/10.1007/978-3-030-58452-8_24

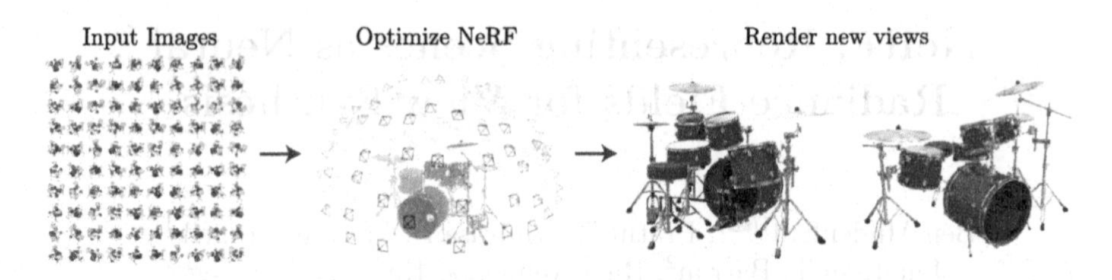

Fig. 1. We present a method that optimizes a continuous 5D neural radiance field representation (volume density and view-dependent color at any continuous location) of a scene from a set of input images. We use techniques from volume rendering to accumulate samples of this scene representation along rays to render the scene from any viewpoint. Here, we visualize the set of 100 input views of the synthetic *Drums* scene randomly captured on a surrounding hemisphere, and we show two novel views rendered from our optimized NeRF representation.

density at each point which acts like a differential opacity controlling how much radiance is accumulated by a ray passing through (x, y, z). Our method optimizes a deep fully-connected neural network without any convolutional layers (often referred to as a multilayer perceptron or MLP) to represent this function by regressing from a single 5D coordinate (x, y, z, θ, ϕ) to a single volume density and view-dependent RGB color. To render this *neural radiance field* (NeRF) from a particular viewpoint we: 1) march camera rays through the scene to generate a sampled set of 3D points, 2) use those points and their corresponding 2D viewing directions as input to the neural network to produce an output set of colors and densities, and 3) use classical volume rendering techniques to accumulate those colors and densities into a 2D image. Because this process is naturally differentiable, we can use gradient descent to optimize this model by minimizing the error between each observed image and the corresponding views rendered from our representation. Minimizing this error across multiple views encourages the network to predict a coherent model of the scene by assigning high volume densities and accurate colors to the locations that contain the true underlying scene content. Figure 2 visualizes this overall pipeline.

We find that the basic implementation of optimizing a neural radiance field representation for a complex scene does not converge to a sufficiently high-resolution representation and is inefficient in the required number of samples per camera ray. We address these issues by transforming input 5D coordinates with a positional encoding that enables the MLP to represent higher frequency functions, and we propose a hierarchical sampling procedure to reduce the number of queries required to adequately sample this high-frequency scene representation.

Our approach inherits the benefits of volumetric representations: both can represent complex real-world geometry and appearance and are well suited for gradient-based optimization using projected images. Crucially, our method overcomes the prohibitive storage costs of *discretized* voxel grids when modeling complex scenes at high-resolutions. In summary, our technical contributions are:

– An approach for representing continuous scenes with complex geometry and materials as 5D neural radiance fields, parameterized as basic MLP networks.

- A differentiable rendering procedure based on classical volume rendering techniques, which we use to optimize these representations from standard RGB images. This includes a hierarchical sampling strategy to allocate the MLP's capacity towards space with visible scene content.
- A positional encoding to map each input 5D coordinate into a higher dimensional space, which enables us to successfully optimize neural radiance fields to represent high-frequency scene content.

We demonstrate that our resulting neural radiance field method quantitatively and qualitatively outperforms state-of-the-art view synthesis methods, including works that fit neural 3D representations to scenes as well as works that train deep convolutional networks to predict sampled volumetric representations. As far as we know, this paper presents the first continuous neural scene representation that is able to render high-resolution photorealistic novel views of real objects and scenes from RGB images captured in natural settings.

2 Related Work

A promising recent direction in computer vision is encoding objects and scenes in the weights of an MLP that directly maps from a 3D spatial location to an implicit representation of the shape, such as the signed distance [5] at that location. However, these methods have so far been unable to reproduce realistic scenes with complex geometry with the same fidelity as techniques that represent scenes using discrete representations such as triangle meshes or voxel grids. In this section, we review these two lines of work and contrast them with our approach, which enhances the capabilities of neural scene representations to produce state-of-the-art results for rendering complex realistic scenes.

A similar approach of using MLPs to map from low-dimensional coordinates to colors has also been used for representing other graphics functions such as images [43], textured materials [11,30,35,36], and indirect illumination values [37].

Neural 3D Shape Representations. Recent work has investigated the implicit representation of continuous 3D shapes as level sets by optimizing deep networks that map xyz coordinates to signed distance functions [14,31] or occupancy fields [10,26]. However, these models are limited by their requirement of access to ground truth 3D geometry, typically obtained from synthetic 3D shape datasets such as ShapeNet [2]. Subsequent work has relaxed this requirement of ground truth 3D shapes by formulating differentiable rendering functions that allow neural implicit shape representations to be optimized using only 2D images. Niemeyer *et al.* [28] represent surfaces as 3D occupancy fields and use a numerical method to find the surface intersection for each ray, then calculate an exact derivative using implicit differentiation. Each ray intersection location is provided as the input to a neural 3D texture field that predicts a diffuse color for that point. Sitzmann *et al.* [41] use a less direct neural 3D representation that simply outputs a feature vector and RGB color at each continuous 3D coordinate, and propose a differentiable rendering function consisting of a recurrent neural network that marches along each ray to decide where the surface is located.

Though these techniques can potentially represent complicated and high-resolution geometry, they have so far been limited to simple shapes with low geometric complexity, resulting in oversmoothed renderings. We show that an alternate strategy of optimizing networks to encode 5D radiance fields (3D volumes with 2D view-dependent appearance) can represent higher-resolution geometry and appearance to render photorealistic novel views of complex scenes.

View Synthesis and Image-Based Rendering. Given a dense sampling of views, photorealistic novel views can be reconstructed by simple light field sample interpolation techniques [4,6,20]. For novel view synthesis with sparser view sampling, the computer vision and graphics communities have made significant progress by predicting traditional geometry and appearance representations from observed images. One popular class of approaches uses mesh-based representations of scenes with either diffuse [47] or view-dependent [1,7,48] appearance. Differentiable rasterizers [3,9,22,24] or pathtracers [21,29] can directly optimize mesh representations to reproduce a set of input images using gradient descent. However, gradient-based mesh optimization based on image reprojection is often difficult, likely because of local minima or poor conditioning of the loss landscape. Furthermore, this strategy requires a template mesh with fixed topology to be provided as an initialization before optimization [21], which is typically unavailable for unconstrained real-world scenes.

Another class of methods use volumetric representations to address the task of high-quality photorealistic view synthesis from a set of input RGB images. Volumetric approaches are able to realistically represent complex shapes and materials, are well-suited for gradient-based optimization, and tend to produce less visually distracting artifacts than mesh-based methods. Early volumetric approaches used observed images to directly color voxel grids [18,39,44]. More recently, several methods [8,12,16,27,32,42,45,51] have used large datasets of multiple scenes to train deep networks that predict a sampled volumetric representation from a set of input images, and then use either alpha-compositing [33] or learned compositing along rays to render novel views at test time. Other works have optimized a combination of convolutional networks (CNNs) and sampled voxel grids for each specific scene, such that the CNN can compensate for discretization artifacts from low resolution voxel grids [40] or allow the predicted voxel grids to vary based on input time or animation controls [23]. While these volumetric techniques have achieved impressive results for novel view synthesis, their ability to scale to higher resolution imagery is fundamentally limited by poor time and space complexity due to their discrete sampling—rendering higher resolution images requires a finer sampling of 3D space. We circumvent this problem by instead encoding a *continuous* volume within the parameters of a deep fully-connected neural network, which not only produces significantly higher quality renderings than prior volumetric approaches, but also requires just a fraction of the storage cost of those *sampled* volumetric representations.

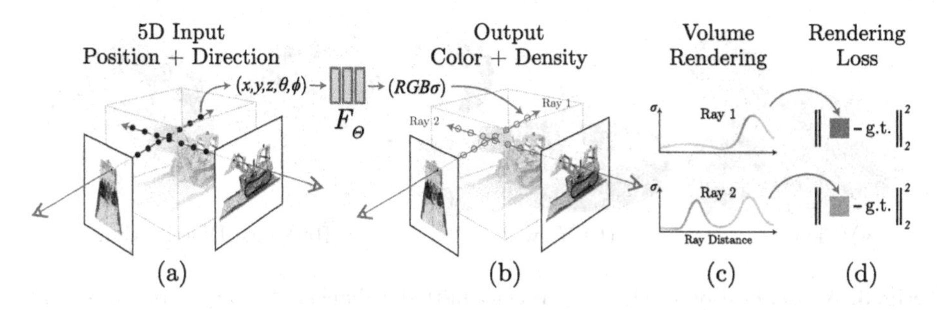

Fig. 2. An overview of our neural radiance field scene representation and differentiable rendering procedure. We synthesize images by sampling 5D coordinates (location and viewing direction) along camera rays (a), feeding those locations into an MLP to produce a color and volume density (b), and using volume rendering techniques to composite these values into an image (c). This rendering function is differentiable, so we can optimize our scene representation by minimizing the residual between synthesized and ground truth observed images (d). (Color figure online)

3 Neural Radiance Field Scene Representation

We represent a continuous scene as a 5D vector-valued function whose input is a 3D location $\mathbf{x} = (x, y, z)$ and 2D viewing direction (θ, ϕ), and whose output is an emitted color $\mathbf{c} = (r, g, b)$ and volume density σ. In practice, we express direction as a 3D Cartesian unit vector $\mathbf{d}$. We approximate this continuous 5D scene representation with an MLP network $F_\Theta : (\mathbf{x}, \mathbf{d}) \rightarrow (\mathbf{c}, \sigma)$ and optimize its weights Θ to map from each input 5D coordinate to its corresponding volume density and directional emitted color.

We encourage the representation to be multiview consistent by restricting the network to predict the volume density σ as a function of only the location $\mathbf{x}$, while allowing the RGB color $\mathbf{c}$ to be predicted as a function of both location and viewing direction. To accomplish this, the MLP F_Θ first processes the input 3D coordinate $\mathbf{x}$ with 8 fully-connected layers (using ReLU activations and 256 channels per layer), and outputs σ and a 256-dimensional feature vector. This feature vector is then concatenated with the camera ray's viewing direction and passed to one additional fully-connected layer (using a ReLU activation and 128 channels) that output the view-dependent RGB color.

See Fig. 3 for an example of how our method uses the input viewing direction to represent non-Lambertian effects. As shown in Fig. 4, a model trained without view dependence (only $\mathbf{x}$ as input) has difficulty representing specularities.

4 Volume Rendering with Radiance Fields

Our 5D neural radiance field represents a scene as the volume density and directional emitted radiance at any point in space. We render the color of any ray passing through the scene using principles from classical volume rendering [15]. The volume density $\sigma(\mathbf{x})$ can be interpreted as the differential probability of a

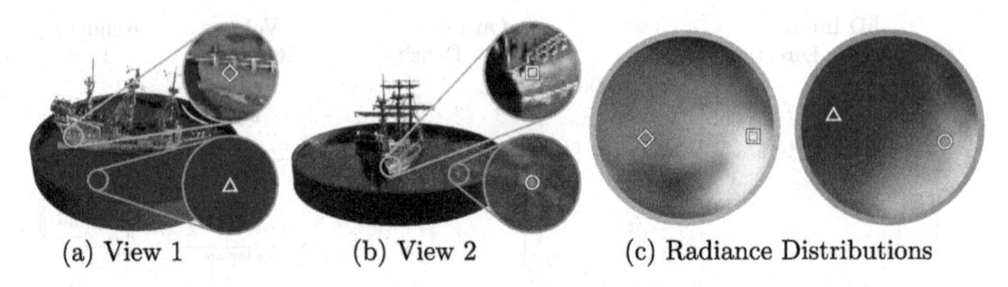

Fig. 3. A visualization of view-dependent emitted radiance. Our neural radiance field representation outputs RGB color as a 5D function of both spatial position $\mathbf{x}$ and viewing direction $\mathbf{d}$. Here, we visualize example directional color distributions for two spatial locations in our neural representation of the *Ship* scene. In (a) and (b), we show the appearance of two fixed 3D points from two different camera positions: one on the side of the ship (orange insets) and one on the surface of the water (blue insets). Our method predicts the changing specular appearance of these two 3D points, and in (c) we show how this behavior generalizes continuously across the whole hemisphere of viewing directions. (Color figure online)

ray terminating at an infinitesimal particle at location $\mathbf{x}$. The expected color $C(\mathbf{r})$ of camera ray $\mathbf{r}(t) = \mathbf{o} + t\mathbf{d}$ with near and far bounds t_n and t_f is:

$$C(\mathbf{r}) = \int_{t_n}^{t_f} T(t)\sigma(\mathbf{r}(t))\mathbf{c}(\mathbf{r}(t), \mathbf{d})dt, \text{ where } T(t) = \exp\left(-\int_{t_n}^{t} \sigma(\mathbf{r}(s))ds\right). \quad (1)$$

The function $T(t)$ denotes the accumulated transmittance along the ray from t_n to t, *i.e.*, the probability that the ray travels from t_n to t without hitting any other particle. Rendering a view from our continuous neural radiance field requires estimating this integral $C(\mathbf{r})$ for a camera ray traced through each pixel of the desired virtual camera.

We numerically estimate this continuous integral using quadrature. Deterministic quadrature, which is typically used for rendering discretized voxel grids, would effectively limit our representation's resolution because the MLP would only be queried at a fixed discrete set of locations. Instead, we use a stratified sampling approach where we partition $[t_n, t_f]$ into N evenly-spaced bins and then draw one sample uniformly at random from within each bin:

$$t_i \sim \mathcal{U}\left[t_n + \frac{i-1}{N}(t_f - t_n),\ t_n + \frac{i}{N}(t_f - t_n)\right]. \quad (2)$$

Although we use a discrete set of samples to estimate the integral, stratified sampling enables us to represent a continuous scene representation because it results in the MLP being evaluated at continuous positions over the course of optimization. We use these samples to estimate $C(\mathbf{r})$ with the quadrature rule discussed in the volume rendering review by Max [25]:

$$\hat{C}(\mathbf{r}) = \sum_{i=1}^{N} T_i(1 - \exp(-\sigma_i\delta_i))\mathbf{c}_i\,, \text{ where } T_i = \exp\left(-\sum_{j=1}^{i-1}\sigma_j\delta_j\right), \quad (3)$$

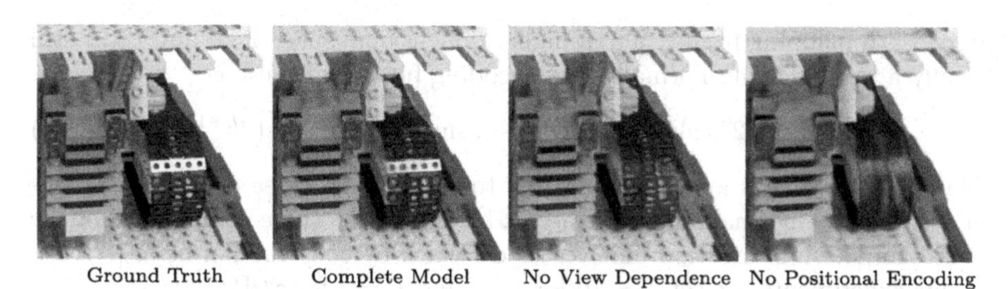

Fig. 4. Here we visualize how our full model benefits from representing view-dependent emitted radiance and from passing our input coordinates through a high-frequency positional encoding. Removing view dependence prevents the model from recreating the specular reflection on the bulldozer tread. Removing the positional encoding drastically decreases the model's ability to represent high frequency geometry and texture, resulting in an oversmoothed appearance.

where $\delta_i = t_{i+1} - t_i$ is the distance between adjacent samples. This function for calculating $\hat{C}(\mathbf{r})$ from the set of $(\mathbf{c}_i, \sigma_i)$ values is trivially differentiable and reduces to traditional alpha compositing with alpha values $\alpha_i = 1 - \exp(-\sigma_i \delta_i)$.

5 Optimizing a Neural Radiance Field

In the previous section we have described the core components necessary for modeling a scene as a neural radiance field and rendering novel views from this representation. However, we observe that these components are not sufficient for achieving state-of-the-art quality, as demonstrated in Sect. 6.4). We introduce two improvements to enable representing high-resolution complex scenes. The first is a positional encoding of the input coordinates that assists the MLP in representing high-frequency functions, and the second is a hierarchical sampling procedure that allows us to efficiently sample this high-frequency representation.

5.1 Positional Encoding

Despite the fact that neural networks are universal function approximators [13], we found that having the network F_Θ directly operate on $xyz\theta\phi$ input coordinates results in renderings that perform poorly at representing high-frequency variation in color and geometry. This is consistent with recent work by Rahaman *et al.* [34], which shows that deep networks are biased towards learning lower frequency functions. They additionally show that mapping the inputs to a higher dimensional space using high frequency functions before passing them to the network enables better fitting of data that contains high frequency variation.

We leverage these findings in the context of neural scene representations, and show that reformulating F_Θ as a composition of two functions $F_\Theta = F'_\Theta \circ \gamma$, one learned and one not, significantly improves performance (see Fig. 4 and Table 2).

Here γ is a mapping from $\mathbb{R}$ into a higher dimensional space $\mathbb{R}^{2L}$, and F'_Θ is still simply a regular MLP. Formally, the encoding function we use is:

$$\gamma(p) = \left(\sin(2^0\pi p), \cos(2^0\pi p), \cdots, \sin(2^{L-1}\pi p), \cos(2^{L-1}\pi p)\right). \quad (4)$$

This function $\gamma(\cdot)$ is applied separately to each of the three coordinate values in $\mathbf{x}$ (which are normalized to lie in $[-1, 1]$) and to the three components of the Cartesian viewing direction unit vector $\mathbf{d}$ (which by construction lie in $[-1, 1]$). In our experiments, we set $L = 10$ for $\gamma(\mathbf{x})$ and $L = 4$ for $\gamma(\mathbf{d})$.

A similar mapping is used in the popular Transformer architecture [46], where it is referred to as a *positional encoding*. However, Transformers use it for a different goal of providing the discrete positions of tokens in a sequence as input to an architecture that does not contain any notion of order. In contrast, we use these functions to map continuous input coordinates into a higher dimensional space to enable our MLP to more easily approximate a higher frequency function. Concurrent work on a related problem of modeling 3D protein structure from projections [50] also utilizes a similar input coordinate mapping.

5.2 Hierarchical Volume Sampling

Our rendering strategy of densely evaluating the neural radiance field network at N query points along each camera ray is inefficient: free space and occluded regions that do not contribute to the rendered image are still sampled repeatedly. We draw inspiration from early work in volume rendering [19] and propose a hierarchical representation that increases rendering efficiency by allocating samples proportionally to their expected effect on the final rendering.

Instead of just using a single network to represent the scene, we simultaneously optimize two networks: one "coarse" and one "fine". We first sample a set of N_c locations using stratified sampling, and evaluate the "coarse" network at these locations as described in Eqs. 2 and 3. Given the output of this "coarse" network, we then produce a more informed sampling of points along each ray where samples are biased towards the relevant parts of the volume. To do this, we first rewrite the alpha composited color from the coarse network $\hat{C}_c(\mathbf{r})$ in Eq. 3 as a weighted sum of all sampled colors c_i along the ray:

$$\hat{C}_c(\mathbf{r}) = \sum_{i=1}^{N_c} w_i c_i\,, \qquad w_i = T_i(1 - \exp(-\sigma_i\delta_i))\,. \quad (5)$$

Normalizing these weights as $\hat{w}_i = {w_i}/{\sum_{j=1}^{N_c} w_j}$ produces a piecewise-constant PDF along the ray. We sample a second set of N_f locations from this distribution using inverse transform sampling, evaluate our "fine" network at the union of the first and second set of samples, and compute the final rendered color of the ray $\hat{C}_f(\mathbf{r})$ using Eq. 3 but using all $N_c + N_f$ samples. This procedure allocates more samples to regions we expect to contain visible content. This addresses a similar goal as importance sampling, but we use the sampled values as a nonuniform discretization of the whole integration domain rather than treating each sample as an independent probabilistic estimate of the entire integral.

5.3 Implementation Details

We optimize a separate neural continuous volume representation network for each scene. This requires only a dataset of captured RGB images of the scene, the corresponding camera poses and intrinsic parameters, and scene bounds (we use ground truth camera poses, intrinsics, and bounds for synthetic data, and use the COLMAP structure-from-motion package [38] to estimate these parameters for real data). At each optimization iteration, we randomly sample a batch of camera rays from the set of all pixels in the dataset, and then follow the hierarchical sampling described in Sect. 5.2 to query N_c samples from the coarse network and $N_c + N_f$ samples from the fine network. We then use the volume rendering procedure described in Sect. 4 to render the color of each ray from both sets of samples. Our loss is simply the total squared error between the rendered and true pixel colors for both the coarse and fine renderings:

$$\mathcal{L} = \sum_{\mathbf{r}\in\mathcal{R}} \left[\left\| \hat{C}_c(\mathbf{r}) - C(\mathbf{r}) \right\|_2^2 + \left\| \hat{C}_f(\mathbf{r}) - C(\mathbf{r}) \right\|_2^2 \right] \tag{6}$$

where $\mathcal{R}$ is the set of rays in each batch, and $C(\mathbf{r})$, $\hat{C}_c(\mathbf{r})$, and $\hat{C}_f(\mathbf{r})$ are the ground truth, coarse volume predicted, and fine volume predicted RGB colors for ray $\mathbf{r}$ respectively. Note that even though the final rendering comes from $\hat{C}_f(\mathbf{r})$, we also minimize the loss of $\hat{C}_c(\mathbf{r})$ so that the weight distribution from the coarse network can be used to allocate samples in the fine network.

In our experiments, we use a batch size of 4096 rays, each sampled at $N_c = 64$ coordinates in the coarse volume and $N_f = 128$ additional coordinates in the fine volume. We use the Adam optimizer [17] with a learning rate that begins at 5×10^{-4} and decays exponentially to 5×10^{-5} over the course of optimization (other Adam hyperparameters are left at default values of $\beta_1 = 0.9$, $\beta_2 = 0.999$, and $\epsilon = 10^{-7}$). The optimization for a single scene typically take around 100–300k iterations to converge on a single NVIDIA V100 GPU (about 1–2 days).

6 Results

We quantitatively (Tables 1) and qualitatively (Figs. 5 and 6) show that our method outperforms prior work, and provide extensive ablation studies to validate our design choices (Table 2). We urge the reader to view our supplementary video to better appreciate our method's significant improvement over baseline methods when rendering smooth paths of novel views.

6.1 Datasets

Synthetic Renderings of Objects. We first show experimental results on two datasets of synthetic renderings of objects (Table 1, "Diffuse Synthetic 360°" and "Realistic Synthetic 360°"). The DeepVoxels [40] dataset contains four Lambertian objects with simple geometry. Each object is rendered at 512 × 512 pixels

Table 1. Our method quantitatively outperforms prior work on datasets of both synthetic and real images. We report PSNR/SSIM (higher is better) and LPIPS [49] (lower is better). The DeepVoxels [40] dataset consists of 4 diffuse objects with simple geometry. Our realistic synthetic dataset consists of pathtraced renderings of 8 geometrically complex objects with complex non-Lambertian materials. The real dataset consists of handheld forward-facing captures of 8 real-world scenes (NV cannot be evaluated on this data because it only reconstructs objects inside a bounded volume). Though LLFF achieves slightly better LPIPS, we urge readers to view our supplementary video where our method achieves better multiview consistency and produces fewer artifacts than all baselines.

Method	Diffuse Synthetic 360° [40]			Realistic Synthetic 360°			Real Forward-Facing [27]		
	PSNR↑	SSIM↑	LPIPS↓	PSNR↑	SSIM↑	LPIPS↓	PSNR↑	SSIM↑	LPIPS↓
SRN [41]	33.20	0.963	0.073	22.26	0.846	0.170	22.84	0.668	0.378
NV [23]	29.62	0.929	0.099	26.05	0.893	0.160	–	–	–
LLFF [27]	34.38	0.985	0.048	24.88	0.911	0.114	24.13	0.798	**0.212**
Ours	**40.15**	**0.991**	**0.023**	**31.01**	**0.947**	**0.081**	**26.50**	**0.811**	0.250

from viewpoints sampled on the upper hemisphere (479 as input and 1000 for testing). We additionally generate our own dataset containing pathtraced images of eight objects that exhibit complicated geometry and realistic non-Lambertian materials. Six are rendered from viewpoints sampled on the upper hemisphere, and two are rendered from viewpoints sampled on a full sphere. We render 100 views of each scene as input and 200 for testing, all at 800×800 pixels.

Real Images of Complex Scenes. We show results on complex real-world scenes captured with roughly forward-facing images (Table 1, "Real Forward-Facing"). This dataset consists of 8 scenes captured with a handheld cellphone (5 taken from the LLFF paper and 3 that we capture), captured with 20 to 62 images, and hold out $1/8$ of these for the test set. All images are 1008×756 pixels.

6.2 Comparisons

To evaluate our model we compare against current top-performing techniques for view synthesis, detailed below. All methods use the same set of input views to train a separate network for each scene except Local Light Field Fusion [27], which trains a single 3D convolutional network on a large dataset, then uses the same trained network to process input images of new scenes at test time.

Neural Volumes (NV). [23] synthesizes novel views of objects that lie entirely within a bounded volume in front of a distinct background (which must be separately captured without the object of interest). It optimizes a deep 3D convolutional network to predict a discretized RGBα voxel grid with 128^3 samples as well as a 3D warp grid with 32^3 samples. The algorithm renders novel views by marching camera rays through the warped voxel grid.

Scene Representation Networks (SRN). [41] represent a continuous scene as an opaque surface, implicitly defined by a MLP that maps each (x, y, z)

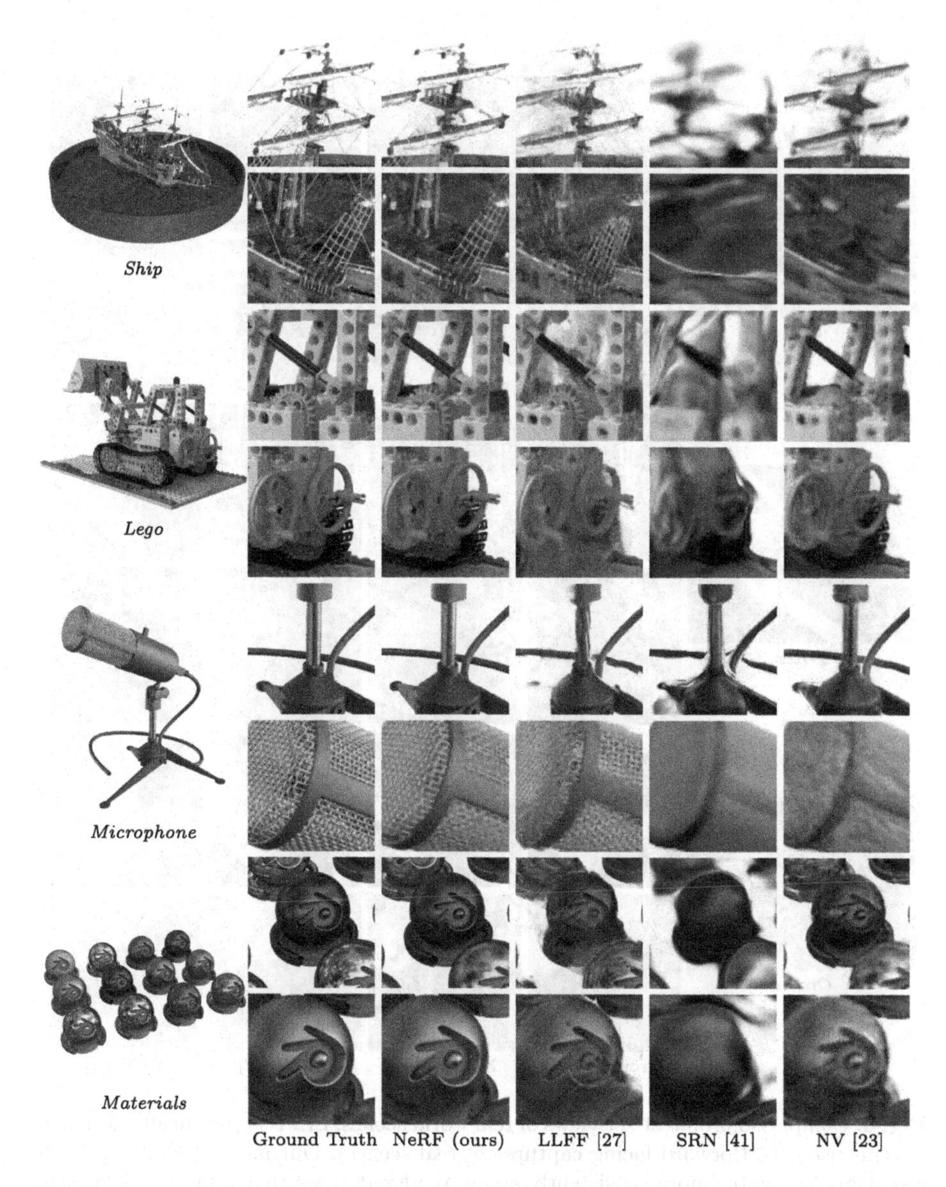

Fig. 5. Comparisons on test-set views for scenes from our new synthetic dataset generated with a physically-based renderer. Our method is able to recover fine details in both geometry and appearance, such as *Ship*'s rigging, *Lego*'s gear and treads, *Microphone*'s shiny stand and mesh grille, and *Material*'s non-Lambertian reflectance. LLFF exhibits banding artifacts on the *Microphone* stand and *Material*'s object edges and ghosting artifacts in *Ship*'s mast and inside the *Lego* object. SRN produces blurry and distorted renderings in every case. Neural Volumes cannot capture the details on the *Microphone*'s grille or *Lego*'s gears, and it completely fails to recover the geometry of *Ship*'s rigging.

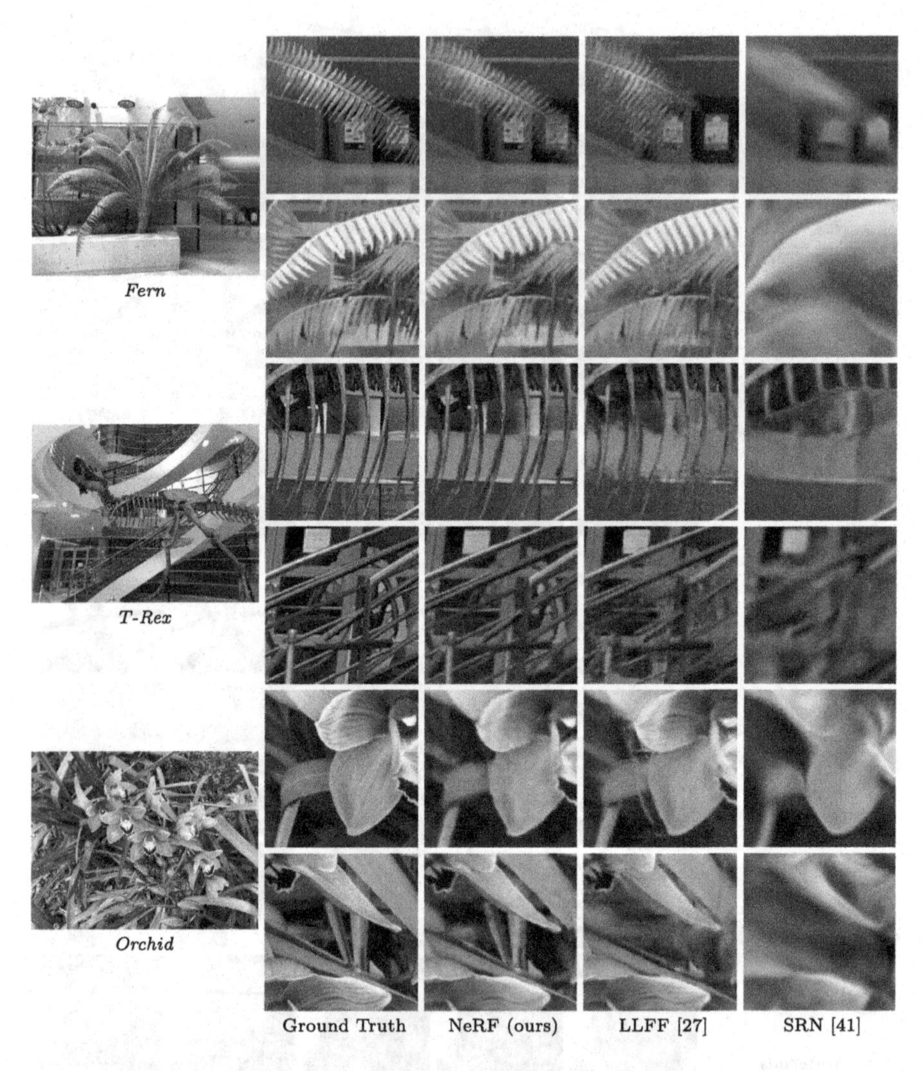

Fig. 6. Comparisons on test-set views of real world scenes. LLFF is specifically designed for this use case (forward-facing captures of real scenes). Our method is able to represent fine geometry more consistently across rendered views than LLFF, as shown in *Fern*'s leaves and the skeleton ribs and railing in *T-rex*. Our method also correctly reconstructs partially occluded regions that LLFF struggles to render cleanly, such as the yellow shelves behind the leaves in the bottom *Fern* crop and green leaves in the background of the bottom *Orchid* crop. Blending between multiples renderings can also cause repeated edges in LLFF, as seen in the top *Orchid* crop. SRN captures the low-frequency geometry and color variation in each scene but is unable to reproduce any fine detail. (Color figure online)

coordinate to a feature vector. They train a recurrent neural network to march along a ray through the scene representation by using the feature vector at any 3D coordinate to predict the next step size along the ray. The feature vector from the final step is decoded into a single color for that point on the surface. Note that SRN is a better-performing followup to DeepVoxels [40] by the same authors, which is why we do not include comparisons to DeepVoxels.

Local Light Field Fusion (LLFF). [27] LLFF is designed for producing photorealistic novel views for well-sampled forward facing scenes. It uses a trained 3D convolutional network to directly predict a discretized frustum-sampled RGBα grid (multiplane image or MPI [51]) for each input view, then renders novel views by alpha compositing and blending nearby MPIs into the novel viewpoint.

6.3 Discussion

We thoroughly outperform both baselines that also optimize a separate network per scene (NV and SRN) in all scenarios. Furthermore, we produce qualitatively and quantitatively superior renderings compared to LLFF (across all except one metric) while using only their input images as our entire training set.

The SRN method produces heavily smoothed geometry and texture, and its representational power for view synthesis is limited by selecting only a single depth and color per camera ray. The NV baseline is able to capture reasonably detailed volumetric geometry and appearance, but its use of an underlying explicit 128^3 voxel grid prevents it from scaling to represent fine details at high resolutions. LLFF specifically provides a "sampling guideline" to not exceed 64 pixels of disparity between input views, so it frequently fails to estimate correct geometry in the synthetic datasets which contain up to 400–500 pixels of disparity between views. Additionally, LLFF blends between different scene representations for rendering different views, resulting in perceptually-distracting inconsistency as is apparent in our supplementary video.

The biggest practical tradeoffs between these methods are time versus space. All compared single scene methods take at least 12 h to train per scene. In contrast, LLFF can process a small input dataset in under 10 min. However, LLFF produces a large 3D voxel grid for every input image, resulting in enormous storage requirements (over 15 GB for one "Realistic Synthetic" scene). Our method requires only 5 MB for the network weights (a relative compression of 3000$\times$ compared to LLFF), which is even less memory than the *input images alone* for a single scene from any of our datasets.

6.4 Ablation Studies

We validate our algorithm's design choices and parameters with an extensive ablation study in Table 2. We present results on our "Realistic Synthetic 360°" scenes. Row 9 shows our complete model as a point of reference. Row 1 shows a minimalist version of our model without positional encoding (PE), view-dependence (VD), or hierarchical sampling (H). In rows 2–4 we remove these

Table 2. An ablation study of our model. Metrics are averaged over the 8 scenes from our realistic synthetic dataset. See Sect. 6.4 for detailed descriptions.

		Input	#Im	L	(N_c, N_f)	PSNR↑	SSIM↑	LPIPS↓
1)	No PE, VD, H	xyz	100	–	(256, –)	26.67	0.906	0.136
2)	No pos. encoding	$xyz\theta\phi$	100	–	(64, 128)	28.77	0.924	0.108
3)	No view dependence	xyz	100	10	(64, 128)	27.66	0.925	0.117
4)	No hierarchical	$xyz\theta\phi$	100	10	(256, –)	30.06	0.938	0.109
5)	Far fewer images	$xyz\theta\phi$	25	10	(64, 128)	27.78	0.925	0.107
6)	Fewer images	$xyz\theta\phi$	50	10	(64, 128)	29.79	0.940	0.096
7)	Fewer frequencies	$xyz\theta\phi$	100	5	(64, 128)	30.59	0.944	0.088
8)	More frequencies	$xyz\theta\phi$	100	15	(64, 128)	30.81	0.946	0.096
9)	Complete model	$xyz\theta\phi$	100	10	(64, 128)	**31.01**	**0.947**	**0.081**

three components one at a time from the full model, observing that positional encoding (row 2) and view-dependence (row 3) provide the largest quantitative benefit followed by hierarchical sampling (row 4). Rows 5–6 show how our performance decreases as the number of input images is reduced. Note that our method's performance using only 25 input images still exceeds NV, SRN, and LLFF across all metrics when they are provided with 100 images (see supplementary material). In rows 7–8 we validate our choice of the maximum frequency L used in our positional encoding for $\mathbf{x}$ (the maximum frequency used for $\mathbf{d}$ is scaled proportionally). Only using 5 frequencies reduces performance, but increasing the number of frequencies from 10 to 15 does not improve performance. We believe the benefit of increasing L is limited once 2^L exceeds the maximum frequency present in the sampled input images (roughly 1024 in our data).

7 Conclusion

Our work directly addresses deficiencies of prior work that uses MLPs to represent objects and scenes as continuous functions. We demonstrate that representing scenes as 5D neural radiance fields (an MLP that outputs volume density and view-dependent emitted radiance as a function of 3D location and 2D viewing direction) produces better renderings than the previously-dominant approach of training deep convolutional networks to output discretized voxel representations.

Although we have proposed a hierarchical sampling strategy to make rendering more sample-efficient (for both training and testing), there is still much more progress to be made in investigating techniques to efficiently optimize and render neural radiance fields. Another direction for future work is interpretability: sampled representations such as voxel grids and meshes admit reasoning about the expected quality of rendered views and failure modes, but it is unclear how to analyze these issues when we encode scenes in the weights of a deep neural network. We believe that this work makes progress towards a graphics pipeline based on real world imagery, where complex scenes could be composed of neural radiance fields optimized from images of actual objects and scenes.

References

1. Buehler, C., Bosse, M., McMillan, L., Gortler, S., Cohen, M.: Unstructured lumigraph rendering. In: SIGGRAPH (2001)
2. Chang, A.X., et al.: ShapeNet: an information-rich 3D model repository. arXiv:1512.03012 (2015)
3. Chen, W., et al.: Learning to predict 3D objects with an interpolation-based differentiable renderer. In: NeurIPS (2019)
4. Cohen, M., Gortler, S.J., Szeliski, R., Grzeszczuk, R., Szeliski, R.: The lumigraph. In: SIGGRAPH (1996)
5. Curless, B., Levoy, M.: A volumetric method for building complex models from range images. In: SIGGRAPH (1996)
6. Davis, A., Levoy, M., Durand, F.: Unstructured light fields. In: Eurographics (2012)
7. Debevec, P., Taylor, C.J., Malik, J.: Modeling and rendering architecture from photographs: a hybrid geometry-and image-based approach. In: SIGGRAPH (1996)
8. Flynn, J., et al.: DeepView: view synthesis with learned gradient descent. In: CVPR (2019)
9. Genova, K., Cole, F., Maschinot, A., Sarna, A., Vlasic, D., Freeman, W.T.: Unsupervised training for 3D morphable model regression. In: CVPR (2018)
10. Genova, K., Cole, F., Sud, A., Sarna, A., Funkhouser, T.: Local deep implicit functions for 3D shape. In: CVPR (2020)
11. Henzler, P., Mitra, N.J., Ritschel, T.: Learning a neural 3D texture space from 2D exemplars. In: CVPR (2020)
12. Henzler, P., Rasche, V., Ropinski, T., Ritschel, T.: Single-image tomography: 3D volumes from 2D cranial X-rays. In: Eurographics (2018)
13. Hornik, K., Stinchcombe, M., White, H.: Multilayer feedforward networks are universal approximators. Neural Netw. **2**(5), 359–366 (1989)
14. Jiang, C., Sud, A., Makadia, A., Huang, J., Nießner, M., Funkhouser, T.: Local implicit grid representations for 3D scenes. In: CVPR (2020)
15. Kajiya, J.T., Herzen, B.P.V.: Ray tracing volume densities. Comput. Graph. (SIGGRAPH) **18**(3), 165–174 (1984)
16. Kar, A., Häne, C., Malik, J.: Learning a multi-view stereo machine. In: NeurIPS (2017)
17. Kingma, D.P., Ba, J.: Adam: a method for stochastic optimization. In: ICLR (2015)
18. Kutulakos, K.N., Seitz, S.M.: A theory of shape by space carving. Int. J. Comput. Vis. **1**, 307–314 (2000)
19. Levoy, M.: Efficient ray tracing of volume data. ACM Trans. Graph. **9**(3), 245–261 (1990)
20. Levoy, M., Hanrahan, P.: Light field rendering. In: SIGGRAPH (1996)
21. Li, T.M., Aittala, M., Durand, F., Lehtinen, J.: Differentiable Monte Carlo ray tracing through edge sampling. ACM Trans. Graph. (SIGGRAPH Asia) **37**(6), 1–11 (2018)
22. Liu, S., Li, T., Chen, W., Li, H.: Soft rasterizer: a differentiable renderer for image-based 3D reasoning. In: ICCV (2019)

23. Lombardi, S., Simon, T., Saragih, J., Schwartz, G., Lehrmann, A., Sheikh, Y.: Neural volumes: learning dynamic renderable volumes from images. ACM Trans. Graph. (SIGGRAPH) (2019)
24. Loper, M.M., Black, M.J.: OpenDR: an approximate differentiable renderer. In: ECCV (2014)
25. Max, N.: Optical models for direct volume rendering. IEEE Trans. Visual. Comput. Graph. **1**(2), 99–108 (1995)
26. Mescheder, L., Oechsle, M., Niemeyer, M., Nowozin, S., Geiger, A.: Occupancy networks: learning 3D reconstruction in function space. In: CVPR (2019)
27. Mildenhall, B., et al.: Local light field fusion: practical view synthesis with prescriptive sampling guidelines. ACM Trans. Graph. (SIGGRAPH) **38**(4), 1–14 (2019)
28. Niemeyer, M., Mescheder, L., Oechsle, M., Geiger, A.: Differentiable volumetric rendering: learning implicit 3D representations without 3D supervision. In: CVPR (2019)
29. Nimier-David, M., Vicini, D., Zeltner, T., Jakob, W.: Mitsuba 2: a retargetable forward and inverse renderer. ACM Trans. Graph. (SIGGRAPH Asia) **38**(6), 1–17 (2019)
30. Oechsle, M., Mescheder, L., Niemeyer, M., Strauss, T., Geiger, A.: Texture fields: learning texture representations in function space. In: ICCV (2019)
31. Park, J.J., Florence, P., Straub, J., Newcombe, R., Lovegrove, S.: DeepSDF: learning continuous signed distance functions for shape representation. In: CVPR (2019)
32. Penner, E., Zhang, L.: Soft 3D reconstruction for view synthesis. ACM Trans. Graph. (SIGGRAPH Asia) **36**(6), 1–11 (2017)
33. Porter, T., Duff, T.: Compositing digital images. Comput. Graph (SIGGRAPH) (1984)
34. Rahaman, N., et al.: On the spectral bias of neural networks. In: ICML (2018)
35. Rainer, G., Ghosh, A., Jakob, W., Weyrich, T.: Unified neural encoding of BTFs. Comput. Graph. Forum (Eurographics) (2020)
36. Rainer, G., Jakob, W., Ghosh, A., Weyrich, T.: Neural BTF compression and interpolation. Comput. Graph. Forum (Eurographics) **38**(2), 235–244 (2019)
37. Ren, P., Wang, J., Gong, M., Lin, S., Tong, X., Guo, B.: Global illumination with radiance regression functions. ACM Trans. Graph. **32**(4), 1–12 (2013)
38. Schönberger, J.L., Frahm, J.M.: Structure-from-motion revisited. In: CVPR (2016)
39. Seitz, S.M., Dyer, C.R.: Photorealistic scene reconstruction by voxel coloring. Int. J. Comput. Vis. **35**, 151–173 (1999). https://doi.org/10.1023/A:1008176507526
40. Sitzmann, V., Thies, J., Heide, F., Nießner, M., Wetzstein, G., Zollhöfer, M.: DeepVoxels: learning persistent 3D feature embeddings. In: CVPR (2019)
41. Sitzmann, V., Zollhoefer, M., Wetzstein, G.: Scene representation networks: continuous 3D-structure-aware neural scene representations. In: NeurIPS (2019)
42. Srinivasan, P.P., Tucker, R., Barron, J.T., Ramamoorthi, R., Ng, R., Snavely, N.: Pushing the boundaries of view extrapolation with multiplane images. In: CVPR (2019)
43. Stanley, K.O.: Compositional pattern producing networks: a novel abstraction of development. Genet. Program. Evolvable Mach. **8**, 131–162 (2007). https://doi.org/10.1007/s10710-007-9028-8
44. Szeliski, R., Golland, P.: Stereo matching with transparency and matting. In: ICCV (1998)
45. Tulsiani, S., Zhou, T., Efros, A.A., Malik, J.: Multi-view supervision for single-view reconstruction via differentiable ray consistency. In: CVPR (2017)
46. Vaswani, A., et al.: Attention is all you need. In: NeurIPS (2017)

47. Waechter, M., Moehrle, N., Goesele, M.: Let there be color! large-scale texturing of 3D reconstructions. In: ECCV (2014)
48. Wood, D.N., et al.: Surface light fields for 3D photography. In: SIGGRAPH (2000)
49. Zhang, R., Isola, P., Efros, A.A., Shechtman, E., Wang, O.: The unreasonable effectiveness of deep features as a perceptual metric. In: CVPR (2018)
50. Zhong, E.D., Bepler, T., Davis, J.H., Berger, B.: Reconstructing continuous distributions of 3D protein structure from cryo-EM images. In: ICLR (2020)
51. Zhou, T., Tucker, R., Flynn, J., Fyffe, G., Snavely, N.: Stereo magnification: learning view synthesis using multiplane images. ACM Trans. Graph. (SIGGRAPH) (2018)

ReferIt3D: Neural Listeners for Fine-Grained 3D Object Identification in Real-World Scenes

Panos Achlioptas[1(✉)], Ahmed Abdelreheem[2], Fei Xia[1], Mohamed Elhoseiny[1,2], and Leonidas Guibas[1]

[1] Stanford University, Stanford, USA
{panos,feixia,elhoseiny,guibas}@cs.stanford.edu
[2] King Abdullah University of Science and Technology, Thuwal, Saudi Arabia
{ahmed.abdelreheem,mohamed.elhoseiny}@kaust.edu.sa

Abstract. In this work we study the problem of using referential language to identify common objects in real-world 3D scenes. We focus on a challenging setup where the referred object belongs to a *fine-grained* object class and the underlying scene contains *multiple* object instances of that class. Due to the scarcity and unsuitability of existent 3D-oriented linguistic resources for this task, we first develop two large-scale and complementary visio-linguistic datasets: i) **Sr3D**, which contains 83.5 K template-based utterances leveraging *spatial relations* among fine-grained object classes to localize a referred object in a scene, and ii) **Nr3D** which contains 41.5K *natural, free-form*, utterances collected by deploying a 2-player object reference game in 3D scenes. Using utterances of either datasets, human listeners can recognize the referred object with high (>86%, 92% resp.) accuracy. By tapping on this data, we develop novel neural listeners that can comprehend object-centric natural language and identify the referred object *directly* in a 3D scene. Our key technical contribution is designing an approach for combining linguistic and geometric information (in the form of 3D point clouds) and creating multi-modal (3D) neural listeners . We also show that architectures which promote object-to-object communication via graph neural networks outperform less context-aware alternatives, and that fine-grained object classification is a bottleneck for language-assisted 3D object identification.

1 Introduction

The progress on connecting language and vision in the past decade has rekindled interest in tasks like visual question answering (e.g., [12,53]), image captioning (e.g., [6,28,41,62,67]), and sentence-to-image similarity (e.g., [28,31]). Recent works have enhanced the accessibility of visual content through language

Electronic supplementary material The online version of this chapter (https://doi.org/10.1007/978-3-030-58452-8_25) contains supplementary material, which is available to authorized users.

A. Vedaldi et al. (Eds.): ECCV 2020, LNCS 12346, pp. 422–440, 2020.
https://doi.org/10.1007/978-3-030-58452-8_25

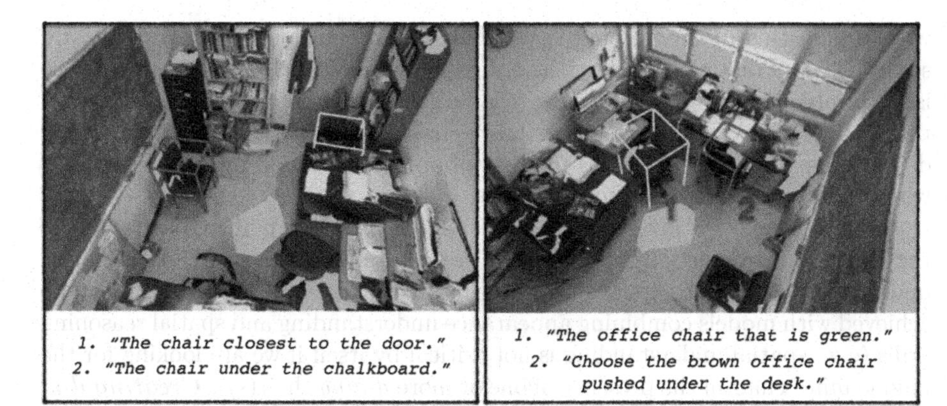

1. "The chair closest to the door."
2. "The chair under the chalkboard."
1. "The office chair that is green."
2. "Choose the brown office chair pushed under the desk."

Fig. 1. Examples of natural free-form utterances. Each color-coded utterance distinguishes the corresponding object (marked with same color) against a distracting object in the underlying scene; contrasting two simple chairs (left) and two office-chairs (right). The use of a specific contrasting context inside a scene (as delineated by the bounding boxes surrounding *all and only those* objects of the same fine-grained class) fosters the production of *discriminative* language and lifts the reference problem beyond fine-grained classification. (Color figure online)

via grounding (e.g., [47,48]), showing strong results in locating linguistically described visual elements in images. However, most of these works focus on developing better models that connect vision to language in images, which express after all only a 2D view of our 3D reality. Even in embodied AI most works (e.g., embodied QA [21], or embodied visual recognition [68]), fine-grained 3D object identification is not explicitly modeled. Fine-grained 3D understanding however can be essential to more 3D-oriented visually-grounded embodied tasks, such as those that need to be performed by autonomous robotic agents [56,65].

Humans possess an astonishing capacity to reason about, describe, and locate 3D objects. Over time we have developed efficient communication protocols to linguistically express such processes – e.g., given the utterance *"the laptop placed on the table next to the main door"*, one can identify the referred object in the room, as long as the reference conveys some unique aspect of that object. Solving such a reference problem *directly* in 3D space – i.e., *without* a camera view dependency – can benefit many downstream robotics applications, including embodied question answering [21], visual- and language-based navigation [9], instruction following [57], and manipulating objects in a scene [37,66]. Despite this, developing *datasets* and *methods* with characteristics that enable machine learning models to perform well on this 3D reference task is far from straightforward; in this work, we examine how to address both.

Leveraging 3D visual understanding for solving vision and language tasks has been recently explored in Visual Question Answering [33] and Visual Grounding [49]. Still, the focus has been on synthetic datasets without 3D understanding going beyond (at best) multiple 2D views. An alternative, yet more direct way to gain this understanding is by analyzing point cloud data of real-world scenes [3,51]. Point clouds carry the entire geometric and appearance characteristics of objects and provide access to a larger spatial context (within a scene) than a single 2D view [13]. This flexibility enables us also to bypass camera view dependency (e.g., having access to parts of a scene occluded by a fixed camera) when we refer linguistically to objects.

In this paper, we investigate object references when multiple instances of the same fine-grained object class are present in a 3D scene. Discriminative understanding of object classes is important at the fine-grained level and can be achieved with models combining appearance understanding and spatial reasoning skills (e.g., spatial understanding is not critical by itself if we are looking for the unique *office* chair in the presence of one or more *dining* chair(s)). ***Creating discriminative linguistic descriptions:*** We focus on designing a data collection strategy that covers *both* spatial and appearance based identification (Sect. 3). As we show in our experiments, this step is critical for progress in 3D visual object identification from free-form language descriptions. Our strategy involves both synthetic and human-based generation of utterances, and has the following characteristics: (i) in every single example there are multiple object instances of the same class referred in the language describing the target 3D object; (ii) in the case of human language utterances, we explicitly ask the human subject to describe a target object in contrast to other instances of the *same* object class. By explicitly contrasting the same fine-grained class instances *and only them,* the resulting utterances are discriminative, even if uttered by crowd-sourced annotators unfamiliar with the environment. Figure 1 illustrates examples of this. ***Developing a 3D neural listener:*** We also design a novel visio-linguistic graph-convolution network that predicts the referred object given a language description, by enabling communication among objects in a 3D visual scene. Our contributions can be summarized below:

1. **Fine-Grained ReferIt3D** task: We introduce the task of language-based identification of specific 3D object instances, where fine-grained object-centric and multi-object understanding is necessary for its completion.
2. **Nr3D** and **Sr3D** datasets: We contribute a new dataset that contains natural and synthetic language descriptions, namely Nr3D and Sr3D respectively. For Sr3D we propose a simple but effective methodology for building template-based and *spatially-oriented* object referential language in 3D scenes. We show that training with Sr3D in addition to natural language data (Nr3D or [18]) improves neural-based pipelines.
3. **ReferIt3DNet**: We explore the task of understanding object references grounded in real-world 3D data (including both language and scenes) by designing a novel visio-linguistic graph neural network, termed *ReferIt3DNet*[1].

[1] The datasets and neural listener code are available at https://referit3d.github.io.

2 Related Work

2D High-Level Vision and Language: Vision & Language, also sometimes called Visual Semantic modeling, has been extensively studied in a variety of 2D tasks. Among early approaches of combining Vision & Semantics are tasks such as zero-shot learning where language/unseen descriptions of an unseen class are provided to describe it (e.g., [7,24–26,35,38,39,54,58,60,69,73]). Similar approaches have been developed to model image-sentence similarity for bidirectional retrieval of images given a sentence (e.g., [28,31]). More recently, the development of a large scale dataset of 2D Visual Question answering (VQA) [12] enabled new approaches on how to best represent question and images for this task. However, a huge language bias was revealed: just by looking at the language and without necessarily understanding the visual content, the predictive performance of the right answer is high [5]. The same bias was shown in tasks such as image-captioning [22]. More balanced VQA benchmarks [27] mitigated some of the biases and motivated the development of better attention mechanisms (e.g., [30]) and modular networks [11,70]. As per the example of the *2D Vision and Language* community, properly modeling 3D visio-lingual tasks requires establishing carefully designed connections between language and the 3D visual data, encoded with point clouds.

2D ReferIt Game and Grounded Vision and Language: Several papers explore connecting referential language to image regions for co-reference resolution (e.g., [15,42,52] - in videos, e.g., [8]), for generating referring expressions [29,42,43], and more. Recent work grounds noun phrases in image captions, such as in Flickr30KEntities [48] and ActivityNetEntities [72] in videos. In [23,42], the authors proposed the use of referring expressions for human-robot interaction and object localization in real-word environments but using primarily 2D images in contrast to our work.

Visual Relationships and Spatial Reasoning: Detecting visual relationships in images such as `<woman, carrying, umbrella>` (e.g., [1,40,71]) has been explored using datasets such as VRD [40] and more recently on the large Visual Genome dataset [32]. Spatial relations have also been studied in 3D by Rosema *et al.* [55]. However, relations in that work are not described in free form and hence are of restricted vocabulary. Also, the goal in [55] is simpler than identifying a target object in a complex 3D environment (our goal).

3D Vision and Language: Connecting 3D vision to natural language is relatively understudied. From a generative angle, [16] presented conditional generation of 3D models from text, which could be useful in augmented reality applications. In a concurrent work [18], Chen *et al.*, collected natural language to localize/discover referred objects in 3D scenes. In contrast, we assume that we are given the segmented object instances in a room and focus on identifying a referred object among instances of *the same* fine-grained category.

ReferIt 3D Game. Our 41,503 human language utterances were collected via a reference game between two humans, as inspired by 2D ReferItGame [29] and ShapeGlot [3]. The basic arrangement of such games can be traced back to the language games explored by Wittgenstein [64] and Lewis [36]. Recently, these approaches have also been adopted as a benchmark for discriminative and context-aware NLP [4,10,19,34,45,46,59,61]. Our paper goes beyond this prior work by grounding language behavior in a reference task containing objects in complex (real-world) 3D scenes, thereby eliciting compositional spatial and color/shape-oriented language.

3 Developing Referential 3D-Centric Data

The problem of language driven disambiguation of common objects in real world 3D scenes is new, and as such, not many datasets exist that are well suited for this task. With this in mind we introduce a two-part dataset: a high quality synthetic dataset of referential utterances (**Sr3D**) and a dataset with natural (human) referential utterances (**Nr3D**). Both Sr3D and Nr3D are built on top of ScanNet [20], a real-world 3D scene dataset with extensive semantic annotations that we utilize to create appropriate contrastive *communication contexts*. We define *communication context* as a `(scene, target, distractor(s))` tuple, where *scene* is one of the 707 unique indoor scenes of ScanNet, *target* is one of 76 fine-grained object classes (e.g., office-chairs, armchairs, etc.), and *distractors* are instances of the same fine-grained object class as the target that are contained in the same scene. We generate a total of 5878 unique tuples. We select the 76 object classes by applying the following intuitive criteria. A class is a valid class for a *target* if: (a) it is contained in at least 5 *scenes*; and (b) each *scene* contains multiple *distractors* but not more than six (to promote a problem beyond fine-grained (FG) classification without making it too hard even for human annotators). We add the constraint of having 5 such *scenes* per class, to foster generalization and make the problem less heavy-tailed (15.26% of all annotated ScanNet classes appear with multiple instances in exactly one *scene*). We also exclude the few classes that are object parts (e.g. a door of a closet) or are structural elements of the scenes (i.e., walls, floors, and ceiling) to ensure that we are working with common objects.

3.1 Creating Template Based Spatial References

We introduce the **Spatial Reference in 3D** (**Sr3D**) dataset, consisting of 83,572 utterances. Each utterance aims to uniquely refer to a *target* object in a ScanNet 3D scene by defining a relationship between the *target* and a surrounding object (*anchor*). *Anchors* are object instances that can belong to a set of 100 object classes in ScanNet, comprising of the 76 mentioned above and an additional 24 that: (a) frequently appear as singletons in a scene; and (b) are large objects (e.g., a fireplace or a TV). However, an *anchor* can never belong to the same class as the *target* and, as such, its *distractors*.

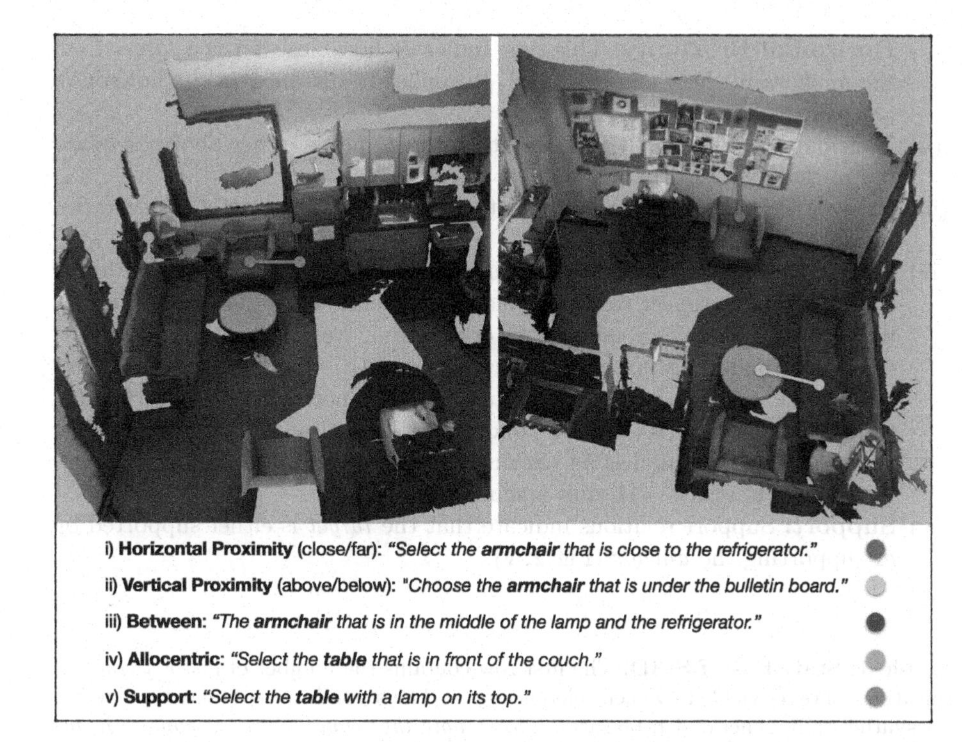

Fig. 2. Examples of spatial reference types of **Sr3D**. In the left image, there are examples of "horizontal proximity", "between", and "support" relations; the target object in the first two is an armchair and the target in the third relation is a table. In the right image, there are examples of "vertical proximity" and "allocentric" relations; the target objects are an armchair and a table respectively. The left and the right images represent a ScanNet scene where there exist two armchairs (one is beside the refrigerator and the other under the bulletin board) and three tables (a black one under the bulletin board, one in front of the couch, and one in the corner of the room). (Color figure online)

Consider, for instance, an underlying 3D scene with a *target* object (e.g. desk) that can be completely disambiguated from its *distractors* with the help of a spatial relation (e.g., closest) to an *anchor* object (e.g., door). We synthesize discriminative **Sr3D** utterances using the following compositional template:

$$< \text{target-class} > \quad < \text{spatial-relation} > \quad < \text{anchor-class(es)} > \tag{1}$$

e.g., *"the desk that is closest to the door"*. Per (1), the **Sr3D** template consists of three placeholders. Our goal is to find combinations of them that can uniquely characterize target objects among their distractors in their scenes.

We define the following five types of spatial object-to-object relations. For more details we refer the reader to Table 1 for a summary of statistics and to the Supplementary Material [2].

(i) **Horizontal Proximity:** This type indicates how close/far is a *target* from the *anchors* in the scene (Fig. 2, i). It applies to distance on the horizontal placement of the objects.
(ii) **Vertical Proximity:** It indicates that the *target* is either above or below the *anchor* (Fig. 2, ii).
(iii) **Between:** Between relations indicate the existence of a *target* between two *anchors* (Fig. 2, iii).
(iv) **Allocentric:** Allocentric relations encode information about the location of the *target* with respect to the intrinsic self-orientation of an *anchor* (Fig. 2, iv). To define the aforementioned orientation, we need to know; (a) the orientated bounding boxes of the *anchor* and (b) whether the *anchor* has an intrinsic front (e.g., a chair with a back) or not (e.g., a stool). For (a) we utilized the Scan2CAD [14] annotations that provide 9DOF alignments between ShapeNet models and ScanNet objects, and for (b) we used a combination of PartNet's [44] and manual annotations.
(v) **Support:** Support relations indicate that the *target* is either supported by or supporting the *anchor* (Fig. 2, v).

Table 1. Statistics of Sr3D. The first row contains the number of *distinct* communication contexts yielded by each reference-type. The second row contains the number of synthetically generated utterances. *Please note that communication context in the Sr3D setup also takes into account spatial relationships and anchors.*

Relationship	Horizontal Prox	Vertical Prox	Support	Allocentric	Between	All
|Context|	34001	1589	747	1880	3569	41786
|Utterances|	68002	3178	1494	3760	7138	83572

Discussion Our protocol for generating Sr3D is *simple* but also **effective**:

- A user study conducted in Amazon Mechanical Turk (AMT) revealed that 86.1% of the time, humans guessed correctly the target when provided with a sampled utterance of **Sr3D** (2K samples, $p < 0.001$).
- As shown in Sect. 5, **Sr3D** allows us to investigate the reference problem in a more controlled manner than **Nr3D**, by providing a homogeneous vocabulary and a specific type of reasoning. For example, it bypasses color- or shape-based reference, and other complicated factual reasoning (e.g., use of brand names or metaphors).

Sr3D+: In addition to the dataset generation described above, we augment Sr3D with more utterances choosing the target object's class among those that do not comply with the criterion of having more than one *distractors* in the scene. Given the synthetic nature of the data, we can generate a large amount of utterances in a cost-free way. We explore the contribution of Sr3D+ to the final performance of our neural listener in Sect. 5. This additional set of data will be particularly useful when comparing our method to the *Unique* setting of [18] (Table 4), since it assumes that the target object is the only instance of that class in the scene.

3.2 Natural Reference in 3D Scenes

The Natural Reference in 3D (**Nr3D**) dataset contains 41,503 human utterances collected by deploying an online reference game in AMT. The game is played between two humans: a 'speaker' who was asked to describe a designated *target* object in a ScanNet 3D scene and a 'listener' who, given the speaker's utterance, was asked to select the referred object among its *distractors*. The game is structured such that both 'speaker' and 'listener' are rewarded when the *target* is successfully selected, hence incentivising descriptions that are most discriminative in the context of a scene and a general audience.

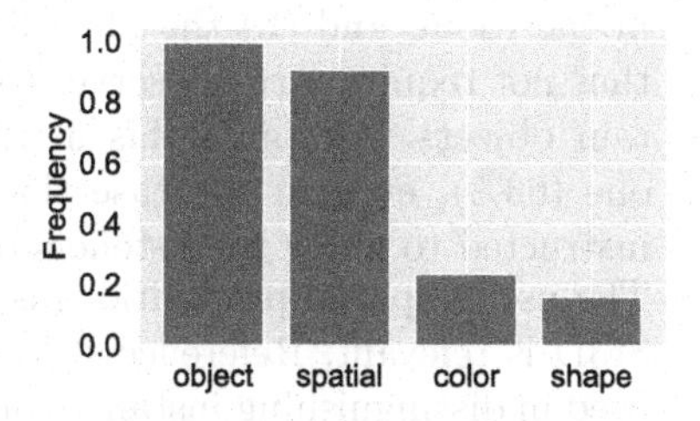

Fig. 3. Vocabulary histogram.

Both players are shown the same 3D scene in the form of a decimated mesh model and can interact with it through a 3D interface. In order to remove any camera view bias, we initialize the 'speaker' and 'listener' 3D interfaces with different randomized camera parameters. Given the specifics of the task and the difficulty of understanding the depicted 3D real-world visual content by the non-expert players, we highlight with bounding boxes (oriented when available) the *target* and *distractors*. We distinguish them for the 'speaker' with red and green color respectively, whereas for the 'listener' there is no distinction among them. To encourage players to explore the scene and familiarize themselves with all highlighted objects, we also provide them with a total count of bounding boxes they should expect in the scene. For an example of the speaker's interface we refer the reader to Fig. 1 in Supplementary Material [2].

We collect at least 7 utterances from different player pairs per target object. During the collection process, we iterate over all object instances with the same fine-grained object class in a scene (e.g., all 6 sofa chairs Fig. 1 in Supplementary Material [2]), providing to the dataset a symmetric property. Among the collected utterances, some originate from games with unsuccessful results; these are not used for training/learning purposes.

Discussion. Before presenting our neural agents, we identify several important properties of Nr3D:

- Performance in the gamified data collection process was high (92.2%), but 'listeners' made significantly more errors in the more challenging "hard" contexts (90.0% vs. 94.7%, z = 17.5, $p < 0.001$). We define "hard" contexts as those 3D scenes that contain *more than* 2 *distractors* (Fig. 5 illustrates examples of "hard" vs. "easy").

- Speakers naturally produced longer utterances on average to describe targets in hard contexts (approximately 12.5 words vs 10.2, $t = -35$, $p < 0.001$). The average number of words across all utterances (ignoring punctuation) is 11.4 and the median is 10.
- Regardless of the context difficulty, we identified two attributes in the descriptive power of the utterances (Fig. 5): (a) the *target* is **scene-discoverable** when it is uniquely distinguishable among objects in the entire scene and not only its *distractors*. The majority of the utterances mention the fine-grained class type or a close synonym of the *target* (91.6%). This *naturally emerging* property of Nr3D allows us to identify the target among *all* objects in the room; and (b) the identification of the *target* is **view-independent** thus not requiring the observer to place themselves into the scene facing certain objects. Although this attribute is not as prominent as the previous one (63%), even in the case that there is view dependency, speakers were instructed to guide the listeners on how to place themselves in the scene.
- The use of spatial prepositions is ubiquitous (90.5%), which exemplifies why Sr3D is relevant. Reference to color and shape properties is drastically less used in distinguishing instances of the same fine-grained object class. Figure 3 is a frequency-histogram of the different types of language use.

4 Developing 3D Neural Listeners

Given a 3D scene $\mathcal{S}$ represented as an RGB-colored point-cloud of N points $\mathcal{S} \in \mathbb{R}^{N \times 6}$ and a word-tokenized utterance $U = \{u_1, \ldots, u_t\}$ we want to build a neural listener that can identify the referred target object $\mathcal{T} \subset \mathcal{S}$. To this end, we assume access to a partition of $\mathcal{S} = \{O_1, \ldots, O_M\}$ that represents the objects (O_i) present in $\mathcal{S}$. While it is feasible to attempt identification (or more precisely, in this case, object localization) by operating directly on the unstructured $\mathcal{S}$ ([18,50]), the problem of instance localization (*especially* for FG classes) remains vastly unsolved. To overcome this and decouple the 3D instance-segmentation problem from our referential setting, we assume access to the instance-level object segmentations of the underlying scene. This choice allows us to cast the 3D reference problem into a classification problem that aims to predict the referred "target" among M segmented 3D instances.

While the above assumption eliminates the need to define each object in $\mathcal{S}$, it still leaves open the problems of: (i) FG object classification; (ii) recognition of the referred object class (per the utterance); and (iii) the original problem of selecting the referred object among the m options. For the first two tasks, we experiment with a neural listener that utilizes two auxiliary cross-entropy losses ($\mathcal{L}_{fg}$, $\mathcal{L}_{text}$) aimed to decouple these intrinsic aspects of the original task. Specifically, the two losses are added to the cross-entropy loss of the main task in hand ($\mathcal{L}_{ref}$) making a final loss that is a weighted sum of these terms:

$$\mathcal{L}_{total} = \alpha_1 \mathcal{L}_{fg} + \alpha_2 \mathcal{L}_{text} + \mathcal{L}_{ref}$$

Contextual Scene Understanding. The above design is object-aware, but our underlying task is also scene-oriented and heavily relies on the *configuration* of the objects present in a scene. Because of this reason it is important to provide a neural listener with a signal that contains explicit information about the scene it operates. A baseline that we explored to this end, is to create a PointNet++ hierarchical scene-feature (based on a large number of points of $\mathcal{S}$) which we fused with every visual representation extracted independently for each object O_i. While the resulting representation is simultaneously object-centric and scene-aware it is not taking into account explicit object-to-object interactions. A more sophisticated approach – which is part of the **ReferIt3DNet** – uses a structured and explicit way of capturing object-to-object interactions to provide information about the scene. Specifically, we use a dynamic graph-convolutional network (DGCN) [63] that operates on the visual features of the objects present in a scene (the object are nodes of a graph). The edges of this graph are computed dynamically at each layer of the DGCN according to the Euclidean similarity among the updated (per-node) visual features. In our experiments we use the

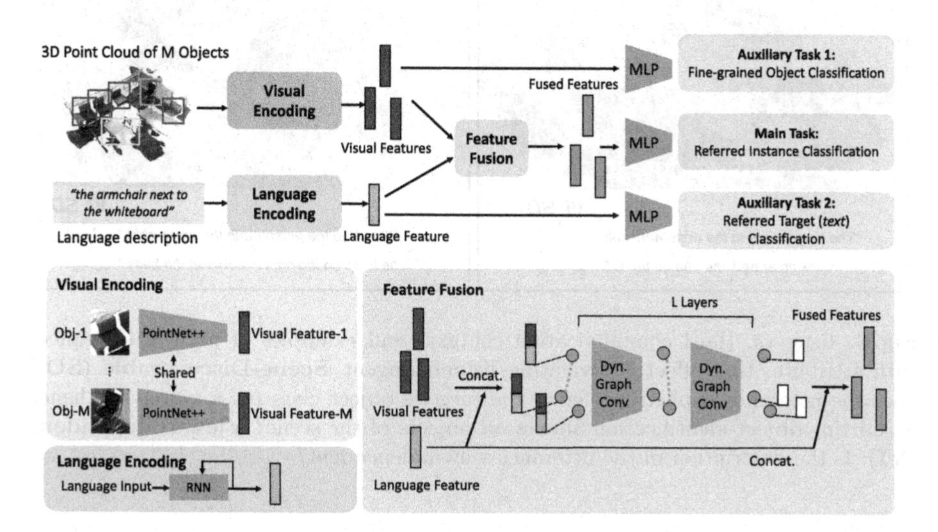

Fig. 4. The *ReferIt3DNet* neural listener. A visual encoder processes (via a shared PointNet++) each 3D object of a given scene that is represented by a 6D point cloud containing its *xyz* coordinates and RGB color. Simultaneously, the utterance describing the referred object (e.g., "*the armchair next to the whiteboard*") is processed by a Recurrent Neural Network (RNN). The resulting visio-linguistic representations are fused together and processed by a Dynamic Graph Convolution Network (DGCN) which creates an object-centric *and* scene- (context-) aware representation per object. The output of the DGCN is processed by an MLP classifier that estimates for every object its likelihood to be the referred one. Two auxiliary losses modulate the visio-linguistic representations before they are processed by the DGCN via an FG object-class classifier and a referential-text classifier ($\mathcal{L}_{fg}$ and $\mathcal{L}_{text}$ – see text for details). (Color figure online)

k-nearest neighbor-graph among the nodes ($k = 7$, chosen per validation). We note that $k = 7$ creates a relatively sparse graph (the 90th percentile of the number of objects in the training scenes is 52). For further details we refer the reader to the Supplementary Material [2].

Incorporation of Language. An important decision regards how one should "fuse" the linguistic signal in a pipeline like the above. Despite a chair being visually different from a door, our graph-network should inspect the relation among these objects, especially when the reference requires it (e.g., "the chair close to the door"). To promote this action we fuse the visual (object) features with the linguistic ones (derived by an RNN) *before* we pass them to the DGCN. We also explore the effect of adding the linguistic features *after* the DGCN and in both places – which is the best performing option. An overview of our pipeline is illustrated in Fig. 4.

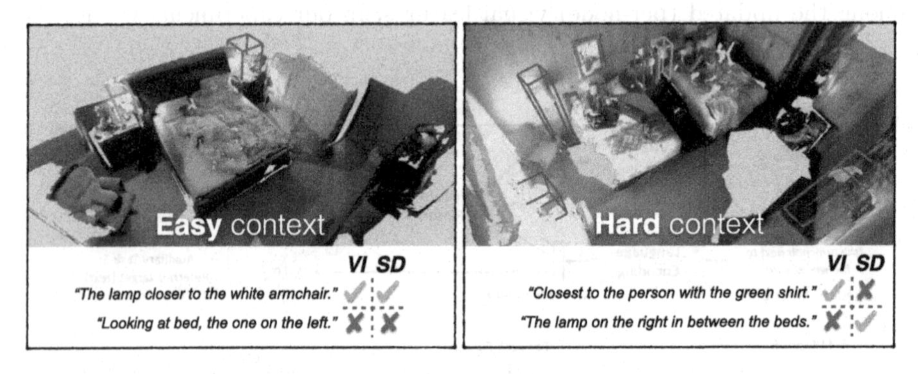

Fig. 5. Easy vs. Hard communication contexts and examples of natural utterances with attributes that affect a navigating/listening agent. **Scene-Discoverable (SD)**: does the utterance explicitly refer to the target's object class (or a synonym), hence permitting object-identification among *all* objects of the scene? **View-Independent (VI)**: Is the description in the utterance view-independent?

5 Experiments and Analysis

We explore different listening architectures[2] and report the listening accuracy; each test utterance receives a binary score (1 if the correct object is predicted as target and 0 otherwise). For all experiments we use the official-ScanNet splits.

1. **Decoupled approach**: This is a baseline listener consisting of a text classifier and an (FG) object classifier that are trained separately. Given an utterance we use the text-clf to predict the referred object-class. Then we select uniformly i.i.d. (and output) an object from $O_i \in \mathcal{S}$ for which the object-clf

[2] Architecture details and hyper-parameters for all the experiments, are provided in the Supplementary Material [2].

Table 2. ReferIt3DNet performance on Nr3D with/out Sr3D. The first row contains the achieved accuracy on the Nr3D testing data for a listener trained solely with the Nr3D training set; the other rows showcase the effect of training simultaneously with the Sr3D/Sr3D+, respectively.

	Overall	Easy	Hard	View-dep	View-indep
Nr3D	35.6% ± 0.7%	43.6% ± 0.8%	27.9% ± 0.7%	32.5% ± 0.7%	37.1% ± 0.8%
w/Sr3D	37.2% ± 0.3%	44.0% ± 0.6%	**30.6% ± 0.3%**	**33.3% ± 0.6%**	39.1% ± 0.2%
w/Sr3D+	**37.6% ± 0.4%**	**45.4% ± 0.6%**	30.0% ± 0.4%	33.1% ± 0.5%	**39.8% ± 0.4%**

Table 3. Listening performance of various ablated models. The first two columns contain the obtained accuracy when no auxiliary losses are used, and the last two the accuracy when these losses are included.

	Nr3D	**Sr3D**	**Nr3D**	**Sr3D**
Aux. classification loss	**No**		**Yes**	
Decoupled	25.45%	31.73%	–	–
V + L	26.12% ± 0.5%	32.61% ± 0.4%	26.62% ± 0.5%	32.98% ± 0.4%
V + L + C	27.45% ± 0.6%	34.7% ± 0.4%	28.51% ± 0.6%	37.2% ± 0.4%
ReferIt3DNet-A	32.3% ± 0.3%	**39.7% ± 0.3%**	33.4% ± 0.3%	**41.0% ± 0.3%**
ReferIt3DNet-B	31.8% ± 0.3%	38.1% ± 0.2%	33.0% ± 0.3%	40.5% ± 0.2%
ReferIt3DNet	**32.4% ± 0.5%**	38.4% ± 0.2%	**35.6% ± 0.7%**	39.8% ± 0.2%

matches the text-based prediction. We note that in Nr3D (Sr3D) test accuracies for the two classifiers are 93.0% (100.0%) and 64.7% (67.4%), indicating a noticeable asymmetry in the difficulty of solving the two tasks.

2. **Vision + Language, no Context (V + L)**: Inspired by context-free listening architectures like those in [4,45], we ground an RNN with the visual feature of each object of $O_i \in \mathcal{S}$, *independently*, and use a shallow classifier to predict the likelihood of each O_i for being the referred target. This baseline can encode visual properties of an object beyond its FG class enabling rich (context-free) distinctions (e.g., "very small, or yellow colored chair").
3. **Vision + Language + Holistic Context (V + L + C)**: Similar to the above, but also fuses a PointNet++ scene-feature with each object's visual feature to ground the RNN. This enables the inspection of non-structured context when solving the reference task (PointNet++ is applied on a non-segmented scene point cloud).
4. **Vision + Language + Graph (structured) Context (*ReferIt3DNet*)**: This is our proposed listener and comes in three variants that differ w.r.t. *where* we fuse the linguistic with the visual information.

Table 4. ScanRefer performance with/out Sr3D. MeanIoU improvements when combining Sr3D data with ScanRefer's data during training.

Dataset	Unique		Multiple		**Overall**	
	P@0.25	P@0.5	P@0.25	P@0.5	P@0.25	P@0.5
ScanRefer	53.75%	37.47%	21.03%	12.83%	26.44%	16.90%
w/Sr3D	59.99%	39.06%	21.69%	14.33%	28.02%	18.42%
w/Sr3D+	**63.55%**	**42.18%**	**24.12%**	**15.75%**	**30.64%**	**20.12%**

Neural Listeners. Comparisons for the above models are presented in Table 3. We observe the following main trends[3]: i) using the visual and linguistic auxiliary classification losses improves performance; ii) Simplified language (Sr3D) makes identification easier; iii) scene context matters a lot, but most importantly how we incorporate the context (e.g., via DGCN, or direct fusion of PointNet++) makes an important difference in performance. As expected, a more structured versus a rudimentary representation favors better results; iv) where we fuse language matters as well: ReferIt3DNet-*A* fuses after the DGCN, ReferIt3DNet-*B* before, and the best performing (for Nr3D) model fuses in both places.

The results shown in Fig. 6 show the neural listener's capability to understand and locate objects in challenging 3D scenarios. For example, the top-right example was successful despite the utterance being long. Referring to this particular trashcan among other similar ones requires both spatial reasoning and visual 3D understanding. Similar capability can be demonstrated in the two examples in the second row about the door and the cabinets. Finally, the last row shows two challenging failure cases of our model. In the bottom-left example, the utterance has wrongly placed the listener in the room (the chair is at the 3 o'clock position instead of 9). The bottom-right example is particularly hard to solve; the scene is almost symmetric and stripped of visual features, making it hard to discriminate among the chairs.

Combining Nr3D and Sr3D. In Table 2, we observe how combining the two datasets provides a consistent boost in performance. This demonstrates the contribution of adding a synthetically generated dataset to a human one. We get a similar outcome when combining Sr3D to the ScanRefer [18] data (see Table 4). We performed this experiment following the implementation in [17]. Going back to Table 2, the results showcase that our neural listener performs better by a margin in "easy" versus "hard" cases. This is expected and solidifies the understanding that more work needs to be done in discriminatively distinguishing objects when there are multiple distractors in the scene. Another important finding is that view-independent utterances are easier to solve than dependent ones. This does not come as a surprise, since the network has naturally more work to do to comprehend nuances related to viewing the scene w.r.t. another object.

[3] In all results mean accuracies and standard errors across 5 random seeds are reported, to control for the point cloud scene sampling.

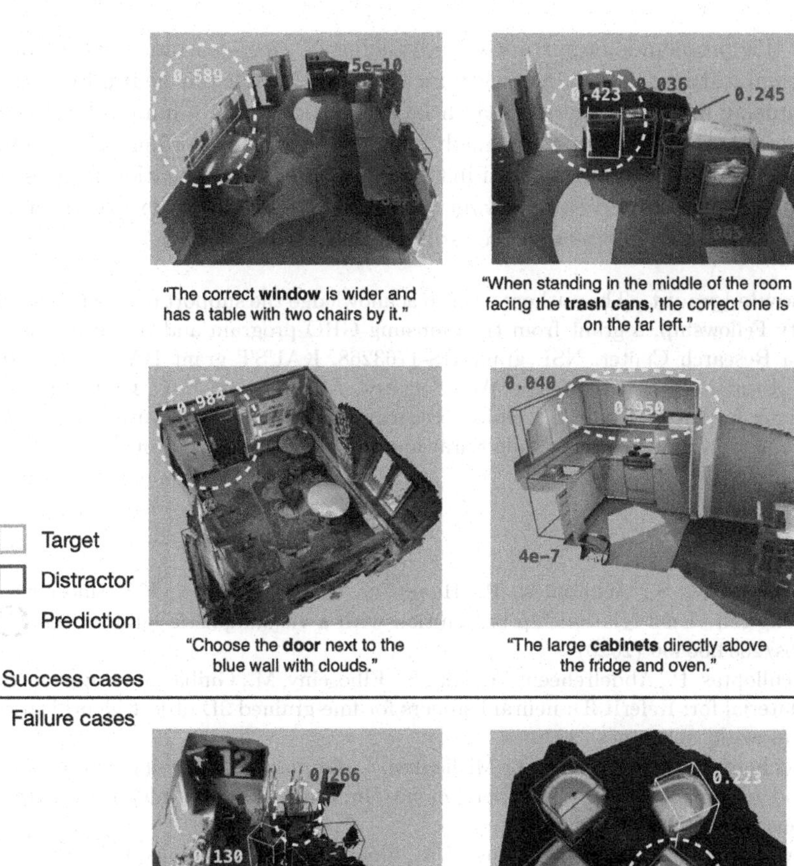

0.589
5e-10
"The correct **window** is wider and has a table with two chairs by it."
0.423
0.036
0.245
"When standing in the middle of the room facing the **trash cans**, the correct one is on the far left."
0.984
"Choose the **door** next to the blue wall with clouds."
0.040
0.950
4e-7
"The large **cabinets** directly above the fridge and oven."
Target
Distractor
Prediction
Success cases
Failure cases

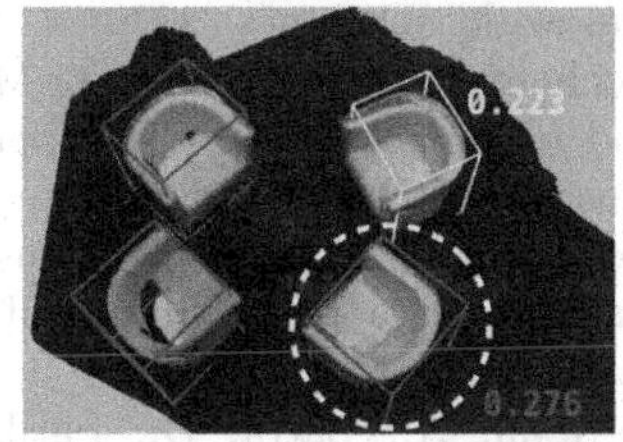

"Looking at the 12 poster, the **chair** in the 9 o'clock position."

"The one **chair** not arranged properly."

Fig. 6. Qualitative results. Success cases are in the top four images and Failure in the bottom two. Targets are shown in green boxes and distractors in red. The network predictions are shown in a dashed yellow circle, along with the predicted probabilities. Please note that probabilities of inter class distractors are not illustrated here. (Color figure online)

6 Conclusion

Language assisted object disambiguation done directly for 3D objects in 3D environments is a novel but very challenging task. This is especially true when one tries to distinguish among multiple instances of the same fine-grained object category. In addition to the intrinsic difficulty of the problem, there is a scarcity of appropriate datasets. Creating relevant visio-linguistic data that allow us to study this problem is important for advancing 3D deep-learning that, similar to 2D visual learning, is a data hungry methodology. While our neural listeners are a promising first step, more research has to be done before human-level performance and generalization is attained. In summary, this paper has (a) introduced the problem of fine-grained multi-instance 3D object identification in real-world scenes; (b) contributed two relevant public datasets; and (c) explored an array of sensible neural architectures for solving the referential task.

Acknowledgment. The authors wish to acknowledge the support of a Vannevar Bush Faculty Fellowship, a grant from the Samsung GRO program and the Stanford SAIL Toyota Research Center, NSF grant IIS-1763268, KAUST grant BAS/1/1685-01-01, and a research gift from Amazon Web Services. Also, they wish to thank Prof. Angel X. Chang for the inspiring discussions regarding the creation of synthetic 3D spatial data, and Iro Armeni and Antonia Saravanou for their help in writing.

References

1. Abdelkarim, S., Achlioptas, P., Huang, J., Li, B., Church, K., Elhoseiny, M.: Long-tail visual relationship recognition with a visiolinguistic hubless loss. CoRR abs/2004.00436 (2020)
2. Achlioptas, P., Abdelreheem, A., Xia, F., Elhoseiny, M., Guibas, L.: Supplementary material for: ReferIt3D: neural listeners for fine-grained 3D object identification in real world 3D scenes (2020)
3. Achlioptas, P., Diamanti, O., Mitliagkas, I., Guibas, L.: Learning representations and generative models for 3D point clouds. In: International Conference on Machine Learning (ICML) (2018)
4. Achlioptas, P., Fan, J., Hawkins, R.X., Goodman, N.D., Guibas, L.J.: ShapeGlot: learning language for shape differentiation. In: International Conference on Computer Vision (ICCV) (2019)
5. Agrawal, A., Batra, D., Parikh, D.: Analyzing the behavior of visual question answering models. In: Empirical Methods in Natural Language Processing (EMNLP) (2016)
6. Agrawal, H., et al.: nocaps: novel object captioning at scale. In: Conference on Computer Vision and Pattern Recognition (CVPR) (2019)
7. Akata, Z., Reed, S., Walter, D., Lee, H., Schiele, B.: Evaluation of output embeddings for fine-grained image classification. In: CVPR (2015)
8. Anayurt, H., Ozyegin, S.A., Cetin, U., Aktas, U., Kalkan, S.: Searching for ambiguous objects in videos using relational referring expressions. CoRR abs/1908.01189 (2019)
9. Anderson, P., et al.: Vision-and-language navigation: interpreting visually-grounded navigation instructions in real environments. In: Conference on Computer Vision and Pattern Recognition (CVPR) (2018)
10. Andreas, J., Klein, D.: Reasoning about pragmatics with neural listeners and speakers. CoRR abs/1604.00562 (2016)
11. Andreas, J., Rohrbach, M., Darrell, T., Klein, D.: Neural module networks. In: Conference on Computer Vision and Pattern Recognition (CVPR) (2016)
12. Antol, S., et al.: VQA: visual question answering. In: International Conference on Computer Vision (ICCV) (2015)

13. Armeni, I., et al.: 3D semantic parsing of large-scale indoor spaces. In: Conference on Computer Vision and Pattern Recognition (CVPR) (2016)
14. Avetisyan, A., Dahnert, M., Dai, A., Savva, M., Chang, A.X., Nießner, M.: Scan2CAD: learning cad model alignment in RGB-D scans. In: Conference on Computer Vision and Pattern Recognition (CVPR) (2019)
15. Kong, C., Lin, D., Bansal, M., Urtasun, R., Fidler, S.: What are you talking about? Text-to-image coreference. In: CVPR (2014)
16. Chen, K., Choy, C.B., Savva, M., Chang, A.X., Funkhouser, T., Savarese, S.: Text2Shape: generating shapes from natural language by learning joint embeddings. CoRR abs/1803.08495 (2018)
17. Chen, Z.D., Chang, A.X., Nießner, M.: https://github.com/daveredrum/ScanRefer. Accessed 17 July 2020
18. Chen, Z.D., Chang, A.X., Nießner, M.: ScanRefer: 3D object localization in RGB-D scans using natural language. CoRR abs/1912.08830 (2019)
19. Cohn-Gordon, R., Goodman, N., Potts, C.: Pragmatically informative image captioning with character-level inference. CoRR abs/1804.05417 (2018)
20. Dai, A., Chang, A.X., Savva, M., Halber, M., Funkhouser, T., Nießner, M.: ScanNet: richly-annotated 3D reconstructions of indoor scenes. In: Conference on Computer Vision and Pattern Recognition (CVPR) (2017)
21. Das, A., Datta, S., Gkioxari, G., Lee, S., Parikh, D., Batra, D.: Embodied question answering. In: Conference on Computer Vision and Pattern Recognition (CVPR) (2018)
22. Devlin, J., Gupta, S., Girshick, R., Mitchell, M., Zitnick, C.L.: Exploring nearest neighbor approaches for image captioning. arXiv preprint arXiv:1505.04467 (2015)
23. Doğan, F.I., Kalkan, S., Leite, I.: Learning to generate unambiguous spatial referring expressions for real-world environments. CoRR (2019)
24. Elhoseiny, M., Elfeki, M.: Creativity inspired zero-shot learning. In: Proceedings of the IEEE International Conference on Computer Vision, pp. 5784–5793 (2019)
25. Elhoseiny, M., Saleh, B., Elgammal, A.: Write a classifier: zero-shot learning using purely textual descriptions. In: ICCV (2013)
26. Elhoseiny, M., Zhu, Y., Zhang, H., Elgammal, A.: Link the head to the "beak": zero shot learning from noisy text description at part precision. In: Conference on Computer Vision and Pattern Recognition (CVPR) (2017)
27. Goyal, Y., Khot, T., Summers-Stay, D., Batra, D., Parikh, D.: Making the V in VQA matter: elevating the role of image understanding in visual question answering. In: Conference on Computer Vision and Pattern Recognition (CVPR) (2017)
28. Karpathy, A., Joulin, A., Li, F.F.F.: Deep fragment embeddings for bidirectional image sentence mapping. In: Advances in Neural Information Processing Systems (NeurIPS) (2014)
29. Kazemzadeh, S., Ordonez, V., Matten, M., Berg, T.: ReferitGame: referring to objects in photographs of natural scenes. In: Empirical Methods in Natural Language Processing (EMNLP) (2014)
30. Kim, J.H., Jun, J., Zhang, B.T.: Bilinear attention networks. In: Advances in Neural Information Processing Systems (NeurIPS) (2018)
31. Kiros, R., Salakhutdinov, R., Zemel, R.S., et al.: Unifying visual-semantic embeddings with multimodal neural language models. Trans. Assoc. Comput. Linguist. (TACL) (2015)
32. Krishna, R., et al.: Visual genome: connecting language and vision using crowdsourced dense image annotations. Int. J. Comput. Vis. **123**(1), 32–73 (2017). https://doi.org/10.1007/S11263-016-0981-7

33. Kulkarni, N., Misra, I., Tulsiani, S., Gupta, A.: 3D-RelNet: joint object and relational network for 3D prediction. In: Conference on Computer Vision and Pattern Recognition (CVPR) (2019)
34. Lazaridou, A., Hermann, K.M., Tuyls, K., Clark, S.: Emergence of linguistic communication from referential games with symbolic and pixel input. arXiv preprint arXiv:1804.03984 (2018)
35. Ba, J.L., Swersky, K., Fidler, S., et al.: Predicting deep zero-shot convolutional neural networks using textual descriptions. In: ICCV (2015)
36. Lewis, D.: Convention: A Philosophical Study. Wiley, Hoboken (2008)
37. Li, C., Xia, F., Martín-Martín, R., Savarese, S.: HRL4IN: hierarchical reinforcement learning for interactive navigation with mobile manipulators. In: Conference on Robot Learning (2020)
38. Long, Y., Shao, L.: Describing unseen classes by exemplars: zero-shot learning using grouped simile ensemble. In: Winter Conference on Applications of Computer Vision (WACV) (2017)
39. Long, Y., Shao, L.: Learning to recognise unseen classes by a few similes. In: Proceedings of the 25th ACM International Conference on Multimedia, pp. 636–644. ACM (2017)
40. Lu, C., Krishna, R., Bernstein, M., Fei-Fei, L.: Visual relationship detection with language priors. In: Leibe, B., Matas, J., Sebe, N., Welling, M. (eds.) ECCV 2016. LNCS, vol. 9905, pp. 852–869. Springer, Cham (2016). https://doi.org/10.1007/978-3-319-46448-0_51
41. Mao, J., Xu, W., Yang, Y., Wang, J., Yuille, A.: Deep captioning with multimodal recurrent neural networks (m-RNN). In: International Conference on Learning Representations (ICLR) (2015)
42. Mauceri, C., Palmer, M., Heckman, C.: SUN-Spot: an RGB-D dataset with spatial referring expressions. In: International Conference on Computer Vision Workshop on Closing the Loop Between Vision and Language (2019)
43. Mitchell, M., van Deemter, K., Reiter, E.: Generating expressions that refer to visible objects. In: North American Chapter of the Association for Computational Linguistics (NAACL) (2013)
44. Mo, K., et al.: A large-scale benchmark for fine-grained and hierarchical part-level 3D object understanding. In: Conference on Computer Vision and Pattern Recognition (CVPR) (2019)
45. Monroe, W., Hawkins, R.X., Goodman, N.D., Potts, C.: Colors in context: a pragmatic neural model for grounded language understanding. Trans. Assoc. Comput. Linguist. (TACL) (2017)
46. Paetzel, M., Racca, D.N., DeVault, D.: A multimodal corpus of rapid dialogue games. In: LREC, pp. 4189–4195 (2014)
47. Plummer, B.A., et al.: Revisiting image-language networks for open-ended phrase detection. CoRR abs/1811.07212 (2018)
48. Plummer, B.A., Wang, L., Cervantes, C.M., Caicedo, J.C., Hockenmaier, J., Lazebnik, S.: Flickr30k entities: collecting region-to-phrase correspondences for richer image-to-sentence models. In: Conference on Computer Vision and Pattern Recognition (CVPR) (2015)
49. Prabhudesai, M., Tung, H.Y.F., Javed, S.A., Sieb, M., Harley, A.W., Fragkiadaki, K.: Embodied language grounding with implicit 3D visual feature representations. CoRR abs/1910.01210 (2019)
50. Qi, C.R., Litany, O., He, K., Guibas, L.: Deep hough voting for 3D object detection in point clouds. In: International Conference on Computer Vision (ICCV) (2019)

51. Qi, C.R., Yi, L., Su, H., Guibas, L.J.: PointNet++: deep hierarchical feature learning on point sets in a metric space. In: Advances in Neural Information Processing Systems (NeurIPS) (2017)
52. Ramanathan, V., Joulin, A., Liang, P., Fei-Fei, L.: Linking people with "their" names using coreference resolution. In: European Conference on Computer Vision (ECCV) (2014)
53. Ren, M., Kiros, R., Zemel, R.: Exploring models and data for image question answering. In: Advances in Neural Information Processing Systems (NeurIPS) (2015)
54. Romera-Paredes, B., Torr, P.: An embarrassingly simple approach to zero-shot learning. In: ICML, pp. 2152–2161 (2015)
55. Rosman, B., Ramamoorthy, S.: Learning spatial relationships between objects. Int. J. Rob. Res. **30**(11), 1328–1342 (2011)
56. Savva, M., et al.: Habitat: a platform for embodied AI research. In: Conference on Computer Vision and Pattern Recognition (CVPR) (2019)
57. Shridhar, M., et al.: ALFRED: a benchmark for interpreting grounded instructions for everyday tasks. CoRR abs/1912.01734 (2019)
58. Socher, R., Ganjoo, M., Manning, C.D., Ng, A.: Zero-shot learning through cross-modal transfer. In: NeurIPS, pp. 935–943 (2013)
59. Su, J.C., Wu, C., Jiang, H., Maji, S.: Reasoning about fine-grained attribute phrases using reference games. In: Proceedings of the IEEE International Conference on Computer Vision, pp. 418–427 (2017)
60. Tsai, Y.H.H., Huang, L.K., Salakhutdinov, R.: Learning robust visual-semantic embeddings. In: ICCV (2017)
61. Vedantam, R., Bengio, S., Murphy, K., Parikh, D., Chechik, G.: Context-aware captions from context-agnostic supervision. In: Proceedings of the IEEE Conference on Computer Vision and Pattern Recognition, pp. 251–260 (2017)
62. Vinyals, O., Toshev, A., Bengio, S., Erhan, D.: Show and tell: a neural image caption generator. In: Conference on Computer Vision and Pattern Recognition (CVPR) (2015)
63. Wang, Y., Sun, Y., Liu, Z., Sarma, S.E., Bronstein, M.M., Solomon, J.M.: Dynamic graph CNN for learning on point clouds. ACM Trans. Graph. (TOG) **38**(5), 1–12 (2019)
64. Wittgenstein., L.: Philosophical Investigations: the English text of the third edition. Wiley, Hoboken (1953)
65. Xia, F., Zamir, A.R., He, Z., Sax, A., Malik, J., Savarese, S.: Gibson Env: real-world perception for embodied agents. In: Conference on Computer Vision and Pattern Recognition (CVPR) (2018)
66. Xiang, F., et al.: SAPIEN: a simulated part-based interactive environment. In: Proceedings of the IEEE/CVF Conference on Computer Vision and Pattern Recognition, pp. 11097–11107 (2020)
67. Xu, K., et al.: Show, attend and tell: neural image caption generation with visual attention. In: International Conference on Machine Learning (ICML) (2015)
68. Yang, J., et al.: Embodied visual recognition. CoRR abs/1904.04404 (2019)
69. Yang, Y., Hospedales, T.M.: A unified perspective on multi-domain and multi-task learning. In: ICLR (2015)
70. Yu, Z., Yu, J., Cui, Y., Tao, D., Tian, Q.: Deep modular co-attention networks for visual question answering. In: Conference on Computer Vision and Pattern Recognition (CVPR) (2019)

71. Zhang, J., Kalantidis, Y., Rohrbach, M., Paluri, M., Elgammal, A., Elhoseiny, M.: Large-scale visual relationship understanding. In: AAAI Conference on Artificial Intelligence (2019)
72. Zhou, L., Kalantidis, Y., Chen, X., Corso, J.J., Rohrbach, M.: Grounded video description. In: Conference on Computer Vision and Pattern Recognition (CVPR) (2019)
73. Zhu, Y., Elhoseiny, M., Liu, B., Peng, X., Elgammal, A.: A generative adversarial approach for zero-shot learning from noisy texts. In: CVPR (2018)

MatryODShka: Real-time 6DoF Video View Synthesis Using Multi-sphere Images

Benjamin Attal[1,2(✉)], Selena Ling[1], Aaron Gokaslan[1], Christian Richardt[3], and James Tompkin[1]

[1] Brown University, Providence, USA
[2] Carnegie Mellon University, Pittsburgh, USA
battal@andrew.cmu.edu
[3] University of Bath, Bath, UK

Abstract. We introduce a method to convert stereo 360° (omnidirectional stereo) imagery into a layered, multi-sphere image representation for six degree-of-freedom (6DoF) rendering. Stereo 360° imagery can be captured from multi-camera systems for virtual reality (VR), but lacks motion parallax and correct-in-all-directions disparity cues. Together, these can quickly lead to VR sickness when viewing content. One solution is to try and generate a format suitable for 6DoF rendering, such as by estimating depth. However, this raises questions as to how to handle disoccluded regions in dynamic scenes. Our approach is to simultaneously learn depth and disocclusions via a multi-sphere image representation, which can be rendered with correct 6DoF disparity and motion parallax in VR. This significantly improves comfort for the viewer, and can be inferred and rendered in real time on modern GPU hardware. Together, these move towards making VR video a more comfortable immersive medium.

1 Introduction

360° imagery is a valuable tool for virtual reality (VR) as the viewer is immersed in a captured real-world environment. Stereo 360° imagery aims to increase this immersion by providing binocular disparity as a depth cue in all directions, and video provides depiction for dynamic scenes. This imagery is usually captured with wide-angle multi-camera systems, arranged in a circle [1,45], with state-of-the-art systems providing high-resolution and high-quality stitched imagery. Stereo 360° cameras are decreasing in cost (≈$5k), which will increase their deployment across many industries. However, there are problems with this format. Motion parallax is missing from stereo 360° imagery, which can cause viewing discomfort. Further, stereo 360° formats have disparity problems:

Electronic supplementary material The online version of this chapter (https://doi.org/10.1007/978-3-030-58452-8_26) contains supplementary material, which is available to authorized users.

A. Vedaldi et al. (Eds.): ECCV 2020, LNCS 12346, pp. 441–459, 2020.
https://doi.org/10.1007/978-3-030-58452-8_26

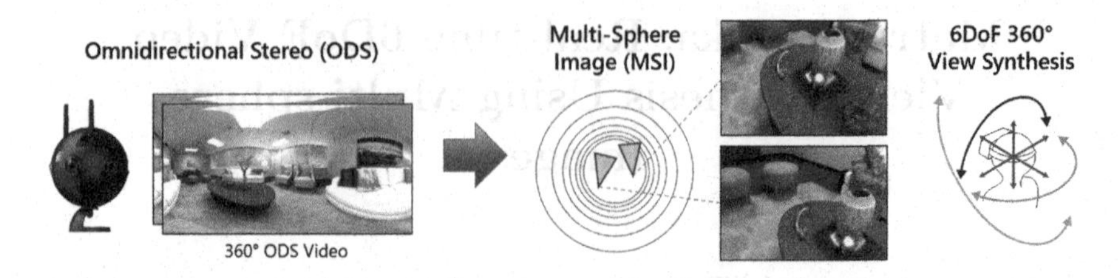

Fig. 1. Our approach takes 360° omnidirectional stereo video as input and predicts multi-sphere images that enable six degree-of-freedom 360° view synthesis in real time. This produces a more comfortable and immersive VR video viewing experience.

side-by-side equirectangular projection (ERP) formats have incorrect disparity everywhere but in one direction [25], and omnidirectional stereo (ODS) formats [21,36] have diminished disparity as the view approaches the zenith or nadir [1]. In short, long-term viewing is difficult as vestibulo-ocular comfort is low [56], which can cause VR sickness [34].

Our goal is to provide six-degree-of-freedom (6DoF) video with accurate motion parallax and disparity cues for a reasonably-sized headbox. Large-baseline camera systems can *interpolate* views to provide this via optical flow or depth-based reprojection, but usually the desired human motion is too large to build practical camera systems that operate in this way. Thus, we must *extrapolate* views beyond the camera system's baseline. This requires estimating content for unseen regions via hallucination or inpainting. Further, for video, we want this view synthesis to happen quickly, preferably in real time when applied to a stereo 360° camera feed, so as to avoid preprocessing and allow live applications.

Our approach is to simultaneously estimate depth and inpaint the holes by using a learning-based approach on a layer-based scene representation. Inspired by recent work on stereo magnification [48,64] and light field fusion [32], we learn to decompose a scene into multi-sphere images (MSI), each with RGB and alpha (RGBA) values. This is created by a network architecture which supports stereo 360° input in the omnidirectional stereo format, uses spherically-aware convolutions and losses, and maintains temporal consistency for video without additional network parameters via spherical single-image transform-inverse regularization [15]. We demonstrate quantitatively and qualitatively that these contributions increase reconstruction quality both spatially and temporally against existing view expansion methods, and that our approach can be applied and rendered in real time to 4 K videos on modern GPUs. Our contributions are:

- A multi-sphere image scene representation for omnidirectional view synthesis.
- A method to recover the MSI representation from ODS imagery via a learning-based soft spherical 3D reconstruction method. This uses an architecture and losses for spherical images, including spherical temporal consistency.
- A real-time inference and VR rendering engine for MSI from ODS input.

These are complemented by an open-source system, with mono (ERP) and stereo 360° (ODS) renderers to generate synthetic training data [18,44,49], TensorFlow models, real-time TensorFlow and TensorRT inference within Unity that outputs to GPU textures, and a real-time multi-sphere video renderer in Unity.

2 Related Work

360° Video Stitching. Builds on seminal work in panorama image stitching [5,53], which automatically aligns and blends multiple photos of a scene into a single, wide field-of-view panorama. Subsequent work on stitching 360° videos [27,38] addresses temporally coherent stitching from multi-view video input, as commonly used in commercial 360° videos. However, monocular 360° videos only provide views for a single center of projection, and hence no depth perception. Omnidirectional stereo (ODS) is a circular projection [21,36,40] that improves depth perception using the disparity between the left- and right-eye panoramic views. ODS has become popular for stereo 360° [1,41,45] as it is a good fit for existing processing, compression and transmission pipelines.

360° Depth Estimation. Aims to recover dense 360° depth maps, which can be used for rendering novel views using a mesh-based renderer. Assuming a moving camera in a static environment, structure-from-motion and multi-view stereo can be used [19,20]. However, the made assumptions are actually violated by most usage scenarios, like stationary cameras or dynamic environments. Learning-based depth estimation approaches have the potential to overcome these limitations by using single-image input to predict 360° depth maps [25,57,65,66]. Nevertheless, view synthesis from RGBD (RGB+Depth) data is fundamentally limited by 3D reconstruction accuracy, and one cannot look behind occlusions [19,25].

360° View Synthesis. Creates new panoramic viewpoints from different input [42]. For example, ODS video can be created from three fisheye cameras [8], two 360° cameras mounted side by side [31], or two rotating line cameras [23]. However, ODS provides only binocular disparity and no motion parallax. Novel-view synthesis with motion parallax can be achieved using depth-augmented ODS [55], flow-based blending [3,30], or a layered scene representation [46]. However, these approaches do not support up/down motion. Parra Pozo et al.'s spherical video camera rig enables high-quality 6DoF view synthesis [35], but not in real time. Serrano et al. similarly propose an offline, optimization-based method for high-quality 6DoF video generation from RGBD equirectangular video input [46]. We create a fast learning-based view synthesis method that is applicable to 'in the wild' ODS videos or streams, e.g., including all YouTube ODS videos.

Perspective View Synthesis. Has made leaps in visual quality using soft 3D reconstructions [10,37] and multi-plane images [17,32,48,64]. MPIs are stacks of semi-transparent layers representing scene appearance without explicit geometry, and can easily be reprojected into novel views. Learning-based approaches optimize MPIs from stereo images [48,64], or 4–5 input images [17,32]. Most approaches optimize RGBA colors per layer; the alpha channel allows for 'soft' reconstructions by blending the layers for perspectives from different input views

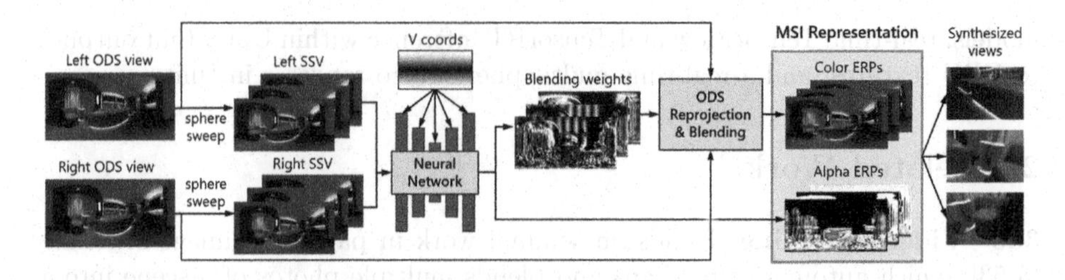

Fig. 2. Given an ODS image, we generate left/right sphere sweep volumes (SSV). These are input to a fully-convolutional neural network, with V-coordinate convolution, that predicts blending weights and alpha ERPs per multi-sphere image (MSI) layer. The final high-res MSI enables real-time 6DoF view synthesis. (Figure inspired by Zhou et al. [64].)

[37]. We extend these ideas to multi-sphere images, a layered spherical scene representation that enables omnidirectional 6DoF view synthesis.

CNNs on Spheres. Need to be adapted to correctly handle the unique distortions of 360° images. Su and Grauman work directly on equirectangular images using wider kernels near the poles [50], but there is no information shared between kernels. 360° images can also be projected into a cubemap, processing all sides as perspective images, and recombined [9]. More principled are distortion-aware convolutions [12,51,54], which also allow transfer of perspectively trained models to equirectangular images. Full rotation-equivariance can be achieved with spherical convolutions [11,16], but this is not necessarily desirable as videos can exploit the fixed gravity vector. Recent work generalizes cubemaps to icosahedra, and the resulting 20 triangles are unwrapped into five rectangles with shared convolution kernels [28,61]. These approaches add expense at inference time, or trade model capacity for spherical awareness; neither of which is desirable.

3 Method

Our goal is to enable real-time 6DoF view synthesis in the vicinity of an input stereo 360° video (Fig. 2). We begin with omnidirectional stereo (ODS) imagery, which has an image for each eye given a position in the world [21,36,40]. Given a database of synthetic ODS image pairs [18], our method trains a network to generate a multi-sphere image (MSI) representation of the scene. Then, in VR, we infer an MSI for each ODS video frame of an input video, and render it from novel headset viewpoints for the left and right eyes of the user.

Equirectangular Projection (ERP): Pixel coordinates in equirectangular images directly map to directions on the unit sphere. A pixel's x-coordinate maps to the azimuth angle $\theta = \pi \times (2x-1) \in [-\pi, \pi]$, and its y-coordinate to the elevation angle $\varphi = \pi \times (\frac{1}{2} - y) \in$

Omnidirectional projections
ERP
ODS (right view)

$[-\frac{\pi}{2}, \frac{\pi}{2}]$. A point $\mathbf{p} = (p_x, p_y, p_z)$ projects to:

$$\theta_{\text{ERP}} = -\arctan(p_z/p_x) \quad \text{and} \quad \varphi_{\text{ERP}} = \arctan\left(p_y/\sqrt{p_x^2 + p_z^2}\right), \tag{1}$$

with x-forward, z-left, and y-down [1]. A major disadvantage of the ERP format for 360° stereo imagery is that disparity is zero along the camera baseline, and so depth cannot be determined for pixels that lie along the baseline [31].

ODS Projection: This is defined by the *viewing circle* to which all camera rays are tangent [21,36]. Without loss of generality, this circle lies in the x-z-plane, is centered at the origin and has radius r = IPD/2 = 31.5 mm [1,13]. A 3D point $\mathbf{p} = (p_x, p_y, p_z)$ projects into the left and right ODS images at:

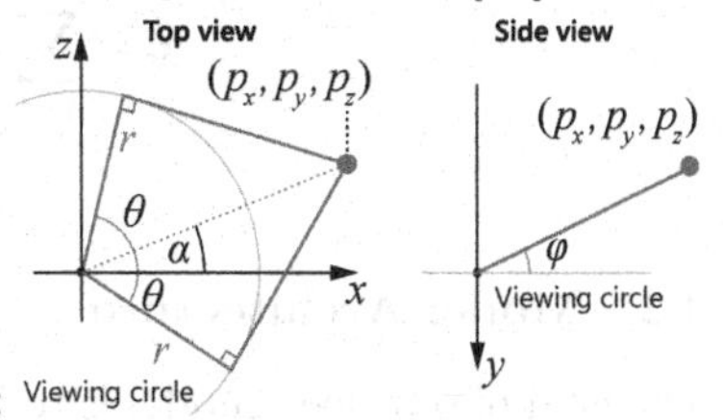

$$\theta_{\text{ODS}}^{\text{L/R}} = \pm \arcsin\left(r/\sqrt{p_x^2 + p_z^2}\right) - \arctan(p_z/p_x), \tag{2}$$

$$\varphi_{\text{ODS}} = \arctan\left(p_y/\sqrt{p_x^2 + p_z^2 - r^2}\right). \tag{3}$$

Unlike ERP, ODS encodes disparity in all azimuth directions, and therefore is richer input for view-synthesis tasks.

3.1 Multi-sphere Image Representation

The core representation for our scene inference and view-synthesis pipeline is the multi-sphere image (MSI)—the spherical equivalent of multi-plane images [64]. MSIs represent a scene as concentric spheres with color and transparency (RGBA) at each point on a sphere. One advantage of MSIs is that they allow for fast and dense rendering of novel views using traditional graphics pipelines (e.g., in Unity). Unlike multi-plane images, multi-sphere images are omnidirectional and thus enable view synthesis for any camera orientation and position within the innermost sphere. During training, inference, and rendering, we store MSIs as a sequence of ERPs parametrized by spherical coordinates $\{(\theta_i, \varphi_i)\}$ and their sphere radii $\{r_i\}$. In practice, we use 32 layers as a trade-off between inference speed and quality, with the smallest radius being 1 m, and the largest being 100 m. We generate the radii for in-between layers by linearly interpolating reciprocal depths. This creates more layers for nearby scene geometry (half the layers closer than 2 m). This sampling pattern is also desirable as it separates layers linearly in disparity when projected into a translated view close to the center of projection. See concurrent work by Broxton et al. [6] for a theoretical analysis.

Differentiable Rendering from MSIs: Novel views can easily be rendered from MSIs for a range of projections, including perspective, ERP and ODS. We use this to render *target* views for supervision during training. Each pixel in the projection defines a ray, whose color is given by repeated over-compositing [39] on the MSI, similar to MPIs [17]. Each such ray is intersected with the concentric spheres of the MSI, producing intersection points $\{\mathbf{p}_i\}_{i=1}^N$ with spherical layers 1 to N, from near to far. Next, each intersection point $\mathbf{p}_i$ is converted from Cartesian to spherical coordinates (θ_i, φ_i), so we can sample the RGB colors $\{\mathbf{c}_i\}_{i=1}^N$ and alphas (opacities) $\{\alpha_i\}_{i=1}^N$ corresponding to these points from MSI layer i. We use bilinear interpolation for sub-pixel precision. The final color is:

$$\mathbf{c} = \sum_{i=1}^{N} \mathbf{c}_i \cdot \underbrace{\alpha_i \cdot \prod_{j=1}^{i-1} (1 - \alpha_j)}_{\text{Net opacity of layer } i} . \tag{4}$$

3.2 Model Architecture

The goal of our deep model is to infer an RGBA MSI from a pair of ODS images (Fig. 2). We use a U-Net-style architecture [43,64] to perform MSI inference, with specific adjustments for the spherical domain. Our approach of inferring MSI alphas via a new ODS reprojection component conceptually corresponds to a soft 3D reconstruction [37], and is similar in idea to depth probability volumes [10].

Beyond alpha, we structure our network to additionally learn a blending weight between the left and right ODS views for each layer of the MSI, to be used within our reprojection method. This allows the network to blend between views as appropriate, and to fill holes with content from at least one input view. This lets us overcome occlusions in one view. In principle, this also handles specular highlights, reflections, and transparent content that does not correspond between views and so does not have a natural disparity. This is also useful during training or when inference is imperfect: any ghosting from combining left/right ODS views is minimized when the inferred alphas are opaque at or beyond the correct scene depth, and when the alphas are transparent in front of the correct scene depth.

ODS Reprojection: We first preprocess a left/right ODS pair I_L and I_R into a pair of *sphere sweep volumes* [20]: these are defined as a set of ERP images, each corresponding to the reprojection of concentric spheres with radii $\{r_i\}_{i=1}^N$ into one of the ODS images. We generate each ERP in the sphere sweep as follows:

1. Back-project ERP pixels (θ_i, φ_i) to points $\mathbf{p}_i$ on the sphere of radius r_i.
2. Project these points into the image I_L or I_R (Eqs. 2 and 3).
3. Look up pixel colors from I_L or I_R (bilinearly interpolated).

Blending: For each MSI layer, our network predicts alpha values and left/right ODS blending weights. Specifically, let $\mathcal{S}^{\mathrm{L/R}} = \{S_i^{\mathrm{L/R}}\}_{i=1}^N$ be the sphere sweep for the left/right ODS image, and let $\{\beta_i\}_{i=1}^N$ be the per-layer blending weights. Then, the colors $\{\mathbf{c}_i\}$ for MSI layer i are calculated via element-wise multiplication ($\odot$):

$$\mathbf{c}_i = \beta_i \odot S_i^{\mathrm{L}} + (1 - \beta_i) \odot S_i^{\mathrm{R}}. \tag{5}$$

Angle-aware Kernels: To create a training and inference approach which is efficient and implicitly aware of the angular distortions within ERP and ODS images, we provide the network with information about each pixel's relative location in the spherical projection using coordinate convolution [29]. Existing approaches provide two additional channels, U and V, to each convolutional layer within the network, with each containing the normalized azimuth and elevation at each pixel position [65]. However, the shape of the image distortion under ERP/ODS projection is independent of azimuth, and is symmetric in elevation around the equator. Thus we only use $V(x, y) = |\sin(\varphi(y))|$.

3.3 Training Losses

During training, we learn to assign alpha values and blending weights to each MSI layer by penalizing the L_2 re-rendering error between a predicted target view $\hat{I} = \text{Render}(\text{MSI}, P)$ for pose P and the ground-truth image I_{GT} at P:

$$\mathcal{L}_{\text{L2}} = \sum_{x,y} \left\| \hat{I}(x, y) - I_{\text{GT}}(x, y) \right\|_2^2. \tag{6}$$

Spherical Weighting: Applying an L_2 loss directly on equirectangular images puts disproportionate weight on regions near the poles as the projection does not conserve area. Instead, we use a spherical weighting scheme which weights pixels by their area on the sphere's surface. We generate a map of area weights A by projecting corner coordinates at pixel (x, y) into spherical coordinates (θ_0, θ_1) and (φ_0, φ_1), and then computing their subtended area on the unit sphere $(r = 1)$:

$$A(x, y) = r^2(\theta_1 - \theta_0)(\cos(\varphi_1) - \cos(\varphi_0)). \tag{7}$$

Given a target image I_{GT} and the predicted image $\hat{I} = \text{Render}(\text{MSI}, P)$ from the MSI at pose P, we then apply the spherical area weighting A to the L_2 loss:

$$\mathcal{L}_{\text{ERP-L2}} = \sum_{x,y} \left\| A(x, y) \cdot \left(\hat{I}(x, y) - I_{GT}(x, y) \right) \right\|_2^2. \tag{8}$$

Transform-inverse MSI Regularization: Per-frame processing can lead to undesirable flickering in videos. We improve the temporal consistency of our model with a spherical 3D procedure derived from the 2D image transform-inverse regularization approach [15]. The motivating idea is that predicting an image under a small transformation of the target view, and then transforming it back to the original target view, should only incur a small difference in appearance. Penalizing this difference then leads to smoothness over time, as small transformations mimic the minor frame-to-frame differences in a video. *Single-frame* temporal consistency methods are more efficient to train and infer than two-frame optical-flow-based methods [4] or recurrent network methods [26].

We develop a new approach to apply this to MPIs and MSIs. The input to our network is 3D rather than 2D, via two sphere sweep volumes, $\mathcal{S}^{\text{L}}$ and $\mathcal{S}^{\text{R}}$.

Likewise, our output is a 3D representation of a scene. As such, let us consider a 3D rigid body transformation on the inputs and outputs $T = [R \mid t]$, and let f be the function to infer our MSI representation. We would like to compute the loss:

$$\mathcal{L}_{\mathrm{TI}} = \left\| f(T(\mathcal{S}^{\mathrm{L}}), T(\mathcal{S}^{\mathrm{R}})) - T(f(\mathcal{S}^{\mathrm{L}}, \mathcal{S}^{\mathrm{R}})) \right\|_2 . \tag{9}$$

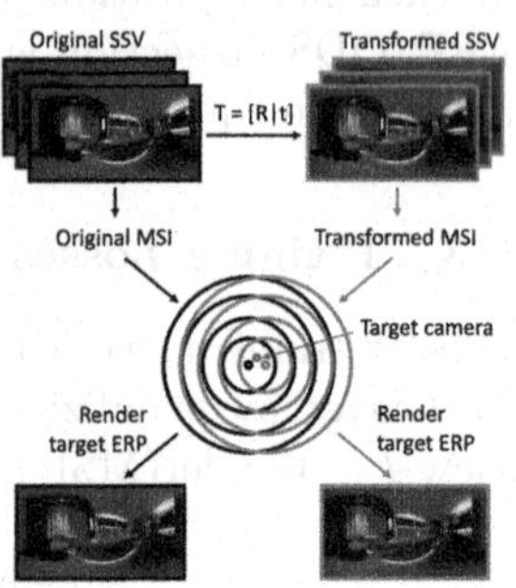

Applying T to the input corresponds to resampling the sphere sweeps for a set of concentric spheres transformed by T. This is easily accomplished by applying T to back-projected points in the sphere sweep volume generation process (Sect. 3.2). However, applying T to an output MSI is less straightforward as, given an MSI, we must determine a new MSI at a pose transformed by T. This requires interpolating alpha values and blending weights for the layers of the new MSI, while still preserving the correct output color along rays, which may blur features of the original MSI. Instead, our approach is to penalize the re-rendering loss for a target view with pose P between the MSI predicted for transformed inputs, and the original MSI:

$$\mathcal{L}_{\mathrm{TI}} = \left\| \mathrm{Render}(f(T(\mathcal{S}^{\mathrm{L}}), T(\mathcal{S}^{\mathrm{R}})), P) - \mathrm{Render}(T(f(\mathcal{S}^{\mathrm{L}}, \mathcal{S}^{\mathrm{R}})), P) \right\|_2 , \tag{10}$$

which is equivalent to

$$\mathcal{L}_{\mathrm{TI}} = \left\| \mathrm{Render}(f(T(\mathcal{S}^{\mathrm{L}}), T(\mathcal{S}^{\mathrm{R}})), P) - \mathrm{Render}(f(\mathcal{S}^{\mathrm{L}}, \mathcal{S}^{\mathrm{R}}), TP) \right\|_2 . \tag{11}$$

This can be computed without explicitly interpolating MSIs, and only requires transforming the target pose. If the re-renderings are close, then this implies the MSIs $f(T(\mathcal{S}_L), T(\mathcal{S}_R))$ and $T(f(\mathcal{S}_L, \mathcal{S}_R))$ are consistent scene representations.

Perceptual Loss: Finally, we experiment with replacing the $\mathcal{L}_{\mathrm{ERP\text{-}L2}}$ loss with a perceptual E-LPIPS [22] loss, which penalizes transformations of the image through comparisons of feature activations of the VGG network:

$$\mathcal{L}_{\mathrm{Per}} = \text{E-LPIPS}(\mathrm{Render}(\mathrm{MSI}, P), I_{\mathrm{GT}}). \tag{12}$$

Thus, our final loss is: $\mathcal{L}_{\mathrm{Final}} = \lambda_{\mathrm{Per}} \mathcal{L}_{\mathrm{Per}} + \lambda_{\mathrm{TI}} \mathcal{L}_{\mathrm{TI}}$, with $\lambda_{\mathrm{Per}} = 1, \lambda_{\mathrm{TI}} = 10$.

3.4 High-resolution Rendering

An advantage of any approach that directly predicts scene structure (MPIs, MSIs, meshes) is that, no matter the resolution of the output representation, it can be textured with a high-resolution image. We combine *multiple* high-resolution images (left and right ODS) using learned blending weights, which allows for high-resolution synthesis of both visible *and* single-view occluded regions, while maintaining real-time frame rates. We render a set of concentric spheres as meshes in Unity, textured with the alpha values described above and RGBs derived from combining the two high-resolution ODS images using the inferred low-resolution blending weights (Eq. 5).

Further, given an inferred MSI, we can apply a GPU-based joint bilateral upsampling filter [24] between the high-resolution 4K×2K blended RGB images at each layer and the lower-resolution 640×320 inferred alpha layers. This offers the following advantages:

1. It allows us to perform inference at low spatial resolution for speed, and then upsample the MSI alphas to match the higher-resolution input RGBs.
2. Training produces checkerboard artifacts through the architecture of the strided transpose convolutions [33]; filtering reduces these, which improves the quality of occlusion edges during view synthesis.

Different architecture choices are also possible to reduce checkerboarding, such as bilinear upsampling followed by convolution [52], but typically these are more expensive on the GPU at inference time ($\approx 2\times$ slower).

4 Experiments

We experiment by training our model on a dataset of indoor scenes with moving cameras. We quantitatively compare our model against ground-truth data with image quality and perceptual metrics. Further, we test our approach on ODS footage from real cameras collected online—please see our supplemental video.

Training Data: Existing view synthesis and expansion approaches rely on large databases of permissively-licensed perspective video [64], which are not available for stereo 360° imagery as the format is still nascent. Instead, we generate synthetic training data using the Replica dataset [49] for supervision from *target views*. Replica contains high-quality scenes with few holes or missing/incorrect geometry, as might be found in real-world scan databases [7,60]. In principle, we can use any projection for the target views during training (Sect. 3). Choosing an omnidirectional projection allows us to back-propagate our loss through all parts of the MSI simultaneously. If we trained on real video data, we would project inferred MSIs into target views by tracking the ODS video over time and selecting frames with desired poses. However, our ultimate goal is to produce views that are correct to each human eye within head-tracked VR. For this, ERP images more closely match our desire than ODS reprojections. Given that our data is synthetic, we exploit this fact and render ERP target views for supervision.

Data Generation: First, we develop a custom ERP and ODS renderer for Replica [49]. Then, we sample floor positions in Replica via uniformly randomly selecting from navigable positions available via the Facebook AI Habitat simulator [44]. For each floor position, we sample a vertical position from a Gaussian distribution over human heights. At each position, we render the camera's left-eye and right-eye ODS images, along with a triplet of ERP target views within the desired headbox for VR rendering. We render one *interpolation* target view

Table 1. Quantitative comparison of baseline methods (top) and ablations of our approach (bottom). We show single-image reconstruction error (E-LPIPS, SSIM, PSNR) on a Replica ODS test set. We report 1000×E-LPIPS [22] as 'mE-LPIPS' (milli-E-LPIPS). For training data, (P) is perspective views and (ODS) is omnidirectional stereo. '▲' means higher is better, '▼' means lower is better. Numbers are mean±standard error. Datasets: RealEstate10K [64], Replica [49], Stanford 2D-3D-S [2]. Best numbers in green. *90 Hz at 256×128 pixels ≈ 15 Hz at 640×320 pixels. †Please see text discussion.

Baseline/Ablation Model	Inference▲	Training Data	mE-LPIPS▼	SSIM▲	PSNR▲
Ground-truth depth+mesh rendering	N/A	(not trained)	3.08±6.17	0.93±0.05	26.75±3.73
Zhou et al. [64] adapted to ODS	2 Hz	RealEstate10K (P)	7.38±4.00	0.78±0.06	20.65±2.15
Double plane sweep (DPS-RE10K)	30 Hz	RealEstate10K (P)	4.28±3.00	0.90±0.04	27.52±3.52
Double plane sweep (DPS-Rep)	30 Hz	Replica (P)	7.46±3.86	0.80±0.06	21.55±2.34
ODS-Net [25]+mesh rendering	15 Hz*	Stanford 2D-3D-S	5.58±3.54	0.82±0.06	21.82±3.06
Ours (real-time)	30 Hz	Replica (ODS)	2.29±2.21	0.92±0.04	29.10±3.91
– CoordConv	30 Hz	Replica (ODS)	2.43±2.10	0.92±0.04	29.14±3.92
– E-LPIPS (using L2)	30 Hz	Replica (ODS)	2.88±2.53	0.92±0.04	29.82±3.91
Ours (with 2 × 32 = 64 MSI layers)	15 Hz	Replica (ODS)	2.45±2.25	0.92±0.04	29.25±3.89
Ours (graph conv. network)	2 Hz	Replica (ODS)	3.06±2.40	0.92±0.04	30.03±4.15
Ours (with background layer)	30 Hz†	Replica (ODS)	2.04±2.14	0.91±0.04	30.03±4.04

randomly within the ODS viewing circle, and two *extrapolation* target views at random offset from the camera. To reduce render aliasing as the synthetic data does not have mipmaps, we render each view at 4K×2K, apply a Gaussian blur with $\sigma = 3.2$ pixels, and downsample to our training resolution of 640×320. We render ERP images from 30,000 positions in total. Finally, we split the data 70%/30% across training and test sets by scene. See our supplement for details.

Image Metrics: To quantitatively compare our results against ground truth and other baselines, we use three evaluation metrics: mean E-LPIPS [22], SSIM [59], and PSNR, along with their standard error over the test set of frames. E-LPIPS is a relatively new metric, built upon LPIPS [62], which computes and compares VGG16 feature responses [47] under different perturbations by self-ensembling through random transformations. As a more advanced, learned 'perceptual' metric, its intent is to be more robust to transformations which are equivalent in PSNR and SSIM but noticeable to the human eye [14].

Baseline Comparisons: To our knowledge, no existing method infers MSIs from 360° imagery; we thus adapt existing baselines to our purpose and compare them in Table 1. Our first baseline creates novel views using textured mesh rendering with ground-truth depth maps. In practice, perfect depth is unattainable, but it provides an upper bound on the performance of depth-based re-rendering.

Zhou et al.'s. Released model [63] was trained on the extensive RealEstate10K dataset of perspective videos. Their network takes as input a *reference* view and a plane sweep volume for the *source* view, and generates an MPI for the reference view. We first naïvely extend their approach by providing the left ODS image as the reference view and a sphere sweep volume of the right

ODS image as the source view. This leads to bad performance across all metrics in Table 1 and highlights the mismatch between Zhou et al.'s architecture and our desire to estimate an MSI at the center of the camera system from ODS imagery.

Double-plane-sweep Baseline. To better represent the nature of ODS imagery, we create an adaption of Zhou et al.'s method that takes as input a plane sweep volume for each of the two views, and produces a multi-plane image. We train this model on perspective views from the synthetic Replica and the natural RealEstate10K dataset. Please see the supplement for details. Both datasets have similar content, but RealEstate10K is more varied and not synthetic. At test time, we apply this model to sphere sweep volumes of the left/right ODS views, exploiting the 'pseudo-perspective' projection in the equatorial region of ODS imagery. Despite the mismatch between perspective training and ODS testing, the model trained on the RealEstate10K dataset performs well, and clearly better than the model that was trained on the less varied Replica dataset.

ODS-net. [25] is a real-time learning-based method specifically aimed at 6DoF video generation. This method predicts a 256×128 depth map per video frame, and synthesizes views by rendering a mesh based on the depth map (without inpainting). Note that we provide double-ERP input, as expected by ODS-Net, while all other comparisons use ODS input. In our experiments, ODS-Net failed to produce metrically accurate depth on our Replica test data, and performed worse than mesh-based rendering with ground-truth depth in Table 1.

Serrano et al. [46] is an offline optimization-based method for 6DoF video generation from RGBD ERP video input. The method produces a set of RGBD+α layers for real-time rendering of novel views. While their results show sharper occlusion boundaries and more accurate parallax, our method can better synthesize occluded content by extrapolating from both the left and right ODS images (see Fig. 3). As we do not assume that depth is available a priori, our method is also applicable to in-the-wild ODS videos without any additional preprocessing steps, such as depth estimation. Our method also performs inference in real time at 30 Hz, while Serrano et al.'s method requires one minute per frame. Further, they assume a static camera for layer computation, while our method can, with sufficient training data, be applied to scenes with arbitrary camera motion.

Ablations: Table 1 also shows quantitative measures for ablations of our approach, which reduce the important perceptual E-LPIPS score.

GCN. Our first ablation replaces convolutions directly on the ODS domain with a graph convolutional network (GCN) on a sphere mesh with approximately as many vertices as image pixels (164K). We project the sphere sweep volumes into the spherical basis via this mesh, which then uses GCN layers (in architecture of Pixel2Mesh [58]) to transform them into per-vertex alpha/blending weights, which are projected back to ERP for the MSI. This approach is (almost) rotation equivariant, and performs well in terms of the metrics in Table 1. However, inference times are slow and currently cannot be accelerated to real-time with TensorRT due to its lack of sparse matrix support for graph convolutions.

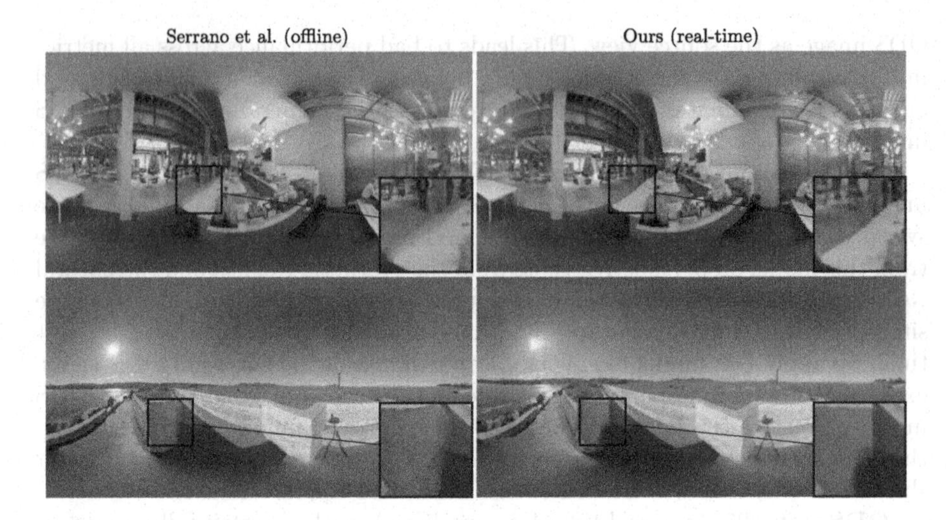

Fig. 3. Qualitative comparison between our results (*right*) and Serrano et al. [46] (*left*). Our method exhibits improved synthesis of occluded content *(first row)*. As their method uses RGBD input, and includes a depth preprocessing step, it predicts disparity with higher accuracy, especially in challenging textureless regions *(second row)*.

With Background Layer. We predict an additional RGB layer which can help to inpaint extrapolated disocclusions [64]. This improves PSNR and E-LPIPS scores in Table 1. However, at runtime, there are two issues to combine this with our high-resolution input videos: 1) the background layer is limited to the inference resolution, and so looks blurry in relation, and 2) we must explicitly find disoccluded regions to blend in this background layer, which complicates and slows virtual view render times. This is an application-level trade off.

Temporal Consistency. We generate five ground-truth video sequences with moving cameras to measure temporal consistency. Then, we apply a low-pass filter (with $\sigma = 11$ pixels) to the output videos, and compute absolute frame-to-frame differences ('f2f' metrics) between consecutive frames and depth maps, which detects temporal inconsistencies such as flickering. As we increase the transformation range for transform-inverse regularization, f2f RGB metrics improve while image metrics degrade slightly due to increased blur in the output MSI RGBs and alphas. Beyond RGB, depth consistency improves more significantly—this is important since we desire accurate and consistent re-rendering and disparity cues for consecutive frames, and for different eye IPDs in VR.

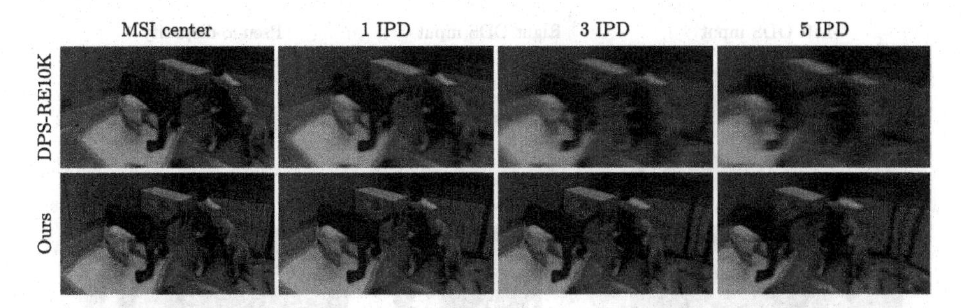

Fig. 4. Perspective reprojections of the *puppy* video in our real-time renderer with our model (*bottom*) and the baseline perspectively-trained model (*top*; row three of Table 1). Our results are significantly less blurry, though both models have difficulty inferring fine features, like the puppy guard wires, as they operate at lower inference resolutions.

Table 2. Quantitative trade-off of transform-inverse regularization on temporal consistency and image quality, at $\lambda_{TI} = 10$. 'f2f-(depth—rgb)' measures the average frame-to-frame difference between consecutive low-pass-filtered depth maps or RGB images as measure of temporal consistency. '▲' means higher is better, '▼' means lower is better. Numbers are mean±standard error. 'Transform range' indicates the scale factor on the transformed pose, where 1× is: translation (x, y, z)±0.01 metres; rotation $(\theta, \phi, \psi) \pm 1.7°$.

Transform range	f2f-depth▼	f2f-rgb▼	mE-LPIPS▼	SSIM▲	PSNR▲
None	3.27±0.70	1.27±0.16	2.88±2.53	0.92±0.04	29.82±3.92
×0.5	1.64±0.37	1.29±0.16	2.37±2.18	0.92±0.04	29.74±4.07
×1.0	1.65±0.41	1.29±0.16	2.51±2.21	0.92±0.04	29.62±4.01
×2.0	1.48±0.28	1.29±0.16	2.59±2.27	0.92±0.04	29.52±4.19
×5.0	1.04±0.27	1.28±0.16	2.87±2.54	0.91±0.04	29.20±4.20

Qualitative Results: First, we show qualitative comparisons for reprojected views against Zhou et al.'s method [64] augmented with a double-plane-sweep input architecture (Fig. 4). These use our real-time pipeline with high-resolution input videos applied to the MSI. Our approach produces results which are visually sharper as we extrapolate the virtual camera away from the baseline at 3× and 5× interpupillary distance (IPD). In Figs. 5 and 6, we show the output of our approach at the inference resolution of 640×320, i.e., without the full real-time pipeline which uses the high-resolution input video. We generate the MSI representation at the center of the ODS camera system, and then generate ODS imagery at 1× IPD to show reconstruction, and at 3× and 5× IPD to show extrapolation (a similar max. distance to Zhou et al. [64]). In our supplement, we show results on synthetic and real ODS video test sequences.

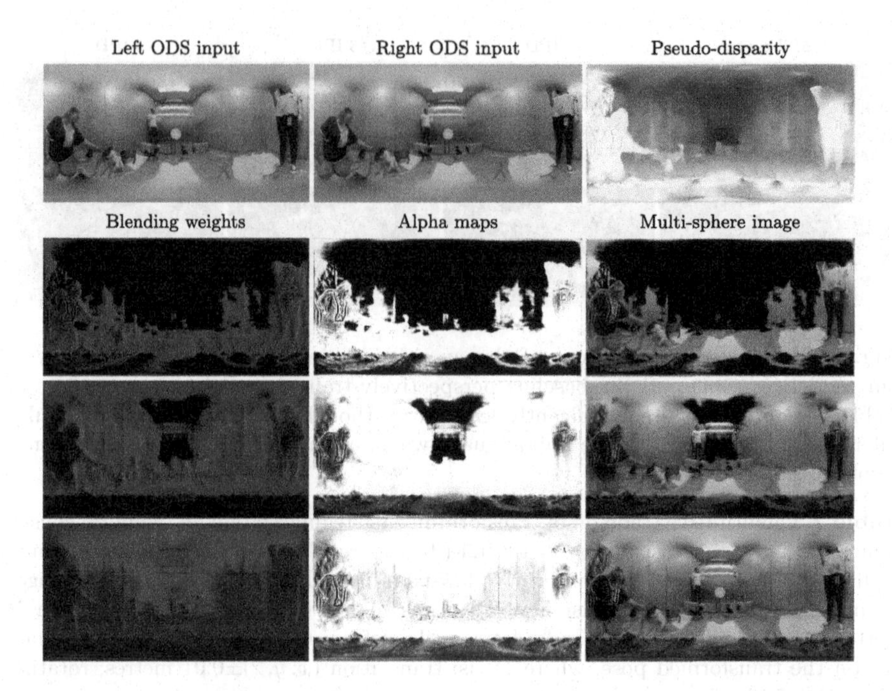

Fig. 5. Inferred MSI representation for the *Puppy* video. Blending weights are red for left ODS and blue for right ODS. Alpha maps are black for transparent and white for opaque. Each row shows a single layer (out of 32) at the near, mid, and far extents of the scene; content exists across all layers to produce the final result. (Color figure online)

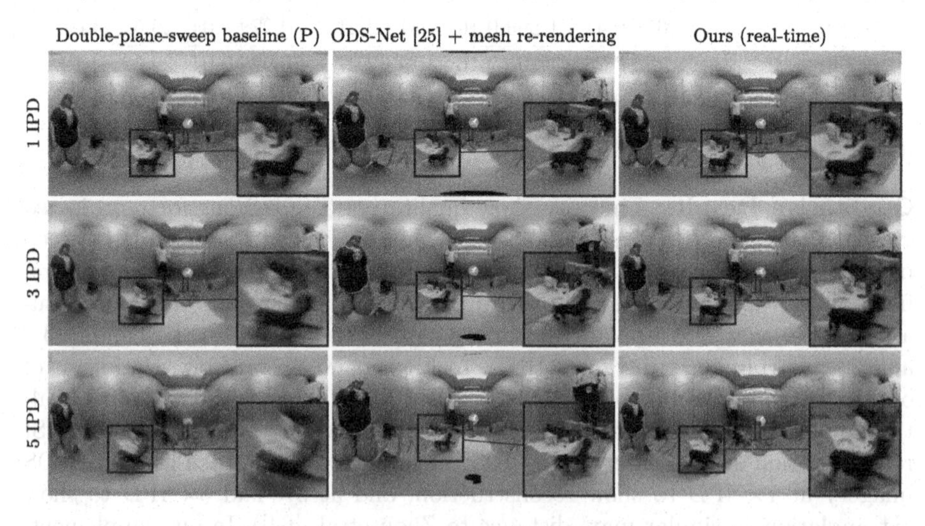

Fig. 6. *Puppy* video results. Inference for low-resolution (640×320) MSI representation for comparison on spherical images (not our high-resolution real-time perspective VR view). We compare our results on the right to the baseline double-plane-sweep baseline method, and to ODS-Net with mesh re-rendering which induces high distortion.

5 Discussion and Conclusion

Via the MSI representation and our learned network, our approach provides real-time inference on ODS video, and stereo VR rendering at 80 Hz. On our test set, we demonstrate results on baselines up to 5.6× larger than ODS interpupillary distance, and produce the highest quantitative performance. In a VR headset, the user's motion is unconstrained and can lead to *much* larger baselines than any current technique can handle [10,48], especially in real-time. The effect is to 'see behind the curtain': our representation still provides correct disparity and motion parallax for scene objects with accurate alpha (which improves comfort), but large disocclusions reveal the layered MSI structure underneath (see supplement).

Some multi-camera systems can see more scene content than is captured in the final ODS projection, and designing a system to start from raw camera feeds would allow this content to contribute to a reconstruction. Our proposed approach could be adapted to work with per-camera feeds, which we leave for future work. Our approach is applicable to online 'in the wild' ODS videos as well as ODS live streams supported by several off-the-shelf cameras.

Limitations: The quality at larger extrapolations is currently limited by computational trade-offs both during training and in real-time application. First, our per-layer ODS blending for re-rendering could be improved by flow-based interpolation methods [3], which would be required during both training and inference. Second, inferring higher-resolution MSI representations is possible at a slower speed with a network with more layers [6,17,48] and so a larger receptive field. Third, an explicit spherical convolution approach may improve quality, but current approaches are too expensive for real-time applications. Further, stereo 360° video is a nascent format, and there is insufficient training data available. Given our training data, the proposed method works best on indoor scenes with moving cameras, and more work must be done for dynamic scene objects.

MSI depth discretization limits the range of non-aliased views, beyond which the layered nature of the MSI becomes noticeable [48]. Increasing only the spatial resolution of the RGB for each MSI layer via the high-resolution input video increases this effect. However, this approach to VR rendering can be a better trade-off in terms of quality and comfort than keeping low-resolution imagery.

Conclusion: Stereo 360° video is only 'half-way' to comfortable (and thus useful) representations of our environments, with the missing piece being 6DoF video. Our work suggests one solution to this problem by learning to create representations which implicitly compute depth and fill disoccluded regions. Our end-to-end system takes images from a stereo 360° camera and converts them into a 6DoF multi-sphere image representation in real time for viewing. This learns how to distribute the scene over different depths per frame, and ensures temporal consistency. We show competitive quantitative metrics for image quality while remaining fast in inference speed—this is important for situations where preprocessing is not an option, like communications and robotic teleoperation.

Overall, we move towards a more natural 6DoF viewing experience for stereo 360° video.

Acknowledgment. We thank Ana Serrano for help with RGB-D comparisons and Eliot Laidlaw for improving the Unity renderer. We thank Frédéric Devernay, Brian Berard, and an Amazon Research Award, and NVIDIA for a GPU donation. This work was supported by a Brown OVPR Seed Award, RCUK grant CAMERA (EP/M023281/1), and an EPSRC-UKRI Innovation Fellowship (EP/S001050/1).

References

1. Anderson, R., et al.: Jump: virtual reality video. ACM Trans. Graph. **35**(6), 198:1–198:13 (2016). https://doi.org/10.1145/2980179.2980257
2. Armeni, I., Sax, S., Zamir, A.R., Savarese, S.: Joint 2D–3D-semantic data for indoor scene understanding (2017). http://arxiv.org/abs/1702.01105arXiv:1702.01105
3. Bertel, T., Campbell, N.D.F., Richardt, C.: MegaParallax: casual 360° panoramas with motion parallax. TVCG **25**(5), 1828–1835 (2019). https://doi.org/10.1109/TVCG.2019.2898799
4. Bonneel, N., Tompkin, J., Sunkavalli, K., Sun, D., Paris, S., Pfister, H.: Blind video temporal consistency. ACM Trans. Graph. **34**(6), 196:1–196:9 (2015). https://doi.org/10.1145/2816795.2818107
5. Brown, M., Lowe, D.G.: Automatic panoramic image stitching using invariant features. IJCV **74**(1), 59–73 (2007). https://doi.org/10.1007/s11263-006-0002-3
6. Broxton, M., et al.: Immersive light field video with a layered mesh representation. ACM Trans. Graph. **39**(4), 86:1–86:15 (2020)
7. Chang, A., et al.: Matterport3D: learning from RGB-D data in indoor environments. In: 3DV, pp. 667–676 (2017). https://doi.org/10.1109/3DV.2017.00081
8. Chapdelaine-Couture, V., Roy, S.: The omnipolar camera: a new approach to stereo immersive capture. In: ICCP (2013). https://doi.org/10.1109/ICCPhot.2013.6528311
9. Cheng, H.T., Chao, C.H., Dong, J.D., Wen, H.K., Liu, T.L., Sun, M.: Cube padding for weakly-supervised saliency prediction in 360° videos. In: CVPR, pp. 1420–1429 (2018). https://doi.org/10.1109/CVPR.2018.00154
10. Choi, I., Gallo, O., Troccoli, A., Kim, M.H., Kautz, J.: Extreme view synthesis. In: ICCV, pp. 7780–7789 (2019). https://doi.org/10.1109/ICCV.2019.00787
11. Cohen, T.S., Geiger, M., Koehler, J., Welling, M.: Spherical CNNs. In: ICLR (2018)
12. Coors, B., Condurache, A.P., Geiger, A.: SphereNet: learning spherical representations for detection and classification in omnidirectional images. In: ECCV, pp. 518–533 (2018). https://doi.org/10.1007/978-3-030-01240-3_32
13. Dodgson, N.A.: Variation and extrema of human interpupillary distance. In: Stereoscopic Displays and Virtual Reality Systems (2004). https://doi.org/10.1117/12.529999
14. Dosselmann, R., Yang, X.D.: A comprehensive assessment of the structural similarity index. Signal Image Video Process. **5**(1), 81–91 (2011). https://doi.org/10.1007/s11760-009-0144-1
15. Eilertsen, G., Mantiuk, R.K., Unger, J.: Single-frame regularization for temporally stable CNNs. In: CVPR, pp. 11168–11177 (2019). https://doi.org/10.1109/CVPR.2019.01143

16. Esteves, C., Allen-Blanchette, C., Makadia, A., Daniilidis, K.: Learning SO(3) equivariant representations with spherical CNNs. In: ECCV, pp. 52–68 (2018). https://doi.org/10.1007/978-3-030-01261-8_4
17. Flynn, J., et al.: DeepView: view synthesis with learned gradient descent. In: CVPR, pp. 2367–2376 (2019). https://doi.org/10.1109/CVPR.2019.00247
18. Google Inc.: Rendering omni-directional stereo content (2015). https://developers.google.com/vr/jump/rendering-ods-content.pdf
19. Huang, J., Chen, Z., Ceylan, D., Jin, H.: 6-DOF VR videos with a single 360-camera. In: IEEE VR, pp. 37–44 (2017). https://doi.org/10.1109/VR.2017.7892229
20. Im, S., Ha, H., Rameau, F., Jeon, H., Choe, G., Kweon, I.: All-around depth from small motion with a spherical panoramic camera. In: Leibe, B., Matas, J., Sebe, N., Welling, M. (eds.) ECCV 2016. LNCS, vol. 9907, pp. 156–172. Springer, Cham (2016). https://doi.org/10.1007/978-3-319-46487-9_10
21. Ishiguro, H., Yamamoto, M., Tsuji, S.: Omni-directional stereo. TPAMI **14**(2), 257–262 (1992). https://doi.org/10.1109/34.121792
22. Kettunen, M., Härkönen, E., Lehtinen, J.: E-LPIPS: robust perceptual image similarity via random transformation ensembles (2019). http://arxiv.org/abs/1906.03973arXiv:1906.03973
23. Konrad, R., Dansereau, D.G., Masood, A., Wetzstein, G.: SpinVR: towards live-streaming 3D virtual reality video. ACM Trans. Graph. **36**(6), 209:1–209:12 (2017). https://doi.org/10.1145/3130800.3130836
24. Kopf, J., Cohen, M.F., Lischinski, D., Uyttendaele, M.: Joint bilateral upsampling. ACM Trans. Graph. **26**(3), 96 (2007). https://doi.org/10.1145/1276377.1276497
25. Lai, P.K., Xie, S., Lang, J., Laganière, R.: Real-time panoramic depth maps from omni-directional stereo images for 6 DoF videos in virtual reality. In: IEEE VR, pp. 405–412 (2019). https://doi.org/10.1109/VR.2019.8798016
26. Lai, W.S., Huang, J.B., Wang, O., Shechtman, E., Yumer, E., Yang, M.H.: Learning blind video temporal consistency. In: ECCV, pp. 170–185 (2018). https://doi.org/10.1007/978-3-030-01267-0_11
27. Lee, J., Kim, B., Kim, K., Kim, Y., Noh, J.: Rich360: optimized spherical representation from structured panoramic camera arrays. ACM Trans. Graph. **35**(4), 63:1–63:11 (2016). https://doi.org/10.1145/2897824.2925983
28. Lee, Y.K., Jeong, J., Yun, J.S., June, C.W., Yoon, K.J.: SpherePHD: applying CNNs on a spherical PolyHeDron representation of 360° images. In: CVPR, pp. 9173–9181 (2019). https://doi.org/10.1109/CVPR.2019.00940
29. Liu, R., et al.: An intriguing failing of convolutional neural networks and the CoordConv solution. In: NeurIPS (2018)
30. Luo, B., Xu, F., Richardt, C., Yong, J.H.: Parallax360: stereoscopic 360° scene representation for head-motion parallax. TVCG **24**(4), 1545–1553 (2018). https://doi.org/10.1109/TVCG.2018.2794071
31. Matzen, K., Cohen, M.F., Evans, B., Kopf, J., Szeliski, R.: Low-cost 360 stereo photography and video capture. ACM Trans. Graph. **36**(4), 148:1–148:12 (2017). https://doi.org/10.1145/3072959.3073645
32. Mildenhall, B., et al.: Local light field fusion: practical view synthesis with prescriptive sampling guidelines. ACM Trans. Graph. **38**(4), 29:1–29:14 (2019). https://doi.org/10.1145/3306346.3322980
33. Odena, A., Dumoulin, V., Olah, C.: Deconvolution and checkerboard artifacts. Distill (2016). 10.23915/distill.00003
34. Padmanaban, N., Ruban, T., Sitzmann, V., Norcia, A.M., Wetzstein, G.: Towards a machine-learning approach for sickness prediction in 360° stereoscopic videos. TVCG **24**(4), 1594–1603 (2018). https://doi.org/10.1109/TVCG.2018.2793560

35. Parra Pozo, A., et al.: An integrated 6DoF video camera and system design. ACM Trans. Graph. **38**(6), 216:1–216:16 (2019). https://doi.org/10.1145/3355089.3356555
36. Peleg, S., Ben-Ezra, M., Pritch, Y.: Omnistereo: panoramic stereo imaging. TPAMI **23**(3), 279–290 (2001). https://doi.org/10.1109/34.910880
37. Penner, E., Zhang, L.: Soft 3D reconstruction for view synthesis. ACM Trans. Graph. **36**(6), 235:1–235:11 (2017). https://doi.org/10.1145/3130800.3130855
38. Perazzi, F., et al.: Panoramic video from unstructured camera arrays. Comput. Graph. Forum **34**(2), 57–68 (2015). https://doi.org/10.1111/cgf.12541
39. Porter, T., Duff, T.: Compositing digital images. Comput. Graph. (Proc. SIGGRAPH) **18**(3), 253–259 (1984). https://doi.org/10.1145/800031.808606
40. Richardt, C.: Omnidirectional stereo. In: Ikeuchi, K. (ed.) Computer Vision: A Reference Guide. Springer, Berlin (2020). https://doi.org/10.1007/978-3-030-03243-2_808-1
41. Richardt, C., Pritch, Y., Zimmer, H., Sorkine-Hornung, A.: Megastereo: constructing high-resolution stereo panoramas. In: CVPR, pp. 1256–1263 (2013). https://doi.org/10.1109/CVPR.2013.166
42. Richardt, C., Tompkin, J., Halsey, J., Hertzmann, A., Starck, J., Wang, O.: Video for virtual reality. In: SIGGRAPH Courses (2017). https://doi.org/10.1145/3084873.3084894
43. Ronneberger, O., Fischer, P., Brox, T.: U-Net: convolutional networks for biomedical image segmentation. In: MICCAI, pp. 234–241 (2015). https://doi.org/10.1007/978-3-319-24574-4_28
44. Savva, M., et al.: Habitat: a platform for embodied AI research. In: ICCV, pp. 9339–9347 (2019). https://doi.org/10.1109/ICCV.2019.00943
45. Schroers, C., Bazin, J.C., Sorkine-Hornung, A.: An omnistereoscopic video pipeline for capture and display of real-world VR. ACM Trans. Graph. **37**(3), 37:1–37:13 (2018). https://doi.org/10.1145/3225150
46. Serrano, A., et al.: Motion parallax for 360° RGBD video. TVCG **25**(5), 1817–1827 (2019). https://doi.org/10.1109/TVCG.2019.2898757
47. Simonyan, K., Zisserman, A.: Very deep convolutional networks for large-scale image recognition. In: ICLR (2015)
48. Srinivasan, P.P., Tucker, R., Barron, J.T., Ramamoorthi, R., Ng, R., Snavely, N.: Pushing the boundaries of view extrapolation with multiplane images. In: CVPR, pp. 175–184 (2019). https://doi.org/10.1109/CVPR.2019.00026
49. Straub, J., et al.: The replica dataset: a digital replica of indoor spaces (2019). http://arxiv.org/abs/1906.05797arXiv:1906.05797
50. Su, Y.C., Grauman, K.: Learning spherical convolution for fast features from 360° imagery. In: NIPS (2017)
51. Su, Y.C., Grauman, K.: Kernel transformer networks for compact spherical convolution. In: CVPR, pp. 9442–9451 (2019). https://doi.org/10.1109/CVPR.2019.00967
52. Sugawara, Y., Shiota, S., Kiya, H.: Super-resolution using convolutional neural networks without any checkerboard artifacts. In: ICIP, pp. 66–70 (2018). https://doi.org/10.1109/ICIP.2018.8451141
53. Szeliski, R.: Image alignment and stitching: a tutorial. Found. Trends Comput. Graph. Vis. **2**(1), 1–104 (2006). https://doi.org/10.1561/0600000009
54. Tateno, K., Navab, N., Tombari, F.: Distortion-aware convolutional filters for dense prediction in panoramic images. In: ECCV, pp. 732–750 (2018). https://doi.org/10.1007/978-3-030-01270-0_43

55. Thatte, J., Boin, J.B., Lakshman, H., Girod, B.: Depth augmented stereo panorama for cinematic virtual reality with head-motion parallax. In: ICME (2016). https://doi.org/10.1109/ICME.2016.7552858
56. Thatte, J., Girod, B.: Towards perceptual evaluation of six degrees of freedom virtual reality rendering from stacked OmniStereo representation. Electron. Imaging **2018**(5), 1-6–352 (2018). https://doi.org/10.2352/ISSN.2470-1173.2018.05.PMII-352
57. Wang, F.E., et al.: Self-supervised learning of depth and camera motion from 360° videos. In: ACCV, pp. 53–68 (2018). https://doi.org/10.1007/978-3-030-20873-8_4
58. Wang, N., Zhang, Y., Li, Z., Fu, Y., Liu, W., Jiang, Y.G.: Pixel2Mesh: generating 3D mesh models from single RGB images. In: ECCV, pp. 52–67 (2018). https://doi.org/10.1007/978-3-030-01252-6_4
59. Wang, Z., Bovik, A.C., Sheikh, H.R., Simoncelli, E.P.: Image quality assessment: from error visibility to structural similarity. IEEE Trans. Image Process. **13**(4), 600–612 (2004). https://doi.org/10.1109/TIP.2003.819861
60. Xia, F., Zamir, A.R., He, Z., Sax, A., Malik, J., Savarese, S.: Gibson Env: real-world perception for embodied agents. In: CVPR, pp. 9068–9079 (2018). https://doi.org/10.1109/CVPR.2018.00945
61. Zhang, C., Liwicki, S., Smith, W., Cipolla, R.: Orientation-aware semantic segmentation on icosahedron spheres. In: ICCV, pp. 3533–3541 (2019). https://doi.org/10.1109/ICCV.2019.00363
62. Zhang, R., Isola, P., Efros, A.A., Shechtman, E., Wang, O.: The unreasonable effectiveness of deep features as a perceptual metric. In: CVPR (2018). https://doi.org/10.1109/CVPR.2018.00068
63. Zhou, H., Ummenhofer, B., Brox, T.: DeepTAM: deep tracking and mapping. In: ECCV, pp. 822–838 (2018). https://doi.org/10.1007/978-3-030-01270-0_50
64. Zhou, T., Tucker, R., Flynn, J., Fyffe, G., Snavely, N.: Stereo magnification: learning view synthesis using multiplane images. ACM Trans. Graph. **37**(4), 65:1–65:12 (2018). https://doi.org/10.1145/3197517.3201323
65. Zioulis, N., Karakottas, A., Zarpalas, D., Alvarez, F., Daras, P.: Spherical view synthesis for self-supervised 360° depth estimation. In: 3DV, pp. 690–699 (2019). https://doi.org/10.1109/3DV.2019.00081
66. Zioulis, N., Karakottas, A., Zarpalas, D., Daras, P.: OmniDepth: dense depth estimation for indoors spherical panoramas. In: ECCV, pp. 448–465 (2018). https://doi.org/10.1007/978-3-030-01231-1_28

Learning and Aggregating Deep Local Descriptors for Instance-Level Recognition

Giorgos Tolias(✉), Tomas Jenicek, and Ondřej Chum

Visual Recognition Group, Faculty of Electrical Engineering, Czech Technical University, Prague, Czech Republic
giorgos.tolias@cmp.felk.cvut.cz

Abstract. We propose an efficient method to learn deep local descriptors for instance-level recognition. The training only requires examples of positive and negative image pairs and is performed as metric learning of sum-pooled global image descriptors. At inference, the local descriptors are provided by the activations of internal components of the network. We demonstrate why such an approach learns local descriptors that work well for image similarity estimation with classical efficient match kernel methods. The experimental validation studies the trade-off between performance and memory requirements of the state-of-the-art image search approach based on match kernels. Compared to existing local descriptors, the proposed ones perform better in two instance-level recognition tasks and keep memory requirements lower. We experimentally show that global descriptors are not effective enough at large scale and that local descriptors are essential. We achieve state-of-the-art performance, in some cases even with a backbone network as small as ResNet18.

Keywords: Deep local descriptors · Deep local features · Efficient match kernel · ASMK · Image retrieval · Instance-level recognition

1 Introduction

Instance-level recognition tasks are dealing with a very large number of classes and relatively small intra-class variability. Typically, even instance-level classification tasks are cast as instance-level search in combination with nearest neighbor classifiers. The first instance-level search approach to achieve good performance, *i.e.* Video Google [42], is based on local features and the Bag-of-Words (BoW) representation. Representing images as collections of vector-quantized descriptors of local features allows for efficient spatial verification [32], which turns out to be a key ingredient in search for small objects. Follow-up approaches improve the BoW paradigm either with finer quantization and better matching schemes [2,20,44] or with compact global descriptors generated through aggregation [1,22]. Good performance is achieved even without spatial verification.

A. Vedaldi et al. (Eds.): ECCV 2020, LNCS 12346, pp. 460–477, 2020.
https://doi.org/10.1007/978-3-030-58452-8_27

Fig. 1. Learned local features and descriptors matched with ASMK. Features assigned to the same visual word (65k words codebook) are shown in the same color; only top 20 common visual words (out of 94) are included. Accurate localization is not required since we do not use spatial verification to perform instance-level recognition.

The advent of deep networks made it easy to generate and train global image descriptors. A variety of approaches exist [3,14,16,28,35,50] that differ in the training data, in the loss function, or in the global pooling operation. However, the performance of global descriptors deteriorates for very large collections of images. Noh *et al.* [29] are the first to exploit the flexibility of global descriptor training in order to obtain local features and descriptors, called DELF, for the task of instance-level recognition. DELF descriptors are later shown [34] to achieve top performance when combined with the state-of-the-art image search approach, *i.e.* the Aggregated Selective Match Kernel (ASMK) [44]. Compared to compact global descriptors, this comes with higher memory requirements and search time cost. In contrast to other learned local feature detectors [13] that use keypoint-level supervision and non-maxima suppression, DELF features are not precisely localized and suffer from redundancy since deep network activations are typically spatially correlated.

In this work[1], we propose a local feature detector and descriptor based on a deep network. It is trained through metric learning of a global descriptor with image-level annotation. We design the architecture and the loss function so that the local features and their descriptors are suitable for matching with ASMK, see Fig. 1. ASMK is known to deliver good performance even without spatial verification, *i.e.* precise feature localization is not crucial, and it deals well with repeated or bursty features. Therefore, the common drawbacks of existing deep local features for instance-level recognition are overcome. Unlike classical local features that attempt to offer precise localization to extract reliable descriptors, multiple nearby locations can give rise to a similar descriptor in our training; multiple similar responses are not suppressed, but averaged.

The main contribution of this work is the proposed combination of deep feature detector and descriptor with ASMK matching, which outperforms existing global and local descriptors on two instance-level recognition tasks, *i.e.* classifi-

[1] https://github.com/gtolias/how.

cation and search, in the domain of landmarks. Our ablation study shows that the proposed components reduce the memory required by ASMK. The learned local descriptors outperform by far deep global descriptors as well as other deep local descriptors combined with ASMK. Finally, we provide insight into why the image-level optimization is relevant for local-descriptors and ASMK matching.

2 Related Work

We review the related work in learning global or local descriptors for instance-level matching task and local descriptors for registration tasks.

Global Descriptors. A common approach to obtain global image descriptors with deep fully-convolutional neural networks is to perform global pooling on 3D feature maps. This approach is applied to activations generated by pre-trained networks [4,23,45] or end-to-end learned networks [14,15,35]. One of the first examples is SPoC descriptor by Babenko and Lempitsky [4] that is generated by simple global sum-pooling. Weighted sum-pooling is performed in CroW by Kalantidis *et al.* [23], where the weights are given by the magnitude of the activation vectors at each spatial location of the feature map. Such a 2D map of weights, seen as an attention map, is related to our approach as discussed in Sect. 4.2. Inspired by classical embeddings, Arandjelović *et al.* [3] extend the VLAD [22] descriptor to NetVLAD. Its contextually re-weighted counterpart, proposed by Kim and Frahm [24], introduces a learned attention map which is generated by a small network.

Local Features and Descriptors for Instance-level Recognition. Numerous classical approaches that are based on hand-crafted local features [25,27] and descriptors [8,26] exist in the literature of instance-level search [30,32,42,44,51]. Inspired by classical feature detection, Simeoni *et al.* [41] perform MSER [25] detection on activation maps. The features detected at one feature channel are used as tentative correspondences, hence no descriptors are required; the approach is applicable to any network and does not require learning. Learning of attentive deep local features (DELF) is introduced in the work of Noh *et al.* [29]. A global descriptor is derived from a network that learns to attend on feature map positions. The global descriptor is optimized with category-level labels and classification loss. At test time, locations with the strongest attention scores are selected while the descriptors are the activation vectors at the selected locations. This approach is highly relevant to ours. We therefore provide a number of different ablation experiments to reveal the key differences. A recent variant shows that it is possible to jointly learn DELF-like descriptors and global descriptors with a single model [11]. A similar achievement appears in the work of Yang *et al.* [49] with a scope that goes beyond instance-level recognition and covers image registration too.

Local Features and Descriptors for Registration. A richer line of work exists in learning local feature detection and description for image registration where denser point correspondences are required. As in the previous tasks, some methods do not require any learning and are applicable on any pre-trained network. This is the case of the work Benbihi *et al.* [9] where activation magnitudes are back-propagated to the input image and local-maxima are detected. Learning is performed with or without labeling at the local level in a number of different approaches [7,10,12,13,38,47]. A large number of features is typically required for good performance. This line of research differentiates from our work; our focus is on large-scale instance-level recognition where memory requirements matter.

3 Background

In this section, the binarized versions of Selective Match Kernel (SMK) and its extension, the Aggregated Selective Match Kernel (ASMK) [44][2], are reviewed as the necessary background. This paper exploits the ASMK indexing and retrieval.

In SMK, an image is represented by a set $\mathcal{X} = \{\mathbf{x} \in \mathbb{R}^d\}$ of $n = |\mathcal{X}|$ d-dimensional local descriptors. The descriptors are quantized by k-means quantizer $q : \mathbb{R}^d \rightarrow \mathcal{C} \subset \mathbb{R}^d$, where $\mathcal{C} = \{\mathbf{c}_1, \ldots, \mathbf{c}_k\}$ is a codebook comprising $|\mathcal{C}|$ vectors (visual words). Descriptor $\mathbf{x}$ is assigned to its nearest visual word $q(\mathbf{x})$. We denote by $\mathcal{X}_\mathbf{c} = \{x \in \mathcal{X} : q(x) = \mathbf{c}\}$ the subset of descriptors in $\mathcal{X}$ that are assigned to visual word $\mathbf{c}$, and by $\mathcal{C}_\mathcal{X}$ the set of all visual words that appear in $\mathcal{X}$. Descriptor $\mathbf{x}$ is mapped to a binary vector through function $b : \mathbb{R}^d \rightarrow \{-1, 1\}^d$ given by $b(\mathbf{x}) = \text{sign}(r(\mathbf{x}))$, where $r(\mathbf{x}) = \mathbf{x} - q(\mathbf{x})$ is the residual vector w.r.t. the nearest visual word and sign is the element-wise sign function.

The SMK similarity of two images, represented by $\mathcal{X}$ and $\mathcal{Y}$ respectively, is estimated by cross-matching all pairs of local descriptors with match kernel

$$S_{\text{SMK}}(\mathcal{X}, \mathcal{Y}) = \gamma(\mathcal{X})\,\gamma(\mathcal{Y}) \sum_{\mathbf{x}\in\mathcal{X}} \sum_{\mathbf{y}\in\mathcal{Y}} [q(\mathbf{x}) = q(\mathbf{y})] k(b(\mathbf{x}), b(\mathbf{y})), \tag{1}$$

where $[\cdot]$ is the Iverson bracket, $\gamma(\mathcal{X})$ is a scalar normalization that ensures unit self-similarity[3], *i.e.* $S_{\text{SMK}}(\mathcal{X}, \mathcal{X}) = 1$. Function $k : \{-1, 1\}^d \times \{-1, 1\}^d \rightarrow [0, 1]$ is given by

$$k(b(\mathbf{x}), b(\mathbf{y})) = \begin{cases} \left(\frac{b(\mathbf{x})^\top b(\mathbf{y})}{d}\right)^\alpha, & \frac{b(\mathbf{x})^\top b(\mathbf{y})}{d} \geq \tau \\ 0, & \text{otherwise,} \end{cases} \tag{2}$$

where $\tau \in [0, 1]$ is a threshold parameter. Only descriptor pairs that are assigned to the same visual word contribute to the image similarity in (1). In practice, not all pairs need to be enumerated and image similarity is equivalently given by

$$S_{\text{SMK}}(\mathcal{X}, \mathcal{Y}) = \gamma(\mathcal{X})\,\gamma(\mathcal{Y}) \sum_{\mathbf{c}\in\mathcal{C}_\mathcal{X}\cap\mathcal{C}_\mathcal{Y}} \sum_{\mathbf{x}\in\mathcal{X}_\mathbf{c}} \sum_{\mathbf{y}\in\mathcal{Y}_\mathbf{c}} k(b(\mathbf{x}), b(\mathbf{y})), \tag{3}$$

[2] The binarized versions are originally [44] referred to as SMK* and ASMK*. Only binarized versions are considered in this work and the asterisk is omitted.

[3] To simplify, we use the same notation, *i.e.* $\gamma(\cdot)$, for the normalization of different similarity measures in the rest of the text. In each case, it ensures unit self-similarity of the corresponding similarity measure.

where cross-matching is only performed within common visual words.

ASMK first *aggregates* the local descriptors assigned to the same visual word into a single binary vector. This is performed by $B(\mathcal{X}_c) = \text{sign}\left(\sum_{x \in \mathcal{X}_c} r(\mathbf{x})\right)$, with $B(\mathcal{X}_c) \in \{-1, 1\}^d$. Image similarity in ASMK is given by

$$S_{\text{ASMK}}(\mathcal{X}, \mathcal{Y}) = \gamma(\mathcal{X})\, \gamma(\mathcal{Y}) \sum_{c \in \mathcal{C}_\mathcal{X} \cap \mathcal{C}_\mathcal{Y}} k(B(\mathcal{X}_c), B(\mathcal{Y}_c)). \tag{4}$$

This is computationally and memory-wise more efficient than SMK. In practice, it is known to perform better due to handling the burstiness phenomenon [18]. Efficient search is performed by using an inverted-file indexing structure.

Simplifications. Compared to the original approach [44], we drop IDF weighting, pre-binarization random projections, and median-value thresholding, as these are found unnecessary.

4 Method

Learning local descriptors with ASMK in an end-to-end manner is challenging and impractical due to the use of large visual codebooks and the hard-assignment of descriptors to visual words. In this section, we first describe a simple framework to generate global descriptors that can be optimized with image-level labels and provide an insight into why this is relevant to optimizing the local representation too. Then, we extend the framework by additional components and discuss their relation to prior work.

4.1 Derivation of the Architecture

In the following, let us assume a deep Fully Convolutional Network (FCN), denoted by function $f : \mathbb{R}^{w \times h \times 3} \rightarrow \mathbb{R}^{W \times H \times D}$, that maps an input image I to a 3D tensor of activations $f(I)$. The FCN is used as an extractor of dense deep local descriptors. The 3D activation tensor can be equivalently seen as a set $\mathcal{U} = \{\mathbf{u} \in \mathbb{R}^D\}$ of $W \times H$ D-dimensional local descriptors[4]. Each local descriptor is associated to a keypoint, also called local feature, that is equivalent to the receptive field, or a fraction of it, of the corresponding activations.

Let us consider global image descriptors constructed by global sum-pooling, known as SPoC [4]. Pairwise image similarity is estimated by the inner product of the corresponding ℓ_2-normalized SPoC descriptors. This can be equivalently seen as an efficient match kernel. Let $\mathcal{U} = f(I)$ and $\mathcal{V} = f(J)$ be sets of dense feature descriptors in images I and J, respectively. Image similarity is given by

$$S_{\text{SPoC}}(\mathcal{U}, \mathcal{V}) = \gamma(\mathcal{U})\, \gamma(\mathcal{V}) \sum_{\mathbf{u} \in \mathcal{U}} \sum_{\mathbf{v} \in \mathcal{V}} \mathbf{u}^\top \mathbf{v} \tag{5}$$

[4] Both $f(I)$ and $\mathcal{U}$ correspond to the same representation seen as a 3D tensor and a set of descriptors, respectively. We write $\mathcal{U} = f(I)$ implying the tensor is transformed into a set of vectors. $\mathcal{U}$ is, in fact, a multi-set, but it is referred to as set in the paper.

$$= \left(\gamma(\mathcal{U})\sum_{\mathbf{u}\in\mathcal{U}}\mathbf{u}\right)^\top\left(\gamma(\mathcal{V})\sum_{\mathbf{v}\in\mathcal{V}}\mathbf{v}\right) = Z_{\text{SPoC}}(\mathcal{U})^\top Z_{\text{SPoC}}(\mathcal{V}), \tag{6}$$

where $\gamma(\mathcal{U}) = 1/||\sum_{\mathbf{u}\in\mathcal{U}}\mathbf{u}||$. The optimization of the network parameters is cast as metric learning.

The interpretation of global descriptor matching in (6) as local descriptor cross-matching in (5) provides some useful insight. Local descriptor similarity is estimated by $\mathbf{u}^\top\mathbf{v} = ||\mathbf{u}|| \cdot ||\mathbf{v}|| \cdot \cos(\mathbf{u}, \mathbf{v})$. Optimizing SPoC with image-level labels and contrastive loss implicitly optimizes the following four cases. First, local descriptors of background features, or image regions, are pushed to have small magnitude, *i.e.* small $||\mathbf{u}||$, so that their contribution in cross-matching is minimal. Similarly, local descriptors of foreground features are pushed to have large magnitude. Additionally, local descriptors of truly-corresponding features, *i.e.* image locations depicting the same object or object part, are pushed closer in the representation space, so that their inner product is maximized. Finally, local descriptors of non-corresponding features are pushed apart, so that their inner product is minimized. Therefore, optimizing global descriptors is a good surrogate to optimize local descriptors that are used to measure similarity via efficient match kernels, such as in ASMK. Importantly, this is possible with image-level labels and no local correspondences are required for training. In the following, we introduce additional components to the presented model, that are designed to amplify these properties.

Feature Strength and Attention. The feature strength, or importance, is estimated as the ℓ_2 norm of the feature descriptor $\mathbf{u}$ by the attention function $w(\mathbf{u}) = ||\mathbf{u}||$. During training, the feature strength is used to weigh the contribution of each feature descriptor in cross-matching. This way, the impact of the weak features is limited during the training, which is motivated by only the strongest features being selected in the test time. The attention function is fixed, without any parameters to be learned.

Local Smoothing. Large activation values tend to appear in multiple channels of the activation tensor on foreground features [41]. Moreover, these activations tend not to be spatially well aligned. We propose to spatially smooth the activations by average pooling in an $M \times M$ neighborhood. The result is denoted by $\bar{\mathcal{U}} = h(f(I))$ or $\bar{\mathcal{U}} = h(\mathcal{U})$. Our experiments show that it is beneficial for the aggregation operation performed in ASMK. It is a fixed function, without any further parameters to be learned, and parameter M is a design choice.

Mean Subtraction. Commonly used FCNs (all that are used in this work) generate non-negative activation tensors since a Rectified Linear Unit (ReLU) constitutes the last layer. Therefore, the inner product for all local descriptor pairs in cross-matching is non-negative and contributes to the image similarity. Mean descriptor subtraction is known to capture negative evidence [19] and allows to better disambiguate non-matching descriptors.

Descriptor Whitening. Local descriptor dimensions are de-correlated by a linear whitening transformation; PCA-whitening improves the discriminability of

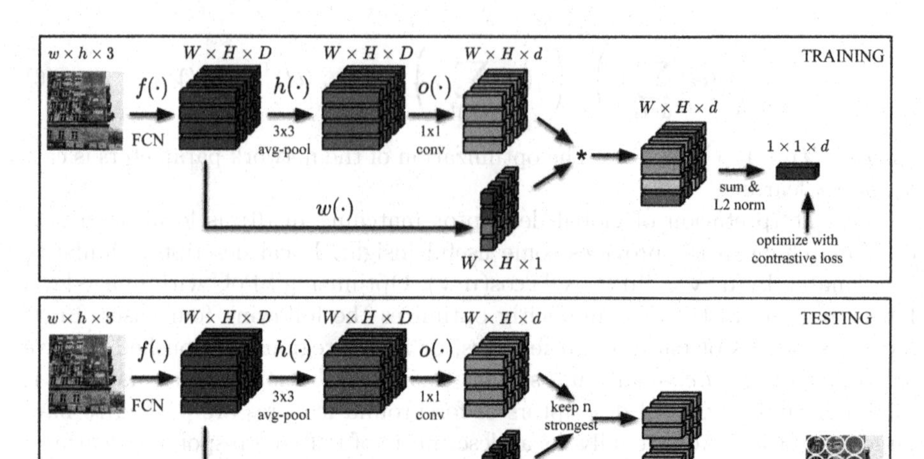

Fig. 2. Training and testing architecture overview for HOW local features and descriptors. A global descriptor is generated for each image during training and optimized with contrastive loss and image-level labels. During testing the strongest local descriptors (features), according to the attention map, are kept to represent the image. Then, these are used with ASMK for image search.

both local [6] and global descriptors [36]. For efficiency, dimensionality reduction is performed jointly with the whitening. Formally, we group mean subtraction, whitening, and dimensionality reduction into function $o : \mathbb{R}^D \rightarrow \mathbb{R}^d$ given by $o(\mathbf{u}) = P(\mathbf{u} - \mathbf{m})$, where $P \in \mathbb{R}^{d \times D}$, $d \leq D$. Function $o(\cdot)$ is implemented by 1×1 convolution with bias. In practice, we initialize P and $\mathbf{m}$ according to the result of PCA whitening on a set of local descriptors from the training set and keep them fixed during the training.

Learning. Let $\bar{\mathcal{V}} = h(\mathcal{V})$ and $\bar{\mathbf{v}} \in \bar{\mathcal{V}}$ denote the activation vector after local average pooling $h(\cdot)$ at the same spatial location as $\mathbf{v} \in \mathcal{V}$. Similarly for $\bar{\mathcal{U}}$ and $\bar{\mathbf{u}}$. The image similarity that is being optimized during learning is expressed as

$$S_{\mathit{how}}(\mathcal{U}, \mathcal{V}) = \gamma(\bar{\mathcal{U}})\, \gamma(\bar{\mathcal{V}}) \sum_{\mathbf{u} \in \mathcal{U}} \sum_{\mathbf{v} \in \mathcal{V}} w(\mathbf{u}) \cdot w(\mathbf{v}) \cdot o(\bar{\mathbf{u}})^\top o(\bar{\mathbf{v}}) \tag{7}$$

$$= \Big(\gamma(\bar{\mathcal{U}}) \sum_{\mathbf{u} \in \mathcal{U}} w(\mathbf{u}) o(\bar{\mathbf{u}})\Big)^{\!\top} \Big(\gamma(\bar{\mathcal{V}}) \sum_{\mathbf{v} \in \mathcal{V}} w(\mathbf{v}) o(\bar{\mathbf{v}})\Big) = Z_{\mathit{how}}(\mathcal{U})^\top Z_{\mathit{how}}(\mathcal{V}). \tag{8}$$

A metric learning approach is used to train the network. In particular, contrastive loss is minimized: $||Z_{\mathit{how}}(\mathcal{U}) - Z_{\mathit{how}}(\mathcal{V})||^2$ if $\mathcal{U}$ and $\mathcal{V}$ originate from matching (positive) image pairs, or $([\mu - ||Z_{\mathit{how}}(\mathcal{U}) - Z_{\mathit{how}}(\mathcal{V})||]_+)^2$ otherwise, where $[\cdot]_+$ denotes the positive part.

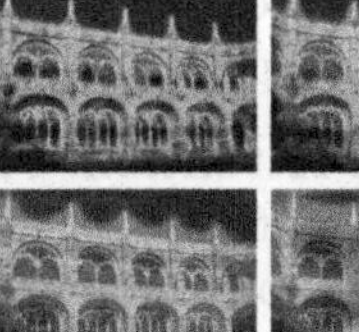
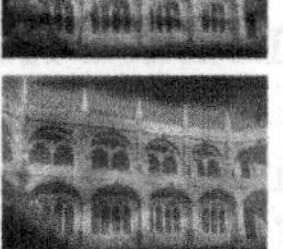

Fig. 3. Example of multi-scale local feature detection. Left: Strongest 1,000 local features with color-coded strength; red is the strongest. Middle: Only the strongest feature per visual word is shown. Right: Attention maps for input images resized by scaling factors 0.25, 0.5, 1, and 2. (Color figure online)

Test-Time Architecture. The architecture in test time is slightly modified; no global descriptor is generated. The n strongest descriptors $o(\bar{\mathbf{u}})$ for $\mathbf{u} \in \mathcal{U}$ are kept according to the importance given by $w(\mathbf{u})$. These are used as the local descriptor set $\mathcal{X}$ in ASMK. Multi-scale extraction is performed by resizing the input image at multiple resolutions. Local features from all resolutions are merged and ranked jointly according to strength. Selection of the strongest features is performed in the merged set. Multi-scale extraction is performed only during testing, and not during training. The training and testing architectures are summarized in Fig. 2, while examples of detected features are shown in Fig. 3.

4.2 Relation to Prior Work

The proposed method has connections to different approaches in the literature which are discussed in this section. The work closest to ours, in terms of training of the local detector and descriptor, is DELF [29]. Following our notation, DELF generates local feature descriptors $Z_{\hat{h}\hat{o}\hat{w}}(\mathcal{U})$, where $\hat{h}(.)$ is identity, *i.e.* no local smoothing, $\hat{o}(.)$ is identity, *i.e.* no mean subtraction, whitening, and dimensionality reduction, and $\hat{w}(.)$ is a learned attention function. In particular, $\hat{w} : \mathbb{R}^D \rightarrow \mathbb{R}_+$ is a 2 layer convolutional network with 1×1 convolutions whose parameters are learned during the training. Dimensionality reduction of the descriptor space in DELF is performed as post-processing and is not a part of the optimization.[5] In terms of optimization, DELF performs the training in a classification manner with cross entropy loss. We show experimentally, that when combined with ASMK, the proposed descriptors are superior to DELF.

Fixed Attention. Function $w(\cdot)$ is previously used to weigh activations and generate global descriptor, in particular by CroW [23]. It is also used in concurrent work for deep local features by Yang *et al.* [49]. The same attention function is used by Iscen *et al.* [17] for feature detection on top of dense SIFT descriptors without any learning. In our case, during learning, background or domain irrele-

[5] The main difference is that we do not follow the two stage training performed in the original work [29]; DELF is trained in a single stage for our ablations.

vant descriptors are pushed to have low ℓ_2 norm in order to contribute less. The corresponding features are consequently not detected during test time.

Learned Attention. Further example of learned attention, apart from DELF, is the contextual re-weighting that is performed to construct global descriptors in the work of Kim and Frahm [24]. The attention function is similar to DELF but with larger spatial kernel; a contextual neighborhood is used. A comparison between learned and fixed attention is included in the experimental ablations in Sect. 5.

Whitening. A common approach to whitening is to apply it as the last step in the pipeline, as post-processing. A similar approach to ours – applying the PCA whitening during training and learning it end-to-end – is followed by Gordo *et al.* [14] in the context of processing and aggregating regional descriptors to construct global descriptors. Comparison between "in-network" and post-processing whitening-reduction is included in Sect. 5.

5 Experiments

We first review the datasets used for training, validation, and testing. Then, we discuss the implementation details for training and for testing with ASMK. Finally, we present our results on two instance-level tasks, namely recognition and search in the domain of landmarks and buildings.

5.1 Datasets

Training. The training dataset *SfM120k* [35] is used. It is the outcome of Stucture-from-Motion (SfM) [40] with 551 3D models for training. Matching pairs (anchor-positive) are formed by images with visual overlap (same 3D model). Non-matching pairs (anchor-negative) come from different 3D models.

Validation. We use the remaining 162 3D models of SfM120k to construct a challenging validation set reflecting the target task; this is different than the validation in [35]. We randomly choose 5 images per 3D model as queries. Then, for each query, images of the same 3D model with enough (more than 3), but not too many (at most 10), common 3D points with the query are marked as positive images to be retrieved. This avoids dominating the evaluation measure by a large number of easy examples. The remaining images of the same 3D model are excluded from evaluation for the specific query [32]. Skipping queries with an empty list of positive images results in 719 queries and 12,441 database images in total. Evaluation on the validation set is performed by instance-level search and performance is measured by mean Average Precision (mAP).

Evaluation on Instance-level Search. We use $\mathcal{R}$Oxford [32] and $\mathcal{R}$Paris [33] to evaluate search performance in the revisited benchmark [34]. They consist of 70 queries each, and 4993 and 6322 database images, respectively, and 1 million distractors called $\mathcal{R}$1M. We measure mAP on the Medium and Hard setups.

Evaluation on Instance-level Classification. We use instance-level classification as another task on which to evaluate the performance of the learned local descriptors. We perform search with ASMK and use k-nearest-neighbors classifiers for class predictions. The *Google Landmarks Dataset – version 2* (GLD_2) [48] is used. It consists of 4,132,914 train/database images with known class labels, and 117,577 test/query images which either correspond to the database landmarks, to other landmarks, or to non-landmark images. The query images are split into testing (private) and validation (public) sets with 76,627 and 40,950 images, respectively. Performance is measured by micro Average Precision (μAP) [31], also known as Global Average Precision (GAP), on the testing split. Note that we do not perform any learning on this dataset. We additionally create a mini version of GLD_2 to use for ablation experiments. It includes 1,000 query images, that are sampled from the testing split, and 10,0000 database images with labels, where the images come from 50 landmarks in total. We denote it by Tiny-GLD_2.

Classification is performed by accumulating the N top-ranked images per class. Prediction is given by the top-ranked class and the corresponding confidence is equal to the accumulated similarity. We use three variants, *i.e.* $N = 1$ (CLS1), $N = 10$ (CLS2), and $N = 10$ with accumulation of square-rooted similarity multiplied by a class weight (CLS3). The class weight is equal to the logarithm of the number of classes divided by the class frequency to down-weigh frequent classes.

5.2 Implementation Details

Network Architecture. We perform experiments with a backbone FCN ResNet18 and ResNet50, initialized by pre-training on ImageNet [39]. Descriptor dimensionality D is equal to 512 and 2048, respectively. We additionally experiment by removing the last block, *i.e.* "conv5_x", where D becomes 256 and 1024, respectively. We set $d = 128$ and $M = 3$ to perform 3×3 average pooling for local smoothing.

Training. We use a batch size of 5 tuples, where a tuple consists of 7 images – an anchor, a positive, and 5 negatives. Training images are restricted to a maximum resolution of 1,024 pixels. For each epoch, we randomly choose 2,000 anchor-positive pairs. The pool of candidate negatives contains 20,000 randomly chosen images, and hard-negative mining is performed before every epoch. We adopt the above choices from the work of Radenovic *et al.* [35], whose public implementation[6] we use to implement our method. To initialize P and $\mathbf{m}$, we use local descriptors from 5,000 training images extracted at a single scale. We use Adam optimizer with weight decay equal to 10^{-4}. The learning rate and margin μ are tuned, per variant, according to the performance on the validation set, by trying values $10^{-6}, 5 \cdot 10^{-6}, 10^{-5}, 5 \cdot 10^{-5}$ for learning rate and 0.5 to 0.9 with step 0.05 for margin μ. Margin μ is set equal to 0.8 for the proposed method, and

[6] https://github.com/filipradenovic/cnnimageretrieval-pytorch.

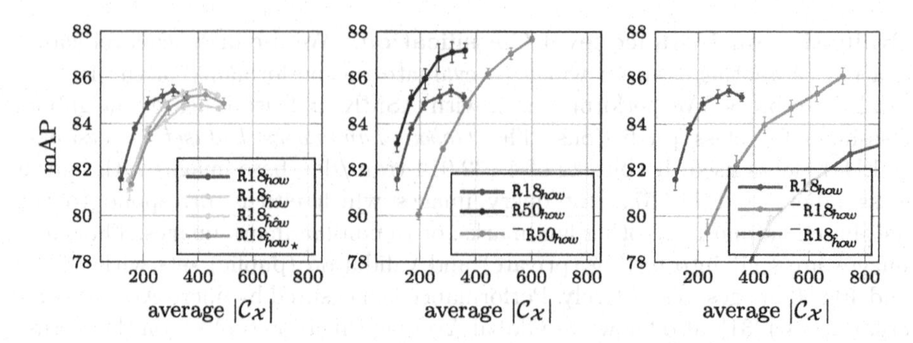

Fig. 4. Ablation study reporting performance versus average number of vectors per image in ASMK on the validation set for varying n (400,600,800,1000,1200,1400). Components used by DELF are denoted with $\hat{}$, while ours without. $\star$: random initialization of $\mathbf{m}$ and P (learned during training). Mean and standard deviation over 5 runs.

learning rate equal to $5 \cdot 10^{-6}$ and 10^{-5} for ResNet18 and ResNet50 without the last block, respectively, according to the tuning process. Training is performed for 20 epochs and 1 epoch takes about 22 min for ResNet50 without the last block on a single NVIDIA Tesla V100 GPU with 32GB of DRAM. Training with cross entropy loss for ablations is performed with a batch size equal to 64 for 10 epochs. We repeat the training of each model and report mean and standard deviation over 5 runs. In large scale experiments, we evaluate a single model, the one with median performance on the validation set.

Validation. Validation performance is measured with ASMK-based search on the validation set. We measure validation performance every 5 epochs during training and the best performing model is kept.

Testing. We use ASMK to perform testing and to evaluate the performance of the learned local descriptors. The default ASMK configuation is as follows. We set threshold $\tau = 0$, $d = 128$, and use a codebook of $\kappa = 65536$ visual words, which is learned on local descriptors from 20,000 training images extracted at a single scale. Images are resized to have maximum resolution of 1024 pixels and multi-scale extraction is performed by re-scaling with factors 0.25, 0.353, 0.5, 0.707, 1.0, 1.414, 2.0. Assignment to multiple, in particular 5, visual words is performed for the query images. The strongest $n = 1000$ local descriptors are kept. The default configuration is used unless otherwise stated. The inverted file is compressed by delta coding.

5.3 Ablation Experiments

We denote ResNet18 and ResNet50 by R18 and R50, and their versions with the last block skipped by $^{-}$R18 and $^{-}$R50, respectively. The local smoothing, whitening with reduction, and the fixed attention are denoted by subscripts h, o and w, respectively. The proposed method is denoted by R18$_{how}$ when the backbone network is ResNet18. Following the same convention, the original DELF

Table 1. Ablation study for performance and average number of descriptors per image in ASMK (mean and standard deviation over 5 runs). 1: our method, 5: DELF variant. CO: contrastive loss, CE: cross entropy. On Tiny-GLD$_2$, classifier CLS3 is used.

Method	Loss	Validation		$\mathcal{R}$Oxford		Tiny-GLD$_2$	
		mAP	$\lvert\mathcal{C}_\mathcal{X}\rvert$	mAP	$\lvert\mathcal{C}_\mathcal{X}\rvert$	μAP	$\lvert\mathcal{C}_\mathcal{X}\rvert$
1: R18$_{how}$	CO	85.2 ± 0.3	263.8 ± 0.1	74.8 ± 0.2	283.6 ± 0.2	81.3 ± 1.0	252.2 ± 0.5
2: R18$_{\hat{h}ow}$	CO	85.3 ± 0.2	344.1 ± 0.7	75.4 ± 0.3	365.9 ± 0.5	80.6 ± 0.3	332.8 ± 1.0
3: R18$_{\hat{h}\hat{o}w}$	CO	84.8 ± 0.2	343.5 ± 2.7	73.1 ± 0.3	365.7 ± 2.8	78.6 ± 1.0	336.2 ± 3.5
4: R18$_{\hat{h}\hat{o}\hat{w}}$	CO	83.7 ± 0.9	354.4 ± 2.0	70.0 ± 1.7	380.7 ± 2.9	74.2 ± 3.6	358.6 ± 5.5
5: R18$_{\hat{h}\hat{o}\hat{w}}$	CE	75.5 ± 1.3	391.0 ± 8.2	63.7 ± 1.6	442.3 ± 9.7	64.0 ± 1.8	427.5 ± 15.6
6: R18$_{\hat{h}\hat{o}w}$	CE	77.1 ± 0.9	375.0 ± 8.5	67.0 ± 1.0	429.0 ± 13.8	67.3 ± 2.1	417.8 ± 11.7
7: R18$_{\hat{h}o\hat{w}}$	CE	78.4 ± 0.8	354.6 ± 10.5	67.8 ± 1.3	402.8 ± 12.2	66.7 ± 1.5	367.2 ± 14.8
8: R18$_{h\hat{o}\hat{w}}$	CE	77.0 ± 0.9	279.6 ± 5.6	65.4 ± 0.5	320.6 ± 6.8	68.6 ± 1.8	300.9 ± 11.4
9: R18$_{how}$	CE	80.4 ± 0.5	308.3 ± 4.0	69.9 ± 1.4	345.4 ± 5.2	71.1 ± 1.8	308.4 ± 4.1

architecture is denoted by $^-$R50$_{\hat{h}\hat{o}\hat{w}}$, where the dimensionality reduction is not part of the network but is performed by PCA whitening as post-processing.

Figure 4 shows the performance on the validation set versus the average number of binary vectors indexed by ASMK for the database images. We perform an ablation by excluding the proposed components and by using different backbone networks. Local smoothing results in larger amount of aggregation in ASMK (red vs pink) and reduces the memory requirements, which are linear in $|\mathcal{C}_\mathcal{X}|$. It additionally improves the performance when the last ResNet block is removed and feature maps have two times larger resolution (green vs magenta). Initializing and fixing $h(\cdot)$ with the result of PCA whitening is a beneficial choice too (orange vs pink). ResNet50 gives a good performance boost compared to ResNet18 (blue vs red), while removing the last block is able to reach higher performance at the cost of larger memory requirements (brown vs blue). More ablation experiments are shown in Table 1. Fixed attention is better than learned attention (3 vs 4). Metric learning with the contrastive loss delivers significantly better performance than cross entropy loss, in a classification manner, as done by Noh *et al.* [29] (4 vs 5). This is a confirmation of results that appear in the literature of instance-level recognition [14,46].

In Fig. 5, we present the evolution of the model during training. We evaluate the performance of the optimized global descriptor for nearest neighbor search with multi-scale global descriptors, *i.e.* aggregation of global descriptors extracted at multiple image resolutions (same set of 7 resolutions as for the local descriptors). We additionally evaluate performance of the corresponding local descriptors with ASMK. The descriptor that is directly optimized in the loss performs worse than the internal local descriptors.

In the following we use two backbone networks – R18, which is fast with low memory footprint, and $^-$R50, which achieves better results at the cost of slower extraction and more memory.

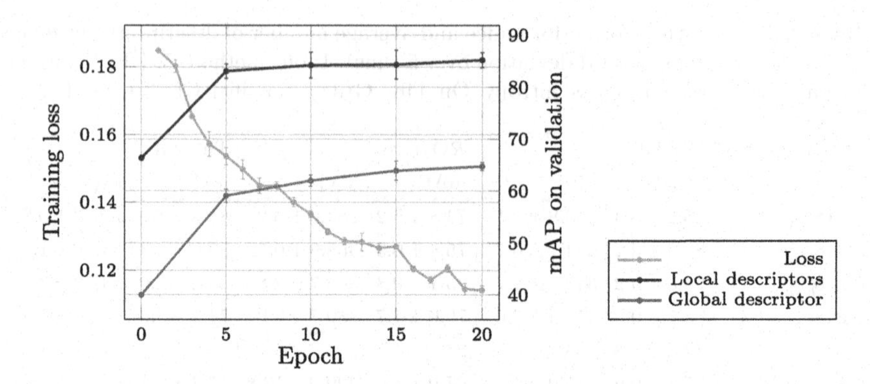

Fig. 5. Training loss and validation performance during the training. Mean and standard deviation (×5 for better visualization) is reported over 5 runs. Both performance curves correspond to the same model, only its inference differs.

5.4 Large-Scale Instance-Level Search

The performance comparison on $\mathcal{R}$Oxford and $\mathcal{R}$Paris is presented in Table 2. We do our best to evaluate the best available variants or models of the state-of-the-art approaches. The proposed local descriptors and DELF descriptors are evaluated with identical implementation and configuration of ASMK. The proposed descriptors outperform all approaches by a large margin at large scale; global descriptors perform well enough at small scale, but at large scale, local representation is essential. Even with a backbone network as small as ResNet18, we achieve the second best performance (ranked after our method with ResNet50) on all cases of $\mathcal{R}$Oxford and at the hard setup of $\mathcal{R}$Paris +$\mathcal{R}$1M. Compared to DELF, our descriptors, named *HOW*, perform better for less memory.

Teichmann *et al.* [43] achieve average performance (mean all in Table 2) equal to 56.0 and 57.3 without and with spatial verification, respectively. They use additional supervision, *i.e.* manually created bounding boxes for 94,000 images, which we do not. The concurrent work of Cao *et al.* [11] achieves average performance equal to 58.3 but requires 485 GB of RAM, and slightly lower performance with 22.6 GB of RAM for binarized descriptors.

In an effort to further compress the memory requirements of competing global descriptors, we evaluate the best variant with dimensionality reduction and with Product Quantization (PQ) [21]. We further improve their performance by concatenating the two best performing ones. Among all these variants, the proposed approach appears to be a good solution in the performance-memory trade-off.

Storing less vectors in ASMK affects memory but also speed. A query of average statistics computes the hamming distance for about $1.2 \cdot 10^6$ (average number of vectors stored per inverted list multiplied by average $|\mathcal{C}_\mathcal{X}|$) 128D binary vector pairs in the case of $^-$R18 with our method at large scale. The same number for $^-$R50 is $3.2 \cdot 10^6$. Search on $\mathcal{R}$Oxford +$\mathcal{R}$1M takes on average 0.75 seconds on a single threaded Python non-optimized CPU implementation.

Table 2. Performance comparison with global and local descriptors for instance-level search on $\mathcal{R}$Oxford ($\mathcal{R}$O) and $\mathcal{R}$Paris ($\mathcal{R}$P). Memory is reported for $\mathcal{R}$1M. Methods marked by † are evaluated by us using the public models for descriptor extraction. The method marked by ‡ is evaluated by us using the public descriptors [34]. GEM$_w$ is a public model that includes a "whitening" (FC) layer. Dimensionality reduction is denoted by ▷ and descriptor concatenation by +. PQ8 and PQ1 denote PQ quantization using 8D and 1D sub-spaces, respectively.

Method	FCN	Mem	Mean		$\mathcal{R}$O		$\mathcal{R}$O +$\mathcal{R}$1M		$\mathcal{R}$Par		$\mathcal{R}$P +$\mathcal{R}$1M	
		(GB)	all	$\mathcal{R}$1M	med	hard	med	hard	med	hard	med	hard
Global descriptors & Euclidean distance search												
R-MAC [14]	R101	7.6	45.8	33.6	60.9	32.4	39.3	12.5	78.9	59.4	54.8	28.0
GeM [35]	R101	7.6	47.4	35.5	64.7	38.5	45.2	19.9	77.2	56.3	52.3	24.7
GeM [35]†	R101	7.6	47.3	35.2	65.4	40.1	45.1	22.7	76.7	55.2	50.8	22.4
GeM$_w$†	R101	7.6	49.0	37.2	67.8	41.7	47.7	23.3	77.6	56.3	52.9	25.0
GeM$_{AP}$[37]	R101	7.6	49.9	37.1	67.5	42.8	47.5	23.2	80.1	60.5	52.5	25.1
GeM$_{AP}$[37]†	R101	7.6	49.7	36.7	67.1	42.3	47.8	22.5	80.3	60.9	51.9	24.6
GeM$_{AP}$[37] PQ1	R101	1.9	49.6	36.7	67.1	42.2	47.7	22.5	80.3	60.8	51.9	24.6
GeM$_{AP}$[37] PQ8	R101	0.2	48.1	35.5	65.1	40.4	46.1	21.6	78.7	58.7	50.7	23.5
GeM$_{AP}$[37]† ▷1024	R101	3.8	49.0	35.8	66.6	41.6	46.7	21.7	80.0	60.3	51.0	23.8
GeM$_{AP}$[37]† ▷512	R101	1.9	47.5	33.7	65.9	40.5	43.9	19.7	79.5	59.4	48.9	22.3
GeM$_{AP}$+ GeM$_w$†	R101	15.3	53.4	41.4	70.5	45.7	52.6	27.1	**81.9**	**63.4**	57.0	29.1
Local descriptors & ASMK												
DELF [29]†	$^{-}$R50	9.2	53.0	43.1	69.0	44.0	54.1	31.1	79.5	58.9	59.3	28.1
DELF [29] ‡	$^{-}$R50	9.2	52.5	42.7	69.2	44.3	54.3	31.3	78.7	57.4	58.2	26.9
DELF [29][34]	$^{-}$R50	9.2	51.5	42.2	67.8	43.1	53.8	31.2	76.9	55.4	57.3	26.4
R18$_{how}$, $n = 1000$	R18	4.6	54.8	43.5	75.1	51.7	55.7	32.0	79.4	58.3	57.4	28.9
$^{-}$R50$_{how}$, $n = 1000$	$^{-}$R50	7.9	58.0	47.4	78.3	55.8	63.6	36.8	80.1	60.1	58.4	30.7
$^{-}$R50$_{how}$, $n = 1200$	$^{-}$R50	9.2	58.8	48.4	78.8	56.7	64.5	37.7	80.6	61.0	59.6	31.7
$^{-}$R50$_{how}$, $n = 1400$	$^{-}$R50	10.6	59.3	49.0	79.1	56.8	64.9	38.2	81.0	61.5	60.4	32.6
$^{-}$R50$_{how}$, $n = 2000$	$^{-}$R50	14.3	**60.1**	**50.1**	**79.4**	**56.9**	**65.8**	**38.9**	81.6	62.4	**61.8**	**33.7**

5.5 Large-Scale Instance-Level Classification

Performance comparison on GLD$_2$ is presented in Table 3. We extract DELF keeping at most top 1,000 local descriptors and reduce dimensionality to 128. The proposed local descriptors and DELF descriptors are evaluated with identical configuration of ASMK. Our method outperforms all other methods with memory requirements that are even lower than raw 2048D global descriptors. DELF, R18$_{how}$, and $^{-}$R50$_{how}$, end up with 347, 252, and 423 vectors to store per image on average, respectively. Due to the strength threshold, DELF extracted 504 local descriptors per image on average which is significantly less than for images of $\mathcal{R}$1M. Multiple visual word assignment is not performed in this experiment, for any of the methods, to reduce the computational cost of search.

Table 3. Performance comparison on instance-level recognition (GLD_2). μAP is reported for classification with 3 different k-nn classifiers. Existing methods are evaluated by us using the public models. Global descriptors are combined with simple nearest neighbor search, and local descriptors are combined with ASMK-based retrieval.

Method	FCN	Training set	Memory (GB)	CLS1	CLS2	CLS3
GEM [35]	R101	SfM-120k	31.5	1.9	18.0	24.1
GEM_w	R101	SfM-120k	31.5	3.7	23.4	28.7
GEM-AP [37]	R101	SfM-120k	31.5	2.8	14.8	20.7
DELF [29]	⁻R50	Landmarks [5]	24.1	2.1	11.9	21.9
$R18_{how}$	R18	SfM-120k	17.5	8.5	20.0	27.0
$^{-}R50_{how}$	⁻R50	SfM-120k	29.3	18.5	33.1	36.5

6 Conclusions

An architecture for extracting deep local features is designed to be combined with ASMK matching. The proposed method consistently outperforms other methods on a number of standard benchmarks, even if a less powerful backbone network is used. Through an extensive ablation study, we show that the SoA performance is achieved by the synergy of the proposed local feature detector with ASMK. We show that methods based on local features outperform global descriptors in large scale problems, and also that the proposed method outperforms other local feature detectors combined with ASMK. We demonstrate why the proposed architecture, despite being trained with image-level supervision only, is effective in learning image similarity based on local features.

Acknowledgement. The authors would like to thank Yannis Kalantidis for valuable discussions. This work was supported by MSMT LL1901 ERC-CZ grant. Tomas Jenicek was supported by CTU student grant SGS20/171/OHK3/3T/13.

References

1. Arandjelović, R., Zisserman, A.: All about VLAD. In: CVPR (2013)
2. Arandjelović, R., Zisserman, A.: DisLocation: scalable descriptor distinctiveness for location recognition. In: Cremers, D., Reid, I., Saito, H., Yang, M.H. (eds.) ACCV 2014. LNCS, vol. 9006, pp. 188–204. Springer, Cham (2015). https://doi.org/10.1007/978-3-319-16817-3_13
3. Arandjelović, R., Gronat, P., Torii, A., Pajdla, T., Sivic, J.: NetVLAD: CNN architecture for weakly supervised place recognition. In: CVPR (2016)
4. Babenko, A., Lempitsky, V.: Aggregating deep convolutional features for image retrieval. In: ICCV (2015)
5. Babenko, A., Slesarev, A., Chigorin, A., Lempitsky, V.: Neural codes for image retrieval. In: Fleet, D., Pajdla, T., Schiele, B., Tuytelaars, T. (eds.) ECCV 2014. LNCS, vol. 8689, pp. 584–599. Springer, Cham (2014). https://doi.org/10.1007/978-3-319-10590-1_38

6. Balntas, V., Lenc, K., Vedaldi, A., Mikolajczyk, K.: Hpatches: a benchmark and evaluation of handcrafted and learned local descriptors. In: CVPR (2017)
7. Barroso Laguna, A., Riba, E., Ponsa, D., Mikolajczyk, K.: Key. net: keypoint detection by handcrafted and learned cnn filters. In: ICCV (2019)
8. Bay, H., Ess, A., Tuytelaars, T., Gool, L.V.: SURF: speeded up robust features. Comput. Vis. Image Underst. **110**(3), 346–359 (2008)
9. Benbihi, A., Geist, M., Pradalier, C.: Elf: embedded localisation of features in pre-trained cnn. In: CVPR (2019)
10. Bhowmik, A., Gumhold, S., Rother, C., Brachmann, E.: Reinforced feature points: optimizing feature detection and description for a high-level task. In: CVPR (2020)
11. Cao, B., Araujo, A., Sim, J.: Unifying deep local and global features for efficient image search. In: arxiv (2020)
12. DeTone, D., Malisiewicz, T., Rabinovich, A.: Superpoint: self-supervised interest point detection and description. In: CVPRW (2018)
13. Dusmanu, M., et al.: D2-net: a trainable cnn for joint detection and description of local features. In: CVPR (2019)
14. Gordo, A., Almazán, J., Revaud, J., Larlus, D.: End-to-End learning of deep visual representations for image retrieval. Int. J. Comput. Vis. **124**(2), 237–254 (2017). https://doi.org/10.1007/s11263-017-1016-8
15. Gu, Y., Li, C., Jiang, Y.G.: Towards optimal cnn descriptors for large-scale image retrieval. In: ACM Multimedia (2019)
16. Husain, S., Bober, M.: Improving large-scale image retrieval through robust aggregation of local descriptors. PAMI **39**(9), 1783–1796 (2016)
17. Iscen, A., Tolias, G., Gosselin, P.H., Jégou, H.: A comparison of dense region detectors for image search and fine-grained classification. IEEE Trans. Image Process. **24**(8), 2369–2381 (2015)
18. Jégou, H., Douze, M., Schmid, C.: On the burstiness of visual elements. In: CVPR, June 2009
19. Jégou, H., Chum, O.: Negative evidences and co-occurences in image retrieval: the benefit of PCA and whitening. In: Fitzgibbon, A., Lazebnik, S., Perona, P., Sato, Y., Schmid, C. (eds.) ECCV 2012. LNCS, vol. 7573, pp. 774–787. Springer, Heidelberg (2012). https://doi.org/10.1007/978-3-642-33709-3_55
20. Jégou, H., Douze, M., Schmid, C.: Improving bag-of-features for large scale image search. IJCV **87**(3), 316–336 (2010)
21. Jégou, H., Douze, M., Schmid, C.: Product quantization for nearest neighbor search. PAMI **33**(1), 117–128 (2011)
22. Jégou, H., Perronnin, F., Douze, M., Sánchez, J., Pérez, P., Schmid, C.: Aggregating local descriptors into compact codes. In: PAMI, Sep 2012
23. Kalantidis, Y., Mellina, C., Osindero, S.: Cross-dimensional weighting for aggregated deep convolutional features. In: Hua, G., Jégou, H. (eds.) ECCV 2016. LNCS, vol. 9913, pp. 685–701. Springer, Cham (2016). https://doi.org/10.1007/978-3-319-46604-0_48
24. Kim, H.J., Dunn, E., Frahm, J.M.: Learned contextual feature reweighting for image geo-localization. In: CVPR (2017)
25. Matas, J., Chum, O., Urban, M., Pajdla, T.: Robust wide-baseline stereo from maximally stable extremal regions. Image Vis. Comput. **22**(10), 761–767 (2004)
26. Mikolajczyk, K., Schmid, C.: A performance evaluation of local descriptors. PAMI **27**(10), 1615–1630 (2005)
27. Mikolajczyk, K., et al.: A comparison of affine region detectors. IJCV **65**(1/2), 43–72 (2005)

28. Mohedano, E., McGuinness, K., O'Connor, N.E., Salvador, A., Marques, F., Giro-i Nieto, X.: Bags of local convolutional features for scalable instance search. In: ICMR (2016)
29. Noh, H., Araujo, A., Sim, J., Weyand, T., Han, B.: Large-scale image retrieval with attentive deep local features. In: ICCV (2017)
30. Perronnin, F., Liu, Y., Sanchez, J., Poirier, H.: Large-scale image retrieval with compressed Fisher vectors. In: CVPR (2010)
31. Perronnin, F., Liu, Y., Renders, J.M.: A family of contextual measures of similarity between distributions with application to image retrieval. In: CVPR, pp. 2358–2365 (2009)
32. Philbin, J., Chum, O., Isard, M., Sivic, J., Zisserman, A.: Object retrieval with large vocabularies and fast spatial matching. In: CVPR (2007)
33. Philbin, J., Chum, O., Isard, M., Sivic, J., Zisserman, A.: Lost in quantization: improving particular object retrieval in large scale image databases. In: CVPR, June 2008
34. Radenović, F., Iscen, A., Tolias, G., Avrithis, Y., Chum, O.: Revisiting oxford and paris: large-scale image retrieval benchmarking. In: CVPR (2018)
35. Radenović, F., Tolias, G., Chum, O.: Fine-tuning CNN image retrieval with no human annotation. PAMI **41**(7), 1655–1668 (2019)
36. Razavian, A.S., Sullivan, J., Carlsson, S., Maki, A.: Visual instance retrieval with deep convolutional networks. ITE Trans. Media Technol. Appl. **4**(3), 251–258 (2016)
37. Revaud, J., Almazán, J., de Rezende, R.S., de Souza, C.R.: Learning with average precision: training image retrieval with a listwise loss. In: ICCV (2019)
38. Revaud, J., et al.: R2d2: repeatable and reliable detector and descriptor. In: NeurIPS (2019)
39. Russakovsky, O., et al.: Imagenet large scale visual recognition challenge. Int. J. Comput. Vis. **115**(3), 211–252 (2015). https://doi.org/10.1007/s11263-015-0816-y
40. Schönberger, J.L., Radenović, F., Chum, O., Frahm, J.M.: From single image query to detailed 3D reconstruction. In: CVPR (2015)
41. Siméoni, O., Avrithis, Y., Chum, O.: Local features and visual words emerge in activations. In: CVPR (2019)
42. Sivic, J., Zisserman, A.: Video google: a text retrieval approach to object matching in videos. In: ICCV (2003)
43. Teichmann, M., Araujo, A., Zhu, M., Sim, J.: Detect-to-retrieve: efficient regional aggregation for image search. In: CVPR (2019)
44. Tolias, G., Avrithis, Y., Jégou, H.: Image search with selective match kernels: aggregation across single and multiple images. IJCV **116**(3), 247–261 (2015). https://doi.org/10.1007/s11263-015-0810-4
45. Tolias, G., Sicre, R., Jégou, H.: Particular object retrieval with integral max-pooling of CNN activations. In: ICLR (2016)
46. Vo, N., Jacobs, N., Hays, J.: Revisiting im2gps in the deep learning era. In: CVPR (2017)
47. Wang, Q., Zhou, X., Hariharan, B., Snavely, N.: Learning feature descriptors using camera pose supervision. In: arXiv (2020)

48. Weyand, T., Araujo, A., Cao, B., Sim, J.: Google landmarks dataset v2-a large-scale benchmark for instance-level recognition and retrieval. In: CVPR (2020)
49. Yang, T., Nguyen, D., Heijnen, H., Balntas, V.: Ur2kid: unifying retrieval, keypoint detection, and keypoint description without local correspondence supervision. In: arxiv (2020)
50. Yue-Hei Ng, J., Yang, F., Davis, L.S.: Exploiting local features from deep networks for image retrieval. In: CVPR (2015)
51. Zhu, C.Z., Jégou, H., ichi Satoh, S.: Query-adaptive asymmetrical dissimilarities for visual object retrieval. In: ICCV (2013)

A Consistently Fast and Globally Optimal Solution to the Perspective-n-Point Problem

George Terzakis[1(✉)] and Manolis Lourakis[2]

[1] Rovco, The Quorum, Bond Street, Bristol BS1 3AE, UK
george.terzakis@rovco.com

[2] Foundation for Research and Technology – Hellas, N. Plastira 100, 70013 Heraklion, Greece
lourakis@ics.forth.gr

Abstract. An approach for estimating the pose of a camera given a set of 3D points and their corresponding 2D image projections is presented. It formulates the problem as a non-linear quadratic program and identifies regions in the parameter space that contain unique minima with guarantees that at least one of them will be the global minimum. Each regional minimum is computed with a sequential quadratic programming scheme. These premises result in an algorithm that always determines the global minima of the perspective-n-point problem for any number of input correspondences, regardless of possible coplanar arrangements of the imaged 3D points. For its implementation, the algorithm merely requires ordinary operations available in any standard off-the-shelf linear algebra library. Comparative evaluation demonstrates that the algorithm achieves state-of-the-art results at a consistently low computational cost.

Keywords: Perspective-n-point problem · Pose estimation · Non-linear quadratic program · Sequential quadratic programming · Global optimality

1 Introduction

The perspective-n-point (PnP) problem concerns the recovery of 6D pose given the central projections of $n \geq 3$ known 3D points on a calibrated camera (cf. Fig. 1). It arises often in vision and robotics applications involving localization, pose tracking and multi-view 3D reconstruction, e.g. [23,32,33,36,39,40,42,44,50].

We begin with the definition of the PnP problem in order to establish notation for the rest of the paper and remind the reader of the most typical cost function formulations associated with the problem. For a set of known Euclidean

---

Electronic supplementary material The online version of this chapter (https://doi.org/10.1007/978-3-030-58452-8_28) contains supplementary material, which is available to authorized users.

A. Vedaldi et al. (Eds.): ECCV 2020, LNCS 12346, pp. 478–494, 2020.
https://doi.org/10.1007/978-3-030-58452-8_28

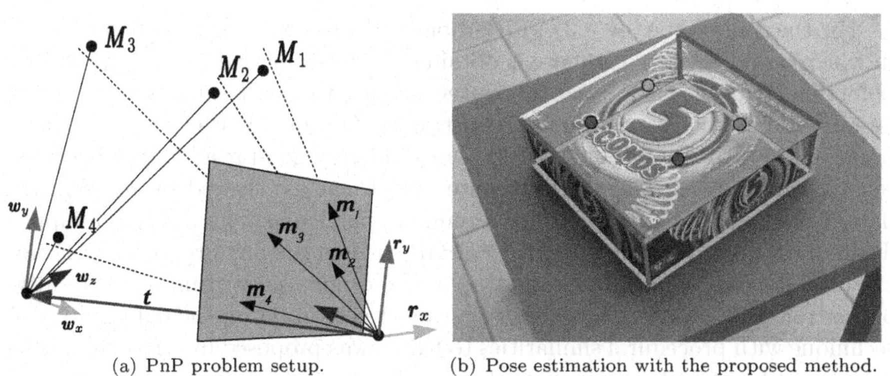

(a) PnP problem setup. (b) Pose estimation with the proposed method.

Fig. 1. (a) The PnP setup for $n = 4$ points. The unknown camera pose comprises the rotation matrix $\boldsymbol{R} = \left[\boldsymbol{r}_x \, \boldsymbol{r}_y \, \boldsymbol{r}_z\right]^T$ and the vector $\boldsymbol{t}$ from the camera center to the world frame. (b) Camera pose estimation with the proposed method for four coplanar points (colored dots). The inserted wireframe model reflects the pose estimate's quality.

world points $\boldsymbol{M}_i \in \mathbb{R}^3$, $i \in \{1, \ldots, n\}$ and their corresponding normalized projections $\boldsymbol{m}_i$ on the $Z = 1$ plane of an unknown camera coordinate frame, we seek to recover the rotation matrix $\boldsymbol{R}$ and translation vector $\boldsymbol{t}$ minimizing the cumulative squared projection error

$$\sum_{i=1}^{n} \left\| \boldsymbol{m}_i - \frac{\boldsymbol{R}\boldsymbol{M}_i + \boldsymbol{t}}{\mathbf{1}_z^T \left(\boldsymbol{R}\boldsymbol{M}_i + \boldsymbol{t}\right)} \right\|^2, \tag{1}$$

where $\mathbf{1}_z = \left[0\;0\;1\right]^T$ and $\left(\boldsymbol{R}\boldsymbol{M}_i + \boldsymbol{t}\right) / \left(\mathbf{1}_z^T \left(\boldsymbol{R}\boldsymbol{M}_i + \boldsymbol{t}\right)\right)$ is the projection of $\boldsymbol{M}_i$ on the Euclidean plane $Z = 1$ in the camera coordinate frame. The cost function (1) is often employed with non-linear least squares to iteratively refine an initial camera pose estimate; see, e.g., the use of the Gauss-Newton algorithm in [28,49].

Instead of comparing image projections, a slightly different formulation introduced in [17] considers the sum of squared differences between the measured direction vectors and those estimated from world points transformed in the camera frame, i.e. $\sum_{i=1}^{n} \left\| \boldsymbol{u}_i - \frac{\left(\boldsymbol{R}\boldsymbol{M}_i + \boldsymbol{t}\right)}{\left\|\boldsymbol{R}\boldsymbol{M}_i + \boldsymbol{t}\right\|} \right\|^2$, where $\boldsymbol{u}_i = \frac{\boldsymbol{m}_i}{\|\boldsymbol{m}_i\|}$ is the unit direction vector associated with the measured Euclidean projection $\boldsymbol{m}_i$ in the camera frame.

1.1 Related Work

Over the last two decades, numerous solutions to the PnP problem have been proposed. In many cases, authors target specializations of the problem for a specific number of points. Among them, of particular interest is the minimal case for $n = 3$, i.e. P3P, which can be solved analytically by assuming noise-free data [14–16,25]. When $n > 3$, multiple P3P solutions are typically used in the context of random sampling schemes [10] to identify mismatches and determine pose [32,42]. This approach is usually effective, however using minimal sets can also produce skewed estimates if the data points are very noisy [43].

For the generic (i.e., $n \geq 3$) PnP problem, Lepetit et al. [28] proposed EPnP, in which an initial pose estimate is obtained by rotating and translating the 3D points in their eigenspace and thereafter, solving the least squares (LS) formulation without the orthonormality constraints. Optionally, the estimate can be improved iteratively. Although EPnP can yield very good results, it nonetheless relies on an unconstrained LS estimate, which can be skewed by noise in the data. It can also get trapped in local minima, particularly for small size inputs. There have been a few extensions to EPnP such as the one proposed by Ferraz et al. [9], whose main improvement upon the original algorithm is an iterative outlier rejection scheme that enforces a rank-1 constraint on the data matrix. A technique with procedural similarities to EPnP was proposed by Urban et al. [48] with a cost function that penalizes the misalignment of a reconstructed ray from the measured bearing and solved with ordinary least squares to obtain an initial solution. Amongst general solvers there is also a purely iterative method proposed by Lu et al. [34], which is initialized with a weak perspective approximation and refines the rotation matrix via successive solutions to the absolute orientation problem [20]. All the aforementioned methods can efficiently solve the general PnP problem, however they are all heuristic in nature and do not account for particular data configurations which yield multiple minima (e.g., P3P, P4P with coplanar points, etc.).

The largest class of PnP solvers in the literature comprises methods that render the problem unconstrained by utilizing a minimal degree of freedom (DoF) parameterization scheme for the rotation matrix [46,47]. The corresponding first order optimality conditions of the unconstrained optimization problem yield a complicated system of (at least cubic) polynomials. Solving polynomial systems with the aid of Gröbner basis solvers is a plausible approach [5,26,35]. Despite that these solvers tend to perform better in practice than their overall exponential rating [2], they cannot provide strict limits on execution time. This can be a serious drawback in applications such as robotic localization, which demand results within tight time constraints. One of the earliest such methods is that of Hesch and Roumeliotis [17], who proposed the Direct Least Squares (DLS) algorithm that employs the Cayley transform [8] to parametrize the rotation matrix and solve with resultants a quartic polynomial system obtained by the first order optimality conditions. DLS inherits a singularity for 180° rotations from its use of the Cayley transform [47]. A very similar solution was later proposed by Zheng et al. [51,52], with improved rotation parameterization for singularity avoidance and a Gröbner polynomial solver. In principle, both methods require a number of elimination steps to recover in the order of 40 solutions which are later substituted in the cost function to determine the global minimum.

Kukelova et al. [27] simplified the use of Gröbner basis solvers by introducing an automatic solver generator suited for computer vision problems. This generator revived the interest in polynomial solvers and instigated a new work cycle on the PnP problem. Bujnak et al. [6,7] used it to propose solutions for the P3P and P4P with unknown focal length problems and reported relatively short execution times. A more recent work that makes use of the automatic generator is that by Nakano [37]. He derives an optimality condition without Lagrange multipliers and proposes a globally optimal DLS method, parameterized by the Cayley representation. It is also worth mentioning here the RPnP method by Li et al. [29] which uses a relaxation of the PnP cost function by partitioning the points in triads so that each triad can minimize the P3P condition under a common unknown parameter. Owing to the partitioning, the first order conditions of RPnP are an eighth-degree polynomial in a single variable that can be readily solved numerically. Despite its efficiency, the method is suboptimal due to the bias associated with the choice of two common points in all triads. This fact is acknowledged by Wang et al. [49], who extend RPnP by incorporating a Gauss-Newton refinement of its estimate.

1.2 The PnP as a Quadratic Program with Quadratic Constraints

The objective function (Eq. (2)) that will be suggested in Sect. 2 along with the constraints on the elements of the rotation matrix can be used to cast the PnP problem as a quadratically constrained quadratic program (QCQP). Equality-constrained QCQPs are generally non-convex, NP-hard problems [21,41] and there has been a substantial number of contributions dealing with various manifestations of these problems in the optimization literature, e.g. [1,3,11,19,22].

To the best of our knowledge, only Schweighofer and Pinz [45] have approached the PnP problem as a QCQP. In that work, the problem is transformed into a semi-definite positive program that is solved with general purpose software. To achieve this, the original problem is substituted with a relaxation that seeks to maximize a lower bound of the global minimum expressed as a sum of squares polynomials. The method is generally effective, yet overly slow and requires slightly different approaches for special cases (e.g. coplanar points) as well as careful parameter tuning to achieve convergence.

1.3 Contributions

In this work we present an algorithm, called SQPnP, which casts PnP as a non-linear quadratic program (NLQP) with a cost function similar to these in [34,45]. However, instead of solving a relaxation of the problem or a polynomial system on the rotation parameters, our approach concentrates on special feasible points, from which it locates a small set of regional minima, guaranteed to contain the global one. Our contribution is two-fold:

1. Present a novel non-polynomial solver that possesses a number of desirable features generally not jointly present in existing methods:
   - It is truly generic and solves the PnP for any number and/or spatial arrangement of points without the need for any special treatment.
   - It is resilient to noise and recovers the global optima with the same or higher accuracy and consistency than those of state-of-the-art polynomial solvers.
   - Its complexity scales linearly with the number of points by virtue of invoking a small number of local searches, each completed in a bounded number of steps.

  - Its implementation is relatively short and simple, requiring only standard linear algebra operations.
2. Establish a novel mathematical framework which fully justifies the efficiency of the solver and provides a walk-through of the search space towards the global minimum.

2 Method

We consider the following cost function stemming from the sum of squared reprojection errors in (1):

$$\mathcal{E}^2 = \sum_{i=1}^{n} \left\| \mathbf{1}_z^T \left(\boldsymbol{R}\boldsymbol{M}_i + \boldsymbol{t} \right) \boldsymbol{m}_i - \left(\boldsymbol{R}\boldsymbol{M}_i + \boldsymbol{t} \right) \right\|^2 \tag{2}$$

The squared terms in Eq. (2) penalize the distances between the reconstructed and the actual 3D points in the camera frame. This type of rearrangement in the cost function that replaces the reprojection error with a back-projection error is common in pertinent literature (e.g., [17,34]) and facilitates algebraic manipulations that lead to the elimination of the unknown translation $\boldsymbol{t}$.

Let now $\boldsymbol{r} \in \mathbb{R}^9$ be the vector formed by stacking the rows of $\boldsymbol{R}$. For future use, we will denote the inverse operation as $\text{mat}(\boldsymbol{r}) = \boldsymbol{R}$. For a world point $\boldsymbol{M}_i$, we denote with $\boldsymbol{A}_i \in \mathbb{R}^{3\times 9}$ the matrix

$$\boldsymbol{A}_i = \begin{bmatrix} \boldsymbol{M}_i^T & \mathbf{0}_3^T & \mathbf{0}_3^T \\ \mathbf{0}_3^T & \boldsymbol{M}_i^T & \mathbf{0}_3^T \\ \mathbf{0}_3^T & \mathbf{0}_3^T & \boldsymbol{M}_i^T \end{bmatrix}, \tag{3}$$

so that $\boldsymbol{R}\boldsymbol{M}_i = \boldsymbol{A}_i\boldsymbol{r}$. Substituting this into Eq. (2) and making use of the fact that $\mathbf{1}_z^T \left(\boldsymbol{R}\boldsymbol{M}_i + \boldsymbol{t} \right) \boldsymbol{m}_i = \boldsymbol{m}_i \mathbf{1}_z^T \left(\boldsymbol{R}\boldsymbol{M}_i + \boldsymbol{t} \right)$, the cost function can be factored into

$$\mathcal{E}^2 = \sum_{i=1}^{n} \left(\boldsymbol{A}_i \boldsymbol{r} + \boldsymbol{t} \right)^T \boldsymbol{Q}_i \left(\boldsymbol{A}_i \boldsymbol{r} + \boldsymbol{t} \right), \tag{4}$$

where $\boldsymbol{Q}_i$ is a symmetric positive semidefinite (PSD) matrix associated with the normalized Euclidean projection $\boldsymbol{m}_i$ via $\boldsymbol{Q}_i = \left(\boldsymbol{m}_i \mathbf{1}_z^T - \boldsymbol{I}_3 \right)^T \left(\boldsymbol{m}_i \mathbf{1}_z^T - \boldsymbol{I}_3 \right)$.

By considering the first order optimality conditions, $\boldsymbol{t}$ can be eliminated from the factorized cost of Eq. (4). Zeroing the derivative of $\mathcal{E}^2$ with respect to $\boldsymbol{t}$ yields:

$$\sum_{i=1}^{n} \boldsymbol{Q}_i \left(\boldsymbol{A}_i \boldsymbol{r} + \boldsymbol{t} \right) = 0 \Leftrightarrow \left(\sum_{i=1}^{n} \boldsymbol{Q}_i \right) \boldsymbol{t} = - \left(\sum_{i=1}^{n} \boldsymbol{Q}_i \boldsymbol{A}_i \right) \boldsymbol{r}$$

Proposition 1. *The matrix* $\sum_{i=1}^{n} \boldsymbol{Q}_i$ *is invertible.*

Proof. Provided in the supplementary material.

Hence, the translation vector can be directly expressed in terms of rotation $\boldsymbol{r}$ as

$$\boldsymbol{t} = \boldsymbol{P}\boldsymbol{r}, \tag{5}$$

where

$$\boldsymbol{P} = -\left(\sum_{i=1}^{n} \boldsymbol{Q}_i\right)^{-1} \left(\sum_{i=1}^{n} \boldsymbol{Q}_i \boldsymbol{A}_i\right). \tag{6}$$

Substituting Eq. (5) in Eq. (4) yields a squared error quadratic expression in terms of the rotation vector $\boldsymbol{r}$ only, i.e. $\mathcal{E}^2 = \boldsymbol{r}^T \boldsymbol{\Omega} \boldsymbol{r}$, where $\boldsymbol{\Omega}$ is the 9×9 PSD matrix

$$\boldsymbol{\Omega} = \sum_{i=1}^{n} (\boldsymbol{A}_i + \boldsymbol{P})^T \boldsymbol{Q}_i (\boldsymbol{A}_i + \boldsymbol{P}). \tag{7}$$

Thus far, $\boldsymbol{r}$ has been assumed to represent a valid rotation matrix. We may now cast the PnP problem as a NLQP[1] over an unknown vector denoted $\boldsymbol{x}$ to emphasize that it can assume values in $\mathbb{R}^9$ that do not correspond to valid rotation matrices. Unless otherwise specified, we reserve the notation $\boldsymbol{r}$ to imply a rotation matrix. The NLQP is then

$$\underset{\boldsymbol{x} \in \mathbb{R}^9}{\text{minimize}} \; \boldsymbol{x}^T \boldsymbol{\Omega} \boldsymbol{x} \; \text{ s.t. } \; \boldsymbol{h}(\boldsymbol{x}) = \boldsymbol{0}_6, \tag{8}$$

where $\boldsymbol{\Omega}$ is given by Eq. (7) and $\boldsymbol{h}(\boldsymbol{x}) \in \mathbb{R}^6$ is the vector of constraints ensuring that when $\boldsymbol{h}(\boldsymbol{x}) = \boldsymbol{0}_6$, vector $\boldsymbol{x}$ represents a rotation matrix:

$$\boldsymbol{h}(\boldsymbol{x}) = \left[\boldsymbol{x}_{1:3}^T \boldsymbol{x}_{1:3} - 1,\; \boldsymbol{x}_{4:6}^T \boldsymbol{x}_{4:6} - 1,\; \boldsymbol{x}_{1:3}^T \boldsymbol{x}_{4:6},\; \boldsymbol{x}_{1:3}^T \boldsymbol{x}_{7:9},\; \boldsymbol{x}_{4:6}^T \boldsymbol{x}_{7:9},\; \det(\text{mat}(\boldsymbol{x})) - 1\right]^T,$$

where $\boldsymbol{x}_{i:j}$ denotes the subvector of $\boldsymbol{x}$ from the ith to the jth component. Note that the unit norm constraint for $\boldsymbol{x}_{7:9}$ is redundant and therefore omitted from the components of $\boldsymbol{h}$. These constraints will be henceforth referred to as proper orthonormality constraints and $\boldsymbol{h}(\boldsymbol{x})$ as the proper orthonormality function.

Proposition 2. *Define* $\boldsymbol{H}_{\mathbf{x}} \equiv \left.\frac{\partial \boldsymbol{h}(\boldsymbol{x})}{\partial \boldsymbol{x}}\right|_{\boldsymbol{x}=\mathbf{x}} \in \mathbb{R}^{6\times 9}$ *to be the Jacobian matrix of the proper orthonormality function at* $\mathbf{x}$. *If* $\text{rank}(\text{mat}(\mathbf{x})) \geq 2$, *then* $\text{rank}(\boldsymbol{H}_{\mathbf{x}}) = 6$.

Proof. Provided in the supplementary material. □

Proposition 2 will be useful in showing that the quadratic programming algorithm adapted for our method will always involve a non-singular system matrix (Sect. 2.2); additionally, it implies that when $\text{rank}(\text{mat}(\mathbf{x})) \geq 2$, the null space of $\boldsymbol{H}_{\mathbf{x}}$ is 3-dimensional, which reflects the fact that rotations have 3 DoF. Note that the null space of $\boldsymbol{H}_{\mathbf{x}}$ is also the tangent space of the 3D rotation group $\mathcal{SO}(3)$ at $\mathbf{x}$, hence the two terms will be interchangeable throughout this text.

[1] For brevity, the determinant constraint we employ here is cubic. Alternatively, it can be imposed with 3 quadratic constraints, thereby the formulation becomes a QCQP.

2.1 Minima on the 8-Sphere

Rather than parametrizing the rotation with a minimal representation, we next consider the problem of finding feasible solutions, i.e. proper orthogonal matrices, as 9-vectors for the NLQP of (8). Clearly, the feasible set of the constrained program lies on the hypersphere of radius $\sqrt{3}$ centered at the origin of $\mathbb{R}^9$, henceforth referred to as the 8-sphere and simply denoted by $\mathbb{S}^8$. For simplicity, we assume that $\boldsymbol{\Omega}$ has exactly nine non-vanishing eigenvalues. However, as will be explained in Sect. 2.3, the results that follow also hold in the general case where $\boldsymbol{\Omega}$ is singular. Consider now a relaxed, more general problem, specifically that of finding the minima of the quadratic function defined on $\mathbb{S}^8$:

$$f(\boldsymbol{x}) = \boldsymbol{x}^T \boldsymbol{\Omega} \boldsymbol{x},\ \boldsymbol{x} \in \mathbb{S}^8. \tag{9}$$

It is a well-known fact that the stationary points of a unit-norm constrained quadratic function are the eigenvectors of $\boldsymbol{\Omega}$ [13, Thm. 8.5], which we will henceforth denote $\boldsymbol{e}_1, \ldots, \boldsymbol{e}_9$, in descending order of the eigenvalues $s_1 > \cdots > s_9$. Thus, the function f in Eq. (9) has exactly 18 distinct stationary points $\boldsymbol{x}_1, \ldots, \boldsymbol{x}_{18} \in \mathbb{S}^8$ that correspond to the 9 eigenvectors of $\boldsymbol{\Omega}$ scaled by $\pm\sqrt{3}$, i.e.

$$\boldsymbol{x}_1 = +\sqrt{3}\boldsymbol{e}_1,\ \ldots,\ \boldsymbol{x}_9 = +\sqrt{3}\boldsymbol{e}_9,\ \ \boldsymbol{x}_{10} = -\sqrt{3}\boldsymbol{e}_1,\ \ldots,\ \boldsymbol{x}_{18} = -\sqrt{3}\boldsymbol{e}_9. \tag{10}$$

In the rest of this section, we establish a close relationship between the feasible local minima and the solutions of the nearest orthogonal matrix problem (NOMP) [18] in a special spherical region around a minimizer of f. Through Propositions 3, 4, 5, we ensure that the aforementioned spherical region always contains feasible solutions of the fully constrained QCQP and, through Proposition 6, that the pertinent minima can be exhaustively traced from the solutions of the NOMP associated with the minimizer of f.

Proposition 3. *The function $f(\boldsymbol{x}) = \boldsymbol{x}^T \boldsymbol{\Omega} \boldsymbol{x}$, $\boldsymbol{x} \in \mathbb{S}^8$, is convex in a region of the 8-sphere of radius $\sqrt{3}$ that contains a local minimum and the spherical points that form angles less than $90°$ with the minimum.*

Proof. If $\boldsymbol{e}$ is a local minimizer of f, then it is an eigenvector of $\boldsymbol{\Omega}$. Since $\boldsymbol{\Omega}$ is a PSD matrix, its eigenvectors are mutually orthogonal. Thus, the nearest inflection point to $\boldsymbol{e}$ must also be an eigenvector of $\boldsymbol{\Omega}$ and forms an angle of at least $90°$ with $\boldsymbol{e}$. □

Proposition 4. *Let $\boldsymbol{e} \in \mathbb{R}^9$ s.t. $\|\boldsymbol{e}\| = 1$. If $\boldsymbol{r}$ represents a rotation minimizing*

$$\boldsymbol{r} = \underset{\text{mat}(\boldsymbol{x}) \in \mathcal{SO}(3)}{\text{argmin}} \left\| \boldsymbol{x} - \sqrt{3}\boldsymbol{e} \right\|^2, \tag{11}$$

then the angle between vectors $\boldsymbol{e}$ and $\boldsymbol{r}$ is strictly less than $71°$.

Proof. Provided in the supplementary material. □

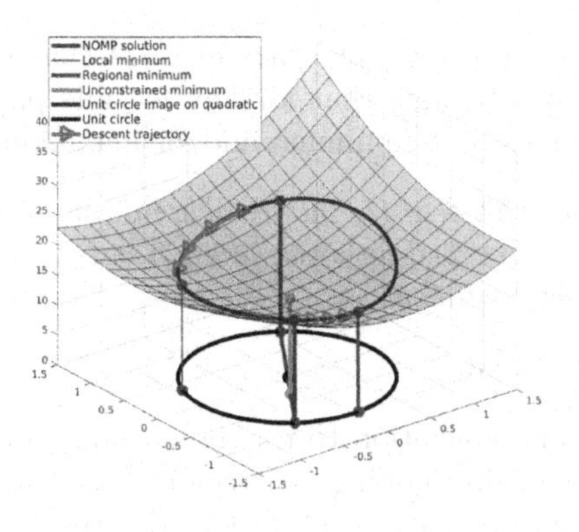

(a) Descent to regional minima.

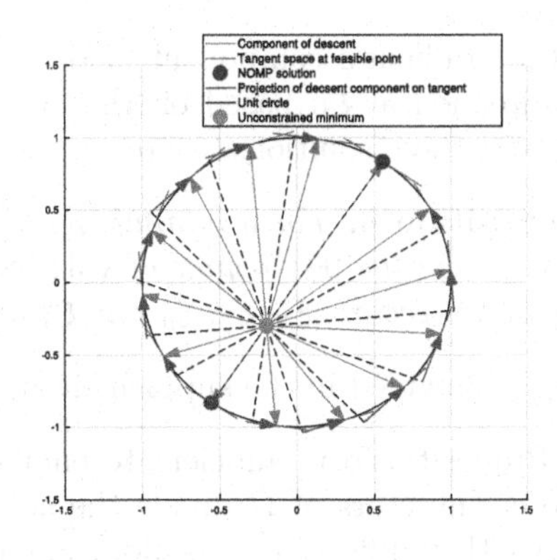

(b) Gradient descent component.

Fig. 2. Illustration of regional search for minima from the solutions of the NOMP (antipodal dark green points on the unit circle) for the case where $\boldsymbol{R} \in \mathcal{SO}(2)$. (a) Descent from the two NOMP solutions leads to the two local minima shown in red. (b) The component of the gradient responsible for descent (light green arrows actually drawn in the direction of ascent for illustration purposes) at a feasible point is the direction vector from the minimum of the parabola to that point; the projection of this component on the tangent at the feasible point changes direction on the NOMP solutions, thus ensuring there can be at most one minimum and/or maximum in a feasible path between any two such solutions.

Propositions 3 and 4 suggest that we can identify 90° regions of convexity of f which are guaranteed to contain a non-empty set of feasible solutions, i.e. rotations. For $\boldsymbol{x} \in \mathbb{R}^9$, the Euclidean norm $\|\boldsymbol{x}\|$ equals the Frobenius norm $\|\text{mat}(\boldsymbol{x})\|_F$. Hence, Eq. (11) amounts to finding the orthogonal matrix minimizing the Frobenius distance from a given matrix, i.e., the NOMP [18].

Proposition 5. *For* $\boldsymbol{e} \in \mathbb{R}^9$ *with* $\|\boldsymbol{e}\|^2 = 1$, *there exist exactly 4 vectors* $\boldsymbol{\xi}_1, \boldsymbol{\xi}_2, \boldsymbol{\xi}_3$ *and* $\boldsymbol{\xi}_4$ *with* $\text{mat}(\boldsymbol{\xi}_i) \in \mathcal{O}(3)$ *in the 90° region of* $\sqrt{3}\boldsymbol{e}$ *for which the vectors* $\sqrt{3}\boldsymbol{e} - \boldsymbol{\xi}_i$ *are orthogonal to the tangent space of* $\mathcal{O}(3)$ *at* $\boldsymbol{\xi}_i$.

Proof. Provided in the supplementary material. □

Note here that $\mathcal{O}(3)$ is the orthogonal group of dimension 3, consisting of all 3×3 orthogonal matrices. These matrices have determinant either 1 or –1. The PnP quadratic program of Eq. (8) can be equivalently cast with orthogonality constraints, owing to the fact that the value of f for an orthogonal matrix $\boldsymbol{\xi}$ with determinant –1 is the same as the one for the rotation $\boldsymbol{r} = -\boldsymbol{\xi}$. Thus, by focusing on simply orthogonal matrices in the region of $\sqrt{3}\boldsymbol{e}$ allows us to study the original program in the 90° region of $\sqrt{3}\boldsymbol{e}$ where the behavior of f is known.

In the special case where $\boldsymbol{e}$ is a minimizing eigenvector, the direction of the projection of $\sqrt{3}\boldsymbol{e} - \boldsymbol{\xi}_i$ on the tangent space of the sphere at $\boldsymbol{\xi}_i$ should be the component of the gradient of f responsible for descent (Fig. 2(b)), owing to the fact that f is convex in the 90° region of $\sqrt{3}\boldsymbol{e}$. It follows that the projection of the gradient's component of descent on the tangent space of $\mathcal{O}(3)$ changes its direction at $\boldsymbol{\xi}_i$; based on the latter we conclude that for any feasible path between these solutions, there can be at most one minimum and/or maximum:

Proposition 6. *For a minimizing eigenvector* $\mathbf{e}$ *of* $\boldsymbol{\Omega}$*, the feasible minimum in* $\mathcal{O}(3)$ *inside the* 90° *region of* $\sqrt{3}\,\mathbf{e}$ *can be reached by descending from at least one of the vectors* $\boldsymbol{\xi}_1$, $\boldsymbol{\xi}_2$, $\boldsymbol{\xi}_3$ *and* $\boldsymbol{\xi}_4$ *of Proposition 5.*

Proof. Provided in the supplementary material. □

Proposition 6 is sufficient to enable safe navigation to the global minimum. The eigenvectors of $\boldsymbol{\Omega}$ divide the sphere in overlapping $90°$ regions associated with either inflection or saddle points on the surface of f. Clearly, the best candidates to begin the search for the global minimum of the fully constrained problem are the global minima of f associated with eigenvector $\boldsymbol{e}_9$. The feasible regional minimizers are located by initiating a descent from the orthogonal matrix vector nearest to the minimum of f. Similar searches may be performed in the regions of eigenvectors that may simply be saddle points of f, owing to the overlap of the pertinent regions with regions associated with minimizers. Figure 2 visualizes this process in 3D for the special PnP case where $\boldsymbol{R} \in \mathcal{SO}(2)$. $\mathcal{SO}(2)$ is effectively a 2D circle and f is a quadratic translated from the origin.

In practice, we do not need to thoroughly examine all eigenvectors as the fact that the corresponding stationary values of f are the eigenvalues of $\boldsymbol{\Omega}$ scaled by 3 can be used to avoid unnecessary regional searches [13]. Thus, we begin from the region of $\sqrt{3}\,\boldsymbol{e}_9$ and, if the recovered regional minimum has a value above $3\,s_8$, we examine the region $\sqrt{3}\,\boldsymbol{e}_8$ and repeat in ascending order of eigenvalues, until one of the new minima is less than the remaining (scaled by 3) eigenvalues, or the set of eigenvectors is exhausted.

2.2 Sequential Quadratic Programming

Sequential quadratic programming (SQP) is an iterative technique for solving non-linear constrained optimization problems [4,12,38]. The core idea in SQP is to approximate the cost function with a quadratic and the constraints with linear functions in order to produce a linearly constrained quadratic program (LCQP) which can be solved analytically with a linear system that comprises the first order conditions of the Lagrangian function and the linearized constraints. The solution of the linearly constrained quadratic program yields a perturbation in the vector of unknowns and the vector of Lagrange multipliers at the solution. The non-linear program is subsequently approximated at the new estimate and the process is repeated until convergence.

In the case of the NLQP for the PnP problem (8), we introduce the SQP approximation at a feasible point $\boldsymbol{r}$ on the 8-sphere of radius $\sqrt{3}$. Since the cost function is already a quadratic, we express it in terms of the difference $\boldsymbol{\delta} = \boldsymbol{x} - \boldsymbol{r}$ and linearize the function $\boldsymbol{h}(\boldsymbol{x})$ using its first order Taylor approximation:

$$\underset{\boldsymbol{\delta}\in\mathbb{R}^9}{\text{minimize}}\; \boldsymbol{\delta}^T \boldsymbol{\Omega}\,\boldsymbol{\delta} + 2\boldsymbol{r}^T\boldsymbol{\Omega}\,\boldsymbol{\delta} \;\text{ s.t. }\; \boldsymbol{H}_{\boldsymbol{r}}\,\boldsymbol{\delta} = -\boldsymbol{h}(\boldsymbol{r}). \tag{12}$$

The first order conditions of the Lagrangian function along with the linear constraints of the LCQP in Eq. (12) yield a linear system, the solution of which is a descent direction that converges towards[2] or stays on a trajectory of feasible solutions. For the solution $\hat{\boldsymbol{\delta}}$ of the linear system, the NLQP is approximated at a new point $\boldsymbol{r}' = \boldsymbol{r} + \hat{\boldsymbol{\delta}}$ and a new descent direction is obtained. The process is repeated until the norm of $\hat{\boldsymbol{\delta}}$ drops below a threshold.

Proposition 7. *Let $\boldsymbol{r} \in \mathbb{R}^9$ be the estimate of the rotation matrix which may not be feasible at some step of the SQP. If $\operatorname{rank}(\boldsymbol{\Omega}) \geq 3$, then the linearly constrained quadratic program of Eq. (12) has a unique solution in $\mathbb{R}^9$.*

Proof. Provided in the supplementary material. □

Proposition 7 ensures that the SQP linear system will have a unique solution in every step of the process. This is because in the presence of non-degenerate data, $\operatorname{rank}(\boldsymbol{\Omega})$ will be at least 3 for the problem to be fully constrained.

To solve the PnP, the aim is to recover the regional minima associated initially with the global minimizer $\boldsymbol{e}_9$ of the quadratic $f(\boldsymbol{x}) = \boldsymbol{x}^T \boldsymbol{\Omega} \boldsymbol{x}$ on the 8-sphere and, if necessary, proceed to repeat the process in the region of the next eigenvector in ascending order of eigenvalues (cf. Sect. 2.1).

As explained in Sect. 2.1, we may recover the regional minimum associated with $\boldsymbol{e}$ by descending along the feasible path from the solutions $\boldsymbol{\xi}_1, \boldsymbol{\xi}_2, \boldsymbol{\xi}_3$ and $\boldsymbol{\xi}_4$ (cf. Proposition 5, 6) of the NOMP associated with $\sqrt{3}\boldsymbol{e}$. We empirically determined that descending only from the two nearest of $\boldsymbol{\xi}_1, \boldsymbol{\xi}_2, \boldsymbol{\xi}_3, \boldsymbol{\xi}_4$ to $\sqrt{3}\boldsymbol{e}$ allows the method to converge to the global minimum. Finding the two nearest solutions of the NOMP related to eigenvector $\boldsymbol{e}$ is equivalent to finding the rotation matrices $\boldsymbol{r}_1, \boldsymbol{r}_2$ nearest to $\sqrt{3}\boldsymbol{e}$, $-\sqrt{3}\boldsymbol{e}$ respectively. Thus, each inspected eigenvector $\boldsymbol{e}$ contributes to the overall search with two minima recovered via SQP descent from the following rotations:

$$\boldsymbol{r}_1 = \underset{\operatorname{mat}(\boldsymbol{x}) \in \mathcal{SO}(3)}{\operatorname{argmin}} \left\| \boldsymbol{x} - \sqrt{3}\boldsymbol{e} \right\|^2, \quad \boldsymbol{r}_2 = \underset{\operatorname{mat}(\boldsymbol{x}) \in \mathcal{SO}(3)}{\operatorname{argmin}} \left\| \boldsymbol{x} + \sqrt{3}\boldsymbol{e} \right\|^2. \tag{13}$$

2.3 The General Case

We have thus far assumed that the PSD data matrix $\boldsymbol{\Omega}$ has an empty null space. However, this is not generally the case, particularly when n is small. In these cases, the intersection of the null space with the sphere of radius $\sqrt{3}$, referred to here as the null sphere, is treated as a generalized minimum. This "minimum" is a flat region for f, which can only contain a finite number of solutions. However, we know that the 90° regions of the null space basis vectors are not entirely flat due to overlap with the corresponding regions of eigenvectors with non-vanishing eigenvalues. Based on this, we generalize the approach of Sect. 2.2 and devise a number of SQPs equal to the number of null space vectors with feasible

[2] During the first few steps of SQP, the solutions may not be entirely feasible due to inaccuracies in the linear approximations of the constraints.

starting points obtained as in Eq. (13). More formally, suppose that the last (in descending eigenvalue order) k eigenvectors of $\boldsymbol{\Omega}$ are the null-space basis, $\text{null}(\boldsymbol{\Omega}) = \langle \boldsymbol{e}_{10-k}, \ldots, \boldsymbol{e}_9 \rangle$, $k \geq 1$. We then perform $2k$ SQPs with starting points

$$\boldsymbol{r}_i = \underset{\text{mat}(\boldsymbol{x}) \in \mathcal{SO}(3)}{\text{argmin}} \left\| \boldsymbol{x} - (-1)^{\lfloor (i-1)/k \rfloor} \sqrt{3}\, \boldsymbol{e}_{9-k+i-\lfloor i/k \rfloor k} \right\|^2, \tag{14}$$

where $i \in \{1, \ldots, 2k\}$. Note here that large null spaces (up to 6 basis vectors) are typically associated with small numbers of points (up to 6) and therefore multiple solutions may exist. In these cases, the typical treatment involving the positive depth test applies. The overall procedure is detailed in Algorithm 1.

3 The SQPnP Algorithm

SQPnP is described in pseudocode as Algorithms 1 and 2 that jointly yield the global minima of the PnP problem. Algorithm 1 solves the PnP problem by delivering a list of minimizers that contains the global minima. The algorithm computes the PSD data matrix $\boldsymbol{\Omega}$, the matrix $\boldsymbol{P}$ required by Eq. (5) for the computation of the translation vector, and the feasible starting points from which it initiates iterative searches using SQP, as detailed by Algorithm 2. SQP typically converges within 10 iterations, hence the recommendation $T \geq 15$. Similarly, we empirically determined that 10^{-5} suffices as the perturbation norm tolerance.

4 Experiments

Based on Matlab implementations, we report next results from the comparison of SQPnP with several prominent PnP methods, namely DLS [17], LHM [34], RPnP [29], OPnP [51], MLPnP [48], REPPnP [9], EPnP [28] and optDLS [37]; both [17,37] include the three rotations by 90° preprocessing to avoid the Cayley singularity. Our investigation focuses primarily on the reprojection error achieved by these methods for increasing numbers of points n and amounts of noise. However, acknowledging that execution time is crucial for many practical applications, we also provide timing measurements with the forewarning that Matlab implementations are less efficient than implementations in compiled languages such as C++. We exclude UPnP [24] from our comparison since, as detailed in [37], it is a suboptimal method worse than optDLS and OPnP.

SQPnP calls for i) the computation of the data matrix $\boldsymbol{\Omega}$ in Eq. (7), which is linear in the number of points and ii) the search for minima via SQPs which are bounded by a finite number of iterations and starting points. Therefore, the execution time of SQPnP has a baseline offset which largely depends on the cost of the linear system solution required in every SQP iteration (cf. Algorithm 2). Our current implementation employs Matlab's general-purpose linear system solver `linsolve`. However, this operation can be considerably accelerated by exploiting the special structure of the system's matrix and $\boldsymbol{H}_{\mathbf{x}}$ in particular.

The nearest orthogonal approximation problem in Eq. (14) is solved without costly matrix factorizations using the FOAM algorithm, as discussed in [30,31].

Algorithm 1. SQPnP: SolvePnP

Require:
Number of points, $n \geq 3$
World points $\boldsymbol{M}_i \in \mathbb{R}^3$, $1 \leq i \leq n$
Projections $\boldsymbol{m}_i = \left[x_i\ y_i\ 1\right]^T$, $1 \leq i \leq n$

Perturbation tolerance, $\epsilon \leq 10^{-5}$
Maximum number of iterations, $T \geq 15$

$\{\boldsymbol{\Omega}, \boldsymbol{P}\} \leftarrow$ eqs. (6), (7)
$\{\boldsymbol{e}_1, \ldots, \boldsymbol{e}_9, s_1, \ldots, s_9\} \leftarrow \mathrm{SVD}(\boldsymbol{\Omega})$
$\{\boldsymbol{e}_{10-k}, \ldots, \boldsymbol{e}_9\} \leftarrow \underset{\boldsymbol{x} \in \mathbb{S}^8}{\operatorname{argmin}}\ \boldsymbol{x}^T \boldsymbol{\Omega} \boldsymbol{x}$
for $i \leftarrow 1$ **to** $2k$ **do**
  $\mu \leftarrow \lfloor (i-1)/k \rfloor$
  $\nu \leftarrow 9 - k + i - \lfloor i/k \rfloor k$
  $\boldsymbol{r}_i \leftarrow \underset{\mathrm{mat}(\boldsymbol{x}) \in \mathcal{SO}(3)}{\operatorname{argmin}} \left\|\boldsymbol{x} - (-1)^{\mu}\sqrt{3}\boldsymbol{e}_{\nu}\right\|^2$
  $\hat{\boldsymbol{r}}_i \leftarrow \mathrm{SolveSQP}(\boldsymbol{r}_i, \boldsymbol{\Omega}, \epsilon, T)$
  $\mathcal{E}_i^2 \leftarrow \hat{\boldsymbol{r}}_i^T \boldsymbol{\Omega} \hat{\boldsymbol{r}}_i$
end for

while $\min\left\{\mathcal{E}_1^2, \ldots, \mathcal{E}_{2k}^2\right\} \geq s_{9-k}$ **do**
  for $i \leftarrow 1$ **to** 2 **do**
    $\boldsymbol{r}_{2k+i} \leftarrow \underset{\mathrm{mat}(\boldsymbol{x}) \in \mathcal{SO}(3)}{\operatorname{argmin}} \left\|\boldsymbol{x} - (-1)^{i}\sqrt{3}\boldsymbol{e}_{9-k}\right\|^2$
    $\hat{\boldsymbol{r}}_{2k+i} \leftarrow \mathrm{SolveSQP}(\boldsymbol{r}_{2k+i}, \boldsymbol{\Omega}, \epsilon, T)$
    $\mathcal{E}_{2k+i}^2 \leftarrow \hat{\boldsymbol{r}}_{2k+i}^T \boldsymbol{\Omega} \hat{\boldsymbol{r}}_{2k+i}$
  end for
  $k \leftarrow k + 1$
end while
return $\hat{\boldsymbol{r}}_1, \ldots, \hat{\boldsymbol{r}}_{2k}, \mathcal{E}_1^2, \ldots, \mathcal{E}_{2k}^2$

Algorithm 2. SQPnP: SolveSQP

Require:
Starting point $\boldsymbol{r} \in \mathbb{R}^9$, s.t. $\mathrm{mat}(\boldsymbol{r}) \in \mathcal{SO}(3)$
Data matrix, $\boldsymbol{\Omega} \in \mathbb{R}^{9\times 9}$, $\boldsymbol{\Omega} \succeq \boldsymbol{0}$
Tolerance in perturbation estimate norm, $\epsilon \leq 10^{-5}$
Maximum number of iterations, $T \geq 15$

step $\leftarrow 0$
$\hat{\boldsymbol{r}} \leftarrow \boldsymbol{r}$
repeat
  $\boldsymbol{H}_{\hat{\boldsymbol{r}}} \leftarrow \left.\frac{\partial \boldsymbol{h}(\boldsymbol{x})}{\partial \boldsymbol{x}}\right|_{\boldsymbol{x}=\hat{\boldsymbol{r}}}$
  $\begin{bmatrix}\hat{\boldsymbol{\delta}} \\ \hat{\boldsymbol{\lambda}}\end{bmatrix} \leftarrow \begin{bmatrix}\boldsymbol{\Omega} & \boldsymbol{H}_{\hat{\boldsymbol{r}}}^T \\ \boldsymbol{H}_{\hat{\boldsymbol{r}}} & \boldsymbol{0}_{6\times 6}\end{bmatrix}^{-1} \begin{bmatrix}-\boldsymbol{\Omega}\hat{\boldsymbol{r}} \\ -\boldsymbol{h}(\hat{\boldsymbol{r}})\end{bmatrix}$
  $\hat{\boldsymbol{r}} \leftarrow \hat{\boldsymbol{r}} + \hat{\boldsymbol{\delta}}$
  step $\leftarrow$ step + 1
until $\left\|\hat{\boldsymbol{\delta}}\right\| < \epsilon$ **or** step $> T$
return $\hat{\boldsymbol{r}}$

4.1 Synthetic Experiments

Procedure. In our experiments, Euclidean quantities are expressed in units of meters. World 3D points were randomly sampled from an isotropic Gaussian distribution with SD 3, i.e. $\boldsymbol{M}_i \sim \mathcal{N}\left(\overline{\mathbf{M}},\ 3^2 \boldsymbol{I}_3\right)$, where $\overline{\mathbf{M}} \equiv \left[3/4\ 3/4\ 12\right]^T$. Similarly, camera poses comprising position $\boldsymbol{b}$ and MRP [47] orientation parameters $\boldsymbol{\psi}$ in the world frame are sampled from a zero-mean 6D Gaussian distribution

$$\begin{bmatrix}\boldsymbol{b} \\ \boldsymbol{\psi}\end{bmatrix} \sim \mathcal{N}\left(\mathbf{0}_6, \begin{bmatrix}\sigma_b^2 \boldsymbol{I}_3 & \mathbf{0}_3 \\ \mathbf{0}_3 & \sigma_{\psi}^2 \boldsymbol{I}_3\end{bmatrix}\right). \tag{15}$$

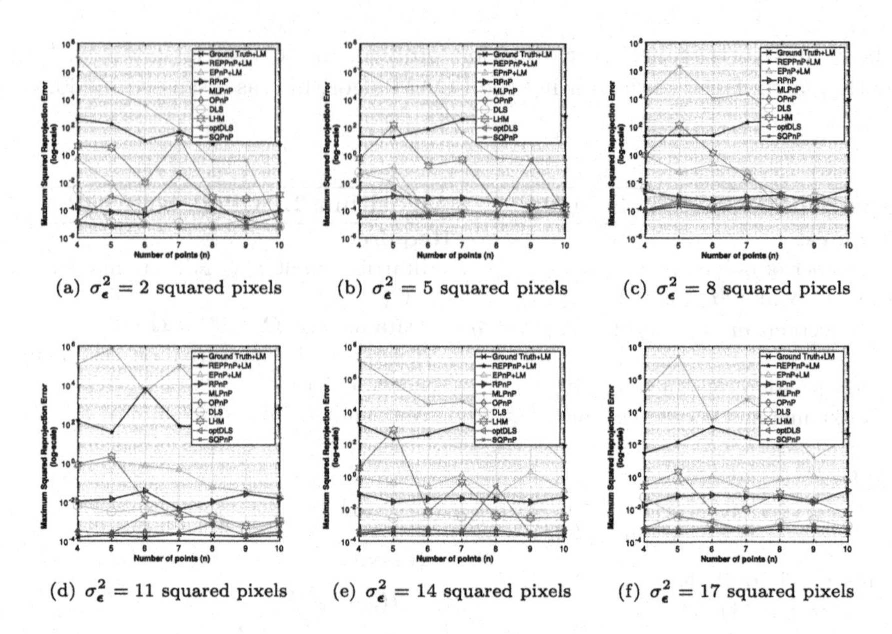

(a) $\sigma_{\epsilon}^2 = 2$ squared pixels (b) $\sigma_{\epsilon}^2 = 5$ squared pixels (c) $\sigma_{\epsilon}^2 = 8$ squared pixels

(d) $\sigma_{\epsilon}^2 = 11$ squared pixels (e) $\sigma_{\epsilon}^2 = 14$ squared pixels (f) $\sigma_{\epsilon}^2 = 17$ squared pixels

Fig. 3. Plots of maximum squared reprojection error for 500 executions of each PnP solver on n random points, $4 \leq n \leq 10$. For each n, points are repeatedly sampled from a previously generated point population contaminated with additive Gaussian noise. Each plot represents the results obtained by points drawn from a different population and whose projections were contaminated with zero-mean Gaussian noise of variance $\sigma_{\epsilon}^2 \in \{2, 5, 8, 11, 14, 17\}$ squared pixels (top left to bottom right). Notice the different scales in the vertical axes of the plots.

The values chosen for the standard deviations were sufficiently small (i.e., $\sigma_{\boldsymbol{b}} = 0.2$ and $\sigma_{\boldsymbol{\psi}} = 0.05$) to ensure that the generated points will always be in front of the simulated camera, assumed to have a focal length $f = 1400$ pixels and image size 1800×1800.

Using six different levels of additive Gaussian noise with $\sigma_{\epsilon}^2 \in \{2, 5, 8, 11, 14, 17\}$ squared pixels, we performed experiments summarized in the plots of Fig. 3. For each noise level, we generated 100 random 3D points. Then, for every $n \in \{4, \ldots, 10\}$ we randomly sampled a population of 500 sets of n 3D points each, generated a camera pose with Eq. (15) for each set, projected all sets on the image plane and perturbed the projections with noise $\boldsymbol{\epsilon}_i \sim \mathcal{N}\left(\mathbf{0}_2, \sigma_{\epsilon}^2 \boldsymbol{I}_2\right)$. For every set in a certain population, we executed all PnP solvers under comparison with default parameters[3] and calculated the maximum squared reprojection error for each solver across these executions. We also determined the reprojection error corresponding to the maximum likelihood estimate, obtained by minimizing the total reprojection error for each set's noisy points with the Levenberg-Marquardt (LM) algorithm initiated at the true pose.

[3] SQPnP employed maximum iterations $T = 15$ and tolerance $\epsilon = 10^{-8}$.

Table 1. Average and median execution times (ms) of several PnP solvers implemented in Matlab, computed across all executions for every $4 \leq n \leq 10$. Time for any additional non-linear refinement was not taken into account.

		(ours)	PnP Solver							
		SQPnP	optDLS	LHM	DLS	OPnP	MLPnP	RPnP	EPnP	REPPnP
Time	Mean	2.7	3.5	4.7	3.5	14.0	2.3	0.7	1.1	2.1
	Median	2.0	3.5	4.7	3.5	14.1	2.4	0.8	1.1	2.1

The maximum was preferred over the average squared reprojection error since we are primarily interested in demonstrating the consistency of our solver in reaching a squared error that is similar to that of the maximum likelihood estimate. Nevertheless, plots of the average squared reprojection errors corresponding to exactly the same experiments can be found in the supplementary material. For completeness, the supplementary material also includes plots of the pose translational and rotational errors for the same experiments.

It is finally noted that our experiments concern relatively small numbers of points since, in practice, these are the typical sample sizes that yield candidate solutions in sampling-based camera resectioning [10].

Results. The plots of Fig. 3 illustrate the maximum total squared reprojection errors (cf. (1)) in terms of n for each PnP solver applied to the samples drawn from the populations generated for a certain level of additive noise. The results obtained by the EPnP and REPPnP were further improved with the LM algorithm. Moreover, the plots incorporate the reprojection error corresponding to the maximum likelihood estimates (labeled "Ground Truth+LM").

Using as metric the infinity norm (i.e., maximum) of the reprojection errors pertaining to the multiple executions of the methods being compared, ensures that the plots will reflect the repeatability of the methods in approaching the minimum for a given n. As expected, we observe that SQPnP does not deviate from the ground truth more than 10^{-3} in any of the 500 executions, regardless of the number of points or levels of additive noise. SQPnP consistently approaches the minimum similarly to the polynomial solvers OPnP, DLS and optDLS that provide strong theoretical guarantees of finding the global minimizer. In doing so, SQPnP attains better accuracy compared to DLS and optDLS, and very similar to that of OPnP. OPnP performs well in all experiments, albeit at a much higher computational cost in comparison to SQPnP (cf. Table 1). In contrast, methods such as EPnP, REPPnP and MLPnP that employ unconstrained LS formulations, tend to perform equally well for small amounts of noise (cf. Figs. 3(a), 3(b)), yet give rise to erratic convergence patterns when the noise increases. The execution times of the various PnP methods are in Table 1, showing that SQPnP is competitive also in terms of computational cost.

5 Conclusion

This paper presented SQPnP, a fast and globally convergent non-polynomial PnP solver. SQPnP casts the PnP problem as a quadratically constrained quadratic program and solves it by conducting local searches in the vicinity of special feasible points from which the global minima are located in a few steps. SQPnP admits a simple implementation that requires standard linear algebra operations and incurs a low computational cost. Comparative experiments confirm that SQPnP performs competitively to state-of-the-art PnP solvers and consistently recovers the true camera pose regardless of the noise and the spatial arrangement of input data. A C++ implementation of SQPnP is available at https://github.com/terzakig/sqpnp.

Acknowledgements. M. Lourakis has been funded by the EU H2020 Programme under Grant Agreement No. 826506 (sustAGE).

References

1. Andersen, E.D., Roos, C., Terlaky, T.: On implementing a primal-dual interior-point method for conic quadratic optimization. Math. Program. **95**(2), 249–277 (2003)
2. Bardet, M., Faugere, J.C., Salvy, B.: On the complexity of Gröbner basis computation of semi-regular overdetermined algebraic equations. In: Proceedings of the International Conference on Polynomial System Solving, pp. 71–74 (2004)
3. Beck, A., Eldar, Y.C.: Strong duality in nonconvex quadratic optimization with two quadratic constraints. SIAM J. Optim. **17**(3), 844–860 (2006)
4. Boggs, P.T., Tolle, J.W.: Sequential quadratic programming. Acta Numerica **4**, 1–51 (1995)
5. Buchberger, B., Kauers, M.: Groebner basis. Scholarpedia **5**(10), 7763 (2010)
6. Bujnak, M., Kukelova, Z., Pajdla, T.: A general solution to the P4P problem for camera with unknown focal length. In: Conference on Computer Vision and Pattern Recognition, pp. 1–8. IEEE (2008)
7. Bujnak, M., Kukelova, Z., Pajdla, T.: New efficient solution to the absolute pose problem for camera with unknown focal length and radial distortion. In: Kimmel, R., Klette, R., Sugimoto, A. (eds.) ACCV 2010. LNCS, vol. 6492, pp. 11–24. Springer, Heidelberg (2011). https://doi.org/10.1007/978-3-642-19315-6_2
8. Cayley, A.: Sur quelques propriétés des déterminants gauches. J. für die reine und angewandte Mathematik **32**, 119–123 (1846)
9. Ferraz, L., Binefa, X., Moreno-Noguer, F.: Very fast solution to the PnP problem with algebraic outlier rejection. In: Proceedings of the IEEE Conference on Computer Vision and Pattern Recognition, pp. 501–508 (2014)
10. Fischler, M.A., Bolles, R.C.: Random sample consensus: a paradigm for model fitting with applications to image analysis and automated cartography. Commun. ACM **24**(6), 381–395 (1981)
11. Floudas, C.A., Visweswaran, V.: Quadratic optimization. In: Horst, R., Pardalos, P.M. (eds.) Handbook of Global Optimization. Nonconvex Optimization and Its Applications, vol. 2, pp. 217–269. Springer, Boston (1995). https://doi.org/10.1007/978-1-4615-2025-2_5

12. Forst, W., Hoffmann, D.: Optimization - Theory and Practice. Springer, Heidelberg (2010). https://doi.org/10.1007/978-0-387-78977-4
13. Fraleigh, J., Beauregard, R.: Linear Algebra. Addison-Wesley, Boston (1995)
14. Gao, X.S., Hou, X.R., Tang, J., Cheng, H.F.: Complete solution classification for the perspective-three-point problem. IEEE Trans. Pattern Anal. Mach. Intell. **25**(8), 930–943 (2003)
15. Grunert, J.: Das pothenotische Problem in erweiterter Gestalt nebst über seine Anwendungen in Geodäsie. Grunerts Archiv für Mathematik und Physik (1841)
16. Haralick, R.M., Joo, H., Lee, C.N., Zhuang, X., Vaidya, V.G., Kim, M.B.: Pose estimation from corresponding point data. IEEE Trans. Syst. Man Cybern. **19**(6), 1426–1446 (1989)
17. Hesch, J.A., Roumeliotis, S.I.: A direct least-squares (DLS) method for PnP. In: International Conference on Computer Vision, pp. 383–390. IEEE (2011)
18. Higham, N.J.: Functions of Matrices: Theory and Computation. Society for Industrial and Applied Mathematics, Philadelphia (2008)
19. Hmam, H.: Quadratic optimisation with one quadratic equality constraint. Technical Report 2416, Defence Science and Technology Organisation, Australia (2010)
20. Horn, B.K., Hilden, H.M., Negahdaripour, S.: Closed-form solution of absolute orientation using orthonormal matrices. JOSA A **5**(7), 1127–1135 (1988)
21. Johnson, D.S., Garey, M.R.: Computers and Intractability: A Guide to the Theory of NP-Completeness. WH Freeman, New York (1979)
22. Kim, S., Kojima, M.: Second order cone programming relaxation of nonconvex quadratic optimization problems. Optim. Methods Softw. **15**(3–4), 201–224 (2001)
23. Klein, G., Murray, D.: Parallel tracking and mapping on a camera phone. In: IEEE International Symposium on Mixed and Augmented Reality, pp. 83–86. IEEE (2009)
24. Kneip, L., Li, H., Seo, Y.: UPnP: an optimal $O(n)$ solution to the absolute pose problem with universal applicability. In: Fleet, D., Pajdla, T., Schiele, B., Tuytelaars, T. (eds.) ECCV 2014. LNCS, vol. 8689, pp. 127–142. Springer, Cham (2014). https://doi.org/10.1007/978-3-319-10590-1_9
25. Kneip, L., Scaramuzza, D., Siegwart, R.: A novel parametrization of the perspective-three-point problem for a direct computation of absolute camera position and orientation. In: Proceedings of the IEEE Conference on Computer Vision and Pattern Recognition, pp. 2969–2976. IEEE (2011)
26. Kukelova, Z., Bujnak, M., Pajdla, T.: Polynomial eigenvalue solutions to minimal problems in computer vision. IEEE Trans. Pattern Anal. Mach. Intell. **34**(7), 1381–1393 (2012)
27. Kukelova, Z., Bujnak, M., Pajdla, T.: Automatic generator of minimal problem solvers. In: Forsyth, D., Torr, P., Zisserman, A. (eds.) ECCV 2008. LNCS, vol. 5304, pp. 302–315. Springer, Heidelberg (2008). https://doi.org/10.1007/978-3-540-88690-7_23
28. Lepetit, V., Moreno-Noguer, F., Fua, P.: EPnP: an accurate O(n) solution to the PnP problem. Int. J. Comput. Vis. **81**(2), 155 (2009)
29. Li, S., Xu, C., Xie, M.: A robust O(n) solution to the perspective-n-point problem. IEEE Trans. Pattern Anal. Mach. Intell. **34**(7), 1444–1450 (2012)
30. Lourakis, M.: An efficient solution to absolute orientation. In: International Conference on Pattern Recognition (ICPR), pp. 3816–3819 (2016)
31. Lourakis, M., Terzakis, G.: Efficient absolute orientation revisited. In: Proceedings of the IEEE/RSJ International Conference on Intelligent Robots and Systems (IROS), pp. 5813–5818 (2018)

32. Lourakis, M., Zabulis, X.: Model-based pose estimation for rigid objects. In: Chen, M., Leibe, B., Neumann, B. (eds.) ICVS 2013. LNCS, vol. 7963, pp. 83–92. Springer, Heidelberg (2013). https://doi.org/10.1007/978-3-642-39402-7_9
33. Lourakis, M.I.A., Argyros, A.A.: Efficient, causal camera tracking in unprepared environments. Comput. Vis. Image Underst. **99**(2), 259–290 (2005)
34. Lu, C.P., Hager, G.D., Mjolsness, E.: Fast and globally convergent pose estimation from video images. IEEE Trans. Pattern Anal. Mach. Intell. **22**(6), 610–622 (2000)
35. Mora, T.: Solving Polynomial Equation Systems II: Macaulay's Paradigm and Gröbner Technology, vol. 2. Cambridge University Press, Cambridge (2003)
36. Mur-Artal, R., Montiel, J.M.M., Tardós, J.D.: ORB-SLAM: a versatile and accurate monocular SLAM system. IEEE Trans. Rob. **31**(5), 1147–1163 (2015)
37. Nakano, G.: Globally optimal DLS method for PnP problem with Cayley parameterization. In: British Machine Vision Conference, pp. 78.1-78.11 (2015)
38. Nocedal, J., Wright, S.J.: Sequential Quadratic Programming, pp. 529–562. Springer, New York (2006). https://doi.org/10.1007/978-0-387-40065-5_18
39. Nousias, S., Lourakis, M., Bergeles, C.: Large-scale, metric structure from motion for unordered light fields. In: Proceedings of the IEEE Conference on Computer Vision and Pattern Recognition, pp. 3292–3301 (2019)
40. Ohayon, S., Rivlin, E.: Robust 3D head tracking using camera pose estimation. In: International Conference on Pattern Recognition (ICPR), vol. 1, pp. 1063–1066. IEEE (2006)
41. Pardalos, P.M., Vavasis, S.A.: Quadratic programming with one negative eigenvalue is NP-hard. J. Glob. Optim. **1**(1), 15–22 (1991)
42. Romea, A.C., Torres, M.M., Srinivasa, S.: The MOPED framework: object recognition and pose estimation for manipulation. Int. J. Rob. Res. **30**(10), 1284–1306 (2011)
43. Rosten, E., Reitmayr, G., Drummond, T.: Improved RANSAC performance using simple, iterative minimal-set solvers. CoRR abs/1007.1432 (2010)
44. Schönberger, J.L., Frahm, J.M.: Structure-from-motion revisited. In: Proceedings of the IEEE Conference on Computer Vision and Pattern Recognition, pp. 4104–4113 (2016)
45. Schweighofer, G., Pinz, A.: Globally optimal O(n) solution to the PnP problem for general camera models. In: British Machine Vision Conference, pp. 1–10 (2008)
46. Shuster, M.: A survey of attitude representations. J. Astronaut. Sci. **41**(4), 439–517 (1993)
47. Terzakis, G., Lourakis, M., Ait-Boudaoud, D.: Modified Rodrigues parameters: an efficient representation of orientation in 3D vision and graphics. J. Math. Imaging Vis. **60**(3), 422–442 (2018)
48. Urban, S., Leitloff, J., Hinz, S.: MLPnP - a real-time maximum likelihood solution to the perspective-n-point problem. ISPRS Ann. Photogram. Remote Sens. Spat. Inf. Sci. **3**, 131–138 (2016)
49. Wang, P., Xu, G., Cheng, Y., Yu, Q.: A simple, robust and fast method for the perspective-n-point problem. Pattern Recogn. Lett. **108**, 31–37 (2018)
50. Zheng, E., Wu, C.: Structure from motion using structure-less resection. In: Proceedings of the IEEE International Conference on Computer Vision, pp. 2075–2083 (2015)
51. Zheng, Y., Kuang, Y., Sugimoto, S., Åström, K., Okutomi, M.: Revisiting the PnP problem: a fast, general and optimal solution. In: Proceedings of the IEEE International Conference on Computer Vision, pp. 2344–2351 (2013)
52. Zheng, Y., Sugimoto, S., Okutomi, M.: ASPnP: an accurate and scalable solution to the perspective-n-point problem. IEICE Trans. Inf. Syst. **96**(7), 1525–1535 (2013)

Learn to Recover Visible Color for Video Surveillance in a Day

Guangming Wu[1], Yinqiang Zheng[2(✉)], Zhiling Guo[1], Zekun Cai[1], Xiaodan Shi[1], Xin Ding[3,4], Yifei Huang[1], Yimin Guo[1], and RyosukeShibasaki[1]

[1] The University of Tokyo, Tokyo 113-8654, Japan
{huster-wgm,guozhilingcc,caizekun,shixiaodan,guo.ym, shiba}@csis.u-tokyo.ac.jp, hyf@iis.u-tokyo.ac.jp
[2] National Institute of Informatics, Tokyo 101-8430, Japan
yqzheng@nii.ac.jp
[3] Wuhan University, Hubei 430072, China
ding-xin@whu.edu.cn
[4] Peng Cheng Laboratory, Shenzhen 518055, China

Abstract. In silicon sensors, the interference between visible and near-infrared (NIR) signals is a crucial problem. For all-day video surveillance, commercial camera systems usually adopt NIR cut filter, and auxiliary NIR LED illumination to selectively block or enhance NIR signal according to the surrounding light conditions. This switching between the daytime and the nighttime mode inevitably involves mechanical parts, and thus requires frequent maintenance. Furthermore, images captured at nighttime mode are in shortage of chrominance, which might hinder human interpretation and high-level computer vision algorithms in succession. In this paper, we present a deep learning based approach that directly generates human-friendly, visible color for video surveillance in a day. To enable training, we capture well-aligned video pairs through a customized optical device and contribute a large-scale dataset, video surveillance in a day (VSIAD). We propose a novel multi-task deep network with state synchronization modules to better utilize texture and chrominance information. Our trained model generates high-quality visible color images and achieves state-of-the-art performance on multiple metrics as well as subjective judgment.

Keywords: Video surveillance in a day · Color recovery · State synchronization network

1 Introduction

In recent years, surveillance cameras have been widely used for security and scientific purposes. Most commercial surveillance cameras are based on silicon

Electronic supplementary material The online version of this chapter (https://doi.org/10.1007/978-3-030-58452-8_29) contains supplementary material, which is available to authorized users.

A. Vedaldi et al. (Eds.): ECCV 2020, LNCS 12346, pp. 495–511, 2020.
https://doi.org/10.1007/978-3-030-58452-8_29

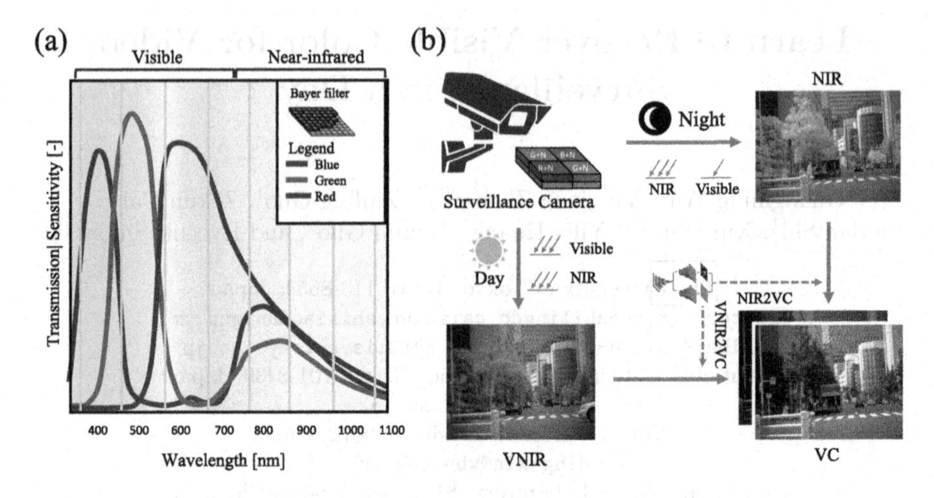

Fig. 1. (a) The spectrum response of the Bayer silicon sensor; and (b) Our integrated pipeline for video surveillance in a day.

sensors, usually equipped with an RGB color filter array, which are sensitive to light with a wavelength from about 400 nm to 1000 nm [37], covering both the visible and near-infrared (NIR) spectrum (as shown in Fig. 1(a)). During daytime, because of the mixture of visible and NIR signal, the captured visible and near-infrared (VNIR) imagery suffers from severe color and contrast degradation [35], as shown in Fig. 1(b). While at nighttime, due to the deficient level of illumination, getting visible color imagery is pretty challenging [44].

For all-day surveillance, the industry practice is to adopt a switchable infrared cut filter (IRCF) to physically block NIR signal at daytime, and to use NIR LEDs, usually centered at 850 nm for illumination during nighttime [40]. This switching mechanism is troubled with frequent maintenance of the mechanical parts. Besides, even NIR shares many properties with visible light, NIR imagery contains less color or texture information, which might hinder human interpretation as well as high-level computer vision applications, *e.g.*, visual tracking [8] and object recognition [19].

To resolve the first drawback, dual-sensor camera systems adopt a beam splitter to split the light and then capture visible and NIR images independently [30]. These systems are free from moving parts, and can directly generate paired visible and NIR images without further image processing steps. Another choice is to use a multispectral filter array (MSFA), which separately records visible and NIR signals in a specially mosaiced sensor [20,28]. The MSFA system can get rid of the mechanical IR cut filter and produce visible as well as NIR images simultaneously through specialized demosaicing [35]. These two solutions capture NIR images independently, which can be used for further visible color image enhancement, such as denoising [10], deblurring [24], and dehazing [38]. However, these two solutions, especially the dual-sensor systems, are relatively expensive and limited to professional usage.

Rather than adding an extra set of imaging sensors or modifying the color filter array, we propose a software solution by training deep convolutional networks (DCN) to realize the automatic translation from the VNIR or NIR input to visible color output. With our proposed model, visible color information can be extracted from the mixed VNIR imagery during the daytime. While at nighttime, NIR images will be colorized into visible color images. To train such a model, it is a big challenge to get sufficient data with ground truth image pairs, *i.e.*, paired VNIR&VC images of daytime, and NIR&VC images of nighttime. Currently, publicly available datasets partially contain either VNIR&VC [34] or NIR&VC [6,7] image pairs. They are limited to small-scale static scenes only[29], thus inappropriate for surveillance usages, for which moving objects like vehicles and pedestrians are of central importance. Besides, because these paired images are captured from different light sources or view angles, there are obvious distortions and misalignments in each pair [11]. To address these problems, we propose a novel optical system with a beam splitter followed by two geometrically aligned and temporally synchronized sensors. We add a NIR cut filter to capture VNIR&VC video pairs and a NIR bandpass filter to capture NIR&VC video pairs. We also introduce large-scale video surveillance in a day (VSIAD) dataset, which is likely to boost other researches.

In order to fully exploit the potential of the dataset, efficient and generalized algorithms are very critical. In recent years, deep convolutional networks (DCNs), including various generative adversarial networks (GANs), have shown promising results for various image-to-image translation tasks [14,48]. Among them, there is a relatively similar topic termed image colorization, which aims to colorize low chrominance images into visible color images. Since texture information is well provided, chrominance recovery is the only issue to be addressed [12,45]. However, in our task, due to the complexity of light sources, only learning chrominance is not sufficient. Hence, we propose a novel multi-task fully convolutional network with state synchronization modules, to learn proper texture and chrominance information from multispectral inputs. To evaluate our approach, we conduct comparison experiments on the newly captured VSIAD dataset. Inspired by the existing researches [21,23], peak signal-to-noise ratio (PSNR), structural similarity (SSIM) [43], learned perceptual image patch similarity (LPIPS) [46] as well as human judgment are used for our image quality evaluation. The results demonstrate that the proposed network achieves considerable translation accuracy in both VNIR2VC and NIR2VC tasks, and outperforms the state-of-the-art image colorization techniques.

The main contributions of this study can be summarized as follows:

- We design a novel optical system to capture well-aligned VNIR&VC and NIR&VC image pairs and contribute a large-scale dataset, video surveillance in a day (VSIAD).
- We demonstrate a software solution of recovering visible color for all-day video surveillance, in contrast to existing hardware solutions that require switchable filters, multispectral filter arrays, or dual sensors.

- We propose a novel multi-task fully convolutional network with state synchronization modules to ensure the consistency of the generated texture and chrominance information.

2 Related Work

To our best knowledge, there is no other research at present trying to handle both VNIR2VC and NIR2VC in a unified network for video surveillance in a day. Instead, several similar studies are working on each topic separately.

VINR2VC. VNIR2VC is a typical imaging problem that aims to extract vivid visible color images from multispectral VNIR images. Zhang *et al.* [47] proposed a dual-camera system using a 45° hot mirror to separate visible from NIR light, and then captured them independently. Additionally, Kise *et al.* [18] designed a triple camera system equipped with interchangeable optical filters. Rather than using two or more camera systems, multispectral filter arrays (MSFA) offer an alternative option. Lu *et al.* [28] presented a customized 4×4 CFA through spatial-domain optimization, which enables the extraction of visible and NIR image pairs from single RAW measurements. Similarly, Chen *et al.* [4] introduces a four-channel bayer pattern (*i.e.*. R, G, B, and NIR) to record visible and NIR signal independently.

Nevertheless, the above-mentioned solutions, which require customized hardware, are relatively expensive and limited to professional usage.

NIR2VC. NIR2VC is slightly different from grayscale image colorization. In grayscale image colorization, the grayscale input and corresponding color output are derived from the same visible color image. Texture information from input and output are almost identical, and chrominance is the only factor to be learned [5,22]. Zhang *et al.* [45] turned grayscale image colorization as classification of chrominance values. Further, Iizuka *et al.* [12] proposed a multi-task network that combines color prediction and scene classification to achieve more natural results.

Different from grayscale image colorization, NIR2VC is also subject to texture recovering from NIR to visible light (as shown in Fig. 1(a) and Fig. 2(c)). Recently, Berg *et al.* [1] utilized an additional structure loss that can minimize the texture difference between thermal infrared (TIR) and grayscale. To avoid misalignments between images pairs, Mehri *et al.* [31] and Nyberg *et al.* [33] adopt modified CycleGANs [48] for unpaired thermal infrared to visual color (TIR2VC) translation. Because of the loose connection between texture and chrominance, their result suffers from severe blurring as well as mismatching of texture and chrominance.

Deep Learning Based Low Light Enhancement. Low light enhancement, which aims to enhance image quality under deficient illuminance condition, is significant in video surveillance. Chen *et al.* [3] built a See-in-the-Dark (SID) dataset captured by various exposure time for training a model to brighten extreme dark images. To see motion in the dark, Chen *et al.* [2] and Jiang *et al.* [15] introduced learning-based pipelines to recover texture and chrominance information from dynamic scenes. Theoretically, these approaches can also be applied for video surveillance in a day without NIR illuminant. However, in practice, RAW images or videos required as input in their systems, are not available in the majority of existing surveillance cameras.

Image-to-Image Translation. Both VNIR2VC & NIR2VC can also be viewed as a specific form of image-to-image translation that enables the mapping between an input image and a corresponding output image. Isola *et al.* [14] built a general image-to-image translation framework using conditional adversarial networks. Zhu *et al.* [48] introduced cycle-consistent adversarial networks (cycleGAN) to get rid of aligned image pairs. For high-resolution image synthesis, Wang *et al.* [42] proposed novel adversarial loss, as well as new multi-scale generator and discriminator architectures. Due to the texture difference between NIR and VC, the NIR2VC task can not be properly addressed with general image-to-image translation methods.

Very recently, Lv *et al.* [29] introduced an integrated enhancement solution for 24-hour colorful imaging. However, their approach is mainly based on indoor static scenes that are inappropriate for surveillance usages, for which moving objects like vehicles and pedestrians are of central importance.

3 Dataset

To enable training, we introduce a novel dataset, video surveillance in a day (VSIAD), which contains ground-truth image pairs of both VNIR&VC and NIR&VC, taken with our co-axis optical imaging system. For data preprocessing, we registered the captured image pairs with feature matching and geometric transformation.

3.1 Data Capturing

The optical imaging system mainly consists of one beam splitter, and two IRCF-free CCD cameras (FLIR GS3-U3-15S4C). A key feature of our system is that we can switch between daytime and nighttime mode easily.

- **Daytime mode.** As shown in Fig. 2(a), light is firstly divided into two branches by a beam splitter. One beam goes to the bayer sensor that yields color imagery containing both visible and NIR information (VNIR). The other one will first pass through the NIR-cut filter to filter out NIR information and then reach the sensor to generate an image of visible color (VC).

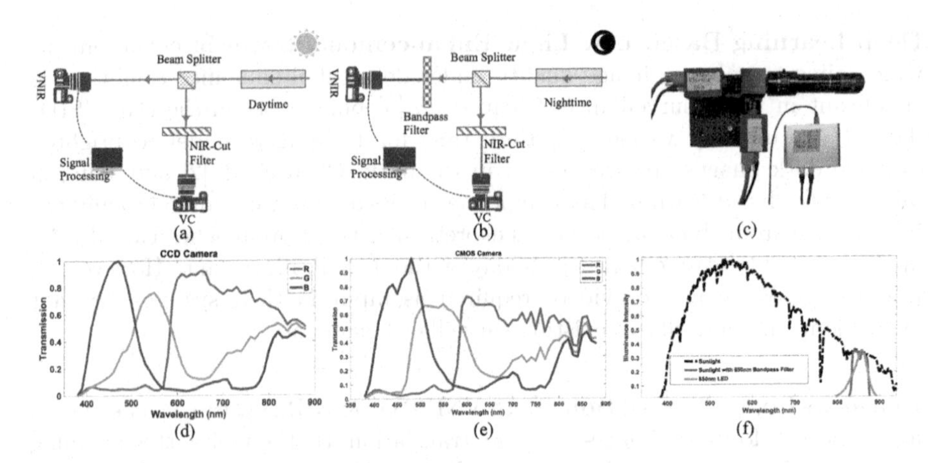

Fig. 2. Overview of the co-axis optical system. (a) The architecture of the daytime mode; (b) The architecture of the nighttime mode; (c) The physical devices; (d) The spectral response of the CCD camera; (e) The spectral response of the CMOS camera; and (f) The spectral distributions of sunlight, sunlight with 850 nm bandpass filter, and 850 nm LED illuminant.

- **Nighttime mode.** Note that it is impossible to capture moving objects in low light condition by an ordinary camera. Therefore, we capture NIR and VC pairs at daylight. As shown in Fig. 2(b), nighttime mode utilizes a similar architecture to daytime. To simulate the NIR image captured by a surveillance camera with 850 nm LED illumination, we use an 850 nm bandpass filter, with an FWHM of 50 nm, to filter out visible information. The rationale of this choice is verified by the similarity between the spectral distribution of the conventional 850 LED illuminants and the daylight spectrum filtered by the aforementioned NIR bandpass filter, as shown in Fig. 2(f). Because of the various distributions of light sources (*i.e.*, sunlight at daytime, and LED illuminant at nighttime), there is a slight difference as compared to real-world conditions.

With the proposed imaging system, 80 video clips (40,000 images) were captured from several street spots. All images are saved in 8-bit BMP format with 1384×1032 pixels. The numbers of video clips of VNIR&VC from daytime and NIR&VC from nighttime are set to be equal for the daytime-nighttime balancing.

3.2 Data Preprocessing

Although the two imaging sensors are well-aligned with similar positions, there are inevitably pixel-level rotation or translation misalignments. To address this issue, we employ projective transformation to wrap the VC image based on scale-invariant feature transform (SIFT) [27] features of corresponding VNIR or NIR images. The window with a size of 1200 × 900 is used to crop the center area of

Fig. 3. A subset of the VSIAD dataset. The upper row is VNIR&VC image pairs taken by the daytime mode, while the remaining row is NIR&VC image pairs taken by nighttime mode.

overlapping registered image pairs to eliminate boundary aliases and artifacts. Later, the cropped images are resized to 640×480 to reduce storage.

Figure 3 shows sample image pairs from our VSIAD dataset. The upper row contains four sets of VNIR&VC image pairs taken by the daytime mode. Because of interference of the NIR signal, VNIR images result in apparent color degradation. Sampled image sets of NIR&VC image pairs taken by the nighttime mode are presented at the bottom row. Due to the NIR spectrum characteristic, captured NIR images present a strong signal on green plants but a weak signal on dyes (*e.g.*, color on clothes, and character of the signpost).

4 State Synchronization Network

Inspired by existing end-to-end fully convolutional networks [25,26], we design a novel state synchronization network (SSN), which utilizes parallel decoders and state synchronization modules (SSMs) to estimate texture and chrominance information separately. Differing with existing colorization methods that use grayscale information as input, our model directly uses RGB values of VNIR or NIR to prevent information loss during the conversion. We note that, although the camera responses of three color channels around 850 nm (see Fig. 2(c)) are quite similar, they are indeed slightly different. Thus, visible color recovery from NIR (*i.e.*, NIR2VC) is less ill-posed than retrieving chrominance from a single-channel grayscale image.

Network Architecture. As shown in Fig. 4, the proposed SSN consists of one encoder and two parallel decoders with four state synchronization modules. The encoder follows the design of the classic ResNet [9] using sequential basic residual blocks [9] and max-pooling layers. The parallel decoders share the identical architecture except for the final prediction layer. Similar to the encoder, the

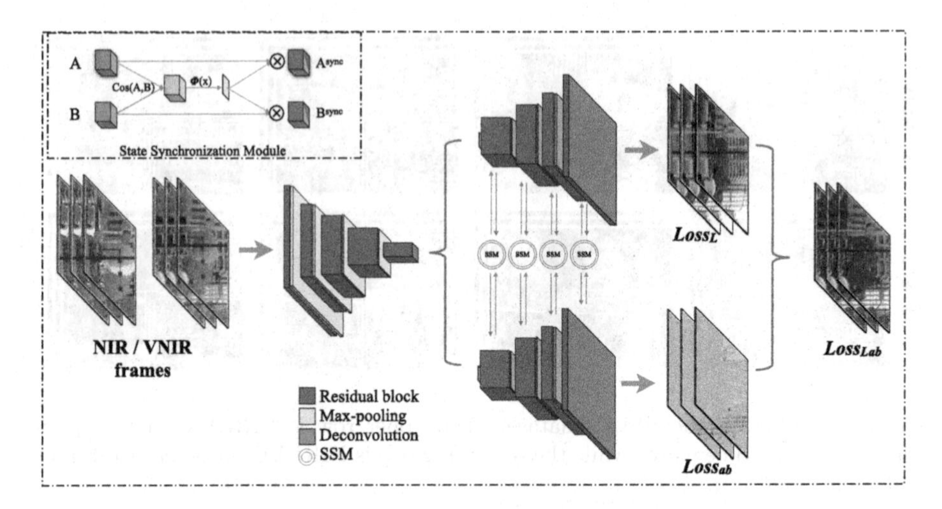

Fig. 4. Architecture of the proposed state synchronization network (SSN). Input VNIR or NIR frames will be translated into visible color images by the network.

decoder applies sequential deconvolutional layers [32], residual blocks, and skip connections [36] to refine to original height and width gradually. To avoid interference within batch samples, batch normalization layers [13] are replaced by instance normalization [41].

State Synchronization Module (SSM). For an image, the texture and chrominance information are highly correlated (*e.g.*, tree texture usually correlates with green color). If we train L and ab independently, consistency of texture and chrominance will be misconducted.

To avoid this, we design the state synchronization module (SSM) to update the state of the parallel decoders continuously. As shown in Fig. 4, features from L or ab decoder are denoted as $\boldsymbol{A}$ and $\boldsymbol{B}$, respectively. Within the SSM, the cosine similarly ($\boldsymbol{CS}$) followed by 2D convolution with $k \times k$ gaussian kernel ($\boldsymbol{G}$) are applied to the generated state map of both features ($\boldsymbol{S_{map}}$). Specifically,

$$\boldsymbol{CS} = \frac{\boldsymbol{A} \bullet \boldsymbol{B}}{||\boldsymbol{A}|| \times ||\boldsymbol{B}||} \tag{1}$$

$$\boldsymbol{S_{map}} = \sum_i \sum_j \boldsymbol{CS}_{i:i+k,j:j+k} \bullet \boldsymbol{G} \tag{2}$$

Then, $\boldsymbol{S_{map}}$ is applied to update feature $\boldsymbol{A}$ and $\boldsymbol{B}$ as $\boldsymbol{A}^{sync}$ and $\boldsymbol{B}^{sync}$ through hadamard product by each channel (c) as follows

$$\boldsymbol{A^{sync}}_c = \boldsymbol{A}_c \odot \boldsymbol{S_{map}} \tag{3}$$

$$\boldsymbol{B^{sync}}_c = \boldsymbol{B}_c \odot \boldsymbol{S_{map}} \tag{4}$$

Objective Function. After parallel decoders with SSMs, predictions of L and ab are generated separately. Structural dissimilarity (DSSIM) [43] and L1 distance between these predictions and ground truths are denoted as $\mathcal{L}_L$ and $\mathcal{L}_{ab}$, respectively.

$$\mathcal{L}_L = \frac{1 - \boldsymbol{SSIM}(L^{pred} - L^{gt})}{2} \tag{5}$$

$$\mathcal{L}_{ab} = |ab^{pred} - ab^{gt}| \tag{6}$$

Finally, independent predictions of L and ab are concatenated as predicted Lab. To evaluate the consistency of texture and corresponding chrominance, a perceptual loss ($\mathcal{L}_{Lab}$) [16] is calculated.

$$\mathcal{L}_{Lab} = \sum_{layer} |\boldsymbol{VGG}(Lab^{pred}) - \boldsymbol{VGG}(Lab^{gt})|_{layer} \tag{7}$$

where layer $\in [relu1\_2, relu2\_2, relu3\_3, relu4\_3]$ of pre-trained VGG16 network [39]. Thus, the final objective function is formulated as:

$$\mathcal{L}_{final} = \alpha \times \mathcal{L}_L + \beta \times \mathcal{L}_{ab} + \lambda \times \mathcal{L}_{Lab} \tag{8}$$

In our experiments, the configuration of the parameters are set as: $k = 3$, $\alpha = 1$, $\beta = 1$, $\lambda = 10$, respectively.

5 Experimental Setup

We split 20,000 image pairs in our VSIAD dataset into training, validation, and testing. Their ratios are 60:20:20 so that there are 12,000 image pairs for training, 4000 for validation, and 4000 for testing. The numbers of VNIR&VC and NIR&VC pairs are set to be equal by each set. We select a batch size of 8 and randomly crop 256×256 patches from a full-resolution VNIR or NIR image as input for training. We implement the proposed networks using PyTorch 1.0 (https://github.com/huster-wgm/VSIAD) and train it with NVIDIA Tesla V100 . The proposed model is trained for 100,000 iterations with 100 validations performed by every 1,000 iterations. In our experiment, parameters are optimized by the Adam optimizer [17] using initial learning rate = $1e^{-4}$, $\beta_1 = 0.9$, $\beta_2 = 0.999$, and $\epsilon = 1e^{-8}$.

5.1 Baselines

For comparison, we choose several representative image colorization methods: Iizuka *et al.* [12], which jointly learns global and local features to exploit classification labels for better colorization; Berg *et al.* [1] that introduces a combination loss for generated texture and chrominance information; Mehri *et al.* [31] and Nyberg *et al.* [33] which apply modified CycleGANs [48] for unpaired thermal infared to visual color (TIR2VC) translation; and pix2pixHD [42], a general high-resolution image-to-image translation framework.

For Iizuka *et al.*'s model, we first try to fine-tune their model on visible images in our VSIAD dataset. However, due to the lack of classification labels, the performance gained from fine-tuning is minimal (± 0.2 for PSNR). Thus, we directly use the pre-trained model for comparisons. As for Berg *et al.*'s model, we carefully implement and train it from scratch using the standard setup in the literature. Thanks to the publicly available training code, we fine-tune Mehri *et al.*'s, Nyberg *et al.*'s, and pix2pixHD models using our VSIAD dataset.

6 Results and Discussions

6.1 Quantitative Evaluation

To evaluate our method and the baselines, evaluation metrics, including pixel-based PSNR, structure-based SSIM, and learning-based LPIPS, are adopted. Note that the lower score of LPIPS indicates better image quality.

The relative performances of different methods over testing data are listed in Table 1. In general, values of PSNR, SSIM, and LPIPS in the VNIR2VC task are higher than those in the NIR2VC task.

Compared with other baselines, pix2pixHD [42] and our SSN present significantly higher scores in PSNR and SSIM as well as lower scores in LPIPS. As for pixel-based metrics, pix2pixHD has a higher PSNR value in the VNIR2VC task. However, in the NIR2VC task, our model performs as good as pix2pixHD. For more generalized metrics, SSIM and LPIPS, our model outperforms all baselines in both VNIR2VC and NIR2VC tasks. These numbers are consistent with our qualitative observation (see details in Sect. 6.2). Besides, comparing with pix2pixHD, our model shows a relatively smaller performance gap between VNIR2VC and NIR2VC tasks. These results indicate that our proposed network can handle both VNIR2VC and NIR2VC tasks efficiently and accurately.

6.2 Qualitative Results

Qualitative comparison of our model against baselines on both VNIR2VC and NIR2VC tasks are shown in Fig. 5 and 6, respectively. The sequential input

Table 1. Performance comparison on both VNIR2VC and NIR2VC tasks. Metric with '↑' means the higher the better image quality, while '↓' means the opposite.

Method	PSNR ↑		SSIM ↑		LPIPS ↓	
	VNIR2VC	NIR2VC	VNIR2VC	NIR2VC	VNIR2VC	NIR2VC
Iizuka *et al.*[12]	14.812	14.465	0.662	0.513	0.321	0.460
Berg *et al.*[1]	20.188	16.543	0.755	0.622	0.236	0.370
Mehri *et al.*[31]	22.359	14.025	0.779	0.491	0.223	0.454
Nyberg *et al.*[33]	21.096	16.474	0.754	0.573	0.174	0.360
pix2pixHD [42]	**25.003**	19.654	0.790	0.641	0.139	0.287
Ours	24.336	**19.690**	**0.836**	**0.698**	**0.109**	**0.248**

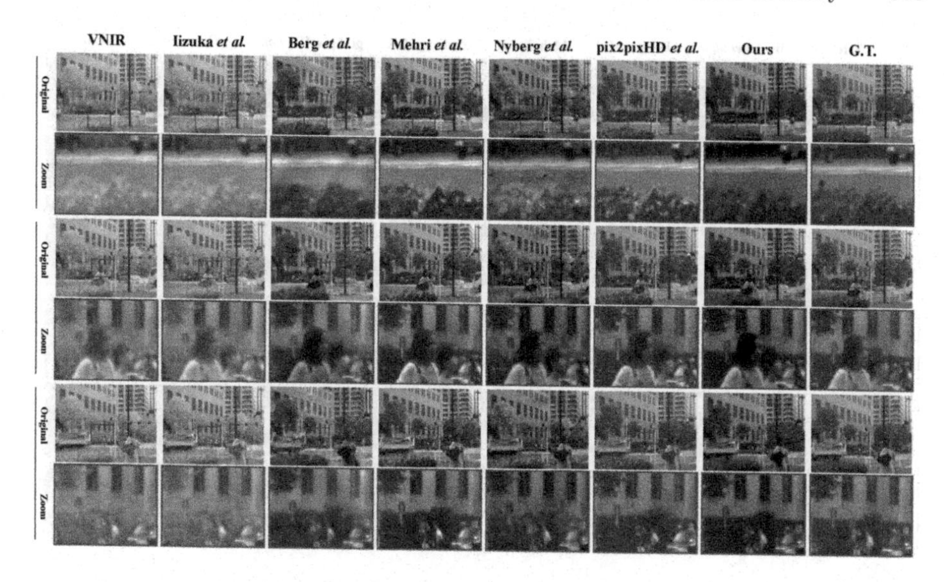

Fig. 5. Qualitative result from the baselines and our SSN model on the VNIR2VC task. The sequential frames are derived from the same location but different viewpoints.

images, including VNIR and NIR, are derived from the same location but different viewpoints. Within these images, moving vehicles and pedestrians are coarsely presented.

Compared to the baselines mentioned above, our method, as well as the fine-tuned pix2pixHD model, generates images with higher color fidelity. On the VNIR2VC task, images generated by pix2pixHD and our model show somewhat similar chrominance and slight sharpness differences of texture. Generally, images from pix2pixHD are sharper than those from us. However, their images are too sharp that perceptually unnatural (*e.g.*, leaves in trees at the 2nd row, Fig. 5).

On the NIR2VC task, even with some artifacts, our method shows significantly better translated images than pix2pixHD (*e.g.*, cars from the 2nd and 6th rows, Fig. 6). Considering both VNIR2VC and NIR2VC, which is critical for video surveillance in a day, our method yields the most consistent visual result using both VNIR and NIR inputs.

6.3 Perceptual Experiments

We evaluate the perceptual quality of the generated images through blind testing. In each inquiry, we present the participant with videos. At every frame, VNIR&VC (or NIR&VC) image pairs and corresponding images generated from ours or a baseline model are organized side by side. The participants are asked to pick up the one that is more close to the original visible color video. In the experiment, 134 feedbacks are collected and listed in Table 2. Videos generated by our SSN achieve a significantly higher preference rate under blind, subjective judgment.

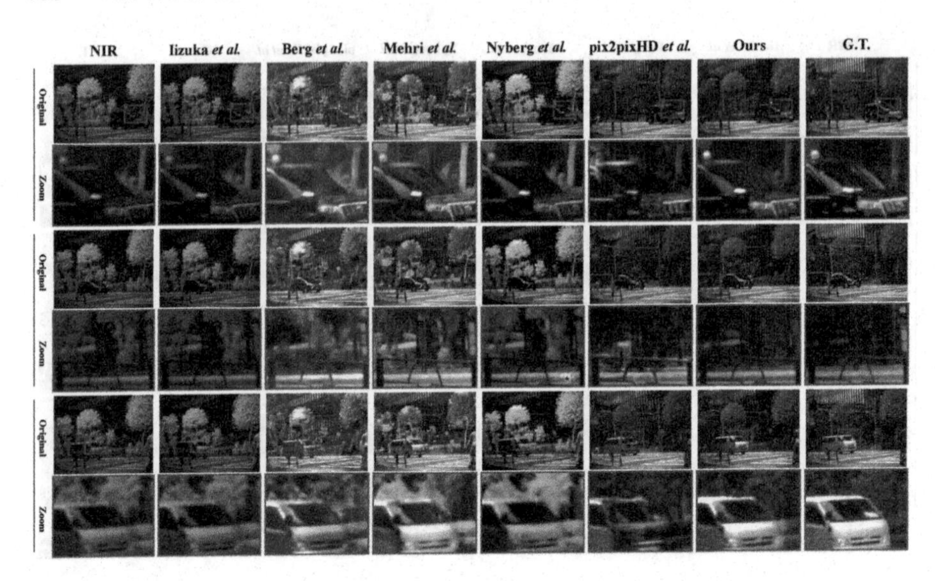

Fig. 6. Qualitative result from the baselines and our SSN model on the NIR2VC task. The sequential frames are derived from the same location but different viewpoints.

Table 2. Preference rates of videos generated by different methods.

Tasks	Preference Rate		
	Ours	pix2pixHD [42]	No preference
VNIR2VC	**73.5%**	4.4%	22.1%
NIR2VC	**82.3%**	1.5%	16.2%

6.4 Generalization Analysis

To evaluate the robustness and generalization capability of the proposed method, we test our trained model on real-world time-elapse images captured from a static viewpoint. We also use a CMOS camera (FLIR BFS-U3-63S4C) and remove its IR-cut filter, which is different from the CCD camera (FLIR GS3-U3-15S4C) in training data capture. Despite their difference, we can see that their spectral response curves are quite similar (*e.g.*, Fig. 2(d) and (e)).

As shown in Fig. 7, even with some artifacts, our model can generate proper visible images from VNIR or NIR images captured by the CCD camera at most times. Because of the difference in spectral response curves, VNIR/NIR images captured by CMOS show significantly different color style when compared with those images taken by CCD (1st vs. 3rd row). Despite this, our model can produce pretty natural visible images during daytime (*e.g.*, at 09:00, 12:00, and 15:00) using the CMOS camera. As for the nighttime (*e.g.*, at 18:00 and 21:00), CMOS results are less satisfactory. We note that cross-camera color recovery at night is extremely challenging, because of the inevitable interference by light

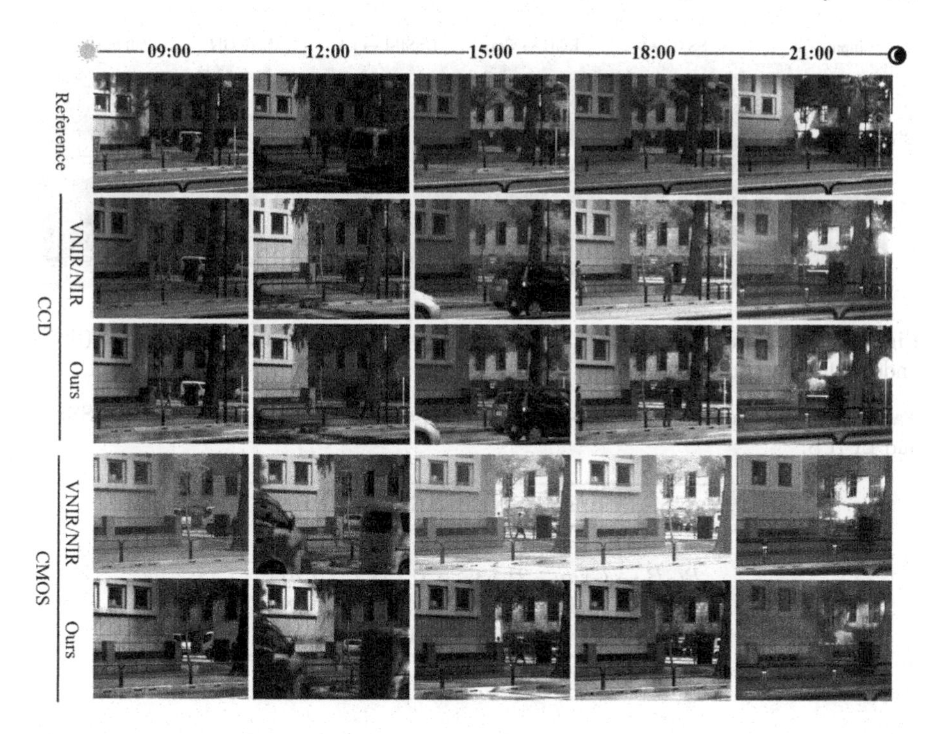

Fig. 7. Time-elapse experiment on both CCD and CMOS cameras. Reference images are captured using a NIR cut filter at daytime and using long exposure at nighttime.

contamination from street light and image noise, which we plan to study further as future work.

6.5 Ablation Experiment

To investigate the effectiveness of different components, we conduct ablation studies on VSIAD. The performance under the different conditions are illustrated in Table 3 and Fig. 8.

State Synchronization Module (SSM). In 1st row, removing SSM leads to significant performance decline (*e.g.*, the value of LPIPS increases about 15.6%), which demonstrates the effectiveness and importance of our SSM.

Color. As shown in 2nd row, changing RGB input to grayscale causes performance losses of 2.2% in PSNR, 1.7% in SSIM, and 14.3% in LPIPS.

Texture. As presented at 3rd row, while replacing structural dissimilarity (DSSIM) with L1 distance of $\mathcal{L}_L$ (Eq. 5), a slight performance degradation can be observed.

Fig. 8. Representative results of the proposed state synchronization network (SSN) under different conditions.

Table 3. Ablation analysis results. The table reports the mean values of PSNR, SSIM, and LPIPS.

Conditions	PSNR ↑	SSIM ↑	LPIPS ↓
$SSM \rightarrow 0$	21.419	0.747	0.212
$RGB \rightarrow Gray$	21.535	0.754	0.209
$DSSIM \rightarrow L1$	21.910	0.759	0.188
$IN \rightarrow GN$	**23.145**	**0.773**	0.182
Full model	22.013	0.767	**0.179**

Normalization. From the 4rd row, if replacing instance normalization (IN) with group normalization (GN), the perceptual performance (*i.e.*, LPIPS) gets worse, while the PSNR and SSIM can be slightly improved.

7 Conclusion

We have demonstrated the effectiveness of our integrated pipeline for video surveillance in a day. Degraded images, including VNIR and NIR, are directly translated into visible color images through a learned model. In contrast to existing hardware solutions that require switchable filters, multispectral filter arrays, or dual sensors, our approach can directly apply to commercial surveillance cameras that are much more cost-efficient. To enable training, we collect a new dataset that contains well-aligned VNIR&VC and NIR&VC image pairs, and introduce a novel parallel network with state synchronization modules that can keep consistency between texture and chrominance information. We also notice that slight performance degradation happened during cross-camera color recovery, which we plan to study further as future work.

Acknowledgements. This work was supported in part by the JSPS KAKENHI under Grant No. 19K20307. A part of this work was finished during Y. Zheng's visit and X. Ding's internship at Peng Cheng Laboratory.

References

1. Berg, A., Ahlberg, J., Felsberg, M.: Generating visible spectrum images from thermal infrared. In: Proceedings of the IEEE Conference on Computer Vision and Pattern Recognition Workshops, pp. 1143–1152 (2018)
2. Chen, C., Chen, Q., Do, M.N., Koltun, V.: Seeing motion in the dark. In: Proceedings of the IEEE International Conference on Computer Vision, pp. 3185–3194 (2019)
3. Chen, C., Chen, Q., Xu, J., Koltun, V.: Learning to see in the dark. In: Proceedings of the IEEE Conference on Computer Vision and Pattern Recognition, pp. 3291–3300 (2018)
4. Chen, Z., Wang, X., Liang, R.: Rgb-nir multispectral camera. Opt. Express **22**(5), 4985–4994 (2014)
5. Cheng, Z., Yang, Q., Sheng, B.: Deep colorization. In: Proceedings of the IEEE International Conference on Computer Vision, pp. 415–423 (2015)
6. Choe, G., Kim, S.H., Im, S., Lee, J.Y., Narasimhan, S.G., Kweon, I.S.: Ranus: RGB and NIR urban scene dataset for deep scene parsing. IEEE Rob. Autom. Lett. **3**(3), 1808–1815 (2018)
7. Fredembach, C., Süsstrunk, S.: Colouring the near-infrared. In: Color and Imaging Conference, vol. 2008, pp. 176–182. Society for Imaging Science and Technology (2008)
8. Gao, S., Cheng, Y., Zhao, Y.: Method of visual and infrared fusion for moving object detection. Opt. Lett. **38**(11), 1981–1983 (2013)
9. He, K., Zhang, X., Ren, S., Sun, J.: Deep residual learning for image recognition. In: Proceedings of the IEEE Conference on Computer Vision and Pattern Recognition, pp. 770–778 (2016)
10. Honda, H., Timofte, R., Van Gool, L.: Make my day-high-fidelity color denoising with near-infrared. In: Proceedings of the IEEE Conference on Computer Vision and Pattern Recognition Workshops, pp. 82–90 (2015)
11. Hwang, S., Park, J., Kim, N., Choi, Y., So Kweon, I.: Multispectral pedestrian detection: Benchmark dataset and baseline. In: Proceedings of the IEEE Conference on Computer Vision and Pattern Recognition, pp. 1037–1045 (2015)
12. Iizuka, S., Simo-Serra, E., Ishikawa, H.: Let there be color!: joint end-to-end learning of global and local image priors for automatic image colorization with simultaneous classification. ACM Trans. Graph. (TOG) **35**(4), 110 (2016)
13. Ioffe, S., Szegedy, C.: Batch normalization: Accelerating deep network training by reducing internal covariate shift. arXiv preprint arXiv:1502.03167 (2015)
14. Isola, P., Zhu, J.Y., Zhou, T., Efros, A.A.: Image-to-image translation with conditional adversarial networks. In: Proceedings of the IEEE Conference on Computer Vision and Pattern Recognition, pp. 1125–1134 (2017)
15. Jiang, H., Zheng, Y.: Learning to see moving objects in the dark. In: Proceedings of the IEEE International Conference on Computer Vision, pp. 7324–7333 (2019)
16. Johnson, J., Alahi, A., Fei-Fei, L.: Perceptual losses for real-time style transfer and super-resolution. In: Leibe, B., Matas, J., Sebe, N., Welling, M. (eds.) ECCV 2016. LNCS, vol. 9906, pp. 694–711. Springer, Cham (2016). https://doi.org/10.1007/978-3-319-46475-6_43
17. Kingma, D.P., Ba, J.: Adam: A method for stochastic optimization. arXiv preprint arXiv:1412.6980 (2014)
18. Kise, M., Park, B., Heitschmidt, G.W., Lawrence, K.C., Windham, W.R.: Multispectral imaging system with interchangeable filter design. Comput. Electron. Agric. **72**(2), 61–68 (2010)

19. Kleynen, O., Leemans, V., Destain, M.F.: Development of a multi-spectral vision system for the detection of defects on apples. J. Food Eng. **69**(1), 41–49 (2005)
20. Koyama, S., Inaba, Y., Kasano, M., Murata, T.: A day and night vision MOS imager with robust photonic-crystal-based RGB-and-IR. IEEE Trans. Electron Devices **55**(3), 754–759 (2008)
21. Lai, W.S., Huang, J.B., Wang, O., Shechtman, E., Yumer, E., Yang, M.H.: Learning blind video temporal consistency. In: Proceedings of the European Conference on Computer Vision (ECCV), pp. 170–185 (2018)
22. Larsson, G., Maire, M., Shakhnarovich, G.: Learning representations for automatic colorization. In: Leibe, B., Matas, J., Sebe, N., Welling, M. (eds.) ECCV 2016. LNCS, vol. 9908, pp. 577–593. Springer, Cham (2016). https://doi.org/10.1007/978-3-319-46493-0_35
23. Lei, C., Chen, Q.: Fully automatic video colorization with self-regularization and diversity. In: Proceedings of the IEEE Conference on Computer Vision and Pattern Recognition, pp. 3753–3761 (2019)
24. Li, W., Zhang, J., Dai, Q.H.: Robust blind motion deblurring using near-infrared flash image. J. Visual Commun. Image Representation **24**(8), 1394–1413 (2013)
25. Lin, T.Y., Dollár, P., Girshick, R., He, K., Hariharan, B., Belongie, S.: Feature pyramid networks for object detection. In: Proceedings of the IEEE Conference on Computer Vision and Pattern Recognition, pp. 2117–2125 (2017)
26. Long, J., Shelhamer, E., Darrell, T.: Fully convolutional networks for semantic segmentation. In: Proceedings of the IEEE Conference on Computer Vision and Pattern Recognition, pp. 3431–3440 (2015)
27. Lowe, D.G., et al.: Object recognition from local scale-invariant features. In: ICCV, vol. 99, pp. 1150–1157 (1999)
28. Lu, Y.M., Fredembach, C., Vetterli, M., Süsstrunk, S.: Designing color filter arrays for the joint capture of visible and near-infrared images. In: 2009 16th IEEE International Conference on Image Processing (ICIP), pp. 3797–3800. IEEE (2009)
29. Lv, F., Zheng, Y., Li, Y., Lu, F.: An integrated enhancement solution for 24-hour colorful imaging. In: AAAI, pp. 11725–11732 (2020)
30. Matsui, S., Okabe, T., Shimano, M., Sato, Y.: Image enhancement of low-light scenes with near-infrared flash images. Inf. Media Technol. **6**(1), 202–210 (2011)
31. Mehri, A., Sappa, A.D.: Colorizing near infrared images through a cyclic adversarial approach of unpaired samples. In: 2019 IEEE/CVF Conference on Computer Vision and Pattern Recognition Workshops (CVPRW), pp. 971–979. IEEE (2019)
32. Noh, H., Hong, S., Han, B.: Learning deconvolution network for semantic segmentation. In: Proceedings of the IEEE International Conference on Computer Vision, pp. 1520–1528 (2015)
33. Nyberg, A., Eldesokey, A., Bergström, D., Gustafsson, D.: Unpaired thermal to visible spectrum transfer using adversarial training. In: Leal-Taixé, L., Roth, S. (eds.) ECCV 2018. LNCS, vol. 11134, pp. 657–669. Springer, Cham (2019). https://doi.org/10.1007/978-3-030-11024-6_49
34. Özkan, K., Işık, Ş., Yavuz, B.T.: Identification of wheat kernels by fusion of RGB, SWIR, VNIR samples over feature and image domain. J. Sci. Food Agricu. **99**, 4977–4984 (2019)
35. Park, C., Kang, M.: Color restoration of RGBn multispectral filter array sensor images based on spectral decomposition. Sensors **16**(5), 719 (2016)
36. Ronneberger, O., Fischer, P., Brox, T.: U-Net: convolutional networks for biomedical image segmentation. In: Navab, N., Hornegger, J., Wells, W.M., Frangi, A.F. (eds.) MICCAI 2015. LNCS, vol. 9351, pp. 234–241. Springer, Cham (2015). https://doi.org/10.1007/978-3-319-24574-4_28

37. Sadeghipoor, Z., Lu, Y.M., Süsstrunk, S.: A novel compressive sensing approach to simultaneously acquire color and near-infrared images on a single sensor. In: 2013 IEEE International Conference on Acoustics, Speech and Signal Processing, pp. 1646–1650. IEEE (2013)
38. Schaul, L., Fredembach, C., Süsstrunk, S.: Color image dehazing using the near-infrared. In: 2009 16th IEEE International Conference on Image Processing (ICIP), pp. 1629–1632. IEEE (2009)
39. Simonyan, K., Zisserman, A.: Very deep convolutional networks for large-scale image recognition. arXiv preprint arXiv:1409.1556 (2014)
40. Tessler, N., Medvedev, V., Kazes, M., Kan, S., Banin, U.: Efficient near-infrared polymer nanocrystal light-emitting diodes. Science **295**(5559), 1506–1508 (2002)
41. Ulyanov, D., Vedaldi, A., Lempitsky, V.: Instance normalization: The missing ingredient for fast stylization. arXiv preprint arXiv:1607.08022 (2016)
42. Wang, T.C., Liu, M.Y., Zhu, J.Y., Tao, A., Kautz, J., Catanzaro, B.: High-resolution image synthesis and semantic manipulation with conditional GANs. In: Proceedings of the IEEE Conference on Computer Vision and Pattern Recognition, pp. 8798–8807 (2018)
43. Wang, Z., Bovik, A.C., Sheikh, H.R., Simoncelli, E.P., et al.: Image quality assessment: from error visibility to structural similarity. IEEE Trans. Image Process. **13**(4), 600–612 (2004)
44. Zafar, I., Zakir, U., Romanenko, I., Jiang, R.M., Edirisinghe, E.: Human silhouette extraction on FPGAs for infrared night vision military surveillance. In: 2010 Second Pacific-Asia Conference on Circuits, Communications and System, vol. 1, pp. 63–66. IEEE (2010)
45. Zhang, R., Isola, P., Efros, A.A.: Colorful image colorization. In: Leibe, B., Matas, J., Sebe, N., Welling, M. (eds.) ECCV 2016. LNCS, vol. 9907, pp. 649–666. Springer, Cham (2016). https://doi.org/10.1007/978-3-319-46487-9_40
46. Zhang, R., Isola, P., Efros, A.A., Shechtman, E., Wang, O.: The unreasonable effectiveness of deep features as a perceptual metric. In: Proceedings of the IEEE Conference on Computer Vision and Pattern Recognition, pp. 586–595 (2018)
47. Zhang, X., Sim, T., Miao, X.: Enhancing photographs with near infra-red images. In: 2008 IEEE Conference on Computer Vision and Pattern Recognition, pp. 1–8. IEEE (2008)
48. Zhu, J.Y., Park, T., Isola, P., Efros, A.A.: Unpaired image-to-image translation using cycle-consistent adversarial networks. In: Proceedings of the IEEE International Conference on Computer Vision, pp. 2223–2232 (2017)

Deep Fashion3D: A Dataset and Benchmark for 3D Garment Reconstruction from Single Images

Heming Zhu[1,2], Yu Cao[1,3], Hang Jin[1], Weikai Chen[4], Dong Du[1,5], Zhangye Wang[2], Shuguang Cui[1], and Xiaoguang Han[1(✉)]

[1] Shenzhen Research Institute of Big Data, The Chinese University of Hong Kong, Shenzhen, China
hanxiaoguang@cuhk.edu.cn
[2] State Key Lab of CAD&CG, Zhejiang University, Hangzhou, China
[3] Xidian University, Xi'an, China
[4] Tencent America, Palo Alto, USA
[5] University of Science and Technology of China, Hefei, China

Abstract. High-fidelity clothing reconstruction is the key to achieving photorealism in a wide range of applications including human digitization, virtual try-on, etc. Recent advances in learning-based approaches have accomplished unprecedented accuracy in recovering unclothed human shape and pose from single images, thanks to the availability of powerful statistical models, e.g. SMPL, learned from a large number of body scans. In contrast, modeling and recovering clothed human and 3D garments remains notoriously difficult, mostly due to the lack of large-scale clothing models available for the research community. We propose to fill this gap by introducing Deep Fashion3D, the largest collection to date of 3D garment models, with the goal of establishing a novel benchmark and dataset for the evaluation of image-based garment reconstruction systems. Deep Fashion3D contains 2078 models reconstructed from real garments, which covers 10 different categories and 563 garment instances. It provides rich annotations including 3D feature lines, 3D body pose and the corresponded multi-view real images. In addition, each garment is randomly posed to enhance the variety of real clothing deformations. To demonstrate the advantage of Deep Fashion3D, we propose a novel baseline approach for single-view garment reconstruction, which leverages the merits of both mesh and implicit representations. A novel adaptable template is proposed to enable the learning of all types of clothing in a single network. Extensive experiments have been conducted on the proposed dataset to verify its significance and usefulness.

H. Zhu, Y. Cao, H. Jin—The first three authors should be considered as joint first authors.

Electronic supplementary material The online version of this chapter (https://doi.org/10.1007/978-3-030-58452-8_30) contains supplementary material, which is available to authorized users.

A. Vedaldi et al. (Eds.): ECCV 2020, LNCS 12346, pp. 512–530, 2020.
https://doi.org/10.1007/978-3-030-58452-8_30

1 Introduction

Human digitization is essential to a variety of applications ranging from visual effects, video gaming, to telepresence in VR/AR. The advent of deep learning techniques has achieved impressive progress in recovering unclothed human shape and pose simply from multiple [30,63] or even single [5,45,57] images. However, these leaps in performance come only when a large amount of labeled training data is available. Such limitation has led to inferior performance of reconstructing clothing – the key element of casting a photorealistic digital human, compared to that of naked human body reconstruction. One primary reason is the scarcity of 3D garment datasets in contrast with large collections of naked body scans, e.g. SMPL [39], SCAPE [6], etc. In addition, the complex surface deformation and large diversity of clothing topologies have introduced additional challenges in modeling realistic 3D garments (Fig. 1).

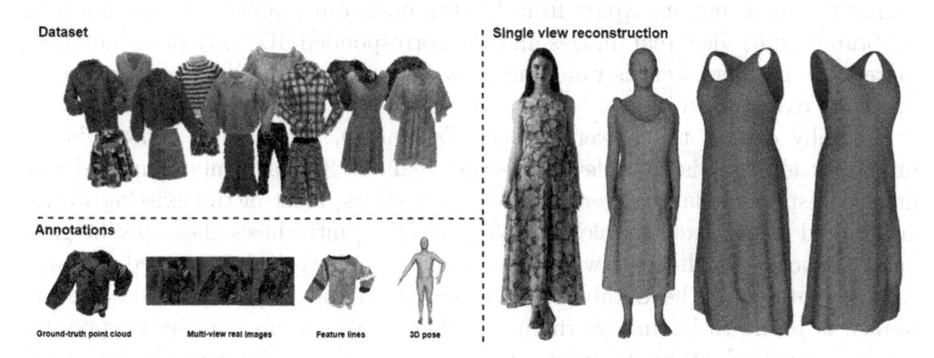

Fig. 1. We present Deep Fashion3D, a large-scale repository of 3D clothing models reconstructed from real garments. It contains over 2000 3D garment models, spanning 10 different cloth categories. Each model is richly labeld with ground-truth point cloud, multi-view real images, 3D body pose and a novel annotation named feature lines. With Deep Fashion3D, inferring the garment geometry from a single image becomes possible.

To address the above issues, there is an increasing need of constructing a high-quality 3D garment database that satisfies the following properties. First of all, it should contain a large-scale repository of 3D garment models that cover a wide range of clothing styles and topologies. Second, it is preferable to have models reconstructed from the real images with physically-correct clothing wrinkles to accommodate the requirement of modeling complicated dynamics and deformations caused by the body motions. Lastly, the dataset should be labeled with sufficient annotations to provide strong supervision for deep generative models.

Multi-Garment Net (MGN) [7] introduces the first dataset specialized for digital clothing obtained from real scans. The proposed digital wardrobe contains 356 digital scans of clothed people which are fitted to pre-defined parametric cloth templates. However, the digital wardrobe only captures 5 garment categories, which is quite limited compared to the large variety of garment styles. Apart from 3D scans, some recent works [26,61] propose to leverage synthetic data obtained from physical simulation. However, the synthetic models lack realism compared to the 3D scans and cannot provide the corresponding real images, which are critical to generalizing the trained model to images in the wild.

In this paper, we address the lack of data by introducing Deep Fashion3D, the largest 3D garment dataset by far, that contains thousands of 3D clothing models with comprehensive annotations. Compared to MGN, the collection of Deep Fashion3D is one order of magnitude larger – including 2078 3D models reconstructed from real garments. It is built from 563 diverse garment instances, covering 10 different clothing categories. Annotation-wise, we introduce a new type of annotation tailored for 3D garment – 3D feature lines. The feature lines denote the most prominent geometrical features on garment surfaces (see Fig. 3), including necklines, cuff contours, hemlines, etc, which provide strong priors for 3D garment reconstruction. Apart from feature lines, our annotations also include calibrated multi-view real images and the corresponded 3D body pose. Furthermore, each garment item is randomly posed to enhance the dataset capacity of modeling dynamic wrinkles.

To fully exploit the power of Deep Fashion3D, we propose a novel baseline approach that is capable of inferring realistic 3D garments from a single image. Despite the large diversity of clothing styles, most of the existing works are limited to one fixed topology [19,33]. MGN [7] introduces class-specific garment network – each deals with a particular topology and is trained by one-category subset of the database. However, given the very limited data, each branch is prone to having overfitting problems. We propose a novel representation, named adaptable template, that can scale to varying topologies during training. It enables our network to be trained using the entire dataset, leading to stronger expressiveness. Another challenge of reconstructing 3D garments is that clothing model is typically a shell structure with open boundaries. Such topology can hardly be handled by the implicit or voxel representation. Yet, the methods based on deep implicit functions [43,48] have shown their ability of modeling fine-scale deformations that the mesh representation is not capable of. We propose to connect the good ends of both worlds by transferring the high-fidelity local details learnt from implicit reconstruction to the template mesh with correct topology and robust global deformations. In addition, since our adaptable template is built upon the SMPL topology, it is convenient to repose or animate the reconstructed results. The proposed framework is implemented in a multi-stage manner with a novel feature line loss to regularize mesh generation.

We have conducted extensive benchmarking and ablation analysis on the proposed dataset. Experimental results demonstrate that the proposed baseline model trained on Deep Fashion3D sets new state of the art on the task of single-view garment reconstruction. Our contributions can be summarized as follows:

- We build Deep Fashion3D, a large-scale, richly annotated 3D clothing dataset reconstructed from real garments. To the best of our knowledge, this is the largest dataset of its kind.
- We introduce a novel baseline approach that combines the merits of mesh and implicit representation and is able to faithfully reconstruct 3D garment from a single image.
- We propose a novel representation, called adaptable template, that enables encoding clothing of various topologies in a single mesh template.
- We first present the feature line annotation specialized for 3D garments, which can provide strong priors for garment reasoning related tasks, e.g., 3D garment reconstruction, classification, retrieval, etc.
- We build a benchmark for single-image garment reconstruction by conducting extensive experiments on evaluating a number of state-of-the-art single-view reconstruction approaches on Deep Fashion3D.

2 Related Work

3D Garment Datasets. While most of existing repositories focus on naked [6,8,9,39] or clothed [68] human body, datasets specially tailored for 3D garment is very limited. BUFF dataset [67] consists of high-resolution 4D scans of clothed human with very limited ammount. In addition, it fails to provide separated models for body and clothing. Segmenting garment models from the 3D scans remains extremely laborious and often leads to corrupted surfaces due to occlusions. To address this issue, Pons-Moll et al. [49] propose an automatic solution to extract the garments and their motion from 4D scans. Recently, a few datasets specialized for 3D garment are proposed. Most of the works [25,61] propose to synthetically generate garment dataset using physical simulation. However, the quality of the synthetic data is not on par with that of real data. In addition, it remains difficult to generalize the trained model to real images as only synthetic images are available. MGN [7] introduces the first garment dataset obtained from 3D scans. However, the dataset only covers 5 cloth categories and is limited to a few hundreds of samples. In contrast, Deep Fashion3D collects more than two thousand clothing models reconstructed from real garments, which covers a much larger diversity of garment styles and topologies. Further, the novel annotation of feature lines provides stronger and more accurate supervision for reconstruction algorithms, which is demonstrated in Sect. 5.

Performance Capture. Over the past decades, progress [42,44,59] has been made to capture cloth surface deformation in motion. Vision-based approaches strive to leverage the easily accessible RGB data and develop frameworks either based on texture pattern tracking [53,62], shading cues [69] or calibrated silhouettes obtained from multi-view videos [11,12,37,55]. However, without dense correspondences or priors, the silhouette-based approaches cannot fully recover the fine details. To improve the reconstruction quality, stronger prior knowledge, including the clothing type [20], pre-scanned template model [27], stereo [10] and photometric [29,58] constraints, has been considered in recent works. With the advances of fusion-based solutions [32,46], template model can be eliminated as the surface geometry can be progressively fused on the fly [18,21] with even a single depth camera [64–66]. Yet, most of the existing works estimate body and clothing jointly and thus cannot obtain a separated cloth surface from the output. Chen et al. [15] propose to model 3D garment from a single depth camera by fitting deformable templates to the initial mesh generated by KinectFusion.

Single-view Garment Reconstruction. Inferring 3D cloth from a single image is highly challenging due to the scarcity of the input and the enormous searching space. Statistical model has been introduced for such ill-posed problem to provide strong priors. However, most models [6,28,34,39,50] are restricted to capturing human body only. Attempts have been made to jointly reconstruct body and clothing from videos [3,4] and multi-view images [30,63]. Recent advances in deep learning based approaches [2,5,14,36,45,51,52,56,57] have achieved single-view clothed body reconstruction. However, for all these methods, tedious manual post-processing is required to extract the clothing surface. And yet, the reconstructed clothing lacks realism. DeepWrinkles [35] synthesizes faithful clothing wrinkles onto a coarse garment mesh following a given pose. Jin et al. [33] leverage similar idea with [31], which encodes detailed geometry deformations in the uv space. However, the method is limited to a fixed topology and cannot scale well to large deformations. Daněřek et al. [19] propose to use physics based simulations as supervision for training a garment shape estimation network. However, the quality of their results is limited to that of the synthetic data and thus cannot achieve high photo-realism. Closer to our work, Multi-Garment Net [7] learns per-category garment reconstruction using scanned data. Nonetheless, their method typically requires 8 frames as input while our approach only consumes a single image. Further, since MGN relies on pre-trained parametric models, it cannot deal with out-of-scope deformations, especially the clothing wrinkles that are dependent on body poses. In contrast, our approach is blendshape-free and is able to faithfully capture multi-scale shape deformations.

3 Dataset Construction

Despite the rapid evolution of 2D garment image datasets from DeepFashion [38] to DeepFashion2 [23] and FashionAI [70], large-scale collection of 3D clothing is very rare. The digital wardrobe released by MGN [7] only contains 356 scans and is limited to only 5 garment categories, which is not sufficient for training an expressive reconstruction model. To fill this gap, we build a more comprehensive dataset named Deep Fashion3D, which is one order larger than MGN, richly annotated and covers a much larger variations of garment styles. We provide more details on data collection and statistics in the following context (Fig. 2).

Cloth Capture. To model the large variety of real-world clothing, we collect a large number of garments, consisting of 563 diverse items that covers 10 clothing categories. The detailed numbers for each category are shown in Table 1. We

Table 1. Statistics of the each clothing categories of Deep Fashion3D.

Type	Number	Type	Number
Long-sleeve coat	157	Long-sleeve dress	18
Short-sleeve coat	98	Short-sleeve dress	34
None-sleeve coat	35	None-sleeve dress	32
Long trousers	29	Long skirt	104
Short trousers	44	Short skirt	48

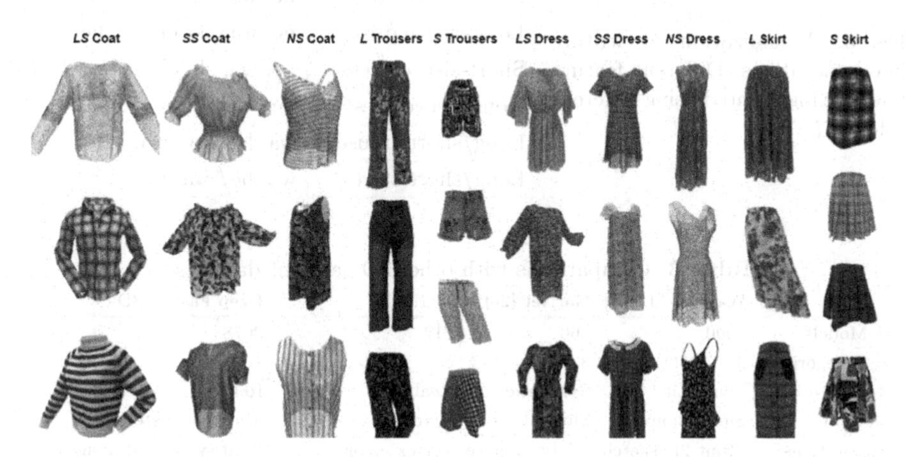

Fig. 2. Example garment models of Deep Fashion3D.

adopt the image-based reconstruction software [1] to reconstruct high-resolution garment models from multi-view images in the form of dense point cloud. In particular, the input images are captured in a multi-view studio with of 50 RGB cameras and controlled lighting. To enhance the expressiveness of the dataset, each garment item is randomly posed on a dummy model or real human to generate a large variety of real deformations caused by body motion. The body parts are manually removed from reconstructed point clouds. With the augmentation of poses, 2078 3D garment models in total are reconstructed from our pipeline.

Annotations. To facilitate future research on 3D garment reasoning, apart from the calibrated multi-view images, we provide additional annotations for Deep Fashion3D. In particular, we introduce *feature line* annotation which is specially tailored for 3D garments. Akin to facial landmarks, the feature lines denote the most prominent features, e.g. the open boundaries, the neckline, cuff, waist, etc, that could provide strong priors for faithful garment reconstruction. The details of feature line annotations are provided in Table 2 and visualized in Fig. 3. We will show in method section that feature line labels can supervise the learning of 3D key lines prediction, which provide explicit constraints for mesh generation.

Fig. 3. Visualization of feature line annotations. Different feature lines are highlighted using different colors.

Table 2. Feature line positions for each cloth category. The meanings for the abbreviations are: 'ne'-neck, 'wa'-waist, 'sh'-shoulder, 'el'-elbow, 'wr'-wrist, 'kn'-knee, 'an'-ankle, 'he'-'hemline'.

Cloth category	Feature line positions
Long-sleeve coat	ne, wa, sh, el, wr
Short-sleeve coat	ne, wa, sh, el
None-sleeve coat	ne, wa, sh
Long-sleeve dress	ne, wa, sh, el, wr, he
Short-sleeve dress	ne, wa, sh, el, he
None-sleeve dress	ne, wa, sh, he
Long/short trousers	wa, kn, an/ wa, kn
Long/short skirt	wa, he/ wa, he

Table 3. Comparisons with other 3D garment datasets.

	Wang et al. [61]	GarNet [26]	MGN [7]	Deep Fashion3D
# Models	2000	600	712	2078
# Categories	3	3	5	10
Real/Synthetic	Synthetic	Synthetic	Real	Real
Method	Simulation	Simulation	Scanning	Multi-view stereo
Annotations	Input 2D sketch	3D body pose	Vertex color 3D body pose	Multi-view real images 3D feature lines 3D body pose

Furthermore, each reconstructed model is labeled with 3D pose represented by SMPL [39] coefficients. The pose is obtained by fitting the SMPL model to the reconstructed dense point cloud. Due to the highly coupled nature between human body and clothing, we believe the labeled 3D pose could be beneficial to infer the global shape and pose-dependent deformations of the garment model.

Data Statistics. To the best of our knowledge, among existing works, there are only three publicly available datasets specialized for 3D garments: Wang et. al [61], GarNet [26] and MGN [7]. In Table 3, we provide detailed comparisons with these datasets in terms of the number of models, categories, data modality, production method and data annotations. Scale-wise, Deep Fashion3D and Wang et al. [61] are one order larger than the other counterparts. However, our dataset covers much more garment categories compared to Wang et al. [61]. Apart from our dataset, only MGN collects models reconstructed from real garments while the other two are fully synthetic. Regarding data annotations, Deep Fashion3D provides the richest data labels. In particular, multi-view real images are only available in our dataset. In addition, we present a new form of garment annotation, the 3D feature lines, which could offer important landmark information for a variety of 3D garment reasoning tasks including garment reconstruction, segmentation, retrieval, etc.

4 A Baseline Approach for Single-View Reconstruction

To demonstrate the usefulness of Deep Fashion3D, we propose a novel baseline approach for single-view garment reconstruction. Specifically, taking a single image I of a garment as input, we aim to reconstruct its 3D shape represented as a triangular mesh. Although recent advances in 3D deep learning techniques have achieved promising progress in single-view reconstruction on general objects, we found all existing approaches have difficulty scaling to cloth reconstruction. The main reasons are threefolds: (1) *Non-closed surfaces.* Unlike the general objects in ShapeNet [13], the garment shape typically appears as a thin layer with open boundary. While implicit representation [43,48] can only model closed surface, voxel based approach [16] is not suited for recovering shell-like structure like the garment surface. (2) *Complex shape topologies.* As all existing mesh-based approaches [24,47,60] rely on deforming a fixed template, they fail to handle the highly diversified topologies introduced by different clothing categories. (3) *Complicated geometric details.* While general man-made objects typically consist of smooth surfaces, the clothing dynamics often introduces intricate high-frequency surface deformations that are challenging to capture.

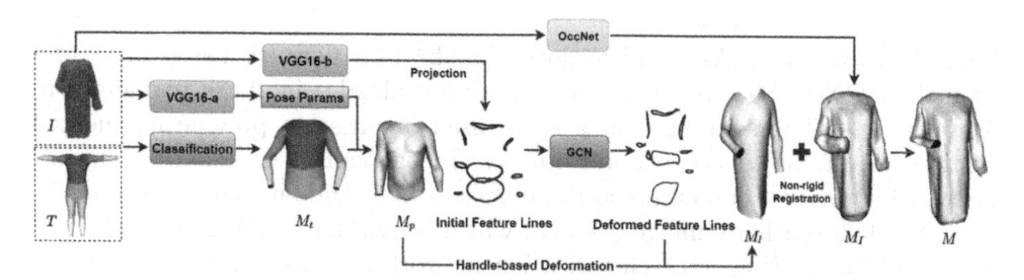

Fig. 4. The pipeline of our proposed approach.

Overview. To address the above issues, we propose to employ a hybrid representation that leverages the merits of each embedding. In particular, we harness both the capability of implicit surface of modeling fine geometric details and the flexibility of mesh representation of handling open surfaces. Our method starts with generating a template mesh M_t which can automatically adapt its topology to fit the target clothing category in the input image. It is then deformed to M_p according to estimated 3D pose. By treating the feature lines as a graph, we then apply image-guided graph convolutional network (GCN) to capture the 3D feature lines, which later trigger handle-based deformation and generates mesh M_l. To exploit the power of implicit representation, we first employ OccNet

[43] to generate a mesh model M_I and then adaptively register M_l to M_I by incorporating the learned fine surface details from M_I while discarding its outliers and noises caused by enforcement of close surface. The proposed pipeline is illustrated in Fig. 4.

4.1 Template Mesh Generation

Adaptable Template. We propose *adaptable template*, a new representation that is scalable to different cloth topologies, enabling the generation of all types of cloth available in the dataset using a single network. The adaptable template is built on the SMPL [39] model by removing the head, hands and feet regions. As seen in Fig. 4, it is then segmented into 6 semantic regions: torso, waist, and upper/lower limbs/legs. During training, the entire adaptable template is fed into the pipeline. However, different semantic regions are activated according to the estimated cloth topology. We denote the template mesh as $M_t = (V, E, B)$, where $V = \{v_i\}$ and E are the set of vertices and edges respectively, and $B = \{b_i\}$ is a per-vertex binary activation mask. v_i will only be activated if $b_i = 1$; otherwise v_i will be detached during the training and removed in the output. The activation mask is determined by the estimated cloth category, where regions of vertices are labeled as a whole. For instance, to model a short-sleeve dress, vertices belonging to the regions of lower limbs and legs are deactivated. Note that in order to adapt the waist region to large deformations for modeling long dresses, we densify its triangulation accordingly using mesh subdivisions.

Cloth Classification. We build a cloth classification network based on a pre-trained VGGNet. The classification network is trained using both real and synthetic images. The synthetic images are used in order to provide augmented lighting conditions to the training images. In particular, we render each garment model under different global illuminations in 5 random views. We generate around 10,000 synthetic images, 90% of which is used for training while the rest is reserved for testing. Our classification network can achieve an accuracy of 99.3%, leading to an appropriate template at both train and test time.

4.2 Learning Surface Reconstruction

To achieve a balanced trade-off between mesh smoothness and accuracy of reconstruction, we propose a multi-stage pipeline to progressively deforming M_t to fit the target shape.

Feature Line-Guided Mesh Generation. It is well understood that, the feature lines, such as necklines, hemlines, etc, play a key role in casting the shape contours of the 3D clothing. Therefore, we propose to first infer the 3D feature lines and then deform M_t by treating the feature lines as deformation handles.

Pose Estimation. Due to the large degrees of freedom of 3D lines, directly regressing their positions is highly challenging. To reduce the searching space, we first estimate the body pose and deform M_t to M_p which provides an initialization $\{l_i^p\}$ of 3D feature lines. Here, the pose of 3D garment is represented with SMPL pose parameters θ [39], which are regressed by a pose estimation network.

GCN-Based Feature Line Regression. We represent the feature lines $\{l_i^p\}$ as polygons during pose estimation. This enables us to treat it as a graph and further employ an image-guided GCN to regress the vertex-wise displacements. We employ another VGG module to extract image features and leverage a similar learning strategy with Pixel2Mesh [60] to infer deformation of feature lines. Note that all of the feature lines predefined on the template are fed into the network, but only the activated subset of the feature lines are adopted to update network parameters.

Handle-Based Deformation. We denote the output feature lines of the above steps as $\{l_i^o\}$. M_l is obtained by deforming M_p so that its feature lines $\{l_i^p\}$ fit our prediction $\{l_i^o\}$. We use the handle-based Laplcacian deformation [54] by setting the alignment between $\{l_i^p\}$ and $\{l_i^o\}$ as hard constrains while optimizing the displacements of the remaining vertices to achieve smooth and visually pleasing deformations. Note that the explicit handle-based deformation can quickly lead to a result that is close to the target surface, which alleviates the difficulty of regressing of a large number of vertices.

Surface Refinement by Fitting Implicit Reconstruction. After obtaining M_l, a straightforward way to obtain surface details is to apply Pixel2Mesh [60] by taking M_l as input. However, as illustrated in Fig. 5, this method fails probably due to the inherent difficulty of learning the high-frequency details while preserving surface smoothness. In contrast, our empirical results indicate that the implicit surface based methods, such as OccNet [43], can faithfully recover the details but only generate closed surface. We therefore perform an adaptive non-rigid registration from M_l to OccNet output for transferring surface details.

Learning Implicit Surface. We directly employ OccNet [43] for learning the implicit surface. Specifically, the input image is first encoded into a latent vector using ResNet-18. For each 3D point in the space, a MLP layer consumes its coordinate and the latent code to predict if the point is inside or outside the surface. Note that we convert all the data into closed meshes using Poisson reconstruction in MeshLab [17]. With the trained network, we first generate an implicit field and then extract the reconstructed surface M_I using marching cube algorithm [40].

Detail Transfer with Adaptive Registration. Though OccNet can synthesize high-quality geometric details, it may also introduce outliers due to its enforcement of generating closed surface. To improve robustness and convergence in conventional non-rigid ICP, we impose normal and distance constraints to filter out wrong correspondences so that only the correct high-frequency details are transferred: (1) the two points of a valid correspondence should have consistent normal direction (i.e., the angle of the two normal directions should be smaller than a threshold which is set as 60°). (2) the bi-directional Chamfer distance between the corresponded points should be less than a preset threshold σ (σ is set as 0.01). The adaptive registrations helps to remove erroneous correspondences and produces our final output M_r.

4.3 Training

There are four sub-networks need to be trained: cloth classification, pose estimation, GCN-based feature line fitting and the implicit reconstruction. Each of the sub-networks is trained independently. In the following subsections, we will provide the details on training data preparation and loss functions.

Training Data Generation

Pose Estimation. We obtain the 3D pose of the garment model by fitting the SMPL model to the reconstructed dense point cloud. The data processing procedures are as follows: 1) for each annotated feature line, we calculate its center point as the its corresponding skeleton joint; 2) we use the joints in the torso region to align all the point clouds to ensure a consistent orientation and scale. 3) lastly, we compute the SMPL pose parameters for each model by fitting the joints and point cloud. The obtained pose parameters will be used for supervising the pose estimation module in Sect. 4.2.

Image Rendering. We augment the input with synthetic images. In particular, for each model, we generate rendered images by randomly sampling 3 viewpoints and 3 different lighting environments, obtaining 9 images in total. Note that we only sample viewpoints from the front viewing angles as we only focus on front-view reconstruction in this work. However, our approach can scale to side or back view prediction by providing corresponding training images.

Loss Functions. The training of cloth classification, pose estimation and implicit reconstruction exactly follows the mainstream protocols. Hence, due to the page limit, we only focus on the part of feature line regression here while leaving other details in the appendix.

Feature Line Regression. Our training goal is to minimize the average distance between the vertices on the obtained feature lines and the ground-truth annotations. Therefore, our loss function is a weighted sum of a distance metric (we use Chamfer distance here) and an edge length regularization loss [60], which helps to smooth the deformed feature lines (more details can be found in supplementals).

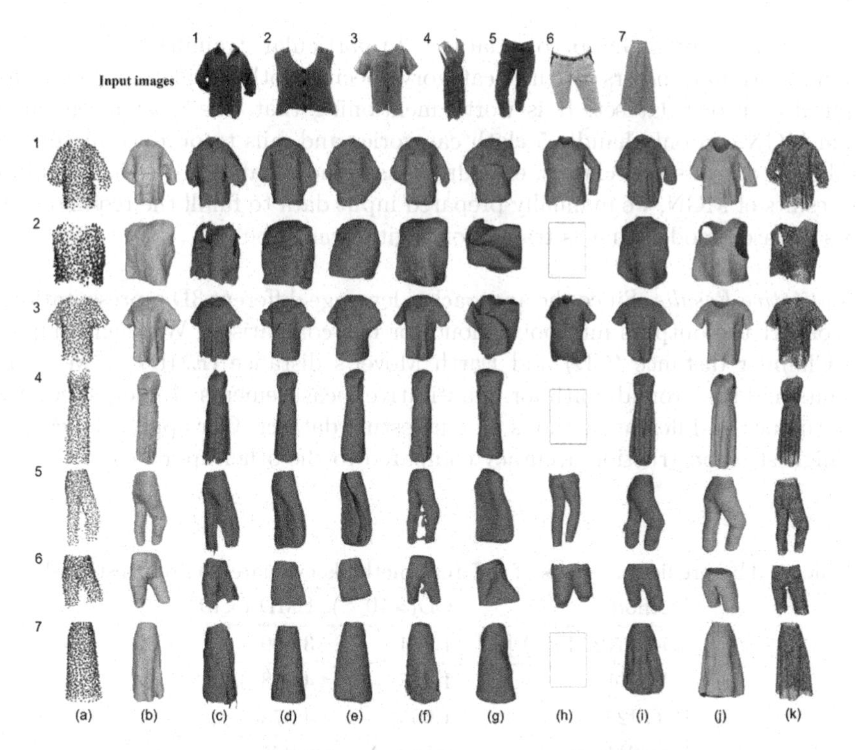

Fig. 5. Experiment results against other methods. Given an image, results are followed with (a) PSG (Point Set Generation) [22]; (b) 3D-R2N2 [16]; (c) AtlasNet [24] with 25 square patches; (d) AtlasNet [24] whith a sphere template; (e) Pixel2Mesh [60]; (f) MVD [41] (multi-view depth generation); (g) TMN [47] (topology modification network); (h) MGN (Multi-Garment Network) [7]; (i) OccNet [43]; (j) Ours; (k) The groundtruth point clouds. The input images on the top. The null means the method fails to generate a result.

5 Experimental Results

Implementation Details. The whole pipeline proposed is implemented using PyTorch. The initialized learning rate is set to 5e–5 and with the batch size of 8. It takes about 30 h to train the whole network using Adam optimization for 50 epochs using a NVIDIA TITAN XP graphics card.

5.1 Benchmarking on Single-View Reconstruction

Methods. We compare our method against six state-of-the-art single-view reconstruction approaches that use different 3D representations: 3D-R2N2 [16], PSG(Point Set Generation) [22], MVD (generating multi-view depth maps) [41], Pixel2Mesh [60], AtlasNet [24], MGN [7] and OccNet [43]. For AtlasNet, we have experimented it using both sphere template and patch template, which are denoted as "Atlas-Sphere" and "Atlas-Patch". To ensure fairness, we train all the algorithms, except MGN, on our dataset. In particular, training MGN requires ground-truth parameters for their category-specific cloth template, which is not applicable in our dataset. It is worth mentioning that, the most recent algorithm MGN can only handle 5 cloth categories and fails to produce reasonable results for out-of-scope classes, e.g., dress, as demonstrated in Fig. 5. To obtain the results of MGN, we manually prepared input data to fulfill the requirements of its released model, that is trained on digital wardrobe [7].

Quantitative Results. Since the approaches leverage different 3D representations, we convert the outputs into point cloud for fair comparison. We then compute the Chamfer distance (CD) and Earth Mover's distance (EMD) between the outputs and the ground-truth for quantitative measurements. Table 4 shows the performance of different methods on our testing dataset. Our approach achieves the highest reconstruction accuracy compared to the other approaches.

Table 4. The prediction errors of different methods evaluated on our testing data.

Method	CD($\times 10^{-3}$)	EMD ($\times 10^{2}$)
3D-R2N2 (128^3) [16]	1.264	3.609
MVD [41]	1.047	4.058
PSG [22]	1.065	4.675
Pixel2Mesh [60]	0.782	9.078
AtlasNet(sphere) [24]	0.855	6.193
AtlasNet(patch) [24]	0.908	9.428
TMN [47]	0.865	8.580
OccNet (256^3) [43]	0.960	3.431
Ours	**0.679**	**2.942**

Qualitative Results. In Fig. 5, we also provide qualitative comparisons by randomly selecting some samples from different garment categories in arbitrary poses. Compared to the other methods, our approach provides more accurate reconstructions that are closer to ground truths. The reasons are: 1) 3D representations like point set [22], voxel [16] or multi-view depth maps [41] are not suitable for generating a clean mesh. 2) Although template-based methods [24,47,60] are designed for mesh generation, it is hard to use a fixed template for fitting diverse shape complexity of clothing. 3) As shown in the results, method based on implicit function [43] is able to synthesis rich details. However, it can only generate closed shapes, making it difficult to handle garment reconstruction, which typically consists of multiple open boundaries. By explicitly combining the merits of template-based methods and implicit ones, the proposed approach can not only capture the global shape but also generate faithful geometric details.

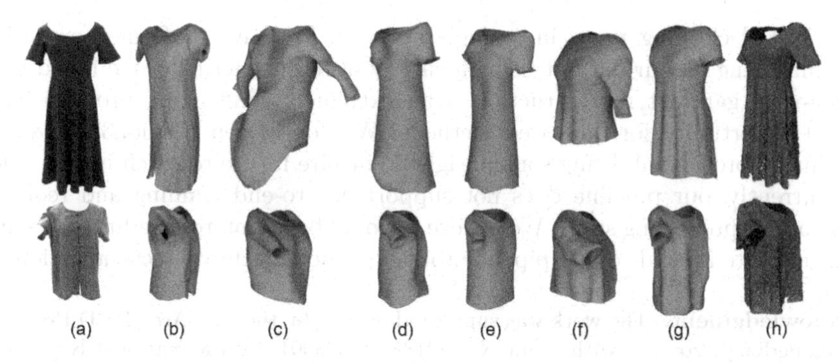

Fig. 6. Results of ablation studies. (a) input images; (b) results of M_t+GCN; (c) results of M_p+GCN; (d) results of M_l+GCN. (e) results of our approach without surface refinement, i.e., M_l. (f) M_t+Regis. (g) results of our full approach. (h) groundtruth point clouds.

5.2 Ablation Analysis

We further validate the effectiveness of each algorithmic component by selectively applying them in different settings: 1) Directly applying GCN on the generated template mesh M_t to fit the target shape, termed as M_t+GCN; 2) Applying GCN on M_p (obtained by deforming M_t with estimated SMPL pose) to fit the target shape, termed as M_p+GCN; 3) Applying GCN on the resulted mesh after feature line-guided deformation, i.e. M_l. This is termed as M_l+GCN; 4) Directly performing registration from M_t to M_I for details transferring, which is termed as M_t+Regis. Figure 6 shows the qualitative comparisons between these settings and the proposed one. As seen, the baseline approach produce the best results.

As observed from the experiments, it is difficult for GCN to learn geometric details. There are two possible reasons: 1) It is inherently difficult to synthesize high-frequency signals while preserving surface smoothness; 2) GCN structure might be not suitable for a fine-grained geometric learning task as graph is a sparse and crude approximation of a surface. We also found that the feature lines are much easier to learn and explicit handle-based deformation works surprisingly well. The deeper study in this regard is left as one of our further works.

6 Conclusions and Discussions

We have proposed a new dataset called Deep Fashion3D for image-based garment reconstruction, which is by far the largest 3D garment collection reconstructed from real clothing images. In particular, it consists of over 2000 highly diversified garment models covering 10 clothing categories and 563 distinct garment items. In addition, each model of Deep Fashion3D is richly labeled with 3D body pose, 3D feature lines and multi-view real images. We also presented a baseline approach for single-view reconstruction to validate the usefulness of the proposed dataset. It uses a novel representation, called adaptable template, to learn a variety of clothing types in a single network. We have performed extensive benchmarking on our dataset using a variety of recent methods. We found that single-view garment reconstruction is an extremely challenging problem with ample opportunity for improved methods. We hope Deep Fashion3D and our baseline approach will bring some insight to inspire future research in this field.

Currently, our pipeline does not support end-to-end training and requires some offline processing steps. We believe it would be an interesting future avenue to investigate an end-to-end pipeline to enable more accurate reconstruction.

Acknowledgment. The work was supported in part by the Key Area R&D Program of Guangdong Province with grant No. 2018B030338001, by the National Key R&D Program of China with grant No. 2018YFB1800800, by Natural Science Foundation of China with grant NSFC-61629101 and 61902334, by Guangdong Research Project No. 2017ZT07X152, and by Shenzhen Key Lab Fund No.ZDSYS201707251409055. The authors would thank Yuan Yu for her early efforts on dataset construction.

References

1. Agisoft: Mentashape (2019). https://www.agisoft.com/
2. Alldieck, T., Magnor, M., Bhatnagar, B.L., Theobalt, C., Pons-Moll, G.: Learning to reconstruct people in clothing from a single RGB camera. In: IEEE Conference on Computer Vision and Pattern Recognition (CVPR) (2019)
3. Alldieck, T., Magnor, M., Xu, W., Theobalt, C., Pons-Moll, G.: Detailed human avatars from monocular video. In: International Conference on 3D Vision (3DV) (2018)
4. Alldieck, T., Magnor, M., Xu, W., Theobalt, C., Pons-Moll, G.: Video based reconstruction of 3d people models. In: IEEE Conference on Computer Vision and Pattern Recognition (CVPR) (2018)
5. Alldieck, T., Pons-Moll, G., Theobalt, C., Magnor, M.: Tex2shape: Detailed full human body geometry from a single image. In: IEEE International Conference on Computer Vision (ICCV). IEEE (2019)
6. Anguelov, D., Srinivasan, P., Koller, D., Thrun, S., Rodgers, J., Davis, J.: SCAPE: shape completion and animation of people. ACM Trans. Graph. **24**(3), 408–416 (2005)
7. Bhatnagar, B.L., Tiwari, G., Theobalt, C., Pons-Moll, G.: Multi-garment net: learning to dress 3D people from images. In: IEEE International Conference on Computer Vision (ICCV). IEEE (2019)
8. Bogo, F., Romero, J., Loper, M., Black, M.J.: FAUST: dataset and evaluation for 3D mesh registration. In: Proceedings IEEE Conference on Computer Vision and Pattern Recognition (CVPR). IEEE, Piscataway (2014)
9. Bogo, F., Romero, J., Pons-Moll, G., Black, M.J.: Dynamic FAUST: registering human bodies in motion. In: Proceedings IEEE Conference on Computer Vision and Pattern Recognition (CVPR). IEEE (2017)
10. Bradley, D., Popa, T., Sheffer, A., Heidrich, W., Boubekeur, T.: Markerless garment capture. In: ACM Transactions on Graphics (TOG), vol. 27, p. 99. ACM (2008)
11. Cagniart, C., Boyer, E., Ilic, S.: Probabilistic deformable surface tracking from multiple videos. In: Daniilidis, K., Maragos, P., Paragios, N. (eds.) ECCV 2010. LNCS, vol. 6314, pp. 326–339. Springer, Heidelberg (2010). https://doi.org/10.1007/978-3-642-15561-1_24

12. Carranza, J., Theobalt, C., Magnor, M.A., Seidel, H.P.: Free-viewpoint video of human actors. ACM Trans. Graph. (TOG) **22**, 569–577 (2003)
13. Chang, A.X., et al.: Shapenet: An information-rich 3D model repository. arXiv preprint arXiv:1512.03012 (2015)
14. Chen, X., Guo, Y., Zhou, B., Zhao, Q.: Deformable model for estimating clothed and naked human shapes from a single image. Visual Comput. **29**(11), 1187–1196 (2013)
15. Chen, X., Zhou, B., Lu, F.X., Wang, L., Bi, L., Tan, P.: Garment modeling with a depth camera. ACM Trans. Graph. **34**(6), 203–2111 (2015)
16. Choy, C.B., Xu, D., Gwak, J., Chen, K., Savarese, S.: 3D–r2n2: a unified approach for single and multi-view 3D object reconstruction. In: Proceedings of the European Conference on Computer Vision (ECCV) (2016)
17. Cignoni, P., Callieri, M., Corsini, M., Dellepiane, M., Ganovelli, F., Ranzuglia, G.: Meshlab: an open-source mesh processing tool. In: Eurographics Italian Chapter Conference, vol. 2008, pp. 129–136. Salerno (2008)
18. Collet, A., et al.: High-quality streamable free-viewpoint video. ACM Trans. Graph. (ToG) **34**(4), 69 (2015)
19. Daněřek, R., Dibra, E., Öztireli, C., Ziegler, R., Gross, M.: Deepgarment: 3D garment shape estimation from a single image. In: Computer Graphics Forum, vol. 36, pp. 269–280. Wiley Online Library (2017)
20. De Aguiar, E., Stoll, C., Theobalt, C., Ahmed, N., Seidel, H.P., Thrun, S.: Performance capture from sparse multi-view video, vol. 27. ACM (2008)
21. Dou, M., et al.: Fusion4d: real-time performance capture of challenging scenes. ACM Trans. Graph. (TOG) **35**(4), 114 (2016)
22. Fan, H., Su, H., Guibas, L.J.: A point set generation network for 3D object reconstruction from a single image. In: The IEEE Conference on Computer Vision and Pattern Recognition (CVPR) (2017)
23. Ge, Y., Zhang, R., Wang, X., Tang, X., Luo, P.: Deepfashion2: a versatile benchmark for detection, pose estimation, segmentation and re-identification of clothing images. In: Proceedings of the IEEE Conference on Computer Vision and Pattern Recognition, pp. 5337–5345 (2019)
24. Groueix, T., Fisher, M., Kim, V.G., Russell, B., Aubry, M.: AtlasNet: a Papier-Mâché Approach to Learning 3D Surface Generation. In: Proceedings IEEE Conf.erenceon Computer Vision and Pattern Recognition (CVPR) (2018)
25. Gundogdu, E., Constantin, V., Seifoddini, A., Dang, M., Salzmann, M., Fua, P.: Garnet: A two-stream network for fast and accurate 3D cloth draping. arXiv preprint arXiv:1811.10983 (2018)
26. Gundogdu, E., Constantin, V., Seifoddini, A., Dang, M., Salzmann, M., Fua, P.: Garnet: A two-stream network for fast and accurate 3D cloth draping. In: Proceedings of the IEEE International Conference on Computer Vision, pp. 8739–8748 (2019)
27. Habermann, M., Xu, W., Zollhoefer, M., Pons-Moll, G., Theobalt, C.: Livecap: real-time human performance capture from monocular video. ACM Trans. Graph. (TOG) **38**(2), 14 (2019)
28. Hasler, N., Stoll, C., Sunkel, M., Rosenhahn, B., Seidel, H.P.: A statistical model of human pose and body shape. In: Computer Graphics Forum, vol. 28, pp. 337–346. Wiley Online Library (2009)
29. Hernández, C., Vogiatzis, G., Brostow, G.J., Stenger, B., Cipolla, R.: Non-rigid photometric stereo with colored lights. In: 2007 IEEE 11th International Conference on Computer Vision, pp. 1–8. IEEE (2007)

30. Huang, Z., et al.: Deep volumetric video from very sparse multi-view performance capture. In: Proceedings of the European Conference on Computer Vision (ECCV), pp. 336–354 (2018)
31. Huynh, L., et al.: Mesoscopic facial geometry inference using deep neural networks. In: The IEEE Conference on Computer Vision and Pattern Recognition (CVPR) (2018)
32. Izadi, S., et al.: Kinectfusion: real-time 3D reconstruction and interaction using a moving depth camera. In: Proceedings of the 24th annual ACM symposium on User interface Software and Technology, pp. 559–568. ACM (2011)
33. Jin, N., Zhu, Y., Geng, Z., Fedkiw, R.: A pixel-based framework for data-driven clothing. arXiv preprint arXiv:1812.01677 (2018)
34. Joo, H., Simon, T., Sheikh, Y.: Total capture: a 3D deformation model for tracking faces, hands, and bodies. In: Proceedings of the IEEE Conference on Computer Vision and Pattern Recognition, pp. 8320–8329 (2018)
35. Lahner, Z., Cremers, D., Tung, T.: Deepwrinkles: accurate and realistic clothing modeling. In: Proceedings of the European Conference on Computer Vision (ECCV), pp. 667–684 (2018)
36. Lazova, V., Insafutdinov, E., Pons-Moll, G.: 360-degree textures of people in clothing from a single image. In: International Conference on 3D Vision (3DV) (2019)
37. Leroy, V., Franco, J.S., Boyer, E.: Multi-view dynamic shape refinement using local temporal integration. In: Proceedings of the IEEE International Conference on Computer Vision, pp. 3094–3103 (2017)
38. Liu, Z., Luo, P., Qiu, S., Wang, X., Tang, X.: Deepfashion: powering robust clothes recognition and retrieval with rich annotations. In: Proceedings of IEEE Conference on Computer Vision and Pattern Recognition (CVPR) (2016)
39. Loper, M., Mahmood, N., Romero, J., Pons-Moll, G., Black, M.J.: SMPL: a skinned multi-person linear model. ACM Trans. Graph. **34**(6), 248:1–248:16 (2015)
40. Lorensen, W.E., Cline, H.E.: Marching cubes: a high resolution 3D surface construction algorithm. ACM Siggraph Comput. Graph. **21**(4), 163–169 (1987)
41. Lun, Z., Gadelha, M., Kalogerakis, E., Maji, S., Wang, R.: 3D shape reconstruction from sketches via multi-view convolutional networks. In: 2017 International Conference on 3D Vision (3DV), pp. 67–77. IEEE (2017)
42. Matsuyama, T., Nobuhara, S., Takai, T., Tung, T.: 3D Video and its Applications. Springer, Heidelberg (2012). https://doi.org/10.1007/978-1-4471-4120-4
43. Mescheder, L., Oechsle, M., Niemeyer, M., Nowozin, S., Geiger, A.: Occupancy networks: learning 3D reconstruction in function space. In: Proceedings of the IEEE Conference on Computer Vision and Pattern Recognition, pp. 4460–4470 (2019)
44. Miguel, E., et al.: Data-driven estimation of cloth simulation models. In: Computer Graphics Forum, vol. 31, pp. 519–528. Wiley Online Library (2012)
45. Natsume, R., et al.: Siclope: silhouette-based clothed people. In: Proceedings of the IEEE Conference on Computer Vision and Pattern Recognition, pp. 4480–4490 (2019)
46. Newcombe, R.A., Fox, D., Seitz, S.M.: Dynamicfusion: reconstruction and tracking of non-rigid scenes in real-time. In: Proceedings of the IEEE Conference on Computer Vision and Pattern Recognition, pp. 343–352 (2015)
47. Pan, J., Han, X., Chen, W., Tang, J., Jia, K.: Deep mesh reconstruction from single RGB images via topology modification networks. In: Proceedings of the IEEE International Conference on Computer Vision, pp. 9964–9973 (2019)

48. Park, J.J., Florence, P., Straub, J., Newcombe, R., Lovegrove, S.: Deepsdf: learning continuous signed distance functions for shape representation. In: Proceedings of the IEEE Conference on Computer Vision and Pattern Recognition, pp. 165–174 (2019)
49. Pons-Moll, G., Pujades, S., Hu, S., Black, M.: ClothCap: seamless 4D clothing capture and retargeting. ACM Trans. Graph. (SIGGRAPH) **36**(4), 1–15 (2017)
50. Pons-Moll, G., Romero, J., Mahmood, N., Black, M.J.: Dyna: a model of dynamic human shape in motion. ACM Trans. Graph. (TOG) **34**(4), 120 (2015)
51. Pumarola, A., Sanchez, J., Choi, G., Sanfeliu, A., Moreno-Noguer, F.: 3DPeople: modeling the geometry of dressed humans. In: International Conference on Computer Vision (ICCV) (2019)
52. Saito, S., Huang, Z., Natsume, R., Morishima, S., Kanazawa, A., Li, H.: Pifu: Pixel-aligned implicit function for high-resolution clothed human digitization. arXiv preprint arXiv:1905.05172 (2019)
53. Scholz, V., Stich, T., Keckeisen, M., Wacker, M., Magnor, M.: Garment motion capture using color-coded patterns. In: Computer Graphics Forum, vol. 24, pp. 439–447. Wiley Online Library (2005)
54. Sorkine, O., Cohen-Or, D., Lipman, Y., Alexa, M., Rössl, C., Seidel, H.P.: Laplacian surface editing. In: Proceedings of the 2004 Eurographics/ACM SIGGRAPH Symposium on Geometry Processing, pp. 175–184. ACM (2004)
55. Starck, J., Hilton, A.: Surface capture for performance-based animation. IEEE Computer Graph. Appl. **27**(3), 21–31 (2007)
56. Tang, S., Tan, F., Cheng, K., Li, Z., Zhu, S., Tan, P.: A neural network for detailed human depth estimation from a single image. In: Proceedings of the IEEE International Conference on Computer Vision, pp. 7750–7759 (2019)
57. Varol, G., et al.: Bodynet: volumetric inference of 3D human body shapes. In: Proceedings of the European Conference on Computer Vision (ECCV), pp. 20–36 (2018)
58. Vlasic, D., et al.: Dynamic shape capture using multi-view photometric stereo. In: ACM Transactions on Graphics (TOG), vol. 28, p. 174. ACM (2009)
59. Wang, H., O'Brien, J.F., Ramamoorthi, R.: Data-driven elastic models for cloth: modeling and measurement. In: ACM Transactions on Graphics (TOG), vol. 30, p. 71. ACM (2011)
60. Wang, N., Zhang, Y., Li, Z., Fu, Y., Liu, W., Jiang, Y.G.: Pixel2mesh: generating 3D mesh models from single RGB images. In: ECCV (2018)
61. Wang, T.Y., Ceylan, D., Popovic, J., Mitra, N.J.: Learning a shared shape space for multimodal garment design. ACM Trans. Graph. **37**(6), 1:1–1:14 (2018). https://doi.org/10.1145/3272127.3275074
62. White, R., Crane, K., Forsyth, D.A.: Capturing and animating occluded cloth. In: ACM Transactions on Graphics (TOG), vol. 26, p. 34. ACM (2007)
63. Xu, Y., Yang, S., Sun, W., Tan, L., Li, K., Zhou, H.: 3D virtual garment modeling from RGB images. arXiv preprint arXiv:1908.00114 (2019)
64. Yu, T., et al.: Bodyfusion: real-time capture of human motion and surface geometry using a single depth camera. In: Proceedings of the IEEE International Conference on Computer Vision, pp. 910–919 (2017)
65. Yu, T., et al.: Doublefusion: real-time capture of human performances with inner body shapes from a single depth sensor. In: Proceedings of the IEEE Conference on Computer Vision and Pattern Recognition, pp. 7287–7296 (2018)
66. Yu, T., et al.: Simulcap: Single-view human performance capture with cloth simulation. arXiv preprint arXiv:1903.06323 (2019)

67. Zhang, C., Pujades, S., Black, M.J., Pons-Moll, G.: Detailed, accurate, human shape estimation from clothed 3D scan sequences. In: Proceedings of the IEEE Conference on Computer Vision and Pattern Recognition, pp. 4191–4200 (2017)
68. Zheng, Z., Yu, T., Wei, Y., Dai, Q., Liu, Y.: Deephuman: 3D human reconstruction from a single image. In: The IEEE International Conference on Computer Vision (ICCV) (2019)
69. Zhou, B., Chen, X., Fu, Q., Guo, K., Tan, P.: Garment modeling from a single image. In: Computer Graphics Forum, vol. 32, pp. 85–91. Wiley Online Library (2013)
70. Zou, X., Kong, X., Wong, W., Wang, C., Liu, Y., Cao, Y.: Fashionai: a hierarchical dataset for fashion understanding. In: Proceedings of the IEEE Conference on Computer Vision and Pattern Recognition Workshops (2019)

Spatially Adaptive Inference with Stochastic Feature Sampling and Interpolation

Zhenda Xie[1,2], Zheng Zhang[2(✉)], Xizhou Zhu[2,3], Gao Huang[4], and Stephen Lin[2]

[1] Tsinghua University, Beijing, China
xzd18@mails.tsinghua.edu.cn
[2] Microsoft Research Asia, Beijing, China
{zhez,stevelin}@microsoft.com
[3] University of Science and Technology of China, Hefei, China
ezra0408@mail.ustc.edu.cn
[4] Department of Automation, Tsinghua University, Beijing, China
gaohuang@tsinghua.edu.cn

Abstract. In the feature maps of CNNs, there commonly exists considerable spatial redundancy that leads to much repetitive processing. Towards reducing this superfluous computation, we propose to compute features only at sparsely sampled locations, which are probabilistically chosen according to activation responses, and then densely reconstruct the feature map with an efficient interpolation procedure. With this sampling-interpolation scheme, our network avoids expending computation on spatial locations that can be effectively interpolated, while being robust to activation prediction errors through broadly distributed sampling. A technical challenge of this sampling-based approach is that the binary decision variables for representing discrete sampling locations are non-differentiable, making them incompatible with backpropagation. To circumvent this issue, we make use of a reparameterization trick based on the Gumbel-Softmax distribution, with which backpropagation can iterate these variables towards binary values. The presented network is experimentally shown to save substantial computation while maintaining accuracy over a variety of computer vision tasks.

Keywords: Sparse convolution · Sparse sampling · Feature interpolation

---

Z. Xie, Z. Zheng and X. Zhu—Equal contribution.
Z. Xie and X. Zhu—This work is done when Zhenda Xie and Xizhou Zhu were interns at Microsoft Research Asia.

Electronic supplementary material The online version of this chapter (https://doi.org/10.1007/978-3-030-58452-8_31) contains supplementary material, which is available to authorized users.

A. Vedaldi et al. (Eds.): ECCV 2020, LNCS 12346, pp. 531–548, 2020.
https://doi.org/10.1007/978-3-030-58452-8_31

1 Introduction

On many computer vision tasks, significant improvements in accuracy have been achieved through increasing model capacity in convolutional neural networks (CNNs) [11,32]. These gains, however, come at a cost of greater processing that can hinder deployment on resource-limited devices. Towards greater practicality of deep models, much attention has been focused on reducing CNN computation.

A common approach to this problem is to prune weights and neurons that are not needed to maintain the network's performance [9,10,13,19,20,28,34,37]. Orthogonal to these architectural changes are methods that eliminate computation at inference time conditioned on the input. These techniques are typically based on feature map sparsity, where the locations of zero-valued activations are predicted so that the computation at those positions can be skipped [1,6,30]. As illustrated in Fig. 1(b), this approach deterministically samples predicted foreground areas while avoiding computational expenditure on the background.

In this paper, we seek a more efficient allocation of computation over a feature map that takes advantage of its spatial redundancy[1]. Rather than exhausting all the computation on areas estimated to have the most significant activation, our approach is to treat the predicted activation map as a probability field, stochastically sample a sparse set of locations based on their probabilities, and then interpolate the features at these samples to reconstruct the rest of the feature map. This strategy avoids wasting computation at locations where the features can simply be interpolated, and it allows feature computation to expand into presumed low-activation regions, which can compensate for errors in activation prediction. With the generated sampling distributions shown in Fig. 1(c) and (d), this sampling-interpolation approach can reduce the computation needed to match the accuracy of feature map sparsity methods, or alternatively increase accuracy through more comprehensive sampling of the feature map while maintaining the same level of overall computation.

To identify sparse points for interpolation, our network trains a content-aware stochastic sampling module that produces a binary sampling mask over the activations. Due to the inability to backpropagate through networks containing binary variables, we employ a reparameterization trick where the non-differentiable mask values are replaced by differentiable samples from a Gumbel-Softmax distribution, which can be smoothly annealed into binary values during training [17]. The module learns to spatially adjust the sampling density according to predicted activations, and the interpolation is performed with a kernel whose parameters are jointly learned with the sampling module. To aid in interpolation of areas far from the content-aware samples, very sparse uniform samples over the feature map are added before interpolation, which we refer to as a grid prior.

With this content-based sampling-interpolation approach, our network obtains appreciable reductions in computation without much loss in accuracy on

[1] We note that CNNs commonly capitalize on spatial redundancy by downsampling input image resolutions or employing strides in convolution.

Fig. 1. **(a)** Input image. **(b)** Deterministic sampling for efficient inference. **(c)** Content-aware stochastic sampling by our method (without *Grid Prior* for better viewing), which yields the same detection accuracy as (b) but with less overall computation. **(d)** Content-aware stochastic sampling with the same overall computation as (b) but with better detection accuracy.

COCO object detection, Cityscapes semantic segmentation, and ImageNet classification. An extensive ablation study validates the sampling and interpolation components of our algorithm, and comparisons to related techniques show that the proposed method provides a superior FLOPs-accuracy tradeoff for object detection and semantic segmentation, and comparable performance for image classification.

2 Related Work

In this section, we briefly review related approaches for reducing computation in convolutional neural networks.

Model Pruning. A widely investigated approach for improving network efficiency is to remove connections or filters that are unimportant for achieving high performance. The importance of these network elements has been approximated in various ways, including by connection weight magnitudes [9,10], filter norms [12,20,34], and filter redundancy within a layer [13]. To reflect network sensitivity to these elements, importance has also been measured based on their effects on the loss [19,28] or the reconstruction error of the final response layer [37] when removing them. Alternatively, sparsity learning techniques identify what to prune in conjunction with network training, through constraints that zero out some filters [34], cause some filters to become redundant and removable [5], scale some filter or block outputs to zero [22], or sparsify batch normalization scaling factors [25,35]. Model pruning techniques as well as other architecture-based acceleration schemes, such as low-rank factorizations of convolutional filters [16] and knowledge distillation of networks [15], are orthogonal to our approach and could potentially be employed in a complementary manner.

Early Stopping. Rather than prune network elements, early stopping techniques reduce computation by skipping the processing at later stages whenever it is deemed to be unnecessary. In [7], an adaptive number of ResNet layers are skipped within a residual block for unimportant regions in object classification.

The skipping mechanism is controlled by halting scores predicted at branches to the output of each residual unit. In [21], a deep model for semantic segmentation is turned into a cascade of sub-models where earlier sub-models handle easy regions and harder cases are progressively fed forward to the next sub-model for further processing. In [18], various predefined downsampling configurations are randomly used during training, and the appropriate configuration is applied according to the computation budget during inference. Like our method, these techniques spatially adapt the processing to the input content. However, they process all spatial positions at least to some degree, which limits the achievable computational savings.

Activation Sparsity. The activations of rectified linear units (ReLUs) are commonly sparse. This property has been exploited for network acceleration by excluding the zero values from subsequent convolutions [29,31]. This approach has been extended by estimating the activation sparsity and skipping the computation for predicted insignificant activations. The sparsity has been predicted from prior knowledge of road and sidewalk locations in autonomous driving applications [30], from model-predicted foreground masks at low resolution [30], from a small auxiliary layer that supplements each convolutional layer [6], and from a highly quantized version of the convolutional layer [1]. Our work instead reconstructs activation maps by interpolation from a sparse set of samples selected in a content-aware fashion, thus avoiding computation at locations where features can be easily reconstructed. Moreover, our probabilistic sampling distributes computation among feature map locations with varying levels of predicted activation, providing greater robustness to activation prediction errors.

Sparse Sampling. To reduce processing cost, PerforatedCNNs compute only sparse samples of a convolutional layer's outputs and interpolate the remaining values [8]. The sampling follows a predefined pattern, and the interpolation is done by nearest neighbors. Our method also takes a sparse sampling and interpolation approach, but in contrast to the input-independent sampling and generic interpolation of PerforatedCNNs, the sampling in our network is adaptively determined from the input such that the sampling density reflects predicted activation values, and the interpolation parameters are learned. As shown later in the experiments, this approach allows for much greater sparsity in the sampling. In [27], high-resolution predictions are generated from low-resolution results. Instead, our method is used for features rather than final outputs.

Gumbel-Based Selection. Random selection based on the Gumbel distribution has been used in making discrete decisions for network acceleration. The Gumbel-Softmax trick was utilized in adaptively choosing network layers to apply on an input image [33] and in selecting channels or layers to skip [14]. In contrast to these techniques which determine computation based on image-level semantics for image classification, our sampling is driven by the spatial

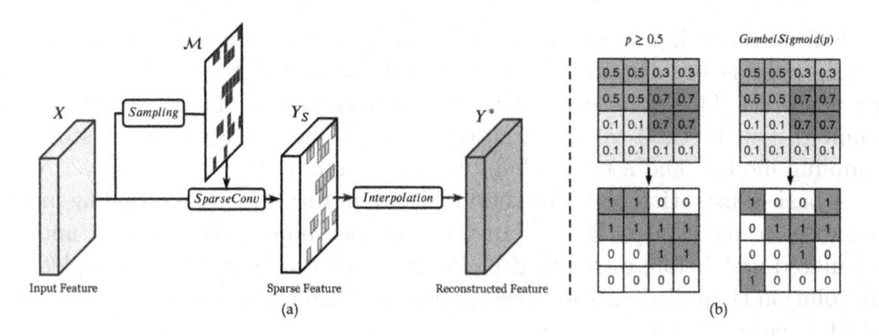

Fig. 2. (a) Stochastic sampling-interpolation network. The stochastic sampling module generates a sparse sampling mask $\mathcal{M}$ based on the input feature map X, and then calculates features only at the sampling points, forming a sparse feature map Y_s. The features of unsampled points are interpolated by the interpolation module to form the output feature map Y^*. (b)(Left) In deterministic sampling, points with the same confidence are either sampled altogether or not sampled at all. (Right) In stochastic sampling, a random subset of the points with the same confidence will be sampled, with a sampling density determined by their confidence.

organization of features and is geared towards accurately reconstructing positional content. As a result, our method is well-suited to spatial understanding tasks such as object detection and semantic segmentation.

3 Methodology

In this section, we first present a general introduction of the stochastic sampling-interpolation network and then describe the stochastic sampling module and interpolation module in detail. Next, we introduce the grid prior which is found to be helpful for interpolation. Finally, we illustrate how to integrate the sampling and interpolation network modules into residual blocks.

3.1 Stochastic Sampling-Interpolation Network

Convolutions in neural networks typically generate an output feature map $Y \in \mathbb{R}^{C_{out} \times H \times W}$ pixel by pixel from an input feature map $X \in \mathbb{R}^{C_{in} \times H \times W}$:

$$Y(p) = \sum_{p' \in R_k} W_c(p') X(p + p'), p \in \Omega, \tag{1}$$

where H and W represent the height and width of the feature map, C_{in} and C_{out} denote the input and output feature dimensions, $\Omega = \{(i,j) | i \leq W, j \leq H, i, j \in \mathbb{Z}^+\}$ represents the spatial domain of the feature map, R_k indicates the support region of kernel offsets with kernel size k (e.g., for a 3×3 convolution, $R_k = \{(-1,-1), (-1,0), ..., (1,1)\}$ and $k = 3$), and $W_c \in \mathbb{R}^{C_{in} \times C_{out} \times k \times k}$ denotes convolution weights.

Spatial redundancy commonly exists in feature maps, such that features at certain points can be approximated by interpolating the features from surrounding positions. Therefore, exhaustive computation across the entire space is not required. Our method takes advantage of this using a content-aware stochastic sampling module and a trainable interpolation module.

A basic illustration of our method is shown in Fig. 2(a). The sampling module generates a mask $\mathcal{M} \in \mathbb{R}^{H \times W}$. In the inference phase, $\mathcal{M}$ is binary and it is calculated first before computing Y. Points masked as 1 in $\mathcal{M}$ are sampled, and convolution is applied only on these points, resulting in a sparse feature map Y_s, which is calculated as:

$$Y_s(p) = \begin{cases} 0 & \mathcal{M}(p) = 0 \\ Y(p) & \mathcal{M}(p) = 1. \end{cases} \tag{2}$$

Then, the features of unsampled points are constructed by the interpolation module C. Together with the features of sampled points, they constitute the reconstructed output feature Y^*:

$$Y^*(p) = \begin{cases} C(Y_s)(p) & \mathcal{M}(p) = 0 \\ Y_s(p) & \mathcal{M}(p) = 1. \end{cases} \tag{3}$$

Since the cost of calculating $\mathcal{M}$ is less than that of Y and C, if $\mathcal{M}$ is sparse, then the computation cost can be reduced. In our experiments, the sparsity of $\mathcal{M}$ can be greater than 70% on average.

A technical challenge of this approach is that the binary sampling mask $\mathcal{M}$ is non-differentiable, making this sampling module incompatible with backpropagation in the training stage. To circumvent this issue, a mask $\mathcal{M}$ that gradually changes from soft to hard during training is used. Therefore, the mask can undergo optimization from the beginning of training and then becomes roughly consistent to a hard mask used in the inference stage by the end of training. During training, with this soft mask, the output Y_s at each point p is calculated as follows:

$$Y_s(p) = \mathcal{M}(p) \odot Y(p), \tag{4}$$

and the full output feature Y^* is calculated by:

$$Y^*(p) = (1 - \mathcal{M}(p)) \odot C(Y_s)(p) + \mathcal{M}(p) \odot Y_s(p), \tag{5}$$

where $\odot$ denotes broadcast multiplication.

3.2 Stochastic Sampling Module

In previous works [1,6,30], sampling is usually done in a deterministic manner, where points with confidence greater than a certain threshold are sampled, as shown in Fig. 2(b)(Left). But in our stochastic sampling, a higher confidence only indicates a higher probability of the point being sampled, as shown in Fig. 2(b)(Right). Due to the spatial redundancy over the feature map, adjacent points may have similar features and confidences, so deterministic sampling typically samples or not samples adjacent points together. However, stochastic sampling can sample a portion of the points, and the features of the other unsampled points can be obtained from the interpolation module. Therefore, sparser sampling can be achieved while maintaining relatively accurate feature maps.

Our sampling is based on the two-class Gumbel-Softmax distribution, which was first introduced in reinforcement learning [17] to simulate stochastic discrete sampling. Thus, the mask $\mathcal{M}$ is defined as:

$$\mathcal{M}(p) = \frac{exp((-log(\pi_1(p)) + g_1(p))/\tau)}{\sum_{j \in \{0,1\}} exp((-log(\pi_j(p)) + g_j(p))/\tau)}, \tag{6}$$

where π denotes confidence map generated from a two-class Softmax activation. g represents noise sampled from a standard Gumbel distribution, and it provides the randomness of Gumbel-Softmax. If the noise g is eliminated, Gumbel-Softmax degenerates into a deterministic function that is approximately equal to the Softmax function with a temperature term. τ is a temperature parameter. When τ approaches 0, $\mathcal{M}(p)$ becomes approximately binary.

In our implementation, π is generated by a 3×3 convolutional layer with a two-class Softmax activation, and τ is exponentially decreased over iterations according to $\tau = \alpha^{iter}\tau_0$, where α is the decay factor, $iter$ is the number of iterations, and τ_0 is the initial temperature. In our experiments, we set $\tau_0 = 1$. Therefore, at the beginning of training, the mask is soft, allowing the sampling module to be trained. At the end of training, τ becomes close to 0, so the mask generated by the sampling module is nearly binary, as desired for our discrete inference. To encourage the network to produce sparse sampling masks, the sparse loss is introduced on the confidence map π_1 for all layers during training:

$$L_{sparse} = \sum_l \|\pi_1^l\|_1 \tag{7}$$

where π_1^l indicates l-th layer's confidence map, and $\|\cdot\|_1$ indicates L1-norm. Different levels of sparsity are achieved by adjusting the weight of the sparse loss. Therefore, the training objective is:

$$L = L_{task} + \gamma L_{sparse}. \tag{8}$$

where L_{task} is the task specific objective and γ is the sparse loss weight.

3.3 Interpolation Module

In previous works [6,7], the features of unsampled points are obtained by reusing the previous features at the corresponding points [7]:

$$Y^*(p) = \mathcal{M}(p) \odot Y_s(p) + (1 - \mathcal{M}(p)) \odot X(p) \tag{9}$$

or just by setting them to zero [6]:

$$Y^*(p) = \mathcal{M}(p) \odot Y_s(p) + (1 - \mathcal{M}(p)) \odot \mathbf{0} = \mathcal{M}(p) \odot Y_s(p). \tag{10}$$

However, these approaches ignore the spatial redundancy of feature maps, which could be used to obtain the features of unsampled points, leading to a more accurate feature map.

Our method capitalizes on this spatial redundancy to generate relatively accurate feature maps. This is done by interpolating the features of unsampled points from those of sampled points as indicated in Eq. (3). The interpolation is formulated as:

$$C(Y_s)(p) = \frac{\sum_{s_i} W_I(p, s_i) Y_s(s_i)}{\sum_{s_i} W_I(p, s_i) + \epsilon}, s_i \in \Omega, \tag{11}$$

where $W_I(p, s_i) \geq 0$ represents interpolation weights, Ω represents the spatial domain like in Eq. (1), and ϵ is a small constant which is set to 10^{-5}. In this formula, the features of the unsampled points are represented by a weighted average of the features at sampled points. However, interpolation by considering all sampled points is costly, even if Y_s is sparse in the inference phase. Fortunately, since neighboring points commonly have stronger feature correlation, we can alleviate this problem by computing the interpolation only using samples that lie within a window centered on the given unsampled point. Specifically, we restrict s to a window of radius r centered on the unsampled point p, defined as $\mathcal{R}_s^r(p) = \{s_i | s_i \in \Omega, ||s_i - p||_\infty \leq r\}$. Thus, the windowed interpolation module is formulated as:

$$C(Y_s)(p) = \frac{\sum_{s_i} W_I(p, s_i) Y_s(s_i)}{\sum_{s_i} W_I(p, s_i)}, s_i \in \mathcal{R}_s^r(p). \tag{12}$$

The best formulation of W_I in irregular spatial sampling is an open problem. We explored different design choices:

RBF Kernel. The radial basis function (RBF) kernel is a commonly used interpolation function, defined as:

$$W_I(p_1, p_2) = exp(-\lambda^2 ||p_1 - p_2||^2), \tag{13}$$

where p_1, p_2 are two points and λ is a learnable parameter. With this kernel, an interpolated point is more greatly affected by closer points. The learnable RBF kernel is adopted by default in our approach.

Plain Convolution. Plain convolution can also be used as a learnable interpolation kernel. However, negative convolution weights may result in a zero denominator for Eq. (12). Therefore, we use the absolute value of the convolution weights in the denominator instead.

Average Pooling. Average pooling is the simplest interpolation kernel. It assigns the same weight to all pixels in the interpolation window. We use this method as a baseline.

3.4 Grid Prior

With our stochastic sampling, the windowed sample set $\mathcal{R}_s^r$ at some unsampled points may be empty. To handle this situation, one way is to fill the features of

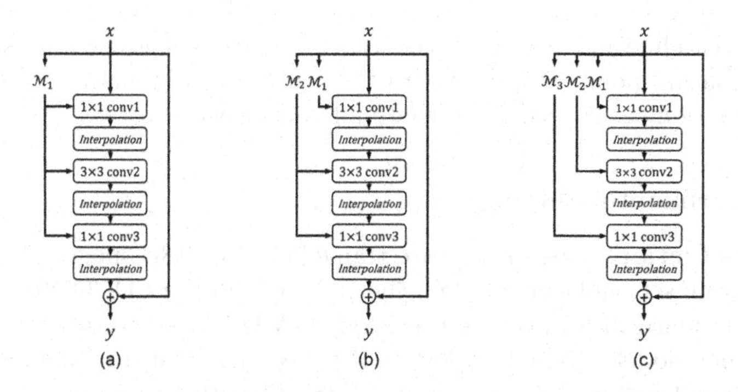

Fig. 3. Integration of our sampling and interpolation modules into a standard residual block. (a) Applying a single mask to all convolutions; (b) Applying a separate mask on conv1, and sharing a mask for conv2 and conv3. This approach is adopted as the default setting in our experiments; (c) Applying three separate masks for each of the convolutions.

these points as zeros. To allow interpolation at these points, we instead build an equal-interval sampling mask $\mathcal{M}_{grid}$ of stride s and combine it with the mask $\mathcal{M}_{sample}$ generated by the sampling module: $\mathcal{M} = \max(\mathcal{M}_{sample}, \mathcal{M}_{grid})$. The combined mask is used in the interpolation module, and we find experimentally that it does not affect the performance but better stabilizes the training process in comparison to zero filling, as shown in Sect. 4.2.

3.5 Integration with Residual Block

Our sampling and interpolation modules can be easily integrated with existing network architectures. Here, we use the residual block [11] as an example to show how these modules can be used. A natural way to insert them is by applying a separate sampling mask for each convolution (Fig. 3(c)). However, since conv3 is a pointwise convolution (kernel size is 1×1), its mask can be shared with conv2 without changing the output (Fig. 3(b)), meanwhile avoiding the computation for generating an extra mask. Another approach is to share a single mask among all convolutions within the residual block (Fig. 3(a)). However, since conv2 is not a pointwise convolution (kernel size is 3×3), sampling a point in conv2 requires the corresponding 3×3 points from conv1 to be sampled, which in turn places strong conditions on the sampling of conv2 if conv1 and conv2 were to share the same mask. Thus, sharing one mask for all convolutions within a block is not an effective solution. We found the two sampling mask approach (Fig. 3(b)) to be slightly better than others in experiments, so it is used by default.

4 Experiments

In this section, we validate our approach on three tasks: object detection, semantic segmentation, and image classification. Comparisons between our approach and other baseline models are conducted in terms of speed-accuracy tradeoff. Average floating point operations (FLOPs) in the backbone network over the whole validation set is used to evaluate inference speed.

4.1 Experimental Settings

Object Detection. Our models are trained on the 118k images of the COCO 2017 [24] train set, and evaluated on the 5k images of the COCO 2017 validation set. The standard mean average precision (mAP) is used to measure accuracy. The baseline model is based on Faster R-CNN with Feature Pyramid Network (FPN) [23] and deformable convolution [4,38]. The other components and hyper-parameters are based on mmdetection [2]. More details are given in Appendix.

For our modules, the sparse sampling module and interpolation module are integrated into all the residual blocks as shown in Fig. 3(b). The window size r for interpolation is set to 7, and the stride s of the grid prior is set to 11. During training, the parameter λ used in the learnable RBF kernel is initialized to 3, and the decay factor α of the stochastic sampling module is set to make the Gumbel-Softmax temperature parameter $\tau = 0.01$ at the end of training. During inference, we use the same τ to produce masks, and points with mask values below than 0.5 are marked as unsampled.

Semantic Segmentation. Our models are trained on the 2975 finely annotated images of the Cityscapes [3] train set and evaluated on the 500 images of the validation set. Accuracy is measured by the standard mean IoU. The baseline model is ResNet-50 based dilated FCN [26] with deformable convolution layers [4]. The implementation and hyper-parameters are based on the open-source implementation of TorchCV [36]. More details are given in Appendix. For our modules, the same hyper-parameters are used as in object detection.

Image Classification. Our models are trained on the ImageNet-1K training set, and follow both the training and inference settings of [6]. We choose ResNet-34 as our baseline for fair comparison. Following [6], our sampling and interpolation modules are incorporated in all residual blocks, except for the first block in each stage. Since ResNet-34 is composed of basic blocks (two 3×3 convolutions), we apply separate masks for each convolution.

Different from the previous two tasks, experiments on ImageNet use a much lower image resolution (i.e. 224×224). Thus, for our modules, we reduce the window size r of the interpolation module to 5, and the stride s of the grid prior to 2, while keeping other hyper-parameters the same as in object detection.

4.2 Ablation Study

We validate several design choices on the COCO2017 object detection benchmark. All the models are trained and evaluated on images with a shorter side of 1000 pixels and with the sparse loss weight set to 0.1.

Table 1. Comparison of different interpolation kernels on COCO2017 validation

	Avg pool	Plain conv	RBF kernel
mAP	41.8	41.9	42.0
GFlops	110.0	109.8	96.6

Table 2. Validation of the interpolation module on COCO2017 validation

	Fill zeros	Reuse features	Ours
mAP	42.0	42.1	42.0
GFlops	164.9	226.1	96.6

Table 3. Comparison of different grid prior settings on COCO2017 validation

	$s = 9$	$s = 11$	$s = 13$	w/o Grid Prior
mAP	41.9	42.0	41.8	41.8
GFlops	95.4	96.6	95.3	95.0

Different Interpolation Kernels. We first compare the different interpolation kernels mentioned in Sect. 3.3: *learnable RBF kernel, plain convolution* and *average pooling.* Results are shown in Table 1, with the sparse loss weight in the training phase set so that the three kernels yield similar accuracy. It can be seen that the *learnable RBF kernel* consumes fewer FLops.

Effect of Removing Interpolation. We further study the effect of removing the interpolation module by replacing it with *reusing features* of the previous layer [7] (see Eq. (9)) and directly *filling zeros* [6] (see Eq. (10)). Results are shown in Table 2. *Reusing features* achieves 42.1 mAP with 226.1 GFLOPs, while *Filling zeros* obtains 39.7 mAP with 121.8 GFLOPs. Both of these methods consume noticeably more computation than our interpolation module with similar or worse accuracy, which indicates that they are inferior to our method in terms of FLOPs-accuracy tradeoff.

Effect of Grid Prior. We compare performance under different grid stride s or by removing the grid prior. Results are shown in Table 3. We found the performance in different settings to be similar, but the training process is sometimes not stable at $s = 13$ and without the grid prior. Therefore, we choose $s = 11$ as our default setting.

4.3 Object Detection

We compared our method to other baselines on the COCO2017 object detection benchmark. To better illustrate the FLOPs-accuracy tradeoff under different

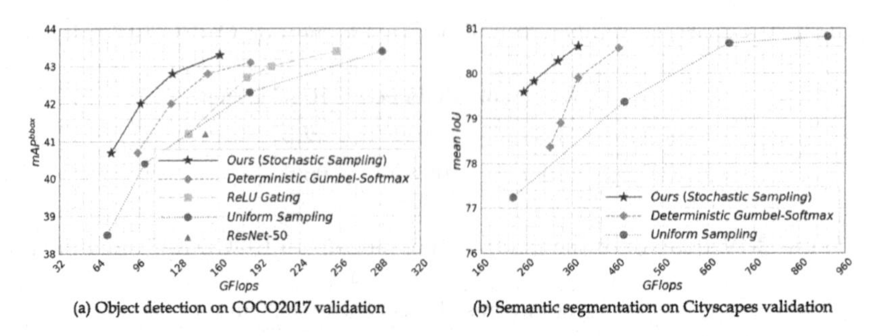

Fig. 4. (a) Tradeoff curves for different sampling methods on object detection (COCO2017 validation). Curve of "Uniform Sampling" is drawn from various resolutions, and others are drawn from various sparse loss weights (with shorter side of 1000 pixels). (b) Experiments on the Cityscapes semantic segmentation benchmark. Except for "Uniform Sampling", all other models are trained and evaluated on images with a shorter side of 1024 pixels.

parameters and settings, we display charts rather than tables and present the numerical results in Appendix.

Uniform Sampling. The direct way to sample uniformly is by using a mask with sampling points at equal intervals. However, since the interval must be integer, the minimal interval would be 2, which limits its feasibility for handling arbitrary sampling ratios. Thus, instead of using an equal-interval sampling mask, we choose to downsample the input images and not include our modules, which can also be seen as a uniform sampling method. We compared these two approaches with a downsampling rate of 2, and experimental results show the performance of these two approaches to be very similar[2].

For uniform sampling, we resize the shorter sides of input images to {1000, 800, 600, 500} to draw the FLOPs-accuracy curve. For our method, the curve is drawn according to different sparse loss weights, i.e. {0.2, 0.1, 0.05, 0.02}. Results are shown in Fig. 4(a). For our baseline without any sampling or interpolation modules, which is trained and evaluated on images with a shorter side of 1000 pixels, it obtains 43.4 mAP with 289.5 GFLOPs. In comparison, our stochastic sampling method performs at 43.3 mAP with only 160.4 GFLOPs, saving nearly half of the computation cost with no accuracy drop.

Deterministic Sampling. We next compare our stochastic sampling strategy to deterministic sampling methods. For fair comparison, we only replace the sampling module, while leaving the other parts unchanged. Two deterministic sampling methods are examined:

- ReLU Gating: Similar to LCCL [6], a ReLU function is applied on sampling confidence to generate a sparse mask, by trimming values smaller than 0

[2] 38.5 mAP for downsampled images vs. 38.7 mAP for equal-interval sampling masks.

during training and inference. The sampling confidence is generated by a 3×3 convolution with a 1-dimensional output. Since the output of ReLU is not binary, Eq. (5) is used to calculate the full output feature.
- Deterministic Gumbel-Softmax: Eliminating the noise g in Gumbel-Softmax naturally results in deterministic sampling, as described in Sect. 3.2, and is approximately equal to the Softmax function with a temperature term.

For our stochastic sampling and the two deterministic sampling methods, the FLOPs-accuracy curve is drawn according to different sparse loss weights, i.e. $\{0.2, 0.1, 0.05, 0.02\}$. Compared to *ReLU Gating*, our stochastic sampling achieves clearly better performance than the ReLU based approach, especially at lower levels of computation. *Deterministic Gumbel-Softmax* performs better than *ReLU Gating*, but is still worse than stochastic sampling by a large margin.

Smaller Backbones. Replacing a large backbone with a smaller backbone is a common method for reducing computation. Thus, we also compare our method (based on ResNet-101) with a baseline using ResNet-50 as the smaller backbone. ResNet-50 achieves 41.2 mAP with 149.2 GFLOPs, which indicates accuracy much worse than that of our method with similar computation costs.

4.4 Semantic Segmentation

We also conduct experiments on the Cityscapes benchmark for semantic segmentation. Unless otherwise specified, all the models are trained and evaluated on images with a shorter side of 1024 pixels. We first compare our content-aware stochastic sampling to *uniform sampling* and deterministic sampling. For deterministic sampling, we choose *deterministic Gumbel-Softmax* for comparison because of its better performance exhibited on the object detection task.

Results are shown in Fig. 4(b) and present the numerical results in Appendix. For our method and *deterministic Gumbel-Softmax*, we draw curves according to different sparse loss weights $\{0.3, 0.2, 0.1, 0.05\}$. For *uniform sampling*, we resize the shorter sides of input images to $\{1024, 896, 736, 512\}$. The original baseline with a shorter side of 1024 pixels obtains 80.8 mean IoU with 920.6 GFLOPs. In comparison, our method obtains 80.6 mean IoU with only 373.2 GFLOPs, saving nearly 60% of the computation cost with almost no accuracy drop. Compared with other sampling methods, i.e. *uniform sampling* and *deterministic Gumbel-Softmax*, our method clearly achieves a better FLOPs-accuracy tradeoff.

4.5 Image Classification

We also compare our method to other state-of-the-art methods [6,12,13,20] for reducing computation on the ImageNet-1K image classification benchmark. Similar to our method, LCCL [6] explores sparsity in the spatial domain, while PFEC [20], SFP [12] and FPGM [13] reduce computation by model pruning. Results are presented in Table 4. Our models are trained under different sparse

Table 4. Performance comparison on the ImageNet validation set. All the methods are based on ResNet-34. Our models are trained with a loss weight of 0.01 and 0.015 to achieve accuracy or FLOPs similar to other methods for fair comparison. "w/o Interp" indicates removing the interpolation module and filling the features of unsampled positions with 0

Method	Type	Top-1/Top-5 Acc Drop(%)	FLOPs	Speedup
LCCL [6]	Spatial	0.43/0.17	2.7×10^9	24.8%
PFEC [20]	Pruning	1.06/-	2.7×10^9	24.2%
SFP [12]	Pruning	2.09/1.29	2.2×10^9	41.1%
FPGM [13]	Pruning	1.29/0.54	2.2×10^9	41.1%
Ours(0.01)	Spatial	0.45/0.20	2.53×10^9	30.8%
Ours(0.015)	Spatial	1.19/0.47	2.16×10^9	42.4%
w/o Interp(0.015)	Spatial	1.07/0.46	2.21×10^9	41.0%

loss weights, 0.01 and 0.015, to reach accuracy or FLOPs similar to other methods for fair comparison. Compared with LCCL [6] and PFEC [20], our approach achieves comparable accuracy with less FLOPs. Compared with SFP [12] and FPGM [13], our approach obtains a smaller accuracy drop with similar FLOPs.

We further remove the interpolation module from our method and fill the features of unsampled points with 0. Results show that removing interpolation does not affect performance on the ImageNet validation set. This is inconsistent with object detection and semantic segmentation. We believe that this is because the classification network is focused on extracting global feature representations. Therefore, as long as the features of certain key points are calculated and preserved, the global features will not be affected and the performance will not be hurt. In other words, in the image classification task, it is not important to reconstruct the features of unsampled points by interpolation.

4.6 Analysis of Sampling and Interpolation Modules

In this section, we further study the relationship between the sampling and interpolation modules. Specifically, we analyze the relationship between sampling sparsity and the parameter λ in the RBF interpolation module. A larger λ indicates a sharper interpolation kernel, and when $\lambda > 3$, the RBF interpolation kernel approximates an identity kernel, for which there is no interpolation.

Results are shown in Fig. 5. For object detection (see Fig. 5(a)) and semantic segmentation (see Fig. 5(b)), sparsity and λ show a strong negative correlation. When sparsity is high, the λ is usually small, which means sampled points far away from the unsampled points also contribute greatly to the interpolation. However, for image classification (see Fig. 5(c)), this correlation has not been observed. λ is quite large in most cases, which means the effect of the interpolation module is limited. This phenomenon is consistent with the experimental results of the "w/o Interp" entry in Table 4, that the results without interpola-

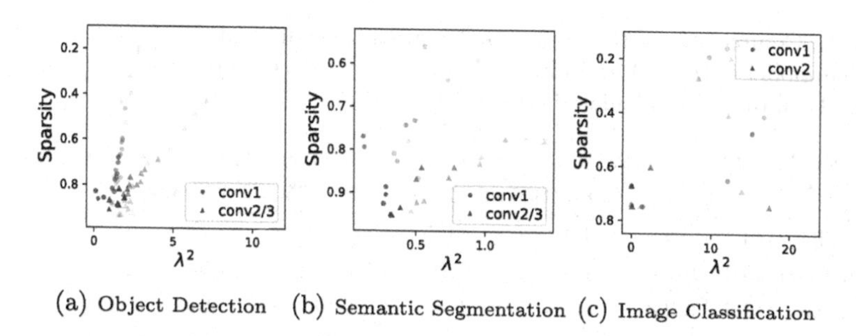

(a) Object Detection (b) Semantic Segmentation (c) Image Classification

Fig. 5. Relationship between interpolation parameter λ and sparsity. The transparency of each point represents its location in the network, where a darker color indicates greater depth. A larger λ yields a sharper interpolation kernel; when $\lambda > 3$, it approximates an identity kernel.

tion are almost identical to our full model, further indicating that interpolation is not important for image classification.

Another observation is that the λ of conv1 are consistently smaller than λ of conv2 and conv3 in object detection and semantic segmentation. The reason is still unclear but we suspect that this phenomenon may be related to the receptive field of operators.

4.7 Realistic Run-Time on CPU

Our method achieves good theoretical speed-accuracy trade-offs. In this section, we present a preliminary evaluation of the realistic run-time of the backbone network on the CPU. According to Eq. (2), the mask $\mathcal{M}$ is calculated before computing Y to decide which points in Y need to calculated. In order to show the realistic speedup of our method under different computing resources, we conducted experiments in two different hardware environments: a workstation (E5-2650 v2, 256G RAM and Ubuntu 16.04 OS) and a laptop (I7-6650U, 16G RAM and Ubuntu 16.04 OS). For fair comparison, we replace all convolutions by our implementation for all models, and enable multi-threading by default (32 threads for E5-2650 v2 and 4 threads for I7-6650U). Results on object detection are shown in Table 5. There is a gap between the theoretical speedup and the realistic speedup, but the gap for laptop (I7-6650U) is smaller than for workstation (E5-2650 v2). The main reason is that the workstation has better computing speed and more cores (E5-2650 v2 has 8 cores and I7-6650U has 2 cores), but has a memory access bottleneck. This suggests that our method is more suitable for devices with less computation speed but relatively faster IO speed, such as mobile or edge computing devices.

In addition, some works [29,30] have developed general techniques to accelerate sparse convolution based methods. For example, SCNN [29] designed a hardware accelerator for sparse convolution and demonstrate that ideal speedup of sparse convolution is achievable in such devices. SBNet [30] is a general method

Table 5. Comparison of theoretical and realistic speedups on E5-2650 and I7-6650U. Baseline model is trained and evaluated on images with a shorter side of 1000 pixels. The CPU run-time is calculated on the COCO2017 validation set

Model	mAP	GFLOPs	Runtime(s/img)		Real speedup		Theo speedup
			E5-2650 v2	I7-6650U	E5-2650 v2	I7-6650U	
Baseline	43.4	289.5	4.9	10.5	–	–	–
Our(0.02)	43.3	160.4	4.0	7.9	1.23	1.33	1.80
Our(0.05)	42.8	122.8	3.5	6.3	1.40	1.67	2.36
Our(0.1)	42.0	96.6	3.2	5.3	1.53	1.98	3.00
Our(0.2)	40.7	73.3	2.7	4.4	1.81	2.39	3.95.

to accelerate sparse convolution in GPU. These general techniques of accelerating sparse convolution are compatible with our method, and can further close the gap between theoretical and actual speedup in real applications.

5 Conclusion

A method for reducing computation in convolutional networks was proposed that exploits the intrinsic sparsity and spatial redundancy in feature maps. We present a stochastic sampling and interpolation scheme to avoid expensive computation at spatial locations that can be effectively interpolated. To overcome the challenge of training binary decision variables for representing discrete sampling locations, Gumbel-Softmax is introduced to our sampling module. The effectiveness of this approach is verified on a variety of computer vision tasks.

Acknowledgment. We would like to thank Jifeng Dai for his early contribution to this work during his work at Microsoft Research Asia. Jifeng later turned to other more exciting projects.

References

1. Cao, S., et al.: Seernet: predicting convolutional neural network feature-map sparsity through low-bit quantization. In: CVPR (2019)
2. Chen, K., et al.: mmdetection (2018). https://github.com/open-mmlab/mmdetection
3. Cordts, M., et al.: The cityscapes dataset for semantic urban scene understanding. In: CVPR (2016)
4. Dai, J., et al.: Deformable convolutional networks. In: CVPR (2017)
5. Ding, X., Ding, G., Guo, Y., Han, J.: Centripetal SGD for pruning very deep convolutional networks with complicated structure. In: CVPR (2019)
6. Dong, X., Huang, J., Yang, Y., Yan, S.: More is less: a more complicated network with less inference complexity. In: CVPR (2017)
7. Figurnov, M., et al.: Spatially adaptive computation time for residual networks. In: CVPR (2017)

8. Figurnov, M., Ibraimova, A., Vetrov, D., Kohli, P.: Perforatedcnns: acceleration through elimination of redundant convolutions. In: NIPS (2016)
9. Guo, Y., Yao, A., Chen, Y.: Dynamic network surgery for efficient DNNs. In: NIPS (2016)
10. Han, S., Pool, J., Tran, J., Dally, W.J.: Learning both weights and connections for efficient neural network. In: NIPS (2015)
11. He, K., Zhang, X., Ren, S., Sun, J.: Deep residual learning for image recognition. In: CVPR (2016)
12. He, Y., Kang, G., Dong, X., Fu, Y., Yang, Y.: Soft filter pruning for accelerating deep convolutional neural networks. arXiv preprint arXiv:1808.06866 (2018)
13. He, Y., Liu, P., Wang, Z., Hu, Z., Yang, Y.: Filter pruning via geometric median for deep convolutional neural networks acceleration. In: CVPR (2019)
14. Herrmann, C., Bowen, R.S., Zabih, R.: An end-to-end approach for speeding up neural network inference. arXiv preprint arXiv:1812.04180v3 (2019)
15. Hinton, G., Vinyals, O., Dean, J.: Distilling the knowledge in a neural network. arXiv preprint arXiv:1503.02531 (2015)
16. Jaderberg, M., Vedaldi, A., Zisserman, A.: Speeding up convolutional neural networks with low rank expansions. In: BMVC (2014)
17. Jang, E., Gu, S., Poole, B.: Categorical reparameterization with gumbel-softmax. In: ICLR (2017)
18. Kuen, J., et al.: Stochastic downsampling for cost-adjustable inference and improved regularization in convolutional networks. In: CVPR (2018)
19. LeCun, Y., Denker, J.S., Solla, S.A.: Optimal brain damage. In: NIPS (1989)
20. Li, H., Kadav, A., Durdanovic, I., Samet, H., Graf, H.P.: Pruning filters for efficient convnets. In: ICLR (2017)
21. Li, X., Liu, Z., Luo, P., Loy, C.C., Tang, X.: Not all pixels are equal: difficulty-aware semantic segmentation via deep layer cascade. In: CVPR (2017)
22. Lin, S., et al.: Towards optimal structured CNN pruning via generative adversarial learning. In: CVPR (2019)
23. Lin, T.Y., Dollár, P., Girshick, R., He, K., Hariharan, B., Belongie, S.: Feature pyramid networks for object detection. In: CVPR (2017)
24. Lin, T.Y., et al.: Microsoft COCO: common objects in context. In: Fleet, D., Pajdla, T., Schiele, B., Tuytelaars, T. (eds.) ECCV 2014. LNCS, vol. 8693, pp. 740–755. Springer, Cham (2014). https://doi.org/10.1007/978-3-319-10602-1_48
25. Liu, Z., Li, J., Shen, Z., Huang, G., Yan, S., Zhang, C.: Learning efficient convolutional networks through network slimming. In: ICCV (2017)
26. Long, J., Shelhamer, E., Darrell, T.: Fully convolutional networks for semantic segmentation. In: CVPR (2015)
27. Mazzini, D., Schettini, R.: Spatial sampling network for fast scene understanding. In: CVPR Workshop (2019)
28. Molchanov, P., Mallya, A., Tyree, S., Frosio, I., Kautz, J.: Importance estimation for neural network pruning. In: CVPR (2019)
29. Parashar, A., et al.: SCNN: an accelerator for compressed-sparse convolutional neural networks. In: International Symposium on Computer Architecture (2017)
30. Ren, M., Pokrovsky, A., Yang, B., Urtasun, R.: Sbnet: sparse blocks network for fast inference. In: CVPR (2018)
31. Shi, S., Chu, X.: Speeding up convolutional neural networks by exploiting the sparsity of rectifier units. arXiv preprint arXiv:1704.07724 (2017)
32. Szegedy, C., et al.: Going deeper with convolutions. In: CVPR (2015)
33. Veit, A., Belongie, S.: Convolutional networks with adaptive inference graphs. In: ECCV (2017)

34. Wen, W., Wu, C., Wang, Y., Chen, Y., Li, H.: Learning structured sparsity in deep neural networks. In: NIPS (2016)
35. Ye, J., Lu, X., Lin, Z., Wang, J.Z.: Rethinking the smaller-norm-less-informative assumption in channel pruning of convolution layers. In: ICLR (2018)
36. You, A., Li, X., Zhu, Z., Tong, Y.: Torchcv: a pytorch-based framework for deep learning in computer vision (2019). https://github.com/donnyyou/torchcv
37. Yu, R., et al.: NISP: pruning networks using neuron importance score propagation. In: CVPR (2018)
38. Zhu, X., Hu, H., Lin, S., Dai, J.: Deformable convnets v2: more deformable, better results. In: CVPR (2019)

BorderDet: Border Feature for Dense Object Detection

Han Qiu[1,2], Yuchen Ma[1(✉)], Zeming Li[1], Songtao Liu[1], and Jian Sun[1]

[1] MEGVII Technology, Beijing, China
{qiuhan,mayuchen,lizeming,liusongtao,sunjian}@megvii.com
[2] Xi'an Jiaotong University, Xi'an, China
qiuhan@stu.xjtu.edu.cn

Abstract. Dense object detectors rely on the sliding-window paradigm that predicts the object over a regular grid of image. Meanwhile, the feature maps on the point of the grid are adopted to generate the bounding box predictions. The point feature is convenient to use but may lack the explicit border information for accurate localization. In this paper, We propose a simple and efficient operator called Border-Align to extract "border features" from the extreme point of the border to enhance the point feature. Based on the BorderAlign, we design a novel detection architecture called BorderDet, which explicitly exploits the border information for stronger classification and more accurate localization. With ResNet-50 backbone, our method improves single-stage detector FCOS by 2.8 AP gains (38.6 v.s. 41.4). With the ResNeXt-101-DCN backbone, our BorderDet obtains 50.3 AP, outperforming the existing state-of-the-art approaches.

Keywords: Dense object detection · Border feature

1 Introduction

Sliding-window object detector [4,11,16–19,21,28], which generates bounding-box predictions over a dense and regular grid, plays an essential role in modern object detection. Most sliding-window object detectors like SSD [17], RetinaNet [16] and FCOS [21] adopt a point-based feature representation of the bounding box, where the bounding box is predicted by the feature on each point of the grid, shown as the "Single Point" in Fig. 1. This single point feature is convenient to be used for object localization and object classification because no additional feature extraction is conducted.

The first two authors contributed equally to this work.

Electronic supplementary material The online version of this chapter (https://doi.org/10.1007/978-3-030-58452-8_32) contains supplementary material, which is available to authorized users.

A. Vedaldi et al. (Eds.): ECCV 2020, LNCS 12346, pp. 549–564, 2020.
https://doi.org/10.1007/978-3-030-58452-8_32

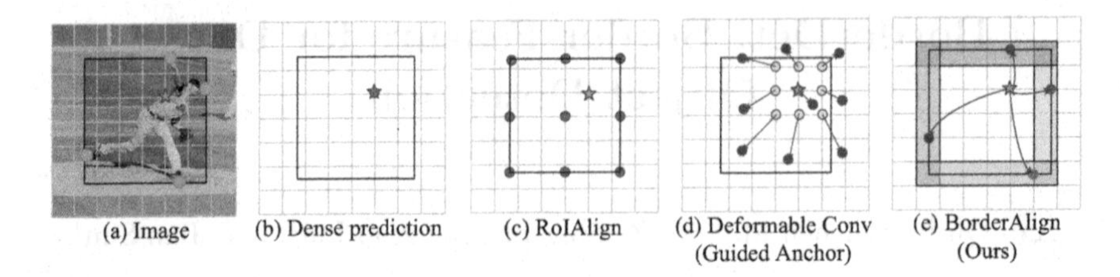

(a) Image (b) Dense prediction (c) RoIAlign (d) Deformable Conv (Guided Anchor) (e) BorderAlign (Ours)

Fig. 1. Different feature extraction strategy. The red pentagram represents the current point that predicts the bounding box. The black rectangle denotes the bounding box predicted on the red pentagrams. And the blue point indicates where the features are extracted. Different from the deformable Convolution and RoIAlign which densely extract the features from the whole bounding box, Our BorderAlign only extracts the features from five points for the current single point and four extreme points of the borders respectively. The orange points in (a) are the extreme points (Color figure online)

However, the point feature may contain insufficient information for representing the full instance with its limited receptive field. Meanwhile, it may also lack the information of the object boundary to precisely regress the bounding box.

Many studies have been focused on the feature representation of the object, such as the GA-RPN [2], RepPoints [26] and Cascade RPN [23], or pooling based methods like RoI pooling [7] and RoIAlign [9]. As shown in Fig. 1, these methods extract more representative features than the point features. However, there are two limitations of implementing these methods for dense object detection: (1) The feature extracted within the whole boxes may involve *unnecessary* computation and easily be affected by the background. (2) These methods extract the border features *implicitly* and *indirectly*. Since the features are discriminated and extracted adaptively within the whole boxes, no specific extraction on the border features is conducted in these methods.

In this work, we propose a powerful feature extraction operator called BorderAlign, which directly utilizes the border features pooled from each boundary to enhance the original point feature. It differs from the other feature extraction operators as shown in Fig. 1, which densely extracts the feature from the whole box. Our proposed BorderAlign focuses on the object border and is designed to adaptively discriminate the representative part of the object border, *e.g.* the extreme point [29], which is shown in Fig. 1(e).

We design BorderDet which utilizes Border Alignment Modules (BAM) to refine the classification score and bounding box regression. Our BorderDet uses less computation than similar feature enhancement methods and achieves better accuracy. Moreover, our method can be easily integrated into any dense object detectors with/without anchors.

To summarize, our contribution is three-fold as follows:

1 We analyze the feature representation for the dense object detector and demonstrate the significance of supplementing the single-point feature representation with the border feature.
2 We propose a novel feature extraction operator called BorderAlign to enhance features by the border features. Based on BorderAlign, we present an efficient and accurate object detector architecture named BorderDet.
3 We achieve state-of-the-art results on COCO dataset without bells and whistles. Our method leads to significant improvements on both single-stage method FCOS and two-stage method FPN, by 2.8 AP and 3.6 AP respectively. Our ResNext-101-DCN based BorderDet yields 50.3 AP, outperforming the existing state-of-the-art approaches.

2 Related Works

Sliding-window Paradigm. Sliding-window Paradigm is widely used in object detection. For the one-stage object detectors, Densebox [11], YOLO [18,19], SSD [17], RetinaNet [16], and FCOS [21] have demonstrated the effectiveness to densely predict the classification and localization scores. For the two-stage object Detectors, R-CNN series [7–10,14,15,20] adopt the region proposal network (RPN) that based on the sliding-window mechanism to generate the initial proposals, and then a refinement stage that consists of a RoIAlign [9] and R-CNN is performed to warp the feature maps of the region-of-interests (RoI) and generate the accurate predictions.

Feature Representation of Object. Typical sliding-window object detectors adopt a point-based feature representation. However, it is difficult for the point feature to maintain the powerful feature representation for both classification and location. Recently, some works [23,24,26] attempt to improve the feature representation of object detection. Guided Anchor [24] is proposed to enhance the single point feature representation by utilizing the deformable convolution. Cascade-RPN [23] presents adaptive convolution to align the features maps to their corresponding object bounding box predictions. Reppoints [26] formulate the object bounding box as a set of representative points and extract the representative point feature by deformable convolution. However, the feature maps proposed in these methods are extracted from the whole object, thus the feature extraction is redundant and easily affected by the background feature maps. In contrast to the above methods, our BorderDet directly enhances the single point feature by the border feature, which enables the feature map to have a high response to the extreme points of the object borders and does not involve the background noise.

Border Localization. There are several methods that search on each row and column of the region or bucket to accurately locate the boundary of the object. LocNet [6] and SABL [25] adopt an additional object localization stage which

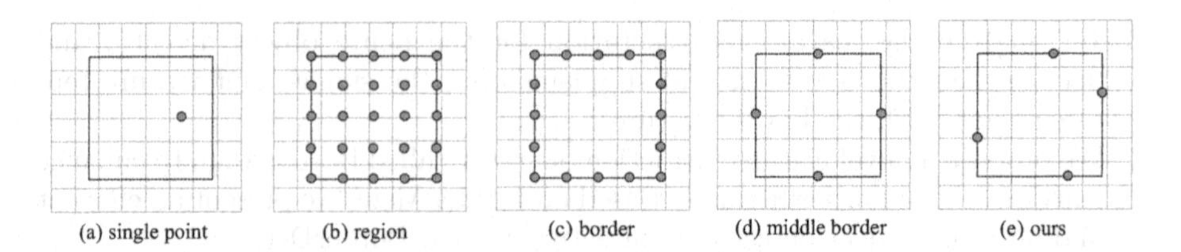

Fig. 2. Different feature representations of the bounding box. (a) denotes *point feature* on each point of the grid. (b) indicates the *region features* extracted from the whole bounding box by RoIAlign [9]. (c) denotes the *border features* extracted from the border of the bounding box. (d) indicates the *border-middle features* which are extracted from the center point of each border. (e) denotes our BorderAlign feature extractor

aggregates the RoI feature maps along with X-axis and Y-axis to locate the object borders and generate the probability for each object border prediction. However, such border localization pipelines rely heavily on the high-resolution RoI feature maps, thus the implementation of these methods in the dense object detector may be restricted. In this work, we aim to efficiently exploit the border feature for accurate object localization.

3 Our Approach

In this section, we first investigate the feature representation of the bounding box in sliding-window object detectors. Then, we propose a new feature extractor called BorderAlign, which extracts the border features to enhance the original point-based feature representation. Based on the BorderAlign, we present the design of BorderDet and discuss its mechanism for extracting boundary features efficiently.

3.1 Motivation

Sliding-window object detectors usually generate bounding box predictions over a dense, regular grid of feature maps. As shown in Fig. 2, the feature on each point of the grid is generally used to predict the category and location of the objects. This point-based feature representation is hard to contain the effective border feature and it may limit the localization ability of the object detectors. As for the two-stage object detectors, the object is described by the region features which are extracted from the whole bounding box, which is shown in Fig. 2 (b). This region-based feature representation is able to provide more abundant features than the point-based feature representation for object classification and localization.

In Table 1, we provide a deeper analysis of the feature representation of the bounding box. Firstly, we adopt a simple dense object detector (FCOS) as our baseline to generate the coarse bounding box predictions. Next, we will re-extract the features as shown in Fig. 2 from the second-to-last feature map of FCOS.

Table 1. Comparison of different feature representation of the bounding box. The first row is the baseline. The F_{point} indicates the feature used in the first prediction. F'_{point}, F_{region}, F_{border} and F_{middle} indicate the features used in the second prediction. The specific illustration of these features are shown in Fig. 2. The final column "N" denotes how many points are sampled to extract feature in the second prediction where "N" equals 5 in these experiments

F_{point}	F'_{point}	F_{region}	F_{border}	F_{middle}	AP	AP_{50}	AP_{75}	AP_S	AP_M	AP_L	N
✓					38.6	57.2	41.7	23.5	42.8	48.9	0
✓	✓				38.9	57.7	42.1	23.7	43.1	49.3	1
✓	✓	✓			**39.9**	58.9	43.4	24.6	44.1	50.8	n^2+1
✓	✓		✓		39.6	58.5	43.2	24.2	43.8	50.4	$4n+1$
✓	✓			✓	**39.9**	58.7	43.4	24.8	44.0	50.4	$4+1$

Then we gradually supplement the single point feature with different features to refine the coarse predictions. We make the following observations on these experiments. (1) Region features are more representative than the point feature. Enhancing the single point feature with the region features leads to an improvement of 1.3 AP. (2) The border features play a major role in the region features when the region features are used to enhance the single point feature. Performance only reduces 0.3 AP if we ignore the inner part of the bounding box and only introduce the border features. (3) Extracting the border features effectively leads to further improvement than densely extracting the border features. The experiment in the fourth column of Table 1 shows that the middle border features is 0.3 AP higher than border features and reaches the same performance to the region features with fewer sample points.

In consequence, for the feature representation in the dense object detector, the point-based feature representation is lack of the explicit feature of the whole object and the feature enhancement is requisite. However, extracting the feature from the whole boxes is unnecessary and redundant. Meanwhile, a more efficient extraction strategy of the border features will lead to better performance. Based on these conceptions, we explore how to boost the dense object detector performance by using border feature enhancement in the next section.

3.2 Border Align

Owing to our observation above, the border features are important in achieving better detection performance. However, it is inefficient to extract features intensively on the borders since there is usually very little foreground and lots of background on the borders of the object (*e.g.* the person in Fig. 1). We thus propose a novel feature extractor, called BorderAlign to effectively exploit the border feature.

The architecture of the BorderAlign is illustrated in Fig. 3. Inspired by the R-FCN [14], our BorderAlign take the *border-sensitive* feature maps I with $(4+1)C$ channels as the input. The $4C$ channels of the feature maps correspond to the four borders (left, top, right, bottom), while the other C channels corresponds to

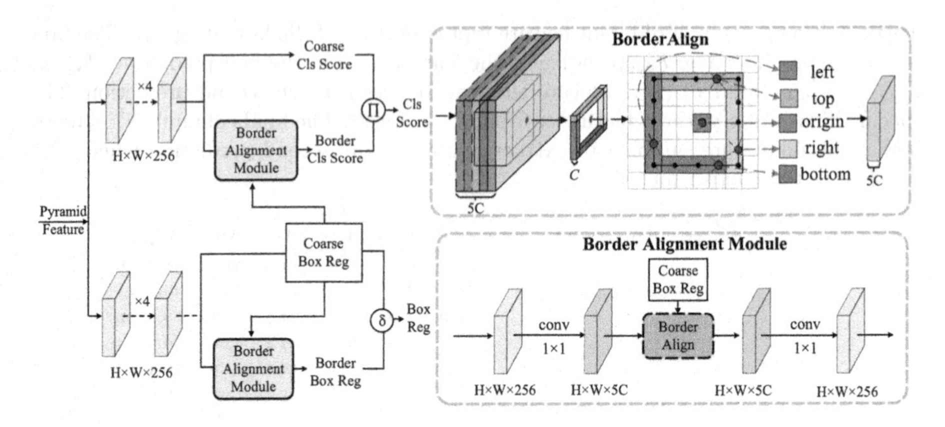

Fig. 3. The architecture of BorderDet. Firstly, we adopt a regular single-stage object detector to generate the coarse predictions of the classification score and bounding box location. Then the Border Alignment Module is applied to refine the coarse predictions with the border features. The π indicates multiplication and the δ denotes the combination of the two bounding box locations (Color figure online)

the original single point features as shown in Fig. 2. Then, each border is evenly subdivided into N points and the feature values of these N points are aggregated by the max-pooling. N denotes the pooling size and is set to 10 in this paper as default. The proposed BorderAlign could adaptively exploit the representative border features from the extreme points of the borders.

It is worth noting that our BorderAlign adopts a channel-wise max-pooling scheme that the four borders are max-pooled independently within each C channels of the input feature maps. Assuming the input feature maps are in the order of (single point, left border, top border, right border and bottom border), the output feature maps $\mathcal{F}$ can be formulated as the following equation:

$$F_c(i,j) = \begin{cases} I_c(i,j) & 0 \leq c < C \\ \max\limits_{0 \leq k \leq N-1} (I_c(x_0, y_0 + \frac{kh}{N})) & C \leq c < 2C \\ \max\limits_{0 \leq k \leq N-1} (I_c(x_0 + \frac{kw}{N}, y_0)) & 2C \leq c < 3C \\ \max\limits_{0 \leq k \leq N-1} (I_c(x_1, y_0 + \frac{kh}{N})) & 3C \leq c < 4C \\ \max\limits_{0 \leq k \leq N-1} (I_c(x_0 + \frac{kw}{N}, y_1)) & 4C \leq c < 5C \end{cases} \quad (1)$$

Here $\mathcal{F}_c(i,j)$ is the feature value on the (i,j)-th point for the c-th channel of the output feature maps $\mathcal{F}$, (x_0, y_0, x_1, y_1) is the bounding box prediction on the point (i,j), w and h are the width and height of (x_0, y_0, x_1, y_1). To avoid the quantization error, the exact value I_c is computed by bilinear interpolation [12] with the nearby feature value on the feature maps.

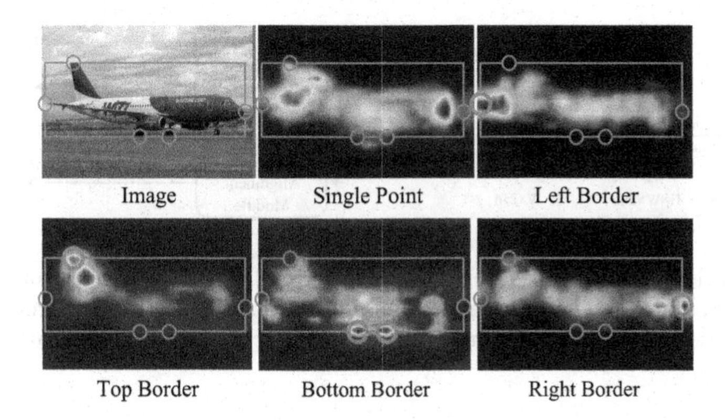

Fig. 4. Visualization of the *border-sensitive* feature maps. The orange circle on the border indicate the extreme points. The feature maps of 'Single Point', 'Left Border', 'Top Border', 'Right Border' and 'Bottom Border' are the maximum feature value on each C channels of the *border-sensitive* feature maps

In Fig. 4, we visualize the maximum value on each C channels of the *border-sensitive* feature maps. It reveals that the bank of $(4 + 1)C$ feature maps are guided to activated in their corresponding location of the object. For example, the first C channels of the show strong response over the whole object. Meanwhile, the second C exhibits a high response near the left border of the object. These *border-sensitive* feature maps facilitate our BorderAlign to extract the border feature in a principle way.

3.3 Network Architecture

BorderDet. We now present the network architecture of our BorderDet. In our experiments, we adopt a simple anchor-free object detector FCOS as our baseline. Since the border extraction procedure in BorderAlign requires border location as input, our BorderDet adopts two prediction stages as shown in Fig. 3. Taken the pyramid feature maps as input, the BorderDet first predicts the coarse classification scores and coarse bounding box locations. Then the coarse bounding box locations and the feature maps are fed into the Border Alignment Module (BAM) to generate the feature maps which contain explicit border information. Finally, we apply a 1×1 convolutional layers to predict the border classification score and border locations. The above two predictions will be unified to form the final predictions. It is to note that the border classification score is category-aware to avoid ambiguous predictions when there is an overlapping among different category boundaries.

It is worth noting that although our BorderDet adopts two extra predictions for both object classification and object localization, the additional computation is negligible due to the effective structure and layer sharing. In addition, the proposed method can be integrated into other object detectors in a plug-and-play manner, including RetinaNet [16], FCOS [21] and so forth.

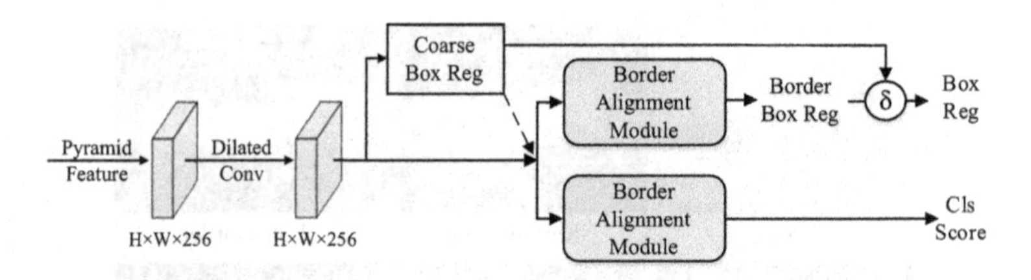

Fig. 5. The architecture of BorderRPN. We use the BAM to enhance the origin feature of the RPN, and combine the coarse bounding box locations and the border locations by δ. Meanwhile, we use the border classification score as the classification score of BorderRPN

Border Alignment Module. The structure of Border Alignment Module (BAM) is illustrated in the green box of Fig. 3. BAM takes the feature maps with C channels as input, followed by a 1×1 convolutional layer with instance normalization to output the *border-sensitive* feature maps. The border-sensitive feature maps composed of five feature maps with C channels for each border and the single point. Thus the channels of the output feature maps have $(4+1)C$ channels. In our experiments, C is set to 256 for the classification branch and to 128 for the regression branch. Finally, we adopt BorderAlign module to extract the border feature from the *border-sensitive* feature maps, and apply a 1×1 convolutional layer to reduce the $(4+1)C$ channels back to C.

BorderRPN. Our method can also be served as a better proposal generator for the typical two-stage detectors. We add the border alignment module to RPN and denote the new structure as BorderRPN. The architecture of BorderRPN is shown in Fig. 5. We remain the regression branch in RPN to predict the coarse bounding box locations. The first 3×3 convolution in RPN is replaced with a 3×3 dilated convolution to increase the effective receptive field.

3.4 Model Training and Inference

Target Assignment. We adopt FCOS [21] as our baseline to predict the coarse classification scores and coarse bounding box prediction (x_0, y_0, x_1, y_1). Then, in the second stage, a coarse bounding box prediction will be assigned to the ground-truth box $(x_0^t, y_0^t, x_1^t, y_1^t)$ by using an intersection-over-union (IoU) threshold of 0.6. And its regression targets $(\delta x_0, \delta y_0, \delta x_1, \delta y_1)$ are computed as following:

$$\delta x_0 = \frac{x_0^t - x_0}{w * \sigma} \quad \delta y_0 = \frac{y_0^t - y_0}{h * \sigma} \quad \delta x_1 = \frac{x_1^t - x_1}{w * \sigma} \quad \delta y_1 = \frac{y_1^t - y_1}{h * \sigma}, \tag{2}$$

where w, h are the width and height of the coarse bounding box prediction, and σ is the variance to improve the effectiveness of multi-task learning (σ equals to 0.5 by default).

Loss Function. The proposed BorderDet is easy to optimize in an end-to-end way using a multi-task loss. Combining the output of the BorderDet, we define our training loss function as follows:

$$\mathcal{L} = \mathcal{L}^C_{cls} + \mathcal{L}^C_{reg} + \frac{1}{\mathcal{N}_{pos}} \sum_{x,y} \mathcal{L}^B_{cls}(\mathcal{P}^B, \mathcal{C}^*) + \mathcal{L}^B_{reg_{\{\mathcal{C}^*>0\}}}(\Delta, \Delta^*), \tag{3}$$

where $\mathcal{L}^C_{cls}$ and $\mathcal{L}^C_{reg}$ are the coarse classification loss and coarse regression loss. In the implementation, focal loss [16] and IoU loss are used as the classification loss and regression loss respectively, which are the same as FCOS [21]. $\mathcal{L}^B_{cls}$ is the focal loss computed between the border classification and its assigned ground truth $\mathcal{C}^*$, the loss is averaged by the number of the positive samples $\mathcal{N}_{pos}$. We use $\mathcal{L}_1$ loss as our corner regression loss. $\mathcal{P}^B$ represents the predicted border classification scores and Δ is the predicted border offset.

Inference. BorderDet predicts classification scores and box locations for each pixel on the feature maps, while the final classification score is obtained by multiplying the coarse score and border score. The bounding boxe location is computed in a simple transformation as illustrated above. Finally, the predictions from all levels are merged and non-maximum suppression (NMS) with a threshold of 0.6.

4 Experiments

4.1 Implementation Details

Following the common practice, our ablation experiments are trained on COCO trainval35k set (115K images) and evaluated on COCO minival set (5K images). To compare with the state-of-art approaches, we report COCO AP on the test-dev set (20K images). We use ResNet-50 with FPN as our backbone network for all the experiments, if not otherwise specified. We use synchronized stochastic gradient descent(SGD) over 8 GPUs with a total of 16 images per minibatch (2 images per GPU) for 90k iterations. With an initial learning rate of 0.01, we decrease it by a factor of 10 after 60k iterations and 80k iterations respectively. We use horizontal image flipping as the only form of data augmentation. Weight decay of 0.0001 and momentum of 0.9 are used. We initialize our backbone network with the weights pre-trained on ImageNet. Unless specified, the input images are resized to ensure their shorter edge being 800 and the longer edge less than 1333.

4.2 Ablation Study

We gradually add the Border Alignment Module (BAM) to the baseline to investigate the effectiveness of our proposed BorderDet. We first apply the BAM on the classification branch. As shown in the second row of Table 2, the BAM leads to a gain of 1.1 AP. It is worth noting that the improvement mainly occurs in the

Table 2. Ablation studies of BorderDet. The 'cls' and 'reg' denote the implementation of the Border Alignment Module (BAM) on the classification branch and the regression branch respectively

Cls-BAM	Reg-BAM	AP	AP_{50}	AP_{60}	AP_{70}	AP_{80}	AP_{90}
		38.6	57.2	53.3	46.7	35.3	16.0
✓		39.7	58.4	54.8	48.5	36.2	15.9
	✓	39.7	57.3	53.3	47.3	36.9	18.6
✓	✓	**41.4**	**59.4**	**55.4**	**49.4**	**38.6**	**19.5**

Table 3. Ablation studies on the pooling size of the BorderAlign. Considering the speed/accuracy trade-off, pooling size equals to 10 in all our experiments

Size	0	2	4	6	8	10	16	32
AP	39.0	40.5	41.2	40.9	41.0	**41.4**	41.3	41.4
fps	18.3	17.0	17.0	16.9	16.9	16.7	16.6	15.9

AP with a low threshold and the improvement decays along with the increase of IoU threshold. The improvement at low IoU threshold is because the BAM can rescore the bounding boxes according to their border features, and maintain the predictions with both high classification score and localization accuracy. And the performance at a high IoU threshold is restricted by a lack of high-quality detected bounding boxes.

As opposed to BAM on the classification branch, the improvement made by the BAM on the regression branch is mainly concentrated on the AP with the high IoU threshold. The third row in Table 2 shows that conducting the BAM on the regression branch boosts the performance from 38.6 to 39.7. The BAM on the regression branch can significantly raise the localization accuracy of the detected bounding boxes, and lead to a gain of 2.6 AP_{90}.

Finally, as shown in the last row in Table 2, the implementation of the BAM on both branches can further improve the performance from 38.6 to 41.4. And the improvements are achieved over all IoU thresholds (from AP_{50} to AP_{90}), with AP_{50} increasing by 2.2 and AP_{90} by 3.5. It is worth mentioning that the AP_{90} has been improved by 20% compared with the baseline. This dramatic performance improvement demonstrates the effectiveness of our proposed BorderDet, especially for the detection with high IoU thresholds.

4.3 Border Align

Pooling Size. As described in Sect. 3.2, the BorderAlign first subdivides each border into several points and then pools over each border to extract the border features. One new hyper-parameter, the pooling size, is introduced during the procedure of BorderAlign. We compare the detection performance of different pooling size in BorderAlign. Results are shown in Table 3. When the pooling size

Table 4. Ablation studies on the border-sensitive feature maps. 'border sensitive' indicates that the extractions of border features and original single point feature are conducted on the different feature maps, while 'border agnostic' means the feature extractions are conducted on the single feature map

	AP	AP_{50}	AP_{75}	AP_S	AP_M	AP_L
Border agnostic	40.8	59.1	44.0	23.7	44.7	52.7
Border sensitive	41.4	59.4	44.5	23.6	45.1	54.6

Table 5. Ablation studies on border feature aggregation strategy in BorderAlign. For the "Border-Wise" strategy, firstly, the feature maps are aggregate along channel dimension by different pooling methods to generate the feature maps with channel equals 1. Then, a max-pooling is conducted on each border of the object to explore the extreme point, and the feature maps on the extreme points are extracted to form the border features. For the "Channel-Wise" strategy, the border feature of each channel is aggregated along the border by average-pooling or max-pooling independently

Aggregation strategy		AP	AP_{50}	AP_{75}	AP_S	AP_M	AP_L
Border-Wise	average-pooling	39.9	58.8	43.0	23.0	43.8	51.9
	max-pooling	39.5	58.2	42.7	22.6	43.3	51.3
Channel-Wise	average-pooling	40.6	58.9	43.8	23.8	44.4	52.7
	max-pooling	**41.4**	59.4	44.5	23.6	45.1	54.6

equals 0, the experiment is equivalent to iteratively predict the bounding box. The experiments show that the results are robust to the value of pooling size in a large range. As a large pooling size expenses extra computation, while a small pooling size leads to unstable results, the pooling size is set to 10 as our default setting.

Border-Sensitive Feature Maps. To analyse the impact of the *border-sensitive* feature maps as described in Sect. 3.2, we also apply the BorderAlign on border-agnostic feature maps with C channels. All the features in BorderAlign will be extracted from the same C feature maps. As shown in Table 4, *border-sensitive* feature maps improve the AP from 40.8 to 41.4. This is because the *border-sensitive* feature maps could be highly activated on the extreme points of different borders on different channels, thus facilitate the border feature extraction.

Border Feature Aggregation Strategy. In BorderAlign, we adopt a channel-wise max-pooling strategy that the border feature of each channel is aggregated along the border independently. We investigate the influence of the aggregation strategy from both the channel-wise and border-wise. As illustrated in Table 5, the channel-wise max-pooling strategy achieves the best performance of 41.4 AP. Compare to the other methods, the proposed channel-wise max-pooling strategy could extract the representative border feature without involving the background noise.

Table 6. Comparison of different feature extraction strategies. All the fps of the extraction strategies are tested on a single NVIDIA 2080Ti GPU

Method	AP	AP_{50}	AP_{75}	AP_S	AP_M	AP_L	fps
FCOS [21]	38.6	57.2	41.7	23.5	42.8	48.9	18.4
w/Iter-Box [5]	39.0	58.0	42.0	21.8	42.9	50.7	18.3
w/Adaptive Conv [23]	39.6	58.5	42.8	22.0	43.5	51.3	16.8
w/Deformable Conv [2]	39.5	58.5	42.9	22.0	43.5	52.0	16.8
w/RoIAlign [9]	40.4	58.6	43.6	22.6	44.1	53.1	12.6
w/**BorderAlign**	**41.4**	**59.4**	**44.5**	**23.6**	**45.1**	**54.6**	16.7

Comparison with Other Feature Extraction Operators. Cascade-RPN [23] and GA-RPN [24] proposed to ease the misalignment between the prediction bounding boxes and their corresponding feature. Both the two methods adopt some irregular convolutions, like deformable convolution [2] and adaptive convolution [23] to extract the feature of the bounding boxes. These irregular convolutions can also extract the border feature implicitly. To further prove the effectiveness of our proposed BorderAlign, we directly replace the BorderAlign and the second convolution in Border Alignment Module (BAM) (Fig. 3) with the adaptive convolution and deformable convolution respectively. For the fair comparison, we remain the first 1×1 convolution with Instance Nornalization [22] in the BorderDet. Meanwhile, we also compare the BorderAlign with the RoIAlign by replacing the BorderAlign with RoIAlign. Table 6 reveals that the BorderAlign outperforms other feature extraction operators by 1.0 AP at least.

Our proposed BorderAlign can concentrate on the representative part of the border, like the extreme points, and extract the border features explicitly and efficiently. On the contrary, the other operators which extract the feature from the whole boxes will introduce the redundant features and limit the detection performance.

4.4 Analysis of BorderDet

Border Feature Representation. BorderAlign is accomplished by a channel-wise max-pooling along the border that guarantees the feature extraction process is conducted around the representative extreme points on the borders. We demonstrate this perspective by a quantitative method. Concretely, we first use the annotations of the instance segmentation to yield the locations of the extreme points (top-most, left-most, bottom-most, right-most). Then, we calculate the counts of the normalized distances from the BorderAlign sample points to the extreme points in each response maps during the training (5k iteration, 30k iterations and 90k iterations), which is shown in Fig. 6. The mean value of the normalized distance is almost equal to zero. Meanwhile, the variance of the distance decreased gradually during the training. It means that our BorderDet can

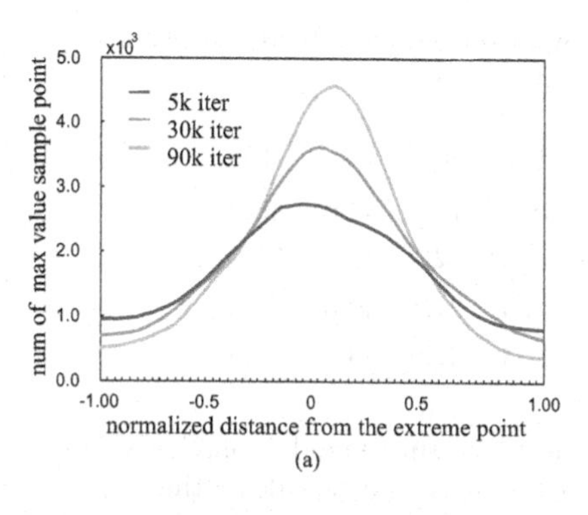

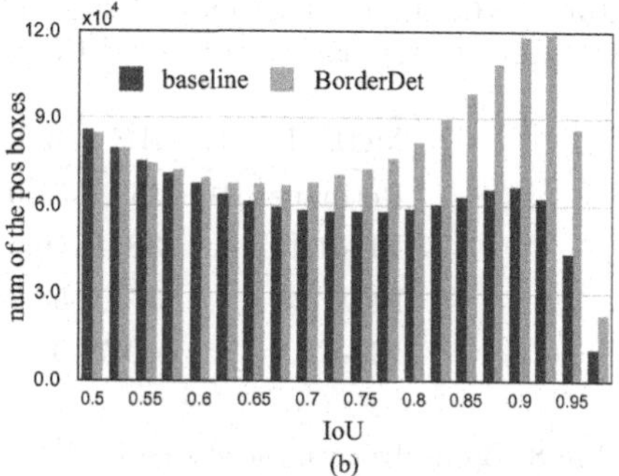

Fig. 6. (a) The statistical analysis of the border extraction. The horizontal axis indicates the normalized distance from the extreme point to the point with max feature value in BorderAlign. (b) The IoU histogram of output bounding boxes. Both the quality and quantity of the output boxes have been greatly improved by the BorderDet

adaptively learn to extract the feature near the extreme point. These results further demonstrate the effectiveness of the proposed BorderDet for border feature extraction. For further visualization results, please refer to Figure S1.

Regression Performance. To further investigate the benefit of the border feature in object localization, we count the number of bounding box predictions with different IoU thresholds separately. Figure 6 shows the comparison of the distributions of the bounding box predictions in the FCOS and BorderDet. We can see the localization accuracy of the bounding boxes is improved significantly. The number of valid prediction boxes (IoU greater than 0.5) increased by about 30%. In particular, the number of boxes with IoU greater than 0.9 has nearly doubled. This observation can also explain the significant improvement in AP_{90} as shown in Table 2.

4.5 Generalization of BorderDet

Our BorderDet can be easily integrated with the many popular object detectors, e.g. RetinaNet and FPN. To prove the generalization of the BorderDet, we first add the proposed border alignment module to the RetinaNet. For a fair comparison, without modifying any setting of RetinaNet, we directly select the one with the highest score from the nine prediction boxes of each pixel to refine. As shown in Table 7, BorderDet can consistently improve the RetinaNet by 2.3 AP. For the two-stage method FPN, our experiments show that the proposed BorderRPN gains 3.6 AP improvement.

4.6 Comparisons with State-of-the-Art Detectors

The BorderDet, based on FCOS and ResNet-101 backbone, is compared to the state-of-the-art methods in Table 8 under standard setting and advanced setting.

Table 7. Results of combining BorderDet with one-stage detector (RetinaNet) and two-stage detector (Faster R-CNN, FPN based)

Method	AP	AP_{50}	AP_{75}	AP_S	AP_M	AP_L
Retinanet [16]	36.1	55.0	38.4	19.1	39.6	48.2
BD-Retinanet	**38.4**	56.5	55.5	22.4	41.6	51.0
FPN [20]	37.1	58.7	40.3	21.1	40.3	48.6
BD-FPN	**40.7**	57.8	44.3	21.9	43.7	54.8

Table 8. BorderDet vs. the state-of-the-art mothods (single model) on COCO test-dev set. "†" indicates multi-scale training. "‡" indicates the multi-scale testing

Method	Backbone	Iter	AP	AP_{50}	AP_{75}	AP_S	AP_M	AP_L
FPN [15]	ResNet-101-FPN	180k	36.2	59.1	39.0	18.2	39.0	48.2
Mask R-CNN [9]	ResNet-101-FPN	180k	38.2	60.3	41.7	20.1	41.1	50.2
Cascade R-CNN [1]	ResNet-101	280k	42.8	62.1	46.3	23.7	45.5	55.2
RefineDet512 [27]	Resnet-101	280k	41.8	62.9	45.7	25.6	45.1	54.1
RetinaNet [16]	ResNet-101-FPN	135k	39.1	59.1	42.3	21.8	42.7	50.2
FSAF [30]	ResNet-101-FPN	135k	40.9	61.5	44.0	24.0	44.2	51.3
FCOS [21]	ResNet-101-FPN	180k	41.5	60.7	45.0	24.4	44.8	51.6
FCOS-imprv [21]	ResNet-101-FPN	180k	43.0	61.7	46.3	26.0	46.8	55.0
CornerNet [13]	Hourglass-104	500k	40.6	56.4	43.2	19.1	42.8	54.3
CenterNet [3]	Hourglass-104	500k	44.9	62.4	48.1	25.6	47.4	57.4
BorderDet	ResNet-101-FPN	90k	43.2	62.1	46.7	24.4	46.3	54.9
BorderDet†	ResNet-101-FPN	180k	45.4	64.1	48.8	26.7	48.3	56.5
BorderDet†	ResNeXt-32x8d-101	180k	45.9	65.1	49.7	28.4	48.6	56.7
BorderDet†	ResNeXt-64x4d-101	180k	46.5	65.7	50.5	29.1	49.4	57.5
BorderDet†	ResNet-101-DCN	180k	47.2	66.1	51.0	28.1	50.2	59.9
BorderDet†	ResNeXt-64x4d-101-DCN	180k	48.0	67.1	52.1	29.4	50.7	60.5
BorderDet‡	ResNeXt-64x4d-101-DCN	180k	50.3	68.9	55.2	32.8	52.8	62.3

The standard setting is the same as the setting in Sect. 4.1. The advanced setting follows the setting that using the jitter over scales {640, 672, 704, 736, 768, 800}, and number the training iterations are doubled to 180K. Table 8 shows the comparison with the state-of-the-art detectors on the MS-COCO test-dev set. With the standard setting, the proposed BorderDet achieves an AP of 43.2. It surpasses the anchor-free approaches including GuidedAnchoring, FSAF and CornerNet. By adopting advanced settings, BorderDet reaches 50.3 AP, the state of the art among existing one-stage methods and two-stage methods.

5 Conclusion

In this work, we present the BorderDet, a simple yet effective network architecture that extracts border features in both the classification and regression procedure to improve the localization ability of the object detector. The introduced border features are extracted by a novel operation called BorderAlign. Through BorderAlign, the object detector is able to adaptively learn to extract the features of the extreme point on each borders. Extensive experiments are conducted to validate the BorderAlign has higher performance than the previous feature refinement operations.

Acknowledgement. This work was supported in part by the National Key Research and Development Program of China under Grant 2017YFA0700800.

References

1. Cai, Z., Vasconcelos, N.: Cascade R-CNN: delving into high quality object detection. In: Proceedings of the IEEE Conference on Computer Vision and Pattern Recognition, pp. 6154–6162 (2018)
2. Dai, J., Qi, H., Xiong, Y., Li, Y., Zhang, G., Hu, H., Wei, Y.: Deformable convolutional networks. In: The IEEE International Conference on Computer Vision (ICCV), October 2017
3. Duan, K., Bai, S., Xie, L., Qi, H., Huang, Q., Tian, Q.: CenterNet: keypoint triplets for object detection. In: Proceedings of the IEEE International Conference on Computer Vision, pp. 6569–6578 (2019)
4. Fu, C.Y., Liu, W., Ranga, A., Tyagi, A., Berg, A.C.: DSSD: deconvolutional single shot detector. arXiv preprint arXiv:1701.06659 (2017)
5. Gidaris, S., Komodakis, N.: Attend refine repeat: active box proposal generation via in-out localization. In: Richard C.,W.E.R.H., Smith, W.A.P. (eds.) Proceedings of the British Machine Vision Conference (BMVC), pp. 90.1–90.13. BMVA Press, September 2016. https://doi.org/10.5244/C.30.90
6. Gidaris, S., Komodakis, N.: LocNet: improving localization accuracy for object detection. In: 2016 IEEE Conference on Computer Vision and Pattern Recognition (CVPR) (2016)
7. Girshick, R.: Fast R-CNN. In: Proceedings of the IEEE International Conference on Computer Vision, pp. 1440–1448 (2015)
8. Girshick, R., Donahue, J., Darrell, T., Malik, J.: Rich feature hierarchies for accurate object detection and semantic segmentation. In: Proceedings of the IEEE Conference on Computer Vision and Pattern Recognition, pp. 580–587 (2014)
9. He, K., Gkioxari, G., Dollar, P., Girshick, R.: Mask R-CNN. In: The IEEE International Conference on Computer Vision (ICCV), October 2017
10. He, K., Zhang, X., Ren, S., Sun, J.: Spatial pyramid pooling in deep convolutional networks for visual recognition. In: Fleet, D., Pajdla, T., Schiele, B., Tuytelaars, T. (eds.) ECCV 2014. LNCS, vol. 8691, pp. 346–361. Springer, Cham (2014). https://doi.org/10.1007/978-3-319-10578-9_23
11. Huang, L., Yang, Y., Deng, Y., Yu, Y.: DenseBox: unifying landmark localization with end to end object detection. arXiv preprint arXiv:1509.04874 (2015)
12. Jaderberg, M., Simonyan, K., Zisserman, A., kavukcuoglu, k.: Spatial transformer networks. In: Cortes, C., Lawrence, N.D., Lee, D.D., Sugiyama, M., Garnett, R. (eds.) Advances in Neural Information Processing Systems 28, pp. 2017–2025. Curran Associates, Inc. (2015). http://papers.nips.cc/paper/5854-spatial-transformer-networks.pdf
13. Law, H., Deng, J.: Cornernet: Detecting objects as paired keypoints. In: The European Conference on Computer Vision (ECCV), September 2018
14. Li, Y., He, K., Sun, J., et al.: R-FCN: Object detection via region-based fully convolutional networks. In: Advances in Neural Information Processing Systems, pp. 379–387 (2016)

15. Lin, T.Y., Dollár, P., Girshick, R.B., He, K., Hariharan, B., Belongie, S.J.: Feature pyramid networks for object detection. In: CVPR, vol. 1, p. 4 (2017)
16. Lin, T.Y., Goyal, P., Girshick, R., He, K., Dollar, P.: Focal loss for dense object detection. In: 2017 IEEE International Conference on Computer Vision (ICCV), pp. 2999–3007. IEEE (2017)
17. Liu, W., et al.: SSD: single shot multibox detector. In: Leibe, B., Matas, J., Sebe, N., Welling, M. (eds.) European Conference on Computer Vision, pp. 21–37. Springer, Cham (2016). https://doi.org/10.1007/978-3-319-46448-0_2
18. Redmon, J., Divvala, S., Girshick, R., Farhadi, A.: You only look once: unified, real-time object detection. In: Proceedings of the IEEE Conference on Computer Vision and Pattern Recognition, pp. 779–788 (2016)
19. Redmon, J., Farhadi, A.: Yolo9000: better, faster, stronger. In: 2017 IEEE Conference on Computer Vision and Pattern Recognition (CVPR), pp. 6517–6525. IEEE (2017)
20. Ren, S., He, K., Girshick, R., Sun, J.: Faster R-CNN: towards real-time object detection with region proposal networks. In: Advances in Neural Information Processing Systems, pp. 91–99 (2015)
21. Tian, Z., Shen, C., Chen, H., He, T.: FCOS: fully convolutional one-stage object detection. arXiv preprint arXiv:1904.01355 (2019)
22. Ulyanov, D., Vedaldi, A., Lempitsky, V.: Instance normalization: the missing ingredient for fast stylization. arXiv e-prints arXiv:1607.08022 (Jul 2016)
23. Vu, T., Jang, H., Pham, T.X., Yoo, C.D.: Cascade RPN: delving into high-quality region proposal network with adaptive convolution. arXiv e-prints arXiv:1909.06720 (Sep 2019)
24. Wang, J., Chen, K., Yang, S., Loy, C.C., Lin, D.: Region proposal by guided anchoring. In: The IEEE Conference on Computer Vision and Pattern Recognition (CVPR), June 2019
25. Wang, J., et al.: Side-aware boundary localization for more precise object detection. arXiv e-prints arXiv:1912.04260, December 2019
26. Yang, Z., Liu, S., Hu, H., Wang, L., Lin, S.: Reppoints: point set representation for object detection. In: The IEEE International Conference on Computer Vision (ICCV), October 2019
27. Zhang, S., Wen, L., Bian, X., Lei, Z., Li, S.Z.: Single-shot refinement neural network for object detection. CoRR abs/1711.06897 (2017), http://arxiv.org/abs/1711.06897
28. Zhang, S., Wen, L., Bian, X., Lei, Z., Li, S.Z.: Single-shot refinement neural network for object detection. In: The IEEE Conference on Computer Vision and Pattern Recognition (CVPR), June 2018
29. Zhou, X., Zhuo, J., Krähenbühl, P.: Bottom-up object detection by grouping extreme and center points. CoRR abs/1901.08043 (2019). http://arxiv.org/abs/1901.08043
30. Zhu, C., He, Y., Savvides, M.: Feature selective anchor-free module for single-shot object detection. In: Proceedings of the IEEE Conference on Computer Vision and Pattern Recognition, pp. 840–849 (2019)

Regularization with Latent Space Virtual Adversarial Training

Genki Osada[1,2,5(✉)], Budrul Ahsan[1,3], Revoti Prasad Bora[4], and TakashiNishide[2]

[1] Philips Co-Creation Center, Tokyo, Japan
homerunrun@hotmail.com
[2] University of Tsukuba, Tsukuba, Japan
[3] The Tokyo Foundation for Policy Research, Tokyo, Japan
[4] Philips India Limited, Bangalore, India
[5] I Dragon Corporation, Tokyo, Japan

Abstract. Virtual Adversarial Training (VAT) has shown impressive results among recently developed regularization methods called consistency regularization. VAT utilizes adversarial samples, generated by injecting perturbation in the input space, for training and thereby enhances the generalization ability of a classifier. However, such adversarial samples can be generated only within a very small area around the input data point, which limits the adversarial effectiveness of such samples. To address this problem we propose LVAT (Latent space VAT), which injects perturbation in the latent space instead of the input space. LVAT can generate adversarial samples flexibly, resulting in more adverse effect and thus more effective regularization. The latent space is built by a generative model, and in this paper we examine two different type of models: variational auto-encoder and normalizing flow, specifically Glow.

We evaluated the performance of our method in both supervised and semi-supervised learning scenarios for an image classification task using SVHN and CIFAR-10 datasets. In our evaluation, we found that our method outperforms VAT and other state-of-the-art methods.

Keywords: Consistency regularization · Adversarial training · Image classification · Semi-supervised learning · And unsupervised learning

1 Introduction

One of the goals of training machine learning models is to avoid overfitting. To address overfitting, we use various regularization techniques. Recently *consistency regularization* has shown remarkable results to overcome overfitting problems in deep neural networks. Consistency regularization is also known as perturbation-based methods, and its basic strategy is to perturb inputs during

Revoti Prasad Bora—Currently working at Lowe's Services India Pvt. Ltd.

A. Vedaldi et al. (Eds.): ECCV 2020, LNCS 12346, pp. 565–581, 2020.
https://doi.org/10.1007/978-3-030-58452-8_33

the learning process and force the model to be robust against them. Perturbation is defined as randomizing operations such as dropout, introducing Gaussian noise, and data augmentation. Consistency regularization is achieved by introducing a regularization term, called consistency cost. Consistency cost is designed to penalize the discrepancy between the model outputs, with and without perturbations. Thus consistency regularization can work without class labels, which leads to two attractive features: 1) it can extend most supervised learning methods to semi-supervised learning methods, which enables us to use unlabeled data for the model training, and 2) like the techniques such as dropout and data augmentation, it can be applied to most existing models without modifying the model architecture.

Among the consistency regularization methods, Virtual Adversarial Training (VAT) [25,26] has shown promising results. The way of generating perturbation in VAT is different from other consistency regularization methods. Perturbation in VAT is not random, instead deliberately generated towards a direction that causes the most adverse effects on the model output. However, perturbations in VAT can work only a very small area around the input data points. It is because those perturbations are generated based on the local sensitivity, i.e., the gradients of the model outputs w.r.t. the tiny shift in the input space (Sect. 3.1). We focus on the fact that such *local constraint* limits the offensive power of adversarial perturbations, and thus hinders the effectiveness of VAT as a consistency regularization. Based on this, we aim to overcome this limitation and develop a method to generate perturbation more flexibly, which would lead to better generalization.

In this paper, we propose VAT based consistency regularization that utilizes the latent space. The key idea underlying our method is transforming the space in which the computation of perturbation is performed. In our method, the input data is mapped to the latent space, and we compute perturbation and inject it into the point in the latent space. Then, adversarial examples are generated by mapping such perturbed latent point back to the point in the input space, and therefore, there is no local constraint which hinders VAT effectiveness. To map to and from the latent space, we use a generative model, which we call *transformer*, and in this paper we examine two models: variational auto-encoder (VAE) [16] and normalizing flow, specifically, Glow [17]. We used SVHN and CIFAR-10 datasets, which are common benchmarking datasets, to demonstrate that our method improves VAT and outperforms other state-of-the-art methods. To the best of our knowledge, this work is the first to introduce the latent space in the context of consistency regularization.

Notation. We consider a classification task. Suppose we have K classes, taking an input $\mathbf{x}_i \in \mathcal{X}$, we want to predict the class label $y_i \in \{1, 2, \ldots, K\}$[1], where $\mathcal{X}$ is a sample space with data points. Our purpose is to learn K-class classifier model $f : \mathcal{X} \mapsto \mathbb{R}^K$ parameterized by θ, where $\mathbb{R}$ denotes real numbers, the k-th element

[1] We write scalars and vectors by non-bold and bold letters, respectively.

$f(\mathbf{x})_k$ is called logit for class k, and $y_{\text{pred}} = \arg\max_k f(\mathbf{x})_k \in \{1, 2, \ldots, K\}$ corresponds to the model prediction of the class. Also, labeled dataset is denoted by $D_{\text{l}} = \{(\mathbf{x}_i, y_i)\}_{i=1}^{N_{\text{l}}}$ with N_{l} samples and unlabeled dataset is denoted by $D_{\text{u}} = \{\mathbf{x}_i\}_{i=1}^{N_{\text{u}}}$ with N_{u} samples.

2 Related Work

We focus mainly on state-of-the-art methods that are related to our approach. These can be broadly categorized as follows: consistency regularization, graph-based methods, and GAN-based methods.

Consistency Regularization. The VAT and our proposed method LVAT belong to this category. The assumption underlying these methods is *local consistency* [37]: nearby points in the input space are likely to have the same output. In general, the model predictions for the points near the decision boundaries are sensitive to perturbations and prone to be misclassified by being perturbed. To mitigate this sensitivity, this type of method employs the regularizing loss function, called consistency cost, which aims to train the model so that its outputs would be consistent for the inputs both with and without perturbation. Because the consistency cost becomes large at the points near the decision boundaries, the regularizing effect works so that the decision boundary would be kept far away from such points, which leads to better generalization in testing time. As the simplest case, the Π-Model [20] employs the following consistency cost $R(\mathbf{x})$:

$$R(\mathbf{x}) := \|f(\tilde{\mathbf{x}}_1, \theta) - f(\tilde{\mathbf{x}}_2, \theta)\|_2^2 \tag{1}$$

$$\tilde{\mathbf{x}}_1 \sim \text{Perturb}(\mathbf{x}),\ \tilde{\mathbf{x}}_2 \sim \text{Perturb}(\mathbf{x}) \tag{2}$$

where $\|\cdot\|_2$ denotes L_2 norm, and Perturb$(\cdot)$ is a function that applies a stochastic deformation (i.e., data augmentation), random noise addition, and dropout, thus outputting different $\tilde{\mathbf{x}}_i$ each time.

Dropout, Gaussian noise, and randomized data augmentation have been chosen as perturbations in [1,2,20,28,31,34]. [36] has shown that the latest data augmentation techniques, AutoAugment and Cutout, were quite effective to use as perturbations. Although in the Π-Model, Perturb$(\cdot)$ is applied to both of $\mathbf{x}$ in $R(\mathbf{x})$, Perturb$(\cdot)$ is applied only to one $\mathbf{x}$ in $R(\mathbf{x})$ in VAT and LVAT. Furthermore, perturbations used in VAT and LVAT are not random but are carefully computed, as we describe it in the next section.

Graph-Based Methods. In contrast to the above consistency regularization methods, graph-based methods assume *global consistency* [37]: all samples that map to the same class label should belong to a single cluster. [14] proposed a method that captures a structure of samples within a mini-batch by means of label propagation, and then forces samples belonging to the same class to form compact clusters in the feature space. Smooth Neighbors on Teacher Graphs

(SNTG) [23] computes unsupervised loss function for each mini-batch, in which attraction force between samples belonging to the same class and repulsion force between samples belonging to the different classes are realized. Consistency regularizations focus on the sensitivity of each data point to perturbations, whereas graph-based methods regularize a whole structure of data points within a mini-batch. Graph-based methods and consistency regularizations are not mutually exclusive, but rather complementary. These two could be implemented at the same time, and in fact [23] reported that combining with SNTG steadily improved the performance of all consistency regularization methods in their experiments.

GAN-Based Methods. Several works have utilized the samples generated in GAN framework as another kind of perturbation injection [8,21,32,33]. Among them, BadGAN [5] presented impressive performance. As opposed to the usual GAN framework, the generator in BadGAN generates unrealistic samples, which can be viewed as a data augmentation targeting the lower density regions in the given data distribution. With such data augmentation, BadGAN aims to draw better decision boundaries, and thus their objective is similar to VAT and our proposed method. We compare our method with these methods in result section.

Adversarial Examples for Generative Models. We finally note that, in the filed of adversarial machine learning, there are studies in regard to the latent space of generative models. However, their objectives are not regularization like ours. [3] studied attacking the generative models, and [9] derived a fundamental upper bound on robustness against adversarial perturbations. Our goal is to achieve better consistency regularization, and to this end, we aim to generate more effective adversarial examples by utilizing the latent space.

3 Background

3.1 Virtual Adversarial Training and Local Constraint

Perturbations in VAT are deliberately generated so that its direction could cause the most adverse effects on the model outputs, i.e., classification predictions. Formally, letting

$$R(\mathbf{x}, \mathbf{r}) := \mathrm{KL}(f(\mathbf{x}, \theta) \parallel f(\mathbf{x}+\mathbf{r}, \theta)), \tag{3}$$

adversarial perturbation $\mathbf{r}_{\mathrm{vat}}$ is defined as

$$\mathbf{r}_{\mathrm{vat}} := \arg\max_{\mathbf{r}}\{R(\mathbf{x}, \mathbf{r}); \|\mathbf{r}\|_2 \leq \epsilon_{\mathrm{vat}}\} \tag{4}$$

where $\mathrm{KL}(p \parallel q)$ denotes Kullback-Leibler (KL) divergence between distributions p and q, and ϵ_{vat} is a hyper-parameter to decide the magnitude of $\mathbf{r}_{\mathrm{vat}}$[2]. Once

[2] We use the suffix of 'vat' to distinguish from the symbols that will be used later in the description of our proposed method.

$\mathbf{r}_{\text{vat}}$ is computed, the consistency cost is given as:

$$\begin{aligned} L_{\text{vat}} &:= R(\mathbf{x}, \mathbf{r}_{\text{vat}}) \\ &= \text{KL}(f(\mathbf{x}, \theta) \parallel f(\mathbf{x} + \mathbf{r}_{\text{vat}}, \theta)). \end{aligned} \tag{5}$$

This regularizing cost encourages the classifier to be trained so that it outputs consistent predictions for the clean input $\mathbf{x}$ and the adversarially perturbed input $\mathbf{x} + \mathbf{r}_{\text{vat}}$.

Equation (4) can be rewritten as $\mathbf{r}_{\text{vat}} = \epsilon_{\text{vat}}\mathbf{u}$, where $\mathbf{u}$ is a unit vector in the same direction as $\mathbf{r}_{\text{vat}}$ and the maximum magnitude of $\mathbf{r}_{\text{vat}}$ is given by ϵ_{vat}. To calculate $\mathbf{u}$, [26] has presented following fast approximation method. Under the assumption that $f(\mathbf{x}, \theta)$ is twice differentiable with respect to θ, the second-order Taylor expansion around the point of $\mathbf{r} = 0$ yields

$$R(\mathbf{x}, \mathbf{r}) \approx R(\mathbf{x}, 0) + R'(\mathbf{x}, 0)\mathbf{r} + \frac{1}{2}\mathbf{r}^T H \mathbf{r} \tag{6}$$

$$= \frac{1}{2}\mathbf{r}^T H \mathbf{r} \tag{7}$$

where H is the Hessian matrix given by $H := R''(\mathbf{x}, 0)$, and $R(\mathbf{x}, 0)$ and $R'(\mathbf{x}, 0)$ in the first line are zeros since $\text{KL}(p \parallel q)$ takes the minimal value zero when $p = q$, i.e., $\mathbf{r} = 0$. Thus, taking the eigenvector of H which has the largest eigenvalue is required to solve Eq. (4). To reduce the computational cost, a finite difference power method is introduced. Given a random unit vector $\mathbf{d}$, the iterative calculation of $\mathbf{d} \leftarrow \overline{H\mathbf{d}}$ where $\overline{H\mathbf{d}} := H\mathbf{d}/\|H\mathbf{d}\|_2$, makes the $\mathbf{d}$ converge to $\mathbf{u}$. With a small constant ξ, finite difference approximation follows as

$$H \approx (R'(\mathbf{x}, 0 + \xi\mathbf{d}) - R'(\mathbf{x}, 0)) / \xi\mathbf{d} \tag{8}$$

$$Hd = R'(\mathbf{x}, \xi\mathbf{d})/\xi \tag{9}$$

where we use the fact that $R'(\mathbf{x}, 0) = 0$. Then the repeated application of $\mathbf{d} \leftarrow \overline{R'(\mathbf{x}, \xi\mathbf{d})}$ yields $\mathbf{u}$. [26] reported that sufficient result was reached by only one iteration. As a result, with a given ϵ_{vat}, Eq. (4) can be computed as:

$$\mathbf{r}_{\text{vat}} = \epsilon_{\text{vat}}\overline{R'(\mathbf{x}, \xi\mathbf{d})}. \tag{10}$$

The pseudo-code describing the computation of L_{vat} defined in Eq. (5) is shown in Algorithm 1, which will be helpful to clarify in which part our proposed method differs from the VAT.

Local Constraint. As we can see in Eq. (6), VAT algorithm works under the assumption that $\mathbf{r}$ is very small such that Taylor expansion is applicable. This means that adversarial examples $\mathbf{x} + \mathbf{r}_{\text{vat}}$ are crafted as only a very small shift from $\mathbf{x}$, which hinders the search for more adverse examples. Our purpose is to remove this constraint to generate adversarial examples flexibly, and to this end, we compute $\mathbf{r}_{\text{vat}}$ in the latent space.

3.2 Transformer

To map to and from the latent space, we use a generative model. In this paper, we examine two types of model, VAE and normalizing flow. The VAE is approximate inference and the dimensionality of the latent space is usually much smaller than that of the input space. On the other hand, the normalizing flow is exact inference and the dimensionality of the latent space is kept equal to that of the input space, i.e., lossless conversion. We will use the term *transformer* as a generic name to refer to these two.

Variational Auto-Encoder consists of two networks: the encoder (Enc) that maps a data sample $\mathbf{x}$ to $\mathbf{z}$ in latent space, and the decoder (Dec) that maps $\mathbf{z}$ back to a point $\hat{\mathbf{x}}$ in the input space as:

$$\mathbf{z} \sim \mathrm{Enc}(\mathbf{x}) = q(\mathbf{z}|\mathbf{x}), \; \hat{\mathbf{x}} \sim \mathrm{Dec}(\mathbf{z}) = p(\mathbf{x}|\mathbf{z}). \tag{11}$$

The VAE regularizes the encoder by imposing a prior over the latent distribution $p(\mathbf{z})$. Typically $p(\mathbf{z})$ is set as a standard normal distribution $\mathcal{N}(0, \mathbf{I})$. The VAE loss is:

$$L_{\text{vae}} = -\mathbb{E}_{q(\mathbf{z}|\mathbf{x})}\left[\log \frac{p(\mathbf{x}|\mathbf{z})p(\mathbf{z})}{q(\mathbf{z}|\mathbf{x})}\right] \tag{12}$$

and it can be written as the sum of the following two terms: the expectation of negative log likelihood, i.e., the reconstruction error, $\mathbb{E}_{q(\mathbf{z}|\mathbf{x})}[\log p(\mathbf{x}|\mathbf{z})]$, and a prior regularization term, $\mathrm{KL}(q(\mathbf{z}|\mathbf{x}) \parallel p(\mathbf{z}))$.

Normalizing Flow. Suppose $g(\cdot)$ is an invertible function and let $\mathbf{h}_0$ and $\mathbf{h}_1$ be random variables of equal dimensionality. Under the change of variables rule, transformation $\mathbf{h}_1 = g(\mathbf{h}_0)$ can be written as the change in the probability density function: $p(\mathbf{h}_0) = p(\mathbf{h}_1)|\det(d\mathbf{h}_1/d\mathbf{h}_0)|$. Stacking this transformation L-times as $\mathbf{h}_1, \mathbf{h}_2, \ldots, \mathbf{h}_L$ yields:

$$p(\mathbf{h}_0) = p(\mathbf{h}_L)\prod_{i=1}^{L} |\det(d\mathbf{h}_i/d\mathbf{h}_{i-1})|, \tag{13}$$

and taking the logarithm results in:

$$\log p(\mathbf{x}) = \log p(\mathbf{z}) + \sum_{i=1}^{L} \log |\det(d\mathbf{h}_i/d\mathbf{h}_{i-1})| \tag{14}$$

where we define $\mathbf{h}_0 := \mathbf{x}$ and $\mathbf{h}_L := \mathbf{z}$. Such a series of transformations can gradually transform $p(\mathbf{x})$ into a target distribution $p(\mathbf{z})$ of any form. Setting $p(\mathbf{z}) = \mathcal{N}(0, \mathbf{I})$ is especially called a normalizing flow [30].

As Eq. (14) is the form of log-likelihood, the learning objective is to maximize $\mathbb{E}_{p(\mathbf{x})}\log p(\mathbf{x})$ by optimizing $g(\cdot)$. The function $g(\cdot)$ must be designed to

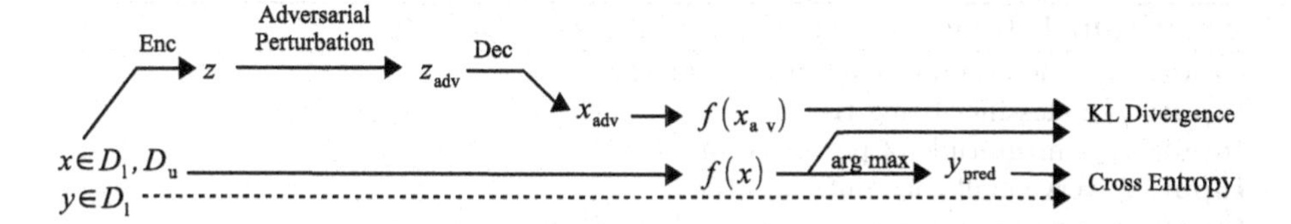

Fig. 1. Overview of our method. Only during training, we place trained transfomer, i.e., Enc() and Dec(), in front of classifier $f()$ being trained. While for $\mathbf{x} \in D_\mathrm{l}, D_\mathrm{u}$ classifier outputs $f(\mathbf{x})$, predictive class label y_pred is produced only for $\mathbf{x} \in D_\mathrm{l}$. KL divergence corresponds to the consistency const L_lvat.

have the tractability to compute its inverse and the determinant of Jacobian matrix $|\det(d\mathbf{h}_i/d\mathbf{h}_{i-1})|$ in Eq. (14), and several methods have been proposed in this regard. *Autoregressive* models [11,18,27] have a powerful expression but are computationally slow due to non-parallelization. Thus, we use *split coupling* models [6,7], specifically, Glow [17]. For brevity, we refer the reader to [17].

Similarly to the case of VAE, we denote transformation $\mathbf{x} \to \mathbf{z}$ by $\mathbf{z} = \mathrm{Enc}(\mathbf{x})$ and $\mathbf{z} \to \mathbf{x}$ by $\mathbf{x} = \mathrm{Dec}(\mathbf{z})$, and we call them just Enc() and Dec() as generic notations, for convenience.

4 Method

Our proposed method applies Eq. (4), i.e., Eq. (10), to the latent space. It means that our method generates perturbations based on the gradients of the model outputs w.r.t. the shift in the latent space, and therefore, the latent space in our method is required to be a continuous distribution. Thus, vanilla Auto-Encoder and Denoising Auto-Encoder [35], which do not construct the latent space as a continuous distribution, are out of our selection. Instead, we choose two different types of generative models, VAE and Glow, and we build those models so that the latent space $p(\mathbf{z})$ forms $\mathcal{N}(0, \mathbf{I})$. We call our proposed method *LVAT* standing for Virtual Adversarial Training in the Latent space, and we refer to LVAT using VAE and LVAT using Glow as *LVAT-VAE* and *LVAT-Glow*, respectively. In Fig. 1, we show the overview of LVAT. We deploy the transfomer in the fore stage of the classifier that we want to train. During training, by mapping the input $\mathbf{x} \in D_\mathrm{l}, D_\mathrm{u}$ to the latent space by Enc(), the latent representation $\mathbf{z} = \mathrm{Enc}(\mathbf{x})$ is computed. It is followed by applying Eq. (4) to $\mathbf{z}$ and computing the adversarial perturbation in the latent space, $\mathbf{r}_\mathrm{lvat}$, and the adversarial latent representation $\mathbf{z}_\mathrm{adv} = \mathbf{z} + \mathbf{r}_\mathrm{lvat}$ is computed. Then, by putting $\mathbf{z}_\mathrm{adv}$ through Dec(), we obtain adversarial samples $\mathbf{x}_\mathrm{adv} = \mathrm{Dec}(\mathbf{z}_\mathrm{adv})$. Here, we define

$$R_\mathrm{lvat}(\mathbf{x}, \mathbf{r}) := \mathrm{KL}(f(\mathbf{x}, \theta) \parallel f(\mathbf{x}', \theta)) \tag{15}$$

$$\mathbf{x}' = \mathrm{Dec}(\mathrm{Enc}(\mathbf{x}) + \mathbf{r}) \tag{16}$$

Algorithm 1. Computation of consistency cost for VAT

Input: X: random mini-batch from dataset
Input: $f()$: classifier being trained
Input: ϵ_{vat}: magnitude of perturbation
Input: ξ: very small constant, e.g., 1e−6
Input: $\mathbf{d}$: random unit vector of same shape of X
Output: L_{vat}: consistency cost of VAT
1: $\mathbf{g} \leftarrow \nabla_{\mathbf{d}}\ \text{KL}(f(X) \parallel f(X + \xi\mathbf{d})$
2: $\mathbf{r}_{\text{vat}} \leftarrow \epsilon_{\text{vat}}\mathbf{g}/\|\mathbf{g}\|_2$
3: $L_{\text{vat}} \leftarrow \text{KL}(f(X) \parallel f(X + \mathbf{r}_{\text{vat}}))$
4: **return** L_{vat}

Algorithm 2. Computation of consistency cost for LVAT

Input: X: random mini-batch from dataset
Input: $f()$: classifier being trained
Input: Enc() and Dec(): encoder and decoder of transfomer
Input: ϵ_{lvat}: magnitude of perturbation
Input: ξ: very small constant, e.g., 1e−6
Input: $\mathbf{d}$: random unit vector of same size as latent space
Output: L_{lvat}: consistency cost of LVAT
1: $Z \leftarrow \text{Enc}(X)$
2: $\mathbf{g} \leftarrow \nabla_{\mathbf{d}}\ \text{KL}(f(X) \parallel f(\text{Dec}(Z + \xi\mathbf{d})))$
3: $\mathbf{r}_{\text{lvat}} \leftarrow \epsilon_{\text{lvat}}\mathbf{g}/\|\mathbf{g}\|_2$
4: $L_{\text{lvat}} \leftarrow \text{KL}(f(X) \parallel f(\text{Dec}(Z + \mathbf{r}_{\text{lvat}})))$
5: **return** L_{lvat}

and the adversarial perturbation $\mathbf{r}_{\text{lvat}}$ and the consistency cost L_{lvat} are defined as:

$$\mathbf{r}_{\text{lvat}} := \arg\max_{\mathbf{r}}\{R_{\text{lvat}}(\mathbf{x}, \mathbf{r}); \|\mathbf{r}\|_2 \leq \epsilon_{\text{lvat}}\} \tag{17}$$

$$\mathbf{x}_{\text{adv}} := \text{Dec}(\text{Enc}(\mathbf{x}) + \mathbf{r}_{\text{lvat}}) \tag{18}$$

$$L_{\text{lvat}} := R_{\text{lvat}}(\mathbf{x}, \mathbf{r}_{\text{lvat}}) \tag{19}$$

$$= \text{KL}(f(\mathbf{x}, \theta) \parallel f(\mathbf{x}_{\text{adv}}, \theta)). \tag{20}$$

where ϵ_{lvat} is a hyper-parameter to decide the magnitude of $\mathbf{r}_{\text{lvat}}$. The ϵ_{vat} in VAT gives the L_2 distance $\|\mathbf{x} - \mathbf{x}_{\text{adv}}\|_2$ in the input space, whereas ϵ_{lvat} in LVAT gives the L_2 distance between $\|\mathbf{z} - \mathbf{z}_{\text{adv}}\|_2$ in the latent space. The pseudo-code to obtain L_{lvat} is shown in Algorithm 2.

The full loss function L is thus given by

$$L = L_{\text{sl}}(D_{\text{l}}, \theta) + \alpha L_{\text{usl}}(D_{\text{l}}, D_{\text{u}}, \theta) \tag{21}$$
$$L_{\text{usl}} = \mathbb{E}_{\mathbf{x} \in D_{\text{l}}, D_{\text{u}}}[L_{\text{lvat}}]$$

where L_{usl} is the unsupervised loss, i.e., consistency cost, L_{sl} is a typical supervised loss (cross-entropy for our task), and α is a coefficient relative to the supervised cost.

Table 1. Architecture of classifier. BNorm stands for batch normalization [12]. Slopes of all Leaky ReLU (lReLU) [24] are set to 0.1.

Input: 32 × 32 RGB image	8: 2 × 2 max-pool, dropout 0.5
1: 3 × 3 conv. 128 same padding, BNorm, lReLU	9: 3 × 3 conv. 512 valid padding, BNorm, lReLU
2: 3 × 3 conv. 128 same padding, BNorm, lReLU	10: 1 × 1 conv. 256 BNorm, lReLU
3: 3 × 3 conv. 128 same padding, BNorm, lReLU	11: 1 × 1 conv. 128 BNorm, lReLU
4: 2 × 2 max-pool, dropout 0.5	12: Global average pool 6 × 6 → 1 × 1
5: 3 × 3 conv. 256 same padding, BNorm, lReLU	13: Fully connected 128 → 10
6: 3 × 3 conv. 256 same padding, BNorm, lReLU	14: BNorm (only for SVHN)
7: 3 × 3 conv. 256 same padding, BNorm, lReLU	15: Softmax

5 Experiments

We evaluate our proposed method in an image classification task using SVHN and CIFAR-10 datasets. Both supervised learning (SL) and semi-supervised learning (SSL) tests are conducted. The experimental code[3] was run with NVIDIA GeForce GTX 1070.

5.1 Datasets

The street view house numbers (SVHN) dataset consists of 32 × 32 pixel RGB images of real-world house numbers, having 10 classes. The CIFAR-10 dataset also consists of 32×32 pixel RGB images in 10 different classes, *airplanes, cars, birds, cats, deer, dogs, frogs, horses, ships*, and *trucks*. The numbers of training/test images are $73,257/26,032$ for SVHN and $50,000/10,000$ for CIFAR-10, respectively.

We also evaluate our method using augmented datasets. We augmented data using random 2×2 translation for both datasets and horizontal flips only for CIFAR-10 same as previous study [25]. These augmentations are dynamically applied for each mini-batch. We denote the datasets with data augmentation by (w/aug.), and our evaluation is conducted with four datasets, SVHN, SVHN (w/aug.), CIFAR-10, and CIFAR-10 (w/aug.).

In tests in SL, all labels in the training dataset are used, and the results are averaged over 3 runs. In tests in SSL, $1,000$ and $4,000$ labeled data points are randomly sampled for SVHN and CIFAR-10, respectively. To evaluate different combinations of labeled data in tests in SSL, we prepared 5 different datasets with 5 different seeds for random sampling of labeled data points, and the results are averaged over them.

5.2 Model Training

The transformer can be modularized in our method, and thus we first train only the transformer for each dataset separately from the classifier. Once we build the transformers, then, training the classifier with LVAT using the trained

[3] https://github.com/geosada/LVAT.

Table 2. Architecture of encoder and decoder of VAE. Dimensionality of latent space is 128. BNorm stands for batch normalization. Slopes of Leaky ReLU (lReLU) are set to 0.1.

Enc	Dec
Input: 32 × 32 × 3 image	Input: 128-dimensional vector
2 × 2 conv. 128 valid padding, BNorm, ReLU	Fully connected 128 → 512 (4 × 4 × 32), lReLU
2 × 2 conv. 256 valid padding, BNorm, ReLU	2 × 2 deconv. 512 same padding, BNorm, ReLU
2 × 2 conv. 512 valid padding, BNorm, tanh	2 × 2 deconv. 256 same padding, BNorm, ReLU
Fully connected 8192 → 128	2 × 2 deconv. 128 same padding, BNorm, ReLU
	1 × 1 conv. 128 valid padding, sigmoid

transformer model follows. We can use the same transformers throughout all the experiments, which benefits us as it reduces experiment time significantly, especially when we have to run the experiments many times (e.g., for grid-searching for hyper-parameters). This can be viewed as a sort of curriculum strategy that is found for example in [22].

Model Architectures. The architecture of the classifier is the same as that of the previous works, and the detail is shown in Table 1. The architecture of the VAE is designed based on DCGAN [29], which is shown in Table 2. The architecture of the Glow mainly consists of two parameters: the depth of flow K and the number of levels L. We set $K = 22$ and $L = 3$, respectively[4]. Refer to our experimental code for more details.

Hyper-parameters. For the classifier with LVAT, we fixed the coefficient $\alpha = 1$ in Eq. (21), like the original VAT. We used the Adam optimizer [15] with the momentum parameters $\beta_1 = 0.9$ and $\beta_2 = 0.999$. The initial learning rate is set to 0.001 and decays linearly with the last 16,000 updates, and β_1 is changed to 0.5 when the learning rate starts decaying. The size of a mini-batch is 32 and 128 for L_{sl} and L_{usl}, respectively for both datasets. We trained each model with 48,000 and 200,000 updates for SVHN and CIFAR-10, respectively. The best hyper-parameter ϵ_{lvat} in Eq. (17) was found through a grid search in the SSL setting. For LVAT-VAE, 1.5 and 1.0 were selected for SVHN and CIFAR-10, respectively, from {0.1, 0.25, 0.5, 0.75, 3.0, 4.0, ..., 15.0}. For LVAT-Glow, 1.0 was selected for both SVHN and CIFAR-10 from {0.5, 1.0, 1.5}.

Also for a fair comparison, we conduct the test in SL for VAT, since the results for these were not reported in the original paper except for the one on CIFAR-10 (w/aug.). According to the code the original authors provide[5], we set ϵ_{vat} to $2.5, 3.5, 10.0$, and 8.0 for SVHN, SVHN (w/aug.), CIFAR-10, and CIFAR-10 (w/aug.), respectively. Although these values are provided for SSL test and we also attempted other values, it was found that the above ones were

[4] We implemented Glow model based on [19].

[5] https://github.com/takerum/vat_tf.

better. Regarding ϵ_{lvat} and ϵ_{vat}, we use these values for LVAT and VAT for all experiments unless otherwise noted.

The VAE and Glow were trained using the same data that was used for training the classifier. For the VAE, we also used the Adam optimizer with $\beta_1 = 0.9$ and $\beta_2 = 0.999$, with batch size 256. The learning rate starts with 0.001 and exponentially decays with rate 0.97 at every 2 epochs after the first 80 epochs, and we trained for 300 epochs. For the Glow, the learning rate starts with 0.0001, and we trained 3,200 iterations for SVHN and CIFAR-10 and 5,200 iterations for SVHN (w/aug.) and CIFAR-10 (w/aug.).

Table 3. Error rates (%) comparing to VAT and other methods. Results with data augmentation are denoted with (w/aug.). SSL indicates semi-supervised learning, i.e., number of labeled data N_l is 1,000 and 4,000 for SVHN and CIFAR-10, respectively. SL indicates supervised learning, i.e., all training data are used with label.

Methods	SVHN		SVHN (w/aug.)		CIFAR-10		CIFAR-10 (w/aug.)	
	SSL	SL	SSL	SL	SSL	SL	SSL	SL
Consistency Regularization								
Sajjadi *et al.* [31]	-	-	-	2.22 (± 0.04)	-	11.29 (± 0.24)	-	
MT [34]	5.21 (± 0.21)	2.77 (± 0.09)	3.95 (± 0.19)	2.50 (± 0.05)	17.74 (± 0.30)	7.21 (± 0.24)	12.31 (± 0.28)	5.94 (± 0.14)
Π-Model [20]	5.43 (± 0.25)	-	4.82 (± 0.17)	2.54 (± 0.04)	16.55 (± 0.29)	-	12.36 (± 0.31)	5.56 (± 0.10)
TempEns [20]	-	-	4.42 (± 0.16)	2.74 (± 0.06)	-	-	12.16 (± 0.24)	5.60 (± 0.14)
VAT [25]	5.77 (± 0.32)	2.34 (± 0.05)[a]	5.42 (± 0.22)	2.22 (± 0.08)[a]	16.92 (± 0.45)[b, c]	8.175[b]	11.36 (± 0.34)	5.81(± 0.02)
Graph-based Methods								
LBA[d] [10]	9.25 (± 0.65)	3.61 (± 0.10)	9.25 (± 0.65)	3.61 (± 0.10)	19.33 (± 0.51)	8.46 (± 0.18)	19.33 (± 0.51)	8.46 (± 0.18)
CCLP [14]	5.69 (± 0.28)	3.04 (± 0.05)	-	-	18.57 (± 0.41)	8.04 (± 0.18)	-	-
GAN-based Methods								
ALI [8]	7.42 (± 0.65)	-	-	-	17.99 (± 1.62)	-	-	-
CatGAN [33]	-	-	-	-	19.58 (± 0.58)	9.38	-	-
TripleGAN [21]	5.77 (± 0.17)	-	-	-	16.99 (± 0.36)	-	-	-
ImprovedGAN [32]	8.11 (± 1.30)	-	-	-	18.63 (± 2.32)	-	-	-
BadGAN [5]	4.25 (± 0.03)	-	-	-	14.41 (± 0.30)	-	-	-
LVAT-VAE (Ours)	4.44 (± 0.36)	**2.26** (± 0.08)	4.20 (± 0.23)	**2.02** (± 0.04)	13.90 (± 0.36)	8.05 (± 0.30)	14.64 (± 0.54)	6.54 (± 0.26)
LVAT-Glow (Ours)	**4.20** (± 0.45)	**2.23** (± 0.07)	**3.83** (± 0.37)	**2.13** (± 0.07)	**9.94** (± 0.22)	**5.24** (± 0.20)	**7.34** (± 0.24)	**3.94** (± 0.05)

[a]Results of our experiments with code [25] provided.
[b]Results of our experiments with code [25] provided without ZCA.
[c]Reported result in [25] is 14.87 (± 0.38) with ZCA.
[d]Results of re-implementation by [14].

Table 4. Error rates (%) comparing to combination methods. Notations are same as those in Table 3.

Methods	SVHN		SVHN (w/aug.)		CIFAR-10		CIFAR-10 (w/aug.)	
	SSL	SL	SSL	SL	SSL	SL	SSL	SL
Combination Methods								
MT + SNTG [23]	-	-	3.86 (± 0.27)	2.42 (± 0.06)	-	-	-	-
MT + fast-SWA [1]	-	-	-	-	-	-	9.05 (± 0.21)	4.73 (± 0.18)
Π-Model + SNTG [23]	4.22 (± 0.16)	-	**3.82** (± 0.25)	2.42 (± 0.05)	13.62 (± 0.17)	-	11.00 (± 0.13)	5.19 (± 0.14)
Π-Model + fast-SWA [1]	-	-	-	-	-	-	10.07 (± 0.27)	4.72 (± 0.04)
TempEns + SNTG [23]	-	-	3.98 (± 0.21)	2.44 (± 0.03)	-	-	10.93 (± 0.14)	5.20 (± 0.14)
VAT + Ent [25]	4.28 (± 0.10)	-	3.86 (± 0.11)	-	13.15 (± 0.21)	-	10.55 (± 0.05)	-
VAT + Ent + SNTG [23]	**4.02** (± 0.20)	-	3.83 (± 0.22)	-	12.49 (± 0.36)	-	9.89 (± 0.34)	-
VAT + Ent + fast-SWA [1]	-	-	-	-	-	-	10.97	-
VAT + LGA [13]	6.58 (± 0.36)	-	-	-	12.06 (± 0.19)	-	-	-
LVAT-VAE (Ours)	4.44 (± 0.36)	**2.26** (± 0.08)	4.20 (± 0.23)	**2.02** (± 0.04)	13.90 (± 0.36)	8.05 (± 0.30)	14.64 (± 0.54)	6.54 (± 0.26)
LVAT-Glow (Ours)	4.20 (± 0.45)	**2.23** (± 0.07)	3.83 (± 0.37)	**2.13** (± 0.07)	**9.94** (± 0.22)	**5.24** (± 0.20)	**7.34** (± 0.24)	**3.94** (± 0.05)

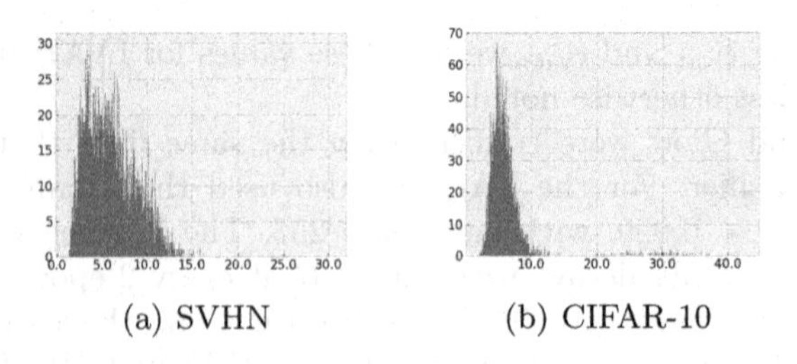

(a) SVHN (b) CIFAR-10

Fig. 2. Histograms of L_2 distance between the original input images and the adversarial images generated by LVAT-Glow with $\epsilon_{\text{lvat}} = 1.0$. x-axis is $\|\mathbf{x} - \text{Dec}(\text{Enc}(\mathbf{x}) + \mathbf{r}_{\text{lvat}})\|_2$ and y-axis is frequency. For each dataset 5,000 samples are randomly sampled. This indicates that LVAT generates various magnitudes of perturbations.

5.3 Results

We show classification accuracies. Note that some methods in Tables 3 and 4, including VAT, performed image pre-processing with ZCA on CIFAR-10, which is not used in our experiments. In terms of the model capacity, it is a fair comparison as all other methods used the same network architecture as ours.

In Table 3, we compared LVAT to VAT, other consistency regularizations, and also other approaches introduced in Sect. 2. We can see that LVAT substantially improved the original VAT, and moreover, outperformed all of other methods in all eight experimental settings.

It has been reported that even better results can be obtained by combining consistency regularizations (MT, Π-Model, TempEns, and VAT) together with other techniques, such as graph-based method (SNTG). We also compared our method to those combinations in Table 4. It is noteworthy that LVAT still surpassed all other methods, except for the result in the SSL testings on SVHN (VAT + Ent + SNTG) and on SVHN (w/aug.) (Π-Model + SNTG). In particular, LVAT-Glow on CIFAR-10 and CIFAR-10 (w/aug.) showed outstanding performance. We believe that combining LVAT with other methods (e.g., LVAT + SNTG) would also achieve further improved performance, but we leave it as future works.

6 Discussions

6.1 Adversarial Examples

In this section, we analyze our results focusing on the adversarial examples that LVAT generates.

Perturbation Magnitude. First, we see the magnitude of perturbation in the input space. Figure 2 shows the histograms of L_2 distance between the original input images and the adversarial images that LVAT generates, i.e., $\|\mathbf{x} - \text{Dec}(\text{Enc}(\mathbf{x}) + \mathbf{r}_{\text{lvat}})\|_2$, which corresponds to ϵ_{vat} in the original VAT. It is

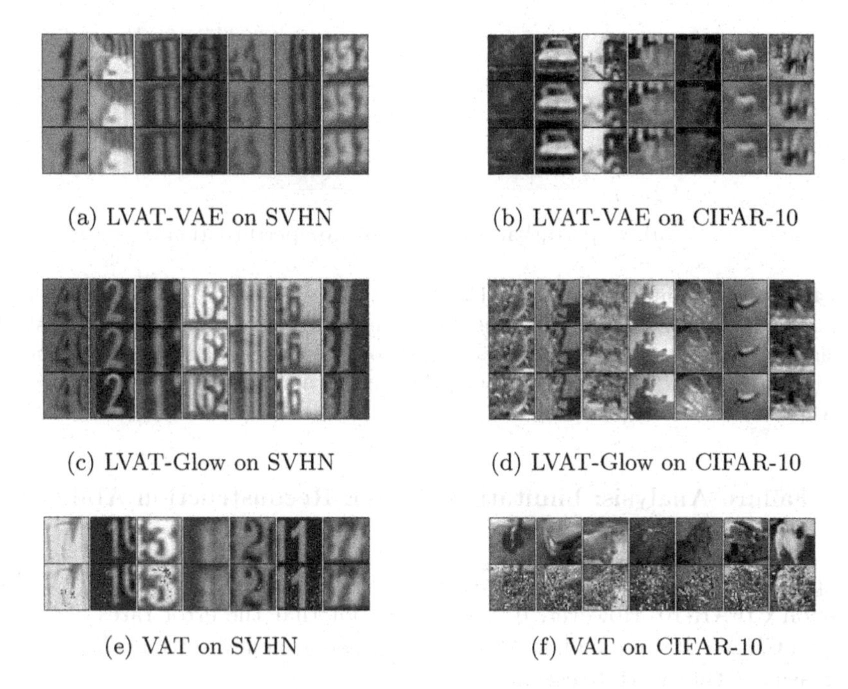

(a) LVAT-VAE on SVHN (b) LVAT-VAE on CIFAR-10

(c) LVAT-Glow on SVHN (d) LVAT-Glow on CIFAR-10

(e) VAT on SVHN (f) VAT on CIFAR-10

Fig. 3. Generated Images. For (a) through (d), first row: original images $\mathbf{x}$, second row: reconstructed images via transformer without perturbation $\hat{\mathbf{x}} = \mathrm{Dec}(\mathrm{Enc}(\mathbf{x}))$, third row: adversarial images $\mathbf{x}_{\mathrm{adv}} = \mathrm{Dec}(\mathrm{Enc}(\mathbf{x}) + \mathbf{r}_{\mathrm{lvat}})$. For (e) and (f), first row: $\mathbf{x}$, second row: $\mathbf{x}_{\mathrm{adv}} = \mathbf{x} + \mathbf{r}_{\mathrm{vat}}$.

shown that LVAT generates adversarial examples in the wide range of magnitude, unlike in the original VAT where every adversarial example is generated with the same given magnitude ϵ_{vat}. In terms of perturbation magnitude, we can see that the LVAT can generate various adversarial examples as we aimed.

Visual Appearance. Next, we see the visual appearance of adversarial examples of LVAT and VAT. Figure 3 shows that adversarial images of VAT are tainted with artifacts, whereas the ones of LVAT look realistic. In both transformer VAE and Glow, the latent space $p(\mathbf{z})$ is constructed so that the points in the high-density area in $p(\mathbf{z})$ correspond to the data used during the model training, i.e., correspond to real images. Thus, unless ϵ_{lvat} is not too large, the perturbed latent representation $\mathbf{z}_{\mathrm{adv}}$ computed in LVAT still should correspond to a realistic image. It has been argued in [36] that these noisy images generated in VAT seem harmful to further performance improvement. Unlike VAT, there is no such concern in LVAT.

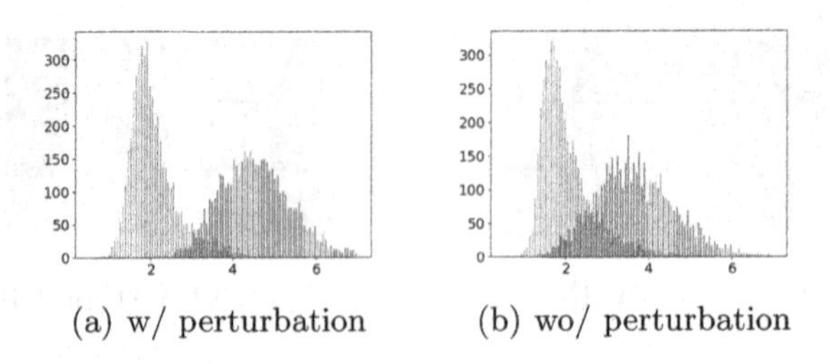

(a) w/ perturbation (b) wo/ perturbation

Fig. 4. Distance from $\mathbf{x}$ (L_2 norm) in LVAT-VAE. (a) is $\|\mathbf{x} - \text{Dec}(\text{Enc}(\mathbf{x}) + \mathbf{r}_{\text{lvat}})\|_2$ and (b) is $\|\mathbf{x} - \text{Dec}(\text{Enc}(\mathbf{x}))\|_2$, i.e., reconstruction error. SVHN is orange and CIFAR-10 is blue. For each dataset 5,000 samples are randomly sampled, and y-axis is frequency. This indicates regardless of perturbation, the reconstruction error of VAE is larger on CIFAR-10 than on SVHN.

6.2 Failure Analysis: Limitation of VAE Reconstruction Ability on CIFAR-10

Our proposed method LVAT achieved good results as we saw, especially LVAT-Glow on CIFAR-10. However, it also turned out that the error rates of LVAT-VAE on CIFAR-10 were higher than the other experimental settings. We analyze the reason of that in this section.

In Fig. 3(b), we can see that the adversarial images (the third row) on CIFAR-10 are blurred, and more importantly, the images just reconstructed without perturbation (the second row) are also blurry. This indicates that regardless of perturbation, just passing Enc() and Dec() of VAE will blur the input image, which can be viewed as the known VAE characteristics [4]. Figure 4 shows the reconstruction error in VAE: L_2 distance between the original images and the decoded images by VAE for both with and without perturbing, i.e., $\|\mathbf{x} - \text{Dec}(\text{Enc}(\mathbf{x}) + \mathbf{r}_{\text{lvat}})\|_2$ and $\|\mathbf{x} - \text{Dec}(\text{Enc}(\mathbf{x}))\|_2$. It can be seen that the reconstruction error of VAE is larger on CIFAR-10 than on SVHN, and we think that it is caused by the difference in the complexity of images contained in each dataset.

Contrary to the VAE, the Glow reconstructs very sharp images (the second row in Fig. 3(d)) and the classification performance of LVAT-Glow was very good on CIFAR-10. Given these observations, we conclude that the reconstruction ability of the transformer is crucial to the quality of regularization, which caused the high error rates of LVAT-VAE on CIFAR-10.

7 Conclusion

We focused on the local constraint of VAT: VAT can generate adversarial perturbation only within a very small area around the input data point. In order to circumvent this constraint, we proposed LVAT in which computing and injecting perturbation are done in the latent space. Since adversarial examples in LVAT are generated via the latent space, they are more flexible than those in the original VAT, which led to more effective consistency regularization and better classification performance as a result. To the best of our knowledge, this work is the first to introduce the latent space in the context of consistency regularization. We compared LVAT with VAT and other state-of-the-art methods in supervised and semi-supervised scenarios for a classification task in SVHN and CIFAR-10 datasets (both with and without data-augmentation). Our evaluation indicates that LVAT outperforms state-of-the-art methods in terms of classification accuracy in different scenarios.

Acknowledgement. This work was supported in part by JSPS KAKENHI Grant Number 20K11807.

References

1. Athiwaratkun, B., Finzi, M., Izmailov, P., Wilson, A.G.: There are many consistent explanations of unlabeled data: Why you should average. In: International Conference on Learning Representations (2019). https://openreview.net/forum?id=rkgKBhA5Y7
2. Berthelot, D., Carlini, N., Goodfellow, I., Papernot, N., Oliver, A., Raffel, C.: MixMatch: a holistic approach to semi-supervised learning. In: NeurIPS (2019)
3. Cao, X., Gong, N.Z.: Mitigating evasion attacks to deep neural networks via region-based classification. In: Proceedings of the 33rd Annual Computer Security Applications Conference, pp. 278–287. ACM (2017)
4. Chen, X., et al.: Variational lossy autoencoder. In: International Conference on Learning Representations (2017)
5. Dai, Z., Yang, Z., Yang, F., Cohen, W.W., Salakhutdinov, R.R.: Good semi-supervised learning that requires a bad GAN. In: Guyon, I., et al. (eds.) Advances in Neural Information Processing Systems 30, pp. 6510–6520. Curran Associates, Inc. (2017). http://papers.nips.cc/paper/7229-good-semi-supervised-learning-that-requires-a-bad-gan.pdf
6. Dinh, L., Krueger, D., Bengio, Y.: Nice: non-linear independent components estimation. In: International Conference on Learning Representations (2015)
7. Dinh, L., Sohl-Dickstein, J., Bengio, S.: Density estimation using real NVP. In: International Conference on Learning Representations (2017)
8. Dumoulin, V., et al.: Adversarially learned inference. In: International Conference on Learning Representations (2017)
9. Fawzi, A., Fawzi, H., Fawzi, O.: Adversarial vulnerability for any classifier. In: Bengio, S., Wallach, H., Larochelle, H., Grauman, K., Cesa-Bianchi, N., Garnett, R. (eds.) Advances in Neural Information Processing Systems, vol. 31, pp. 1178–1187. Curran Associates, Inc. (2018). http://papers.nips.cc/paper/7394-adversarial-vulnerability-for-any-classifier.pdf
10. Haeusser, P., Mordvintsev, A., Cremers, D.: Learning by association - a versatile semi-supervised training method for neural networks. In: The IEEE Conference on Computer Vision and Pattern Recognition (CVPR), July 2017
11. Huang, C.W., Krueger, D., Lacoste, A., Courville, A.: Neural autoregressive flows. In: Dy, J., Krause, A. (eds.) Proceedings of the 35th International Conference on Machine Learning. Proceedings of Machine Learning Research, vol. 80, pp. 2078–2087. PMLR, Stockholmsmässan, Stockholm Sweden, 10–15 Jul 2018. http://proceedings.mlr.press/v80/huang18d.html

12. Ioffe, S., Szegedy, C.: Batch normalization: Accelerating deep network training by reducing internal covariate shift. In: Bach, F., Blei, D. (eds.) Proceedings of the 32nd International Conference on Machine Learning. Proceedings of Machine Learning Research, vol. 37, pp. 448–456. PMLR, Lille, France, 07–09 July 2015. http://proceedings.mlr.press/v37/ioffe15.html
13. Jackson, J., Schulman, J.: Semi-supervised learning by label gradient alignment. arXiv preprint arXiv:1902.02336 (2019)
14. Kamnitsas, K., et al.: Semi-supervised learning via compact latent space clustering. In: Dy, J., Krause, A. (eds.) Proceedings of the 35th International Conference on Machine Learning. Proceedings of Machine Learning Research, vol. 80, pp. 2459–2468. PMLR, Stockholmsmässan, Stockholm Sweden, 10–15 July 2018. http://proceedings.mlr.press/v80/kamnitsas18a.html
15. Kingma, D.P., Ba, J.: Adam: a method for stochastic optimization. In: International Conference on Learning Representations (2015)
16. Kingma, D.P., Welling, M.: Auto-encoding variational Bayes. In: International Conference on Learning Representations (2014)
17. Kingma, D.P., Dhariwal, P.: Glow: Generative flow with invertible 1 × 1 convolutions. In: Bengio, S., Wallach, H., Larochelle, H., Grauman, K., Cesa-Bianchi, N., Garnett, R. (eds.) Advances in Neural Information Processing Systems 31, pp. 10215–10224. Curran Associates, Inc. (2018). http://papers.nips.cc/paper/8224-glow-generative-flow-with-invertible-1x1-convolutions.pdf
18. Kingma, D.P., Salimans, T., Jozefowicz, R., Chen, X., Sutskever, I., Welling, M.: Improved variational inference with inverse autoregressive flow. In: Lee, D.D., Sugiyama, M., Luxburg, U.V., Guyon, I., Garnett, R. (eds.) Advances in Neural Information Processing Systems 29, pp. 4743–4751. Curran Associates, Inc. (2016), http://papers.nips.cc/paper/6581-improved-variational-inference-with-inverse-autoregressive-flow.pdf
19. Kolasinski, K.: An implementation of the GLOW paper and simple normalizing flows lib (2018). https://github.com/kmkolasinski/deep-learning-notes/tree/master/seminars/2018-10-Normalizing-Flows-NICE-RealNVP-GLOW
20. Laine, S., Aila, T.: Temporal ensembling for semi-supervised learning. In: International Conference on Learning Representations (2017)
21. Li, C., Xu, T., Zhu, J., Zhang, B.: Triple generative adversarial nets. In: Guyon, I., et al. (eds.) Advances in Neural Information Processing Systems 30, pp. 4088–4098. Curran Associates, Inc. (2017). http://papers.nips.cc/paper/6997-triple-generative-adversarial-nets.pdf
22. Li, Y., Liu, S., Yang, J., Yang, M.H.: Generative face completion. In: The IEEE Conference on Computer Vision and Pattern Recognition (CVPR), July 2017
23. Luo, Y., Zhu, J., Li, M., Ren, Y., Zhang, B.: Smooth neighbors on teacher graphs for semi-supervised learning. In: The IEEE Conference on Computer Vision and Pattern Recognition (CVPR), June 2018
24. Maas, A.L., Hannun, A.Y., Ng, A.Y.: Rectifier nonlinearities improve neural network acoustic models. In: ICML Workshop on Deep Learning for Audio, Speech and Language Processing (2013)
25. Miyato, T., Maeda, S.i., Koyama, M., Ishii, S.: Virtual adversarial training: a regularization method for supervised and semi-supervised learning. IEEE Trans. Pattern Anal. Mach. Intell. **41**(8), 1979–1993 (2018)
26. Miyato, T., Maeda, S.i., Koyama, M., Nakae, K., Ishii, S.: Distributional smoothing with virtual adversarial training. In: International Conference on Learning Representations (2016)

27. Papamakarios, G., Pavlakou, T., Murray, I.: Masked autoregressive flow for density estimation. In: Guyon, I., et al. (eds.) Advances in Neural Information Processing Systems 30, pp. 2338–2347. Curran Associates, Inc. (2017). http://papers.nips.cc/paper/6828-masked-autoregressive-flow-for-density-estimation.pdf
28. Park, S., Park, J., Shin, S.J., Moon, I.C.: Adversarial dropout for supervised and semi-supervised learning. In: Thirty-Second AAAI Conference on Artificial Intelligence (2018)
29. Radford, A., Metz, L., Chintala, S.: Unsupervised representation learning with deep convolutional generative adversarial networks. In: International Conference on Learning Representations (2016)
30. Rezende, D., Mohamed, S.: Variational inference with normalizing flows. In: Bach, F., Blei, D. (eds.) Proceedings of the 32nd International Conference on Machine Learning. Proceedings of Machine Learning Research, vol. 37, pp. 1530–1538. PMLR, Lille, France, 07–09 July 2015. http://proceedings.mlr.press/v37/rezende15.html
31. Sajjadi, M., Javanmardi, M., Tasdizen, T.: Regularization with stochastic transformations and perturbations for deep semi-supervised learning. In: Lee, D.D., Sugiyama, M., Luxburg, U.V., Guyon, I., Garnett, R. (eds.) Advances in Neural Information Processing Systems 29, pp. 1163–1171. Curran Associates, Inc. (2016). http://papers.nips.cc/paper/6333-regularization-with-stochastic-transformations-and-perturbations-for-deep-semi-supervised-learning.pdf
32. Salimans, T., Goodfellow, I., Zaremba, W., Cheung, V., Radford, A., Chen, X.: Improved techniques for training GANs. In: Advances in Neural information Processing Systems, pp. 2234–2242 (2016)
33. Springenberg, J.T.: Unsupervised and semi-supervised learning with categorical generative adversarial networks. In: International Conference on Learning Representations (2016)
34. Tarvainen, A., Valpola, H.: Mean teachers are better role models: Weight-averaged consistency targets improve semi-supervised deep learning results. In: Guyon, I., et al. (eds.) Advances in Neural Information Processing Systems 30, pp. 1195–1204. Curran Associates, Inc. (2017). http://papers.nips.cc/paper/6719-mean-teachers-are-better-role-models-weight-averaged-consistency-targets-improve-semi-supervised-deep-learning-results.pdf
35. Vincent, P., Larochelle, H., Lajoie, I., Bengio, Y., Manzagol, P.A.: Stacked denoising autoencoders: Learning useful representations in a deep network with a local denoising criterion. J. Mach. Learn. Res. **11**, 3371–3408 (2010)
36. Xie, Q., Dai, Z., Hovy, E., Luong, M.T., Le, Q.V.: Unsupervised data augmentation for consistency training (2019)
37. Zhou, D., Bousquet, O., Lal, T.N., Weston, J., Schölkopf, B.: Learning with local and global consistency. In: Thrun, S., Saul, L.K., Schölkopf, B. (eds.) Advances in Neural Information Processing Systems 16, pp. 321–328. MIT Press (2004). http://papers.nips.cc/paper/2506-learning-with-local-and-global-consistency.pdf

Du^{2}Net: Learning Depth Estimation from Dual-Cameras and Dual-Pixels

Yinda Zhang(✉), Neal Wadhwa, Sergio Orts-Escolano, Christian Häne, Sean Fanello, and Rahul Garg

Google Research, Mountain View, USA
yindaz@google.com

Abstract. Computational stereo has reached a high level of accuracy, but degrades in the presence of occlusions, repeated textures, and correspondence errors along edges. We present a novel approach based on neural networks for depth estimation that combines stereo from dual cameras with stereo from a dual-pixel sensor, which is increasingly common on consumer cameras. Our network uses a novel architecture to fuse these two sources of information and can overcome the above-mentioned limitations of pure binocular stereo matching. Our method provides a dense depth map with sharp edges, which is crucial for computational photography applications like synthetic shallow-depth-of-field or 3D Photos. Additionally, we avoid the inherent ambiguity due to the aperture problem in stereo cameras by designing the stereo baseline to be orthogonal to the dual-pixel baseline. We present experiments and comparisons with state-of-the-art approaches to show that our method offers a substantial improvement over previous works.

Keywords: Dual-pixels · Stereo matching · Depth estimation · Computational photography

1 Introduction

Despite their maturity, modern stereo depth estimation techniques still suffer from artifacts in occluded areas, around object boundaries and in regions containing edges parallel to the baseline (the so-called aperture problem). These errors are especially problematic for applications requiring a depth map that is accurate near object boundaries, such as synthetic shallow depth-of-field or 3D photos.

Electronic supplementary material The online version of this chapter (https://doi.org/10.1007/978-3-030-58452-8_34) contains supplementary material, which is available to authorized users.

A. Vedaldi et al. (Eds.): ECCV 2020, LNCS 12346, pp. 582–598, 2020.
https://doi.org/10.1007/978-3-030-58452-8_34

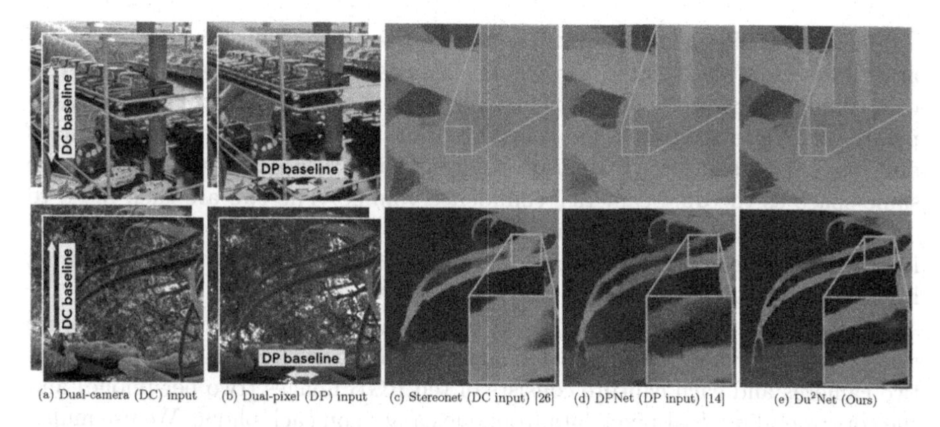

Fig. 1. Du²Net combines dual-camera (DC) and dual-pixel (DP) images to produce edge-aware disparities with high precision even near occlusion boundaries. The large vertical DC baseline complements the small horizontal DP baseline to mitigate the aperture problem (top) and occlusions (bottom).

While these problems can be mitigated by using more than two cameras, a recent improvement to consumer camera sensors allows us to alleviate them without any extra hardware. Specifically, camera manufacturers have added dual-pixel (DP) sensors to DSLR and smartphone cameras to assist with focusing. These sensors work by capturing two views of a scene through the camera's single lens, thereby creating a tiny baseline binocular stereo pair (Fig. 2). Recent work has shown that it is possible to estimate depth from these dense dual-pixel sensors [14,47]. Due to the tiny baseline, there are fewer occluded areas between the views, and as a result the depth from dual-pixels is more accurate near object boundaries than the depth from binocular stereo. However, the tiny baseline also means that the depth quality is worse than stereo at farther distances due to the quadratic increase in depth error in triangulation-based systems [45].

In this work, we consider a dual-camera (DC) system where one camera has a dual-pixel sensor, a common setup on recently-released flagship smartphones. We propose a deep learning solution to estimate depth from both dual-pixels and dual-cameras. Because depth from dual-cameras and depth from dual-pixels have complementary errors, such a setup promises to have accurate depth at both near and far distances and around object boundaries. In addition, in our setup, the dual-pixel baseline is orthogonal to the dual-camera baseline. This allows us to estimate depth even in regions where image texture is parallel to one of the two baselines (Fig. 1). This is usually difficult due to the well known aperture problem [35].

One key problem that prevents the trivial solution of multi-view stereo matching from working is a fundamental affine ambiguity in the depth estimated from dual-pixels [14]. This is because disparity is related to inverse depth via an affine transformation that depends on the camera's focus distance, focal length and aperture size, which are often unknown or inaccurately recorded.

To address this issue, we propose an end-to-end solution that uses two separate low resolution learned *confidence volumes*, i.e. the softmax of a negative cost volume, to compute disparity maps from dual-cameras and dual-pixels independently. We then fit an affine transformation between the two disparity maps and use it to resample the dual-pixels' confidence volume, so that it is in the same space as the dual-cameras' confidence volume. The two are then fused to estimate a low resolution disparity map. A final edge-aware refinement [26] that leverages features computed from dual-pixels is then used to obtain the final high resolution disparity map.

To train and evaluate our approach, we capture a new dataset using a capture rig containing five synchronized Google Pixel 4 smartphones. Each phone has two cameras and each capture consists of ten RGB images (two per phone) and the corresponding dual-pixel data from one camera on each phone. We use multi-view stereo techniques to estimate ground truth disparity using all ten views. We plan to release this dataset.

Via extensive experiments and comparisons with state-of-the-art approaches, we show that our solution effectively leverages both dual-cameras and dual-pixels. We additionally show applications in computational photography where precise edges and dense disparity maps are the key for compelling results.

2 Related Work

Stereo matching is a fundamental problem in computer vision and is often used in triangulation systems to estimate depth for various applications such as computational photography [47], autonomous driving [34], robotics [11], augmented and virtual reality [38] and volumetric capture [16]. Traditionally, stereo matching pipelines [42] follow these main steps: matching cost computation, cost aggregation, and disparity optimization, often followed by a disparity refinement (post-processing) step. This problem has been studied for over four decades [31] and we refer the reader to [17,42,43] for a survey of traditional techniques.

Recent classical approaches aim at improving the disparity correspondence search by using either global [3,12,28,29] or local [4,9,46] optimization schemes. These methods usually rely on hand-crafted descriptors [4,46] or learned shallow binary features [9,10] followed by sequential propagation steps [4,10] or fast parallel approximated CRF inference [9,46]. However, these methods cannot compete with recent deep learning-based methods that use end-to-end training [5,6,8,25,26,30,44,48]. Such approaches were introduced by [23,32], who used encoder-decoder networks for the problems of disparity and flow estimation.

Kendall et al. [25], inspired by classical methods, employed a model architecture that constructs a full cost-volume with 3D convolutions as an intermediate stage and infers the final disparity through a soft-arg min function. Khamis et al. [26] extended this concept by using a learned edge aware refinement step as the final stage of the model to reduce computational cost. More recently, PSMNet [6] used a multi-scale pooling approach to improve the accuracy of the predicted disparities. Finally [49], inspired by [21], used a semi-global matching approach to replace the expensive 3D convolutions.

Other end-to-end approaches use multiple iterative refinements to converge to a final disparity solution. Gidaris et al. [15] propose a generic architecture for labeling problems, such as depth estimation, that is trained end-to-end to predict and refine the output. Pang et al. [39] propose a cascaded approach to learn the depth residual from an initial estimate. Despite this progress, stereo depth estimation systems still suffer from limited precision in occlusion boundaries, imprecise edges, errors in areas with repeated textures, and the aperture problem. The aperture problem can be addressed with two orthogonal dual camera pairs [33]. Occlusions can be reduced by using trinocular stereo [36]. Both of these approaches require additional hardware and more complex calibration. In this work, we combine bincocular stereo with a dual-pixel sensor, a hardware available in most modern smartphone and DSLR cameras where they are used for autofocus.

Recently, a handful of techniques have been proposed to recover depth from a single camera using dual-pixels [14,24,40,47]. Dual-pixels are essentially a two-view light field [37], providing two slightly different views of the scene. These two views can be approximated as a stereo pair except for a fundamental ambiguity identified by [14] discussed in the introduction. In addition to depth estimation, dual-pixels have been used for dereflection [41].

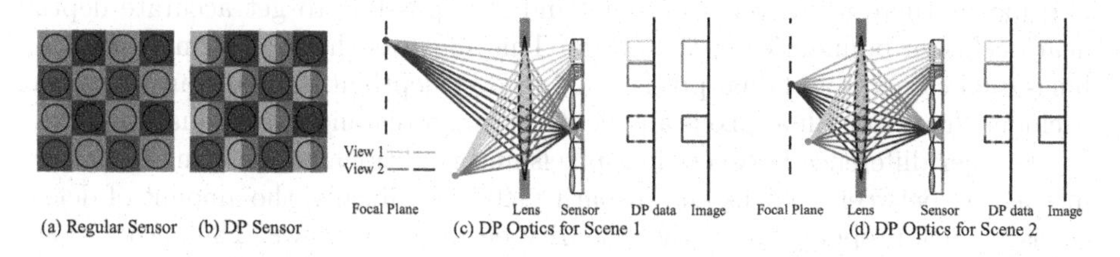

Fig. 2. In a regular Bayer sensor, each pixel has a microlens on top to collect more light (a). Dual-pixel sensors split some of the pixels underneath the microlens into two halves; the green pixels in (b). The two dual-pixel views get their light from different halves of the aperture, resulting in a slight depth-dependent disparity between the views (c). Different scenes can produce the same dual-pixel images if the focus distance changes ((c) vs. (d)). This is a fundamental ambiguity of dual-pixel sensors. (Reproduced with permission from Garg et al. [14].)

3 Dual-Pixel Sensors

Dual-pixel sensors work by splitting each pixel in half, such that the left half integrates light over the right half of the aperture and the right half integrates light over the left half of the aperture. Because the two half pixels see light from different halves of the aperture, they form a kind of "stereo pair", whose centers of projection are in the centers of each half aperture. Since the two half pixel images account for all the light going through the aperture, when they are added together, the full normal image is recovered.

These sensors are becoming increasingly common in smartphone and DSLR cameras because they assist in auto-focus. The reason for this is that the zero-disparity distance corresponds exactly to the image being in-focus (e.g. the blue point in Fig. 2(c)) and disparity is exactly proportional to how much the lens needs to be moved to make the image in-focus. This property also implies that unlike rectified stereo image pairs, the range of disparities can be both negative and positive for DP data.

In addition to this dependence on focus distance, dual-pixels have a number of other key differences from stereo cameras. On the positive side, the dual-pixel views are perfectly synchronized and have the same white balance, exposure and focus, making matching easier. In addition, they are perfectly rectified. This means that the baseline is perfectly horizontal for a sensor whose dual-pixels are split horizontally as in Fig. 2(b). Another advantage of this perfect rectification is that, dual-pixel sensors are not affected by rolling shutter or optical image stabilization [7], which shifts the principal point and center of projection of a camera. While we need to calibrate for this with stereo cameras, it does not cause problems for dual-pixel images.

Like in a stereo pair, the small baseline of dual-pixel images means that depth estimation at large distances is difficult. Please see the supplementary material for a visualization of the tiny parallax between the DP images. However, it also means there are fewer occlusions and it is possible to get accurate depth near occlusion boundaries in the image. This suggests that a system that combines dual-cameras and dual-pixels could recover depth at short distances and in occluded areas from dual-pixels and depth at larger distances from dual-cameras.

Another difference between dual-pixels and traditional stereo cameras is the interaction between defocus and disparity [40]. Specifically, the amount of defocus is exactly proportional to the disparity between the views. This means that a learned model that makes use of dual-pixels could make use of defocus as well to resolve ambiguities that typically fool matching-based approaches, such as repeated textures.

Finally, we elaborate on the affine ambiguity of depth predictions discussed in the introduction (Fig. 2(c–d)). This happens because the mapping between disparity and depth depends on focus distance which is often inaccurate or unknown in cheap smartphone camera modules. Garg et al. [14] used the paraxial and thin-lens approximation to show that $D_{DP}(x,y) = \alpha + \beta/Z(x,y)$, where $Z(x,y)$ is the depth for pixel x, y; $D_{DP}(x,y)$ is the dual-pixel disparity at (x,y), and α and β are constants that depend on the aperture, point spread function, and the focus distance of the lens. Because these can be difficult to determine, inverse depth can be estimated only up to an unknown affine transform.

If there is a second camera in addition to the camera with dual-pixels, such as in our setup, the stereo disparity D_{DC} of the dual-cameras is bf/Z [45] where b is the baseline and f is the focal length. From this, it follows that D_{DC} and D_{DP} are also related via an affine transform

$$D_{DC}(x,y) = \alpha' + \beta' D_{DP}(x,y) \tag{1}$$

We use this observation in our network architecture to effectively integrate stereo and dual-pixel cues. Note that no further rectification is required between dual-pixel and stereo input since the dual-pixels naturally align with one of the camera in stereo.

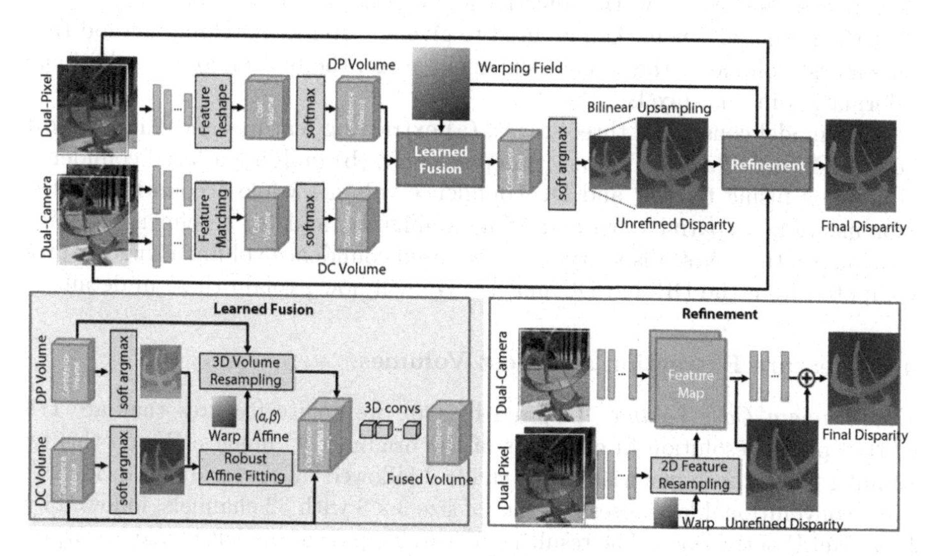

Fig. 3. Overview of Du^{2}Net. Top: two disparity maps are separately inferred from the dual-camera and the dual-pixel branches. An affine transformation is fit between them and used to resample the dual-pixel confidence volume. It is then fused with the dual-camera volume and they are together used to infer the unrefined disparity. An edge-aware refinement step uses the dual-pixel features to predict the final disparity. Bottom: details of the volume fusion and refinement step. See text for more details.

4 Fusing Dual-Pixels and Dual-Cameras

We describe our deep learning model to predict disparity from both dual-camera and dual-pixel data (Fig. 3). The input to our system is a pair of rectified dual-camera (DC) images, I_l and I_r corresponding to the left and the right cameras, and a pair of dual-pixel (DP) images from the right camera sensor, I_t^{DP} and I_b^{DP} corresponding to the top and bottom half-pixels on the sensor.

At capture time, the DP images are perfectly aligned with the right DC image. However, after stereo rectification, the left and right dual-camera images are respectively warped by spatial homography transformations $\mathbf{W}_l(x, y)$ and $\mathbf{W}_r(x, y)$, which remaps every pixel to new coordinates in both images. As a result, the right image is no longer aligned with the dual-pixel images. In addition, as explained in Sect. 3, the two disparity maps coming from DC and DP respectively, are related via an affine transformation. Our method takes both of these issues into account when fusing information from the two sources.

Our model uses two building blocks from other state-of-the-art stereo matching architectures. Specifically, a cost volume [6,25,26] and refinement stages [26,39]. Our main contributions are a method to fuse the confidence volumes (the softmax of the negative cost volume) computed from dual-cameras and dual-pixels, and to show the effectiveness of dual-pixels for refinement. Note that the proposed scheme can be used to give any stereo matching method that uses a cost-volume or that has a refinement stage, the benefits of the additional information in dual-pixels.

Our model consists of three stages, (a) extracting features and building cost volumes from DP and DC inputs independently, (b) building a fused confidence volume by fusing the DP and DC confidence volumes while accounting for the aforementioned spatial warp and affine ambiguity, and (c) a refinement stage that refines the coarse disparity from the fused confidence volume using features computed from the DP and DC images. We will now explain these in detail.

4.1 Feature Extraction and Cost Volumes

Dual-Camera Cost Volume. Inspired by [26], we create features that are 1/8 of the spatial resolution in each axis of the original DC images. We do this by running three 2D convolutions with stride 2 followed by six residual blocks [19]. Each convolutional layer uses a kernel of size 3×3 with 32 channels, followed by leaky ReLU activations. The resulting feature map from the left image is warped to the right feature map using multiple disparity hypotheses $d \in [0, 16]$. For each hypothesis, we calculate the distance between the two feature maps. Unlike [26], we use the ℓ_1 distance instead of subtracting the features since this increases the stability of training.

Dual-Pixel Cost Volume. While we can align the dual-camera features from the two cameras by warping them according to disparity, we cannot do the same for dual-pixel images due to the interaction between dual-pixel disparity and defocus [40]. The two DP images will not align even after warping due to the different defocus blurs. Hence, instead of warping and subtracting the dual-pixel features, we concatenate the two DP images and implicitly produce a cost volume. This also allows the model to use the depth cues from defocus blur since the amount of defocus is proportional to depth. To do so, we feed the DP images into a 2D network with six residual blocks. Each layer consists of convolutions of size 3×3 with 32 channels followed by leaky ReLU activations. A 2D convolution is attached to the end to produce a feature map with N_d channels, where $N_d = 17$ is the number of desired disparity hypotheses. We then reshape the feature map and convert it to a 3D volume by expanding the final dimension.

4.2 Fused Confidence Volume

It is not straightforward to merge the DP and DC cost volumes due to the rectification warp $\mathbf{W}_r$ between the DP and DC images and the affine ambiguity in DP disparity (Sect. 3). In addition, the costs in the two volumes may be scaled differently since they are predicted from different network layers.

We first normalize the two cost volumes to the range $[0, 1]$ by applying soft-max to the negative cost volume along the disparity dimension. We call the resulting tensors confidence volumes. To handle the affine ambiguity, we first predict disparity maps D_{DP} and D_{DC} from the two confidence volumes using a soft–arg max operator [25], and fit an affine transformation between the two by solving a Tikhonov-regularized least squares problem that biases the solution to be close to $\alpha = 0, \beta = 1$:

$$\hat{\alpha}, \hat{\beta} = \underset{\alpha,\beta}{argmin} \left\|(\alpha + \beta \cdot D_{DP}) - D_{DC}\right\|^2 + \gamma \left\|\beta - 1\right\|^2 + \gamma \left\|\alpha\right\|^2, \quad (2)$$

where $\gamma = 0.1$ is a regularization constant. Then, we use the known rectification warp $\mathbf{W}_r(x, y) = [W_r^x(x, y), W_r^y(x, y)]$ and the estimated affine transformation to warp the DP confidence volume into the DC space:

$$C_{warp}^{DP}(x, y, z) = C^{DP}\left(W_r^x(x, y), W_r^y(x, y), (z - \hat{\alpha})/\hat{\beta}\right), \quad (3)$$

where C^{DP} is the confidence volume built from DP images. The 3D warping uses differentiable bilinear interpolation and zero padding.

The warped DP confidence volume C_{warp}^{DP} is now aligned with the DC confidence volume. We stack these together to form a 4D tensor that is fused into a 3D confidence volume by a shallow network consisting of three layers of 3D convolutions with leaky ReLU activations and a softmax at the end. Finally, the fused volume is converted into a disparity map D_{unref} using a soft–arg max.

4.3 Disparity Refinement

The next step is to refine the low resolution disparity D_{unref}. Khamis et al. [26] use the RGB image as the guide image to upsample the disparity while applying a learned residual to improve edges and minimize the final error. A straightforward extension of [26] is to warp the DP images using the rectification warp $\mathbf{W}_r$ and use them along with the RGB image as the guide image. However, we find that this yields inferior results compared to our method, presumably because warping and resampling makes it harder to extract disparity cues from DP images.

Instead, we extract features from the input DP images and then warp them using $\mathbf{W}_r$. This way, the model can easily extract disparity cues from the perfectly rectified DP image pair. The warped features are concatenated with the features extracted from the right RGB image and the unrefined disparity D_{unref}. These are fed into six residual blocks, with 3×3 convolutions followed by a leaky ReLU activation, to predict a residual R. The final output is set to $D_{ref} = D_{unref} + R$.

4.4 Loss Function

To train our network we use a weighted Huber loss [22]:

$$L(D) = \frac{\sum_{\mathbf{p}} \mathcal{H}(D(\mathbf{p}) - D^{gt}(\mathbf{p}), \delta) \cdot C^{gt}(\mathbf{p})}{\sum_{\mathbf{p}} C^{gt}(\mathbf{p})}, \tag{4}$$

where D^{gt} is the ground truth disparity, C^{gt} is the per-pixel confidence of the ground truth, and δ is the switching point between the quadratic and the linear function, which is set to 1 for disparity in range $[0, 128]$. The overall loss is a weighted sum of four terms:

$$L_{total} = \lambda_{DP} L(\hat{\alpha} + \hat{\beta} \cdot D_{DP}) + \lambda_{DC} L(D_{DC}) + \lambda_{unref} L(D_{unref}) + \lambda_{ref} L(D_{ref}), \tag{5}$$

where λ_{DC} is set to 10 and the other weights are set to 1.

5 Evaluation

In this section, we perform extensive experiments to evaluate our model. We conduct an ablation study to show the effectiveness of our design choices. We also compare to other stereo and dual-pixel methods. We focus our experiments on thin structures, edges and occlusion boundaries, to show the effectiveness of the complementary information coming from DP and DC data. For quantitative evaluations we report MAE, RMSE and the *bad* δ metric, i.e. the percentage of pixels with disparity error greater than δ. These are weighted by the ground-truth confidence. See the supplementary material for details.

5.1 Data Collection

We collect a new data set using the Google Pixel 4 smartphone, which has a dual camera system consisting of a main camera with a dual-pixel sensor and a regular telephoto camera. We refer to the main camera as the right camera and the telephoto camera as the left camera. We use a data acquisition set up similar to [14], i.e., a capture rig consisting of 5 phones (Fig. 4a) synchronized with [2]. Structure from motion [18] and multi-view stereo techniques are used to generate depth maps (Supplementary Material Sec. 1.2). Similar to [14], we also compute a per-pixel confidence for the depth by checking for depth coherence with neighboring views. The center phone in the rig is used for training and evaluation since its depth quality is higher than other views especially for occluded regions.

We rectify the stereo images from the center phone using estimated camera poses [13]. The estimated depth is converted to disparity and then rectified along with confidences to yield D_l^{gt}, C_l^{gt} and D_r^{gt}, C_r^{gt} for the left and right images respectively. As described in Sect. 4, dual-pixel images from the main (right) camera are not rectified or warped. Instead, we store the warp map $\mathbf{W}_r$ that is needed to warp and align the DP images with the rectified right camera image

(a) Capture Rig (b) DP Views (c) Right View (d) Left View (e) GT (f) GT (Occ.)

Fig. 4. Our capture rig (a) similar to [14] but with phones that can capture both dual-pixel (b) and dual-camera (c, d) data. The left and right views are rectified, and the ground truth disparity (e) corresponding to the right view is computed using multi-view stereo techniques on all 10 views captured by the rig. Low confidence depth samples are rendered in black. The multitude of views ensures that we have good quality depth in regions that are occluded in the left view. (f) shows the GT depth masked to regions that are occluded in the left view.

Table 1. Ablation Study. Left: We compare different ways of fusing DP with the DC confidence volume. '(2D)' indicates fusion of the 2D disparity maps extracted from the two confidence volumes. '(C)' indicates fusing cost volumes instead of confidence volumes. Right: We compare the different ways of using DP to refine the best unrefined disparity from the left and show evaluation on the final disparity. '(I)' indicates that input DP images are warped before computing features for refinement.

DP in Confidence volume					DP in refinement				
Conf. volume	MAE	RMSE	$\delta>2$	$\delta>3$	Refinement	MAE	RMSE	$\delta>2$	$\delta>3$
DC	1.023	2.502	10.65	6.32	RGB	0.838	2.197	7.74	4.55
DP+DC (2D)	0.969	2.423	9.72	5.79	DP (I)	0.835	2.173	7.75	4.54
DP+DC (C)	0.964	2.372	9.79	5.80	DP	0.829	2.184	7.51	4.45
DP+DC (Ours)	**0.889**	**2.263**	**8.78**	**5.18**	RGB+DP (Ours)	**0.802**	**2.147**	**7.17**	**4.25**

(see Sect. 4). In addition, to evaluate the quality of the estimated disparity in regions that are visible in only one of the cameras in the stereo pair, we also compute C_i^{occ} for $i \in l, r$, i.e., a per-pixel confidence indicating that the ground-truth disparity is correct but the pixel is occluded in the other camera. In total, we collect 3308 training examples and 1077 testing examples. Please refer to supplementary materials for details of the calculation of C_i^{occ}, data collection and ground truth calculation.

5.2 Training Scheme

We use Tensorflow [1] to implement the network and train using Adam [27] for 2 million iterations with a batch size of 1. The learning rate is set to 3×10^{-5} and then reduced to 3×10^{-6} after 1.5 million iterations. Training takes roughly 16 hours using 8 Tesla V100 GPUs. Inputs to the network I_l, I_r, and $\mathbf{W_r}$ are resized to match the resolution of the predicted and the ground truth disparity, i.e., 448×560. DP images I_t^{DP} and I_b^{DP} are of size 1000×1250.

5.3 Ablation Study

We evaluate the effect of each component of the model. In particular we focus on the impact of dual-pixels on the fused volume and the refinement stage. We provide quantitative comparisons (Table 1) and qualitative comparisons (Fig. 5).

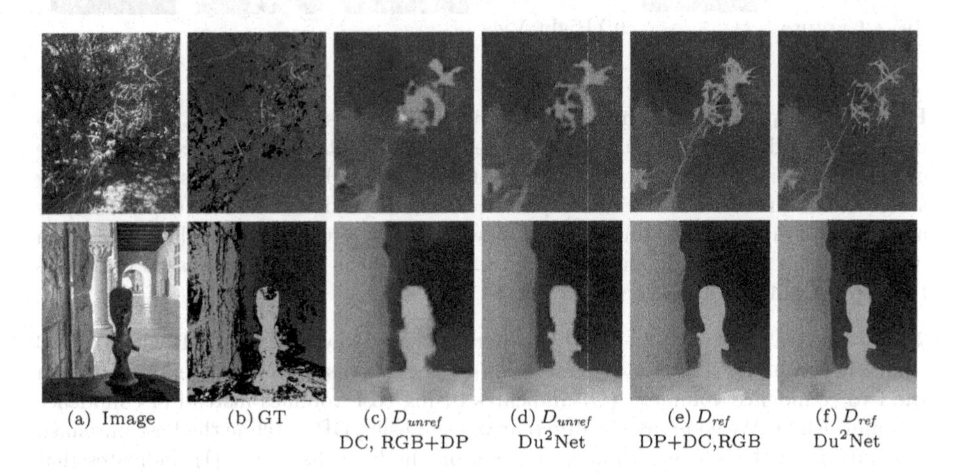

Fig. 5. Ablations of our method. The right camera image (a), ground truth disparity (b) with low confidence disparity in black, D_{unref} (c) from an ablation where only the DC input is used for the confidence volume, D_{unref} (d) from Du^{2}Net, D_{ref} (e) from an ablation where only the RGB image is used for refinement, and D_{ref} (f) from Du^{2}Net. DP input is useful for both the confidence volume and refinement stages to recover accurate depth for fine structures and occluded regions.

Dual-Pixels in the Confidence Volume. Table 1 (left) shows the error of the unrefined disparity D_{unref} using different fusing strategies for the cost volume. Our method of merging the DP and DC volumes (DP+DC) significantly outperforms using only the DC cost volume. We also compare to fusing the 2D disparity maps instead of the the 3D confidence volumes. Specifically, we concatenate D_{DP} and D_{DC} after the affine transformation and use a 2D neural network with six residual blocks to predict D_{unref}. This is worse than our method according to all metrics (DP+DC (2D) vs DP+DC). Finally, we evaluate fusing cost volumes instead of confidence volumes. This, DP+DC (C), is also inferior to DP+DC.

Qualitative comparisons are shown in Fig. 5(c) and (d). Compared to a DC only cost volume (c), fusing DP into the cost volume (d) adds more details to the unrefined disparity and prevents errors at object boundaries. This is critical for getting high quality disparities since the refinement is only able to make local adjustments to the disparity and cannot fix large errors.

Table 2. Quantitative comparisons to the state-of-the-art. Note how the proposed approach substantially outperforms all the competitors in all and occluded regions. '*' indicates a final affine transformation applied to the output disparity and the best results are highlighted separately (see text for details).

Method	Input	All Pixels				Occluded Pixels			
		MAE	RMSE	$\delta>2$	$\delta>3$	MAE	RMSE	$\delta>2$	$\delta>3$
GA-Net [49]	DC	1.001	2.425	9.31	6.09	6.068	8.386	69.81	60.07
PSM-Net [6]	DC	0.815	2.289	7.98	4.97	2.799	5.188	34.73	26.78
StereoNet [26]	DC	0.935	2.432	9.07	5.41	3.123	5.632	38.13	28.49
Du^{2}Net(ours)	DC + DP	**0.802**	**2.147**	**7.17**	**4.25**	**2.396**	**4.543**	**30.62**	**21.91**
DPNet* [14]	DP	1.090	1.989	12.60	5.80	2.594	**4.307**	38.14	26.18
Du^{2}Net* (ours)	DC + DP	**0.746**	**1.825**	**6.63**	**3.63**	**2.352**	4.373	**32.65**	**21.77**

Dual-Pixels in Refinement. We show that our method of using dual-pixels in the refinement stage is better than several baselines (Table 1 (right)). We use the fused DP + DC confidence volume for all cases and extract RGB and DP features for refinement using networks with the same capacity for a fair comparison. Using DP for refinement is better than just using the right RGB image. However, results are best when both are used (RGB+DP). Notably, if the DP is warped before feature extraction (instead of after), performance (DP (I) in Table 1) is not better than using only the RGB image. This suggests that DP cues are not effective after warping, and it is important to extract features before warping.

Qualitative comparisons are shown in Fig. 5(e) and (f). While both methods use DP and DC to compute the unrefined disparity, (e) uses only the right RGB image for refinement, and (f) uses RGB and DP for refinement. Even though DP increases the quality of the unrefined disparity, using it during refinement further improves depth quality at thin structures and near object boundaries. This indicates that it is important to use DP for both the cost volume and the refinement to achieve the best performance.

5.4 Comparison to State-of-the-Art Methods

We compare to other stereo and dual-pixel depth estimation methods (Table 2). All methods are trained on our data using code provided by the authors. We modified the original loss functions to use the confidence maps C_r^{gt} from our dataset in the same way that these are used in our method (as a per pixel weight). Additionally, for the stereo methods we compare to, we set the maximum disparity range to 128. The baseline methods were trained until convergence on the test set error. We used the hyper-parameters and training strategies (including annealing schedules for learning rates) provided in the implementations.

For stereo baselines, we compare to StereoNet [26] which is similar to our model with only the DC input, PSMNet [6] and GANet [49]. PSMNet [6] uses multi-scale feature extraction and cost volumes, and GANet [49] uses a sophisticated semi-global aggregation [49]. In Table 2, we report quantitative results. Our model uses a low-resolution cost volume and a refinement stage, with a run-time

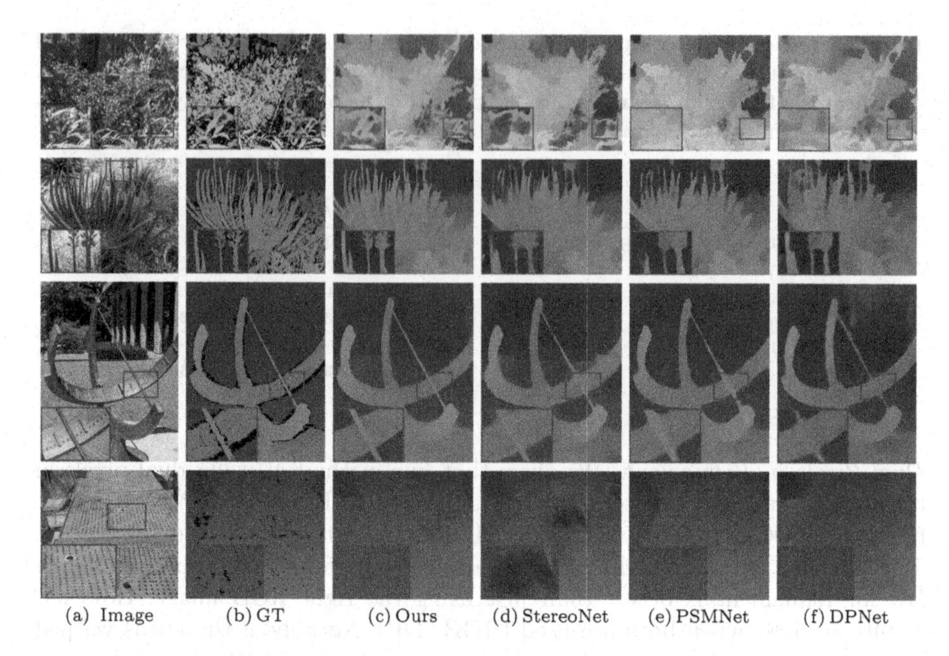

Fig. 6. Qualitative comparison to the state-of-the-art. Right camera image (a) from our test set, ground truth disparity with low confidence disparity in black (b), and results from our method (c), stereo only methods StereoNet [26] (d) and PSMNet [6] (e), and DP only method DPNet [14] (f). Stereo only methods fail for vertical structures due to the aperture problem, e.g., in the second image. They also fail in regions with fine structures and occlusions, e.g., in the first three images. StereoNet fails on the last image potentially due to repeated texture. DPNet's accuracy falls quickly with distance due to the small dual-pixel baseline. Our method overcomes these problems by fusing the two cues.

comparable with StereoNet, while achieving accuracy higher than more computationally expensive models, such as PSMNet (which uses *25* 3D convolutional layers as opposed to our 8) consistently over all the evaluation metrics.

For DP input baselines, DPNet [14] predicts disparity *up to an unknown affine transformation*. To handle this, like [14], we find the best fit (according to MSE) affine transformation between the prediction and the ground truth and transform the prediction to compute the metrics. For a fair comparison, we apply the same post processing to our method (Du^{2}Net* in Table 2), showing it consistently outperforms DPNet.

As mentioned in Sect. 5.1, we also compute an occlusion mask C_r^{occ} and evaluate the methods only on these pixels. The results are reported in Table 2 under "Occluded pixels". Our method outperforms the other competitors by a substantial margin in those areas showing the advantage of small baseline DP data.

Qualitative comparisons are provided in Fig. 6. Note how we better capture fine details and small structures, while correctly inferring disparity near occlusion boundaries. The orthogonal baselines of the dual-pixels and dual-cameras also helps mitigate the aperture problem and issues due to repeated textures. Additional results are available in the supplementary material.

5.5 Applications in Computational Photography

Predicting accurate disparities, hence depth, is crucial for many applications in computational photography. These applications usually require accurate depth for fine structures and near occlusion boundaries. Figure 7 and 8 show how our more accurate depth leads to fewer artifacts when used to produce synthetic shallow depth-of-field images and 3D photos [20] respectively.

Fig. 7. Synthetic shallow depth-of-field results for different methods. Top: Accurate depth near occlusion boundaries is critical for avoiding artifacts near the subject boundary. Bottom: DPNet [14] is unable to resolve the small depth difference between the flower and the twigs in the background. As a result, parts of the background are incorrectly sharp.

6 Discussion

We presented the first method to combine dual-camera and dual-pixel data. The inherent affine ambiguity of disparity computed from dual-pixel images prevents a straightforward integration of the two modalities. Therefore, we proposed a novel solution that resamples the confidence volume computed from dual-pixels and concatenates it with the dual-camera volume. A refinement stage leverages dual-pixels to infer the final disparity map. We show the effectiveness of the proposed solution with experiments, comparisons to the state-of-the-art and applications. Our dataset will be released publicly and we hope it can advance the field. While the orthogonality of the baselines allows us to avoid the aperture problem, our method doesn't work on textureless regions. Perhaps this could be

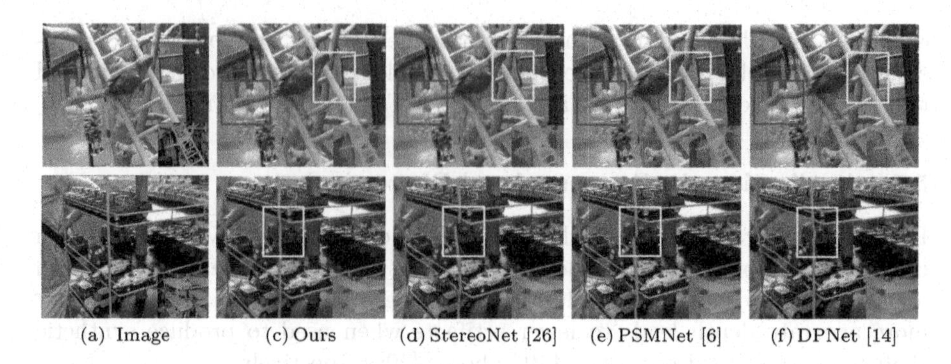

Fig. 8. 3D photos results [20]. Novel views of the scene are rendered by warping the image according to the estimated depth to new camera positions. Depth errors lead to unnatural distortion of rigid scene structures in the novel views.

handled by combining information from additional modalities like active depth sensors. Another interesting direction for future work would be to consider dual-camera pairs where both cameras have dual-pixels.

Acknowledgements. We thank photographers Michael Milne and Andrew Radin for collecting the data, Jon Barron and Marc Levoy for their comments on the text, Yael Pritch, Sameer Ansari, Christoph Rhemann and Shahram Izadi for the help, useful discussions and support for this work.

References

1. Abadi, M., Agarwal, A., Barham, P., Brevdo, E., et al.: TensorFlow: large-scale machine learning on heterogeneous systems (2015). https://www.tensorflow.org/
2. Ansari, S., Wadhwa, N., Garg, R., Chen, J.: Wireless software synchronization of multiple distributed cameras. In: ICCP (2019)
3. Besse, F., Rother, C., Fitzgibbon, A., Kautz, J.: PMBP: patchmatch belief propagation for correspondence field estimation. IJCV **110**(1), 2–13 (2014)
4. Bleyer, M., Rhemann, C., Rother, C.: Patchmatch stereo-stereo matching with slanted support windows. In: BMVC (2011)
5. Chabra, R., Straub, J., Sweeney, C., Newcombe, R., Fuchs, H.: StereoDRNet: dilated residual stereonet. In: CVPR (2019)
6. Chang, J., Chen, Y.: Pyramid stereo matching network. In: CVPR (2018)
7. DiVerdi, S., Barron, J.T.: Geometric calibration for mobile, stereo, autofocus cameras. In: WACV (2016)
8. Duggal, S., Wang, S., Ma, W.C., Hu, R., Urtasun, R.: DeepPruner: learning efficient stereo matching via differentiable patchmatch. In: ICCV (2019)
9. Fanello, S.R., et al.: Low compute and fully parallel computer vision with hashmatch. In: ICCV (2017)
10. Fanello, S.R., et al.: UltraStereo: efficient learning-based matching for active stereo systems. In: CVPR (2017)
11. Fanello, S., et al.: 3D stereo estimation and fully automated learning of eye-hand coordination in humanoid robots. In: Humanoids (2014)

12. Felzenszwalb, P.F., Huttenlocher, D.P.: Efficient belief propagation for early vision. IJCV **19**, 57–92 (2006)
13. Fusiello, A., Trucco, E., Verri, A.: A compact algorithm for rectification of stereo pairs. Mach. Vis. Appl. **12**, 16–22 (2000)
14. Garg, R., Wadhwa, N., Ansari, S., Barron, J.T.: Learning single camera depth estimation using dual-pixels. In: ICCV (2019)
15. Gidaris, S., Komodakis, N.: Detect, replace, refine: deep structured prediction for pixel wise labeling. In: CVPR (2017)
16. Guo, K., et al.: The relightables: volumetric performance capture of humans with realistic relighting. TOG (2019)
17. Hamzah, R.A., Ibrahim, H.: Literature survey on stereo vision disparity map algorithms. J. Sens. **2016**, 1–23 (2016)
18. Hartley, R., Zisserman, A.: Multiple View Geometry in Computer Vision. Cambridge University Press, Cambridge (2003)
19. He, K., Zhang, X., Ren, S., Sun, J.: Deep residual learning for image recognition. In: CVPR (2016)
20. Hedman, P., Kopf, J.: Instant 3D photography. In: SIGGRAPH (2018)
21. Hirschmuller, H.: Stereo processing by semiglobal matching and mutual information. TPAMI **30**, 328–341 (2008)
22. Huber, P.J.: Robust estimation of a location parameter. In: Kotz, S., Johnson, N.L. (eds.) Breakthroughs in Statistics, pp. 492–518. Springer, New York (1992). https://doi.org/10.1007/978-1-4612-4380-9_35
23. Ilg, E., Mayer, N., Saikia, T., Keuper, M., Dosovitskiy, A., Brox, T.: Flownet 2.0: evolution of optical flow estimation with deep networks. In: CVPR (2017)
24. Jang, J., Park, S., Jo, J., Paik, J.: Depth map generation using a single image sensor with phase masks. Optics express (2016)
25. Kendall, A., et al.: End-to-end learning of geometry and context for deep stereo regression. In: CVPR (2017)
26. Khamis, S., Fanello, S., Rhemann, C., Kowdle, A., Valentin, J., Izadi, S.: StereoNet: guided hierarchical refinement for real-time edge-aware depth prediction. In: Ferrari, V., Hebert, M., Sminchisescu, C., Weiss, Y. (eds.) ECCV 2018. LNCS, vol. 11219, pp. 596–613. Springer, Cham (2018). https://doi.org/10.1007/978-3-030-01267-0_35
27. Kingma, D.P., Ba, J.: Adam: a method for stochastic optimization. In: ICLR (2015)
28. Klaus, A., Sormann, M., Karner, K.: Segment-based stereo matching using belief propagation and a self-adapting dissimilarity measure. In: ICPR (2006)
29. Kolmogorov, V., Zabih, R.: Computing visual correspondence with occlusions using graph cuts. In: ICCV (2001)
30. Liang, Z., et al.: Learning for disparity estimation through feature constancy. In: CVPR (2018)
31. Marr, D., Poggio, T.: Cooperative computation of stereo disparity. Science **194**, 283–287 (1976)
32. Mayer, N., et al.: A large dataset to train convolutional networks for disparity, optical flow, and scene flow estimation. In: CVPR (2016)
33. Meier, L., Honegger, D., Vilhjalmsson, V., Pollefeys, M.: Real-time stereo matching failure prediction and resolution using orthogonal stereo setups. In: ICRA (2017)
34. Menze, M., Geiger, A.: Object scene flow for autonomous vehicles. In: CVPR (2015)
35. Morgan, M., Castet, E.: The aperture problem in stereopsis. Vis. Res. **37**, 2737–2744 (1997)
36. Mulligan, J., Isler, V., Daniilidis, K.: Trinocular stereo: a real-time algorithm and its evaluation. IJCV **47**, 51–61 (2002)

37. Ng, R., Levoy, M., Brédif, M., Duval, G., Horowitz, M., Hanrahan, P.: Light field photography with a hand-held plenoptic camera. Technical report, Stanford University (2005)
38. Orts-Escolano, S., et al.: Holoportation: virtual 3D teleportation in real-time. In: UIST (2016)
39. Pang, J., Sun, W., Ren, J., Yang, C., Yan, Q.: Cascade residual learning: a two-stage convolutional neural network for stereo matching. In: ICCV Workshop (2017)
40. Punnappurath, A., Abuolaim, A., Afifi, M., Brown, M.S.: Modeling defocus-disparity in dual-pixel sensors. In: ICCP (2020)
41. Punnappurath, A., Brown, M.S.: Reflection removal using a dual-pixel sensor. In: CVPR (2019)
42. Scharstein, D., Szeliski, R.: A taxonomy and evaluation of dense two-frame stereo correspondence algorithms. IJCV **47**, 7–42 (2002). https://doi.org/10.1023/A:1014573219977
43. Sinha, S.N., Scharstein, D., Szeliski, R.: Efficient high-resolution stereo matching using local plane sweeps. In: CVPR (2014)
44. Song, X., Zhao, X., Hu, H., Fang, L.: EdgeStereo: a context integrated residual pyramid network for stereo matching. In: ACCV (2018)
45. Szeliski, R.: Computer Vision: Algorithms and Applications, 1st edn. Springer, Heidelberg (2010)
46. Tankovich, V., et al.: Sos: stereo matching in O(1) with slanted support windows. In: IROS (2018)
47. Wadhwa, N., et al.: Synthetic depth-of-field with a single-camera mobile phone. In: SIGGRAPH (2018)
48. Yang, G., Zhao, H., Shi, J., Deng, Z., Jia, J.: SegStereo: exploiting semantic information for disparity estimation. In: Ferrari, V., Hebert, M., Sminchisescu, C., Weiss, Y. (eds.) ECCV 2018. LNCS, vol. 11211, pp. 660–676. Springer, Cham (2018). https://doi.org/10.1007/978-3-030-01234-2_39
49. Zhang, F., Prisacariu, V.A., Yang, R., Torr, P.H.S.: GA-Net: guided aggregation net for end-to-end stereo matching. In: CVPR (2019)

Model-Agnostic Boundary-Adversarial Sampling for Test-Time Generalization in Few-Shot Learning

Jaekyeom Kim(✉), Hyoungseok Kim, and Gunhee Kim

Computer Science and Engineering, Seoul National University, Seoul, Korea
{jaekyeom,harry2636,gunhee}@snu.ac.kr
http://vision.snu.ac.kr/projects/mabas

Abstract. Few-shot learning is an important research problem that tackles one of the greatest challenges of machine learning: learning a new task from a limited amount of labeled data. We propose a model-agnostic method that improves the test-time performance of any few-shot learning models with no additional training, and thus is free from the training-test domain gap. Based on only the few support samples in a meta-test task, our method generates the samples adversarial to the base few-shot classifier's boundaries and fine-tunes its embedding function in the direction that increases the classification margins of the adversarial samples. Consequently, the embedding space becomes denser around the labeled samples which makes the classifier robust to query samples. Experimenting on miniImageNet, CIFAR-FS, and FC100, we demonstrate that our method brings significant performance improvement to three different base methods with various properties, and achieves the state-of-the-art performance in a number of few-shot learning tasks.

Keywords: Few-shot learning · Meta-learning · Adversarial learning

1 Introduction

One of the greatest challenges for machine intelligence to meet human intelligence is the ability to quickly adapt to novel tasks. Humans learn how to solve new tasks with only a small amount of training, by taking advantage of the prior information they have learned for lives. Few-shot learning [11,28] is a research problem to make the most of knowledge gained during training to deal with novel tasks with a limited number of labeled samples. Its core difficulty lies in the data deficiency for novel tasks.

In the few-shot learning problem, the base classes in training and novel classes in test are disjoint. The test phase consists of multiple tasks where each contains a small labeled *support* set and an unlabeled *query* set. The goal of each task is to predict the labels of the query data based on the support data. The meta-learning scheme [12,36], which forms batches of tasks for the training as well,

A. Vedaldi et al. (Eds.): ECCV 2020, LNCS 12346, pp. 599–617, 2020.
https://doi.org/10.1007/978-3-030-58452-8_35

has become dominant in this field of research. Thus, the training and test phases are often called the *meta-training* and *meta-test* phases. Also, many of modern few-shot learning methods use embedding functions (or feature extractors) such as ResNet-12 and handle the data in the embedding space.

With a recent surge of interest in few-shot learning, there have been various approaches proposed, including distance metric methods [40,45], meta-learning methods [12,38], and data augmentation methods [17,39]. Among them, the data augmentation (or hallucination) methods augment the support set by generating fake labeled data for few-shot learning methods (referred to as *base models* or *base classifiers*) [7,17,39,46,51]. The key merit of this approach is applicability to a wide range of base classifiers since it can directly generate fake labeled data. However, most previous approaches have some limitations. (i) Such methods *learn* to generate additional examples with the meta-training set, and thus may not be effective if meta-test domains are far from the meta-training domain. (ii) Since they do not update the trained parameters of the base classifier models at test time, they have no chance to correct the errors that exist in the base classifiers (*e.g.* overfitting of the embedding functions to the meta-training set). (iii) These methods need to be re-trained for each base classifier to generate fake labeled data optimal for the base model.

In this work, we propose a novel model-agnostic sample generation approach for few-shot learning that does not suffer from the aforementioned limitations. The keys to our method named MABAS (*Model-Agnostic Boundary-Adversarial Sampling*) are to perform *no training* for data generation and to generate samples for embedding function fine-tuning. Given only the few labeled data (*i.e.* support samples) in a meta-test task, it creates samples adversarial to the classification boundaries of the base model targeting every meta-test class using each support sample. It then updates the embedding function in the direction that increases the classification margins of the adversarial samples; as a result, the embedding space becomes denser around the labeled samples, which makes the classifier robust to query samples. For sample generation in the embedding space is more advantageous rather than in image space for generalization to unseen classes, since adversarial gradients in the embedding space can directly attack the classification margins.

Finally, we can summarize the main contributions of this work as follows.

1. To the best of our knowledge, our approach is the first pure test-time method for few-shot learning that generates samples for embedding function fine-tuning without learning how to create samples. It simply creates samples adversarial to the classification boundaries of the base few-shot model and fine-tunes the embedding function using the new samples to improve the few-shot generalization performance.
2. Our approach is free from the training-test domain gap and integrable with any base classifier models. We apply our approach to three representative few-shot learning methods, including MetaOptNet [22], Few-Shot without Forgetting (FSwF) [16] and standard transfer learning (STL) [6,42].

3. Our experiments demonstrate that MABAS provides all of the three few-shot learners with significant performance gains and achieves the state-of-the-art performance in a number of tasks on three benchmarks: miniImageNet [45], CIFAR-FS [3] and FC100 [34].

2 Related Work

2.1 Few-Shot Learning

We review a large body of existing few-shot learning methods in three categories: distance metric, meta-learning, and data augmentation methods as follows.

Distance metric methods tackle the few-shot learning problem by learning distance metrics to measure more similar images closer. Matching networks [45] and Prototypical networks [40] predict the labels of the query data based on their learned distances to the support samples in the embedding space. Relational networks [43] propose the embedding and relation module to learning to compare query data with support samples.

Meta-learning methods deal with the problem using the *learning to learn* paradigm, in which an outer loop optimizes meta-variables that controls the optimization of model parameters in an inner loop. MAML [12] proposes the objective of the outer-loop that can learn a good initialization for the inner-loop few-shot learners. LEO [38] introduces latent meta-variables for neural network parameters and take gradient steps within the low-dimensional latent space instead of the high-dimensional parameter space. MetaOptNet [22] uses a convex learner such as multi-class SVMs in the inner loop and update the embedding function in the outer loop to be optimal for the inner loop. MTL [42] utilizes scaling and shifting parameters to adapt the learned embedding parameters to each task differently. LGM-Net [23] encodes prior knowledge of tasks into the context encoder to generate task-specific function weights of embedding networks.

Data augmentation methods learn to augment data to resolve the problem of few-shot learning that lacks enough labeled data. Hariharan and Girshick [17] use the modes of intra-class variation of the base training classes to generate additional samples for the novel classes. Δ-encoder [39] employs auto-encoders that learn to extract transferable intra-class deformations from training data. Zhang *et al.* [51] introduce a saliency map extractor that separates foregrounds with backgrounds to hallucinate datapoints. Chen *et al.* [7] propose a deformation network that fuses few-shot images with unlabeled images. Wang *et al.* [46] combine a meta-learner with a generative model to produce imaginary examples from an anchor example. The existing data augmentation methods share the limitations introduced in Sect. 1. Our method is free from the issues since it does not rely on learning from training data to generate samples. That our method generates samples for embedding function fine-tuning instead of simply enlarging support sets, is also a fundamental difference.

2.2 Adversarial Learning

Since neural network classifiers were known to be vulnerable to even small input perturbations [44], adversarial learning has been actively studied. Adversarial attack methods aim at generating adversarial examples by adding perturbations to samples to fail the classifier [4,5,21,25,31,41,48]. Few-shot adversarial learning methods exploit adversarial signals from discriminators and generators to augment the few-shot classes [1,2,8,10,27,30,35,47,53]. Mottian *et al.* [32] design a multi-class adversarial discriminator to address the supervised adaptation problem in the few-shot domain. Zhang *et al.* [52] employ a generative adversarial network (GAN) that produces fake samples to make sharper decision boundaries. Gao *et al.* [14] model the latent distribution of novel classes with adversarial networks by preserving the covariance information.

Compared to the existing few-shot adversarial learning approaches, we propose to generate adversarial samples purely in test time with no training of additional models. Consequently, our method is orthogonal and easily adaptable to any other few-shot learning methods.

3 MABAS: Boundary-Adversarial Sample Generation

We briefly review the formulation of the few-shot classification (Sect. 3.1) and the idea of test-time fine-tuning of embedding functions (Sect. 3.2). We then propose MABAS as an adversarial learning approach to adaptively fine-tuning the embedding function to each meta-test task (Sect. 3.3).

3.1 The Few-Shot Classification Problem

We begin with the formulation of few-shot classification following previous work [22,36]. The meta-test phase of the few-shot classification problem is comprised of I tasks (*i.e.* episodes): $\mathcal{D}^{\text{test}} = \{\mathcal{T}_i^{\text{test}}\}_{i=1}^{I}$. The i-th task $\mathcal{T}_i^{\text{test}} = (\mathcal{S}_i^{\text{test}}, \mathcal{Q}_i^{\text{test}})$ consists of two sets of data: the support set $\mathcal{S}_i^{\text{test}}$ and the query set $\mathcal{Q}_i^{\text{test}}$. Each task is a K-way M-shot classification problem; the support $\mathcal{S}_i^{\text{test}}$ consists of K different classes, each of which contains M labeled samples (*i.e.* $|\mathcal{S}_i^{\text{test}}| = KM$). The meta-training dataset $\mathcal{D}^{\text{train}}$ consists of the classes disjoint with the meta-test set $\mathcal{D}^{\text{test}}$. For a classifier $\mathcal{C}$ trained with $\mathcal{D}^{\text{train}}$, the meta-test accuracy is defined as $\sum_{i=1}^{I} \sum_{(\boldsymbol{x},y)\in\mathcal{Q}_i^{\text{test}}} \mathbb{1}\left(\mathcal{C}\left(\boldsymbol{x}|\mathcal{S}_i^{\text{test}}\right) = y\right)$.

3.2 Test-Time Fine-Tuning of Embedding Functions

Since each meta-test task consists of the classes that are never seen during meta-training, it is a common approach in few-shot learning to fine-tuning the learned parameters using the support samples of the novel task. For instance, MAML [12] and LEO [38] use the trained model as initialization and fine-tuning it to meta-test tasks, and MTL [42] applies scaling and shifting to the learned parameters for each meta-test task differently, which could be better than direct update of parameters to reduce the overfitting to a small number of samples.

Likewise, our approach aims at fine-tuning the learned parameters of the base few-shot learner adaptively to novel tasks. However, we limit to update the parameters of the *embedding function* (or the feature extractor). It is a universally applicable idea since most recent few-shot learning models employ CNNs as their embedding functions that extract features from images [22,29].

We formulate the iterative fine-tuning procedure of the embedding function f_ϕ with parameter $\phi \in \psi$ for the classifier $\mathcal{C}$ defined by $\mathcal{C}\left(\boldsymbol{x}|\mathcal{S}^{\text{test}}\right) = \text{argmax}_k\, h\left(f_\phi(\boldsymbol{x}), k|\psi, \mathcal{S}^{\text{test}}\right)$, as follows. For a given fine-tuning loss function $\mathcal{L}^{\text{fine-tune}}(\mathcal{S}, \phi)$, we update the classifier $\mathcal{C}$ as

$$\mathcal{C}\left(\boldsymbol{x}|\mathcal{S}^{\text{test}}\right) = \underset{k}{\text{argmax}}\, h\left(f_{\phi'}(\boldsymbol{x}), k|\psi', \mathcal{S}^{\text{test}}\right) \tag{1}$$

where $\phi' \in \psi'$. The fine-tuning loss $\mathcal{L}^{\text{fine-tune}}(\mathcal{S}, \phi)$ and the score function h differ according to the base models, and their definitions will be described in Sect. 4. For simplicity, we denote $\boldsymbol{z} = f_\phi(\boldsymbol{x})$. The new parameter ϕ' is obtained by multiple updates via gradient descent:

$$\phi^i = \phi^{i-1} - \beta \cdot \nabla_{\phi^{i-1}}\left(\mathcal{L}^{\text{fine-tune}}(\mathcal{S}^{\text{test}}, \phi^{i-1})\right), \tag{2}$$

for $i = 1, \ldots, U$ where U is the number of updates and β is a fine-tuning step size. We initialize $\phi^0 = \phi$ with the parameter learned during meta-training and finally set $\phi' = \phi^U$.

3.3 Fine-Tuning by Boundary-Adversarial Samples

We assume that a base few-shot classifier is chosen and trained with training data. Our approach solely focuses on meta-test time; it first generates samples adversarial to the classification boundaries defined by the few-shot classifier in the embedding space, and use them to fine-tune only the parameter ϕ of the embedding function. Figure 1 intuitively visualizes how our approach works. For the success of few-shot learning, it is important to transfer the embedding function to the domain that lacks labeled samples. We generate boundary-adversarial samples by moving every support sample toward each of the classification boundaries. The embedding function is fine-tuned in the direction that increases the margins of the adversarial samples. After the update, the data embeddings are denser around the support embeddings, and thus the recomputed classification boundaries better separate the queries from different classes.

Generation of adversarial samples. We first define the classification margin for sample $\boldsymbol{z}$ between classes k and k' as

$$m\left(\boldsymbol{z}, k, k'|\psi, \mathcal{S}\right) := h\left(\boldsymbol{z}, k|\psi, \mathcal{S}\right) - h\left(\boldsymbol{z}, k'|\psi, \mathcal{S}\right). \tag{3}$$

For every support sample $(\boldsymbol{x}, y) \in \mathcal{S}$ and each attack target class $k' \in \{1, \ldots, K\} \setminus y$, we create a boundary-adversarial sample $\boldsymbol{z}^{\text{adv}}_{y,k'}$ by moving $\boldsymbol{z} = f_\phi(\boldsymbol{x})$ in the direction that minimizes its margin against k' in the embedding space:

$$\boldsymbol{z}^{\text{adv}}_{y,k'} := \boldsymbol{z} - \delta \cdot \nabla_{\boldsymbol{z}}\; m\left(\boldsymbol{z}, y, k'|\psi, \mathcal{S}\right) \tag{4}$$

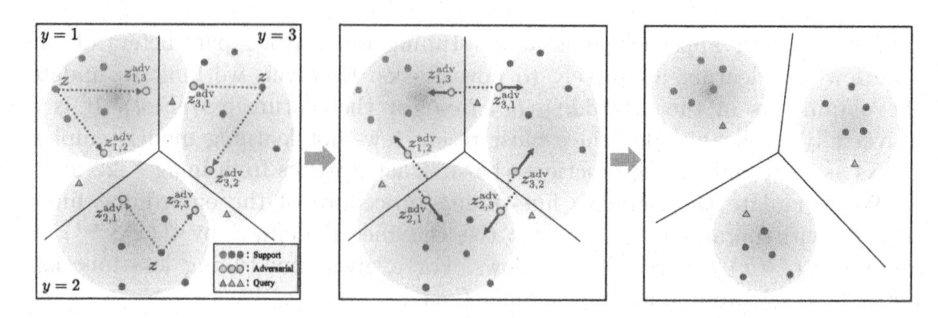

Fig. 1. Conceptual visualization of our MABAS approach in the embedding space. The solid circles, triangles, and yellow circles are the support, query, and boundary-adversarial samples, respectively. Each shaded area represents a region to which most samples from the class are mapped by the embedding function. The black lines are the classification boundaries computed from the supports. *Left*: The boundary-adversarial samples are generated by moving each support sample toward the classification boundaries (Eq. (4)). *Middle*: The embedding function is updated in the direction that increases the margins of the adversarial samples (Eq. (6)) while holding the support embeddings (Eq. (5)). *Right*: After the fine-tuning of the embedding function, data embeddings are denser around the support embeddings, and the classification boundaries are updated to better separate queries from different classes.

where δ is a step size. $\boldsymbol{z}_{y,k'}^{\text{adv}}$ is obtained by applying single-step gradient descent to $\boldsymbol{z}$ in the direction that minimizes the margin (or score gap) of the embedding between y and k'. The single-step update is sufficient for the generation as the fine-tuning process is alternation between the adversarial sample generation and the embedding function update. The resulting $\boldsymbol{z}_{y,k'}^{\text{adv}}$ is an adversarial sample based on $\boldsymbol{z}$ against the target class k'. It can be regarded as an augmented data of class y located near to the classification boundary between y and k'. When deriving the adversarial gradient in Eq. (4), we fix h even in the case of differentiable base learners. We will elaborate it with some examples of differentiable base models in Sect. 4.

In the meta-test task, for every $(\boldsymbol{x}, y) \in \mathcal{S}$, $K-1$ adversarial samples are generated one for each target class $(\{1, \ldots, K\} \setminus y)$, and thus $|\mathcal{S}|(K-1) = MK(K-1)$ adversarial samples are created in total at each fine-tuning step.

Fine-tuning of the embedding function. With the adversarial samples, we update the parameter ϕ of the embedding function via the gradient descent in Eq. (2). The fine-tuning loss $\mathcal{L}^{\text{fine-tune}}$ is defined as

$$\mathcal{L}^{\text{fine-tune}}(\mathcal{S}, \phi) := \mathcal{L}^{\text{adv}}(\mathcal{S}, \phi) + \eta \cdot \frac{1}{|\mathcal{S}|} \sum_{(\boldsymbol{x},y)\in\mathcal{S}} \|f_\phi(\boldsymbol{x})\|^2, \tag{5}$$

whose first term is the adversarial loss term and second term is the regularizer with a coefficient η. We define $\mathcal{L}^{\text{adv}}$ as

$$\mathcal{L}^{\text{adv}}(\mathcal{S}, \phi) := \frac{1}{|\mathcal{S}|} \sum_{(\boldsymbol{x},y)\in\mathcal{S}} \left[\frac{1}{K-1} \sum_{k'\neq y} \left\{ \alpha_{\boldsymbol{x},y} - \min_{k\neq y} m\left(\boldsymbol{z}^{\text{adv}}_{y,k'}, y, k|\psi, \mathcal{S}\right) \right\}_{+} \right] \quad (6)$$

where $\alpha_{\boldsymbol{x},y} = \frac{1}{K-1}\sum_{k\neq y} m(\boldsymbol{z}, y, k|\psi, \mathcal{S})$. The objective $\mathcal{L}^{\text{adv}}$ chooses the minimum margin per adversarial sample and increases the margin but not larger than the anchor $\alpha_{\boldsymbol{x},y}$. The anchor $\alpha_{\boldsymbol{x},y}$ is the average margin for $(\boldsymbol{x}, y)$ and provides a reasonable upper limit for the adversarial samples' margins. We treat $\alpha_{\boldsymbol{x},y}$ as a constant threshold rather than letting fine-tuning gradients flow through $\alpha_{\boldsymbol{x},y}$, since otherwise it might dominate the objective and disturb the pushing of the adversarial samples against the boundaries.

The second regularization term in Eq. (5) not only prevents the excessive expansion of the supports' embedding space but also stabilizes the updates of the embedding space.

4 Application to Various Few-Shot Methods

To show the flexibility and generality of our MABAS approach, we apply it to three representative few-shot learning methods, including MetaOptNet [22], Few-Shot without Forgetting (FSwF) [16] and the standard transfer learning (STL) method [6,42]. These methods show diverse characteristics; MetaOptNet and FSwF have differentiable base learners while the STL does not. Also, the classifiers of MetaOptNet and STL are linear, whereas FSwF is not. In this section, we present the key ideas of each base method and how our approach is integrated with them.

4.1 MetaOptNet

Original methodology. MetaOptNet [22] uses a differentiable SVM solver for few-shot classification. In each meta-training or meta-test task, MetaOptNet solves the multi-class SVM problem for the support data to make predictions for the query data. More specifically, given a task with support $\mathcal{S}$ and query $\mathcal{Q}$, it solves the K-class SVM problem [9] for support samples $(\boldsymbol{x}_n, y_n) \in \mathcal{S}$, $n = 1, \ldots, N$, whose objective is defined by

$$\begin{aligned} &\underset{\boldsymbol{w}_1,\ldots,\boldsymbol{w}_K,\xi_1,\ldots,\xi_N}{\text{minimize}} \ \frac{1}{2}\sum_k \|\boldsymbol{w}_k\|_2^2 + C\sum_n \xi_n \\ &\text{s.t.} \quad (\boldsymbol{w}_{y_n} - \boldsymbol{w}_k)^\top f_\phi(\boldsymbol{x}_n) \geq 1 - \mathbb{1}(y_n = k) - \xi_n, \quad \forall n, k. \end{aligned} \quad (7)$$

The score function for the task is defined using the SVM solution $\boldsymbol{w}_1, \ldots, \boldsymbol{w}_K$:

$$h\left(f_\phi(\boldsymbol{x}), k|\psi, \mathcal{S}\right) := \boldsymbol{w}_k^\top f_\phi(\boldsymbol{x}). \quad (8)$$

The parameter ϕ of the embedding function f_ϕ is trained with the classification loss in the meta-training phase, and not updated in the meta-test time.

Boundary-adversarial fine-tuning. By plugging Eq. (8) into Eqs. (4) and (6), we obtain $z^{\mathrm{adv}}_{y,k'}$ and $\mathcal{L}^{\mathrm{adv}}$ for MetaOptNet as

$$z^{\mathrm{adv}}_{y,k'} = z - \delta \cdot \nabla_z \left((w_y - w_{k'})^\top z \right) = z - \delta \cdot (w_y - w_{k'}), \tag{9}$$

$$\mathcal{L}^{\mathrm{adv}}(\mathcal{S}, \phi) = \frac{1}{|\mathcal{S}|} \sum_{(x,y) \in \mathcal{S}} \left[\frac{1}{K-1} \sum_{k' \neq y} \left\{ \alpha_{x,y} - \min_{k \neq y} \left((w_y - w_k)^\top z^{\mathrm{adv}}_{y,k'} \right) \right\}_+ \right].$$

As mentioned in Sect. 3.3, although $w_1, \ldots, w_K$ are differentiable with respect to z, we assume they are fixed constants to derive the attack gradient as $\nabla_z((w_y - w_{k'})^\top z) = w_y - w_{k'}$ in Eq. (9). Only the embedding function parameter ϕ is fine-tuned iteratively, as described in Sect. 3.

Variation. We also apply our approach to a modified version of MetaOptNet, which uses the SVM solution for $\{(\frac{1}{M} \sum_{n: y_n = k} f_\phi(x_n), k) | \text{ for } k = 1, \ldots, K\}$ instead of $\{(f_\phi(x), y) | (x, y) \in \mathcal{S}\}$. The difference is that K class prototypes (*i.e.* per-class average embeddings of the supports) are used instead of KM support embeddings as input to the multi-class SVM solver, inspired by [40]. Except it, the derivation is the same. This variation solves the multi-class SVM problem with fewer samples but still shows competitive results. We refer to this variation as MetaOptNet-Proto for the rest of this paper.

Setting of δ. Although fixing an adversarial step size δ in Eq. 9 works well with MetaOptNet(-Proto), we can formulate δ so that the change of the margin per adversarial step is fixed at λ, using the fact that the multi-class SVM is linear and the maximum margin is 1:

$$z^{\mathrm{adv}}_{y,k'} = z - \delta_{y,k'} \cdot (w_y - w_{k'}), \quad \text{where } \delta_{y,k'} = \frac{\lambda}{\|w_y - w_{k'}\|^2}. \tag{10}$$

With this definition, $m(z, y, k' | \psi, \mathcal{S}) - m(z^{\mathrm{adv}}_{y,k'}, y, k' | \psi, \mathcal{S}) = \lambda$ regardless of y and k' during all fine-tuning updates.

4.2 Few-Shot Without Forgetting

Original methodology. The Few-Shot without Forgetting (FSwF) [16] learns not only the embedding function f_ϕ but also the attention-based classification weight generator. The role of the weight generator G with parameter θ is to compute the classification weight vector w_k for the novel class k using two types of input: (i) the classification weights for the B base classes (*i.e.* $v_1, \ldots, v_B$) trained from $\mathcal{D}^{\mathrm{train}}$ and (ii) the support of novel class k (*i.e.* $S_k = \{(x, y) | y = k, \forall (x, y) \in \mathcal{S}\}$) in each few-shot task:

$$G_\theta(\mathcal{S}, k, \boldsymbol{v}_1, \ldots, \boldsymbol{v}_B) = \theta^{avg} \odot \boldsymbol{w}_k^{avg} + \theta^{att} \odot \boldsymbol{w}_k^{att}, \tag{11}$$

$$\boldsymbol{w}_k^{avg} = \frac{1}{|S_k|} \sum_{(\boldsymbol{x},y) \in S_k} \frac{f_\phi(\boldsymbol{x})}{\|f_\phi(\boldsymbol{x})\|}, \tag{12}$$

$$\boldsymbol{w}_k^{att} = \frac{1}{|S_k|} \sum_{(\boldsymbol{x},y) \in S_k} \sum_{b=1}^{B} \text{Att}(\theta^{att} \frac{f_\phi(\boldsymbol{x})}{\|f_\phi(\boldsymbol{x})\|}, \boldsymbol{\theta}_b^{key}) \cdot \frac{\boldsymbol{v}_b}{\|\boldsymbol{v}_b\|}, \tag{13}$$

where $\odot$ is the element-wise product and Att() is the attention kernel. The learnable parameters include ϕ and $\theta = \{\theta^{avg}, \theta^{att}, \boldsymbol{\theta}_1^{key}, \ldots, \boldsymbol{\theta}_B^{key}\}$, where θ is trained on meta-training tasks and not updated in the meta-test phase. Finally, the novel class weight vector is $\boldsymbol{w}_k = G_\theta(\mathcal{S}, k, \boldsymbol{v}_1, \ldots, \boldsymbol{v}_B)$.

The score function of FSwF for the task becomes

$$h\left(f_\phi(\boldsymbol{x}), k | \psi, \mathcal{S}\right) := \frac{\boldsymbol{w}_k^\top f_\phi(\boldsymbol{x})}{\|\boldsymbol{w}_k\| \|f_\phi(\boldsymbol{x})\|}. \tag{14}$$

Boundary-adversarial fine-tuning. By applying the definition of h from Eq. (14) to Eqs. (4) and (6), $\boldsymbol{z}_{y,k'}^{\text{adv}}$ for FSwF is defined by

$$\begin{aligned} \boldsymbol{z}_{y,k'}^{\text{adv}} &= \boldsymbol{z} - \delta \cdot \nabla_{\boldsymbol{z}} \left(\left(\frac{\boldsymbol{w}_y}{\|\boldsymbol{w}_y\|} - \frac{\boldsymbol{w}_{k'}}{\|\boldsymbol{w}_{k'}\|} \right)^\top \frac{\boldsymbol{z}}{\|\boldsymbol{z}\|} \right) \\ &= \boldsymbol{z} - \delta \cdot \left(\frac{I_d}{\|\boldsymbol{z}\|} - \frac{\boldsymbol{z}\boldsymbol{z}^\top}{\|\boldsymbol{z}\|^3} \right) \left(\frac{\boldsymbol{w}_y}{\|\boldsymbol{w}_y\|} - \frac{\boldsymbol{w}_{k'}}{\|\boldsymbol{w}_{k'}\|} \right) \end{aligned} \tag{15}$$

where $\boldsymbol{z} \in \mathbb{R}^d$, and the adversarial loss $\mathcal{L}^{\text{adv}}$ becomes

$$\begin{aligned} \mathcal{L}^{\text{adv}}(\mathcal{S}, \phi) = \frac{1}{|\mathcal{S}|} \sum_{(\boldsymbol{x},y) \in \mathcal{S}} \Bigg[\frac{1}{K-1} \sum_{k' \neq y} \\ \left\{ \alpha_{\boldsymbol{x},y} - \min_{k \neq y} \left(\frac{\boldsymbol{w}_y}{\|\boldsymbol{w}_y\|} - \frac{\boldsymbol{w}_k}{\|\boldsymbol{w}_k\|} \right)^\top \frac{\boldsymbol{z}_{y,k'}^{\text{adv}}}{\|\boldsymbol{z}_{y,k'}^{\text{adv}}\|} \right\}_+ \Bigg]. \end{aligned} \tag{16}$$

Similarly to the derivation for MetaOptNet in Sect. 4.1, the adversarial gradient in Eq. 15 is derived while fixing the classification weights of $\boldsymbol{w}_y$ and $\boldsymbol{w}_{k'}$ with respect to $\boldsymbol{z}$.

4.3 Standard Transfer Learning

Original methodology. STL [6,42] is a standard transfer learning approach to the few-shot classification problem. It learns the embedding function f_ϕ during meta-training, and obtains the linear classification weight matrix $\mathbf{W} = [\boldsymbol{w}_1; \ldots; \boldsymbol{w}_K] \in \mathbb{R}^{d \times K}$ per meta-test task using f_ϕ for $\boldsymbol{z} \in \mathbb{R}^d$:

$$[\boldsymbol{w}_1; \ldots; \boldsymbol{w}_K] = \underset{[\boldsymbol{w}'_1; \ldots; \boldsymbol{w}'_K]}{\text{argmin}} \frac{1}{|\mathcal{S}|} \sum_{(\boldsymbol{x},y) \in \mathcal{S}} -\log \left(\frac{\exp(\boldsymbol{w}'^\top_y f_\phi(\boldsymbol{x}))}{\sum_k \exp(\boldsymbol{w}'^\top_k f_\phi(\boldsymbol{x}))} \right) \tag{17}$$

for the support $\mathcal{S}$. The STL computes $[\boldsymbol{w}_1; \ldots; \boldsymbol{w}_K]$ using the gradient descent on Eq. (17), and the procedure is not differentiable.

Finally, the score function h of STL for the given task becomes

$$h\left(f_\phi(\boldsymbol{x}), k|\psi, \mathcal{S}\right) := \boldsymbol{w}_k^\top f_\phi(\boldsymbol{x}). \tag{18}$$

Boundary-adversarial fine-tuning. By the definition of h from Eq. (18), the boundary-adversarial sample generation for STL is derived as

$$\boldsymbol{z}_{y,k'}^{\text{adv}} = \boldsymbol{z} - \delta \cdot \nabla_{\boldsymbol{z}}\left((\boldsymbol{w}_y - \boldsymbol{w}_{k'})^\top \boldsymbol{z}\right) = \boldsymbol{z} - \delta \cdot (\boldsymbol{w}_y - \boldsymbol{w}_{k'}), \tag{19}$$

and $\mathcal{L}^{\text{adv}}$ is

$$\mathcal{L}^{\text{adv}}(\mathcal{S}, \phi) = \frac{1}{|\mathcal{S}|} \sum_{(\boldsymbol{x},y)\in\mathcal{S}} \left[\frac{1}{K-1} \sum_{k'\neq y} \left\{\alpha_{\boldsymbol{x},y} - \min_{k\neq y}\left((\boldsymbol{w}_y - \boldsymbol{w}_k)^\top \boldsymbol{z}_{y,k'}^{\text{adv}}\right)\right\}_+\right].$$

5 Experiments

We conduct experiments to evaluate the few-shot classification performance of our MABAS approach. We first present experimental setup (Sect. 5.1) and discuss the quantitative and qualitative results (Sects. 5.2 and 5.3). Please refer to Appendix for additional experimental results including evaluation on tieredImageNet [37].

5.1 Experimental Setup

Datasets. We use three benchmark datasets for evaluation of few-shot classification. (1) miniImageNet [45] consists of 100 classes each of which has 600 images with a size of $84 \times 84 \times 3$. We adopt the same class split used by [22,36]: 64, 16 and 20 classes for training, validation and the test, respectively. (2) CIFAR-FS [3] splits all of the classes in CIFAR100 [20] into 64 training, 16 validation and 20 test sets, respectively. (3) FC100 [34] is another CIFAR100-based dataset. Classes are split into 60, 20 and 20 for training, validation and test, respectively. This class split is designed to minimize the overlap of information between all three subsets, to be more challenging than CIFAR-FS for few-shot learning.

Embedding functions. We employ ResNet-12, which is one of the popular choices for few-shot learning research [22,29,34]. For the experiments of MetaOptNet [22] and STL [6,42], we use the same architecture of ResNet-12 as [22]. The only architectural difference for STL from MetaOptNet is that an average pooling is applied to the last residual block. For FSwF experiments, we use the architecture in [29] following [16].

Meta-training and meta-validation phase. For the MetaOptNet, FSwF and STL models, we follow the training and validation protocol in the original papers [6,16,22] with some minor modifications as follows. For MetaOptNet-Proto, we use a learning decay rate of 0.1 with a decay period of 15 epochs for simplicity. For the STL models, we train for 100 epochs with a batch size of 256 and a learning rate of 0.001 using the cosine annealing decay [18].

Meta-test phase. We test all the models on the 5-way 5-shot and 5-way 1-shot classification tasks. For a fair comparison, all of the meta-test results are obtained using the same setup with the previous works [12,16,42]. Each meta-test run consists of 600 tasks sampled from $\mathcal{D}^{\text{test}}$, and a single task contains 15 query samples per each of the 5 classes.

Boundary-adversarial fine-tuning. For all the four base methods, we fine-tune only the last (*i.e.* the fourth) block of the ResNet-12 embedding function, since it preserves most of the representation power of the embedding function. We use Adam [19] optimizer for fine-tuning, and maintain a single set of hyper-parameters per base method across all datasets. In all experiments, we update the embedding function for 150 steps and use the step-based learning rate decay with a decay rate of 0.8 and a period of 5. For MetaOptNet(-Proto), we use $\eta = 0.0005$ as its regularization coefficient and use an initial learning rate of 0.000025. We set the adversarial step size δ to fix the change of margin to $\lambda = 1$, as in Sect. 4.1. Since FSwF uses the cosine similarity, which is a measure invariant to scaling of inputs, we set $\eta = 0$. Its initial learning rate is 0.0003 and the adversarial step size is $\delta = 10$. In the STL experiments, we perform the fine-tuning with an initial learning rate of 0.0001, $\eta = 0.01$ and $\delta = 10$.

5.2 Quantitative Evaluation

Table 1 reports the test-time accuracies of the four base methods with or without MABAS on the six tasks. We also summarize the accuracy gains by our method. Table 2 compares our method with the state-of-the-art models from the original papers. Here are some important observations to emphasize:

1. Our method achieves the new state-of-the-art performance on four of the six tasks (miniImageNet 5-shot, CIFAR-FS 1-shot and 5-shot, and FC100 5-shot settings) if we exclude the methods with WRN-28-10 [50], which consumes about three times more parameters than ResNet-12 [22].
2. Our method improves the accuracy in every experiment of the four base methods on three datasets in Table 1, where only a single set of fine-tuning hyperparameters is used for all experiments per base method.
3. The largest accuracy improvement brought by our method is $8.34\%p$, and the average improvement is $\approx 2.80\%p$.
4. These results demonstrate that our approach is effective for solving the labeled data scarcity problem of few-shot learning. Also, our method is universal enough to improve various base methods with different properties.

Effects of embedding function architectures. FSwF [16] is tested with three ConvNet and one ResNet-12 embedding functions in the original paper. In its experiments on 5-way miniImageNet tasks, the accuracies on $\mathcal{D}^{\text{train}}$ are

Table 1. Meta-test accuracies (%) of the four base methods before and after applying our MABAS method with the 95% confidence interval. We also report the accuracy gains (%*p*) by MABAS. We obtain all results using the source codes provided by the original authors to precisely measure the accuracy improvements by our method.

Method	miniImageNet, 5-way		CIFAR-FS, 5-way		FC100, 5-way	
	1-shot	5-shot	1-shot	5-shot	1-shot	5-shot
STL [6,42]	55.98 ± 0.78	76.19 ± 0.60	64.32 ± 0.92	83.62 ± 0.61	38.83 ± 0.70	54.54 ± 0.72
+ **Ours**	60.19 ± 0.79	79.34 ± 0.57	67.41 ± 0.91	84.29 ± 0.65	40.76 ± 0.68	58.16 ± 0.78
Gain (%*p*)	4.21	3.15	3.09	0.67	1.93	3.62
FSwF [16]	55.64 ± 0.82	69.94 ± 0.68	69.23 ± 0.90	82.52 ± 0.68	37.91 ± 0.77	49.75 ± 0.72
+ **Ours**	60.45 ± 0.82	78.28 ± 0.61	70.71 ± 0.89	85.25 ± 0.65	40.63 ± 0.75	54.95 ± 0.75
Gain (%*p*)	4.81	8.34	1.48	2.73	2.72	5.20
MetaOptNet [22]	62.25 ± 0.82	78.55 ± 0.58	72.11 ± 0.96	84.32 ± 0.65	40.15 ± 0.71	54.92 ± 0.75
+ **Ours**	64.21 ± 0.82	81.01 ± 0.57	73.24 ± 0.95	85.65 ± 0.65	41.74 ± 0.73	57.11 ± 0.75
Gain (%*p*)	1.96	2.46	1.13	1.33	1.59	2.19
MetaOptNet-Proto	61.68 ± 0.85	78.36 ± 0.59	72.46 ± 0.91	84.02 ± 0.67	40.60 ± 0.75	55.04 ± 0.75
+ **Ours**	65.08 ± 0.86	82.70 ± 0.54	73.51 ± 0.92	85.49 ± 0.68	42.31 ± 0.75	57.56 ± 0.78
Gain (%*p*)	3.40	4.34	1.05	1.47	1.71	2.52

the highest with ResNet-12, while its accuracies on $\mathcal{D}^{\text{test}}$ are worse than those of the ConvNet architectures. Inspired by this observation, we experiment the fine-tuning accuracy of our approach with different embedding functions.

Table 3 shows that our method increases the fine-tuning accuracy of ResNet-12 much larger than ConvNet architectures. The accuracy gains with ConvNets are less than 1%*p*, while the gain becomes 8.34%*p* with the ResNet-12 on the 5-way 5-shot miniImageNet task. This result hints that ResNet-12 has high capacity and thus suffers from overfitting, more seriously in few-shot tasks. Our method helps each embedding function to maximize its representation ability even with only a few examples of novel classes.

Comparison with naive fine-tuning. To highlight the effectiveness of MABAS, we compare MABAS with the naive fine-tuning of the embedding function, which is fine-tuning using support samples only (*i.e.* no adversarial sample). Table 4 shows that the naive fine-tuning provides only small performance gains whereas MABAS brings significant improvements.

Table 2. Comparison with the state-of-the-art few-shot learning methods. We present the average meta-test accuracy with its 95% confidence interval.[‡] denotes the results from [22,36,42], while all the other baseline scores are referred to their original papers. The numbered superscripts denote the architecture of f_ϕ: [1]Conv-4, [2]Conv-4+MetaNet, [3]ResNet-12, [4]WRN-28-10, [5]Conv4+MetaGAN. Note that WRN-28-10 involves three times more parameters than ResNet-12.

Method	miniImageNet, 5-way		CIFAR-FS, 5-way		FC100, 5-way	
	1-shot	5-shot	1-shot	5-shot	1-shot	5-shot
Meta-LSTM[1] [36]	43.44 ± 0.77	60.60 ± 0.71	-	-	-	-
MatchingNets[1‡] [45]	43.56 ± 0.84	55.31 ± 0.73	-	-	-	-
MAML[1‡] [12]	48.70 ± 1.84	63.11 ± 0.92	58.9 ± 1.9	71.5 ± 1.0	38.1 ± 1.7	50.4 ± 1.0
ProtoNets[1‡] [40]	49.42 ± 0.78	68.20 ± 0.66	55.5 ± 0.7	72.0 ± 0.6	35.3 ± 0.6	48.6 ± 0.6
RelationNets[1‡] [43]	50.44 ± 0.82	65.32 ± 0.70	55.0 ± 1.0	69.3 ± 0.8	-	-
R2D2[1] [3]	51.20 ± 0.60	68.80 ± 0.10	65.30 ± 0.20	79.40 ± 0.10	-	-
FSwF[1] [16]	56.20 ± 0.86	73.00 ± 0.64	-	-	-	-
FSwF[3] [16]	55.45 ± 0.89	70.13 ± 0.68	-	-	-	-
Bilevel Program[3] [13]	50.54 ± 0.85	64.53 ± 0.68	-	-	-	-
MetaGAN[5] [52]	52.71 ± 0.64	68.63 ± 0.67	-	-	-	-
SNAIL[3] [29]	55.71 ± 0.99	68.88 ± 0.92	-	-	-	-
AdaResNet[3] [33]	56.88 ± 0.62	71.94 ± 0.57	-	-	-	-
TADAM[3] [34]	58.5 ± 0.3	76.7 ± 0.3	-	-	40.1 ± 0.4	56.1 ± 0.4
ProtoNets[3‡] [40]	59.25 ± 0.64	75.60 ± 0.48	72.2 ± 0.7	83.5 ± 0.5	37.5 ± 0.6	52.5 ± 0.6
MTL[3] [42]	61.2 ± 1.8	75.5 ± 0.8	-	-	**45.1±1.8**	57.6 ± 0.9
TapNet[3] [49]	61.65 ± 0.15	76.36 ± 0.10	-	-	-	-
MetaOptNet[3] [22]	62.64 ± 0.61	78.63 ± 0.46	72.0 ± 0.7	84.2 ± 0.5	41.1 ± 0.6	55.5 ± 0.6
LEO[4] [38]	61.76 ± 0.08	77.59 ± 0.12	-	-	-	-
CC+rot[4] [15]	62.93 ± 0.45	79.87 ± 0.33	**76.09±0.30**	**87.83±0.21**	-	-
$S2M2_R$[4] [26]	64.93 ± 0.18	**83.18±0.11**	74.81 ± 0.19	87.47 ± 0.13	-	-
LGM-Net[2] [23]	**69.13±0.35**	71.18 ± 0.68	-	-	-	-
STL + **Ours**[3]	60.19 ± 0.79	79.34 ± 0.57	67.41 ± 0.91	84.29 ± 0.65	40.76 ± 0.68	**58.16±0.78**
MetaOptNet +**Ours**[3]	64.21 ± 0.82	81.01 ± 0.57	73.24 ± 0.95	**85.65±0.65**	41.74 ± 0.73	57.11 ± 0.75
–Proto + **Ours**[3]	65.08 ± 0.86	**82.70±0.54**	**73.51±0.92**	85.49 ± 0.68	42.31 ± 0.75	57.56 ± 0.78

Table 3. Meta-test accuracy of FSwF with different embedding functions with or without MABAS. All ConvNets share the same hyperparameters.

f_ϕ	FSwF	+ MABAS
Conv32	70.01 ± 0.66	70.05 ± 0.65
Conv64	71.89 ± 0.67	72.15 ± 0.66
Conv128	72.59 ± 0.65	72.79 ± 0.63
ResNet-12	69.94 ± 0.68	78.28 ± 0.61

Table 4. Comparison of meta-test accuracy between naive fine-tuning and MABAS using FSwF on the miniImageNet dataset.

Method	miniImageNet, 5-way	
	1-shot	5-shot
FSwF	55.64 ± 0.82	69.94 ± 0.68
+ Naive FT	55.73 ± 0.82	70.14 ± 0.67
+ Ours	60.45 ± 0.82	78.28 ± 0.61

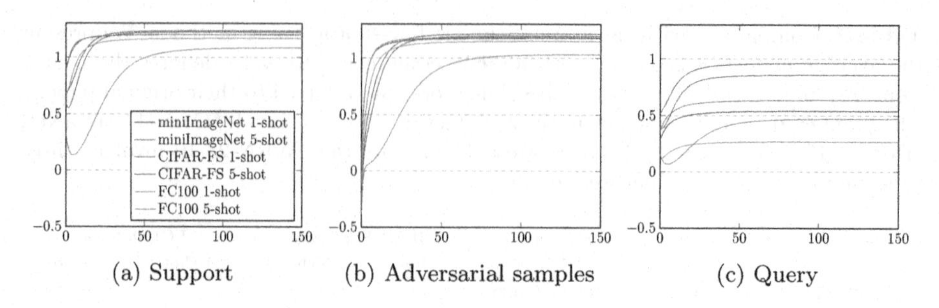

(a) Support (b) Adversarial samples (c) Query

Fig. 2. Average classification margin with FSwF + Ours for support (*left*), adversarial (*middle*) and query (*right*) samples on 5-way tasks. The *x-axis* and *y-axis* denote the number of fine-tuning updates and the average classification margin, respectively. The margin values are averaged over all the meta-test tasks.

Evolution of support, adversarial, and query sample margins. Fig. 2 shows the evolution of average classification margin for support, adversarial and query samples with FSwF on 5-way meta-test tasks. In each fine-tuning update step, MABAS generates adversarial samples and update the embedding space by increasing the classification margins for those samples. As the fine-tuning progresses, the embedding space becomes denser around the support samples and the margins for novel query samples increase as in (c), due to the changes in the embedding space. It indicates that their classification confidences increase too, which results in accuracy improvement.

5.3 Qualitative Evaluation

Figure 3 illustrates how the embeddings of the support and query change according to fine-tuning updates. Using t-SNE [24], we visualize the embeddings computed by MetaOptNet-Proto and FSwF for the 5-way 5-shot and 1-shot miniImageNet problems. As FSwF uses the cosine similarity in its score function, its ℓ_2-normalized embeddings are taken as input to t-SNE. Before applying our method, the embeddings from different classes are distributed in a mixed way with no clear class separation. As the boundary-adversarial fine-tuning proceeds, not only the support embeddings but also most query embeddings are condensed, and samples from distinct classes become distant.

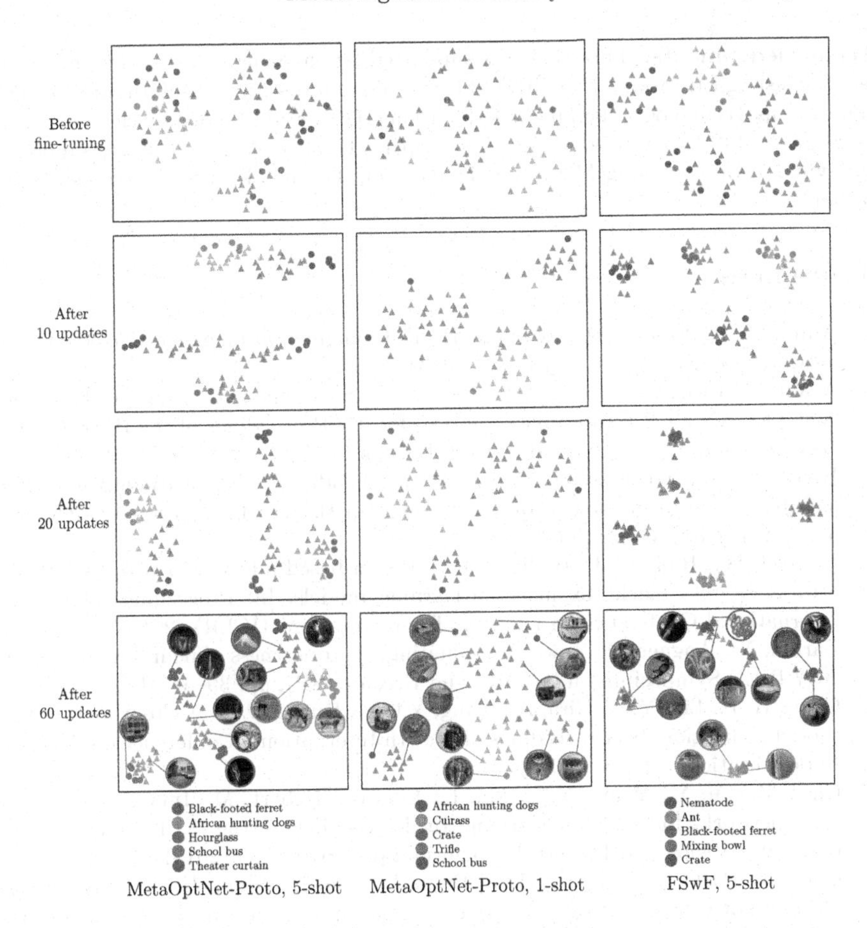

Fig. 3. t-SNE [24] visualization of the support embeddings (*circles*) and query embeddings (*triangles*) obtained by MetaOptNet-Proto and FSwF before and after fine-tuning updates at meta-test time for 5-way miniImageNet tasks.

6 Conclusion

We presented MABAS, a novel model-agnostic approach to generating adversarial samples in the embedding space at test time for few-shot generalization. MABAS is a practical method that works with no additional training and is integrable with any few-shot learning methods. Our results on three few-shot benchmark datasets – miniImageNet, CIFAR-FS, and FC100 – showed that MABAS significantly enhanced the performance of the base methods with various characteristics, and consequently achieved the state-of-the-art performance in several tasks. We believe this work provides a low-effort add-on method for performance enhancement with existing and future few-shot learning methods.

Acknowledgements. This work was supported by Samsung Research Funding Center of Samsung Electronics (No. SRFC-IT1502-51) and Institute of Information & communications Technology Planning & Evaluation (IITP) grant funded by the Korea government (MSIT) (No. 2017-0-01772, Video Turing Test). Jaekyeom Kim was supported by Hyundai Motor Chung Mong-Koo Foundation. Gunhee Kim is the corresponding author.

References

1. Antoniou, A., Storkey, A., Edwards, H.: Data augmentation generative adversarial networks. arXiv preprint arXiv:1711.04340 (2017)
2. Azadi, S., Fisher, M., Kim, V.G., Wang, Z., Shechtman, E., Darrell, T.: Multi-content GAN for few-shot font style transfer. In: Proceedings of the IEEE Conference on Computer Vision and Pattern Recognition, pp. 7564–7573 (2018)
3. Bertinetto, L., Henriques, J.F., Torr, P.H., Vedaldi, A.: Meta-learning with differentiable closed-form solvers. In: Proceedings of the 7th International Conference on Learning Representations (ICLR) (2019)
4. Brendel, W., Rauber, J., Bethge, M.: Decision-based adversarial attacks: reliable attacks against black-box machine learning models. In: Proceedings of the 6th International Conference on Learning Representations (ICLR) (2018)
5. Carlini, N., Wagner, D.: Towards evaluating the robustness of neural networks. In: 2017 IEEE Symposium on Security and Privacy (SP), pp. 39–57. IEEE (2017)
6. Chen, W.Y., Liu, Y.C., Kira, Z., Wang, Y.C.F., Huang, J.B.: A closer look at few-shot classification. In: Proceedings of the 7th International Conference on Learning Representations (ICLR) (2019)
7. Chen, Z., Fu, Y., Wang, Y.X., Ma, L., Liu, W., Hebert, M.: Image deformation meta-networks for one-shot learning. In: Proceedings of the IEEE Conference on Computer Vision and Pattern Recognition, pp. 8680–8689 (2019)
8. Choe, J., Park, S., Kim, K., Hyun Park, J., Kim, D., Shim, H.: Face generation for low-shot learning using generative adversarial networks. In: Proceedings of the IEEE International Conference on Computer Vision, pp. 1940–1948 (2017)
9. Crammer, K., Singer, Y.: On the algorithmic implementation of multiclass kernel-based vector machines. J. Mach. Learn. Res. **2**, 265–292 (2001)
10. Dong, N., Xing, E.P.: Domain adaption in one-shot learning. In: Joint European Conference on Machine Learning and Knowledge Discovery in Databases, pp. 573–588. Springer (2018)
11. Fei-Fei, L., Fergus, R., Perona, P.: One-shot learning of object categories. IEEE Trans. Pattern Anal. Mach. Intell. **28**(4), 594–611 (2006)
12. Finn, C., Abbeel, P., Levine, S.: Model-agnostic meta-learning for fast adaptation of deep networks. In: Proceedings of the 34th International Conference on Machine Learning-Volume 70, pp. 1126–1135. JMLR. org (2017)
13. Franceschi, L., Frasconi, P., Salzo, S., Grazzi, R., Pontil, M.: Bilevel programming for hyperparameter optimization and meta-learning. In: International Conference on Machine Learning, pp. 1563–1572 (2018)
14. Gao, H., Shou, Z., Zareian, A., Zhang, H., Chang, S.F.: Low-shot learning via covariance-preserving adversarial augmentation networks. In: Advances in Neural Information Processing Systems, pp. 975–985 (2018)
15. Gidaris, S., Bursuc, A., Komodakis, N., Perez, P., Cord, M.: Boosting few-shot visual learning with self-supervision. In: The IEEE International Conference on Computer Vision (ICCV), October 2019

16. Gidaris, S., Komodakis, N.: Dynamic few-shot visual learning without forgetting. In: Proceedings of the IEEE Conference on Computer Vision and Pattern Recognition, pp. 4367–4375 (2018)
17. Hariharan, B., Girshick, R.: Low-shot visual recognition by shrinking and hallucinating features. In: Proceedings of the IEEE International Conference on Computer Vision, pp. 3018–3027 (2017)
18. He, T., Zhang, Z., Zhang, H., Zhang, Z., Xie, J., Li, M.: Bag of tricks for image classification with convolutional neural networks. In: Proceedings of the IEEE Conference on Computer Vision and Pattern Recognition, pp. 558–567 (2019)
19. Kingma, D.P., Ba, J.L.: Adam: a method for stochastic optimization. In: Proceedings of the 3rd International Conference on Learning Representations (ICLR) (2015)
20. Krizhevsky, A., Hinton, G., et al.: Learning multiple layers of features from tiny images. Technical report, Citeseer (2009)
21. Kurakin, A., Goodfellow, I., Bengio, S.: Adversarial machine learning at scale. In: Proceedings of the 5th International Conference on Learning Representations (ICLR) (2017)
22. Lee, K., Maji, S., Ravichandran, A., Soatto, S.: Meta-learning with differentiable convex optimization. In: Proceedings of the IEEE Conference on Computer Vision and Pattern Recognition, pp. 10657–10665 (2019)
23. Li, H., Dong, W., Mei, X., Ma, C., Huang, F., Hu, B.G.: LGM-NET: learning to generate matching networks for few-shot learning. In: International Conference on Machine Learning, pp. 3825–3834 (2019)
24. van der Maaten, L., Hinton, G.: Visualizing data using t-SNE. J. Mach. Learn. Res. **9**, 2579–2605 (2008)
25. Madry, A., Makelov, A., Schmidt, L., Tsipras, D., Vladu, A.: Towards deep learning models resistant to adversarial attacks. In: Proceedings of the 6rd International Conference on Learning Representations (ICLR) (2018)
26. Mangla, P., Kumari, N., Sinha, A., Singh, M., Krishnamurthy, B., Balasubramanian, V.N.: Charting the right manifold: Manifold mixup for few-shot learning. In: Proceedings of the IEEE/CVF Winter Conference on Applications of Computer Vision (WACV), March 2020
27. Mehrotra, A., Dukkipati, A.: Generative adversarial residual pairwise networks for one shot learning. arXiv preprint arXiv:1703.08033 (2017)
28. Miller, E.G., Matsakis, N.E., Viola, P.A.: Learning from one example through shared densities on transforms. In: Proceedings IEEE Conference on Computer Vision and Pattern Recognition. CVPR 2000 (Cat. No. PR00662), vol. 1, pp. 464–471. IEEE (2000)
29. Mishra, N., Rohaninejad, M., Chen, X., Abbeel, P.: A simple neural attentive meta-learner. In: Proceedings of the 6th International Conference on Learning Representations (ICLR) (2018)
30. Mondal, A.K., Dolz, J., Desrosiers, C.: Few-shot 3D multi-modal medical image segmentation using generative adversarial learning. arXiv preprint arXiv:1810.12241 (2018)
31. Moosavi-Dezfooli, S.M., Fawzi, A., Fawzi, O., Frossard, P.: Universal adversarial perturbations. In: Proceedings of the IEEE Conference on Computer Vision and Pattern Recognition, pp. 1765–1773 (2017)
32. Motiian, S., Jones, Q., Iranmanesh, S., Doretto, G.: Few-shot adversarial domain adaptation. In: Advances in Neural Information Processing Systems, pp. 6670–6680 (2017)

33. Munkhdalai, T., Yuan, X., Mehri, S., Trischler, A.: Rapid adaptation with conditionally shifted neurons. In: International Conference on Machine Learning, pp. 3661–3670 (2018)
34. Oreshkin, B., López, P.R., Lacoste, A.: Tadam: task dependent adaptive metric for improved few-shot learning. In: Advances in Neural Information Processing Systems, pp. 721–731 (2018)
35. Pahde, F., Ostapenko, O., Hnichen, P.J., Klein, T., Nabi, M.: Self-paced adversarial training for multimodal few-shot learning. In: 2019 IEEE Winter Conference on Applications of Computer Vision (WACV), pp. 218–226. IEEE (2019)
36. Ravi, S., Larochelle, H.: Optimization as a model for few-shot learning. In: Proceedings of the 5th International Conference on Learning Representations (ICLR) (2017)
37. Ren, M., Triantafillou, E., Ravi, S., Snell, J., Swersky, K., Tenenbaum, J.B., Larochelle, H., Zemel, R.S.: Meta-learning for semi-supervised few-shot classification. In: Proceedings of the 6th International Conference on Learning Representations (ICLR) (2018)
38. Rusu, A.A., et al.: Meta-learning with latent embedding optimization. In: Proceedings of the 7th International Conference on Learning Representations (ICLR) (2019)
39. Schwartz, E., et al.: Delta-encoder: an effective sample synthesis method for few-shot object recognition. In: Advances in Neural Information Processing Systems, pp. 2845–2855 (2018)
40. Snell, J., Swersky, K., Zemel, R.: Prototypical networks for few-shot learning. In: Advances in Neural Information Processing Systems, pp. 4077–4087 (2017)
41. Su, J., Vargas, D.V., Sakurai, K.: One pixel attack for fooling deep neural networks. IEEE Trans. Evol. Comput. **23**, 828–841 (2019)
42. Sun, Q., Liu, Y., Chua, T.S., Schiele, B.: Meta-transfer learning for few-shot learning. In: Proceedings of the IEEE Conference on Computer Vision and Pattern Recognition, pp. 403–412 (2019)
43. Sung, F., Yang, Y., Zhang, L., Xiang, T., Torr, P.H., Hospedales, T.M.: Learning to compare: Relation network for few-shot learning. In: Proceedings of the IEEE Conference on Computer Vision and Pattern Recognition, pp. 1199–1208 (2018)
44. Szegedy, C., et al.: Intriguing properties of neural networks. In: Proceedings of the 2nd International Conference on Learning Representations (ICLR) (2014)
45. Vinyals, O., Blundell, C., Lillicrap, T., Wierstra, D., et al.: Matching networks for one shot learning. In: Advances in Neural Information Processing Systems, pp. 3630–3638 (2016)
46. Wang, Y.X., Girshick, R., Hebert, M., Hariharan, B.: Low-shot learning from imaginary data. In: Proceedings of the IEEE Conference on Computer Vision and Pattern Recognition, pp. 7278–7286 (2018)
47. Wu, L., Wang, Y., Yin, H., Wang, M., Shao, L., Lovell, B.C.: Few-shot deep adversarial learning for video-based person re-identification. arXiv preprint arXiv:1903.12395 (2019)
48. Xiao, C., Li, B., Zhu, J.Y., He, W., Liu, M., Song, D.: Generating adversarial examples with adversarial networks. In: 27th International Joint Conference on Artificial Intelligence, IJCAI 2018, pp. 3905–3911. International Joint Conferences on Artificial Intelligence (2018)
49. Yoon, S.W., Seo, J., Moon, J.: Tapnet: Neural network augmented with task-adaptive projection for few-shot learning. In: International Conference on Machine Learning, pp. 7115–7123 (2019)

50. Zagoruyko, S., Komodakis, N.: Wide residual networks. arXiv preprint arXiv:1605.07146 (2016)
51. Zhang, H., Zhang, J., Koniusz, P.: Few-shot learning via saliency-guided hallucination of samples. In: Proceedings of the IEEE Conference on Computer Vision and Pattern Recognition, pp. 2770–2779 (2019)
52. Zhang, R., Che, T., Ghahramani, Z., Bengio, Y., Song, Y.: Metagan: an adversarial approach to few-shot learning. In: Advances in Neural Information Processing Systems, pp. 2365–2374 (2018)
53. Zou, H., Zhou, Y., Yang, J., Liu, H., Das, H.P., Spanos, C.J.: Consensus adversarial domain adaptation. Proceedings of the AAAI Conference on Artificial Intelligence, vol. 33, pp. 5997–6004 (2019)

Targeted Attack for Deep Hashing Based Retrieval

Jiawang Bai[1,2], Bin Chen[1,2(✉)], Yiming Li[1], Dongxian Wu[1,2], Weiwei Guo[3], Shu-Tao Xia[1,2], and En-Hui Yang[4]

[1] Tsinghua Shenzhen International Graduate School, Tsinghua University, Shenzhen, China
cb17@mails.tsinghua.edu.cn
[2] Peng Cheng Laboratory, PCL Research Center of Networks and Communications, Shenzhen, China
[3] vivo AI Lab, Shenzhen, China
[4] Department of Electrical and Computer Engineering, University of Waterloo, Waterloo, Canada

Abstract. The deep hashing based retrieval method is widely adopted in large-scale image and video retrieval. However, there is little investigation on its security. In this paper, we propose a novel method, dubbed deep hashing targeted attack (DHTA), to study the targeted attack on such retrieval. Specifically, we first formulate the targeted attack as a *point-to-set* optimization, which minimizes the average distance between the hash code of an adversarial example and those of a set of objects with the target label. Then we design a novel *component-voting scheme* to obtain an *anchor code* as the representative of the set of hash codes of objects with the target label, whose optimality guarantee is also theoretically derived. To balance the performance and perceptibility, we propose to minimize the Hamming distance between the hash code of the adversarial example and the anchor code under the ℓ^∞ restriction on the perturbation. Extensive experiments verify that DHTA is effective in attacking both deep hashing based image retrieval and video retrieval.

Keywords: Targeted attack · Deep hashing · Adversarial attack · Similarity retrieval

1 Introduction

High-dimension and large-scale data approximate nearest neighbor (ANN) retrieval has been widely adopted in online search engines, *e.g.*, Google or

J. Bai, B. Chen and Y. Li—Equal contribution.

Electronic supplementary material The online version of this chapter (https://doi.org/10.1007/978-3-030-58452-8_36) contains supplementary material, which is available to authorized users.

A. Vedaldi et al. (Eds.): ECCV 2020, LNCS 12346, pp. 618–634, 2020.
https://doi.org/10.1007/978-3-030-58452-8_36

Bing, due to its efficiency and effectiveness. Within all ANN retrieval methods, hashing-based methods [42] have attracted a lot of attentions due to their compact binary representations and rapid similarity computation between hash codes with Hamming distance. In particular, deep learning based hashing methods [2,8,19,24,28,35] have shown their superiority in performance since they generally learn more meaningful semantic hash codes through learnable hashing functions with deep neural networks (DNNs).

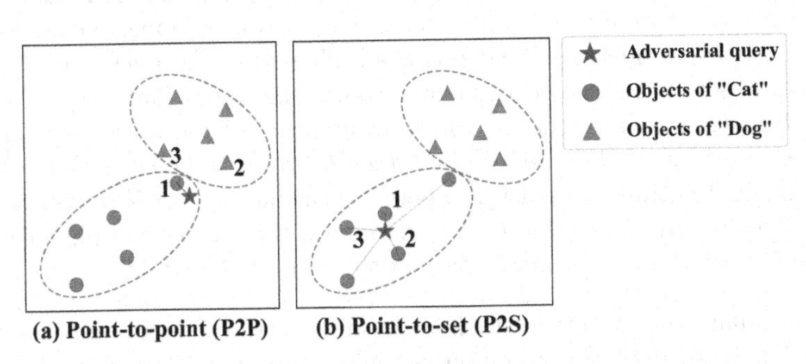

Fig. 1. The comparison between the P2P attack paradigm and proposed P2S paradigm. There are two object classes (*i.e.* 'Cat' and 'Dog') as shown above, where the target label being attack is 'Cat'. In the P2P paradigm, a object with the target label is randomly selected as the reference to generate the adversarial query. But if the selected object is close to the category boundary (dotted lines in the figure) or is an outlier, the attack performance will be poor. In this example, the 'targeted attack success rate' of P2P and P2S is 33.3% and 100%, respectively.

Recent studies [1,6,14,16,39,48] revealed that DNNs are vulnerable to adversarial examples, which are crafted by adding intentionally small perturbations to benign examples and fool DNNs to confidently make incorrect predictions. While deep retrieval systems take advantage of the power of DNNs, they also inherit the vulnerability to adversarial examples [15,25,40,51]. Previous research [51] only paid attention to design a non-targeted attack in deep hashing based retrieval, *i.e.*, returning retrieval objects with incorrect labels. Compared with non-targeted attacks, targeted attacks are more malicious since they make the adversarial examples misidentified as a predefined label and can be used to achieve some malicious purposes [5,13,32]. For example, a hashing based retrieval system may return violent images when a child queries with an intentionally perturbed cartoon image by the adversary. Accordingly, it is desirable to study the targeted attacks on deep hashing models and address their security concerns.

This paper focuses on the targeted attack in hashing based retrieval. Different from classification, retrieval aims at returning multiple relevant objects instead of one result, which indicates that the query has more important relationship with the set of relevant objects than with other objects. Motivated by this fact, we formulate the targeted attack as a *point-to-set* (P2S) optimization, which minimizes the average distance between the compressed representations (*e.g.*, hash codes in Hamming space) of the adversarial example and those of a set of objects with the target label. Compared with the *point-to-point* (P2P) paradigm [40] which directs the adversarial example to generate a representation similar to that of a randomly chosen object with the target label, our proposed point-to-set attack paradigm is more stable and efficient. The detailed comparison between P2S and P2P attack paradigm is shown in Fig. 1. In particular, when minimizing the average Hamming distances between a hash code and those of an object set, we prove that the globally optimal solution (dubbed *anchor code*) can be achieved through a simple component-voting scheme, which is a gift from the nature of hashing-based retrieval. Therefore, the anchor code can be naturally chosen as a targeted hash code to direct the generation of adversarial query. To further balance the attack performance and the imperceptibility, we propose a novel attack method, dubbed ***d****eep* ***h****ashing* ***t****argeted* ***a****ttack* (DHTA), by minimizing the Hamming distance between the hash code of adversarial query and the anchor code under the ℓ^∞ restriction on the adversarial perturbations.

In summary, the main contribution of this work is four-fold:

- We formulate the targeted attack on hashing retrieval as a point-to-set optimization instead of the common point-to-point paradigm considering the characteristics of retrieval tasks.
- We propose a novel component-voting scheme to obtain an anchor code as the representative of the set of hash codes of objects with the target label, whose theoretical optimality of proposed attack paradigm with average-case point-to-set metric is discussed.
- We develop a simple yet effective targeted attack, the DHTA, which efficiently balances the attack performance and the perceptibility. This is the first attempt to design a targeted attack on hashing based retrieval.
- Extensive experiments verify that DHTA is effective in attacking both deep hashing based image retrieval and video retrieval.

2 Related Work

2.1 Deep Hashing Based Similarity Retrieval

Hashing methods can map semantically similar objects to similar compact binary codes in Hamming space, which are widely adopted to accelerate the ANN retrieval [42]. The classical version of data-dependent hashing consists of two parts, including hash function learning and binary inference [26,29,34,43].

Recently, more and more deep learning techniques were introduced to the traditional hashing-based retrieval methods and reach state-of-the-art performance, thanks to the powerful feature extraction of deep neural networks. The first deep hashing method was proposed in [47] focusing on image retrieval. Recent works showed that learning hashing mapping in an end-to-end manner can greatly improve the quality of the binary codes [2,3,23,28]. The above-mentioned methods can be easily extended to multi-label image retrieval, *e.g.*, [44,53]. Depending on the availability of unlabeled images, other researchers devoted to design novel hashing methods to cope with the lack of labeled images, *e.g.*, unsupervised deep hashing method [50], and semi-supervised one [49]. Different from deep image hashing methods, deep video hashing usually first extract frame features by a convolutional neural network (CNN), then fuse them to learn global hashing function. Among various kinds of fusion methods, recurrent neural network (RNN) architecture is the most common choice, which can well model the temporal structure of videos [17]. Moreover, some of the unsupervised video hashing methods were also proposed [27,46], which organize the hash code learning in a self-taught manner to reduce the time and labor consuming labeling.

2.2 Adversarial Attack

DNNs can be easily fooled to confidently make incorrect predictions by intentional and human-imperceptible perturbations. The process of generating adversarial examples is called *adversarial attack*, which was initially proposed by Szegedy *et al.* [39] in the image classification task. To achieve such adversarial examples, the fast gradient sign method (FGSM) [16] aims to maximize the loss along the gradient direction. After that, projected gradient descent (PGD) [22] was proposed to reach better performance. Deepfool finds the smallest perturbation by exploring the nearest decision boundary [31]. Except for the aforementioned attacks, many other methods [4,10,45,52] have also been developed to find the adversarial perturbation in the image classification problem.

Besides, there are also other DNN based tasks that inherit the vulnerability to adversarial examples [11,12,30,48]. Especially for the deep learning based similarity retrieval, it raises wide concerns on its security issues. For feature-based retrieval, Li *et al.* [25] focused on non-targeted attack by adding universal adversarial perturbations (UAPs), while targeted mismatch adversarial attack was explored in [40]. In [15], adversarial queries for deep product quantization network are generated by perturbing the overall soft-quantized distributions. However, for hashing based retrieval, one of the most important retrieval methods, its robustness analysis is left far behind. There is only one previous work in attacking deep hashing based retrieval [51], which paid attention to the non-targeted attack, *i.e.*, returning retrieval objects with the incorrect label. The targeted attack in such retrieval a system remains unaddressed.

3 The Proposed Method

3.1 Preliminaries

In this section, we briefly review the process of deep hashing based retrieval. Suppose $\boldsymbol{X} = \{(\boldsymbol{x}_i, \boldsymbol{y}_i)\}_{i=1}^{N}$ indicates a set of N sample collection labeled with C classes, where $\boldsymbol{x}_i$ indicates the retrieval object, *e.g.*, a image or a video, and $\boldsymbol{y}_i \in \{0,1\}^C$ corresponds to a label vector. The c-th component of indicator vector $\boldsymbol{y}_i^c = 1$ means that the sample $\boldsymbol{x}_i$ belongs to class c. Let $\boldsymbol{X}^{(t)} = \{(\boldsymbol{x}, \boldsymbol{y}) \in \boldsymbol{X} \mid \boldsymbol{y} = \boldsymbol{y}_t\}$ be a subset of $\boldsymbol{X}$ consisting of those objects with label $\boldsymbol{y}_t$.

Deep Hashing Model. The hash code of a query object $\boldsymbol{x}$ of deep hashing model is generated as follows:

$$\boldsymbol{h} = F(\boldsymbol{x}) = \text{sign}\left(f_\theta(\boldsymbol{x})\right), \tag{1}$$

where $f_\theta(\cdot)$ is a DNN. In general, $f_\theta(\cdot)$ consists of a feature extractor followed by the fully-connected layers. Specifically, the feature extractor is usually specified by a CNN for image retrieval [2,3,7], while CNN stacked with RNN is widely adopted for video retrieval [17,27,37]. In particular, the sign$(\cdot)$ function is approximated by the tanh$(\cdot)$ function during the training process in deep hashing based retrieval methods to alleviate the *gradient vanishing* problem [3].

Similarity-Based Retrieval. Given a deep hashing model $F(\cdot)$, a query object $\boldsymbol{x}$ and a object database $\{\boldsymbol{x}_i\}_{i=1}^M$, the retrieval process is as follows. Firstly, the query $\boldsymbol{x}$ is fed into the deep hashing model and binary code $F(\boldsymbol{x})$ can be obtained through Eq. (1). Secondly, the Hamming distance between the hash code of query $\boldsymbol{x}$ and that of each object $\boldsymbol{x}_i$ in the database is calculated, denoted as $d_H(F(\boldsymbol{x}), F(\boldsymbol{x}_i))$. Finally, the retrieval system returns a list of objects, which is produced by sorting their Hamming distances.

3.2 Deep Hashing Targeted Attack

Problem Formulation. In general, given a benign query $\boldsymbol{x}$, the objective of targeted attack in retrieval is to generate an attacked version $\boldsymbol{x}'$ of $\boldsymbol{x}$, which would cause the targeted model to retrieve objects with the target label $\boldsymbol{y}_t$. This objective can be achieved through minimizing the distance between the hash code of the attacked sample $\boldsymbol{x}'$ and those of the object subset $\boldsymbol{X}^{(t)}$ with the target label $\boldsymbol{y}_t$, *i.e.*,

$$\min_{\boldsymbol{x}'} d\left(F(\boldsymbol{x}'), F(\boldsymbol{X}^{(t)})\right), \tag{2}$$

where $F(\boldsymbol{X}^{(t)}) = \{F(\boldsymbol{x})|\boldsymbol{x} \in \boldsymbol{X}^{(t)})\}$, and $d(\cdot,\cdot)$ denotes a point-to-set metric.

Once the problem is formulated as objective (2), the remaining problem is how to define the point-to-set metric. In this paper, we use the most widely used point-to-set metric, the *average-case metric*, as shown in Definition 1.

Definition 1. *Given a point* $\boldsymbol{h}_0 \in \{-1,+1\}^K$ *and a set of points* $\mathcal{A}$ *in* $\{-1,+1\}^K$ *and point-to-point metric* d_H*, the average-case point-to-set metric is defined as follows:*

$$d_{Ave}\left(\boldsymbol{h}_0, \mathcal{A}\right) \triangleq \frac{1}{|\mathcal{A}|} \sum_{\boldsymbol{h}\in\mathcal{A}} d_H(\boldsymbol{h}_0, \boldsymbol{h}). \tag{3}$$

Remark 1. If average-case point-to-set metric is adopted, the objective function (2) is specified as

$$\min_{\boldsymbol{h}'} \frac{1}{|\mathcal{A}|} \sum_{\boldsymbol{h}\in F(\boldsymbol{X}^{(t)})} d_H(\boldsymbol{h}', \boldsymbol{h}), \tag{4}$$

where $\boldsymbol{h}'$ is the hash code corresponding to the adversarial example $\boldsymbol{x}'$.

In particular, there exists an analytical optimal solution (dubbed *anchor code*) of the optimization problem (4) obtained through a *component-voting scheme*, which is a property arising from the nature of Hamming distance of hashing-based retrieval. The *component-voting scheme* is shown in Algorithm 1, and the optimality of anchor code is verified in Theorem 1. The proof is shown in the **Appendix**.

Theorem 1. *Anchor code* $\boldsymbol{h}_a$ *calculated by Algorithm 1 is the binary code achieving the minimal sum of Hamming distances with respect to* $\boldsymbol{h}_i$, $i = 1, \ldots, n_t$, *i.e.*,

$$\boldsymbol{h}_a = \arg \min_{\boldsymbol{h} \in \{+1,-1\}^K} \sum_{i=1}^{n_t} d_H(\boldsymbol{h}, \boldsymbol{h}_i). \tag{5}$$

Algorithm 1. Component-voting Scheme

Input: K-bits hash codes $\{\boldsymbol{h}_i\}_{i=1}^{n_t}$ of objects with the target label t.
Output: Anchor code $\boldsymbol{h}_a$.
1: **for** $j = 1 : K$ **do**
2: Conduct voting process through counting up the number of $+1$ and -1, denoted by N_{+1}^j and N_{-1}^j, respectively. For the j-th component among $\{\boldsymbol{h}_i\}_{i=1}^{n_t}$, *i.e.*,

$$N_{+1}^j = \sum_{i}^{n_t} \mathbb{I}(\boldsymbol{h}_i^j = +1), \qquad N_{-1}^j = \sum_{i=1}^{n_t} \mathbb{I}(\boldsymbol{h}_i^j = -1), \tag{6}$$

where $\mathbb{I}(\cdot)$ is an indicator function.
3: Determine the j-th component of anchor code $\boldsymbol{h}_a^j$ as

$$\boldsymbol{h}_a^j = \begin{cases} +1, & \text{if } N_{+1}^j \geqslant N_{-1}^j \\ -1, & \text{otherwise} \end{cases}. \tag{7}$$

4: **end for**
5: **return** Anchor code $\boldsymbol{h}_a$.

Overall Objective Function. Due to the optimal representative property of anchor code for the set of hash codes of objects with the target label (Theorem 1), we can naturally choose the anchor code as a targeted hash code to direct the generation of the adversarial query. However, the attacked object corresponding to the anchor code may be far different from the original one visually, which would cause the attacked object to be easily detectable. To solve this problem, we introduce the ℓ^∞ restriction on the adversarial perturbations by minimizing the Hamming distance between the hash code of attacked object and that of the anchor code as follows:

$$\min_{\boldsymbol{x}'} d_H(\text{sign}(f_\theta(\boldsymbol{x}')), \boldsymbol{h}_a) \quad s.t. \ \|\boldsymbol{x}' - \boldsymbol{x}\|_\infty \le \epsilon, \tag{8}$$

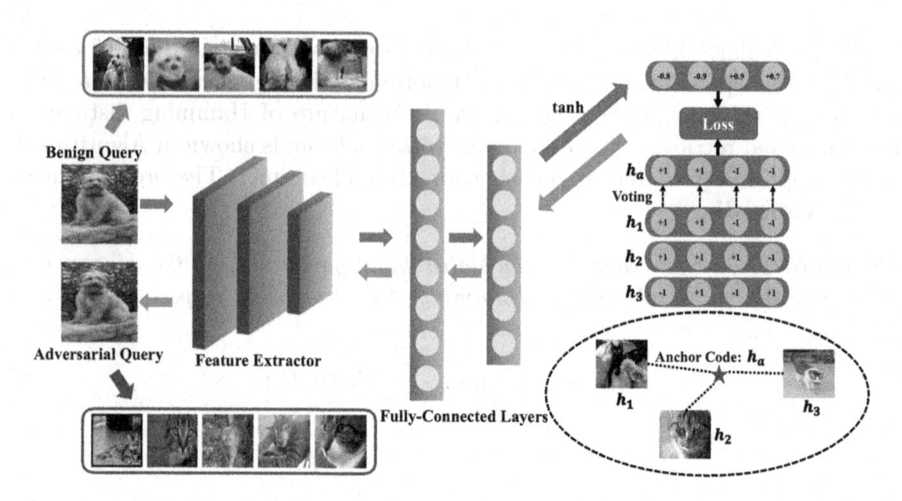

Fig. 2. The pipeline of proposed DHTA method, where the gray and orange arrows indicate forward and backward propagation, respectively. The adversarial query is generated through minimizing the loss calculated by its hash code and the anchor code of the set of objects with the target label. The anchor code h_a is calculated through the component-voting scheme (*i.e.* an entry-wise voting process). In this toy example, $\boldsymbol{h}_1$, $\boldsymbol{h}_2$ and $\boldsymbol{h}_3$ are three 4 bits hash codes of objects with the target label "Cat".

where ϵ denotes the maximum perturbation strength, $\boldsymbol{h}_a$ is the anchor code of object set with the target label.

Besides, given a pair of binary codes $\boldsymbol{h}_i$ and $\boldsymbol{h}_j$, since $d_H(\boldsymbol{h}_i, \boldsymbol{h}_j) = \frac{1}{2}(K - \boldsymbol{h}_i^\top \boldsymbol{h}_j)$, we can equivalently replace Hamming distance with inner product in the objective function. In particular, similar to deep hashing methods [3], we adopt the hyperbolic tangent (tanh) function to approximate sign function for the adversarial generation. Similar to [51], we also introduce the factor α to address the *gradient vanishing* problem. In summary, the overall optimization objective of proposed method is as follows:

$$\min_{\boldsymbol{x}'} \ -\frac{1}{K}\boldsymbol{h}_a^\top \tanh(\alpha f_\theta(\boldsymbol{x}')) \quad s.t. \ \ ||\boldsymbol{x}' - \boldsymbol{x}||_\infty \leq \epsilon, \tag{9}$$

where the hyper-parameter $\alpha \in [0, 1]$, $\boldsymbol{h}_a$ is the anchor code.

The overall process of proposed DHTA is shown in Fig. 2.

4 Experiments

4.1 Benchmark Datasets and Evaluation Metrics

Four benchmark datasets, including ImageNet [33], NUS-WIDE [9], JHMDB [20], and UCF-101 [38], are adopted in our experiments. The first two datasets are used for image retrieval, while the last two are used for video retrieval. More details about datasets are described in the **Appendix**.

For the evaluation of targeted attacks, we define the targeted mean average precision (t-MAP) as the evaluation metric, which is similar to mean average precision (MAP) widely used in information retrieval [54]. Specifically, the referenced label of t-MAP is the targeted label instead of the original one of the query object in MAP. The higher the t-MAP, the better the targeted attack performance. In image hashing, we evaluate t-MAP on top 5,000 and 1,000 retrieved images on NUS-WIDE and ImageNet, respectively. We evaluate t-MAP on all retrieved videos in video hashing. Besides, we also present the precision-recall curves (PR curves) of different methods for more comprehensive comparison.

Table 1. t-MAP (%) of targeted attack methods and MAP (%) of query with benign objects ('Original') with various code lengths on two image datasets.

Method	Metric	ImageNet				NUS-WIDE			
		16 bits	32 bits	48 bits	64 bits	16 bits	32 bits	48 bits	64 bits
Original	t-MAP	3.80	1.36	1.64	1.98	37.62	36.03	38.32	38.69
Noise	t-MAP	3.29	1.24	1.89	2.10	37.34	36.15	38.25	38.57
P2P	t-MAP	44.35	58.32	62.50	65.61	75.45	78.59	81.40	81.28
DHTA	t-MAP	**63.68**	**77.76**	**82.31**	**82.10**	**82.35**	**85.66**	**86.80**	**88.84**
Original	MAP	51.02	62.70	67.80	70.11	76.93	80.37	82.06	81.62

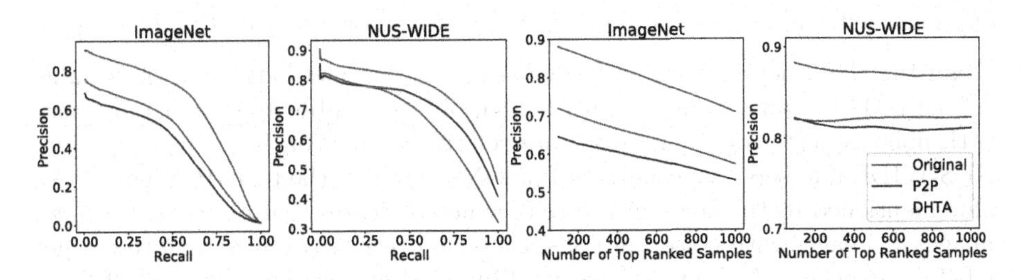

Fig. 3. Precision-recall and precision curves under 48 bits code length in image retrieval. P2P attack and DHTA are evaluated based on the target label, while the result of 'Original' is calculated based on the label of query object.

4.2 Overall Results on Image Retrieval

Evaluation Setup. For image hashing, we adopt VGG-11 [36] as the backbone network pre-trained on ImageNet to extract features, then replace the last fully-connected layer of softmax classifier with the hashing layer. The detailed settings of training image hashing models are illustrated in the **Appendix**. For each dataset, we randomly select 100 samples from the query set as benign queries

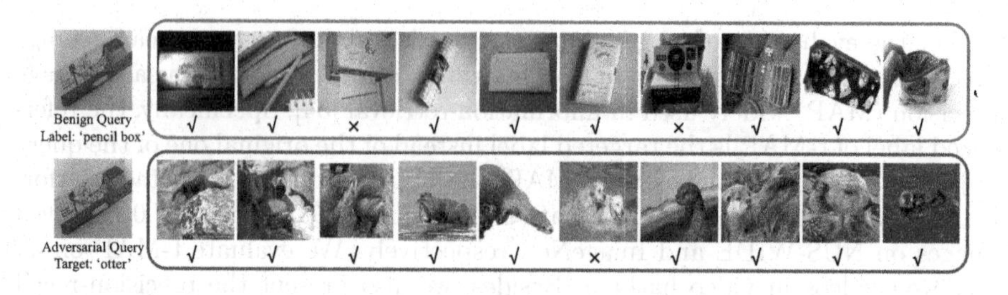

Fig. 4. An example of image retrieval with benign query and its adversarial query on ImageNet. Retrieved objects with top-10 similarity are shown in the box. The tick and cross indicate whether the retrieved object is consistent with the desired label (the original label for benign query and the target label for adversarial query).

to evaluate the performance of attack. For each generation, we randomly select a label as the target label different from the label of query. When generating an anchor code, we randomly sample images from objects in the database with the target label to form the hash code set. For all adversarial examples, the perturbation magnitude ϵ of normalized data and n_t is set to 0.032 and 9, respectively. We adopt PGD [22] to optimize the proposed attack. We attack image hashing models with learning rate 1 and the number of iterations is set to 2,000. Following [51], the parameter α is set as 0.1 during the first 1,000 iterations, and is updated every 200 iterations according to [0.2, 0.3, 0.5, 0.7, 1.0] during the last 1,000 iterations. We compare DHTA with targeted attack with P2P paradigm [40], which is specified as DHTA with $n_t = 1$. We also show the t-MAP results of images with additive noise sampled from the uniform distribution $U(-\epsilon, +\epsilon)$.

Results. The general attack performance of different methods is shown in Table 1. The t-MAP values of query with benign objects (dubbed *Original*) or query with noisy objects (dubbed *Noise*) are relatively small on both ImageNet and NUS-WIDE datasets. Especially on ImageNet dataset, the t-MAP values of two aforementioned methods are closed to 0, which indicates that query with benign images or images with noise can not successfully retrieve objects with the target labels as expected. In contrast, designed targeted attack methods (*i.e.* P2P and DHTA) can significantly improve the t-MAP values. For example, compared with the t-MAP of benign query on ImageNet dataset, the improvement of P2P methods is over 40% in all cases. Especially under the relatively large code length (64 bits), the improvement even goes to 63%. Among two targeted attack methods, the proposed DHTA method achieves the best performance. Compared with P2P, the t-MAP improvement of DHTA is over 16% (usually over 19%) in all cases on the ImageNet dataset. Moreover, the t-MAP values of targeted attacks increase as the number of bits, which is probably caused by the extra information introduced in the longer code length. In particular, an interesting phenomenon is that the t-MAP value of DHTA is even significantly higher than the MAP value of 'Original', which suggests that the attack performance of DHTA is not hindered by the performance of the original hashing model (*i.e.* threat model) to some extent. An example of the results of query with a benign image and an adversarial image is displayed in Fig. 4.

Furthermore, we also provide the precision-recall and precision curves for a more comprehensive comparison. As shown in Fig. 3, the curves of DHTA are always above all other curves, which demonstrates that the performance of DHTA does better than all other methods.

Table 2. t-MAP (%) of targeted attack methods and MAP (%) of query with benign objects ('Original') with various code lengths on two video datasets.

Method	Metric	JHMDB				UCF-101			
		16 bits	32 bits	48 bits	64 bits	16 bits	32 bits	48 bits	64 bits
Original	t-MAP	6.73	6.26	6.48	6.89	1.69	1.67	1.79	1.86
Noise	t-MAP	6.67	6.13	6.50	6.94	1.69	1.72	1.87	1.85
P2P	t-MAP	39.67	42.37	44.78	44.38	55.57	53.49	55.27	51.88
DHTA	t-MAP	**56.47**	**62.04**	**63.02**	**66.06**	**67.84**	**66.18**	**69.72**	**67.83**
Original	MAP	35.18	42.46	45.80	45.50	55.16	55.25	56.56	56.79

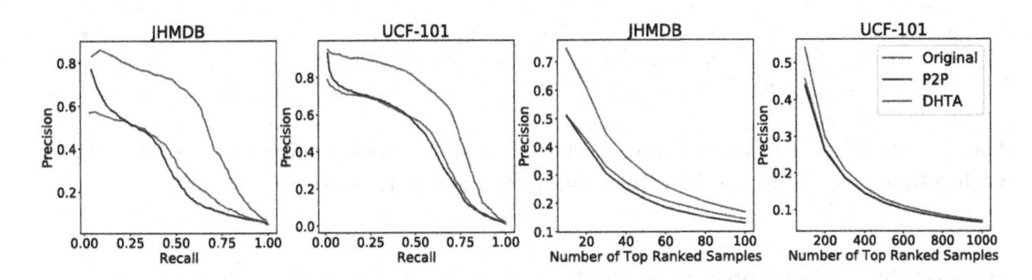

Fig. 5. Precision-recall and precision curves under 48 bits code length in video retrieval. P2P attack and DHTA are evaluated based on the target label, while the result of 'Original' method is calculated based on the label of query object.

4.3 Overall Results on Video Retrieval

Evaluation Setup. According to model architectures of the state-of-the-art deep video retrieval methods [17,27,37], we adopt AlexNet [21] to extract spatial features and LSTM [18] to fuse the temporal information. The detailed settings of training video hashing model are presented in the **Appendix**. For attacking video hashing, the number of iterations is 500, and the parameter α is fixed at 0.1. Other settings are the same as those used in Sect. 4.2.

Results. The attack performance in video retrieval is shown in Table 2. Similar to the image scenario, query with benign videos or videos with noise can not

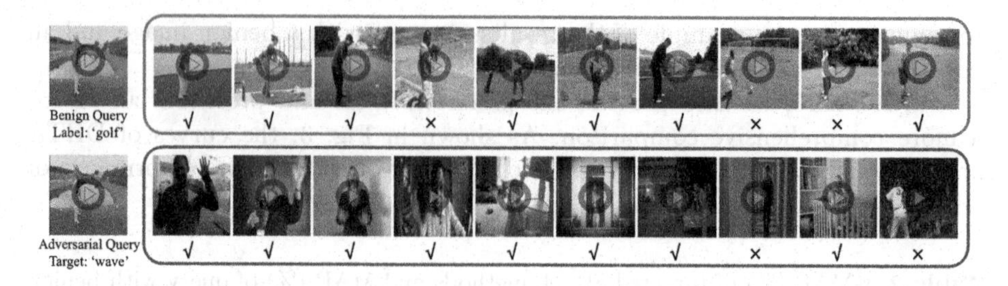

Fig. 6. An example of video retrieval with benign query and its adversarial query on JHMDB. Retrieved objects with top-10 similarity are shown in the box. The tick and cross indicate whether the retrieved object is consistent with the desired label (the original label for benign query and the target label for adversarial query).

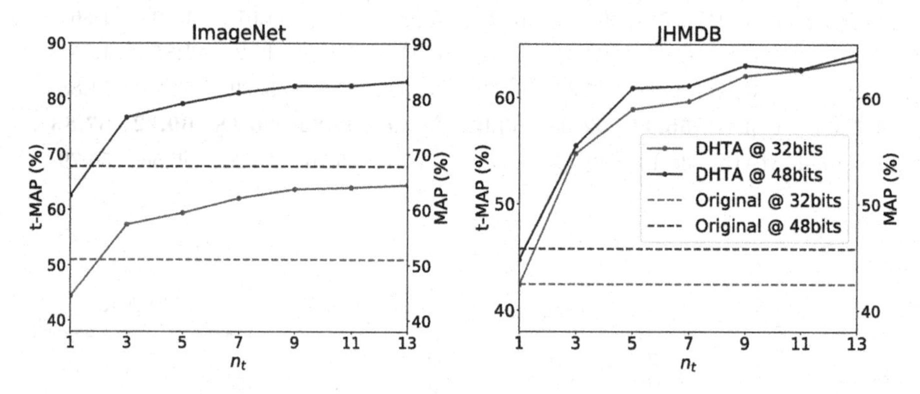

Fig. 7. t-MAP (%) of DHTA and MAP (%) of query with benign objects ('Original') with different n_t and code length on ImageNet and JHMDB.

successfully retrieve objects with the target label, thus fails to attack the deep hashing based retrieval. In contrast, deep hashing based video retrieval can be easily attacked by designed targeted attacks, especially the DHTA proposed in this paper. For example, the t-MAP value of DHTA is 59% over query with benign videos, and 21% over P2P attack paradigm on the JHMDB dataset with code length 64 bits. The precision-recall and the precision curves also verify the superiority of DHTA over other methods, as shown in Fig. 5. Especially on JHMDB dataset, there exists a significantly large gap between the PR curve of DHTA and those of other methods. In addition, the t-MAP value of DHTA is again significantly larger than the MAP of the benign query (the 'Original'). An example of the results of query with a benign video and an adversarial video is displayed in Fig. 6.

4.4 Discussion

Effect of n_t. To analyze the effect of the size of object set for generating the anchor code (*i.e.*, n_t), we discuss the t-MAP of DHTA under different values

Table 3. t-MAP (%) of DHTA with different iterations on ImageNet.

Iteration	16 bits	32 bits	48 bits	64 bits
100	52.99	66.29	70.65	72.43
500	55.18	68.30	74.47	76.15
1000	56.96	68.36	75.03	76.25
1500	62.81	74.11	79.28	78.71
2000	63.68	77.76	82.31	82.10

Table 4. t-MAP (%) of DHTA with different iterations on JHMDB.

Iteration	16 bits	32 bits	48 bits	64 bits
10	28.51	23.88	22.84	23.21
50	48.69	48.18	47.01	48.97
100	53.21	54.91	55.94	58.28
500	56.47	62.04	63.02	66.06

of $n_t \in \{1, 3, 5, 7, 9, 11, 13\}$. Other settings are the same as those used in Sect. 4.2–4.3. We use ImageNet and JHMDB as the representative for analysis.

As shown in Fig. 7, the t-MAP value increase as the increase of n_t under different code lengths. The MAP of corresponding query with benign objects (*i.e.* the 'Original') can be regarded as the reference of the retrieval performance. We observe that the t-MAP is higher than the MAP of its corresponding 'Original' method in all cases when $n_t \geq 3$. In other words, DHTA can still have satisfying performance with relatively small n_t. This advantage is critical for attackers, since the bigger the n_t, the higher the cost of data collection and adversarial generation for an attack. It is worth noting that the attack performance degrades significantly when $n_t = 1$, which exactly corresponds to the P2P attack paradigm.

Effect of the Number of Iterations. Table 3–4 present the t-MAP of DHTA with different iterations on ImageNet and JHMDB datasets. Except for the iterations, other settings are the same as those used in Sect. 4.2–4.3.

As expected, the t-MAP values increase with the number of iterations. Even with relatively few iterations, the proposed DHTA can still achieve satisfying performance. For example, with 100 iterations, the t-MAP values are over 50% under all code lengths. Especially on the ImageNet dataset, the t-MAP is over 70% with relatively larger code length (≥48 bits). These results consistently verify the high-efficiency of our DHTA method.

Evaluation from the Perspective of Non-targeted Attack. Targeted attack can be regarded as a special non-targeted attack, since the target label is usually different from the one of query object. In this part, we compare the targeted attacks (P2P and DHTA) with other methods, including additive noise and HAG [51] (which is the state-of-the-art non-targeted attack), in the non-targeted attack scenario.

The MAP results of different methods are reported in Table 5. The lower the MAP, the better the non-targeted attack performance. As shown in the table, although targeted attacks are not designed for the non-targeted scenario, they still have competitive performance. For example, the MAP values of DHTA are 50% smaller than those of 'Original' under all code length on ImageNet.

Table 5. MAP (%) of different methods on ImageNet and JHMDB. The best results are marked with boldface, while the second best results are marked with underline.

Method	ImageNet				JHMDB			
	16 bits	32 bits	48 bits	64 bits	16 bits	32 bits	48 bits	64 bits
Original	51.02	62.70	67.80	70.11	35.18	42.46	45.80	45.50
Noise	50.94	62.52	66.69	69.85	35.04	42.15	45.67	45.63
P2P	3.36	**2.48**	2.45	3.93	7.71	8.20	8.14	10.19
HAG	1.88	4.96	3.89	2.34	**3.52**	**3.58**	**3.42**	**3.34**
DHTA	**0.54**	5.64	**2.30**	**1.70**	6.76	7.23	6.56	7.55

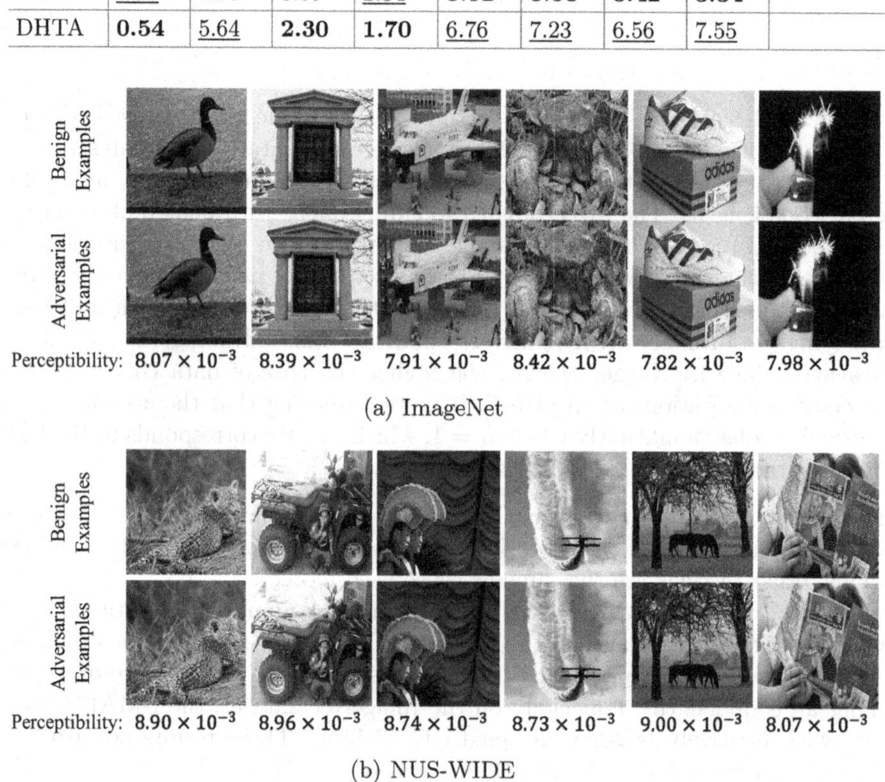

(a) ImageNet

(b) NUS-WIDE

Fig. 8. Visualization examples of generated adversarial examples in image hashing.

Especially for the proposed DHTA, it even has better non-targeted attack performance (*i.e.* smaller MAP) compared with HAG on ImageNet in most cases.

Perceptibility. Except for the attack performance, the *perceptibility* of adversarial perturbations is also important. Following the setting suggested in [39,41], given a benign query $\boldsymbol{x}$, the perceptibility of its corresponding adversarial query $\boldsymbol{x}'$ is defined as $\sqrt{\frac{1}{n}\|\boldsymbol{x}'-\boldsymbol{x}\|_2^2}$, where n is the size of the object and pixel values are scaled to be in the range [0, 1].

For each dataset, we calculate the average perceptibility over all generated adversarial objects. The perceptibility value of ImageNet and NUS-WIDE datasets is 8.35×10^{-3} and 9.07×10^{-3}, respectively. In video retrieval tasks, the value is 5.81×10^{-3} and 7.72×10^{-3} on JHMDB and UCF-101 datasets, respectively. These results indicate that the adversarial queries are very similar to their original versions. Some adversarial images are shown in Fig. 8, while examples of video retrieval are shown in the **Appendix**.

4.5 Open-Set Targeted Attack

Evaluation Setup. In the above experiments, the target label is selected from those of training set. In this section, we use ImageNet dataset as an example to further evaluate the proposed DHTA under a tougher open-set scenario, where the out-of-sample class will be assigned as the target label. This setting is more realistic since the attacker may probably not be able to access the training set of the attacked deep hashing model. For example, the deep hashing model may be downloaded from a third-party open-source platform where the training set is unavailable.

Table 6. t-MAP (%) of DHTA with out-of-sample target label on ImageNet.

Method	16 bits	32 bits	48 bits	64 bits
DHTA ($n_t = 5$)	33.67	46.34	48.91	48.27
DHTA ($n_t = 7$)	34.77	50.92	51.68	49.18
DHTA ($n_t = 9$)	37.34	54.13	55.12	52.17
DHTA ($n_t = 11$)	38.00	54.05	56.93	54.12

Specifically, we randomly select 10 additional classes different from those used for training a deep hashing model in Sect. 4.1. These selected images from 10 additional classes will be treated as an open set for our evaluation. When generating the anchor code of objects with the target label (within the open set), we remain our deep hashing model trained on the previous 100 classes.

Results. As shown in Table 6, DHTA still has a certain attack effect even if the target label is out-of-sample. Especially when the n_t and the code length tend larger, the t-MAP values of DHTA are over 50%. This phenomenon may reveal that the learned feature extractor did learn some useful low-level features, which represents those objects with the same class in some similar locations in Hamming space, no matter the class is learned or not. In addition, the attack performance is also increasing with the n_t and code length.

5 Conclusion and Future Work

In this paper, we explore the landscape of the targeted attack for deep hashing based retrieval. Based on the characteristics of the retrieval task, we formulate the attack as a point-to-set optimization, which minimizes the average distance between the hash code of the adversarial example and those of a set of objects with the target label. Theoretically, we propose a component-voting scheme to obtain the optimal representative, the anchor code, for the code set of point-to-set optimization. Based on the anchor code, we propose a novel targeted attack method, the DHTA, to balance the performance and perceptibility through minimizing the Hamming distance between the hash code of adversarial example and the anchor code under the ℓ^∞ restriction on the adversarial perturbation. Extensive experiments are conducted, which verifies the effectiveness of DHTA in attacking both deep hashing based image retrieval and video retrieval. To alleviate the proposed threat, we will discuss how to generalize existing adversarial training based methods from P2P to the P2S scheme for the defense. The specific approaches will be further demonstrated in our future works.

Acknowledgments. This work is supported in part by the National Key Research and Development Program of China under Grant 2018YFB1800204, the National Natural Science Foundation of China under Grant 61771273, the R&D Program of Shenzhen under Grant JCYJ20180508152204044, the project "PCL Future Greater-Bay Area Network Facilities for Large-scale Experiments and Applications (LZC0019)", the Natutal Sciences and Engineering Research Council of Canada under Grant RGPIN203035-16, and the Canada Research Chairs Program. We also thank vivo and Rejoice Sport Tech. co., LTD. for their GPUs.

References

1. Bai, Y., Zeng, Y., Jiang, Y., Wang, Y., Xia, S.T., Guo, W.: Improving query efficiency of black-box adversarial attack. In: ECCV 2020 (2020)
2. Cao, Y., Long, M., Liu, B., Wang, J.: Deep Cauchy Hashing for hamming space retrieval. In: CVPR (2018)
3. Cao, Z., Long, M., Wang, J., Yu, P.S.: HashNet: deep learning to hash by continuation. In: ICCV (2017)
4. Carlini, N., Wagner, D.: Towards evaluating the robustness of neural networks. In: IEEE S&P (2017)
5. Carlini, N., Wagner, D.: Audio adversarial examples: targeted attacks on speech-to-text. In: IEEE S&P Workshops (2018)
6. Chen, W., Zhang, Z., Hu, X., Wu, B.: Boosting decision-based black-box adversarial attacks with random sign flip. In: ECCV (2020)
7. Chen, Y., Lai, Z., Ding, Y., Lin, K., Wong, W.K.: Deep supervised hashing with anchor graph. In: CVPR (2019)
8. Chen, Z., Yuan, X., Lu, J., Tian, Q., Zhou, J.: Deep hashing via discrepancy minimization. In: CVPR (2018)
9. Chua, T.S., Tang, J., Hong, R., Li, H., Luo, Z., Zheng, Y.: NUS-WIDE: a real-world web image database from national university of Singapore. In: ICMR (2009)
10. Dong, Y., et al.: Boosting adversarial attacks with momentum. In: CVPR (2018)
11. Dong, Y., et al.: Efficient decision-based black-box adversarial attacks on face recognition. In: CVPR (2019)
12. Duan, R., Ma, X., Wang, Y., Bailey, J., Qin, A.K., Yang, Y.: Adversarial camouflage: hiding physical-world attacks with natural styles. In: CVPR (2020)

13. Eykholt, K., et al.: Robust physical-world attacks on deep learning visual classification. In: CVPR (2018)
14. Fan, Y., Wu, B., Li, T., Zhang, Y., Li, M., Li, Z., Yang, Y.: Sparse adversarial attack via perturbation factorization. In: ECCV (2020)
15. Feng, Y., Chen, B., Dai, T., Xia, S.T.: Adversarial attack on deep product quantization network for image retrieval. In: AAAI (2020)
16. Goodfellow, I.J., Shlens, J., Szegedy, C.: Explaining and harnessing adversarial examples. In: ICLR (2015)
17. Gu, Y., Ma, C., Yang, J.: Supervised recurrent hashing for large scale video retrieval. In: ACM MM (2016)
18. Hochreiter, S., Schmidhuber, J.: Long short-term memory. Neural Comput. **9**(8), 1735–1780 (1997)
19. Hu, D., Nie, F., Li, X.: Deep binary reconstruction for cross-modal hashing. IEEE Trans. Multimedia **21**(4), 973–985 (2018)
20. Jhuang, H., Gall, J., Zuffi, S., Schmid, C., Black, M.J.: Towards understanding action recognition. In: ICCV (2013)
21. Krizhevsky, A., Sutskever, I., Hinton, G.E.: ImageNet classification with deep convolutional neural networks. In: NeurIPS, pp. 1097–1105 (2012)
22. Kurakin, A., Goodfellow, I., Bengio, S.: Adversarial examples in the physical world. In: ICLR (2017)
23. Lai, H., Pan, Y., Liu, Y., Yan, S.: Simultaneous feature learning and hash coding with deep neural networks. In: CVPR (2015)
24. Li, C., Deng, C., Li, N., Liu, W., Gao, X., Tao, D.: Self-supervised adversarial hashing networks for cross-modal retrieval. In: CVPR (2018)
25. Li, J., Ji, R., Liu, H., Hong, X., Gao, Y., Tian, Q.: Universal perturbation attack against image retrieval. In: ICCV (2019)
26. Li, P., Wang, M., Cheng, J., Xu, C., Lu, H.: Spectral hashing with semantically consistent graph for image indexing. IEEE Trans. Multimedia **15**(1), 141–152 (2012)
27. Li, S., Chen, Z., Lu, J., Li, X., Zhou, J.: Neighborhood preserving hashing for scalable video retrieval. In: ICCV, pp. 8212–8221 (2019)
28. Liu, H., Wang, R., Shan, S., Chen, X.: Deep supervised hashing for fast image retrieval. In: CVPR (2016)
29. Liu, W., Wang, J., Ji, R., Jiang, Y.G., Chang, S.F.: Supervised hashing with kernels. In: CVPR (2012)
30. Ma, X., et al.: Understanding adversarial attacks on deep learning based medical image analysis systems. Pattern Recogn. (2019)
31. Moosavi-Dezfooli, S.M., Fawzi, A., Frossard, P.: DeepFool: a simple and accurate method to fool deep neural networks. In: CVPR (2016)
32. Qin, Y., Carlini, N., Cottrell, G., Goodfellow, I., Raffel, C.: Imperceptible, robust, and targeted adversarial examples for automatic speech recognition. In: ICML (2019)
33. Russakovsky, O., et al.: ImageNet large scale visual recognition challenge. Int. J. Comput. Vision **115**(3), 211–252 (2015)
34. Shen, F., Shen, C., Liu, W., Tao Shen, H.: Supervised discrete hashing. In: CVPR (2015)
35. Shen, F., Xu, Y., Liu, L., Yang, Y., Huang, Z., Shen, H.T.: Unsupervised deep hashing with similarity-adaptive and discrete optimization. IEEE Trans. Pattern Anal. Mach. Intell. **40**(12), 3034–3044 (2018)
36. Simonyan, K., Zisserman, A.: Very deep convolutional networks for large-scale image recognition. In: ICLR (2015)

37. Song, J., Zhang, H., Li, X., Gao, L., Wang, M., Hong, R.: Self-supervised video hashing with hierarchical binary auto-encoder. IEEE Trans. Image Process. **27**(7), 3210–3221 (2018)
38. Soomro, K., Zamir, A.R., Shah, M.: Ucf101: a dataset of 101 human actions classes from videos in the wild. arXiv preprint arXiv:1212.0402 (2012)
39. Szegedy, C., et al.: Intriguing properties of neural networks. In: ICLR (2014)
40. Tolias, G., Radenovic, F., Chum, O.: Targeted mismatch adversarial attack: query with a flower to retrieve the tower. In: ICCV (2019)
41. Tramèr, F., Kurakin, A., Papernot, N., Goodfellow, I., Boneh, D., McDaniel, P.: Ensemble adversarial training: attacks and defenses. In: ICLR (2018)
42. Wang, J., Zhang, T., Sebe, N., Shen, H.T., et al.: A survey on learning to hash. IEEE Trans. Pattern Anal. Mach. Intell. **40**(4), 769–790 (2017)
43. Wang, J., Liu, W., Kumar, S., Chang, S.F.: Learning to hash for indexing big data-a survey. Proc. IEEE **104**(1), 34–57 (2015)
44. Wu, D., Lin, Z., Li, B., Ye, M., Wang, W.: Deep supervised hashing for multi-label and large-scale image retrieval. In: ICMR (2017)
45. Wu, D., Wang, Y., Xia, S.T., Bailey, J., Ma, X.: Skip connections matter: on the transferability of adversarial examples generated with ResNets. In: ICLR (2020)
46. Wu, G., et al.: Unsupervised deep video hashing via balanced code for large-scale video retrieval. IEEE Trans. Image Process. **28**(4), 1993–2007 (2018)
47. Xia, R., Pan, Y., Lai, H., Liu, C., Yan, S.: Supervised hashing for image retrieval via image representation learning. In: AAAI (2014)
48. Xu, Y., et al.: Exact adversarial attack to image captioning via structured output learning with latent variables. In: CVPR (2019)
49. Yan, X., Zhang, L., Li, W.J.: Semi-supervised deep hashing with a bipartite graph. In: IJCAI (2017)
50. Yang, E., Liu, T., Deng, C., Liu, W., Tao, D.: DistillHash: unsupervised deep hashing by distilling data pairs. In: CVPR (2019)
51. Yang, E., Liu, T., Deng, C., Tao, D.: Adversarial examples for hamming space search. IEEE Trans. Cybern. **50**(4), 1473–1484 (2018)
52. Yao, Z., Gholami, A., Xu, P., Keutzer, K., Mahoney, M.W.: Trust region based adversarial attack on neural networks. In: CVPR (2019)
53. Zhao, F., Huang, Y., Wang, L., Tan, T.: Deep semantic ranking based hashing for multi-label image retrieval. In: CVPR (2015)
54. Zuva, K., Zuva, T.: Evaluation of information retrieval systems. Int. J. Comput. Sci. Inf. Technol. **4**(3), 35 (2012)

Gradient Centralization: A New Optimization Technique for Deep Neural Networks

Hongwei Yong[1,2], Jianqiang Huang[2], Xiansheng Hua[2], and Lei Zhang[1,2](✉)

[1] Department of Computing, The Hong Kong Polytechnic University, Kowloon, Hong Kong
{cshyong,cslzhang}@comp.polyu.edu.hk

[2] DAMO Academy, Alibaba Group, Hangzhou, China
jianqiang.jqh@gmail.com, huaxiansheng@gmail.com

Abstract. Optimization techniques are of great importance to effectively and efficiently train a deep neural network (DNN). It has been shown that using the first and second order statistics (e.g., mean and variance) to perform Z-score standardization on network activations or weight vectors, such as batch normalization (BN) and weight standardization (WS), can improve the training performance. Different from these existing methods that mostly operate on activations or weights, we present a new optimization technique, namely gradient centralization (GC), which operates directly on gradients by centralizing the gradient vectors to have zero mean. GC can be viewed as a projected gradient descent method with a constrained loss function. We show that GC can regularize both the weight space and output feature space so that it can boost the generalization performance of DNNs. Moreover, GC improves the Lipschitzness of the loss function and its gradient so that the training process becomes more efficient and stable. GC is very simple to implement and it can be embedded into existing gradient based DNN optimizers with only one line of code. Our experiments on various applications, including general image classification, fine-grained image classification, detection and segmentation, demonstrate that GC can consistently improve the performance of DNN learning. The code of GC can be found at https://github.com/Yonghongwei/Gradient-Centralization.

Keywords: Deep network optimization · Gradient descent

1 Introduction

The broad success of deep learning largely owes to the recent advances on large-scale datasets [41], powerful computing resources (e.g., GPUs and TPUs), sophisticated network architectures [14,15] and optimization algorithms [3,22]. Among

Electronic supplementary material The online version of this chapter (https://doi.org/10.1007/978-3-030-58452-8_37) contains supplementary material, which is available to authorized users.

A. Vedaldi et al. (Eds.): ECCV 2020, LNCS 12346, pp. 635–652, 2020.
https://doi.org/10.1007/978-3-030-58452-8_37

these factors, the efficient optimization techniques, such as stochastic gradient descent (SGD) with momentum [36], Adagrad [9] and Adam [22], make it possible to train very deep neural networks (DNNs) with a large-scale dataset, and consequently deliver more powerful and robust DNN models in practice. The generalization performance of the trained DNN models as well as the efficiency of training process depend essentially on the employed optimization techniques.

There are two major goals for a good DNN optimizer: accelerating the training process (i.e., spending less time and cost to reach a good local minima) and improving the model generalization capability (i.e., making accurate predictions on test data). A variety of optimization algorithms [9,22,36] have been proposed to achieve these goals. SGD [3,4] and its extension SGD with momentum (SGDM) [36] are among the most commonly used ones. To overcome the gradient vanishing problems, a few successful techniques have been proposed, such as weight initialization strategies [10,13], efficient activation functions (e.g., ReLU [33]), gradient clipping [34,35], adaptive learning rate [9,22], and so on.

In addition to the above techniques, the sample/feature statistics such as mean and variance can also be used to normalize the network activations or weights to make the training process more stable. The representative methods operating on activations include batch normalization (BN) [18], instance normalization (IN) [17,45], layer normalization (LN) [27] and group normalization (GN) [49]. Among them, BN is the most widely used optimization technique which normalizes the features along the sample dimension in a mini-batch for training. BN smooths the optimization landscape [43] and it can speed up the training process and boost model generalization performance with a proper batch size [14,51]. Another line of statistics based methods operate on weights, such as weight normalization (WN) [16,42] and weight standardization (WS) [37]. These methods re-parameterize weights to restrict weight vectors during training. For example, WN decouples the length of weight vectors from their direction to accelerate the training of DNNs. WS uses the weight vectors' mean and variance to standardize them to have zero mean and unit variance. Similar to BN, WS can also smooth the loss landscape and speed up training. Nevertheless, such methods operating on weight vectors cannot directly adopt the pre-trained models (e.g., on ImageNet) because their weights may not meet the condition of zero mean and unit variance.

Different from the above techniques, we propose a very simple yet effective DNN optimization technique operating on the gradient of weight, namely gradient centralization (GC). As illustrated in Fig. 1(a), GC simply centralizes the gradient vectors to have zero mean. It can be easily embedded into the current gradient based optimization algorithms (e.g., SGDM [36], Adam [22]) using only one line of code. Though simple, GC demonstrates various desired properties, such as accelerating the training process, improving the generalization performance, and the compatibility for fine-tuning pre-trained models. The main contributions of this paper are highlighted as follows:

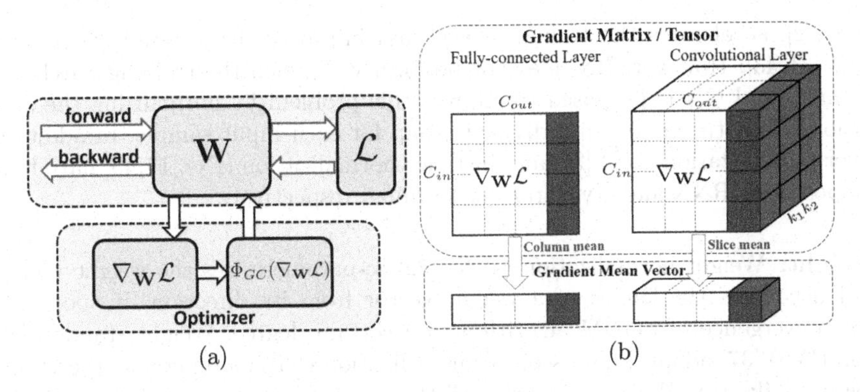

Fig. 1. (a) Sketch map for using gradient centralization (GC). $\mathbf{W}$ is the weight, $\mathcal{L}$ is the loss function, $\nabla_{\mathbf{W}}\mathcal{L}$ is the gradient of weight, and $\Phi_{GC}(\nabla_{\mathbf{W}}\mathcal{L})$ is the centralized gradient. GC is simple to be embedded into existing network optimizers by replacing $\nabla_{\mathbf{W}}\mathcal{L}$ with $\Phi_{GC}(\nabla_{\mathbf{W}}\mathcal{L})$. (b) Illustration of the GC on gradient matrix/tensor of weights in the fully-connected layer (left) and convolutional layer (right). GC computes the column/slice mean of gradient matrix/tensor and centralizes each column/slice.

- We propose a new general network optimization technique, namely gradient centralization (GC), which can not only smooth and accelerate the training process of DNN but also improve the model generalization performance.
- We analyze the theoretical properties of GC, and show that GC constrains the loss function by introducing a new constraint on weight vector, which regularizes both the weight space and output feature space so that it can boost model generalization performance. Besides, the constrained loss function has better Lipschitzness than the original one, which makes the training process more stable and efficient.

Finally, we perform comprehensive experiments on various applications, including general image classification, fine-grained image classification, object detection and instance segmentation. The results demonstrate that GC can consistently improve the performance of learned DNN models in different applications. It is a simple, general and effective network optimization method.

2 Related Work

In order to accelerate the training and boost the generalization performance of DNNs, a variety of optimization techniques [18,34,36,37,42,49] have been proposed from three perspectives: activation, weight and gradient.

Activation: The activation normalization layer has become a common setting in DNN, such as batch normalization (BN) [18] and group normalization (GN) [49]. BN was originally introduced to solve the internal covariate shift by normalizing the activations along the sample dimension. It allows higher learning rates [2], accelerates the training speed and improves the generalization accuracy [31,43]. However, BN does not perform well when the training batch size is small, and GN is proposed to address this problem by normalizing the activations or feature maps in a divided group for each input sample. In addition, layer normalization (LN) [27] and instance normalization (IN) [17,45] have been proposed for RNN and style transfer learning, respectively.

Weight: Weight normalization (WN) [42] re-parameterizes the weight vectors and decouples the length of a weight vector from its direction. It speeds up the convergence of SGDM algorithm to a certain degree. Weight standardization (WS) [37] adopts the Z-score standardization to re-parameterize the weight vectors. Like BN, WS can also smooth the loss landscape and improve training speed. Besides, binarized DNN [7,8,38] quantifies the weight into binary values, which may also improve the generalization capability. However, those methods operating on weights cannot be directly used to fine-tune pre-trained models since the pre-trained weight may not meet their constraints.

Gradient: A commonly used technique on gradient is the momentum [36]. With the momentum of gradient, SGDM accelerates SGD in the relevant direction and dampens oscillations. Besides, L_2 regularization based weight decay, which introduces a term into the gradient of weight, has long been a standard trick to improve the generalization performance of DNNs [25,52]. To make DNN training more stable and avoid gradient explosion, gradient clipping [1,21,34,35] has been proposed to train a very deep DNNs. In addition, the projected gradient methods and Riemannian approach [6,47] project the gradient on a subspace or a Riemannian manifold to regularize the learning of weights.

3 Gradient Centralization

3.1 Motivation

BN [18] is a powerful DNN optimization technique, which uses the first and second order statistics to perform Z-score standardization on activations. It has been shown in [43] that BN reduces the Lipschitz constant of loss function and makes the gradients more Lipschitz smooth. WS [37] can also reduce the Lipschitzness of loss function and smooth the optimization landscape through Z-score standardization on weight vectors.

Apart from operating on activation (e.g. BN) and weight (e.g. WS), can we directly operate on gradient to make the training process more effective and stable? One intuitive idea is to perform Z-score standardization on gradient, like what has been done by BN and WS on activation and weight. Unfortunately, we found that normalizing gradient cannot improve the stability of training. Instead, we propose to compute the mean of gradient vectors and centralize the gradients to have zero mean. As we will see in the following development, the so called gradient centralization (GC) method can have good Lipschitz property, smooth the DNN training and improve the model generalization performance.

Algorithm 1. SGDM with Gradient Centralization

Input: Weight vector $\mathbf{w}^0$, step size α, momentum factor β, $\mathbf{m}^0$

Training step:

1: **for** $t = 1, ...T$ **do**
2: $\mathbf{g}^t = \nabla_{\mathbf{w}^t}\mathcal{L}$
3: $\widehat{\mathbf{g}}^t = \Phi_{GC}(\mathbf{g}^t)$
4: $\mathbf{m}^t = \beta\mathbf{m}^{t-1} + (1-\beta)\widehat{\mathbf{g}}^t$
5: $\mathbf{w}^{t+1} = \mathbf{w}^t - \alpha\mathbf{m}_t$
6: **end for**

Algorithm 2. Adam with Gradient Centralization

Input: Weight vector $\mathbf{w}^0$, step size α, β_1, β_2, ϵ, $\mathbf{m}^0$,$\mathbf{v}^0$

Training step:

1: **for** $t = 1, ...T$ **do**
2: $\mathbf{g}^t = \nabla_{\mathbf{w}^t}\mathcal{L}$
3: $\widehat{\mathbf{g}}^t = \Phi_{GC}(\mathbf{g}^t)$
4: $\mathbf{m}^t = \beta_1\mathbf{m}^{t-1} + (1-\beta_1)\widehat{\mathbf{g}}^t$
5: $\mathbf{v}^t = \beta_2\mathbf{v}^{t-1} + (1-\beta_2)\widehat{\mathbf{g}}^t \odot \widehat{\mathbf{g}}^t$
6: $\widehat{\mathbf{m}}^t = \mathbf{m}^t/(1-(\beta_1)^t)$
7: $\widehat{\mathbf{v}}^t = \mathbf{v}^t/(1-(\beta_2)^t)$
8: $\mathbf{w}^{t+1} = \mathbf{w}^{t+1} = \mathbf{w}^t - \alpha\frac{\widehat{\mathbf{m}}^t}{\sqrt{\widehat{\mathbf{v}}^t}+\epsilon}$
9: **end for**

3.2 Notations

We define some basic notations. For fully connected layers (FC layers), the weight matrix is denoted as $\mathbf{W}_{fc} \in \mathbb{R}^{C_{in}\times C_{out}}$, and for convolutional layers (Conv layers) the weight tensor is denoted as $\mathbf{W}_{conv} \in \mathbb{R}^{C_{in}\times C_{out}\times(k_1k_2)}$, where C_{in} is the number of input channels, C_{out} is the number of output channels, and k_1, k_2 are the kernel size of convolution layers. For the convenience of expression, we unfold the weight tensor of Conv layer into a matrix and use a unified notation $\mathbf{W} \in \mathbb{R}^{M\times N}$ for weight matrix in FC layer ($\mathbf{W} \in \mathbb{R}^{C_{in}\times C_{out}}$) and Conv layers ($\mathbf{W} \in \mathbb{R}^{(C_{in}k_1k_2)\times C_{out}}$). Denote by $\mathbf{w}_i \in \mathbb{R}^M$ $(i = 1, 2, ..., N)$ the i-th column vector of weight matrix $\mathbf{W}$ and $\mathcal{L}$ the objective function. $\nabla_{\mathbf{W}}\mathcal{L}$ and $\nabla_{\mathbf{w}_i}\mathcal{L}$ denote the gradient of $\mathcal{L}$ w.r.t. $\mathbf{W}$ and $\mathbf{w}_i$, respectively. Let $\mathbf{X}$ be the input activations for this layer and $\mathbf{W}^T\mathbf{X}$ be its output activations. $\mathbf{e} = \frac{1}{\sqrt{M}}\mathbf{1}$ denotes an M dimensional unit vector and $\mathbf{I} \in \mathbb{R}^{M\times M}$ denotes an identity matrix.

3.3 Formulation of GC

For a FC layer or a Conv layer, suppose that we have obtained the gradient through backward propagation, then for a weight vector $\mathbf{w}_i$ whose gradient is $\nabla_{\mathbf{w}_i}\mathcal{L}$ $(i = 1, 2, ..., N)$, the GC operator, denoted by Φ_{GC}, is defined as follows:

$$\Phi_{GC}(\nabla_{\mathbf{w}_i}\mathcal{L}) = \nabla_{\mathbf{w}_i}\mathcal{L} - \frac{1}{M}\sum_{j=1}^{M}\nabla_{w_{i,j}}\mathcal{L} \tag{1}$$

The formulation of GC is very simple. As shown in Fig. 1(b), we only need to compute the mean of the gradient vectors, and then remove the mean from them. We can also have a matrix formulation of Eq. (1):

$$\Phi_{GC}(\nabla_{\mathbf{W}}\mathcal{L}) = \mathbf{P}\nabla_{\mathbf{W}}\mathcal{L}, \qquad \mathbf{P} = \mathbf{I} - \mathbf{e}\mathbf{e}^T \tag{2}$$

The physical meaning of $\mathbf{P}$ will be explained later in Sect. 4.1. In practical implementation, we can directly remove the mean value from each weight vector to accomplish the GC operation. The computation is very simple and efficient.

3.4 Embedding of GC to SGDM/Adam

GC can be easily embedded into the current DNN optimization algorithms such as SGDM [4,36] and Adam [22]. After obtaining the centralized gradient $\Phi_{GC}(\nabla_{\mathbf{w}}\mathcal{L})$, we can directly use it to update the weight. Algorithm 1 and Algorithm 2 show how to embed GC into the two most popular optimization algorithms, SGDM and Adam, respectively. Moreover, if we want to use weight decay, we can set $\widehat{\mathbf{g}}^t = \mathbf{P}(\mathbf{g}^t + \lambda\mathbf{w})$, where λ is the weight decay factor. It only needs to add one line of code to execute GC with negligible additional computation. For example, it costs only 0.6 s extra training time in one epoch on CIFAR100 with ResNet50 model in our experiments (71 s for one epoch).

4 Properties of GC

As we will see in the section of experimental result, GC can accelerate the training process and improve the generalization performance of DNNs. In this section, we perform theoretical analysis to explain why GC works.

4.1 Improving Generalization Performance

One important advantage of GC is that it can improve the generalization performance of DNNs. We explain this advantage from two aspects: weight space regularization and output feature space regularization.

Weight Space Regularization: Let's first explain the physical meaning of $\mathbf{P}$ in Eq.(2). Actually, it is easy to prove that: $\mathbf{P}^2 = \mathbf{P} = \mathbf{P}^T$, $\mathbf{e}^T\mathbf{P}\nabla_{\mathbf{w}}\mathcal{L} = 0$. The above equations show that $\mathbf{P}$ is the projection matrix for the hyperplane with normal vector $\mathbf{e}$ in weight space, and $\mathbf{P}\nabla_{\mathbf{w}}\mathcal{L}$ is the projected gradient.

The property of projected gradient has been investigated in some previous works [6,11,26,47], which indicate that projecting the gradient of weight will constrict the weight space in a hyperplane or a Riemannian manifold. Similarly, the role of GC can also be viewed from the perspective of projected gradient descent. We give a geometric illustration of SGD with GC in Fig. 2. As shown in Fig. 2, in the t-th step of SGD with GC, the gradient is first projected on the hyperplane determined by $\mathbf{e}^T(\mathbf{w} - \mathbf{w}^t) = 0$, where $\mathbf{w}^t$ is the weight vector in the t-th iteration, and then the weight is updated along the direction of projected gradient $-\mathbf{P}\nabla_{\mathbf{w}^t}\mathcal{L}$. From $\mathbf{e}^T(\mathbf{w} - \mathbf{w}^t) = 0$, we have $\mathbf{e}^T\mathbf{w}^{t+1} = \mathbf{e}^T\mathbf{w}^t = ... = \mathbf{e}^T\mathbf{w}^0$, i.e., $\mathbf{e}^T\mathbf{w}$ is a constant during training. Mathematically, the latent objective function w.r.t. one weight vector $\mathbf{w}$ can be written as follows:

$$\min_{\mathbf{w}} \mathcal{L}(\mathbf{w}), \quad s.t. \quad \mathbf{e}^T(\mathbf{w} - \mathbf{w}^0) = 0 \tag{3}$$

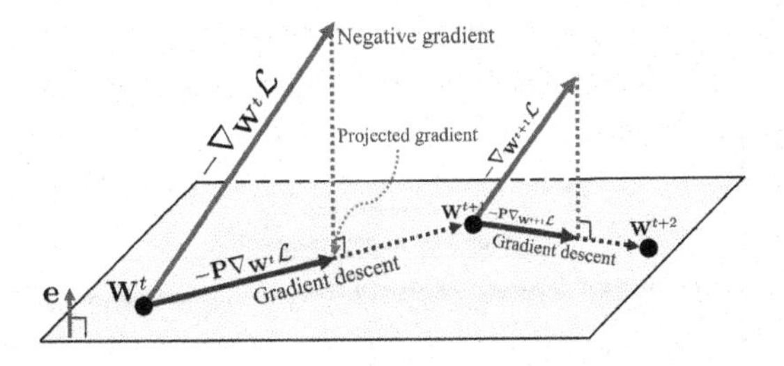

Fig. 2. The geometrical interpretation of GC. The gradient is projected on a hyperplane $\mathbf{e}^T(\mathbf{w}-\mathbf{w}^t)=0$, where the projected gradient is used to update the weight.

Clearly, this is a constrained optimization problem on weight vector $\mathbf{w}$. It regularizes the solution space of $\mathbf{w}$, reducing the possibility of over-fitting on training data. As a result, GC can improve the generalization capability of trained DNN models, especially when the number of training samples is limited.

It is noted that WS [37] uses a constraint $\mathbf{e}^T\mathbf{w}=0$ for weight optimization. It reparameterizes weights to meet this constraint. However, this constraint largely limits its practical applications because the initialized weight may not satisfy this constraint. For example, a pretrained DNN on ImageNet usually cannot meet $\mathbf{e}^T\mathbf{w}^0=0$ for its initialized weight vectors. If we use WS to fine-tune this DNN, the advantages of pretrained models will disappear. Therefore, we have to retrain the DNN on ImageNet with WS before we fine-tune it. This is very cumbersome. Fortunately the weight constraint of GC in Eq. (3) fits any initialization of weight, since $\mathbf{e}^T(\mathbf{w}^0-\mathbf{w}^0)=0$ is always true. Moreover, GC also works well for FC layer, while WS only performs well on Conv layer.

Output Feature Space Regularization: For SGD based algorithms, we have $\mathbf{w}^{t+1}=\mathbf{w}^t-\alpha^t\mathbf{P}\nabla_{\mathbf{w}^t}\mathcal{L}$. It can be derived that $\mathbf{w}^t=\mathbf{w}^0-\mathbf{P}\sum_{i=0}^{t-1}\alpha^{(i)}\nabla_{\mathbf{w}^{(i)}}\mathcal{L}$. For any input feature vector $\mathbf{x}$, we have the following theorem:

Corollary 4.1: *Suppose that SGD (or SGDM) with GC is used to update the weight vector* $\mathbf{w}$*, for any input feature vectors* $\mathbf{x}$ *and* $\mathbf{x}+\gamma\mathbf{1}$*, we have*

$$(\mathbf{w}^t)^T\mathbf{x}-(\mathbf{w}^t)^T(\mathbf{x}+\gamma\mathbf{1})=\gamma\mathbf{1}^T\mathbf{w}^0 \tag{4}$$

where $\mathbf{w}^0$ *is the initial weight vector and* γ *is a scalar.*

Please find the proof in the **Supplementary Materials**. Corollary 4.1 indicates that a constant intensity change (i.e., $\gamma\mathbf{1}$) of an input feature causes a change of output activation; interestingly, this change is only related to γ and $\mathbf{1}^T\mathbf{w}^0$ but not the current weight vector $\mathbf{w}^t$. $\mathbf{1}^T\mathbf{w}^0$ is the scaled mean of the initial weight vector $\mathbf{w}^0$. In particular, if the mean of $\mathbf{w}^0$ is close to zero, then the output activation is not sensitive to the intensity change of input features, and the output feature space becomes more robust to training sample variations.

Indeed, the mean of $\mathbf{w}^0$ is small by the commonly used weight initialization strategies, such as Kaiming initialization [13] and even ImageNet pre-trained

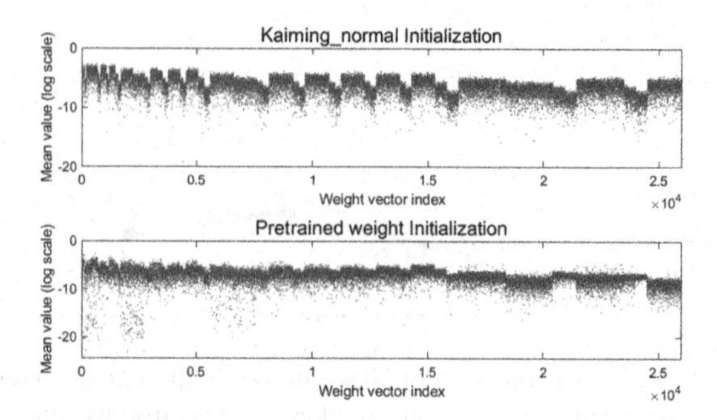

Fig. 3. The absolute value (log scale) of the mean of weight vectors for convolution layers in ResNet50. The x-axis is the weight vector index. We plot the mean value of different convolution layers from left to right with the order from sallow to deep layers. Kaiming normal initialization [13] (top) and ImageNet pre-trained weight initialization (bottom) are employed here. We can see that the mean values are usually very small (less than e^{-7}) for most of the weight vectors.

weight initialization. Figure 3 shows the absolute value (log scale) of the mean of weight vectors for Conv layers in ResNet50 with Kaiming normal and ImageNet pre-trained weight initialization. We can see that the mean values of most weight vectors are very small (less than e^{-7}). This ensures that the output features will not be sensitive to the variation of the intensity of input features. Such a property regularizes the output feature space and further boosts the generalization performance. It should be stressed that although the weight means are small, they are not equal to zero and still keep useful information of pre-trained models. If we force the weight means in pre-trained models strictly to zero (e.g., WS), the performance will drop a lot.

4.2 Accelerating Training Process

Optimization Landscape Smoothing: It has been shown in [37,43] that both BN and WS smooth the optimization landscape. Although BN and WS operate on activations and weights, they implicitly constrict the gradient of weights, making the gradient of weight more predictive and stable for fast training. Specifically, BN and WS use the gradient magnitude $||\nabla f(\mathbf{x})||_2$ to capture the Lipschitzness of function $f(\mathbf{x})$. For the loss and its gradients, $f(\mathbf{x})$ will be $\mathcal{L}$ and $\nabla_{\mathbf{w}}\mathcal{L}$, respectively, and $\mathbf{x}$ will be $\mathbf{w}$. The upper bounds of $||\nabla_{\mathbf{w}}\mathcal{L}||_2$ and $||\nabla^2_{\mathbf{w}}\mathcal{L}||_2$ ($\nabla^2_{\mathbf{w}}\mathcal{L}$ is the Hessian matrix of $\mathbf{w}$) have been given in [37,43] to illustrate the optimization landscape smoothing property of BN and WS. Similar conclusion can be made for our proposed GC by comparing the Lipschitzness of original loss function $\mathcal{L}(\mathbf{w})$ with the constrained loss function in Eq. (3) and the Lipschitzness of their gradients. We have the following theorem:

Proposition 4.2: *Suppose $\nabla_{\mathbf{w}}\mathcal{L}$ is the gradient of loss function $\mathcal{L}$ w.r.t. weight vector $\mathbf{w}$. With the $\Phi_{GC}(\nabla_{\mathbf{w}}\mathcal{L})$ defined in Eq. (2), we have the following conclusion for the loss function and its gradient, respectively:*

$$\begin{cases} ||\Phi_{GC}(\nabla_{\mathbf{w}}\mathcal{L})||_2 \leq ||\nabla_{\mathbf{w}}\mathcal{L}||_2, \\ ||\nabla_{\mathbf{w}}\Phi_{GC}(\nabla_{\mathbf{w}}\mathcal{L})||_2 \leq ||\nabla^2_{\mathbf{w}}\mathcal{L}||_2. \end{cases} \tag{5}$$

The proof of Proposition 4.2 can be found in the **Supplementary Materials**. Proposition 4.2 shows that for the loss function $\mathcal{L}$ and its gradient $\nabla_{\mathbf{w}}\mathcal{L}$, the constrained loss function in Eq. (3) by GC leads to a better Lipschitzness than the original loss function so that the optimization landscape becomes smoother. This means that GC has similar advantages to BN and WS on accelerating training. A good Lipschitzness on gradient implies that the gradients used in training are more predictive and well-behaved so that the optimization landscape can be smoother for faster and more effective training.

Gradient Explosion Suppression: Another benefit of GC for DNN training is that GC can avoid gradient explosion and make training more stable. This property is similar to gradient clipping [1,21,34,35]. Too large gradients will make the weights change abruptly during training so that the loss may severely oscillate and hard to converge. It has been shown that gradient clipping can suppress large gradient so that the training can be more stable and faster [34, 35]. There are two popular gradient clipping approaches: element-wise value clipping [21,34] and norm clipping [1,35]. In Fig. 4 we plot the max value and L_2 norm of gradient matrix of the first convolutional layer and the fully-connected layer in ResNet50 (trained on CIFAR100) with and without GC. It can be seen that both the max value and the L_2 norm of the gradient matrix become smaller by using GC in training. This is in accordance to our conclusion in Proposition 4.2 that GC can make training process smoother and faster.

5 Experimental Results

5.1 Setup of Experiments

Extensive experiments are performed to validate the effectiveness of GC. To make the results as fair and clear, we arrange the experiments as follows:

- We start from experiments on the Mini-ImageNet dataset [46] to demonstrate that GC can accelerate the DNN training process and improve the model generalization performance. We also evaluate the combinations of GC with BN and WS to show that GC can improve them for DNN optimization.
- We then use the CIFAR100 dataset [24] to evaluate GC with various DNN optimizers (e.g., SGDM, Adam) and DNN architectures (e.g., ResNet, DenseNet).
- We then perform experiments on ImageNet [41] to demonstrate that GC also works well on large scale image classification, and show that GC can also work well with normalization methods other than BN, such as GN.

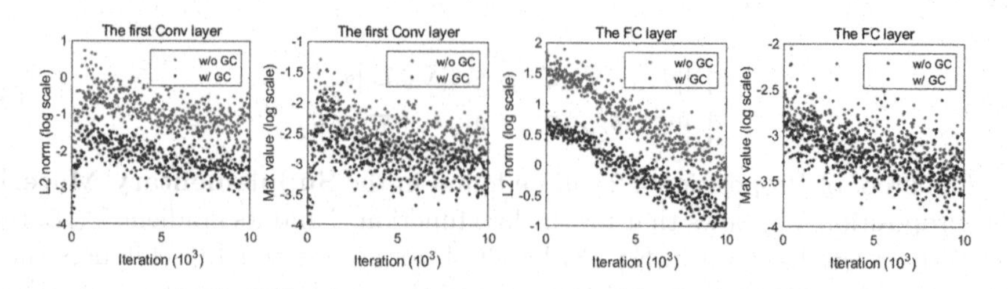

Fig. 4. The L_2 norm (log scale) and max value (log scale) of gradient matrix or tensor vs. iterations. ResNet50 trained on CIFAR100 is used as the DNN model here. The left two sub-figures show the results on the first Conv layer and the right two show the FC layer. The red and blue points represent the results of training w/o and w/ GC, respectively. We can see that GC reduces the L_2 norm and max value of gradient. (Color figure online)

- We consequently perform experiments on four fine-grained image classification datasets (FGVC Aircraft [32], Stanford Cars [23], Stanford Dogs [20] and CUB-200-2011 [48]) to show that GC can be directly adopted to fine-tune the pre-trained DNN models and improve them.
- At last, we perform experiments on the COCO dataset [29] to show that GC also works well on other tasks such as objection detection and segmentation.

GC can be applied to either Conv layer or FC layer, or both of them. In all of our following experiments, if not specified, we always apply GC to both Conv and FC layers. Except for Sect. 5.3 where we embed GC into different DNN optimizers for test, in all other sections we embed GC into SGDM for experiments, and the momentum is set to 0.9. All experiments are conducted under the Pytorch 1.3 framework and the GPUs are NVIDIA Tesla P100. We would like to stress that no additional hyper-parameter is introduced in our GC method. We compare the performances of DNN models trained w/ and w/o GC to validate the effectiveness of GC.

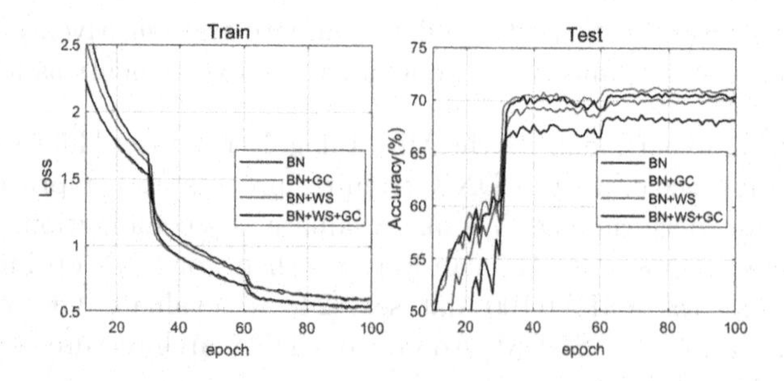

Fig. 5. Training loss (left) and testing accuracy (right) curves vs. training epoch on the Mini-ImageNet. The ResNet50 is used as the DNN model. The compared optimization techniques include BN, BN+GC, BN+WS and BN+WS+GC.

Table 1. Testing accuracies of different DNN models on CIFAR100

Model	R18	R101	X29	V11	D121
w/o GC	76.87 ± 0.26	78.82 ± 0.42	79.70 ± 0.30	70.94 ± 0.34	79.31 ± 0.33
w/ GC	$\mathbf{78.82 \pm 0.31}$	$\mathbf{80.21 \pm 0.31}$	$\mathbf{80.53 \pm 0.33}$	$\mathbf{71.69 \pm 0.37}$	$\mathbf{79.68 \pm 0.40}$

Table 2. Testing accuracies of different optimizers on CIFAR100

Algorithm	SGDM	Adam	Adagrad	SGDW	AdamW
w/o GC	78.23 ± 0.42	71.64 ± 0.56	70.34 ± 0.31	74.02 ± 0.27	74.12 ± 0.42
w/ GC	$\mathbf{79.14 \pm 0.33}$	$\mathbf{72.80 \pm 0.62}$	$\mathbf{71.58 \pm 0.37}$	$\mathbf{76.82 \pm 0.29}$	$\mathbf{75.07 \pm 0.37}$

5.2 Results on Mini-Imagenet

Mini-ImageNet [46] is a subset of the ImageNet dataset [41]. We use the train/test splits provided by [19,39]. It consists of 100 classes and each class has 500 images for training and 100 images for testing. The image resolution is 84×84. We resize the images into standard ImageNet training input size 224×224. The DNN we used here is ResNet50, which is trained on 4 GPUs with batch size 128. Other settings are the same as training ImageNet. We repeat the experiments for 10 times and report the average results over the 10 runs.

BN, WS and GC operate on activations, weights and gradients, respectively, and they can be used together to train DNNs. Actually, it is necessary to normalize the feature space by methods such as BN; otherwise, the model is hard to be well trained. Therefore, we evaluate four combinations here: BN, BN+GC, BN+WS and BN+WS+GC. The optimizer is SGDM with momentum 0.9. Figure 5 presents the training loss and testing accuracy curves of these four combinations. Compared with BN, the training loss of BN+GC decreases much faster and the testing accuracy increases more rapidly. For both BN and BN+WS, GC can further speed up their training speed. Moreover, we can see that BN+GC achieves the highest testing accuracy, validating that GC can accelerate training and enhance the generalization performance simultaneously.

5.3 Experiments on CIFAR100

CIFAR100 [24] consists of 50K training images and 10K testing images from 100 classes. The size of input image is 32×32. Since the image resolution is small, we found that applying GC to the Conv layer is good enough on this dataset. All DNN models are trained for 200 epochs using one GPU with batch size 128. The results are reported in mean $\pm$ std format with repeating 10 times.

Different Networks: We testify GC on different DNN architectures, including ResNet18 (R18), ResNet101 (R101) [14], ResNeXt29 4x64d (X29) [50], VGG11 (V11) [44] and DenseNet121 (D121) [15]. SGDM is used as the network optimizer. The weight decay is set to 0.0005. The initial learning rate is 0.1 and it

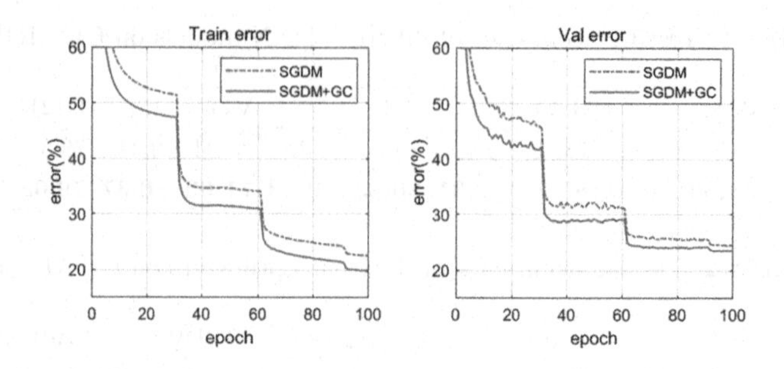

Fig. 6. Training error (left) and validation error (right) curves vs. training epoch on ImageNet. The DNN model is ResNet50 with GN.

Table 3. Top-1 error rates on ImageNet w/o GC and w/ GC.

Model	R50BN	R50GN	R101BN	R101GN
w/o GC	23.71	24.50	22.37	23.34
w/ GC	**23.21**	**23.53**	**21.82**	**22.14**

is multiplied by 0.1 for every 60 epochs. Table 1 reports the their testing accuracies. It shows that the performance of all DNNs is improved by GC, which verifies GC is a general optimization technique for different DNN architectures.

Different Optimizers: We embed GC into different DNN optimizers, including SGDM [36], Adagrad [9], Adam [22], SGDW and AdamW [30], to test their performance. SGDW and AdamW optimizers directly apply weight decay on weight, instead of using L_2 weight decay regularization. Weight decay is set to 0.001 for SGDW and AdamW, and 0.0005 for other optimizers. The initial learning rate is set to 0.1, 0.01 and 0.001 for SGDM/SGDW, Adagrad, Adam/AdamW, respectively, and the learning rate is multiplied by 0.1 for every 60 epochs. The other hyper-parameters are set by their default settings on Pytorch. The DNN model used here is ResNet50. Table 2 shows the testing accuracies. It can be seen that GC boosts the generalization performance of all the five optimizers. It is also found that adaptive learning rate based algorithms Adagrad and Adam have poor generalization performance on CIFAR100, while GC can improve their performance by $> 0.9\%$.

5.4 Results on ImageNet

We then evaluate GC on the large-scale ImageNet dataset [41] which consists of 1.28 million images for training and 50K images for validation from 1000 categories. We use the common training settings and embed GC to SGDM on Conv layer. The ResNet50 and ResNet101 are used as the backbone networks. For the former, we use 4 GPUs with batch size 64 per GPU, and for the latter, we use 8 GPUs with batch size 32 per GPU. We evaluate four models here:

ResNet50 with BN (R50BN), ResNet50 with GN (R50GN), ResNet101 with BN (R101BN) and ResNet101 with GN (R101GN). Table 3 shows the final Top-1 errors of these four DNN models trained with and without GC. We can see that GC can improve the performance by 0.5%–1.2% on ImageNet. Figure 6 plots the training and validation error curves of ResNet50 (GN is used for feature normalization). We can see that GC can largely speed up the training with GN.

5.5 Results on Fine-Grained Image Classification

In order to show that GC can also work with the pre-trained models, we conduct experiments on four challenging fine-grained image classification datasets, including FGVC Aircraft [32], Stanford Cars [23], Stanford Dogs [20] and CUB-200-2011 [48]. We use the official pre-trained ResNet50 provided by Pytorch as the baseline DNN for all these four datasets. The original images are resized into 512×512 and we crop the center region with 448×448 as input for both training and testing. We use SGDM with momentum of 0.9 to fine-tune ResNet50 for 100 epochs on 4 GPUs with batch size 256. The initial learning rate is 0.1 for the last FC layer and 0.01 for all pre-trained Conv layers. The learning rate is multiplied by 0.1 at the 50th and 80th epochs. We repeat the experiments for 10 times and report the result in mean $\pm$ std format.

Table 4. Testing accuracies on the four fine-grained image classification datasets.

Datesets	FGVC Aircraft	Stanford Cars	Stanford Dogs	CUB-200-2011
w/o GC	86.62 ± 0.31	88.66 ± 0.22	76.16 ± 0.25	82.07 ± 0.26
w/ GC	**87.77 ± 0.27**	**90.03 ± 0.26**	**78.23 ± 0.24**	**83.40 ± 0.30**

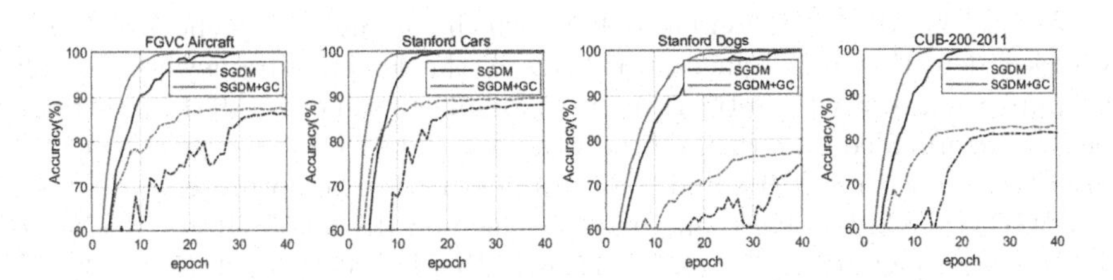

Fig. 7. Training accuracy (solid line) and testing accuracy (dotted line) curves vs. training epoch on four fine-grained image classification datasets.

Figure 7 shows the training and testing accuracies of SGDM and SGDM+GC for the first 40 epochs on the four fine-grained image classification datasets. Table 4 shows the final testing accuracies. We can see that both the training and testing accuracies of SGDM are improved by GC. For the final classification accuracy, GC improves SGDM by 1.1%–2.1% on these four datasets. This sufficiently demonstrates the effectiveness of GC on fine-tuning pre-trained models.

Table 5. Detection results on COCO by using Faster-RCNN and FPN with various backbone models.

Method	Backbone	AP	$AP_{.5}$	$AP_{.75}$	Backbone	AP	$AP_{.5}$	$AP_{.75}$
w/o GC	R50	36.4	58.4	39.1	X101-32x4d	40.1	62.0	43.8
w/ GC	R50	37.0	59.0	40.2	X101-32x4d	40.7	62.7	43.9
w/o GC	R101	38.5	60.3	41.6	X101-64x4d	41.3	63.3	45.2
w/ GC	R101	38.9	60.8	42.2	X101-64x4d	41.6	63.8	45.4

Table 6. Detection and segmentation results on COCO by using Mask-RCNN and FPN with various backbone models.

Method	Backbone	AP^b	$AP^b_{.5}$	$AP^b_{.75}$	AP^m	$AP^m_{.5}$	$AP^m_{.75}$	Backbone	AP^b	$AP^b_{.5}$	$AP^b_{.75}$	AP^m	$AP^m_{.5}$	$AP^m_{.75}$
w/o GC	R50	37.4	59.0	40.6	34.1	55.5	36.1	R50 (4c1f)	37.5	58.2	41.0	33.9	55.0	36.1
w/ GC	R50	37.9	59.6	41.2	34.7	56.1	37.0	R50 (4c1f)	38.4	59.5	41.8	34.6	55.9	36.7
w/o GC	R101	39.4	60.9	43.3	35.9	57.7	38.4	R101GN	41.1	61.7	44.9	36.9	58.7	39.3
w/ GC	R101	40.0	61.5	43.7	36.2	58.1	38.7	R101GN	41.7	62.3	45.3	37.4	59.3	40.3
w/o GC	X101-32x4d	41.1	62.8	45.0	37.1	59.4	39.8	R50GN+WS	40.0	60.7	43.6	36.1	57.8	38.6
w/ GC	X101-32x4d	41.6	63.1	45.5	37.4	59.8	39.9	R50GN+WS	40.6	61.3	43.9	36.6	58.2	39.1
w/o GC	X101-64x4d	42.1	63.8	46.3	38.0	60.6	40.9							
w/ GC	X101-64x4d	42.8	64.5	46.8	38.4	61.0	41.1							

5.6 Objection Detection and Segmentation

Finally, we evaluate GC on object detection and segmentation tasks to show that GC can also be applied to more tasks beyond image classification. The models are pre-trained on ImageNet, and trained on COCO *train*2017 dataset (118K images) and evaluated on COCO *val*2017 dataset (40K images) [29]. COCO dataset can be used for multiple tasks, including image classification, object detection, semantic segmentation and instance segmentation.

We use the MMDetection [5] toolbox, which contains comprehensive models on object detection and segmentation tasks, as the detection framework. The official implementations and settings are used for all experiments. All the pre-trained models are provided from their official websites, and we fine-tune them on COCO *train*2017 set with 8 GPUs and 2 images per GPU. The optimizers are SGDM and SGDM+GC. The backbone networks include ResNet50 (R50), ResNet101 (R101), ResNeXt101-32x4d (X101-32x4d), ResNeXt101-64x4d (X101-32x4d). The Feature Pyramid Network (FPN) [28] is also used. The learning rate schedule is $1X$ for both Faster R-CNN [40] and Mask R-CNN [12], except R50 with GN and R101 with GN, which use $2X$ learning rate schedule.

Table 5 shows the Average Precision (AP) results of Faster R-CNN. We can see that all the backbone networks trained with GC can achieve a performance gain about 0.3%–0.6% on object detection. Table 6 presents the Average Precision for bounding box (AP^b) and instance segmentation (AP^m). It can be seen that the AP^b increases by 0.5%–0.9% for object detection task and the AP^m increases by 0.3%–0.7% for instance segmentation task. Moreover, we find that if 4conv1fc bounding box head, like R50 (4c1f), is used, the performance can increase more by GC. And GC can also boost the performance of GN (see R101GN) and improve the performance of WS (see R50GN+WS). Overall, we can see that GC boosts the generalization performance of all evaluated models.

6 Conclusions

How to efficiently and effectively optimize a DNN is one of the key issues in deep learning research. Previous methods such as batch normalization (BN) and weight standardization (WS) mostly operate on network activations or weights to improve DNN training. We proposed a different approach which operates directly on gradients. Specifically, we removed the mean from the gradient vectors and centralized them to have zero mean. The so-called Gradient Centralization (GC) method demonstrated several desired properties of deep network optimization. We showed that GC actually improves the loss function with a constraint on weight vectors, which regularizes both weight space and output feature space. We also showed that this constrained loss function has better Lipschitzness than the original one so that it has a smoother optimization landscape. Comprehensive experiments were performed and the results demonstrated that GC can be well applied to different tasks with different optimizers and network architectures.

Acknowledgements. This research is supported by the Hong Kong RGC GRF grant (PolyU 152216/18E).

References

1. Abadi, M., et al.: Deep learning with differential privacy. In: Proceedings of the 2016 ACM SIGSAC Conference on Computer and Communications Security, pp. 308–318. ACM (2016)
2. Bjorck, J., Gomes, C., Selman, B., Weinberger, K.Q.: Understanding batch normalization, pp. 7694–7705 (2018)
3. Bottou, L.: Stochastic gradient learning in neural networks. Proc. Neuro-Nımes **91**(8), 12 (1991)
4. Bottou, L.: Large-scale machine learning with stochastic gradient descent. In: Lechevallier, Y., Saporta, G. (eds.) Proceedings of COMPSTAT 2010, pp. 177–186. Springer, Heidelberg (2010). https://doi.org/10.1007/978-3-7908-2604-3_16
5. Chen, K., et al.: MMDetection: open MMLab detection toolbox and benchmark. arXiv preprint arXiv:1906.07155 (2019)
6. Cho, M., Lee, J.: Riemannian approach to batch normalization. In: Advances in Neural Information Processing Systems, pp. 5225–5235 (2017)
7. Courbariaux, M., Bengio, Y., David, J.P.: BinaryConnect: training deep neural networks with binary weights during propagations. In: Advances in Neural Information Processing Systems, pp. 3123–3131 (2015)
8. Courbariaux, M., Hubara, I., Soudry, D., El-Yaniv, R., Bengio, Y.: Binarized neural networks: training deep neural networks with weights and activations constrained to +1 or −1. arXiv preprint arXiv:1602.02830 (2016)
9. Duchi, J., Hazan, E., Singer, Y.: Adaptive subgradient methods for online learning and stochastic optimization. J. Mach. Learn. Res. **12**, 2121–2159 (2011)

10. Glorot, X., Bengio, Y.: Understanding the difficulty of training deep feedforward neural networks. In: Proceedings of the Thirteenth International Conference on Artificial Intelligence and Statistics, pp. 249–256 (2010)
11. Gupta, H., Jin, K.H., Nguyen, H.Q., McCann, M.T., Unser, M.: CNN-based projected gradient descent for consistent CT image reconstruction. IEEE Trans. Med. Imaging **37**(6), 1440–1453 (2018)
12. He, K., Gkioxari, G., Dollár, P., Girshick, R.: Mask R-CNN. In: Proceedings of the IEEE International Conference on Computer Vision, pp. 2961–2969 (2017)
13. He, K., Zhang, X., Ren, S., Sun, J.: Delving deep into rectifiers: surpassing human-level performance on ImageNet classification. In: Proceedings of the IEEE International Conference on Computer Vision, pp. 1026–1034 (2015)
14. He, K., Zhang, X., Ren, S., Sun, J.: Deep residual learning for image recognition. In: Proceedings of the IEEE Conference on Computer Vision and Pattern Recognition, pp. 770–778 (2016)
15. Huang, G., Liu, Z., Van Der Maaten, L., Weinberger, K.Q.: Densely connected convolutional networks. In: Proceedings of the IEEE Conference on Computer Vision and Pattern Recognition, pp. 4700–4708 (2017)
16. Huang, L., Liu, X., Liu, Y., Lang, B., Tao, D.: Centered weight normalization in accelerating training of deep neural networks. In: Proceedings of the IEEE International Conference on Computer Vision, pp. 2803–2811 (2017)
17. Huang, X., Belongie, S.: Arbitrary style transfer in real-time with adaptive instance normalization. In: Proceedings of the IEEE International Conference on Computer Vision, pp. 1501–1510 (2017)
18. Ioffe, S., Szegedy, C.: Batch normalization: accelerating deep network training by reducing internal covariate shift. arXiv preprint arXiv:1502.03167 (2015)
19. Iscen, A., Tolias, G., Avrithis, Y., Chum, O.: Label propagation for deep semi-supervised learning. In: Proceedings of the IEEE Conference on Computer Vision and Pattern Recognition, pp. 5070–5079 (2019)
20. Khosla, A., Jayadevaprakash, N., Yao, B., Li, F.F.: Novel dataset for FGVC: stanford dogs. In: CVPR Workshop on FGVC, San Diego, vol. 1 (2011)
21. Kim, J., Kwon Lee, J., Mu Lee, K.: Accurate image super-resolution using very deep convolutional networks. In: Proceedings of the IEEE Conference on Computer Vision and Pattern Recognition, pp. 1646–1654 (2016)
22. Kingma, D.P., Ba, J.: Adam: a method for stochastic optimization. arXiv preprint arXiv:1412.6980 (2014)
23. Krause, J., Stark, M., Deng, J., Fei-Fei, L.: 3D object representations for fine-grained categorization. In: Proceedings of the IEEE International Conference on Computer Vision Workshops, pp. 554–561 (2013)
24. Krizhevsky, A., Hinton, G., et al.: Learning multiple layers of features from tiny images. Technical report, Citeseer (2009)
25. Krogh, A., Hertz, J.A.: A simple weight decay can improve generalization. In: Advances in Neural Information Processing Systems, pp. 950–957 (1992)
26. Larsson, M., Arnab, A., Kahl, F., Zheng, S., Torr, P.: A projected gradient descent method for CRF inference allowing end-to-end training of arbitrary pairwise potentials. In: Pelillo, M., Hancock, E. (eds.) EMMCVPR 2017. LNCS, vol. 10746, pp. 564–579. Springer, Cham (2018). https://doi.org/10.1007/978-3-319-78199-0_37
27. Lei Ba, J., Kiros, J.R., Hinton, G.E.: Layer normalization. arXiv preprint arXiv:1607.06450 (2016)
28. Lin, T.Y., Dollár, P., Girshick, R., He, K., Hariharan, B., Belongie, S.: Feature pyramid networks for object detection. In: Proceedings of the IEEE Conference on Computer Vision and Pattern Recognition, pp. 2117–2125 (2017)

29. Lin, T.-Y., et al.: Microsoft COCO: common objects in context. In: Fleet, D., Pajdla, T., Schiele, B., Tuytelaars, T. (eds.) ECCV 2014. LNCS, vol. 8693, pp. 740–755. Springer, Cham (2014). https://doi.org/10.1007/978-3-319-10602-1_48
30. Loshchilov, I., Hutter, F.: Decoupled weight decay regularization. arXiv preprint arXiv:1711.05101 (2017)
31. Luo, P., Wang, X., Shao, W., Peng, Z.: Towards understanding regularization in batch normalization (2018)
32. Maji, S., Rahtu, E., Kannala, J., Blaschko, M., Vedaldi, A.: Fine-grained visual classification of aircraft. arXiv preprint arXiv:1306.5151 (2013)
33. Nair, V., Hinton, G.E.: Rectified linear units improve restricted Boltzmann machines. In: Proceedings of the 27th International Conference on Machine Learning (ICML 2010), pp. 807–814 (2010)
34. Pascanu, R., Mikolov, T., Bengio, Y.: Understanding the exploding gradient problem. CoRR abs/1211.5063 (2012)
35. Pascanu, R., Mikolov, T., Bengio, Y.: On the difficulty of training recurrent neural networks. In: International Conference on Machine Learning, pp. 1310–1318 (2013)
36. Qian, N.: On the momentum term in gradient descent learning algorithms. Neural Netw. **12**(1), 145–151 (1999)
37. Qiao, S., Wang, H., Liu, C., Shen, W., Yuille, A.: Weight standardization. arXiv preprint arXiv:1903.10520 (2019)
38. Rastegari, M., Ordonez, V., Redmon, J., Farhadi, A.: XNOR-Net: ImageNet classification using binary convolutional neural networks. In: Leibe, B., Matas, J., Sebe, N., Welling, M. (eds.) ECCV 2016. LNCS, vol. 9908, pp. 525–542. Springer, Cham (2016). https://doi.org/10.1007/978-3-319-46493-0_32
39. Ravi, S., Larochelle, H.: Optimization as a model for few-shot learning (2016)
40. Ren, S., He, K., Girshick, R., Sun, J.: Faster R-CNN: towards real-time object detection with region proposal networks. In: Advances in Neural Information Processing Systems, pp. 91–99 (2015)
41. Russakovsky, O., et al.: ImageNet large scale visual recognition challenge. Int. J. Comput. Vision **115**(3), 211–252 (2015)
42. Salimans, T., Kingma, D.P.: Weight normalization: a simple reparameterization to accelerate training of deep neural networks. In: Advances in Neural Information Processing Systems, pp. 901–909 (2016)
43. Santurkar, S., Tsipras, D., Ilyas, A., Madry, A.: How does batch normalization help optimization? (no, it is not about internal covariate shift), pp. 2483–2493 (2018)
44. Simonyan, K., Zisserman, A.: Very deep convolutional networks for large-scale image recognition. arXiv preprint arXiv:1409.1556 (2014)
45. Ulyanov, D., Vedaldi, A., Lempitsky, V.: Instance normalization: the missing ingredient for fast stylization. arXiv preprint arXiv:1607.08022 (2016)
46. Vinyals, O., Blundell, C., Lillicrap, T., Wierstra, D., et al.: Matching networks for one shot learning. In: Advances in Neural Information Processing Systems, pp. 3630–3638 (2016)
47. Vorontsov, E., Trabelsi, C., Kadoury, S., Pal, C.: On orthogonality and learning recurrent networks with long term dependencies. In: Proceedings of the 34th International Conference on Machine Learning-Volume 70, pp. 3570–3578. JMLR. org (2017)
48. Wah, C., Branson, S., Welinder, P., Perona, P., Belongie, S.: The caltech-UCSD birds-200-2011 dataset (2011)
49. Wu, Y., He, K.: Group normalization. In: Ferrari, V., Hebert, M., Sminchisescu, C., Weiss, Y. (eds.) ECCV 2018. LNCS, vol. 11217, pp. 3–19. Springer, Cham (2018). https://doi.org/10.1007/978-3-030-01261-8_1

50. Xie, S., Girshick, R., Dollár, P., Tu, Z., He, K.: Aggregated residual transformations for deep neural networks. In: Proceedings of the IEEE Conference on Computer Vision and Pattern Recognition, pp. 1492–1500 (2017)
51. Zhang, C., Bengio, S., Hardt, M., Recht, B., Vinyals, O.: Understanding deep learning requires rethinking generalization. arXiv preprint arXiv:1611.03530 (2016)
52. Zhang, G., Wang, C., Xu, B., Grosse, R.: Three mechanisms of weight decay regularization. arXiv preprint arXiv:1810.12281 (2018)

Content-Aware Unsupervised Deep Homography Estimation

Jirong Zhang[1,2], Chuan Wang[2], Shuaicheng Liu[1(✉)], Lanpeng Jia[2], Nianjin Ye[2], Jue Wang[2], Ji Zhou[1], and Jian Sun[2]

[1] University of Electronic Science and Technology of China, Chengdu, China
zhangjirong@std.uestc.edu.cn, {liushuaicheng,jzhou233}@uestc.edu.cn
[2] Megvii Technology, Beijing, China
{wangchuan,jialanpeng,yenianjin,wangjue,sunjian}@megvii.com
https://github.com/JirongZhang/DeepHomography

Abstract. Homography estimation is a basic image alignment method in many applications. It is usually conducted by extracting and matching sparse feature points, which are error-prone in low-light and low-texture images. On the other hand, previous deep homography approaches use either synthetic images for supervised learning or aerial images for unsupervised learning, both ignoring the importance of handling depth disparities and moving objects in real world applications. To overcome these problems, in this work we propose an unsupervised deep homography method with a new architecture design. In the spirit of the RANSAC procedure in traditional methods, we specifically learn an outlier mask to only select reliable regions for homography estimation. We calculate loss with respect to our learned deep features instead of directly comparing image content as did previously. To achieve the unsupervised training, we also formulate a novel triplet loss customized for our network. We verify our method by conducting comprehensive comparisons on a new dataset that covers a wide range of scenes with varying degrees of difficulties for the task. Experimental results reveal that our method outperforms the state-of-the-art including deep solutions and feature-based solutions.

Keywords: Homography · Deep homography · Image alignment · RANSAC

1 Introduction

Homography can align images taken from different perspectives if they approximately undergo a rotational motion or the scene is close to a planar surface [13].

J. Zhang and C. Wang—Joint First Author.

Electronic supplementary material The online version of this chapter (https://doi.org/10.1007/978-3-030-58452-8_38) contains supplementary material, which is available to authorized users.

A. Vedaldi et al. (Eds.): ECCV 2020, LNCS 12346, pp. 653–669, 2020.
https://doi.org/10.1007/978-3-030-58452-8_38

Fig. 1. Our deep homography estimation on challenging cases, compared with one traditional feature-based, i.e. SIFT [23] + RANSAC and one unsupervised DNN-based method [27]. (a) An example with dominate moving foreground. (b) A low texture example. (c) A low light example. We mix the blue and green channels of the warped image and the red channel of the target image to obtain the visualization results as above, where the misaligned pixels appear as red or green ghosts. The same visualization method is applied for the rest of this paper. (Color figure online)

For scenes that satisfy the constraints, a homography can align them directly. For scenes that violate the constraints, e.g., a scene that consists of multiple planes or contains moving objects, homography usually serves as an initial alignment model before more advanced models such as mesh flow [20] and optical flow [16]. Most of the time, such a pre-alignment is crucial for the final quality. As a result, the homography has been widely applied in vision tasks such as multi-frame HDR imaging [10], multi-frame image super resolution [34], burst image denoising [22], video stabilization [21], image/video stitching [12,36], SLAM [26,42], augmented reality [30] and camera calibration [40].

Homography estimation by traditional approaches generally requires matched image feature points such as SIFT [23]. Specifically, after a set of feature correspondences are obtained, a homography matrix is estimated by Direct Linear Transformation (DLT) [13] with RANSAC outlier rejection [9]. Feature-based methods commonly could achieve good performance while they highly rely on the quality of image features. Estimation could be inaccurate due to insufficient number of matched points or poor distribution of the features, which is a common case due to the existence of textureless regions (e.g., blue sky and white wall), repetitive patterns or illumination variations. Moreover, the rejection of outlier points, e.g., point matches that located on the non-dominate planes or dynamic objects, is also important for high quality results. Consequently, feature-based homography estimation is usually a challenging task for these non-regular scenes.

Due to the development of deep neural networks (DNN) in recent years, DNN-based solutions to homography estimation are gradually proposed such as supervised [7] and unsupervised [27] ones. For the former solution, it requires homography as ground truth (GT) to supervise the training, so that only synthetic target images warped by the GT homography could be generated. Although the synthetic image pairs can be produced in arbitrary scale, they are far from real cases because real depth disparities are unavailable in the training data. As such, this method suffers from bad generalization to real images. To tackle this issue, Nguyen *et al.* proposed the latter unsupervised solution [27], which minimizes the photometric loss on real image pairs. However, this method has two main problems. One is that the loss calculated with respect to image intensity is less effective than that in the feature space, and the loss is calculated uniformly in the entire image ignoring the RANSAC-like process. As a result, this method cannot exclude the moving or non-planar objects to contribute the final loss, so as to potentially decrease the estimation accuracy. To avoid the above phenomenons, Nguyen *et al.* [27] has to work on aerial images that are far away from the camera to minimize the influence of depth variations of parallax.

To tackle the aforementioned issues, we propose an unsupervised solution to homography estimation by a new architecture with content-awareness learning. It is designed specially for image pairs with a **small baseline**, as this case is commonly applicable for consecutive video frames, burst image capturing or photos captured by a dual-camera cellphone. In particular, to robustly optimize a homography, our network implicitly learns a deep feature for alignment and a content-aware mask to reject outlier regions simultaneously. The learned feature is used for loss calculation instead of using photometric loss as in [7], and learning a content-aware mask makes the network concentrate on the important and registrable regions. We further formulate a novel triplet loss to optimize the network so that the unsupervised learning could be achieved. Experimental results demonstrate the effectiveness of all the newly involved techniques for our network, and qualitative and quantitative evaluations also show that our network outperforms the state-of-the-art as shown in Figs. 1, 6 and 7. We also introduce a comprehensive image pair dataset, which contains 5 categories of scenes as well as human-labeled GT point correspondences for quantitative evaluation of its validation set (Fig. 5). To summarize, our main contributions are:

- A novel network structure that enables content-aware robust homography estimation from two images with small baseline.
- A triplet loss designed for unsupervised training, so that an optimal homography matrix could be produced as an output, together with a deep feature map for alignment and a mask highlighting the alignment inliers being implicitly learned as intermediate results.
- A comprehensive dataset covers various scenes for unsupervised training of image alignment models, including but not limited to homography, mesh warps or optical flow.

2 Related Work

Traditional Homography. A homography is a 3×3 matrix which compensates plane motions between two images. It consists of 8 degree of freedom (DOF), with each 2 for scale, translation, rotation and perspective [13] respectively. To solve a homography, traditional approaches often detect and match image features, such as SIFT [23], SURF [4], ORB [29], LPM [25], GMS [5], SOSNet [32], LIFT [35] and OAN [38]. Two sets of correspondences were established between two images, following which robust estimation is adopted, such as the classic RANSAC [9], IRLS [15] and MAGSAC [3], for the outlier rejection during the model estimation. A homogrpahy can also be solved directly without image features. The direct methods, such as seminal Lucas-Kanade algorithm [24], calculates sum of squared differences (SSD) between two images. The differences guide the shift of the images, yielding homography updates. A random initialized homography is optimized in this way iteratively [2]. Moreover, the SSD can be replaced with enhanced correlation coefficient (ECC) for the robustness [8].

Deep Homography. Following the success of various deep image alignment methods such as optical flow [16,33], dense matching [28], learned descriptors [32] and deep features [1], a deep homography solution was first proposed by [7] in 2016. The network takes source and target images as input and produces 4 corner displacement vectors of source image, so as to yield the homography. It used GT homography to supervise the training. However, the training images with GT homography is generated without depth disparity. To overcome such issue, Nguyen *et al.* [27] proposed an unsupervised approach that computed photometric loss between two images and adopted Spatial Transform Network (STN) [17] for image warping. However, they calculated loss directly on the intensity and uniformly on the image plane. In contrast, we learn a content-aware mask. Notebaly, predicting mask for effective estimation has been attempted in other tasks, such as monocular depth estimation [11,41]. Here, it is introduced for the unsupervised homography learning.

Image Stitching. Image stitching methods [19,36] are traditional methods that focus on stitching images under large baselines [37] for the purpose of constructing the panorama [6]. The stitched images were often captured with dramatic viewpoint differences. In this work, we focus on images with small baselines for the purpose of multi-frame applications.

3 Algorithm

3.1 Network Structure

Our method is built upon convolutional neural networks. It takes two grayscale image patches I_a and I_b as input, and produces a homography matrix $\mathcal{H}_{ab}$ from I_a to I_b as output. The entire structure could be divided into three modules: a

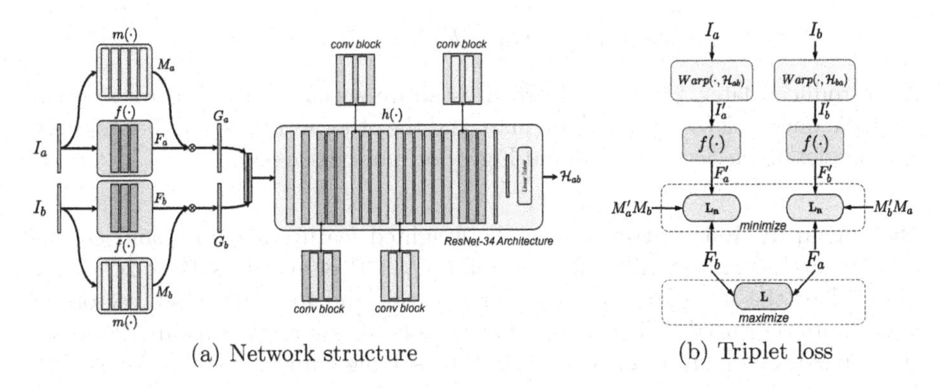

(a) Network structure (b) Triplet loss

Fig. 2. The overall structure of our deep homography estimation network (a) and the triplet loss we design to train the network (b).

feature extractor $f(\cdot)$, a mask predictor $m(\cdot)$ and a homography estimator $h(\cdot)$. $f(\cdot)$ and $m(\cdot)$ are fully convolutional networks which accepts input of arbitrary sizes, and the $h(\cdot)$ utilizes a backbone of ResNet-34 [14] and produces 8 values. Figure 2(a) illustrates the network structure.

Feature Extractor. Unlike previous DNN based methods that directly utilizes the pixel intensity values as the feature, here our network automatically learns a deep feature from the input for robust feature alignment. To this end, we build a fully convolutional network (FCN) that takes an input of size $H \times W \times 1$, and produces a feature map of size $H \times W \times C$. For inputs I_a and I_b, the feature extractor shares weights and produces feature maps F_a and F_b, i.e.

$$F_\beta = f(I_\beta), \quad \beta \in \{a, b\} \tag{1}$$

The learned feature is more robust than pixel intensity when applied to loss calculation. Especially for the images with luminance variations, the learned feature is pretty robust when compared to the pixel intensity. See Sect. 4.3 and Fig. 3 for a detailed verification of the effectiveness of this module.

Mask Predictor. In non-planar scenes, especially those including moving objects, there exists no single homography that can align the two views. In traditional algorithm, RANSAC is widely applied to find the inliers for homography estimation, so as to solve the most approximate matrix for the scene alignment. Following the similar idea, we build a sub-network to automatically learn the positions of inliers. Specifically, a sub-network $m(\cdot)$ learns to produce an inlier probability map or mask, highlighting the content in the feature maps that contribute much for the homography estimation. The size of the mask is the same as the size of the feature maps F_a and F_b. With the masks, we further weight the features extracted by f before feeding them to the homography estimator, obtaining two weighted feature maps G_a and G_b as,

$$M_\beta = m(I_\beta), \quad G_\beta = F_\beta M_\beta, \quad \beta \in \{a, b\} \tag{2}$$

As introduced later, the mask learned as above actually play two roles in the network, one works as an attention map, and the other works as a outlier rejecter. See the details in Sect. 3.2, 4.3 and Fig. 4 for more discussion.

Homography Estimator. Given the weighted feature maps G_a and G_b, we concatenate them to build a feature map $[G_a, G_b]$ of size $H \times W \times 2C$. Then it is fed to the homography estimator network and four 2D offset vectors (8 values) are produced. With the 4 offset vectors, it is straight-forward to obtain the homography matrix $\mathcal{H}_{ab}$ with 8 DOF by solving a linear system. We use $h(\cdot)$ to represent the whole process, i.e.

$$\mathcal{H}_{ab} = h([G_a, G_b]) \tag{3}$$

The backbone of $h(\cdot)$ follows a ResNet-34 structure. It contains 34 layers of strided convolutions followed by a global average pooling layer, which generates fixed size (8 in our case) of feature vectors regardless of the input feature dimensions. Please refer to the project page for more details.

3.2 Triplet Loss for Robust Homography Estimation

With the homography matrix $\mathcal{H}_{ab}$ estimated, we warp image I_a to I'_a and then further extracts its feature map as F'_a. Intuitively, if the homography matrix $\mathcal{H}_{ab}$ is accurate enough, F'_a should be well aligned with F_b, causing a low l_1 loss between them. Considering in real scenes, a single homography matrix normally cannot satisfy the transformation between the two views, we also normalize the l_1 loss by M'_a and M_b. Here M'_a is the warped version of M_a. So the loss between the warped I_a and I_b is as follows,

$$\mathbf{L_n}(I'_a, I_b) = \frac{\sum_i M'_a M_b \cdot ||F'_a - F_b||_1}{\sum_i M'_a M_b} \tag{4}$$

where $F'_a = f(I'_a)$ and $I'_a = Warp(I_a, \mathcal{H}_{ab})$. Index i indicates pixel locations in the masks and feature maps. STN [17] is used to achieve the warping operation.

Directly minimizing Eq. 4 may easily cause trivial solutions, where the feature extractor only produces all zero maps, i.e. $F'_a = F_b = 0$. In this case, the features learned indeed describe the fact that I'_a and I_b are "well aligned", but it fails to reflect the fact that the original images I_a and I_b are mis-aligned. To this end, we involve another loss between F_a and F_b, i.e.

$$\mathbf{L}(I_a, I_b) = ||F_a - F_b||_1 \tag{5}$$

and further maximize it when minimizing Eq. 4. This strategy avoids the trivial all-zero solutions, and enables the network to learn a discriminative feature map.

In practise, we swap the features of I_a and I_b and produce another homography matrix $\mathcal{H}_{ba}$. Following Eq. 4, we involve a loss $\mathbf{L_n}(I'_b, I_a)$ between the warped

Fig. 3. Ablation study on the effectiveness of our feature extractor, demonstrated by examples with illuminance change, displayed separately in the left and right two columns. For each example, the input and target GT images are in Row 1, followed by the results by disabling the feature extractor $f(\cdot)$ (Row 2) and by ours (Row 3), including the learned masks and the aligned results in odd and even columns. As seen, our results are obviously stable for such a case.

I_b and I_a. We also add a constraint that enforces $\mathcal{H}_{ab}$ and $\mathcal{H}_{ba}$ to be inverse. So, the optimization procedure of the network is written as follows,

$$\min_{m,f,h} \mathbf{L_n}(I'_a, I_b) + \mathbf{L_n}(I'_b, I_a) - \lambda \mathbf{L}(I_a, I_b) + \mu||\mathcal{H}_{ab}\mathcal{H}_{ba} - \mathcal{I}||_2^2 \tag{6}$$

where λ and μ are balancing hyper-parameters, and $\mathcal{I}$ is a 3-order identity matrix. We set $\lambda = 2.0$ and $\mu = 0.01$ in our experiments. We show the loss formulations in Fig. 2(b), and validate its effectiveness by an ablation study detailed in Sect. 4.3, which shows that it decreases the error at least 50% in average.

3.3 Unsupervised Content-Awareness Learning

As mentioned above, our network contains a sub-network $m(\cdot)$ to predict an inlier probability mask. It is such designed that our network can be of content-awareness by the two-fold roles. First, we use the masks M_a, M_b to explicitly weight the features F_a, F_b, so that only highlighted features could be fully fed into homography estimator $h(\cdot)$. The masks actually serve as attention maps for the feature maps. Second, they are also implicitly involved into the normalized loss Eq. 4, working as a weighting item. By doing this, only those regions that are really fit for alignment would be taken into account. For those areas containing low texture or moving foreground, because they are non-distinguishable or misleading for alignment, they are naturally removed for homography estimation during optimizing the triplet loss as proposed. Such a content-awareness is achieved fully by an unsupervised learning scheme, without any GT mask data as supervision. To demonstrate the effectiveness of the mask as the two roles, we

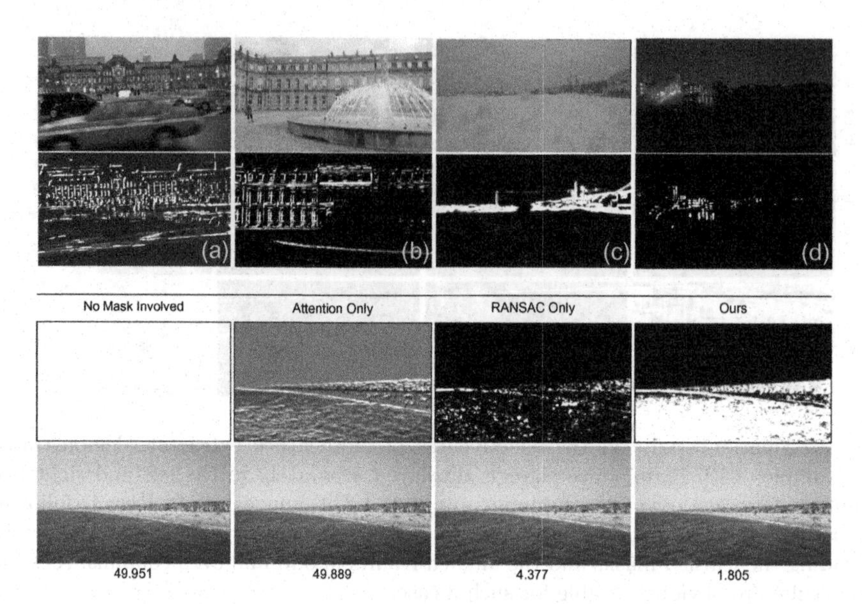

Fig. 4. Row 1 and 2: Our predicted masks for various of scenes. (a) and (b) contains large dynamic foreground. (c) contains few textures and (d) is a night example. Row 3 and 4: Ablation study on the content-aware mask. We disable both or either role of the mask for comparisons. Errors are shown at the bottom.

conduct an ablation study by disabling the effect of mask working as an attention map or as a loss weighting item. As seen in Table 1(c), the accuracy has a significant decrease when mask is removed in either case.

We also illustrate several examples in Fig. 4 to show the mask effectiveness. For example, in Fig. 4(a)(b) where the scenes contain large dynamic foregrounds, our network successfully rejects moving objects, even if the movements are inapparent as the fountain in (b), or the objects occupy a large space as in (a). These cases are very difficult for RANSAC to find robust inliers. Figure 4(c) is a low-textured example, in which the sky and snow ground occupies almost the entire image. It is challenging for traditional methods because not enough feature matches can be provided. Our predicted mask concentrates on the horizon for the alignment. Last, Fig. 4(d) is a low light example, where only visible areas contain weights as seen. We also illustrate an example to show the two effects by the mask as separate roles in the bottom 2 rows of Fig. 4. Details about this ablation study are introduced later in Sect. 4.3.

We adopt a two-stage strategy to train our network. Specifically, we first train the network by disabling the attention map role of the mask, i.e. $G_\beta = F_\beta,\ \beta \in \{a, b\}$. After about 60k iterations, we finetune the network by involving the attention map role of the mask as Eq. 2. We validate this training strategy by another ablation study detailed in Sect. 4.3, where we train the network totally from scratch. This two-stage training strategy reduces the error by 4.40% in average, as shown in Row 10 of Table 1(c).

4 Experimental Results

4.1 Dataset and Implementation Details

We propose our dataset for comprehensive homography evaluation considering there lacks dedicated dataset for this task. Our dataset contains 5 categories of totally 80k image pairs, including regular (**RE**), low-texture (**LT**), low-light (**LL**), small-foregrounds (**SF**), and large-foregrounds (**LF**) scenes, with each category ≈16k image pairs, as shown in Fig. 5. For the test data, 4.2k image pairs are randomly chosen from all categories. For each pair, we manually marked 6–8 equally distributed matching points for the purpose of quantitative comparisons, as illustrated in the rightmost column of Fig. 5. The category partition is based on the understanding and property of traditional homography registration. Experimental results demonstrate our method is robust over all categories as seen in Figs. 1, 6, 7 and the supplementary materials, which also contain a detailed introduction to each category.

Our network is trained with 120k iterations by an Adam optimizer [18], with parameters being set as $l_r = 1.0 \times 10^{-4}$, $\beta_1 = 0.9$, $\beta_2 = 0.999$, $\varepsilon = 1.0 \times 10^{-8}$. The batch size is 64, and for every 12k iterations, the learning rate l_r is reduced by 20%. Each iteration costs about 1.2 s and it takes nearly 40 h to complete the entire training. The implementation is based on PyTorch and the network training is performed on 4 NVIDIA RTX 2080 Ti. To augment the training data and avoid black boundaries appearing in the warped image, we randomly crop patches of size 315 × 560 from the original image to form I_a and I_b. Code is available at https://github.com/JirongZhang/DeepHomography.

4.2 Comparisons with Existing Methods

Qualitative Comparison. We first compare our method with the existing two deep homography methods, the supervised [7] and the unsupervised [27] approaches, as illustrated in Fig. 6. Figure 6(a) shows an synthesized example with no disparities. In this case, the supervised solution [7] performs well enough as ours. However, it fails in the case that real consecutive frames of the same footage are applied (Fig. 6(b)), because it is unable to handle large disparities and moving objects of the scene. Figure 6(c) shows an example that contains a dominate planar building surface, where all methods work well. However, if the image pair involves illumination variation caused by camera flash, the unsupervised method [27] fails due to its alignment metric being pixel intensity value difference instead of semantic feature difference, as seen in Fig. 6(d). Figure 6(e) and (f) contain near-range objects and two dominate planes with moving objects at corners respectively, and Fig. 6(g) and (h) are low texture and low light examples separately. Similarly, in all of these scenarios, our method produces warped images with more pixels aligned, so as to obviously outperform the other two DNN-based methods.

We also compare our method with some feature-based solutions. Specially, we choose SIFT [23], ORB [29], LIFT [35] and SOSNet [32] as the feature descriptors

Fig. 5. A glace of our dataset. For left 6 columns, from top to bottom are the 5 categories of the dataset. The rightmost column shows two examples of human labeled point correspondences for quantitative evaluation.

and choose RANSAC [9] and MAGSAC [3] as the outlier rejection algorithms, obtaining 8 combinations. We show 3 examples in Fig. 7, where (a)(b) show the 8 combinations produce reasonable but low quality results, and (c) shows one that most of them fail thoroughly. Note that the failure cases caused by low texture or low light condition frequently appears in our dataset, and it may lead to unstable results in real applications such as video stabilization or multi-frame image fusion. In comparison, our method is robust against these challenges.

Quantitative Comparison. We demonstrate the performance of our method by comparing it with all of the other methods quantitatively. The comparison is based on our dataset and the average l_2 distances between the warped points and the human-labeled GT points are evaluated as the error metric. We report the errors for each category and the overall averaged error in Table 1, where $\mathcal{I}_{3\times3}$ refers to a 3×3 identity matrix as a "no-warping" homography for reference. As seen, our method outperforms the others for all categories, except for regular (RE) scenes if compared with feature-based methods. This result is reasonable because in RE scenes, rich texture delivers sufficient high quality features so that it is naturally friendly for the feature-based solutions. Even though, our error is only 5.85% higher than the best solution in this case, i.e. SIFT [23] + MAGSAC [3]. For the rest scenes, our method consistently beats the others, especially for the low texture (LT) and low light (LL) scenes, where our error is lower than the 2nd best by 25.78% and 7.62% respectively. For the scenes containing small (SF) and large (LF) foreground, although the 2nd best method SOSNet [32] + MAGSAC [3] only loses to ours very slightly (0.57% and 2.82%), it cannot well handle the LT and LL scenes, where its errors are higher than the 2nd best by 100.78% and 109.05% separately. It is worth noting that the two solutions involving LIFT [35] feature produce rather stable results for all scenes, but their average errors are higher than ours by at least 12.08%. As for

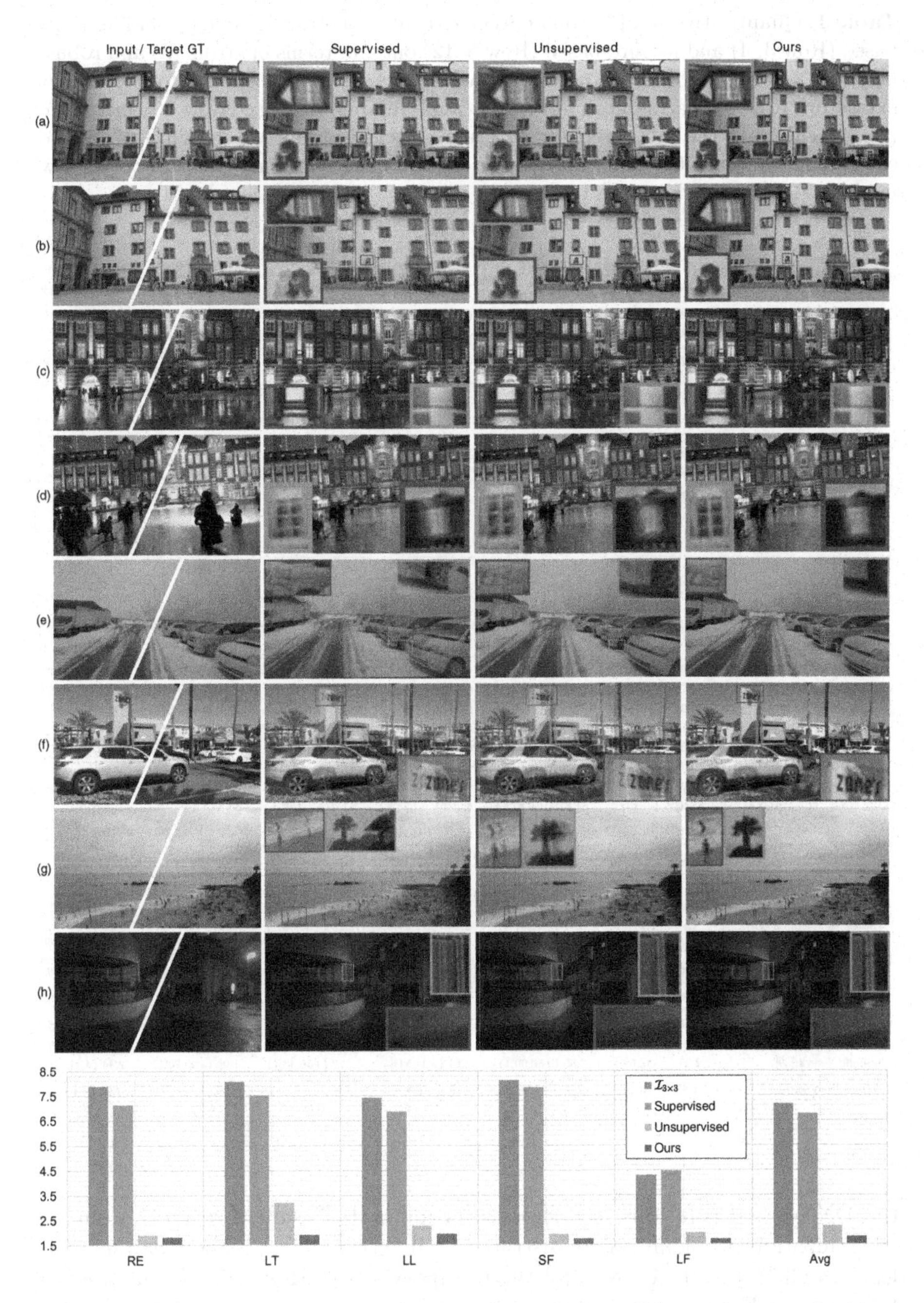

Fig. 6. Comparison with existing DNN-based approaches. Column 1 shows the input and GT target images, columns 2 to 4 are results by the supervised [7], the unsupervised [27] and our method. The errors by all the DNN-based methods are displayed by a bar chart at the bottom.

Table 1. Quantitative comparison between ours and all other methods including DNN-based (Row 3, 4) and feature-based (Row 5–12) ones, in terms of errors (a) and robustness (b), as well as ablation studies on mask (Rows 2–4), triplet loss (Row 5), feature extractor (Row 6), backbones (Rows 7–9) and training strategy (Row 10) in (c). For (b), we calculate the inlier percentage when matched points are within 3 pixels. For each scene, we mark the best solution in red. For the scenes ours beats the others, we mark the 2nd best solution in blue.

(a) Errors

1)	RE	LT	LL	SF	LF	Avg
2)$\mathcal{I}_{3\times3}$	7.88 (+360.82%)	8.07 (+215.23%)	7.41 (+252.86%)	8.11 (+360.80%)	4.29 (+142.37%)	7.15 (+245.41%)
3)Supervised [7]	7.12 (+316.37%)	7.53 (+194.14%)	6.86 (+226.67%)	7.83 (+344.89%)	4.46 (+151.98%)	6.76 (+226.57%)
4)Unsupervised [27]	1.88 (+9.94%)	3.21 (+25.39%)	2.27 (+8.10%)	1.93 (+9.66%)	1.97 (+11.30%)	2.25 (+8.70%)
5)SIFT [23] + RANSAC [9]	1.72 (+0.58%)	2.56 (+0.00%)	4.97 (+136.67%)	1.82 (+3.41%)	1.84 (+3.95%)	2.58 (+24.64%)
6)SIFT [23] + MAGSAC [3]	1.71 (+0.00%)	3.15 (+23.05%)	4.91 (+133.81%)	1.88 (+6.82%)	1.79 (+1.13%)	3.20 (+54.59%)
7)ORB [29] + RANSAC [9]	1.85 (+8.19%)	3.76 (+46.88%)	2.56 (+21.90%)	2.00 (+13.64%)	2.29 (+29.38%)	2.49 (+20.29%)
8)ORB [29] + MAGSAC [3]	2.02 (+18.13%)	5.18 (+102.34%)	2.78 (+32.38%)	1.92 (+9.09%)	2.25 (+27.12%)	2.83 (+36.71%)
9)LIFT [35] + RANSAC [9]	1.76 (+2.92%)	3.04 (+18.75%)	2.14 (+1.90%)	1.82 (+3.41%)	1.92 (+8.47%)	2.14 (+3.38%)
10)LIFT [35] + MAGSAC [3]	1.73 (+1.17%)	2.92 (+14.06%)	2.10 (+0.00%)	1.79 (+1.70%)	1.79 (+1.13%)	2.07 (+0.00%)
11)SOSNet [32] + RANSAC [9]	1.72 (+0.58%)	3.70 (+44.53%)	4.58 (+118.09%)	1.84 (+4.54%)	1.83 (+3.39%)	2.73 (+31.88%)
12)SOSNet [32] + MAGSAC [3]	1.73 (+1.17%)	5.14 (+100.78%)	4.39 (+109.05%)	1.76 (+0.00%)	1.77 (+0.00%)	2.99 (+44.44%)
13)Ours	**1.81 (+5.85%)**	1.90 (-25.78%)	1.94 (-7.62%)	1.75 (-0.57%)	1.72 (-2.82%)	1.82 (-12.08%)

(b) Robustness: Inlier Percentage When Matched Points Are within 3 Pixels

1)	RE	LT	LL	SF	LF	Avg
2)$\mathcal{I}_{3\times3}$	12.75% (-85.35%)	37.83% (-54.13%)	36.68% (-55.32%)	48.46% (-42.43%)	64.30% (-25.15%)	38.76% (-53.79%)
3)Supervised [7]	16.17% (-81.42%)	42.76% (-48.16%)	40.73% (-50.38%)	48.24% (-42.69%)	61.29% (-28.65%)	40.89% (-51.25%)
4)Unsupervised [27]	85.57% (-1.69%)	71.41% (-13.42%)	79.45% (-3.22%)	82.52% (-1.96%)	83.65% (-2.62%)	79.80% (-4.86%)
5)SIFT [23]+RANSAC [9]	86.95% (-0.10%)	81.98% (-0.61%)	80.79% (-1.58%)	84.17% (+0.00%)	85.36% (-0.63%)	83.77% (-0.13%)
6)SIFT [23]+MAGSAC [3]	86.70% (-0.39%)	82.48% (+0.00%)	80.67% (-1.73%)	83.69% (-0.57%)	85.90% (+0.00%)	83.88% (+0.00%)
7)ORB [29]+RANSAC [9]	85.31% (-1.99%)	77.21% (-6.39%)	81.44% (-0.79%)	83.55% (-0.74%)	79.70% (-7.22%)	81.00% (-3.43%)
8)ORB [29]+MAGSAC [3]	83.55% (-4.01%)	75.15% (-8.89%)	80.77% (-1.61%)	81.25% (-3.47%)	79.80% (-7.10%)	79.70% (-4.98%)
9)LIFT [35]+RANSAC [9]	86.50% (-0.62%)	72.58% (-12.00%)	80.89% (-1.46%)	83.22% (-1.13%)	83.42% (-2.89%)	80.63% (-3.87%)
10)LIFT [35]+MAGSAC [3]	87.04% (+0.00%)	74.53% (-9.64%)	82.09% (+0.00%)	83.84% (-0.39%)	85.61% (-0.34%)	82.03% (-2.21%)
11)SOSNet [32]+RANSAC [9]	87.03% (-0.01%)	81.44% (-1.26%)	80.69% (-1.71%)	84.10% (-0.08%)	85.48% (-0.49%)	83.63% (-0.30%)
12)SOSNet [32]+MAGSAC [3]	86.93% (-0.13%)	81.81% (-0.81%)	80.63% (-1.78%)	83.29% (-1.05%)	85.84% (-0.07%)	83.69% (-0.23%)
13)Ours	**86.12% (-1.06%)**	83.58% (+1.33%)	83.63% (+1.88%)	85.23% (+1.26%)	87.36% (+1.70%)	85.10% (+1.45%)

(c) Ablation Studies

1)	RE	LT	LL	SF	LF	Avg
2)No mask involved	2.10 (+16.02%)	2.51 (+32.11%)	2.48 (+27.84%)	3.02 (+72.57%)	1.78 (+3.49%)	2.38 (+30.77%)
3)Mask as attention only	1.85 (+2.21%)	3.37 (+77.37%)	2.16 (+11.34%)	2.29 (+30.86%)	1.75 (+1.74%)	2.27 (+24.73%)
4)Mask as RANSAC only	1.85 (+2.21%)	2.16 (+13.68%)	2.17 (+11.86%)	2.04 (+16.57%)	2.16 (+25.58%)	2.07 (+13.74%)
5)w/o. Triple loss	2.16 (+19.34%)	4.15 (+118.42%)	3.30 (+70.10%)	2.49 (+42.29%)	2.09 (+21.51%)	2.84 (+56.04%)
6)w/o. Feature extractor	1.89 (+4.42%)	2.54 (+33.68%)	2.13 (+9.79%)	1.80 (+2.86%)	1.79 (+4.07%)	2.03 (+11.54%)
7)VGG [31]	1.91 (+5.52%)	2.89 (+52.11%)	2.05 (+5.67%)	2.14 (+22.29%)	1.88 (+9.30%)	2.17 (+19.23%)
8)ResNet-18 [14]	1.84 (+1.66%)	2.30 (+21.05%)	2.05 (+5.67%)	2.28 (+30.29%)	1.85 (+7.56%)	2.06 (+13.19%)
9)ShuffleNet-v2 [39]	2.05 (+13.26%)	2.85 (+50.00%)	2.61 (+34.54%)	2.72 (+55.43%)	1.99 (+15.70%)	2.44 (+34.07%)
10)Train from scratch	1.87 (+3.31%)	2.00 (+5.26%)	1.98 (+2.06%)	1.90 (+8.57%)	1.77 (+2.91%)	1.90 (+4.40%)
11)Ours	1.81	1.90	1.94	1.75	1.72	1.82

the DNN-based solutions, the supervised method [7] suffers severely from the generalization problem as demonstrated by its errors being higher than us by at least 142.37% for all scenes, and the unsupervised method [27] also apparently fails in the LT scene, causing over 50% higher error than ours in this case.

To further evaluate the robustness, a threshold (3 pixels) is used to count the percentage of inliners. Matches that beyond the threshold are considered as outliers. Table 1(b) shows the inlier percentage on different scene categories

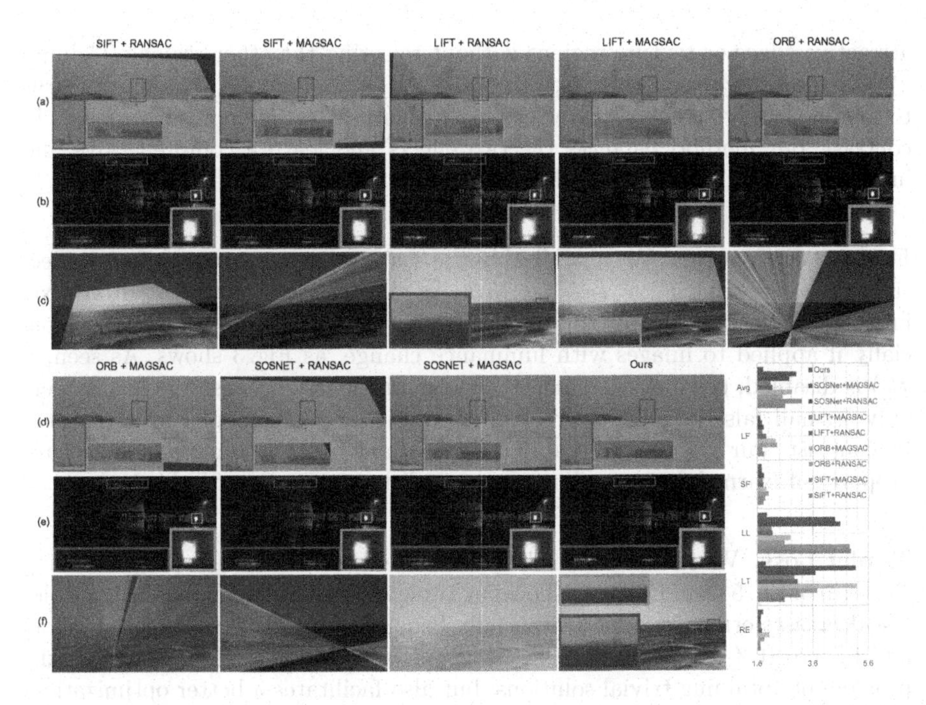

Fig. 7. Comparison with 8 feature-based solutions on 3 examples, shown in (a)(d), (b)(e) and (c)(f). For the first 2 examples, our method produces more accurate results, while for the last one but not the least, most of the feature-based solutions fail extremely, which happens frequently for the low texture or low light scenes. We also display the errors by all the methods in bar chart.

of various methods. As seen, for tough cases, our method achieves the highest robustness compared with other competitors while for regular cases, our performance is on par with the others, which draws similar conclusion as by Table 1(a). Please see Table 1 and bar charts in Figs. 6 and 7 for the detailed comparisons.

4.3 Ablation Studies

Content-Aware Mask. As mentioned in Sect. 3.3, the content-aware mask takes effects in two-folds, working as an attention for the feature map, or as a weighting map to reject the outliers. We verify its effectiveness by evaluating the performance in the case of disabling both or either effect and report the errors in Row 2, 3, 4 of Table 1(c). Specifically, for Row 3 "Mask as attention only", Eq. 4 is modified as $\mathbf{L_n}(I'_a, I_b) = \mathbf{L}(I'_a, I_b) = ||F'_a - F_b||_1$. On the contrary, for Row 4 "Mask as RANSAC only", Eq. 2 is modified as $G_\beta = F_\beta,\ \beta \in \{a, b\}$. As the errors indicate, for most scenes the mask takes effect increasingly by the two roles, except for the scenes LT and LF where disabling one role only may cause the worst result. We also illustrate one example in Row 3, 4 of Fig. 4, where in the case of "Mask as attention only" the mask learns to highlight the most attractive edges or texture regions without rejecting the other regions (Column 2). On the contrary, in the case of "Mask as RANSAC only", the mask learns to highlight only sparse texture regions (Column 3) as inliers for alignment. In contrast, our method balances the two effects and learn a comprehensive and informative weighting map as shown in Column 4.

Feature Extractor. We also disable the feature extractor to verify its effectiveness, i.e. setting $F_\beta = I_\beta$, $\beta \in \{a, b\}$ so that the loss is evaluated on pixel intensity values instead. In this case, the network loses some robustness, especially if applied to images with luminance change, as Fig. 3 shows. As seen, if $f(\cdot)$ is disabled, the masks would be abnormally sparse because the loss reflects only a small falsely "aligned" region, causing a wrong homography estimated. In contrast, our results are stable enough thanks to the luminance invariant property of learned features. The errors are listed in Row 6 of Table 1(c).

Triplet Loss. We further exam the effectiveness of our triplet loss by removing the term of Eq. 5 from Eq. 6. As shown in Table 1(c) "w/o. triplet loss", the triplet loss decreases errors over 50%, especially beneficial in LT (118.42% lower error) and LL (70.10% lower error) scenes, demonstrating that it not only avoids the problem of obtaining trivial solutions, but also facilitates a better optimization.

Backbone. We also exam several popular backbones, including VGG [31], ResNet-18 [14], and ShuffleNet [39] for $h(\cdot)$. As seen in Rows 7–9 of Table 1(c), the ResNet-18 achieves similar performance as ours (ResNet-34). The VGG backbone is slightly worse than ResNet-18 and ResNet-34. Interestingly, the lightweight backbone ShuffleNet achieves similar performance with other large ones, indicating the potential application to portable systems of our method.

Training Strategy. As aforementioned, we use a two-stage strategy to train the network. To validate this strategy, we conduct an ablation study to train the network from scratch. As Row 10 and 11 of Table 1(c) reveal, our training strategy brings a 4.40% lower error in average, demonstrating its usefulness.

5 Conclusions

We have presented a new architecture for unsupervised deep homography estimation with content-aware capability, for small baseline scenarios. Unlike traditional feature base methods that heavily rely on the quality of image features so as to be vulnerable to low-texture and low-light scenes, or previous DNN-based solutions that pay less attention to the depth disparity issue, our network learns a content-aware mask during the estimation to reject outliers, such that the network can concentrate on the regions that can be aligned by a homography. To achieve it, we have designed a novel triplet loss to enable unsupervised training of our network. Moreover, we present a comprehensive dataset for image alignment. The dataset is divided into 5 categories of scenes, which can be used for the future research of image alignment models, including but not limited to homography, mesh alignment and optical flow. Extensive experiments and ablation studies demonstrate the effectiveness of our network as well as the triplet loss design, and reveal the superiority of our method over the state-of-the-art.

Acknowledgment. This research was supported in part by National Key Research and Development Program of China under Grant 2017YFA0700800, in part by National Natural Science Foundation of China under Grants (NSFC, No. 61872067 and No. 61720106004) and in part by Research Programs of Science and Technology in Sichuan Province under Grant 2019YFH0016.

References

1. Altwaijry, H., Veit, A., Belongie, S.J., Tech, C.: Learning to detect and match keypoints with deep architectures. In: Proceedings of BMVC (2016)
2. Baker, S., Matthews, I.: Lucas-kanade 20 years on: a unifying framework. Int. J. Comput. Vis. **56**(3), 221–255 (2004)
3. Barath, D., Matas, J., Noskova, J.: MAGSAC: marginalizing sample consensus. In: Proceedings of CVPR, pp. 10197–10205 (2019)
4. Bay, H., Tuytelaars, T., Van Gool, L.: SURF: speeded up robust features. In: Leonardis, A., Bischof, H., Pinz, A. (eds.) ECCV 2006. LNCS, vol. 3951, pp. 404–417. Springer, Heidelberg (2006). https://doi.org/10.1007/11744023_32
5. Bian, J., Lin, W.Y., Matsushita, Y., Yeung, S.K., Nguyen, T.D., Cheng, M.M.: GMS: grid-based motion statistics for fast, ultra-robust feature correspondence. In: Proceedings of CVPR, pp. 4181–4190 (2017)
6. Brown, M., Lowe, D.: Recognising panoramas. In: Proceedings of ICCV, p. 1218 (2003)
7. DeTone, D., Malisiewicz, T., Rabinovich, A.: Deep image homography estimation. arXiv preprint arXiv:1606.03798 (2016)
8. Evangelidis, G.D., Psarakis, E.Z.: Parametric image alignment using enhanced correlation coefficient maximization. IEEE Trans. Pattern Anal. Mach. Intell. **30**(10), 1858–1865 (2008)
9. Fischler, M.A., Bolles, R.C.: Random sample consensus: a paradigm for model fitting with applications to image analysis and automated cartography. Commun. ACM **24**(6), 381–395 (1981)
10. Gelfand, N., Adams, A., Park, S.H., Pulli, K.: Multi-exposure imaging on mobile devices. In: Proceedings of ACM Multimedia, pp. 823–826 (2010)
11. Godard, C., Mac, O., Firman, M., Brostow, G.J.: Digging into self-supervised monocular depth estimation. In: Proceedings of ICCV, pp. 3828–3838 (2019)
12. Guo, H., Liu, S., He, T., Zhu, S., Zeng, B., Gabbouj, M.: Joint video stitching and stabilization from moving cameras. IEEE Trans. Image Process. **25**(11), 5491–5503 (2016)
13. Hartley, R., Zisserman, A.: Multiple View Geometry in Computer Vision. Cambridge University Press, Cambridge (2003)
14. He, K., Zhang, X., Ren, S., Sun, J.: Deep residual learning for image recognition. In: Proceedings of CVPR, pp. 770–778 (2016)

15. Holland, P.W., Welsch, R.E.: Robust regression using iteratively reweighted least-squares. Commun. Stat. Theo. Methods **6**(9), 813–827 (1977)
16. Ilg, E., Mayer, N., Saikia, T., Keuper, M., Dosovitskiy, A., Brox, T.: FlowNet 2.0: evolution of optical flow estimation with deep networks. In: Proceedings of CVPR, pp. 2462–2470 (2017)
17. Jaderberg, M., Simonyan, K., Zisserman, A., et al.: Spatial transformer networks. In: Advances in Neural Information Processing Systems, pp. 2017–2025 (2015)
18. Kingma, D.P., Ba, J.: Adam: a method for stochastic optimization. arXiv preprint arXiv:1412.6980 (2014)
19. Lin, K., Jiang, N., Liu, S., Cheong, L.F., Do, M., Lu, J.: Direct photometric alignment by mesh deformation. In: Proceedings of CVPR, pp. 2405–2413 (2017)
20. Liu, S., Tan, P., Yuan, L., Sun, J., Zeng, B.: MeshFlow: minimum latency online video stabilization. In: Leibe, B., Matas, J., Sebe, N., Welling, M. (eds.) ECCV 2016. LNCS, vol. 9910, pp. 800–815. Springer, Cham (2016). https://doi.org/10.1007/978-3-319-46466-4_48
21. Liu, S., Yuan, L., Tan, P., Sun, J.: Bundled camera paths for video stabilization. ACM Trans. Graph. **32**(4), 78 (2013)
22. Liu, Z., Yuan, L., Tang, X., Uyttendaele, M., Sun, J.: Fast burst images denoising. ACM Trans. Graph. **33**(6), 1–9 (2014)
23. Lowe, D.G.: Distinctive image features from scale-invariant keypoints. Int. J. Comput. Vis. **60**(2), 91–110 (2004)
24. Lucas, B.D., Kanade, T., et al.: An iterative image registration technique with an application to stereo vision. In: Proceedings of IJCAI (1981)
25. Ma, J., Zhao, J., Jiang, J., Zhou, H., Guo, X.: Locality preserving matching. Int. J. Comput. Vis. **127**(5), 512–531 (2019)
26. Mur-Artal, R., Montiel, J.M.M., Tardos, J.D.: ORB-SLAM: a versatile and accurate monocular slam system. IEEE Trans. Robot. **31**(5), 1147–1163 (2015)
27. Nguyen, T., Chen, S.W., Shivakumar, S.S., Taylor, C.J., Kumar, V.: Unsupervised deep homography: a fast and robust homography estimation model. IEEE Robot. Autom. Lett. **3**(3), 2346–2353 (2018)
28. Revaud, J., Weinzaepfel, P., Harchaoui, Z., Schmid, C.: Deepmatching: hierarchical deformable dense matching. Int. J. Comput. Vis. **120**(3), 300–323 (2016)
29. Rublee, E., Rabaud, V., Konolige, K., Bradski, G.R.: ORB: an efficient alternative to SIFT or SURF. In: Proceedings of ICCV, vol. 11, pp. 2564–2571 (2011)
30. Simon, G., Fitzgibbon, A.W., Zisserman, A.: Markerless tracking using planar structures in the scene. In: Proceedings of International Symposium on Augmented Reality, pp. 120–128 (2000)
31. Simonyan, K., Zisserman, A.: Very deep convolutional networks for large-scale image recognition. arXiv preprint arXiv:1409.1556 (2014)
32. Tian, Y., Yu, X., Fan, B., Wu, F., Heijnen, H., Balntas, V.: SOSNet: second order similarity regularization for local descriptor learning. In: Proceedings of CVPR, pp. 11016–11025 (2019)
33. Weinzaepfel, P., Revaud, J., Harchaoui, Z., Schmid, C.: Deepflow: large displacement optical flow with deep matching. In: Proceedings of CVPR, pp. 1385–1392 (2013)
34. Wronski, B., et al.: Handheld multi-frame super-resolution. ACM Trans. Graph. **38**(4), 1–18 (2019)
35. Yi, K.M., Trulls, E., Lepetit, V., Fua, P.: LIFT: learned invariant feature transform. In: Leibe, B., Matas, J., Sebe, N., Welling, M. (eds.) ECCV 2016. LNCS, vol. 9910, pp. 467–483. Springer, Cham (2016). https://doi.org/10.1007/978-3-319-46466-4_28

36. Zaragoza, J., Chin, T.J., Brown, M.S., Suter, D.: As-projective-as-possible image stitching with moving DLT. In: Proceedings of CVPR, pp. 2339–2346 (2013)
37. Zhang, F., Liu, F.: Parallax-tolerant image stitching. In: Proceedings of CVPR, pp. 3262–3269 (2014)
38. Zhang, J., et al.: Learning two-view correspondences and geometry using order-aware network. In: Proceedings of ICCV, pp. 5845–5854 (2019)
39. Zhang, X., Zhou, X., Lin, M., Sun, J.: ShuffleNet: an extremely efficient convolutional neural network for mobile devices. In: Proceedings of CVPR, pp. 6848–6856 (2018)
40. Zhang, Z.: A flexible new technique for camera calibration. IEEE Trans. Pattern Anal. Mach. Intell. **22**(11), 1330–1334 (2000)
41. Zhou, T., Brown, M., Snavely, N., Lowe, D.G.: Unsupervised learning of depth and ego-motion from video. In: Proceedings of CVPR, pp. 1851–1858 (2017)
42. Zou, D., Tan, P.: CoSLAM: collaborative visual slam in dynamic environments. IEEE Trans. Pattern Anal. Mach. Intell. **35**(2), 354–366 (2012)

Multi-view Optimization of Local Feature Geometry

Mihai Dusmanu[1(✉)], Johannes L. Schönberger[2], and Marc Pollefeys[1,2]

[1] Department of Computer Science, ETH Zürich, Zürich, Switzerland
mihai.dusmanu@inf.ethz.ch
[2] Microsoft, Zürich, Switzerland

Abstract. In this work, we address the problem of refining the geometry of local image features from multiple views without known scene or camera geometry. Current approaches to local feature detection are inherently limited in their keypoint localization accuracy because they only operate on a single view. This limitation has a negative impact on downstream tasks such as Structure-from-Motion, where inaccurate keypoints lead to large errors in triangulation and camera localization. Our proposed method naturally complements the traditional feature extraction and matching paradigm. We first estimate local geometric transformations between tentative matches and then optimize the keypoint locations over multiple views jointly according to a non-linear least squares formulation. Throughout a variety of experiments, we show that our method consistently improves the triangulation and camera localization performance for both hand-crafted and learned local features.

Keywords: 3D reconstruction · Local features

1 Introduction

Local image features are one of the central blocks of many computer vision systems with numerous applications ranging from image matching and retrieval to visual localization and mapping. Predominantly, local feature extraction and matching are the first stages in these systems with high impact on their final performance in terms of accuracy and completeness [37]. The main advantages of local features are their robustness, scalability, and efficient matching, thereby enabling large-scale 3D reconstruction [18] and localization [21].

Handcrafted local feature approaches generally focus on low-level structures for detection [16,23]. Despite the typically accurate keypoint localization of these methods, they are easily perturbed by appearance variations such as day-to-night or seasonal changes, as shown by Sattler *et al.* [35]. To achieve a better robustness against viewpoint and appearance changes, recent methods turned

Electronic supplementary material The online version of this chapter (https://doi.org/10.1007/978-3-030-58452-8_39) contains supplementary material, which is available to authorized users.

A. Vedaldi et al. (Eds.): ECCV 2020, LNCS 12346, pp. 670–686, 2020.
https://doi.org/10.1007/978-3-030-58452-8_39

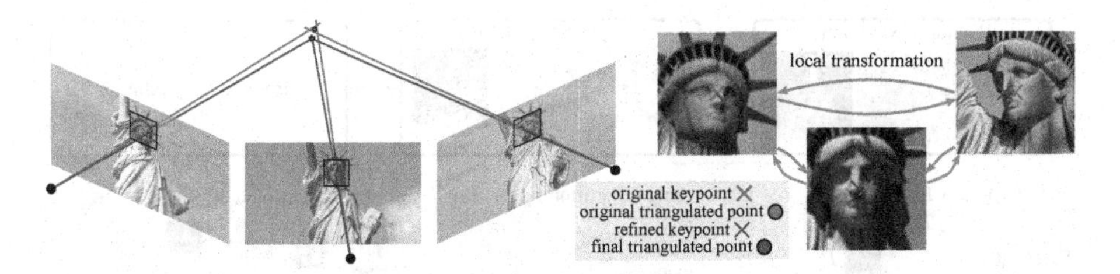

Fig. 1. Multi-view keypoint refinement. The proposed method estimates local transformations between multiple tentative views of a same feature and uses them to refine the 2D keypoint location, yielding more accurate and complete point clouds.

to convolutional neural networks (CNNs) for local feature detection and description [10,12,27,30]. However, this comes at the cost of a poorer keypoint localization, mainly caused by relying on larger receptive fields and feature map down-sampling through pooling or strided convolutions (Fig. 1).

Moreover, both traditional and CNN-based methods only exploit a single view, as feature detection and description is run independently on each image. Even for low-level detectors, there is no inherent reason why detections would be consistent across multiple views, especially under strong viewpoint or appearance changes. While some recent works [15,17,44] consider multiple views to improve the feature matching step, to the best of our knowledge, no prior work exploits multiple views to improve the feature detection stage for more accurate keypoints.

In this paper, we propose a method for optimizing the geometry of local features by exploiting multiple views without any prior knowledge about the camera geometry or scene structure. Our proposed approach first uses a patch-alignment CNN between tentative matches to obtain an accurate two-view refinement of the feature geometry. The second stage aggregates all the two-view refinements in a multi-view graph of relative feature geometry constraints and then globally optimizes them jointly to obtain the refined geometry of the features. The proposed two-stage approach is agnostic to the type of local features and easily integrates into any application relying on local feature matching. Numerous experiments demonstrate the superior performance of our approach for various local features on the tasks of image matching, triangulation, camera localization, and end-to-end 3D reconstruction from unstructured imagery. The source code of our entire method and of the evaluation pipeline will be released as open source.

2 Related Work

Our method is directly related to local features as well as patch description and matching. The two-view alignment network borrows concepts from recent advances in the field of image alignment and visual flow. In this section, we provide an overview of the state of the art in these research directions.

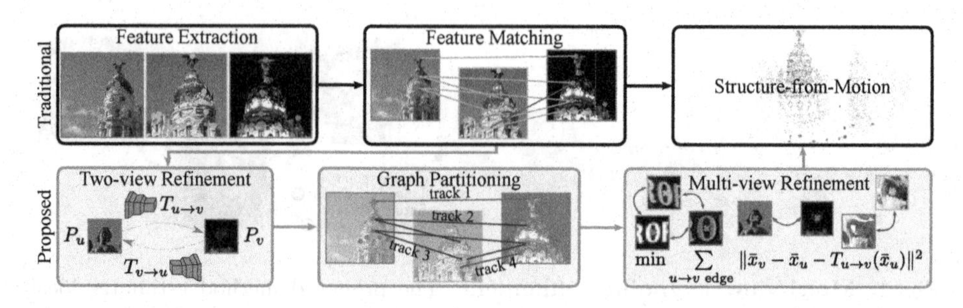

Fig. 2. Overview of the proposed method. Our method operates on the tentative matches graph (with patches as nodes P_u, P_v and matches as edges) without knowledge of scene and camera geometry. A neural network is used to annotate the edges of this graph with local geometric transformations $(T_{u\to v}, T_{v\to u})$. Next, the graph is partitioned into tracks, each track containing at most one patch from each image. Finally, the keypoint locations x_u, x_v are refined using a global optimization over all edges.

Local Features. Traditional local feature extractors can be split into two main stages: first, feature detection finds interesting regions in the image using low-level statistics (*e.g.*, Difference-of-Gaussians [23] or the Harris score [16]), typically followed by the estimation of local feature geometry (*e.g.*, scale, orientation, affine shape) for the detected interest points to achieve viewpoint invariance. Second, feature description then normalizes the local image region around interest points to a canonical frame using the detected feature geometry and finally extracts an illumination invariant, compact numerical representation from the normalized patch (*e.g.*, SIFT [23], Root-SIFT [3], BRIEF [8]). More recently, researchers have developed trainable counterparts that either replace individual parts of the pipeline – learned detectors [6,36,43] and descriptors [5,26] – or reformulate the entire pipeline in an end-to-end trainable manner [29,45].

Lately, methods have moved away from the detect-then-describe methodology to a describe-then-detect approach, mainly due to the sensitivity of detections to changes in image statistics. These methods start by using a CNN as a dense feature extractor and afterwards either train a classifier for detection on top [27], use it as a shared encoder that splits into two decoders for detection and description respectively [10,30], or directly use non-maxima suppression on the deep feature maps [12]. However, these approaches have another issue: due to their large receptive field and feature map down-sampling, the obtained keypoints are generally not well localized when compared to their hand-crafted, low-level counterparts. This is the case even for methods [10] explicitly trained to detect corners. In this paper, we address the limited accuracy of feature detections for both hand-crafted as well as learned features. Our approach only requires images as input and achieves superior detection accuracy by considering multiple views jointly, which is in contrast to existing local feature approaches.

Patch Description and Matching. CNNs have been successfully used to learn local descriptors offering better robustness to viewpoint and illumination changes using different triplet losses [5], hard-negative mining techniques [26], and geometric similarity for training [25]. Likewise, in Multi-View Stereo, hand-crafted similarity metrics [14] and descriptors [42] traditionally used for patch matching were replaced by learned counterparts [15,17,24,44]. Closer to our approach are methods bypassing description and directly considering multiple views to decide whether two points correspond [15,46,47]. While these approaches focus on the second part of the local feature pipeline, we focus our attention on the detection stage. However, the intrinsic motivation is the same: exploiting multiple views facilitates a more informed decision for better results.

Geometric Alignment and Visual Flow. Recent advances in semantic alignment [31,33] and image matching [33] as well as flow estimation [11] use a Siamese network followed by a feature matching layer. Our patch alignment network uses the correlation normalization introduced in [31]. The matching results are processed by a sequence of convolutional and fully connected layers for prediction. Contrary to visual flow, which is generally targeted at temporally adjacent video frames, where pixel displacements remain relatively low and appearance is similar, our method must handle large deformations and drastic illumination changes.

Refinement from Known Geometry or Poses. Closer to our method, Eichhardt *et al.* [13] recently introduced an approach for local affine frame refinement. While they similarly formulate the problem as a constrained, multi-view least squares optimization, their method assumes known two-view camera geometries and does not consider visual cues from two views jointly to compute the patch alignment. Furthermore, they need access to ground-truth feature tracks (computed by an initial Structure-from-Motion process). In contrast, not requiring known camera geometry and feature tracks makes our approach amenable to a much wider range of practical applications, *e.g.*, Structure-from-Motion or visual localization. Moreover, the two methods are in fact complementary - our procedure can improve the quality of Structure-from-Motion, which can then be further refined using their approach.

3 Method

The generic pipeline for multi-view geometry estimation, illustrated in Fig. 2, starts from a set of input images $\mathcal{I} = \{I_1, \ldots, I_N\}$ and first runs feature extraction on each image I_i independently yielding keypoints p_i with associated local descriptors d_i. Feature matching next computes tentative feature correspondences $\mathcal{M}_{i,j} = \{(k,l)$ such that $d_{i,k}$ matches $d_{j,l}\}$ between image pairs (I_i, I_j) based on nearest neighbors search in descriptor space (usually alongside filtering techniques). The output of this step can be interpreted as a tentative matches graph $G = (V, E)$ with keypoints as nodes ($V = \cup_i p_i$) and matches as edges ($E = \cup_{i,j} \mathcal{M}_{i,j}$), optionally weighted (*e.g.*, by the cosine similarity of descriptors). In the last step, the specific application (*e.g.*, a Structure-from-Motion [38] or visual localization pipeline [34]) takes the tentative matches graph as input and estimates camera or scene geometry as the final output.

In this paper, we propose a further geometric refinement of the nodes V in the tentative matches graph, as shown in the bottom part of Fig. 2. This intermediate processing step naturally fits into any generic multi-view geometry pipeline. As demonstrated in experiments, our method significantly improves the geometric accuracy of the keypoints and thereby also the later processing steps, such as triangulation and camera pose estimation.

3.1 Overview

Our proposed method operates in a two-stage approach. First, for each edge, we perform a two-view refinement using a patch alignment network that, given local patches P_u, P_v around the corresponding initial keypoint locations $u, v \in \mathbb{R}^2$, predicts the flow $d_{u \to v}$ of the central pixel from one patch in the other and vice versa as $d_{v \to u}$. This network is used to annotate the edges of the tentative matches graph with geometric transformations $T_{u \to v}$, $T_{v \to u}$. In the second step, we partition the graph into components (*i.e.*, features tracks) and find a global consensus by optimizing a non-linear least squares problem over the keypoint locations, given the estimated two-view transformations.

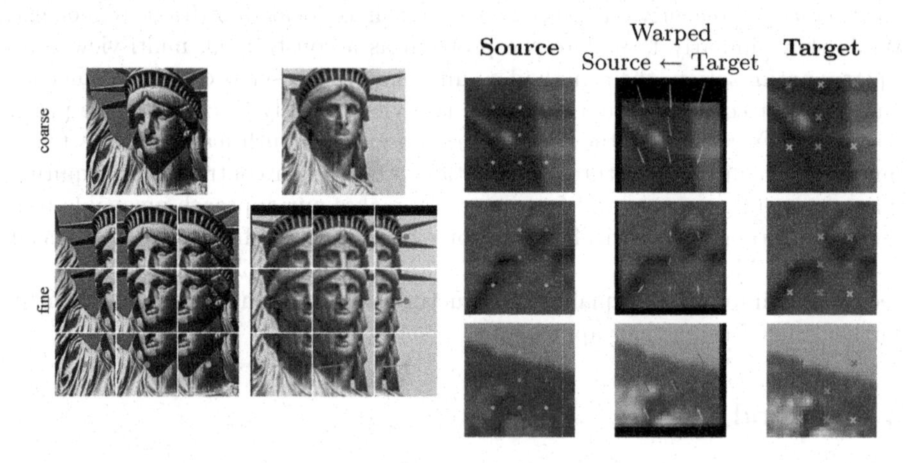

Fig. 3. Coarse-to-fine refinement and qualitative examples. *Left:* We start by a coarse alignment at feature extraction resolution taking into account only the central point, followed by a fine refinement on sub-patches corresponding to each grid point. *Right:* The first and last columns show the source and the target patch, respectively. The 3×3 regular grid is plotted with circles. For the target patch, we plot the deformed grid predicted by the coarse-to-fine refinement with crosses. The middle column shows the warped target patch using bisquare interpolation in between grid locations.

3.2 Two-View Refinement

Our method starts by computing a two-view refinement for every edge in the graph. Similarly to previous works in the field of CNNs for semantic alignment

[31,32], image matching [33], and visual flow [11], we employ a Siamese architecture for feature extraction followed by a correlation layer. The final flow is predicted by a succession of convolutional and fully connected layers.

Feature Extraction and Correlation. The architecture first densely extracts features in both patches (P_u, P_v) with a standard CNN architecture. The output is two 3D tensors $F_u, F_v \in \mathbb{R}^{h \times w \times d}$, each of which can be interpreted as a set of d-dimensional descriptors associated to a $h \times w$ spatial grid in their corresponding patches $\mathbf{f}_u(i,j), \mathbf{f}_v(i,j) \in \mathbb{R}^d$. Before matching the descriptors using dot-product correlation, we perform L2-normalization as $\hat{\mathbf{f}}(i,j) = \frac{\mathbf{f}(i,j)}{\|\mathbf{f}(i,j)\|_2}$.

Dense matching can be implemented using a correlation layer yielding a 4D tensor $c \in \mathbb{R}^{h \times w \times h \times w}$ defined by $c(i_1, j_1, i_2, j_2) = \hat{\mathbf{f}}_u(i_1, j_1)^T \hat{\mathbf{f}}_v(i_2, j_2)$. This volume can be interpreted as a 3D tensor $m \in \mathbb{R}^{h \times w \times (h \cdot w)}$, where each channel is associated to a different grid position in the opposite patch: $m(i_1, j_1)_k = c(i_1, j_1, i_2, j_2)$ where $k = i_2 \cdot w + j_2$.

Following the methodology proposed by [31], we use L2-normalization across the channel dimension to lower the values of ambiguous matches

$$\hat{m}(i,j) = \frac{\mathrm{ReLU}(m(i,j))}{\|\mathrm{ReLU}(m(i,j))\|_2}, \tag{1}$$

when the opposite patch contains more than one similar descriptor.

Regression. The final matching result $\hat{m}$ is post-processed by a CNN to aggregate local information. Finally, to enforce a patch-level consistency, a sequence of fully connected layers predicts the final output $d_{u \to v}$. Please refer to the supplementary material for more details regarding the architecture.

3.3 Multi-view Refinement

In a two-view scenario, the network described in the previous section is sufficient: (u and $v + d_{u \to v}$) or ($u + d_{v \to u}$ and v) can directly be used as the refined keypoint locations. However, given that our final goal is to perform optimization over multiple views, there are several challenges we need to overcome.

Firstly, since corresponding features are generally observed from different viewpoints and looking at non-planar scene structures, the computed displacement vector is only valid for the central pixel and not constant within the patch (*i.e.*, $\frac{\delta}{\delta u} d_{u \to v} \neq \mathbf{0}_{2,2}$). Thus, when refining keypoint locations $u, v, w, \ldots$ over multiple views, consistent results can only be produced by forming displacement chains (*e.g.*, $d_{u \to v} + d_{(v + d_{u \to v}) \to w} + \ldots$) without loops. However, such an approach does not consider all possible edges in the graph and quickly accumulate errors along the chain. Another possible way to perform the refinement is to predict new displacements every time the keypoint locations are updated during the multi-view optimization. The main downside of this approach is its run-time, since the two-view network would have to be run for each edge after each optimization step. Therefore, to refine the keypoints over the entire graph and also achieve practical run-times, we use the two-view network to estimate local flow fields $T_{u \to v}$ prior to multi-view refinement and then efficiently interpolate displacements within the patch during the optimization. Some qualitative examples are shown in Fig. 3 (right).

Secondly, the connected components of G generally contain feature tracks of different scene points, as the graph topology is purely based on appearance and feature matching is imperfect despite various filtering constraints – a single incorrect match can merge two tracks. As such, we partition the connected components into smaller, more reliable subsets based on the descriptor cosine similarity $s_{u,v}$ between patch pairs (u, v).

Thirdly, predicting the reverse flow or loops in the graph does not necessarily produce a consistent result (*e.g.*, $T_{v \to u} \circ T_{u \to v} \neq \mathbf{id}$, $T_{w \to u} \circ T_{v \to w} \circ T_{u \to v} \neq \mathbf{id}$) due to wrong matches or noisy network predictions. We tackle this by formulating a joint robust optimization of all tentatively matching keypoint locations considering all the edges over multiple views, analogous to Pose Graph Optimization [28]. In the following paragraphs, we detail our solutions to the issues mentioned above.

Flow Field Prediction. To facilitate the multi-view optimization of the keypoint locations, we use repeated forward passes of the central flow network to predict a local flow field $T_{u \to v}$ around the initial keypoint location u. Note that this prediction is directionally biased and, as such, we always also predict the inverse flow field $T_{v \to u}$. For further space and time efficiency considerations, we approximate the full flow field between two patches by a 3×3 displacement grid and use bi-square interpolation with replicate padding in between the grid points. Assuming locally smooth flow fields, we can efficiently chain the transformations from any node u to another node w without any additional forward-passes of the two-view network. To obtain correspondences for all points of the 3×3 grid, we first predict a coarse alignment $d^c_{u \to v}$ using patches around matched features u, v at original keypoint extraction resolution. Subsequently, we further refine the coarse flow at a finer resolution using sub-patches around each 3×3 grid position g, $d^f_{u+g \to v+d^c_{u \to v}+g}$. The final transformation is given by: $T_{u \to v}(g) = d^c_{u \to v} + d^f_{u+g \to v+d^c_{u \to v}+g}$. This process is illustrated in Fig. 3 (left).

Match Graph Partitioning. To address the second issue, our multi-view refinement starts by partitioning the tentative matches graph into disjoint components called tracks. A track is defined as a subset of the nodes V containing at most one node (patch) from each image. This is similar to a 3D feature track (*i.e.*, the set of 2D keypoints corresponding to the same 3D point). For each node $u \in V$, we denote t_u the track containing u. For a subset S of V, we define $\mathbf{I}_S$ as the set of images in which the features (nodes) of S were extracted (*i.e.*, $\mathbf{I}_S = \{I \in \mathcal{I} | \exists u \in S \text{ s.t. } u \in I\}$).

The proposed algorithm for track separation follows a greedy strategy and is closely related to Kruskal's minimum-spanning-tree algorithm [20]. The edges $(u \to v) \in E$ are processed in decreasing order of their descriptor similarity $s_{u \to v}$. Given an edge $u \to v$ linking two nodes from different tracks (*i.e.*, $t_u \neq t_v$), the two tracks are joined only if their patches come from different images (*i.e.*, $\mathbf{I}_{t_u} \cap \mathbf{I}_{t_v} = \emptyset$). The pseudo-code of this algorithm is defined in Fig. 4 (left).

Another challenge commonly arising due to repetitive scene structures are very large connected components in the tentative matches graph. These large components are generally caused by a small number of low-similarity edges and lead to excessively large optimization problems. To prevent these large components from slowing down the optimization, we use recursive normalized graph-cuts (GC) on the meta-graph of tracks $\mathcal{G} = (\mathcal{V}, \mathcal{E})$ until each remaining connected component has fewer nodes than the number of images N. The nodes of $\mathcal{G}$ correspond to tracks ($\mathcal{V} = \{t_u | u \in V\}$) and its edges aggregate over the edges of G, $\mathcal{E} = \{(t_u, t_v, w_{t_u,t_v}) | (u \to v) \in E, w_{t_u,t_v} = \sum_{(u' \to v') \in E \text{ s.t. } t_{u'} = t_u, t_{v'} = t_v} s_{u' \to v'}\}$. The G-cardinality of a subset $\mathcal{A} \subseteq \mathcal{V}$ is defined as: $|\mathcal{A}|_G = |\{u \in V | t_u \in \mathcal{A}\}|$. The pseudo-code is detailed in Fig. 4 (right). This step returns a pair-wise disjoint family of sets $\mathcal{S}$ corresponding to the final connected components of $\mathcal{G}$.

Given the track assignments and a set of tracks $\mathcal{A} \in \mathcal{S}$, we define the set of intra-edges connecting nodes within a track as $E^{\mathcal{A}}_{\text{intra}} = \{(u \to v) \in E | t_u = t_v, t_u \in \mathcal{A}\}$ and the set of inter-edges connecting nodes of different tracks as $E^{\mathcal{A}}_{\text{inter}} = \{(u \to v) \in E | t_u \neq t_v, t_u \in \mathcal{A}, t_v \in \mathcal{A}\}$. In the subsequent optimization step, the intra-edges are considered more reliable and prioritized, since they correspond to more confident matches.

```
Input: Graph G = (V, E)
Output: Track assignments t_u, ∀u ∈ V
for u ∈ V do
 | t_u ← new track {u};
end
F ← E sorted by decreasing similarity;
for (u, v) ∈ F do
 | if I_{t_u} ∩ I_{t_v} = ∅ then
 | | merge t_u and t_v;
 | end
end
```

```
Input: Meta-graph 𝒢 = (𝒱, ℰ)
Output: Family of sets 𝒮
𝒮 ← {};
for 𝒞 connected component of 𝒢 do
 | RecursiveGraphCut(𝒞);
end
Function RecursiveGraphCut(𝒞)
 | if |𝒞|_G > N then
 | | 𝒜, ℬ ← NormalizedGC(𝒞);
 | | RecursiveGraphCut(𝒜);
 | | RecursiveGraphCut(ℬ);
 | else
 | | 𝒮 ← 𝒮 ∪ {𝒞};
 | end
```

Fig. 4. Algorithms. *Left – track separation algorithm:* the tentative matches graph is partitioned into tracks following a greedy strategy. Each track contains at most one patch from each image. *Right – recursive graph cut:* we remove edges until having connected components of size at most N - the number of images. This algorithm yields a pair-wise disjoint family of sets $\mathcal{S}$, each set representing an ensemble of tracks.

Graph Optimization. Given the tentative matches graph augmented by differentiable flow fields T for all edges, the problem of optimizing the keypoint locations x_p can be formulated independently for each set of tracks $\mathcal{A} \in \mathcal{S}$ as the bounded non-linear least squares problem

$$\begin{aligned}\min_{\{x_p|t_p\in\mathcal{A}\}} &\sum_{(u\to v)\in E^{\mathcal{A}}_{\text{intra}}} s_{u\to v}\rho(\|\bar{x}_v-\bar{x}_u-T_{u\to v}(\bar{x}_u)\|^2)\\ &+\sum_{(u\to v)\in E^{\mathcal{A}}_{\text{inter}}} s_{u\to v}\psi(\|\bar{x}_v-\bar{x}_u-T_{u\to v}(\bar{x}_u)\|^2)\\ &\text{s.t.}\|\bar{x}_p\|_1=\|x_p-x^0_p\|_1\leq K,\forall p,\end{aligned} \tag{2}$$

where $x^0_\cdot$ are the initial keypoint locations, ρ is a soft, unbounded robust function for intra-edges, ψ is a stronger, bounded robust function for inter-edges, and K is the degree of liberty of each keypoint (in pixels). Finally, $s_{u\to v}$ is the cosine similarity between descriptors of nodes u and v; thus, closer matches in descriptor space are given more confidence during the optimization.

The inter-edges are essential since most features detectors in the literature sometimes fire multiple times for the same visual feature despite non-max suppression (at multiple scales or with different orientations). Without inter-edges, given our definition of a track as only containing at most one feature from each image, these detections would be optimized separately. With inter-edges, the optimization can merge different tracks for higher estimation redundancy if the deviations from the intra-track solutions are not too high.

Note that this problem can have multiple local minima corresponding to different scene points observed in all the patches of a track. For robust convergence of the optimization to a good local minimum, we fix the keypoint location of the node r_τ with the highest connectivity score[1] in each track τ, $r_\tau = \arg\max_{\{u|t_u=\tau\}} \gamma(u)$.

4 Implementation Details

This section describes the loss and dataset used for training the patch alignment network in a supervised manner, as well as details regarding the graph optimization algorithm, hyperparameters, and runtime.

Training Loss. For training the network, we use a squared L2 loss: $\mathcal{L} = \sum_{P_1,P_2}\|d_{1\to 2} - d^{\text{gt}}_{1\to 2}\|_2^2$, where d and d^{gt} are the predicted and ground-truth displacements for the central pixel from patch 1 to patch 2, respectively.

Training Dataset. We use the MegaDepth dataset [22] consisting of 196 different scenes reconstructed from internet images using COLMAP [38,39] to generate training data. Given the camera intrinsics, extrinsics, and depth maps of each image, a random triangulated SIFT keypoint is selected as reference and reprojected to a matching image to generate a corresponding patch pair. We enforce depth consistency to ensure that the reference pixel is not occluded in the other view. We discarded 16 scenes due to inconsistencies between sparse

[1] The connectivity score of a node u is defined as the similarity-weighted degree of the intra-edges $\gamma(u) = \sum_{\{(u\to v)|t_u=t_v\}} s_{u\to v}$.

and dense reconstructions. The extracted patch pairs are centered around the SIFT keypoint in the reference view and its reprojected correspondence in the target view respectively (*i.e.*, the ground-truth flow is $\mathbf{0}$). Random homographies are used on the target view to obtain varied ground-truth central point flow. While the MegaDepth dataset provides training data across a large variety of viewpoint and illumination conditions, the ground-truth flow is sometimes not perfectly sub-pixel accurate due to errors in the dense reconstruction. Therefore, we synthesize same-condition patch pairs with perfect geometric flow annotation using random warping of reference patches to generate a synthetic counterpart.

Feature Extraction CNN. As the backbone architecture for feature extraction, we use the first two blocks of VGG16 [41] (up to `conv2_2`) pretrained on ImageNet [9]. To keep the features aligned with input patch pixels, we replace the 2×2 max-pooling with stride 2 by a 3×3 max-pooling with stride 2 and zero padding.

Training Methodology. We start by training the regression head for 5 epochs. Afterwards, the entire network is trained end-to-end for 30 epochs, with the learning rate divided by 10 every 10 epochs. Adam [19] serves as the optimizer with an initial learning rate of 10^{-3} and a batch size of 32. To counter scene imbalance, 100 patch pairs are sampled from every scene during each epoch.

Graph Optimization. During the optimization, keypoints are allowed to move a maximum of $K = 16$ pixels in any direction. We initialize x_p to the initial keypoint locations x_p^0. Empirically, we model the soft robust function ρ as Cauchy scaled at 4 pixels, and the strong one ψ as Tukey scaled at 1 pixel. We solve the problems from Eq. 2 for each connected component $\mathcal{A} \in \mathcal{S}$ independently using Ceres [1] with sparse Cholesky factorization on the normal equations.

Runtime. The coarse-to-fine patch transformation prediction processes 1–4 image pairs per second on a modern GPU depending on the number of matches. The average runtime of the graph optimization across all methods on the ETH3D scenes is 3.0 s (median runtime 1.0 s) on a CPU with 16 logical processors.

5 Experimental Evaluation

Despite being trained on SIFT keypoints, our method can be used with a variety of different feature detectors. To validate this, we evaluate our approach in conjunction with two well-known hand-crafted features (SIFT [23] and SURF [7]), one learned detector combined with a learned descriptor (Key.Net [6] with HardNet [26]), and three learned ones (SuperPoint [10] denoted SP, D2-Net [12], and R2D2 [30]). For all methods, we resize the images before feature extraction such that the longest edge is at most 1600 pixels (lower resolution images are kept unchanged). We use the default parameters as released by their authors in the associated public code repositories. Our refinement protocol takes exactly the same input as the feature extraction. The main objective is not to compare these methods against each other, but rather to show that each of them independently significantly improves when coupled with our refinement procedure.

First, we evaluate the performance with and without refinement on a standard image matching task containing sequences with illumination and viewpoint changes. Then, we present results in the more complex setting of Structure-from-Motion. In particular, we demonstrate large improvements on the tasks of multi-view triangulation, camera localization, as well as their combination in an end-to-end image-based 3D reconstruction scenario.

For the Structure-from-Motion evaluations, we use the following matching protocol: for SIFT and SURF, we use a symmetric second nearest neighbor ratio test (with the standard threshold of 0.8) and mutual nearest neighbors filtering. For Key.Net+HardNet, we use the same protocol with a threshold of 0.9. For the remaining methods, we use mutual nearest neighbors filtering with different similarity thresholds - 0.755 for SuperPoint, 0.8 for D2-Net, and 0.9 for R2D2.[2]

5.1 Image Matching

In this experiment, we evaluate the effect of our refinement procedure on the full image sequences from the well-known HPatches dataset [4]. This dataset consists of 116 sequences of 6 images with changes in either illumination or viewpoint. We follow the standard evaluation protocol introduced by [12] that discards 8 of the sequences due to resolution considerations. The protocol reports the mean matching accuracy per image pair of a mutual nearest neighbors matcher while varying the pixel threshold up to which a match is considered to be correct.

Figure 5 shows the results for illumination-only, viewpoint-only, as well as overall for features with and without refinement. As expected, our method greatly improves upon learned features under either condition. Note that the evaluated learned methods represent the state of the art on this benchmark already and we further improve their results. For SIFT [23], while the performance remains roughly the same under viewpoint changes, our method significantly improves the results under illumination sequences, where low-level changes in image statistics perturb the feature detector. It is also worth noting that, especially in the viewpoint sequences for learned features, our refinement procedure improves the results for coarse thresholds by correcting wrong, far-away correspondences.

5.2 Triangulation

Next, we evaluate the triangulation quality with known ground-truth camera poses and intrinsics on the ETH3D benchmark [40]. Originally, this benchmark was proposed for multi-view stereo methods and provides highly accurate ground-truth camera poses and dense 3D point-clouds. Nevertheless, the same evaluation protocol also applies to our scenario – we want to evaluate the impact of refined keypoint locations on the completeness and accuracy of sparse multi-view triangulation. For each method, we run the multi-view triangulator of COLMAP [38]

[2] The thresholds for the learned methods were determined following the methodology of [23]. Please refer to the supplementary material for more details.

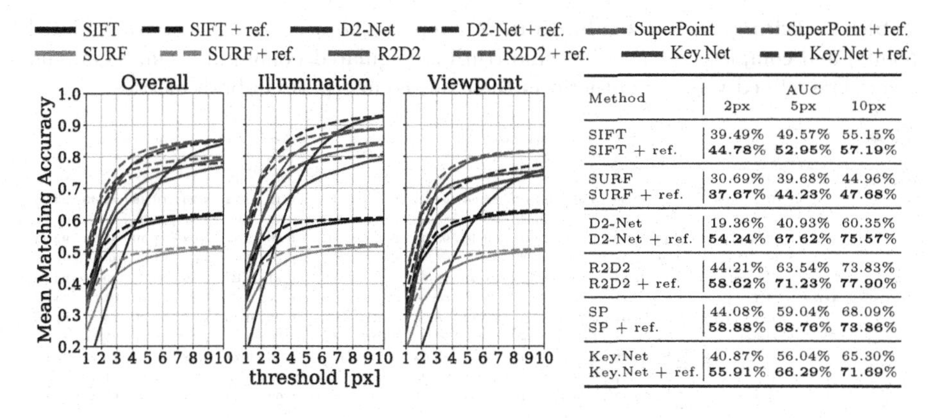

Method	AUC 2px	5px	10px
SIFT	39.49%	49.57%	55.15%
SIFT + ref.	**44.78%**	**52.95%**	**57.19%**
SURF	30.69%	39.68%	44.96%
SURF + ref.	**37.67%**	**44.23%**	**47.68%**
D2-Net	19.36%	40.93%	60.35%
D2-Net + ref.	**54.24%**	**67.62%**	**75.57%**
R2D2	44.21%	63.54%	73.83%
R2D2 + ref.	**58.62%**	**71.23%**	**77.90%**
SP	44.08%	59.04%	68.09%
SP + ref.	**58.88%**	**68.76%**	**73.86%**
Key.Net	40.87%	56.04%	65.30%
Key.Net + ref.	**55.91%**	**66.29%**	**71.69%**

Fig. 5. Matching evaluation. We plot the mean matching accuracy on HPatches Sequences at different thresholds for illumination and viewpoint sequences, as well as overall. We also report the area under the overall curve (AUC) up to 2, 5, and 10 pixels. All methods have their performance improved by the proposed refinement procedure.

with fixed camera intrinsics and extrinsics. Given the sparse point cloud, we run the ETH3D evaluation code to report the accuracy (% of triangulated points) and completeness (% of ground-truth triangulated points) at different real-world thresholds. We refer to the original paper for more details about the evaluation.

Table 1 compares the different local feature approaches with their refined counterparts. Our proposed keypoint refinement procedure improves the results across the board for all methods. Once again, the learned keypoints that suffer from poor localization due to downsampling and large receptive field are drastically improved for both indoor and outdoor scenarios. Even though the performance gain is smaller in the case of SIFT, this experiment shows that exploiting multi-view information is beneficial for very well localized features as well. The increase in completeness for all local features shows that our approach does not trim the 3D models to only contain accurate points, but rather improves the overall quality by yielding more triangulated points which are also more precise. Please refer to the supplementary material for results on each dataset.

5.3 Camera Localization

We also evaluate the camera localization performance under strict thresholds on the ETH3D dataset [40]. For each scene, we randomly sample 10 images that will be treated as queries (130 query images in total). For each query, a partial 3D model is built without the query image and its 2 closest neighbors in terms of co-visibility in the reference model (released with the dataset); 2D-3D correspondences are inferred from the tentative matches between the query image and all (partial) 3D model images; finally, absolute pose estimation with non-linear refinement from COLMAP is used to obtain the camera pose. The partial models are built independently, *i.e.*, multi-view optimization is only run on the

Table 1. Triangulation evaluation. We report the accuracy (% of triangulated points) and completeness (% of ground-truth triangulated points) at 1 cm, 2 cm, and 5 cm. The refined versions outperform their raw counterparts in both metrics.

Dataset	Method	Comp. (%)			Accuracy (%)			Method	Comp. (%)			Accuracy (%)		
		1 cm	2 cm	5 cm	1 cm	2 cm	5 cm		1 cm	2 cm	5 cm	1 cm	2 cm	5 cm
Indoors 7 scenes	SIFT	0.20	0.86	3.61	75.74	84.77	92.26	SURF	0.08	0.41	1.97	66.37	79.05	89.61
	SIFT + ref.	**0.24**	**0.96**	**3.88**	**81.06**	**88.64**	**94.61**	SURF + ref.	**0.12**	**0.52**	**2.26**	**76.28**	**85.30**	**92.36**
	D2-Net	0.46	1.83	7.00	46.95	64.91	83.25	R2D2	0.53	2.04	8.53	66.70	79.26	90.04
	D2-Net + ref.	**1.44**	**4.53**	**12.97**	**78.53**	**86.46**	**93.05**	R2D2 + ref.	**0.66**	**2.32**	**9.08**	**77.56**	**85.74**	**92.54**
	SP	0.59	2.21	8.86	75.26	85.27	93.30	Key.Net	0.16	0.68	3.01	66.51	80.44	91.61
	SP + ref	**0.71**	**2.51**	**9.55**	**86.03**	**91.91**	**95.83**	Key.Net + ref	**0.21**	**0.81**	**3.36**	**80.51**	**89.24**	**94.73**
Outdoors 6 scenes	SIFT	0.06	0.34	2.44	58.31	73.13	86.24	SURF	0.03	0.17	1.22	44.21	63.11	79.71
	SIFT + ref.	**0.07**	**0.41**	**2.75**	**61.61**	**76.89**	**88.96**	SURF + ref.	**0.05**	**0.26**	**1.68**	**62.88**	**74.67**	**87.10**
	D2-Net	0.03	0.19	1.80	21.35	35.08	56.75	R2D2	0.11	0.55	3.61	48.75	65.74	82.81
	D2-Net + ref.	**0.21**	**1.09**	**6.13**	**59.07**	**72.34**	**85.62**	R2D2 + ref.	**0.16**	**0.71**	**4.08**	**63.85**	**78.10**	**90.09**
	SP	0.09	0.54	3.86	49.67	64.57	80.79	Key.Net	0.01	0.09	0.75	39.25	54.57	72.30
	SP + ref.	**0.15**	**0.77**	**4.91**	**65.23**	**77.50**	**88.37**	Key.Net + ref.	**0.02**	**0.13**	**0.91**	**55.62**	**69.41**	**85.56**

views that are part of each partial model (without the query and holdout images). For the query keypoints, central point flow is predicted from the reprojected locations of 3D scene points in the matching views to the query view. To obtain a single 2D coordinate for each matching 3D point, we compute the similarity-weighted average of the flow for each track, which is equivalent to solving Eq. 2, where nodes of keypoints in the 3D model are connected through a single edge to matching query keypoints.

The results of this experiment are presented in Fig. 6. The performance of SIFT [23] after refinement is on par with the unrefined version despite the increase in point-cloud accuracy and completeness; this suggests that the method has nearly saturated on this localization task. All the other features have their performance greatly improved by the proposed refinement. It is worth noting that the refined versions of SuperPoint [10] and R2D2 [30] drastically outperform SIFT especially on the finer thresholds (1 mm and 1 cm).

5.4 Structure-from-Motion

Finally, we evaluate our refinement procedure on the scenario of end-to-end 3D reconstruction from unstructured imagery on the benchmark introduced in [37]. For the internet datasets (Madrid Metropolis, Gendarmenmarkt, and Tower of London), instead of exhaustively matching all images, we use NetVLAD [2] to retrieve top 20 related views for each image and only match against these. Due to the wide range of resolutions in internet images, we impose the use of multi-scale features if available and not active by default (*i.e.*, for D2-Net [12]).

After matching and feature refinement, we run COLMAP [38] to obtain sparse 3D reconstructions. Finally, different reconstruction statistics taking into account only the images registered both with and without refinement are reported in Table 2. For independent results, please refer to the supplementary material.

Overall, the results with refined keypoints achieve significantly better statistics than their original counterparts. On the small datasets all refined methods

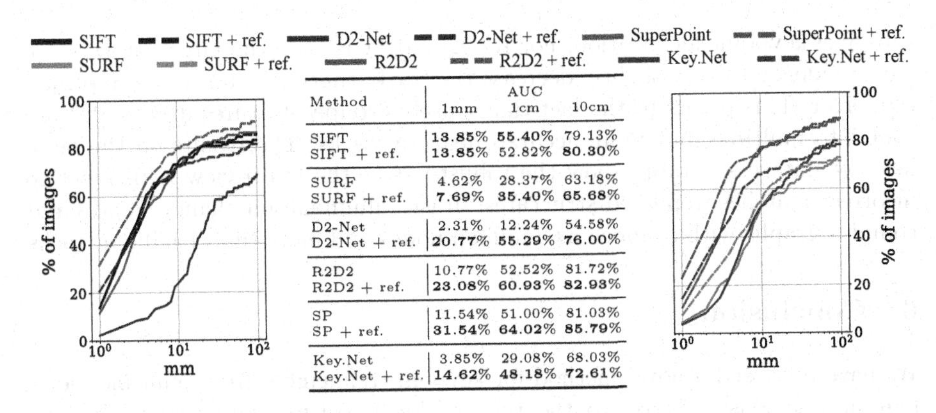

Method	AUC 1mm	AUC 1cm	AUC 10cm
SIFT	**13.85%**	**55.40%**	79.13%
SIFT + ref.	**13.85%**	52.82%	**80.30%**
SURF	4.62%	28.37%	63.18%
SURF + ref.	**7.69%**	**35.40%**	**65.68%**
D2-Net	2.31%	12.24%	54.58%
D2-Net + ref.	**20.77%**	**55.29%**	**76.00%**
R2D2	10.77%	52.52%	81.72%
R2D2 + ref.	**23.08%**	**60.93%**	**82.93%**
SP	11.54%	51.00%	81.03%
SP + ref.	**31.54%**	**64.02%**	**85.79%**
Key.Net	3.85%	29.08%	68.03%
Key.Net + ref.	**14.62%**	**48.18%**	**72.61%**

Fig. 6. Camera localization evaluation. We report the percentage of localized images at different camera position error thresholds as well as the area under the curve (AUC) up to 1 mm, 1 cm and 10 cm. The performance of SIFT remains similar on this task. All other features show greatly improved camera pose accuracy after refinement.

Table 2. Local Feature Evaluation Benchmark. A 3D model is built for each method and different reconstruction statistics are reported. For the large datasets, we report the statistics on the common images only.

Dataset	Method	Reg. images	Num. obs.	Track length	Reproj. error	Method	Reg. images	Num. obs.	Track length	Reproj. error
Herzjesu 8 images	SIFT	8	15.9K	4.10	0.59	SURF	8	5.0K	3.64	0.70
	SIFT + ref.		**16.2K**	**4.16**	**0.29**	SURF + ref.		**5.2K**	**3.70**	**0.30**
	D2-Net	8	38.5K	3.36	1.32	R2D2	8	21.1K	5.84	1.08
	D2-Net + ref.		**47.7K**	**4.06**	**0.41**	R2D2 + ref.		**21.6K**	**6.04**	**0.57**
	SP	8	17.2K	4.54	1.00	Key.Net	8	5.0K	4.29	1.00
	SP + ref		**17.9K**	**4.72**	**0.36**	Key.Net + ref		**5.3K**	**4.46**	**0.42**
Fountain 11 images	SIFT	11	27.0K	4.51	0.55	SURF	11	5.6K	3.91	0.64
	SIFT + ref.		**27.4K**	**4.56**	**0.26**	SURF + ref.		**5.7K**	**3.95**	**0.30**
	D2-Net	11	62.0K	3.51	1.36	R2D2	11	33.0K	7.11	1.10
	D2-Net + ref.		**77.4K**	**4.47**	**0.40**	R2D2 + ref.		**33.6K**	**7.47**	**0.62**
	SP	11	21.5K	4.93	1.06	Key.Net	11	8.4K	5.53	1.00
	SP + ref		**22.4K**	**5.19**	**0.43**	Key.Net + ref		**8.7K**	**5.70**	**0.44**
Madrid Metropolis 1344 images	SIFT	379	187.2K	6.83	0.70	SURF	268	**116.0K**	**6.25**	0.76
	SIFT + ref.		**187.7K**	**6.86**	**0.66**	SURF + ref.		115.2K	**6.25**	**0.66**
	D2-Net	372	668.8K	6.00	1.47	R2D2	410	355.2K	**10.20**	0.90
	D2-Net + ref.		**752.5K**	**7.28**	**0.96**	R2D2 + ref.		**356.8K**	10.17	**0.76**
	SP	414	269.7K	7.64	0.98	Key.Net	304	111.9K	9.18	0.94
	SP + ref		**277.7K**	**8.20**	**0.72**	Key.Net + ref		**114.5K**	**9.31**	**0.75**
Gendarmenmarkt 1463 images	SIFT	874	440.3K	6.33	0.82	SURF	472	163.9K	**5.45**	0.90
	SIFT + ref.		**441.4K**	**6.42**	**0.75**	SURF + ref.		**164.8K**	5.43	**0.78**
	D2-Net	858	1.479M	5.33	1.44	R2D2	929	**1.043M**	**10.09**	0.99
	D2-Net + ref.		**1.665M**	**6.37**	**1.04**	R2D2 + ref.		**1.043M**	10.05	**0.89**
	SP	911	626.9K	6.84	1.05	Key.Net	810	253.3K	7.08	0.99
	SP + ref		**648.0K**	**7.10**	**0.89**	Key.Net + ref		**258.6K**	**7.25**	**0.86**
Tower of London 1576 images	SIFT	561	447.8K	7.90	0.69	SURF	430	212.0K	**5.94**	0.70
	SIFT + ref.		**449.0K**	**7.96**	**0.59**	SURF + ref.		**212.7K**	5.92	**0.58**
	D2-Net	635	1.408M	5.96	1.48	R2D2	689	758.0K	13.44	0.92
	D2-Net + ref.		**1.561M**	**7.63**	**0.91**	R2D2 + ref.		**759.2K**	**13.74**	**0.76**
	SP	621	442.9K	8.06	0.95	Key.Net	495	186.5K	9.02	0.85
	SP + ref		**457.6K**	**8.55**	**0.69**	Key.Net + ref		**190.8K**	**9.18**	**0.65**

apart from R2D2 have sub-pixel keypoint accuracy (*i.e.*, a reprojection error lower than 0.5). SuperPoint and Key.Net, despite being targeted at low-level features, are still largely behind SIFT in terms of reprojection error without refinement. The refinement lowers this gap while also improving their already significant track length. For SIFT, the main improvement is in terms of reprojection error showing that it is possible to refine even features with accurate, sub-pixel keypoint localization. For R2D2 and SURF, on the large datasets, we see a tendency to very slightly decrease the track length to improve the reprojection error. This points to the fact that loosely grouped features during SfM are split into multiple, but more accurate feature tracks. The results on the large internet datasets notably show the robustness of the multi-view refinement to incorrect matches, repeated structures, drastic illumination changes, and large, complex graphs with as much as 5 million nodes and more than 1 million tracks.

6 Conclusion

We have proposed a novel method for keypoint refinement from multiple views. Our approach is agnostic to the type of local features and seamlessly integrates into the standard feature extraction and matching paradigm. We use a patch alignment neural network for two-view flow prediction and formulate the multi-view refinement as a non-linear least squares optimization problem. The experimental evaluation demonstrates drastically improved performance on the Structure-from-Motion tasks of triangulation and camera localization. Throughout our experiments, we have shown that our refinement cannot only address the poor keypoint localization of recent learned feature approaches, but it can also improve upon SIFT – the arguably most well-known handcrafted local feature with accurate sub-pixel keypoint refinement.

Acknowledgements. This work was supported by the Microsoft Mixed Reality & AI Zürich Lab PhD scholarship.

References

1. Agarwal, S., Mierle, K., et al.: Ceres solver. http://ceres-solver.org
2. Arandjelovic, R., Gronat, P., Torii, A., Pajdla, T., Sivic, J.: NetVLAD: CNN architecture for weakly supervised place recognition. In: Proceedings CVPR (2016)
3. Arandjelovic, R., Zisserman, A.: Three things everyone should know to improve object retrieval. In: Proceedings of CVPR (2012)
4. Balntas, V., Lenc, K., Vedaldi, A., Mikolajczyk, K.: HPatches: a benchmark and evaluation of handcrafted and learned local descriptors. In: Proceedings of CVPR (2017)
5. Balntas, V., Riba, E., Ponsa, D., Mikolajczyk, K.: Learning local feature descriptors with triplets and shallow convolutional neural networks. In: Proceedings of BMVC (2016)
6. Barroso-Laguna, A., Riba, E., Ponsa, D., Mikolajczyk, K.: Key. Net: Keypoint detection by handcrafted and learned CNN filters. In: Proceedings ICCV (2019)
7. Bay, H., Tuytelaars, T., Van Gool, L.: SURF: speeded up robust features. In: Leonardis, A., Bischof, H., Pinz, A. (eds.) ECCV 2006. LNCS, vol. 3951, pp. 404–417. Springer, Heidelberg (2006). https://doi.org/10.1007/11744023_32
8. Calonder, M., Lepetit, V., Strecha, C., Fua, P.: BRIEF: binary robust independent elementary features. In: Daniilidis, K., Maragos, P., Paragios, N. (eds.) ECCV 2010. LNCS, vol. 6314, pp. 778–792. Springer, Heidelberg (2010). https://doi.org/10.1007/978-3-642-15561-1_56

9. Deng, J., Dong, W., Socher, R., Li, L.J., Li, K., Fei-Fei, L.: ImageNet: a large-scale hierarchical image database. In: Proceedings of CVPR (2009)
10. DeTone, D., Malisiewicz, T., Rabinovich, A.: SuperPoint: self-supervised interest point detection and description. In: CVPR Workshops (2018)
11. Dosovitskiy, A., et al.: FlowNet: learning optical flow with convolutional networks. In: Proceedings of ICCV (2015)
12. Dusmanu, M., et al.: D2-Net: a trainable CNN for joint detection and description of local features. In: Proceedings of CVPR (2019)
13. Eichhardt, I., Barath, D.: Optimal multi-view correction of local affine frames. In: Proceedings of BMVC (2019)
14. Goesele, M., Curless, B., Seitz, S.M.: Multi-view stereo revisited. In: Proceedings of CVPR (2006)
15. Han, X., Leung, T., Jia, Y., Sukthankar, R., Berg, A.C.: MatchNet: unifying feature and metric learning for patch-based matching. In: Proceedings of CVPR (2015)
16. Harris, C., Stephens, M.: A combined corner and edge detector. In: Proceedings of Alvey Vision Conference (1988)
17. Hartmann, W., Galliani, S., Havlena, M., Van Gool, L., Schindler, K.: Learned multi-patch similarity. In: Proceedings of ICCV (2017)
18. Heinly, J., Schönberger, J.L., Dunn, E., Frahm, J.M.: Reconstructing the world* in six days *(as captured by the Yahoo 100 million image dataset). In: Proceedings of CVPR (2015)
19. Kingma, D.P., Ba, J.: Adam: a method for stochastic optimization. In: Proceedings of ICLR (2015)
20. Kruskal, J.B.: On the shortest spanning subtree of a graph and the traveling salesman problem. Proc. Am. Math. Soc. **7**(1), 48–50 (1956)
21. Li, Y., Snavely, N., Huttenlocher, D., Fua, P.: Worldwide pose estimation using 3D point clouds. In: Fitzgibbon, A., Lazebnik, S., Perona, P., Sato, Y., Schmid, C. (eds.) ECCV 2012. LNCS, vol. 7572, pp. 15–29. Springer, Heidelberg (2012). https://doi.org/10.1007/978-3-642-33718-5_2
22. Li, Z., Snavely, N.: MegaDepth: learning single-view depth prediction from internet photos. In: Proceedings of CVPR (2018)
23. Lowe, D.G.: Distinctive image features from scale-invariant keypoints. Int. J. Comput. Vis. **60**, 91–110 (2004)
24. Luo, W., Schwing, A.G., Urtasun, R.: Efficient deep learning for stereo matching. In: Proceedings of CVPR (2016)
25. Luo, Z., et al.: GeoDesc: learning local descriptors by integrating geometry constraints. In: Ferrari, V., Hebert, M., Sminchisescu, C., Weiss, Y. (eds.) ECCV 2018. LNCS, vol. 11213, pp. 170–185. Springer, Cham (2018). https://doi.org/10.1007/978-3-030-01240-3_11
26. Mishchuk, A., Mishkin, D., Radenovic, F., Matas, J.: Working hard to know your neighbor's margins: local descriptor learning loss. In: Advances in NeurIPS (2017)
27. Noh, H., Araujo, A., Sim, J., Weyand, T., Han, B.: Largescale image retrieval with attentive deep local features. In: Proceedings of ICCV (2017)
28. Olson, E., Leonard, J., Teller, S.: Fast iterative optimization of pose graphs with poor initial estimates. In: Proceedings of ICRA (2006)
29. Ono, Y., Trulls, E., Fua, P., Yi, K.M.: LF-Net: learning local features from images. In: Advances in NeurIPS (2019)
30. Revaud, J., Weinzaepfel, P., de Souza, C.R., Humenberger, M.: R2D2: repeatable and reliable detector and descriptor. In: Advances in NeurIPS (2019)
31. Rocco, I., Arandjelović, R., Sivic, J.: Convolutional neural network architecture for geometric matching. In: Proceedings of CVPR (2017)

32. Rocco, I., Arandjelović, R., Sivic, J.: End-to-end weakly-supervised semantic alignment. In: Proceedings of CVPR (2018)
33. Rocco, I., Cimpoi, M., Arandjelović, R., Torii, A., Pajdla, T., Sivic, J.: Neighbourhood consensus networks. In: Advances in NeurIPS (2018)
34. Sattler, T., Leibe, B., Kobbelt, L.: Fast image-based localization using direct 2D-to-3D matching. In: Proceedings of ICCV (2011)
35. Sattler, T., et al.: Benchmarking 6DoF outdoor visual localization in changing conditions. In: Proceedings of CVPR (2018)
36. Savinov, N., Seki, A., Ladicky, L., Sattler, T., Pollefeys, M.: Quad-networks: unsupervised learning to rank for interest point detection. In: Proceedings of CVPR (2017)
37. Schönberger, J.L., Hardmeier, H., Sattler, T., Pollefeys, M.: Comparative evaluation of hand-crafted and learned local features. In: Proceedings of CVPR (2017)
38. Schönberger, J.L., Frahm, J.M.: Structure-from-motion revisited. In: Proceedings of CVPR (2016)
39. Schönberger, J.L., Zheng, E., Frahm, J.-M., Pollefeys, M.: Pixelwise view selection for unstructured multi-view stereo. In: Leibe, B., Matas, J., Sebe, N., Welling, M. (eds.) ECCV 2016. LNCS, vol. 9907, pp. 501–518. Springer, Cham (2016). https://doi.org/10.1007/978-3-319-46487-9_31
40. Schöps, T., et al.: A multi-view stereo benchmark with high-resolution images and multi-camera videos. In: Proceedings of CVPR (2017)
41. Simonyan, K., Zisserman, A.: Very deep convolutional networks for large-scale image recognition. In: Proceedings of ICLR (2015)
42. Tola, E., Lepetit, V., Fua, P.: Daisy: an efficient dense descriptor applied to wide-baseline stereo. IEEE PAMI **32**(5), 815–830 (2009)
43. Verdie, Y., Yi, K., Fua, P., Lepetit, V.: TILDE: a temporally invariant learned detector. In: Proceedings of CVPR (2015)
44. Yao, Y., Luo, Z., Li, S., Fang, T., Quan, L.: MVSNet: depth inference for unstructured multi-view stereo. In: Ferrari, V., Hebert, M., Sminchisescu, C., Weiss, Y. (eds.) ECCV 2018. LNCS, vol. 11212, pp. 785–801. Springer, Cham (2018). https://doi.org/10.1007/978-3-030-01237-3_47
45. Yi, K.M., Trulls, E., Lepetit, V., Fua, P.: LIFT: learned invariant feature transform. In: Leibe, B., Matas, J., Sebe, N., Welling, M. (eds.) ECCV 2016. LNCS, vol. 9910, pp. 467–483. Springer, Cham (2016). https://doi.org/10.1007/978-3-319-46466-4_28
46. Zagoruyko, S., Komodakis, N.: Learning to compare image patches via convolutional neural networks. In: Proceedings of CVPR (2015)
47. Zbontar, J., LeCun, Y.: Stereo matching by training a convolutional neural network to compare image patches. J. Mach. Learn. Res. **17**(1), 2287–2318 (2016)

The Phong Surface: Efficient 3D Model Fitting Using Lifted Optimization

Jingjing Shen(✉), Thomas J. Cashman, Qi Ye, Tim Hutton, Toby Sharp, Federica Bogo, Andrew Fitzgibbon, and Jamie Shotton

Microsoft Mixed Reality & AI Labs, Cambridge, UK
{jinshen,tcashman,yeqi,tihutt,tsharp,febogo,awf,jamiesho}@microsoft.com

Abstract. Realtime perceptual and interaction capabilities in mixed reality require a range of 3D tracking problems to be solved at low latency on resource-constrained hardware such as head-mounted devices. Indeed, for devices such as HoloLens 2 where the CPU and GPU are left available for applications, multiple tracking subsystems are required to run on a continuous, real-time basis while sharing a single Digital Signal Processor. To solve model-fitting problems for HoloLens 2 hand tracking, where the computational budget is approximately 100 times smaller than an iPhone 7, we introduce a new surface model: the 'Phong surface'. Using ideas from computer graphics, the Phong surface describes the same 3D shape as a triangulated mesh model, but with continuous surface normals which enable the use of lifting-based optimization, providing significant efficiency gains over ICP-based methods. We show that Phong surfaces retain the convergence benefits of smoother surface models, while triangle meshes do not.

Keywords: Model-fitting · Optimization · Hand tracking · Pose estimation

1 Introduction

As computer vision systems are increasingly deployed on wearable or mobile computing platforms, they are required to operate with low power and limited computational resources. In this context, the problem of pose estimation (as applied, for example, to tracking hands [15,28], human bodies [2,17,31], or faces [10]) is often tackled with a hybrid architecture that combines discriminative machine-learnt models with generative model fitting to explain the observed data [12,15,27–29]. For the purposes of this paper, we define 'model fitting' as the registration of a 3D surface model to a point set observation. Model fitters can benefit from powerful priors learned from data, and recent work even shows the benefits of including model fitting in the training loop [9,30]. An optimizer

Electronic supplementary material The online version of this chapter (https://doi.org/10.1007/978-3-030-58452-8_40) contains supplementary material, which is available to authorized users.

A. Vedaldi et al. (Eds.): ECCV 2020, LNCS 12346, pp. 687–703, 2020.
https://doi.org/10.1007/978-3-030-58452-8_40

(a) Subdiv. surface

(b) Phong surface

(c) Triangular mesh

Surface type	eval.	with $\partial/\partial u$
Subdiv. surface	0.329s	1.241s
Phong surface	0.049s	0.279s
Triangular mesh	0.047s	0.196s

(d) Timings on 10^6 evaluations

Fig. 1. A hand model represented by a Loop subdivision surface [11] (a), a Phong surface (b) and a triangle mesh (c), with surface normals visualized by mapping x, y, z coordinates to red, green and blue components respectively. (d) shows timings in seconds on PC: *eval.* refers to evaluation of 10^6 surface point positions and normals, and *with* $\partial/\partial u$ includes the cost of derivative calculations w.r.t. surface coordinate as well. (Color figure online)

with fast convergence is critical for building real-time systems that operate with low compute. However, the *correspondences* between the observed data and the model are often unknown, and need to be discovered in the course of the optimization.

Two main optimization alternatives have been proposed for solving this problem. **Iterative Closest Point (ICP)** algorithms [1,7,20,25] solve for model pose via 'block coordinate descent': first finding closest points on the model surface, and then fixing those correspondences while solving for model pose alone.

The alternative approach is **'lifted' optimization**: to solve for correspondences and model pose *simultaneously*, using a lifted objective function that explicitly parametrizes the unknown correspondences. Taylor et al. [26,27] demonstrate hand tracking systems using this approach, and claim that the smoothness of the model surface is an important prerequisite for a smooth energy landscape that allows lifted optimization to converge efficiently. However, the complex surface representation they propose comes at a cost, entailing as much as 58% of the per-iteration model-fitting time [27, supplementary material].

In this paper, we show that much simpler surface representations are sufficient, thus making it cheaper for vision systems to access the convergence benefits of lifted optimization. In particular, since it is often beneficial to include surface orientation properties in the objective function, one might assume that a normal field with continuous first derivatives (second-order surface smoothness) is a necessary minimum for a gradient-based optimizer. This implies that a simple triangle mesh cannot suffice and we show this is true: a simple triangle mesh is insufficient, but it *is* sufficient to pair a triangle mesh with a normal field that is simply linearly interpolated over each triangle. We call the resulting representation a 'Phong surface', after the Phong shading [19] technique in computer graphics, which evaluates the lighting equation using similarly interpolated normals. Figure 1 shows the normal field evaluated inside the optimizer for different surface representations, illustrating that the Phong surface (b) leads to surface evaluations that are a close approximation to evaluations on the smooth subdivision surface (a). Informally, our contribution to 3D model fitting is a surface representation which is designed to be "as simple as possible, but no simpler".

We evaluate Phong surfaces in comparison to two other model representations; the smooth subdivision surfaces used by prior work [4,26,27] and simple triangle meshes with a piecewise-constant normal field. Our experiments in rigid pose alignment and in the example application of hand tracking compare these alternative models in the context of different optimizers, and confirm earlier results that lifted optimization converges faster and with a wider basin of convergence than ICP [27].

We have successfully applied lifted optimization with Phong surfaces to implement a fully articulated hand tracker on HoloLens 2 [14], a self-contained head-mounted holographic computer. The tracker needs to run on an embedded Digital Signal Processor (DSP) with a compute budget of 4 GFLOPS, which is 100 times smaller than an iPhone 7, or 1000 times smaller than a high-end laptop. This application requires real-time continuous tracking under tight thermal and power constraints, motivating us to find the simplest possible computation that would allow the optimizer to converge to an accurate hand pose estimate. The Phong surface representation is one of the key innovations that make this hand tracker run in realtime on the HoloLens 2.

In summary, our contributions are that we

- introduce Phong surfaces by transferring the concept of Phong shading from computer graphics to model fitting in computer vision;
- show that Phong surfaces combine the convergence benefits of lifted methods with the computational cost of planar mesh models;
- demonstrate that fitting Phong surface models allows us to implement real-time hand tracking under a computational budget of 4 GFLOPS.

1.1 Related Work

ICP has a long history in surface reconstruction and point set registration, starting from its first descriptions by Besl and McKay [1] and Chen and Medioni [5]. The simplicity of ICP means that it is easy to implement and was broadly adopted, spawning a host of variations on the two fundamental steps of closest point finding and error minimization, as described by Rusinkiewicz and Levoy [22]. Neugebauer [16] and Fitzgibbon [6] formulate ICP as an instance of a non-linear least squares problem, solved by general optimizers such as the Levenberg-Marquardt algorithm [13]. This opens up new possibilities for the objective function, such as allowing a robust kernel to be included directly in the objective as a way to smoothly handle outliers [6].

Meanwhile, ICP was generalized to articulated models by Pellegrini et al. [18], and demonstrated in this form for fitting models of human bodies [3,32] and hands [25]. Another key advantage of ICP is the broad range of geometric representations that can be fitted: any point set or surface that could support the chosen 'closest point' query is included in a straightforward manner. However, a disadvantage of the alternating coordinate descent is that ICP algorithms *converge slowly*, reducing the objective only *linearly* as optimization proceeds [28]. This is particularly apparent when a model needs to slide relative to a data set, as this requires that the model's pose is updated in harmony with data correspondences. Point-to-plane ICP [5] attempts to address this for common cases, but thereby limits the set of objectives which can be minimized, losing the freedom introduced by Neugebauer and Fitzgibbon. Recently Rusinkiewicz [21] presents a symmetric objective function for ICP which is particularly suited to the original task of point set to point set alignment, but not of point set to parametric surface alignment, which is the focus of this paper. Note that we are explicitly in a non-symmetric scenario. An extension of [21] could be considered by sampling the model, but this would lose the structure that the model provides, and which is available in many real scenarios.

Several authors have attempted to address the shortcomings of ICP by estimating unknown correspondences simultaneously with model parameters. Sullivan and Ponce [24] present one of the first systems to do so in a model-fitting context, and Cashman and Fitzgibbon [4] elaborate on this idea to also fit smooth surface models to silhouette constraints. Subsequently, Taylor et al. [26,27] and Khamis et al. [8] demonstrate the benefits of lifted optimization in the area of articulated hand tracking. However, all of these methods required complicated smooth surface constructions, in stark contrast to the freedom available when using ICP. This paper addresses this disadvantage, by showing that lifted optimization is equally applicable to models with extremely simple geometry.

Another line of work performs model fitting without estimating any data correspondences at all. Taylor et al. [28] follow the same approach originally proposed by Fitzgibbon [6] by using articulated distance fields to find correspondences; this is conceptually similar to an ICP closest-point search, but can be implemented extremely efficiently using graphics hardware. Mueller et al. [15] use a discriminative network to perform dense correspondence regression, thus allowing the model-fitting stage to proceed under the assumption that the correspondences are fixed. Neither of these approaches are currently suitable for deployment on low-power devices.

2 Method

We describe model-fitting problems in a common framework with the following notation:

- A 3D surface model $S(\theta) \subset \mathbb{R}^3$ parameterized by a vector θ, which might for example specify rigid pose, shape variations or joint angles,
- A list of sampled data points $\{\mathbf{x}_i\}_{i=1}^D$ with estimated data normals $\{\mathbf{x}_i^\perp\}_{i=1}^D$,
- An objective function $E(\theta)$ that penalizes differences between the parameterized model and the observed data,
- An optimizer that iteratively updates the current hypothesis for θ to locally minimize the objective function.

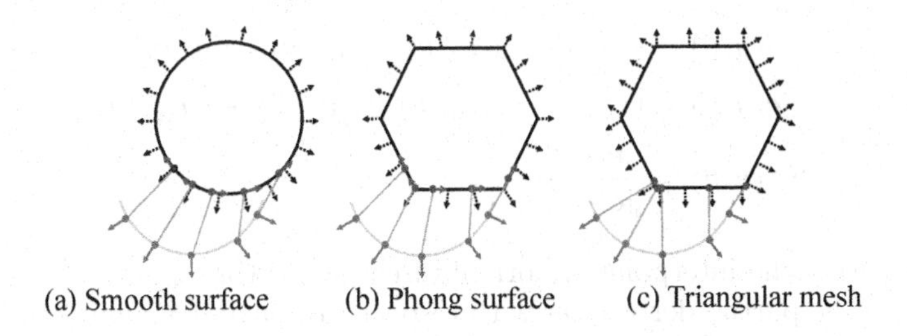

Fig. 2. Illustration of the cross section of a surface model (black) with normals (dashed black) close to some target data (green). With a continuous normal field, either from a smooth model (a) or from a Phong surface (b), the normal term forces shown in blue drive the red correspondences towards their correct location, improving convergence. These forces do not exist for a triangular mesh (c) where the normals are constant on the facet and so the derivatives are all zero. (Color figure online)

We follow the model fitting work of Taylor et al. [27] which optimizes a differentiable lifted objective function with a Levenberg optimizer. The smoothness of the subdivision surface model used in [27] encourages good convergence properties, but at the cost of expensive function evaluations for the surface positions, normals and their derivatives. We focus on efficiency and investigate the requirements for surface and normal smoothness in a lifted optimizer.

2.1 Phong Surface Model

The key idea is to generate surface normal vectors by linearly interpolating vertex normals over each planar triangle. This is the same approximation of a smooth surface used in the Phong shading method for computer graphics rendering [19], motivating our use of the 'Phong surface' moniker.

The surface model is defined by a triangle control mesh containing N control vertices each with a position and normal vector, $V(\theta) \in \mathbb{R}^{6\times N}$. The model also has a fixed triangulation of the vertices, where each triangle in the mesh corresponds to a parameterized triangular patch of the surface.

As illustrated in Fig. 3a, let $u = \{p, v, w\}$ be a surface coordinate where $p \in \mathbb{N}$ is the index of a triangular patch, and $v \in [0, 1], w \in [0, 1 - v]$ parameterize the unit triangle. $S(u, \theta) \in \mathbb{R}^3$ denotes the surface position and $S^{\perp}(u, \theta) \in \mathbb{R}^3$ the unit-length normal to the surface, both evaluated at the given coordinate u. Let $\boldsymbol{v}_1(\theta), \boldsymbol{v}_2(\theta), \boldsymbol{v}_3(\theta)$ be the control vertex positions and $\boldsymbol{v}_1^{\perp}(\theta), \boldsymbol{v}_2^{\perp}(\theta), \boldsymbol{v}_3^{\perp}(\theta)$ the control vertex normals of the pth triangular patch as specified by u, where $\boldsymbol{v}_i(\theta)$ and $\boldsymbol{v}_i^{\perp}(\theta)$ are determined by the pose and/or shape parameter vector θ. Then the Phong surface evaluation is defined simply as a linear interpolation of the control vertices:

$$S(u, \theta) = (1 - v - w)\boldsymbol{v}_1(\theta) + v\boldsymbol{v}_2(\theta) + w\boldsymbol{v}_3(\theta), \tag{1}$$

$$\boldsymbol{c}(u, \theta) = (1 - v - w)\boldsymbol{v}_1^\perp(\theta) + v\boldsymbol{v}_2^\perp(\theta) + w\boldsymbol{v}_3^\perp(\theta), \tag{2}$$

$$S^\perp(u, \theta) = \frac{\boldsymbol{c}(u, \theta)}{\|\boldsymbol{c}(u, \theta)\|}, \tag{3}$$

where $\boldsymbol{c}(u, \theta)$ is the interpolated normal direction vector.

We give the partial derivatives with respect to v, w and θ compactly in terms of the total differentials:

$$\mathrm{d}S(u, \theta) = \mathrm{d}v(\boldsymbol{v}_2 - \boldsymbol{v}_1) + \mathrm{d}w(\boldsymbol{v}_3 - \boldsymbol{v}_1) + (1 - v - w)\mathrm{d}\boldsymbol{v}_1 + v\mathrm{d}\boldsymbol{v}_2 + w\mathrm{d}\boldsymbol{v}_3,$$

$$\mathrm{d}\boldsymbol{c}(u, \theta) = \mathrm{d}v(\boldsymbol{v}_2^\perp - \boldsymbol{v}_1^\perp) + \mathrm{d}w(\boldsymbol{v}_3^\perp - \boldsymbol{v}_1^\perp) + (1 - v - w)\mathrm{d}\boldsymbol{v}_1^\perp + v\mathrm{d}\boldsymbol{v}_2^\perp + w\mathrm{d}\boldsymbol{v}_3^\perp,$$

$$\mathrm{d}S^\perp(u, \theta) = \frac{1}{\|\boldsymbol{c}\|}\left(I_3 - \frac{\boldsymbol{c}\boldsymbol{c}^T}{\|\boldsymbol{c}\|^2}\right)\mathrm{d}\boldsymbol{c}.$$

The partials can be read off from this notation as the coefficient of the relevant differential, e.g. $\frac{\partial S}{\partial v} = \boldsymbol{v}_2 - \boldsymbol{v}_1$ by setting $\mathrm{d}v = 1, \mathrm{d}w = 0, \mathrm{d}\boldsymbol{v}_i = \boldsymbol{0}, \mathrm{d}\boldsymbol{v}_i^\perp = \boldsymbol{0}$ in $\mathrm{d}S$.

2.2 Lifted Optimization with the Phong Surface

The objective function of the optimization in model fitting typically includes data terms and problem-dependent prior terms or regularization terms. Here we briefly describe the data terms.

We penalize the distance from each data point $\mathbf{x}_i$ to its corresponding surface point defined by the surface coordinate u_i, and the difference in surface orientation to the associated data normal $\mathbf{x}_i^\perp$, using

$$E(\theta, U) = \frac{1}{D}\sum_{i=1}^{D}\left(\|S(u_i, \theta) - \mathbf{x}_i\|^2 + \lambda_n\|S^\perp(u_i, \theta) - \mathbf{x}_i^\perp\|^2\right) \tag{4}$$

where D is the number of data points, λ_n is the contribution weight for normals, and $U = \{u_i\}_{i=1}^D$ are the surface coordinates corresponding to each data point.

Note that the surface coordinates U are optimized jointly with θ by the lifted optimizer, whereas the ICP optimizer alternates between updating θ and U. This means that the lifted optimizer can choose to slide these coordinates on the surface to better match the data points and data normals, while simultaneously updating the shape or pose hypothesis. On the other hand, the ICP optimizer operates on only one set of variables each iteration, without access to gradient information in the other variables. U can be viewed as latent variables as we eventually retain only the final θ estimate. As in [27], the lifted optimizer uses Levenberg steps with damping.

Figure 2 illustrates how the continuous surface normals for the subdivision surface and Phong surface cause the data normal term to 'pull' the coordinates in the directions of the blue arrows, leading to faster convergence than for the triangular mesh.

Note that *lifted optimization* has a close relation to *point-to-plane ICP* but is mathematically richer while being more efficient. Both optimizers allow a model to slide against the data in each iteration, improving convergence. However, *point-to-plane ICP* addresses this for only one energy formulation (point-to-model distances) whereas a lifted optimizer generalizes to arbitrary differentiable objectives. For example, we have a normal disparity term in our energy (Eq. 4), which is critical for faster convergence as shown in Fig. 7 and Table 1; point-to-plane ICP cannot minimize this objective. See our supplementary material for more illustrations.

2.3 Correspondence Update on Triangles

The surface types *Subdiv.*, *Phong* and *Tri. mesh* are all defined by a triangular control mesh. As mentioned in Sect. 2.1, we write correspondences as surface coordinates $u = \{p, v, w\}$. The entire surface can be viewed as a collection of triangular patches indexed by p, with each patch parameterized by the unit triangle.

After a Levenberg step in lifted optimization, we apply an update $u := u + \delta u$, which may involve walking across adjacent triangles. We follow the same triangle-walking scheme as in [4, 26], as illustrated in Fig. 3b.

Figure 3b (right) shows a single transition of a correspondence from one triangle to its neighbor; u and δu are denoted as a 2D point and a 2D vector respectively, in the domain space of the current patch p (colored in light blue). When $u + \delta u$ leaves the domain of triangle p, we calculate the partial update such that $u + r\delta u$ lies on the boundary of that triangle, and then map the remaining update $(1 - r)\delta u$ to the adjacent patch q (colored in light green). For *Subdiv.*, this results in a path that is tangent-continuous on the surface because this surface has C^1 continuity across patches. Note that this C^1 assumption does not hold for *Phong* and *Tri. mesh*; however, we implement the walking in exactly the same way for all surface types and find that it works well in practice.

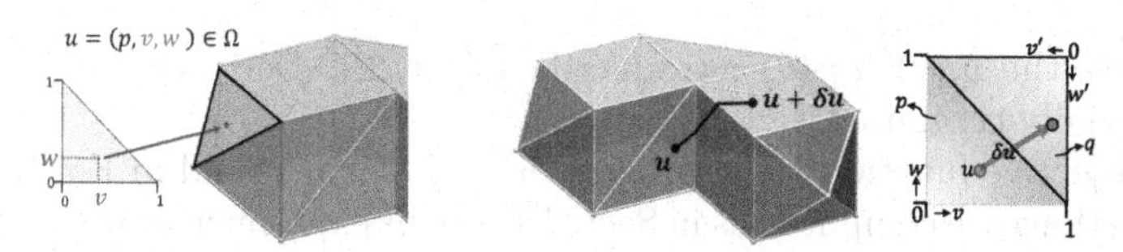

(a) Surface parameterization (b) Correspondence update across triangles.

Fig. 3. (a) A surface correspondence u that lies on the p-th triangle patch, and its coordinate (v, w) in the unit-triangle domain Ω of this patch. (b) Walking on a triangle mesh to apply an update δu and walking in the domain space (figure from [26]). (Color figure online)

3 Experiments

We compare the three surface types mentioned above: Loop subdivision surface (*Subdiv*), Phong surface (*Phong*) and triangular mesh surface (*Tri. mesh*), in both *Lifted* and *ICP* optimization frameworks. First we analyze their efficiency and convergence properties on rigid pose estimation of an ellipsoid. We further extend them to a more challenging scenario: fully articulated hand tracking. Specifically, we implement and evaluate these methods in the hand tracker by Taylor et al. [27] and in the hand tracker on HoloLens 2 [14].

In each experiment, we define the control parameters of each surface as follows: for each control vertex of *Subdiv*, we compute its position and normal on the limit Loop subdivision surface; we use these limit positions as the vertices of *Tri. mesh*, and use both limit positions and normals to define control vertices and normals of *Phong*. Note that this definition is purely to give surface models that are comparable for *Phong*, *Subdiv* and *Tri. mesh*, and we are free to define the Phong Surface model (i.e. control vertex positions and normals) in the way that best represents the target geometry.

We run most experiments on a desktop machine equipped with an Intel® Xeon® W-2155 CPU and 32 GB RAM, except for Fig. 12b, which is on a Microsoft HoloLens 2.

3.1 Rigid Pose Alignment of an Ellipsoid

Problem. We parametrize the ellipsoid pose as a 6D vector θ, storing translation and (axis-angle) rotation $[t_x, t_y, t_z, r_x, r_y, r_z]$. It defines a translation vector $\mathbf{t}(\theta) \in \mathbb{R}^3$ and a rotation matrix $R(\theta) \in \mathbb{R}^{3\times3}$. Given a template mesh, each control vertex position $\boldsymbol{v}_i$ is posed according to θ:

$$\boldsymbol{v}_i(\theta) = R(\theta)\boldsymbol{v}_i + \boldsymbol{t}(\theta). \tag{5}$$

For Phong surfaces, we express also control vertex normals as a function of θ (this is more efficient than re-computing the normals from the posed vertices):

$$\boldsymbol{v}_i^{\perp}(\theta) = R(\theta)\boldsymbol{v}_i^{\perp}. \tag{6}$$

This defines the $\boldsymbol{v}_i(\theta)$, $\boldsymbol{v}_i^{\perp}(\theta)$ introduced in Sect. 2.1. Derivatives $\frac{\partial \boldsymbol{v}_i(\theta)}{\partial \theta}, \frac{\partial \boldsymbol{v}_i^{\perp}(\theta)}{\partial \theta}$ can be trivially computed.

The objective function includes only the data term stated in Eq. 4. The surface evaluations are computed as in Sect. 2.1. In our experiments, we set $D = 200$. For fair comparisons among various surfaces and especially *Tri. mesh*, we did parameter sweeping on λ_n in the range $[0.0, 1.0]$ for fitting 400 random rotations. Based on the average fitting error, we find optimal $\lambda_n = 1.0$ for *Subdiv* and *Phong*, and $\lambda_n = 0.05$ for *Tri. mesh*. For the following experiments, for each surface type, we use its optimal λ_n. Note that we have included 0.0 in our λ_n sweeping.

Figure 4 shows an example fitting result. See supplementary material for more qualitative comparisons.

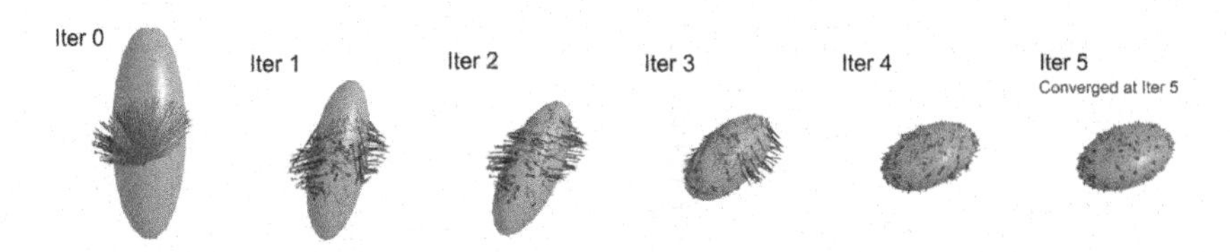

Fig. 4. Lifted optimization using a Phong surface ellipsoid model. Green points are the target data, with black lines joining them to their corresponding surface points in red. (Color figure online)

Input Data. Starting with the axis-aligned ellipsoid (with radii 1, 2, 3) centered at the origin (referred to as *neutral pose*), we apply a target rigid transform to obtain a ground-truth (target) pose. We then sample randomly 200 data points and normals on the subdivision surface defined by this mesh. For quantitative convergence analysis, we only sample from the triangle patches facing the positive z-axis: this gives us an incomplete set of data points. We add random noise to the data points and normals by sampling uniform random distributions with range [0.0, 0.1] in each dimension. The initial starting pose for model fitting is always the neutral pose.

Metrics. For the quantitative analysis, we focus on rotations. We compute a pose estimation error by applying the fitted and ground-truth rigid transformations to a fixed vector (e.g., $[1, 0, 0]$), and measuring the angle (in degrees) between the two transformed vectors. To allow for the 180° symmetry of an ellipsoid, we define the fitting error as the minimum of two angles, one computed using $[1, 0, 0]$ as the fixed vector and the other computed using $[-1, 0, 0]$ as the fixed vector.

Quantitative Analysis. We run 400 trials targeting ground-truth poses $[0, 0, 0, y, y, y]$, where y is uniformly sampled from $(-\pi, \pi)$. Figure 5 and 6 show the performance obtained for the rigid alignment of an ellipsoid model with 320 facets. Figure 5 shows that *Phong* performs as well as *Subdiv.* in accuracy, and much better than *Tri. mesh*, with either *ICP* or *lifted* optimization. In Fig. 6 (left), we see that *lifted* optimization converges much faster than *ICP* for both *Phong* and *Subdiv.*

Figure 6 (right) further plots accuracy (on the x-axis) against speed (on the y-axis, measured in milliseconds). It shows that *Phong* achieves the same level of accuracy as *Subdiv.* and runs as fast as *Tri. mesh* for both *lifted* (solid lines) and *ICP* (dashed lines). It also shows that lifted optimization converges much faster than ICP, e.g. *Lifted Phong* can achieve an average rotation error $< 10^\circ$ within 8 iters, while *ICP Phong* needs 30 iterations.

Note that our runtime for *ICP* is slightly slower than *lifted* here. While a faster ICP might be achievable by further code optimization, we emphasize that the per-iteration cost of lifted is actually comparable to ICP when counting the theoretical FLOP computations. See our supplementary material (part 3) for details.

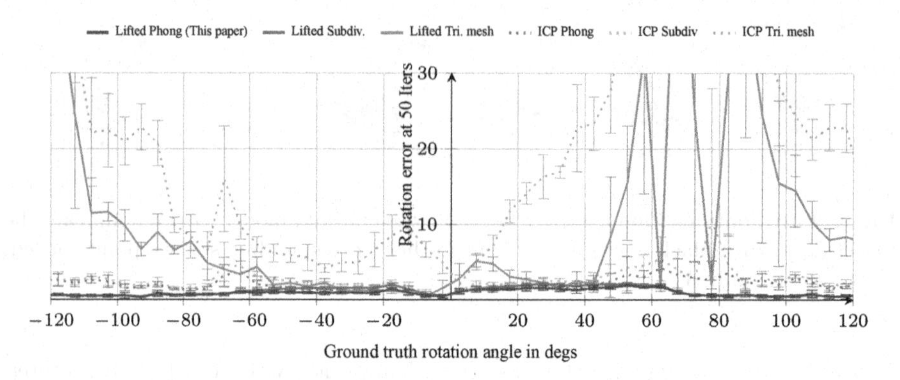

Fig. 5. Accuracy results for the rigid pose alignment of an ellipsoid with *320* facets. Optimizer ran for max. 50 iterations.

Table 1. Average rotation error after 10 and 50 lifted optimization iterations, with and without data normal term.

Surface type	Avg. rot. err. at 10 iters	Avg. rot. err. at 50 iters
Subdiv.	9.89°	1.21°
Subdiv. w/o normal	14.81°	2.57°
Phong	**8.13°**	**0.99°**
Phong w/o normal	23.24°	3.54°
Tri. mesh	17.47°	11.07°
Tri. mesh w/o normal	23.24°	3.54°

Data Normal Term. Here we assess the importance of the data normal term in Eq. 4. In particular, we demonstrate that this term is critical for fast convergence in both lifted and ICP optimizations, and that a continuous normal field improves both the basin of convergence and the accuracy of pose estimation.

As shown in Fig. 7, *Phong* and *Subdiv.*, which have a continuous normal field, converge much faster when the normal term is included, and achieve better accuracy after convergence (see also Table 1). Note that Phong w/o normal (solid bright red) coincides with Tri. mesh w/o normal (dashed black) for both lifted and ICP optimization.

For *Tri. mesh* (dashed lines in Fig. 7 (left)), we observe faster convergence with the data normal term within 10 iterations, but the accuracy reached after convergence is worse. We believe this is because the piecewise-constant surface normals introduce local minima for the correspondences, where reassignment to a different triangle causes discrete jumps in the energy, and it is difficult for the optimizer to find a global minimum for these correspondences as the partial derivatives $\frac{\partial S^{\perp}}{\partial u}$ are zero. Note that the accuracy of *Tri. mesh w/o normal* is still 3 times worse than the *Phong* and *Subdiv.* models when including the normal term (see Table 1, third column).

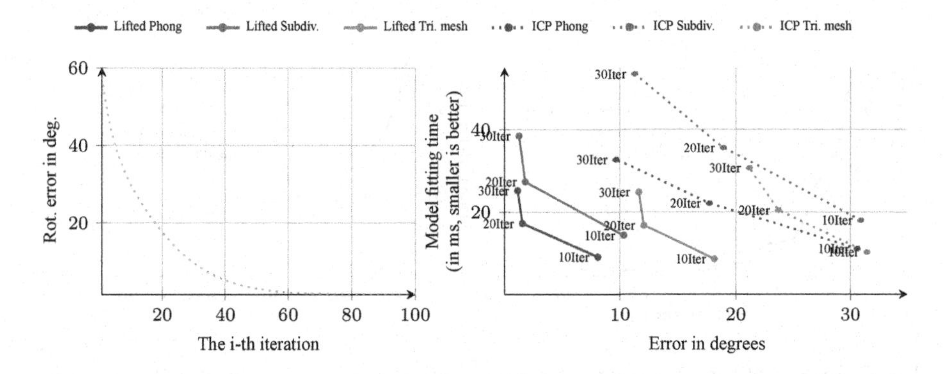

Fig. 6. Left: Convergence results for the rigid pose alignment of an ellipsoid with 320 facets. Right: Speed (model-fitting time in milliseconds) vs. accuracy (avg. error) comparisons on lifted and ICP optimizations, for max. 10, 20, and 30 iters. Closer to the origin is better.

3.2 Performance on Hand Tracking

The *lifted* optimizer used in the hand tracker by Taylor et al. [27] uses Loop subdivision surfaces. In this experiment, we simply replace their subdivision surface with our Phong surface. We leave everything else unchanged. Figure 8 shows an example result obtained fitting our Phong surface with lifted optimization. Figure 9 compares qualitatively the hand model fitting results at the 2nd and 4th iterations of lifted optimization using various surface types, given the same starting point.

Problem. We optimize a set of hand pose parameters $\theta \in \mathbb{R}^{28}$; θ stores hand orientation and translation, plus the 22 joint angles of the hand skeleton. From θ, the hand surface vertices are computed by Linear Blend Skinning [8].

For Phong surface, the 3D mesh vertex normals are deformed in the same way as vertex positions according to their LBS weights. This computation for the control normals gives a close approximation of the normals that we would obtain by rederiving normals from the posed positions of local vertices, with far greater efficiency.

We refer the reader to [27] for details on the objective function, data and experimental setup and evaluation metrics.

Performance on Dexter. In Fig. 10, 11 and 12a, we compare the accuracy and the computational efficiency of the tracker with different surface types on the Dexter dataset [23]. We adopt the same experimental setup as in [27].

The tracking settings for Fig. 10 are 192 data points, 10 starting points and 10 iterations. Figure 10a shows the max and average error of per-frame fitting, i.e. fitting each frame independently with 10 new starting points. Figure 10b shows the max and average error of tracking, i.e. using the tracked pose in the previous frame (if available) as one of the 10 starting points for current frame.

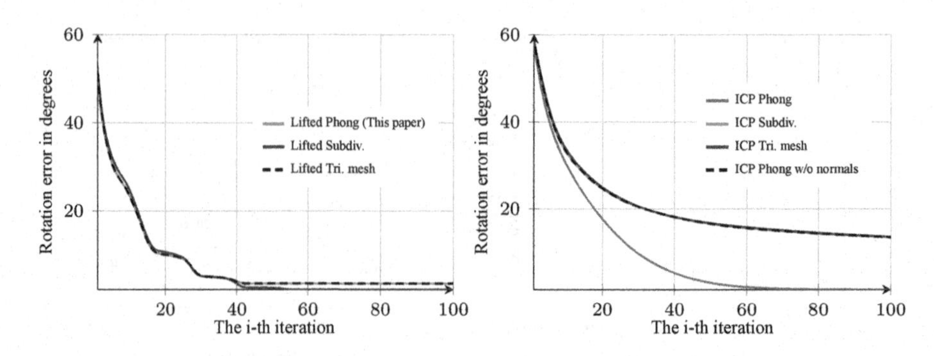

Fig. 7. Impact of the data normal term on optimization convergence, for the rigid pose alignment of an ellipsoid (320 facets). Left: Lifted optimization. Right: ICP optimization.

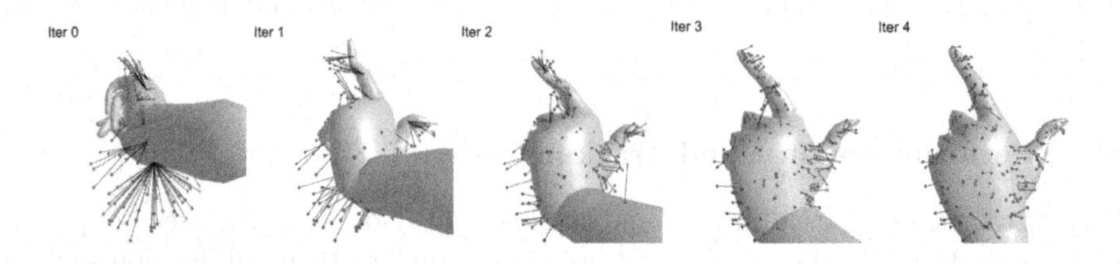

Fig. 8. A starting point and 4 iterations of lifted optimization using a Phong surface hand model. Green points are the target data, with black lines joining them to their corresponding surface points in red. (Color figure online)

Both show that on this dataset, *Phong* performs as well as *Subdiv* and achieves higher accuracy than *Tri. mesh*. *Lifted* optimization is slightly better than *ICP*.

Robustness to Initialization. Model-fitting methods are often sensitive to initialization, due to the non-convex objectives. This is why the tracking accuracy (Fig. 10b) is better than the per-frame fitting (Fig. 10a), where the optimization starts from scratch each time. The gap between *Phong* and *Tri. mesh* is larger in the per-frame fitting (Fig. 10a), which means that the smooth normal field is the key for faster convergence when the starting point is poorer. We emphasize the importance of fast convergence in live experience, as tracking failures often occur when hands are out of the field of view, or in the presence of self- and object-occlusions.

The computational cost of model fitting is dominated by 3 variables: (i) the number of data points in the data term; (ii) the number of starting points used to initialize the optimizer; and (iii) the number of iterations for each starting point. Figure 11 shows the impact of these variables on accuracy and convergence. Again, *Phong* behaves similarly to *Subdiv.*, and much better than *Tri. mesh*. Figure 11b shows how *Lifted* exhibits better convergence than *ICP*, confirming the conclusion in [27].

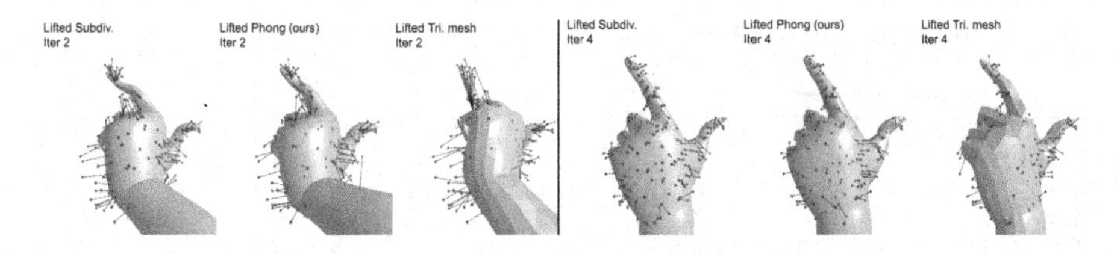

Fig. 9. Fitted hand model result at the 2nd and 4th iterations of lifted optimization using various surface types.

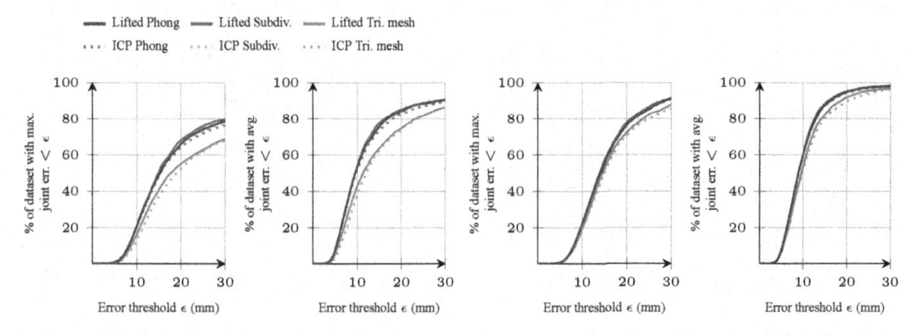

(a) Max (left) and Avg. (right) joint error on *per-frame fitting.*

(b) Max (left) and Avg. (right) joint error on *tracking.*

Fig. 10. Accuracy comparison of surface types with lifted optimizer and ICP on Dexter.

These tests show that the smooth normal field of the model is important for faster convergence in lifted optimization; it is not actually relevant whether the mesh geometry is smooth or not.

Figure 12a shows the speed vs. accuracy plot in the per-frame fitting case (192 data points, 10 starting points). For each surface type, the number of iterations varies from 2 to 10. The x-axis reports the accuracy, measured as the percentage of dataset frames that have average joint error <20 mm. The y-axis reports the speed in FPS (per starting point) of the model-fitting stage, i.e. not including the preprocessing time. For example, in the *lifted* case (solid lines), if we require to run the fitting at 50fps, we can perform 6 iterations for *Phong* and *Tri. mesh*, but only 4 iterations for *Subdiv.*, and *Phong* provides the highest accuracy at this speed. Alternatively, if we require the fitter to reach near 80% accuracy, we can run 4 iterations with *Phong* and *Subdiv.*, but *Phong* is 20% faster. So *Phong* achieves almost the same level of accuracy as *Subdiv*, while being as cheap as *Tri. mesh* in terms of efficiency.

Similar conclusions can be drawn for ICP optimizations (dashed lines in Fig. 12a). As pointed out earlier at the end of Sect. 3.1, our runtime for *ICP* is slightly slower than *lifted*, but we show that per-iteration cost of lifted is actually comparable with ICP in our supplementary material, part 3.

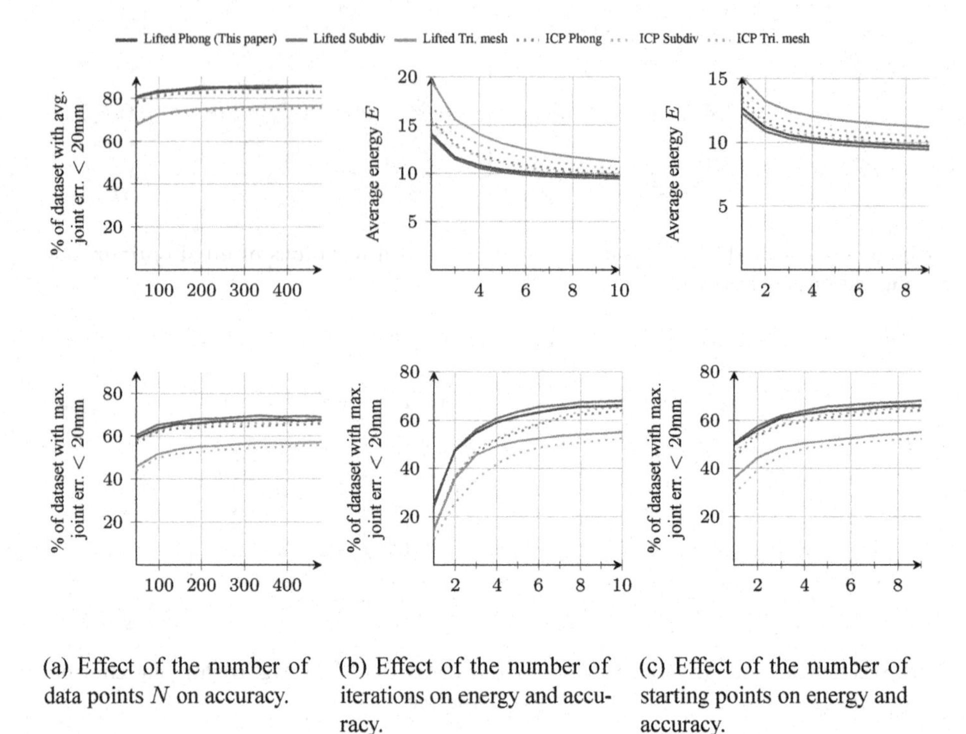

(a) Effect of the number of data points N on accuracy.

(b) Effect of the number of iterations on energy and accuracy.

(c) Effect of the number of starting points on energy and accuracy.

Fig. 11. Effect of alternative optimization configurations when fitting to Dexter. The results shown are for per-frame model fitting.

Note that the accuracy achieved by *Lifted Subdiv.* after 8 iterations coincides with that achieved by *Lifted Phong* after 10 iterations. This is because *Lifted Phong* already converged after 8 iterations, and further iterations do not improve accuracy further (see also Fig. 11b).

Performance on HoloLens 2. Starting from the work of [27], we made many improvements to enable us to run a hand tracker in real time on the HoloLens 2 [14], a mobile device with very limited computational and power resources. The Phong surface model presented here was one of the key efficiency improvements that was required. Figure 12b shows the speed vs. accuracy of various surface types with lifted optimization in this hand tracker on HoloLens 2, evaluated on a captured depth dataset. The *Phong* surface (red dot) can be evaluated twice as fast as the subdivision surface (green dot), and gives the same level of accuracy.

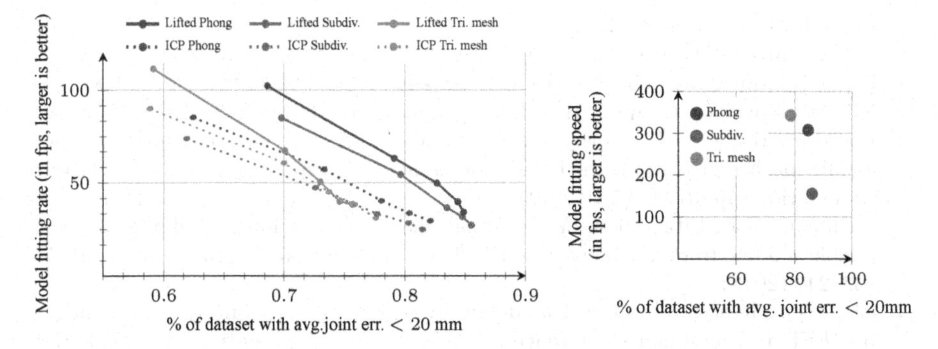

(a) Application on the hand tracker by Taylor et al. [27]. (b) Our hand tracker on HoloLens 2.

Fig. 12. (a) The speed (model-fitting speed in fps for one starting point) vs. accuracy (percentage of frames with avg. joint error less than 20 mm) for per-frame fitting using the hand tracker by Taylor et al. [27] on Dexter. Upwards and to the right is better. The dots from top to bottom on each line denote the number of iterations being 2, 4, 6, 8, 10. (b) The speed vs. accuracy for our hand tracker on HoloLens 2. (Color figure online)

4 Conclusions

In this paper we demonstrated that the convergence benefits of lifted optimization are available to a wider range of surface models than was previously thought. We introduced Phong surfaces, and showed that they provide sufficient information about the local model geometry to allow a model-fitting optimizer to converge fast, whilst requiring a fraction of the compute of expensive smooth surface models. Beside rigid pose alignment and hand tracking, the proposed method can be applied to various 3D surface model-fitting applications, for example, 3D pose and shape estimation of SMPL body [2], face [10] and animals [33,34], and is particularly valuable when computational budget is limited.

Given the generalization we show in this paper, a natural question is 'what are the set of requirements in order for lifted optimization to work effectively?' We hypothesize that the only requirement is for a model to provide sufficient approximations to the energy tangent space $\partial E/\partial \theta$ for a gradient-based optimizer to take efficient steps in each iteration, with an implied freedom on global topology and connectivity, as well as the form taken by those approximations. We intend to explore this hypothesis more fully in future work.

References

1. Besl, P.J., McKay, N.D.: A method for registration of 3-D shapes. IEEE Trans. Pattern Anal. Mach. Intell. **14**(2), 239–256 (1992)

2. Bogo, F., Kanazawa, A., Lassner, C., Gehler, P., Romero, J., Black, M.J.: Keep it SMPL: automatic estimation of 3D human pose and shape from a single image. In: Leibe, B., Matas, J., Sebe, N., Welling, M. (eds.) ECCV 2016. LNCS, vol. 9909, pp. 561–578. Springer, Cham (2016). https://doi.org/10.1007/978-3-319-46454-1_34
3. Bogo, F., Romero, J., Loper, M., Black, M.J.: FAUST: dataset and evaluation for 3D mesh registration. In: IEEE Conference on Computer Vision and Pattern Recognition, pp. 3794–3801 (2014)
4. Cashman, T.J., Fitzgibbon, A.W.: What shape are dolphins? Building 3D morphable models from 2D images. IEEE Trans. Pattern Anal. Mach. Intell. **35**(1), 232–244 (2013)
5. Chen, Y., Medioni, G.: Object modeling by registration of multiple range images. In: IEEE International Conference on Robotics and Automation, pp. 2724–2729 (1991)
6. Fitzgibbon, A.: Robust registration of 2D and 3D point sets. In: Proceedings of the British Machine Vision Conference, pp. 411–420 (2001)
7. Hirshberg, D.A., Loper, M., Rachlin, E., Black, M.J.: Coregistration: simultaneous alignment and modeling of articulated 3D shape. In: Fitzgibbon, A., Lazebnik, S., Perona, P., Sato, Y., Schmid, C. (eds.) ECCV 2012. LNCS, vol. 7577, pp. 242–255. Springer, Heidelberg (2012). https://doi.org/10.1007/978-3-642-33783-3_18
8. Khamis, S., Taylor, J., Shotton, J., Keskin, C., Izadi, S., Fitzgibbon, A.: Learning an efficient model of hand shape variation from depth images. In: IEEE Conference on Computer Vision and Pattern Recognition, pp. 2540–2548 (2015)
9. Kolotouros, N., Pavlakos, G., Black, M., Daniilidis, K.: Learning to reconstruct 3D human pose and shape via model-fitting in the loop. In: IEEE Conference on Computer Vision and Pattern Recognition, pp. 2252–2261 (2019)
10. Li, T., Bolkart, T., Black, M.J., Li, H., Romero, J.: Learning a model of facial shape and expression from 4D scans. ACM Trans. Graph. **36**(6), 194:1–194:17 (2017)
11. Loop, C.T.: Smooth subdivision surfaces based on triangles. Master's thesis, University of Utah (1987)
12. Magic Leap Inc.: Perception at Magic Leap (2019). https://sites.google.com/view/perceptionatmagicleap/
13. Marquardt, D.W.: An algorithm for least-squares estimation of nonlinear parameters. J. Soc. Ind. Appl. Math. **11**(2), 431–441 (1963)
14. Microsoft: HoloLens 2 (2019). https://blogs.microsoft.com/blog/2019/02/24/microsoft-at-mwc-barcelona-introducing-microsoft-hololens-2
15. Mueller, F., et al.: Real-time pose and shape reconstruction of two interacting hands with a single depth camera. ACM Trans. Graph. **38**(4), 49:1–49:13 (2019)
16. Neugebauer, P.J.: Geometrical cloning of 3D objects via simultaneous registration of multiple range images. In: International Conference on Shape Modeling and Applications, pp. 130–139 (1997)
17. Pavlakos, G., et al.: Expressive body capture: 3D hands, face, and body from a single image. In: IEEE Conference on Computer Vision and Pattern Recognition, pp. 10975–10985 (2019)
18. Pellegrini, S., Schindler, K., Nardi, D.: A generalisation of the ICP algorithm for articulated bodies. In: Proceedings of the British Machine Vision Conference, pp. 87.1–87.10 (2008)
19. Phong, B.T.: Illumination for computer generated pictures. Commun. ACM **18**(6), 311–317 (1975)
20. Qian, C., Sun, X., Wei, Y., Tang, X., Sun, J.: Realtime and robust hand tracking from depth. In: IEEE Conference on Computer Vision and Pattern Recognition, pp. 1106–1113 (2014)

21. Rusinkiewicz, S.: A symmetric objective function for ICP. ACM Trans. Graph. **38**(4), 85:1–85:7 (2019)
22. Rusinkiewicz, S., Levoy, M.: Efficient variants of the ICP algorithm. In: International Conference on 3D Digital Imaging and Modeling, pp. 145–152 (2001)
23. Sridhar, S., Oulasvirta, A., Theobalt, C.: Interactive markerless articulated hand motion tracking using RGB and depth data. In: International Conference on Computer Vision, pp. 2456–2463 (2013)
24. Sullivan, S., Ponce, J.: Automatic model construction and pose estimation from photographs using triangular splines. IEEE Trans. Pattern Anal. Mach. Intell. **20**(10), 1091–1097 (1998)
25. Tagliasacchi, A., Schröder, M., Tkach, A., Bouaziz, S., Botsch, M., Pauly, M.: Robust articulated-ICP for real-time hand tracking. Comput. Graph. Forum **34**(5), 101–114 (2015)
26. Taylor, J., et al.: User-specific hand modeling from monocular depth sequences. In: IEEE Conference on Computer Vision and Pattern Recognition, pp. 644–651 (2014)
27. Taylor, J., et al.: Efficient and precise interactive hand tracking through joint, continuous optimization of pose and correspondences. ACM Trans. Graph. **35**(4), 143:1–143:12 (2016)
28. Taylor, J., et al.: Articulated distance fields for ultra-fast tracking of hands interacting. ACM Trans. Graph. **36**(6), 244:1–244:12 (2017)
29. Tkach, A., Pauly, M., Tagliasacchi, A.: Sphere-meshes for real-time hand modeling and tracking. ACM Trans. Graph. **35**(6), 222:1–222:11 (2016)
30. Wan, C., Probst, T., Gool, L.V., Yao, A.: Self-supervised 3D hand pose estimation through training by fitting. In: IEEE Conference on Computer Vision and Pattern Recognition, pp. 10845–10854 (2019)
31. Xiang, D., Joo, H., Sheikh, Y.: Monocular total capture: posing face, body, and hands in the wild. In: IEEE Conference on Computer Vision and Pattern Recognition, pp. 10957–10966 (2018)
32. Zheng, J., Zeng, M., Cheng, X., Liu, X.: SCAPE-based human performance reconstruction. Comput. Graph. **38**, 191–198 (2014)
33. Zuffi, S., Kanazawa, A., Black, M.J.: Lions and tigers and bears: capturing non-rigid, 3D, articulated shape from images. In: IEEE Conference on Computer Vision and Pattern Recognition, pp. 3955–3963 (2018)
34. Zuffi, S., Kanazawa, A., Jacobs, D.W., Black, M.J.: 3D menagerie: modeling the 3D shape and pose of animals. In: IEEE Conference on Computer Vision and Pattern Recognition, pp. 5524–5532 (2017)

Forecasting Human-Object Interaction: Joint Prediction of Motor Attention and Actions in First Person Video

Miao Liu[1](✉), Siyu Tang[3], Yin Li[2], and James M. Rehg[1]

[1] Georgia Institute of Technology, Atlanta, USA
mliu328@gatech.edu
[2] University of Wisconsin-Madison, Madison, USA
[3] ETH Zürich, Zürich, Switzerland

Abstract. We address the challenging task of anticipating human-object interaction in first person videos. Most existing methods either ignore how the camera wearer interacts with objects, or simply considers body motion as a separate modality. In contrast, we observe that the intentional hand movement reveals critical information about the future activity. Motivated by this observation, we adopt intentional hand movement as a feature representation, and propose a novel deep network that jointly models and predicts the egocentric hand motion, interaction hotspots and future action. Specifically, we consider the future hand motion as the motor attention, and model this attention using probabilistic variables in our deep model. The predicted motor attention is further used to select the discriminative spatial-temporal visual features for predicting actions and interaction hotspots. We present extensive experiments demonstrating the benefit of the proposed joint model. Importantly, our model produces new state-of-the-art results for action anticipation on both EGTEA Gaze+ and the EPIC-Kitchens datasets. Our project page is available at https://aptx4869lm.github.io/ForecastingHOI/.

Keywords: First Person Vision · Action anticipation · Motor attention

1 Introduction

The human ability of "looking into the near future" remains a key challenge for computer vision. Consider the example in Fig. 1, given a video shortly before the start of an action, we can easily predict what will happen next, e.g., the person will take the canister of salt. Even without seeing any future frames, we can vividly imagine how the person will perform the action, e.g., the trajectory of the hand when reaching for the canister or the location on the canister that will be grasped.

Electronic supplementary material The online version of this chapter (https://doi.org/10.1007/978-3-030-58452-8_41) contains supplementary material, which is available to authorized users.

A. Vedaldi et al. (Eds.): ECCV 2020, LNCS 12346, pp. 704–721, 2020.
https://doi.org/10.1007/978-3-030-58452-8_41

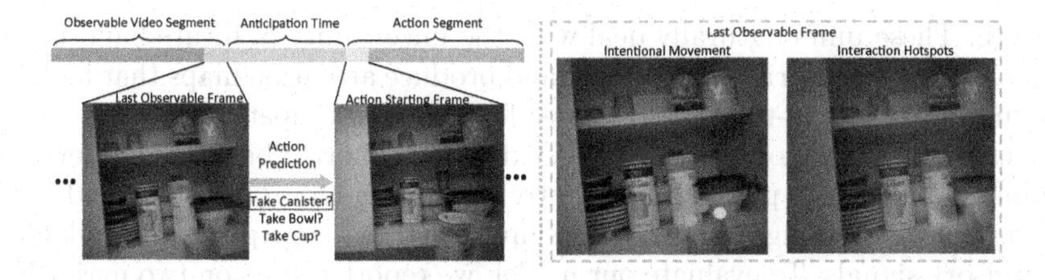

Fig. 1. *What is the most likely future interaction?* Our model takes advantage of the connection between motor attention and visual perception. In addition to future action label, our model also predicts the interaction hotspots on the last observable frame and hand trajectory (in the order of yellow, green, cyan, and magenta) between the last observable time step to action starting point. Visualizations of hand trajectory are projected to the last observable frame (best viewed in color). (Color figure online)

There is convincing evidence that our remarkable ability to forecast other individuals' actions depends critically upon our perception and interpretation of their body motion. The investigation of this anticipatory mechanism dates back to 19th century, when William James argued that future expectations are intrinsically related to purposive body movements [25]. Additional evidence for a link between perceiving and performing actions was provided by the discovery of mirror neurons [8,20]. The observation of others' actions activates our motor cortex, the same brain regions that are in charge of the planning and control of intentional body motion. This activation can happen even before the onset of the action and is highly correlated with the anticipation accuracy [1]. A compelling explanation from [45] suggests that *motor attention*, i.e., the active prediction of meaningful future body movements, serves as a key representation for anticipation. A goal of this work is to develop a computational model for motor attention that can enable more accurate action prediction.

Despite these relevant findings in cognitive neuroscience, the role of intentional body motion in action anticipation is largely ignored by the existing literature [11,13,15,16,27,28,38,56]. In this work, we focus on the problem of forecasting human-object interactions in First Person Vision (FPV). Interactions consist of a single verb and one or more nouns, with "take bowl" as an example. FPV videos capture complex hand movements during a rich set of interactions, thus providing a powerful vehicle for studying the connection between motor attention and future representation. Several previous works have investigated the problems of FPV activity anticipation [13,15] and body movement prediction [2,12,19,57]. We believe we are the first to utilize a motor attention model for FPV action anticipation.

To this end, we propose a novel deep model that predicts "motor attention"—the future trajectory of the hands, as an anticipatory representation of actions. Based on motor attention, our model further localizes the future contact region of the interaction, i.e., interaction hotspots [39] and recognizes the type of future interactions. Importantly, we characterize motor attention and interaction hotspots as probabilistic variables modeled by stochastic units in a deep network. These units naturally deal with the uncertainty of future hand motion and contact region during interaction, and produce attention maps that highlight discriminative spatial-temporal features for action anticipation.

During inference, our model takes video clips shortly before the interaction as inputs, and jointly predicts motor attention, interaction hotspots, and action labels. During training, our model assumes that these outputs are available as supervisory signals. To evaluate our model, we report results on two major FPV benchmarks: EGTEA Gaze+ and EPIC-Kitchens. Our approach outperforms prior state-of-the-art methods by a significant margin. In addition, we conduct extensive ablation studies to verify the design of our model and evaluate our model for motor attention prediction and interaction hotspots estimation. Our model demonstrates strong results for both tasks. We believe our model provides a solid step towards the challenge of FPV visual anticipation.

2 Related Works

There has recently been substantial interest in learning to forecast future events in videos. The most relevant works to ours are those investigations on FPV action anticipation. Our work is also related to previous studies on third person action anticipation, other visual prediction tasks, and visual affordance.

FPV Action Anticipation. Action anticipation aims at predicting an action before it happens. We refer the readers to a recent survey [30] for a distinction between action recognition and anticipation. FPV action recognition has been studied extensively [10,32,34,36,41,42,46,63], while fewer works have targeted egocentric action anticipation. Shen et al. [49] investigated how different egocentric modalities affect the action anticipation performance. Soran et al. [52] adopted Hidden Markov Model to compute the transition probability among sequences of actions. A similar idea was explored in [38]. Furnari et al. [13] considered the task of predicting the next-active objects. Their recent work [15] proposed to factorize the anticipation model into a "Rolling" LSTM that summarizes the past activity and an "Unrolling" LSTM that makes hypotheses of the future activity. Ke et al. [28] proposed a time-conditioned skip connection operation to extract relevant information for action anticipation. In contrast to our proposed method, these prior works did not exploit the connection between human motor attention and visual perception, and did not explicitly model the contact region during human-object interaction.

Third Person Action Anticipation. Several previous efforts seek to address the task of action anticipation in third person vision. Kris et al. [29] combined semantic scene labeling with a Markov decision process to forecast the behavior and trajectory of a subject. Vondrick et al. [56] proposed to predict the future video representation from large scale unlabeled video data. Gao et al. [16] proposed a Reinforced Encoder-Decoder network to create a summary representation of past frames and produce a hypothesis of future action. Kataoka et al. [27] introduced a subtle motion descriptor to identify the difference between an on-going action and a transitional action, and thereby facilitate future anticipation. Our work shares the same goal of future forecasting, but we focus on leveraging abundant visual cues from egocentric videos for action anticipation.

Other Prediction Tasks. Anticipation has been studied under other vision tasks. In particular, human body motion prediction has been extensively studied [12,19,40,55,57,58], including recent work in the setting of FPV. Rhinehart et al. [44] proposed an online learning algorithm to forecast the first-person trajectory. Park et al. [51] proposed a deep network to infer possible human trajectories from egocentric stereo images. Wei et al. [61] utilized a probabilistic model to infer 3D human attention and intention. Tagi et al. [62] addressed a novel task of predicting the future locations of an observed subject in egocentric videos. Ryoo et al. [47] proposed a novel method to summarize pre-activity observations for robot-centric activity prediction. However, none of these previous work considered modeling body movement for action anticipation.

Visual Affordance. The problem of predicting visual affordances has attracted growing interest in computer vision. Affordance can be helpful for scene understanding [7,18,59], human-object interaction recognition [53], and action analysis [31,43]. Several recent works have focused on estimating visual affordances that are grounded on human object interaction. Chen et al. [5] proposed to estimate likely object interaction regions by learning the connection between subject and object. Fang et al. [9] proposed to estimate interaction regions by learning from demonstration videos. However, none of these previous works considered future prediction. More recently, Tushar et al. [39] introduced an unsupervised learning method that uses the backward attention map to approximate the interaction hotspots grounded on a future action. However, their method did not model the presence of objects and thus can not be used to anticipate human-object interactions. However, we compare to their results for interaction hotspot estimation in our experiments.

3 Method

We consider the setting of action anticipation from [6]. Denote an input video segment as $x : [\tau_a - \Delta\tau_o, \tau_a]$. x starts at $\tau_a - \Delta\tau_o$ and ends at τ_a with duration $\Delta\tau_o > 0$ as the "observation time". Our goal is to predict the label y of an immediate future interaction starting at $\tau_s = \tau_a + \Delta\tau_a$, where $\Delta\tau_a > 0$ is a fixed interval known as the "anticipation time." Moreover, we seek to estimate future hand trajectories $\mathcal{M}$ within $[\tau_a, \tau_s]$ (projected back to the last observable frame at τ_a), and to localize interaction hotspots $\mathcal{A}$ at τ_a (the last observable frame). Figure 1 illustrates our setting.

To summarize, our model seeks to anticipate the future action y by jointly predicting the future hand trajectory $\mathcal{M}$ and interaction hotspots $\mathcal{A}$ at the last observable frame. Predicting the future is fundamentally ambiguous, since the observation of future interaction only represents one of the many possibilities characterized by an underlying distribution. Our key idea is thus to model motor

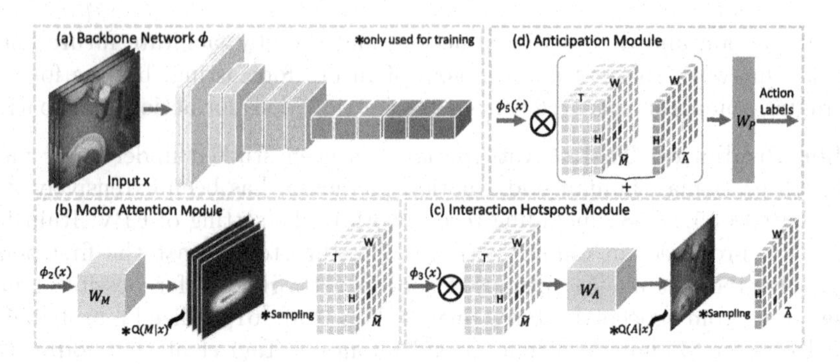

Fig. 2. Overview of our model. A 3D convolutional network $\phi(x)$ is used as our backbone network, with features from its ith convolution block as $\phi_i(x)$ (a). A motor attention module (b) makes use of stochastic units to generate sampled future hand trajectories $\tilde{\mathcal{M}}$ used to guide interaction hotspots estimation in module (c). Module (c) further generates sampled interaction hotspots $\tilde{\mathcal{A}}$ with a similar stochastic units as in module (b). Both $\tilde{\mathcal{M}}$ and $\tilde{\mathcal{A}}$ are used to guide action anticipation in anticipation module (d). During testing, our model takes only video clips as inputs, and predicts motor attention, interaction hotspots, and action labels. Note that $\otimes$ represents element-wise multiplication for weighted pooling.

attention and interaction hotspots as probabilistic variables in order to account for their uncertainty. We present an overview of our model in Fig. 2.

Specifically, we make use of a 3D backbone network $\phi(x)$ for video representation learning. Following the approach in [21,50], we utilize 5 convolutional blocks, and denote the features from the ith convolution block as $\phi_i(x)$. Based on $\phi(x)$, our motor attention module (b) predicts future hand trajectories as motor attention $\mathcal{M}$ and uses stochastic units to sample from $\mathcal{M}$. The sampled motor attention $\tilde{\mathcal{M}}$ is an indicator of important spatial-temporal features for interaction hotspot estimation. Our interaction hotspot module (c) further produces an interaction hotspot distribution $\mathcal{A}$ and its sample $\tilde{\mathcal{A}}$. Finally, our anticipation module (d) makes use of both $\tilde{\mathcal{M}}$ and $\tilde{\mathcal{A}}$ to aggregate network features, and predicts the future interaction y.

3.1 Joint Modeling of Human-Object Interaction

Formally, we consider motor attention $\mathcal{M}$ and interaction hotspots $\mathcal{A}$ as probabilistic variables, and model the conditional probability of the future action label y given the input video x as a latent variable model, where

$$p(y|x) = \int_{\mathcal{M}} \int_{\mathcal{A}} p(y|\mathcal{A}, \mathcal{M}, x) p(\mathcal{A}|\mathcal{M}, x) p(\mathcal{M}|x) \, d\mathcal{A} \, d\mathcal{M}, \quad (1)$$

$p(\mathcal{M}|x)$ first estimates motor attention from video input x. $\mathcal{M}$ is further used to estimate interaction hotspots A ($p(\mathcal{A}|\mathcal{M}, x)$). Given x, $\mathcal{M}$ and $\mathcal{A}$, the action label y is determined by $p(y|\mathcal{A}, \mathcal{M}, x)$. Our model thus consists of three main components.

Motor Attention Module tackles $p(\mathcal{M}|x)$. Given the network features $\phi_2(x)$, our model uses a function F_M to predict motor attention $\mathcal{M}$. $\mathcal{M}$ is represented as a 3D tensor of size $T_m \times H_m \times W_m$. Moreover, $\mathcal{M}$ is normalized within each temporal slice, i.e., $\sum_{w,h} \mathcal{M}(t, w, h) = 1$.

Interaction Hotspots Module targets at $p(\mathcal{A}|\mathcal{M}, x)$. Our model uses a function F_A to estimate the interaction hotspots $\mathcal{A}$ based on the network feature $\phi_3(x)$ and sampled motor attention $\tilde{\mathcal{M}}$. $\mathcal{A}$ is represented as a 2D attention map of size $H_a \times W_a$. A further normalization constrained that $\sum_{w,h} \mathcal{A}(w, h) = 1$.

Anticipation Module makes use of the predicted motor attention and interaction hotspots for action anticipation. Specifically, sampled motor attention $\tilde{\mathcal{M}}$ and sampled interaction hotspots $\tilde{\mathcal{A}}$ are used to aggregate feature $\phi_5(x)$ via weighted pooling. An action anticipation function F_P further maps the aggregated features to future action label y.

3.2 Motor Attention Module

Motor Attention Generation. The motor attention prediction function F_M is composed of a linear function with parameter W_M on top of network features $\phi_2(x)$. The linear function is realized by a 3D convolution and a softmax function is used to normalized the attention map. This is given by $\psi = softmax(W_M^T \phi_2(x))$, where the output ψ is a 3D tensor of size $T_m \times H_m \times W_m$. We further model $p(\mathcal{M}|x)$ by normalizing ψ within each temporal slice:

$$\mathcal{M}_{m,n,t} = \frac{\psi_{m,n,t}}{\sum_{m,n} \psi_{m,n,t}}, \tag{2}$$

where $\psi_{m,n,t}$ is the value at location (m, n) and time step t in the 3D tensor of ψ. And $\mathcal{M}$ can be considered as the expectation of $p(\mathcal{M}|x)$.

Stochastic Modeling. Modeling motor attention in the context of forecasting human-object interaction requires a mechanism for addressing the stochastic nature of motor attention in developing the joint model. Here, we propose to use stochastic units to model the uncertainty. The key idea is to sample from the motor attention distribution. We follow the Gumbel-Softmax and reparameterization trick introduced in [26,37] to design a differentiable sampling mechanism:

$$\tilde{\mathcal{M}}_{m,n,t} \sim \frac{\exp((\log \psi_{m,n,t} + G_{m,n,t})/\theta)}{\sum_{m,n} \exp((\log \psi_{m,n,t} + G_{m,n,t})/\theta)}, \tag{3}$$

where G is a Gumbel Distribution used to sample from discrete distribution. This Gumbel-Softmax trick produces a "soft" sampling step that allows the direct back-propagation of gradients to ψ. θ is the temperature parameter that controls the "sharpness" of the distribution. We set $\theta = 2$ for all of our experiments.

3.3 Interaction Hotspots Module

The predicted motor attention $\mathcal{M}$ is further used to guide interaction hotspots estimation $p(\mathcal{A}|x)$ by considering the conditional probability

$$p(\mathcal{A}|x) = \int_{\mathcal{M}} p(\mathcal{A}|\mathcal{M}, x) p(\mathcal{M}|x) d\mathcal{M}. \tag{4}$$

In practice, $p(\mathcal{A}|x)$ is estimated using sampled motor attention $\tilde{\mathcal{M}}$ based on $p(\mathcal{A}|\tilde{\mathcal{M}}, x)$ and $p(\tilde{\mathcal{M}}|x)$. For each sample $\tilde{\mathcal{M}}$, $p(\mathcal{A}|\tilde{\mathcal{M}}, x)$ is defined by the interaction hotspots estimation function F_A. F_A takes the input of a motor attention map $\tilde{\mathcal{M}}$ and $\phi_3(x)$, and has the form of a linear 2D convolution parameterzied by W_A followed by a softmax function.

$$p(\mathcal{A}|\tilde{\mathcal{M}}, x) = softmax\left(W_A^T(\tilde{\mathcal{M}} \otimes \phi_3(x))\right), \tag{5}$$

where $\otimes$ is the Hadamard product (element-wise multiplication). The result $p(\mathcal{A}|\mathcal{M}, x)$ is a 2D map of size $H_a \times W_a$. Intuitively, $\tilde{\mathcal{M}}$ presents a spatial-temporal saliency map to highlight feature representation $\phi_3(x)$. F_A thus normalizes (using softmax) the output of a linear model on the selected features $\tilde{\mathcal{M}} \otimes \phi_3(x)$, and is a convex function. Finally, a similar sampling mechanism as in Eq. 3 can be used to sample $\tilde{\mathcal{A}}$ from $p(\mathcal{A}|x)$.

3.4 Anticipation Module

We now present the last piece of our model—the action anticipation module. The action anticipation function $p(y|\mathcal{A}, \mathcal{M}, x) = F_P(\mathcal{A}, \mathcal{M}, x)$ is defined as a function of the sampled motor attention map (3D) $\tilde{\mathcal{M}}$, sampled interaction heatmap (2D) $\tilde{\mathcal{A}}$ and the network feature $\phi_5(x)$. This is given by

$$p(y|\tilde{\mathcal{A}}, \tilde{\mathcal{M}}, x) = softmax\left(W_P^T \Sigma\left(\tilde{\mathcal{M}} \otimes \phi_5(x)\right) + W_P^T \Sigma\left(\tilde{\mathcal{A}} \odot \phi_5(x)\right)\right), \tag{6}$$

where $\otimes$ is again the Hadamard product. Σ is the global average pooling operation that pools a vector representation from a 2D or 3D feature map. $\odot$ is to use a 2D map ($\tilde{\mathcal{A}}$) to conduct Hadamard product to the last temporal slice of a 3D tensor $\phi_5(x)$. This is because the interaction hotspots $\tilde{\mathcal{A}}$ is only defined on the last observable frame. W_P is a linear function that maps the features into prediction logits. F_P is a combination of linear operations followed by a softmax function, and thus remains a convex function.

3.5 Training and Inference

Training our proposed joint model is challenging, as $p(\mathcal{M}|x)$ and $p(\mathcal{A}|\mathcal{M}, x)$ are intractable. Fortunately, variational inference comes to the rescue.

Prior Distribution. During training, we assume that reference distributions of future hand position $Q(\mathcal{M}|x)$ and interaction hotspots $Q(\mathcal{A}|x)$ are known in prior. These distributions can be derived from manual annotation of 2D fingertips and interaction hotspots, as we will describe in Sect. 4.1. A 2D isotropic Gaussian is further applied to the annotated 2D points, leading to the distributions of $Q(\mathcal{M}|x)$ and $Q(\mathcal{A}|x)$. If annotations are not available, we adopt uniform distributions for both $Q(\mathcal{M}|x)$ and $Q(\mathcal{A}|x)$.

Variational Learning. Our proposed model seeks to jointly predict motor attention $\mathcal{M}$, interaction hotspots $\mathcal{A}$, and the action label y. Therefore, we inject the posterior $p(\mathcal{A}, \mathcal{M}|x)$ into $p(y|x)$. We further assume $p(\mathcal{A}, \mathcal{M}|x)$ can be factorized into $p(\mathcal{A}|x)$ and $p(\mathcal{M}|x)$ (see supplementary materials for details). Our model thereby optimizes the resulting latent variable model by maximizing the Evidence Lower Bound (ELBO), given by[1]

$$\begin{aligned} \log p(y|x) \geq & E_{p(\mathcal{A},\mathcal{M}|x)}[\log p(y|\mathcal{A},\mathcal{M},x)] - log(p(\mathcal{A},\mathcal{M}|x))] \\ = & \sum_{\mathcal{A},\mathcal{M}} \log p(y|\mathcal{A},\mathcal{M},x) - KL[p(\mathcal{A}|x)||Q(\mathcal{A}|x)] - KL[p(\mathcal{M}|x)||Q(\mathcal{M}|x)]. \end{aligned} \tag{7}$$

Therefore, the loss function $\mathcal{L}$ is given by

$$\mathcal{L} = -\sum_{\mathcal{A},\mathcal{M}} \log p(y|\mathcal{A},\mathcal{M},x) + KL[p(\mathcal{A}|x)||Q(\mathcal{A}|x)] + KL[p(\mathcal{M}|x)||Q(\mathcal{M}|x)]. \tag{8}$$

The first term in the loss function is the cross entropy loss for action anticipation. The last two terms use KL-Divergence to align the predicted distributions of motor attention $p(\mathcal{M}|x)$ and interaction hotspots $p(\mathcal{A}|x)$ to their reference distributions ($Q(\mathcal{M}|x)$ and $Q(\mathcal{A}|x)$). To make the training practical, we draw a single sample for each input within a mini-batch similar to [26,37]. Multiple samples of the same input will be drawn at different iterations.

Approximate Inference. At inference time, our model could have drawn many samples of motor attention $\tilde{\mathcal{M}}$ and interaction hotspots $\tilde{\mathcal{A}}$ for the anticipation. However, the sampling and averaging is computationally expensive. We choose to feed deterministic $\mathcal{M}$ and $\mathcal{A}$ into Eq. 5 and Eq. 6 at inference time. Note that F_A and F_P are convex, since they are composed of linear mapping function and softmax function. By Jensen's inequality, we have

$$E[F_A(\tilde{\mathcal{M}}, x)] \geq F_A(E[\tilde{\mathcal{M}}], x) = F_A(\mathcal{M}, x), \tag{9}$$

$$E[F_P(\tilde{\mathcal{A}}, \tilde{\mathcal{M}}, x)] \geq F_P(E[\tilde{\mathcal{A}}], E[\tilde{\mathcal{M}}], x) = F_P(\mathcal{A}, \mathcal{M}, x) \tag{10}$$

Therefore, such approximation provides a valid lower bound of $E[F_P(\tilde{\mathcal{A}}, \tilde{\mathcal{M}}, x)]$ and $E[F_A(\tilde{\mathcal{M}}, x)]$, and serves as a shortcut to avoid sampling during testing.

3.6 Network Architecture

We consider two different backbone networks for our model, including lightweight I3D-Res50 network [4,60] pre-trained on Kinetics and heavy CSN-152 [54] network pre-trained on IG-65M [17]. We use I3D-Res50 for our ablation study on

[1] See supplementary material for the derivation.

EGTEA and EPIC-Kitchens, and report results using CSN-152 backbone when competing on the EPIC-Kitchens dataset. Both networks have five convolutional blocks. The motor attention module, the interaction hotspots module and the recognition module are attached to the 2nd, the 3rd and the 5th block, respectively. We use 3D max pooling to match the size of attention map to the size of the feature map in Eq. 5 and Eq. 6. For training, our model takes an input of 32 frames (every other frame from a 64-frame chunk) with a resolution of 224×224. For inference, our model samples 30 clips from a video (3 along width of frame and 10 in time). Each clip has 32 frames with a resolution of 256×256. We average the scores of all sampled clips for video level prediction. Other implementation details will be discussed in the experiments.

4 Experiments

We now present our experiments and results. We briefly introduce our implementation details and describe the datasets and annotations. Moreover, we present our results on EPIC-Kitchens action anticipation challenge, followed by ablation studies that further evaluate our model on interaction hotspot estimation and motor attention prediction. Finally, we provide a discussion of our method.

Implementation Details. Our model is trained using SGD with momentum 0.9 and batch size 64 on 4 GPUs. The initial learning rate is 2.5e−4 with cosine decay. We set weight decay to 1e−4 and enable batch norm [24]. We downsample all frames to 320×256 (24 fps) for EGTEA, and 512×288 (30 fps) for EPIC-Kitchens. We apply several data augmentation techniques, including random flipping, rotation, cropping and color jittering to avoid overfitting.

4.1 Datasets and Annotations

Datasets. We make use of two FPV datasets: EGTEA Gaze+ [32,33] and Epic-Kitchens [6]. EGTEA comes with 10,321 action instances from 19/53/106 verb/noun/action classes. We report results on the first split of the dataset. EPIC-Kitchens contains 39,596 instances from 125 verbs and 352 nouns. We follow [15] to split the public training set into training and validation sets with 2513 action classes. We conduct ablation studies on this train/val split, and present the action anticipation results on the testing sets. We set the anticipation time as 0.5 s for EGTEA and 1 s [6] for EPIC-Kitchens.

Annotations. Our model requires supervisory signals of interaction hotspots and hand trajectories during training. We provide extra annotations for both EGTEA and EPIC-Kitchens datasets. These annotations will be made publicly available. Specifically, we manually annotated interaction hotspots as 2D points on the last observable frames for all instances on EGTEA and a subset of instances on EPIC-Kitchens. This is because many noun labels in Epic-Kitchens have very few instances, hence we focus on interaction hotspots of action instances that include many-shot nouns [6] in the training set.

Moreover, we explore different approaches to generate the pseudo ground truth of future hand trajectories. On EGTEA, we trained a hand segmentation model ([35] using hand masks from the dataset). The motor attention was approximated by segmenting hands at every frame and tracking the fingertip closest to an active object. To mitigate ego-motion, we used optical flow and RANSAC to compute a homography transform, and project the motor attention to the last observable frame. As EPIC-Kitchens does not provide hand masks, we instead annotated the fingertip closest to an interaction hotspots on the last observable frame. A linear interpolation of 2D motion between the fingertip and the interaction hotspots was used to approximate the motor attention.

4.2 FPV Action Anticipation on EPIC-Kitchens

We highlight our results for FPV action anticipation on EPIC-Kitchens dataset.

Table 1. Action anticipation results on Epic-Kitchens. Ours+Obj model outperforms state-of-the-art by a notable margin. See discussions of Ours+Obj in Sect. 4.2.

	Method	Top1/Top5 accuracy		
		Verb	Noun	Action
s1	2SCNN [6]	29.76/76.03	15.15/38.65	4.32/15.21
	TSN [6]	31.81/76.56	16.22/42.15	6.00/18.21
	TSN+MCE [14]	27.92/73.59	16.09/39.32	10.76/25.28
	Trans R(2+1)D [38]	30.74/76.21	16.47/42.72	9.74/25.44
	RULSTM [15]	33.04/**79.55**	22.78/50.95	14.39/33.73
	Ours	34.99/77.05	20.86/46.45	14.04/31.29
	Ours+Obj	**36.25**/79.15	**23.83/51.98**	**15.42/34.29**
s2	2SCNN [6]	25.23/68.66	9.97/27.38	2.29/9.35
	TSN [6]	25.30/68.32	10.41/29.50	2.39/9.63
	TSN+MCE [14]	21.27/63.66	9.90/25.50	5.57/25.28
	Trans R(2+1)D [38]	28.37/69.96	12.43/32.20	7.24/19.29
	RULSTM [15]	27.01/69.55	15.19/34.38	8.16/21.20
	Ours	28.27/70.67	14.07/34.35	8.64/22.91
	Ours+Obj	**29.87/71.77**	**16.80/38.96**	**9.94/23.69**

Experiment Setup. To compete for EPIC-Kitchens anticipation challenge, we used the backbone network CSN152. We trained our model on the public training set and report results using top-1/5 accuracy as in [6].

Results. Table 1 compares our results to latest methods on EPIC-Kitchens. Our model outperforms strong baselines (TSN and 2SCNN) reported in [6] by a large margin. Compared to previous best results from RULSTM [15], our model archives +2%/−1.9%/−0.3% for verb/noun/action on seen set, and +1.3%/−1.1%/+0.6% on unseen set of EPIC-Kitchens. Our results are better for verb, worse for noun and comparable or better for actions. Notably, RULSTM requires object boxes & optical flow for training and object features & optical flow for testing. In contrast, our method uses hand trajectories and interaction hotspots for training and needs *only RGB frames* for testing.

To further improve the performance, we fuse the object stream from RULSTM with our model (Ours+Obj). Compared to RULSTM, Ours+Obj has a performance gain of +3.2%/+2.9% for verb, +1.1%/+1.6% for noun, and +1.0%/+1.8% for action (seen/unseen). It is worthy pointing out that RULSTM benefits from an extra flow network, while ours+Obj model takes additional supervisory signals of hands and hotspots. Note that our performance boost does not simply come from those extra annotations. In a subsequent ablation study, we have shown that simply training with these extra annotations has minor improvement, when used without our proposed probabilistic deep model.

We note that it is not possible to make a direct apples-to-apples comparison between our model and RULSTM [15], as the two models used vastly different training signals. We refer readers to the supplementary materials for a detailed experiment setup comparison. In terms of performance, our model is comparable to RULSTM without using any side information for inference. When using additional object stream during inference as in RULSTM, our model outperforms RULSTM by a relative improvement of **7%/22%** on seen/unseen set. More

Table 2. Ablation study for action anticipation. We compare our model with backbone I3D network, and further analyze the role of motor attention prediction, interaction hotspots estimation, and stochastic units in joint modelling. See discussions in Sect. 4.3.

Method	EGTEA			Epic-Kitchens		
	Top1 accuracy/Mean Cls accuracy			Top1 accuracy/Top5 accuracy		
	Verb	Noun	Action	Verb	Noun	Action
I3D-Res50	48.01/31.25	42.11/30.01	34.82/23.20	30.06/76.86	16.07/41.67	9.60/24.29
JointDet	48.58/32.21	43.95/31.26	35.69/23.59	30.16/**76.86**	16.25/41.71	9.76/24.40
Hotspots Only	47.95/31.94	44.02/32.53	35.50/23.82	30.21/75.93	16.57/42.28	9.66/24.33
Motor Only	**49.35**/32.34	**45.69/33.93**	36.49/25.13	30.63/76.69	17.28/42.56	10.21/25.32
Ours	48.96/**32.48**	45.50/32.73	**36.60/25.30**	**30.65**/76.53	**17.40/42.60**	**10.38/25.48**

Table 3. Ablation study for interaction hotspots estimation. Jointly modeling motor attention with stochastic units can greatly benefit the performance of interaction hotspots estimation. (↑/↓ indicates higher/lower is better) See discussions in Sect. 4.3.

Method	EGTEA				Epic-Kitchens			
	Prec ↑	Recall ↑	F1 ↑	KLD ↓	Prec ↑	Recall ↑	F1 ↑	KLD ↓
I3DHeatmap	12.82	37.53	19.11	2.66	17.20	77.39	28.15	3.07
JointDet	16.11	41.82	23.26	1.84	17.32	85.79	28.83	2.21
Ours	**17.43**	**48.81**	**25.69**	**1.62**	**17.86**	**86.59**	**29.60**	**1.99**

importantly, our model also provides the additional capabilities of predicting future hand trajectories and estimating interaction hotspots.

4.3 Ablation Study

We present ablation studies of our model. We introduce our experiment setup, evaluate each component of our model, and then contrast our method to a series of baselines on motor attention prediction and interaction hotspot estimation

Experiment Setup. For all of our ablation studies, we adopt the lightweight I3D-Res50 [60] as backbone network to reduce computational cost. Our model is evaluated for action anticipation, motor attention prediction and interaction hotspots estimation across EGTEA (using split1) and EPIC-Kitchens (using the train/val split from [15]). Specifically, we consider the following metrics.

- **Action Anticipation**. We report Top1/Mean Class accuracy on EGTEA as in [34] and Top1/Top5 accuracy as on EPIC-Kitchens following [15].
- **Interaction Hotspots Estimation**. We report F1 score as in [32] and KL-Divergence (KLD) as in [39] using a downsampled heatmap (32x) at the last observable frame.
- **Motor Attention Prediction**. We report the average and final displacement errors between the most confident location on a predicted attention map and the ground-truth hand points, similar to previous work on trajectory prediction [3]. Note that the motor attention maps is downsampled by a factor of 32/8 in space/time. Hence, we report displacement errors normalized in spatial and temporal dimension.

Benefits of Joint Modeling. As a starting point, we compare our model with a backbone I3D-Res50 model. We present the results of action anticipation in Table 2. In comparison to I3D-Res50, our model improves noun and action prediction by +3.4%/1.8% on EGTEA and +1.3%/0.8% on EPIC-Kitchens. Moreover, we show that our model improves the performance of interaction hotspots estimation. We consider the baseline I3D model that only estimates interaction region with interaction hotspots module as I3DHeatmap. As shown in Table 3, our model improves the F1 score by 6.6%/1.5% on EGTEA/EPIC-Kitchens.

Stochastic Modeling vs. Deterministic Modeling. We further evaluate the benefits of probabilistic modeling of motor attention and interaction hotspots. To this end, we compare our model with a deterministic joint model (*JointDet*). JointDet has the same architecture as our model, except for the stochastic units. As shown in Table 2, JointDet slightly improve the I3D baseline for action anticipation (+0.87% on EGTEA and +0.16% on EPIC-Kitchens), yet lags behind our probabilistic model. Specifically, our model outperforms JointDet by 0.91% and 0.62% on EGTEA and EPIC-Kitchens. Moreover, in comparison to JointDet, our model has better performance for interaction hotspots estimation (+2.4%/+0.8% in F1 scores on EGTEA/EPIC-Kitchens). These results suggest that simply training with extra annotations might fail to capture the uncertainty of visual anticipation. In contrast, our design choice of probabilistic modeling can effectively deal with those uncertainty, therefore helps to improve the performance of joint modeling.

Motor Attention vs. Interaction Hotspots. Furthermore, we evaluate the contributions of motor attention and interaction hotspots for FPV action anticipation. We consider two baseline models in Table 3: I3D model equipped with only motor attention module (*Motor Only*), and I3D model equipped with only interaction hotspots module (*Hotspots Only*). Both models underperform the full model across the two datasets, yet the gap between *Motor Only* and the full model is smaller. These results suggest that both components contribute to the performance boost of action anticipation, yet the modeling of motor attention weights more than the modeling of interaction hotspots.

Interaction Hotspots Estimation. We present additional results on interaction hotspots estimation. We compare our results to the following baselines.

- **Center Prior** represents a Gaussian Distribution at the center of the image.
- **Grad-Cam** uses the same I3D backbone network as our model, and produces a saliency map via Grad-Cam [48].
- **EgoGaze** considers possible gaze position as salient region of a given image. This model is trained on eye fixation annotation from EGTEA-Gaze+ [23]. The assumption is that the person is likely to look at the interaction hotspots.
- **DSS Saliency** predicts salient region during human object interaction. This model is trained on pixel-level saliency annotation from [22].
- **EgoHotspots** is the latest work [39] for estimating interaction hotspots.

Our results are shown in Table 4. Our model outperforms the best baselines (EgoGaze and EgoHotspots) by 5.4% on EGTEA and 3.6% on EPIC-Kitchens in F1 scores. These results suggest that our proposed joint model can effectively identify future interaction region. Another observation is that our model performs better on EPIC-Kitchens than EGTEA. This is probably due to the larger number of available training samples.

Table 4. Interaction hotspots estimation results on EGTEA and EPIC-Kitchens. Our model outperforms a set of strong baselines. (↑/↓ indicates higher/lower is better)

Method	EGTEA				Epic-Kitchens			
	Prec↑	Recall↑	F1↑	KLD↓	Prec↑	Recall↑	F1↑	KLD↓
Center prior	10.87	17.65	13.45	10.64	11.66	16.97	13.82	10.27
Grad-Cam [48]	9.98	22.13	13.76	8.73	10.85	20.01	14.07	8.06
DSS [22]	9.02	39.49	14.69	6.12	12.03	33.75	17.74	5.21
EgoGaze [23]	15.02	31.34	20.31	3.20	11.30	27.65	16.05	3.37
EgoHotspots [39]	16.51	24.07	19.59	3.36	**22.26**	31.37	26.04	2.84
Ours	**17.43**	**48.81**	**25.69**	**1.62**	17.86	**86.5**	**29.6**	**1.99**

Table 5. Motor attention prediction results on EGTEA. Our model compares favourably to strong baselines. (↑/↓ indicates higher/lower is better)

Method	Avg. Disp. Error ↓	Final Disp. Error ↓
Kalman filter	0.32	0.48
GPR	0.29	0.37
LSTM	**0.22**	**0.35**
Ours	0.23	0.36

Motor Attention Prediction. We report our results on motor attention prediction. We consider the following baselines and only report results on EGTEA, as the future hand position on EPIC-Kitchens is not accurate (see Sect. 4.1).

- **Kalman Filter** describes the hand trajectory prediction problem with state-space model, and assumes linear acceleration during update step.
- **Gaussian Process Regression (GPR)** iteratively predicts the future hand position using Gaussian Process Regression.
- **LSTM** adopts a vanilla LSTM network for trajectory forecasting. We use the implementation from [3].

The results are presented in Table 5. Our model outperforms Kalman filter and GPR, yet is slightly worse than LSTM model (+0.01 in both errors). Note that all baseline methods need the coordinate of the first observed hand for prediction. This simplifies trajectory prediction into a less challenging regression problem. In contrast, our model does not need hand coordinates for inference. A model that relies on the observation of hand positions will encounter failure cases when the hand has not been observed, while our model is still capable of "imagining" the possible hand trajectory. See "Operate Microwave" and "Wash Coffee Cup" in Fig. 3 for example results from our model.

Fig. 3. Visualization of motor attention (left image), interaction hotspots (right image), and action labels (captions above the images) on sample frames from EGTEA (first row) and EPIC-Kitchens (second row). Both successful (green label) and failure cases (red label) are shown. Future hands position are predicted at every 8 frames and plotted on the last observable frame with the order of yellow, green, cyan, and magenta. (Color figure online)

Visualization of Motor Attention and Interaction Hotspots. Finally, we visualize the predicted motor attention, interaction hotspots, and action labels from our model in Fig. 3. The predicted motor attention almost always attends to the predicted objects and corresponding interaction hotspots. Hence, our model can address challenging cases where next-active objects are ambiguous. Take the first example of "Operate Stove" in Fig. 3. Our model successfully predicted the future objects and estimated the interaction hotspots as the stove control knob.

4.4 Remarks and Discussion

We must also point out that our method has certain limitations, which point to exciting future research directions. For example, our model requires additional annotations for training, which might bring scalability issues when analyzing other datasets. These dense annotations can indeed be approximated using sparsely annotated frames as discussed in Sect. 4.1. We speculate that more advanced hand tracking and object segmentation models can be explored to generating the pseudo ground truth of motor attention and interaction hotspots. Moreover, our model shares a similar conundrum faced by previous work on anticipation. Our model is likely to fail when future active objects are not observed. See "Close Fridge Drawer" and "Put Coffee Maker" in Fig. 3. We conjecture that these cases requires incorporating logical reasoning into learning based methods—an active research topic in our community.

5 Conclusions

We presented the first deep model that jointly predicts motor attention, interaction hotspots, and future action labels in FPV. Importantly, we demonstrated that motor attention plays an important role in forecasting human-object interactions. Another key insight is that characterizing motor attention and interaction hotspots as probabilistic variables can account for the stochastic pattern of human intentional movement. We believe that our model provides a solid step towards the challenging problem of visual anticipation.

Acknowledgments. Portions of this research were supported in part by National Science Foundation Award 1936970 and a gift from Facebook. YL acknowledges the support from the Wisconsin Alumni Research Foundation.

References

1. Aglioti, S.M., Cesari, P., Romani, M., Urgesi, C.: Action anticipation and motor resonance in elite basketball players. Nat. Neurosci. **11**(9), 1109 (2008)
2. Aksan, E., Kaufmann, M., Hilliges, O.: Structured prediction helps 3D human motion modelling. In: ICCV (2019)
3. Alahi, A., Goel, K., Ramanathan, V., Robicquet, A., Fei-Fei, L., Savarese, S.: Social LSTM: human trajectory prediction in crowded spaces. In: CVPR (2016)

4. Carreira, J., Zisserman, A.: Quo vadis, action recognition? A new model and the kinetics dataset. In: CVPR (2017)
5. Chen, C.Y., Grauman, K.: Subjects and their objects: localizing interactees for a person-centric view of importance. Int. J. Comput. Vision **126**(2–4), 292–313 (2018). https://doi.org/10.1007/s11263-016-0958-6
6. Damen, D., et al.: Scaling egocentric vision: the epic-kitchens dataset. In: Ferrari, V., Hebert, M., Sminchisescu, C., Weiss, Y. (eds.) ECCV 2018. LNCS, vol. 11208, pp. 753–771. Springer, Cham (2018). https://doi.org/10.1007/978-3-030-01225-0_44
7. Delaitre, V., Fouhey, D.F., Laptev, I., Sivic, J., Gupta, A., Efros, A.A.: Scene semantics from long-term observation of people. In: Fitzgibbon, A., Lazebnik, S., Perona, P., Sato, Y., Schmid, C. (eds.) ECCV 2012. LNCS, vol. 7577, pp. 284–298. Springer, Heidelberg (2012). https://doi.org/10.1007/978-3-642-33783-3_21
8. di Pellegrino, G., Fadiga, L., Fogassi, L., Gallese, V., Rizzolatti, G.: Understanding motor events: a neurophysiological study. Exp. Brain Res. **91**, 176–180 (1992). https://doi.org/10.1007/BF00230027
9. Fang, K., Wu, T.L., Yang, D., Savarese, S., Lim, J.J.: Demo2Vec: reasoning object affordances from online videos. In: CVPR (2018)
10. Fathi, A., Farhadi, A., Rehg, J.M.: Understanding egocentric activities. In: ICCV (2011)
11. Felsen, P., Agrawal, P., Malik, J.: What will happen next? Forecasting player moves in sports videos. In: ICCV (2017)
12. Fragkiadaki, K., Levine, S., Felsen, P., Malik, J.: Recurrent network models for human dynamics. In: ICCV (2015)
13. Furnari, A., Battiato, S., Grauman, K., Farinella, G.M.: Next-active-object prediction from egocentric videos. J. Vis. Commun. Image Represent. **49**, 401–411 (2017)
14. Furnari, A., Battiato, S., Farinella, G.M.: Leveraging uncertainty to rethink loss functions and evaluation measures for egocentric action anticipation. In: Leal-Taixé, L., Roth, S. (eds.) ECCV 2018. LNCS, vol. 11133, pp. 389–405. Springer, Cham (2019). https://doi.org/10.1007/978-3-030-11021-5_24
15. Furnari, A., Farinella, G.M.: What would you expect? Anticipating egocentric actions with rolling-unrolling LSTMs and modality attention. In: ICCV (2019)
16. Gao, J., Yang, Z., Nevatia, R.: Red: reinforced encoder-decoder networks for action anticipation. In: BMVC (2017)
17. Ghadiyaram, D., Tran, D., Mahajan, D.: Large-scale weakly-supervised pre-training for video action recognition. In: CVPR (2019)
18. Grabner, H., Gall, J., Van Gool, L.: What makes a chair a chair? In: CVPR (2011)
19. Gui, L.-Y., Wang, Y.-X., Liang, X., Moura, J.M.F.: Adversarial geometry-aware human motion prediction. In: Ferrari, V., Hebert, M., Sminchisescu, C., Weiss, Y. (eds.) ECCV 2018. LNCS, vol. 11208, pp. 823–842. Springer, Cham (2018). https://doi.org/10.1007/978-3-030-01225-0_48
20. Hari, R., Forss, N., Avikainen, S., Kirveskari, E., Salenius, S., Rizzolatti, G.: Activation of human primary motor cortex during action observation: a neuromagnetic study. Proc. Natl. Acad. Sci. **95**(25), 15061–15065 (1998)
21. He, K., Zhang, X., Ren, S., Sun, J.: Deep residual learning for image recognition. In: CVPR (2016)
22. Hou, Q., Cheng, M.M., Hu, X., Borji, A., Tu, Z., Torr, P.: Deeply supervised salient object detection with short connections. In: CVPR (2017)

23. Huang, Y., Cai, M., Li, Z., Sato, Y.: Predicting gaze in egocentric video by learning task-dependent attention transition. In: Ferrari, V., Hebert, M., Sminchisescu, C., Weiss, Y. (eds.) ECCV 2018. LNCS, vol. 11208, pp. 789–804. Springer, Cham (2018). https://doi.org/10.1007/978-3-030-01225-0_46
24. Ioffe, S., Szegedy, C.: Batch normalization: accelerating deep network training by reducing internal covariate shift. In: ICML (2015)
25. James, W., Burkhardt, F., Bowers, F., Skrupskelis, I.K.: The Principles of Psychology, vol. 1. Macmillan, London (1890)
26. Jang, E., Gu, S., Poole, B.: Categorical reparameterization with Gumbel-Softmax. In: ICLR (2017)
27. Kataoka, H., Miyashita, Y., Hayashi, M., Iwata, K., Satoh, Y.: Recognition of transitional action for short-term action prediction using discriminative temporal CNN feature. In: BMVC (2016)
28. Ke, Q., Fritz, M., Schiele, B.: Time-conditioned action anticipation in one shot. In: CVPR (2019)
29. Kitani, K.M., Ziebart, B.D., Bagnell, J.A., Hebert, M.: Activity forecasting. In: Fitzgibbon, A., Lazebnik, S., Perona, P., Sato, Y., Schmid, C. (eds.) ECCV 2012. LNCS, vol. 7575, pp. 201–214. Springer, Heidelberg (2012). https://doi.org/10.1007/978-3-642-33765-9_15
30. Kong, Y., Fu, Y.: Human action recognition and prediction: a survey. arXiv preprint arXiv:1806.11230 (2018)
31. Koppula, H.S., Saxena, A.: Anticipating human activities using object affordances for reactive robotic response. IEEE Trans. Pattern Anal. Mach. Intell. **38**(1), 14–29 (2015)
32. Li, Y., Liu, M., Rehg, J.M.: In the eye of beholder: joint learning of gaze and actions in first person video. In: Ferrari, V., Hebert, M., Sminchisescu, C., Weiss, Y. (eds.) ECCV 2018. LNCS, vol. 11209, pp. 639–655. Springer, Cham (2018). https://doi.org/10.1007/978-3-030-01228-1_38
33. Li, Y., Liu, M., Rehg, J.M.: In the eye of the beholder: gaze and actions in first person video. arXiv preprint arXiv:2006.00626 (2020)
34. Li, Y., Ye, Z., Rehg, J.M.: Delving into egocentric actions. In: CVPR (2015)
35. Long, J., Shelhamer, E., Darrell, T.: Fully convolutional networks for semantic segmentation. In: CVPR (2015)
36. Ma, M., Fan, H., Kitani, K.M.: Going deeper into first-person activity recognition. In: CVPR (2016)
37. Maddison, C.J., Mnih, A., Teh, Y.W.: The concrete distribution: a continuous relaxation of discrete random variables. In: ICLR (2017)
38. Miech, A., Laptev, I., Sivic, J., Wang, H., Torresani, L., Tran, D.: Leveraging the present to anticipate the future in videos. In: CVPR Workshops (2019)
39. Nagarajan, T., Feichtenhofer, C., Grauman, K.: Grounded human-object interaction hotspots from video. In: ICCV (2019)
40. Pavlovic, V., Rehg, J.M., MacCormick, J.: Learning switching linear models of human motion. In: Leen, T.K., Dietterich, T.G., Tresp, V. (eds.) NeurIPS, pp. 981–987. MIT Press, Cambridge (2001)
41. Pirsiavash, H., Ramanan, D.: Detecting activities of daily living in first-person camera views. In: CVPR (2012)
42. Poleg, Y., Ephrat, A., Peleg, S., Arora, C.: Compact CNN for indexing egocentric videos. In: WACV (2016)
43. Rhinehart, N., Kitani, K.M.: Learning action maps of large environments via first-person vision. In: CVPR (2016)

44. Rhinehart, N., Kitani, K.M.: First-person activity forecasting with online inverse reinforcement learning. In: ICCV (2017)
45. Rushworth, M., Johansen-Berg, H., Göbel, S.M., Devlin, J.: The left parietal and premotor cortices: motor attention and selection. Neuroimage **20**, S89–S100 (2003)
46. Ryoo, M.S., Rothrock, B., Matthies, L.: Pooled motion features for first-person videos. In: CVPR (2015)
47. Ryoo, M., Fuchs, T.J., Xia, L., Aggarwal, J.K., Matthies, L.: Robot-centric activity prediction from first-person videos: what will they do to me? In: HRI (2015)
48. Selvaraju, R.R., Cogswell, M., Das, A., Vedantam, R., Parikh, D., Batra, D.: Grad-CAM: visual explanations from deep networks via gradient-based localization. In: ICCV (2017)
49. Shen, Y., Ni, B., Li, Z., Zhuang, N.: Egocentric activity prediction via event modulated attention. In: Ferrari, V., Hebert, M., Sminchisescu, C., Weiss, Y. (eds.) ECCV 2018. LNCS, vol. 11206, pp. 202–217. Springer, Cham (2018). https://doi.org/10.1007/978-3-030-01216-8_13
50. Simonyan, K., Zisserman, A.: Very deep convolutional networks for large-scale image recognition. In: ICLR (2015)
51. Soo Park, H., Hwang, J.J., Niu, Y., Shi, J.: Egocentric future localization. In: CVPR (2016)
52. Soran, B., Farhadi, A., Shapiro, L.: Generating notifications for missing actions: don't forget to turn the lights off! In: ICCV (2015)
53. Thermos, S., Papadopoulos, G.T., Daras, P., Potamianos, G.: Deep affordance-grounded sensorimotor object recognition. In: CVPR (2017)
54. Tran, D., Wang, H., Torresani, L., Feiszli, M.: Video classification with channel-separated convolutional networks. In: ICCV (2019)
55. Urtasun, R., Fleet, D.J., Geiger, A., Popović, J., Darrell, T.J., Lawrence, N.D.: Topologically-constrained latent variable models. In: ICML (2008)
56. Vondrick, C., Pirsiavash, H., Torralba, A.: Anticipating visual representations from unlabeled video. In: CVPR (2016)
57. Walker, J., Marino, K., Gupta, A., Hebert, M.: The pose knows: video forecasting by generating pose futures. In: ICCV (2017)
58. Wang, J.M., Fleet, D.J., Hertzmann, A.: Gaussian process dynamical models for human motion. IEEE Trans. Pattern Anal. Mach. Intell. **30**(2), 283–298 (2007)
59. Wang, X., Girdhar, R., Gupta, A.: Binge watching: scaling affordance learning from sitcoms. In: CVPR (2017)
60. Wang, X., Girshick, R., Gupta, A., He, K.: Non-local neural networks. In: CVPR (2018)
61. Wei, P., Xie, D., Zheng, N., Zhu, S.C.: Inferring human attention by learning latent intentions. In: IJCAI (2017)
62. Yagi, T., Mangalam, K., Yonetani, R., Sato, Y.: Future person localization in first-person videos. In: CVPR (2018)
63. Zhou, Y., Ni, B., Hong, R., Yang, X., Tian, Q.: Cascaded interactional targeting network for egocentric video analysis. In: CVPR (2016)

Learning Stereo from Single Images

Jamie Watson[1(✉)], Oisin Mac Aodha[2], Daniyar Turmukhambetov[1], Gabriel J. Brostow[1,3], and Michael Firman[1]

[1] Niantic, San Francisco, USA
jwatson@nianticlabs.com
[2] University of Edinburgh, Edinburgh, UK
[3] UCL, London, UK

Abstract. Supervised deep networks are among the best methods for finding correspondences in stereo image pairs. Like all supervised approaches, these networks require ground truth data during training. However, collecting large quantities of accurate dense correspondence data is very challenging. We propose that it is unnecessary to have such a high reliance on ground truth depths or even corresponding stereo pairs. Inspired by recent progress in monocular depth estimation, we generate plausible disparity maps from single images. In turn, we use those flawed disparity maps in a carefully designed pipeline to generate stereo training pairs. Training in this manner makes it possible to convert any collection of single RGB images into stereo training data. This results in a significant reduction in human effort, with no need to collect real depths or to hand-design synthetic data. We can consequently train a stereo matching network from scratch on datasets like COCO, which were previously hard to exploit for stereo. Through extensive experiments we show that our approach outperforms stereo networks trained with standard synthetic datasets, when evaluated on KITTI, ETH3D, and Middlebury. Code to reproduce our results is available at https://github.com/nianticlabs/stereo-from-mono/.

Keywords: Stereo matching · Correspondence training data

1 Introduction

Given a pair of scanline rectified images, the goal of stereo matching is to estimate the per-pixel horizontal displacement (i.e. disparity) between the corresponding location of every pixel from the first view to the second, or vice versa. This is typically framed as a matching problem, where the current best performance is achieved by deep stereo networks e.g. [5,25,73,76]. These networks rely on expensive to obtain ground truth correspondence training data, for example captured with LiDAR scanners [20,21,52]. Synthetic data is a common alternative; given

Electronic supplementary material The online version of this chapter (https://doi.org/10.1007/978-3-030-58452-8_42) contains supplementary material, which is available to authorized users.

A. Vedaldi et al. (Eds.): ECCV 2020, LNCS 12346, pp. 722–740, 2020.
https://doi.org/10.1007/978-3-030-58452-8_42

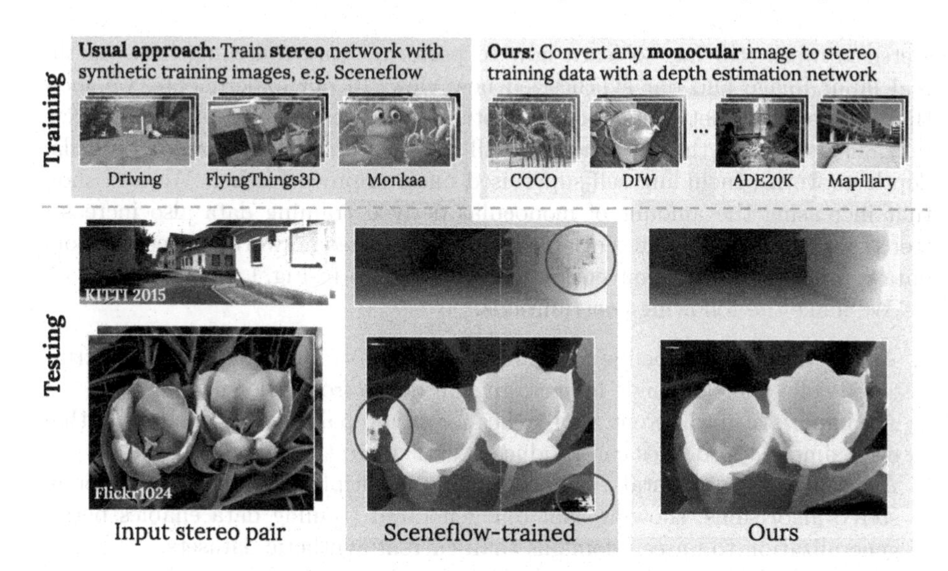

Fig. 1. Training stereo correspondence from monocular data. Deep stereo networks are typically trained on synthetic data, with per-dataset finetuning when ground truth disparity is available. We instead train on real images, which we convert to tuples of stereo training data using off-the-shelf monocular depth estimation. Our approach offers better generalization to new datasets, fewer visual artifacts, and improved scores.

a collection of 3D scenes, large quantities of artificially rendered stereo pairs can be created with ground truth disparities [4,14,18,43]. Deployment in novel *real* scenes currently hinges on finetuning with additional correspondence data from that target domain. Assuming access to stereo pairs, a promising alternative could be to do self-supervised finetuning using image reconstruction losses and no ground truth correspondence. Unfortunately, self-supervised performance is still short of supervised methods for real domains [57,59,80].

We tackle the problem of generating training data for deep stereo matching networks. Recent works have shown that pretraining on large amounts of noisily labeled data improves performance on image classification [41,55,62,71]. A similar approach has not, to the best of our knowledge, been applied to training stereo matching networks. We introduce a fully automatic stereo data synthesis approach which only requires a collection of single images as input. In doing so, we open up the data available for deep stereo matching to any 2D image. Our approach uses a monocular depth estimation network to predict a disparity map for each training image. We refine this initial prediction before using it to synthesize a corresponding stereo pair from a new viewpoint, taking care to address issues associated with occlusions and collisions. Using natural images, as opposed to synthetic ones, ensures that the stereo network sees *realistic* textures and layouts, with *plausible* disparity maps during training. We show that networks trained from monocular data, or even low quality depth, provide an informative signal when generating training data. This is because the stereo network still has to learn how to perform dense matching between the real input image and our synthesized new view. In our experiments, we show that progressively better monocular depth yields gains in stereo matching accuracy. This suggests that our method will benefit 'for free' as future monocular depth networks, including self-supervised ones, improve further. We also show that increasing the amount of monocular derived training data also increases stereo performance. We present an example of a stereo network trained with our data compared to using conventional synthetic data in Fig. 1.

We make the following contributions:

1. A fully automatic pipeline for generating stereo training data from unstructured collections of single images, given a depth-from-color model.
2. A comparison to different approaches for stereo pair synthesis, showing that even simple approaches can produce useful stereo training data.
3. A detailed experimental evaluation across multiple domains with different stereo algorithms, showing that our generated training data enables better generalization to unseen domains than current synthetic datasets.

2 Related Work

Stereo Matching—Traditionally, matching between stereo image pairs was posed as an energy minimization problem consisting of separate, hand-designed, appearance matching and data smoothness terms e.g. [3,26,69]. [33,75] instead learned the appearance matching term while still retaining conventional stereo methods for the additional processing. Subsequently, fully end-to-end approaches replaced the entire conventional stereo matching pipeline [31,43]. Recent advances have focused on improving stereo network architectures and training losses. This has included innovations like spatial pyramids [5] and more sophisticated approaches for regularizing the generated cost volume [8,76,77]. In this work, we propose an approach for generating training data for deep stereo matching algorithms that is agnostic to the choice of stereo network. Our experiments across multiple different stereo networks show consistent improvements, independent of the underlying architecture used.

Just as recent works have adapted stereo models to help improve monocular depth estimation [24,40,60,68], in our work we leverage monocular models to improve stereo. This is related to concurrent work [1], where a monocular completion network is utilized to help self-supervised stereo. However, in contrast to these approaches, we do not require stereo pairs or ground truth disparity for data synthesis, but instead use monocular depth networks to generate training data for stereo networks.

Stereo Training Data—While deep stereo algorithms have proven very successful, the best performing methods are still fully supervised and thus require large quantities of stereo pairs with *ground truth* correspondence data at training time [21,52]. Acquiring this data for real world scenes typically involves specially designed depth sensing hardware, coupled with conventional intensity cameras for capturing scene appearance e.g. [20,21,52]. In addition to currently being laborious and expensive, these commonly used laser-based depth sensors are limited in the types of scenes they can operate in e.g. no direct sunlight.

Other potential sources of stereo data include static stereo cameras [70], internet videos [63], 3D movies [48], and multi-view images [37]. Crucially however, these images do not come with ground truth correspondence. Estimating disparity for these images is especially challenging due to issues such as negative disparity values in the case of 3D movies and extreme appearance changes over time in multi-view image collections. These issues make it difficult to apply conventional stereo algorithms to process the data. It also poses a 'chicken and egg' problem, as we first need accurate stereo to extract reliable correspondences.

Synthetically rendered correspondence data is an attractive alternative as it leverages 3D graphics advances in place of specialized hardware. One of the first large-scale synthetic datasets for stereo matching consisted of simple 3D primitives floating in space [43]. Other synthetic datasets have been proposed to better match the statistics of real world domains, such as driving scenes [4,14,18] or indoor man-made environments [36,65]. With synthetic data, it is also possible to vary properties of the cameras viewing the scene (e.g. the focal length) so the trained models can be more robust to these variations at test time [77].

While dense correspondence networks can be trained from unrealistic synthetic data [13,42], the best performance is still achieved when synthetic training data is similar in appearance and depth to the target domain [77], and most networks still finetune on stereo pairs from the target domain for state-of-the-art performance e.g. [5,76,77]. This suggests that the goal of general stereo matching by training with just synthetic data has not yet been reached. It is possible to simulate stereo training data that matches the target domain, e.g. [14,60,77]. However, constructing varied scenes with realistic shape and appearance requires significant manual effort, and can be expensive to render. We perform experiments that show that our approach to generating training data generalizes better to new scenes, while requiring no manual effort to create.

Image Augmentation Based Training Data—An alternative to using computer generated synthetic data is to synthesize images by augmenting natural training images with pre-defined and known pixel-level transformations. The advantage of this approach is that it mostly retains the realistic appearance information of the input images. One simple method is to apply a random homography to an input image to create its corresponding matching pair [12]. Alternative methods include more sophisticated transformations such as pasting one image, or segmented foreground object, on top of another background image [13,29,42], random non-rigid deformations of foreground objects [30,34,56], and using supervised depth and normals to augment individual objects [2].

Augmentation-based approaches are largely heuristic and do not produce realistic correspondence data due to their inability to model plausible occlusions and depth. In contrast, we use off-the-shelf networks trained for monocular depth estimation to guide our image synthesis, so depths and occlusions are plausible. We perform experimental comparisons with existing image augmentation approaches and show superior performance with our method.

Domain Adaptation for Stereo—There is a large body of work that explores the problem of domain adaptation in stereo by addressing the domain gap between synthetic training data and real images. Existing approaches use additional supervision in the form of highly confident correspondences [57], apply iterative refinement [46], meta-learning [58], consistency checks [78,80], occlusion reasoning [35], or temporal information [59,66]. In the context of fully supervised adaptation, [77] proposed a novel domain normalization layer to account for statistical differences between different data domains (Fig. 2).

To address the domain adaptation problem we propose a method for automatically synthesizing large quantities of diverse stereo training data from single image collections. We show that our approach results in a large performance increase when testing on domains that have not been seen at all during training.

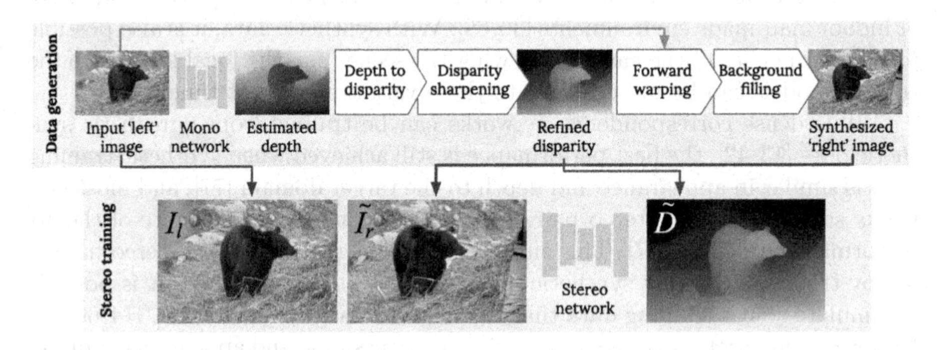

Fig. 2. Overview of our data generation approach. We use off-the-shelf monocular depth estimation to synthesize stereo training pairs from single images. We convert estimated depth to a sharpened disparity map, and then forward warp pixels from the input left image I_l to create a plausible and geometrically accurate right image $\tilde{I}_r$.

3 Method

Supervised training of deep stereo networks requires rectified pairs of color images I_l and I_r, together with a ground truth disparity map D.[1] We propose a method to create plausible tuples of training data $(I_l, \tilde{I}_r, \tilde{D})$ from collections of individual and unrelated input images. We use a pretrained monocular depth network to estimate a plausible depth map Z for an image I_l, which we then use to synthesize $\tilde{I}_r$ and $\tilde{D}$. This section describes in detail the important design decisions required at each step of this process. Our goal here is similar to works that synthesize novel views of a single image e.g. [9,82]. However, our aim is to

[1] For simplicity of notation we assume that disparity and depth maps are all aligned to the left input image I_l.

synthesize a new view together with a geometrically correct depth map aligned with the input image, which is non-trivial to obtain from existing methods. We choose a simple and efficient algorithm for this based on pixel manipulation, which we found to be surprisingly effective.

3.1 Stereo Training Data from Monocular Depth

At test time, a monocular depth estimation network g takes as input a single color image, and for each pixel estimates the depth Z to the camera,

$$Z = g(I). \tag{1}$$

These monocular networks can be parameterized as deep neural networks that are trained with ground truth depth [16] or via self-supervision [19,22,81]. To train our stereo network we need to covert the estimated depth Z to a disparity map $\tilde{D}$. To achieve our aim of generalizable stereo matching, we want to simulate stereo pairs with a wide range of baselines and focal lengths. We achieve this by converting depth to disparity with $\tilde{D} = \frac{s Z_{\max}}{Z}$, where s is a randomly sampled (uniformly from $[d_{\min}, d_{\max}]$) scaling factor which ensures the generated disparities are in a plausible range.

Our goal is to synthesize a right image $\tilde{I}_r$ from the input 'left' image I_l and the predicted disparity $\tilde{D}$. The disparity map and the color image are both aligned to the left image I_l, so backward warping [19,53] is not appropriate. Instead, we use $\tilde{D}$ to synthesize a stereo pair $\tilde{I}_r$ via *forward* warping [53]. For each pixel i in I_l, we translate it $\tilde{D}^i$ pixels to the left, and perform interpolation of the warped pixel values to obtain $\tilde{I}_r$. Due to the nature of forward warping, $\tilde{I}_r$ will contain missing pixel artifacts due to occluded regions, and in some places collisions will occur when multiple pixels will land at the same location. Additionally, monocular networks often incorrectly predict depths around depth discontinuities, resulting in unrealistic depth maps. Handling these issues is key to synthesising realistic stereo pairs, as we demonstrate in our ablation experiments.

3.2 Handling Occlusion and Collisions

Naively synthesizing images based on the estimated disparity will result in two main sources of artifacts: occlusion holes and collisions. Occluded regions are pixels in one image in a stereo pair which are not visible in the corresponding image [64]. In the forward warping process, pixels in $\tilde{I}_r$ with no match in I_l manifest themselves as *holes* in the reconstructed image $\tilde{I}_r$. While the synthesized image gives an accurate stereo pair for I_l, the blank regions result in unnatural artifacts. We can increase the realism of the synthesized image by filling the missing regions with a texture from a randomly selected image I_b from our training set, following [15]. We perform color transfer between I_l and I_b using [49] to obtain $\hat{I}_b$. We subsequently set all missing pixels in $\tilde{I}_r$ to the values of $\hat{I}_b$ at the corresponding positions. This results in textures from natural images being copied into the missing regions of $\tilde{I}_r$. Separately, *collisions* are pixels that appear in I_l but are not in $\tilde{I}_r$, and thus multiple pixels from I_l can map to the same point in $\tilde{I}_r$. For collisions, we select the pixel from I_l which has the greater disparity value since these pixels are closer and should be visible in both images.

3.3 Depth Sharpening

As opposed to realistic sharp discontinuities, monocular depth estimation networks typically produce blurry edges at depth discontinuities [47]. When forward warping during image synthesis this leads to 'flying pixels' [50], where individual pixels are warped to empty regions between two depth surfaces. In Fig. 3 we observe this phenomenon for the monocular depth network of [48]. Not correcting this problem would result in a collection of synthesized training images where the stereo matching algorithm is expected to match pixels in I_l to individual, isolated pixels in $\tilde{I}_r$; a highly unrealistic test-time scenario. To address flying pixels in our synthesized $\tilde{I}_r$ images we perform *depth sharpening*. We identify flying pixels as those for which the disparity map has a Sobel edge filter response of greater than 3 [27,54]. These pixels are then assigned the disparity of the nearest 'non-flying' pixel in image space. We use this corrected disparity map for both image sampling and when training our stereo networks.

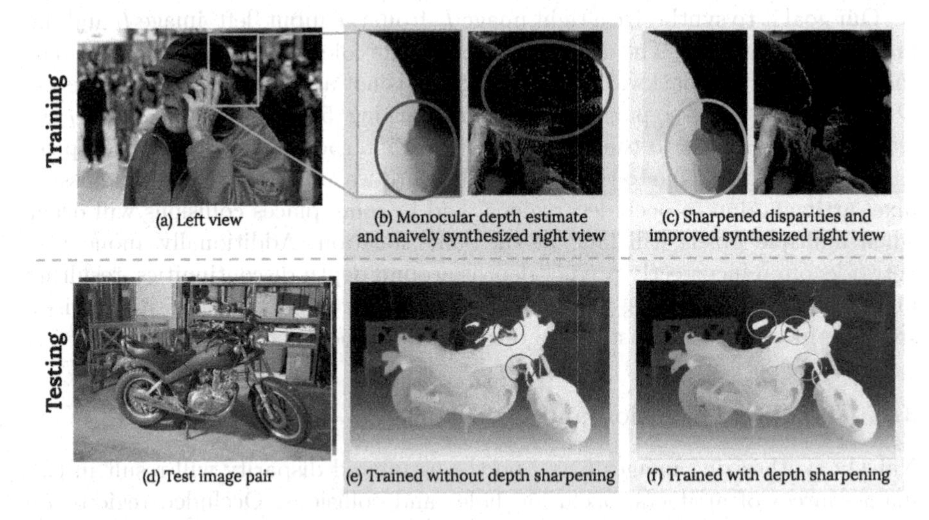

Fig. 3. Depth map sharpening. Depth maps from monocular depth networks typically have 'blurry' edges, which manifest themselves as flying pixels in $\tilde{I}_r$ (b). Our depth sharpening method reduces the effect of flying pixels (c). Models trained with depth sharpening have better scores (Table 4) and fewer visual artifacts; (e) versus (f).

3.4 Implementation Details

Training Details—Unless otherwise stated, we use the widely-used PSMNet hourglass network variant [5] for stereo matching, using the official architecture and training code. We train all models with a batch size of 2 for 250,000 steps, equivalent to 7 epochs of the Sceneflow dataset [43]. With the exception of some baselines, we do not use any synthetic data when training our models. We follow [5] in using a learning rate of 0.001 for all experiments. Unless otherwise stated, we perform monocular depth estimation using MiDaS [48], a network with a ResNeXt-101 [72] backbone, trained on a mixture of monocular and stereo datasets including 3D movies. Their training depths were generated by running a supervised optical flow network [28] on their stereo training pairs.

Our networks are set to predict a maximum disparity of 192. We set $d_{\min} = 50$ and $d_{\max} = 225$, which produces training images with a diverse range of baselines and focal lengths, including stereo pairs with disparity ranges outside the network's modeling range for further robustness to test time scenarios. During training, we take crops of size 608×320. If images are smaller than these dimensions (or significantly larger, e.g. Mapillary [44]), we isotropically resize to match the constraining dimension. Additionally, we follow PSMNet [5] by performing ImageNet [11] normalization on all images.

Training Images—Our method allows us to train using any color images. To maximise the ability of our algorithm to learn to adapt to different test domains, we train models on a combination of varied single image datasets which we call the 'Mono for Stereo' dataset, or **MfS**. MfS comprises all the training images (without depth or semantic labels) from COCO 2017 [39], Mapillary Vistas [44], ADE20K [79], Depth in the Wild [6], and DIODE [61] (see Fig. 1 for examples). This results in 597,727 training images, an order of magnitude more than Sceneflow's 35,454 training pairs, from which we use a random subset of 500,000, unless stated otherwise.

Color Augmentation—Cameras in a stereo pair are typically nominally identical. However, each image in a pair may have slightly different white balancing, lens geometry, glare, and motion blur. To account for these differences when training stereo networks we augment $\tilde{I}_r$ with pixel-level noise with standard deviation 0.05, and we randomly adjust contrast, brightness, saturation, and hue with ranges of ± 0.2, ± 0.2, ± 0.2, and ± 0.01. Finally, with 50% probability, we add Gaussian blur to $\tilde{I}_r$ with kernel size $\sigma \sim \text{Unif}[0, 1]$. Interestingly, we observe that these simple augmentations also significantly improve the performance of the baseline stereo models trained on Sceneflow [43] alone.

4 Experiments

In this section we present experimental results to validate our approach to stereo data generation. Our experiments show that:

1. Our approach produces better and more general stereo matching networks compared to training on alternative sources of data, such as synthetic scenes
2. We are robust to errors in monocular depth predictions, and we perform well regardless of the source of monocular training data
3. Our design decisions are sensible, validated through ablation of our method
4. Our performance gains are agnostic to the stereo architecture used

5. As the amount of generated data increases, so does stereo performance
6. Finetuning models trained with our approach compared to synthetic data with ground truth disparity results in better performance.

4.1 Evaluation Datasets and Metrics

We evaluate multiple stereo models on a variety of datasets to demonstrate generality of our approach. Our test datasets are: (i) **KITTI 2012** and **2015** [20,21]: real-world driving scenes with LiDAR ground truth, (ii) **ETH3D** [52]: grayscale stereo pairs from indoor and outdoor environments, again with LiDAR ground truth, and (iii) **Middlebury** [51]: high-resolution stereo pairs with structured light ground truth. Middlebury and ETH3D are particularly compelling for demonstrating general-purpose stereo matching as their training data is very limited, with just 15 and 27 pairs respectively. Note that we never train on any of these datasets, with the exception of KITTI finetuning experiments in Sect. 4.6. For Middlebury, we evaluate on the 15 training images provided for the official stereo benchmark. We report results on the full training set for the two view benchmark for ETH3D. For both KITTI 2012 and 2015 we evaluate on the validation set from [5], and provide the image indices in the supplementary material. For all datasets we report end-point-error (EPE) and the % of pixels with predicted disparity within τ pixels of the ground truth e.g. <3px. Unless otherwise stated, we evaluate every pixel for which there is a valid ground truth disparity value (i.e. we include occluded pixels). We only report scores with the most commonly reported value of τ for each dataset. Finally, we show qualitative results on **Flickr1024** [67], a diverse dataset of stereo pairs originally captured by internet users for stereoscopic viewing.

Table 1. Our approach outperforms alternative stereo training data generation methods. Each row represents a single PSMNet trained with a different data synthesis approach, without any dataset-specific finetuning. We can see that simple synthesis methods still produce valuable training data. However, our approach in the bottom row, which incorporates a monocular depth network [48], outperforms even synthetic Sceneflow data. Numbers here are directly comparable to Table 3, showing we beat the baselines no matter what depth network or training data we use.

Synthesis approach	Training data	KITTI '12		KITTI '15		Middlebury		ETH3D	
		EPE	<3px	EPE	<3px	EPE	<2px	EPE	<1px
Affine warps (e.g. [12])	MfS	3.74	14.78	2.33	14.94	21.50	61.19	1.28	24.21
Random pasted shapes (e.g. [42])	MfS	2.45	11.89	1.49	7.57	11.38	40.37	0.77	14.19
Random superpixels	MfS	1.46	6.58	1.31	6.28	11.87	32.70	2.21	16.34
Synthetic	Sceneflow [43]	1.04	5.76	1.19	5.80	9.42	34.99	1.25	13.04
SVSM selection module [40]	MfS	1.04	5.66	1.10	5.47	8.45	31.34	0.57	11.61
Ours	MfS	**0.95**	**4.43**	**1.04**	**4.75**	**6.33**	**26.42**	**0.53**	**8.17**

4.2 Comparison to Alternative Data Generation Methods

In our first experiment, we compare our approach to alternative baseline methods for stereo data generation. To do this, we train a PSMNet stereo network using data generated by each of the baselines and compare the network's stereo estimation performance on three held out datasets. We compare to: (i) **Affine warps**: warps of single images similar to [12], (ii) **Random pasted shapes**: our implementation of the optical flow data generation method of [42], adapted for stereo training, (iii) **Random superpixels**: where we initialise a disparity map as a plane, and assign disparity values drawn uniformly from $[d_{\min}, d_{\max}]$ to randomly selected image superpixels [17], (iv) **Synthetic**: 35K synthetic image pairs from Sceneflow [43], and (v) **SVSM selection module**: our implementation of the selection module from [40], where $\tilde{I}_r$ is the weighted sum of shifted left images, with weightings derived from the depth output of [48]. With the exception of the synthetic baseline, all results use the MfS dataset to synthesize stereo pairs. Detailed descriptions are in the supplementary material. Table 1 shows that our fully automatic approach outperforms all of these methods. Qualitative results are in Figs. 1, 6 and 5.

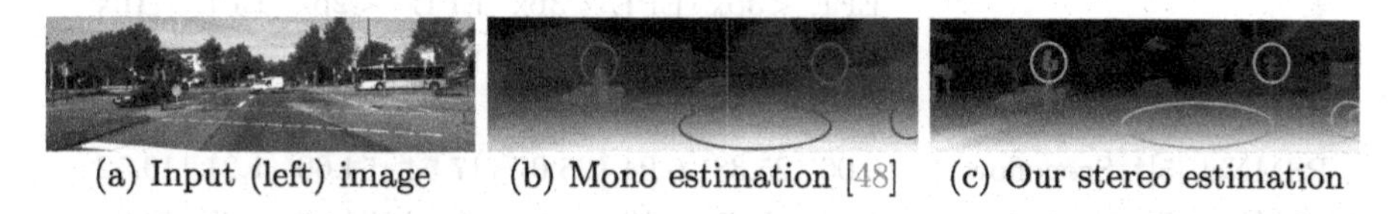

(a) Input (left) image (b) Mono estimation [48] (c) Our stereo estimation

Fig. 4. We can recover from errors in monocular depth estimation. Problems present in monocular depths (b), such as missing objects and uneven ground do not transfer to our eventual stereo predictions (c).

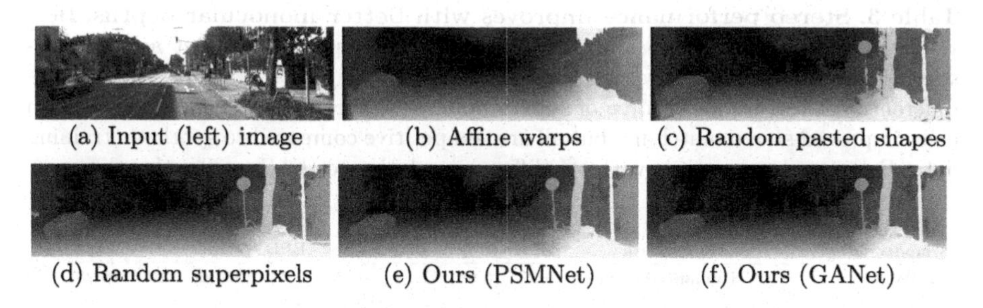

(a) Input (left) image (b) Affine warps (c) Random pasted shapes
(d) Random superpixels (e) Ours (PSMNet) (f) Ours (GANet)

Fig. 5. Comparison to baseline data generation methods. We show that simple image synthesis can be a surprisingly effective method for generating stereo training data e.g. (c) and (d). However, our approach produces sharper and cleaner disparities, whether using PSMNet (e) or the larger GANet (f).

4.3 Model Architecture Ablation

In Table 2 we use our data synthesis approach to train three different stereo matching networks. With the exception of iResNet [38], we used the same architectures and training code provided by the original authors, and trained each model for 250,000 iterations from scratch. For fair comparison, we also trained the same models without our data but instead with Sceneflow using the same color augmentations described in Sect. 3.4. Interestingly, we observe that this results in an improvement over the publicly released model snapshots. We see that our approach consistently outperforms Sceneflow training irrespective of the stereo network used.

Table 2. Our approach is agnostic to stereo model architecture. We compare three diverse stereo architectures, from the lightweight iResnet [38], through to the computationally expensive, but state-of-the-art, GANet [76]. Rows marked with † use the publicly released model weights. We see that our synthesized data (MfS) consistently performs better than training with the synthetic Sceneflow dataset.

Architecture	Training data	KITTI '12		KITTI '15		Middlebury		ETH3D	
		EPE	<3px	EPE	<3px	EPE	<2px	EPE	<1px
iResnet [38]	Sceneflow	1.07	6.40	**1.15**	5.98	12.13	41.65	1.57	17.90
iResnet [38]	MfS	**0.93**	**5.38**	**1.15**	**5.75**	**8.52**	**33.36**	**0.61**	**11.65**
PSMNet [5]†	Sceneflow	6.06	25.38	6.04	25.07	17.54	54.60	1.21	19.52
PSMNet [5]	Sceneflow	1.04	5.76	1.19	5.80	9.42	34.99	1.25	13.04
PSMNet [5]	MfS	**0.95**	**4.43**	**1.04**	**4.75**	**6.33**	**26.42**	**0.53**	**8.17**
GANet [76]†	Sceneflow	1.30	7.97	1.51	9.52	11.88	37.34	0.48	8.19
GANet [76]	Sceneflow	0.96	5.24	1.14	5.43	9.81	32.20	0.48	9.45
GANet [76]	MfS	**0.81**	**4.31**	**1.02**	**4.56**	**5.66**	**24.41**	**0.43**	**6.62**

Table 3. Stereo performance improves with better monocular depths. Here we train PSMNet with MfS and our data synthesis approach without any dataset-specific finetuning. Each row uses a different monocular depth network, each with a different architecture and source of supervision. As expected, better monocular depth gives improved stereo matching, but all are competitive compared to synthetic training data. Below each row we show the % difference relative to the Sceneflow baseline.

Monocular model	Monocular training data	KITTI '12		KITTI '15		Middlebury		ETH3D	
		EPE	<3px	EPE	<3px	EPE	<2px	EPE	<1px
DiW [6]	Human labelling	0.93	5.08	1.13	6.05	9.96	32.96	0.63	12.23
		−10.6	*−11.8*	*−5.0*	***4.3***	***5.7***	*−5.8*	*−49.6*	*−6.2*
Monodepth2 (M) [23]	KITTI Monocular	0.92	5.01	1.10	5.11	9.41	30.76	0.60	12.56
		−11.5	*−13.0*	*−7.6*	*−11.9*	*−0.1*	*−12.1*	*−52.0*	*−3.7*
Megadepth [37]	SfM reconstructions	**0.89**	**4.29**	1.07	4.96	7.66	28.86	0.54	10.41
		−14.4	*−25.5*	*−10.1*	*−14.5*	*−18.7*	*−17.5*	*−56.8*	*−20.2*
MiDaS [48]	3D Movies + others	0.95	4.43	**1.04**	**4.75**	**6.33**	**26.42**	**0.53**	**8.17**
		−8.7	*−23.1*	*−12.6*	*−18.1*	*−32.8*	*−24.5*	*−57.6*	*−37.3*
Sceneflow baseline		1.04	5.76	1.19	5.80	9.42	34.99	1.25	13.04

4.4 Comparing Different Monocular Depth Networks

The majority of our experiments are performed using monocular depth from MiDaS [48], which itself is trained with a variety of different datasets e.g. flow derived pseudo ground truth and structured light depth. In Table 3 we show that our approach still produces competitive stereo performance when we use alternative monocular networks that have been trained using much *weaker* supervision. This includes using monocular video only self-supervision on the KITTI dataset (Monodepth2 (M) [23]), pairwise ordinal human annotations (DiW [6]) and multi-view image collections (Megadepth [37]). It is worth noting that even though Megadepth [37] training depth is computed from unstructured image collections, without using any trained correspondence matching, using it to synthesize training data still outperforms the synthetic Sceneflow baseline. This is likely due to the fact that stereo matching is fundamentally different from monocular depth estimation, and even potentially unrealistic disparity information can help the stereo network learn how to better match pixels. While the recent MiDaS [48] performs best overall, we observe a consistent performance increase in stereo estimation with better monocular depth. This is very encouraging, as it indicates that our performance may only get better in future as monocular depth networks improve. We also show qualitatively in Fig. 4 how our stereo predictions can recover from errors made by monocular models.

4.5 Ablating Components of Our Method

In Table 4 we train stereo networks with data synthesized with and without depth sharpening and background texture filling (i.e. I_b). For the case where we do not use background texture filling we simply leave black pixels. We observe that disabling these components results in worse stereo performance compared to the full pipeline, with Fig. 3 demonstrating the qualitative impact of depth sharpening. This highlights the importance of realistic image synthesis. A more sophisticated warping approach [45] or inpainting with a neural network [74] should increase the realism and is left as future work. We instead opt for a simple warping and background filling approach for computational efficiency, which allows us to generate new views online during training.

Table 4. Depth sharpening and background filling helps. Disabling depth sharpening and background filling hurts PSMNet stereo performance when trained on MfS.

		KITTI '12		KITTI '15		Middlebury		ETH3D	
Sharpening	Background filling	EPE	<3px	EPE	<3px	EPE	<2px	EPE	<1px
✗	✓	1.02	4.75	1.06	4.76	7.57	28.16	**0.47**	8.21
✓	✗	0.96	4.81	1.10	5.27	7.11	27.22	0.68	8.70
✓	✓	**0.95**	**4.43**	**1.04**	**4.75**	**6.33**	**26.42**	0.53	**8.17**

Table 5. KITTI 2015 finetuning. All trained by us on PSMNet. Our data generation approach, when finetuned on KITTI LiDAR data (last row), results in superior performance when compared to Sceneflow pretraining and KITTI finetuning (4th row). †Sceneflow models cannot be finetuned on single KITTI images without our method.

Pretraining	Finetuning method	EPE Noc	<3px Noc	EPE All	<3px All
Sceneflow	None	1.18	5.62	1.19	5.80
Sceneflow	KITTI left images	†	†	†	†
Sceneflow	KITTI self-supervised	1.03	4.90	1.05	5.07
Sceneflow	KITTI LiDAR	0.73	2.21	0.74	2.36
Ours with MfS	None	1.02	4.57	1.04	4.75
Ours with MfS	KITTI left images	1.00	4.44	1.01	4.58
Ours with MfS	KITTI self-supervised	1.00	4.57	1.01	4.71
Ours with MfS	KITTI LiDAR	0.68	2.05	0.69	2.15

4.6 Adapting to the Target Domain

Our method allows us to train a generalizable stereo network which outperforms baselines. However, in cases where we have data from the target domain, e.g. in the form of single unpaired images, stereo pairs, or stereo pairs with ground truth LiDAR. We compare adaptation with each of these data sources in Table 5. First, we train on stereo pairs generated from the **KITTI left images** with our method, using the 29K KITTI images from the 33 scenes not included in the KITTI 2015 stereo training set and monocular predictions from [48]. For **KITTI self-supervised** finetuning, we follow the method proposed in [24] using the stereo image pairs from the same 33 scenes. Finally, for **KITTI LiDAR** we finetune on 160 training images and point clouds (following [5]) for 200 epochs. We give results for unoccluded pixels (*Noc*) and all pixels (*All*); see [5] for more details of the metrics. We outperform Sceneflow pretraining in all cases.

4.7 Varying the Amount of Training Data

In Table 6 we train a PSMNet stereo network by varying the amount of training images from our MfS dataset while keeping the monocular network (MiDaS [48]) fixed. We observe that as the number of training images increases so does stereo performance. Even in the smallest data regime (35K images), which is comparable in size to Sceneflow [43], we still outperform training on the synthetic data alone in all metrics. This indicates that it is not just the power of the monocular network that drives our performance, but the large and varied image set that we make available to stereo training.

Table 6. Performance improves with more synthesized images. Here we train PSMNet with varying numbers of images from our MfS dataset using our synthesis approach, without any finetuning to the test datasets. As expected, more training data improves the stereo network, even outperforming synthetic data. Note that all models were trained for 250K steps. Below each results row we show the % difference relative to the Sceneflow baseline.

Monocular model	Dataset	# of images	KITTI '12		KITTI '15		Middlebury		ETH3D	
			EPE	<3px	EPE	<3px	EPE	<2px	EPE	<1px
Megadepth [37]	MfS	35,000	0.88	4.47	1.25	5.52	9.31	28.13	0.59	12.24
			−15.4	*−22.4*	*−3.4*	*−4.8*	*−1.2*	*−19.6*	*−52.8*	*−6.1*
Megadepth [37]	MfS	500,000	0.89	**4.29**	1.07	4.96	7.66	28.86	0.54	10.41
			−14.4	*−25.5*	*−10.1*	*−14.5*	*−18.7*	*−17.5*	*−56.8*	*−20.2*
MiDaS [48]	MfS	35,000	**0.87**	4.35	1.09	5.06	6.77	28.86	0.56	8.18
			−16.3	*−24.5*	*−8.4*	*−12.8*	*−28.1*	*−17.5*	*−55.2*	*−29.6*
MiDaS [48]	MfS	500,000	0.95	4.43	**1.04**	**4.75**	**6.33**	**26.42**	**0.53**	**8.17**
			−8.7	*−23.1*	*−12.6*	*−18.1*	*−32.8*	*−24.5*	*−57.6*	*−37.3*
Sceneflow baseline		35,454	1.04	5.76	1.19	5.80	9.42	34.99	1.25	13.04

5 Discussion

Many of our results used the monocular depth estimation network from [48], which was trained on a mixture of datasets including stereo pairs, which in turn had their depth estimated using an optical flow network trained on synthetic data [28]. However, experiments in Table 3 show we perform well regardless of the source of data used to train the underlying monocular estimator, with good predictions with depth estimation networks trained on SfM reconstructions [37] or monocular video sequences [23]. In fact, even the naive 'disparity estimation' methods in Table 1 generate reasonable stereo training data. Furthermore, we observe in Table 6 that as we increase the amount of training images our stereo performance improves. This indicates the power of the new additional training data that our approach makes available for stereo training.

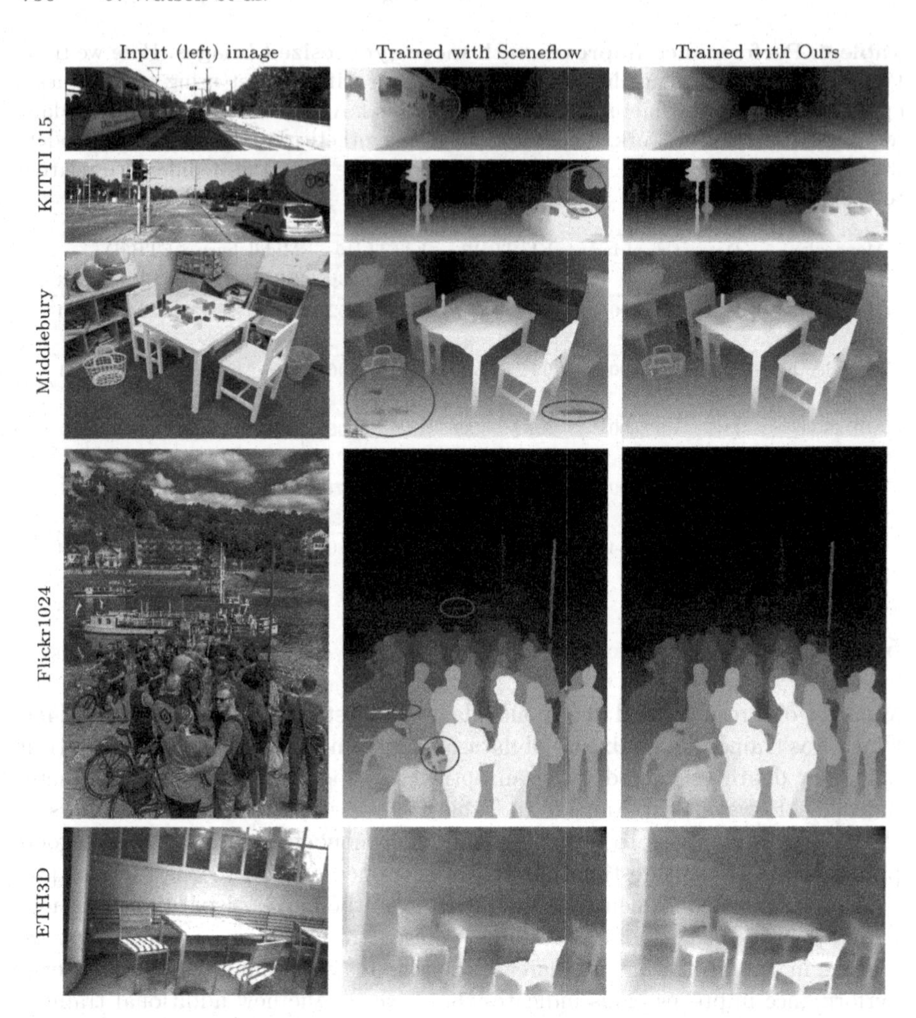

Fig. 6. Qualitative Results. Comparing our synthetic data generation approach with Sceneflow, using a PSMNet model trained on both. Our predictions benefit from semantically plausible depths at training time and have fewer artifacts.

6 Conclusion

We presented a fully automatic pipeline for synthesizing stereo training data from single monocular images. We showed that our approach is robust to the source of monocular depth and results in a significant increase in test-time generalization performance for multiple different stereo algorithms. Importantly, our approach enables the possibility of converting any monocular image into stereo training data, opening up the door to using existing large image collections to improve stereo estimation. In future we intend to explore different ways to automatically estimate the quality of our synthesized data [7] and approaches for performing additional augmentations during training [10]. Another interesting question is if we can use our improved stereo networks to train better monocular depth networks, e.g. by providing sparse depth cues [32,60,68], which in turn could be used to further improve stereo with our method.

Acknowledgements. Large thanks to Aron Monszpart for help with baseline code, to Grace Tsai for feedback, and Sara Vicente for suggestions for baseline implementations. Also thanks to the authors who shared models and images.

References

1. Aleotti, F., Tosi, F., Zhang, L., Poggi, M., Mattoccia, S.: Reversing the cycle: self-supervised deep stereo. In: ECCV (2020)
2. Abu Alhaija, H., Mustikovela, S.K., Geiger, A., Rother, C.: Geometric image synthesis. In: Jawahar, C.V., Li, H., Mori, G., Schindler, K. (eds.) ACCV 2018. LNCS, vol. 11366, pp. 85–100. Springer, Cham (2019). https://doi.org/10.1007/978-3-030-20876-9_6
3. Bleyer, M., Rhemann, C., Rother, C.: PatchMatch stereo - stereo matching with slanted support windows. In: BMVC (2011)
4. Cabon, Y., Murray, N., Humenberger, M.: Virtual KITTI 2. arXiv:2001.10773 (2020)
5. Chang, J.R., Chen, Y.S.: Pyramid stereo matching network. In: CVPR (2018)
6. Chen, W., Fu, Z., Yang, D., Deng, J.: Single-image depth perception in the wild. In: NeurIPS (2016)
7. Chen, W., Qian, S., Deng, J.: Learning single-image depth from videos using quality assessment networks. In: CVPR (2019)
8. Cheng, X., Wang, P., Yang, R.: Learning depth with convolutional spatial propagation network. PAMI **42**(10), 2361–2379 (2019)
9. Choi, I., Gallo, O., Troccoli, A., Kim, M.H., Kautz, J.: Extreme view synthesis. In: ICCV (2019)
10. Cubuk, E.D., Zoph, B., Mane, D., Vasudevan, V., Le, Q.V.: AutoAugment: learning augmentation strategies from data. In: CVPR (2019)
11. Deng, J., Dong, W., Socher, R., Li, L.J., Li, K., Fei-Fei, L.: ImageNet: a large-scale hierarchical image database. In: CVPR (2009)
12. DeTone, D., Malisiewicz, T., Rabinovich, A.: SuperPoint: self-supervised interest point detection and description. In: CVPR Deep Learning for Visual SLAM Workshop (2018)
13. Dosovitskiy, A., et al.: FlowNet: learning optical flow with convolutional networks. In: ICCV (2015)
14. Dosovitskiy, A., Ros, G., Codevilla, F., Lopez, A., Koltun, V.: CARLA: an open urban driving simulator. In: CoRL (2017)
15. Dwibedi, D., Misra, I., Hebert, M.: Cut, paste and learn: surprisingly easy synthesis for instance detection. In: ICCV (2017)
16. Eigen, D., Puhrsch, C., Fergus, R.: Depth map prediction from a single image using a multi-scale deep network. In: NeurIPS (2014)
17. Felzenszwalb, P.F., Huttenlocher, D.P.: Efficient graph-based image segmentation. Int. J. Comput. Vision **59**, 167–181 (2004). https://doi.org/10.1023/B:VISI.0000022288.19776.77

18. Gaidon, A., Wang, Q., Cabon, Y., Vig, E.: Virtual worlds as proxy for multi-object tracking analysis. In: CVPR (2016)
19. Garg, R., Vijay Kumar, B.G., Carneiro, G., Reid, I.: Unsupervised CNN for single view depth estimation: geometry to the rescue. In: Leibe, B., Matas, J., Sebe, N., Welling, M. (eds.) ECCV 2016. LNCS, vol. 9912, pp. 740–756. Springer, Cham (2016). https://doi.org/10.1007/978-3-319-46484-8_45
20. Geiger, A., Lenz, P., Stiller, C., Urtasun, R.: Vision meets robotics: the KITTI dataset. Int. J. Robot. Res. **1**, 6 (2013)
21. Geiger, A., Lenz, P., Urtasun, R.: Are we ready for autonomous driving? The KITTI vision benchmark suite. In: CVPR (2012)
22. Godard, C., Mac Aodha, O., Brostow, G.J.: Unsupervised monocular depth estimation with left-right consistency. In: CVPR (2017)
23. Godard, C., Mac Aodha, O., Firman, M., Brostow, G.J.: Digging into self-supervised monocular depth estimation. In: ICCV (2019)
24. Guo, X., Li, H., Yi, S., Ren, J., Wang, X.: Learning monocular depth by distilling cross-domain stereo networks. In: Ferrari, V., Hebert, M., Sminchisescu, C., Weiss, Y. (eds.) ECCV 2018. LNCS, vol. 11215, pp. 506–523. Springer, Cham (2018). https://doi.org/10.1007/978-3-030-01252-6_30
25. Guo, X., Yang, K., Yang, W., Wang, X., Li, H.: Group-wise correlation stereo network. In: CVPR (2019)
26. Hirschmuller, H.: Stereo processing by semiglobal matching and mutual information. PAMI **30**(2), 328–341 (2007)
27. Hu, J., Ozay, M., Zhang, Y., Okatani, T.: Revisiting single image depth estimation: toward higher resolution maps with accurate object boundaries. In: WACV (2019)
28. Ilg, E., Mayer, N., Saikia, T., Keuper, M., Dosovitskiy, A., Brox, T.: FlowNet 2.0: evolution of optical flow estimation with deep networks. In: CVPR (2017)
29. Janai, J., Güney, F., Ranjan, A., Black, M., Geiger, A.: Unsupervised learning of multi-frame optical flow with occlusions. In: Ferrari, V., Hebert, M., Sminchisescu, C., Weiss, Y. (eds.) ECCV 2018. LNCS, vol. 11220, pp. 713–731. Springer, Cham (2018). https://doi.org/10.1007/978-3-030-01270-0_42
30. Kanazawa, A., Jacobs, D.W., Chandraker, M.: WarpNet: weakly supervised matching for single-view reconstruction. In: CVPR (2016)
31. Kendall, A., et al.: End-to-end learning of geometry and context for deep stereo regression. In: ICCV (2017)
32. Klodt, M., Vedaldi, A.: Supervising the new with the old: learning SFM from SFM. In: Ferrari, V., Hebert, M., Sminchisescu, C., Weiss, Y. (eds.) ECCV 2018. LNCS, vol. 11214, pp. 713–728. Springer, Cham (2018). https://doi.org/10.1007/978-3-030-01249-6_43
33. Ladickỳ, L., Häne, C., Pollefeys, M.: Learning the matching function. arXiv:1502.00652 (2015)
34. Le, H.A., Nimbhorkar, T., Mensink, T., Baslamisli, A.S., Karaoglu, S., Gevers, T.: Unsupervised generation of optical flow datasets. arXiv:1812.01946 (2018)
35. Li, A., Yuan, Z.: Occlusion aware stereo matching via cooperative unsupervised learning. In: Jawahar, C.V., Li, H., Mori, G., Schindler, K. (eds.) ACCV 2018. LNCS, vol. 11366, pp. 197–213. Springer, Cham (2019). https://doi.org/10.1007/978-3-030-20876-9_13
36. Li, W., et al.: InteriorNet: mega-scale multi-sensor photo-realistic indoor scenes dataset. In: BMVC (2018)
37. Li, Z., Snavely, N.: MegaDepth: learning single-view depth prediction from internet photos. In: CVPR (2018)

38. Liang, Z., et al.: Learning for disparity estimation through feature constancy. In: CVPR (2018)
39. Lin, T.-Y., et al.: Microsoft COCO: common objects in context. In: Fleet, D., Pajdla, T., Schiele, B., Tuytelaars, T. (eds.) ECCV 2014. LNCS, vol. 8693, pp. 740–755. Springer, Cham (2014). https://doi.org/10.1007/978-3-319-10602-1_48
40. Luo, Y., et al.: Single view stereo matching. In: CVPR (2018)
41. Mahajan, D., et al.: Exploring the limits of weakly supervised pretraining. In: Ferrari, V., Hebert, M., Sminchisescu, C., Weiss, Y. (eds.) ECCV 2018. LNCS, vol. 11206, pp. 185–201. Springer, Cham (2018). https://doi.org/10.1007/978-3-030-01216-8_12
42. Mayer, N., et al.: What makes good synthetic training data for learning disparity and optical flow estimation? Int. J. Comput. Vision **126**(9), 942–960 (2018). https://doi.org/10.1007/s11263-018-1082-6
43. Mayer, N., et al.: A large dataset to train convolutional networks for disparity, optical flow, and scene flow estimation. In: CVPR (2016)
44. Neuhold, G., Ollmann, T., Rota Bulò, S., Kontschieder, P.: The Mapillary Vistas Dataset for semantic understanding of street scenes. In: ICCV (2017)
45. Niklaus, S., Liu, F.: Softmax splatting for video frame interpolation. In: CVPR (2020)
46. Pang, J., et al.: Zoom and learn: generalizing deep stereo matching to novel domains. In: CVPR (2018)
47. Ramamonjisoa, M., Lepetit, V.: SharpNet: fast and accurate recovery of occluding contours in monocular depth estimation. In: ICCV Workshops (2019)
48. Ranftl, R., Lasinger, K., Hafner, D., Schindler, K., Koltun, V.: Towards robust monocular depth estimation: mixing datasets for zero-shot cross-dataset transfer. arXiv:1907.01341 (2019)
49. Reinhard, E., Adhikhmin, M., Gooch, B., Shirley, P.: Color transfer between images. IEEE Comput. Graphics Appl. **21**(5), 34–41 (2001)
50. Reynolds, M., Doboš, J., Peel, L., Weyrich, T., Brostow, G.J.: Capturing time-of-flight data with confidence. In: CVPR (2011)
51. Scharstein, D., et al.: High-resolution stereo datasets with subpixel-accurate ground truth. In: Jiang, X., Hornegger, J., Koch, R. (eds.) GCPR 2014. LNCS, vol. 8753, pp. 31–42. Springer, Cham (2014). https://doi.org/10.1007/978-3-319-11752-2_3
52. Schops, T., et al.: A multi-view stereo benchmark with high-resolution images and multi-camera videos. In: CVPR (2017)
53. Schwarz, L.A.: Non-rigid registration using free-form deformations. Technische Universität München (2007)
54. Sobel, I., Feldman, G.: A 3x3 isotropic gradient operator for image processing. A talk at the Stanford Artificial Project (1968)
55. Sun, C., Shrivastava, A., Singh, S., Gupta, A.: Revisiting unreasonable effectiveness of data in deep learning era. In: ICCV (2017)
56. Thewlis, J., Bilen, H., Vedaldi, A.: Unsupervised learning of object landmarks by factorized spatial embeddings. In: ICCV (2017)
57. Tonioni, A., Poggi, M., Mattoccia, S., Di Stefano, L.: Unsupervised adaptation for deep stereo. In: ICCV (2017)
58. Tonioni, A., Rahnama, O., Joy, T., di Stefano, L., Ajanthan, T., Torr, P.H.S.: Learning to adapt for stereo. In: CVPR (2019)
59. Tonioni, A., Tosi, F., Poggi, M., Mattoccia, S., Stefano, L.D.: Real-time self-adaptive deep stereo. In: ICCV (2019)
60. Tosi, F., Aleotti, F., Poggi, M., Mattoccia, S.: Learning monocular depth estimation infusing traditional stereo knowledge. In: CVPR (2019)

61. Vasiljevic, I., et al.: DIODE: a dense indoor and outdoor depth dataset. arXiv:1908.00463 (2019)
62. Veit, A., Alldrin, N., Chechik, G., Krasin, I., Gupta, A., Belongie, S.: Learning from noisy large-scale datasets with minimal supervision. In: CVPR (2017)
63. Wang, C., Lucey, S., Perazzi, F., Wang, O.: Web stereo video supervision for depth prediction from dynamic scenes. In: 3DV (2019)
64. Wang, J., Zickler, T.: Local detection of stereo occlusion boundaries. In: CVPR (2019)
65. Wang, Q., Zheng, S., Yan, Q., Deng, F., Zhao, K., Chu, X.: IRS: a large synthetic indoor robotics stereo dataset for disparity and surface normal estimation. arXiv:1912.09678 (2019)
66. Wang, Y., Wang, P., Yang, Z., Luo, C., Yang, Y., Xu, W.: UnOS: unified unsupervised optical-flow and stereo-depth estimation by watching videos. In: CVPR (2019)
67. Wang, Y., Wang, L., Yang, J., An, W., Guo, Y.: Flickr1024: a large-scale dataset for stereo image super-resolution. In: ICCV Workshops (2019)
68. Watson, J., Firman, M., Brostow, G.J., Turmukhambetov, D.: Self-supervised monocular depth hints. In: ICCV (2019)
69. Woodford, O., Torr, P., Reid, I., Fitzgibbon, A.: Global stereo reconstruction under second-order smoothness priors. PAMI **31**(12), 2115–2128 (2009)
70. Xian, K., et al.: Monocular relative depth perception with web stereo data supervision. In: CVPR (2018)
71. Xiao, T., Xia, T., Yang, Y., Huang, C., Wang, X.: Learning from massive noisy labeled data for image classification. In: CVPR (2015)
72. Xie, S., Girshick, R.B., Dollár, P., Tu, Z., He, K.: Aggregated residual transformations for deep neural networks. In: CVPR (2016)
73. Yin, Z., Darrell, T., Yu, F.: Hierarchical discrete distribution decomposition for match density estimation. In: CVPR (2019)
74. Yu, J., Lin, Z., Yang, J., Shen, X., Lu, X., Huang, T.S.: Free-form image inpainting with gated convolution. In: ICCV (2019)
75. Žbontar, J., LeCun, Y.: Stereo matching by training a convolutional neural network to compare image patches. JMLR **17**(1), 2287–2318 (2016)
76. Zhang, F., Prisacariu, V., Yang, R., Torr, P.H.: GA-Net: guided aggregation net for end-to-end stereo matching. In: CVPR (2019)
77. Zhang, F., Qi, X., Yang, R., Prisacariu, V., Wah, B., Torr, P.: Domain-invariant stereo matching networks. In: ECCV (2020)
78. Zhong, Y., Dai, Y., Li, H.: Self-supervised learning for stereo matching with self-improving ability. arXiv:1709.00930 (2017)
79. Zhou, B., Zhao, H., Puig, X., Fidler, S., Barriuso, A., Torralba, A.: Scene parsing through ADE20K dataset. In: CVPR (2017)
80. Zhou, C., Zhang, H., Shen, X., Jia, J.: Unsupervised learning of stereo matching. In: ICCV (2017)
81. Zhou, T., Brown, M., Snavely, N., Lowe, D.G.: Unsupervised learning of depth and ego-motion from video. In: CVPR (2017)
82. Zhou, T., Tulsiani, S., Sun, W., Malik, J., Efros, A.A.: View synthesis by appearance flow. In: Leibe, B., Matas, J., Sebe, N., Welling, M. (eds.) ECCV 2016. LNCS, vol. 9908, pp. 286–301. Springer, Cham (2016). https://doi.org/10.1007/978-3-319-46493-0_18

Prototype Rectification for Few-Shot Learning

Jinlu Liu, Liang Song, and Yongqiang Qin(✉)

AInnovation Technology Co., Ltd., Beijing, China
{liujinlu,songliang,qinyongqiang}@ainnovation.com

Abstract. Few-shot learning requires to recognize novel classes with scarce labeled data. Prototypical network is useful in existing researches, however, training on narrow-size distribution of scarce data usually tends to get biased prototypes. In this paper, we figure out two key influencing factors of the process: *the intra-class bias* and *the cross-class bias*. We then propose a simple yet effective approach for prototype rectification in transductive setting. The approach utilizes label propagation to diminish the intra-class bias and feature shifting to diminish the cross-class bias. We also conduct theoretical analysis to derive its rationality as well as the lower bound of the performance. Effectiveness is shown on three few-shot benchmarks. Notably, our approach achieves state-of-the-art performance on both miniImageNet (70.31% on 1-shot and 81.89% on 5-shot) and tieredImageNet (78.74% on 1-shot and 86.92% on 5-shot).

Keywords: Few-shot learning · Prototype rectification · Intra-class bias · Cross-class bias

1 Introduction

Many deep learning based methods have achieved significant performance on object recognition tasks with abundant labeled data provided [9,12,27]. However, these methods generally perform unsatisfactorily if the labeled data is scarce. To reduce the dependency of data annotation, more researchers make efforts to develop powerful methods to learn new concepts from very few samples, which is so-called Few-Shot Learning (FSL) [5,17,33]. In FSL, we aim to learn prior knowledge on base classes with large amounts of labeled data and utilize the knowledge to recognize few-shot classes with scarce labeled data. It is usually formed as N-way K-shot few-shot tasks where each task consists of N few-shot classes with K labeled samples per class (the support set) and some unlabeled samples (the query set) for test.

Classifying test samples by matching them to the nearest class prototype [30] is a common practice in FSL. It is supposed that an expected prototype has the

Electronic supplementary material The online version of this chapter (https://doi.org/10.1007/978-3-030-58452-8_43) contains supplementary material, which is available to authorized users.

A. Vedaldi et al. (Eds.): ECCV 2020, LNCS 12346, pp. 741–756, 2020.
https://doi.org/10.1007/978-3-030-58452-8_43

minimal distance to all samples within the same class. However, the prototypes they get are always biased due to the data scarcity in few-shot scenarios. The internal factors that restrict the representation ability of the prototypes should be identified for performance improvement. Hence, we figure out the bias in prototype computation and accordingly propose the diminishing methods for rectification.

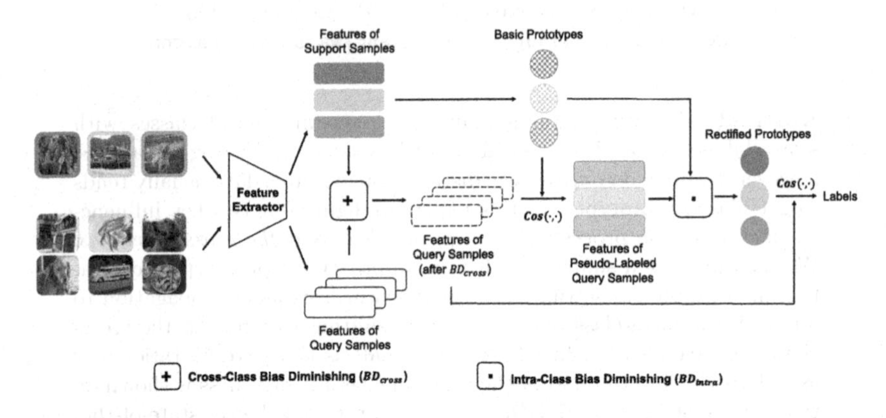

Fig. 1. Framework of our proposed method for prototype rectification. *The cross-class bias diminishing module* reduces the bias between the support set and the query set while *the intra-class bias diminishing module* reduces the bias between the actually computed prototypes and the expected prototypes.

In this paper, we target to find the expected prototypes which have the maximum cosine similarity to all data points within the same class. The cosine similarity based prototypical network (CSPN) is firstly proposed to extract discriminative features and compute basic prototypes from the limited samples. In CSPN, we firstly train a feature extractor with a cosine classifier on the base classes. The cosine classifier has a strong capability of driving the feature extractor to learn discriminative features. It learns an embedding space where features belonging to the same class cluster more tightly. At the inference stage, we use class means as the basic prototypes of few-shot classes. Classification can be directly performed by nearest prototype matching based on cosine similarity.

Since the basic prototypes are biased due to data scarcity, we import a bias diminishing module into the network for prototype rectification, which is called BD-CSPN in this paper. We figure out two key factors: the intra-class bias and the cross-class bias, which influence the representativeness of class prototypes. The approach to reduce the bias is accordingly proposed as shown in Fig. 1. The *intra-class bias* refers to the distance between the expectedly unbiased prototype and the prototype actually computed from the available data. To reduce it, we adopt the pseudo-labeling strategy to add unlabeled samples with high prediction confidence into the support set in transductive setting. Considering that some of the pseudo-labeled samples are possibly misclassified, we use the weighted sum as the modified prototypes instead of simple averaging. It avoids bringing larger bias into prototype computation. The *cross-class bias* refers to the distance between the representative vectors of training and test datasets, which are commonly represented as the mean vectors. We reduce it by importing a shifting term ξ to the query samples, driving them to distribute closely to the support samples.

To verify the rationality of our bias diminishing method, we give the theoretical analysis in Sect. 4. The derivation of the expected performance of cosine-similarity based prototypical network is firstly given. It shows that the lower bound of the expected accuracy is positively correlated with the number of samples. *We demonstrate the effectiveness and simplicity of our pseudo-labeling strategy in raising the lower bound, which leads to significant improvement as shown in experiments.* Then we give the derivation of shifting term ξ in cross-class bias diminishing. In conclusion, we argue that our method is simpler yet more efficient than many complicated few-shot learning methods. Also, it is mathematically rigorous with the theoretical analysis.

Our contributions are summarized as:

1) We figure out the internal factors: the intra-class bias and the cross-class bias which restrict the representational ability of class prototypes in few-shot learning.
2) We propose the bias diminishing module for prototype rectification, which is mainly conducted by pseudo-labeling and feature shifting. It is conceptually simple but practically effective to improve the performance.
3) To verify the rationality of the intra-class bias diminishing method, we theoretically analyze the correlation between the number of sample and the lower bound of the expected performance. Furthermore, we give the derivation of the shifting term in cross-class bias diminishing.
4) We conduct extensive experiments on three popular few-shot benchmarks and achieve the state-of-the-art performance. The experiment results demonstrate that our proposed bias diminishing module can bring in significant improvement by a large margin.

2 Related Works

Few-Shot Learning. Few-shot learning methods can be divided into two groups: *gradient based methods* and *metric learning based methods*. Gradient based methods focus on fast adapting model parameters to new tasks through gradient descent [2,6,13,15,20,28]. Typical methods such as MAML [6] and Reptile [20] aim to learn a good parameter initialization that enables the model easy to fine-tune. In this section, we focus on metric learning based methods which are more closely to our approach. Metric learning based methods learn an informative metric to indicate the similarity relationship in the embedding space [1,30,31,33]. Relation network [31] learns a distance metric to construct the relation of samples within an episode. The unlabeled samples thus can be classified according to the computed relation scores. Prototypical Networks (PN)

[30] views the mean feature as the class prototype and assigns the points to the nearest class prototype based on Euclidean distance in the embedding space. It is indicated in [13] that PN shows limited performance in the high-dimensional embedding space. In some recent works, models trained with a cosine-similarity based classifier are more effective in learning discriminative features [3,8]. In this paper, we use cosine classifier to learn a discriminative embedding space and compute the cosine distance to the class prototype (mean) for classification. The prototype computed in the discriminative feature space is more robust to represent a class.

According to the test setting, FSL can be divided into two branches: *inductive few-shot learning* and *transductive few-shot learning*. The former predicts the test samples one by one while the latter predicts the test samples as a whole. Early proven in [10,36], transductive inference outperforms inductive inference especially when training data is scarce. Some literatures recently attack few-shot learning problem in transductive setting. In [20], the shared information between test samples via normalization is used to improve classification accuracy. Different from [20], TPN [16] adopts transductive inference to alleviate low-data problem in few-shot learning. It constructs a graph using the union of the support set and the query set, where labels are propagated from support to query. Under transductive inference, the edge-labeling graph neural network (EGNN) proposed in [11] learns more accurate edge-labels through exploring the intra-cluster similarity and the inter-cluster dissimilarity. Our method takes the advantage of transductive inference that samples with higher prediction confidence can be obtained when the test samples are predicted as a whole.

Semi-supervised Few-Shot Learning. In semi-supervised few-shot learning, an extra unlabeled set not contained in current episode is used to improve classification accuracy [14,25,29]. In [25], the extended versions of Prototypical Networks [30] are proposed to use unlabeled data to create class prototypes by Soft k-Means. LST [14] employs pseudo-labeling strategy to the unlabeled set, then it re-trains and fine-tunes the base model based on the pseudo-labeled data. For recognizing the novel classes, it utilizes dynamically sampled data which is not contained in the current episode. Different from these methods, the unlabeled data in our method comes from the query set and we requires no extra datasets besides the support and query set.

3 Methodology

We firstly use cosine similarity based prototypical network (CSPN) to learn a discriminative feature space and get the basic prototypes of few-shot classes. Then we figure out two influencing factors in prototype computation: the intra-class bias and the cross-class bias. Accordingly, we propose the bias diminishing (BD) method for prototype rectification in transductive setting.

3.1 Denotation

At the training stage, a labeled dataset $\mathcal{D}$ of base classes $\mathcal{C}_{base}$ is given to train the feature extractor $F_\theta(\cdot)$ and the cosine classifier $C(\cdot|W)$. At the inference stage, we aim to recognize few-shot classes $\mathcal{C}_{few}$ with K labeled images per class. Episodic sampling is adopted to form such N-way K-shot tasks. Each episode consists of a support set $\mathcal{S}$ and a query set $\mathcal{Q}$. In the support set, all samples x are labeled and we use the extracted features $X = F_\theta(x)$ to compute the prototypes P of few-shot classes. The samples in the query set are unlabeled for test.

3.2 Cosine Similarity Based Prototypical Network

We propose a metric learning based method: cosine similarity based prototypical network (CSPN) to compute the basic prototypes of few-shot classes. Training a good feature extractor that can extract discriminative features is of great importance. Thus, we firstly train a feature extractor $F_\theta(\cdot)$ with a cosine similarity based classifier $C(\cdot|W)$ on the base classes. The cosine classifier $C(\cdot|W)$ is:

$$C(F_\theta(x) \mid W) = Softmax(\tau \cdot Cos(F_\theta(x), W)) \tag{1}$$

where W is the learnable weight of the base classes and τ is a scalar parameter. We target to minimize the negative log-likelihood loss on the supervised classification task:

$$L(\theta, W \mid \mathcal{D}) = \mathbb{E}[-logC(F_\theta(x) \mid W)] \tag{2}$$

At the inference stage, retraining $F_\theta(\cdot)$ and classification weights on the scarce data of $\mathcal{C}_{few}$ classes is likely to run into overfitting. To avoid it, we directly compute the basic prototype P_n of class n as follows:

$$P_n = \frac{1}{K}\sum_{i=1}^{K} \overline{X}_{i,n} \tag{3}$$

where $\overline{X}$ is the normalized feature of support samples. The query samples can be classified by finding the nearest prototype based on cosine similarity.

3.3 Bias Diminishing for Prototype Rectification

In CSPN, we can obtain the basic prototypes by simply averaging the features of support samples. However, the prototypes computed in such low-data regimes are biased against the expected prototypes we want to find. Therefore, we identify two influencing factors: the intra-class bias and the cross-class bias, and accordingly propose the bias diminishing approach.

The intra-class bias within a class is defined by Eq. (4):

$$B_{intra} = \mathbb{E}_{X' \sim p_{X'}}[X'] - \mathbb{E}_{X \sim p_X}[X] \tag{4}$$

where $p_{X'}$ is the distribution of all data belonging to a certain class and p_X is the distribution of the available labeled data of this class. It is easy to observe the difference between the expectations of the two distributions. The difference becomes more significant in low-data regimes. Since the prototype is computed by feature averaging, the intra-class bias also can be understood as the difference between the expected prototype and the actually computed prototype. The expected prototype is supposed to be represented by the mean feature of all samples within a class. In practice, only a part of samples are available for training which is to say that, it is almost impossible to get the expected prototype. In few-shot scenario, we merely have K samples per few-shot class. The number of available samples are far less than the expected amount. Computed from scarce samples, the prototypes obviously tend to be biased.

To reduce the bias, we adopt the pseudo-labeling strategy to augment the support set, which assigns temporary labels to the unlabeled data according to their prediction confidence [14]. Pseudo-labeled samples can be augmented into the support set such that we can compute new prototypes in a 'higher-data' regime. We can simply select top Z confidently predicted query samples per class to augment the support set $\mathcal{S}$ with their pseudo labels. We use CSPN as recognition model to get the prediction scores. Then we have an augmented support set with confidently predicted query samples: $\mathcal{S}' = \mathcal{S} \cup \mathcal{Q}^Z_{pseudo}$. Since some pseudo-labeled samples are likely to be misclassified, simple averaging with the same weights is possible to result in larger bias in prototype computation. To compute the new prototypes in a more reasonable way, we use the weighted sum of $X^{'}$ as the rectified prototype. We note that $X^{'}$ refers to the feature of the sample in $\mathcal{S}'$ including both original support samples and pseudo-labeled query samples. The rectified prototype of a class is thus computed from the normalized features $\overline{X}^{'}$:

$$P_n^{'} = \sum_{i=1}^{Z+K} w_{i,n} \cdot \overline{X}^{'}_{i,n} \tag{5}$$

where $w_{i,n}$ is the weight indicating the relation of the augmented support samples and the basic prototypes. The weight is computed by:

$$w_{i,n} = \frac{exp(\varepsilon \cdot Cos(X^{'}_{i,n}, P_n))}{\sum_{j=1}^{K+Z} exp(\varepsilon \cdot Cos(X^{'}_{j,n}, P_n))} \tag{6}$$

ε is a scalar parameter and P_n is the basic prototype obtained in Sect. 3.2. Samples with larger cosine similarity to the basic prototypes hold larger proportions in prototype rectification. Compared with the basic prototype P_n, the rectified prototype $P_n^{'}$ distributes closer to the expected prototype.

The cross-class bias refers to the distance between the mean vectors of support and query datasets. It is derived from the domain adaptation problem where the mean value is used as a type of the first order statistic information to represent a dataset [34]. Minimizing the distance between different domains is a typical method of mitigating domain gaps. Since the support set and the query set are assumed to distribute in the same domain, the distance between them is the distribution bias rather than the domain gap. The cross-class bias B_{cross} is formulated as:

$$B_{cross} = \mathbb{E}_{X_s \sim p_S}[X_s] - \mathbb{E}_{X_q \sim p_Q}[X_q] \tag{7}$$

where $p_{\mathcal{S}}$ and $p_{\mathcal{Q}}$ respectively represent the distributions of support and query sets. Notably, the support set $\mathcal{S}$ and the query set $\mathcal{Q}$ include N few-shot classes in Eq. (7). To diminish B_{cross}, we can shift the query set towards the support set. In practice, we add a shifting term ξ to each normalized query feature $\overline{X}_q$ and ξ is defined as:

$$\xi = \frac{1}{|\mathcal{S}|}\sum_{i=1}^{|\mathcal{S}|}\overline{X}_{i,s} - \frac{1}{|\mathcal{Q}|}\sum_{j=1}^{|\mathcal{Q}|}\overline{X}_{j,q} \tag{8}$$

The detailed derivation of ξ is given in the next section.

4 Theoretical Analysis

We give the theoretical analysis to show the rationality of our proposed bias diminishing method.

4.1 Lower Bound of the Expected Performance

We derive the formulation of our expected performance in theory and point out what factors influence the final result. We use X to represent the feature of a class. For clear illustration, the formulation of class prototype we use in this section is given:

$$P = \frac{\sum_i^T \overline{X}_i^{'}}{T} \tag{9}$$

where $T = K + Z$, $\overline{X}_i^{'} \in \mathcal{S}^{'}$ and $\mathcal{S}^{'}$ is a subset sampled from X. $\overline{X}$ is the normalized feature and $\overline{P}$ is the normalized prototype. For cosine similarity based prototypical network, an expected prototype should have the largest cosine similarity to all samples within its class. Our objective is to maximize the expected cosine similarity which is positively correlated with the classification accuracy. It is formulated as:

$$\max \mathbb{E}_P[\mathbb{E}_X[Cos(P, X)]] \tag{10}$$

And we derive it as:

$$\begin{aligned} \mathbb{E}_P[\mathbb{E}_X[Cos(P, X)]] &= \mathbb{E}_{P,X}[\overline{P} \cdot \overline{X}] \\ &= \mathbb{E}[\overline{X}] \cdot \mathbb{E}[\frac{P}{\|P\|_2}] \end{aligned} \tag{11}$$

From previous works [21,26], we know that:

$$\mathbb{E}[\frac{A}{B}] = \frac{\mathbb{E}[A]}{\mathbb{E}[B]} + O(n^{-1}) \quad (first\ order) \tag{12}$$

where A and B are random variables. In Eq. (12), $\frac{\mathbb{E}[A]}{\mathbb{E}[B]}$ is the first order estimator of $\mathbb{E}[\frac{A}{B}]$. Thus, Eq. (11) is approximate to:

$$\mathbb{E}_P[\mathbb{E}_X[Cos(P, X)]] \approx \frac{\mathbb{E}[\overline{X}] \cdot \mathbb{E}[P]}{\mathbb{E}[\|P\|_2]} \tag{13}$$

Based on Cauchy-Schwarz inequality, we have:

$$\mathbb{E}[\|P\|_2] \leq \sqrt{\mathbb{E}[\|P\|_2^2]} \tag{14}$$

P and $\overline{X}$ are D-dimensional vectors which can be denoted as $P = [p_1, p_2, ..., p_D]$ and $\overline{X} = [\overline{x}_1, \overline{x}_2, ..., \overline{x}_D]$ respectively. In our method, we assume that each dimension of a vector is independent from each other. Then, we can derive that:

$$\begin{aligned} \mathbb{E}[\|P\|_2^2] = \mathbb{E}[\sum_{i=1}^{D} p_i^2] &= \sum_{i=1}^{D} [Var[p_i] + \mathbb{E}[p_i]^2] \\ &= \sum_{i=1}^{D} [Var[p_i] + \mathbb{E}[\overline{x}_i]^2] \\ &= \sum_{i=1}^{D} [\frac{1}{T} Var[\overline{x}_i] + \mathbb{E}[\overline{x}_i]^2] \end{aligned} \tag{15}$$

Thus, the lower bound of the expected cosine similarity is formulated as:

$$\begin{aligned} \mathbb{E}_P[\mathbb{E}_X[Cos(P, X)]] &\geq \frac{\mathbb{E}[\overline{X}] \cdot \mathbb{E}[P]}{\sqrt{\mathbb{E}[\|P\|_2^2]}} \\ &= \frac{\sum_{i=1}^{D} \mathbb{E}[\overline{x}_i]^2}{\sqrt{\frac{1}{T}\sum_{i=1}^{D} Var[\overline{x}_i] + \sum_{i=1}^{D} \mathbb{E}[\overline{x}_i]^2}} \end{aligned} \tag{16}$$

Maximizing the expected accuracy is approximate to maximize its lower bound of the cosine similarity as shown in Eq. (16). It can be seen that the number T of the sample is positively correlated with the lower bound of the expected performance. Thus, we import more pseudo-labeled samples into prototype computation. *The rationality of the pseudo-labeling strategy in improving few-shot accuracy is that, it can effectively raise the lower bound of the expected performance.*

4.2 Derivation of Shifting Term ξ

We propose to reduce the cross-class bias by feature shifting and the derivation of shifting term ξ is provided as follows. In N-way K-shot Q-query tasks, the accuracy can be formalized as:

$$Acc = \frac{1}{NQ} \sum_{i}^{N} \sum_{q}^{Q} \mathbb{1}(y_{i,q} == i) \tag{17}$$

$$= \frac{1}{NQ} \sum_{i}^{N} \sum_{q}^{Q} \mathbb{1}(Cos(P_i, X_{i,q}) > \max_{j \neq i}\{Cos(P_j, X_{i,q})\}) \tag{18}$$

where $y_{i,q}$ is the predicted label and i is the true class label. $\mathbb{1}(b)$ is an indicator function. $\mathbb{1}(b) = 1$ if b is true and 0 otherwise. P_i is the prototype of class i and $X_{i,q}$ is the q-th query feature of class i. Based on Eq. (18), the accuracy formulation can be further rewritten as:

$$Acc = \frac{1}{NQ}\sum_{i}^{N}\sum_{q}^{Q}\mathbb{1}(Cos(P_i, X_{i,q}) > t_i) \tag{19}$$

where t_i denotes the cosine similarity threshold of the i-th class. Improving the accuracy is equal to maximize the cosine similarity $Cos(\cdot)$.

As mentioned above, there is a bias between the support and query set of a class i. We assume that the bias can be diminished by adding a shifting term ξ_i to the query samples. Since the class labels are unknown, we approximately add the same term ξ to all query samples. The term ξ should follow the objective:

$$\arg\max_{\xi} \frac{1}{NQ}\sum_{i}^{N}\sum_{q}^{Q} Cos(P_i, X_{i,q} + \xi) \tag{20}$$

We assume that each feature X can be represented as $X = P + \epsilon$. Equation (20) can be further formalized as:

$$\arg\max_{\xi} \frac{1}{NQ}\sum_{i}^{N}\sum_{q}^{Q} Cos(P_i, P_i + \epsilon_{i,q} + \xi) \tag{21}$$

To maximize the cosine similarity, we should minimize the following objective:

$$\min \frac{1}{NQ}\sum_{i}^{N}\sum_{q}^{Q}(\epsilon_{i,q} + \xi) \tag{22}$$

The term ξ is thus computed:

$$\xi = -E[\epsilon] \tag{23}$$

$$= \frac{1}{NQ}\sum_{i}^{N}\sum_{q}^{Q}(P_i - X_{i,q}) \tag{24}$$

We can see that Eq. (24) is in line with Eq. (8). For cosine similarity computation, the shifting term is calculated from the normalized features as displayed in Sect. 3.3.

5 Experiments

5.1 Datasets

miniImageNet consists of 100 randomly chosen classes from ILSVRC-2012 [27]. We adopt the split proposed in [24] where the 100 classes are split into 64 training classes, 16 validation classes and 20 test classes. Each class contains 600 images of size 84 × 84. ***tieredImageNet*** [25] is also a derivative of ILSVRC-2012 [27] containing 608 low-level categories, which are split into 351, 97, 160 categories for training, validation, test with image size of 84 × 84. ***Meta-Dataset*** [32] is a new benchmark that is large-scale and consists of diverse datasets for training and evaluating models.

5.2 Implementation Details

We train the base recognition model CSPN in the supervised way with SGD optimizer and test the validation set on 5-way 5-shot tasks for model selection. WRN-28-10 [35] is used as the main backbone. ConvNets [8] and ResNet-12 [13] are used for ablation. The results are averaged from 600 randomly sampled episodes. Each episode contains 15 query samples per class. The initial value of τ is 10 and ε is fixed at 10. More details are shown in the supplementary materials.

Table 1. Average accuracy (%) comparison on miniImageNet. ‡ Training set and validation set are used for training.

Setting	Methods	Backbone	miniImageNet	
			1-shot	5-shot
Inductive	Matching Network [33]	ConvNet-64	43.56 ± 0.84	55.31 ± 0.73
	MAML [6]	ConvNet-32	48.70 ± 1.84	63.11 ± 0.92
	Prototypical Networks‡ [30]	ConvNet-64	49.42 ± 0.78	68.20 ± 0.66
	Relation Net [31]	ConvNet-256	50.44 ± 0.82	65.32 ± 0.70
	SNAIL [18]	ResNet-12	55.71 ± 0.99	68.88 ± 0.92
	LwoF [8]	ConvNet-128	56.20 ± 0.86	73.00 ± 0.64
	AdaResNet [19]	ResNet-12	56.88 ± 0.62	71.94 ± 0.57
	TADAM [22]	ResNet-12	58.50 ± 0.30	76.70 ± 0.30
	Activation to Parameter‡ [23]	WRN-28-10	59.60 ± 0.41	73.74 ± 0.19
	LEO‡ [28]	WRN-28-10	61.76 ± 0.08	77.59 ± 0.12
	MetaOptNet-SVM [13]	ResNet-12	62.64 ± 0.61	78.63 ± 0.46
	BFSL [7]	WRN-28-10	62.93 ± 0.45	79.87 ± 0.33
Semi-Supervised	ML [25]	ConvNet-128	49.04 ± 0.31	62.96 ± 0.14
	LST [14]	ResNet-12	70.1 ± 1.9	78.7 ± 0.8
Transductive	TPN [16]	ConvNet-64	55.51 ± 0.86	69.86 ± 0.65
	EGNN [11]	ConvNet-256	–	76.37
	Transductive Fine-Tuning [4]	WRN-28-10	65.73 ± 0.68	78.40 ± 0.52
	BD-CSPN (ours)	WRN-28-10	$\mathbf{70.31 \pm 0.93}$	$\mathbf{81.89 \pm 0.60}$

5.3 Results on MiniImageNet and TieredImageNet

The results on miniImageNet and tieredImageNet are shown in Table 1 and Table 2 respectively. It can be seen that we achieve state-of-the-art performance in all cases. Compared with existing transductive methods [4,11,16], our proposed

Table 2. Average accuracy (%) comparison on tieredImageNet. * Results by our implementation. ‡ Training set and validation set are used for training.

Setting	Methods	Backbone	tieredImageNet	
			1-shot	5-shot
Inductive	MAML [6]	ConvNet-32	51.67 ± 1.81	70.30 ± 1.75
	Prototypical Networks‡ [30]	ConvNet-64	53.31 ± 0.89	72.69 ± 0.74
	Relation Net [31]	ConvNet-256	54.48 ± 0.93	71.32 ± 0.78
	LwoF [8]	ConvNet-128	60.35 ± 0.88*	77.24 ± 0.72*
	LEO‡ [28]	WRN-28-10	66.33 ± 0.05	81.44 ± 0.09
	MetaOptNet-SVM [13]	ResNet-12	65.99 ± 0.72	81.56 ± 0.53
Semi-Supervised	ML [25]	ConvNet-128	51.38 ± 0.38	69.08 ± 0.25
	LST [14]	ResNet-12	77.7 ± 1.6	85.2 ± 0.8
Transductive	TPN [16]	ConvNet-64	59.91 ± 0.94	73.30 ± 0.75
	EGNN [11]	ConvNet-256	–	80.15
	Transductive Fine-Tuning [4]	WRN-28-10	73.34 ± 0.71	85.50 ± 0.50
	BD-CSPN (ours)	WRN-28-10	$\mathbf{78.74 \pm 0.95}$	$\mathbf{86.92 \pm 0.63}$

BD-CSPN consistently achieves the best performance on both datasets. EGNN [11] transductively learns edge-labels through exploring the intra-cluster similarity and the inter-cluster dissimilarity. Transductive Fine-Tuning [4] is newly published, providing a strong baseline by simple fine-tuning techniques. In comparison with TPN [16], we achieve better results with a simpler implementation of label propagation technique. Given the similar backbone ConvNet-128 on miniImageNet, BD-CSPN produces good results of 61.74% and 76.12% on 1-shot and 5-shot tasks respectively, surpassing TPN by large margins.

Our method also shows superiority compared with existing semi-supervised methods [14,25]. Note that LST [14] uses extra unlabeled data as auxiliary information in evaluation, which is not contained in current episode. It re-trains and fine-tunes the model on each novel task. We have a simpler technique without re-training and fine-tuning which is more efficient in computation.

5.4 Results on Meta-Dataset

To further illustrate the effectiveness of our method, we show 5-shot results on the newly proposed Meta-Dataset [32] in Table 3. The average rank of our 5-shot model is **1.9**. More details are provided in our supplementary materials.

Table 3. 5-shot results on Meta-Dataset: the model is trained on ILSVRC-2012 only and test on the listed test sources.

Test	5-shot	Test	5-shot	Test	5-shot	Test	5-shot	Test	5-shot
ILSVRC	59.80	Omniglot	78.29	Aircraft	43.42	Birds	67.22	Textures	54.82
Quick Draw	58.80	Fungi	61.56	VGG Flower	83.88	Traffic Signs	68.68	MSCOCO	52.69

5.5 Ablation Study

The ablation results are shown in Table 4. We display the results of CSPN as baselines which are obtained in inductive setting. The network is trained on traditional supervised tasks (64-way), following the setting in [7,8]. It achieves better performance than some complicated meta-trained methods [23,28] with the same backbone, as shown in Table 1. Based on CSPN, our BD module makes an improvement by large margins up to 9% and 3% on 1-shot and 5-shot tasks respectively. It leads to relatively minor improvements in 5-shot scenarios.

Table 4. Ablative results of bias diminishing module. CSPN: without bias diminishing modules; BD_c-CSPN: with cross-class bias diminishing module; BD_i-CSPN: with intra-class bias diminishing module; BD-CSPN: with both modules.

Dataset	Method	1-shot	5-shot	Dataset	Method	1-shot	5-shot
miniImageNet	CSPN	61.84	78.64	tieredImageNet	CSPN	69.20	84.31
	BD_c-CSPN	62.54	79.32		BD_c-CSPN	70.84	84.99
	BD_i-CSPN	69.81	81.58		BD_i-CSPN	78.12	86.67
	BD-CSPN	**70.31**	**81.89**		BD-CSPN	**78.74**	**86.92**

Ablation of Intra-Class Bias Diminishing. It can be seen in Table 4 that BD_i-CSPN brings in significant improvements on both datasets. The intra-class bias diminishing module especially shows its merit in 1-shot scenarios. With intra-class bias diminished, the accuracy on 1-shot miniImageNet increases from 61.84% to 69.81% and the accuracy on 1-shot tieredImageNet raises to 78.12% from 69.20%.

Furthermore, to intuitively demonstrate the influence of our proposed intra-class bias diminishing module, we display the 5-way accuracy in Fig. 2(a)–2(b). The results are reported without using cross-class bias diminishing module. It shows a coincident tendency that with more pseudo-labeled samples, there is an obvious growth of classification accuracy. We use the validation set to determine the value of Z and set it to 8 for accuracy comparison in Table 1 and Table 2.

Theoretical Value. As we know, the expected accuracy $Acc(P, X)$ has a positive correlation with the expected cosine similarity. Then we derive the first-order estimation of $Acc(P, X)$ from Eq. (16) which is formulated as:

$$Acc(P, X) \approx \eta \cdot \frac{\alpha}{\sqrt{\lambda \cdot \frac{1}{K+Z} + \alpha}} \tag{25}$$

where η is a coefficient and $K + Z = T$. λ and α are values correlated with the variance term and the expectation term in Eq. (16). The theoretical values of λ and α can be approximately computed from the extracted features. Furthermore, we can compute the value of η by 1-shot and 5-shot accuracies of CSPN. Thus, the number Z is the only variable in Eq. (25). The theoretical curves are displayed as the dashed lines in Fig. 2(c) to show the impact of Z on classification

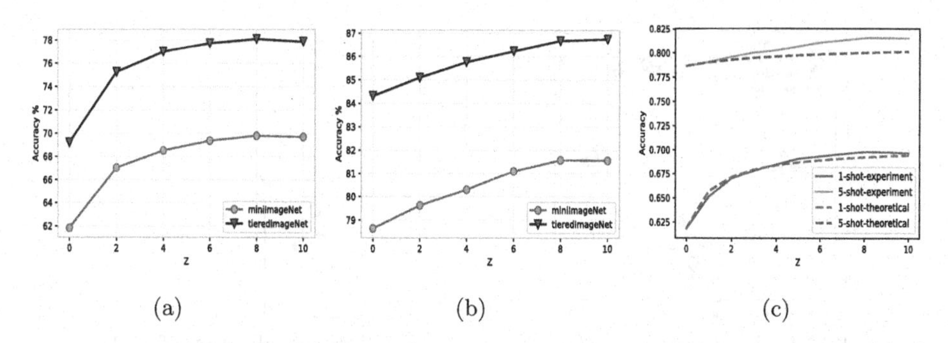

Fig. 2. Effectiveness of intra-bias diminishing. Z: the number of pseudo-labeled samples. (a) 5-way 1-shot results. (b) 5-way 5-shot results. (c) Theoretical value on miniImageNet. The experiment results (solid lines) show a consistent tendency with the theoretical results (dashed lines).

accuracy. The dashed lines, showing the theoretical lower bound of the expected accuracy, have a consistent tendency with our experiment results in Fig. 2(a)–2(b). Since the cosine similarity is continuous and the accuracy is discrete, the accuracy stops increasing when the cosine similarity grows to a certain value.

T-SNE Visualization. We show t-SNE visualization of our intra-bias diminishing method in Fig. 3(a) for intuitive illustration. The basic prototype of each class is computed from the support set while the rectified prototype is computed from the augmented support set. In this section, the expected prototype refers to the first term in Eq. (4) which is represented by the average vector of all samples (both support and query samples) of a class in an episode. Due to the scarcity of labeled samples, there is a large bias between the basic prototype and the expected prototype. The bias can be reflected by the distance between the stars and the triangles in Fig. 3(a).

Ablation of Cross-Class Bias Diminishing. Table 4 shows the ablative results of the cross-class bias diminishing module. It illustrates an overall improvement as a result of diminishing the cross-class bias. Moving the whole query set towards the support set center by importing the shifting term ξ is an effective approach to reduce the bias between the two datasets. For example, the accuracy increases by 1.64% on 1-shot tieredImageNet.

T-SNE Visualization. In few-shot learning, the support set includes far less samples compared with the query set in an episode. There exists a large distance between the two mean vectors of the datasets. We aim to decrease the distance by shifting the query samples towards the center of the support set as shown in Fig. 3(b). It depicts the spatial changing of the query samples, before and after cross-class bias diminishing. The significant part is zoomed in for clear visualization, where the query samples with BD_{cross} (marked in green) distribute more closely to the center of support set.

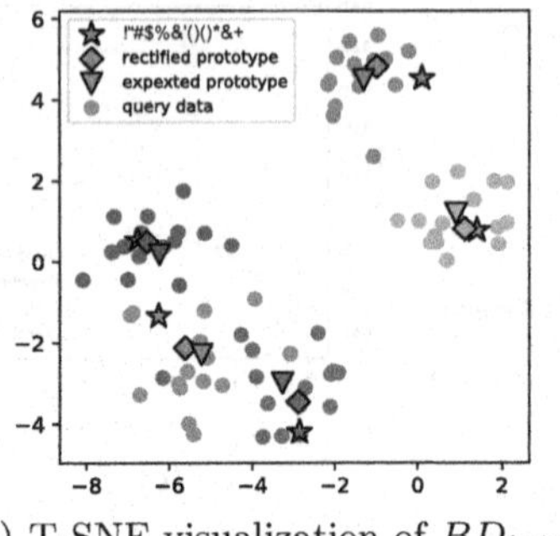

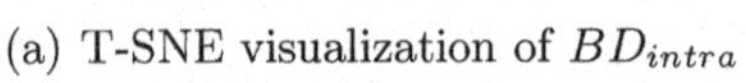

(a) T-SNE visualization of BD_{intra}

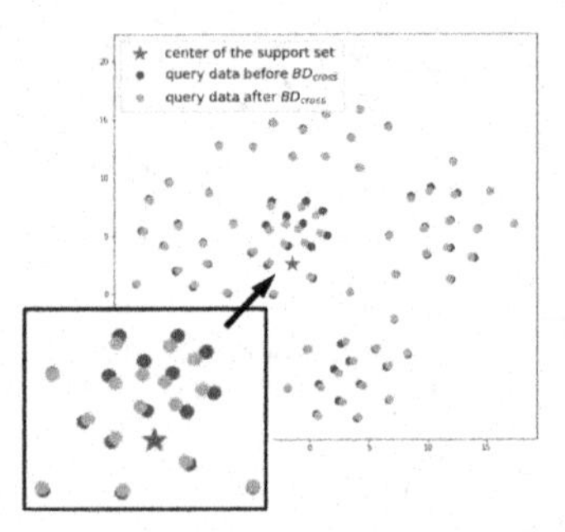

(b) T-SNE visualization of BD_{cross}

Fig. 3. We randomly sample a 5-way 1-shot episode on tieredImageNet. Different classes are marked in different colors. Best viewed in color with zoom in. (Color figure online)

Table 5. Ablation of **backbones** and result **comparison with TFT** (Transductive Fine-Tuning [4]) on miniImageNet. * The backbone is ConvNet-64.

1-shot	CSPN	BD-CSPN	TFT	5-shot	CSPN	BD-CSPN	TFT
ConvNet-128	55.62	**61.74**	50.46*	ConvNet-128	72.57	**76.12**	66.68*
ResNet-12	59.14	**65.94**	62.35	ResNet-12	76.26	**79.23**	74.53
WRN-28-10	61.84	**70.31**	65.73	WRN-28-10	78.64	**81.89**	78.40

Ablation of Backbone. The results on miniImageNet are displayed in Table 5 and more ablation results are given in the supplementary materials. Our method also shows good performance based on ConvNet-128 and ResNet-12, which is better than most approaches in Table 1. For example, with ResNet-12, we achieve 79.23% in 5-shot scenario, outperforming the strongest baselines: 78.7% [14] and 78.63% [13].

5.6 Comparison with Transductive Fine-Tuning

We compare our method with TFT [4] in Table 5, which is recently proposed as a new baseline for few-shot image classification. BD-CSPN outperforms it given different backbones. For example, we achieve better results which are higher than TFT by 3% to 5% given ResNet-12. Since BD-CSPN and TFT conduct experiments in the same transductive setting, the comparison between these two methods is more persuasive to demonstrate the effectiveness of the approach.

6 Conclusions

In this paper, we propose a powerful method of prototype rectification in few-shot learning, which is to diminish the intra-class bias and the cross-class bias of class prototypes. Our theoretical analysis verifies that, the proposed bias diminishing method is effective in raising the lower bound of the expected performance. Extensive experiments on three few-shot benchmarks demonstrate the effectiveness of our method. The proposed bias diminishing method achieves significant improvements in transductive setting by large margins (e.g. 8.47% on 1-shot miniImageNet and 9.54% on 1-shot tieredImageNet).

References

1. Allen, K., Shelhamer, E., Shin, H., Tenenbaum, J.: Infinite mixture prototypes for few-shot learning. In: ICML, pp. 232–241 (2019)
2. Andrychowicz, M., et al.: Learning to learn by gradient descent by gradient descent. In: NIPS, pp. 3981–3989 (2016)
3. Chen, W.Y., Liu, Y.C., Kira, Z., Wang, Y.C.F., Huang, J.B.: A closer look at few-shot classification. In: ICLR (2019)
4. Dhillon, G.S., Chaudhari, P., Ravichandran, A., Soatto, S.: A baseline for few-shot image classification. In: ICLR (2020)
5. Fei-Fei, L., Fergus, R., Perona, P.: One-shot learning of object categories **28**, 594–611 (2006)
6. Finn, C., Abbeel, P., Levine, S.: Model-agnostic meta-learning for fast adaptation of deep networks. In: ICML, pp. 1126–1135 (2017)
7. Gidaris, S., Bursuc, A., Komodakis, N., Perez, P.P., Cord, M.: Boosting few-shot visual learning with self-supervision. In: ICCV, pp. 8058–8067 (2019)
8. Gidaris, S., Komodakis, N.: Dynamic few-shot visual learning without forgetting. In: CVPR, pp. 4367–4375 (2018)
9. He, K., Zhang, X., Ren, S., Sun, J.: Deep residual learning for image recognition. In: CVPR, pp. 770–778 (2016)
10. Joachims, T.: Transductive inference for text classification using support vector machines. In: ICML, pp. 200–209 (1999)
11. Kim, J., Kim, T., Kim, S., Yoo, C.D.: Edge-labeling graph neural network for few-shot learning. In: CVPR, pp. 11–20 (2019)
12. Krizhevsky, A., Sutskever, I., Hinton, G.E.: ImageNet classification with deep convolutional neural networks **141**, 1097–1105 (2012)
13. Lee, K., Maji, S., Ravichandran, A., Soatto, S.: Meta-learning with differentiable convex optimization. In: CVPR, pp. 10657–10665 (2019)
14. Li, X., et al.: Learning to self-train for semi-supervised few-shot classification. In: NeurIPS (2019)
15. Li, Z., Zhou, F., Chen, F., Li, H.: Meta-SGD: learning to learn quickly for few shot learning (2017)
16. Liu, Y., et al.: Learning to propagate labels: transductive propagation network for few-shot learning. In: ICLR (2019)
17. Miller, E., Matsakis, N., Viola, P.: Learning from one example through shared densities on transforms. In: CVPR, vol. 1, pp. 464–471 (2000)
18. Mishra, N., Rohaninejad, M., Chen, X., Abbeel, P.: A simple neural attentive meta-learner. In: ICLR (2018)
19. Munkhdalai, T., Yuan, X., Mehri, S., Trischler, A.: Rapid adaptation with conditionally shifted neurons. In: ICML, pp. 3661–3670 (2018)
20. Nichol, A., Achiam, J., Schulman, J.: On first-order meta-learning algorithms (2018)

21. Nowozin, S.: Optimal decisions from probabilistic models: the intersection-over-union case. In: CVPR, pp. 548–555 (2014)
22. Oreshkin, B.N., Lpez, P.R., Lacoste, A.: TADAM: task dependent adaptive metric for improved few-shot learning. In: NIPS, pp. 721–731 (2018)
23. Qiao, S., Liu, C., Shen, W., Yuille, A.L.: Few-shot image recognition by predicting parameters from activations. In: CVPR, pp. 7229–7238 (2018)
24. Ravi, S., Larochelle, H.: Optimization as a model for few-shot learning. In: ICLR (2017)
25. Ren, M., et al.: Meta-learning for semi-supervised few-shot classification. In: ICLR (2018)
26. Rice, S.H.: The expected value of the ratio of correlated random variables. Texas Tech University (2015)
27. Russakovsky, O., et al.: ImageNet large scale visual recognition challenge **115**, 211–252 (2015)
28. Rusu, A.A., et al.: Meta-learning with latent embedding optimization. In: ICLR (2019)
29. Satorras, V.G., Estrach, J.B.: Few-shot learning with graph neural networks. In: ICLR (2018)
30. Snell, J., Swersky, K., Zemel, R.S.: Prototypical networks for few-shot learning. In: NIPS, pp. 4077–4087 (2017)
31. Sung, F., Yang, Y., Zhang, L., Xiang, T., Torr, P.H., Hospedales, T.M.: Learning to compare: relation network for few-shot learning. In: CVPR, pp. 1199–1208 (2018)
32. Triantafillou, E., et al.: Meta-dataset: a dataset of datasets for learning to learn from few examples. In: ICLR (2020)
33. Vinyals, O., Blundell, C., Lillicrap, T.P., Kavukcuoglu, K., Wierstra, D.: Matching networks for one shot learning. In: NIPS, pp. 3637–3645 (2016)
34. Wang, Y., Li, W., Dai, D., Gool, L.V.: Deep domain adaptation by geodesic distance minimization. In: ICCVW, pp. 2651–2657 (2017)
35. Zagoruyko, S., Komodakis, N.: Wide residual networks. In: BMVC (2016)
36. Zhou, D., Bousquet, O., Lal, T.N., Weston, J., Schlkopf, B.: Learning with local and global consistency. In: NIPS, pp. 321–328 (2003)

Learning Feature Descriptors Using Camera Pose Supervision

Qianqian Wang[1,2](✉), Xiaowei Zhou[3], Bharath Hariharan[1], and Noah Snavely[1,2]

[1] Cornell University, Ithaca, USA
[2] Cornell Tech, New York, USA
qw246@cornell.edu
[3] Zhejiang University, Hangzhou, China

Abstract. Recent research on learned visual descriptors has shown promising improvements in correspondence estimation, a key component of many 3D vision tasks. However, existing descriptor learning frameworks typically require ground-truth correspondences between feature points for training, which are challenging to acquire at scale. In this paper we propose a novel *weakly-supervised* framework that can learn feature descriptors *solely* from relative camera poses between images. To do so, we devise both a new loss function that exploits the epipolar constraint given by camera poses, and a new model architecture that makes the whole pipeline differentiable and efficient. Because we no longer need pixel-level ground-truth correspondences, our framework opens up the possibility of training on much larger and more diverse datasets for better and unbiased descriptors. We call the resulting descriptors **CA**mera **P**ose **S**upervised, or **CAPS**, descriptors. Though trained with weak supervision, CAPS descriptors outperform even prior *fully-supervised* descriptors and achieve state-of-the-art performance on a variety of geometric tasks. (Project page: https://qianqianwang68.github.io/CAPS/.)

Keywords: Local features · Feature descriptors · Correspondence · Image matching · Camera pose

1 Introduction

Finding local feature correspondence is a fundamental component of many computer vision tasks, such as structure from motion (SfM) [56] and visual localization [54]. Recently, learned feature descriptors [42,58,65] have shown significant improvements over hand-crafted ones [4,26,37] on standard benchmarks. However, other recent work has observed that, when applied to real-world unseen scenarios, learned descriptors do not always generalize well [39,57].

Electronic supplementary material The online version of this chapter (https://doi.org/10.1007/978-3-030-58452-8_44) contains supplementary material, which is available to authorized users.

A. Vedaldi et al. (Eds.): ECCV 2020, LNCS 12346, pp. 757–774, 2020.
https://doi.org/10.1007/978-3-030-58452-8_44

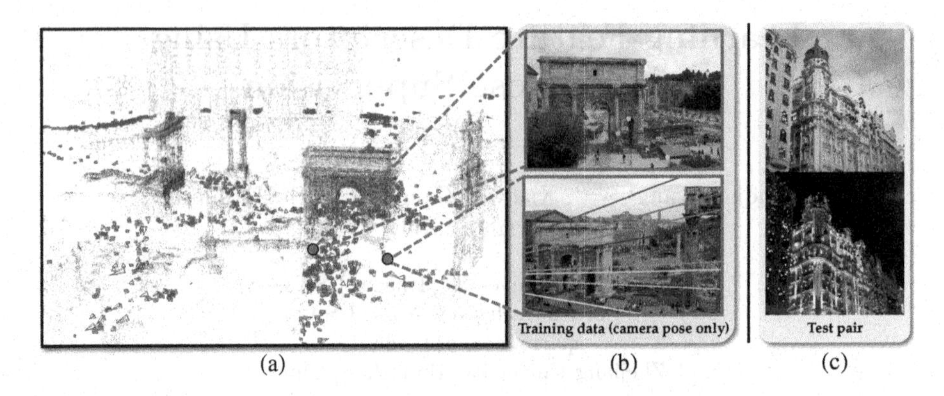

Fig. 1. Overview of our method. Our model can learn descriptor using only relative camera poses (e.g., from SfM reconstructions **(a)**). Knowing camera poses, we obtain epipolar constraints illustrated in **(b)**, where points in the first image correspond to the epipolar lines in same color in the second image. We utilize such epipolar constraints as our supervision signal (see Fig. 2). **(c)** shows that at inference, our descriptors establish reliable correspondences even for challenging image pairs.

One potential cause of such limited generalization is the insufficiency of high-quality training data in both quantity and diversity [57]. Ideally, one would train descriptors on fully accurate, dense ground-truth correspondence between image pairs. However, it is hard to collect such data for real imagery, and only a few datasets of this form exist [6,10]. As an alternative, many previous methods resort to SfM datasets that provide pseudo ground-truth correspondences given by matched and reconstructed feature points [39,42,48,65], but these correspondences are sparse and potentially biased by the keypoints used in the SfM pipeline. Another option for obtaining correspondence annotations is synthetic image pairs warped by homographies [13,40]. However, homographies do not capture the full range of geometric and photometric variations observed in real images.

In this paper, we address the challenge of limited training data in descriptor learning by relaxing this requirement of ground-truth pixel-level correspondences. We propose to learn descriptors solely from relative camera poses between pairs of images. Camera poses can be obtained via a variety of non-vision-based sensors, such as IMUs and GPS, and can also be estimated reliably using SfM pipelines [56]. By reducing the supervision requirement to camera poses, it becomes possible to learn better descriptors on much larger and more diverse datasets.

However, existing metric learning based methods for learning descriptors cannot utilize camera poses as supervision, as the triplet or contrastive losses used in such methods cannot be defined with respect to camera poses. Hence, we propose a novel framework to leverage camera pose supervision. Specifically, we translate the relative camera pose between an image pair into an epipolar constraint on pixel locations of matched points as our supervision signal (Fig. 2).

The remaining challenge is to make the locations of matched points differentiable with respect to descriptors for training, for which we introduce a new differentiable matching layer (Fig. 3(a)). To further reduce the computation cost and accelerate training, we use a coarse-to-fine matching scheme (Fig. 3(b)) that computes the correspondence at a lower resolution, then locally refines at a finer scale.

Once trained, our system can generate dense feature descriptors for an arbitrary input image, which can then be combined with existing keypoint detectors for downstream tasks. Despite the fact that we only train with *weak* camera pose supervision, our learned descriptors are on par with or even outperform prior *fully-supervised* state-of-the-art methods that train with ground-truth correspondence annotations. Furthermore, while enabling training with solely camera poses, our framework can also be trained with ground-truth correspondences, yielding even better results.

Figure 1 summarizes our approach. To conclude, our main contributions are:

- We show that camera poses alone suffice to learn good descriptors, which has not been explored in the literature to our knowledge.
- To enable learning from camera poses, we depart from existing metric learning-based approaches and design a novel loss function as well as a new, efficient network architecture.
- We achieve state-of-the-art performance across a range of geometric tasks.

2 Related Work

Descriptor Learning. The dominant paradigm for learning feature descriptors is essentially deep metric learning [8], which encourages matching points to be close whereas non-matching points to be far away in the feature space. Various loss functions (e.g., pairwise and triplet loss [3,8,13,31,66], structured loss [42,47,59,61]) have been developed. Based on the input type, current descriptor learning approaches roughly fall into two categories, *patch-based* and *dense* descriptor methods. Patch-based methods [3,15,21,27,39,42–44,48,58,61,65] produce a feature descriptor for each *patch* defined by a keypoint detector, which can be viewed as direct counterparts for hand-crafted feature descriptors [4,5,37,53]. Dense descriptor methods [9,13,14,16,34,50,55] instead use fully-convolutional neural networks [35] to extract dense feature descriptors for the whole image in one forward pass. Our method gives dense descriptors, and unlike the prior work that requires ground-truth correspondence annotations to train, we are able to learn descriptors from the weak supervision of camera pose.

Correspondence Learning. Our differentiable matching layer is related to the correlation layer and cost volume that are widely used to compute stereo correspondences [7,28] or optical flow [17,22,60] in a differentiable manner. However, the search space in these problems is limited to either a single scanline or a local patch, while in wide-baseline matching we must search for matches over

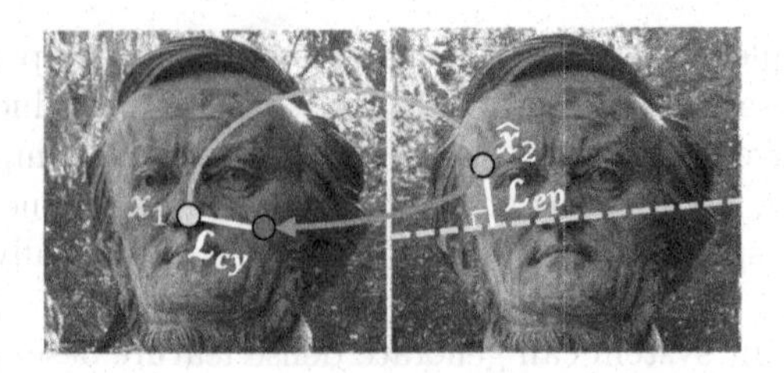

Fig. 2. Epipolar loss and cycle consistency loss. $\mathbf{x}_1$ (yellow) is the query point, and $\hat{\mathbf{x}}_2$ (orange) is the predicted correspondence. The epipolar loss $\mathcal{L}_{ep}$ is the distance between $\hat{\mathbf{x}}_2$ and ground-truth epipolar line $\mathbf{F}\mathbf{x}_1$. The cycle consistency loss $\mathcal{L}_{cy}$ is the L_2 distance between $\mathbf{x}_1$ and its forward-backward corresponding point (green). (Color figure online)

the whole image. This necessitates the efficient coarse-to-fine architecture we use. Our method is also related to weakly-supervised semantic correspondence approaches [24,29,46,51,69]. However, they usually assume a simpler parametric transformation between images and tolerate much coarser correspondences than what is required for geometric tasks. Recent work [40,52] explores dense geometric correspondence, but focuses on global optimization of the estimated correspondences rather than the descriptors themselves. In contrast to these prior work, we propose a new architecture that is more suitable for descriptor learning.

Epipolar Constraint. Epipolar constraint has been shown to be useful for learning local features [23,64] and optical flow [67]. MONET [23] proposes the epipolar divergence for learning semantic keypoints, but this loss does not apply to dense descriptor learning. [64] leverages epipolar constraints to generate pseudo-groundtruth correspondences but this process is non-differentiable. In contrast, we enable differentiable training of dense descriptors using the epipolar constraint.

3 Method

Given only image pairs with camera pose, standard deep metric learning methods do not apply. Therefore, we devise a new method to exploit the geometric information of camera pose for descriptor learning. Specifically, we translate relative camera pose into an epipolar constraint between image pairs, and enforce the predicted matches to obey this constraint (Sect. 3.1). Since this constraint is imposed on pixel coordinates, we must make the coordinates of correspondences differentiable with respect to the feature descriptors. For this we devise a differentiable matching layer (Sect. 3.2). To further improve efficiency, we introduce a coarse-to-fine architecture (Sect. 3.3) to accelerate training, which also boosts the descriptor performance. We elaborate on our method below.

3.1 Loss Formulation

Our training data consists of image pairs with relative camera poses. To train our correspondence system with such data, we propose to use two complimentary loss terms: a novel epipolar loss, and a cycle consistency loss (Fig. 2).

Given the relative pose and camera intrinsics for a pair of images $\mathbf{I}_1$ and $\mathbf{I}_2$, one can compute the fundamental matrix $\mathbf{F}$. The epipolar constraint states that $\mathbf{x}_2^T\mathbf{F}\mathbf{x}_1 = 0$ holds if $\mathbf{x}_1$ and $\mathbf{x}_2$ is a true match, where $\mathbf{F}\mathbf{x}_1$ can be interpreted as the epipolar line corresponding to $\mathbf{x}_1$ in $\mathbf{I}_2$.[1] We treat $\mathbf{x}_1$ as the query point and re-fashion this constraint into an epipolar loss based on the distance between the predicted correspondence location and the ground-truth epipolar line:

$$\mathcal{L}_{ep}(\mathbf{x}_1) = \text{dist}(h_{1\rightarrow 2}(\mathbf{x}_1), \mathbf{F}\mathbf{x}_1), \tag{1}$$

where $h_{1\rightarrow 2}(\mathbf{x}_1)$ is the predicted correspondence in $\mathbf{I}_2$ for the point $\mathbf{x}_1$ in $\mathbf{I}_1$, and $\text{dist}(\cdot,\cdot)$ is the distance between a point and a line.

The epipolar loss alone only encourages a predicted match to lie on the epipolar line, rather than near the ground-truth correspondence location (which is at an unknown location on the line). To provide additional supervision, we additionally introduce a cycle consistency loss. This loss encourages the forward-backward mapping of a point to be spatially close to itself [63]:

$$\mathcal{L}_{cy}(\mathbf{x}_1) = ||h_{2\rightarrow 1}(h_{1\rightarrow 2}(\mathbf{x}_1)) - \mathbf{x}_1||_2. \tag{2}$$

This term encourages the network to find true correspondences and suppress other outputs, especially those that satisfy the epipolar constraint alone.

Full Training Objective. For each image pair, our total objective is a weighted sum of epipolar and cycle consistency losses, totaled over n sampled query points:

$$\mathcal{L}(\mathbf{I}_1, \mathbf{I}_2) = \sum_{i=1}^{n} [\mathcal{L}_{ep}(\mathbf{x}_1^i) + \lambda\mathcal{L}_{cy}(\mathbf{x}_1^i)], \tag{3}$$

where $\mathbf{x}_1^i$ is the i-th training point in $\mathbf{I}_1$, and λ is a weight for the cycle consistency loss term. At the end of Sect. 3.2, we further show how we can reweight individual training instances in Eq. (3) to improve training.

3.2 Differentiable Matching Layer

The objective defined above is a simple function of the pixel locations of the predicted correspondences. Minimizing this objective through gradient descent therefore requires these locations to be differentiable with respect to the network parameters. Many prior methods establish correspondence by identifying nearest neighbor matches, which unfortunately is a non-differentiable operation.

To address this challenge, we propose a differentiable matching layer, illustrated in Fig. 3(a). Given a pair of images, we first use convolutional networks

[1] For simplicity, we use the same symbols for homogeneous and Cartesian coordinates.

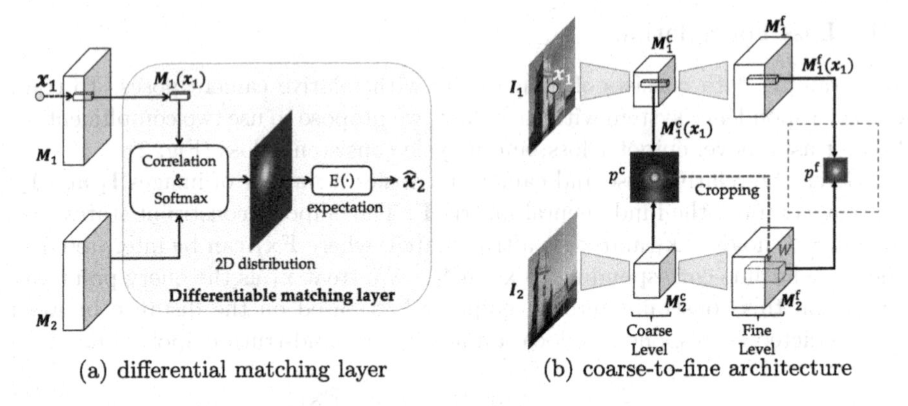

(a) differential matching layer (b) coarse-to-fine architecture

Fig. 3. Network architecture design. (a) differentiable matching layer. For a query point, its correspondence location is represented as the expectation of a distribution computed from the correlation between feature descriptors. **(b) The coarse-to-fine module.** We use the location of highest probability at coarse level (red circle) to determine the location of a local window W at the fine level. During training, we compute the correspondence locations at both coarse and fine level from distribution p^c and p^f, respectively, and impose our loss functions on both. This allows us to train both coarse- and fine-level features simultaneously. (Color figure online)

with shared weights to extract dense feature descriptors $\mathbf{M}_1$ and $\mathbf{M}_2$. To compute the correspondence for a query point $\mathbf{x}_1$ in $\mathbf{I}_1$, we correlate the feature descriptor at $\mathbf{x}_1$, denoted by $\mathbf{M}_1(\mathbf{x}_1)$, with all of $\mathbf{M}_2$. Following a 2D softmax operation [19], we obtain a distribution over 2D pixel locations of $\mathbf{I}_2$, indicating the probability of each location being the correspondence of $\mathbf{x}_1$. We denote this probability distribution as $p(\mathbf{x}|\mathbf{x}_1, \mathbf{M}_1, \mathbf{M}_2)$:

$$p(\mathbf{x}|\mathbf{x}_1, \mathbf{M}_1, \mathbf{M}_2) = \frac{\exp\left(\mathbf{M}_1(\mathbf{x}_1)^{\mathrm{T}}\mathbf{M}_2(\mathbf{x})\right)}{\sum_{\mathbf{y}\in\mathbf{I}_2}\exp\left(\mathbf{M}_1(\mathbf{x}_1)^{\mathrm{T}}\mathbf{M}_2(\mathbf{y})\right)}, \quad (4)$$

where $\mathbf{y}$ varies over the pixel grid of $\mathbf{I}_2$. A single 2D match can then be computed as the expectation of this distribution:

$$\hat{\mathbf{x}}_2 = h_{1\to 2}(\mathbf{x}_1) = \sum_{\mathbf{x}\in\mathbf{I}_2} \mathbf{x}\cdot p(\mathbf{x}|\mathbf{x}_1, \mathbf{M}_1, \mathbf{M}_2). \quad (5)$$

This makes the entire system end-to-end trainable. Since the correspondence location is computed from the correlation between feature descriptors, enforcing it to be correct would facilitate descriptor learning.

Leveraging Uncertainty During Training. This differentiable matching also provides an interpretable measure of *uncertainty.* For each query point $\mathbf{x}_1$, we can calculate the total variance $\sigma^2(\mathbf{x}_1)$ as an uncertainty measure, which is defined as the trace of the covariance matrix of the 2D distribution $p(\mathbf{x}|\mathbf{x}_1, \mathbf{M}_1, \mathbf{M}_2)$. High variance indicates multiple or diffuse modes, signifying an unreliable prediction.

This uncertainty can help identify unreliable correspondences and improve training. In particular, due to the lack of ground-truth correspondence annotations, it is unknown if a query point has a true correspondence in the other image during training (which could be missing due to occlusion or truncation). Minimizing the loss for such points can lead to incorrect training signals. To alleviate this issue, we reweight the losses for each individual point using the total variance defined above, resulting in the final weighted loss function:

$$\mathcal{L}(\mathbf{I}_1, \mathbf{I}_2) = \sum_{i=1}^{n} \frac{1}{\sigma(\mathbf{x}_1^i)} [\mathcal{L}_{ep}(\mathbf{x}_1^i) + \lambda \mathcal{L}_{cy}(\mathbf{x}_1^i)], \tag{6}$$

where the weight $1/\sigma(\mathbf{x}_1^i)$ are normalized so that they sum up to one. This weighting strategy weakens the effect of infeasible and non-discriminative training points, which we find to be critical for rapid convergence. Prior work [25,46] on semantic correspondence leverages the uncertainty in a similar way, but their uncertainty is predicted using extra network parameters whereas ours is directly derived from the learned descriptors.

3.3 Coarse-to-Fine Architecture

During training, we impose supervision only on sparsely sampled query points for each pair of images. While the computational cost is made manageable in this way, having to search correspondence over the entire image space is still costly. To overcome this issue, we propose a coarse-to-fine architecture that significantly improves computational efficiency, while preserving the resolution of learned descriptors. Figure 3(b) illustrates the coarse-to-fine module. Instead of generating a flat feature descriptor map, we produce both coarse-level feature descriptors $\mathbf{M}_1^c, \mathbf{M}_2^c$ and fine-level feature descriptors $\mathbf{M}_1^f, \mathbf{M}_2^f$.

Coarse-to-fine matching works as follows. Given a query point $\mathbf{x}_1$, we first compute the distribution $p^c(\mathbf{x}|\mathbf{x}_1, \mathbf{M}_1^c, \mathbf{M}_2^c)$ over *all locations* of the coarse feature map. At the fine level, on the contrary, we compute the fine-level distribution only in a *local window* W centered at the highest probability location in the coarse-level distribution (with coordinates rescaled appropriately). Given coarse- and fine-level distributions, correspondences at both levels can be computed. We then impose our loss function (Eq. (6)) on correspondences at both levels, which allows us to train both coarse and fine features descriptors simultaneously.

This architecture allows us to learn high-resolution descriptors without evaluating full correlation between large feature maps, significantly reducing computational cost. In addition, as observed by Liu et al. [33], we find that coarse-to-fine reasoning not only improves efficiency but also boosts matching accuracy (Sect. 4.3). By concatenating both coarse- and fine-level descriptors, we obtain the final hierarchical descriptors [16] that capture both abstract and detailed information.

3.4 Discussion

Effectiveness of Epipolar Constraint. The seemingly weak epipolar constraint actually provides empirically sufficient supervision for descriptor learning, as suggested by results in Sect. 4. One key reason is that the epipolar constraint suppresses a large number of incorrect correspondence—i.e., every point not on the epipolar line. Moreover, among all valid predictions that satisfy the epipolar constraint, true correspondences are most likely to have similar feature encodings given their local appearance similarity. Therefore, by aggregating such a geometric constraint over all training data, the network learns to encode the similarity between true correspondences, leading to effective learned descriptors.

Training with Ground-Truth Correspondence Annotations. Although the focus of this paper is on learning from camera poses alone, our system can also be trained with ground-truth correspondence annotations when such data is available. In this case, we can replace our loss functions with an L_2 distance between the pixel locations of the predicted and ground-truth correspondence. As shown in Fig. 7, our method trained with groundtruth correspondences achieves even better performance than our method trained with camera poses, with both outperforming prior fully supervised methods.

Matching at Test Time. The descriptors learned by our system can be integrated in standard feature matching pipelines. Given a detected keypoint, feature vectors in the coarse and fine feature maps are extracted by interpolation and concatenated to form the final descriptor. We then match features using the standard Euclidean distance between them.

3.5 Implementation Details

Architecture. We use a ImageNet-pretrained ResNet-50 [11,20,49] architecture, truncated after `layer3`, as our backbone. With an additional convolutional layer we obtain the coarse-level feature map. The fine-level feature map is obtained by further convolutional layers along with up-sampling and skip-connections. The sizes of the coarse- and fine-level feature map are 1/16 and 1/4 of the original image size, respectively. They both have a feature dimensionality of 128. The size of the local window W at fine level is $1/8\times$ the size of the fine-level feature map.

Training Data. We train using the MegaDepth dataset [32], which consists of 196 different scenes reconstructed from over 1M internet photos using COLMAP [56]. 130 out of 196 scenes are used for training and the rest are for validation and testing. This gives us millions of training pairs with known camera poses. We train our system on these pairs using only the provided camera poses and intrinsics.

Training Details. We train the network using Adam [30] with a base learning rate of 10^{-4}. The weight λ for the cycle consistency term is set to 0.1. $n = 500$

query points are used in each training image pair due to memory constraints. These query points consist of 90% SIFT [37] keypoints and 10% random points.

For more implementation details, please refer to the supplementary material.

4 Experimental Results

To evaluate our descriptors, referred to as CAPS, we conduct three sets of experiments:

1. **Feature matching experiments:** The most direct evaluation of CAPS is in terms of how accurately they can be matched between images. We evaluate both sparse and dense feature matching on the HPatches dataset [2].
2. **Experiments on downstream tasks:** Feature matches are rarely the end-goal. Instead, they form a core part of many 3D reconstruction tasks. We evaluate the impact of CAPS on downstream tasks (homography estimation on HPatches as well as relative pose estimation on MegaDepth [32] and ScanNet [10]) and 3D reconstruction (as part of an SfM pipeline in the ETH local feature benchmark [57]).
3. **Ablation study:** We evaluate the impact of each proposed contribution using the HPatches dataset.

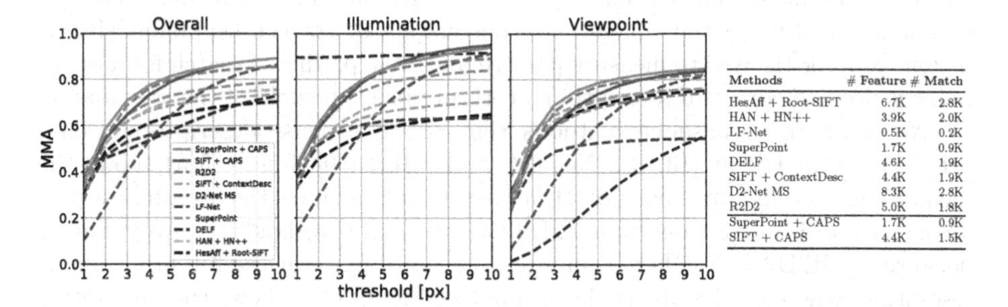

Methods	# Feature	# Match
HesAff + Root-SIFT	6.7K	2.8K
HAN + HN++	3.9K	2.0K
LF-Net	0.5K	0.2K
SuperPoint	1.7K	0.9K
DELF	4.6K	1.9K
SIFT + ContextDesc	4.4K	1.9K
D2-Net MS	8.3K	2.8K
R2D2	5.0K	1.8K
SuperPoint + CAPS	1.7K	0.9K
SIFT + CAPS	4.4K	1.5K

Fig. 4. Mean matching accuracy (MMA) on HPatches [2]. For each method, we show the MMA with varying pixel error thresholds. We also report the mean number of detected features and mutual nearest neighbor matches. With SuperPoint [13] keypoints, our approach achieves the best overall performance after 2px.

4.1 Feature Matching Results

We evaluate our descriptors on both sparse and dense feature matching on the HPatches dataset [2]. HPatches is a homography dataset containing 116 sequences, where 57 sequences have illumination changes and 59 have viewpoint changes.

Sparse Feature Matching. Given a pair of images, we extract keypoints in both images and match them using feature descriptors. We follow the same evaluation protocol as in D2-Net [14] and use the mean matching accuracy (MMA) as the evaluation metric. The MMA score is defined as the average percentage of correct matches per image pair under a certain pixel error threshold. Only mutual nearest neighbor matches are considered.

We combine CAPS with SIFT [37] and SuperPoint [13] keypoints which are representative of hand-crafted and learned keypoints, respectively. We compare to several baselines: Hessian affine detector [41] with RootSIFT descriptor [1,37] (HesAff + RootSIFT), HesAffNet [43] regions with HardNet++ descriptors [42] (HAN + HN++), DELF [45], SuperPoint [13], LF-Net [48], multi-scale D2-Net [14] (D2-Net MS), SIFT detector with ContextDesc descriptors [38] (SIFT + ContextDesc), as well as R2D2 [50].

Figure 4 shows MMA results on the HPatches dataset. We report results for the whole dataset, as well as for subsets corresponding to illumination and viewpoint changes. Following D2-Net [14], we additionally present the mean number of detected features per image and mutual nearest neighbor matches per pair. With SuperPoint keypoints CAPS achieves the best overall performance, and with SIFT keypoints CAPS also achieves competitive performance. In addition, with the same detectors, CAPS shows clear improvements over its counterparts ("SIFT + CAPS" vs. "SIFT + ContextDesc", "SuperPoint + CAPS" vs. "SuperPoint").

Dense Feature Matching. To evaluate our dense matching capability, we extract keypoints on image grids in the first image and find their nearest neighbor match in the full second image. The percentage of correct keypoints (PCK) metric [8,36,68] is used to measure performance: the predicted match for a query point is deemed correct if it is within a certain pixel threshold of the true match.

We compare to baseline methods that produce dense descriptors: Dense SIFT [37], SuperPoint [13], D2-Net [14] and R2D2 [50]. Figure 5(a) shows the mean PCK over all image pairs on HPatches. CAPS achieves the overall best performance and is only worse than R2D2 [50] at small thresholds ($\leq$ 4px). This is because the R2D2 we use here computes descriptor maps at the full input image resolution, whereas ours are 4x downsampled. Figure 5(b) shows the qualitative performance of our dense correspondence.

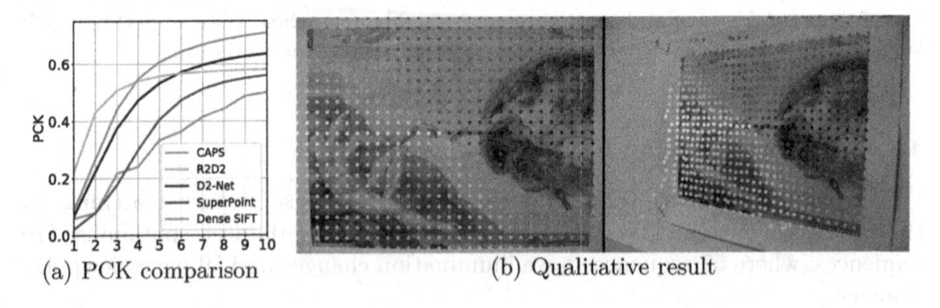

(a) PCK comparison (b) Qualitative result

Fig. 5. Dense feature matching on HPatches. (a) PCK comparison. CAPS outperforms other methods at larger pixel thresholds (> 4px). **(b)** Qualitative result of dense feature matching. Color indicates correspondence. (Color figure online)

4.2 Results on Downstream Tasks

Next, we evaluate how well CAPS facilitates downstream tasks. We focus on two tasks related to two-view geometry estimation: homography estimation and relative camera pose estimation, and a third task related to 3D reconstruction.

Homography Estimation. We use the same HPatches dataset as in Sect. 4.1 for the homography estimation task. We follow the corner correctness metric used in SuperPoint [12,13]. The four corners of one image are transformed to the other image using the estimated homography and compared with the corners computed using the groundtruth homography. The estimated homography is deemed correct if the average error of the four corners is less than ϵ pixels.

Table 1. Homography estimation accuracy [%] at 1, 3, 5 pixels on HPatches. CAPS with SuperPoint keypoints achieves the overall best performance.

Methods	$\epsilon = 1$	$\epsilon = 3$	$\epsilon = 5$
SIFT [37]	40.5	68.1	77.6
LF-Net [48]	34.8	62.9	73.8
SuperPoint [13]	37.4	73.1	82.8
D2-Net [14]	16.7	61.0	75.9
ContextDesc [38]	41.0	73.1	82.2
R2D2 [50]	40.0	**75.0**	84.7
CAPS w/ SIFT kp.	34.6	72.2	81.7
CAPS w/ SuperPoint kp.	**44.8**	74.5	**85.7**

Table 2. Relative pose estimation accuracy on ScanNet [10] **and Mega Depth** [32]. Each cell shows the accuracy of estimated rotations and translations (as *rotation accuracy/translation accuracy*). Each value shown is the percentage of pairs with relative pose error under a certain threshold (5° for ScanNet and 10° for MegaDepth). **Higher is better.** d_{frame} represents the interval between frames. Larger frame intervals imply harder pairs for matching.

Methods	Accuracy on ScanNet [%]			Accuracy on MegaDepth [%]		
	$d_{\text{frame}} = 10$	$d_{\text{frame}} = 30$	$d_{\text{frame}} = 60$	*easy*	*moderate*	*hard*
SIFT [37]	91.0/14.1	65.1/15.6	41.4/11.9	58.9/20.2	26.9/11.8	13.6/9.6
SIFT w/ ratio test [37]	91.2/15.9	67.1/19.8	44.3/15.9	63.9/25.6	36.5/17.0	20.8/13.2
SuperPoint [13]	94.4/17.5	75.9/26.3	53.4/22.1	67.2/27.1	38.7/18.8	24.5/14.1
HardNet [42]	95.8/18.2	79.0/24.7	55.6/21.8	66.3/26.7	39.3/18.8	22.5/12.3
LF-Net [48]	93.6/17.4	76.0/22.4	49.9/18.0	52.3/18.6	25.5/13.2	15.4/11.1
D2-Net [14]	91.6/13.3	68.4/19.5	42.0/14.6	61.8/23.6	35.2/19.2	19.1/12.2
ContextDesc [38]	91.5/16.3	73.8/21.8	51.4/18.5	68.9/27.1	43.1/21.5	27.5/14.1
R2D2 [50]	**97.4/22.3**	**86.1/31.7**	**62.9/28.8**	69.4/30.3	48.3/23.9	32.6/17.4
CAPS w/ SIFT kp.	92.3/16.3	74.8/22.5	50.8/20.9	70.0/30.5	50.2/24.8	36.8/16.1
CAPS w/ SuperPoint kp.	96.1/17.1	79.5/27.2	59.3/26.1	**72.9/30.5**	**53.5/27.9**	**38.1/19.2**

Following SuperPoint [13], we extract a maximum of 1,000 keypoints from each image, and robustly estimate the homography from mutual nearest neighbor matches. The comparison of homography accuracy between CAPS and other methods is shown in Table 1. As can be seen, CAPS improves over both SIFT and SuperPoint descriptors. With SuperPoint keypoints, CAPS achieves the overall best performance even without training on annotated correspondences.

Relative Pose Estimation. We also evaluate the performance of CAPS on the task of relative camera pose estimation. Note that we train only on MegaDepth [32] but test on both MegaDepth and ScanNet [10], an indoor dataset that we use to test the generalization of CAPS. For MegaDepth, we generate overlapping image pairs from test scenes, and sort them into three subsets according to relative rotation angle: *easy* ($[0°, 15°]$), *moderate* ($[15°, 30°]$) and *hard* ($[30°, 60°]$). For ScanNet, we follow LF-Net [48] and randomly sample image pairs at three different frame intervals, 10, 30, and 60. Each subset in MegaDepth and ScanNet consists of 1,000 image pairs.

To estimate relative pose, we first estimate the essential matrix from mutual nearest neighbor matches (RANSAC [18] is applied), and then decompose it to get the relative pose. For SIFT [37] we additionally prune matches using the ratio test [37], since that is the common practice for camera pose estimation (i.e, we report results of both plain SIFT and SIFT with a carefully-tuned ratio test).

Following UCN [8], we evaluate the estimated camera pose using angular deviation for both rotation and translation. We consider a rotation or translation to be correct if the angular deviation is less than a threshold, and report the average accuracy for that threshold. We set a threshold of 5° for ScanNet and 10° for MegaDepth, as MegaDepth is harder due to larger illumination changes. Results for all methods are reported in Table 2. CAPS improves performance over SIFT and SuperPoint descriptors, and “CAPS w/SuperPoint keypoints” outperforms all other methods but is outperformed by R2D2 [50] on ScanNet. Qualitative results on MegaDepth test images are shown in Fig. 6.

3D Reconstruction. Finally, we evaluate the effectiveness of CAPS descriptors in the context of 3D reconstruction using the ETH local features benchmark [57]. We extract CAPS descriptors at keypoint locations provided by [57] and feed them into the protocol. Following [39], we do not conduct the ratio test, in order to investigate the direct matching performance of the descriptors. To quantify the quality of SfM, we report the number of registered images (#Registered), sparse 3D points (#Sparse Points) and image observations (#Obs), the mean track lengths (Track Len.), and the mean reprojection error (Reproj. Err.).

We use SIFT [37], GeoDesc [39], D2-Net [14] and SOSNet [62] as baselines and show the results in Table 3. CAPS is comparable to or even outperforms our baselines in terms of the completeness of the sparse reconstruction (i.e., the number of registered images, sparse points and observations). However, we do not achieve the lowest reprojection error. A similar situation is observed in [39,62], which can be explained by the trade-off between completeness of reconstruction and low reprojection error: fewer matches tend to lead to lower reprojection error. Taking all metrics into consideration, the performance of CAPS for SfM

Fig. 6. Sparse feature matching results after RANSAC. The test image pairs are from MegaDepth [32]. Green lines indicate correspondences. Our method works well even under challenging illumination and viewpoint changes. (Color figure online)

Table 3. Evaluation on the ETH local features benchmark [57]**.** Note that SIFT and D2-Net [14] apply ratio test but other methods do not. Overall, CAPS performs on par with state-of-the-art local features on this task.

		#Registered	#Sparse points	#Obs.	Track Len.	Reproj. Err.
Madrid Metropolis 1,344 images	SIFT [37]	500	116K	734K	6.32	**0.61**px
	GeoDesc [39]	809	307K	1,200K	3.91	0.66px
	D2-Net [14]	501	84K	-	**6.33**	1.28px
	SOSNet [62]	844	**335K**	1,411K	4.21	0.70px
	CAPS	**851**	242K	**1,489K**	6.16	1.03px
Gendarmen-markt 1,463 images	SIFT	1,035	339K	1,872K	**5.52**	**0.70**px
	GeoDesc	1,208	780K	2,903K	3.72	0.74px
	D2-Net	1,053	250K	-	5.08	1.19px
	SOSNet	**1,201**	**816K**	3,255K	3.98	0.77px
	CAPS	1,179	627K	**3,330K**	5.31	1.00px
Tower of London 1,576 images	SIFT	804	240K	1,863K	**7.77**	**0.62**px
	GeoDesc	1,081	**622K**	**2,852K**	4.58	0.69px
	D2-Net	785	180K	-	5.32	1.24px
	CAPS	**1,104**	452K	2,627K	5.81	0.98px

is competitive, indicating the advantages of CAPS even trained with only weak pose supervision.

4.3 Ablation Analysis

In this section, we conduct ablation analysis to demonstrate the effectiveness of our proposed camera pose supervision and architectural designs. We follow the evaluation protocol in Sect. 4.1 and report MMA and PCK score over all image pairs in the HPatches dataset [2]. For sparse feature matching, we combine our descriptors with SIFT [37] keypoints. The variants of our default method (*Ours*) are introduced below. For fair comparison, we train each variant on the same training data (~20K image pairs) from MegaDepth [32] for 10 epochs.

Variants. *Ours from scratch* is trained from scratch instead of using ImageNet [11] pretrained weights. *Ours supervised* is trained on sparse ground-truth correspondences provided by the SfM models of MegaDepth [32]. We simply change the epipolar loss to a L_2 loss between predicted and groundtruth correspondence locations. *Triplet Loss* is also trained on sparse ground-truth correspondences, but using a standard triplet loss and a hard negative mining strategy [8]. *Ours w/o c2f* is a single-scale version of our method, where the coarse-level feature maps are removed and only the fine-level feature maps are trained and used as descriptors. *Ours w/o cycle* does not use the cycle consistency loss term ($\lambda = 0$), and *Ours w/o reweighting* does not use the uncertainty re-weighting strategy, but uses uniform weights during training. Below we provide a detailed analysis based on these variants. The results are shown in Fig. 7.

Analysis of Supervision Signal. Both *Ours supervised* and *Ours* outperform the plain version of *Triplet Loss*, where *Ours supervised* and *Triplet Loss* share the same correspondence annotations but *Ours* uses only camera pose. *Ours supervised* outperforms *Triplet Loss* because of the geometric distance-based losses (as opposed to metric learning) and the coarse-to-fine architecture. Compared to *Ours supervised*, the gains of *Ours* decrease a bit, but our epipolar loss still leverages the rich information in the epipolar constraint and allows us to outperform *Triplet Loss* and other past fully supervised work in Sect. 4. In terms of loss functions, cycle consistency only provides marginal improvement, and training with only cycle consistency loss fails. This validates the importance of epipolar constraint. *Ours from scratch* shows that even with randomly initialized weights, our network still succeeds to converge and learn descriptors, further validating the effectiveness of our loss functions.

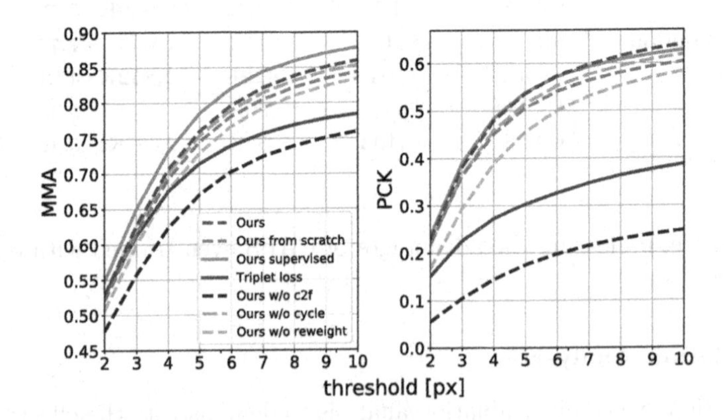

Fig. 7. Ablation study on HPatches. Solid lines indicate methods trained with ground-truth correspondence; dashed lines indicate ones trained with only camera pose.

Analysis of Architecture Design. As shown in Fig. 7, the coarse-to-fine module significantly improves performance. Two explanations for this improvement include: 1) At the fine level, correspondence is computed within a local window, which may reduce issues arising from multi-modal distributions compared to a flat model that computes expectations over the whole image; and 2) The coarse-to-fine module produces hierarchical feature descriptors that capture both global and local information, which may be beneficial for feature matching.

5 Conclusion

In this paper, we propose a novel descriptor learning framework that can be trained using only camera pose supervision. We present both new loss functions that exploit the epipolar constraints, and a new efficient architectural design that enables learning by making the correspondence differentiable. Experiments showed that our method achieves state-of-the-art performance across a range of geometric tasks, outperforming fully supervised counterparts without using any correspondence annotations for training. In future work, we will study how to further improve invariance of the learned descriptors to large geometric transformations. It is also worth investigating if the pose supervision and traditional metric learning losses are complementary to each other, and if their combination can lead to even better performance.

Acknowledgements. We thank Kai Zhang, Zixin Luo, Zhengqi Li for helpful discussion and comments. This work was partly supported by a DARPA LwLL grant, and in part by the generosity of Eric and Wendy Schmidt by recommendation of the Schmidt Futures program.

References

1. Arandjelović, R., Zisserman, A.: Three things everyone should know to improve object retrieval. In: CVPR (2012)
2. Balntas, V., Lenc, K., Vedaldi, A., Mikolajczyk, K.: HPatches: a benchmark and evaluation of handcrafted and learned local descriptors. In: CVPR (2017)
3. Balntas, V., Riba, E., Ponsa, D., Mikolajczyk, K.: Learning local feature descriptors with triplets and shallow convolutional neural networks. In: Proceedings of the British Machine Vision Conference (BMVC), p. 3 (2016)
4. Bay, H., Tuytelaars, T., Van Gool, L.: SURF: speeded up robust features. In: Leonardis, A., Bischof, H., Pinz, A. (eds.) ECCV 2006. LNCS, vol. 3951, pp. 404–417. Springer, Heidelberg (2006). https://doi.org/10.1007/11744023_32
5. Calonder, M., Lepetit, V., Strecha, C., Fua, P.: BRIEF: binary robust independent elementary features. In: Daniilidis, K., Maragos, P., Paragios, N. (eds.) ECCV 2010. LNCS, vol. 6314, pp. 778–792. Springer, Heidelberg (2010). https://doi.org/10.1007/978-3-642-15561-1_56
6. Chang, A., et al.: Matterport3D: learning from RGB-D data in indoor environments. arXiv preprint arXiv:1709.06158 (2017)
7. Chang, J.R., Chen, Y.S.: Pyramid stereo matching network. In: CVPR (2018)

8. Choy, C.B., Gwak, J., Savarese, S., Chandraker, M.: Universal correspondence network. In: NeurIPS (2016)
9. Christiansen, P.H., Kragh, M.F., Brodskiy, Y., Karstoft, H.: UnsuperPoint: end-to-end unsupervised interest point detector and descriptor. arXiv preprint arXiv:1907.04011 (2019)
10. Dai, A., Chang, A.X., Savva, M., Halber, M., Funkhouser, T., Nießner, M.: ScanNet: richly-annotated 3D reconstructions of indoor scenes. In: CVPR (2017)
11. Deng, J., Dong, W., Socher, R., Li, L.J., Li, K., Fei-Fei, L.: ImageNet: a large-scale hierarchical image database. In: CVPR (2009)
12. DeTone, D., Malisiewicz, T., Rabinovich, A.: Deep image homography estimation. arXiv preprint arXiv:1606.03798 (2016)
13. DeTone, D., Malisiewicz, T., Rabinovich, A.: SuperPoint: self-supervised interest point detection and description. In: CVPR Workshops (2018)
14. Dusmanu, M., et al.: D2-Net: a trainable CNN for joint detection and description of local features. arXiv preprint arXiv:1905.03561 (2019)
15. Ebel, P., Mishchuk, A., Yi, K.M., Fua, P., Trulls, E.: Beyond cartesian representations for local descriptors. In: ICCV (2019)
16. Fathy, M.E., Tran, Q.-H., Zia, M.Z., Vernaza, P., Chandraker, M.: Hierarchical metric learning and matching for 2D and 3D geometric correspondences. In: Ferrari, V., Hebert, M., Sminchisescu, C., Weiss, Y. (eds.) ECCV 2018. LNCS, vol. 11219, pp. 832–850. Springer, Cham (2018). https://doi.org/10.1007/978-3-030-01267-0_49
17. Fischer, P., et al.: FlowNet: learning optical flow with convolutional networks. arXiv preprint arXiv:1504.06852 (2015)
18. Fischler, M.A., Bolles, R.C.: Random sample consensus: a paradigm for model fitting with applications to image analysis and automated cartography. Commun. ACM **24**(6), 381–395 (1981)
19. Goodfellow, I., Bengio, Y., Courville, A.: Deep Learning. MIT Press, Cambridge (2016). http://www.deeplearningbook.org
20. He, K., Zhang, X., Ren, S., Sun, J.: Deep residual learning for image recognition. In: CVPR (2016)
21. He, K., Lu, Y., Sclaroff, S.: Local descriptors optimized for average precision. In: CVPR (2018)
22. Ilg, E., Mayer, N., Saikia, T., Keuper, M., Dosovitskiy, A., Brox, T.: FlowNet 2.0: evolution of optical flow estimation with deep networks. In: CVPR (2017)
23. Jafarian, Y., Yao, Y., Park, H.S.: MONET: multiview semi-supervised keypoint via epipolar divergence. arXiv preprint arXiv:1806.00104 (2018)
24. Jeon, S., Kim, S., Min, D., Sohn, K.: PARN: pyramidal affine regression networks for dense semantic correspondence. In: Ferrari, V., Hebert, M., Sminchisescu, C., Weiss, Y. (eds.) ECCV 2018. LNCS, vol. 11210, pp. 355–371. Springer, Cham (2018). https://doi.org/10.1007/978-3-030-01231-1_22
25. Jeon, S., Min, D., Kim, S., Sohn, K.: Joint learning of semantic alignment and object landmark detection. In: ICCV (2019)
26. Ke, Y., Sukthankar, R.: PCA-SIFT: a more distinctive representation for local image descriptors. In: CVPR (2004)
27. Keller, M., Chen, Z., Maffra, F., Schmuck, P., Chli, M.: Learning deep descriptors with scale-aware triplet networks. In: CVPR (2018)
28. Kendall, A., et al.: End-to-end learning of geometry and context for deep stereo regression. In: ICCV (2017)
29. Kim, S., Lin, S., Jeon, S.R., Min, D., Sohn, K.: Recurrent transformer networks for semantic correspondence. In: NeurIPS (2018)

30. Kingma, D.P., Ba, J.: Adam: a method for stochastic optimization. arXiv preprint arXiv:1412.6980 (2014)
31. Kumar, B., Carneiro, G., Reid, I., et al.: Learning local image descriptors with deep siamese and triplet convolutional networks by minimising global loss functions. In: CVPR (2016)
32. Li, Z., Snavely, N.: MegaDepth: learning single-view depth prediction from internet photos. In: CVPR (2018)
33. Liu, C., Yuen, J., Torralba, A., Sivic, J., Freeman, W.T.: SIFT Flow: dense correspondence across different scenes. In: Forsyth, D., Torr, P., Zisserman, A. (eds.) ECCV 2008. LNCS, vol. 5304, pp. 28–42. Springer, Heidelberg (2008). https://doi.org/10.1007/978-3-540-88690-7_3
34. Liu, Y., Shen, Z., Lin, Z., Peng, S., Bao, H., Zhou, X.: Gift: learning transformation-invariant dense visual descriptors via group CNNs. In: NeurIPS (2019)
35. Long, J., Shelhamer, E., Darrell, T.: Fully convolutional networks for semantic segmentation. In: CVPR (2015)
36. Long, J.L., Zhang, N., Darrell, T.: Do convnets learn correspondence? In: NeurIPS (2014)
37. Lowe, D.G.: Distinctive image features from scale-invariant keypoints. IJCV **60**(2), 91–110 (2004)
38. Luo, Z., et al.: ContextDesc: local descriptor augmentation with cross-modality context. In: CVPR (2019)
39. Luo, Z., et al.: GeoDesc: learning local descriptors by integrating geometry constraints. In: Ferrari, V., Hebert, M., Sminchisescu, C., Weiss, Y. (eds.) ECCV 2018. LNCS, vol. 11213, pp. 170–185. Springer, Cham (2018). https://doi.org/10.1007/978-3-030-01240-3_11
40. Melekhov, I., Tiulpin, A., Sattler, T., Pollefeys, M., Rahtu, E., Kannala, J.: DGC-Net: dense geometric correspondence network. In: WACV (2019)
41. Mikolajczyk, K., Schmid, C.: Scale & affine invariant interest point detectors. IJCV **60**(1), 63–86 (2004). https://doi.org/10.1023/B:VISI.0000027790.02288.f2
42. Mishchuk, A., Mishkin, D., Radenovic, F., Matas, J.: Working hard to know your neighbor's margins: local descriptor learning loss. In: NeurIPS (2017)
43. Mishkin, D., Radenović, F., Matas, J.: Repeatability is not enough: learning affine regions via discriminability. In: Ferrari, V., Hebert, M., Sminchisescu, C., Weiss, Y. (eds.) ECCV 2018. LNCS, vol. 11213, pp. 287–304. Springer, Cham (2018). https://doi.org/10.1007/978-3-030-01240-3_18
44. Mukundan, A., Tolias, G., Chum, O.: Explicit spatial encoding for deep local descriptors. In: CVPR (2019)
45. Noh, H., Araujo, A., Sim, J., Weyand, T., Han, B.: Large-scale image retrieval with attentive deep local features. In: ICCV (2017)
46. Novotny, D., Albanie, S., Larlus, D., Vedaldi, A.: Self-supervised learning of geometrically stable features through probabilistic introspection. In: CVPR (2018)
47. Oh Song, H., Xiang, Y., Jegelka, S., Savarese, S.: Deep metric learning via lifted structured feature embedding. In: CVPR (2016)
48. Ono, Y., Trulls, E., Fua, P., Yi, K.M.: LF-Net: learning local features from images. In: NeurIPS (2018)
49. Paszke, A., et al.: Automatic differentiation in PyTorch. In: NIPS Autodiff Workshop (2017)
50. Revaud, J., Weinzaepfel, P., de Souza, C.R., Humenberger, M.: R2D2: repeatable and reliable detector and descriptor. In: NeurIPS (2019)
51. Rocco, I., Arandjelovic, R., Sivic, J.: Convolutional neural network architecture for geometric matching. In: CVPR (2017)

52. Rocco, I., Cimpoi, M., Arandjelović, R., Torii, A., Pajdla, T., Sivic, J.: Neighbourhood consensus networks. In: NeurIPS (2018)
53. Rublee, E., Rabaud, V., Konolige, K., Bradski, G.R.: ORB: an efficient alternative to SIFT or SURF. In: Proceedings of the International Conference on Computer Vision (ICCV). Citeseer (2011)
54. Sattler, T., et al.: Benchmarking 6dof outdoor visual localization in changing conditions. In: CVPR (2018)
55. Schmidt, T., Newcombe, R., Fox, D.: Self-supervised visual descriptor learning for dense correspondence. IEEE Robot. Autom. Lett. **2**(2), 420–427 (2016)
56. Schonberger, J.L., Frahm, J.M.: Structure-from-motion revisited. In: CVPR (2016)
57. Schönberger, J.L., Hardmeier, H., Sattler, T., Pollefeys, M.: Comparative evaluation of hand-crafted and learned local features. In: CVPR (2017)
58. Simo-Serra, E., Trulls, E., Ferraz, L., Kokkinos, I., Fua, P., Moreno-Noguer, F.: Discriminative learning of deep convolutional feature point descriptors. In: ICCV (2015)
59. Sohn, K.: Improved deep metric learning with multi-class N-pair loss objective. In: NeurIPS (2016)
60. Sun, D., Yang, X., Liu, M.Y., Kautz, J.: PWC-Net: CNNs for optical flow using pyramid, warping, and cost volume. In: CVPR (2018)
61. Tian, Y., Fan, B., Wu, F.: L2-Net: deep learning of discriminative patch descriptor in Euclidean space. In: CVPR (2017)
62. Tian, Y., Yu, X., Fan, B., Wu, F., Heijnen, H., Balntas, V.: SOSNet: second order similarity regularization for local descriptor learning. In: CVPR (2019)
63. Wang, X., Jabri, A., Efros, A.A.: Learning correspondence from the cycle-consistency of time. In: CVPR (2019)
64. Yang, G., et al.: Learning data-adaptive interest points through epipolar adaptation. In: CVPR Workshops (2019)
65. Yi, K.M., Trulls, E., Lepetit, V., Fua, P.: LIFT: learned invariant feature transform. In: Leibe, B., Matas, J., Sebe, N., Welling, M. (eds.) ECCV 2016. LNCS, vol. 9910, pp. 467–483. Springer, Cham (2016). https://doi.org/10.1007/978-3-319-46466-4_28
66. Zhang, L., Rusinkiewicz, S.: Learning local descriptors with a CDF-based dynamic soft margin. In: ICCV (2019)
67. Zhong, Y., Ji, P., Wang, J., Dai, Y., Li, H.: Unsupervised deep epipolar flow for stationary or dynamic scenes. In: CVPR (2019)
68. Zhou, T., Jae Lee, Y., Yu, S.X., Efros, A.A.: FlowWeb: joint image set alignment by weaving consistent, pixel-wise correspondences. In: CVPR (2015)
69. Zhou, T., Krahenbuhl, P., Aubry, M., Huang, Q., Efros, A.A.: Learning dense correspondence via 3D-guided cycle consistency. In: CVPR (2016)

Semantic Flow for Fast and Accurate Scene Parsing

Xiangtai Li[1], Ansheng You[1], Zhen Zhu[2], Houlong Zhao[3], Maoke Yang[3], Kuiyuan Yang[3], Shaohua Tan[1], and Yunhai Tong[1(✉)]

[1] Key Laboratory of Machine Perception, MOE, School of EECS, Peking University, Beijing, China
yhtong@pku.edu.cn
[2] Huazhong University of Science and Technology, Wuhan, China
[3] DeepMotion, Wuhan, China

Abstract. In this paper, we focus on designing effective method for fast and accurate scene parsing. A common practice to improve the performance is to attain high resolution feature maps with strong semantic representation. Two strategies are widely used—atrous convolutions and feature pyramid fusion, are either computation intensive or ineffective. Inspired by the Optical Flow for motion alignment between adjacent video frames, we propose a Flow Alignment Module (FAM) to learn Semantic Flow between feature maps of adjacent levels, and broadcast high-level features to high resolution features effectively and efficiently. Furthermore, integrating our module to a common feature pyramid structure exhibits superior performance over other real-time methods even on light-weight backbone networks, such as ResNet-18. Extensive experiments are conducted on several challenging datasets, including Cityscapes, PASCAL Context, ADE20K and CamVid. Especially, our network is the first to achieve 80.4% mIoU on Cityscapes with a frame rate of 26 FPS. The code is available at https://github.com/donnyyou/torchcv.

Keywords: Scene parsing · Semantic flow · Flow alignment module

1 Introduction

Scene parsing or semantic segmentation is a fundamental vision task which aims to classify each pixel in the images correctly. Two important factors that are highly influential to the performance are: detailed information [46] and strong semantics representation [6,64]. The seminal work of Long *et. al.* [33] built a

X. Li and A. You—Equal contribution.

Electronic supplementary material The online version of this chapter (https://doi.org/10.1007/978-3-030-58452-8_45) contains supplementary material, which is available to authorized users.

A. Vedaldi et al. (Eds.): ECCV 2020, LNCS 12346, pp. 775–793, 2020.
https://doi.org/10.1007/978-3-030-58452-8_45

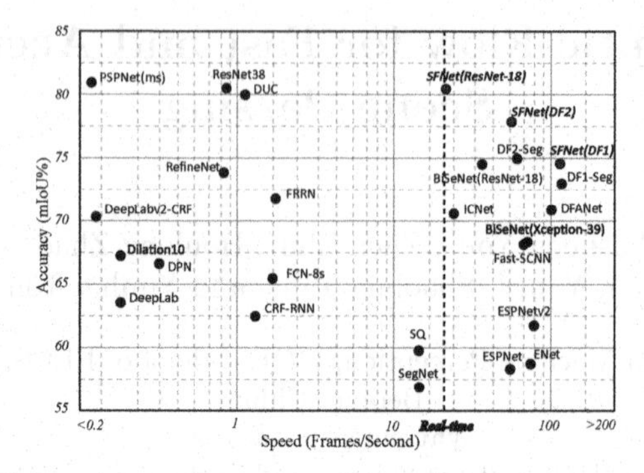

Fig. 1. Inference speed versus mIoU performance on test set of Cityscapes. Previous models are marked as red points, and our models are shown in blue points which achieve the best speed/accuracy trade-off. Note that our method with ResNet-18 as backbone even achieves comparable accuracy with all accurate models at much faster speed. (Color figure online)

deep Fully Convolutional Network (FCN), which is mainly composed from convolutional layers, in order to carve strong semantic representation. However, detailed object boundary information, which is also crucial to the performance, is usually missing due to the use of the down-sampling layers. To alleviate this problem, state-of-the-art methods [15,64,65,68] apply atrous convolutions [55] at the last several stages of their networks to yield feature maps with strong semantic representation while at the same time maintaining the high resolution.

Nevertheless, doing so inevitably requires intensive extra computation since the feature maps in the last several layers can reach up to 64 times bigger than those in FCNs. Given that the FCN using ResNet-18 [19] as the backbone network has a frame rate of 57.2 FPS for a 1024 × 2048 image, after applying atrous convolutions [55] to the network as done in [64,65], the modified network only has a frame rate of 8.7 FPS. Moreover, under a single GTX 1080Ti GPU with no other ongoing programs, the previous state-of-the-art model PSPNet [64] has a frame rate of only 1.6 FPS for 1024 × 2048 input images. As a consequence, this is very problematic to many advanced real-world applications, such as self-driving cars and robots navigation, which desperately demand real-time online data processing.

In order to not only maintain detailed resolution information but also get features that exhibit strong semantic representation, another direction is to build FPN-like [23,32,46] models which leverage the lateral path to fuse feature maps in a top-down manner. In this way, the deep features of the last several layers strengthen the shallow features with high resolution and therefore, the refined features are possible to satisfy the above two factors and beneficial to the accuracy improvement. However, the accuracy of these methods [1,46] is still unsatisfactory when compared to those networks who hold large feature maps in the last several stages. We suspect the low accuracy problem arises from the ineffective propagation of semantics from deep layers to shallow layers.

To mitigate this issue, we propose to learn the **Semantic Flow** between two network layers of different resolutions. The concept of Semantic Flow is inspired from optical flow, which is widely used in video processing task [67] to represent the pattern of apparent motion of objects, surfaces, and edges in a visual scene caused by relative motion. In a flash of inspiration, we feel the relationship between two feature maps of arbitrary resolutions from the same image can also be represented with the "motion" of every pixel from one feature map to the other one. In this case, once precise Semantic Flow is obtained, the network is able to propagate semantic features with minimal information loss. It should be noted that Semantic Flow is apparently different from optical flow, since Semantic Flow takes feature maps from different levels as input and assesses the discrepancy within them to find a suitable flow field that will give dynamic indication about how to align these two feature maps effectively.

Based on the concept of Semantic Flow, we design a novel network module called Flow Alignment Module (FAM) to utilize Semantic Flow in the scene parsing task. Feature maps after FAM are embodied with both rich semantics and abundant spatial information. Because FAM can effectively transmit the semantic information from deep layers to shallow layers through very simple operations, it shows superior efficacy in both improving the accuracy and keeping superior efficiency. Moreover, FAM is end-to-end trainable, and can be plugged into any backbone networks to improve the results with a minor computational overhead. For simplicity, we call the networks that all incorporate FAM but have different backbones as **SFNet(backbone)**. As depicted in Fig. 1, SFNet with different backbone networks outperforms other competitors by a large margin under the same speed. In particular, our method adopting ResNet-18 as backbone achieves **80.4%** mIoU on the Cityscapes test server with a frame rate of **26 FPS**. When adopting DF2 [29] as backbone, our method achieves 77.8% mIoU with 61 FPS and 74.5% mIoU with 121 FPS when equipped with the DF1 backbone. Moreover, when using deeper backbone networks, such as ResNet-101, SFNet achieves better results(81.8% mIoU) than the previous state-of-the-art model DANet [15](81.5% mIoU), and only requires **33%** computation of DANet during the inference. Besides, the consistent superior efficacy of SFNet across various datasets also clearly demonstrates its broad applicability.

To conclude, our main contributions are three-fold:

- We introduce the concept of Semantic Flow in the field of scene parsing and propose a novel flow-based align module (FAM) to learn the Semantic Flow between feature maps of adjacent levels and broadcast high-level features to high resolution features more effectively and efficiently.
- We insert FAMs into the feature pyramid framework and build a feature pyramid aligned network called SFNet for fast and accurate scene parsing.
- Detailed experiments and analysis indicate the efficacy of our proposed module in both improving the accuracy and keeping light-weight. We achieve state-of-the-art results on Cityscapes, Pascal Context, Camvid datasets and a considerable gain on ADE20K.

2 Related Work

For scene parsing, there are mainly two paradigms for high-resolution semantic map prediction. One paradigm tries to keep both spatial and semantic information along the main network pathway, while the other paradigm distributes spatial and semantic information to different parts in a network, then merges them back via different strategies.

The first paradigm mostly relies on some network operations to retain high-resolution feature maps in the latter network stages. Many state-of-the-art accurate methods [15,64,68] follow this paradigm to design sophisticated head networks to capture contextual information. PSPNet [64] proposes to leverage pyramid pooling module (PPM) to model multi-scale contexts, whilst DeepLab series [5–7,52] uses astrous spatial pyramid pooling (ASPP). In [15,17,18,20,27,56,69], non-local operator [50] and self-attention mechanism [49] are adopted to harvest pixel-wise context from the whole image. Meanwhile, several works [22,26,30,59,60] use graph convolutional neural networks to propagate information over the image by projecting features into an interaction space.

The second paradigm contains several state-of-the-art fast methods, where high-level semantics are represented by low-resolution feature maps. A common strategy is to fuse multi-level feature maps for high-resolution spatiality and strong semantics [1,28,33,46,51]. ICNet [63] uses multi-scale images as input and a cascade network to be more efficient. DFANet [25] utilizes a light-weight backbone to speed up its network and proposes a cross-level feature aggregation to boost accuracy, while SwiftNet [42] uses lateral connections as the cost-effective solution to restore the prediction resolution while maintaining the speed. To further speed up, low-resolution images are used as input for high-level semantics [35,63] which reduce features into low resolution and then upsample them back by a large factor. The direct consequence of using a large upsample factor is performance degradation, especially for small objects and object boundaries. Guided upsampling [35] is related to our method, where the semantic map is upsampled back to the input image size guided by the feature map from an early layer. However, this guidance is still insufficient for some cases due to the information gap between the semantics and resolution. In contrast, our method aligns feature maps from adjacent levels and further enhances the feature maps using a feature pyramid framework towards both high resolution and strong semantics, consequently resulting in the state-of-the-art performance considering the trade-off between high accuracy and fast speed.

There is another set of works focusing on designing light-weight backbone networks to achieve real-time performances. ESPNets [36,37] save computation by decomposing standard convolution into point-wise convolution and spatial pyramid of dilated convolutions. BiSeNet [53] introduces spatial path and semantic path to reduce computation. Recently, several methods [29,39,62] use AutoML techniques to search efficient architectures for scene parsing. Our method is complementary to some of these works, which further boosts their accuracy. Since our proposed semantic flow is inspired by optical flow [13], which is used in video

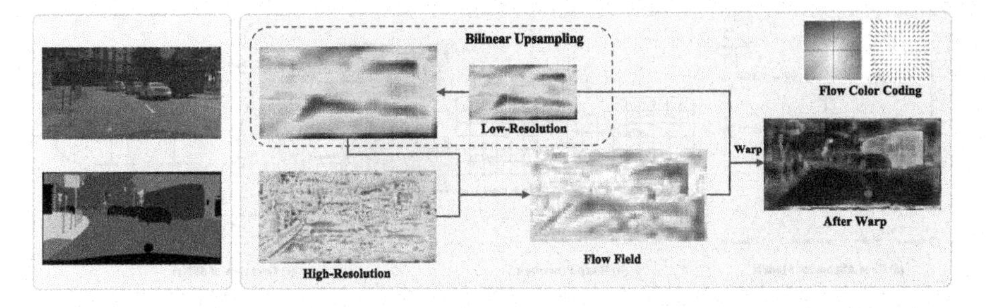

Fig. 2. Visualization of feature maps and semantic flow field in FAM. Feature maps are visualized by averaging along the channel dimension. Larger values are denoted by hot colors and vice versa. We use the color code proposed in [2] to visualize the Semantic Flow field. The orientation and magnitude of flow vectors are represented by hue and saturation respectively.

semantic segmentation, we also discuss several works in video semantic segmentation. For accurate results, temporal information is exceedingly exploited by using optical flow. Gadde *et. al.* [16] warps internal feature maps and Nilsson *et. al.* [41] warps final semantic maps from nearby frame predictions to the current map. To pursue faster speed, optical flow is used to bypass the low-level feature computation of some frames by warping features from their preceding frames [31,67]. Our work is different from theirs by propagating information hierarchically in another dimension, which is orthogonal to the temporal propagation for videos.

3 Method

In this section, we will first give some preliminary knowledge about scene parsing and introduce the misalignment problem therein. Then, we propose the Flow Alignment Module (FAM) to resolve the misalignment issue by learning Semantic Flow and warping top-layer feature maps accordingly. Finally, we present the whole network architecture equipped with FAMs based on the FPN framework [32] for fast and accurate scene parsing.

3.1 Preliminary

The task of scene parsing is to map an RGB image $\mathbf{X} \in \mathbb{R}^{H \times W \times 3}$ to a semantic map $\mathbf{Y} \in \mathbb{R}^{H \times W \times C}$ with the same spatial resolution $H \times W$, where C is the number of predefined semantic categories. Following the setting of FPN [32], the input image $\mathbf{X}$ is firstly mapped to a set of feature maps $\{\mathbf{F}_l\}_{l=2,...,5}$ from each network stage, where $\mathbf{F}_l \in \mathbb{R}^{H_l \times W_l \times C_l}$ is a C_l-dimensional feature map defined on a spatial grid Ω_l with size of $H_l \times W_l$, $H_l = \frac{H}{2^l}$, $W_l = \frac{W}{2^l}$. The coarsest feature map $\mathbf{F}_5$ comes from the deepest layer with strongest semantics. FCN-32s directly predicts upon $\mathbf{F}_5$ and achieves over-smoothed results without fine details. However, some improvements can be achieved by fusing predictions from

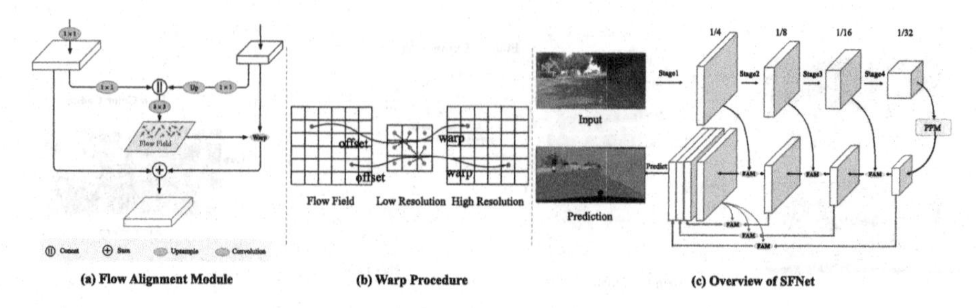

Fig. 3. (a) The details of Flow Alignment Module. We combine the transformed high-resolution feature map and low-resolution feature map to generate the semantic flow field, which is utilized to warp the low-resolution feature map to high-resolution feature map. (b) Warp procedure of Flow Alignment Module. The value of the high-resolution feature map is the bilinear interpolation of the neighboring pixels in low-resolution feature map, where the neighborhoods are defined according learned semantic flow field. (c) Overview of our proposed SFNet. ResNet-18 backbone with four stages is used for exemplar illustration. FAM: Flow Alignment Module. PPM: Pyramid Pooling Module [64]. Best view it in color and zoom in. (Color figure online)

lower levels [33]. FPN takes a step further to gradually fuse high-level feature maps with low-level feature maps in a top-down pathway through 2× bi-linear upsampling, which was originally proposed for object detection [32] and recently introduced for scene parsing [23,51]. The whole FPN framework highly relies on upsampling operator to upsample the spatially smaller but semantically stronger feature map to be larger in spatial size. However, the bilinear upsampling recovers the resolution of downsampled feature maps by interpolating a set of uniformly sampled positions (i.e., it can only handle one kind of fixed and predefined misalignment), while the misalignment between feature maps caused by a residual connection, repeated downsampling and upsampling, is far more complex. Therefore, position correspondence between feature maps needs to be explicitly and dynamically established to resolve their actual misalignment.

3.2 Flow Alignment Module

Design Motivation. For more flexible and dynamic alignment, we thoroughly investigate the idea of optical flow, which is very effective and flexible to align two adjacent video frame features in the video processing task [4,67]. The idea of optical flow motivates us to design a *flow-based alignment module* (**FAM**) to align feature maps of two adjacent levels by predicting a flow field inside the network. We define such flow field as *Semantic Flow*, which is generated between different levels in a feature pyramid. For efficiency, while designing our network, we adopt an efficient backbone network—FlowNet-S [13].

Module Details. FAM is built within the FPN framework, where feature map of each level is compressed into the same channel depth through two 1×1 convolution layers before entering the next level. Given two adjacent feature maps

$\mathbf{F}_l$ and $\mathbf{F}_{l-1}$ with the same channel number, we up-sample $\mathbf{F}_l$ to the same size as $\mathbf{F}_{l-1}$ via a bi-linear interpolation layer. Then, we concatenate them together and take the concatenated feature map as input for a sub-network that contains two convolutional layers with the kernel size of 3×3. The output of the sub-network is the prediction of the semantic flow field $\Delta_{l-1} \in \mathbb{R}^{H_{l-1} \times W_{l-1} \times 2}$. Mathematically, the aforementioned steps can be written as:

$$\Delta_{l-1} = \mathrm{conv}_l(\mathrm{cat}(\mathbf{F}_l, \mathbf{F}_{l-1})), \tag{1}$$

where $\mathrm{cat}(\cdot)$ represents the concatenation operation and $\mathrm{conv}_l(\cdot)$ is the 3×3 convolutional layer. Since our network adopts strided convolutions, which could lead to very low resolution, for most cases, the respective field of the 3×3 convolution conv_l is sufficient to cover most large objects of that feature map. Note that, we discard the correlation layer proposed in FlowNet-C [13], where positional correspondence is calculated explicitly. Because there exists a huge semantic gap between higher-level layer and lower-level layer, explicit correspondence calculation on such features is difficult and tends to fail for offset prediction. Moreover, adopting such a correlation layer introduces heavy computation cost, which violates our goal for the network to be fast and accurate.

After having computed Δ_{l-1}, each position p_{l-1} on the spatial grid Ω_{l-1} is then mapped to a point p_l on the upper level l via a simple addition operation. Since there exists a resolution gap between features and flow field shown in Fig. 3(b), the warped grid and its offset should be halved as Eq. 2,

$$p_l = \frac{p_{l-1} + \Delta_{l-1}(p_{l-1})}{2}. \tag{2}$$

We then use the differentiable bi-linear sampling mechanism proposed in the spatial transformer networks [21], which linearly interpolates the values of the 4-neighbors (top-left, top-right, bottom-left, and bottom-right) of p_l to approximate the final output of the FAM, denoted by $\widetilde{\mathbf{F}}_l(p_{l-1})$. Mathematically,

$$\widetilde{\mathbf{F}}_l(p_{l-1}) = \mathbf{F}_l(p_l) = \sum_{p \in \mathcal{N}(p_l)} w_p \mathbf{F}_l(p), \tag{3}$$

where $\mathcal{N}(p_l)$ represents neighbors of the warped points p_l in $\mathbf{F}_l$ and w_p denotes the bi-linear kernel weights estimated by the distance of warped grid. This warping procedure may look similar to the convolution operation of the deformable kernels in deformable convolution network (DCN) [10]. However, our method has a lot of noticeable difference from DCN. First, our predicted offset field incorporates both higher-level and lower-level features to *align the positions* between high-level and low-level feature maps, while the offset field of DCN moves the positions of the kernels according to the predicted location offsets in order to *possess larger and more adaptive respective fields*. Second, our module focuses on aligning features while DCN works more like an attention mechanism that attends to the salient parts of the objects. More detailed comparison can be found in the experiment part.

On the whole, the proposed FAM module is light-weight and end-to-end trainable because it only contains one 3×3 convolution layer and one parameter-free warping operation in total. Besides these merits, it can be plugged into networks multiple times with only a minor extra computation cost overhead. Figure 3(a) gives the detailed settings of the proposed module while Fig. 3(b) shows the warping process. Figure 2 visualizes feature maps of two adjacent levels, their learned semantic flow and the finally warped feature map. As shown in Fig. 2, the warped feature is more structurally neat than normal bi-linear upsampled feature and leads to more consistent representation of objects, such as the bus and car.

3.3 Network Architectures

Figure 3(c) illustrates the whole network architecture, which contains a bottom-up pathway as the encoder and a top-down pathway as the decoder. While the encoder has a backbone network offering feature representations of different levels, the decoder can be seen as a FPN equipped with several FAMs.

Encoder Part. We choose standard networks pre-trained on ImageNet [47] for image classification as our backbone network by removing the last fully connected layer. Specifically, ResNet series [19], ShuffleNet v2 [34] and DF series [29] are used and compared in our experiments. All backbones have 4 stages with residual blocks, and each stage has a convolutional layer with stride 2 in the first place to downsample the feature map chasing for both computational efficiency and larger receptive fields. We additionally adopt the Pyramid Pooling Module (PPM) [64] for its superior power to capture contextual information. In our setting, the output of PPM shares the same resolution as that of the last residual module. In this situation, we treat PPM and the last residual module together as the last stage for the upcoming FPN. Other modules like ASPP [6] can also be plugged into our network, which are also experimentally ablated in Sect. 4.1.

Aligned FPN Decoder takes feature maps from the encoder and uses the aligned feature pyramid for final scene parsing. By replacing normal bi-linear up-sampling with FAM in the top-down pathway of FPN [32], $\{\mathbf{F}_l\}_{l=2}^4$ is refined to $\{\widetilde{\mathbf{F}}_l\}_{l=2}^4$, where top-level feature maps are aligned and fused into their bottom levels via element-wise addition and l represents the range of feature pyramid level. For scene parsing, $\{\widetilde{\mathbf{F}}_l\}_{l=2}^4 \cup \{\mathbf{F}_5\}$ are up-sampled to the same resolution (*i.e.*, 1/4 of input image) and concatenated together for prediction. Considering there are still misalignments during the previous step, we also replace these up-sampling operations with the proposed FAM.

Cascaded Deeply Supervised Learning. We use deeply supervised loss [64] to supervise intermediate outputs of the decoder for easier optimization. In addition, following [53], online hard example mining [48] is also used by only training on the 10% hardest pixels sorted by cross-entropy loss.

4 Experiments

We first carry out experiments on the Cityscapes [9] dataset, which is comprised of a large set of high-resolution (2048×1024) images in street scenes. This dataset has 5,000 images with high quality pixel-wise annotations for 19 classes, which is further divided into 2975, 500, and 1525 images for training, validation and testing. To be noted, coarse data are not used in this work. Besides, more experiments on Pascal Context [14], ADE20K [66] and CamVid [3] are summarised to further prove the generality of our method.

4.1 Experiments on Cityscapes

Implementation Details: We use PyTorch [44] framework to carry out following experiments. All networks are trained with the same setting, where stochastic gradient descent (SGD) with batch size of 16 is used as optimizer, with momentum of 0.9 and weight decay of 5e−4. All models are trained 50K iterations with an initial learning rate of 0.01. As a common practice, the "poly" learning rate policy is adopted to decay the initial learning rate by multiplying $(1 - \frac{\text{iter}}{\text{total\_iter}})^{0.9}$ during training. Data augmentation contains random horizontal flip, random resizing with scale range of [0.75, 2.0], and random cropping with crop size of 1024 × 1024. During inference, we use the whole picture as input to report performance unless explicitly mentioned. For quantitative evaluation, mean of class-wise intersection-over-union (mIoU) is used for accurate comparison, and number of float-point operations (FLOPs) and frames per second (FPS) are adopted for speed comparison.

Comparison with Baseline Methods: Table 1(a) reports the comparison results against baselines on the validation set of Cityscapes [9], where ResNet-18 [19] serves as the backbone. Comparing with the naive FCN, dilated FCN improves mIoU by 1.1%. By appending the FPN decoder to the naive FCN, we get 74.8% mIoU by an improvement of 3.2%. By replacing bilinear upsampling with the proposed FAM, mIoU is boosted to 77.2%, which improves the naive FCN and FPN decoder by 5.7% and 2.4% respectively. Finally, we append PPM (Pyramid Pooling Module) [64] to capture global contextual information, which achieves the best mIoU of 78.7% together with FAM. Meanwhile, FAM is complementary to PPM by observing FAM improves PPM from 76.6% to 78.7%.

Positions to Insert FAM: We insert FAM to different stage positions in the FPN decoder and report the results as Table 1(b). From the first three rows, FAM improves all stages and gets the greatest improvement at the last stage, which demonstrate that misalignment exists in all stages on FPN and is more severe in coarse layers. This is consistent with the fact that coarse layers containing stronger semantics but with lower resolution, and can greatly boost segmentation performance when they are appropriately upsampled to high resolution. The best result is achieved by adding FAM to all stages in the last row.

Table 1. Experiments results on network design using Cityscapes validation set.

Method	Stride	mIoU (%)	Δa(%)
FCN	32	71.5	-
Dilated FCN	8	72.6	1.1 ↑
+FPN	32	74.8	3.3 ↑
+FAM	32	77.2	5.7 ↑
+FPN + PPM	32	76.6	5.1 ↑
+FAM + PPM	32	**78.7**	7.2 ↑

(a) Ablation study on baseline model.

Method	$\mathbf{F}_3$	$\mathbf{F}_4$	$\mathbf{F}_5$	mIoU(%)	Δa(%)
FPN+PPM				76.6	-
	✓			76.9	0.3 ↑
		✓		77.0	0.4 ↑
			✓	77.5	0.9 ↑
		✓	✓	77.8	1.2 ↑
	✓	✓	✓	78.3	1.7 ↑

(b) Ablation study on insertion position.

Method	mIoU(%)	Δa(%)	#GFLOPs
FAM	76.4	-	-
+PPM [64]	78.3	1.9↑	123.5
+NL [50]	76.8	0.4↑	148.0
+ASPP [6]	77.6	1.2↑	138.6
+DenseASPP [52]	77.5	1.1↑	141.5

(c) Ablation study on context module.

Backbone	mIoU(%)	Δa(%)	#GFLOPs
ResNet-50 [19]	76.8	-	332.6
w/ FAM	79.2	2.4 ↑	337.1
ResNet-101 [19]	77.6	-	412.7
w/ FAM	79.8	2.2↑	417.5
ShuffleNetv2 [34]	69.8	-	17.8
w/ FAM	72.1	2.3 ↑	18.1
DF1 [29]	72.1	-	18.6
w/ FAM	74.3	2.2 ↑	18.7
DF2 [29]	73.2	-	48.2
w/ FAM	75.8	2.6 ↑	48.5

(d) Ablation on study on various backbones.

Table 2. Experiments results on FAM design using Cityscapes validation set.

Method	mIoU (%)
bilinear upsampling	78.3
deconvolution	77.9
nearest neighbor	78.2

(a) Ablation study on Upsampling operation in FAM.

Method	mIoU (%)	Gflops
$k = 1$	77.8	120.4
$k = 3$	78.3	123.5
$k = 5$	78.1	131.6
$k = 7$	78.0	140.5

(b) Ablation study on kernel size k in FAM where 3 FAMs are involved.

Method	mIoU (%)	Δa(%)
FPN +PPM	76.6	-
correlation [13]	77.2	0.6 ↑
Ours	77.5	0.9 ↑

(l) Ablation with FlowNet-C [13] in FAM.

Method	$\mathbf{F}_3$	$\mathbf{F}_4$	$\mathbf{F}_5$	mIoU(%)	Δa(%)
FPN +PPM	-	-	-	76.6	-
DCN			✓	76.9	0.3 ↑
Ours			✓	77.5	0.9 ↑
DCN	✓	✓	✓	77.2	0.6 ↑
Ours	✓	✓	✓	78.3	1.7 ↑

(d) Comparison with DCN [10].

Ablation Study on Network Architecture Design: Considering current state-of-the-art contextual modules are used as heads on dilated backbone networks [6,15,52,58,64,65], we further try different contextual heads in our methods where coarse feature map is used for contextual modeling. Table 1(c) reports the comparison results, where PPM [64] delivers the best result, while more recently proposed methods such as Non-Local based heads [50] perform worse. Therefore, we choose PPM as our contextual head considering its better performance with lower computational cost. We further carry out experiments with different backbone networks including both deep and light-weight networks, where FPN decoder with PPM head is used as a strong baseline in Table 1(d). For heavy networks, we choose ResNet-50 and ResNet-101 [19] as representation. For light-weight networks, ShuffleNetv2 [34] and DF1/DF2 [29] are employed. FAM significantly achieves better mIoU on all backbones with slightly extra computational cost.

Ablation Study on FAM Design: We first explore the effect of upsampling in FAM in Table 2(a). Replacing the bilinear upsampling with deconvolution and nearest neighbor upsampling achieves 77.9 mIoU and 78.2 mIoU, respectively, which are similar to the 78.3 mIoU achieved by bilinear upsampling. We also try the various kernel size in Table 2(b). Larger kernel size of 5×5 is also tried which results in a similar (78.2) but introduces more computation cost. In Table 2(c), replacing FlowNet-S with correlation in FlowNet-C also leads to slightly worse results (77.2) but increases the inference time. The results show that it is enough to use lightweight FlowNet-S for aligning feature maps in FPN. In Table 2(d), we compare our results with DCN [10]. We apply DCN on the concatenated feature map of bilinear upsampled feature map and the feature map of next level. We first insert one DCN in higher layers $\mathbf{F}_5$ where our FAM is better than it. After applying DCN to all layers, the performance gap is much larger. This denotes our method can also align low level edges for better boundaries and edges in lower layers, which will be shown in visualization part.

Aligned Feature Representation: In this part, we give more visualization on aligned feature representation as shown in Fig. 4. We visualize the upsampled feature in the final stage of ResNet-18. It shows that compared with DCN [10], our FAM feature is more structural and has much more precise objects boundaries which is consistent with the results in Table 2(d). That indicates FAM is **not** an attention effect on feature similar to DCN, but actually aligns feature towards more precise shape as compared in red boxes.

Visualization of Semantic Flow: Figure 5 visualizes semantic flow from FAM in different stages. Similar with optical flow, semantic flow is visualized by color coding and is bilinearly interpolated to image size for quick overview. Besides, vector fields are also visualized for detailed inspection. From the visualization, we observe that semantic flow tends to diffuse out from some positions inside objects, where these positions are generally near object centers and have better receptive fields to activate top-level features with pure, strong semantics. Top-level features at these positions are then propagated to appropriate

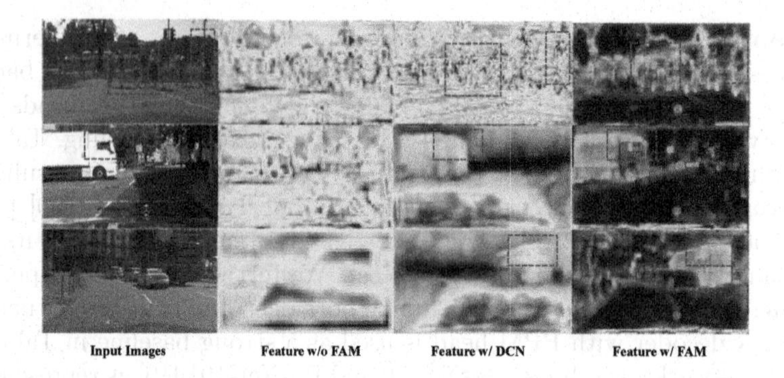

Fig. 4. Visualization of the aligned feature. Compared with DCN, our module outputs more structural feature representation. (Color figure online)

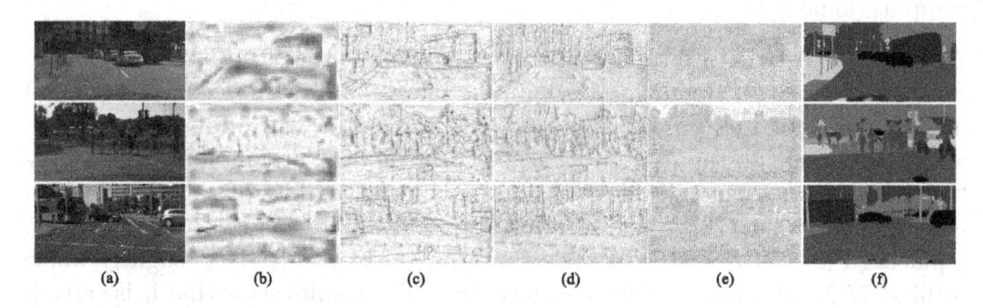

Fig. 5. Visualization of the learned semantic flow fields. Column (a) lists three exemplary images. Column (b)–(d) show the semantic flow of the three FAMs in an ascending order of resolution during the decoding process, following the same color coding of Fig. 2. Column (e) is the arrowhead visualization of flow fields in column (d). Column (f) contains the segmentation results.

high-resolution positions following the guidance of semantic flow. In addition, semantic flows also have coarse-to-fine trends from top level to bottom level, which phenomenon is consistent with the fact that semantic flows gradually describe offsets between gradually smaller patterns.

Visual Improvement Analysis: Figure 6(a) visualizes the prediction errors by both methods, where FAM considerably resolves ambiguities inside large objects (e.g., truck) and produces more precise boundaries for small and thin objects (e.g., poles, edges of wall). Figure 6 (b) shows our model can better handle the small objects with shaper boundaries than dilated PSPNet due to the alignment on lower layers.

Comparison with Real-Time Models: All compared methods are evaluated by single-scale inference and input sizes are also listed for fair comparison. Our speed is tested on one GTX 1080Ti GPU with full image resolution 1024×2048 as input, and we report speed of two versions, i.e., without and with TensorRT acceleration. As shown in Table 3, our method based on DF1 achieves a more

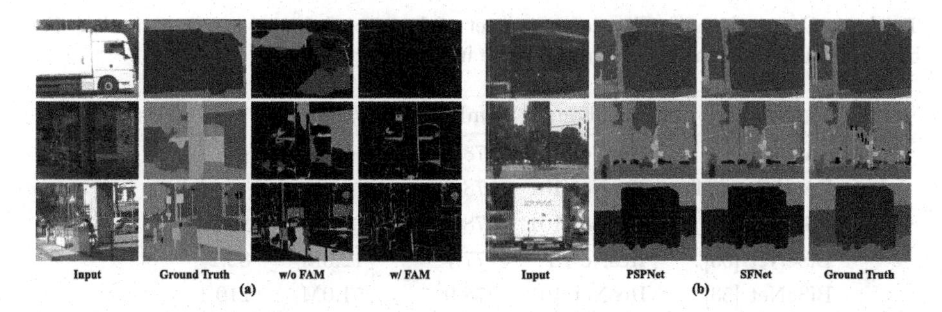

Fig. 6. (a), Qualitative comparison in terms of errors in predictions, where correctly predicted pixels are shown as black background while wrongly predicted pixels are colored with their groundtruth label color codes. (b), Scene parsing results comparison against PSPNet [64], where significantly improved regions are marked with red dashed boxes. Our method performs better on both small scale and large scale objects. (Color figure online)

Table 3. Comparison on Cityscapes *test* set with state-of-the-art real-time models. For fair comparison, input size is also considered, and all models use single scale inference.

Method	InputSize	mIoU (%)	#FPS	#Params
ENet [43]	640×360	58.3	60	0.4M
ESPNet [36]	512×1024	60.3	132	0.4M
ESPNetv2 [37]	512×1024	62.1	80	0.8M
ERFNet [45]	512×1024	69.7	41.9	–
BiSeNet(ResNet-18) [53]	768×1536	74.6	43	12.9M
BiSeNet(Xception-39) [53]	768×1536	68.4	72	5.8M
ICNet [63]	1024×2048	69.5	34	26.5M
DF1-Seg [29]	1024×2048	73.0	80	8.55M
DF2-Seg [29]	1024×2048	74.8	55	8.55M
SwiftNet [42]	1024×2048	75.5	39.9	11.80M
SwiftNet-ens [42]	1024×2048	76.5	18.4	24.7M
DFANet [25]	1024×1024	71.3	100	7.8M
CellNet [62]	768×1536	70.5	108	–
SFNet(DF1)	1024×2048	**74.5**	74/<u>121</u>	9.03M
SFNet(DF2)	1024×2048	**77.8**	53/<u>61</u>	10.53M
SFNet(ResNet-18)	1024×2048	**78.9**	18/<u>26</u>	12.87M
SFNet(ResNet-18)†	1024×2048	**80.4**	18/<u>26</u>	12.87M

† Mapillary dataset used for pretraining.

Table 4. Comparison on Cityscapes *test* set with state-of-the-art accurate models. For better accuracy, all models use multi-scale inference.

Method	Backbone	mIoU (%)	#Params	#GFLOPs†
SAC [61]	ResNet-101	78.1	–	–
DepthSeg [24]	ResNet-101	78.2	–	–
PSPNet [64]	ResNet-101	78.4	65.7M	1065.4
BiSeNet [53]	ResNe-18	77.7	12.3M	82.2
BiSeNet [53]	ResNet-101	78.9	51.0M	219.1
DFN [54]	ResNet-101	79.3	90.7M	1121.0
PSANet [65]	ResNet-101	80.1	85.6M	1182.6
DenseASPP [52]	DenseNet-161	80.6	35.7M	632.9
SPGNet [8]	2×ResNet-50	81.1	–	–
ANNet [69]	ResNet-101	81.3	63.0M	1089.8
CCNet [20]	ResNet-101	81.4	66.5M	1153.9
DANet [15]	ResNet-101	81.5	66.6M	1298.8
SFNet	ResNet-18	**79.5**	**12.87M**	**123.5**
SFNet	ResNet-101	**81.8**	**50.32M**	**417.5**

† #GFLOPs calculation adopts 1024×1024 image as input.

accurate result(74.5%) than all methods faster than it. With DF2, our method outperforms all previous methods while running at 60 FPS. With ResNet-18 as backbone, our method achieves 78.9% mIoU and even reaches performance of accurate models which will be discussed in the next experiment. By additionally using Mapillary [40] dataset for pretraining, our ResNet-18 based model achieves 26 FPS with 80.4% mIoU, which sets the new state-of-the-art record on accuracy and speed trade-off on Cityscapes benchmark. More detailed information are in the supplementary file.

Comparison with Accurate Models: State-of-the-art accurate models [15, 52,64,68] perform multi-scale and horizontal flip inference to achieve better results on the Cityscapes test server. For fair comparison, we also report multi-scale with flip testing results following previous methods [15,64]. Model parameters and computation FLOPs are also listed for comparison. Table 4 summarizes the results, where our models achieve state-of-the-art accuracy while costs much less computation. In particular, our method based on ResNet-18 is 1.1% mIoU higher than PSPNet [64] while only requiring **11%** of its computation. Our ResNet-101 based model achieves better results than DAnet [15] by 0.3% mIoU and only requires **30%** of its computation.

4.2 Experiment on More Datasets

We also perform more experiments on other three data-sets including Pascal Context [38], ADE20K [66] and CamVid [3] to further prove the effectiveness of our method. More detailed setting can be found in the supplemental file.

Table 5. Experiments results on Pascal Context and ADE20k(Multi scale inference). #GFLOPs calculation adopts 480 × 480 image as input.

Method	Backbone	mIoU (%)	#GFLOPs
Ding *et al.* [12]	ResNet-101	51.6	-
EncNet [57]	ResNet-50	49.2	-
EncNet [57]	ResNet-101	51.7	-
DANet [15]	ResNet-50	50.1	186.4
DANet [15]	ResNet-101	52.6	257.1
ANNet [69]	ResNet-101	52.8	243.8
BAFPNet [11]	ResNet-101	53.6	-
EMANet [27]	ResNet-101	53.1	209.3
w/o FAM	ResNet-50	49.0	74.5
SFNet	ResNet-50	**50.7**(1.7 ↑)	75.4
w/o FAM	ResNet-101	51.1	92.7
SFNet	ResNet-101	**53.8**(2.7 ↑)	93.6

(a) Results on Pascal Context. Evaluated on 60 classes.

Method	Backbone	mIoU (%)	#GFLOPs
PSPNet [64]	ResNet-50	42.78	167.6
PSPNet [64]	ResNet-101	43.29	238.4
PSANet [65]	ResNet-101	43.77	264.9
EncNet [57]	ResNet-101	44.65	-
CFNet [58]	ResNet101	44.82	-
w/o FAM	ResNet-50	41.12	74.8
SFNet	ResNet-50	42.81(1.69 ↑)	75.7
w/o FAM	ResNet-101	43.08	93.1
SFNet	ResNet-101	44.67(1.59 ↑)	94.0

(b) Results on ADE20K.

PASCAL Context: The results are illustrated as Table 5(a), our method outperforms corresponding baselines by 1.7% mIoU and 2.6% mIoU with ResNet-50 and ResNet-101 as backbones respectively. In addition, our method on both ResNet-50 and ResNet-101 outperforms their existing counterparts by large margins with significantly lower computational cost.

ADE20K: is a challenging scene parsing dataset. Images in this dataset are from different scenes with more scale variations. Table 5(b) reports the performance comparisons, our method improves the baselines by 1.69% mIoU and 1.59% mIoU respectively, and outperforms previous state-of-the-art methods [64,65] with much less computation.

CamVid: is another road scene dataset. This dataset involves 367 training images, 101 validation images and 233 testing images with resolution of 960×720. We apply our method with different light-weight backbones on this dataset and report comparison results in Table 6. With DF2 as backbone, FAM improves its baseline by 3.2% mIoU. Our method based on ResNet-18 performs best with 73.8% mIoU while running at 35.5 FPS.

Table 6. Accuracy and Speed comparison with previous state-of-the-art real-time models on CamVid [3] test set where the input size is 960 × 720 with single scale inference.

Method	Backbone	mIoU (%)	FPS
ICNet [63]	ResNet-50	67.1	34.5
BiSegNet [53]	Xception-39	65.6	–
BiSegNet [53]	ResNet-18	68.7	–
DFANet A [25]	–	64.7	120
DFANet B [25]	–	59.3	160
w/o FAM	DF2	67.2	139.8
SFNet	DF2	**70.4** (3.2 ↑)	134.1
SFNet	ResNet-18	**73.8**	35.5

5 Conclusion

In this paper, we devise to use the learned **Semantic Flow** to align multi-level feature maps generated by a feature pyramid to the task of scene parsing. With the proposed flow alignment module, high-level features are well fused into low-level feature maps with high resolution. By discarding atrous convolutions to reduce computation overhead and employing the flow alignment module to enrich the semantic representation of low-level features, our network achieves the best trade-off between semantic segmentation accuracy and running time efficiency. Experiments on multiple challenging datasets illustrate the efficacy of our method.

References

1. Badrinarayanan, V., Kendall, A., Cipolla, R.: SegNet: a deep convolutional encoder-decoder architecture for image segmentation. PAMI **39**, 2481–2495 (2017)
2. Baker, S., Scharstein, D., Lewis, J.P., Roth, S., Black, M.J., Szeliski, R.: A database and evaluation methodology for optical flow. Int. J. Comput. Vis. **92**(1), 1–31 (2011). https://doi.org/10.1007/s11263-010-0390-2
3. Brostow, G.J., Fauqueur, J., Cipolla, R.: Semantic object classes in video: a high-definition ground truth database. Pattern Recogn. Lett. (2008)
4. Brox, T., Bruhn, A., Papenberg, N., Weickert, J.: High accuracy optical flow estimation based on a theory for warping. In: Pajdla, T., Matas, J. (eds.) ECCV 2004. LNCS, vol. 3024, pp. 25–36. Springer, Heidelberg (2004). https://doi.org/10.1007/978-3-540-24673-2_3
5. Chen, L.C., Papandreou, G., Kokkinos, I., Murphy, K., Yuille, A.L.: DeepLab: semantic image segmentation with deep convolutional nets, atrous convolution, and fully connected CRFS. PAMI (2018)
6. Chen, L.C., Papandreou, G., Schroff, F., Adam, H.: Rethinking atrous convolution for semantic image segmentation. arXiv preprint arXiv:1706.05587 (2017)

7. Chen, L.-C., Zhu, Y., Papandreou, G., Schroff, F., Adam, H.: Encoder-decoder with atrous separable convolution for semantic image segmentation. In: Ferrari, V., Hebert, M., Sminchisescu, C., Weiss, Y. (eds.) ECCV 2018. LNCS, vol. 11211, pp. 833–851. Springer, Cham (2018). https://doi.org/10.1007/978-3-030-01234-2_49
8. Cheng, B., et al.: SPGNet: semantic prediction guidance for scene parsing. In: ICCV, October 2019
9. Cordts, M., et al.: The cityscapes dataset for semantic urban scene understanding. In: CVPR (2016)
10. Dai, J., et al.: Deformable convolutional networks. In: ICCV (2017)
11. Ding, H., Jiang, X., Liu, A.Q., Magnenat-Thalmann, N., Wang, G.: Boundary-aware feature propagation for scene segmentation (2019)
12. Ding, H., Jiang, X., Shuai, B., Qun Liu, A., Wang, G.: Context contrasted feature and gated multi-scale aggregation for scene segmentation. In: CVPR (2018)
13. Dosovitskiy, A., et al.: FlowNet: learning optical flow with convolutional networks. In: CVPR (2015)
14. Everingham, M., Van Gool, L., Williams, C.K., Winn, J., Zisserman, A.: The pascal visual object classes (VOC) challenge. IJCV **88**, 303–338 (2010). https://doi.org/10.1007/s11263-009-0275-4
15. Fu, J., Liu, J., Tian, H., Fang, Z., Lu, H.: Dual attention network for scene segmentation. arXiv preprint arXiv:1809.02983 (2018)
16. Gadde, R., Jampani, V., Gehler, P.V.: Semantic video CNNs through representation warping. In: ICCV, October 2017
17. He, J., Deng, Z., Qiao, Y.: Dynamic multi-scale filters for semantic segmentation. In: ICCV, October 2019
18. He, J., Deng, Z., Zhou, L., Wang, Y., Qiao, Y.: Adaptive pyramid context network for semantic segmentation. In: CVPR, June 2019
19. He, K., Zhang, X., Ren, S., Sun, J.: Deep residual learning for image recognition. In: CVPR (2016)
20. Huang, Z., Wang, X., Huang, L., Huang, C., Wei, Y., Liu, W.: CCNet: Criss-cross attention for semantic segmentation (2019)
21. Jaderberg, M., Simonyan, K., Zisserman, A., Kavukcuoglu, K.: Spatial transformer networks. ArXiv abs/1506.02025 (2015)
22. Kipf, T.N., Welling, M.: Semi-supervised classification with graph convolutional networks. ArXiv abs/1609.02907 (2016)
23. Kirillov, A., Girshick, R., He, K., Dollar, P.: Panoptic feature pyramid networks. In: CVPR, June 2019
24. Kong, S., Fowlkes, C.C.: Recurrent scene parsing with perspective understanding in the loop. In: CVPR (2018)
25. Li, H., Xiong, P., Fan, H., Sun, J.: DFANet: deep feature aggregation for real-time semantic segmentation. In: CVPR, June 2019
26. Li, X., Yang, Y., Zhao, Q., Shen, T., Lin, Z., Liu, H.: Spatial pyramid based graph reasoning for semantic segmentation. In: CVPR (2020)
27. Li, X., Zhong, Z., Wu, J., Yang, Y., Lin, Z., Liu, H.: Expectation-maximization attention networks for semantic segmentation. In: ICCV (2019)
28. Li, X., Houlong, Z., Lei, H., Yunhai, T., Kuiyuan, Y.: GFF: gated fully fusion for semantic segmentation. In: AAAI (2020)
29. Li, X., Zhou, Y., Pan, Z., Feng, J.: Partial order pruning: for best speed/accuracy trade-off in neural architecture search. In: CVPR (2019)
30. Li, Y., Gupta, A.: Beyond grids: learning graph representations for visual recognition. In: NIPS (2018)

31. Li, Y., Shi, J., Lin, D.: Low-latency video semantic segmentation. In: CVPR, June 2018
32. Lin, T.Y., Dollár, P., Girshick, R.B., He, K., Hariharan, B., Belongie, S.J.: Feature pyramid networks for object detection. In: CVPR (2017)
33. Long, J., Shelhamer, E., Darrell, T.: Fully convolutional networks for semantic segmentation. In: CVPR (2015)
34. Ma, N., Zhang, X., Zheng, H.-T., Sun, J.: ShuffleNet V2: practical guidelines for efficient CNN architecture design. In: Ferrari, V., Hebert, M., Sminchisescu, C., Weiss, Y. (eds.) Computer Vision – ECCV 2018. LNCS, vol. 11218, pp. 122–138. Springer, Cham (2018). https://doi.org/10.1007/978-3-030-01264-9_8
35. Mazzini, D.: Guided upsampling network for real-time semantic segmentation. In: BMVC (2018)
36. Mehta, S., Rastegari, M., Caspi, A., Shapiro, L., Hajishirzi, H.: ESPNet: efficient spatial pyramid of dilated convolutions for semantic segmentation. In: Ferrari, V., Hebert, M., Sminchisescu, C., Weiss, Y. (eds.) ECCV 2018. LNCS, vol. 11214, pp. 561–580. Springer, Cham (2018). https://doi.org/10.1007/978-3-030-01249-6_34
37. Mehta, S., Rastegari, M., Shapiro, L., Hajishirzi, H.: ESPNetv2: a light-weight, power efficient, and general purpose convolutional neural network. In: CVPR, June 2019
38. Mottaghi, R., et al.: The role of context for object detection and semantic segmentation in the wild. In: CVPR (2014)
39. Nekrasov, V., Chen, H., Shen, C., Reid, I.: Fast neural architecture search of compact semantic segmentation models via auxiliary cells. In: CVPR, June 2019
40. Neuhold, G., Ollmann, T., Rota Bulo, S., Kontschieder, P.: The mapillary vistas dataset for semantic understanding of street scenes. In: ICCV (2017)
41. Nilsson, D., Sminchisescu, C.: Semantic video segmentation by gated recurrent flow propagation. In: CVPR, June 2018
42. Orsic, M., Kreso, I., Bevandic, P., Segvic, S.: In defense of pre-trained ImageNet architectures for real-time semantic segmentation of road-driving images. In: CVPR, June 2019
43. Paszke, A., Chaurasia, A., Kim, S., Culurciello, E.: ENet: a deep neural network architecture for real-time semantic segmentation. http://arxiv.org/abs/1606.02147
44. Paszke, A., et al.: Automatic differentiation in PyTorch. In: NIPS-W (2017)
45. Romera, E., Alvarez, J.M., Bergasa, L.M., Arroyo, R.: ERFNet: efficient residual factorized convnet for real-time semantic segmentation. IEEE Trans. Intell. Transp. Syst. **19**, 263–272 (2018)
46. Ronneberger, O., Fischer, P., Brox, T.: U-Net: convolutional networks for biomedical image segmentation. In: Navab, N., Hornegger, J., Wells, W.M., Frangi, A.F. (eds.) MICCAI 2015. LNCS, vol. 9351, pp. 234–241. Springer, Cham (2015). https://doi.org/10.1007/978-3-319-24574-4_28
47. Russakovsky, O., et al.: ImageNet large scale visual recognition challenge. IJCV **115**, 211–252 (2015). https://doi.org/10.1007/s11263-015-0816-y
48. Shrivastava, A., Gupta, A., Girshick, R.: Training region-based object detectors with online hard example mining. In: CVPR (2016)
49. Vaswani, A., et al.: Attention is all you need. In: NIPS (2017)
50. Wang, X., Girshick, R., Gupta, A., He, K.: Non-local neural networks. In: CVPR, June 2018
51. Xiao, T., Liu, Y., Zhou, B., Jiang, Y., Sun, J.: Unified perceptual parsing for scene understanding. In: Ferrari, V., Hebert, M., Sminchisescu, C., Weiss, Y. (eds.) ECCV 2018. LNCS, vol. 11209, pp. 432–448. Springer, Cham (2018). https://doi.org/10.1007/978-3-030-01228-1_26

52. Yang, M., Yu, K., Zhang, C., Li, Z., Yang, K.: DenseASPP for semantic segmentation in street scenes. In: CVPR (2018)
53. Yu, C., Wang, J., Peng, C., Gao, C., Yu, G., Sang, N.: BiSeNet: bilateral segmentation network for real-time semantic segmentation. In: Ferrari, V., Hebert, M., Sminchisescu, C., Weiss, Y. (eds.) ECCV 2018. LNCS, vol. 11217, pp. 334–349. Springer, Cham (2018). https://doi.org/10.1007/978-3-030-01261-8_20
54. Yu, C., Wang, J., Peng, C., Gao, C., Yu, G., Sang, N.: Learning a discriminative feature network for semantic segmentation. In: CVPR (2018)
55. Yu, F., Koltun, V.: Multi-scale context aggregation by dilated convolutions. In: ICLR (2016)
56. Yuan, Y., Wang, J.: OCNet: object context network for scene parsing. arXiv preprint arXiv:1809.00916 (2018)
57. Zhang, H., et al.: Context encoding for semantic segmentation. In: CVPR (2018)
58. Zhang, H., Zhang, H., Wang, C., Xie, J.: Co-occurrent features in semantic segmentation. In: CVPR, June 2019
59. Zhang, L., Li, X., Arnab, A., Yang, K., Tong, Y., Torr, P.H.: Dual graph convolutional network for semantic segmentation. In: BMVC (2019)
60. Zhang, L., Xu, D., Arnab, A., Torr, P.H.: Dynamic graph message passing networks. In: CVPR (2020)
61. Zhang, R., Tang, S., Zhang, Y., Li, J., Yan, S.: Scale-adaptive convolutions for scene parsing. In: ICCV (2017)
62. Zhang, Y., Qiu, Z., Liu, J., Yao, T., Liu, D., Mei, T.: Customizable architecture search for semantic segmentation. In: CVPR, June 2019
63. Zhao, H., Qi, X., Shen, X., Shi, J., Jia, J.: ICNet for real-time semantic segmentation on high-resolution images. In: Ferrari, V., Hebert, M., Sminchisescu, C., Weiss, Y. (eds.) ECCV 2018. LNCS, vol. 11207, pp. 418–434. Springer, Cham (2018). https://doi.org/10.1007/978-3-030-01219-9_25
64. Zhao, H., Shi, J., Qi, X., Wang, X., Jia, J.: Pyramid scene parsing network. In: CVPR (2017)
65. Zhao, H., et al.: PSANet: point-wise spatial attention network for scene parsing. In: Ferrari, V., Hebert, M., Sminchisescu, C., Weiss, Y. (eds.) ECCV 2018. LNCS, vol. 11213, pp. 270–286. Springer, Cham (2018). https://doi.org/10.1007/978-3-030-01240-3_17
66. Zhou, B., Zhao, H., Puig, X., Fidler, S., Barriuso, A., Torralba, A.: Semantic understanding of scenes through the ADE20K dataset. arXiv preprint arXiv:1608.05442 (2016)
67. Zhu, X., Xiong, Y., Dai, J., Yuan, L., Wei, Y.: Deep feature flow for video recognition. In: CVPR, July 2017
68. Zhu, Y., et al.: Improving semantic segmentation via video propagation and label relaxation. In: CVPR, June 2019
69. Zhu, Z., Xu, M., Bai, S., Huang, T., Bai, X.: Asymmetric non-local neural networks for semantic segmentation. In: ICCV (2019)

Appearance Consensus Driven Self-supervised Human Mesh Recovery

Jogendra Nath Kundu[1(✉)], Mugalodi Rakesh[1], Varun Jampani[2], Rahul Mysore Venkatesh[1], and R. Venkatesh Babu[1]

[1] Indian Institute of Science, Bangalore, India
jogendrak@iisc.ac.in
[2] Google Research, Cambridge, USA

Abstract. We present a self-supervised human mesh recovery framework to infer human pose and shape from monocular images in the absence of any paired supervision. Recent advances have shifted the interest towards directly regressing parameters of a parametric human model by supervising them on large-scale datasets with 2D landmark annotations. This limits the generalizability of such approaches to operate on images from unlabeled wild environments. Acknowledging this we propose a novel appearance consensus driven self-supervised objective. To effectively disentangle the foreground (FG) human we rely on image pairs depicting the same person (consistent FG) in varied pose and background (BG) which are obtained from unlabeled wild videos. The proposed FG appearance consistency objective makes use of a novel, differentiable *Color-recovery* module to obtain vertex colors without the need for any appearance network; via efficient realization of color-picking and reflectional symmetry. We achieve state-of-the-art results on the standard model-based 3D pose estimation benchmarks at comparable supervision levels. Furthermore, the resulting colored mesh prediction opens up the usage of our framework for a variety of appearance-related tasks beyond the pose and shape estimation, thus establishing our superior generalizability.

1 Introduction

Inferring highly deformable 3D human pose and shape from in-the-wild monocular images has been a longstanding goal in the vision community [12]. This is considered as a key step for a wide range of downstream applications such as robot interaction, rehabilitation guidance, animation industry, etc. Being one of

J. N. Kundu and M. Rakesh—Equal contribution.
Webpage: https://sites.google.com/view/ss-human-mesh.

Electronic supplementary material The online version of this chapter (https://doi.org/10.1007/978-3-030-58452-8_46) contains supplementary material, which is available to authorized users.

A. Vedaldi et al. (Eds.): ECCV 2020, LNCS 12346, pp. 794–812, 2020.
https://doi.org/10.1007/978-3-030-58452-8_46

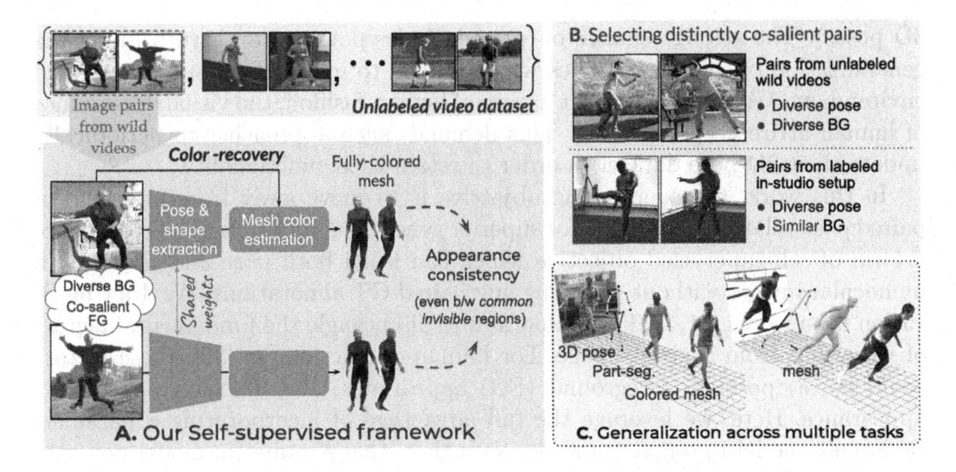

Fig. 1. Our framework disentangles the co-salient FG human from input image pairs. The resulting colored mesh prediction opens up its usage for a variety of tasks.

the important subtasks, human pose estimation has gained considerable performance improvements in recent years [44,56,60], but in a fully-supervised setting. Such approaches heavily rely on large-scale 2D or 3D pose annotations. Following this, the parametric models of human body, such as SCAPE [3], SMPL [39], SMPL(-X) [48,57] lead the way for a full 3D pose and shape estimation. Additionally, to suppress the inherent 2D-to-3D ambiguity, researchers have also utilized auxiliary cues of supervision such as temporal consistency [4,61], multi-view image pairs [14,19,52], or even alternate sensor data from Kinect [66] or IMUs [43]. However, estimating 3D human pose and shape from a single RGB image without relying on any direct supervision remains a very challenging problem.

Early approaches [5,8,34] adopt iterative optimization techniques to fit a parametric human model (*e.g.* SMPL) to a given image observation. These works attempt to iteratively estimate the body pose and shape that best describe the available 2D observation, which is most often the 2D landmark annotations. Though these works usually get good body fits, such approaches are slow and heavily rely on the 2D landmark annotations [2,18,27] or predictions of an off-the-shelf, fully-supervised Image-to-2D pose networks. However, the recent advances in deep learning has shifted the interest towards data-driven regression based methods [20,63], where a deep network directly regresses parameters of the human model for a given input image [47,50,68] in a single-shot computation. This is a promising direction as the network can utilize the full image information instead of just the sparse landmarks to estimate human body shape and pose. In the absence of datasets having images with 3D pose and shape ground-truth (GT), several recent works leverage a variety of available paired 2D annotations [49,62] such as 2D landmarks or silhouettes [50]; alongside the unpaired 3D pose samples to instill the 3D pose priors [20] (*i.e.* to assure recovery of valid

3D poses). The strong reliance on paired 2D keypoint ground-truth limits the generalization of such approaches when applied to images from an unseen wild environment. Given the transient nature of human fashion, the visual appearance of human attire keeps evolving. This demands such approaches to periodically update their 2D pose dataset in order to retain their functionality.

In this work, the overarching objective is to move away from any kind of paired pose-related supervision for superior generalizability. Our aim is to explore a form of self-supervised objective which can learn both pose and shape from monocular images without accessing any paired GT annotations. We draw motivation from works [31,40,45,55] that aim to disentangle the fundamental factors of variations from a given image. For human-centric images [32], these factors could be; a) pose, b) foreground (FG) appearance, and c) background (BG) appearance. Here, we leverage the full advantage of incorporating a parametric human model in our framework. Note that, this parametric model not only encapsulates the pose but also segregates the FG region from the BG, which is enabled by projecting the 3D mesh onto the image plane. Thus, the problem boils down to a faithful registration of the 3D mesh onto the image plane or in other words disentanglement of FG from BG. To achieve this disentanglement, we rely on image pairs depicting consistent FG appearance but varied 3D poses. Such image pairs can be obtained from videos depicting actions of a single person, which are abundantly available on the internet. Our idea stems from the concept of co-saliency detection [13,69] where the objective is to segment out the common, salient FG from a set of two or more images. Surprisingly, this idea works the best for image pairs sampled from wild videos as compared to videos captured in a constrained in-studio setup (static homogeneous background). This is because in wild scenarios, the commonness of FG is distinctly salient in relatively diverse BGs as a result of substantial camera movements (see Fig. 1B). Thus, in contrast to prior self-supervised approaches that either rely on videos with static BG [53] or operate under the assumption of BG commonness between temporally close frames [16]; our approach is more favorable to learn from wild videos hence better generalizable.

Table 1. Characteristic comparison against prior-arts.

Model-based methods	2D keypoint supervision	Temporal supervision	Colored mesh prediction
[20,25,26,47,50]	Yes	No	No
[4,22,61]	Yes	Yes	No
Ours(self-sup.)	**No**	**No**	**Yes**

In the proposed framework, we first employ a CNN regressor to obtain the parameters (both pose and shape) of the SMPL model for a given input image. The human mesh model uses these parameters to output the mesh vertex locations. In contrast to the general trend [1,21], we propose a novel way of inferring mesh texture where the network's burden to regress vertex color or any sort of appearance representation (such as UV map) is entirely taken away. This is realized via a differentiable *Color-recovery* module which aims to assign color to the mesh vertices via spatial registration of the mesh over the image plane while effectively accounting for the challenges of mesh-vertex visibility like self and inter-part occlusions. To obtain a fully-colored mesh, we use a predefined, 4-way symmetry grouping knowledge (front-back and left-right) to propagate the color from camera visible vertices to the non-visible ones in a fully differentiable fashion.

For a given image pair, we pass them through two parallel pathways of our colored mesh prediction framework (see Fig. 1A). The commonness of FG appearance allows us to impose an appearance consistency loss between the predicted mesh representations. In the absence of any paired supervision, this appearance consistency not only helps us to segregate the common FG human from their respective wild BGs but also discovers the required pose deformation in a fully self-supervised manner. The proposed reflectional symmetry module brings in a substantial advantage in our self-supervised framework by allowing us to impose appearance consistency even between body parts which are "*commonly invisible*" in both the images. Recognizing the unreliability of consistent raw color intensities which can easily be violated as result of illumination changes, we propose a *part-prototype* consistency objective. This aims to match a higher level appearance representation beyond the raw color intensities which is enabled by operating the *Color-recovery* module on convolutional feature maps instead of the raw image. Additionally, to regularize the self-supervised framework, we also impose a shape consistency loss alongside the imposition of 3D pose prior learned from a set of unpaired MoCap samples. Note that at test time, we perform single image inference to estimate 3D human pose and shape.

In summary, we make the following main contributions:

- We propose a self-supervised learning technique to perform simultaneous pose and shape estimation which uses image pairs sampled from in-the-wild videos in the absence of any paired supervision.
- The proposed *Color-recovery* module completely eliminates the network's burden to regress any appearance-related representation via efficient realization of color-picking and reflectional symmetry. This best suits our self-supervised framework which relies on FG appearance consistency.
- We demonstrate generalizability of our framework to operate on *unseen* wild datasets. We achieve *state-of-the-art* results against the prior model-based pose estimation approaches when tested at comparable supervision levels.

2 Related Work

Vertex-Color Reconstruction. In literature, we find different ways to infer textured 3D mesh from a monocular RGB image. Certain approaches [33,59] train a deep network to directly regress 3D features (RGB colors) for individual vertices. In the second kind, a fully convolutional deep network is trained to map the location of each pixel to the corresponding continuous UV-map coordinate parameterization [1]. In the third kind, the deep model is trained to directly regress the UV-image [21]. Note that, the spatial structure of the UV image is much different from that of the input image which prevents employing a fully-convolutional network for the same. Recently proposed, Soft-Rasterizer [37] uses a color-selection and color-sampling network whose outputs are processed to obtain the final vertex colors. All the above approaches adopt a learnable way to obtain the mesh color (*i.e.* obtained as neural output). In such cases, the deep network requires substantial training iterations to instill the knowledge of pre-defined UV mapping conventions. We believe this is an additional burden for the network specifically in absence of any auxiliary paired supervisions.

Model-Based Human Mesh Estimation. Recently, parametric human models [3,39] have been used as the output target for the simultaneous pose and shape estimation task. Such a well-defined mesh model with ordered vertices provides a direct mapping to the corresponding 3D pose and part segments. Both optimization [5,34,67] and regression [20,47,50,68] based approaches estimate the body pose and shape that best describes the available 2D observations such as 2D keypoints [20], silhouettes [50], body/part segmentation [47] etc. Due to the lack of datasets having wild images with 3D pose and shape GT, most of the above approaches fully rely on the availability of 2D keypoint annotations [2,36] followed by different variants of a 2D reprojection loss [62,63] (see Table 1).

Use of Auxiliary Supervision. In the absence of any shape supervision, certain prior works also leverage full mesh supervision available from synthetically rendered human images [65] or images with fairly successful body fits [34]. Furthermore, multi-view image pairs have also been used for 3D pose [53] and shape estimation [11,35] via enforcing consistency of canonical 3D pose across multiple views. Liang *et al.* [35] use a multi-stage regressor for multi-view images to further reduce the projection ambiguity in order to obtain a better performance for 3D human body under clothing. To inculcate strong 3D pose prior, Zhou *et al.* [70] makes use of left-right symmetric bone-length constraint for the skeleton based 3D pose estimation task. Further, to assure recovery of valid 3D poses for the model-based pose estimation task, Kanazawa *et al.* [20] enforce learning based human pose and shape prior via adversarial networks using unpaired sample of plausible 3D pose and shape. With the advent of differentiable renderers [10,23] certain methods supervise 3D shape and pose estimation through a textured mesh prediction network to encourage matching of the rendered texture image with the image FG [21], alongside the 2D keypoint supervision [49].

3 Approach

We aim to discover the 3D human pose and shape from unlabeled image pairs of consistent FG appearance. During training, we assume access to a parametric human mesh model to aid our self-supervised paradigm. The mesh model provides a low dimensional parametric representation of variations in human shape and pose deformations. However, by design, this model is unaware of the plausibility restrictions of human pose and shape. Thus, it is prone to implausible poses and self-penetrations specifically in the absence of paired 3D supervision [20]. Therefore, to constrain the pose predictions, we assume access to a pool of human 3D pose samples to learn a 3D pose prior.

Figure 2 shows an overview of our training approach. For a given image pair, two parallel pathways of shared CNN regressors predict the human shape and pose parameters alongside the required camera settings to segregate the co-salient FG human. Moreover, to realize a colored mesh representation, we develop a differentiable *Color-recovery* module which infers mesh vertex colors directly from the given image without employing any explicit appearance extraction network.

3.1 Representation and Notations

Human Mesh Model. We employ the widely used SMPL body model [39] which parameterizes a triangulated human mesh of $K = 6890$ vertices. This model factorizes the mesh deformations into shape $\beta \in \mathbb{R}^{10}$ and pose $\theta \in \mathbb{R}^{3J}$ with $J = 23$ skeleton joints [20]. We use the first 10 PCA coefficients of the shape space as a compact shape representation inline with [20]. And, the pose is parameterized as parent-relative rotations in the axis-angle representation. This differentiable SMPL function outputs mesh vertex locations in a canonical 3D space which is represented as $V \in \mathbb{R}^{K\times 3} = \mathcal{M}(\theta, \beta)$. Here, the corresponding 3D pose (*i.e.* 3D location of J joints) is obtained using a pre-trained linear regressor, *i.e.* $Y \in \mathbb{R}^{J\times 3} = W_p V$ parameterized by $W_p \in \mathbb{R}^{J\times K}$. RGB color corresponding to the mesh vertices, V is denoted as $C \in \mathbb{R}^{3\times K} = \mathit{CRM}(V, I)$, where CRM is the *Color-recovery* module. For each vertex id k, $C^{(k)}$ stores the corresponding RGB color intensities. As shown in Fig. 2, we use subscripts a and b to associate the terms with the respective input images, I_a and I_b.

Camera Model. We define a weak perspective camera model using a global orientation $R \in \mathbb{R}^{3\times 3}$ in axis-angle representation (3 angle parameters), a translation $t \in \mathbb{R}^2$ and a scale $s \in \mathbb{R}$. Given these parameters, the 2D camera space coordinates of the 3D mesh vertices with vertex index k is obtained as $v^{(k)} = \pi(V^{(k)}) = s\Pi(RV^{(k)}) + t$; $v^{(k)} \in \mathcal{U}$, where Π denotes orthographic projection and $\mathcal{U} \subset \mathbb{R}^2$ denotes the space of image coordinates. Similarly, the camera projected 2D joint locations (2D pose) is expressed as $y \in \mathbb{R}^{J\times 2} = \pi(Y)$.

3.2 Mesh Estimation Architecture

For a given monocular image, I as input, we first employ a CNN regressor to predict the SMPL parameters (*i.e.* θ and β) alongside the camera parameters, (R, s, t). This is followed by the *Color-recovery* module. The prime functionality of this module is to assign color to the 3D mesh vertices, $C^{(k)}$; $k = 1, 2, ...K$ based on the corresponding image space coordinates obtained via camera projection. However, a reliable color assignment requires us to segregate the vertices based on the following two important criteria.

a) **Non-camera-facing vertices**: First, the camera-facing vertices are separated from the non-camera-facing ones using the mesh vertex normals. Here,

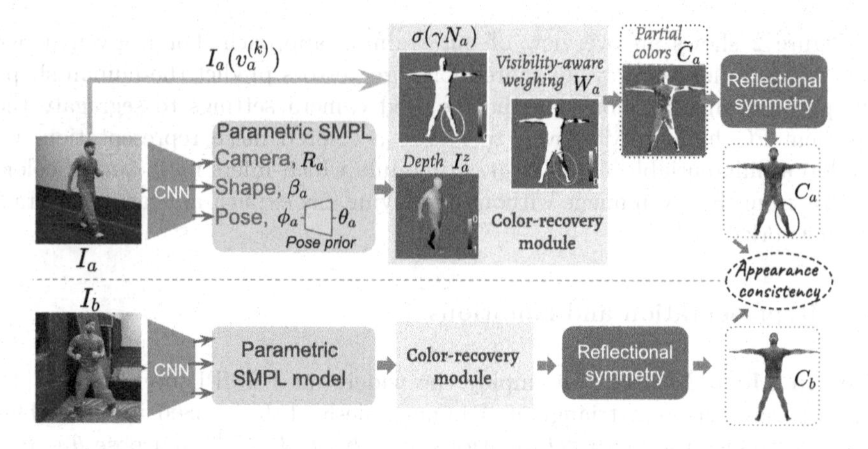

Fig. 2. The proposed self-supervised framework makes use of a differentiable *Color-recovery* module to recover the fully colored mesh vertices. *Yellow-circle*: camera-facing vertices does not account for inter-part occlusion. *Green-circle*: W_a accounts for the inter-part occlusion. *Blue-circle*: Fully colored mesh vertices via reflectional symmetry. (Color figure online)

the vertex normal is computed as the normalized average of the surface normals of the faces connected to a given vertex. We first transform these normals from the default canonical system to the camera coordinate system. Following this, Z-component of the *camera-space-normals*, $N^{(k)} \in \mathbb{R}$ are used to segregate the non-camera-facing vertices via a *sigmoid* operation, as shown in Fig. 2.

b) **Camera-facing, self-occluded vertices**: Note that, $N^{(k)}$ can not be used to select all the camera-visible vertices in presence of inter-part occlusions (see Fig. 2). As, in such scenario, there exist mesh vertices which face the camera but are obscured by other camera-facing vertices which are closer to the camera in 3D. This calls for modeling the relative depth of mesh-vertices as the second criteria to reliably select the vertices which are closer to the camera among all the camera-facing vertices projected to a certain spatial region. To realize this, we utilize *camera-space-depths*, $Z^{(k)} \in \mathbb{R}$ which stores the Z-component (or depth) of the vertex location in the camera transformed space.

3.2.1 Color-Recovery Module

In absence of any appearance related features, we plan to realize a spatial depth map using a fast differentiable renderer [10] where the camera-space-depth of the mesh vertices, Z is treated as the color intensities for the rendering pipeline. The resultant depth-map is represented as $I^z(u)$, where u spans the space of spatial indices. The general idea is to use this depth-map as a margin. More concretely, for effective color assignment, one must select the spatially modulated mesh vertices which have the least absolute depth difference with respect to the above defined depth margin. To realize this, we compute a depth difference $D^{(k)}$ as $|I^z(v^{(k)}) - Z^{(k)}|$, where $I^z(v^{(k)})$ is computed by performing bilinear sampling on $I^z(u)$. In accordance with the above discussion, we formulate a *visibility-aware-weighing* which takes into account both the above mentioned criteria required for an effective mesh vertex selection.

$$W^{(k)} \in [0,1] = \exp(-\alpha D^{(k)})\ \sigma(\gamma N^{(k)}), \text{ where } D^{(k)} = |I^z(v^{(k)}) - Z^{(k)}|$$

Here, $\exp(-\alpha D^{(k)})$ performs a soft selection by assigning a higher weight value (close to 1) for mesh vertices, k whose *camera-space-depth* $Z^{(k)}$ is in agreement with $I^z(v^{(k)})$ and vice-versa. In the second term, σ denotes a sigmoid function with a higher steepness γ to reject the non-camera-facing mesh vertices by attributing a low (close to 0) weighing value. Refer Fig. 2 for visual illustration.

Intermediate Vertex Color Assignment. The above defined *visibility-aware-weighing* is employed to realize a primary vertex color assignment. We denote $\tilde{C} \in \mathbb{R}^{3\times K}$ as the intermediate vertex color, where $\tilde{C}^{(k)}$ stores the corresponding RGB color intensities acquired from the given input image I. Thus, the primary vertex colors are obtained as, $\tilde{C}^{(k)} = I(v^{(k)})\ (2W^{(k)} - 1)$, where $I(v^{(k)})$ stores the RGB color intensities at the spatial coordinates $v^{(k)}$ realized via performing bilinear sampling on the input RGB image I. The scaled weighing function $(2W^{(k)} - 1)$ assigns negative weight to the vertices having low visibility. This assigns a negative color intensity for the corresponding vertices thereby allowing a distinction between the *less-bright* (near-black) colors versus *unassigned* vertices.

3.2.2 Vertex Color Assignment via Reflectional Symmetry

Here, the prime objective is to propagate the reliable color intensities from the assigned vertices to the unreliable/unassigned ones. The idea is to use reflectional symmetry as a prior knowledge by accessing a predefined set of reflectional groups. For each group-id $g = 1, 2, ...G$, a set of 4 vertices are identified according to left-right and front-back symmetry which would have the same color property (except the vertices belonging to the head where only left-right symmetry is used). This symmetry knowledge is stored as a multi-hot encoding denoted as $S^{(g)} \in \{0,1\}^K$ which constitutes of four ones indicating vertex members in the symmetry group g. All the symmetry groups are combined in a symmetry-encoding matrix represented as $S \in \{0,1\}^{G\times K}$. This multi-hot symmetry group representation helps us to perform a fully-differentiable vertex color assignment for all the vertices including the occluded and non-camera facing ones.

To realize the final vertex colors C, we first estimate a group-color for each group g which is denoted by $\mathcal{C}^{(g)} \in \mathbb{R}^3 = (S^{(g)} \circ \text{ReLU}(\tilde{C}))/(S^{(g)} \circ \text{ReLU}(2W - 1))$. Here, $\circ$ denotes dot product between the K-dimensional vectors. The group color can be interpreted as a combination of the intermediate vertex colors weighted by their visibility weighing W. This effectively handles the cases when only one or more of the vertices in a group are initially colored (visible). That is, when visibility is active only for a single vertex among the four vertices in a symmetry set; and when visibility is active for all the 4 vertices in a symmetry set; and also the intermediate cases. Finally, the group color is directly propagated to all the mesh vertices using the following matrix multiplication operation, *i.e.* $C = S^T * \mathcal{C}$, where $\mathcal{C} \in \mathbb{R}^{G\times 3} = [\mathcal{C}^{(1)}, \mathcal{C}^{(2)}, ...\mathcal{C}^{(G)}]$ (see Suppl for more details).

3.3 Self-supervised Learning Objectives

For a given image pair, denoted as I_a and I_b (depicting the same person in diverse pose and BGs), we forward them through two parallel pathways of our colored mesh estimation architecture (see Fig. 2). The commonness of FG appearance allows us to impose an appearance consistency loss between the predicted fully colored mesh representations.

a) **Color consistency.** First, we impose the following consistency loss,

$$\mathcal{L}_{CC} = \mathcal{L}_C + \lambda\mathcal{L}_{\tilde{C}}, \text{ where } \mathcal{L}_C = \|C_a - C_b\| \text{ and } \mathcal{L}_{\tilde{C}} = \|W_a \odot W_b \odot (\tilde{C}_a - \tilde{C}_b)\|$$

Here, $\odot$ denotes element-wise multiplication. Note that, $\mathcal{L}_{\tilde{C}}$ enforces a vertex-color consistency on the co-visible mesh vertices (computed as $(W_a \odot W_b)$), *i.e.* the vertices which are visible in both the mesh representations obtained from the image pair, (I_a, I_b). However, $\mathcal{L}_C$ enforces full vertex color consistency. Here, $\mathcal{L}_{CC}$ combines both of the losses thereby providing a higher weightage to the co-visible vertex colors as compared to the approximate full color representation, considering the approximate nature of the symmetry assumption.

b) **Part-prototype consistency.** The proposed *Color-recovery* module can also be applied on the convolutional feature maps. For a given vertex k and a convolutional feature map $H \in \mathbb{R}^{\tilde{w}\times\tilde{h}\times\tilde{d}}$, we sample $\mathcal{H}^{(k)} \in \mathbb{R}^{\tilde{d}} = H(v^{(k)})$. Note that, we define a fixed vertex to part-segmentation mapping represented as $Q^{(l)}$, which stores a set of vertex indices for each part $l = 1, 2, ...L$. Now, one can use the vertex visibility weighing $W^{(k)}$ to obtain a prototype appearance feature for each body-part l, which is computed as; $\mathcal{F}^{(l)} = (\Sigma_{k\in Q^{(l)}} W^{(k)}\mathcal{H}^{(k)})/(\Sigma_{k\in Q^{(l)}} W^{(k)})$ Following this, we enforce a prototype consistency loss between the image pairs as $\mathcal{L}_P = \Sigma_l\|\mathcal{F}_a^{(l)} - \mathcal{F}_b^{(l)}\|/L$. Note that, the prototype feature computation is inherently aware of the inter-part occlusions as a result of incorporating the visibility weighing $W^{(k)}$. As compared to enforcing vertex-color consistency, $\mathcal{L}_{CC}$ (*i.e.* the raw color intensities), the part-prototype consistency aims to match a higher-level semantic abstraction (*e.g.* checkered regular patterns versus just plain individual colors) of the part appearances extracted from the image pairs. This also helps us to overcome the unreliability of raw vertex colors which could arise due to illumination differences. Motivated by the perceptual loss idea [17], we obtain H_a and H_b as the *Conv2-1* features corresponding to I_a and I_b from an ImageNet trained (frozen) VGG-16 network [58].

c) **Shape-consistency.** We also enforce a shape consistency loss between the shape parameters obtained from the image pair, *i.e.* $\mathcal{L}_\beta = |\beta_a - \beta_b|$. Almost all the prior works [20,49,50] utilize an *unpaired* human shape dataset to enforce plausibility of the shape predictions via adversarial prior. However, in the proposed self-supervised framework we do not access any human shape dataset. To regularize the shape parameters during the initial training iterations we enforce a loss on shape predictions with respect to a fixed mean shape as a regularization. However, after gaining a decent mesh estimation performance we gradually reduce weightage of this loss by allowing shape variations beyond the mean shape driven by the proposed appearance and shape consistency objectives.

d) **Enforcing validity of pose predictions.** Additionally, to assure validity of the predicted pose parameters we train an adversarial auto-encoder [41] to realize a continuous human pose manifold [28,29] mapped from a latent pose representation, $\phi \in [-1,1]^{32}$. This is trained using an unpaired 3D human pose dataset. The frozen pose decoder obtained from this generative framework is directly employed as a module, with instilled human 3D pose prior. More concretely, a *tanh* non-linearity on the pose-prediction head of the CNN regressor (inline with the latent pose ϕ) followed by the frozen pose decoder prevents implausible pose predictions during our self-supervised training. In contrast to enforcing an adversarial pose prior objective [20,49], the proposed setup greatly simplifies our training procedure (devoid of discriminator training).

In absence of paired supervision, parameters of the shared CNN regressor is trained by directly enforcing the above consistency losses, *i.e.* $\mathcal{L}_{CC}$, $\mathcal{L}_P$, and $\mathcal{L}_\beta$.

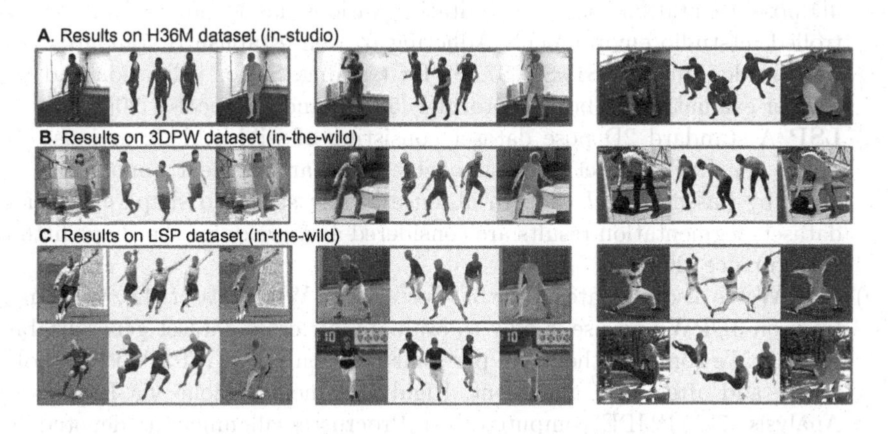

Fig. 3. Qualitative results. In each panel, 1st column depicts the input image, 2nd column depicts our colored mesh prediction, and 3rd column shows the model-based part segments. Our model fails (in magenta) in presence of complex inter-part occlusions.

4 Experiments

We perform thorough experimental analysis to demonstrate the generalizability of our framework across several datasets on a variety of tasks.

Implementation Details. We use Resnet-50 [9] initialized from ImageNet as the base CNN network. The average pooled last layer features are forwarded through a series of fully-connected layers to regress the pose (latent pose encoding ϕ), shape and camera parameters. Note that, the series of differentiable operations post the CNN regressor do not include any trainable parameters even to estimate the vertex colors. During training, we optimize individual loss terms at alternate training iteration using Adam optimizer [24]. We enforce prediction of the mean shape for initial 100k training iterations. We also impose a silhouette loss on the predicted human mesh with respect to a pseudo silhouette ground-truth obtained either by using an unsupervised saliency detection method [71] or by using a background estimate as favourable for static camera scenarios [53].

Datasets. We sample image pairs with diverse BG (pairs with large *L2* distance) from the following standard datasets, *i.e.* Human3.6M [15], MPII [2], MPI-INF-3DHP [46] and an in-house collection of wild YouTube videos. In contrast to the in-studio datasets with hardly any camera movement implying static BG [15], the videos collected from YouTube have diverse camera movements (*e.g.* Parkour and Free-running videos). We prune the raw video samples using a person-detector [51] to obtain reliable human-centric crops as required for the mesh estimation pipeline (see Suppl). The unpaired 3D pose dataset required to train the 3D pose prior is obtained from CMU-MoCap (also used in MoSh [38]).

a) **Human3.6M** This is a widely used dataset consisting of paired image with 3D pose annotations of actors imitating various day-to-day tasks in a controlled in-studio environment. Adhering to well established standards [20] we consider subjects S1, S6, S7, S8 for training, S5 for validation and S9, S11 for evaluation, in both Protocol-1 [53,54] and Protocol-2 [20].
b) **LSP** A standard 2D pose dataset consisting of wild athletic actions. We access the LSP test-set with silhouette and part segment annotations as given by Lassner *et al.* [34]. In absence of any standard shape evaluation dataset, segmentation results are considered as a proxy for the shape fitting performance [20,26].
c) **3DPW** We also evaluate on the 3D Poses in the Wild dataset [42]. We do not train on 3DPW and use it only to evaluate our cross-dataset generalizability [30]. We compute the mean per joint position error (MPJPE) [15], both before and after rigid alignment. Rigid alignment is done via Procrustes Analysis [7]. MPJPE computed post Procrustes alignment is denoted by PA-MPJPE.

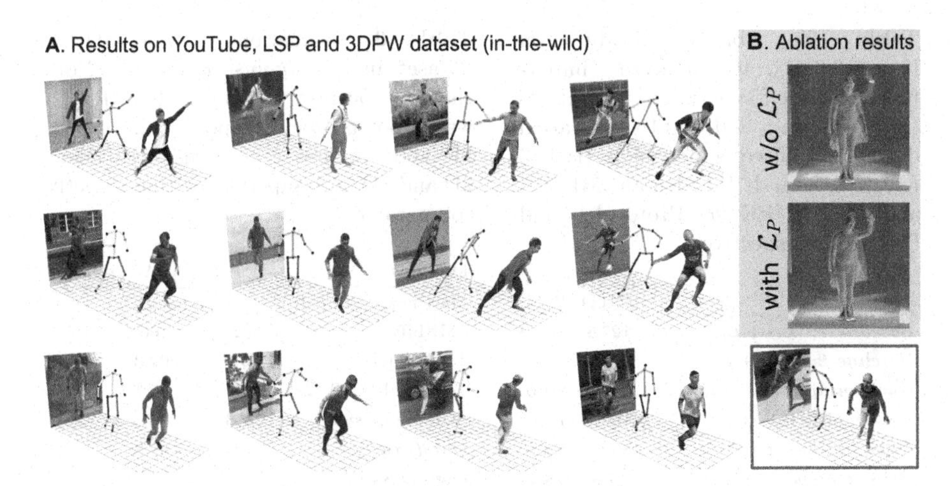

Fig. 4. A. Qualitative results on single image colored human mesh recovery. The model fails in presence of complex inter-limb occlusions (in magenta box). **B.** Qualitative analysis demonstrating importance of incorporating $\mathcal{L}_P$ to extract relevant part-semantics.

4.1 Ablative Study

To analyze effectiveness of individual self-supervised consistency objectives, we perform ablations by removing certain losses as shown in Table 2. First, we train *Baseline-1* by enforcing $\mathcal{L}_C$ and $\mathcal{L}_\beta$. Following this, in *Baseline-2* we enforce $\mathcal{L}_{CC}$ by incorporating $\mathcal{L}_{\tilde{C}}$ which further penalizes color inconsistency between the vertices which are commonly visible in both the mesh representations. This results in marginal improvement of performance. Moving forward, we recognize a clear limitation in our assumption of FG color consistency (raw RGB intensities) which can easily be violated by illumination differences. Further, the assumption of left-right and front-back symmetry in apparel color can also be violated specifically for asymmetric upper body apparel. As a solution, the proposed part-prototype consistency objective, $\mathcal{L}_P$ tries to match a higher level appearance representation beyond just raw color intensities (see Fig. 4B), thus resulting in a significant performance gain (*Ours(unsup)* in Table 2). Note that, $\mathcal{L}_P$ is possible as a consequence of the proposed differentiable *Color-recovery* module.

Further, maintaining a fair comparison ground against the prior weakly supervised approaches, we train 3 variants of the proposed framework by utilizing increasing level of paired supervisions alongside our self-supervised objectives.

a) ***Ours(multi-view-sup)*** Under multi-view supervision, we impose additional consistency loss on the canonically aligned (view-invariant) 3D mesh vertices (*i.e.* $\|V_a - V_b\|$) and the 3D pose (*i.e.* $\|Y_a - Y_b\|$) for the time synchronized multi-view pairs, (I_a, I_b). Inline with Rhodin *et al.* [53], we also use full 3D pose supervision only for S1 while evaluating on the standard

Table 2. Ablative study (on Human3.6M) to analyze importance of self-supervised objectives (first 3 rows), and results at varied degree of paired supervision (last 3 rows). P1 and P2 denote MPJPE and PA-MPJPE in Protocol-1 and Protocol-2 respectively.

Methods	P1(↓)	P2(↓)
Baseline-1; ($\mathcal{L}_C + \mathcal{L}_\beta$)	127.1	101.2
Baseline-2; ($\mathcal{L}_{CC} + \mathcal{L}_\beta$)	119.6	97.4
Ours(unsup.); ($\mathcal{L}_{CC}+\mathcal{L}_\beta+\mathcal{L}_p$)	**110.8**	**90.5**
Ours(multi-view-sup)	102.1	74.1
Ours(weakly-sup)	86.4	58.2
Ours(semi-sup)	**73.8**	**48.1**

Table 3. Evaluation on wild 3DPW dataset in a *fully-unseen* setting. Note that, in contrast to Temporal-HMR [22] we do not use any temporal supervision. Methods in first 5 rows use equivalent 2D and 3D pose supervision, thus directly comparable.

Methods	MPJPE(↓)	PA-MPJPE(↓)
Martinez *et al.* [44]	–	157.0
SMPLify [5]	199.2	106.1
TP-Net [6]	163.7	92.3
Temporal-HMR [22]	127.1	80.1
Ours(semi-sup)	**125.8**	**78.2**
Ours(weakly-sup)	153.4	89.8
Ours(unsup)	187.1	102.7

Human3.6M dataset. We outperform Rhodin *et al.* [53] by a significant margin as reported in the Table 4. This is beyond the usual trend of weaker performance in non-parametric approaches against the model-based parametric ones. Thus, we attribute this performance gain to the proposed appearance consensus driven self-supervised objectives.

b) ***Ours(weakly-sup)*** In this setting, we access image datasets with paired 2D landmark annotations, inline with the supervision setting of prior model-based approaches [20]. Alongside the proposed self-supervised objectives, we impose a direct 2D landmark supervision loss (*i.e.* $\|y - y_{gt}\|$) with respect to the corresponding ground-truths but only on samples from specific datasets, such as LSP, LSP-extended [18] and MPII [2]. Certain prior arts, such as HMR [20], use even more images with paired 2D landmark annotations from COCO [36].

c) ***Ours(semi-sup)*** In this variant, we access paired 3D pose supervision on the widely used in-studio Human3.6M [15] dataset alongside the 2D landmark supervision as used in *Ours(weakly-sup)*. Note that, a better performance on Human3.6M (with limited BG and FG diversity as a result of the in-studio data collection setup) does not translate to the same on wild images as a result of the significant domain gap. As we impose the above supervisions alongside the proposed self-supervised objective on unlabeled wild images, such a training is expected to deliver improved performance by successfully overcoming the domain-shift issue. We evaluate this on the wild 3DPW dataset.

Table 4. Evaluation on Human3.6M (Protocol-2). Methods in first 9 rows use equivalent 2D and 3D pose supervision hence are directly comparable. Same analogy applies for the rows 10–11 and 12–13.

No	Methods	PA-MPJPE(↓)
1	Lassner *et al.* [34]	93.9
2	Pavlakos *et al.* [50]	75.9
3	Omran *et al.* [47]	59.9
4	HMR [20]	56.8
5	Temporal HMR [22]	56.9
6	Arnab *et al.* [4]	54.3
7	Kolotouros *et al.* [26]	50.1
8	TexturePose [49]	49.7
9	*Ours(semi-sup)*	**48.1**
10	HMR unpaired [20]	66.5
11	*Ours(weakly-sup)*	**58.2**
12	Rhodin *et al.* [53]	98.2
13	*Ours(multi-view-sup)*	**74.1**

Table 5. Evaluation of FG-BG and 6-part segmentation on LSP test set. It reports accuracy (Acc.) and F1 score values of ours against the prior-arts. **First group**: Iterative, *optimization-based* approaches. **Last 3 groups**: *Regression-based* methods grouped based on comparable supervision levels.

Methods	**FG-BG Seg.**		**Part Seg.**	
	Acc.(↑)	F1(↑)	Acc.(↑)	F1(↑)
SMPLify *oracle* [5]	92.17	0.88	88.82	0.67
SMPLify [5]	91.89	0.88	87.71	0.64
SMPLify on [50]	92.17	0.88	88.24	0.64
Bodynet [64]	92.75	0.84	–	–
HMR [20]	91.67	0.87	87.12	0.60
Kolotouros *et al.* [26]	91.46	0.87	88.69	0.66
TexturePose [49]	91.82	0.87	89.00	0.67
Ours(semi-sup)	**91.84**	0.87	**89.08**	0.67
HMR unpaired [20]	91.30	0.86	87.00	0.59
Ours(weakly-sup)	**91.70**	**0.87**	**87.12**	**0.60**
Ours(unsup)	91.46	0.86	87.26	0.64

4.2 Comparison with the State-of-the-Art

Evaluation on Human3.6M. Table 4 shows a comparison of different variants of the proposed framework against the prior-arts which are grouped based on the respective supervision levels. We clearly outperform in all the three groups *i.e.* while accessing comparable a) 3D pose supervision, b) 2D landmark supervision, and c) multi-view supervision. Except Rhodin *et al.* [53] all the prior works mentioned in Table 4 use parametric human model for the human mesh estimation task. Note the significant performance gain specifically in absence of any 3D pose supervision, *i.e.* for *Ours(weakly-sup)* and *Ours(multi-view-sup)* against the relevant counterparts as reported in the last 4 rows.

Evaluation on 3DPW. Table 3 reports a comparison of different variants of the proposed framework against the prior-arts which use comparable pose supervision as used in *Ours(semi-sup)* (except certain methods, such as HMR [20] which use even more supervision on 3D pose from the MPI-INF-3DHP [46] dataset). It is worth noting that none of our model variants is trained on the samples from 3DPW dataset (not even in self-supervised paradigm). A better performance in such *unseen* setting highlights our superior cross-dataset generalizability.

Evaluation of Part-Segmentation. We also evaluate our performance on FG-BG segmentation and body part-segmentation tasks which are considered as a proxy to quantify the shape fitting performance. In presence of 2D landmark annotation, iterative model fitting approaches have a clear advantage over

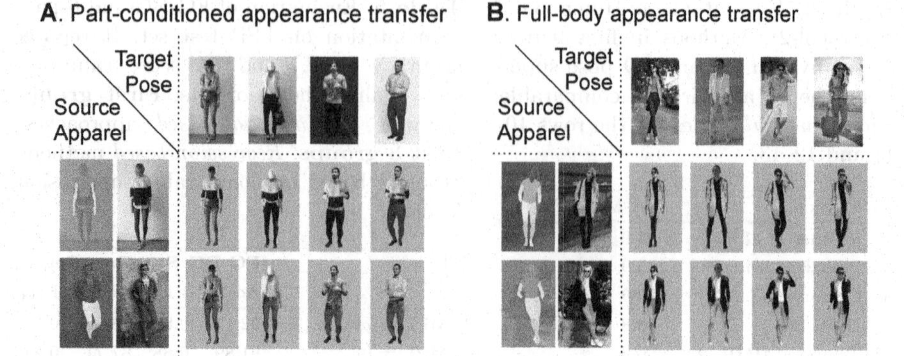

Fig. 5. Qualitative results on **A.** Part-conditioned, and **B.** Full-body appearance transfer. This is enabled as a result of our ability to infer the colored mesh representation.

the single-shot regressor based approaches as shown in Table 5. At comparable supervision, *Ours(semi-sup)* not only outperforms the relevant regression based prior arts but also performs competitive to the iterative model fitting based approaches with a significant advantage on inference time (1 min vs 0.04 s). We also report performance of our self-supervised variant such as *Ours(unsup)* which performs competitive to the prior supervised regression-based approaches, thus establishing the importance of FG appearance consistency for accurate shape recovery. See Suppl for qualitative results.

4.3 Qualitative Results

The proposed mesh recovery model not only infers pose and shape but also outputs a colored mesh representation as a result of the proposed *reflectional-symmetry* procedure. To evaluate effectiveness of the recovered part appearance we perform 2 different tasks a) part-conditioned appearance transfer, and b) full-body appearance transfer as shown in Fig. 5. On the top, we show the target images whose pose and shape (network predicted) is combined with part appearances recovered from the source image (only for the highlighted parts) shown on left, to realize a novel synthesized image. Note that, in case of *part-conditioned* appearance transfer, appearance of the non-highlighted parts are taken from the target image shown on the top. For instance, in the first row, the synthesized image depicts upper-body apparel of the person in the source image combined with the lower-body apparel from the target (and in the target image pose). Qualitative results of *Ours(semi-sup)* model on other primary tasks are shown in Fig. 3 and Fig. 4 with highlighted failure scenarios (see Suppl).

5 Conclusion

We introduce a self-supervised framework for model-based human pose and shape recovery. The proposed appearance consistency not only helps us to segregate the common FG human from their respective wild BGs but also discovers the required pose deformation in a fully self-supervised manner. However, extending such a framework for human centric images with occlusion by external objects or truncated human visibility, remains to be explored in future.

Acknowledgements. We thank Qualcomm Innovation Fellowship India 2020.

References

1. Alp Güler, R., Neverova, N., Kokkinos, I.: DensePose: dense human pose estimation in the wild. In: CVPR (2018)
2. Andriluka, M., Pishchulin, L., Gehler, P., Schiele, B.: 2D human pose estimation: new benchmark and state of the art analysis. In: CVPR (2014)
3. Anguelov, D., Srinivasan, P., Koller, D., Thrun, S., Rodgers, J., Davis, J.: SCAPE: shape completion and animation of people. In: ACM SIGGRAPH (2005)
4. Arnab, A., Doersch, C., Zisserman, A.: Exploiting temporal context for 3D human pose estimation in the wild. In: CVPR (2019)
5. Bogo, F., Kanazawa, A., Lassner, C., Gehler, P., Romero, J., Black, M.J.: Keep it SMPL: automatic estimation of 3D human pose and shape from a single image. In: Leibe, B., Matas, J., Sebe, N., Welling, M. (eds.) ECCV 2016. LNCS, vol. 9909, pp. 561–578. Springer, Cham (2016). https://doi.org/10.1007/978-3-319-46454-1_34
6. Dabral, R., Mundhada, A., Kusupati, U., Afaque, S., Sharma, A., Jain, A.: Learning 3D human pose from structure and motion. In: Ferrari, V., Hebert, M., Sminchisescu, C., Weiss, Y. (eds.) ECCV 2018. LNCS, vol. 11213, pp. 679–696. Springer, Cham (2018). https://doi.org/10.1007/978-3-030-01240-3_41
7. Gower, J.C.: Generalized procrustes analysis. Psychometrika **40**(1), 33–51 (1975)
8. Guan, P., Weiss, A., Balan, A.O., Black, M.J.: Estimating human shape and pose from a single image. In: ICCV (2009)
9. He, K., Zhang, X., Ren, S., Sun, J.: Identity mappings in deep residual networks. In: Leibe, B., Matas, J., Sebe, N., Welling, M. (eds.) ECCV 2016. LNCS, vol. 9908, pp. 630–645. Springer, Cham (2016). https://doi.org/10.1007/978-3-319-46493-0_38
10. Henderson, P., Ferrari, V.: Learning single-image 3D reconstruction by generative modelling of shape, pose and shading. Int. J. Comput. Vision **128**(4), 835–854 (2019). https://doi.org/10.1007/s11263-019-01219-8
11. Hofmann, M., Gavrila, D.M.: Multi-view 3D human pose estimation combining single-frame recovery, temporal integration and model adaptation. In: CVPR (2009)
12. Hogg, D.: Model-based vision: a program to see a walking person. Image Vis. Comput. **1**(1), 5–20 (1983)
13. Hsu, K.-J., Tsai, C.-C., Lin, Y.-Y., Qian, X., Chuang, Y.-Y.: Unsupervised CNN-based co-saliency detection with graphical optimization. In: Ferrari, V., Hebert, M., Sminchisescu, C., Weiss, Y. (eds.) ECCV 2018. LNCS, vol. 11209, pp. 502–518. Springer, Cham (2018). https://doi.org/10.1007/978-3-030-01228-1_30
14. Huang, Y., et al.: Towards accurate marker-less human shape and pose estimation over time. In: 3DV (2017)

15. Ionescu, C., Papava, D., Olaru, V., Sminchisescu, C.: Human3.6M: large scale datasets and predictive methods for 3D human sensing in natural environments. IEEE Trans. Pattern Anal. Mach. Intell. **36**, 1325–1339 (2013)
16. Jakab, T., Gupta, A., Bilen, H., Vedaldi, A.: Unsupervised learning of object landmarks through conditional image generation. In: NeurIPS (2018)
17. Johnson, J., Alahi, A., Fei-Fei, L.: Perceptual losses for real-time style transfer and super-resolution. In: Leibe, B., Matas, J., Sebe, N., Welling, M. (eds.) ECCV 2016. LNCS, vol. 9906, pp. 694–711. Springer, Cham (2016). https://doi.org/10.1007/978-3-319-46475-6_43
18. Johnson, S., Everingham, M.: Clustered pose and nonlinear appearance models for human pose estimation. In: BMVC (2010)
19. Joo, H., Simon, T., Sheikh, Y.: Total capture: a 3D deformation model for tracking faces, hands, and bodies. In: CVPR (2018)
20. Kanazawa, A., Black, M.J., Jacobs, D.W., Malik, J.: End-to-end recovery of human shape and pose. In: CVPR (2018)
21. Kanazawa, A., Tulsiani, S., Efros, A.A., Malik, J.: Learning category-specific mesh reconstruction from image collections. In: Ferrari, V., Hebert, M., Sminchisescu, C., Weiss, Y. (eds.) ECCV 2018. LNCS, vol. 11219, pp. 386–402. Springer, Cham (2018). https://doi.org/10.1007/978-3-030-01267-0_23
22. Kanazawa, A., Zhang, J.Y., Felsen, P., Malik, J.: Learning 3D human dynamics from video. In: CVPR (2019)
23. Kato, H., Ushiku, Y., Harada, T.: Neural 3D mesh renderer. In: CVPR (2018)
24. Kingma, D.P., Ba, J.: Adam: a method for stochastic optimization. arXiv preprint arXiv:1412.6980 (2014)
25. Kolotouros, N., Pavlakos, G., Black, M.J., Daniilidis, K.: Learning to reconstruct 3D human pose and shape via model-fitting in the loop. In: ICCV (2019)
26. Kolotouros, N., Pavlakos, G., Daniilidis, K.: Convolutional mesh regression for single-image human shape reconstruction. In: CVPR (2019)
27. Kundu, J.N., Ganeshan, A., MV, R., Prakash, A., Babu, R.V.: iSPA-Net: iterative semantic pose alignment network. In: ACM Multimedia (2018)
28. Kundu, J.N., Gor, M., Babu, R.V.: BiHMP-GAN: bidirectional 3D human motion prediction GAN. In: AAAI (2019)
29. Kundu, J.N., Gor, M., Uppala, P.K., Babu, R.V.: Unsupervised feature learning of human actions as trajectories in pose embedding manifold. In: WACV (2019)
30. Kundu, J.N., Patravali, J., Babu, R.V.: Unsupervised cross-dataset adaptation via probabilistic amodal 3D human pose completion. In: WACV (2020)
31. Kundu, J.N., Seth, S., Jampani, V., Rakesh, M., Babu, R.V., Chakraborty, A.: Self-supervised 3D human pose estimation via part guided novel image synthesis. In: CVPR (2020)
32. Kundu, J.N., Seth, S., Rahul, M., Rakesh, M., Babu, R.V., Chakraborty, A.: Kinematic-structure-preserved representation for unsupervised 3D human pose estimation. In: AAAI (2020)
33. Navaneet, K.L., Mandikal, P., Jampani, V., Babu, V.: DIFFER: moving beyond 3D reconstruction with differentiable feature rendering. In: CVPR Workshops (2019)
34. Lassner, C., Romero, J., Kiefel, M., Bogo, F., Black, M.J., Gehler, P.V.: Unite the people: closing the loop between 3D and 2D human representations. In: CVPR (2017)
35. Liang, J., Lin, M.C.: Shape-aware human pose and shape reconstruction using multi-view images. In: ICCV (2019)

36. Lin, T.-Y., et al.: Microsoft COCO: common objects in context. In: Fleet, D., Pajdla, T., Schiele, B., Tuytelaars, T. (eds.) ECCV 2014. LNCS, vol. 8693, pp. 740–755. Springer, Cham (2014). https://doi.org/10.1007/978-3-319-10602-1_48
37. Liu, S., Li, T., Chen, W., Li, H.: Soft rasterizer: a differentiable renderer for image-based 3D reasoning. In: ICCV (2019)
38. Loper, M., Mahmood, N., Black, M.J.: MoSh: motion and shape capture from sparse markers. ACM Trans. Graph. (TOG) **33**(6), 220 (2014)
39. Loper, M., Mahmood, N., Romero, J., Pons-Moll, G., Black, M.J.: SMPL: a skinned multi-person linear model. ACM Trans. Graph. (TOG) **34**(6), 1–16 (2015)
40. Ma, L., Sun, Q., Georgoulis, S., Van Gool, L., Schiele, B., Fritz, M.: Disentangled person image generation. In: CVPR (2018)
41. Makhzani, A., Shlens, J., Jaitly, N., Goodfellow, I., Frey, B.: Adversarial autoencoders. arXiv preprint arXiv:1511.05644 (2015)
42. von Marcard, T., Henschel, R., Black, M.J., Rosenhahn, B., Pons-Moll, G.: Recovering accurate 3D human pose in the wild using IMUs and a moving camera. In: Ferrari, V., Hebert, M., Sminchisescu, C., Weiss, Y. (eds.) ECCV 2018. LNCS, vol. 11214, pp. 614–631. Springer, Cham (2018). https://doi.org/10.1007/978-3-030-01249-6_37
43. von Marcard, T., Rosenhahn, B., Black, M.J., Pons-Moll, G.: Sparse inertial poser: automatic 3D human pose estimation from sparse IMUs. In: Computer Graphics Forum, vol. 36, pp. 349–360. Wiley Online Library (2017)
44. Martinez, J., Hossain, R., Romero, J., Little, J.J.: A simple yet effective baseline for 3D human pose estimation. In: ICCV (2017)
45. Mathieu, M.F., Zhao, J.J., Zhao, J., Ramesh, A., Sprechmann, P., LeCun, Y.: Disentangling factors of variation in deep representation using adversarial training. In: NeurIPS, pp. 5040–5048 (2016)
46. Mehta, D., et al.: Monocular 3D human pose estimation in the wild using improved CNN supervision. In: 3DV (2017)
47. Omran, M., Lassner, C., Pons-Moll, G., Gehler, P., Schiele, B.: Neural body fitting: unifying deep learning and model based human pose and shape estimation. In: 3DV (2018)
48. Pavlakos, G., et al.: Expressive body capture: 3D hands, face, and body from a single image. In: CVPR (2019)
49. Pavlakos, G., Kolotouros, N., Daniilidis, K.: TexturePose: supervising human mesh estimation with texture consistency. In: ICCV (2019)
50. Pavlakos, G., Zhu, L., Zhou, X., Daniilidis, K.: Learning to estimate 3D human pose and shape from a single color image. In: CVPR (2018)
51. Ren, S., He, K., Girshick, R., Sun, J.: Faster R-CNN: towards real-time object detection with region proposal networks. In: NeurIPS, pp. 91–99 (2015)
52. Rhodin, H., Robertini, N., Casas, D., Richardt, C., Seidel, H.-P., Theobalt, C.: General automatic human shape and motion capture using volumetric contour cues. In: Leibe, B., Matas, J., Sebe, N., Welling, M. (eds.) ECCV 2016. LNCS, vol. 9909, pp. 509–526. Springer, Cham (2016). https://doi.org/10.1007/978-3-319-46454-1_31
53. Rhodin, H., Salzmann, M., Fua, P.: Unsupervised geometry-aware representation for 3D human pose estimation. In: Ferrari, V., Hebert, M., Sminchisescu, C., Weiss, Y. (eds.) ECCV 2018. LNCS, vol. 11214, pp. 765–782. Springer, Cham (2018). https://doi.org/10.1007/978-3-030-01249-6_46
54. Rhodin, H., et al.: Learning monocular 3D human pose estimation from multi-view images. In: CVPR (2018)

55. Rifai, S., Bengio, Y., Courville, A., Vincent, P., Mirza, M.: Disentangling factors of variation for facial expression recognition. In: Fitzgibbon, A., Lazebnik, S., Perona, P., Sato, Y., Schmid, C. (eds.) ECCV 2012. LNCS, vol. 7577, pp. 808–822. Springer, Heidelberg (2012). https://doi.org/10.1007/978-3-642-33783-3_58
56. Rogez, G., Weinzaepfel, P., Schmid, C.: LCR-Net: localization-classification-regression for human pose. In: CVPR (2017)
57. Romero, J., Tzionas, D., Black, M.J.: Embodied hands: modeling and capturing hands and bodies together. ACM Trans. Graph. (ToG) **36**(6), 245 (2017)
58. Simonyan, K., Zisserman, A.: Very deep convolutional networks for large-scale image recognition. arXiv preprint arXiv:1409.1556 (2014)
59. Song, S., Yu, F., Zeng, A., Chang, A.X., Savva, M., Funkhouser, T.: Semantic scene completion from a single depth image. In: CVPR (2017)
60. Sun, X., Xiao, B., Wei, F., Liang, S., Wei, Y.: Integral human pose regression. In: Ferrari, V., Hebert, M., Sminchisescu, C., Weiss, Y. (eds.) ECCV 2018. LNCS, vol. 11210, pp. 536–553. Springer, Cham (2018). https://doi.org/10.1007/978-3-030-01231-1_33
61. Sun, Y., Ye, Y., Liu, W., Gao, W., Fu, Y., Mei, T.: Human mesh recovery from monocular images via a skeleton-disentangled representation. In: ICCV (2019)
62. Tan, V., Budvytis, I., Cipolla, R.: Indirect deep structured learning for 3D human body shape and pose prediction. In: BMVC (2017)
63. Tung, H.Y., Tung, H.W., Yumer, E., Fragkiadaki, K.: Self-supervised learning of motion capture. In: NIPS (2017)
64. Varol, G., et al.: BodyNet: volumetric inference of 3D human body shapes. In: Ferrari, V., Hebert, M., Sminchisescu, C., Weiss, Y. (eds.) ECCV 2018. LNCS, vol. 11211, pp. 20–38. Springer, Cham (2018). https://doi.org/10.1007/978-3-030-01234-2_2
65. Varol, G., et al.: Learning from synthetic humans. In: CVPR (2017)
66. Weiss, A., Hirshberg, D., Black, M.J.: Home 3D body scans from noisy image and range data. In: ICCV (2011)
67. Zanfir, A., Marinoiu, E., Sminchisescu, C.: Monocular 3D pose and shape estimation of multiple people in natural scenes-the importance of multiple scene constraints. In: CVPR (2018)
68. Zanfir, A., Marinoiu, E., Zanfir, M., Popa, A.I., Sminchisescu, C.: Deep network for the integrated 3D sensing of multiple people in natural images. In: NIPS (2018)
69. Zhang, D., Meng, D., Han, J.: Co-saliency detection via a self-paced multiple-instance learning framework. IEEE Trans. Pattern Anal. Mach. Intell. **39**(5), 865–878 (2016)
70. Zhou, X., Huang, Q., Sun, X., Xue, X., Wei, Y.: Towards 3D human pose estimation in the wild: a weakly-supervised approach. In: CVPR (2017)
71. Zhu, W., Liang, S., Wei, Y., Sun, J.: Saliency optimization from robust background detection. In: CVPR (2014)

Author Index